Magnetic Core Selection for Transformers and Inductors

ELECTRICAL ENGINEERING AND ELECTRONICS

A Series of Reference Books and Textbooks

1. Rational Fault Analysis, *edited by Richard Saeks and S. R. Liberty*
2. Nonparametric Methods in Communications, *edited by P. Papantoni-Kazakos and Dimitri Kazakos*
3. Interactive Pattern Recognition, *Yi-tzuu Chien*
4. Solid-State Electronics, *Lawrence E. Murr*
5. Electronic, Magnetic, and Thermal Properties of Solid Materials, *Klaus Schröder*
6. Magnetic-Bubble Memory Technology, *Hsu Chang*
7. Transformer and Inductor Design Handbook, *Colonel Wm. T. McLyman*
8. Electromagnetics: Classical and Modern Theory and Applications, *Samuel Seely and Alexander D. Poularikas*
9. One-Dimensional Digital Signal Processing, *Chi-Tsong Chen*
10. Interconnected Dynamical Systems, *Raymond A. DeCarlo and Richard Saeks*
11. Modern Digital Control Systems, *Raymond G. Jacquot*
12. Hybrid Circuit Design and Manufacture, *Roydn D. Jones*
13. Magnetic Core Selection for Transformers and Inductors: A User's Guide to Practice and Specification, *Colonel Wm. T. McLyman*

14. Static and Rotating Electromagnetic Devices, *Richard H. Engelmann*
15. Energy-Efficient Electric Motors: Selection and Application, *John C. Andreas*
16. Electromagnetic Compossibility, *Heinz M. Schlicke*
17. Electronics: Models, Analysis, and Systems, *James G. Gottling*
18. Digital Filter Design Handbook, *Fred J. Taylor*
19. Multivariable Control: An Introduction, *P. K. Sinha*
20. Flexible Circuits: Design and Applications, *Steve Gurley, with contributions by Carl A. Edstrom, Jr., Ray D. Greenway, and William P. Kelly*
21. Circuit Interruption: Theory and Techniques, *Thomas E. Browne, Jr.*
22. Switch Mode Power Conversion: Basic Theory and Design, *K. Kit Sum*
23. Pattern Recognition: Applications to Large Data-Set Problems, *Sing-Tze Bow*
24. Custom-Specific Integrated Circuits: Design and Fabrication, *Stanley L. Hurst*
25. Digital Circuits: Logic and Design, *Ronald C. Emery*
26. Large-Scale Control Systems: Theories and Techniques, *Magdi S. Mahmoud, Mohamed F. Hassan, and Mohamed G. Darwish*
27. Microprocessor Software Project Management, *Eli T. Fathi and Cedric V. W. Armstrong (Sponsored by Ontario Centre for Microelectronics)*
28. Low Frequency Electromagnetic Design, *Michael P. Perry*
29. Multidimensional Systems: Techniques and Applications, *edited by Spyros G. Tzafestas*
30. AC Motors for High-Performance Applications: Analysis and Control, *Sakae Yamamura*
31. Ceramic Motors for Electronics: Processing, Properties, and Applications, *edited by Relva C. Buchanan*
32. Microcomputer Bus Structures and Bus Interface Design, *Arthur L. Dexter*
33. End User's Guide to Innovative Flexible Circuit Packaging, *Jay J. Miniet*
34. Reliability Engineering for Electronic Design, *Norman B. Fuqua*
35. Design Fundamentals for Low-Voltage Distribution and Control, *Frank W. Kussy and Jack L. Warren*
36. Encapsulation of Electronic Devices and Components, *Edward R. Salmon*
37. Protective Relaying: Principles and Applications, *J. Lewis Blackburn*
38. Testing Active and Passive Electronic Components, *Richard F. Powell*
39. Adaptive Control Systems: Techniques and Applications, *V. V. Chalam*
40. Computer-Aided Analysis of Power Electronic Systems, *Venkatachari Rajagopalan*
41. Integrated Circuit Quality and Reliability, *Eugene R. Hnatek*
42. Systolic Signal Processing Systems, *edited by Earl E. Swartzlander, Jr.*
43. Adaptive Digital Filters and Signal Analysis, *Maurice G. Bellanger*
44. Electronic Ceramics: Properties, Configuration, and Applications, *edited by Lionel M. Levinson*
45. Computer Systems Engineering Management, *Robert S. Alford*
46. Systems Modeling and Computer Simulation, *edited by Naim A. Kheir*
47. Rigid-Flex Printed Wiring Design for Production Readiness, *Walter S. Rigling*
48. Analog Methods for Computer-Aided Circuit Analysis and Diagnosis, *edited by Takao Ozawa*
49. Transformer and Inductor Design Handbook: Second Edition, Revised and Expanded, *Colonel Wm. T. McLyman*

50. Power System Grounding and Transients: An Introduction, *A. P. Sakis Meliopoulos*
51. Signal Processing Handbook, *edited by C. H. Chen*
52. Electronic Product Design for Automated Manufacturing, *H. Richard Stillwell*
53. Dynamic Models and Discrete Event Simulation, *William Delaney and Erminia Vaccari*
54. FET Technology and Application: An Introduction, *Edwin S. Oxner*
55. Digital Speech Processing, Synthesis, and Recognition, *Sadaoki Furui*
56. VLSI RISC Architecture and Organization, *Stephen B. Furber*
57. Surface Mount and Related Technologies, *Gerald Ginsberg*
58. Uninterruptible Power Supplies: Power Conditioners for Critical Equipment, *David C. Griffith*
59. Polyphase Induction Motors: Analysis, Design, and Application, *Paul L. Cochran*
60. Battery Technology Handbook, *edited by H. A. Kiehne*
61. Network Modeling, Simulation, and Analysis, *edited by Ricardo F. Garzia and Mario R. Garzia*
62. Linear Circuits, Systems, and Signal Processing: Advanced Theory and Applications, *edited by Nobuo Nagai*
63. High-Voltage Engineering: Theory and Practice, *edited by M. Khalifa*
64. Large-Scale Systems Control and Decision Making, *edited by Hiroyuki Tamura and Tsuneo Yoshikawa*
65. Industrial Power Distribution and Illuminating Systems, *Kao Chen*
66. Distributed Computer Control for Industrial Automation, *Dobrivoje Popovic and Vijay P. Bhatkar*
67. Computer-Aided Analysis of Active Circuits, *Adrian Ioinovici*
68. Designing with Analog Switches, *Steve Moore*
69. Contamination Effects on Electronic Products, *Carl J. Tautscher*
70. Computer-Operated Systems Control, *Magdi S. Mahmoud*
71. Integrated Microwave Circuits, *edited by Yoshihiro Konishi*
72. Ceramic Materials for Electronics: Processing, Properties, and Applications, Second Edition, Revised and Expanded, *edited by Relva C. Buchanan*
73. Electromagnetic Compatibility: Principles and Applications, *David A. Weston*
74. Intelligent Robotic Systems, *edited by Spyros G. Tzafestas*
75. Switching Phenomena in High-Voltage Circuit Breakers, *edited by Kunio Nakanishi*
76. Advances in Speech Signal Processing, *edited by Sadaoki Furui and M. Mohan Sondhi*
77. Pattern Recognition and Image Preprocessing, *Sing-Tze Bow*
78. Energy-Efficient Electric Motors: Selection and Application, Second Edition, *John C. Andreas*
79. Stochastic Large-Scale Engineering Systems, *edited by Spyros G. Tzafestas and Keigo Watanabe*
80. Two-Dimensional Digital Filters, *Wu-Sheng Lu and Andreas Antoniou*
81. Computer-Aided Analysis and Design of Switch-Mode Power Supplies, *Yim-Shu Lee*
82. Placement and Routing of Electronic Modules, *edited by Michael Pecht*

83. Applied Control: Current Trends and Modern Methodologies, *edited by Spyros G. Tzafestas*
84. Algorithms for Computer-Aided Design of Multivariable Control Systems, *Stanoje Bingulac and Hugh F. VanLandingham*
85. Symmetrical Components for Power Systems Engineering, *J. Lewis Blackburn*
86. Advanced Digital Signal Processing: Theory and Applications, *Glenn Zelniker and Fred J. Taylor*
87. Neural Networks and Simulation Methods, *Jian-Kang Wu*
88. Power Distribution Engineering: Fundamentals and Applications, *James J. Burke*
89. Modern Digital Control Systems: Second Edition, *Raymond G. Jacquot*
90. Adaptive IIR Filtering in Signal Processing and Control, *Phillip A. Regalia*
91. Integrated Circuit Quality and Reliability: Second Edition, Revised and Expanded, *Eugene R. Hnatek*
92. Handbook of Electric Motors, *edited by Richard H. Engelmann and William H. Middendorf*
93. Power-Switching Converters, *Simon S. Ang*
94. Systems Modeling and Computer Simulation: Second Edition, *Naim A. Kheir*
95. EMI Filter Design, *Richard Lee Ozenbaugh*
96. Power Hybrid Circuit Design and Manufacture, *Haim Taraseiskey*
97. Robust Control System Design: Advanced State Space Techniques, *Chia-Chi Tsui*
98. Spatial Electric Load Forecasting, *H. Lee Willis*
99. Permanent Magnet Motor Technology: Design and Applications, *Jacek F. Gieras and Mitchell Wing*
100. High Voltage Circuit Breakers: Design and Applications, *Ruben D. Garzon*
101. Integrating Electrical Heating Elements in Appliance Design, *Thor Hegbom*
102. Magnetic Core Selection for Transformers and Inductors: A User's Guide to Practice and Specification, Second Edition, *Colonel Wm. T. McLyman*

Additional Volumes in Preparation

Statistical Methods in Control and Signal Processing, *edited by Tohru Katayama and Sueo Sugimoto*

Magnetic Core Selection for Transformers and Inductors

A User's Guide to Practice and Specification

Second Edition

Colonel Wm. T. McLyman

Jet Propulsion Laboratory
California Institute of Technology
Pasadena, California

CRC Press
Taylor & Francis Group
Boca Raton London New York

CRC Press is an imprint of the
Taylor & Francis Group, an **informa** business

CRC Press
Taylor & Francis Group
6000 Broken Sound Parkway NW, Suite 300
Boca Raton, FL 33487-2742

First issued in paperback 2019

ISBN-13: 978-0-8247-9841-3 (hbk)
ISBN-13: 978-0-367-40096-5 (pbk)

Library of Congress Cataloging-in-Publication Data

McLyman, Colonel William T.,
Magnetic core selection for transformers and inductors: a user's guide to practice and specification / Colonel Wm. T. McLyman. – 2nd ed.
p. cm. – (Electrical engineering and electronics : 102)
Includes index.
ISBN 0-8247-9841-4 (alk. paper)
1. Magnetic cores. 2. Electronic transformers–Design and construction. 3. Electric inductors–Design and construction.
I. Title. II. Series.
TK7872.M25M36 1997
621.31′4–dc21 97-1909
CIP

Visit the Taylor & Francis Web site at
http://www.taylorandfrancis.com

and the CRC Press Web site at
http://www.crcpress.com

FOREWORD

Magnetism first became apparent in the Ancient world as an attractive force that exists between two bodies such as lodestone or iron. Since that time, magnetism and later magnetic design, became viewed upon as an art or black magic. Colonel McLyman's first book entitled *Transformer and Inductor Design Handbook* (Marcel Dekker, Inc.) took a lot of the art and black magic out of magnetic design. That book took an easy to use step by step analytical approach that yielded accurate reliable magnetic component designs.

The present volume is an excellent supplement to the *Transformer and Inductor Design Handbook*. This new edition includes step by step design examples of transformers and inductors using powdered iron, molypermalloy, ferrites, cut cores, and so forth. It also has an up-to-date list of the manufacturers of all types of magnetic cores as well as coil forms, mounting hardware, magnet wire, and all the accessories necessary to manufacture magnetic components. This book also includes complete mechanical, electrical, and magnetic data on cut cores, strip wound cores, various powder cores, and ferrites supplied by all the core manufacturers. With this book, the magnetic designer literally has all the core manufacturers catalogs and data books at his finger tips. Many thanks, Colonel, for another valuable tool in the hands of the magnetic component designer.

Robert G. Noah
Magnetics, Division of Spang & Company
Butler, Pennsylvania

PREFACE

This book was written as a supplement to the *Transformer and Inductor Design Handbook: Second Edition, Revised and Expanded* (Volume 49). The idea was to bring the majority of the cores available in the industry in line with standard units of measurement to help the engineer pick the configuration best suited to a specific design.

The aforementioned handbook has new equations and procedures that simplify the design of magnetic components. The equations used in the handbook are in cgs units although magnetic core manufacturers supply data in mixed units and in no standard format. Most of the material in this book is in tabular form to assist the designer in making the trade-offs best suited for his particular application in a minimum amount of time. Approximately 20 core manufacturers are represented, with core types such as

1. Laminations
   (a) EI and EE
   (b) L and DU
   (c) UI
   (d) Three phase
2. C cores, 1, 2, 4, and 12 mil
   (a) EE cores, 4 and 12 mil
3. Tape toroidal core
   (a) Caseless, 1, 2, 4, and 12 mil
4. Ferrites
   (a) Toroids
   (b) EE, EI, and U
   (c) Pot cores
5. Powder cores

When the designer has established the area product A_p or the core geometry coefficient K_g, he can then look in this book for that particular core to obtain the following data:

1. Strip width (cm)
2. Buildup (cm)
3. Window width (cm)
4. Window length (cm)
5. Magnetic path length (cm)
6. Finished transformer height (cm)
7. Finished transformer width (cm)
8. Finished transformer length (cm)
9. Iron weight (gram)
10. Copper weight (gram)
11. Mean length turn (cm)
12. Iron area, A_c (cm^2)
13. Window area, W_a (cm^2)
14. Area product, A_p (cm^4)
15. Core geometry, K_g (cm^5)
16. Transformer surface area $(cm)^2$

Over 12,000 cores have been tabulated for the engineer. The engineer will find that some cores will have the same area product (A_p) or core geometry (K_g) coefficient but will have different size configurations. With these data, the engineer can tell at a glance if that particular design or core configuration will work, or what changes will have to be made.

Possibly for the first time, many manufacturers' core-loss data curves have been organized with the same units for all core losses. The data were digitized right from the manufacturers' data sheets. Then the data were modified to put it in metric units—gauss to tesla and watts per pound to watts per kilogram. This data was then put into the computer to develop a new first-order approximation in the form

$$w = kf^{m}B^{n},$$

where w is the calculated core loss density in watts/kilogram, f is the frequency in hertz, B is the flux density in tesla, and k, m, and n are coefficients derived using a three-dimensional least-square-fit law from the digitized data. These curves include silicon, nickel–iron, ferrites, powdered iron, and metglas. This book can now be used as a new tool to simplify and standardize the process of transformer design.

This new edition is an update using magnetic manufacturers latest catalogs. We also have broadened our database by adding more magnetic manufacturers. We have increased core data to include the latest magnetic materials such as Kool Mμ, Metglas, and Ferrite materials. We have also updated the laminations, C cores, tape toroidal cores, and ferrites.

This handbook provides the latest information on magnetic manufacturers. The data will assist the designer in making the trade-offs best suited to a particular application in a minimum amount of time. This new edition is a valuable tool for the engineer in magnetics design.

Colonel Wm. T. McLyman

ACKNOWLEDGMENTS

In gathering the material for this second edition and preparing it for publication, I have been fortunate in having the assistance and cooperation of a number of people and organizations and wish to express my gratitude to all of them. Too many have helped to mention them all, but I would like to list the following.

Jet Propulsion Laboratory
  Michael Hasbach
  Robert Detwiler
  Dr. John Klein
  Dr. Gene Wester
Magnetics
  Robert Noah
  Harry Savisky (Ret.)
Micrometals
  Jim Cox
  Walt Lewis
  Dale Nicol
Wilorco Inc.
  Kurt Willner
Sherwood Associates
  Ed Sherwood
  Erik Sherwood
and not forgetting . . .
  Charles (CT) Kleiner
  Robert Yahiro
  Laura Young
  Hung Ta

Thanks again!

CONTENTS

MANUFACTURERS

MAGNETIC MATERIALS

NOTES

Allegheny Ludium Steel Corp.
Dept. G, Oliver Bldg.
Pittsburgh, PA 15222
Phone (412) 562-4301

AlliedSignal Inc.
6 Eastmans Road
Parsippany, NJ 07054
Phone (201) 581-7653, Fax (201) 581-7717

Alpha-Core Inc.
915 Pembroke Street
Bridgeport, CT 06608
Phone (203) 335-6805, Fax (203) 384-8120

Armco Steel Corp.
P.O. Box 832
Butler, PA 16003-0832
Phone (412) 284-2000, Fax (412) 284-3064

Arnold Engineering Co.
P.O. Box G
Marengo, IL 60152
Phone (815) 568-2000, Fax (815) 568-2238

Ceramic Magnetics, Inc.
16 Law Drive
Fairfield, NJ 07004
Phone (201) 227-4222, Fax (201) 227-6735

Fair-Rite Products Corp.
1 Commercial Row
P.O. Box J
Wallkill, NY 12589
Phone (914) 895-2055, Fax (914) 895-2629

Ferrite International Co.
39105 North Magnetics Boulevard
Wadsworth, IL 60083
Phone (708) 249-4900, Fax (708) 249-4988

Ferronics, Inc.
45 O'Connor Road
Fairport, NY 14450
Phone (716) 388-1020, Fax (716) 388-0036

NOTES

Magnetics
900 East Butler Road
P.O. Box 391
Butler, PA 16003
Phone (412) 282-8282, Fax (412) 282-6955

Magnetic Metals Corp.
Hayes Ave. at 21 Street
Camden, NJ 08101
Phone (609) 964-7842, Fax (609) 963-8569

Micrometals
1190 N. Hawk Circle
Anaheim, CA 92807
Phone (714) 630-7420, Fax (714) 630-4562

MMG North America
126 Pennsylvania Avenue
Paterson, NJ 07503
Phone (201) 345-8900, Fax (201) 345-1172

National Arnold Corp.
17030 Muskrat Avenue
Adelanto, CA 92301
Phone (619) 246-3020, Fax (619) 246-3870

National Lamination Company
555 Santa Rosa Drive
Des Plaines, IL 60018
Phone (847) 298-7676, Fax (847) 635-8624

Philips Components
1033 Kings Highway
Saugerties, NY 12477
Phone (914) 246-2811, Fax (914) 246-0486

Siemens Corp.
186 Wood Avenue South
Iselin, NJ 08830
Phone (908) 494-1000, Fax (908) 603-5994

Steward, Inc.
East 36th Street
P.O. Box 510
Chattanooga, TN 37401-0510
Phone (423) 867-4100, Fax (423) 867-4102

NOTES

TDK Corp.
MH and W International Corp.
P.O. Box 251 Wyckoff
Saddle Brook, NJ 07481
Phone (201) 891-8800, Fax (201) 423-3716

Temple Steel Co.
5990 West Touhy Avenue
Niles, IL 60648
Phone (312) 282-9400, Fax (708) 647-0731

Thomas and Skinner Inc.
P.O. Box 150-B
1120 East 23rd Street
Indianapolis, IN 46206
Phone (317) 923-2501, Fax (317) 923-5919

Thomson LCC
2211-H Distribution Center
Charlotte, NC 28269
Phone (704) 597-0766, Fax (704) 597-0553

Tokin America, Inc.
155 Nicholson Lane
San Jose, CA 95134
Phone (408) 432-8020, Fax (408) 434-0375

TSC-Pyroferric International, Inc.
200 Madison Street
Toledo, IL 62468
Phone (217) 849-3300, Fax (217) 849-2544

MAGNET WIRE

California Fine Wire Co.
P.O. Box 446
Grover Beach, CA 93483
Phone (805) 489-5144, Fax (805) 489-5352

Essex Magnet Wire
1510 Wall Street
Fort Wayne, IN 46802
Phone (219) 461-4000

Phelps Dodge Magnet Wire
2131 South Coliseum Boulevard
Fort Wayne, IN 46803
Phone (219) 421-5400, Fax (219) 426-4875

MWS Wire Industries
31200 Cedar Valley Drive
Westlake Village, CA 91362
Phone (818) 991-8553, Fax (818) 706-0911

Cooner Wire Company
9265 Owens Mouth Avenue
Chatsworth, CA 91311
Phone (818) 882-8311, Fax (818) 709-8281

New England Wire Corp.
365 Main Street
Lisbon, NH 03585
Phone (603) 838-6624, Fax (603) 838-6160

BOBBIN/PLATFORMS

Cosmo Plastics Co.
30201 Averoea Road
Cleveland, OH 44139-2745
Phone (216) 498-7500

Dorco Electronics
15533 Vermont Avenue
Paramount, CA 90723
Phone (310) 633-4786, Fax (310) 633-0651

Loadestone Pacific
4769 Wesley Drive
Anaheim, CA 92807
Phone (714) 970-0900, Fax (714) 970-0800

Ny-Glass Plastics Inc.
1255 Railroad Street
Corona, CA 91720
Phone (909) 737-9110, Fax (909) 737-9815

HARDWARE

Bahrs Die and Stamping Co.
4375 Rossplain Road
Cincinnati, OH 45236
Phone (513) 793-8100, Fax (513) 793-0620

NOTES

Delbert Blinn Co.
P.O. Box 2007
Pomona, CA 91769
Phone (909) 623-1257, Fax (909) 622-6218

Ram Sales and Distributing
14823 Aetna Street
Van Nuys, CA 91411
Phone (818) 997-8057, Fax (818) 997-7084

Team Sales, Inc.
12564 Palos Tierra Road
Valley Center, CA 92082
Phone (619) 749-2033, Fax (619) 749-6062

Hallmark Metals
600 West Foothill Boulevard
Glendora, CA 91741
Phone (818) 335-1263, Fax (818) 963-1912

NOTES

SYMBOLS

α	Regulation, %
A_c	Effective iron area, cm^2
A_p	Area product, $W_a A_c$, cm^4
AT	Ampere-turns
A_t	Surface area, cm^2
A_w	Wire area, cm^2
$A_{w(B)}$	Wire area bare, cm^2
AWG	American Wire Gauge
B_{ac}	Alternating current flux density, T
B_{dc}	Direct current flux density, T
B_m	Operating flux density, T
B_r	Residual flux density, T
B_s	Saturation flux density, T
cir-mil	Area of a circle whose diameter = 0.001 in.
E	Voltage
Eng	Energy, W s
η	Efficiency
f	Frequency, Hz
F	Fringing flux factor
H	Magnetizing force, amp-turns/cm
H	Magnetizing force, Oer
H_c	Magnetizing force to saturate
I	Current, A
I_{in}	Input current, A
I_m	Excitation current, A

I_o	Load current, A
I_p	Primary current, A
I_s	Secondary current, A
J	Current density, A/cm^2
K	Constant
K_e	Electrical coefficient
K_f	Waveform coefficient
K_g	Core geometry, cm^5
K_i	Gap loss coefficient
K_j	Current density coefficient
K_s	Surface area coefficient
K_u	Window utilization factor
K_v	Volume coefficient
K_w	Weight coefficient
L	Inductance, H
l_g	Gap length, cm
l_m	Magnetic path length, cm
l	Linear dimension, cm
MLT	Mean length turn, cm
MPL	Magnetic path length, cm
μ_Δ	Effective permeability
μ_m	Core material permeability
μ_o	Absolute permeability
μ_r	Relative permeability
n	Turns ratio
N	Turns

N_p	Primary turns
N_s	Secondary turns
P	Power, W
ϕ	Flux, Wb
P_{cu}	Copper loss, W
P_{fe}	Core loss, W
P_{in}	Input power, W
P_g	Gap loss, W
P_o	Output power, W
ψ	Heat flux density, W/cm^2
P_p	Primary loss, W
P_s	Secondary loss, W
P_{Σ}	Total loss (core and copper), W
P_t	Apparent power, W
R	Resistance, Ω
R_{cu}	Copper resistance, Ω
R_o	Load resistance, Ω
R_p	Primary resistance, Ω
R_s	Secondary resistance, Ω
R_t	Total resistance, Ω
S_1	Conductor area/wire area
S_2	Wound area/usable window area
S_3	Usable window area/window area
S_4	Usable window area/usable window area + insulation area
SF	Stacking factor
T	Flux density, T

V A	volt amps, watts
V_d	Diode voltage drop
V_{in}	Input voltage, V
V_o	Load voltage, V
V_p	Primary voltage, V
V_s	Secondary voltage, V
Vol	Volume, cm^3
W	watts
W_a	Window area, cm^2
W s	watt seconds
W_t	weight, grams
W_{tcu}	Copper weight, grams
W_{tfe}	Core weight, grams
ζ	Zeta resistance correction factor for temperature

USEFUL TRANSFORMER AND INDUCTOR DESIGN EQUATIONS

1. $A_t = K_s A_p^{0.5}$ .... [cm²]

2. $W_t = K_w A_p^{0.75}$ .... [grams]

3. $Vol = K_v A_p^{0.75}$ .... [cm³]

4. $J = K_j A_p^{0.125}$ .... [amps/cm²]

5. $A_p = W_a A_c$ .... [cm⁴]

6. $A_p = \dfrac{P_t 10^4}{K_f K_u f B_m J}$ .... [cm⁴]

7. $P_t = P_{in} + P_o$ .... [watts]

8. $K_f = 4.44$ .... [sine wave]

9. $K_f = 4.00$ .... [square wave]

10. $K_u = S_1 S_2 S_3 S_4$ .... [ratio]

11. $A_p = \left(\dfrac{P_t \text{ x } 10^4}{K_f K_u K_j f B_m}\right)^{(Table 1, x)}$ .... [cm⁴]

12. $N_p = \dfrac{E_p \text{ x } 10^4}{K_f B_m f A_c}$ .... [turns]

13. $I_p = \dfrac{P_o}{E_p \eta}$ .... [amps]

14. $A_{w(B)} = \dfrac{I_p}{J}$ .... [cm²]

15. $R_p = MLT \text{ x } N \text{ x } \left(\dfrac{\mu\Omega}{cm}\right) \text{ x } 10^{-6}$ .... [ohms]

16. $P_{cu} = I_p^2 R_p$ .... [watts]

17. $$N_s = \frac{N_p E_s}{E_p}\left(1 + \frac{\alpha}{100}\right)$$ .... [turns]

18. $$P_\Sigma = P_{cu} + P_{fe}$$ .... [watts]

19. $$\psi = \frac{P_\Sigma}{A_t}$$ .... [watts/cm^2]

20. $$Energy = \frac{LI^2}{2}$$ .... [w-s]

21. $$A_p = \frac{2(Eng) \text{ x } 10^4}{B_m K_u J}$$ .... [cm^4]

22. $$A_p = \left(\frac{2(Eng) \text{ x } 10^4}{B_m K_u J}\right)^{(Table1,x)}$$ .... [cm^4]

23. $$B_{max} = B_{dc} + B_{ac}$$ .... [tesla]

24. $$B_{dc} = \frac{0.4\pi N I_{dc} \text{ x } 10^{-4}}{l_g}$$ .... [tesla]

25. $$B_{ac} = \frac{0.4\pi N \frac{\Delta I}{2} \text{ x } 10^{-4}}{l_g}$$ .... [tesla]

26. $$L = \frac{0.4\pi N^2 A_c \text{ x } 10^{-8}}{l_g}$$ .... [henrys]

27. $$F = \left(1 + \frac{l_g}{\sqrt{A_c}} \ln \frac{2G}{l_g}\right)$$, (fringing flux) .... [factor]

28. $$L' = \frac{0.4\pi N^2 A_c F \text{ x } 10^{-8}}{l_g}$$ .... [henrys]

29. $$A_{w(B)} = \frac{I_{dc} + \frac{\Delta I}{2}}{J}$$ .... [cm^2]

30. $W_{a(eff)} = W_a S_3$ .... [cm²]

31. $N = \dfrac{W_{a(eff)} S_2}{A_w}$ .... [turns]

32. $N = \sqrt{\dfrac{l_g L}{0.4\pi A_c F \text{ x } 10^{-8}}}$ .... [turns]

33. $H = \dfrac{0.4\pi NI}{l_m}$ .... [oersteds]

34. $H = \dfrac{NI}{l_m}$ .... [A-T/cm]

35. $NI = 0.84 H l_m$ .... [A-T]

36. $\mu_\Delta = \dfrac{B_m l_m \text{ x } 10^4}{0.4\pi W_a J K_u}$ .... [permeability]

37. $B_m = \dfrac{0.4\pi N\left(I_{dc} + \dfrac{\Delta I}{2}\right) \text{x } 10^{-4}}{l_g}$ .... [tesla]

38. $P_\Sigma = \dfrac{P_o}{\eta} - P_o$ .... [watts]

39. $\alpha = \dfrac{P_{cu}}{P_o + P_{cu}} \text{ x } 100$ .... [regulation]

40. $P_{cu} = \dfrac{P_o \alpha}{(100 - \alpha)}$ .... [watts]

41. $\eta = \dfrac{P_o}{P_o + P_\Sigma}$ .... [efficiency]

42. $VA = K_g K_e \alpha$ .... [watts]

43. $K_g = \dfrac{W_a A_c^2 K_u}{\text{MLT}}$ .... [cm⁵]

44. $K_g = \dfrac{A_p A_c K_u}{\text{MLT}}$ .... [cm⁵]

45. $K_e = 0.145 K_f^{\,2} f^2 B_m^2 \times 10^{-4}$ .... [transformers]

46. $K_g = \dfrac{P_t}{2K_e \alpha}$ .... [cm⁵]

47. $(Energy)^2 = K_g K_e \alpha$ .... [(w-s)²]

48. $K_e = 0.145 P_o B_{dc}^{\,2} \times 10^{-4}$ .... [inductors]

49. $K_g = \dfrac{(Energy)^2}{K_e a}$ .... [cm⁵]

50. $P_o = VA$ .... [watts]

51. $P_g = K_i D l_g f B_m^2$ .... [watts]

52. $A_p = \dfrac{3}{2} W_a A_c$,(3 phase transformers) .... [cm⁴]

53. $P_{fe} = \dfrac{milliwatts}{grams} \times W_{tfe} \times 10^{-3}$ .... [watts]

54. $L = \dfrac{0.4\pi N^2 A_c \times 10^{-8}}{l_g + \dfrac{l_m}{\mu_r}}$ .... [henrys]

55. $J = \dfrac{P_t \times 10^4}{K_f K_u f B_m A_p}$ .... [amps/cm²]

56. $L = \dfrac{0.4\pi N^2 A_c \mu_\Delta 10^{-8}}{l_m}$ .... [henrys]

57. $N = 1000\sqrt{\dfrac{L}{L_{1000}}}$ .... [turns]

58. $L_n = L_{1000} N^2 \times ^{10^{-6}}$ .... [millihenrys]

59. $$P_t = P_o\left(\frac{1}{\eta}+1\right) \text{ [watts]}$$

60. $$P_t = P_o\left(\frac{1}{\eta}+\sqrt{2}\right) \text{ [watts]}$$

61. $$P_t = P_o\left(\frac{\sqrt{2}}{\eta}+\sqrt{2}\right) \text{ [watts]}$$

62. $$I_{rms} = \sqrt{\left(I_{pk}^2 - I_{pk}\Delta I + \frac{\Delta I^2}{3}\right)\left(\frac{t_{on}}{T}\right)} \text{ [amps]}$$

63. $$I_{rms} = \sqrt{\left(I_{pk}^2 - I_{pk}\Delta I + \frac{\Delta I^2}{3}\right)\left(\frac{t_{on}+t_{off}}{T}\right)} \text{ [amps]}$$

Buck Converter

64. $$D_{(\min)} = \frac{V_o}{\left(V_{in(\max)}\eta\right)} \text{ [duty ratio]}$$

65. $$D_{(\max)} = \frac{V_o}{\left(V_{in(\min)}\eta\right)} \text{ [duty ratio]}$$

66. $$L_{(\min)} = \frac{R_{o(\max)}T\left(1-D_{(\min)}\right)}{2} \text{ [henrys]}$$

67. $$\Delta I = \frac{TV_{in(\max)}D_{(\min)}\left(1-D_{(\min)}\right)}{L} \text{ [amps]}$$

68. $$I_{pk} = I_o + \frac{V_{in}TD_{(\min)}\left(1-D_{(\min)}\right)}{2L} \text{ [amps]}$$

Boost Converter

69. $$D_{(\min)} = \left(1-\frac{\eta V_{in(\max)}}{V_o}\right) \text{ [duty ratio]}$$

70. $$D_{(\max)} = \left(1 - \frac{\eta V_{in(\min)}}{V_o}\right)$$ .... [duty ratio]

71. $$L_{(\min)} = \frac{R_{o(\max)} T D_{(\min)} \left(1 - D_{(\min)}\right)^2}{2}$$ .... [henrys]

72. $$\Delta I = \frac{V_{in(\min)} D_{(\max)} T}{L}$$ .... [amps]

73. $$I_{pk} = \frac{I_o}{(1 - D_{(\max)})} + \frac{V_{in} T D_{(\max)}}{2L}$$ .... [amps]

Buck Boost Converter

74. $$D_{(\min)} = \frac{V_o}{\left(\eta V_{in(\max)} + V_o\right)}$$ .... [duty ratio]

75. $$D_{(\max)} = \frac{V_o}{\left(\eta V_{in(\min)} + V_o\right)}$$ .... [duty ratio]

76. $$L_{(\min)} = \frac{R_{o(\max)} T \left(1 - D_{(\min)}\right)^2}{2}$$ .... [henrys]

77. $$\Delta I = \frac{V_{in(\min)} D_{(\max)} T}{L}$$ .... [amps]

78. $$I_{pk} = \frac{I_o}{\left(1 - D_{(\max)}\right)} + \frac{V_{in(\min)} D_{(\max)} T}{2L}$$ .... [amps]

79. $$\mu_\Delta = \frac{\mu_m}{1 + \frac{l_g}{l_m}\mu_m}$$ .... [permeability]

80. $$l_g = \left(\frac{\mu_m}{\mu_\Delta} - 1\right)\frac{l_m}{\mu_m}$$ .... [cm]

81. $$\mu_m = \frac{\mu_\Delta l_m}{\mu_\Delta l_g - l_m}$$ .... [permeability]

DECIMAL AND MILLIMETER EQUIVALENTS

Fraction	DECIMALS	MILLIMETERS
1/64	0.015625	0.397
1/32	.03125	0.794
3/64	.046875	1.191
1/16	.0625	1.588
5/64	.078125	1.984
3/32	.09375	2.381
7/64	.109375	2.778
1/8	.1250	3.175
9/64	.140625	3.572
5/32	.15625	3.969
11/64	.171875	4.366
3/16	.1875	4.763
13/64	.203125	5.159
7/32	.21875	5.556
15/64	.234375	5.953
1/4	.2500	6.350
17/64	.265625	6.747
9/32	.28125	7.144
19/64	.296875	7.541
5/16	.3125	7.938
21/64	.328125	8.334
11/32	.34375	8.731
23/64	.359375	9.128
3/8	.3750	9.525
25/64	.390625	9.922
13/32	.40625	10.319
27/64	.421875	10.716
7/16	.4375	11.113
29/64	.453125	11.509
15/32	.46875	11.906
31/64	.484375	12.303
1/2	.5000	12.700

1 mm = .03937″

Fraction	DECIMALS	MILLIMETERS
33/64	0.515625	13.097
17/32	.53125	13.494
35/64	.546875	13.891
9/16	.5625	14.288
37/64	.578125	14.684
19/32	.59375	15.081
39/64	.609375	15.478
5/8	.6250	15.875
41/64	.640625	16.272
21/32	.65625	16.669
43/64	.671875	17.066
11/16	.6875	17.463
45/64	.703125	17.859
23/32	.71875	18.256
47/64	.734375	18.653
3/4	.7500	19.050
49/64	.765625	19.447
25/32	.78125	19.844
51/64	.796875	20.241
13/16	.8125	20.638
53/64	.828125	21.034
27/32	.84375	21.431
55/64	.859375	21.828
7/8	.8750	22.225
57/64	.890625	22.622
29/32	.90625	23.019
59/64	.921875	23.416
15/16	.9375	23.813
61/64	.953125	24.209
31/32	.96875	24.606
63/64	.984375	25.003
1	1.000	25.400

.001″ = .0254 mm

MM	INCHES	MM	INCHES
.1	.0039	46	1.8110
.2	.0079	47	1.8504
.3	.0118	48	1.8898
.4	.0157	49	1.9291
.5	.0197	50	1.9685
.6	.0236	51	2.0079
.7	.0276	52	2.0472
.8	.0315	53	2.0866
.9	.0354	54	2.1260
1	.0394	55	2.1654
2	.0787	56	2.2047
3	.1181	57	2.2441
4	.1575	58	2.2835
5	.1969	59	2.3228
6	.2362	60	2.3622
7	.2756	61	2.4016
8	.3150	62	2.4409
9	.3543	63	2.4803
10	.3937	64	2.5197
11	.4331	65	2.5591
12	.4724	66	2.5984
13	.5118	67	2.6378
14	.5512	68	2.6772
15	.5906	69	2.7165
16	.6299	70	2.7559
17	.6693	71	2.7953
18	.7087	72	2.8346
19	.7480	73	2.8740
20	.7874	74	2.9134
21	.8268	75	2.9528
22	.8661	76	2.9921
23	.9055	77	3.0315
24	.9449	78	3.0709
25	.9843	79	3.1102
26	1.0236	80	3.1496
27	1.0630	81	3.1890
28	1.1024	82	3.2283
29	1.1417	83	3.2677
30	1.1811	84	3.3071
31	1.2205	85	3.3465
32	1.2598	86	3.3858
33	1.2992	87	3.4252
34	1.3386	88	3.4646
35	1.3780	89	3.5039
36	1.4173	90	3.5433
37	1.4567	91	3.5827
38	1.4961	92	3.6220
39	1.5354	93	3.6614
40	1.5748	94	3.7008
41	1.6142	95	3.7402
42	1.6535	96	3.7795
43	1.6929	97	3.8189
44	1.7323	98	3.8583
45	1.7717	99	3.8976
		100	3.9370

1

MAGNETIC-CORE POWER-HANDLING ABILITY

INTRODUCTION

The conversion process in power electronics requires the use of transformers, components which frequently are the heaviest and bulkiest item in the conversion circuits. They also have a significant effect on the overall performance and efficiency of the system. Accordingly, the design of such transformers has an important influence on overall system weight, power conversion efficiency, and cost. Because of the interdependence and interaction of parameters, judicious trade-offs are necessary to achieve design optimization.

POWER-HANDLING ABILITY

Manufacturers have for years assigned numeric codes to their cores; these codes represent the power-handling ability. This method assigns to each core a number which is the product of its window area (W_a) and core cross-section areas (A_c) and is called "area product," A_P.

These numbers are used by core suppliers to summarize dimensional and electrical properties in their catalogs. They are available for laminations, C-cores, pot cores, powder cores, ferrite toroids, and toroidal tape-wound cores.

The regulation and power-handling ability of a core is related to the core geometry K_g; every core has its own inherent K_g. The core geometry K_g is relatively new and magnetic-core manufacturers do not list this coefficient.

Because of their significance, the area product A_p and core geometry K_g are treated extensively. A great deal of other information is also presented for the convenience of the designer. Much of the material is in tabular form to assist the designer in making the trade-offs best suited for his particular application in a minimum amount of time.

These relationships can now be used as new tools to simplify and standardize the process of transformer design. They make it possible to design transformers of lighter weight and smaller volume or to optimize efficiency without going through a cut-and-try design procedure. Although developed specifically for aerospace applications, the information has wider utility and can be used for the design of nonaerospace transformers as well.

TRANSFORMER DESIGN

The Design Problem Generally

The designer is faced with a set of constraints which must be observed in the design of any transformer. One of these is the output power, P_o (operating voltage multiplied by maximum current demand), which the secondary winding must be capable of delivering to the load within specified regulation limits. Another relates to minimum efficiency of operation which is dependent on the maximum power loss which can be allowed in the transformer. Still another defines the maximum permissible temperature rise for the transformer when used in a specified temperature environment.

Other constraints relate to the volume occupied by the transformer and particularly in aerospace applications. Weight, as weight minimization is an important goal in the design of space flight electronics. Finally, cost-effectiveness is always an important consideration.

Output Power Versus Input Power Versus Apparent Power Capability

Output power (P_o) is of greatest interest to the user. To the transformer designer it is the apparent power (P_t) which is associated with the geometry of the transformer that is of greater importance.

Assume, for the sake of simplicity, that the core of an isolation transformer has but two windings in the window area (W_a): a primary and a secondary. Also, assume that the window area (W_a) is divided up in proportion to the power-handling capability of the windings using equal current density: the primary winding handles P_{in} and the secondary handles P_o to the load. Because the power transformer has to be designed to accommodate the primary P_{in} and P_o, then:

$$P_t = P_{in} + P_o \quad \text{(watts)}$$

$$P_{in} = \frac{P_o}{\eta}$$

$$P_t = \frac{P_o}{\eta} + P_o \quad \text{(watts)}$$

$$P_t = P_o\left(\frac{1}{\eta} + 1\right)$$

The designer must be concerned with the apparent power-handling capability, P_t, of the transformer core and windings. P_t may vary by a factor ranging from 2 to 2.828 times the input power, P_{in}, depending on the type of circuit in which the transformer is used. If the current in the rectifier transformer becomes interrupted, its effective root mean squared (rms) value changes. Transformer size, thus, is not only determined by the load demand but also, by application,because of the different copper losses incurred due to current waveform.

For example, for a load of 1 watt, compare the power-handling capabilities required for each winding (neglecting transformer and diode losses so that $P_{in} = P_o$) for the full-wave bridge circuit of Fig. 1, the full-wave center-tapped secondary circuit of Fig. 2, and the push–pull center-

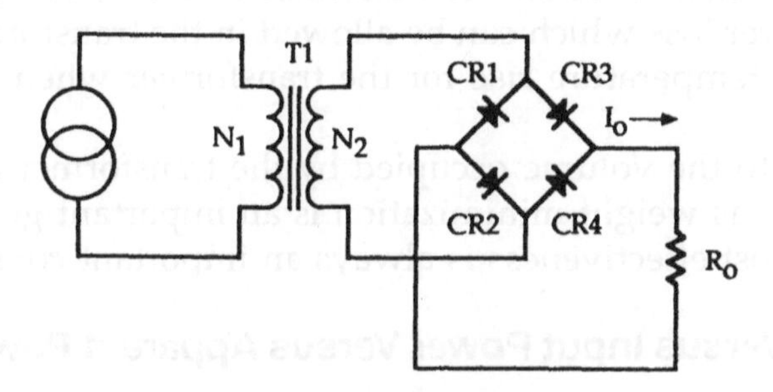

Figure 1 Full-wave bridge circuit.

tapped full-wave circuit in Fig. 3, where all windings have the same number of turns (N). The total apparent power P_t for the circuit shown in Fig. 1 is 2 watts. This is shown in the following equation:

$$P_t = 2P_{in} \quad \text{(watts)}$$

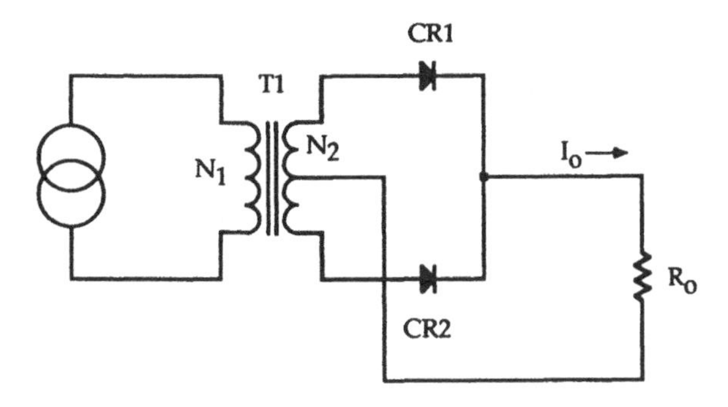

Figure 2 Full-wave, center-tapped circuit.

The total power P_t for the circuit shown in Fig. 2 increased 20.7% due to the distorted waveform of the interrupted current flowing in the secondary winding. This is shown in the following equation:

$$P_t = P_o\left(\frac{1}{\eta} + \sqrt{2}\right) \quad \text{(watts)}$$

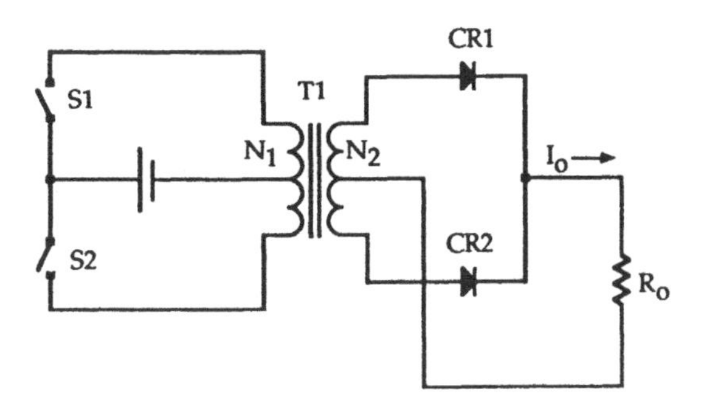

Figure 3 Push–pull, full-wave, center-tapped circuit.

The total power P_t for the circuit shown in Fig. 3, which is typical of a dc to dc converter, increases to 2.828 times P_{in} because of the interrupted current flowing in both the primary and secondary windings, as

$$P_t = P_o\left(\frac{\sqrt{2}}{\eta} + \sqrt{2}\right) \quad \text{(watts)}$$

Relationship of K_g to Power Transformer Regulation Capability

Transformers

Although most transformers are designed for a given temperature rise, they can also be designed for a given regulation. The regulation and power-handling ability of a core is related to two constants:

$$^*\alpha = \frac{P_t}{2K_gK_e} \quad (\%)$$

$$\alpha = \text{Regulation, } (\%)$$

The constant K_g is determined by the core geometry which may be related by the following equation:

$$^*K_g = \frac{W_aA_c^2K_u}{\text{MLT}} \quad (\text{cm}^5)$$

The constant K_e is determined by the magnetic and electric operating conditions which may be related by the following equation:

$$^*K_e = 0.145K_f^2f^2B_m^2\text{x}10^{-4}$$

where

K_f = waveform coefficient
= 4.0 square wave
= 4.44 sine wave

From the above, it can be seen that factors such as flux density, frequency of operation, and waveform coefficient all have an influence on the transformer size.

---

*The derivation for these equations is set forth in detail by the author in the book *Transformer and Inductor Design Handbook,* Marcel Dekker Inc., New York.

Relationship of A_p to Transformer Power-Handling Capability

The customary equation for the power handling capability is:

$$A_p = \frac{P_t(\text{CIR - MIL / AMP})\text{x}10^4}{K_f B_m f K_u} \quad (\text{cm}^4)$$

Various transformer designers have used different approaches in arriving at suitable designs. For example, in many cases, a rule of thumb is used for dealing with current density. Typically, an assumption is made that a good working level is 1000 circular mils per ampere (cir-millamp). This will work in many instances but the wire size needed to meet this requirement may produce a heavier and bulkier transformer than desired or required. The information presented herein makes it possible to avoid the use of this and other rules of thumb and to develop a more economical design with great accuracy.

According to the newly developed approach, the power-handling capability of a core is related to its area product by an equation which may be stated as:

$$A_p = \left(\frac{P_t \text{ x } 10^4}{K_f B_m f K_u K_j}\right)^{(\text{Table 1,x})} \quad (\text{cm}^4)$$

where

K_f = waveform coefficient
= 4.0 square wave
= 4.44 sine wave

From the above, it can be seen that factors such as flux density, frequency of operation, window utilization factor K_u, which defines the maximum space which may be occupied by the copper in the window, and the constant K_j, which is related to temperature rise, all have an influence on the transformer area product. The constant K_j is a new parameter that gives the designer control of the copper loss.

The Area Product A_p and Its Relationships

The A_p of a core is the product of the available window area W_a of the core in square centimeters (cm^2) multiplied by the effective cross-sectional area A_c in square centimeters (cm^2) which may be stated as:

$$A_p = W_a A_c \quad (\text{cm}^4)$$

---

*The derivation for this equation is set forth in detail by the author in the book *Transformer and Inductor Design Handbook*, Marcel Dekker Inc., New York.

Figures 4–8 show in outline form five transformer core types that are typical of those shown in the catalogs of suppliers.

There is a unique relationship between the area product A_p characteristic number for transformer cores and several other important parameters which must be considered in transformer design (see Table 1).

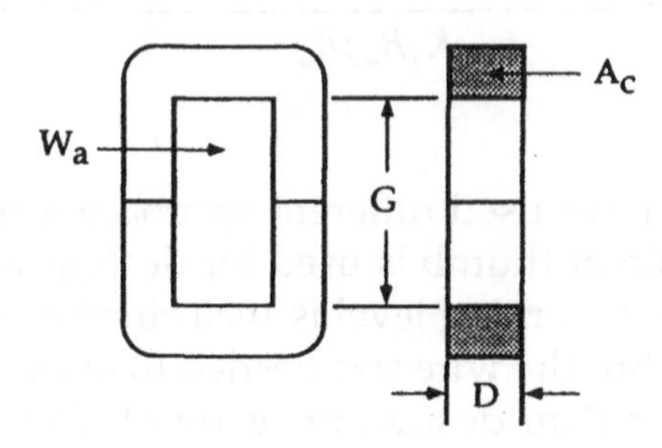

Figure 4 C-core.

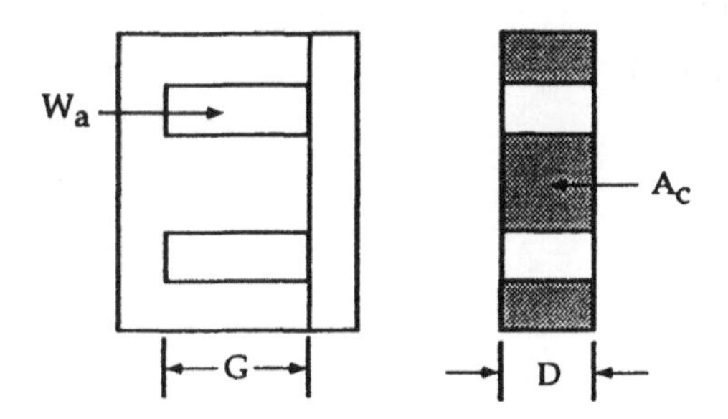

Figure 5 EI lamination.

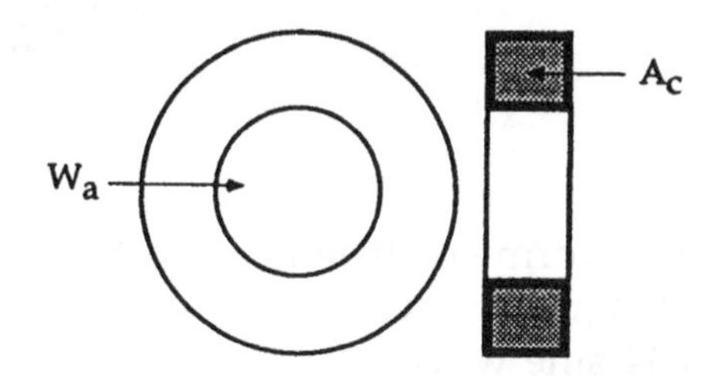

Figure 6 Tape-wound toroidal core.

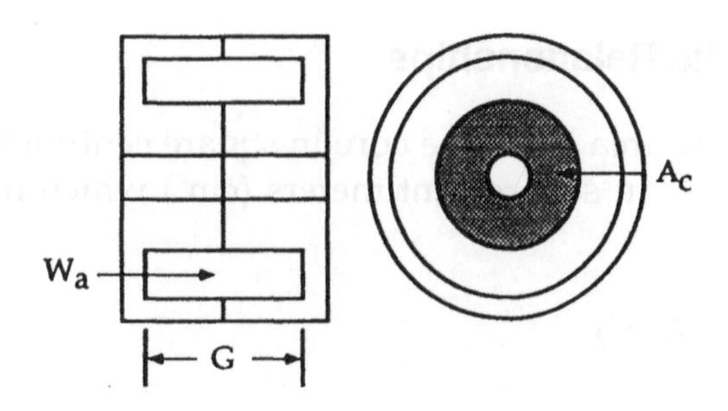

Figure 7 Pot core.

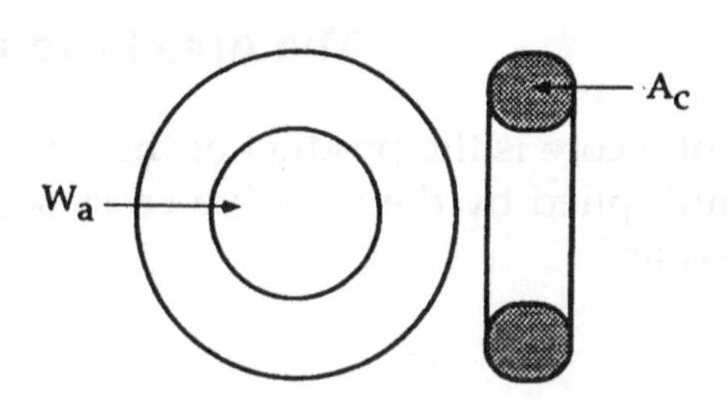

Figure 8 Powder core and ferrites.

Table 1 Core Configuration Constants

Core	Losses	K_j @ (25°C)	K_j @ (50°C)	(x)	K_s	K_w	K_v
Pot Core	$P_{cu} = P_{fe}$	433	632	1.20	33.8	48	14.5
Powder Core	$P_{cu} >> P_{fe}$	403	590	1.14	32.5	58.8	13.1
Laminations	$P_{cu} = P_{fe}$	366	534	1.14	41.3	68.2	19.7
C Core	$P_{cu} = P_{fe}$	323	468	1.16	39.2	66.6	17.9
Single Coil	$P_{cu} >> P_{fe}$	395	569	1.16	44.5	76.6	25.6
Tape Toroid	$P_{cu} = P_{fe}$	250	365	1.15	50.9	82.3	25

$$A_p = \left(\frac{P_t \times 10^4}{K_f B_m f K_u K_j} \right)^{(\text{Table 1},x)} \quad (\text{cm}^4)$$

Transformer Volume

The volume of a transformer can be related to the area product A_p of a transformer. Volume varies in accordance with the cube of any linear dimension where the area product A_p varies as the fourth power.

The volume/area product relationship is:

$$\text{Vol} = K_v A_p^{0.75} \quad (\text{cm}^3)$$

in which K_v is a constant related to the core configuration. These values are given in Table 2.

Table 2 Constant K_v

Core Type	K_v
Pot Core	14.5
Powder Core	13.1
Laminations	19.7
C Cores	17.9
Single Coil	25.6
Tape Toroids	25.0

Transformer Weight

The total weight W_t of a transformer can be related to the area product A_p. The weight W_t varies in accordance with the cube of any linear dimension l^3, whereas the area product A_p varies as the fourth power.

The weight/area product relationship

$$W_t = K_w A_p^{0.75} \quad \text{(grams)}$$

in which K_w is a constant related to core configuration is shown in Table 3.

Table 3 Transformer Weight Constant K_w

Core Type	K_w
Pot Core	48.0
Powder Core	58.8
Laminations	68.2
C Cores	66.6
Single Coil	76.6
Tape Toroids	82.3

Transformer Surface Area

The surface area A_t of a transformer can be related to the area product A_p of a transformer. The surface area varies in accordance with the square of any linear dimension where area product varies as the fourth power.

The surface area/area product relationship

$$A_t = K_s A_p^{0.5} \quad (\text{cm}^2)$$

in which K_s is a constant related to core configuration is shown in Table 4.

Transformer Current Density

The current density J of a transformer can be related to the area product A_p of a transformer for a given temperature rise.

Table 4 Transformer Surface Area Constant K_s

Core Type	K_s
Pot Core	33.8
Powder Core	32.5
Laminations	41.3
C Cores	39.2
Single Coil	44.5
Tape Toroids	50.9

The relationship of current density J to the area product A_p for a given temperature rise is

$$J = K_j A_p^{-0.125} \quad (\text{amps/cm}^2)$$

in which K_j is a constant related to core configuration. The values for K_j shown in Table 5 are empirical.

Table 5 Transformer Current Density Constant K_j

Core Type	K_j @(Δ25°C)	K_j @ (Δ50°C)
Pot Core	433	632
Powder Core	403	590
Laminations	366	534
C Cores	322	468
Single Coil	395	569
Tape Toroids	250	365

INDUCTOR DESIGN

The Design Problem Generally

The designer is faced with a set of constraints which must be observed in the design of any inductor. One of these is copper loss; the winding must be capable of delivering current to the load within specified regulation limits. Another relates to minimum efficiency of operation which is dependent on the maximum power loss which can be allowed in the inductor. Still another defines the maximum permissible temperature rise for the inductor when used in a specified temperature environment. The gapped inductor has three loss components: copper loss P_{cu}, core loss P_{fe}, and gap loss P_g. Maximum efficiency is reached in an inductor, as in a transformer, when the copper loss P_{cu} and the iron loss P_{fe} are equal but only when the core gap is zero. The loss does not occur in the air gap itself but is caused by magnetic flux fringing around the gap and reentering the core in a direction of high loss. As the air gap increases, the fringing flux increase more and more, and some of the fringing flux strikes the core perpendicular to the lamination and sets up eddy currents which cause additional loss. Designing with a molypermalloy powder core, the gap loss is minimized because the powder is insulated with a ceramic material which provides a uniformly distributed air gap. Also, designing with ferrites the gap loss is minimized because ferrite materials have such high resistivity.

Other constraints relate to volume occupied by the inductor, such as weight as weight minimization is an important goal in the design of space flight electronics. Finally, cost-effectiveness is always an important consideration.

Fundamental Conditions in Designing Inductors

The design of a linear reactor depends on four related factors:

1. Desired inductance
2. Direct current
3. Alternating current ΔI
4. Power loss and temperature rise

With these requirements established, the designer must determine the maximum values for B_{dc} and for B_{ac} which will not produce magnetic saturation and must make trade-offs which will yield the highest inductance for a given volume. The core material which is chosen dictates the maximum flux density which can be tolerated for a given design:

$$B_m = \frac{0.4\pi N\left(I_{dc} + \frac{\Delta I}{2}\right) \times 10^{-4}}{l_g} \quad \text{(tesla)}$$

The inductance of an iron-core inductor carrying dc and having an air gap may be expressed as:

$$L = \frac{0.4\pi N^2 A_c \times 10^{-8}}{l_g + \frac{l_m}{\mu_r}} \quad \text{(henry)}$$

Inductance is dependent on the effective length of the magnetic path which is the sum of the air gap length (l_g) and the ratio of the core mean length to relative permeability (l_m / μ_r).

When the core air gap (l_g) is large compared to relative permeability (l_m / μ_r), because of the high relative permeability (μ_r) the variations in μ_r do not substantially affect the total effective magnetic path length or the inductance. The inductance equation then reduces to:

$$L = \frac{0.4\pi N^2 A_c \times 10^{-8}}{l_g} \quad \text{(henry)}$$

Final determination of the air gap size requires consideration of the effect of fringing flux which is a function of gap dimension, the shape of the pole faces, and the shape, size, and location of the winding. Its net effect is to shorten the air gap.

Fringing flux decreases the total reluctance of the magnetic path and therefore increases the inductance by a factor F to a value greater than that calculated from equation 26. Fringing flux is a larger percentage of the total for larger gaps. The fringing flux factor is:

$$F = \left(1 + \frac{l_g}{\sqrt{A_c}} \ln \frac{G}{l_g}\right)$$

The inductance L computed in equation 26 does not include the effect of fringing flux. The value of inductance L' corrected for fringing flux is:

$$L' = \frac{0.4\pi N^2 A_c F \times 10^{-8}}{l_g} \quad \text{(henry)}$$

Distribution of fringing flux is also affected by another aspect of core geometry, the proximity of coil turns to the core, and whether there are turns on both legs.

Accurate prediction of gap loss P_g depends on the amount of fringing flux. This equation for gap loss P_g seems to be a good approximation. Table 6 gives Gap Loss Coefficient for different core configurations

$$P_g = K_i D l_g f B_m^2 \quad \text{(watts)}$$

D = Strip or tongue width (cm)

Table 6 Gap Loss Coefficient

Core Configuration	K_i
Two Coil C Core	0.0388
Single Coil C Core	0.0775
Laminations	0.1550

The fringing flux is around the gap and reenters the core in a direction of high loss as shown in Fig. 9.

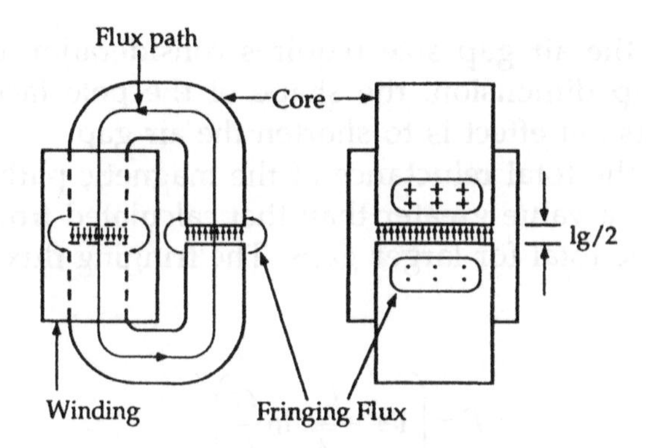

Figure 9 Fringing flux around the gap of an inductor.

Effective permeability may be calculated from the following expression:

$$\mu_\Delta = \frac{\mu_m}{1 + \frac{l_g}{l_m}\mu_m}$$

After establishing the required inductance and the dc bias current which will be encountered, dimensions can be determined. This requires consideration of the energy-handling capability which is controlled by the size of the inductor.

The energy handling capability of a core is:

$$\text{Eng} = \frac{LI^2}{2} \quad \text{(watt seconds)}$$

Toroidal Powder Core Selection

The design of an inductor also frequently involves considerations of the effect of its magnetic field on other devices near where it is placed. This is especially true in the design of high-current inductors for converters and switching regulators used in spacecraft, which may also employ sensitive magnetic field detectors. For this type of design problem it is frequently imperative that a toroidal core be used. The magnetic flux in a moly-permalloy toroid (core) can be contained inside the core more readily than in a lamination or C-type core, as the winding covers the core along the whole magnetic path length.

The author has developed a simplified method of designing optimum dc carrying inductors with moly-permalloy powder cores. This method allows the correct core permeability to be determined without relying on trial and error.

With these requirements established, the designer must determine the maximum values for B_{dc} and B_{ac} which will not produce magnetic saturation and must make trade-offs which will yield the highest inductance for a given volume. The core permeability chosen dictates the maximum dc flux density which can be tolerated for a given design. Permeability values for different powder cores are shown in Table 7.

Table 7 Different Powder Core Permeabilities

Core Permeability	DC Magnetizing Force L < 80%	
	Oersted	Amp-Turn
14	202	253
26	112	140
60	45	56
125	22	28
147	18	23
160	16	20
173	15	19
200	13	16
300	9	11
555	3	4

If an inductance is to be constant with increasing direct current, there must be a negligible drop in inductance over the operating current range. The maximum H, then, is an indication of a core's capability. In terms of ampere-turns and mean magnetic path length l_m,

$$H = \frac{NI}{l_m} \quad \text{(amp-turn/cm)}$$

$$NI = 0.8Hl_m \quad \text{(amp-turn)}$$

Inductance decreases with increasing flux density and magnetizing force for various materials of different values of permeability μ_Δ. The selection of the correct permeability for a given design is made after solving for the energy handling capability:

$$\mu_\Delta = \frac{B_m l_m \times 10^4}{0.4\pi W_a J K_u}$$

It should be remembered that maximum flux density depends on $B_{dc} + B_{ac}$ in the manner shown in Fig. 10:

$$B_{max} = B_{dc} + \frac{B_{ac}}{2} \quad \text{(tesla)}$$

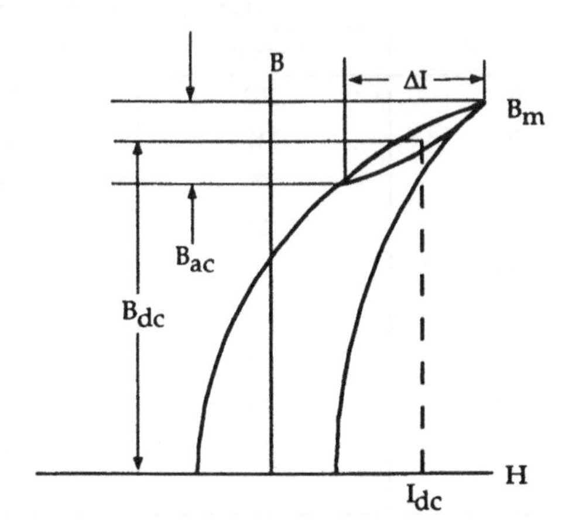

Figure 10 Flux density versus $I_{dc} + \Delta I$.

Moly-permalloy powder cores operating with a dc bias of 0.3 tesla (T) have only about 80% of their original inductance, with very rapid fall off at higher densities.

The flux density for the initial design for moly-permalloy powder cores should be limited to 0.3 T maximum for B_{dc} plus B_{ac}.

The losses in a moly-permalloy inductor due to ac flux density are very low compared to the steady-state dc copper loss. It is then assumed that the majority of the losses are copper:

$$P_{cu} >> P_{fe}$$

Relationship of K_g to Inductor Energy-Handling Capability

Inductors

Inductors, like transformers, are designed for a given temperature rise. They can also be designed for a given regulation. The regulation and energy handling ability of a core is related to two constants:

$$*\alpha = \frac{(\text{Eng})^2}{K_g K_e} \quad (\%)$$

$$\alpha = \text{Regulation, } (\%)$$

The constant K_g is determined by the core geometry:

$$*K_g = \frac{W_a A_c^2 K_u}{\text{MLT}} \quad (\text{cm}^5)$$

The constant K_e is determined by the magnetic and electric operating conditions:

$$K_e = 0.145 P_o B_m^2 \times 10^{-4}$$

where:

$$P_o = \text{output power}$$

$$B_{max} = B_{dc} + \frac{B_{ac}}{2}$$

From the above, it can be seen that flux density is the predominant factor governing size.

Relationship of A_p to Inductor Energy-Handling Capability

According to the newly developed approach, the energy-handling capability of a core is related to its area product A_p by an equation which may be stated as follows:

$$*A_p = \left(\frac{2(\text{Eng}) \times 10^4}{B_m K_u K_j}\right)^{(\text{Table 1, } x)} \quad (\text{cm}^4)$$

*The derivation for these equations are set forth in detail by the author in the book *Transformer and Inductor Design Handbook*, Marcel Dekker Inc., New York.

where:

B_m = flux density, tesla
K_u = window utilization factor
K_j = current density coefficient
Eng = energy, watt seconds

From the above it can be seen that factors such as flux density, window utilization factor K_u (which defines the maximum space which may be occupied by the copper in the window), and the constant K_j (which is related to temperature rise) all have an influence on the inductor area product. The constant K_j is a new parameter that gives the designer control of the copper loss.

Window Utilization Factor K_u

The window utilization factor is the amount of copper that appears in the window area of the transformer or inductor. The window utilization factor is influenced by four factors: (1) wire insulation, (2) wire lay (fill factor), (3) bobbin area (or, when using a toroid, the clearance hole for passage of the shuttle), and (4) insulation required for multilayer windings or between windings. In the design of high-current or low-current transformers, the ratio of conductor area over total wire area can vary from 0.941 to 0.673, depending on the wire size. The wire lay or fill factor can vary from 0.7 to 0.55, depending on the winding technique. The amount and the type of insulation are dependent on the voltage.

The fraction K_u of the available core window space which will be occupied by the winding (copper) is calculated from areas S_1, S_2, S_3, and S_4:

$$K_u = S_1 \times S_2 \times S_3 \times S_4$$

where:

$$S_1 = \frac{\text{Conductor area}}{\text{Wire area}}$$

$$S_2 = \frac{\text{Wound area}}{\text{Usable window area}}$$

$$S_3 = \frac{\text{Usable window area}}{\text{Window area}}$$

$$S_4 = \frac{\text{Usable window area}}{\text{Usable window area} + \text{Insulation area}}$$

in which:

Conductor area = copper area
Wire area = copper area + insulation area

Wound area = number of turns times the wire area of one turn
Usable window area = available window area minus residual area which results from the particular winding technique used
Window area = available window area
Insulation area = area usable for winding insulation

S_1 is dependent on wire size:

$$\text{AWG 10} = \frac{52.61 \text{ cm}^2}{52.90 \text{ cm}^2} = 0.941$$

$$\text{AWG 20} = \frac{5.188 \text{ cm}^2}{6.065 \text{ cm}^2} = 0.855$$

$$\text{AWG 30} = \frac{0.5067 \text{ cm}^2}{0.6785 \text{ cm}^2} = 0.747$$

$$\text{AWG 40} = \frac{0.04869 \text{ cm}^2}{0.0723 \text{ cm}^2} = 0.673$$

When designing low-current transformers, it is advisable to reevaluate S_1 because of the increased amount of insulation.

S_2 is the fill factor for the usable window area. It can be shown that for circular cross-section wire wound on a flat form, the ratio of wire area to the area required for the turns can never be greater than 0.91. In practice, the actual maximum value is dependent on the tightness of winding, variations in insulation thickness, and wire lay. Consequently, the fill factor is always less than the theoretical maximum.

As a typical working value for copper wire with a heavy synthetic film insulation, a ratio of 0.60 may be safely used.

S_3 defines how much of the available window space may actually be used for the winding. The winding area available to the designer depends on the bobbin configuration. A single-bobbin design offers an effective area W_a between 0.835 and 0.929, whereas a two-bobbin configuration offers an effective area W_a between 0.687 and 0.872. A good value to use for both configurations is 0.75.

S_4 defines how much of the usable window space is actually being used for insulation. If the transformer has multiple secondaries having significant amounts of insulation, S_4 should be reduced by 10% for each additional secondary winding because of the added space occupied by insulation and partly due to poorer space factor.

A typical value for the copper fraction in the window area is about 0.40. For example, for AWG 20 wire, $S_1S_2S_3S_4 = (0.855)(0.60)(0.75)(1.0) = 0.385$, which is very close to 0.4.

This may be stated somewhat differently:

$$0.4 = \underbrace{\frac{A_w(\text{bare})}{A_w(\text{total})}}_{(S_1)} \times \underbrace{\text{Fill factor}}_{(S_2)} \times \underbrace{\frac{W_{a(\text{eff})}}{W_a}}_{(S_3)} \times \underbrace{\text{Insulation factor}}_{(S_4)}$$

When a design requires a multitude of windings, all of which have to be insulated, then the insulation factor (S_4) becomes very important in the window utilization factor (K_u). For example, a low-current toroidal transformer with insulation has a significant influence on the window utilization factor as shown below:

S_1 = #40 AWG

$$K_u = S_1 \times S_2 \times S_3 \times S_4$$
$$K_u = 0.673 \times 0.60 \times 0.75 \times 0.80$$
$$K_u = 0.242.$$

The core geometry coefficient K_g in this book has been calculated with a window utilization factor K_u of 0.4. When designing with small wire, rescale the core geometry coefficient K_g required:

$$K_g\left(\frac{K_u}{K_{u(new)}}\right)$$

Efficiency Regulation and Temperature Rise

Transformer efficiency, regulation, and temperature rise are all interrelated. Not all of the input power to the transformer is delivered to the load. The difference between the input power and output power is converted into heat. This power loss can be broken down into two components: core loss and copper loss. The core loss is a fixed loss, and the copper loss is a variable loss which is related to the current demand of the load. Copper loss goes up by the square of the current and is termed quadratic loss. Maximum efficiency is achieved when the fixed loss is equal to the quadratic at rated load. Transformer regulation is copper loss P_{cu} divided by the output power P_o.

Transformer Efficiency

The efficiency of a transformer is a good way to measure the effectiveness of the design. Efficiency is defined as the ratio of the output power P_o to the input power P_{in}. The difference between the P_o and the P_{in} is due to losses. The total power loss in a transformer is determined by the fixed losses in the core and the quadratic losses in the windings or copper. Thus,

$$P_\Sigma = P_{fe} + P_{cu} \quad \text{(watts)}$$

P_{fe} represents the core loss and P_{cu} represents the copper loss.

Maximum efficiency is achieved when the fixed loss is made equal to the quadratic loss:

$$P_{fe} = P_{cu}$$

then

$$P_{in} = P_o + P_\Sigma \quad \text{(watts)}$$

and the efficiency is

$$\eta = \frac{P_o}{P_o + P_\Sigma}$$

Temperature Rise

Not all of the P_{in} to the transformer is delivered to the load as P_o. Some of the input power is converted to heat by hysteresis and eddy currents induced in the core material, and by the resistance of the windings. The first is a fixed loss arising from core excitation and is termed "core loss." The second is a variable loss in the windings which is related to the current demand of the load and thus varies as I^2R. This is termed the quadratic or copper loss.

The heat generated produces a temperature rise which must be controlled to prevent damage to or failure of the windings by breakdown of the wire insulation at elevated temperatures. This heat is dissipated from the exposed surfaces of the transformer by a combination of radiation and convection. The dissipation is therefore dependent on the total exposed surface area of the core and windings.

Ideally, maximum efficiency is achieved when the fixed and quadratic losses are equal. Thus,

$$P_\Sigma = P_{fe} + P_{cu} \quad \text{(watts)}$$

and

$$P_{cu} = \frac{P_\Sigma}{2}$$

Temperature rise in a transformer winding cannot be predicted with complete precision, despite the fact that many different techniques are described in the literature for its calculation. One reasonably accurate method for open core and winding construction is based on the assumption that core and winding losses may be lumped together as

$$P_\Sigma = P_{fe} + P_{cu} \quad \text{(watts)}$$

and the assumption is made that thermal energy is dissipated uniformly throughout the surface area of the core and winding assembly.

The effective surface area A_t required to dissipate heat (expressed as watts dissipated per unit area) is:

$$A_t = \frac{P_\Sigma}{\psi} \quad (\text{cm}^2)$$

in which ψ is the power density of the average power dissipated per unit area from the surface of the transformer and P_Σ is the total power lost or dissipated.

The surface area A_t of a transformer can be related to the area product A_p of a transformer. From this, the following relationship evolves:

$$A_t = K_s A_p^{0.5} = \frac{P_\Sigma}{\psi} \quad (\text{cm}^2)$$

$$\psi = 0.03\ \text{W/cm}^2\ @\ 25°\text{C rise}$$

$$\psi = 0.07\ \text{W/cm}^2\ @\ 50°\text{C rise}$$

as shown in Fig. 11.

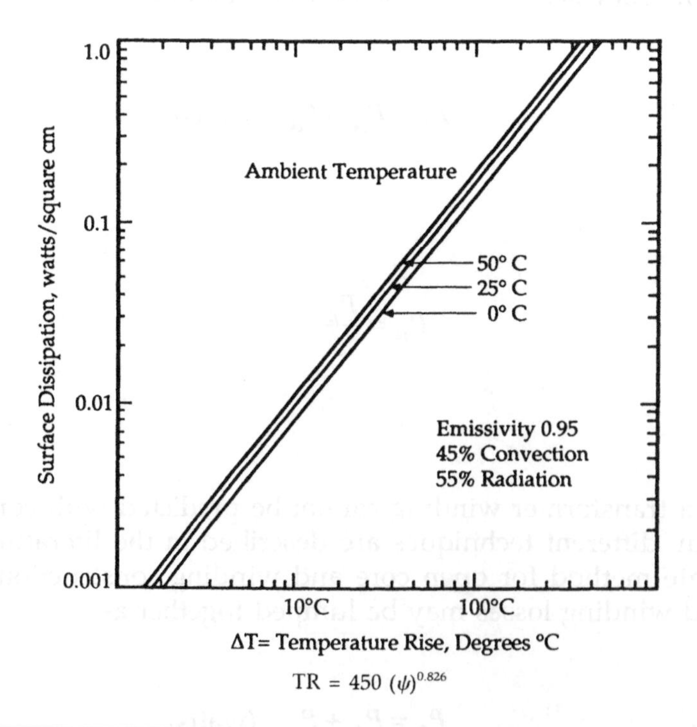

Figure 11 Temperature rise versus surface dissipation.

Regulation

The minimum size of a transformer is usually determined either by a temperature rise limit or by allowable voltage regulation, assuming that size and weight are to be minimized; then

$$\alpha = \left(\frac{P_{cu}}{P_o + P_{cu}} \right) \text{x } 100 \quad (\%)$$

$$\alpha = \text{Regulation } (\%)$$

By definition

$$P_{cu} = P_{fe}$$

then

$$\alpha = \left(\frac{1 - \eta}{1 + \eta} \right) \text{x } 100$$

then

$$\eta = \frac{100 - \alpha}{100 + \alpha}$$

2

TRANSFORMER OUTPUT CIRCUITS

TRANSFORMER RECTIFIER CIRCUITS

The rectifier circuit for a particular application depends on the dc voltage and current requirements and the maximum amount of ripple that can be tolerated in the circuit.

Most Common Circuits

The single-phase full-wave center tap and the full-wave bridge are the most common and are essentially the same except for the additional rectifier and transformer winding. Figure 12 shows the full-wave center tap and the full-wave bridge rectifier configuration.

Both circuits have similar output characteristics, but the transformer ratings and rectifier peak inverse voltage are different.

The output voltage waveforms of the center-tap rectifier and the full-wave bridge rectifier are shown in Fig. 13. The current and voltage wave forms will be in phase and have the same characteristics, providing the transformer reactance is negligible.

Both of the circuits have their weak and strong points. In the center-tap full-wave rectifier circuit, two windings are required, and each of these conduct for only half the time. Also, the peak inverse voltage across the rectifier is twice that of the bridge circuit. The full-wave bridge circuit requires two additional rectifiers, increasing cost and power loss but reducing the peak inverse voltage across each rectifier to only half that of the push–pull circuit.

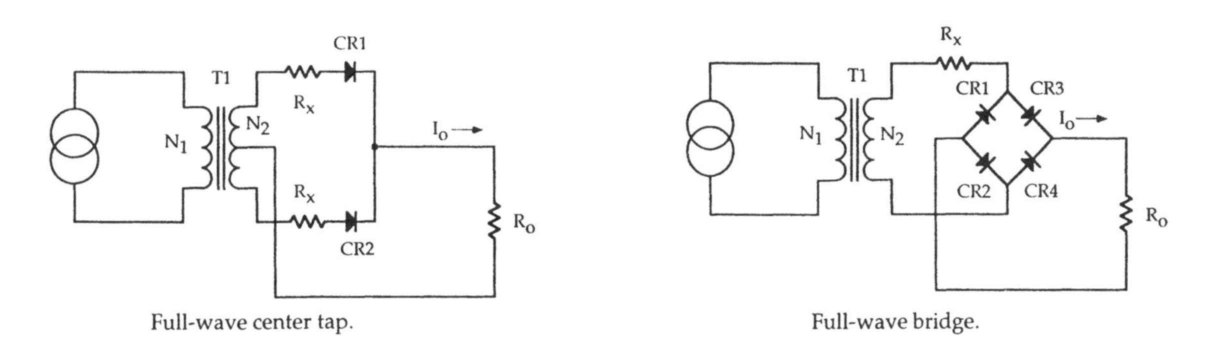

Figure 12 Common center tap and bridge rectifier circuits.

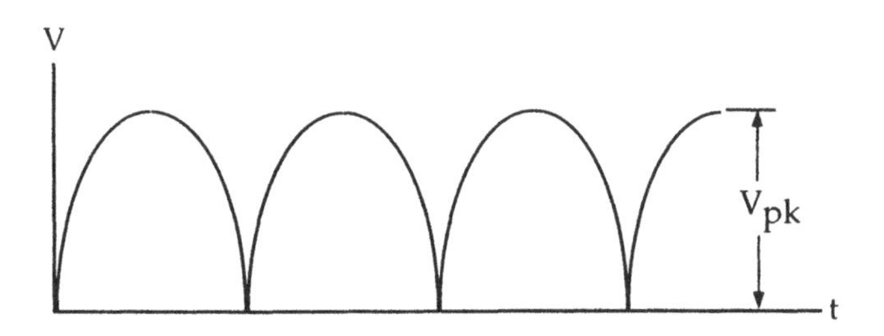

Figure 13 Full-wave rectified sine wave.

Output Filter Networks

Filter networks are used with rectifier circuits to smooth out the ripple in the dc output. Filters consist of two basic types: capacitive input and inductive (choke) input.

The simplest of these filter circuits is the capacitive input types shown in Fig. 14. For this type of filter, the capacitor charges up to approximately the peak of the input voltage on each half-cycle that a rectifier conducts. The output waveform for this circuit are shown in Fig. 15. This simple filter provides poor voltage regulation at high load currents. Although the ripple content can be reduced and the regulation can be improved by use of a larger filter capacitor, the charging current may become excessive and cause damage to the rectifier if the capacitor is made too large.

When power supplies using capacitor input filters are initially energized, a high surge current flows through the rectifier to charge the capacitor. The source resistance R_x and the rectifier resistance limit this current surge whose duration is only a portion of a half-cycle. Surge resistors are inserted in series with the rectifiers in such supplies to limit the in-rush current or the transformer is designed to furnish this resistance.

The value of the surge resistor depends on supply voltage, input filter capacitance, and the rectifier surge rating. The surge rating of the rectifier for one cycle or less is expressed as the maximum amount of heating energy, or I^2T, that the rectifier will generate due to forward current without failing. Figure 16 shows the relationship among these variables for a supply frequency of 60 Hz.

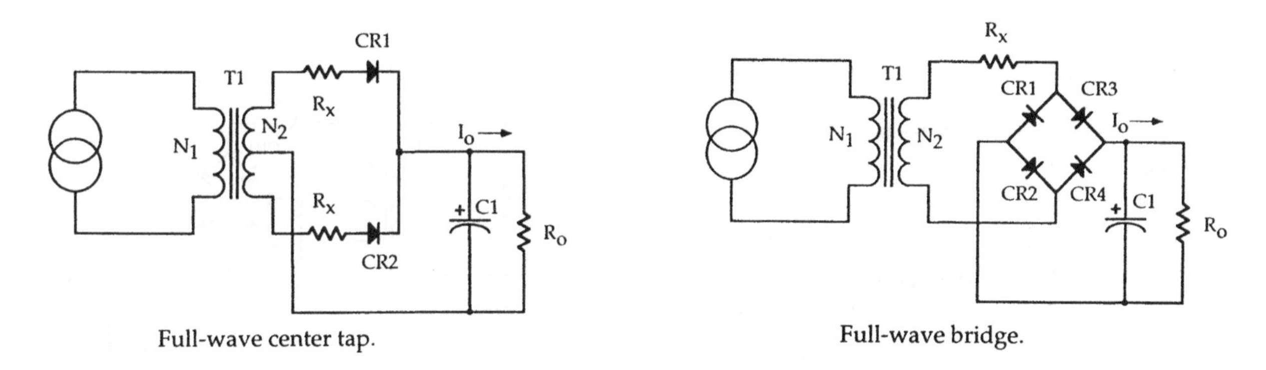

Figure 14 Capacitive input filter.

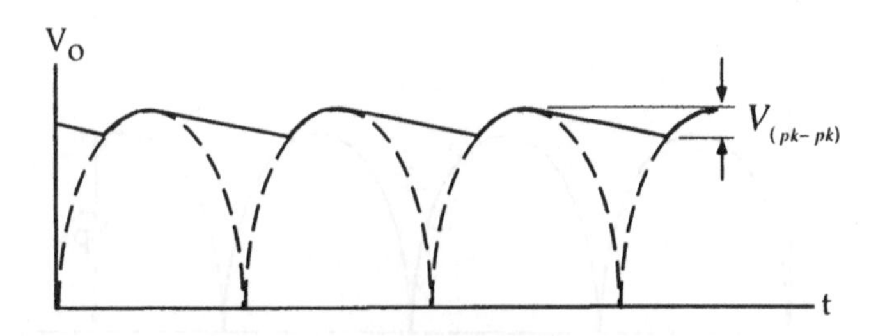

Figure 15 Output ripple waveform.

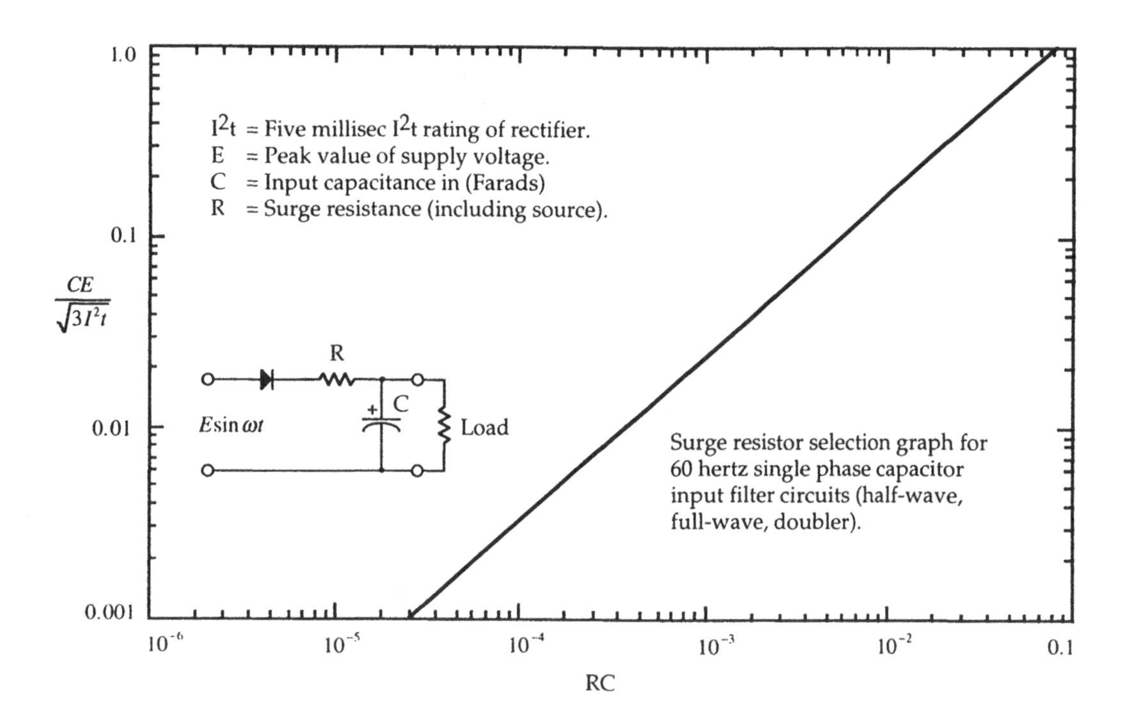

Figure 16 Surge resistor selection graph.

The design of rectifier circuits having capacitive loads often requires the determination of rectifier voltage and current waveforms in terms of average, rms, and peak values; see Fig. 17.

These values are needed for the calculation of circuit parameters, the selection of components, and the matching of circuit parameters with rectifier ratings. Although the actual calculation of rectified voltage is a rather lengthy process, the voltage-relationship graphs shown in Fig. 18 can be used to determine the peak voltage if the average voltage is known.

given value	to get . . .		
	peak	rms	average
peak		0.707(peak)	0.637(peak)
rms	1.41(rms)		0.9(rms)
average	1.57(average)	1.11(average)	

Figure 17 Peak, rms, and average ac.

The ratio of average-to-peak voltage ($V_o / V_{s(pk)}$) are shown in Fig. 18 as functions of the circuit constants ωCR_o and R_x / R_o. The quantity ωCR_o is the resistive-to-capacitive reactance in the load, and the quantity R_x / R_o is the ratio of the limiting resistance to the load resistance. Figure 19 is a plot of the rms ripple as a function of the parameter ωCR_o. Figures 20 and 21 show capacitance versus load current with different values of peak-to-peak ripple voltage at 50 Hz and 400 Hz.

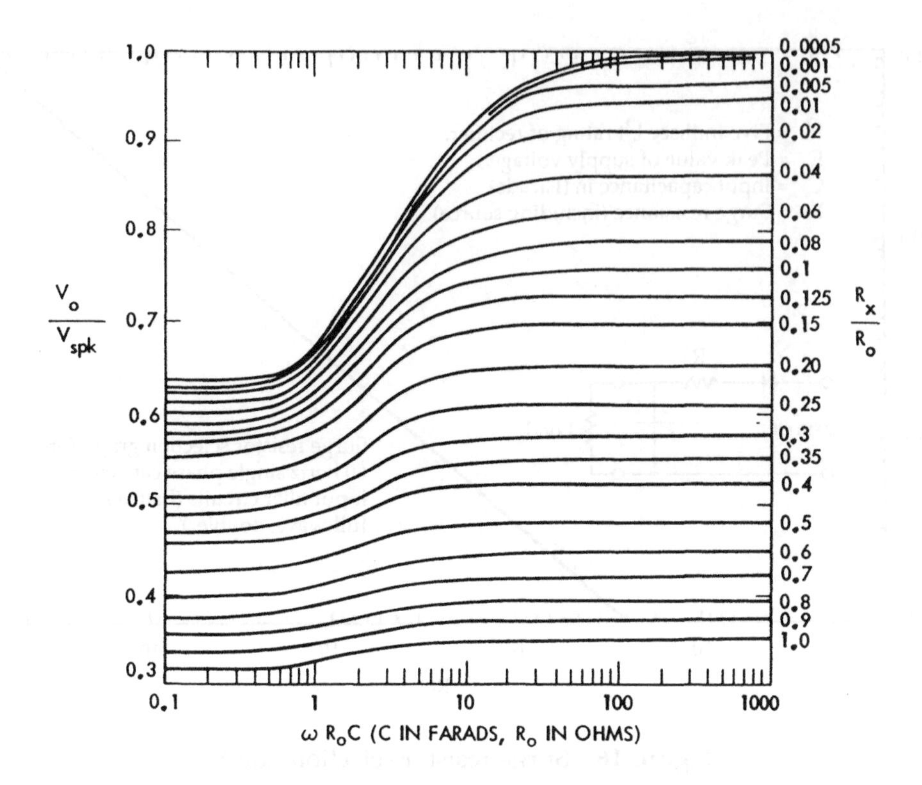

Figure 18 Peak voltage to dc output voltage in full-wave capacitor input circuit.

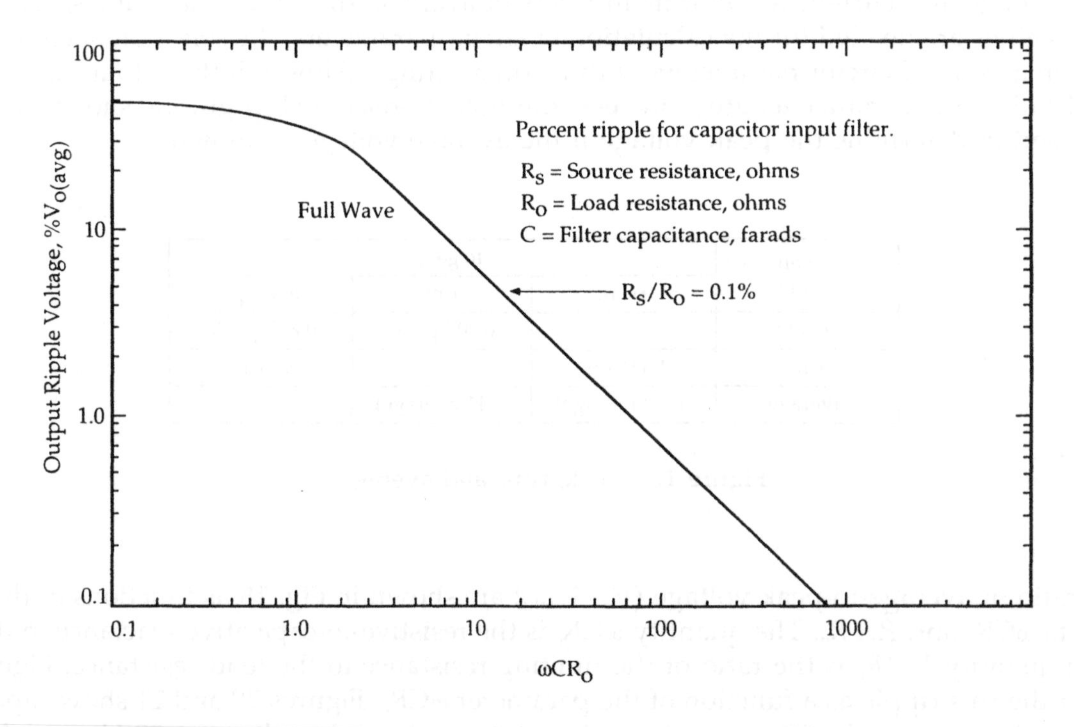

Figure 19 The rms ripple versus ωCR_o.

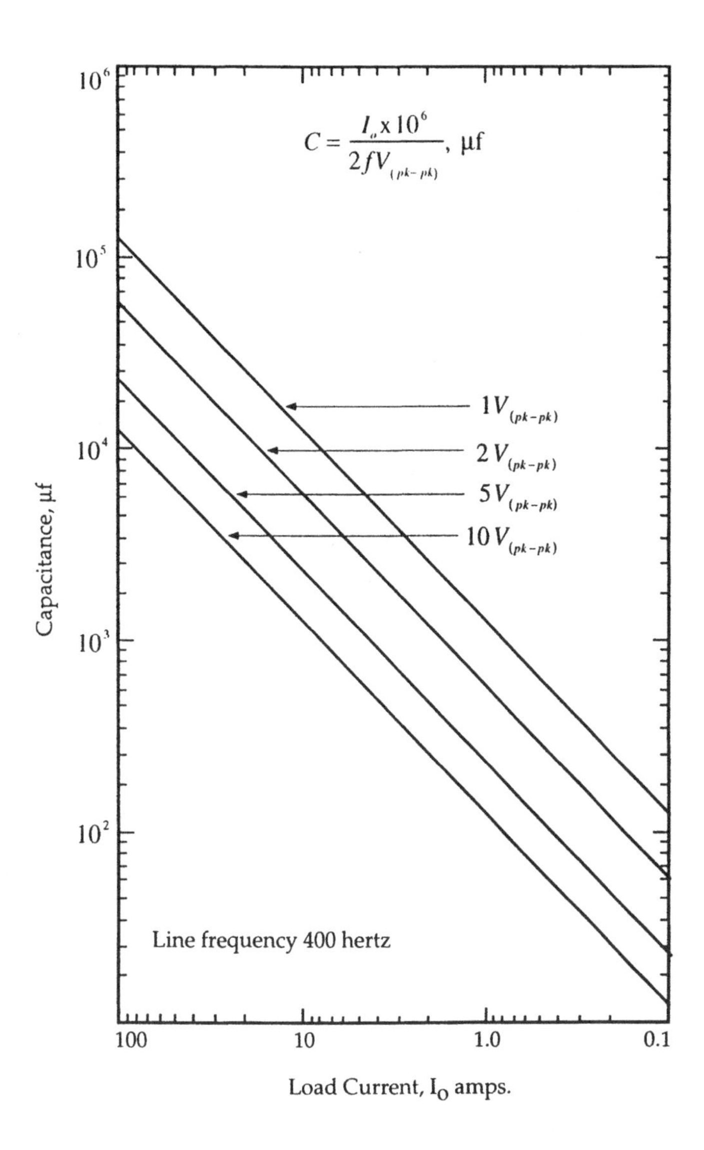

Figure 20 Capacitance versus output current I_o for different $P\text{-}P_{\text{ripple}}$ voltage.

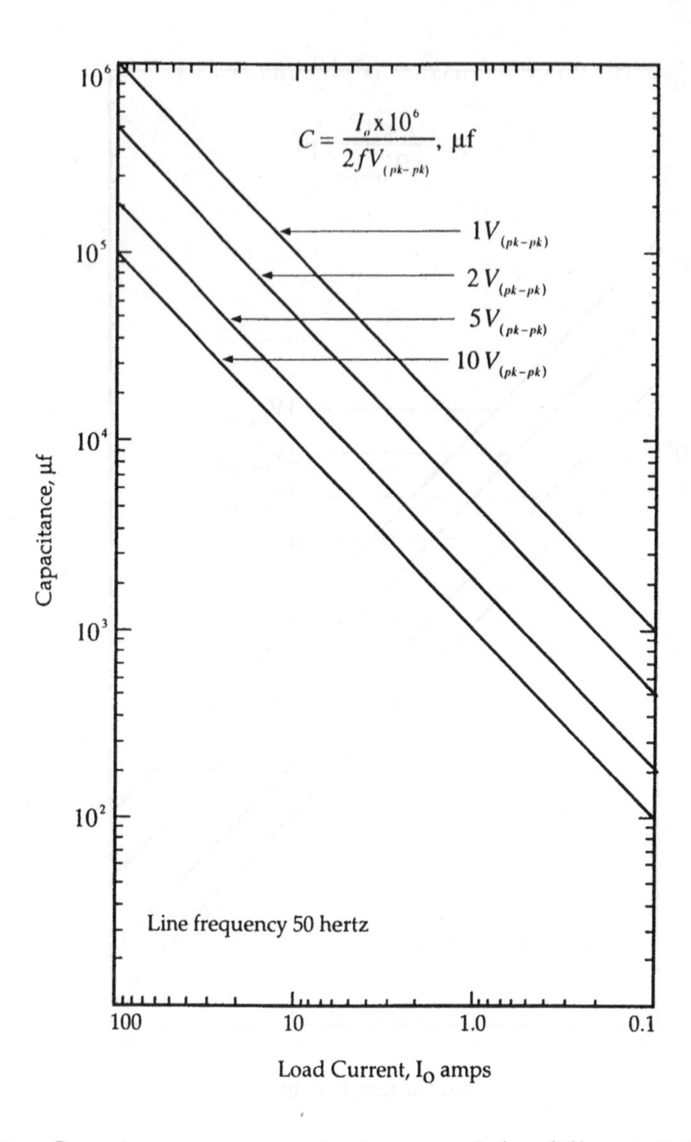

Figure 21 Capacitance versus output current I_o for different $P\text{-}P_{ripple}$ voltage.

Capacitor Input

Using a capacitor input filter, the rectifier current is not continuous throughout each cycle, and the rectified waveform changes. During the voltage peaks of each cycle, the capacitor charges and draws current from the rectifier. During the rest of the time, no current is drawn from the rectifier and the capacitor discharges into the load as shown in Fig. 22.

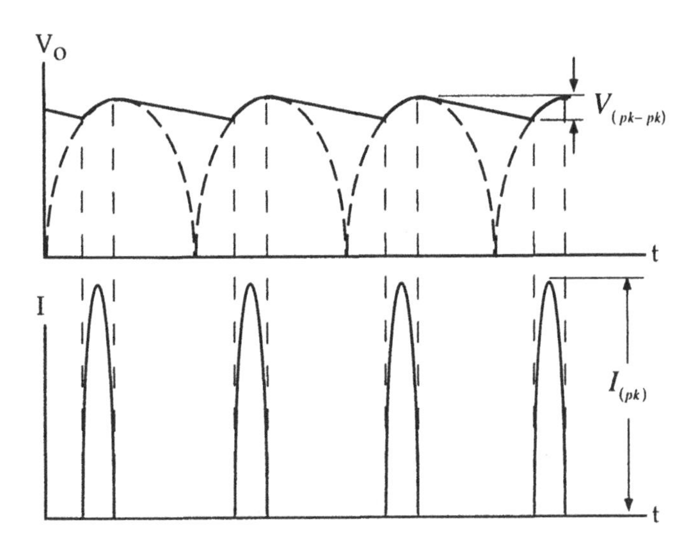

Figure 22 Voltage–current relationship.

The high peak currents flows only while the rectifier voltage is higher than the capacitor dc voltage. The ΔV will look like a triangular waveshape (See Fig. 22). The superimposed ripple voltage value of ΔV will naturally depend on the value of capacitance. Below are the rms ripple equations for voltage and current where:

I_o = dc load current

f = ripple frequency

C = capacitance in farads

$$\Delta V_{rms} = \frac{I_o}{fC2\sqrt{3}}$$

$$I_{rms} = \Delta V_{rms} 2\pi fC$$

As one can see, the capacitor input has peaking current and its rms value has to be taken into consideration when designing these components.

The aluminum electrolytic deviates the most from the theoretically ideal type. In electrolytics, dielectric losses are relatively high. Losses result from direct current leakage (I_l), which together with equivalent series resistance (ESR). Capacitor losses are expressed in terms of the dissipation factor (DF):

$$\mathrm{DF} = (2\pi f C)(\mathrm{ESR}) \times 10^{-4}$$

$$C = \mu\mathrm{f}$$

The dissipation factor (DF) and the equivalent series resistance (ESR) found in the following equations are based on 120-Hz formulas in MIL-C-62B:

$$\mathrm{DF} = 0.006\sqrt{C}$$

$$\mathrm{ESR} = 1326\left(\frac{\mathrm{DF}}{C}\right)$$

The ripple current flowing through the ESR heats the capacitor. The hot spot is the center of the capacitor. The temperature rise in the center of the capacitor depends on the thermal resistance from the center to the ambient. The heat generated in a capacitor is dissipated by the outside surface area. For high values of capacitance, where the ratio of capacitor outside surface area to volume is significantly lower, internal heating becomes a problem. The ripple current rating may be the determining factor in capacitor selection, rather than ripple voltage. In many cases, capacitor size will have to be increased to prevent excessive internal heating.

CAPACITANCE INPUT TRANSFORMER CONSIDERATIONS

The actual dc rectified power output P_o delivered to the load is not the same volt-ampere capacity of the transformer winding.

Using the capacitor input filter, the designer of the transformer must make allowances for high charging currents. The waveshapes of the current that flows through the winding of the transformer are not sinusoidal; the heating of the transformer winding is greater for a capacitance input filter because of the high-current spikes.

Because the transformer apparent power P_t rating varies with the rms content of the rectifier circuit, the rms current must be calculated for each capacitive–input filter design. High R_x/R_o ratios help reduce transformer size but also decrease the regulation.

Inductor Input

The *LC* section filter is one method of reducing ripple levels without the need for single, large-value filter components. The basic circuit is shown in Fig. 23. The capacitive reactance should always be less than 10% of the load resistance at the second harmonic of the incoming frequency.

To achieve normal inductor–input operation, it is necessary that there be a continuous flow of current through the input inductor. The peak value of the alternating current flowing through

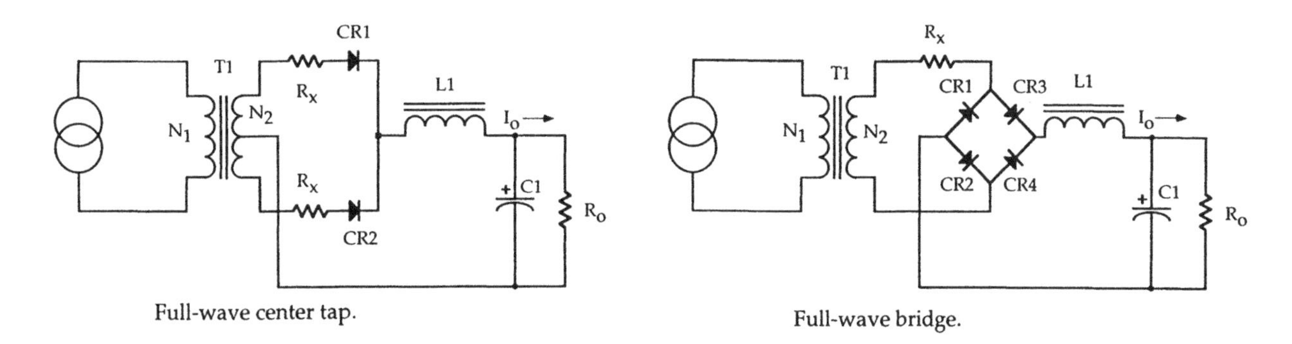

Figure 23 *LC* filter.

the input inductor must, hence, be less than the dc output current of the rectifier. The value for minimum inductance, called critical inductance L_c is

$$L_c = \frac{R_o}{3\omega}$$

where

$$\omega = 2\pi f$$
$$f = \text{line frequency}$$

The higher the load resistance R_o (i.e., the lower the dc load current), the more difficult it is to maintain a continuous flow of current. The filter circuit operates in the following manner: When R_o approaches ∞, under unloaded conditions (no bleeder resistor) $I_o = 0$, the filter capacitor will charge to V_{spk}, the peak voltage. The output voltage will therefore be equal to the peak value of the input voltage as shown in Fig. 24. The actual ripple output of a full-wave single-phase rectifier circuit is shown in Fig. 25 for minimum value of ω^2LC to avoid resonance.

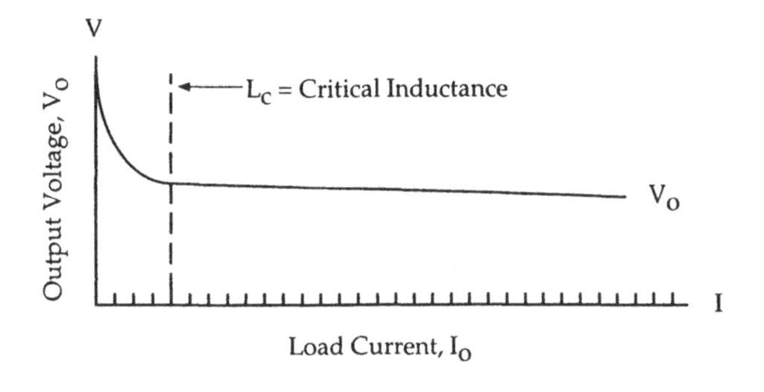

Figure 24 Critical inductance point *LC* filter.

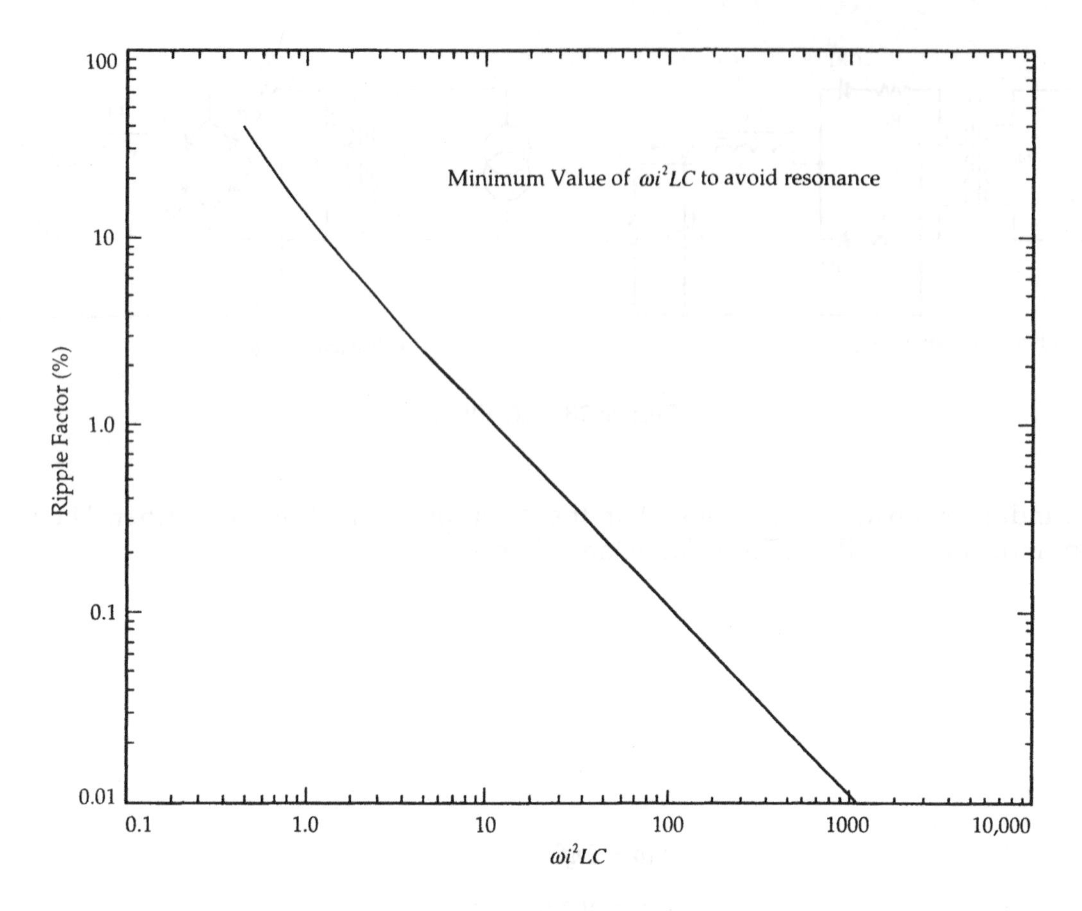

Figure 25 Ripple versus ω^2LC.

Voltage Regulation with Inductor Input

Given ideal components, the voltage regulation of a rectifier–filter system employing input inductance greater than the critical value would be perfect (i.e., voltage output would be independent of load current). In practice, the output voltage falls off with increasing load as a result of resistance in the diodes, filter, and transformer and as a result of the leakage reactance of the supply transformer. All must be taken into consideration when designing for regulation.

SWITCHING MAGNETIC COMPONENTS

The usage of switched-mode power supplies (SMPS) has showed a significant growth. The design of the magnetic components used in switched-mode power supplies takes on added importance because of the trade-offs the engineer has to make.

The engineer has a variety of circuits that are available to convert the dc input voltage to the required dc output voltage. The type circuit chosen by the engineer will be the result of a trade-off study associated with the application for the converter and the characteristics of the various converter circuits. These characteristics are what influence the design on how easy or how hard it is for that application. Some of the circuits available to the engineer are as follows:

1. Push–pull
2. Four-transistor bridge
3. Two-transistor half-bridge
*4. Buck
*5. Buck-boost
*6. Boost
*7. Feed forward

SMPS Fundamental Requirements and Definition

The principle behind flyback converters is based on the storage of energy in the inductor during the charging on-time period t_{on}, and the discharge of the energy to the load during the off-time period t_{off} as shown in Fig. 26, where

$$T = t_{on} + t_{off} = \frac{1}{f}$$

The duty ratio can be expressed as:

$$D \equiv \frac{t_{on}}{T}$$

The minimum duty ratio is:

$$D_{(min)} \equiv \frac{V_o}{V_{in(max)}}$$

The maximum duty ratio is:

$$D_{(max)} \equiv \frac{V_o}{V_{in(min)}}$$

*These circuits can be operated single ended or push–pull.

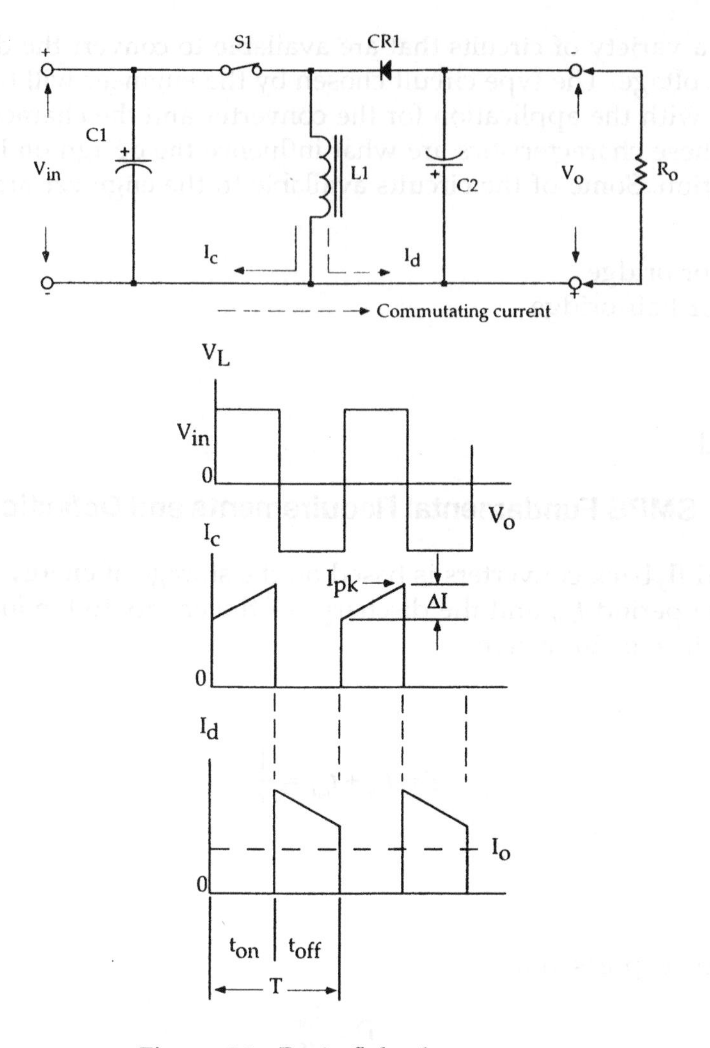

Figure 26 Basic flyback converter.

For the engineer to effectively design a (SMPS) converter, there are seven circuit parameters that must be known:

1. Input voltage range, V_{in}
2. Output voltage, V_o
3. Output current range, I_o
4. Circuit efficiency, η
5. Frequency, f
6. Regulation (copper loss), α
7. Temperature rise, °C

For ease of discussion, assume constant input voltage and assume that the switch is operating at a 50–50 duty cycle where the on time is equal to the off time, and the components are ideal (no losses). These assumptions will be used throughout the discussion of the flyback converter except where noted.

Energy Transfer

Two distinct nodes of operation are possible for the flyback converter:

Discontinuous mode, where all the energy stored in the flyback inductor is transferred to the output capacitor and load circuit before the switch (S) is closed.

Continuous mode, where the energy stored in the flyback converter is not completely transferred to the output capacitor and load circuit before the switch (S) is closed.

In the discontinuous mode, a smaller inductance is required which results in higher peak current in the switch (S) as shown in Figs. 27 and 28. As a consequence, winding losses are increased by 15% comparing the rms values of a square wave and clipped sawtooth. This also results in a higher ripple current in the input capacitor and a higher peak current to the switching device. The advantage of this circuit, other than having a smaller inductor, is that when the switching device is turned on, the initial current is zero. This means the output diode CRI has completely recovered and the switching device does not turn on momentarily into a short. This reduces the EMI interference.

In the continuous mode, a larger inductor is required, resulting in a lower peak current at the end of the cycle than experienced when using a discontinuous system of equivalent output power. On the other hand, the continuous mode demands a high current flowing through the switch during turn on and can lead to higher switch dissipation.

Triangular and Trapezoidal Current

Operating the converter in the discontinuous mode produces a triangular current, whereas operating in the continuous mode produces a trapezoidal current; see Figs. 27 and 28. The current waveshapes in Figs. 27 and 28 have different, effective, or rms values. The larger the ΔI, the larger the rms current for the same output power. The designer of a magnetic component must know the effective or rms value of the current to properly determine the wire size for a given temperature rise or regulation.

The square and triangular waveform would be the extremes for the rms coefficients shown in Fig. 29. The equation for the rms current of a trapezoidal waveform is:

$$I_{rms} = \sqrt{\left(I_{pk}^2 - \left(I_{pk}\Delta I\right) + \frac{\Delta I^2}{3}\right)\frac{t_{on}}{T}}$$

Controlling the Flyback Converter

The flyback converter and its associated waveforms are shown in Fig. 30A. When the switch (S) is closed, the supply is connected across the inductor and the output diode (CR1) is nonconducting. The current rises linearly in the inductor until the switch (S) is opened. At this point, the voltage across the inductor reverses and its stored energy is transferred into the output capacitor and the load circuit. It can be seen from Fig. 30B–D that keeping the switch (S) on and off times equal and keeping the frequency constant results in the output voltage being equal to the input voltage. The waveforms shown in Fig. 30E–G are the effects of halving the on time of the switch and keeping the operating frequency and load resistance constant. This results in the output voltage being halved. This demonstrates the method of controlling the output voltage of a switch-mode power supply (SMPS).

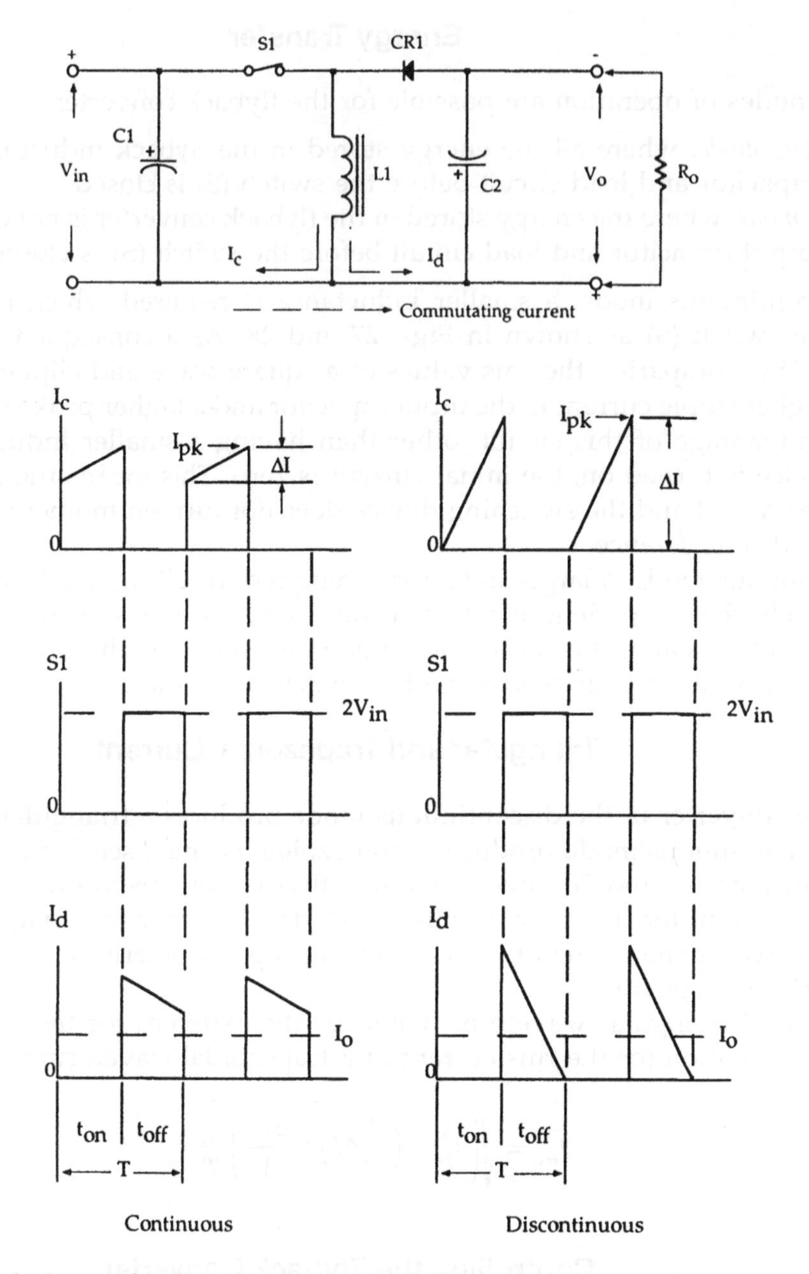

Figure 27 Buck–boost showing continuous and discontinuous waveform comparisons.

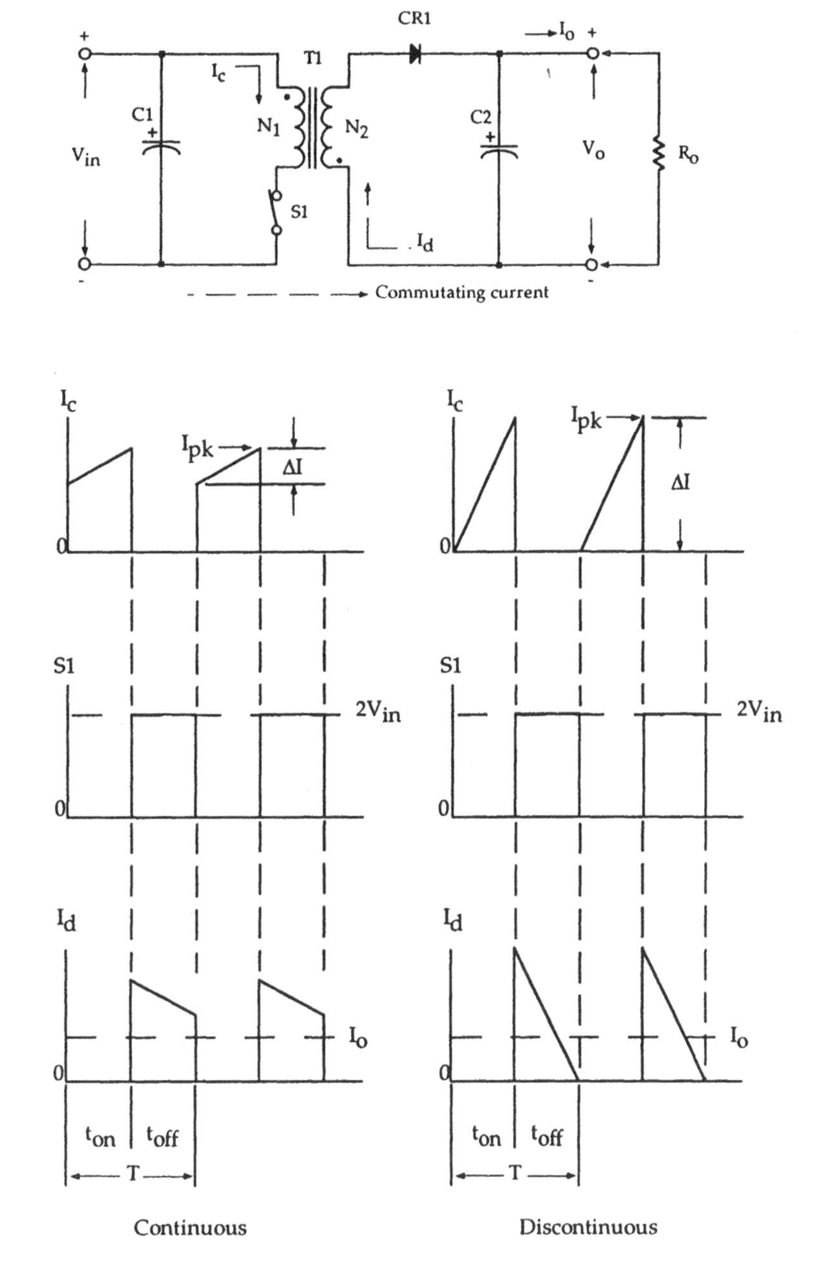

Figure 28 Buck–boost with isolation showing continuous and discontinuous waveforms comparison.

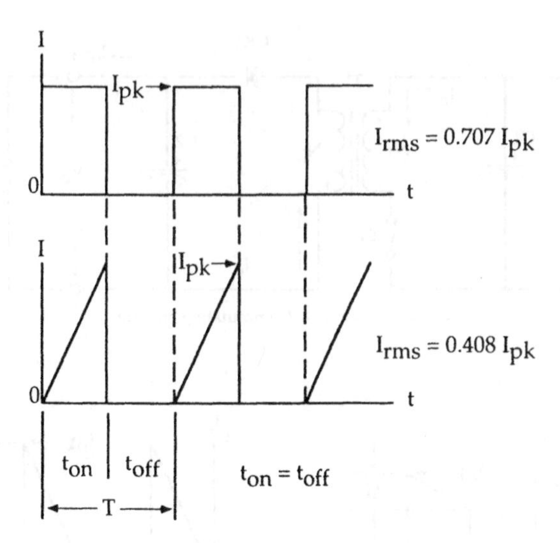

Figure 29 Waveform comparison.

THE FORWARD CONVERTER

The push–pull switching converter is probably the most widely used type of power conversion circuit. In the push–pull converter design, the primary of the transformer can be connected in several ways, as shown in Fig. 31, depending on how you drive the transformer. The push–pull converter is, in fact, an arrangement of two forward converters. The push–pull converters reduces output voltage ripple by doubling the ripple current frequency to the output filter. A further advantage of push–pull operation is that magnetization is applied to the transformer core in both directions. The push–pull converter transformer when subjected to small amounts of dc imbalance can lead to core saturation.

Push–Pull Transformer Consideration

The apparent power P_t of the center-tapped push–pull converter shown in Fig. 31a is

$$P_t = 1.41 P_o \left(\frac{1}{\eta} + 1 \right)$$

If the output circuit was changed to a full-wave bridge, the apparent power P_t would change to

$$P_t = P_o \left(\frac{\sqrt{2}}{\eta} + 1 \right)$$

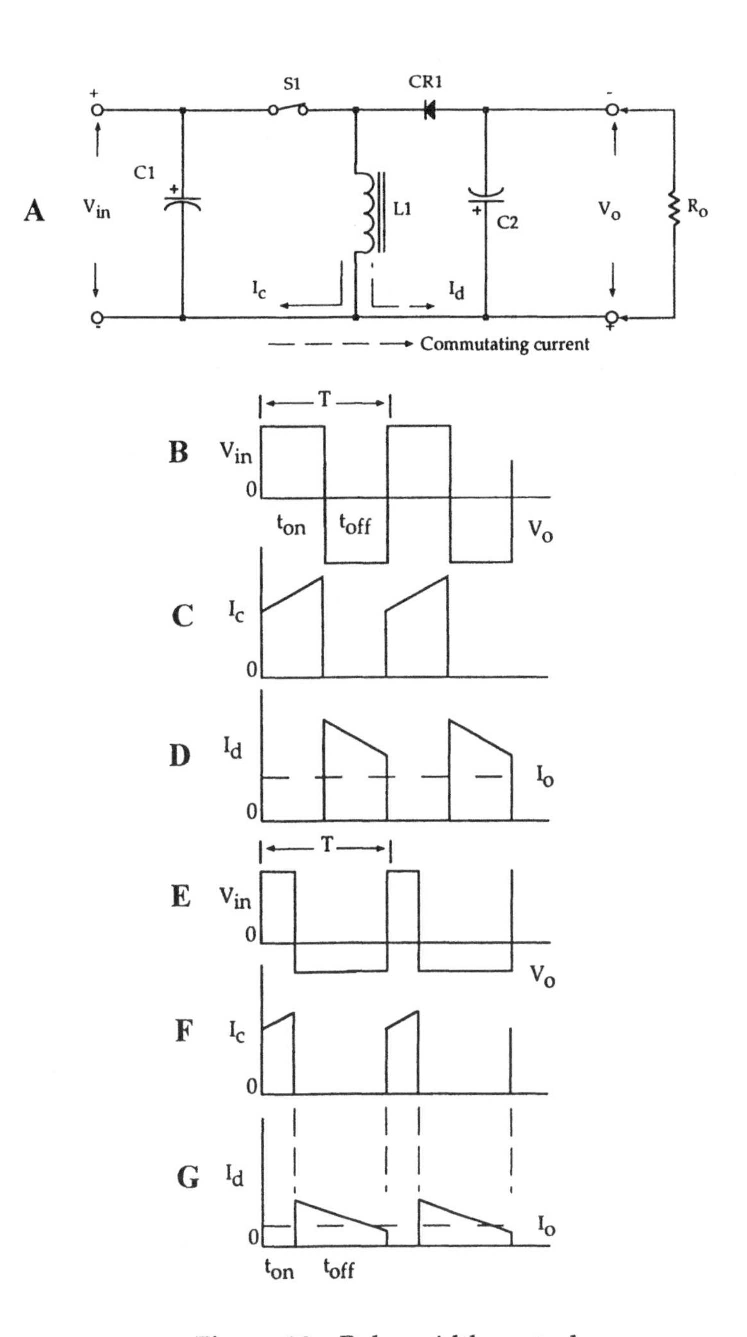

Figure 30 Pulse width control.

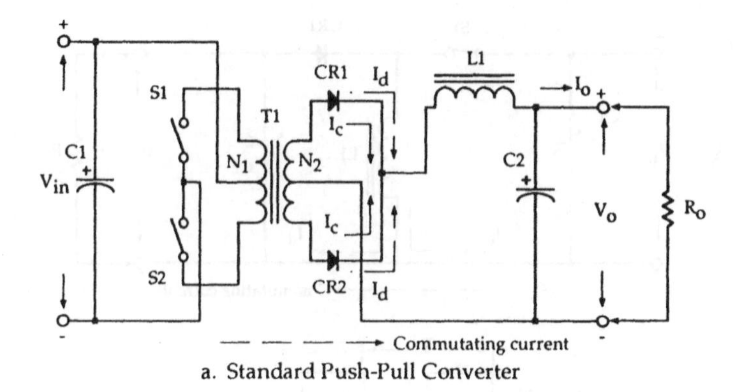

a. Standard Push-Pull Converter

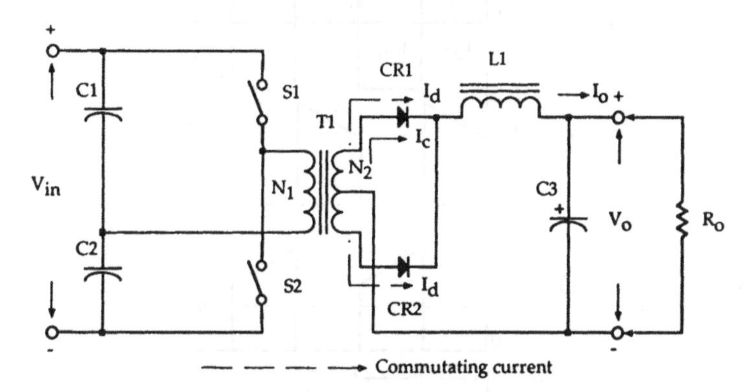

b. Two Transistor Half Bridge Converter

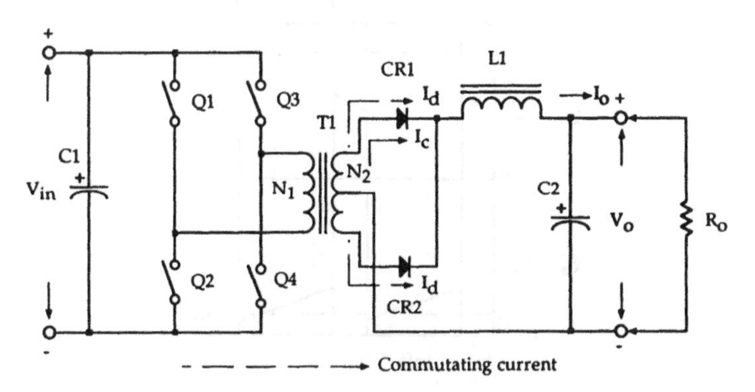

c. Four Transistor Full Bridge Converter

Figure 31 Comparison of push–pull converter circuits.

The voltage across the switch $S_{1\text{-}2}$ is twice the supply voltage V_{in} because of transformer action.

The apparent power P_t of the half-bridge converter shown in Fig. 31b is:

$$P_t = P_o\left(\frac{1}{\eta}+\sqrt{2}\right)$$

If the output circuit was changed to a full-wave bridge, the apparent power P_t would change to:

$$P_t = P_o\left(\frac{1}{\eta}+1\right)$$

The dc voltage at the junction of the two capacitors equals half the supply voltage; therefore, the across the transformer primary is only half in the input voltage V_{in}. The peak voltage across the switch $S_{1\text{-}2}$ equals the supply voltage V_{in}.

The apparent power P_t of the full bridge is the same as the half-bridge. The only difference is that the voltage across the transformer primary is equal to the supply voltage V_{in}. The peak voltage across the switch $S_{1\text{-}4}$ is equal to the supply voltage V_{in}.

SINGLE FORWARD CONVERTER

The forward converter is almost as simple as the one-transistor flyback inductor supply. But its ripple and its output capability are comparable to those of the push–pull converter. The single forward converter does not have the push–pull converter's problem of dc unbalance in the transformer core.

The forward converter and its associated waveforms are shown in Fig. 32. When the switch (S) is closed, the supply is connected across the primary and the rectifier D_1 starts to conduct, and current I_c passes through the output inductor L to the load. During this time, t_{on}, the inductor current ΔI is rising. At the same time, the magnetizing current I_m begins to build up in the transformer primary. The current rises linearly in the transformer primary until the switch (S) is opened. At this point, the voltage across the output inductor reverses and its stored energy is transferred into the load through flyback diode D_2. During this time, t_{off}, the magnetizing current continues to flow through the demagnetizing winding and D_3 back to the source.

Before the switch (S) is turned on again, the magnetizing current must have reached zero, or transformer saturation will occur. If the primary winding to demagnetizing winding turns ratio is assumed to be unity, the maximum duty ratio must not exceed:

$$\mathrm{D} = \frac{t_{on}}{T} = 0.5$$

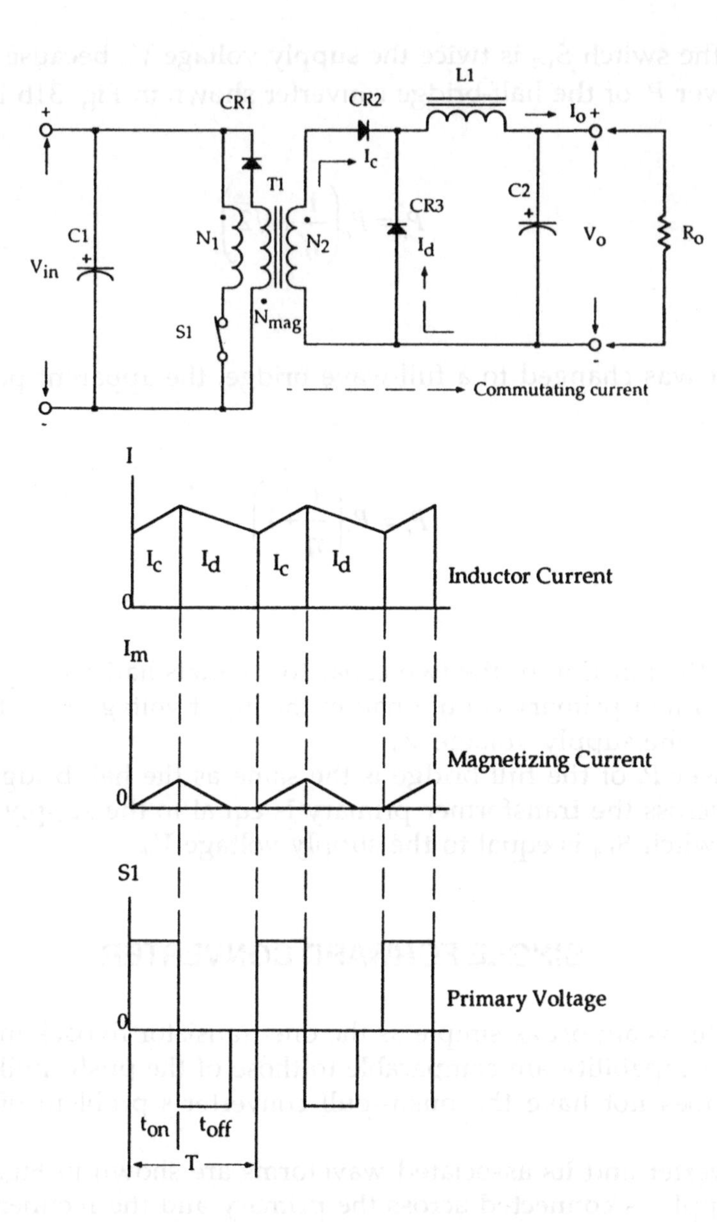

Figure 32 Forward converter.

To ensure smooth transfer of the magnetizing current I_m, the primary and demagnetizing windings must be wound bifilar. The forward converters primary to secondary turns ratio is:

$$n = \frac{DV_{in}}{V_o}$$

If the current in the primary approaches that of a square wave, the apparent power P_t of the forward converter is:

$$P_t = 1.41P_o\left(\frac{1}{\eta} + 1\right)$$

Tapped Inductor

The tapped inductor is used to eliminate premature failure of the switching transistors used in switch regulator circuits. The cause of such failures has been undue stress arising from large voltage spikes. These spikes are attributed by transformer core saturation and the commutating current in the rectifier diodes.

A schematic showing the basic pulse with modulator (PWM) converter circuit is shown in Fig. 33. This circuit combines a converter and switching regulator in one package using push–pull regulator arrangements:

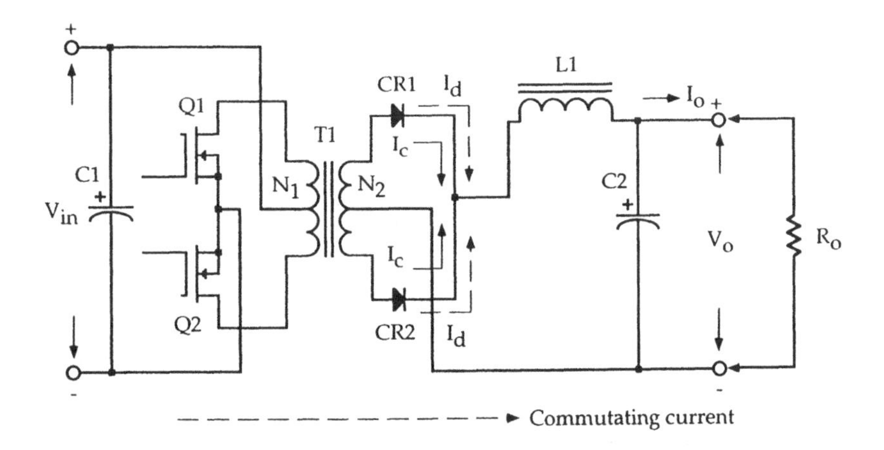

Figure 33 Push–pull two-transistor converter.

In order to simply the discussion, it is assumed that the transistors operate at a 50% duty cycle, and the waveforms have been idealized. Figure 34 shows the load current I_o and the

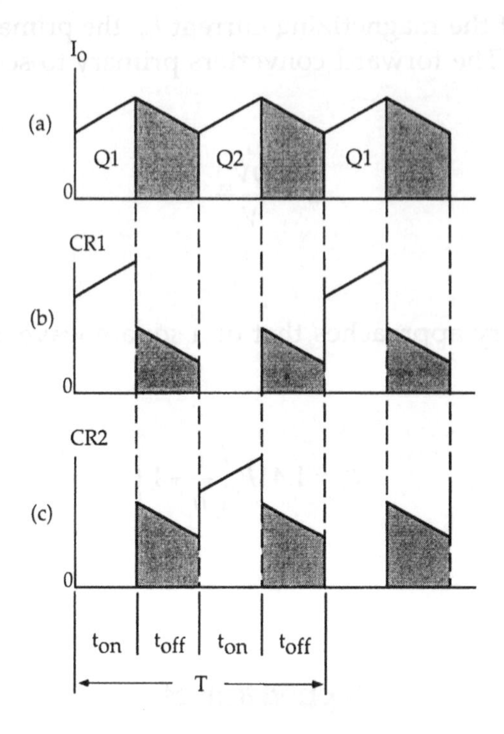

Figure 34 Secondary current waveforms.

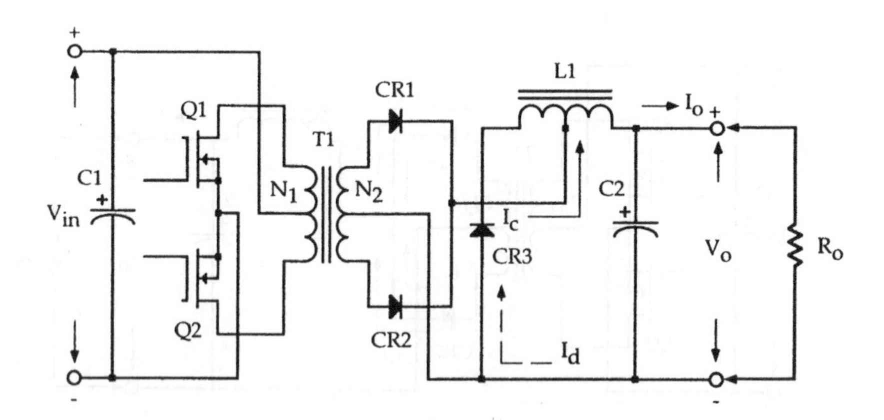

Figure 35 Push–pull converter with tapped inductor.

rectifying diode currents through CR1 and CR2. The shaded portions of Figs. 34b and 34c occur when both power transistors are turned off. The load current I_o of Fig. 34a is the sum of the diode currents shown in Figs. 34b and 34c. The spiking problems of the power transistors Q1 and Q2 are related to the commutating current. When transistor Q1 is turned on, diode CR2 is turned on and is supplying current to the load through L. Now, when transistor Q1 turns off, diode CR2 is still conducting, providing the commutating current for the inductor. Due to transformer action caused by the conduction of diode CR2, diode CR1 starts to conduct to balance the amp-turns in the transformer. This poses two problems. One, the currents are not always balanced in CR1 and CR2, which could cause the core to ratchet to one end or the other of the *B–H* loop and saturate. Second, if CR1 and CR2 are conducting, then when Q2 turns on, it is looking into a momentary short that causes large leading-edge current spikes that could cause premature failure.

The new arrangement is shown in Fig. 35. The difference between the new converter and a conventional PWM converter is that the new converter has a tapped inductor, with overwind N1 and an added diode CR3. The overwind back biases the diodes CR1 and CR2, causing them to turn off and causing current to flow through diode CR3. When the diode CR3 is turned off, the in-rush current is limited by the inductance of the overwind L1. The spiking due to diode turn off is eliminated with the proper amount of overwind. Because the diodes CR1 and CR2 are turned off, a current imbalance does not exist in the secondary, and the core will not saturate.

The amount of overwind needed to turn off the two diodes CR1 and CR2 is very dependent on the leakage inductance of the output transformer. By varying the turns ratio of the two windings of the tapped inductor L, control of the time in which the leakage current goes to zero can be affected. Thus, the higher the reverse voltage to diodes CR1 and CR2, the faster the leakage current falls to zero. The voltage across the leakage inductance of the transformer is:

$$V = L\frac{\Delta I}{\Delta t}$$

$$V = \frac{N_1}{N_2}(V_o + V_d)$$

$$\frac{N_1}{N_2} = \frac{LI_o/\Delta t}{(V_o + V_d)}$$

where:

$$I \cong I_o$$

$N1$ = number of turns on the overwind part of the tapped inductor.

$N2$ = total number of turns.

L = leakage inductance on the secondary of the transformer.

Δt = time for the current to decay to zero.

Vd = 1V assumed drop of one diode.

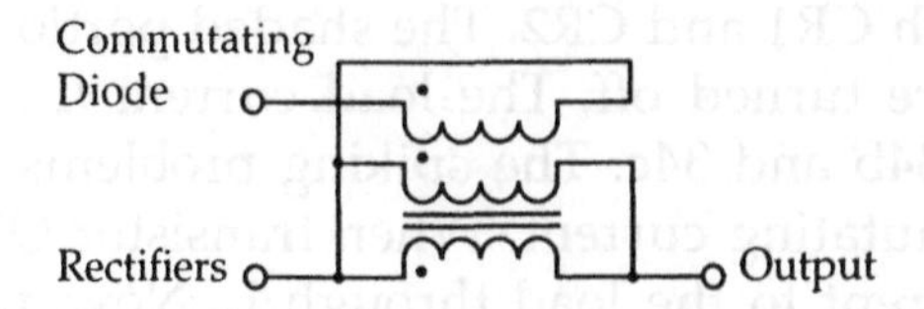

Figure 36 Tapped inductor lead breakout.

The first effort in sizing the inductor for the required watt-second is to multiply the core geometry K_g by 1.5 to make room for the overwinding. Then trifilar wind the inductor and split the windings as shown in Fig. 36.

Output Circuits for Switched-Mode Power Supplies

The converter choice depends on application and performance requirements. The flyback converter is the simplest and least expensive; it is recommended for multioutput supplies with low

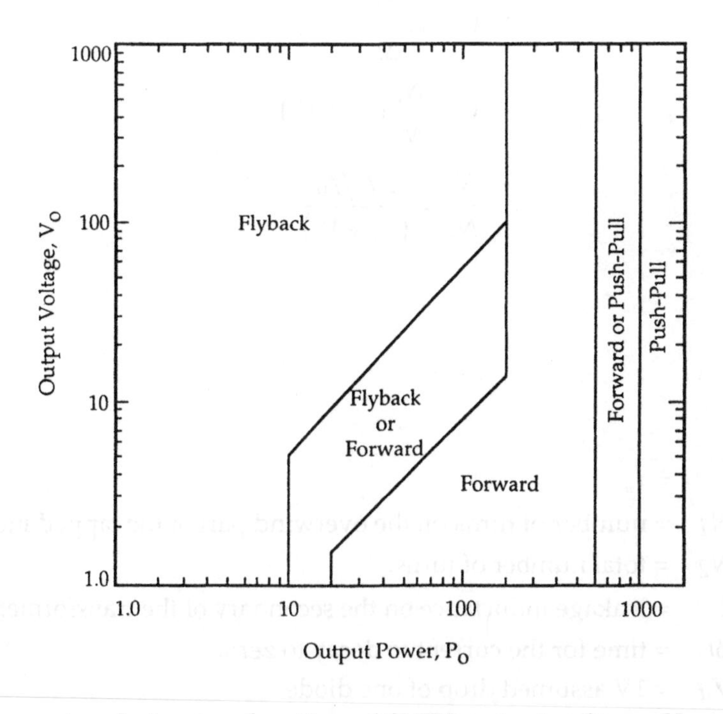

Figure 37 Converter choice as a function of SMPS output voltage, V_o, and output power P_o.

power because each output requires only one diode and one capacitor. For high-performance high power, where the ripple requirement is well below 1%, the push–pull design is the obvious choice.

To select the converter for your switching supply, start by using Fig. 37. Here, converters are typed by their two most distinguishing performance differences: output voltage V_o and output power P_o capability. In the case of the flyback converter, it becomes more difficult to keep the percentage output ripple below an acceptable level as the output power increases and the output voltage decreases.

There are many switched-mode converter circuits. The number of circuits is so large that there is no way to present them all. A few of the most standard circuits are shown in Figs. 38–62, to enable the designer to choose the best circuit for his application. Flyback converter design equations are presented in Table 8.

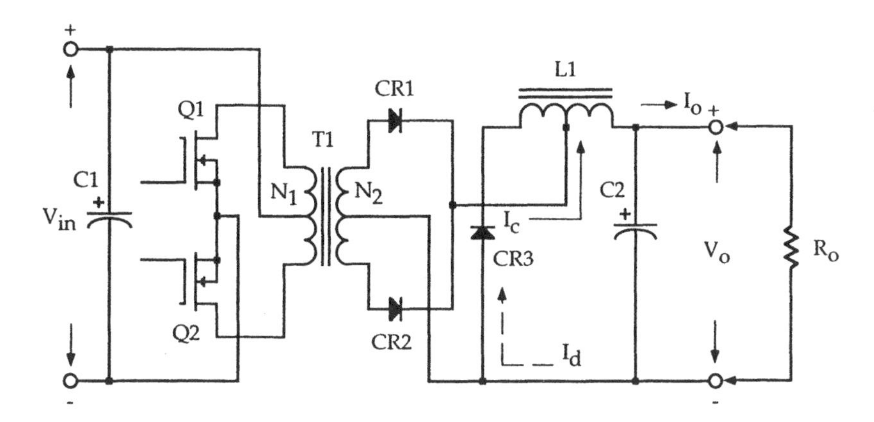

Figure 38 Push–pull converter with tapped inductor.

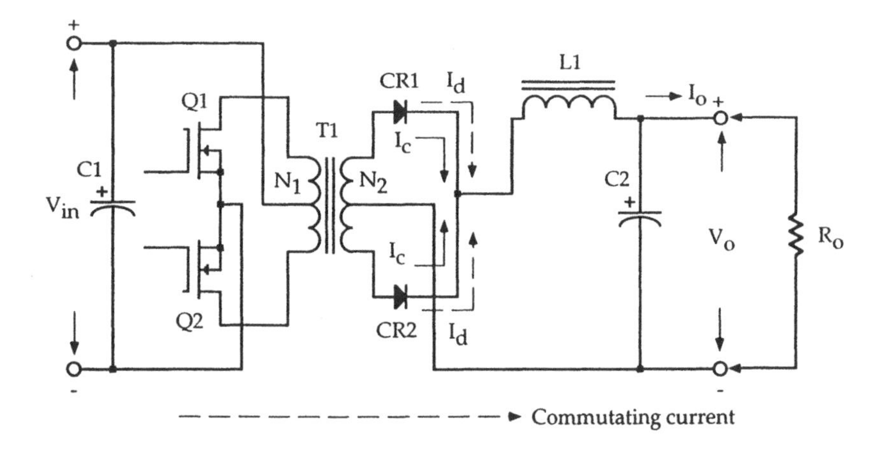

Figure 39 Push–pull two-transistor converter.

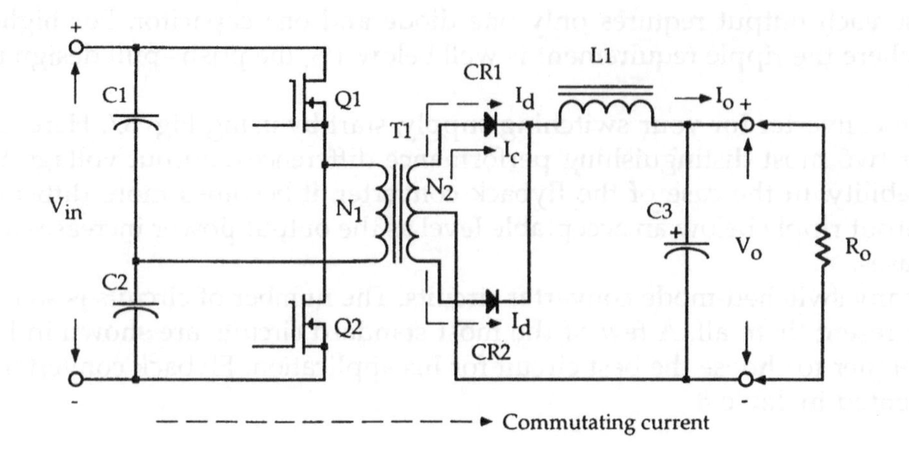

Figure 40 Half-bridge push–pull converter.

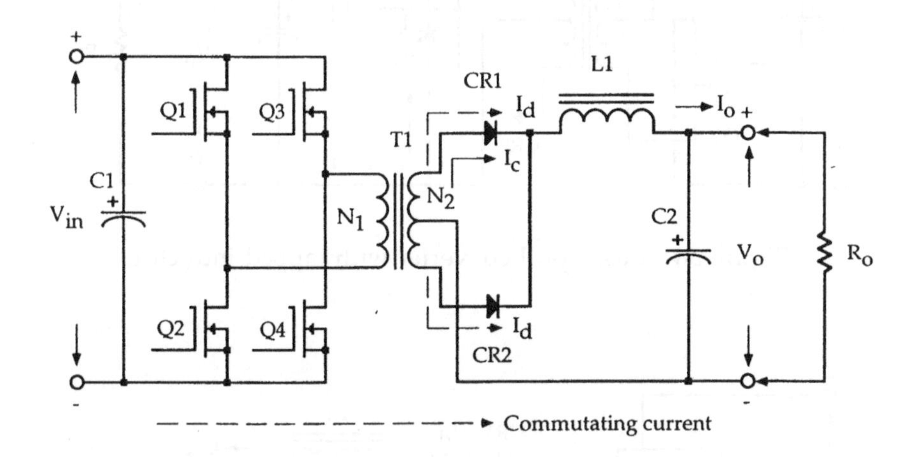

Figure 41 Push–pull four-transistor bridge converter.

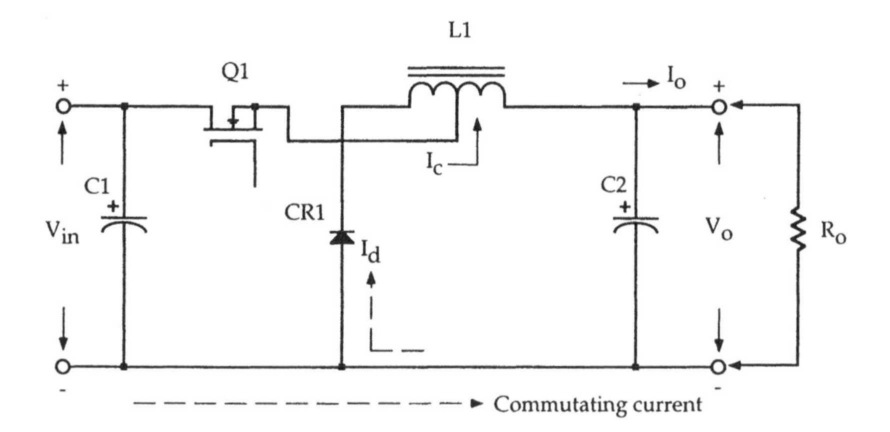

Figure 42 Single-ended buck converter with tapped inductor.

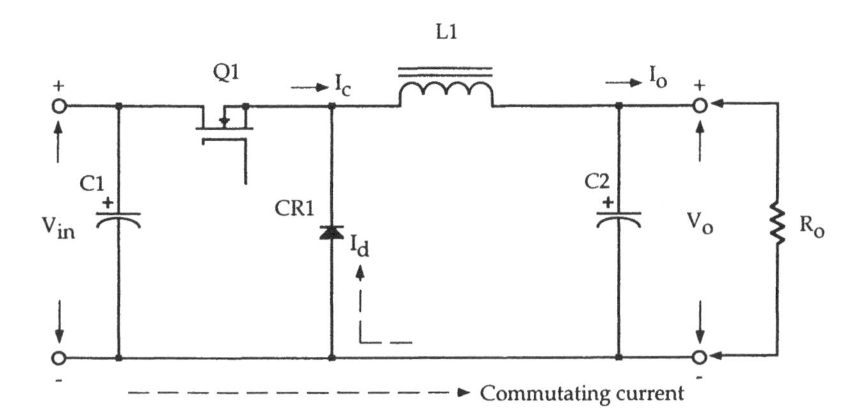

Figure 43 Single-ended buck converter.

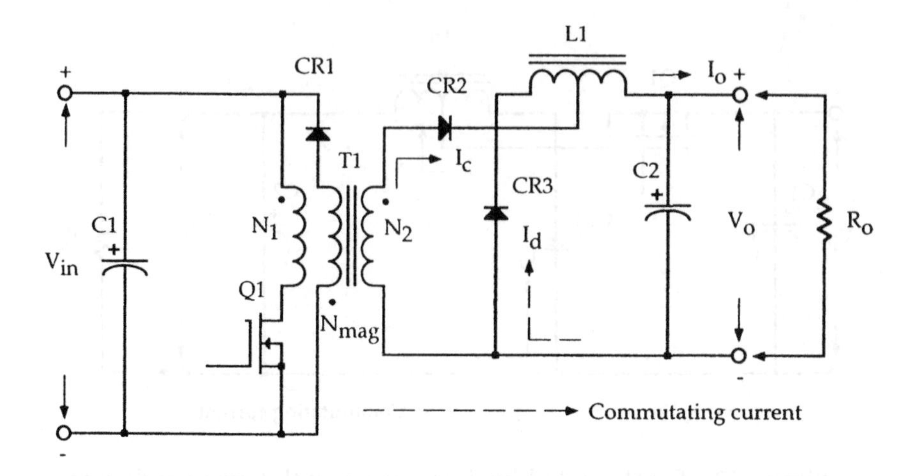

Figure 44 Single-ended forward converter with tapped inductor.

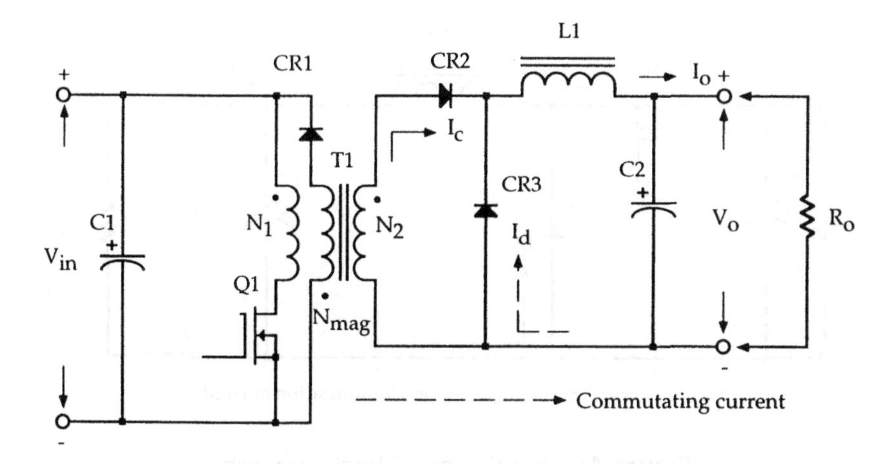

Figure 45 Single-ended forward converter.

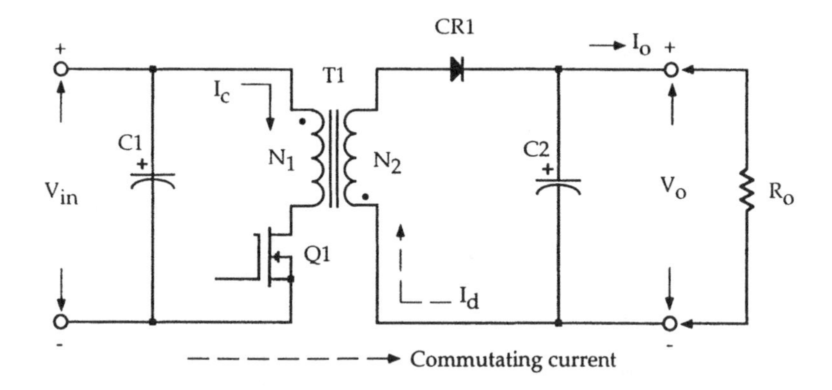

Figure 46 Single-ended buck–boost flyback converter.

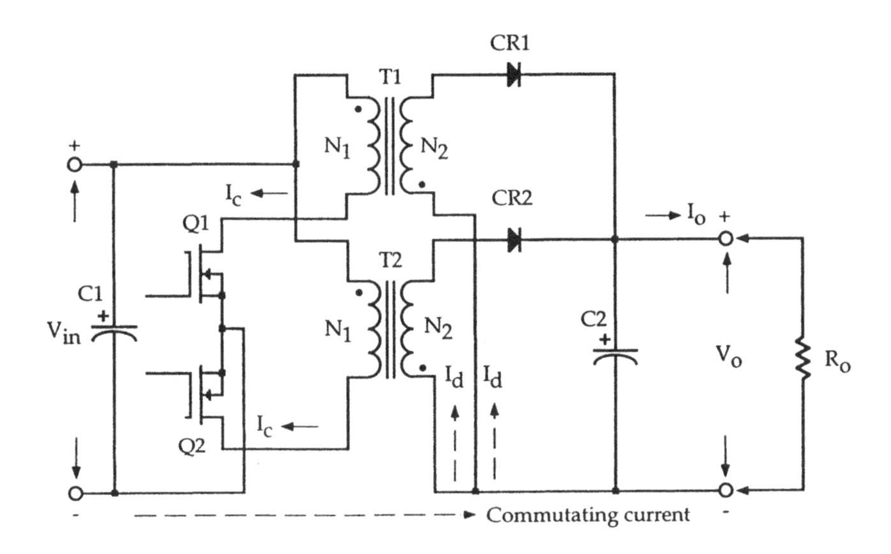

Figure 47 Push–pull buck–boost flyback converter.

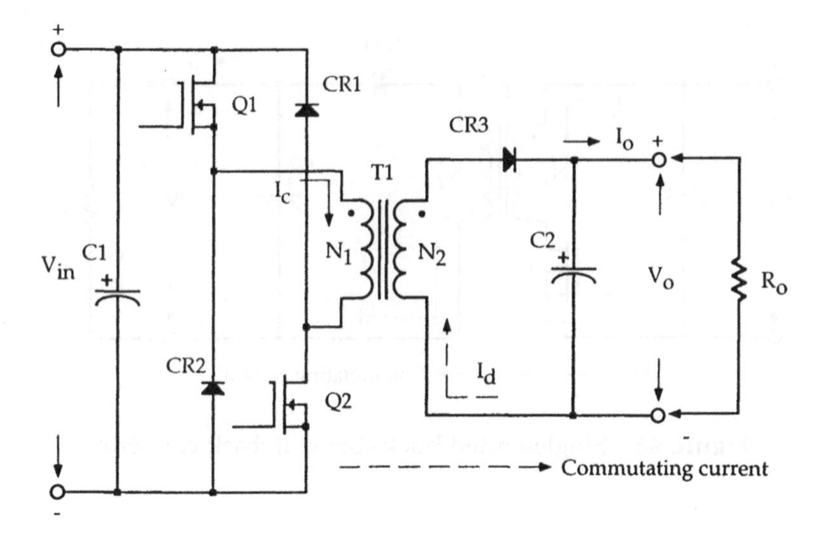

Figure 48 Single-ended two-transistor buck–boost flyback converter with diode clamp.

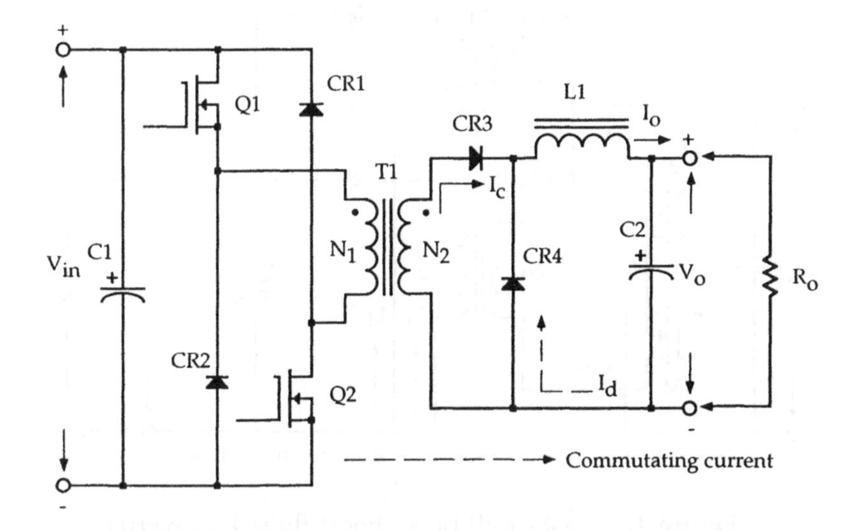

Figure 49 Single-ended two-transistor forward converter with diode clamp.

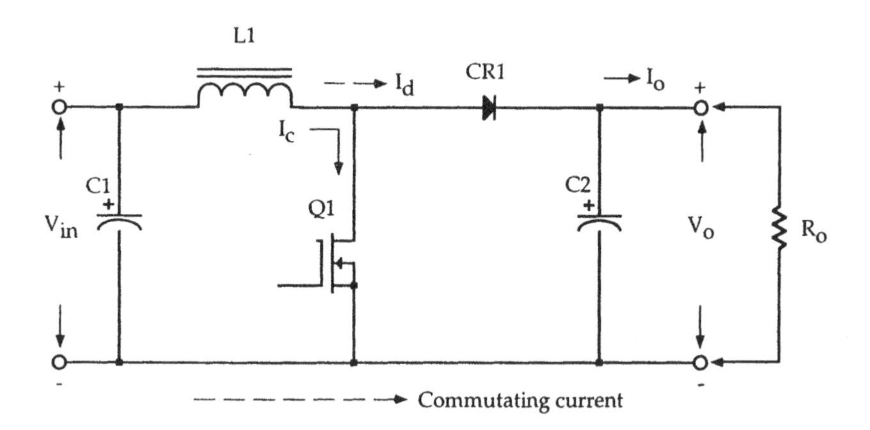

Figure 50 Single-ended boost flyback converter.

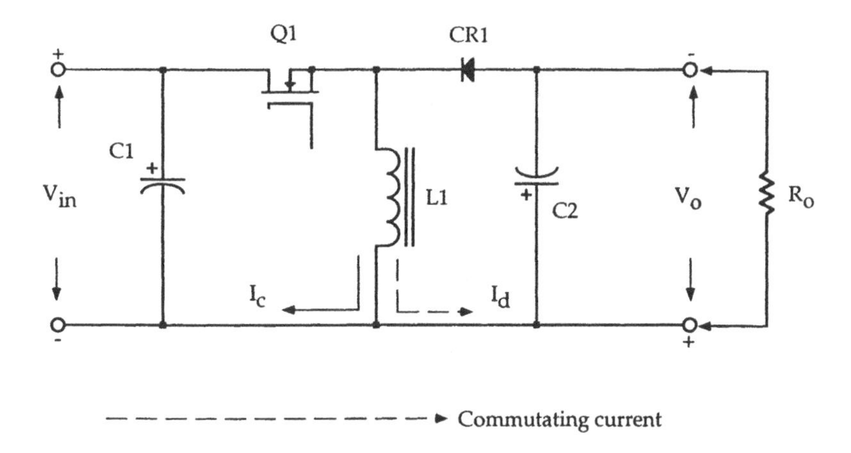

Figure 51 Single-ended flyback converter.

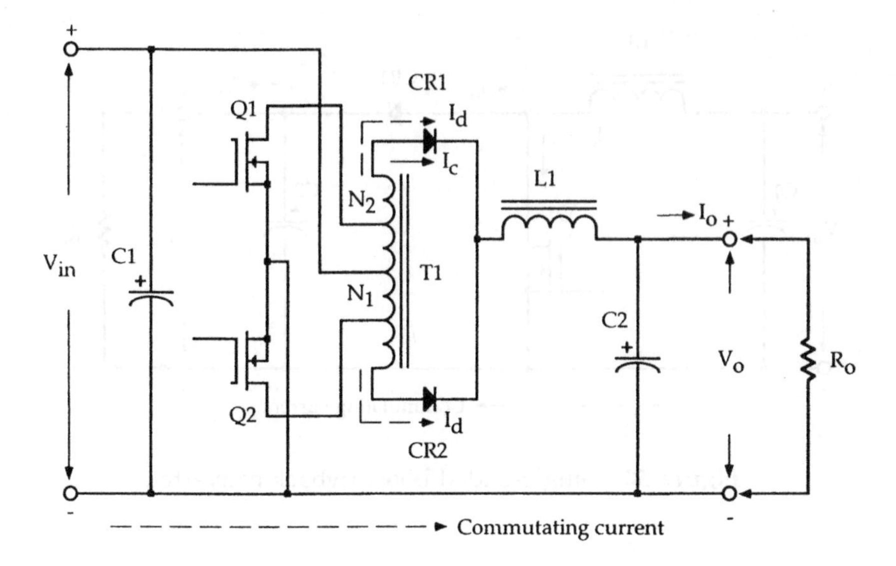

Figure 52 Push–pull boost converter.

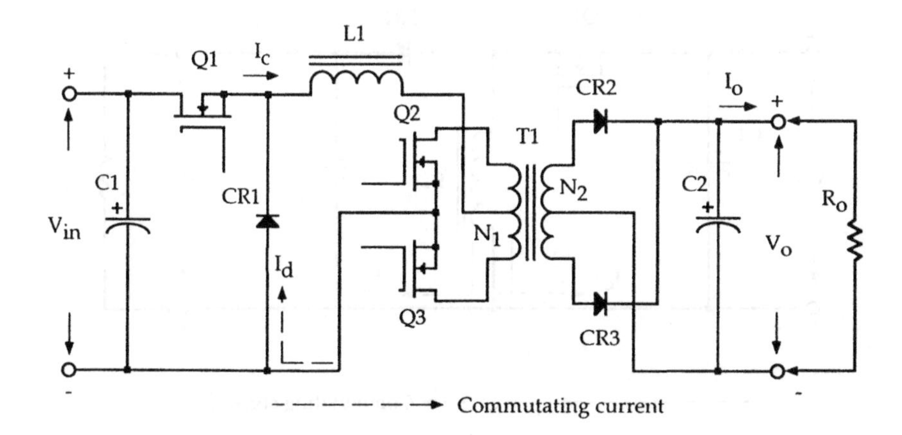

Figure 53 Current-fed dc-to-dc converter.

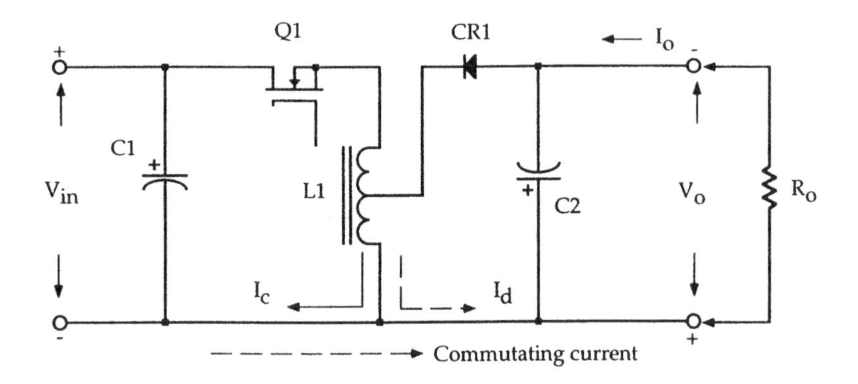

Figure 54 Single-ended step-down flyback converter.

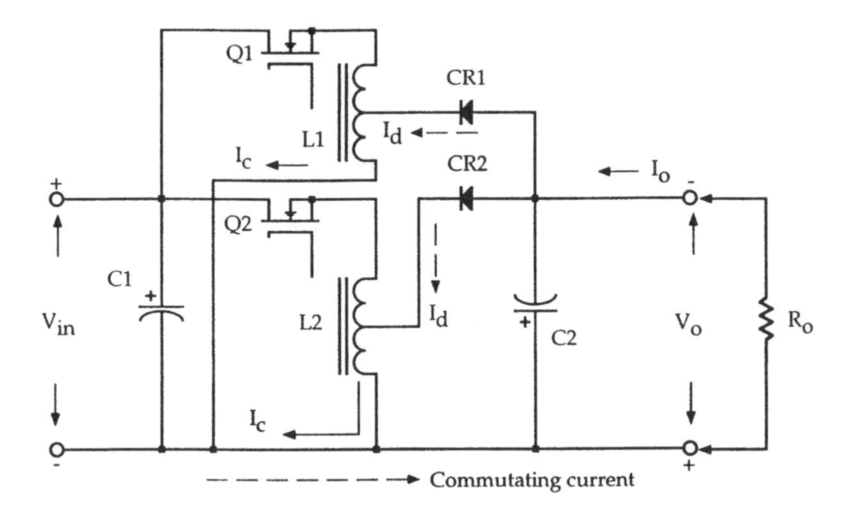

Figure 55 Push–pull step-down flyback converter.

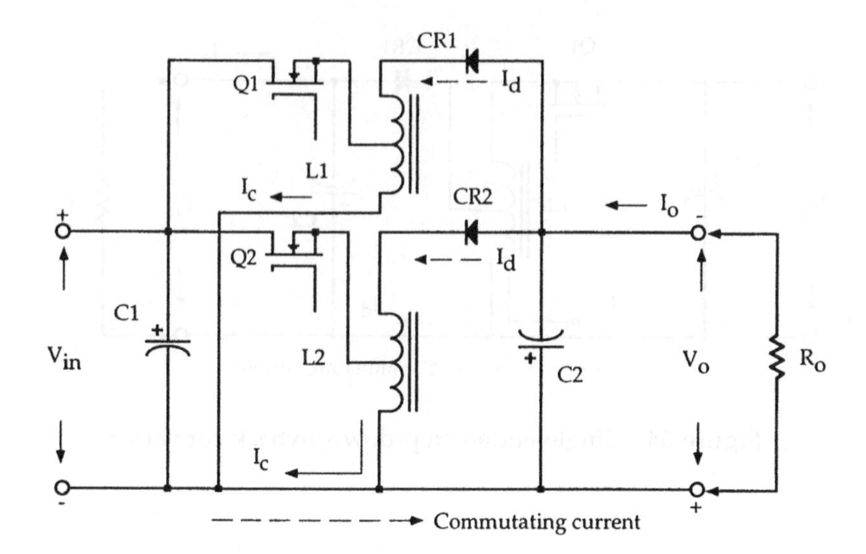

Figure 56 Push–pull step-up flyback converter.

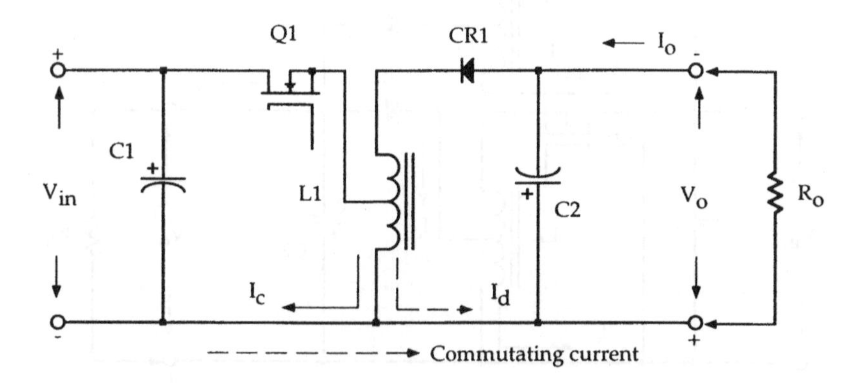

Figure 57 Single-ended step-up flyback converter.

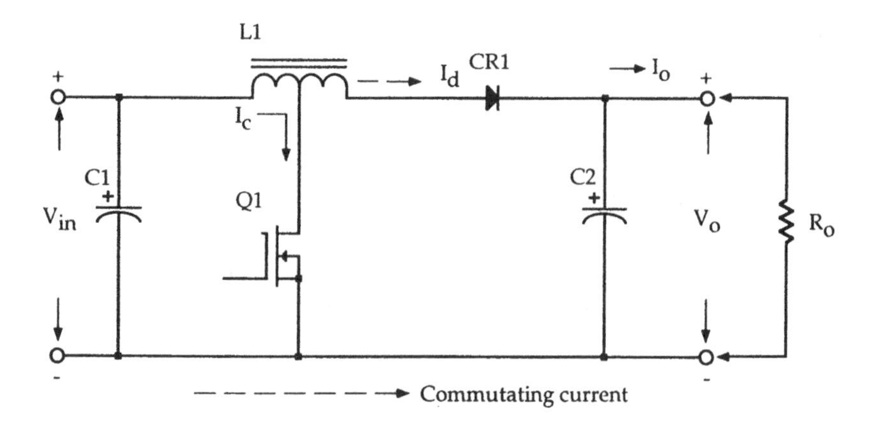

Figure 58 Single-ended boost flyback converter with overwind.

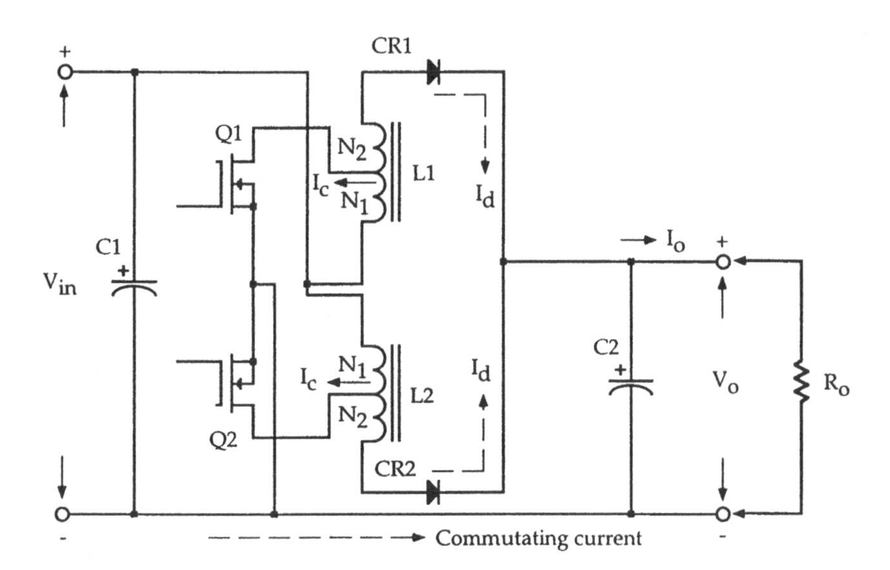

Figure 59 Push–pull boost flyback converter.

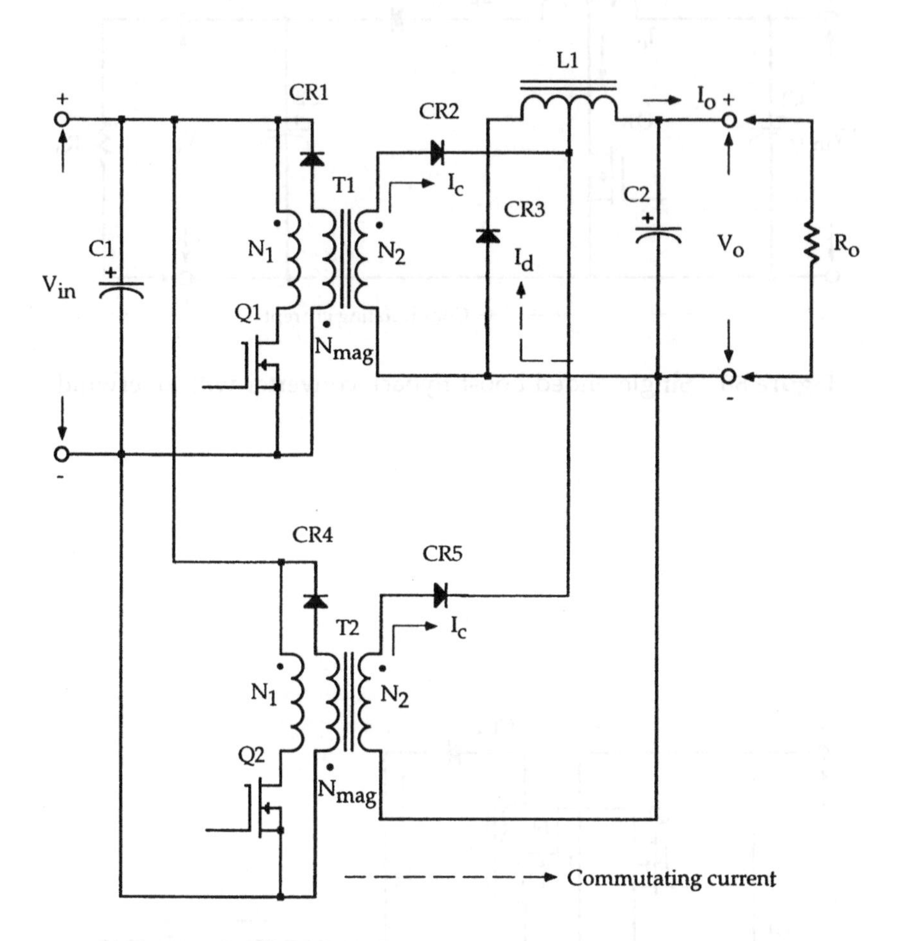

Figure 60 Push–pull forward converter with tapped inductor.

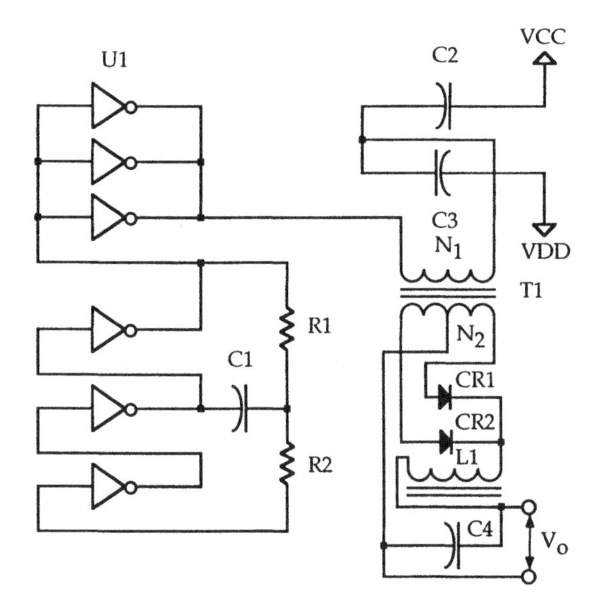

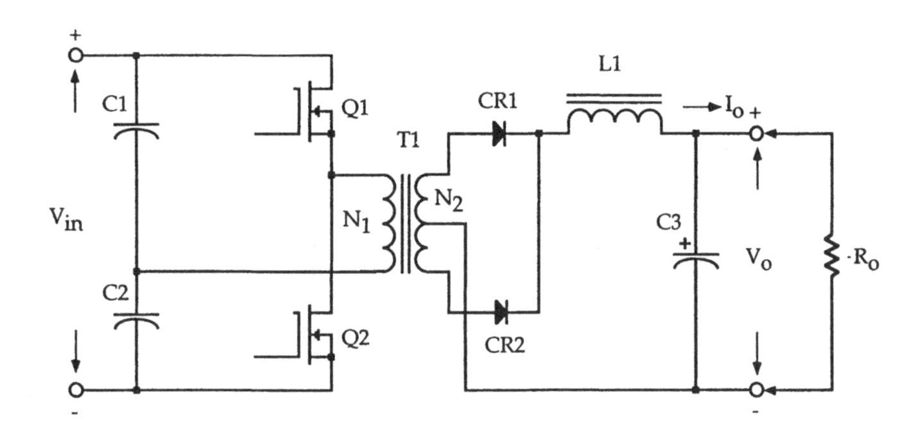

Figure 61 Milliwatt converter.

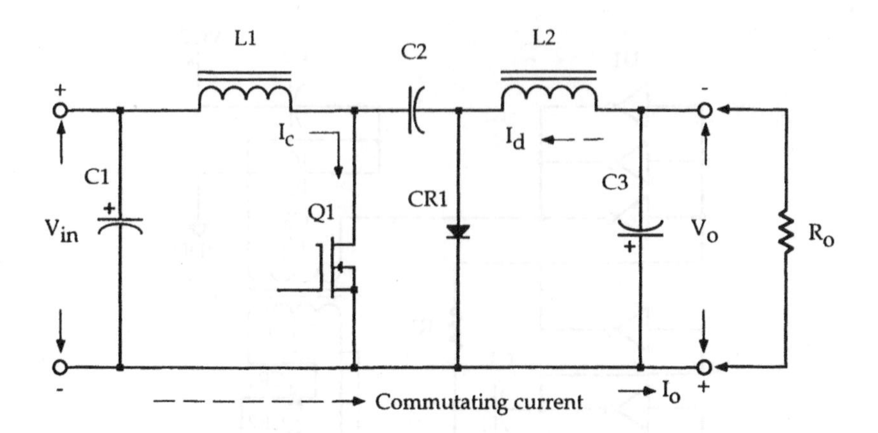

Figure 62 Ćuk converter.

Table 8 Flyback Converter Design Equation

	Buck	Boost	Buck-Boost
D	$\frac{V_o}{(V_{in}\eta)}$	$1-\frac{(V_{in}\eta)}{V_o}$	$\frac{V_o}{(V_{in}\eta+V_o)}$
$L_{(min)}$	$\frac{R_oT(1-D)}{2}$	$\frac{R_oTD(1-D)^2}{2}$	$\frac{R_oT(1-D)^2}{2}$
ΔI	$\frac{V_{in}TD(1-D)}{L}$	$\frac{V_{in}TD}{L}$	$\frac{V_{in}TD}{L}$
I_L	I_o	$\frac{I_o}{(1-D)}$	$\frac{I_o}{(1-D)}$
I_{pk}	$I_o+\frac{V_{in}TD(1-D)}{2L}$	$\frac{I_o}{(1-D)}+\frac{V_{in}TD}{2L}$	$\frac{I_o}{(1-D)}+\frac{V_{in}TD}{2L}$

3

TRANSFORMER AND INDUCTOR DESIGN EXAMPLES

INTRODUCTION

This chapter of the book describes a new method for designing magnetic components for electronic equipment. Detailed design procedures using this method and design examples for transformers and inductors are given. The new method is based on the geometry of the magnetic core. The geometry is symbolized by the constant K_g. The design is an improvement over the A_p (area product) method popularly used today and first introduced in 1927.

The area product method is described and compared to a new K_g method. Design examples using the K_g method are given. Using the core geometry K_g and this handbook, the engineer after one iteration will know the magnetic component's weight, volume, and regulation, and the component will be optimized for its operating frequency and temperature rise.

60-Hz TRANSFORMER DESIGN EXAMPLE USING THE AREA PRODUCT A_p APPROACH

For a typical design example, assume a step-down isolation transformer with the following specification:

1. Input voltage, V_{in} = 230 V
2. Output voltage, V_o = 115 V
3. Output current, I_o = 8.7 A
4. Frequency, f = 60 Hz
5. Temperature rise, 50°C
6. Efficiency, η = 95%
7. Core, lamination 14 mil

Step No. 1. Calculate the output power P_o:

$$P_o = VA, \quad \text{(watts)}$$

$$P_o = (115)(8.7)$$

$$P_o = 1000, \quad \text{(watts)}$$

Step No. 2. Calculate the apparent power P_t:

$$P_t = P_o\left(\frac{1}{\eta}+1\right), \quad \text{(watts)}$$

$$P_t = (1000)\left(\frac{1}{0.95}+1\right)$$

$$P_t = 2052, \quad \text{(watts)}$$

Step No. 3. Calculate the area product A_p:

$$A_p = \left(\frac{P_t x 10^4}{K_f K_u K_j f B_m} \right)^{1.14}, \quad (\text{cm}^4)$$

Assuming:

$K_f = 4.44$, sine wave

$K_u = 0.4$, window utilization factor

$K_j = 534$, see Table 1

$B_m = 1.3$, (tesla)

$$A_p = \left(\frac{2052 x 10^4}{(4.44)(0.4)(534)(60)(1.3)} \right)^{1.14}$$

$$A_p = 609, \quad (\text{cm}^4)$$

Step No. 4. Select a comparable area product A_p from the lamination section in column 12 that is approximately 10% larger than necessary because the area product A_p is listed in gross terms. The area product will have to be multiplied by the stacking factor SF found in Table 10. Also record the gross iron weight W_{tfe} column 7, MLT column 9, gross iron cross section A_c column 10, surface area A_t column 14.

Lamination, 19EI - 1

$A_{p(\text{gross})} = 669, (\text{cm}^4)$

$W_{tfe(\text{gross})} = 4889, (\text{grams})$

$A_{c(\text{gross})} = 19.76, (\text{cm}^2)$

$\text{MLT} = 32.6, (\text{cm})$

$A_t = 1067, (\text{cm}^2)$

$SF = 0.9$

page =

Step No. 5. Calculate the effective area product A_p recorded in Step 4:

$$A_p = (SF)\left(A_{p(gross)}\right), \quad \left(cm^4\right)$$
$$A_p = (0.90)(669)$$
$$A_p = 602, \quad \left(cm^4\right)$$

After the area product A_p has been determined, the geometry of the transformer can be evaluated, such as volume and weight.

Step No. 6. Calculate the effective iron area A_c recorded in Step 4:

$$A_c = (SF)\left(A_{c(gross)}\right), \quad \left(cm^2\right)$$
$$A_c = (0.90)(19.76)$$
$$A_c = 17.8, \quad \left(cm^2\right)$$

Step No. 7. Calculate the number of primary turns:

$$N_p = \left(\frac{V_p x10^4}{K_f B_m f A_c}\right), \quad \text{(turns)}$$
$$N_p = \left(\frac{230x10^4}{(4.44)(1.3)(60)(17.8)}\right)$$
$$N_p = 373, \quad \text{(turns)}$$

Step No. 8. Calculate the primary current I_p:

$$I_p = \frac{P_o}{V_p \eta}, \quad \text{(amps)}$$
$$I_p = \frac{(1000)}{(230)(0.95)}$$
$$I_p = 4.57, \quad \text{(amps)}$$

Step No. 9. Calculate the current density J:

$$J = \left(\frac{P_t x10^4}{K_f K_u f B_m A_p}\right), \quad \text{(amps/cm}^2\text{)}$$

$$J = \left(\frac{2052x10^4}{(4.44)(0.4)(60)(1.3)(602)}\right)$$

$$J = 246, \quad \text{(amps/cm}^2\text{)}$$

Step No. 10. Calculate the bare wire size $A_{w(B)}$ for the primary:

$$A_{w(B)} = \frac{I_p}{J}, \quad \text{(cm}^2\text{)}$$

$$A_{w(B)} = \frac{(4.57)}{(246)}$$

$$A_{w(B)} = 0.0186, \quad \text{(cm}^2\text{)}$$

Step No. 11. Select a wire size from the wire table column 2. Remember, if the wire area is not within 10%, take the next smallest size. Also record micro-ohm per centimeter found in column 4:

$$\text{AWG \#15 with } 0.0165, \quad \text{(cm}^2\text{)}$$

$$\frac{\mu\Omega}{\text{cm}} = 104$$

Step No. 12. Calculate the primary winding resistance. Use MLT from Step 4 and micro-ohm per centimeter from Step 11:

$$R_p = (\text{MLT})(N)\left(\frac{\mu\Omega}{\text{cm}}\right)x10^{-6}, \quad \text{(ohms)}$$

$$R_p = (32.6)(373)(104)x10^{-6}$$

$$R_p = 1.26, \quad \text{(ohms)}$$

Step No. 13. Calculate the primary copper loss P_p:

$$P_p = (I_p)^2 R_p, \quad \text{(watts)}$$
$$P_p = (4.57)^2(1.26)$$
$$P_p = 26.3, \quad \text{(watts)}$$

Step No. 14. Calculate the secondary turns:

$$N_s = \left(\frac{N_p V_s}{V_p}\right), \quad \text{(turns)}$$
$$N_s = \frac{(373)(115)}{(230)}$$
$$N_s = 186, \quad \text{(turns)}$$

Step No. 15. Calculate the bare wire size $A_{sw(B)}$ for the secondary:

$$A_{sw(B)} = \frac{I_o}{J}, \quad (\text{cm}^2)$$
$$A_{sw(B)} = \frac{(8.7)}{(246)}$$
$$A_{sw(B)} = 0.0354, \quad (\text{cm}^2)$$

Step No. 16. Select a wire size from the wire table, column 2. Remember, if the wire area is not within 10%, take the next smallest size. Also, record micro-ohm per centimeter found in column 4:

$$\text{AWG \#12 with } 0.033, \quad (\text{cm}^2)$$
$$\frac{\mu\Omega}{\text{cm}} = 52.1$$

Step No. 17. Calculate the secondary winding resistance. Use MLT from Step 4 and micro-ohm per centimeter found in Step 16:

$$R_s = (\text{MLT})(N)\left(\frac{\mu\Omega}{\text{cm}}\right)\text{x}10^{-6}, \quad \text{(ohms)}$$

$$R_s = (32.6)(186)(52.1)\text{x}10^{-6}$$

$$R_s = 0.316, \quad \text{(ohms)}$$

Step No. 18. Calculate the secondary copper loss P_s:

$$P_s = (I_o)^2 R_s, \quad \text{(watts)}$$

$$P_s = (8.7)^2(0.316)$$

$$P_s = 23.9, \quad \text{(watts)}$$

Step No. 19. Calculate the effective core weight. The gross W_{tfe} and SF are found in Step 4:

$$W_{tfe(\text{eff})} = (\text{SF})W_{tfe(\text{gross})}, \quad \text{(grams)}$$

$$W_{tfe(\text{eff})} = (0.90)(4889)$$

$$W_{tfe(\text{eff})} = 4400, \quad \text{(grams)}$$

Step No. 20. Calculate core loss P_{fe} using the appropriate core loss curves for 14-mil lamination at 1.3 tesla:

$$P_{fe} = \left(\frac{\text{milliwatts}}{\text{gram}}\right)W_{tfe(\text{gross})}\text{x}10^{-3}, \quad \text{(watts)}$$

$$P_{fe} = (0.85)(4400)\text{x}10^{-3}$$

$$P_{fe} = 3.74, \quad \text{(watts)}$$

Step No. 21. Summarize the losses and compare with the total losses P_ε:

$$P_p = 26.1, \text{ Primary}, \quad \text{(watts)}$$
$$P_s = 23.6, \text{ Secondary}, \quad \text{(watts)}$$
$$P_{fe} = 3.84, \text{ Core}, \quad \text{(watts)}$$
$$P_\Sigma = 53.5, \text{ Total}, \quad \text{(watts)}$$

The total power loss in the transformer is 53.5 W, which will effectively meet the required 95% efficiency.

Step No. 22. Calculate the watts per unit area. The surface area A_t is found in Step 4:

$$\psi = \frac{P_\Sigma}{A_t}, \quad \left(\text{watts/cm}^2\right)$$
$$\psi = \frac{(53.5)}{(1067)}$$
$$\psi = 0.0501, \quad \left(\text{watts/cm}^2\right)$$

where:

$$\psi = 0.07 \text{ W/cm}^2 \text{ at } 50°\text{C rise}$$

Step No. 23. Calculate the transformer regulation:

$$\alpha = \frac{P_{cu}}{P_o + P_{cu}}(100), \quad (\%)$$
$$P_{cu} = P_p + P_s, \quad \text{(watts)}$$
$$P_{cu} = (26.3) + (23.9)$$
$$P_{cu} = 50.2, \quad \text{(watts)}$$
$$\alpha = \frac{50.2}{(1000) + (50.2)}(100)$$
$$\alpha = 4.78, \quad (\%)$$

60-Hz TRANSFORMER DESIGN EXAMPLE USING THE CORE GEOMETRY K_g APPROACH

For a typical design example, assume a full-wave center-tapped circuit shown in Fig. 23 with the following specification:

1. Input voltage, V_{in} = 115 V
2. Output voltage, V_o = 28 + V_d
3. Output current, I_o = 5 A
4. Frequency, f = 60 Hz
5. Temperature rise, 25°C
6. Efficiency, η = 95%
7. Regulation, α = 1%
8. Core, lamination 14 mil

Step No. 1. Calculate the transformer output power P_o allowing for a 1.0 V diode drop (V_d) assumed:

$$P_o = VA, \quad \text{(watts)}$$
$$V = 28 + 1, \quad \text{(volts)}$$
$$P_o = (29)(5)$$
$$P_o = 145, \quad \text{(watts)}$$

Step No. 2. Calculate the apparent power P_t:

$$P_t = P_o\left(\frac{1}{\eta} + \sqrt{2}\right), \quad \text{(watts)}$$
$$P_t = (145)\left(\frac{1}{(0.95)} + (1.41)\right)$$
$$P_t = 357, \quad \text{(watts)}$$

Step No. 3. Calculate the electrical conditions:

$$K_e = 0.145 K_f^2 f^2 B_m^2 \text{x}10^{-4}$$
$$K_f = 4.44, \text{ sine wave}$$
$$K_e = (0.145)(4.44)^2(60)^2(1.3)^2 \text{x}10^{-4}$$
$$K_e = 1.74$$

Step No. 4. Calculate the core geometry K_g:

$$K_g = \frac{P_t}{2K_e\alpha}, \quad (\text{cm}^5)$$

$$K_g = \frac{(357)}{2(1.74)(1)}$$

$$K_g = 102, \quad (\text{cm}^5)$$

Step No. 5. Select a comparable core geometry K_g from the lamination section in column 13 that is approximately 20% larger than necessary because the core geometry is listed in gross terms. The core geometry will have to be multiplied by the stacking factor squared $(\text{SF})^2$ found in Table 10. Also record the gross iron weight W_{tfe} column 7, MLT column 9, gross iron cross section A_c column 10, surface area A_t column 14, area product A_p column 12:

Lamination, 150EI - 2

$$K_{g(\text{gross})} = 125, \quad (\text{cm}^5)$$

$$A_{p(\text{gross})} = 316, \quad (\text{cm}^4)$$

$$W_{tfe(\text{gross})} = 5064, \quad (\text{grams})$$

$$A_{c(\text{gross})} = 29, \quad (\text{cm}^2)$$

$$\text{MLT} = 30.1, \quad (\text{cm})$$

$$A_t = 639, \quad (\text{cm}^2)$$

$$\text{SF} = 0.90$$

$$(\text{SF})^2 = 0.81$$

Step No. 6. Calculate the effective core geometry K_g:

$$K_g = (\text{SF})^2 K_{g(\text{gross})}, \quad (\text{cm}^5)$$

$$K_g = (0.81)(125)$$

$$K_g = 101, \quad (\text{cm}^5)$$

After the core geometry K_g has been determined, the geometry of the transformer can be evaluated, such as volume and weight.

Step No. 7. Calculate the effective iron area A_c recorded in Step 5; the staking factor is found in Table 10:

$$A_c = (\text{SF})A_{c(gross)}, \quad (\text{cm}^2)$$
$$A_c = (0.90)(29)$$
$$A_c = 26.1, \quad (\text{cm}^2)$$

Step No. 8. Calculate the number of primary turns:

$$N_p = \left(\frac{V_p \text{x} 10^4}{K_f B_m f A_c}\right), \quad (\text{turns})$$
$$N_p = \left(\frac{115\text{x}10^4}{(4.44)(1.3)(60)(26.1)}\right)$$
$$N_p = 127, \quad (\text{turns})$$

Step No. 9. Calculate the primary current I_p:

$$I_p = \frac{P_o}{V_p \eta}, \quad (\text{amps})$$
$$I_p = \frac{(145)}{(115)(0.95)}$$
$$I_p = 1.33, \quad (\text{amps})$$

Step No. 10. Calculate the current density J:

$$J = \left(\frac{P_t \text{x} 10^4}{K_f K_u f B_m A_p} \right), \quad (\text{amps/cm}^2)$$

$$A_p = (\text{SF}) A_{p(gross)}, \quad (\text{cm}^4)$$

$$A_p = (0.90)(316)$$

$$A_p = 284, \quad (\text{cm}^4)$$

$$K_u = 0.40, \text{ window utilization factor}$$

$$J = \left(\frac{357 \text{x} 10^4}{(4.44)(0.4)(60)(1.3)(284)} \right)$$

$$J = 90.7, \quad (\text{amps/cm}^2)$$

Step No. 11. Calculate the bare wire size $A_{w(B)}$ for the primary:

$$A_{w(B)} = \frac{I_p}{J}, \quad (\text{cm}^2)$$

$$A_{w(B)} = \frac{(1.33)}{(90.7)}$$

$$A_{w(B)} = 0.0146, \quad (\text{cm}^2)$$

Step No. 12. Select a wire size from the wire table column 2. Remember, if the wire area is not within 10%, take the next smallest size. Also, record micro-ohm per centimeter found in column 4.

$$\text{AWG \#16 with } 0.01307, \quad (\text{cm}^2)$$

$$\frac{\mu\Omega}{\text{cm}} = 132$$

Step No. 13. Calculate the primary winding resistance. Use MLT from Step 5 and micro-ohm per centimeter found in Step 12:

$$R_p = (\text{MLT})(N)\left(\frac{\mu\Omega}{\text{cm}}\right)\text{x}10^{-6}, \quad (\text{ohms})$$

$$R_p = (30.1)(127)(132)\text{x}10^{-6}$$

$$R_p = 0.504, \quad (\text{ohms})$$

Step No. 14. Calculate the primary copper loss P_p:

$$P_p = \left(I_p\right)^2 R_p, \quad (\text{watts})$$

$$P_p = (1.33)^2(0.504)$$

$$P_p = 0.891, \quad (\text{watts})$$

Step No. 15. Calculate the secondary turns each side of the center tap:

$$N_s = \left(\frac{N_p V_s}{V_p}\right), \quad (\text{turns})$$

$$N_s = \frac{(127)(29)}{(115)}$$

$$N_s = 32, \quad (\text{turns})$$

Step No. 16. Calculate the bare wire size $A_{sw(\text{B})}$ for the secondary. When using a center-tap configuration, I_o has to be multiplied by (0.707):

$$A_{sw(\text{B})} = \frac{I_o(0.707)}{J}, \quad \left(\text{cm}^2\right)$$

$$A_{sw(\text{B})} = \frac{(5)(0.707)}{(90.7)}$$

$$A_{sw(\text{B})} = 0.0389, \quad \left(\text{cm}^2\right)$$

Step No. 17. Select a wire size from the wire table column 2. Remember, if the wire area is not within 10%, take the next smallest size. Also record micro-ohm per centimeter found in column 4:

$$\text{AWG \#11 with } 0.0416, \quad \left(\text{cm}^2\right)$$

$$\frac{\mu\Omega}{\text{cm}} = 41.4$$

Step No. 18. Calculate the secondary winding resistance. Use MLT from Step 5 and micro-ohm per centimeter from Step 17:

$$R_s = (\text{MLT})(N)\left(\frac{\mu\Omega}{\text{cm}}\right)\text{x}10^{-6}, \quad (\text{ohms})$$

$$R_s = (30.1)(32)(41.4)\text{x}10^{-6}$$

$$R_s = 0.0398, \quad (\text{ohms})$$

Step No. 19. Calculate the secondary copper loss P_s:

$$P_s = \left(I_o\right)^2 R_s, \quad (\text{watts})$$

$$P_s = (5)^2(0.0398)$$

$$P_s = 0.995, \quad (\text{watts})$$

Step No. 20. Calculate the transformer regulation:

$$\alpha = \frac{P_{cu}}{P_o + P_{cu}}(100), \quad (\%)$$

$$P_{cu} = P_p + P_s, \quad (\text{watts})$$

$$P_{cu} = (0.891) + (0.995)$$

$$P_{cu} = 1.88, \quad (\text{watts})$$

$$\alpha = \frac{1.88}{(145) + (1.88)}(100)$$

$$\alpha = 1.28, \quad (\%)$$

Step No. 21. Calculate the effective core weight. The gross W_t is found in Step 5 and the stacking factor is found in Table 10:

$$W_{tfe(eff)} = (SF)W_{tfe(gross)}, \quad \text{(grams)}$$

$$W_{tfe(eff)} = (0.90)(4914)$$

$$W_{tfe(eff)} = 4423, \quad \text{(grams)}$$

Step No. 22. Calculate core loss P_{fe} using the core loss curves for 12-mil material at 1.3 tesla.

$$P_{fe} = \left(\frac{\text{milliwatts}}{\text{gram}}\right) W_{tfe(gross)} \text{x} 10^{-3}, \quad \text{(watts)}$$

$$P_{fe} = (0.85)(4423)\text{x}10^{-3}$$

$$P_{fe} = 3.76, \quad \text{(watts)}$$

Step No. 23. Summarize the losses and compare with the total losses P_{ε}:

$$P_p = 0.891, \text{ Primary}, \quad \text{(watts)}$$

$$P_s = 0.995, \text{ Secondary}, \quad \text{(watts)}$$

$$P_{fe} = 3.76, \text{ Core}, \quad \text{(watts)}$$

$$P_{\Sigma} = 5.65, \text{ Total}, \quad \text{(watts)}$$

Step No. 24. Calculate the efficiency:

$$\eta = \frac{P_o}{P_o + P_{\Sigma}}(100), \quad (\%)$$

$$\eta = \frac{145}{(145) + (5.65)}(100)$$

$$\eta = 96.2, \quad (\%)$$

Step No. 25. Calculate the watts per unit area. The surface area A_t is found in Step 5:

$$\psi = \frac{P_\Sigma}{A_t}, \quad \left(\text{watts/cm}^2\right)$$

$$\psi = \frac{(5.65)}{(639)}$$

$$\psi = 0.009, \quad \left(\text{watts/cm}^2\right)$$

where:

$$\psi = 0.01, \quad \left(\text{watts/cm}^2\right) \ @ \ 10°\text{C rise}$$

TOROIDAL INDUCTOR DESIGN EXAMPLE USING THE AREA PRODUCT A_p APPROACH

For a typical design example circuit shown in Fig. 39, assume the following:

1. Inductance, L = 0.003 H
2. dc current, I_o = 3 A
3. ac current, ΔI = 0.2 A
4. 25°C rise $P_{cu} >> P_{fe}$
5. Molypermalloy powder core (MPP)

Step No. 1. Calculate the energy-handling capability:

$$\text{Energy} = \frac{LI_{pk}^2}{2}, \quad (\text{watt-seconds})$$

$$I_{pk} = I_o + \frac{\Delta I}{2}, \quad (\text{amps})$$

$$I_{pk} = (3) + \frac{(0.2)}{2}$$

$$I_{pk} = 3.1, \quad (\text{amps})$$

$$\text{Energy} = \frac{(0.003)(3.1)^2}{2}$$

$$\text{Energy} = 0.0144, \quad (\text{watt-seconds})$$

Step No. 2. Calculate the area product A_p required:

$$A_p = \left(\frac{2(Eng)x10^4}{B_m K_u K_j} \right)^{1.14}, \quad (cm^4)$$

Assuming

$$B_m = 0.3, \quad (tesla)$$
$$K_u = 0.4, \text{ window utilization factor}$$
$$K_j = 403, \text{ see Table 1}$$

$$A_p = \left(\frac{2(0.0144)x10^4}{(0.3)(0.4)(403)} \right)^{1.14}$$
$$A_p = 7.6, \quad (cm^4)$$

Step No. 3. Select a comparable area product A_p from the powder core section in column 12. Also record the MLT column 9, surface area A_t column 14, window area W_a column 11, magnetic path length MPL column 6.

$$\text{MPP Core} = 55441$$
$$A_p = 8.48, \quad (cm^4)$$
$$MLT = 6.8, \quad (cm)$$
$$A_t = 113, \quad (cm^2)$$
$$W_a = 4.26, \quad (cm^2)$$
$$MPL = 10.7, \quad (cm)$$

After the area product A_p has been determined, the geometry of the inductor can be evaluated, such as volume and weight.

Step No. 4. Calculate the current density J:

$$J = \frac{2(Eng)x10^4}{B_m K_u A_p}, \quad (amps/cm^2)$$
$$J = \frac{2(0.0144)x10^4}{(0.3)(0.4)(8.48)}$$
$$J = 283, \quad (amps/cm^2)$$

Step No. 5. Calculate the permeability of the core required:

$$\mu_\Delta = \frac{B_m(\text{MPL})\text{x}10^4}{(0.4\pi)W_a J K_u}, \quad \text{(perm)}$$

$$\mu_\Delta = \frac{(0.3)(10.7)\text{x}10^4}{(1.26)(4.26)(283)(0.4)}$$

$$\mu_\Delta = 53, \quad \text{(perm)}$$

From the manufacturer's catalog, the core that has the same size but has a permeability closer to the one calculated is the core 55439, with a permeability of 60. This particular core has 135 millihenry per 1000 turns.

Step No. 6. Calculate the number of turns required for 3 millihenry:

$$N = 1000\sqrt{\frac{L}{L_{1000}}}, \quad \text{(turns)}$$

$$N = 1000\sqrt{\frac{(3)}{(135)}}$$

$$N = 149, \quad \text{(turns)}$$

Step No. 7. Calculate the amp-turn per centimeter. Use magnetic path length (MPL) found in Step 3:

$$H = \frac{NI}{(\text{MPL})}, \quad \text{(amp-turns/cm)}$$

$$H = \frac{(149)(3.1)}{(10.7)}$$

$$H = 43, \quad \text{(amp-turns/cm)}$$

Check Table 7 and verify the dc bias inductance.

Step No. 8. Calculate the bare wire size $A_{w(B)}$:

$$A_{w(B)} = \frac{I_{rms}}{J}, \quad (\text{cm}^2)$$

$$I_{rms} = \sqrt{I_o^2 + \frac{\Delta I^2}{12}}, \quad (\text{amps})$$

$$I_{rms} = \sqrt{(3.0)^2 + \frac{(0.2)^2}{12}}$$

$$I_{rms} = 3.0, \quad (\text{amps})$$

$$A_{w(B)} = \frac{(3.0)}{(283)}$$

$$A_{w(B)} = 0.011, \quad (\text{cm}^2)$$

Step No. 9. Select a wire size from the Wire Table, column 2. Remember, if the wire area is not within 10%, take the next smallest size. Also, record micro-ohm per centimeter found in column 4.

$$\text{AWG \#17 with } 0.0104, \quad (\text{cm}^2)$$

$$\frac{\mu\Omega}{\text{cm}} = 166$$

Step No. 10. Calculate the winding resistance. Use MLT found in Step 3 and micro-ohm per centimeter found in Step 9:

$$R = (\text{MLT})(N)\left(\frac{\mu\Omega}{\text{cm}}\right)\text{x}10^{-6}, \quad (\text{ohms})$$

$$R = (6.8)(149)(166)\text{x}10^{-6}$$

$$R = 0.168, \quad (\text{ohms})$$

Step No. 11. Calculate the copper loss P_{cu}:

$$P_{cu} = (I_{rms})^2 R, \quad \text{(watts)}$$
$$P_{cu} = (3.0)^2 (0.168)$$
$$P_{cu} = 1.51, \quad \text{(watts)}$$

Step No. 12. Calculate the watts per unit area. The surface area A_t is found in Step 3:

$$\psi = \frac{P_{cu}}{A_t}, \quad \text{(watts/cm}^2\text{)}$$
$$\psi = \frac{(1.51)}{(113)}$$
$$\psi = 0.0133, \quad \text{(watts/cm}^2\text{)}$$

where:

$$\psi = 0.01, \quad \text{(watts/cm}^2\text{)} \text{ @ } 10°\text{C rise}$$

TOROIDAL INDUCTOR DESIGN EXAMPLE USING THE CORE GEOMETRY K_g APPROACH

For a typical design example circuit shown in Fig. 39, assume the following:

1. Inductance, $L = 0.0025$ H
2. dc current, $I_o = 1.5$ A
3. ac current, $\Delta I = 0.2$ A
4. Output power, $P_o = 100$ W
5. Regulation, $\alpha = 1\%$
6. MPP powder core

Step No. 1. Calculate the energy-handling capability:

$$\text{Energy} = \frac{LI_{pk}^2}{2}, \quad \text{(watt-seconds)}$$

$$I_{pk} = I_o + \frac{\Delta I}{2}, \quad \text{(amps)}$$

$$I_{pk} = (1.5) + \frac{(0.2)}{2}$$

$$I_{pk} = 1.6, \quad \text{(amps)}$$

$$\text{Energy} = \frac{(0.0025)(1.6)^2}{2}$$

$$\text{Energy} = 0.0032, \quad \text{(watt-seconds)}$$

Step No. 2. Calculate the electrical conditions K_e:

$$K_e = 0.145 P_o B_m^2 \text{x} 10^{-4}$$

$$B_m = 0.3, \quad \text{(tesla)}$$

$$K_e = 0.145(100)(0.3)^2 \text{x} 10^{-4}$$

$$K_e = 0.00013$$

Step No. 3. Calculate the core geometry K_g:

$$K_g = \frac{(\text{Energy})^2}{K_e \alpha}, \quad (\text{cm}^5)$$

$$K_g = \frac{(0.0032)^2}{(0.00013)(1)}$$

$$K_g = 0.0787, \quad (\text{cm}^5)$$

Step No. 4. Select a comparable core geometry K_g from the powder core section in column 13. Also record the area product A_p column 12, MLT column 9, surface area A_t column 14, window area W_a column 11, magnetic path length MPL column 6.

$$\text{MPP Core} = 55580$$
$$K_g = 0.0753, \quad (\text{cm}^5)$$
$$A_p = 1.82, \quad (\text{cm}^4)$$
$$\text{MLT} = 4.4, \quad (\text{cm})$$
$$A_t = 64, \quad (\text{cm}^2)$$
$$W_a = 4.01, \quad (\text{cm}^2)$$
$$\text{MPL} = 9.0, \quad (\text{cm})$$

After the core geometry K_g has been determined, the geometry of the inductor can be evaluated, such as volume and weight.

Step No. 5. Calculate the current density J:

$$J = \frac{2(\text{Eng})\text{x}10^4}{B_m K_u A_p}, \quad (\text{amps/cm}^2)$$

$$K_u = 0.4, \text{ window utilization factor}$$

$$J = \frac{2(0.0032)\text{x}10^4}{(0.3)(0.4)(1.82)}$$
$$J = 293, \quad (\text{amps/cm}^2)$$

Step No. 6. Calculate the permeability of the core required:

$$\mu_\Delta = \frac{B_m(\text{MPL})\text{x}10^4}{(0.4\pi)W_a J K_u}, \quad (\text{perm})$$
$$\mu_\Delta = \frac{(0.3)(9.0)\text{x}10^4}{(1.26)(4.01)(293)(0.4)}$$
$$\mu_\Delta = 45.6, \quad (\text{perm})$$

From the manufacturer's catalog, the core that has the same size but has a permeability closer to the one calculated is the core 55586, with a permeability of 60. This particular core has 38 millihenry per 1000 turns.

Step No. 7. Calculate the number of turns required for 2.5 millihenry:

$$N = 1000\sqrt{\frac{L}{L_{1000}}}, \quad \text{(turns)}$$

$$N = 1000\sqrt{\frac{(2.5)}{(38)}}$$

$$N = 256, \quad \text{(turns)}$$

Step No. 8. Calculate the amp-turn per centimeter:

$$H = \frac{NI}{(\text{MPL})}, \quad \text{(amp - turns/cm)}$$

$$H = \frac{(256)(1.6)}{(8.95)}$$

$$H = 45.7, \quad \text{(amp - turns/cm)}$$

Check Table 7 and verify the dc bias inductance.

Step No. 9. Calculate the bare wire size $A_{w(\text{B})}$:

$$A_{w(\text{B})} = \frac{I_{rms}}{J}, \quad \left(\text{cm}^2\right)$$

$$I_{rms} = \sqrt{I_o^2 + \frac{\Delta I^2}{12}}, \quad \text{(amps)}$$

$$I_{rms} = \sqrt{(1.5)^2 + \frac{(0.2)^2}{12}}$$

$$I_{rms} = 1.5, \quad \text{(amps)}$$

$$A_{w(\text{B})} = \frac{(1.5)}{(293)}$$

$$A_{w(\text{B})} = 0.00512, \quad \left(\text{cm}^2\right)$$

Step No. 10. Select a wire size from the wire table column 2. Remember, if the area is not within 10%, take the next smallest size. Also, record micro-ohm per centimeter found in column 4.

$$\text{AWG \#20 with } 0.00519, \quad \left(\text{cm}^2\right)$$

$$\frac{\mu\Omega}{\text{cm}} = 332$$

Step No. 11. Calculate the winding resistance using equation 15. Use MLT found in Step 4 and micro-ohm per-centimeter found in Step 9:

$$R = (\text{MLT})(N)\left(\frac{\mu\Omega}{\text{cm}}\right)\text{x}10^{-6}, \quad \text{(ohms)}$$

$$R = (4.4)(256)(332)\text{x}10^{-6}$$

$$R = 0.374, \quad \text{(ohms)}$$

Step No. 12. Calculate the copper loss P_{cu}:

$$P_{cu} = \left(I_{rms}\right)^2 R, \quad \text{(watts)}$$

$$P_{cu} = (1.5)^2(0.374)$$

$$P_{cu} = 0.842, \quad \text{(watts)}$$

Step No. 13. Calculate the regulation α:

$$\alpha = \frac{P_{cu}}{P_o + P_{cu}}(100), \quad (\%)$$

$$\alpha = \frac{(0.842)}{(100) + (0.842)}(100)$$

$$\alpha = 0.839, \quad (\%)$$

Step No. 14. Calculate the watts per unit area. The surface area A_t is found in Step 4:

$$\psi = \frac{P_{cu}}{A_t}, \quad \left(\text{watts/cm}^2\right)$$

$$\psi = \frac{(0.842)}{(64)}$$

$$\psi = 0.0132, \quad \left(\text{watts/cm}^2\right)$$

where

$$\psi = 0.01, \quad \left(\text{watts/cm}^2\right) \text{ @ } 10^\circ\text{C rise}$$

A C-CORE INDUCTOR DESIGN EXAMPLE USING THE CORE GEOMETRY K_g APPROACH

For a typical design example circuit shown in Fig. 39, assume the following:

1. Inductance, L = 0.002 H
2. dc current, I_o = 5.0 A
3. ac current, ΔI = 0.5 A
4. Output power, P_o = 140 W
5. Regulation, α = 1%
6. Frequency, f = 20 kHz
7. 2-mil silicon C-cores

Step No. 1. Calculate the energy-handling capability:

$$\text{Energy} = \frac{LI_{pk}^2}{2}, \quad (\text{watt-seconds})$$

$$I_{pk} = I_o + \frac{\Delta I}{2}, \quad (\text{amps})$$

$$I_{pk} = (5) + \frac{(0.5)}{2}$$

$$I_{pk} = 5.25, \quad (\text{amps})$$

$$\text{Energy} = \frac{(0.002)(5.25)^2}{2}$$

$$\text{Energy} = 0.0276, \quad (\text{watt-seconds})$$

Step No. 2. Calculate the electrical conditions K_e:

$$K_e = 0.145 P_o B_m^2 \text{x} 10^{-4}$$
$$B_m = 1.4, \quad \text{(tesla)}$$
$$K_e = 0.145(140)(1.4)^2 \text{x} 10^{-4}$$
$$K_e = 0.00398$$

Step No. 3. Calculate the core geometry K_g:

$$K_g = \frac{(\text{Energy})^2}{K_e \alpha}, \quad \left(\text{cm}^5\right)$$
$$K_g = \frac{(0.0276)^2}{(0.00398)(1)}$$
$$K_g = 0.191, \quad \left(\text{cm}^5\right)$$

Step No. 4. Select a comparable core geometry K_g from the 2-mil C-core section in column 15. Also record the G dimension column 4, iron weight column 9, MLT column 11, iron area A_c column 12, window area W_a column 13, area product A_p column 14, and surface area A_t column 16.

$$\text{C Core} = \text{CL-9}$$
$$K_g = 0.2037, \quad \left(\text{cm}^5\right)$$
$$\text{G} = 3.02, \quad \text{(cm)}$$
$$W_{tfe} = 96.6, \quad \text{(grams)}$$
$$\text{MLT} = 6.6, \quad \text{(cm)}$$
$$A_c = 1.08, \quad \left(\text{cm}^2\right)$$
$$W_a = 2.87, \quad \left(\text{cm}^2\right)$$
$$A_p = 3.09, \quad \left(\text{cm}^4\right)$$
$$A_t = 69, \quad \left(\text{cm}^2\right)$$

After the core geometry K_g has been determined, the geometry of the inductor can be evaluated, such as volume and weight.

Step No. 5. Calculate the current density J. Use the area product A_p found in Step 4:

$$J = \frac{2(\text{Eng})\text{x}10^4}{B_m K_u A_p}, \quad (\text{amps/cm}^2)$$

$$K_u = 0.4, \text{ window utilization factor}$$

$$J = \frac{2(0.0276)\text{x}10^4}{(1.4)(0.4)(3.09)}$$

$$J = 319, \quad (\text{amps/cm}^2)$$

Step No. 6. Calculate the bare wire size $A_{w(B)}$:

$$A_{w(B)} = \frac{I_{rms}}{J}, \quad (\text{cm}^2)$$

$$I_{rms} = \sqrt{I_o^2 + \frac{\Delta I^2}{12}}, \quad (\text{amps})$$

$$I_{rms} = \sqrt{(5.0)^2 + \frac{(0.5)^2}{12}}$$

$$I_{rms} = 5.0, \quad (\text{amps})$$

$$A_{w(B)} = \frac{(5.0)}{(319)}$$

$$A_{w(B)} = 0.0157, \quad (\text{cm}^2)$$

Step No. 7. Select a wire size from the Wire Table, column 2. Remember, if the area is not within 10%, take the next smallest size. Also, record micro-ohm per centimeter found in column 4 and the wire area with insulation A_w in column 5.

$$\text{AWG \#15 with } 0.0165, \quad (\text{cm}^2)$$

$$\frac{\mu\Omega}{\text{cm}} = 104$$

$$A_w = 0.01837, \quad (\text{cm}^2)$$

Step No. 8. Calculate the effective window area $W_{a(eff)}$. Use the window area W_a found in Step 4:

$$W_{a(eff)} = W_a S_3, \quad (\text{cm}^2)$$

A typical value for S_3 is 0.75 as shown in the window utilization factor K_u section.

$$W_{a(eff)} = (2.87)(0.75)$$
$$W_{a(eff)} = 2.15, \quad (\text{cm}^2)$$

Step No. 9. Calculate the number of turns. Use the wire area A_w found in Step 7:

$$N = \frac{W_{a(eff)} S_2}{A_w}, \quad (\text{turns})$$

A typical value for S_2 is 0.6 as shown in the window utilization factor K_u section.

$$N = \frac{(2.15)(0.6)}{(0.01837)}$$
$$N = 70, \quad (\text{turns})$$

Step No. 10. Calculate the gap from the inductance. Use the iron area A_c found in Step 4:

$$l_g = \frac{(0.4\pi) N^2 A_c \text{x} 10^{-8}}{L}, \quad (\text{cm})$$
$$l_g = \frac{(1.26)(70)^2 (1.08) \text{x} 10^{-8}}{0.002}$$
$$l_g = 0.0333, \quad (\text{cm})$$

Gap spacing is usually maintained by inserting fish paper. However, this paper is available only in mil thickness. As l_g has been determined in centimeters, it is necessary to convert as follows:

$$\text{cm x } 393.7 = \text{mils (inch system)}$$

Substituting values:

$$0.0333 \text{ x } 393.7 = 13.1, \quad \text{(mils) use 14}$$

An available size of paper is a 7-mil sheet. A single thickness would therefore be used, giving equal gaps in both inside and outside legs.

The equivalent in metric is:

$$0.014 \text{ x } 2.54 = 0.0355, \quad \text{(cm)}$$

Step No. 11. Calculate the amount of fringing flux. Use the G dimension found in Step 4:

$$F = \left(1 + \frac{l_g}{\sqrt{A_c}} \ln \frac{2G}{l_g}\right)$$

$$F = \left(1 + \frac{(0.0355)}{\sqrt{1.08}} \ln \frac{2(3.016)}{(0.0355)}\right)$$

$$F = 1.17$$

Step No. 12. Calculate the new turns by inserting the fringing flux:

$$N = \sqrt{\frac{l_g L}{(0.4\pi) A_c F \text{x} 10^{-8}}}, \quad \text{(turns)}$$

$$N = \sqrt{\frac{(0.0355)(0.002)}{(1.26)(1.08)(1.17)\text{x}10^{-8}}}$$

$$N = 67, \quad \text{(turns)}$$

Step No. 13. Calculate the winding resistance. Use the MLT from Step 4 and micro-ohm per centimeter from Step 7.

$$R = (\text{MLT})(N)\left(\frac{\mu\Omega}{\text{cm}}\right)\text{x}10^{-6}, \quad \text{(ohms)}$$

$$R = (6.6)(67)(104)\text{x}10^{-6}$$

$$R = 0.046, \quad \text{(ohms)}$$

Step No. 14. Calculate the copper loss P_{cu}:

$$P_{cu} = (I_{rms})^2 R, \quad \text{(watts)}$$

$$P_{cu} = (5.0)^2(0.046)$$

$$P_{cu} = 1.15, \quad \text{(watts)}$$

Step No. 15. Calculate the regulation α:

$$\alpha = \frac{P_{cu}}{P_o + P_{cu}}(100), \quad (\%)$$

$$\alpha = \frac{(1.15)}{(140) + (1.15)}(100)$$

$$\alpha = 0.815, \quad (\%)$$

Step No. 16. Calculate the total ac plus dc flux density:

$$B_{max} = \frac{(0.4\pi)N\left(I_{dc} + \frac{\Delta I}{2}\right)}{l_g}\text{x}10^{-4}, \quad \text{(tesla)}$$

$$B_{max} = \frac{(1.26)(67)\left(5 + \frac{0.5}{2}\right)}{(0.0355)}\text{x}10^{-4}$$

$$B_{max} = 1.29, \quad \text{(tesla)}$$

Step No. 17. Calculate the ac flux density:

$$B_{ac} = \frac{(0.4\pi)N\left(\frac{\Delta I}{2}\right)}{l_g} \text{x}10^{-4}, \quad \text{(tesla)}$$

$$B_{ac} = \frac{(1.26)(67)\left(\frac{0.5}{2}\right)}{(0.0355)} \text{x}10^{-4}$$

$$B_{ac} = 0.0615, \quad \text{(tesla)}$$

Step No. 18. Calculate core loss P_{fe} using the appropriate core loss curves for 2-mil silicon at 0.0615 tesla. Iron weight is found in Step 4:

$$P_{fe} = \left(\frac{\text{milliwatts}}{\text{gram}}\right) W_{tfe(\text{gross})} \text{x}10^{-3}, \quad \text{(watts)}$$

$$P_{fe} = (10)(88)\text{x}10^{-3}$$

$$P_{fe} = 0.88, \quad \text{(watts)}$$

Step No. 19. Calculate the total losses P_{ε}:

$$P_{\Sigma} = P_{cu} + P_{fe}, \quad \text{(watts)}$$

$$P_{\Sigma} = (1.15) + (0.88)$$

$$P_{\Sigma} = 2.03, \quad \text{(watts)}$$

Step No. 20. Calculate the efficiency of the inductor:

$$\eta = \frac{P_o}{P_o + P_{\Sigma}}(100), \quad (\%)$$

$$\eta = \frac{140}{(140) + (2.03)}(100)$$

$$\eta = 98.6, \quad (\%)$$

Step No. 21. Calculate the watts per unit area. The surface A_t is found in Step. 4

$$\psi = \frac{P_\Sigma}{A_t}, \quad \left(\text{watts/cm}^2\right)$$

$$\psi = \frac{(2.03)}{(69)}$$

$$\psi = 0.0294, \quad \left(\text{watts/cm}^2\right)$$

where

$$\psi = 0.03, \quad \left(\text{watts/cm}^2\right) \text{ @ } 25°\text{C rise}$$

A POT CORE INDUCTOR DESIGN EXAMPLE USING THE CORE GEOMETRY K_g APPROACH

For a typical design example circuit shown in Fig. 39, assume the following:

1. Inductance, L = 550 μH
2. dc current, I_o = 3.5 A
3. ac current, I = 0.3 A
4. Output power, P_o = 98 W
5. Regulation, α = 1%
6. Frequency, f = 50 kHz
7. Ferrite pot core

Step No. 1. Calculate the energy-handling capability:

$$\text{Energy} = \frac{LI_{pk}^2}{2}, \quad (\text{watt-seconds})$$

$$I_{pk} = I_o + \frac{\Delta I}{2}, \quad (\text{amps})$$

$$I_{pk} = (3.5) + \frac{(0.3)}{2}$$

$$I_{pk} = 3.65, \quad (\text{amps})$$

$$\text{Energy} = \frac{(0.00055)(3.65)^2}{2}$$

$$\text{Energy} = 0.00366, \quad (\text{watt-seconds})$$

Step No. 2. Calculate the electrical conditions K_e:

$$K_e = 0.145 P_o B_m^2 \text{x} 10^{-4}$$

$$B_m = 0.2, \quad \text{(tesla)}$$

$$K_e = 0.145(98)(0.2)^2 \text{x} 10^{-4}$$

$$K_e = 56.8 \text{x} 10^{-6}$$

Step No. 3. Calculate the core geometry K_g:

$$K_g = \frac{(\text{Energy})^2}{K_e \alpha}, \quad \left(\text{cm}^5\right)$$

$$K_g = \frac{(0.00366)^2}{\left(56.8 \text{x} 10^{-6}\right)(1)}$$

$$K_g = 0.236, \quad \left(\text{cm}^5\right)$$

Step No. 4. Select a comparable core geometry K_g from the pot core section in column 12. Also record the G dimension column 4, iron weight column 6, MLT column 8, iron area A_c column 9, window area W_a column 10, area product A_p column 11, surface area A_t and column 13.

Pot Core = B65611, 36x22

$$K_g = 0.218, \quad \left(\text{cm}^5\right)$$

$$\text{G} = 1.46, \quad \text{(cm)}$$

$$W_{tfe} = 59.5, \quad \text{(grams)}$$

$$\text{MLT} = 7.5, \quad \text{(cm)}$$

$$A_c = 2.02, \quad \left(\text{cm}^2\right)$$

$$W_a = 1.00, \quad \left(\text{cm}^2\right)$$

$$A_p = 2.02, \quad \left(\text{cm}^4\right)$$

$$A_t = 45.2, \quad \left(\text{cm}^2\right)$$

After the core geometry has been determined, the geometry of the inductor can be evaluated, such as volume and weight.

Step No. 5. Calculate the current density J. Use the area product A_p found in Step 4:

$$J = \frac{2(\text{Eng})\text{x}10^4}{B_m K_u A_p}, \quad (\text{amps/cm}^2)$$

$$K_u = 0.4, \text{ window utilization factor}$$

$$J = \frac{2(0.00366)\text{x}10^4}{(0.2)(0.4)(2.02)}$$

$$J = 453, \quad (\text{amps/cm}^2)$$

Step No. 6. Calculate the bare wire size $A_{w(B)}$:

$$A_{w(B)} = \frac{I_{rms}}{J}, \quad (\text{cm}^2)$$

$$I_{rms} = \sqrt{I_o^2 + \frac{\Delta I^2}{12}}, \quad (\text{amps})$$

$$I_{rms} = \sqrt{(3.5)^2 + \frac{(0.3)^2}{12}}$$

$$I_{rms} = 3.5, \quad (\text{amps})$$

$$A_{w(B)} = \frac{(3.5)}{(453)}$$

$$A_{w(B)} = 0.00773, \quad (\text{cm}^2)$$

Step No. 7. Select a wire size from the Wire Table, column 2. Remember, if the area is not within 10%, take the next smallest size. Also record micro-ohm per centimeter found in column 4 and wire area with insulation A_w column 5.

$$\text{AWG \#18 with } 0.00822, \quad (\text{cm}^2)$$

$$\frac{\mu\Omega}{\text{cm}} = 209$$

$$A_w = 0.00933, \quad (\text{cm}^2)$$

Step No. 8. Calculate the effective window area $W_{a(eff)}$. Use the window area W_a found in Step 4:

$$W_{a(eff)} = W_a S_3, \quad (cm^2)$$

A typical value for S_3 when designing with pot cores is 0.75 for a single-section bobbin.

$$W_{a(eff)} = (1.00)(0.75)$$

$$W_{a(eff)} = 0.75, \quad (cm^2)$$

Step No. 9. Calculate the number of turns. Use the wire area A_w found in Step 7:

$$N = \frac{W_{a(eff)} S_2}{A_w}, \quad (turns)$$

A typical value for S_2 is 0.6 as shown in the window utilization factor K_u section.

$$N = \frac{(0.75)(0.6)}{(0.00933)}$$

$$N = 48, \quad (turns)$$

Step No. 10. Calculate the gap. Use the iron area A_c found in Step 4:

$$l_g = \frac{(0.4\pi) N^2 A_c x10^{-8}}{L}, \quad (cm)$$

$$l_g = \frac{(1.26)(48)^2 (2.02)x10^{-8}}{(550x10^{-6})}$$

$$l_g = 0.107, \quad (cm)$$

Gap spacing is usually maintained by inserting fish paper. However, this paper is available only in mil thickness. Because l_g has been determined in centimeters, it is necessary to convert as follows:

$$\text{cm x } 393.7 = \text{mils (inch system)}$$

Substituting values:

$$0.107 \text{ x } 393.7 = 42, \quad \text{(mils) use } 40$$

An available size of paper is a 20-mil sheet. Single thickness would therefore be used, giving equal gaps in both the inside and outside.

The equivalent in metric is:

$$0.040 \text{ x } 2.54 = 0.102, \quad \text{(cm)}$$

Step No. 11. Calculate the amount of fringing flux. Use the G dimension found in Step 4:

$$F = \left(1 + \frac{l_g}{\sqrt{A_c}} \ln \frac{2G}{l_g}\right)$$

$$F = \left(1 + \frac{(0.102)}{\sqrt{2.02}} \ln \frac{2(1.46)}{(0.102)}\right)$$

$$F = 1.24$$

Step No. 12. Calculate the new turns by inserting the fringing flux:

$$N = \sqrt{\frac{l_g L}{(0.4\pi) A_c F \text{x} 10^{-8}}}, \quad \text{(turns)}$$

$$N = \sqrt{\frac{(0.102)(550\text{x}10^{-6})}{(1.26)(2.01)(1.24)\text{x}10^{-8}}}$$

$$N = 42, \quad \text{(turns)}$$

Step No. 13. Calculate the winding resistance. Use MLT from Step 4 and micro-ohm per centimeter from Step 7:

$$R = (\text{MLT})(N)\left(\frac{\mu\Omega}{\text{cm}}\right)\text{x}10^{-6}, \quad \text{(ohms)}$$

$$R = (7.5)(42)(209)\text{x}10^{-6}$$

$$R = 0.066, \quad \text{(ohms)}$$

Step No. 14. Calculate the copper loss P_{cu}:

$$P_{cu} = \left(I_{rms}\right)^2 R, \quad \text{(watts)}$$

$$P_{cu} = (3.5)^2(0.066)$$

$$P_{cu} = 0.809, \quad \text{(watts)}$$

Step No. 15. Calculate the regulation α:

$$\alpha = \frac{P_{cu}}{P_o + P_{cu}}(100), \quad (\%)$$

$$\alpha = \frac{(0.809)}{(98)+(0.809)}(100)$$

$$\alpha = 0.819, \quad (\%)$$

Step No. 16. Calculate the total ac plus dc flux density:

$$B_{\max} = \frac{(0.4\pi)N\left(I_{dc} + \frac{\Delta I}{2}\right)}{l_g}\text{x}10^{-4}, \quad \text{(tesla)}$$

$$B_{\max} = \frac{(1.26)(42)\left(3.5 + \frac{0.3}{2}\right)}{(0.102)}\text{x}10^{-4}$$

$$B_{\max} = 0.189, \quad \text{(tesla)}$$

Step No. 17. Calculate the ac flux density:

$$B_{ac} = \frac{(0.4\pi)N\left(\frac{\Delta I}{2}\right)}{l_g} \text{x}10^{-4}, \quad \text{(tesla)}$$

$$B_{ac} = \frac{(1.26)(42)\left(\frac{0.3}{2}\right)}{(0.102)} \text{x}10^{-4}$$

$$B_{ac} = 0.00778, \quad \text{(tesla)}$$

Step No. 18. Calculate core loss P_{fe}. Use the appropriate core loss curves for ferrite at 0.00797 tesla. The iron weight is found in Step 4.

$$P_{fe} = \left(\frac{\text{milliwatts}}{\text{gram}}\right) W_{tfe(\text{gross})} \text{x}10^{-3}, \quad \text{(watts)}$$

$$P_{fe} = (.01)(59.5)\text{x}10^{-3}$$

$$P_{fe} = 0.595\text{x}10^{-3}, \quad \text{(watts)}$$

Step No. 19. Calculate the total losses P_{ε}:

$$P_{\Sigma} = P_{cu} + P_{fe}, \quad \text{(watts)}$$

$$P_{\Sigma} = (0.819) + \left(0.595\text{x}10^{-3}\right)$$

$$P_{\Sigma} = 0.819, \quad \text{(watts)}$$

Step No. 20. Calculate the efficiency of the inductor:

$$\eta = \frac{P_o}{P_o + P_{\Sigma}}(100), \quad (\%)$$

$$\eta = \frac{98}{(98) + (0.819)}(100)$$

$$\eta = 99.2, \quad (\%)$$

Step No. 21. Calculate the watts per unit area. The surface area A_t is found in Step 4:

$$\psi = \frac{P_\Sigma}{A_t}, \quad \text{(watts/cm}^2\text{)}$$

$$\psi = \frac{(0.819)}{(45.2)}$$

$$\psi = 0.0181, \quad \text{(watts/cm}^2\text{)}$$

where

$$\psi = 0.03, \quad \text{(watts/cm}^2\text{)} \text{ @ } 25°\text{C rise}$$

10-kHz dc-TO-dc TRANSFORMER DESIGN EXAMPLE USING THE CORE GEOMETRY K_g APPROACH

For a typical design example, assume a push-pull, full-wave, center-tapped circuit shown in Fig. 39 with the following specification:

1. Input voltage, V_{in} = 22 V
2. Output voltage, V_o = 10 V
3. Output current, I_o = 10 A
4. Frequency, f = 10 kHz
5. Temperature rise, 25°C
6. Efficiency, η = 98%
7. Regulation, α = 1%
8. Core, 2-mil C-core

Step No. 1. Calculate the transformer output power P_o allowing for a 1.0-V diode drop (V_d) assumed:

$$P_o = VI_o, \quad \text{(watts)}$$

$$V = V_o + V_d, \quad \text{(volts)}$$

$$V = 10 + 1$$

$$P_o = (11)(10)$$

$$P_o = 110, \quad \text{(watts)}$$

Step No. 2. Calculate the apparent power P_t:

$$P_t = P_o\left(\frac{\sqrt{2}}{\eta} + \sqrt{2}\right), \quad \text{(watts)}$$

$$P_t = (110)\left(\frac{(1.41)}{(0.98)} + (1.41)\right)$$

$$P_t = 313, \quad \text{(watts)}$$

Step No. 3. Calculate the electrical conditions:

$$K_e = 0.145 K_f^2 f^2 B_m^2 \text{x} 10^{-4}$$

$$K_f = 4.0, \text{ square wave}$$

$$\text{B}_\text{m} = 0.3, \quad \text{(tesla)}$$

$$K_e = (0.145)(4.0)^2(10{,}000)^2(0.3)^2 \text{x} 10^{-4}$$

$$K_e = 2088$$

Step No. 4. Calculate the core geometry K_g:

$$K_g = \frac{P_t}{2K_e\alpha}, \quad \left(\text{cm}^5\right)$$

$$K_g = \frac{(313)}{2(2088)(1)}$$

$$K_g = 0.0749, \quad \left(\text{cm}^5\right)$$

Step No. 5. Select a comparable core geometry K_g from the C-core section in column 15. As this is a high-frequency transformer, the window area W_a should be at least a factor of three larger than the iron area A_c. Also, record the iron weight W_{tfe} column 9, MLT column 11, iron cross section A_c column 12, surface area A_t column 16, area product A_p and column 14.

$$\text{C Core, CL-121}$$
$$K_g = 0.088, \quad (\text{cm}^5)$$
$$A_p = 2.28, \quad (\text{cm}^4)$$
$$W_{tfe} = 48, \quad (\text{grams})$$
$$A_c = 0.54, \quad (\text{cm}^2)$$
$$\text{MLT} = 5.6, \quad (\text{cm})$$
$$A_t = 66, \quad (\text{cm}^2)$$

Step No. 6. Calculate the number of primary turns on each side of center tap:

$$N_p = \left(\frac{V_p \text{x} 10^4}{K_f B_m f A_c}\right)\left(1 + \frac{\alpha}{100}\right), \quad (\text{turns})$$
$$N_p = \left(\frac{22 \text{x} 10^4}{(4.0)(0.3)(10{,}000)(0.54)}\right)\left(1 + \frac{1}{100}\right)$$
$$N_p = 34, \quad (\text{turns})$$

Step No. 7. Calculate the primary current I_p:

$$I_p = \frac{P_o}{V_p \eta}, \quad (\text{amps})$$
$$I_p = \frac{(110)}{(22)(0.98)}$$
$$I_p = 5.10, \quad (\text{amps})$$

Step No. 8. Calculate the current density J:

$$J = \left(\frac{P_t \text{x} 10^4}{K_f K_u f B_m A_p}\right), \quad (\text{amps/cm}^2)$$

$$K_u = 0.40, \text{ window utilization factor}$$

$$J = \left(\frac{313 \text{x} 10^4}{(4.0)(0.4)(10{,}000)(0.3)(2.28)}\right)$$

$$J = 286, \quad (\text{amps/cm}^2)$$

Step No. 9. Calculate the bare wire size $A_{w(B)}$ for the primary. When using a center-tap configuration, I_p has to be multiplied by (0.707):

$$A_{w(B)} = \frac{I_p(0.707)}{J}, \quad (\text{cm}^2)$$

$$A_{w(B)} = \frac{(5.10)(0.707)}{(286)}$$

$$A_{w(B)} = 0.0126, \quad (\text{cm}^2)$$

Step No. 10. Select a wire size from the wire table column 2. Remember, if the wire area is not within 10%, take the next smallest size. Also record micro-ohm per centimeter found in column 4.

AWG #16

$$A_{w(B)} = 0.01307, \quad (\text{cm}^2)$$

$$\frac{\mu\Omega}{\text{cm}} = 132$$

Step No. 11. Calculate the primary winding resistance. Use MLT from Step 5 and micro-ohm per centimeter found in Step 10:

$$R_p = (\text{MLT})(N)\left(\frac{\mu\Omega}{\text{cm}}\right)\text{x}10^{-6}, \quad (\text{ohms})$$

$$R_p = (5.6)(34)(132)\text{x}10^{-6}$$

$$R_p = 0.0251, \quad (\text{ohms})$$

Step No. 12. Calculate the primary copper loss P_p:

$$P_p = (I_p)^2 R_p, \quad \text{(watts)}$$

$$P_p = (5.1)^2 (0.0251)$$

$$P_p = 0.653, \quad \text{(watts)}$$

Step No. 13. Calculate the secondary turns each side of center tap:

$$N_s = \left(\frac{N_p V_s}{V_p}\right), \quad \text{(turns)}$$

$$N_s = \frac{(34)(11)}{(22)}$$

$$N_s = 17, \quad \text{(turns)}$$

Step No. 14. Calculate the bare wire size $A_{w(B)}$ for the secondary. When using a center tap configuration, I_o has to be multiplied by (0.707):

$$A_{w(B)} = \frac{I_o(0.707)}{J}, \quad \left(\text{cm}^2\right)$$

$$A_{w(B)} = \frac{(10)(0.707)}{(286)}$$

$$A_{w(B)} = 0.0247, \quad \left(\text{cm}^2\right)$$

Step No. 15. Select a wire size from the wire table column 2. Remember, if the wire area is not within 10%, take the next smallest size. Also record micro-ohm per centimeter found in column 4.

AWG #13

$$\frac{\mu\Omega}{\text{cm}} = 65.6$$

On small transformers of this type it becomes very clumsy to wind with such large wire because of the wire stiffness and poor lay, which results in a very poor fill factor. Reduce the wire size by three and wind two in parallel. This will give approximately the same cross section but better lay and fill factor.

$$A_{w(B)} = 0.0262, \quad (\text{cm}^2)$$

Step No. 16. Calculate the secondary winding resistance. Use MLT from Step 5 and micro-ohm per centimeter from Step 15.

$$R_p = (\text{MLT})(N)\left(\frac{\mu\Omega}{\text{cm}}\right)\text{x}10^{-6}, \quad (\text{ohms})$$

$$R_p = (5.6)(17)(65.6)\text{x}10^{-6}$$

$$R_p = 0.00625, \quad (\text{ohms})$$

Step No. 17. Calculate the secondary copper loss P_s:

$$P_s = (I_o)^2 R_s, \quad (\text{watts})$$

$$P_s = (10)^2(0.00625)$$

$$P_s = 0.625, \quad (\text{watts})$$

Step No. 18. Calculate the transformer regulation:

$$\alpha = \frac{P_{cu}}{P_o + P_{cu}}(100), \quad (\%)$$

$$P_{cu} = P_p + P_s, \quad (\text{watts})$$

$$P_{cu} = (0.653) + (0.625)$$

$$P_{cu} = 1.28, \quad (\text{watts})$$

$$\alpha = \frac{1.28}{(110) + (1.28)}(100)$$

$$\alpha = 1.15, \quad (\%)$$

Step No. 19. Calculate the combined loss P_Σ:

$$P_\Sigma = \frac{P_o}{\eta} - P_o, \quad \text{(watts)}$$

$$P_\Sigma = \frac{110}{0.98} - 110$$

$$P_\Sigma = 2.24, \quad \text{(watts)}$$

Step No. 20. Calculate the iron loss P_{fe}:

$$P_{fe} = P_\Sigma - P_{cu}, \quad \text{(watts)}$$

$$P_{fe} = (2.24) - (1.28)$$

$$P_{fe} = 0.960, \quad \text{(watts)}$$

Step No. 21. Calculate the core loss P_{fe} in milliwatts per gram:

$$\left(\frac{\text{milliwatts}}{\text{gram}}\right) = \frac{P_{fe}}{W_{tfe} \text{x} 10^{-3}}$$

$$\left(\frac{\text{milliwatts}}{\text{gram}}\right) = \frac{0.960}{(48.3)\text{x}10^{-3}}$$

$$\left(\frac{\text{milliwatts}}{\text{gram}}\right) = 19.8$$

Step No. 22. Find the magnetic material using the appropriate core loss curves for 2-mil material that comes closest to the core loss P_{fe} in Step 21.

$$B_m = 0.3, \quad \text{(tesla)}$$

19.8, (watts/kilogram)

or

19.8, (milliwatts/gram)

In this case, the material that comes closest is 80–20 nickel–iron.

BOOST SWITCHING INDUCTOR DESIGN EXAMPLE USING THE CORE GEOMETRY K_g APPROACH

For a typical design example, assume the circuit shown in Fig. 50 with the following specification:

1. Input voltage, $V_{in} = 28 \pm 6$ V
2. Output voltage, $V_o = 40$ V
3. Output current, $I_o = 2.5–0.25$ A
4. Frequency, $f = 20$ kHz
5. Switching efficiency, $\eta = 90\%$
6. Regulation, $\alpha = 1.0\%$
7. C-core, 1-mil silicon

Step No. 1. Calculate the time period T of operation:

$$T = t_{on} + t_{off}$$

$$T = \frac{1}{f}, \quad \text{(seconds)}$$

$$T = \frac{1}{20{,}000}$$

$$T = 50\text{x}10^{-6}, \quad \text{(seconds)}$$

Step No. 2. Calculate the minimum duty ratio:

$$D_{(\min)} = \left(1 - \frac{V_{in(\max)}\eta}{V_o}\right)$$

$$D_{(\min)} = \left(1 - \frac{(34)(0.9)}{40}\right)$$

$$D_{(\min)} = 0.235$$

Step No. 3. Calculate the maximum duty ratio:

$$D_{(\max)} = \left(1 - \frac{V_{in(\min)}\eta}{V_o}\right)$$

$$D_{(\max)} = \left(1 - \frac{(22)(0.9)}{40}\right)$$

$$D_{(\max)} = 0.505$$

Step No. 4. Calculate the load resistance at minimum load current:

$$R_o = \frac{V_o}{I_{o(\min)}}, \quad \text{(ohms)}$$

$$R_o = \frac{40}{0.25}$$

$$R_o = 160, \quad \text{(ohms)}$$

Step No. 5. Calculate the minimum required inductance:

$$L_{(\min)} = \frac{R_{o(\min)} T D_{(\min)} \left(1 - D_{(\min)}\right)^2}{2}, \quad \text{(henrys)}$$

$$L_{(\min)} = \frac{(160)\left(50\text{x}10^{-6}\right)(0.235)(1-0.235)^2}{2}$$

$$L_{(\min)} = 550\text{x}10^{-6}, \quad \text{(henrys)}$$

Step No. 6. Calculate the peak current I_{pk}:

$$I_{pk} = \frac{I_o}{\left(1 - D_{(\max)}\right)} + \frac{\Delta I}{2}, \quad \text{(amps)}$$

$$\Delta I = \frac{T V_{in(\min)} D_{(\max)}}{L}, \quad \text{(amps)}$$

$$\Delta I = \frac{\left(50\text{x}10^{-6}\right)(22)(0.505)}{550\text{x}10^{-6}}$$

$$\Delta I = 1.01, \quad \text{(amps)}$$

$$I_{pk} = \frac{2.5}{(1-0.505)} + \left(\frac{1.01}{2}\right)$$

$$I_{pk} = 5.55, \quad \text{(amps)}$$

Step No. 7. Calculate the energy-handling capability:

$$\text{Energy} = \frac{LI^2}{2}, \quad \text{(watt - seconds)}$$

$$\text{Energy} = \frac{(0.00055)(5.55)^2}{2}$$

$$\text{Energy} = 0.00847, \quad \text{(watt - seconds)}$$

Step No. 8. Calculate the electrical conditions K_e:

$$K_e = 0.145 P_o B_m^2 \text{x} 10^{-4}$$

$$B_m = 1.5, \quad \text{(tesla)}$$

$$P_o = V_o I_o, \quad \text{(watts)}$$

$$P_o = (40)(2.5)$$

$$P_o = 100, \quad \text{(watts)}$$

$$K_e = 0.145(100)(1.5)^2 \text{x} 10^{-4}$$

$$K_e = 0.00326$$

Step No. 9. Calculate the core geometry K_g:

$$K_g = \frac{(\text{Energy})^2}{K_e \alpha}, \quad (\text{cm}^5)$$

$$K_g = \frac{(0.00847)^2}{(0.00326)(1)}$$

$$K_g = 0.022, \quad (\text{cm}^5)$$

Step No. 10. Select a comparable core geometry K_g from the C-core section in column 15. Also record the D dimension column 1, G dimension column 4, iron weight W_{tf} column 9, MLT column 11, iron area A_c column 12, window area W_a column 13, area product A_p column 14, and surface area A_t column 16.

$$\text{Core} = \text{C Core CL-4}$$
$$K_g = 0.0184, \quad (\text{cm}^5)$$
$$\text{G} = 2.22, \quad (\text{cm})$$
$$W_{tfe} = 22.6, \quad (\text{grams})$$
$$\text{MLT} = 3.9, \quad (\text{cm})$$
$$A_c = 0.36, \quad (\text{cm}^2)$$
$$W_a = 1.41, \quad (\text{cm}^2)$$
$$A_p = 0.51, \quad (\text{cm}^4)$$
$$A_t = 29.8, \quad (\text{cm}^2)$$

Step No. 11. Calculate the current density J. Use the area product A_p found in Step 10:

$$J = \frac{2(\text{Eng})\text{x}10^4}{B_m K_u A_p}, \quad (\text{amps/cm}^2)$$
$$K_u = 0.4, \text{ window utilization factor}$$
$$J = \frac{2(0.00847)\text{x}10^4}{(1.5)(0.4)(0.51)}$$
$$J = 553, \quad (\text{amps/cm}^2)$$

Step No. 12. Calculate the rms current I_{rms} in the inductor:

$$I_{(rms)} = \sqrt{(I_o)^2 + \left(\frac{\Delta I}{12}\right)^2}, \quad (\text{amps})$$
$$I_{(rms)} = \sqrt{(5)^2 + \frac{(0.5)^2}{12}}$$
$$I_{(rms)} = 5.0, \quad (\text{amps})$$

Step No. 13. Calculate the bare wire size $A_{w(B)}$ using the rms current I_{rms}:

$$A_{w(B)} = \frac{I_{(rms)}}{J} \quad (\text{cm}^2)$$

$$A_{w(B)} = \frac{5.0}{553}$$

$$A_{w(B)} = 0.00904 \quad (\text{cm}^2)$$

Step No. 14. Select a wire size from the wire table column 2. Remember, if the wire area is not within 10%, take the next smallest size. Also record micro-ohm per centimeter found in column 4 and wire area with insulation A_w column 5.

AWG #17

$$A_{w(B)} = 0.0104, \quad (\text{cm}^2)$$

$$\frac{\mu\Omega}{\text{cm}} = 65.6$$

$$A_w = 0.0117, \quad (\text{cm}^2)$$

Step No. 15. Calculate the effective window area $W_{a(\text{eff})}$. Use the window area W_a found in Step 10:

$$W_{a(\text{eff})} = W_a S_3, \quad (\text{cm}^2)$$

A typical value for S_3 is 0.75 as shown in the window utilization factor K_u section.

$$W_{a(\text{eff})} = (1.41)(0.75)$$

$$W_{a(\text{eff})} = 1.06, \quad (\text{cm}^2)$$

Step No. 16. Calculate the number of turns. Use the wire area A_w found in Step 14:

$$N = \frac{W_{a(eff)}S_2}{A_w}, \quad \text{(turns)}$$

A typical value for S_2 is 0.6 as shown in the window utilization factor K_u section.

$$N = \frac{(1.06)(0.6)}{(0.0117)}$$

$$N = 54, \quad \text{(turns)}$$

Step No. 17. Calculate the gap required. Use the iron area A_c found in Step 10:

$$l_g = \frac{(0.4\pi)N^2 A_c \text{x} 10^{-8}}{L}, \quad \text{(cm)}$$

$$l_g = \frac{(1.26)(54)^2(0.36)\text{x}10^{-8}}{0.00055}$$

$$l_g = 0.024, \quad \text{(cm)}$$

The gap spacing is usually maintained by inserting fish paper. However, this paper is available only in mil thickness. Because l_g has been determined in centimeters, it is necessary to convert as follows:

cm x 393.7 = mils (inch system)

Substituting values:

0.024 x 393.7 = 9.45, (mils) use 10

An available size of paper is in 5-mil sheets. Single thickness would therefore be used giving 5 mils in each leg.

0.01 x 2.54 = 0.0254, (cm)

Step No. 18. Calculate the amount of fringing flux. Use the G dimension found in Step 10:

$$F = \left(1 + \frac{l_g}{\sqrt{A_c}} \ln \frac{2G}{l_g}\right)$$

$$F = \left(1 + \frac{(0.0254)}{\sqrt{0.36}} \ln \frac{2(2.20)}{(0.0254)}\right)$$

$$F = 1.21$$

Step No. 19. Calculate the new turns by inserting the fringing flux:

$$N = \sqrt{\frac{l_g L}{(0.4\pi) A_c F \text{x} 10^{-8}}}, \quad \text{(turns)}$$

$$N = \sqrt{\frac{(0.0254)(0.00055)}{(1.26)(0.36)(1.21)\text{x}10^{-8}}}$$

$$N = 50, \quad \text{(turns)}$$

Step No. 20. Calculate the winding resistance. Use MLT from Step 10 and micro-ohm per centimeter from Step 14:

$$R = (\text{MLT})(N)\left(\frac{\mu\Omega}{\text{cm}}\right)\text{x}10^{-6}, \quad \text{(ohms)}$$

$$R = (3.9)(50)(166)\text{x}10^{-6}$$

$$R = 0.0323, \quad \text{(ohms)}$$

Step No. 21. Calculate the copper loss P_{cu} using the rms current I_{rms} in Step 12:

$$P_{cu} = (I_{rms})^2 R, \quad \text{(watts)}$$

$$P_{cu} = (5.0)^2(0.0323)$$

$$P_{cu} = 0.808, \quad \text{(watts)}$$

Step No. 22. Calculate the regulation α:

$$\alpha = \frac{P_{cu}}{P_o + P_{cu}}(100), \quad (\%)$$

$$\alpha = \frac{(0.808)}{(100) + (0.808)}(100)$$

$$\alpha = 0.80, \quad (\%)$$

Step No. 23. Calculate the total ac plus dc flux density using the currents from Step 6:

$$B_{max} = \frac{(0.4\pi)N\left(I_{dc} + \frac{\Delta I}{2}\right)}{l_g} \text{x}10^{-4}, \quad \text{(tesla)}$$

$$B_{max} = \frac{(1.26)(50)\left(5 + \frac{1.01}{2}\right)}{(0.0254)} \text{x}10^{-4}$$

$$B_{max} = 1.38, \quad \text{(tesla)}$$

Step No. 24. Calculate the ac flux density:

$$B_{ac} = \frac{(0.4\pi)N\left(\frac{\Delta I}{2}\right)}{l_g} \text{x}10^{-4}, \quad \text{(tesla)}$$

$$B_{ac} = \frac{(1.26)(50)\left(\frac{1.01}{2}\right)}{(0.0254)} \text{x}10^{-4}$$

$$B_{ac} = 0.124, \quad \text{(tesla)}$$

Step No. 25. Calculate core loss P_{fe}. Use the appropriate core loss curves for silicon 1 mil and 0.125 tesla. The iron weight is found in Step 10:

$$P_{fe} = \left(\frac{\text{milliwatts}}{\text{gram}}\right) W_{tfe} \text{x} 10^{-3}, \quad \text{(watts)}$$

$$P_{fe} = (15)(22.6)\text{x}10^{-3}$$

$$P_{fe} = 0.339, \quad \text{(watts)}$$

Step No. 26. Calculate the gap loss P_g:

$$P_g = K_i D l_g f B_m, \quad \text{(watts)}$$

$$P_g = (0.0775)(0.952)(0.254)(0.124)$$

$$P_g = 0.586, \quad \text{(watts)}$$

Step No. 27. Calculate the total loss:

$$P_\Sigma = P_{cu} + P_{fe} + P_g, \quad \text{(watts)}$$

$$P_\Sigma = (0.808) + (0.339) + (0.586)$$

$$P_\Sigma = 1.73, \quad \text{(watts)}$$

Step No. 28. Calculate the efficiency of the inductor:

$$\eta = \frac{P_o}{P_o + P_\Sigma}(100), \quad (\%)$$

$$\eta = \frac{100}{(100) + (1.73)}(100)$$

$$\eta = 98.3, \quad (\%)$$

Step No. 29. Calculate the watts per unit area. The surface area A_t is found in Step 10:

$$\psi = \frac{P_\Sigma}{A_t}, \quad \left(\text{watts/cm}^2\right)$$

$$\psi = \frac{(1.73)}{(29.8)}$$

$$\psi = 0.0581, \quad \left(\text{watts/cm}^2\right)$$

where

$$\psi = 0.07, \quad \left(\text{watts/cm}^2\right) \text{ @ } 50°\text{C rise}$$

BUCK SWITCHING INDUCTOR DESIGN EXAMPLE USING THE CORE GEOMETRY K_g APPROACH

For a typical design example, assume the circuit shown in Fig. 43 with the following specification:

1. Input voltage, V_{in} = 28 ± 6 V
2. Output voltage, V_o = 20 V
3. Output current range, I_o = 5–0.5 A
4. Frequency, f = 20 kHz
5. Switching efficiency, η = 98%
6. Regulation, α = 1.0%
7. Ferrite pot core

Step No. 1. Calculate the time period T of operation:

$$T = t_{on} + t_{off}$$

$$T = \frac{1}{f}, \quad \text{(seconds)}$$

$$T = \frac{1}{20{,}000}$$

$$T = 50\text{x}10^{-6}, \quad \text{(seconds)}$$

Step No. 2. Calculate the minimum duty ratio:

$$D_{(min)} = \frac{V_o}{V_{in(max)}\eta}$$

$$D_{(min)} = \frac{20}{(34)(0.98)}$$

$$D_{(min)} = 0.60$$

Step No. 3. Calculate the maximum duty ratio:

$$D_{(max)} = \frac{V_o}{V_{in(min)}\eta}$$

$$D_{(max)} = \frac{20}{(22)(0.98)}$$

$$D_{(max)} = 0.927$$

Step No. 4. Calculate the load resistance at minimum load current:

$$R_o = \frac{V_o}{I_{o(min)}}, \quad \text{(ohms)}$$

$$R_o = \frac{20}{0.5}$$

$$R_o = 40, \quad \text{(ohms)}$$

Step No. 5. Calculate the minimum required inductance:

$$L_{(min)} = \frac{R_{o(min)}T\left(1 - D_{(min)}\right)}{2}, \quad \text{(henrys)}$$

$$L_{(min)} = \frac{(40)\left(50\text{x}10^{-6}\right)(1 - 0.60)}{2}$$

$$L_{(min)} = 400\text{x}10^{-6}, \quad \text{(henrys)}$$

Step No. 6. Calculate the ΔI in the inductor:

$$\Delta I = \frac{TV_{in(min)}D_{(min)}(1-D_{min})}{L}, \quad \text{(amps)}$$

$$\Delta I = \frac{(50x10^{-6})(34)(0.6)(1-0.6)}{400x10^{-6}}$$

$$\Delta I = 1.0 = 2I_{o(min)}, \quad \text{(amps)}$$

Step No. 7. Calculate the energy-handling capability:

$$\text{Energy} = \frac{LI_{pk}^{2}}{2}, \quad \text{(watt-seconds)}$$

$$I_{pk} = I_o + \frac{\Delta I}{2}, \quad \text{(amps)}$$

$$I_{pk} = (5.0) + \frac{(1.0)}{2}$$

$$I_{pk} = 5.5, \quad \text{(amps)}$$

$$\text{Energy} = \frac{(0.0004)(5.5)^2}{2}$$

$$\text{Energy} = 0.00605, \quad \text{(watt-seconds)}$$

Step No. 8. Calculate the electrical conditions K_e:

$$K_e = 0.145 P_o B_m^2 x10^{-4}$$

$$B_m = 0.3, \quad \text{(tesla)}$$

$$P_o = V_o I_o, \quad \text{(watts)}$$

$$P_o = (20)(5)$$

$$P_o = 100, \quad \text{(watts)}$$

$$K_e = 0.145(100)(0.3)^2 x10^{-4}$$

$$K_e = 0.00013$$

Step No. 9. Calculate the core geometry K_g:

$$K_g = \frac{(\text{Energy})^2}{K_e \alpha}, \quad (\text{cm}^5)$$

$$K_g = \frac{(0.00605)^2}{(0.00013)(1)}$$

$$K_g = 0.282, \quad (\text{cm}^5)$$

Step No. 10. Select a comparable core geometry K_g from the pot core section in column 12. Also record the G dimension column 4, iron weight column 6, MLT column 8, iron area A_c column 9, window area W_a column 10, area product A_p column 11, and surface area A_t column 13.

$$\text{Core} = \text{TDK, ETD-44}$$

$$K_g = 0.395, \quad (\text{cm}^5)$$

$$\text{G} = 3.3, \quad (\text{cm})$$

$$W_{tfe} = 94, \quad (\text{grams})$$

$$\text{MLT} = 9.5, \quad (\text{cm})$$

$$A_c = 1.75, \quad (\text{cm}^2)$$

$$W_a = 3.05, \quad (\text{cm}^2)$$

$$A_p = 5.34, \quad (\text{cm}^4)$$

$$A_t = 87.5, \quad (\text{cm}^2)$$

Step No. 11. Calculate the current density J. Use the product A_p found in Step 10:

$$J = \frac{2(\text{Eng})\text{x}10^4}{B_m K_u A_p}, \quad (\text{amps/cm}^2)$$

$$K_u = 0.4, \text{ window utilization factor}$$

$$J = \frac{2(0.00605)\text{x}10^4}{(0.3)(0.4)(5.34)}$$

$$J = 189, \quad (\text{amps/cm}^2)$$

Step No. 12. Calculate the bare wire size $A_{w(B)}$:

$$A_{w(B)} = \frac{I_{rms}}{J}, \quad (\text{cm}^2)$$

$$I_{rms} = \sqrt{I_o^2 + \frac{\Delta I^2}{12}}, \quad (\text{amps})$$

$$I_{rms} = \sqrt{(5.0)^2 + \frac{(0.5)^2}{12}}$$

$$I_{rms} = 5.0, \quad (\text{amps})$$

$$A_{w(B)} = \frac{(5.0)}{(189)}$$

$$A_{w(B)} = 0.0265, \quad (\text{cm}^2)$$

Step No. 13. Select a wire size from the wire table column 2. Remember, if the area is not within 10%, take the next smallest size. Also record micro-ohm per centimeter found in column 4 and wire area with insulation A_w column 5.

$$\text{AWG } \#13$$

$$A_{w(B)} = 0.02626, \quad (\text{cm}^2)$$

$$\frac{\mu\Omega}{\text{cm}} = 65.6$$

Step No. 14. Calculate the effective window area $W_{a(eff)}$. Use the window area W_a found in Step 10:

$$W_{a(\text{eff})} = W_a S_3, \quad (\text{cm}^2)$$

A typical value for S_3 when designing with pot cores is 0.75 for a single-section bobbin.

$$W_{a(\text{eff})} = (3.05)(0.75)$$

$$W_{a(\text{eff})} = 2.29, \quad (\text{cm}^2)$$

Step No. 15. Calculate the number of turns. Use the wire area A_w found in Step 13:

$$N = \frac{W_{a(eff)}S_2}{A_w}, \quad \text{(turns)}$$

A typical value for S_2 is 0.6 as shown in the window utilization factor K_u section.

$$N = \frac{(2.29)(0.6)}{(0.02626)}$$

$$N = 52, \quad \text{(turns)}$$

Step No. 16. Calculate the gap required. Use the iron area A_c found in Step 10:

$$l_g = \frac{(0.4\pi)N^2 A_c \text{x} 10^{-8}}{L}, \quad \text{(cm)}$$

$$l_g = \frac{(1.26)(52)^2(1.75)\text{x}10^{-8}}{0.0004}$$

$$l_g = 0.149, \quad \text{(cm)}$$

The gap spacing is usually maintained by inserting fish paper. However, this paper is available only in mil thickness. Because l_g has been determined in centimeters, it is necessary to convert as follows:

$$\text{cm x } 393.7 = \text{mils (inch system)}$$

Substituting values

$$0.149 \text{ x } 393.7 = 59.1, \quad \text{(mils) use 60}$$

An available size of paper is 10 and 7 mil sheet. Single thickness of each would therefore be used, giving equal gaps in both inside and outside.

$$0.060 \text{ x } 2.54 = 0.152, \quad \text{(cm)}$$

Step No. 17. Calculate the amount of fringing flux. Use the G dimension found in Step 10:

$$F = \left(1 + \frac{l_g}{\sqrt{A_c}} \ln \frac{2G}{l_g}\right)$$

$$F = \left(1 + \frac{(0.152)}{\sqrt{1.75}} \ln \frac{2(3.30)}{(0.152)}\right)$$

$$F = 1.43$$

Step No. 18. Calculate the new turns by inserting the fringing flux:

$$N = \sqrt{\frac{l_g L}{(0.4\pi) A_c F \text{x} 10^{-8}}}, \quad \text{(turns)}$$

$$N = \sqrt{\frac{(0.152)(0.0004)}{(1.26)(1.75)(1.43)\text{x}10^{-8}}}$$

$$N = 44, \quad \text{(turns)}$$

Step No. 19. Calculate the winding resistance. Use MLT from Step 10 and micro-ohm per centimeter from Step 13:

$$R = (\text{MLT})(N)\left(\frac{\mu\Omega}{\text{cm}}\right)\text{x}10^{-6}, \quad \text{(ohms)}$$

$$R = (9.5)(44)(65.6)\text{x}10^{-6}$$

$$R = 0.0274, \quad \text{(ohms)}$$

Step No. 20. Calculate the copper loss P_{cu}:

$$P_{cu} = \left(I_{rms}\right)^2 R, \quad \text{(watts)}$$

$$P_{cu} = (5.0)^2(0.0274)$$

$$P_{cu} = 0.685, \quad \text{(watts)}$$

Step No. 21. Calculate the regulation α:

$$\alpha = \frac{P_{cu}}{P_o + P_{cu}}(100), \quad (\%)$$

$$\alpha = \frac{(0.685)}{(100)+(0.685)}(100)$$

$$\alpha = 0.680, \quad (\%)$$

Step No. 22. Calculate the total ac plus dc flux density:

$$B_{\max} = \frac{(0.4\pi)N\left(I_{dc} + \frac{\Delta I}{2}\right)}{l_g} \text{x}10^{-4}, \quad \text{(tesla)}$$

$$B_{\max} = \frac{(1.26)(44)\left(5 + \frac{1.0}{2}\right)}{(0.152)} \text{x}10^{-4}$$

$$B_{\max} = 0.20, \quad \text{(tesla)}$$

Step No. 23. Calculate the ac flux density:

$$B_{ac} = \frac{(0.4\pi)N\left(\frac{\Delta I}{2}\right)}{l_g} \text{x}10^{-4}, \quad \text{(tesla)}$$

$$B_{ac} = \frac{(1.26)(44)\left(\frac{1.0}{2}\right)}{(0.152)} \text{x}10^{-4}$$

$$B_{ac} = 0.0182, \quad \text{(tesla)}$$

Step No. 24. Calculate core loss P_{fe}. Use the appropriate core loss curves for ferrite at 0.0248 tesla. The iron weight is found in Step 10:

$$P_{fe} = \left(\frac{\text{milliwatts}}{\text{gram}}\right) W_{tfe} \text{x}10^{-3}, \quad \text{(watts)}$$

$$P_{fe} = (10)(94)\text{x}10^{-3}$$

$$P_{fe} = 0.94, \quad \text{(watts)}$$

Step No. 25. Calculate the total loss P_ε:

$$P_\Sigma = P_{cu} + P_{fe}, \quad \text{(watts)}$$
$$P_\Sigma = (0.685) + (0.94)$$
$$P_\Sigma = 1.63, \quad \text{(watts)}$$

Step No. 26. Calculate the efficiency of the inductor:

$$\eta = \frac{P_o}{P_o + P_\Sigma}(100), \quad (\%)$$
$$\eta = \frac{100}{(100) + (1.63)}(100)$$
$$\eta = 98.4, \quad (\%)$$

Step No. 27. Calculate the watts per unit area. The surface area A_t is found in Step 10:

$$\psi = \frac{P_\Sigma}{A_t}, \quad \left(\text{watts/cm}^2\right)$$
$$\psi = \frac{(1.63)}{(87.5)}$$
$$\psi = 0.0186, \quad \left(\text{watts/cm}^2\right)$$

where

$$\psi = 0.03, \quad \left(\text{watts/cm}^2\right) \text{ @ } 25°\text{C rise}$$

INVERTING BUCK–BOOST SWITCHING INDUCTOR DESIGN EXAMPLE USING THE CORE GEOMETRY K_g APPROACH

For a typical design example, assume the circuit shown in Fig. 51 with the following specification:

1. Input voltage, $V_{in} = 28 \pm 6$ V
2. Output voltage, $V_o = 20$ V
3. Output current, $I_o = 2.22–0.5$ A
4. Frequency, $f = 20$ kHz
5. Switching efficiency, $\eta = 90\%$
6. Regulation, $\alpha = 2.0\%$
7. Moly-permalloy powder core

Step No. 1. Calculate the timed period T of operation:

$$T = t_{on} + t_{off}$$

$$T = \frac{1}{f}, \quad \text{(seconds)}$$

$$T = \frac{1}{20{,}000}$$

$$T = 50\text{x}10^{-6}, \quad \text{(seconds)}$$

Step No. 2. Calculate the minimum duty ratio:

$$D_{(\min)} = \left(\frac{V_o}{\eta V_{in(\max)} + V_o}\right)$$

$$D_{(\min)} = \left(\frac{20}{(0.9)(31) + 20}\right)$$

$$D_{(\min)} = 0.395$$

Step No. 3. Calculate the maximum duty ratio:

$$D_{(max)} = \left(\frac{V_o}{\eta V_{in(min)} + V_o} \right)$$

$$D_{(max)} = \left(\frac{20}{(0.9)(22) + 20} \right)$$

$$D_{(max)} = 0.52$$

Step No. 4. Calculate the load resistance at minimum load current:

$$R_o = \frac{V_o}{I_{o(min)}}, \quad \text{(ohms)}$$

$$R_o = \frac{20}{0.5}$$

$$R_o = 40, \quad \text{(ohms)}$$

Step No. 5. Calculate the minimum required inductance:

$$L_{(min)} = \frac{R_{o(min)} T \left(1 - D_{(min)}\right)^2}{2}, \quad \text{(henrys)}$$

$$L_{(min)} = \frac{(40)(50\text{x}10^{-6})(1 - 0.395)^2}{2}$$

$$L_{(min)} = 366\text{x}10^{-6}, \quad \text{(henrys)}$$

Step No. 6. Calculate the peak current I_{pk}:

$$I_{pk} = \frac{I_o}{\left(1 - D_{(max)}\right)} + \frac{\Delta I}{2}, \quad \text{(amps)}$$

$$\Delta I = \frac{T V_{in(min)} D_{(max)}}{L}, \quad \text{(amps)}$$

$$\Delta I = \frac{(50\text{x}10^{-6})(22)(0.52)}{366\text{x}10^{-6}}$$

$$\Delta I = 1.56, \quad \text{(amps)}$$

$$I_{pk} = \frac{2.22}{(0.48)} + \left(\frac{1.56}{2}\right)$$

$$I_{pk} = 5.40, \quad \text{(amps)}$$

Step No. 7. Calculate the energy-handling capability:

$$\text{Energy} = \frac{LI_{pk}^2}{2}, \quad \text{(watt - seconds)}$$

$$\text{Energy} = \frac{(0.000366)(5.4)^2}{2}$$

$$\text{Energy} = 0.00534, \quad \text{(watt - seconds)}$$

Step No. 8. Calculate the electrical conditions K_e:

$$K_e = 0.145 P_o B_m^2 \text{x}10^{-4}$$

$$B_m = 0.3, \quad \text{(tesla)}$$

$$P_o = V_o I_o, \quad \text{(watts)}$$

$$P_o = (20)(2)$$

$$P_o = 40, \quad \text{(watts)}$$

$$K_e = 0.145(40)(0.3)^2 \text{x}10^{-4}$$

$$K_e = 56(10^{-6})$$

Step No. 9. Calculate the core geometry K_g:

$$K_g = \frac{(\text{Energy})^2}{K_e \alpha}, \quad (\text{cm}^5)$$

$$K_g = \frac{(0.00534)^2}{(56)(2)(10^{-6})}$$

$$K_g = 0.274, \quad (\text{cm}^5)$$

Step No. 10. Select a comparable core geometry K_g from the moly-permalloy powder section in column 13. Also record the magnetic path length MPL column 6, MLT column 9, window area W_a column 11, area product A_p column 12, and surface area A_t column 14.

$$\text{MPP Core} = 55252$$
$$K_g = 0.343, \quad (\text{cm}^5)$$
$$\text{MLT} = 5.7, \quad (\text{cm})$$
$$\text{MPL} = 0.36, \quad (\text{cm})$$
$$W_a = 4.26, \quad (\text{cm}^2)$$
$$A_p = 4.57, \quad (\text{cm}^4)$$
$$A_t = 87.2, \quad (\text{cm}^2)$$

Step No. 11. Calculate the current density J. Use the area product A_p found in Step 10:

$$J = \frac{2(\text{Eng})\text{x}10^4}{B_m K_u A_p}, \quad (\text{amps/cm}^2)$$
$$K_u = 0.4, \text{ window utilization factor}$$
$$J = \frac{2(0.00534)\text{x}10^4}{(0.3)(0.4)(4.57)}$$
$$J = 195, \quad (\text{amps/cm}^2)$$

Step No. 12. Calculate the rms current I_{rms} in the inductor:

$$I_{(rms)} = \sqrt{\left(I_{pk}^2 - I_{pk}\Delta I + \frac{\Delta I^2}{3}\right)\frac{t_{on} + t_{off}}{T}}, \quad (\text{amps})$$
$$I_{(rms)} = \sqrt{(5.4)^2 - (5.4)(1.56) + \frac{(1.56)^2}{3}}$$
$$I_{(rms)} = 4.64, \quad (\text{amps})$$

When the ac current ΔI is small compared to the dc current I_{dc} in the inductor, then this equation is a good approximation for the rms current:

$$I_{(rms)} = \frac{I_o}{\left(1 - D_{(max)}\right)}, \quad \text{(amps)}$$

$$I_{(rms)} = \frac{2.22}{0.48}$$

$$I_{(rms)} = 4.62, \quad \text{(amps)}$$

Step No. 13. Calculate the permeability of the core required:

$$\mu_\Delta = \frac{B_m \text{MPL}\left(10^{-4}\right)}{0.4\pi W_a J K_u}, \quad \text{(perm)}$$

$$\mu_\Delta = \frac{(0.3)(5.7)\left(10^{-4}\right)}{(1.26)(4.26)(195)(0.4)}$$

$$\mu_\Delta = 70.2, \quad \text{(perm)}$$

From the moly-permalloy powder core section, the core that has the same size but has a permeability closer to the one calculated is the core 55083, with a permeability of 60. This particular core has 81 millihenry per 1000 turns.

Step No. 14. Calculate the number of turns required for 0.230 millihenry:

$$N = 1000\sqrt{\frac{L_{(new)}}{L_{(1000)}}}, \quad \text{(turns)}$$

$$N = 1000\sqrt{\frac{0.366}{81}}, \quad \text{(turns)}$$

$$N = 67, \quad \text{(turns)}$$

Step No. 15. Calculate the amp-turn per centimeter using

$$H=\frac{NI_{pk}}{\text{MPL}},\quad \text{(amp-turns/cm)}$$

$$H=\frac{(67)(5.4)}{9.8}$$

$$H=36.9,\quad \text{(amp-turns/cm)}$$

Check Table 7 and verify the dc bias inductance.

Step No. 16. Calculate the bare wire size $A_{w(B)}$:

$$A_{w(B)}=\frac{I_{(rms)}}{J}\quad \left(\text{cm}^2\right)$$

$$A_{w(B)}=\frac{4.64}{195}$$

$$A_{w(B)}=0.0238\quad \left(\text{cm}^2\right)$$

Step No. 17. Select a wire size from the wire table column 2. Remember, if the area is not within 10%, take the next smallest size. Also record micro-ohm per centimeter found in column 4.

AWG #14

$$A_{w(B)}=0.0227,\quad \left(\text{cm}^2\right)$$

$$\frac{\mu\Omega}{\text{cm}}=82.4$$

use (2)#17 or (4)#20

Step No. 18. Calculate the winding resistance. Use MLT found in Step 10 and micro-ohm per centimeter found in Step 17:

$$R=(\text{MLT})(N)\left(\frac{\mu\Omega}{\text{cm}}\right)\text{x}10^{-6},\quad \text{(ohms)}$$

$$R=(5.7)(67)(82.4)\text{x}10^{-6}$$

$$R=0.0315,\quad \text{(ohms)}$$

Step No. 19. Calculate the copper loss P_{cu}:

$$P_{cu} = (I_{rms})^2 R, \quad \text{(watts)}$$

$$P_{cu} = (4.64)^2 (0.0315)$$

$$P_{cu} = 0.678, \quad \text{(watts)}$$

Step No. 20. Calculate the regulation α:

$$\alpha = \frac{P_{cu}}{P_o + P_{cu}}(100), \quad (\%)$$

$$\alpha = \frac{(0.678)}{(40) + (0.678)}(100)$$

$$\alpha = 1.67, \quad (\%)$$

Step No. 21. Calculate the watts per unit area. The surface area A_t is found in Step 10:

$$\psi = \frac{P_{cu}}{A_t}, \quad (\text{watts/cm}^2)$$

$$\psi = \frac{(0.678)}{(87.2)}$$

$$\psi = 0.0078, \quad (\text{watts/cm}^2)$$

where

$$\psi = 0.01, \quad (\text{watts/cm}^2) \text{ @ } 25°\text{C rise}$$

SINGLE-ENDED FORWARD CONVERTER DESIGN EXAMPLE USING THE CORE GEOMETRY K_g APPROACH

For a typical design example, assume the circuit shown in Fig. 45 with the following specification:

1. Input voltage, V_{in} = 140 V
2. Ouput voltage, V_o = 10 V
3. Output current, I_o = 5.0 A
4. Frequency, f = 20 kHz
5. Operating flux, ΔB = 0.1 T
6. Duty ratio, D_{max} = 0.45
7. Regulation, α = 1.0%
8. Switching efficiency, η = 90%
9. Ferrite PQ Core

Step No. 1. Calculate the total period, T:

$$T = t_{on} + t_{off}$$

$$T = \frac{1}{f}, \quad \text{(seconds)}$$

$$T = \frac{1}{20{,}000}$$

$$T = 50\text{x}10^{-6}, \quad \text{(seconds)}$$

Step No. 2. Calculate the maximum transistor on time, t_{on}:

$$t_{on} = TD_{(max)}, \quad \text{(seconds)}$$

$$t_{on} = (50\text{x}10^{-6})(0.45)$$

$$t_{on} = 22.5\text{x}10^{-6}, \quad \text{(seconds)}$$

Step No. 3. Calculate the transformer output power P_o allowing for 1.0-V diode drop (V_d):

$$P_o = VI_o, \quad \text{(watts)}$$
$$V = V_o + V_d, \quad \text{(volts)}$$
$$V = 10 + 1$$
$$P_o = (11)(5)$$
$$P_o = 55, \quad \text{(watts)}$$

Step No. 4. Calculate the input power, P_{in}:

$$P_{in} = \frac{P_o}{\eta}, \quad \text{(watts)}$$
$$P_{in} = \frac{55}{0.90}$$
$$P_{in} = 61, \quad \text{(watts)}$$

Step No. 5. Calculate the electrical conditions, K_e:

$$K_e = 0.145 f^2 \Delta B_m^2 \text{x}10^{-4}$$
$$K_e = 0.145(20{,}000)^2 (0.1)^2 \text{x}10^{-4}$$
$$K_e = 58$$

Step No. 6. Calculate the gore geometry, K_g, adding 10% to the input power P_{in} for the demag winding:

$$K_g = \frac{1.1 P_{in} \text{D}_{(\max)}}{K_e \alpha}, \quad \left(\text{cm}^5\right)$$
$$K_g = \frac{(1.1)(61)(0.45)}{(58)(1)}$$
$$K_g = 0.521, \quad \left(\text{cm}^5\right)$$

Step No. 7. Select a comparable core K_g from the PQ core section in column 12. Also record the iron weight W_{tfe} column 6, MLT column 8, iron cross section A_c column 9, window area W_a column 10, area product A_p column 11, surface area A_t column 13.

PQ Core, 44040

$$K_g = 0.414, \quad (\text{cm}^5)$$

$$A_p = 5.17, \quad (\text{cm}^4)$$

$$W_{tfe} = 95, \quad (\text{grams})$$

$$A_c = 1.67, \quad (\text{cm}^2)$$

$$W_a = 3.1, \quad (\text{cm}^2)$$

$$\text{MLT} = 8.3, \quad (\text{cm})$$

$$A_t = 78.4, \quad (\text{cm}^2)$$

Step No. 8. Calculate the primary rms current, $I_{p(rms)}$:

$$I_{p(rms)} = \frac{P_{in}}{V_{in}\sqrt{D_{(max)}}}, \quad (\text{amps})$$

$$I_{p(rms)} = \frac{(61)}{(140)\sqrt{0.45}}$$

$$I_{p(rms)} = 0.649, \quad (\text{amps})$$

Step No. 9. Calculate the number of primary turns, N_p:

$$N_p = \left(\frac{V_{in} D_{(max)} \text{x} 10^4}{\Delta B_m f A_c}\right), \quad (\text{turns})$$

$$N_p = \left(\frac{(140)(0.45)\text{x}10^4}{(0.1)(20{,}000)(1.67)}\right)$$

$$N_p = 189, \quad (\text{turns})$$

Step No. 10. Calculate current density J using a window utilization $K_u = 0.32$:

$$J = \left(\frac{2P_t \sqrt{D_{(max)}} \mathrm{x} 10^4}{K_u f \Delta B_m A_p} \right), \quad (\text{amps/cm}^2)$$

$$P_t = 1.1 P_{in}, \quad (\text{watts})$$

$$P_t = 1.1(61)$$

$$P_t = 67, \quad (\text{watts})$$

$$K_u = 0.32, \text{ window utilization factor}$$

$$J = \left(\frac{2(67)\sqrt{0.45} \mathrm{x} 10^4}{(0.32)(20{,}000)(0.1)(5.17)} \right)$$

$$J = 272, \quad (\text{amps/cm}^2)$$

Step No. 11. Calculate the required primary bare wire area, $A_{w(B)}$:

$$A_{w(B)} = \frac{I_{p(rms)}}{J}, \quad (\text{cm}^2)$$

$$A_{w(B)} = \frac{(0.649)}{(272)}$$

$$A_{w(B)} = 0.00239, \quad (\text{cm}^2)$$

Step No. 12. Select a wire size with the required area from the wire table column 2. Remember, if the wire area is not within 10%, take the next smallest size. Also record micro-ohm per centimeter found in column 4.

Primary, AWG #23

$$A_{w(B)} = 0.002588, \quad (\text{cm}^2)$$

$$\frac{\mu\Omega}{\text{cm}} = 666$$

Step No. 13. Calculate the primary winding resistance R_p. Use the MLT from Step 7 and micro-ohm per centimeter found in Step 12.

$$R_p = (\text{MLT})(N)\left(\frac{\mu\Omega}{\text{cm}}\right)\text{x}10^{-6}, \quad \text{(ohms)}$$

$$R_p = (8.3)(189)(666)\text{x}10^{-6}$$

$$R_p = 1.04, \quad \text{(ohms)}$$

Step No. 14. Calculate the primary copper loss, P_p:

$$P_p = \left(I_{p(rms)}\right)^2 R_p, \quad \text{(watts)}$$

$$P_p = (0.649)^2(1.04)$$

$$P_p = 0.438, \quad \text{(watts)}$$

Step No. 15. Calculate the transformer secondary voltage, V_s:

$$V_s = \frac{V_o + V_d}{\text{D}_{(\text{max})}}, \quad \text{(volts)}$$

$$V_s = \frac{(10)+(1)}{(0.45)}$$

$$V_s = 24.4, \quad \text{(volts)}$$

Step No. 16. Calculate the number of secondary turns, N_s:

$$N_s = \left(\frac{N_p V_s}{V_p}\right)\left(1+\frac{\alpha}{100}\right), \quad \text{(turns)}$$

$$N_s = \left(\frac{(189)(24.4)}{(140)}\right)\left(1+\frac{1}{100}\right)$$

$$N_s = 33, \quad \text{(turns)}$$

Step No. 17. Calculate the secondary rms current $I_{s(rms)}$:

$$I_{s(rms)} = I_o\sqrt{D_{(max)}}, \quad \text{(amps)}$$
$$I_{s(rms)} = 5\sqrt{0.45}$$
$$I_{s(rms)} = 3.35, \quad \text{(amps)}$$

Step No. 18. Calculate the secondary bare wire area, $A_{w(B)}$:

$$A_{w(B)} = \frac{I_{s(rms)}}{J}, \quad \left(\text{cm}^2\right)$$
$$A_{w(B)} = \frac{(3.35)}{(272)}$$
$$A_{w(B)} = 0.0123, \quad \left(\text{cm}^2\right)$$

Step No. 19. Calculate the skin depth, ε @ 20°C:

$$\varepsilon = \frac{6.62}{\sqrt{f}}, \quad \text{(cm)}$$
$$\varepsilon = \frac{6.62}{\sqrt{20000}}$$
$$\varepsilon = 0.0468, \quad \text{(cm)}$$

let ε = radius

new wire area is:

$$A_{w(new)} = \pi\varepsilon^2, \quad \left(\text{cm}^2\right)$$
$$A_{w(new)} = (3.14)(0.0468)^2$$
$$A_{w(new)} = 0.00689, \quad \left(\text{cm}^2\right)$$

AWG #19 = 0.00653, $\mu\Omega$/cm = 264

AWG #20 = 0.00519, $\mu\Omega$/cm = 332

Step No. 20. Calculate the number of strands, NS:

$$NS = \frac{A_{ws}}{A_{w(19)}}$$

$$NS = \frac{0.0123}{0.00653}$$

$$NS = 1.88, \text{ use } 2$$

$$\text{new}, \frac{\mu\Omega}{\text{cm}} = \frac{\mu\Omega/\text{cm}}{2} = (\#19)\frac{264}{2}$$

$$\text{new}, \frac{\mu\Omega}{\text{cm}} = 132$$

Step No. 21. Calculate the secondary winding resistance R_s:

$$R_s = (\text{MLT})(N)\left(\frac{\mu\Omega}{\text{cm}}\right)\text{x}10^{-6}, \quad (\text{ohms})$$

$$R_s = (8.3)(33)(132)\text{x}10^{-6}$$

$$R_s = 0.0362, \quad (\text{ohms})$$

Step No. 22. Calculate the secondary copper loss, P_s:

$$P_s = \left(I_{s(rms)}\right)^2 R_s, \quad (\text{watts})$$

$$P_s = (3.35)^2(0.0362)$$

$$P_s = 0.406, \quad (\text{watts})$$

Step No. 23. Calculate the transformer regulation, α:

$$\alpha = \frac{P_{cu}}{P_o + P_{cu}}(100), \quad (\%)$$

$$P_{cu} = P_p + P_s, \quad (\text{watts})$$

$$P_{cu} = (0.438) + (0.406)$$

$$P_{cu} = 0.844, \quad (\text{watts})$$

$$\alpha = \frac{0.844}{(55) + (0.844)}(100)$$

$$\alpha = 1.5, \quad (\%)$$

TRANSFORMER RECTIFIER WITH CAPACITANCE INPUT. DESIGN EXAMPLE USING THE CORE GEOMETRY K_g APPROACH

For a typical design example, assume the following using a full-wave bridge as shown in Fig. 14.

1. Input voltage, V_{in} = 115 V
2. Output voltage, V_o = 10 V
3. Output current, I_o = 1–0.1 A
4. Ripple, ΔV_{rms} = 0.5 V
5. Frequency, f = 60 Hz
6. Temperature rise, 25°C
7. Efficiency, η = 95%
8. Regulation, α = 2% (copper)
9. Core, lamination

Step No. 1. Calculate the capacitance necessary for the required ripple at rated current:

$$C = \frac{I_{o(\max)}}{2\sqrt{3} f \Delta V_{rms}}, \quad \text{(farad)}$$

$$C = \frac{(1)}{2(1.73)(120)(0.5)}$$

$$C = 0.00481, \quad \text{(farad)}$$

standard value 5,000 μf

Step No. 2. Calculate the rms current I_{rms} through the capacitor:

$$I_{s(\text{rms})} = \Delta V_{(\text{rms})}(2\pi f C), \quad \text{(amps)}$$

$$I_{s(\text{rms})} = (0.5)(6.28)(120)(0.005)$$

$$I_{s(\text{rms})} = 1.88, \quad \text{(amps)}$$

Step No. 3. Calculate the filter product $\omega R_o C$:

$$\omega = (2\pi f) = 754$$

$$R_o = \frac{V_o}{I_o}, \quad \text{(ohms)}$$

$$R_o = \frac{10}{1}$$

$$R_o = 10, \quad \text{(ohms)}$$

then

$$\omega R_o C = (754)(10)(0.005) \cong 40$$

Step No. 4. Calculate the source resistance R_x for the required regulation α:

$$R_x = \frac{\alpha R_o}{100}, \quad \text{(ohms)}$$

$$\alpha = \text{Regulation \%}$$

$$R_x = \frac{(2)(10)}{100}$$

$$R_x = 0.2, \quad \text{(ohms)}$$

Step No. 5. Calculate the ratio of the source resistance R_x and the load resistance R_o:

$$R_{(source)} = \frac{R_x}{R_o}, \quad \text{(ohms)}$$

$$R_{(source)} = \frac{0.2}{10}$$

$$R_{(source)} = 0.02, \quad \text{(ohms)}$$

Step No. 6. Using the filter product $\omega R_o C$ in Step 3 and the ratio of the source resistance R_x and load resistance R_o in Step 5, find the ratio of the output V_o to the peak voltage V_{spk} required at the intersect in Fig. 18.

$$\gamma = \frac{V_o}{V_{spk}} \cong 0.9$$

Step No. 7. Calculate the peak secondary voltage V_{spk} and the V_{rms} voltage:

$$V_{spk} = \frac{V_o}{\gamma}, \quad \text{(volts)}$$

$$V_{spk} = \frac{10}{0.9}$$

$$V_{spk} = 11.1, \quad \text{(volts)}$$

$$V_s = (0.707)V_{spk}, \quad \text{(volts)}$$

$$V_s = (0.707)(11.1)$$

$$V_s = 7.85, \quad \text{(volts)}$$

Step No. 8. Calculate the secondary output power:

$$P_o = I_{s(\text{rms})} V_s, \quad \text{(watts)}$$

$$P_o = (1.88)(7.85)$$

$$P_o = 14.7, \quad \text{(watts)}$$

Step No. 9. Calculate the apparent power P_t:

$$P_t = P_o\left(\frac{1}{\eta} + 1\right), \quad \text{(watts)}$$

$$P_t = 14.7\left(\frac{1}{0.95} + 1\right)$$

$$P_t = 30.2, \quad \text{(watts)}$$

Step No. 10. Calculate the electrical conditions K_e:

$$K_e = (0.145)K_f^2 f^2 B_m^2 \text{x}10^{-4}$$

$$K_f = 4.44, \text{ sine wave}$$

$$K_e = (0.145)(4.44)^2(60)^2(1.5)^2\text{x}10^{-4}$$

$$K_e = 2.31$$

Step No. 11. Calculate the core geometry K_g:

$$K_g = \frac{P_t}{2K_e\alpha}, \quad (\text{cm}^5)$$

$$K_g = \frac{30.2}{2(2.31)(2)}$$

$$K_g = 3.27, \quad (\text{cm}^5)$$

Step No. 12. Select a comparable core geometry K_g from the lamination section in column 15 that is approximately 20% larger than necessary because the core geometry is listed in gross terms. The core geometry will have to be multiplied by the stacking factor squared $(\text{SF})^2$ found in Table 10. Also record gross iron weight W_{tfe} column 7, MLT column 9, gross iron cross section A_c column 10, surface area A_t column 14, area product A_p column 12.

Lamination, 100EI-1

$$K_{g(\text{gross})} = 5.46, \quad (\text{cm}^5)$$

$$A_{p(\text{gross})} = 31.2, \quad (\text{cm}^4)$$

$$W_{tfe(\text{gross})} = 712, \quad (\text{grams})$$

$$A_{c(\text{gross})} = 6.45, \quad (\text{cm}^2)$$

$$\text{MLT} = 14.8, \quad (\text{cm})$$

$$A_t = 213, \quad (\text{cm}^2)$$

$$\text{SF} = 0.90$$

$$(\text{SF})^2 = 0.81$$

Step No. 13. Calculate the effective core geometry K_g:

$$K_g = (\text{SF})^2 K_{g(\text{gross})}, \quad (\text{cm}^5)$$
$$K_g = (0.81)(5.46)$$
$$K_g = 4.42, \quad (\text{cm}^5)$$

After the core geometry K_g has been determined, the geometry of the transformer can be evaluated, such as volume and weight.

Step No. 14. Calculate the effective iron area A_c recorded in Step 12. The stacking factor is found in Table 10:

$$A_c = (\text{SF})A_{c(gross)}, \quad (\text{cm}^2)$$
$$A_c = (0.90)(6.45)$$
$$A_c = 5.81, \quad (\text{cm}^2)$$

Step No. 15. Calculate the number of primary turns:

$$N_p = \left(\frac{V_p \text{x} 10^4}{K_f B_m f A_c}\right), \quad (\text{turns})$$
$$N_p = \left(\frac{115 \text{x} 10^4}{(4.44)(1.5)(60)(5.81)}\right)$$
$$N_p = 495, \quad (\text{turns})$$

Step No. 16. Calculate the secondary turns:

$$N_s = \left(\frac{N_p V_s}{V_p}\right)\left(1 + \frac{\alpha}{100}\right), \quad (\text{turns})$$
$$N_s = \frac{(495)(7.85)}{(115)}\left(1 + \frac{2}{100}\right)$$
$$N_s = 34, \quad (\text{turns})$$

Step No. 17. Calculate the current density J:

$$J = \left(\frac{P_t \text{x} 10^4}{K_f K_u f B_m A_p} \right), \quad (\text{amps/cm}^2)$$

$$A_p = (\text{SF}) A_{p(gross)}, \quad (\text{cm}^4)$$

$$A_p = (0.90)(31.2)$$

$$A_p = 28.1, \quad (\text{cm}^4)$$

$$K_u = 0.40, \text{ window utilization factor}$$

$$J = \left(\frac{30.2 \text{x} 10^4}{(4.44)(0.4)(60)(1.5)(28.1)} \right)$$

$$J = 67, \quad (\text{amps/cm}^2)$$

Step No. 18. Calculate the primary current I_p:

$$I_p = \frac{N_s I_{s(\text{rms})}}{N_p \eta}, \quad (\text{amps})$$

n = turns ratio
η = efficiency

$$I_p = \frac{(34)(1.88)}{(495)(0.9)}$$

$$I_p = 0.143, \quad (\text{amps})$$

Step No. 19. Calculate the bare wire size $A_{w(B)}$ for the primary:

$$A_{w(\text{B})} = \frac{I_p}{J}, \quad (\text{cm}^2)$$

$$A_{w(\text{B})} = \frac{(0.143)}{(67)}$$

$$A_{w(\text{B})} = 0.00213, \quad (\text{cm}^2)$$

Step No. 20. Select a wire size from the wire table column 2. Remember, if the wire area is not within 10%, take the next smallest size. Also record micro-ohm per centimeter found in column 4.

$$\text{AWG \#24 with } 0.00205, \quad (\text{cm}^2)$$

$$\frac{\mu\Omega}{\text{cm}} = 842$$

Step No. 21. Calculate the primary winding resistance. Use the MLT from Step 12 and micro-ohm per centimeter found in Step 20:

$$R_p = (\text{MLT})(N)\left(\frac{\mu\Omega}{\text{cm}}\right)\text{x}10^{-6}, \quad (\text{ohms})$$

$$R_p = (14.8)(495)(842)\text{x}10^{-6}$$

$$R_p = 6.17, \quad (\text{ohms})$$

Step No. 22. Calculate the primary copper loss P_p:

$$P_p = (I_p)^2 R_p, \quad (\text{watts})$$

$$P_p = (0.143)^2(6.17)$$

$$P_p = 0.126, \quad (\text{watts})$$

Step No. 23. Calculate the bare wire size $A_{w(B)}$ for the secondary:

$$A_{w(B)} = \frac{I_s}{J}, \quad (\text{cm}^2)$$

$$A_{w(B)} = \frac{(1.88)}{(67)}$$

$$A_{w(B)} = 0.0280, \quad (\text{cm}^2)$$

Step No. 24. Select a wire size from the wire table column 2. Remember, if the wire area is not within 10%, take the next smallest size. Also record micro-ohm per centimeter found in column 4.

$$\text{AWG \#13 with } 0.0263, \quad \left(\text{cm}^2\right)$$

$$\frac{\mu\Omega}{\text{cm}} = 65.6$$

Step No. 25. Calculate the secondary winding resistance. Use MLT from Step 12 and micro-ohm per centimeter from Step 24:

$$R_s = (\text{MLT})(N)\left(\frac{\mu\Omega}{\text{cm}}\right)\text{x}10^{-6}, \quad \text{(ohms)}$$

$$R_s = (14.8)(34)(65.6)\text{x}10^{-6}$$

$$R_s = 0.033, \quad \text{(ohms)}$$

Step No. 26. Calculate the secondary copper loss P_s:

$$P_s = \left(I_s\right)^2 R_s, \quad \text{(watts)}$$

$$P_s = (1.88)^2(0.033)$$

$$P_s = 0.116, \quad \text{(watts)}$$

Step No. 27. Calculate the transformer regulation:

$$\alpha = \frac{P_{cu}}{P_o + P_{cu}}(100) \quad (\%)$$

$$P_{cu} = P_p + P_s, \quad \text{(watts)}$$

$$P_{cu} = (0.126) + (0.116)$$

$$P_{cu} = 0.242, \quad \text{(watts)}$$

$$\alpha = \frac{0.242}{(10) + (0.242)}(100)$$

$$\alpha = 2.3 \quad (\%)$$

4

MAGNETIC-CORE MATERIAL TRADE-OFF

INTRODUCTION

The area product A_p and core geometry K_g are associated only with such geometric properties as surface area and volume, weight, and the factors affecting temperature rise such as current density. The area product A_p and core geometry K_g has no relevance to the magnetic-core materials used; however, the designer often must make trade-offs between such goals as efficiency and size which are influenced by core material selection.

Usually in articles written about inverter and converter transformer design, recommendations with respect to choice of core material are a compromise of material characteristics such as those tabulated in Table 9, and graphically displayed in Figure 63. The characteristics shown here are those typical of commercially available core materials. As can be seen, the core material which provides the highest flux density is supermendur. It also produces the smallest component size. If size is the most important consideration, this should determine the choice of materials. On the other hand, the type 78 Supermalloy material (see the 5/78 curve in Figure 63) has the lowest flux density and this material would result in the largest size transformer. However, this material has the lowest coercive force and lowest core loss of any of the available materials. These factors might well be decisive in other applications.

The standard stacking factors for tape cores, wound cut cores, and laminations are shown in Table 10.

Table 9 Magnetic Core Material Characteristics

Material Trade Names	Material Composition	Saturated Flux Density tesla	DC Coercive Force Oersteds	Squareness Ratio	Material Density grams/cm 3rd	Resistivity (ohm-cm)	Curie Temperature Degrees °C	Weight Factor
Supermendur Permendur	49% Co 49% Fe 2% V	1.9-2.2	0.18-0.44	0.90-1.0	8.15	$70x10^{-6}$	930	1.066
Magnesil Silectron Microsil Supersil	3% Si 97% Fe	1.5-1.8	0.5-0.75	0.75-0.85	7.63	$50x10^{-6}$	750	1
Deltamax Orthonol 49 Sq Mu	50% Ni 50% Fe	1.4-1.6	0.125-0.25	0.94-1.0	8.24	$40x10^{-6}$	500	1.079
Allegheny 4750 48 Alloy Carpenter 49	48% Ni 52% Fe	1.15-1.4	0.062-0.187	0.80-0.92	8.19	45x10	480	1.073
4-79 Permalloy Sq Permalloy 80-20 Sq Perm	79% Ni 17% Fe	0.66-0.82	0.025-0.82	0.80-90	8.73	$55x10^{-6}$	460	1.144
Supermalloy	78% Ni 17% Fe 5% Mo	0.65-0.82	0.003-0.008	0.40-0.70	8.76	$65x10^{-6}$	400	1.148
Metglas 2605SC	81% Fe 3.5% Si 13.5% B 2% C	1.5-1.6	0.03-0.08	0.90-1.0	7.32	$135x10^{-6}$	370	1
Metglas 2714B	66% Co 15% Mo 4% Fe 1% Ni	0.5-0.65	.008-0.02	0.90-1.0	7.59	$142x10^{-6}$	205	1
Ferrite	Mn Zn	0.36-0.5	0.06-0.35	0.5-0.75	4.6-4.9	10-100	150-250	0.63
Ferrite	Ni Zn	0.3-0.41	0.17-3	0.5-0.85	4.3-5.2	$1x10^{6}$	120-500	0.63

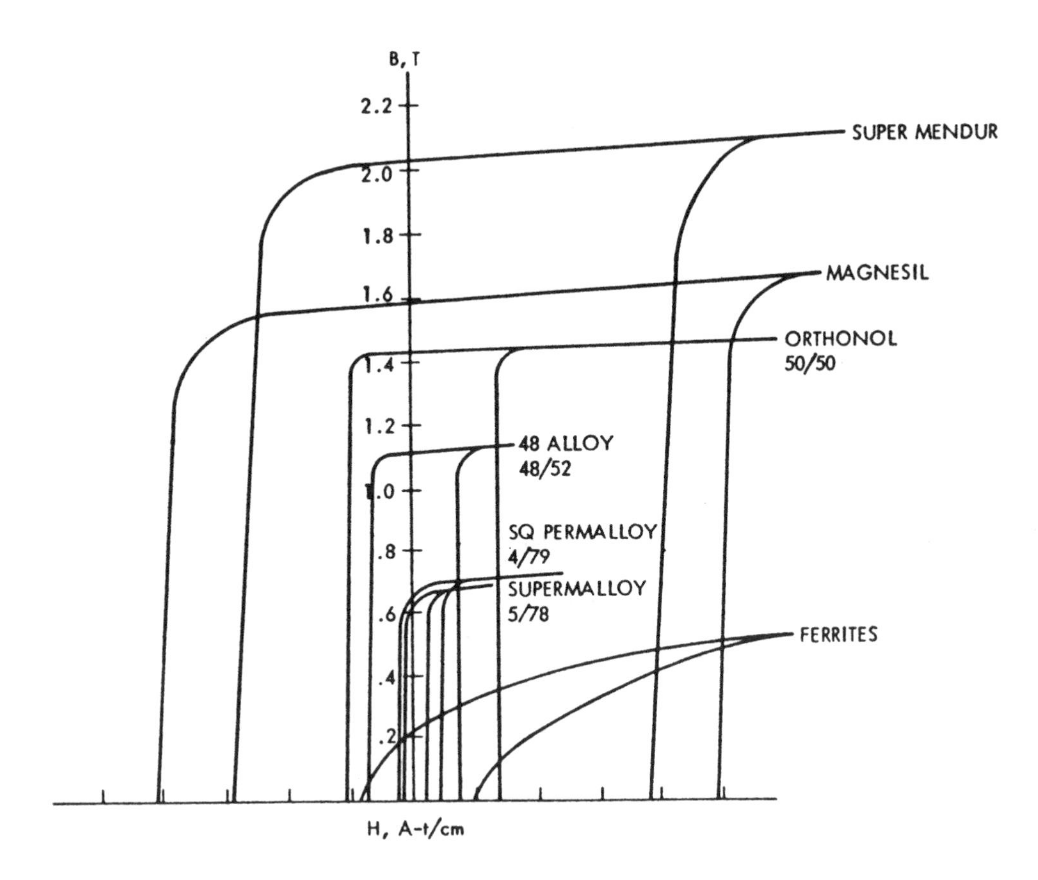

Figure 63 The typical dc B–H loops of magnetic material. (Courtesy of Magnetics.)

Table 10 Standard Stacking Factors for Tape Cores and Laminations

Thickness (mils)	Stacking Factor SF	Stacking Factor $(SF)^2$
	Tape Cores	
0.125	0.250	0.063
0.25	0.375	0.141
0.5	0.500	0.250
1	0.750	0.562
2	0.850	0.722
4	0.900	0.810
12	0.950	0.902
Metglas	0.750	0.563
	C and E Cores	
0.5	0.600	0.360
1	0.800	0.640
2	0.850	0.723
4	0.900	0.810
6	0.900	0.810
12	0.950	0.903
Metglas	0.750	0.563
	Laminations Butt Jointed	
4	0.900	0.810
6	0.900	0.810
14	0.950	0.902
18	0.950	0.902
25	0.950	0.902
	Interleaved One By One	
4	0.800	0.640
6	0.850	0.722
14	0.900	0.810
18	0.900	0.810
25	0.920	0.846

FORTY-NINE CORE-LOSS CURVES

Choice of core material is thus based on achieving the best characteristic for the most critical or important design parameter, with acceptable compromises on all other parameters. Figures 64 through 112 compare the core loss of different magnetic materials as a function of flux density, frequency, and material thickness.

Possibly for the first time, manufacturers core-loss data curves have been organized with the same units for all core losses. The data were digitized right from the manufacturers data sheets. Then the data were modified to put it in metric units gauss to tesla and watts per pound to watts per kilogram. These data were then put into the computer to develop a new first-order approximation in the form of

$$w = kf^m B^n$$

where w is the calculated core-loss density in watts/kilogram, f is the frequency in hertz, B is the flux density in tesla, and k, m, and n are coefficients derived using a three-dimensional least-square-fit law from the digitized data. The new equations were plotted onto the same graphs as the digitized data with a surprisingly close fit considering the nonlinearity of the original curves.

The core-loss curves were created from manufacturers' data. The equations below the core-loss data are a fair approximation and should be used knowing this. The manufacturers' data should be compared in the final design.

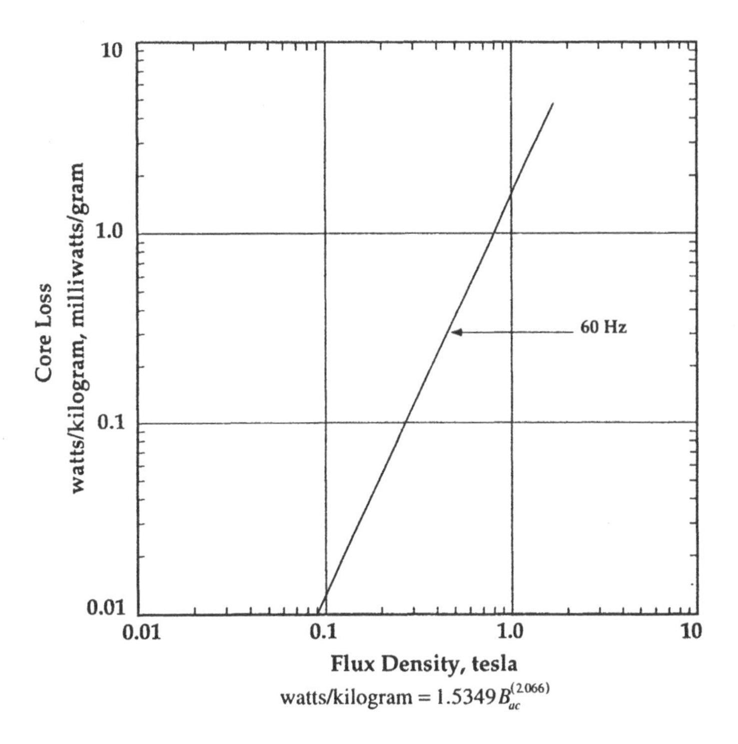

Figure 64 Temple Steel core-loss curve for Silicon 26G M19.

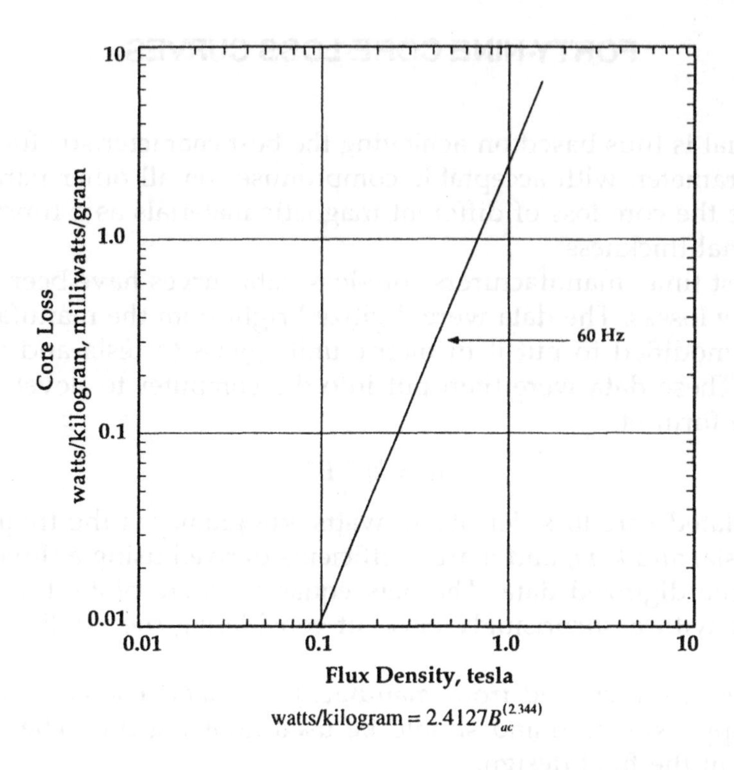

Figure 65 Temple Steel core-loss curve for Silicon 26G M50.

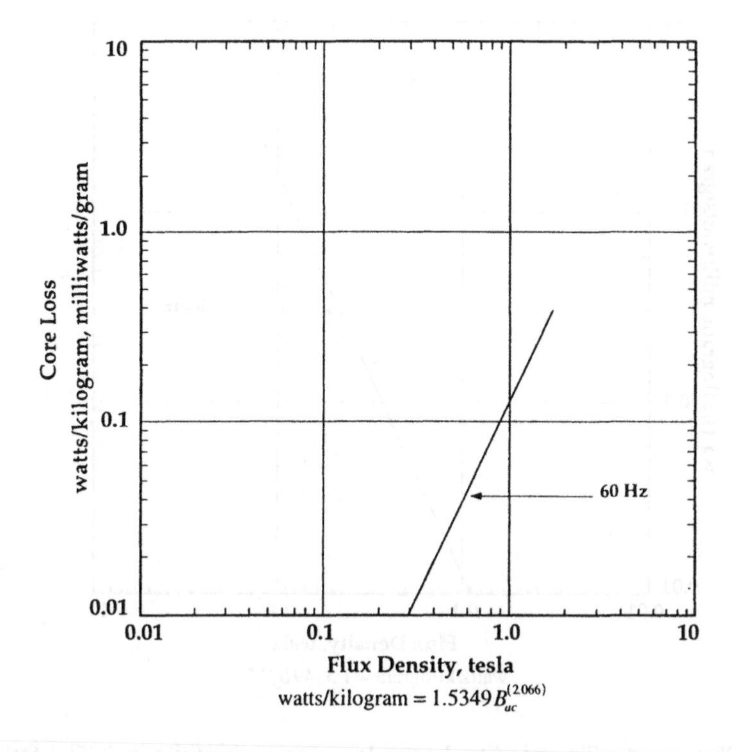

Figure 66 60-Hz core-loss curve for Metglas® Alloy 2605SA1.

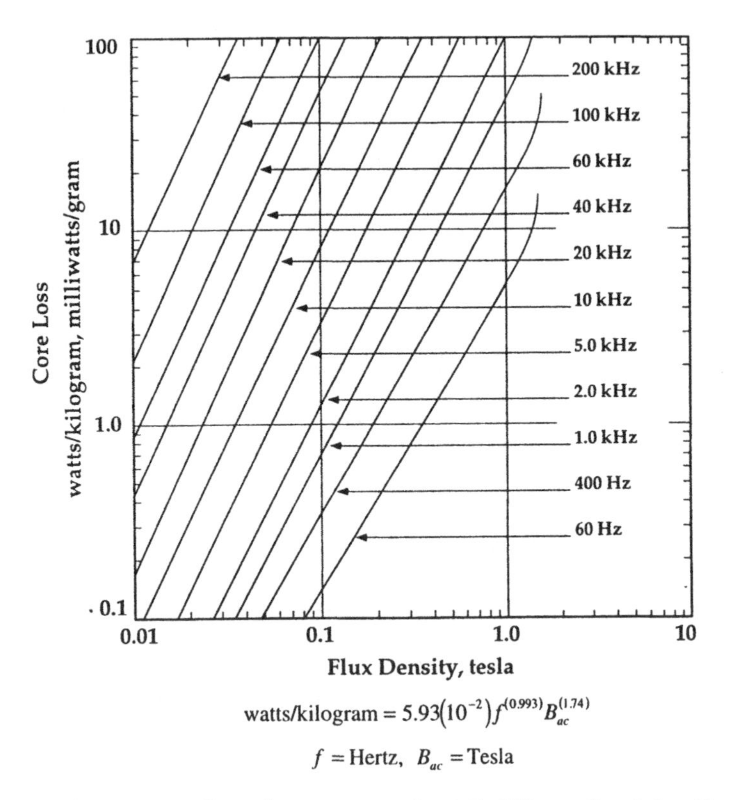

$$\text{watts/kilogram} = 5.93(10^{-2})f^{(0.993)}B_{ac}^{(1.74)}$$

$$f = \text{Hertz}, \quad B_{ac} = \text{Tesla}$$

Figure 67 Core-loss curves for 3% Silicon Steel 1 mil.

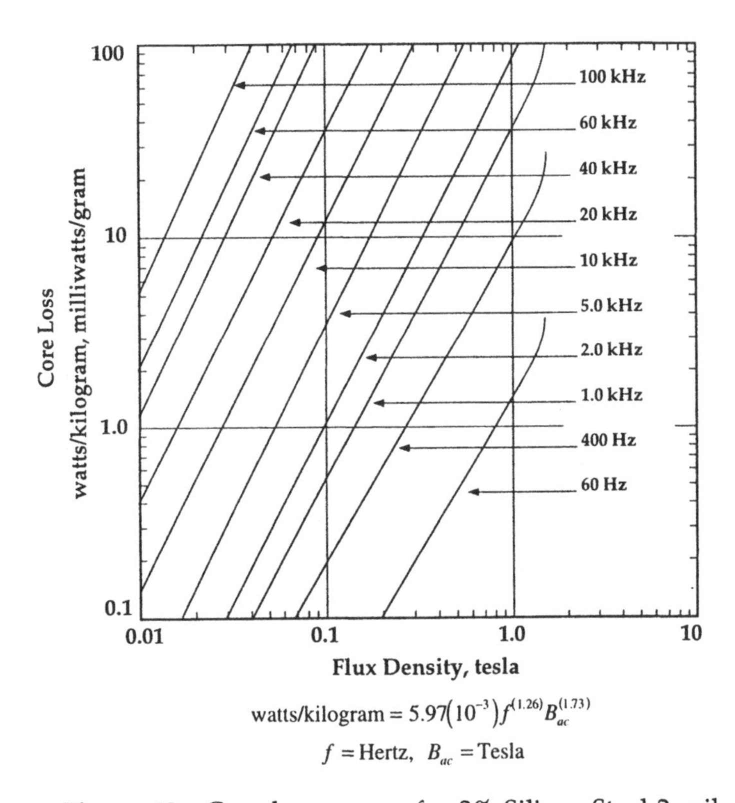

$$\text{watts/kilogram} = 5.97(10^{-3})f^{(1.26)}B_{ac}^{(1.73)}$$

$$f = \text{Hertz}, \quad B_{ac} = \text{Tesla}$$

Figure 68 Core-loss curves for 3% Silicon Steel 2 mil.

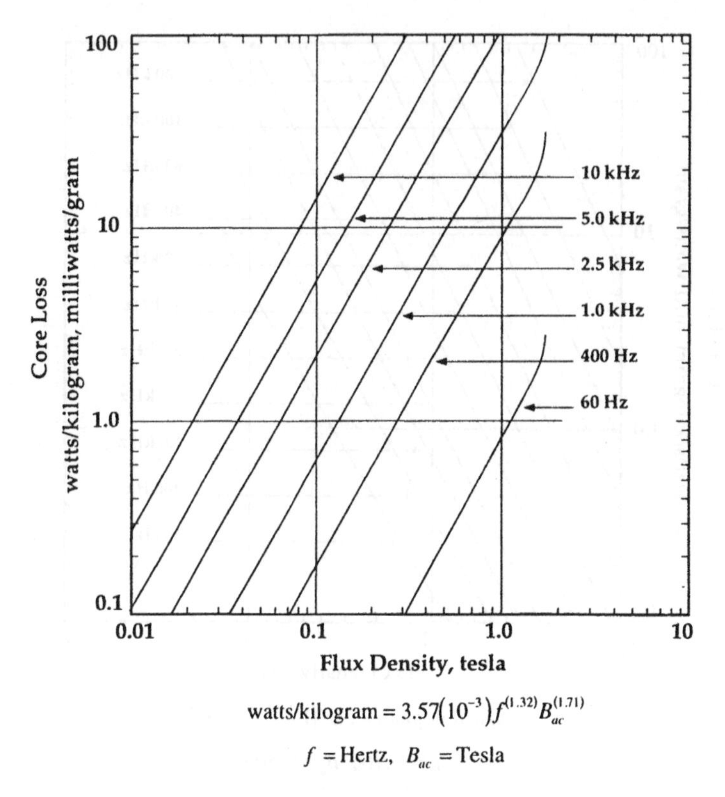

$$\text{watts/kilogram} = 3.57\left(10^{-3}\right) f^{(1.32)} B_{ac}^{(1.71)}$$

$$f = \text{Hertz}, \quad B_{ac} = \text{Tesla}$$

Figure 69 Core-loss curves for 3% Silicon Steel 4 mil.

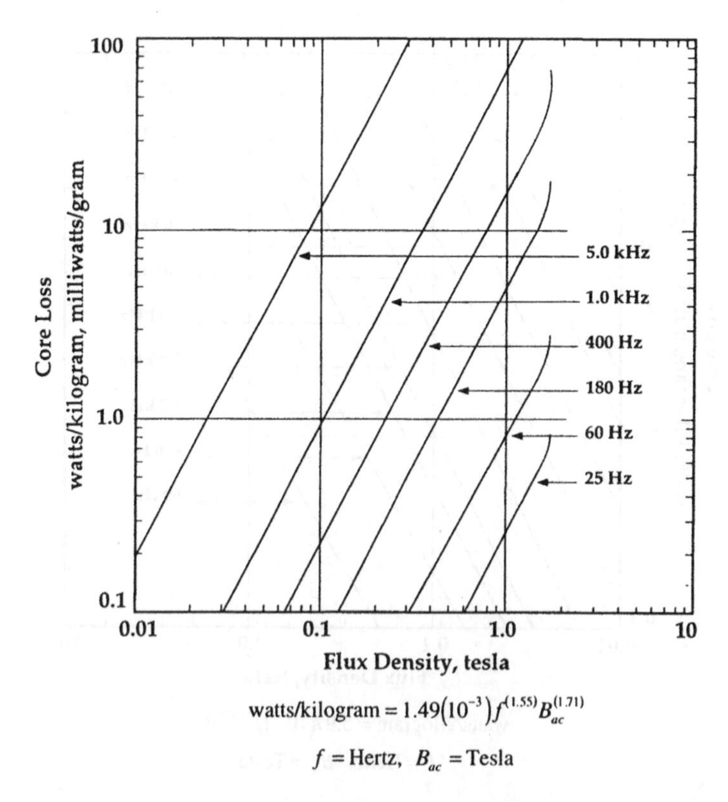

$$\text{watts/kilogram} = 1.49\left(10^{-3}\right) f^{(1.55)} B_{ac}^{(1.71)}$$

$$f = \text{Hertz}, \quad B_{ac} = \text{Tesla}$$

Figure 70 Core-loss curves for 3% Silicon Steel 12 mil.

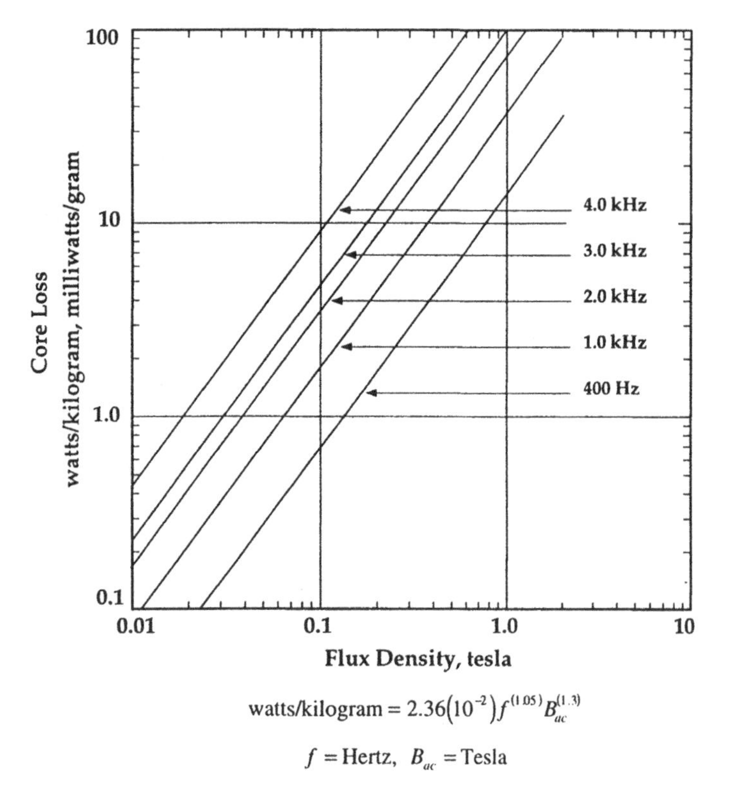

$$\text{watts/kilogram} = 2.36\left(10^{-2}\right)f^{(1.05)}B_{ac}^{(1.3)}$$

$$f = \text{Hertz},\quad B_{ac} = \text{Tesla}$$

Figure 71 Core-loss curves for Supermendur 2 mil.

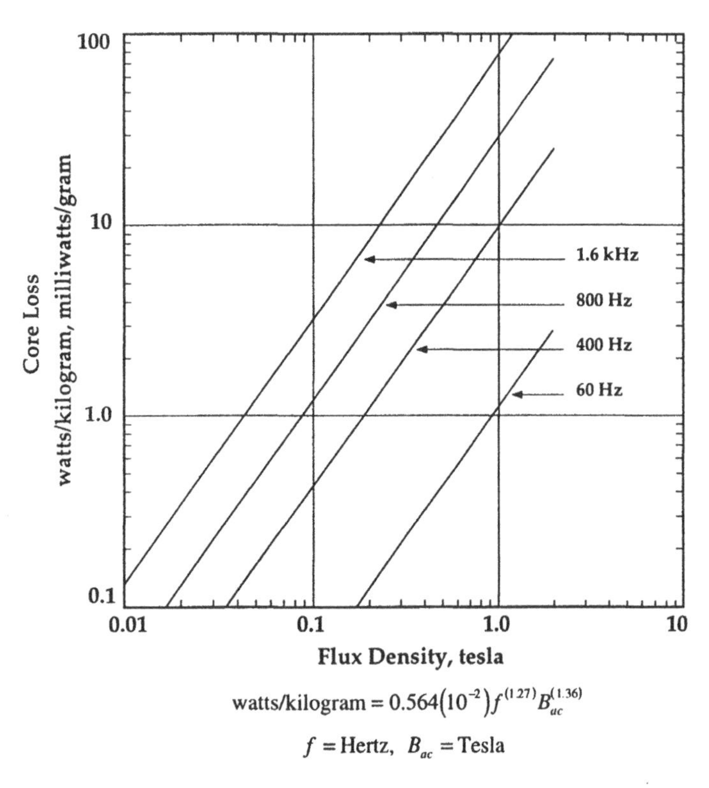

$$\text{watts/kilogram} = 0.564\left(10^{-2}\right)f^{(1.27)}B_{ac}^{(1.36)}$$

$$f = \text{Hertz},\quad B_{ac} = \text{Tesla}$$

Figure 72 Core-loss curves for Supermendur 4 mil.

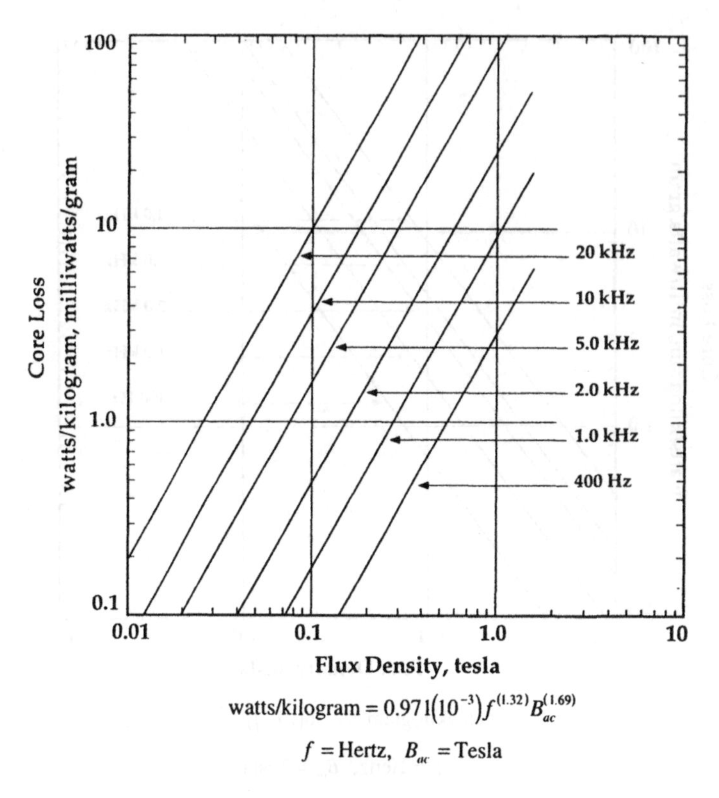

$$\text{watts/kilogram} = 0.971(10^{-3})f^{(1.32)}B_{ac}^{(1.69)}$$

$$f = \text{Hertz},\quad B_{ac} = \text{Tesla}$$

Figure 73 Core-loss curves for 50–50 Ni–Fe 1 mil.

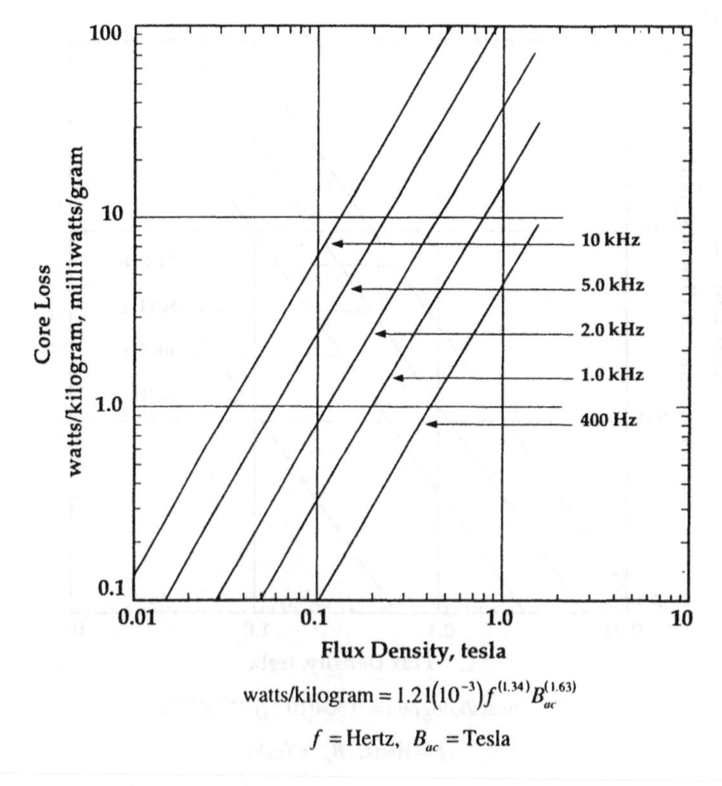

$$\text{watts/kilogram} = 1.21(10^{-3})f^{(1.34)}B_{ac}^{(1.63)}$$

$$f = \text{Hertz},\quad B_{ac} = \text{Tesla}$$

Figure 74 Core-loss curves for 50–50 Ni–Fe 2 mil.

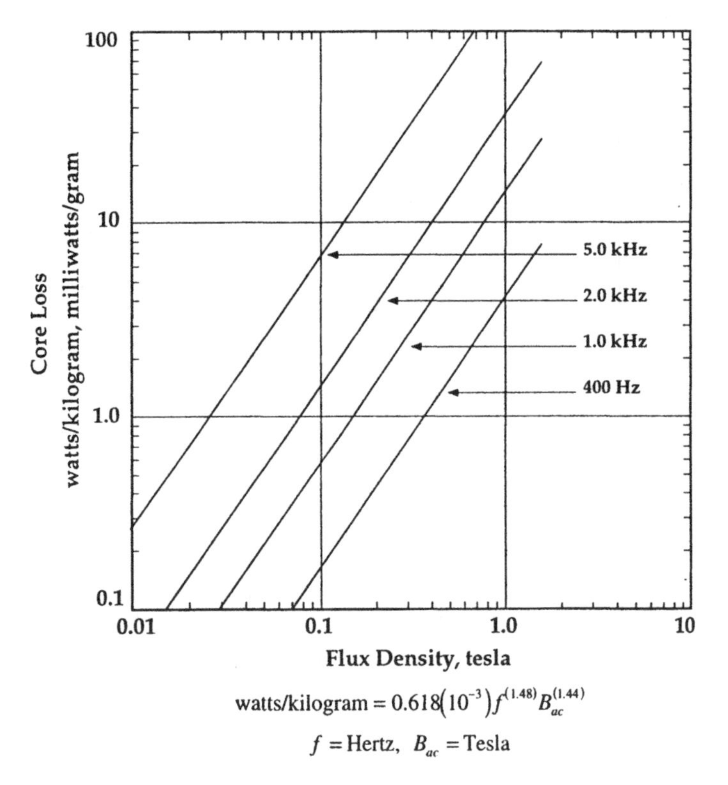

Figure 75 Core-loss curves for 50–50 Ni–Fe 4 mil.

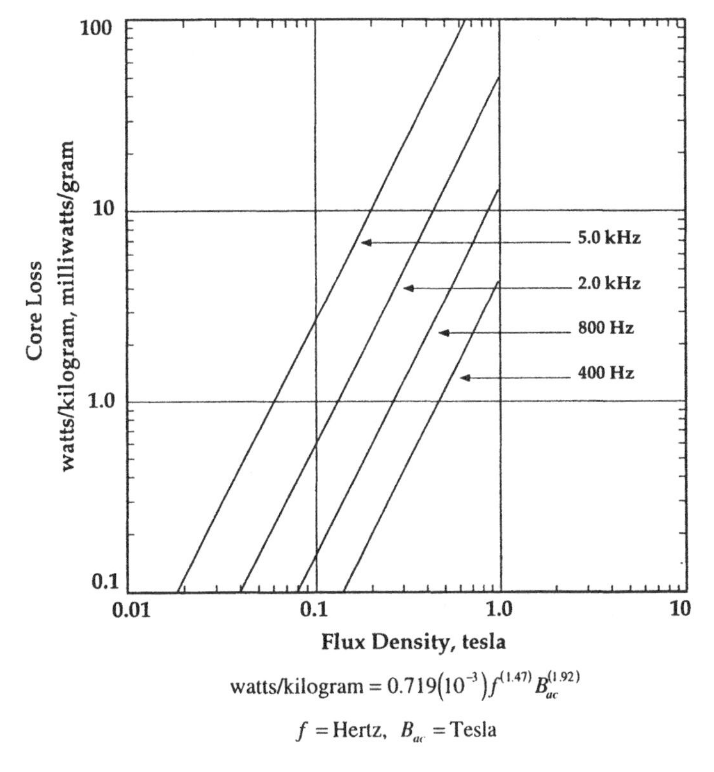

Figure 76 Core-loss curves for 48 Alloy 6 mil.

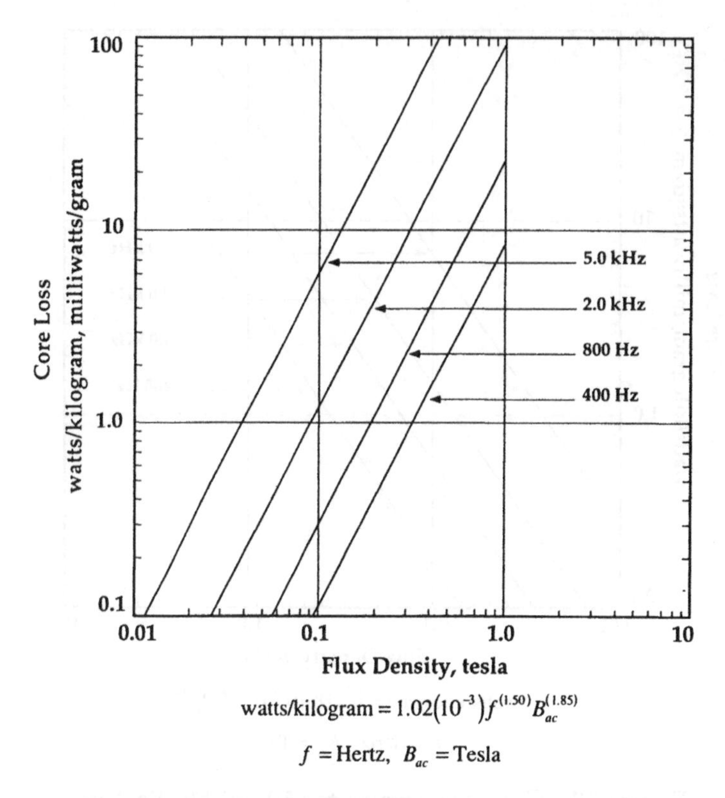

$$\text{watts/kilogram} = 1.02\left(10^{-3}\right) f^{(1.50)} B_{ac}^{(1.85)}$$

$$f = \text{Hertz}, \quad B_{ac} = \text{Tesla}$$

Figure 77 Core-loss curves for 48 Alloy 14 mil.

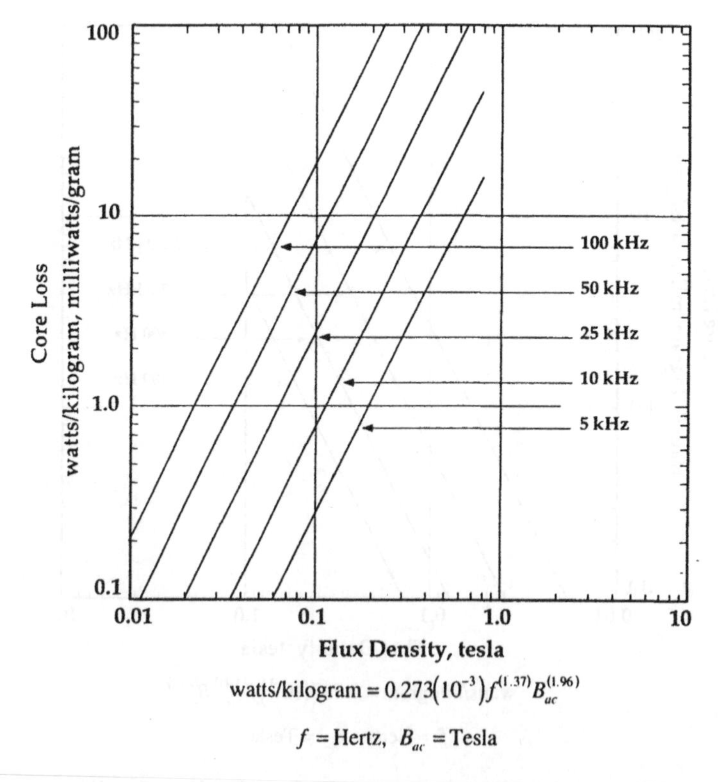

$$\text{watts/kilogram} = 0.273\left(10^{-3}\right) f^{(1.37)} B_{ac}^{(1.96)}$$

$$f = \text{Hertz}, \quad B_{ac} = \text{Tesla}$$

Figure 78 Core-loss curves for Permalloy 80 0.5 mil.

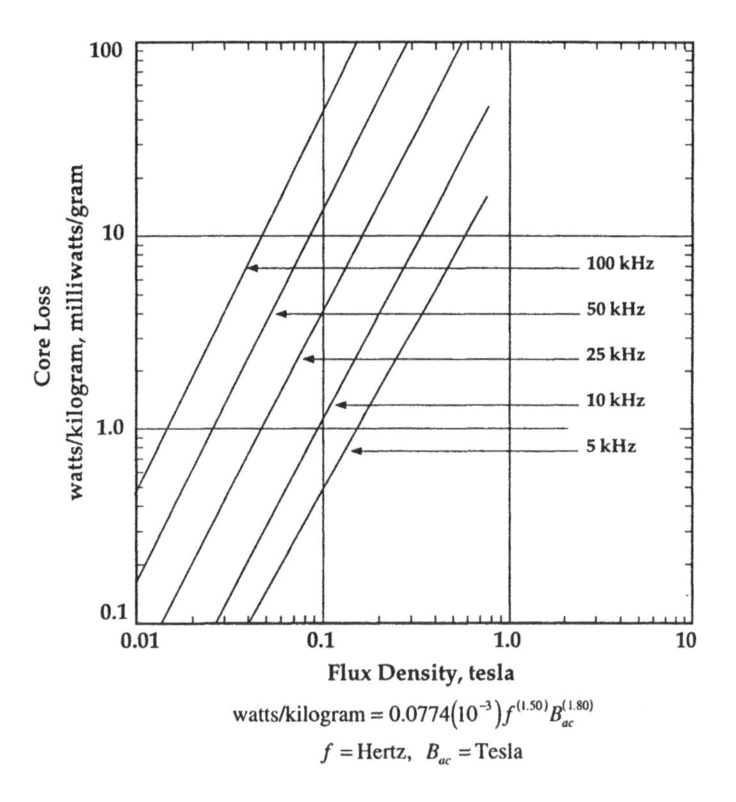

$$\text{watts/kilogram} = 0.0774(10^{-3})f^{(1.50)}B_{ac}^{(1.80)}$$

$$f = \text{Hertz}, \quad B_{ac} = \text{Tesla}$$

Figure 79 Core-loss curves for Permalloy 80 1 mil.

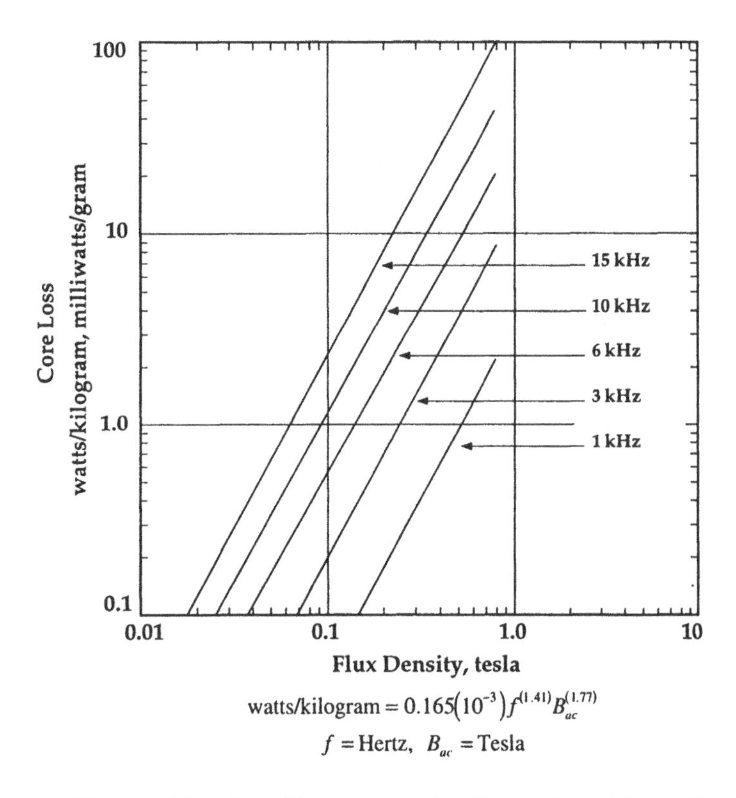

$$\text{watts/kilogram} = 0.165(10^{-3})f^{(1.41)}B_{ac}^{(1.77)}$$

$$f = \text{Hertz}, \quad B_{ac} = \text{Tesla}$$

Figure 80 Core-loss curves for Permalloy 80 2 mil.

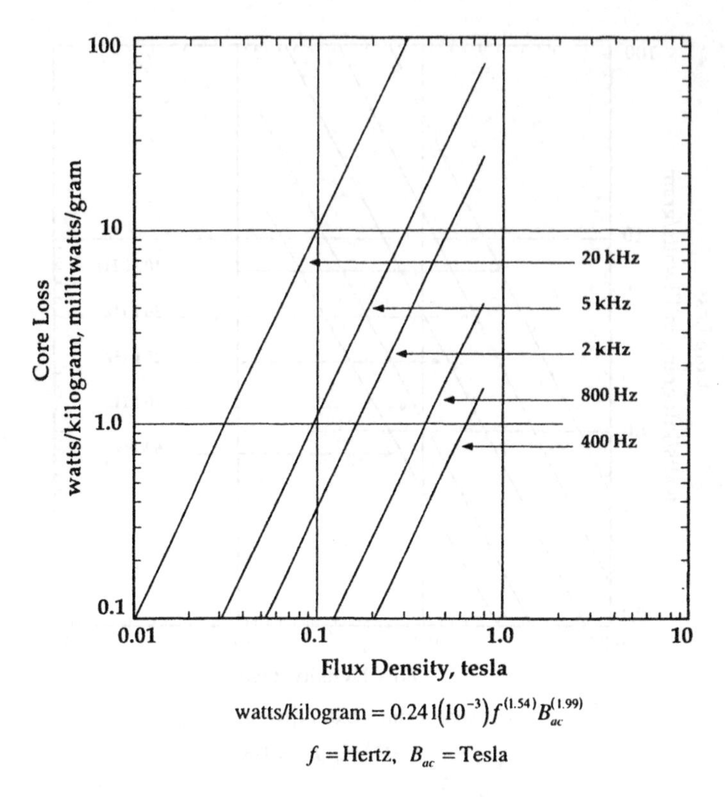

$$\text{watts/kilogram} = 0.241\left(10^{-3}\right) f^{(1.54)} B_{ac}^{(1.99)}$$

$$f = \text{Hertz},\ \ B_{ac} = \text{Tesla}$$

Figure 81 Core-loss curves for Permalloy 80 4 mil.

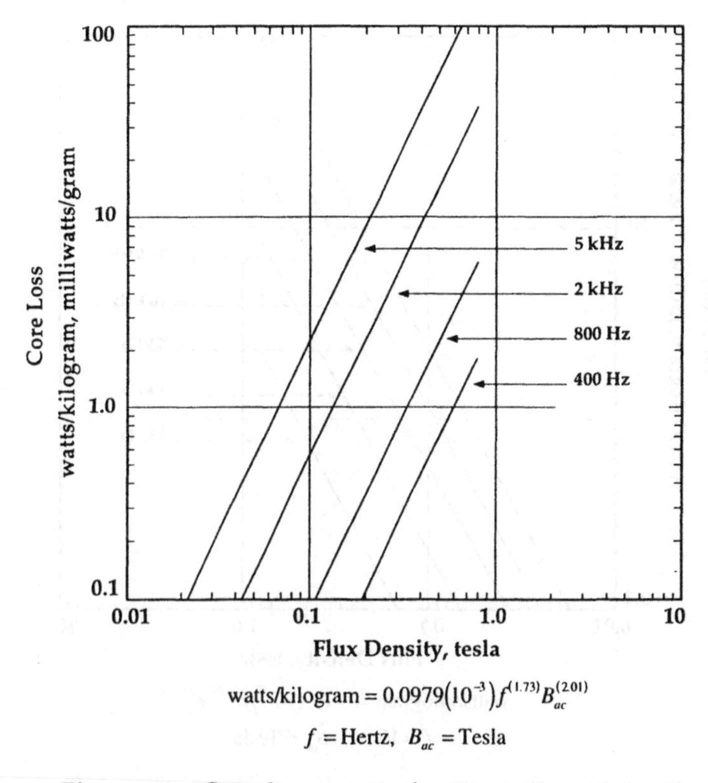

$$\text{watts/kilogram} = 0.0979\left(10^{-3}\right) f^{(1.73)} B_{ac}^{(2.01)}$$

$$f = \text{Hertz},\ \ B_{ac} = \text{Tesla}$$

Figure 82 Core-loss curves for Permalloy 80 6 mil.

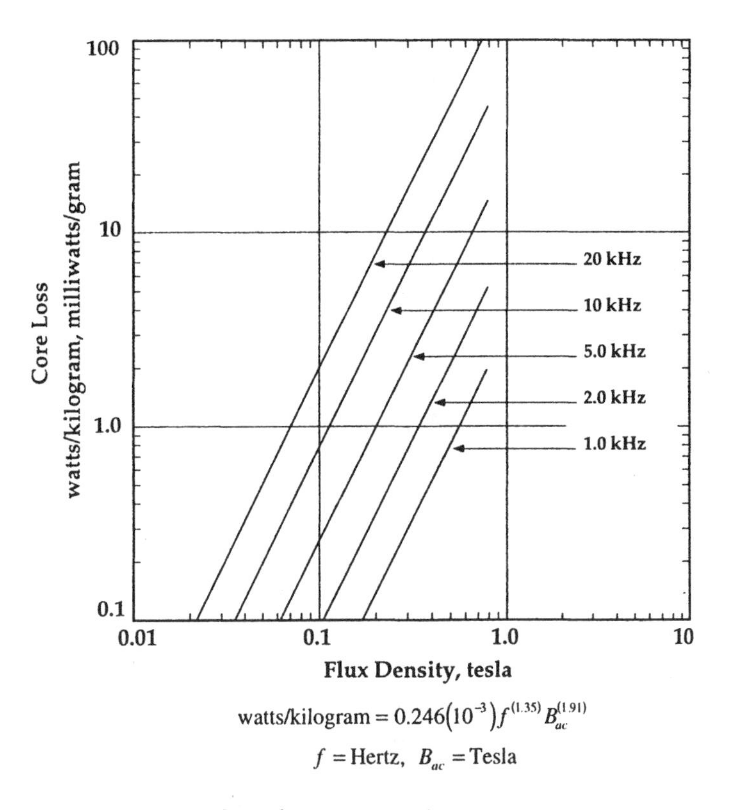

$$\text{watts/kilogram} = 0.246\left(10^{-3}\right) f^{(1.35)} B_{ac}^{(1.91)}$$

$$f = \text{Hertz}, \quad B_{ac} = \text{Tesla}$$

Figure 83 Core-loss curves for Supermalloy 1 mil.

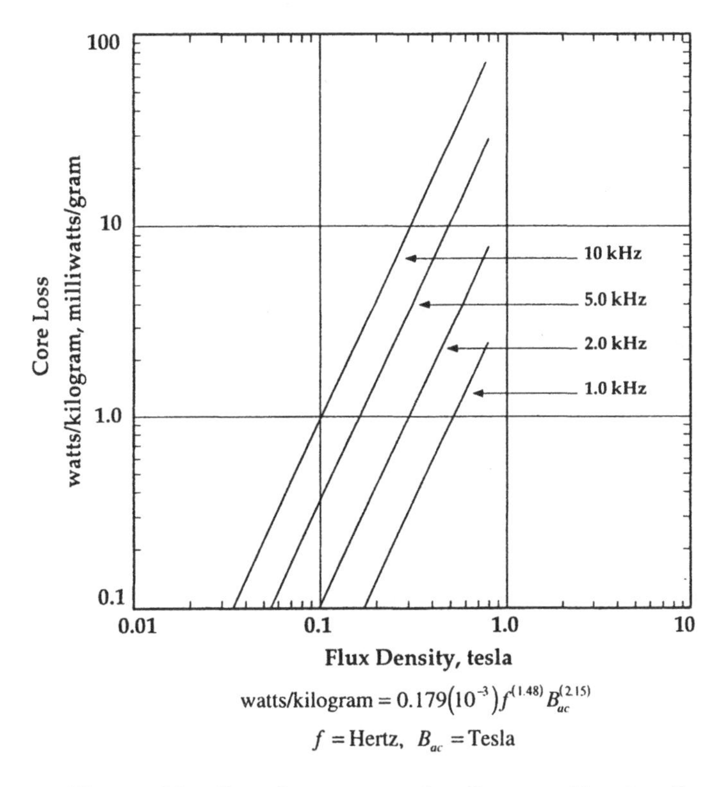

$$\text{watts/kilogram} = 0.179\left(10^{-3}\right) f^{(1.48)} B_{ac}^{(2.15)}$$

$$f = \text{Hertz}, \quad B_{ac} = \text{Tesla}$$

Figure 84 Core-loss curves for Supermalloy 2 mil.

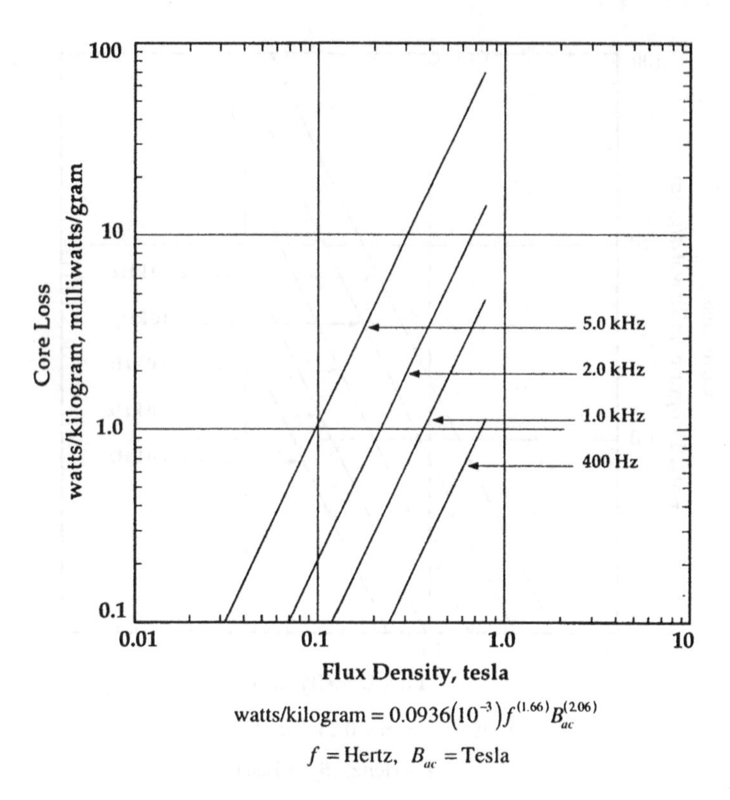

$$\text{watts/kilogram} = 0.0936\left(10^{-3}\right)f^{(1.66)}B_{ac}^{(2.06)}$$

$$f = \text{Hertz},\ B_{ac} = \text{Tesla}$$

Figure 85 Core-loss curves for Supermalloy 4 mil.

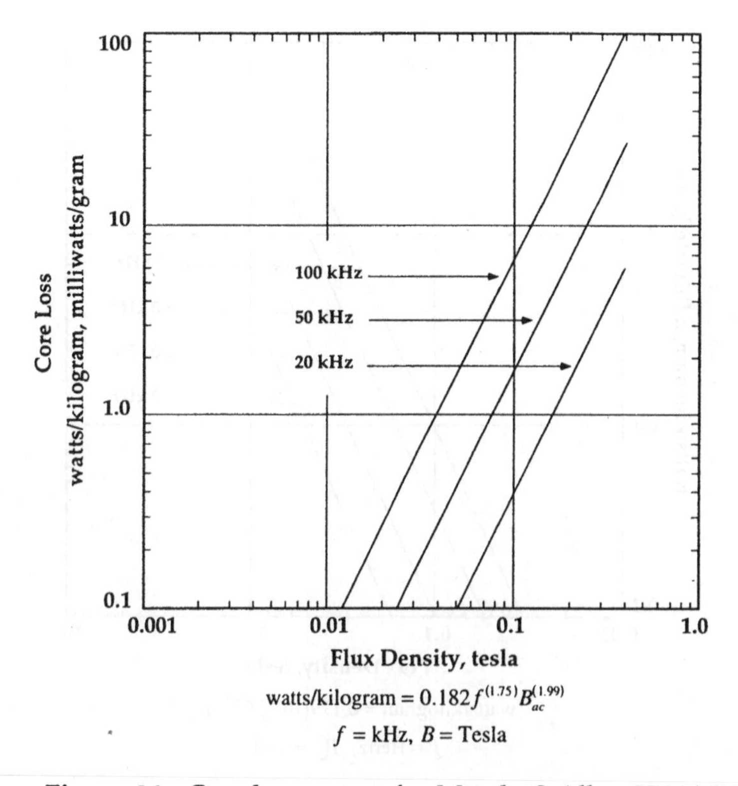

$$\text{watts/kilogram} = 0.182 f^{(1.75)}B_{ac}^{(1.99)}$$

$$f = \text{kHz},\ B = \text{Tesla}$$

Figure 86 Core-loss curves for Metglas® Alloy 2714AF.

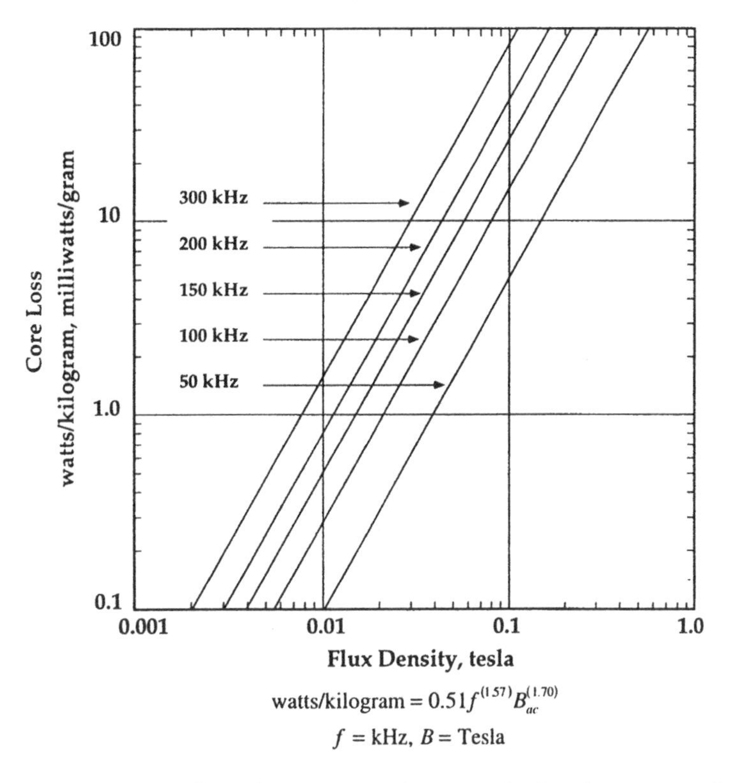

$$\text{watts/kilogram} = 0.51 f^{(1.57)} B_{ac}^{(1.70)}$$
$$f = \text{kHz},\ B = \text{Tesla}$$

Figure 87 Core-loss curves for Metglas® Alloy 2714AS.

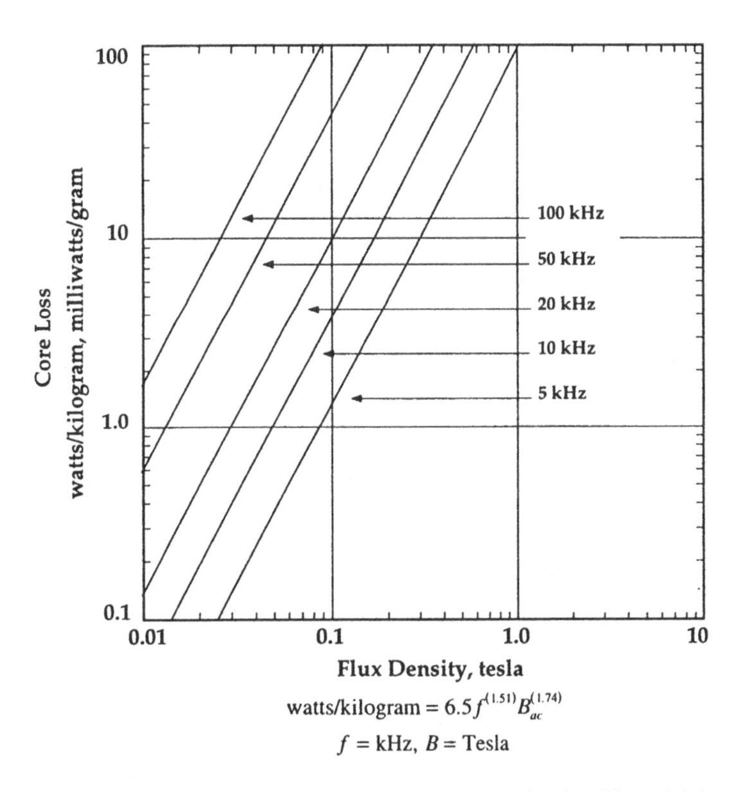

$$\text{watts/kilogram} = 6.5 f^{(1.51)} B_{ac}^{(1.74)}$$
$$f = \text{kHz},\ B = \text{Tesla}$$

Figure 88 Core-loss curves for Metglas® Alloy SA1.

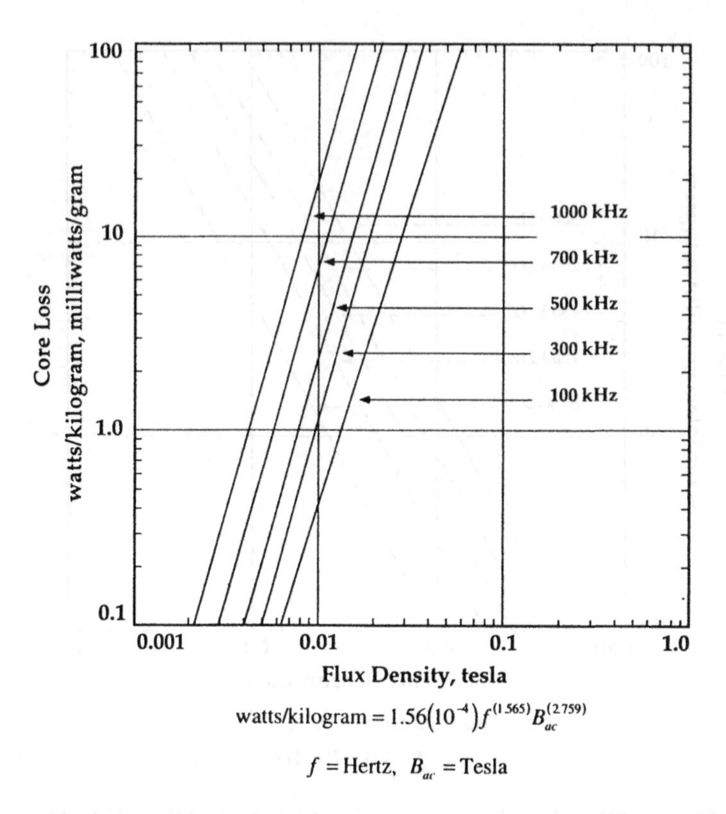

$$\text{watts/kilogram} = 1.56\left(10^{-4}\right)f^{(1.565)}B_{ac}^{(2.759)}$$

f = Hertz, B_{ac} = Tesla

Figure 89 Core-loss curves for Magnetics Ferrite K 1500μ.

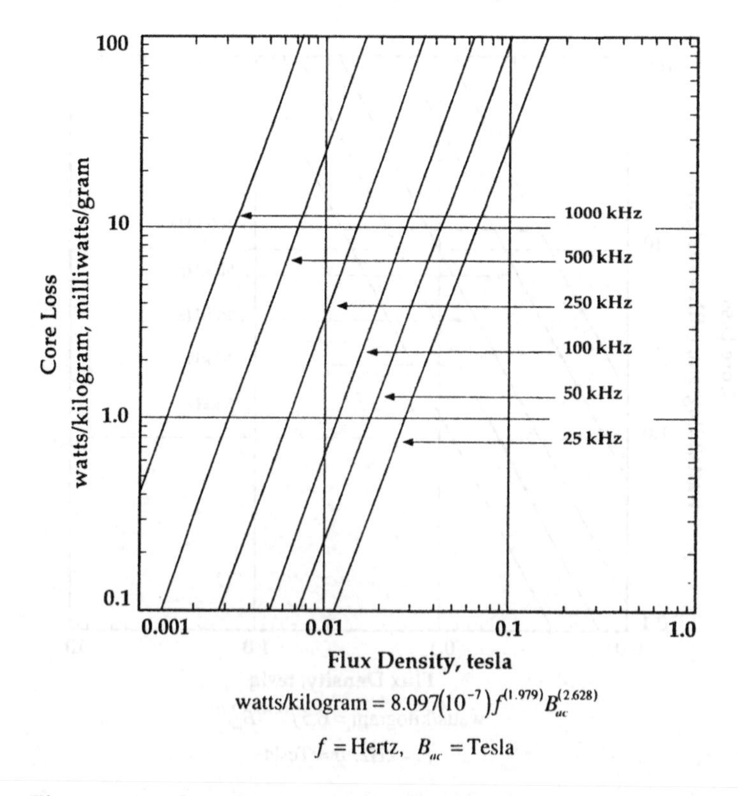

$$\text{watts/kilogram} = 8.097\left(10^{-7}\right)f^{(1.979)}B_{ac}^{(2.628)}$$

f = Hertz, B_{ac} = Tesla

Figure 90 Core-loss curves for Magnetics Ferrite R 2300μ.

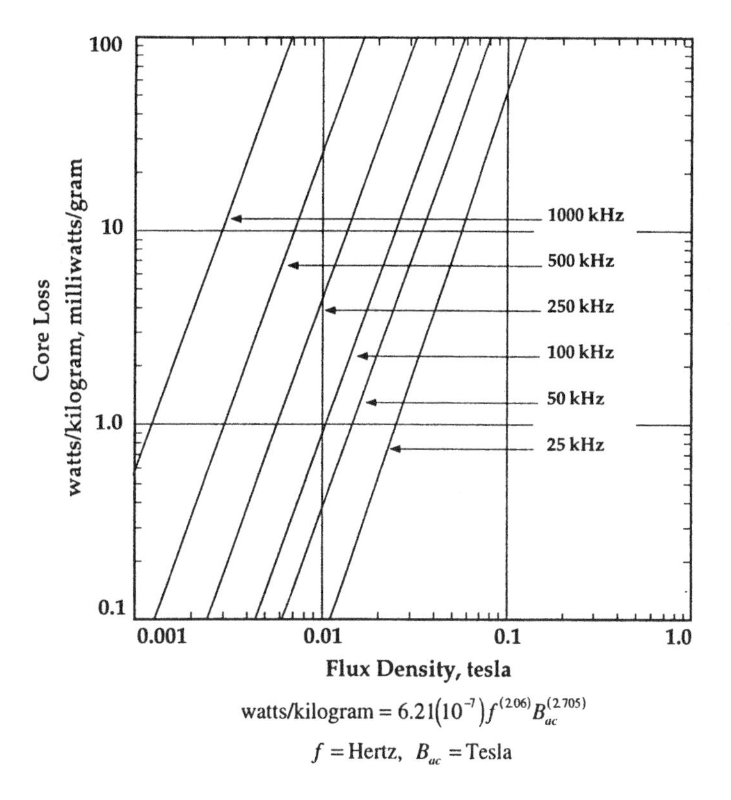

$$\text{watts/kilogram} = 6.21\left(10^{-7}\right)f^{(2.06)}B_{ac}^{(2.705)}$$

$$f = \text{Hertz}, \quad B_{ac} = \text{Tesla}$$

Figure 91 Core-loss curves for Magnetics Ferrite P 2500μ.

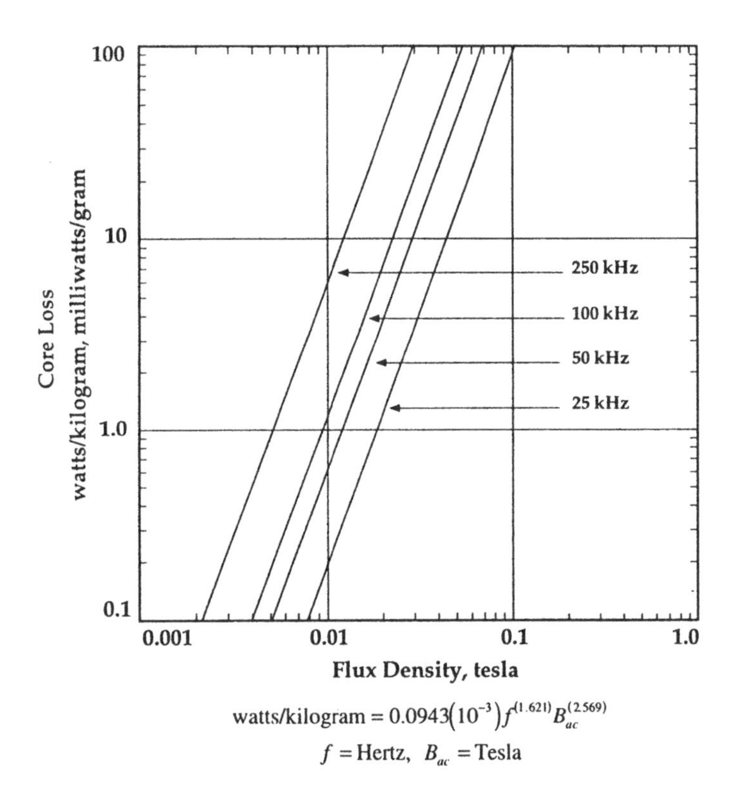

$$\text{watts/kilogram} = 0.0943\left(10^{-3}\right)f^{(1.621)}B_{ac}^{(2.569)}$$

$$f = \text{Hertz}, \quad B_{ac} = \text{Tesla}$$

Figure 92 Core-loss curves for Magnetics Ferrite F 3000μ.

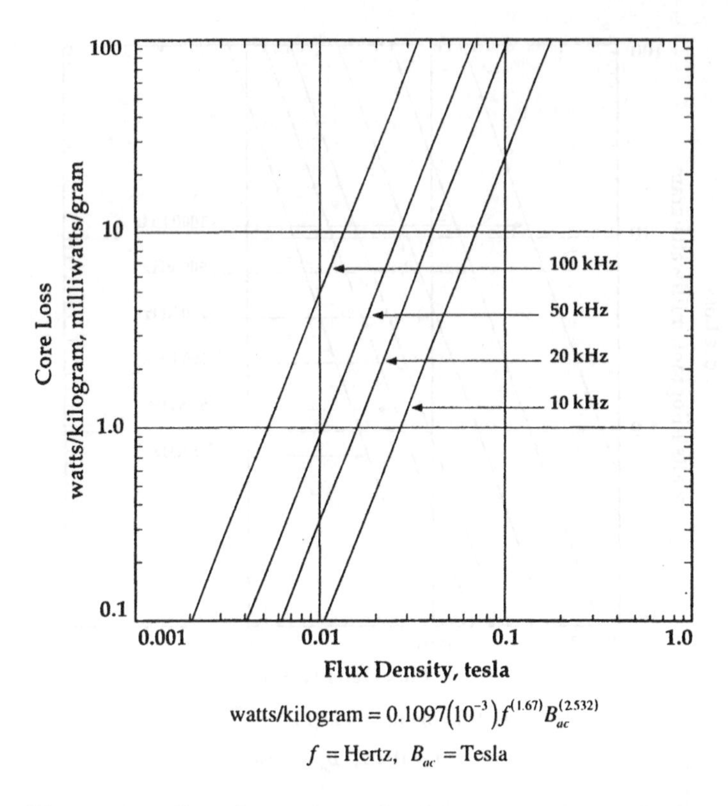

$$\text{watts/kilogram} = 0.1097(10^{-3}) f^{(1.67)} B_{ac}^{(2.532)}$$

$$f = \text{Hertz}, \quad B_{ac} = \text{Tesla}$$

Figure 93 Core-loss curves for Magnetics Ferrite J 5000μ.

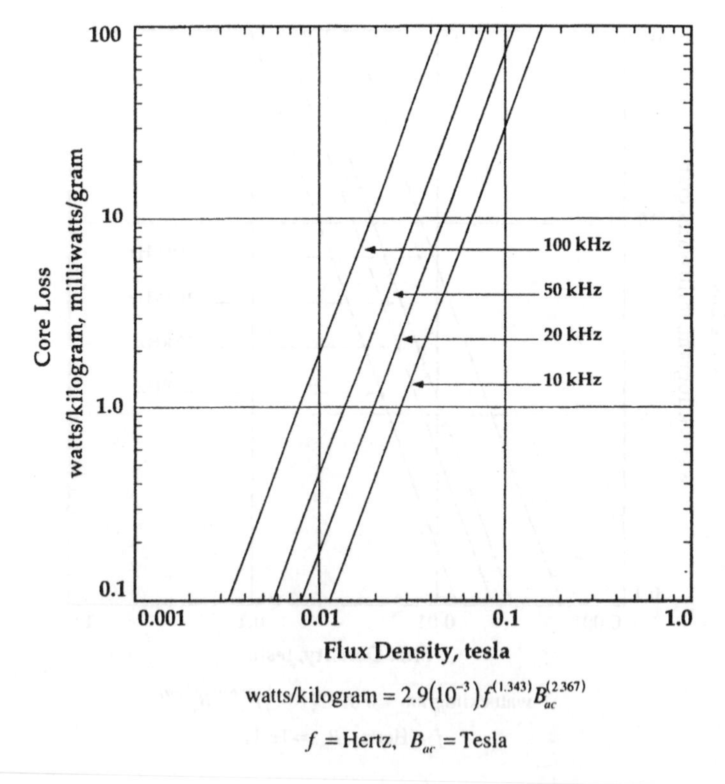

$$\text{watts/kilogram} = 2.9(10^{-3}) f^{(1.343)} B_{ac}^{(2.367)}$$

$$f = \text{Hertz}, \quad B_{ac} = \text{Tesla}$$

Figure 94 Core-loss curves for Magnetics Ferrite W 10,000μ.

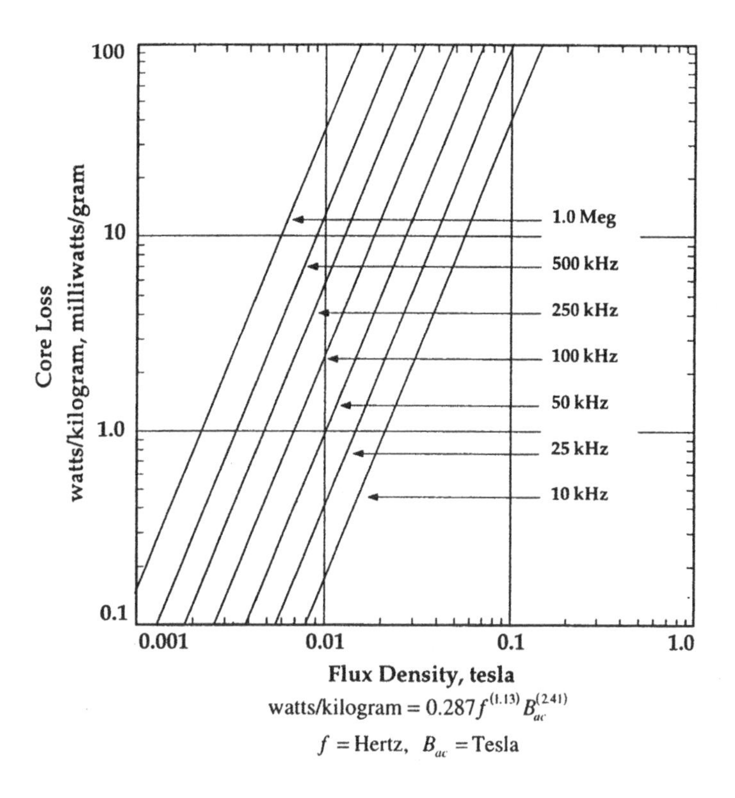

Figure 95 Micrometals core-loss curves for iron powder-08 35μ.

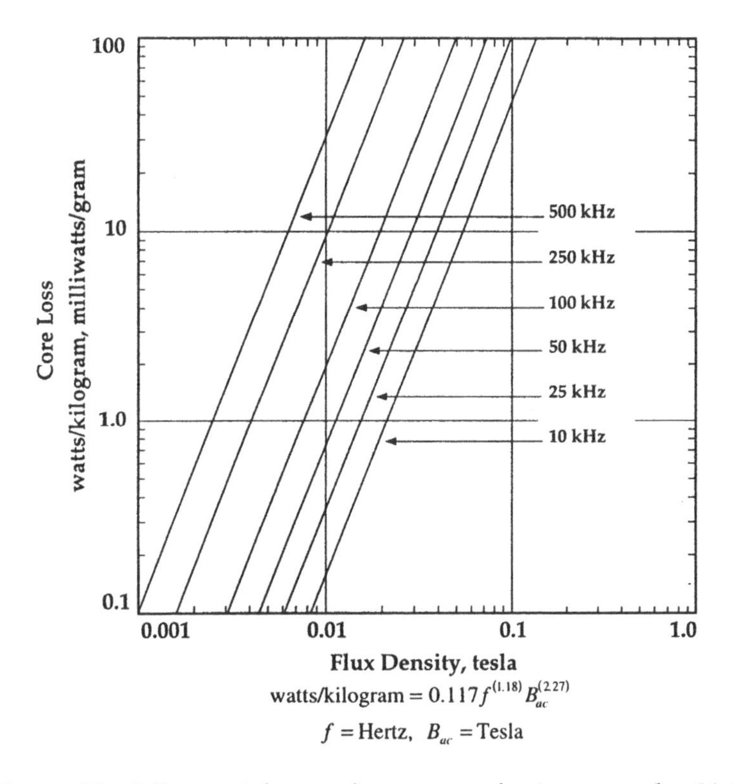

Figure 96 Micrometals core-loss curves for iron powder-18 55μ.

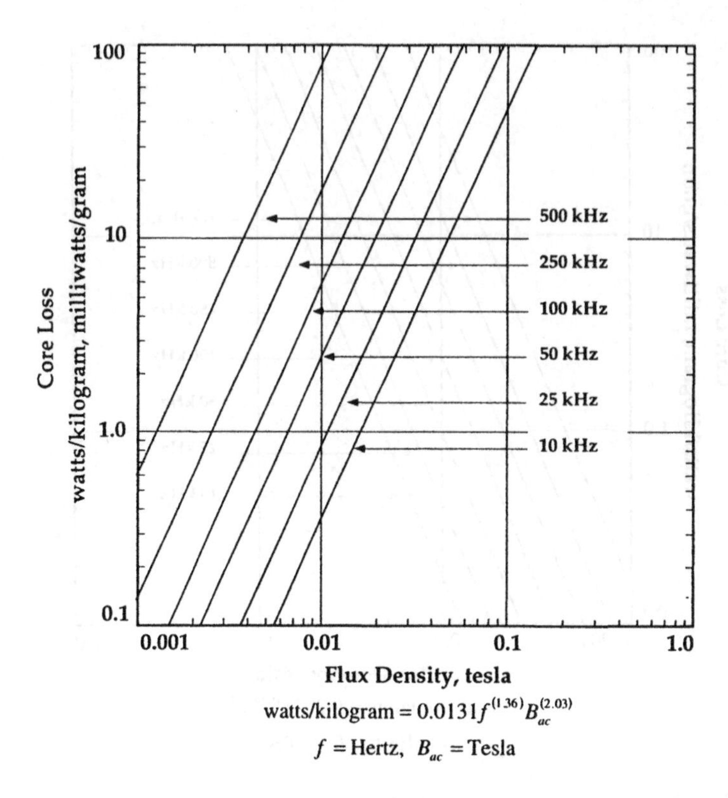

Figure 97 Micrometals core-loss curves for iron powder-26 75μ.

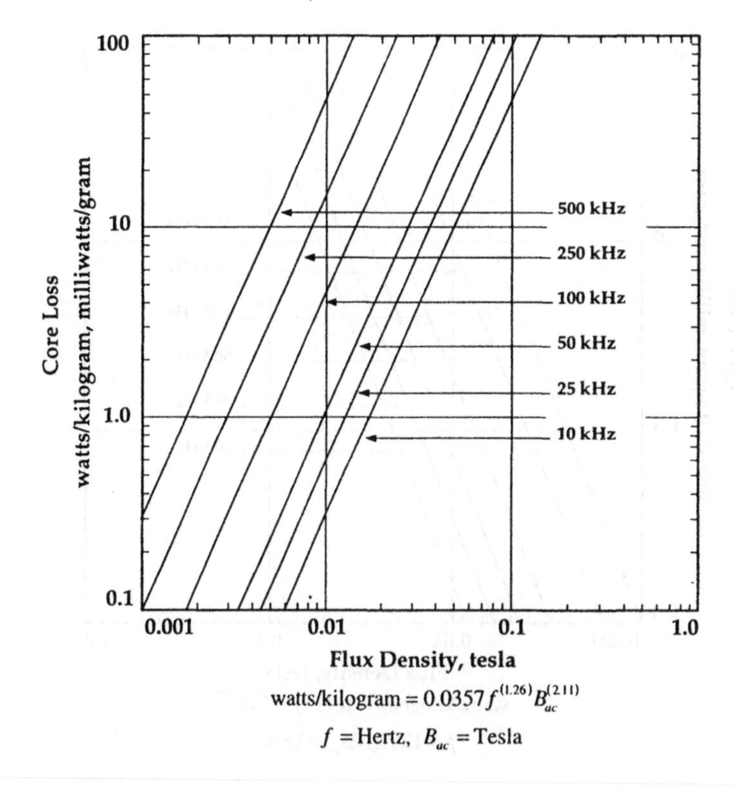

Figure 98 Micrometals core-loss curves for iron powder-52 75μ.

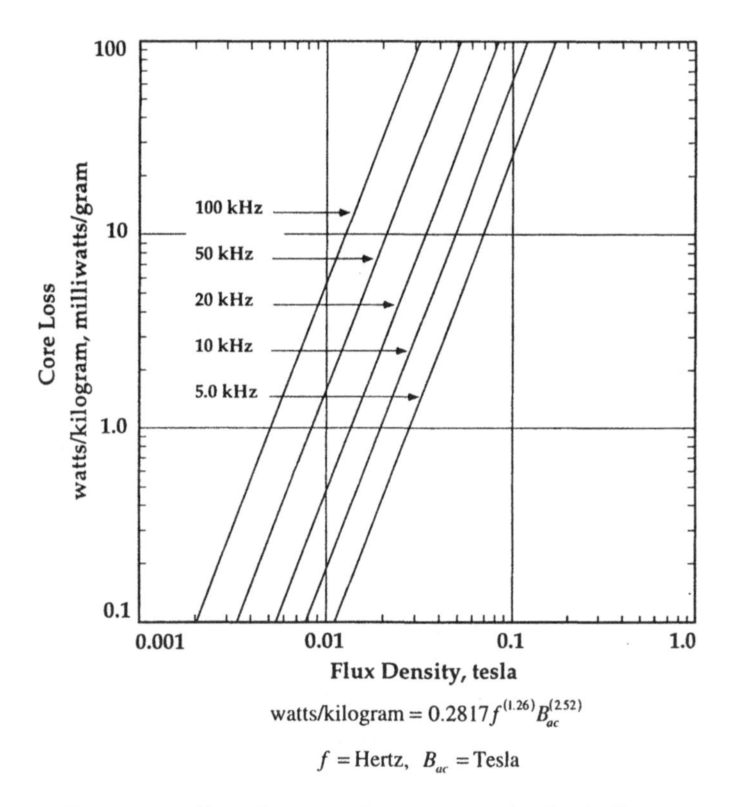

Figure 99 Powder core-loss curves for high-flux 14μ.

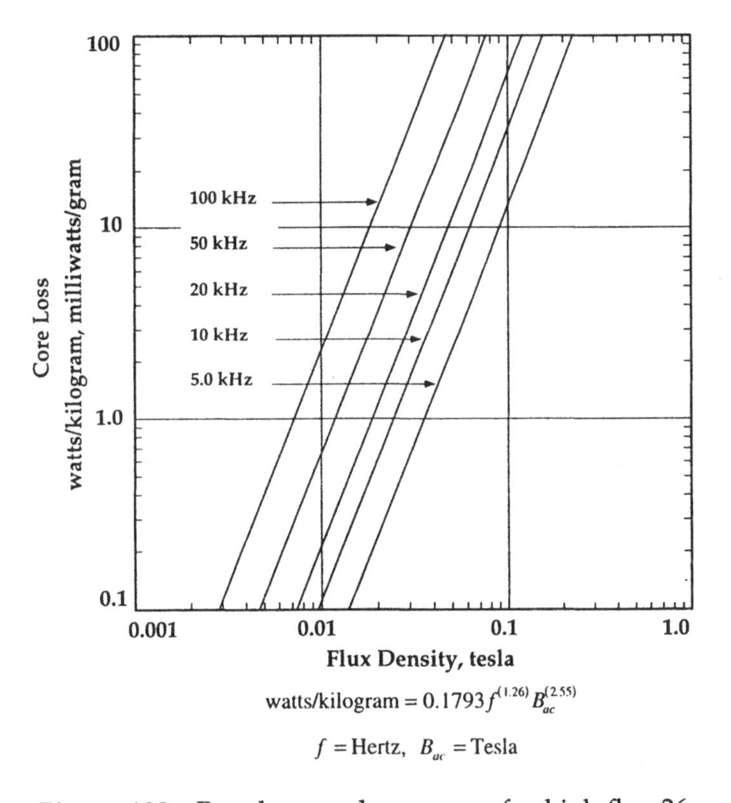

Figure 100 Powder core-loss curves for high-flux 26μ.

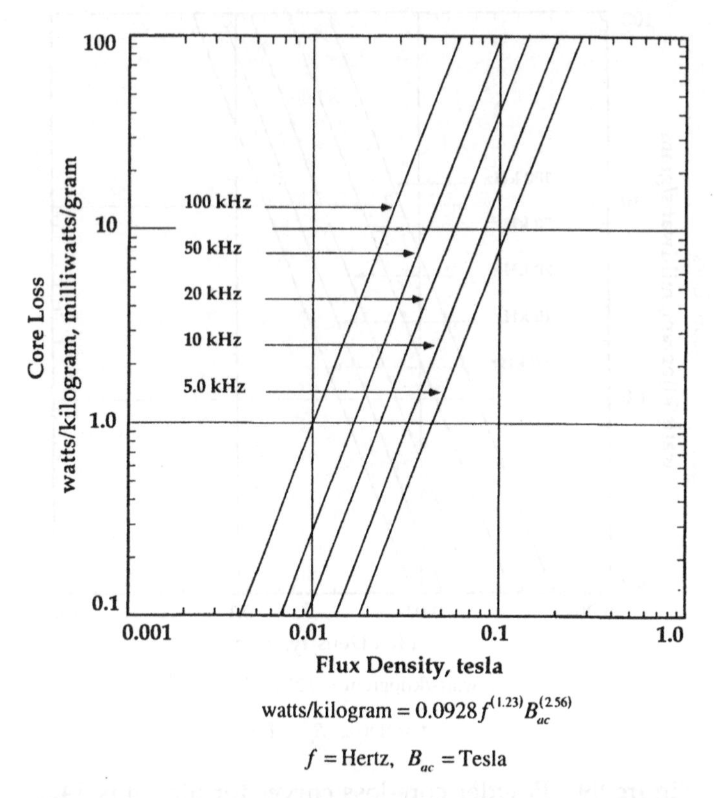

$$\text{watts/kilogram} = 0.0928 f^{(1.23)} B_{ac}^{(2.56)}$$

f = Hertz, B_{ac} = Tesla

Figure 101 Powder core-loss curves for high-flux 60μ.

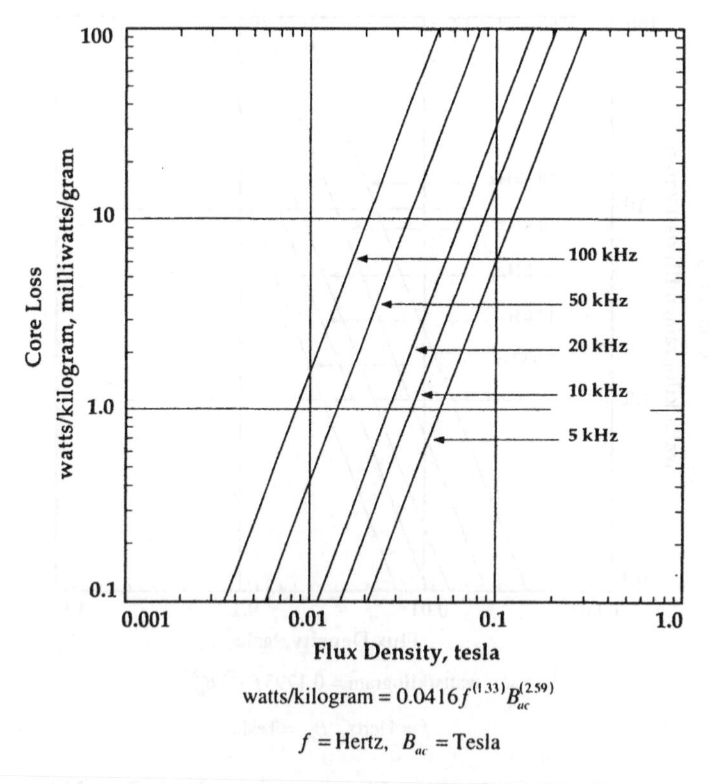

$$\text{watts/kilogram} = 0.0416 f^{(1.33)} B_{ac}^{(2.59)}$$

f = Hertz, B_{ac} = Tesla

Figure 102 Powder core-loss curves for high-flux 125μ.

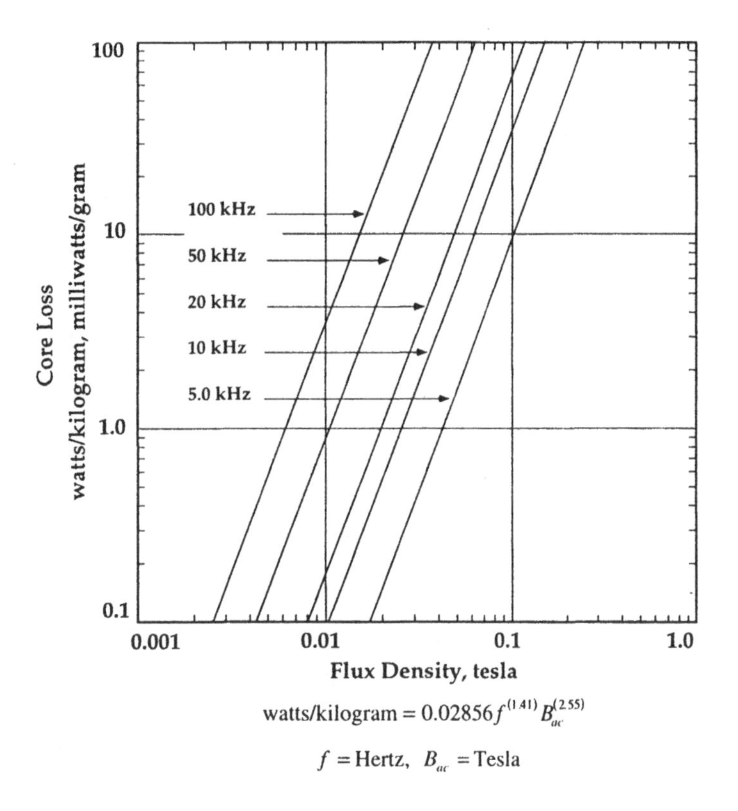

Figure 103 Powder core-loss curves for high-flux 147–160μ.

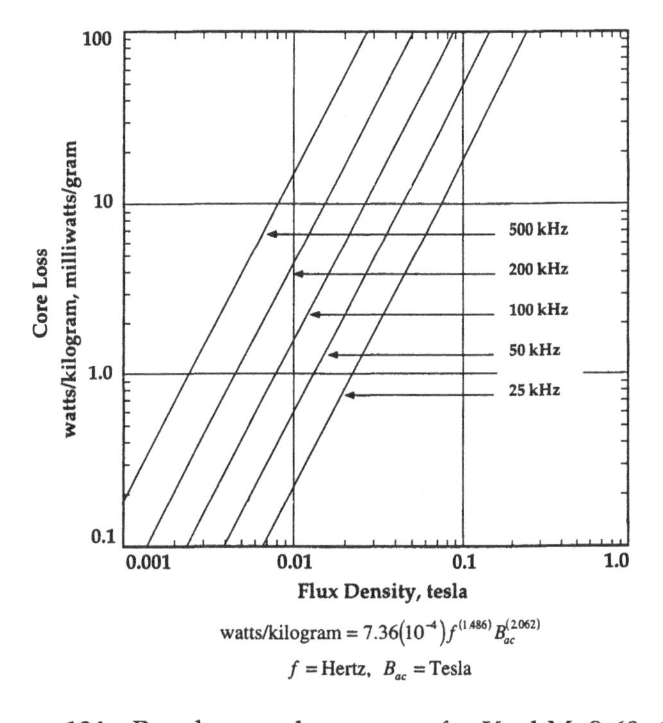

Figure 104 Powder core-loss curves for Kool Mμ® 60–125μ.

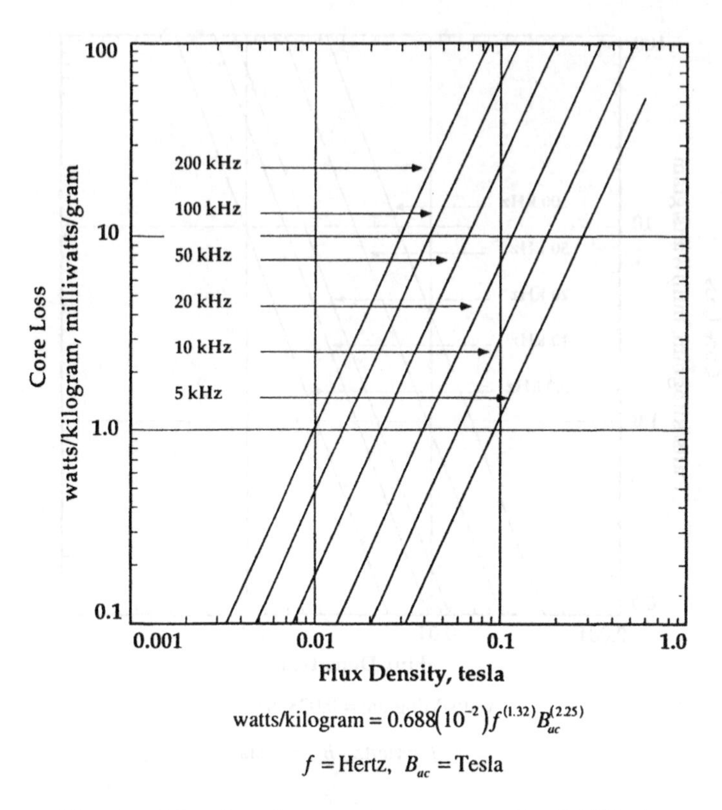

$$\text{watts/kilogram} = 0.688\left(10^{-2}\right)f^{(1.32)}B_{ac}^{(2.25)}$$

$$f = \text{Hertz}, \quad B_{ac} = \text{Tesla}$$

Figure 105 Powder core-loss curves for MPP 14μ.

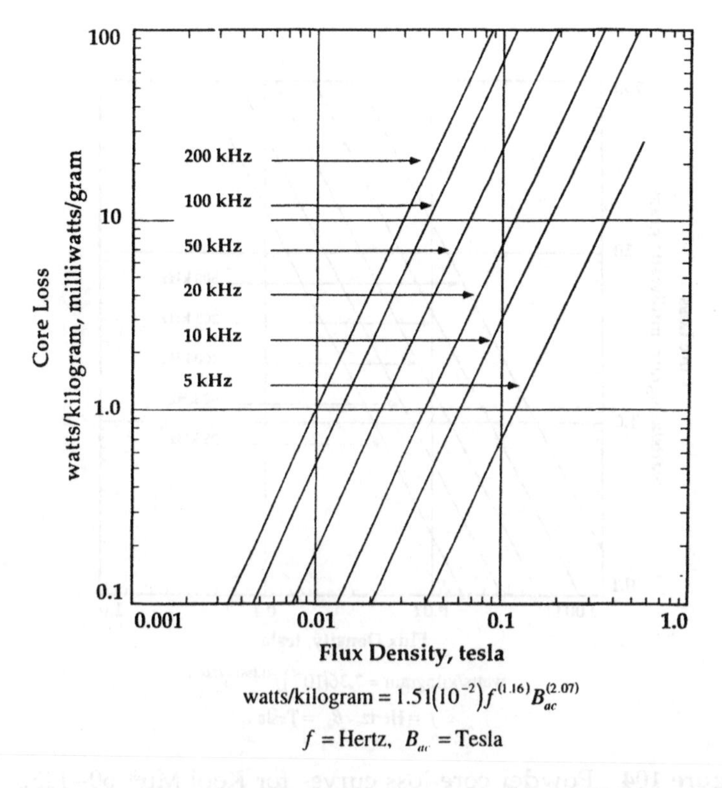

$$\text{watts/kilogram} = 1.51\left(10^{-2}\right)f^{(1.16)}B_{ac}^{(2.07)}$$

$$f = \text{Hertz}, \quad B_{ac} = \text{Tesla}$$

Figure 106 Powder core-loss curves for MPP 26μ.

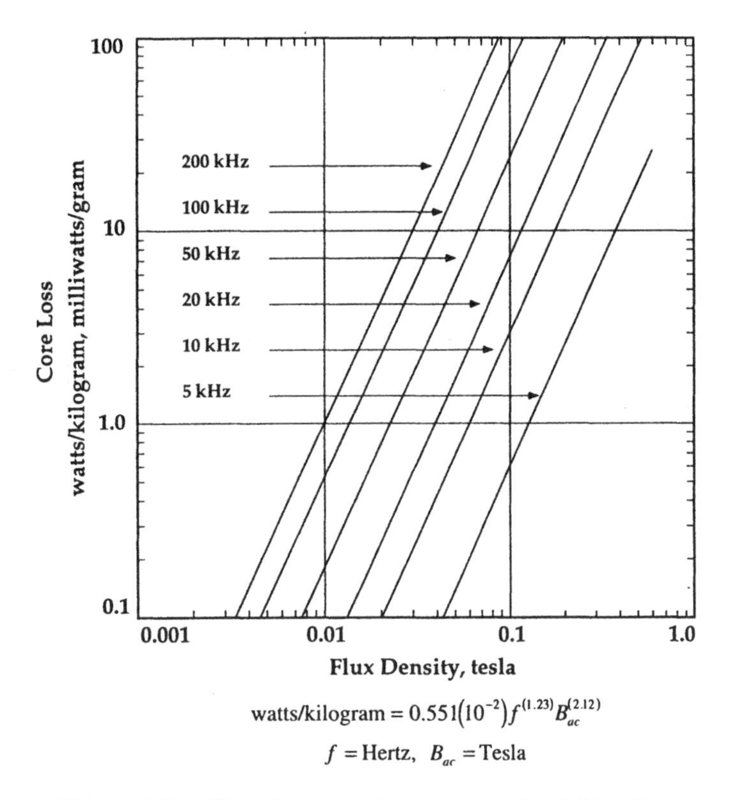

$$\text{watts/kilogram} = 0.551\left(10^{-2}\right)f^{(1.23)}B_{ac}^{(2.12)}$$

$$f = \text{Hertz}, \quad B_{ac} = \text{Tesla}$$

Figure 107 Powder core-loss curves for MPP 60μ.

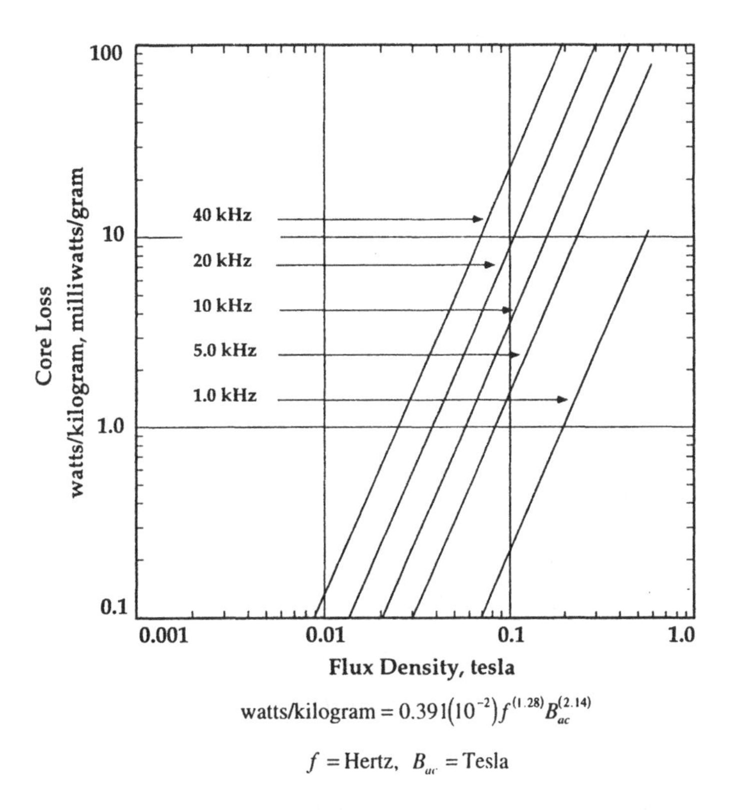

$$\text{watts/kilogram} = 0.391\left(10^{-2}\right)f^{(1.28)}B_{ac}^{(2.14)}$$

$$f = \text{Hertz}, \quad B_{ac} = \text{Tesla}$$

Figure 108 Powder core-loss curves for MPP 125μ.

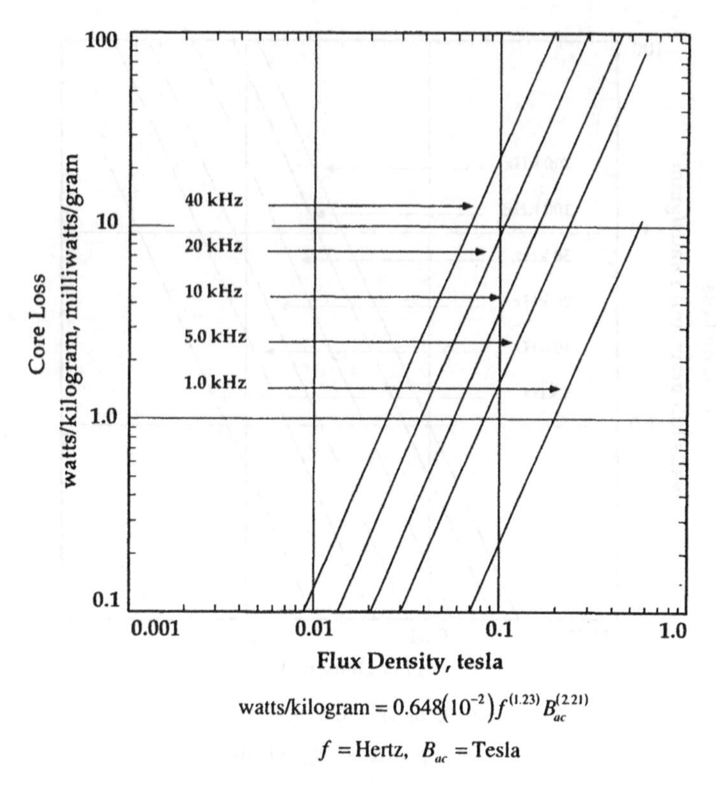

$$\text{watts/kilogram} = 0.648\left(10^{-2}\right)f^{(1.23)}B_{ac}^{(2.21)}$$

$$f = \text{Hertz}, \quad B_{ac} = \text{Tesla}$$

Figure 109 Powder core-loss curves for MPP 147–173μ.

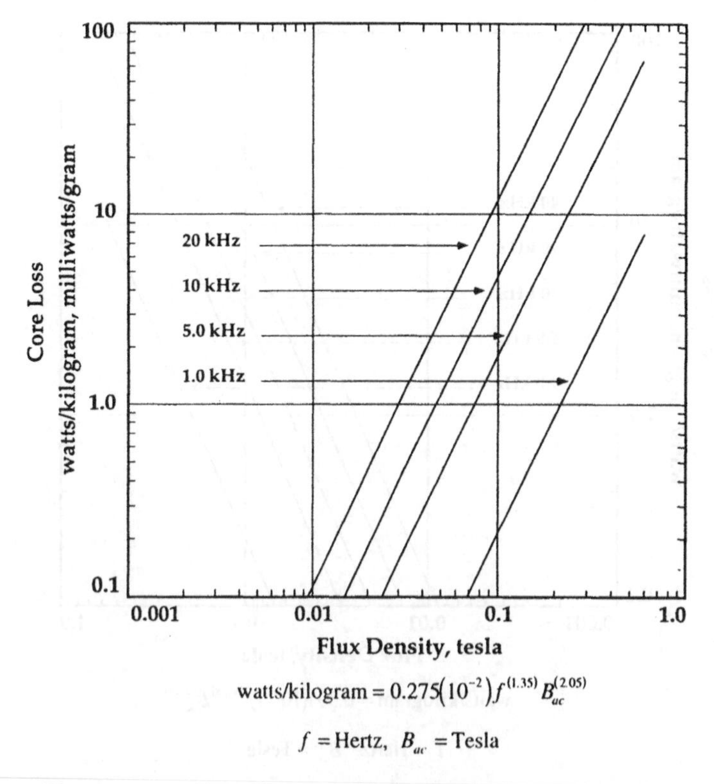

$$\text{watts/kilogram} = 0.275\left(10^{-2}\right)f^{(1.35)}B_{ac}^{(2.05)}$$

$$f = \text{Hertz}, \quad B_{ac} = \text{Tesla}$$

Figure 110 Powder core-loss curves for MPP 200μ.

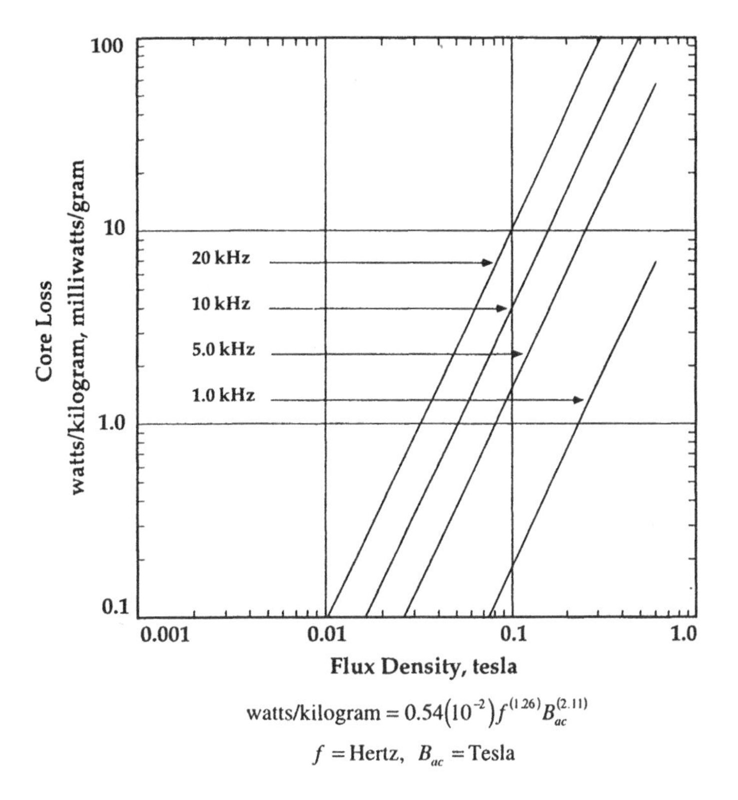

$$\text{watts/kilogram} = 0.54\left(10^{-2}\right)f^{(1.26)}B_{ac}^{(2.11)}$$

$$f = \text{Hertz},\quad B_{ac} = \text{Tesla}$$

Figure 111 Powder core-loss curves for MPP 300μ.

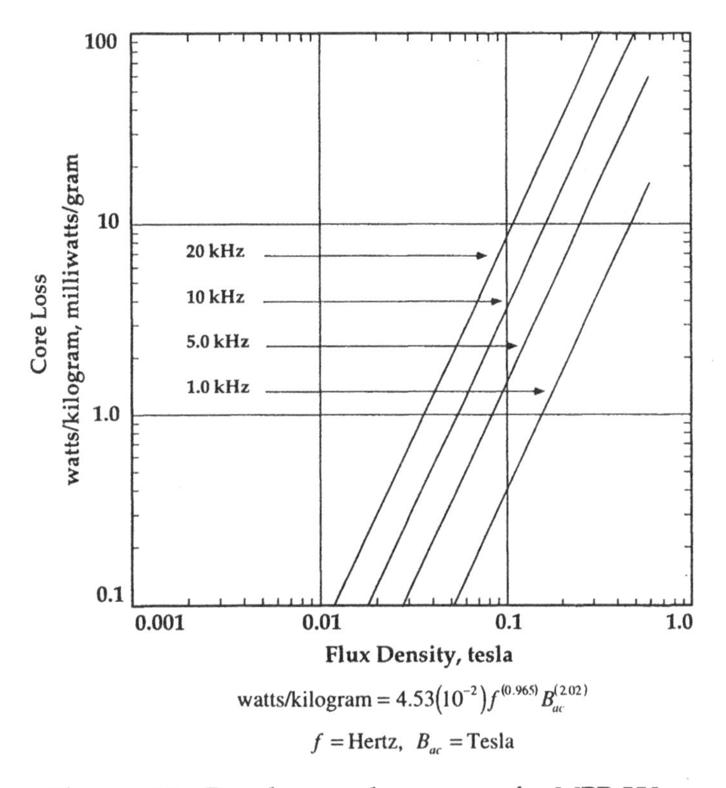

$$\text{watts/kilogram} = 4.53\left(10^{-2}\right)f^{(0.965)}B_{ac}^{(2.02)}$$

$$f = \text{Hertz},\quad B_{ac} = \text{Tesla}$$

Figure 112 Powder core-loss curves for MPP 550μ.

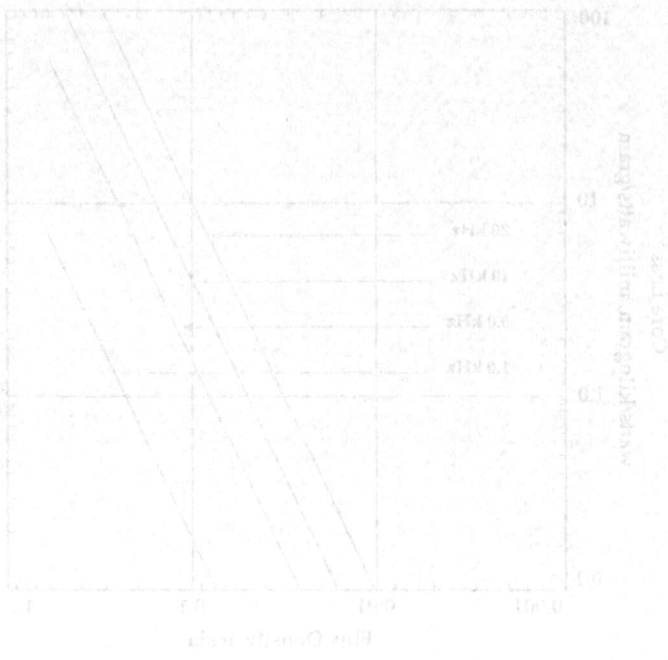

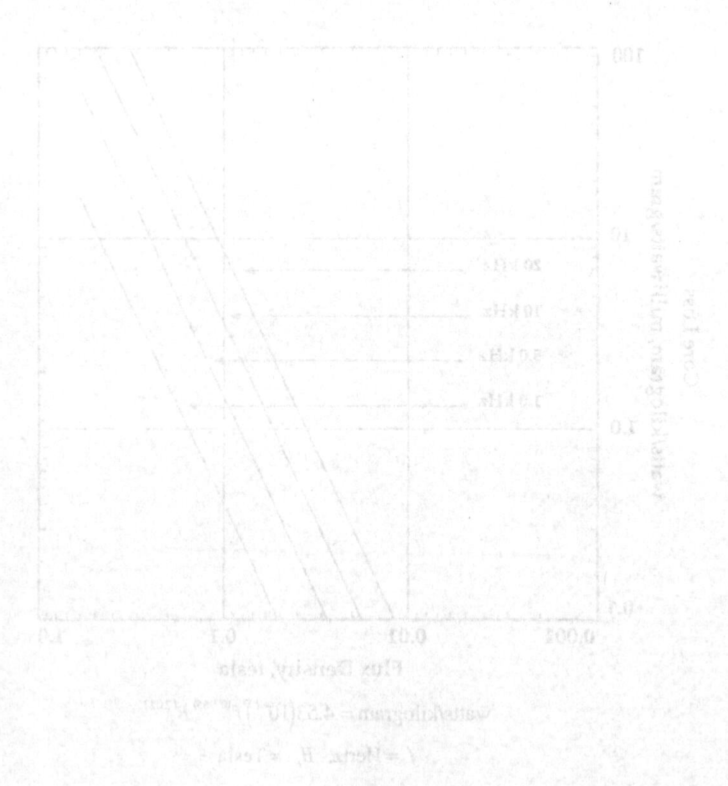

5

CONVERSION DATA FOR MAGNET WIRE

INTRODUCTION

Column 1 in Tables 11, and 12 gives the AWG wire number for the most commonly used sizes. Columns 2 and 3 give the bare area in the commonly used circular mils notation and in the metric equivalent. Column 4 gives the equivalent resistance in microhm per centimeter. Columns 5 and 6 give the area of the wire with heavy synthetic insulation for both circular mils and metric equivalent. Columns 7 and 8 give the nominal diameters including insulation in both inch and metric equivalent. Columns 9 and 10 are turns per linear length in inches and centimeters. Columns 11 and 12 give turns per square inch and turns per square centimeter based on a fill factor ratio of 0.60. Column 13 gives the weight of the wire insulation in grams per centimeter.

MAGNET WIRE MANUFACTURERS

California Fine Wire Co.
P.O. Box 446
Grover Beach, Ca. 93483
Phone (805) 489-5144, Fax (805) 489-5352

Essex Magnet Wire
1510 Wall Street
Fort Wayne, In. 46802
Phone (219) 461-4000

Phelps Dodge Magnet Wire
2131 South Coliseum Blvd.
Fort Wayne, In. 46803
Phone (219) 421-5400, Fax (219) 426-4875

MWS Wire Industries
31200 Cedar Valley Drive
Westlake Village, Ca. 91362
Phone (818) 991-8553, Fax (818) 706-0911

Cooner Wire Company
9265 Owens Mouth Ave.
Chatsworth, CA. 91311
Phone (818) 882-8311, Fax (818) 709-8281

New England Wire Corp.
365 Main Street
Lisbon, NH 03585
Phone (603) 838-6624, Fax (603) 838-6160

TABLES FOR MAGNET WIRE

Definitions for Tables 11 and 12

Information given is listed by column as:

1. AWG, American wire gauge .... AWG
2. Wire area bare .... cm^2
3. Wire area bare .... cir-mil
4. Microhms—per centimeter .... $\mu\Omega$
5. Wire area with insulation .... cm^2
6. Wire area with insulation .... cir-mil
7. Wire diameter .... cm
8. Wire diameter .... inch
9. Turns—per centimeter .... turns
10. Turns—per inch .... turns
11. Turns—per square centimeter .... turns
12. Turns—per square inch .... turns
13. Weight grams—per centimeter .... grams

Table 11 Round Wire

AWG	Bare Area		Resistance μΩ/cm	Heavy Synthetics								
				Area		Diameter		Turns-Per		Turns-Per		Weight
	cm²(10⁻³)	cir-mil	20° C	cm²(10⁻³)	cir-mil	cm	Inch	cm	Inch	cm²	Inch²	gm/cm
1	2	3	4	5	6	7	8	9	10	11	12	13
4	211.4000	41712.00	8.152	220.4000	43510.00	0.53000	0.2086	1.88	4.79	2.72	17.50	1.886000
5	167.6000	33070.00	10.280	175.4000	34630.00	0.47300	0.1861	2.11	5.37	3.42	22.10	1.498100
6	133.0000	26242.00	11.550	139.8000	27590.00	0.42200	0.1661	2.37	6.02	4.29	27.70	1.189000
7	105.5000	20817.00	16.340	111.4000	21990.00	0.37700	0.1483	2.65	6.74	5.39	34.70	0.944000
8	83.6000	16496.00	20.610	88.8000	17530.00	0.33600	0.1324	2.97	7.55	6.75	43.60	0.749100
9	63.3000	12490.00	26.000	70.8000	13970.00	0.30000	0.1182	3.33	8.46	8.47	54.70	0.594300
10	52.6100	10384.00	32.700	55.9000	11046.00	0.26700	0.1051	3.87	9.50	10.73	69.20	0.468000
11	41.6800	8226.00	41.370	44.5000	8798.00	0.23800	0.0938	4.36	10.70	13.48	89.95	0.375000
12	33.0800	6529.00	52.090	35.6400	7022.00	0.21300	0.0838	4.85	11.90	16.81	108.40	0.297700
13	26.2600	5184.00	65.640	28.3600	5610.00	0.19000	0.0749	5.47	13.40	21.15	136.40	0.236700
14	20.8200	4109.00	82.800	22.9500	4556.00	0.17100	0.0675	6.04	14.80	26.14	168.60	0.187900
15	16.5100	3260.00	104.300	18.3700	3624.00	0.15300	0.0602	6.77	16.60	32.66	210.60	0.149200
16	13.0700	2581.00	131.800	14.7300	2905.00	0.13700	0.0539	7.32	18.60	40.73	262.70	0.118400
17	10.3900	2052.00	165.800	11.6800	2323.00	0.12200	0.0482	8.18	20.80	51.36	331.20	0.094300
18	8.2280	1624.00	209.500	9.3260	1857.00	0.10900	0.0431	9.13	23.20	64.33	414.90	0.074740
19	6.5310	1289.00	263.900	7.5390	1490.00	0.09800	0.0386	10.19	25.90	79.85	515.00	0.059400
20	5.1880	1024.00	332.300	6.0650	1197.00	0.08790	0.0346	11.37	28.90	98.93	638.10	0.047260
21	4.1160	812.30	418.900	4.8370	954.80	0.07850	0.0309	12.75	32.40	124.00	799.80	0.037570
22	3.2430	640.10	531.400	3.8570	761.70	0.07010	0.0276	14.25	36.20	155.50	1003.00	0.029650
23	2.5880	510.80	666.000	3.1350	620.00	0.06320	0.0249	15.82	40.20	191.30	1234.00	0.023720
24	2.0470	404.00	842.100	2.5140	497.30	0.05660	0.0223	17.63	44.80	238.60	1539.00	0.018840
25	1.6230	320.40	1062.000	2.0020	396.00	0.05050	0.0199	19.80	50.30	299.70	1933.00	0.014980
26	1.2800	252.80	1345.000	1.6030	316.80	0.04520	0.0178	22.12	56.20	374.20	2414.00	0.011850
27	1.0210	201.60	1687.000	1.3130	259.20	0.04090	0.0161	24.44	62.10	456.90	2947.00	0.009450
28	0.8046	158.80	2142.000	1.0515	207.30	0.03660	0.0144	27.32	69.40	570.60	3680.00	0.007470
29	0.6470	127.70	2664.000	0.8548	169.00	0.03300	0.0130	30.27	76.90	701.90	4527.00	0.006020
30	0.5067	100.00	3402.000	0.6785	134.50	0.02940	0.0116	33.93	86.20	884.30	5703.00	0.004720
31	0.4013	79.21	4294.000	0.5596	110.20	0.02670	0.0105	37.48	95.20	1072.00	6914.00	0.003720
32	0.3242	64.00	5315.000	0.4559	90.25	0.02410	0.0095	41.45	105.30	1316.00	8488.00	0.003050
33	0.2554	50.41	6748.000	0.3662	72.25	0.02160	0.0085	46.33	117.70	1638.00	10565.00	0.002410
34	0.2011	39.69	8572.000	0.2863	56.25	0.01910	0.0075	52.48	133.30	2095.00	13512.00	0.001890
35	0.1589	31.36	10849.000	0.2268	44.89	0.01700	0.0067	58.77	149.30	2645.00	17060.00	0.001500
36	0.1266	25.00	13608.000	0.1813	36.00	0.01520	0.0060	62.52	166.70	3309.00	21343.00	0.001190
37	0.1026	20.25	16801.000	0.1538	30.25	0.01400	0.0055	71.57	181.80	3901.00	25161.00	0.000977
38	0.0811	16.00	21266.000	0.1207	24.01	0.01240	0.0049	80.35	204.10	4971.00	32062.00	0.000773
39	0.0621	12.25	27775.000	0.0932	18.49	0.01090	0.0043	91.57	232.60	6437.00	41518.00	0.000593
40	0.0487	9.61	35400.000	0.0723	14.44	0.00960	0.0038	103.60	263.20	8298.00	53522.00	0.000464
41	0.0397	7.84	43405.000	0.0584	11.56	0.00863	0.0034	115.70	294.10	10273.00	66260.00	0.000379
42	0.0317	6.25	54429.000	0.0456	9.00	0.00762	0.0030	131.20	333.30	13163.00	84901.00	0.000299
43	0.0245	4.84	70308.000	0.0368	7.29	0.00685	0.0027	145.80	370.40	16291.00	105076.00	0.000233
44	0.0202	4.00	85072.000	0.0316	6.25	0.00635	0.0025	157.40	400.00	18957.00	122272.00	0.000195

Table 12 Square Wire

AWG	Bare Area		Resistance µΩ/cm	Heavy Synthetics								
				Area		Diameter		Turns-Per		Turns-Per		Weight
	cm²(10⁻³)	cir-mil	20° C	cm²(10⁻³)	cir-mil	cm	Inch	cm	Inch	cm²	Inch²	gm/cm
1	2	3	4	5	6	7	8	9	10	11	12	13
4	269.00	53100	6.710	278.00	54993	0.530	0.2086	1.88	4.79	0.00	0.00	2.2900
5	213.00	42138	8.570	222.00	43775	0.473	0.1861	2.11	5.37	0.00	0.00	1.8000
6	169.00	33423	10.520	1777.00	34883	0.422	0.1661	2.37	6.02	0.00	0.00	1.4600
7	134.00	26518	13.370	141.00	27820	0.377	0.1483	2.65	6.74	0.00	0.00	1.1500
8	107.00	21029	17.050	112.00	21190	0.336	0.1324	2.97	7.55	0.00	0.00	0.9050
9	84.40	16667	20.750	89.70	17702	0.300	0.1182	3.33	8.46	0.00	0.00	0.7440
10	67.00	13224	26.260	72.40	14282	0.267	0.1051	3.87	9.50	0.00	0.00	0.5880
11	53.10	10477	33.460	57.30	11301	0.238	0.0938	4.36	10.70	0.00	0.00	0.4640
12	42.10	8314	42.320	45.80	9050	0.213	0.0838	4.85	11.90	0.00	0.00	0.3640
13	33.40	6602	52.820	36.80	7259	0.190	0.0749	5.47	13.40	0.00	0.00	0.2930
14	26.50	5233	66.930	29.50	5820	0.171	0.0675	6.04	14.80	0.00	0.00	0.2320

RANDOM WOUND COILS

The toroidal magnetic component has found wide use in industry and aerospace because of its high-frequency capability.

Design Manual TWC-300 of Magnetics, Inc. indicates that random wound cores can be produced with fill factors (S_2) as high as 0.7, but that progressive sector wound cores can be produced with fill factors of only up to 0.55. As a typical working value for copper wire with a heavy synthetic film insulation, a ratio for $S_2 = 0.60$ may be used safely.

The term "usable window/window" (S_3) defines how much of the available window space may actually be used for the winding. For a toroid, assume one-half of the inside diameter (I.D.) will remain to allow free passage of the shuttle. This would mean that if 75% of the window is available for winding, then $S_3 = 0.75$. Figures 113 and 114 are to be used to make a good and fast approximation on wire fit. Figure 113 is based on a fill factor ratio $S_2 = 0.6$ and with one-half of the inside diameter remaining, $S_3 = 0.75$. In this book, all of the toroidal cores tape, ferrites, and powder iron list the bare inside diameter right after the part number. The left column of Fig. 113 shows the inside diameter of the toroid in centimeters. Figure 114 is to be used with magnetic components using bobbins. It gives the maximum number of turns for given wire size with a fill factor of $S_2 = 0.6$. The effective window area is the winding area of the bobbin.

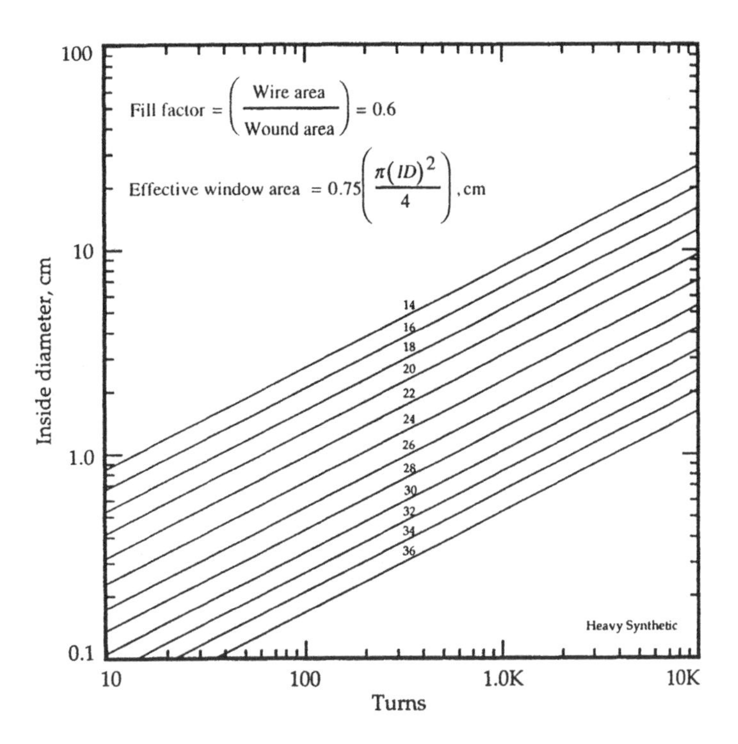

Figure 113 Toroid inside diameter versus turns.

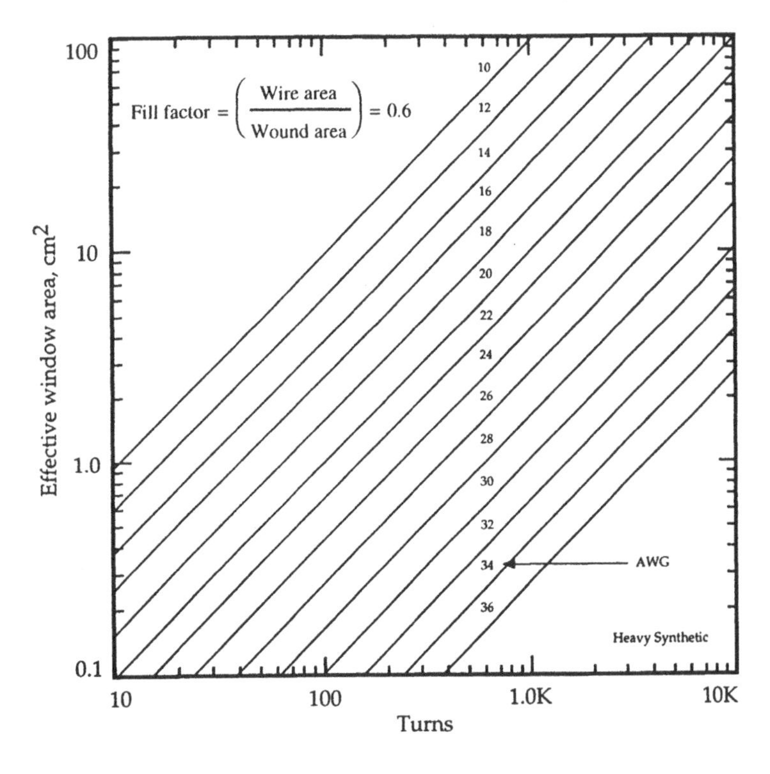

Figure 114 Bobbin winding area versus turns.

SKIN EFFECT

It is now common practice to operate dc-to-dc converters at frequencies up to 200 kHz. At higher frequencies, skin effect alters the predicted efficiency because the current carried by a conductor is distributed uniformly across the conductor cross section only at dc and at low frequencies. The concentration of current near the wire surface at higher frequencies is termed the skin effect. This is the result of magnetic flux lines which circle only part of the conductor. Those portions of the cross section which are circled by the largest number of flux lines exhibit greater reactance.

Skin effects accounts for the fact that the effective alternating current resistance to direct current ratio is greater than unity. The magnitudes of the effects due to increased frequency on conductivity, magnetic permeability, and inductance are sufficient to require further consideration of the size of the conductor. The depth of the skin effect is expressed by

$$\text{Skin depth} = \left(\frac{6.62}{\sqrt{f}}\right) K \text{ (cm)} \tag{1}$$

in which K is a constant according to the relationship

$$K = \sqrt{\left(\frac{1}{\mu_r}\right)\left(\frac{\rho}{\rho_c}\right)} \tag{2}$$

in which

μ_r = relative permeability of conductor material (μ_r = 1 for copper and other nonmagnetic materials)
ρ = resistivity of conductor material at any temperature
ρ_c = resistivity of copper at 20°C = 1.724 μΩ cm
K = unity for copper

Figures 115 and 116, show, respectively, skin depth as a function of frequency according to Eq. (2), and as related to the AWG radius, or as $R_{ac}/R_{dc} = 1$ versus frequency.*

*The data presented are for sine-wave excitation. The author could not find any data for square-wave excitation.

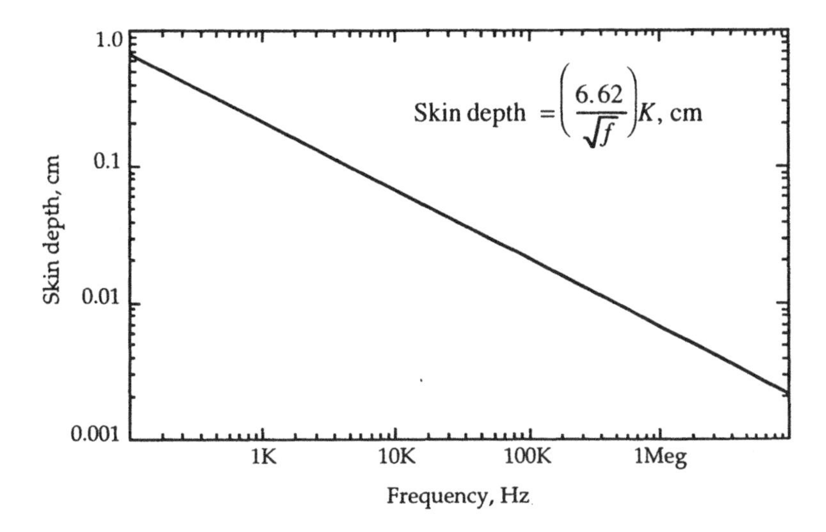

Figure 115 Skin depth versus frequency.

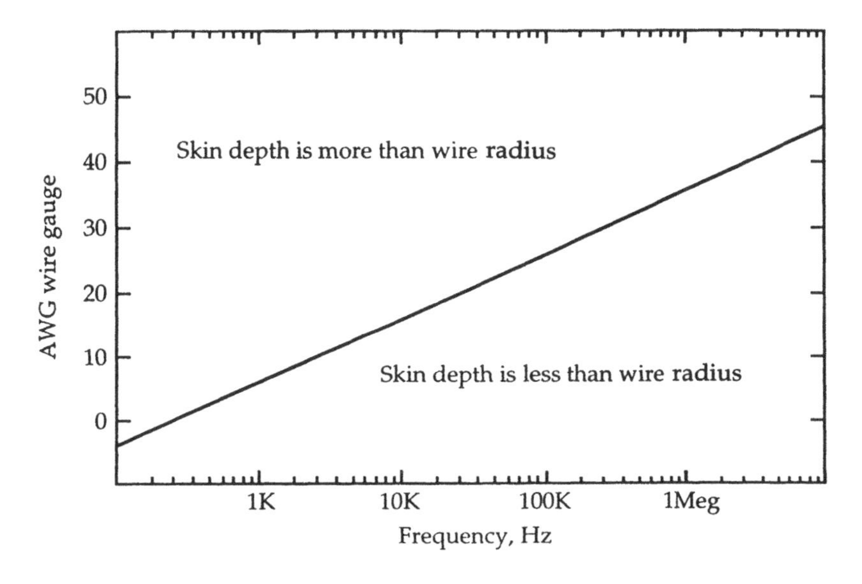

Figure 116 Skin depth equal to AWG radius versus frequency.

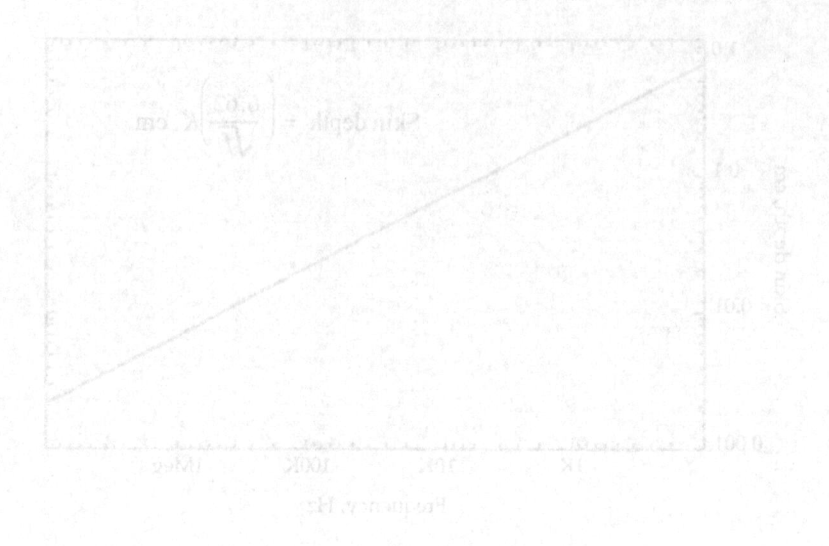

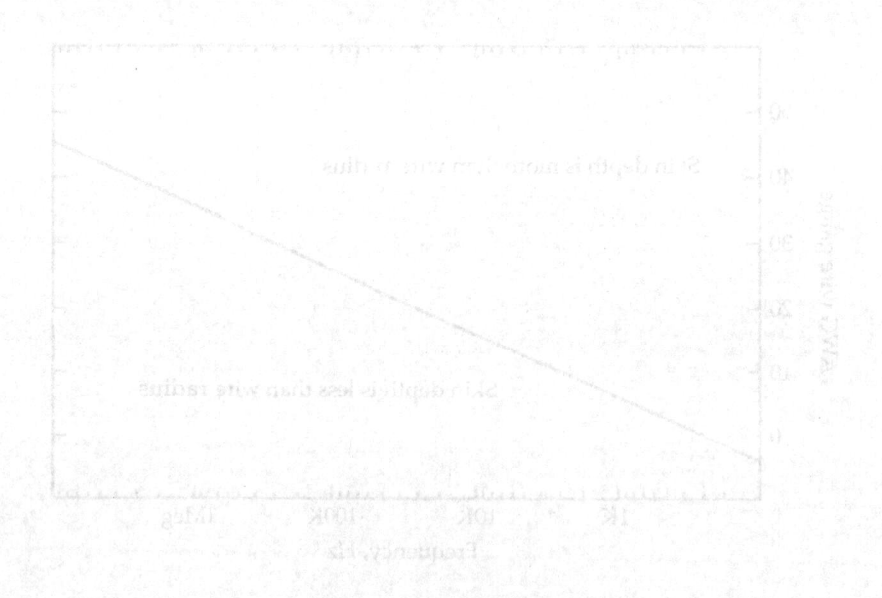

6

CASED TOROIDS

CASED TOROIDAL CORE MANUFACTURERS

Manufacturers address and phone number can be found from the Table of Contents.

Notes

CASED TOROIDAL CORE TABLE DESCRIPTION

Definitions for the Following Tables

Information given is listed by column as

Col.	Dim.	Description	Units
1.	OD	Case plus paint outside diameter	cm
2.	ID	Case plus paint inside diameter	cm
3.	HT	Case plus paint height	cm
4.	OD	Finished wound ($K_u = 0.4$)	cm
5.	HT	Finished wound ($K_u = 0.4$)	cm
6.	MPL	Magnetic path length [MPL = 3.14(OD + ID)(0.5)]	cm
7.	W_{tfe}	Core weight [W_{tfe} = (MPL)(A_c)(7.63)SF]	gr
8.	W_{tcu}	Copper weight W_{tcu} = (MLT)(W_a)(K_u)(8.89)	gr
9.	MLT	Mean length turn [MLT = (OD + 2HT)(0.8), core]	cm
10.	A_c	Effective iron area [A_c = (OD − ID)(HT)(SF)(0.5), core]	cm sq
11.	W_a	Window area [$W_a = 3.14(\text{ID})^2(0.25)$]	cm sq
12.	A_p	Area product [$A_p = (A_c)(W_a)$]	cm 4th
13.	K_g	Core geometry [$K_g = (A_p)(A_c)(K_u)/\text{MLT}$]	cm 5th
14.	A_t	Surface area	cm sq

For mechnical outline see next page

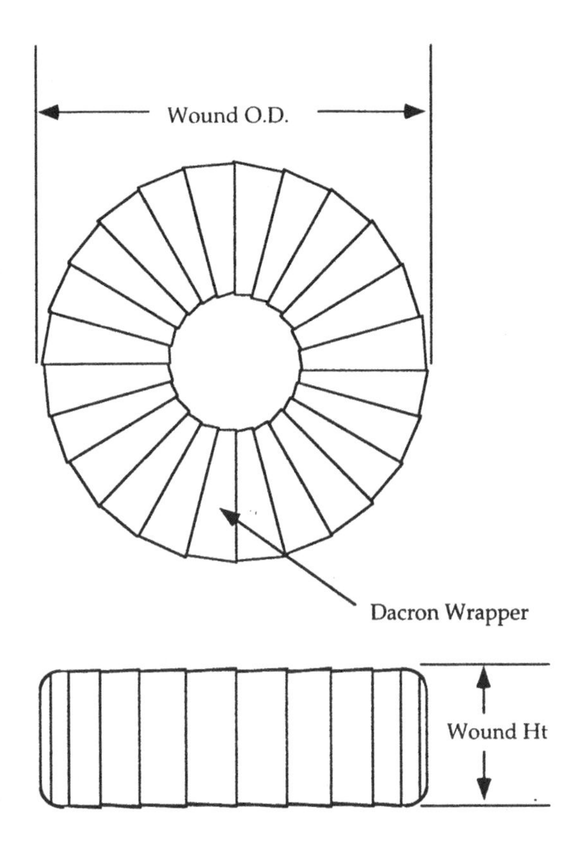

Figure 117 Tape toroids, dimensional outline.

Magnetics Cased Toroids 0.5 mil Iron Alloy														
Part No.	OD cm 1	ID cm 2	HT cm 3	Wound OD cm 4	Wound HT cm 5	MPL cm 6	Wtfe grams 7	Wtcu grams 8	MLT cm 9	Ac cm sq 10	Wa cm sq 11	Ap cm 4th 12	Kg cm 5th 13	At cm sq 14
52402-05	1.384	0.686	0.661	1.732	1.004	3.3	0.3	2.8	2.2	0.0129	0.369	0.005	0.000011	10.2
52107-05	1.689	1.003	0.661	2.184	1.162	4.2	0.4	6.8	2.4	0.0129	0.790	0.010	0.000022	15.5
52403-05	1.854	1.169	0.661	2.429	1.245	4.7	0.5	9.7	2.5	0.0129	1.073	0.014	0.000028	18.8
52485-05	2.019	1.321	0.661	2.669	1.321	5.2	0.5	13.0	2.7	0.0129	1.370	0.018	0.000034	22.3
52153-05	1.537	0.686	0.661	1.895	1.004	3.5	0.7	3.0	2.3	0.0253	0.369	0.009	0.000041	11.6
52356-05	2.172	1.486	0.661	2.903	1.404	5.7	0.6	17.2	2.8	0.0129	1.733	0.022	0.000042	26.0
52154-05	1.689	0.851	0.661	2.119	1.086	4.0	0.8	4.9	2.4	0.0253	0.568	0.014	0.000060	14.3
52056-05	1.854	1.003	0.661	2.355	1.162	4.5	0.9	7.1	2.5	0.0253	0.790	0.020	0.000080	17.3
52057-05	2.172	1.321	0.661	2.823	1.321	5.5	1.1	13.6	2.8	0.0253	1.370	0.035	0.000125	24.2
52143-05	2.489	1.626	0.661	3.289	1.474	6.5	1.3	22.5	3.0	0.0253	2.075	0.052	0.000174	32.2
52155-05	1.697	0.846	0.978	2.126	1.401	4.0	1.5	5.8	2.9	0.0506	0.562	0.028	0.000197	16.4
52374-05	2.819	1.943	0.661	3.776	1.632	7.5	1.4	34.9	3.3	0.0253	2.963	0.075	0.000229	41.7
52086-05	3.137	2.261	0.661	4.253	1.791	8.5	1.6	50.9	3.6	0.0253	4.013	0.101	0.000288	52.3
52000-05	2.172	1.003	0.661	2.690	1.162	5.0	1.9	7.8	2.8	0.0506	0.790	0.040	0.000289	21.2
52063-05	2.489	1.321	0.661	3.150	1.321	6.0	2.3	14.8	3.0	0.0506	1.370	0.069	0.000460	28.7
52002-05	2.553	1.384	0.661	3.243	1.353	6.2	2.4	16.6	3.1	0.0506	1.503	0.076	0.000497	30.3
52632-05	2.172	1.003	0.813	2.690	1.314	5.0	2.9	8.5	3.0	0.0759	0.790	0.060	0.000599	22.5
52134-05	2.819	1.626	0.661	3.624	1.474	7.0	2.7	24.4	3.3	0.0506	2.075	0.105	0.000641	37.4
52459-05	3.137	1.943	0.661	4.094	1.632	8.0	3.1	37.6	3.6	0.0506	2.963	0.150	0.000850	47.3
52176-05	2.172	1.003	0.978	2.690	1.479	5.0	3.8	9.3	3.3	0.1006	0.790	0.079	0.000968	23.9
52011-05	3.467	2.248	0.661	4.573	1.785	9.0	3.5	54.0	3.8	0.0506	3.967	0.201	0.001060	58.5
52666-05	3.774	2.576	0.661	5.042	1.949	10.0	3.8	75.5	4.1	0.0506	5.209	0.264	0.001308	70.8
52033-05	2.489	1.321	0.978	3.150	1.638	6.0	4.6	17.3	3.6	0.1006	1.370	0.138	0.001559	31.8
52076-05	2.832	1.296	0.813	3.503	1.461	6.5	5.6	16.7	3.6	0.1135	1.318	0.150	0.001906	35.3
52296-05	2.553	1.257	0.978	3.192	1.606	6.0	5.5	15.9	3.6	0.1212	1.240	0.150	0.002020	32.1
52454-05	4.747	3.503	0.661	6.479	2.412	13.0	5.0	166.3	4.9	0.0506	9.632	0.487	0.002031	115.0
52061-05	2.819	1.626	0.978	3.624	1.791	7.0	5.4	28.2	3.8	0.1006	2.075	0.209	0.002199	41.0
52691-05	5.070	3.820	0.661	6.962	2.571	14.0	5.4	208.3	5.1	0.0506	11.454	0.579	0.002293	132.3
52654-05	5.387	4.138	0.661	7.441	2.730	15.0	5.8	256.5	5.4	0.0506	13.441	0.680	0.002564	150.7
52106-05	3.137	1.626	0.813	3.954	1.626	7.5	6.5	28.1	3.8	0.1135	2.075	0.236	0.002808	44.7

Magnetics Cased Toroids 0.5 mil Iron Alloy

Part No.	OD cm 1	ID cm 2	HT cm 3	Wound OD cm 4	Wound HT cm 5	MPL cm 6	Wtfe grams 7	Wtcu grams 8	MLT cm 9	Ac cm sq 10	Wa cm sq 11	Ap cm 4th 12	Kg cm 5th 13	At cm sq 14
52007-05	2.832	1.296	0.978	3.503	1.626	6.5	7.5	18.0	3.8	0.1512	1.318	0.199	0.003147	37.2
52004-05	3.467	2.248	0.978	4.573	2.102	9.0	6.9	61.2	4.3	0.1006	3.967	0.399	0.003701	63.0
52748-05	7.328	6.007	0.661	10.335	3.664	20.9	8.1	697.0	6.9	0.0506	28.325	1.433	0.004190	286.6
52168-05	2.819	1.626	1.296	3.624	2.109	7.0	8.1	31.9	4.3	0.1512	2.075	0.314	0.004383	44.6
52115-05	3.772	2.578	0.978	5.041	2.267	10.0	7.7	85.0	4.6	0.1006	5.217	0.525	0.004608	75.8
52084-05	3.137	1.626	0.978	3.954	1.791	7.5	8.6	30.1	4.1	0.1512	2.075	0.314	0.004657	46.8
52167-05	3.137	1.308	0.978	3.836	1.632	7.0	10.7	19.5	4.1	0.2018	1.343	0.271	0.005367	42.8
52039-05	4.089	2.896	0.978	5.516	2.426	11.0	8.4	113.2	4.8	0.1006	6.583	0.662	0.005510	89.8
52392-05	4.089	2.578	0.813	5.358	2.102	10.5	9.1	84.8	4.6	0.1135	5.217	0.592	0.005883	80.4
52285-05	3.467	1.931	0.978	4.427	1.943	8.5	9.8	45.1	4.3	0.1512	2.927	0.442	0.006168	57.8
52094-05	2.819	1.321	1.296	3.499	1.956	6.5	11.2	21.1	4.3	0.2271	1.370	0.311	0.006526	40.7
52318-05	3.454	1.626	0.978	4.290	1.791	8.0	12.3	31.9	4.3	0.2018	2.075	0.419	0.007808	53.0
52029-05	3.785	2.248	0.978	4.895	2.102	9.5	10.9	64.8	4.6	0.1512	3.967	0.600	0.007896	69.9
52034-05	3.137	1.626	1.296	3.954	2.109	7.5	13.0	33.8	4.6	0.2271	2.075	0.471	0.009338	50.7
52228-05	3.772	2.578	1.296	5.041	2.585	10.0	11.5	94.4	5.1	0.1512	5.217	0.789	0.009368	80.8
52393-05	5.448	3.759	0.826	7.299	2.705	14.5	12.5	224.0	5.7	0.1135	11.092	1.259	0.010068	145.6
52133-05	3.188	1.384	1.296	3.917	1.988	7.2	16.6	24.7	4.6	0.3024	1.503	0.455	0.011890	48.5
52391-05	3.452	2.263	1.677	4.565	2.808	9.0	13.8	77.8	5.4	0.2018	4.020	0.811	0.012023	73.0
52181-05	3.467	1.931	1.296	4.427	2.261	8.5	14.7	50.4	4.8	0.2271	2.927	0.665	0.012453	62.2
52032-05	4.102	2.248	0.978	5.222	2.102	10.0	15.3	68.4	4.8	0.2018	3.967	0.800	0.013329	77.3
52188-05	3.467	1.613	1.296	4.299	2.102	8.0	18.4	35.2	4.8	0.3024	2.042	0.617	0.015407	57.4
52553-05	3.785	2.248	1.296	4.895	2.420	9.5	16.4	72.0	5.1	0.2271	3.967	0.901	0.016036	74.8
52018-05	5.448	3.759	0.991	7.299	2.870	14.5	16.7	234.4	5.9	0.1512	11.092	1.677	0.017060	149.4
52030-05	4.750	2.870	0.991	6.165	2.426	12.0	18.4	123.8	5.4	0.2018	6.466	1.305	0.019551	106.6
52504-05	4.102	2.566	1.296	5.365	2.579	10.5	18.1	98.4	5.4	0.2271	5.168	1.174	0.019905	88.6
52383-05	3.785	1.931	1.296	4.759	2.261	9.0	20.7	53.1	5.1	0.3024	2.927	0.885	0.020980	69.3
52323-05	7.455	6.007	0.978	10.454	3.981	21.1	19.5	758.3	7.5	0.1212	28.325	3.432	0.022099	302.3
52481-05	3.467	1.626	1.639	4.303	2.452	8.0	24.6	39.8	5.4	0.4035	2.075	0.837	0.025052	62.2
52026-05	4.102	2.248	1.296	5.222	2.420	10.0	23.0	75.5	5.4	0.3024	3.967	1.199	0.027088	82.5
52315-05	4.750	3.188	1.308	6.318	2.902	12.5	21.6	167.2	5.9	0.2271	7.978	1.811	0.027921	120.2

Magnetics Cased Toroids 0.5 mil Iron Alloy														
Part No.	OD cm 1	ID cm 2	HT cm 3	Wound OD cm 4	Wound HT cm 5	MPL cm 6	Wtfe grams 7	Wtcu grams 8	MLT cm 9	Ac cm sq 10	Wa cm sq 11	Ap cm 4th 12	Kg cm 5th 13	At cm sq 14
52514-05	4.102	1.613	1.321	4.983	2.127	9.0	31.0	39.2	5.4	0.4535	2.042	0.926	0.031145	72.3
52098-05	4.750	2.870	1.308	6.165	2.743	12.0	27.6	135.5	5.9	0.3024	6.466	1.955	0.040124	112.8
52038-05	4.102	2.248	1.639	5.222	2.763	10.0	30.7	83.3	5.9	0.4035	3.967	1.601	0.043765	88.1
52091-05	4.432	2.553	1.651	5.696	2.927	11.0	33.8	112.6	6.2	0.4035	5.116	2.065	0.053863	103.3
52139-05	5.397	3.493	1.308	7.115	3.054	14.0	32.2	218.3	6.4	0.3024	9.577	2.896	0.054637	147.7
52035-05	4.750	2.870	1.651	6.165	3.086	12.0	36.8	148.1	6.4	0.4035	6.466	2.609	0.065382	119.4
52559-05	5.067	3.188	1.664	6.636	3.258	13.0	39.9	190.5	6.7	0.4035	7.978	3.219	0.077378	137.0
52425-05	5.385	2.870	1.321	6.821	2.756	13.0	44.8	147.6	6.4	0.4535	6.466	2.932	0.082845	132.1
52190-05	4.762	2.223	1.664	5.907	2.775	11.0	50.6	89.3	6.5	0.6047	3.879	2.346	0.087671	106.3
52055-05	5.410	3.480	1.664	7.122	3.404	14.0	43.0	236.3	7.0	0.4035	9.506	3.836	0.088583	155.7
52345-05	6.045	4.115	1.677	8.070	3.734	16.0	49.1	355.4	7.5	0.4035	13.292	5.364	0.115150	196.9
52451-05	4.102	2.248	2.909	5.222	4.033	10.0	61.4	111.9	7.9	0.8065	3.967	3.199	0.130045	108.9
52169-05	5.436	2.845	1.664	6.864	3.086	13.0	59.8	158.4	7.0	0.6047	6.353	3.842	0.132555	140.5
52017-05	6.693	4.737	1.677	9.028	4.045	18.0	55.3	503.4	8.0	0.4035	17.614	7.108	0.142750	242.6
52066-05	7.341	4.725	1.359	9.665	3.721	18.9	65.5	501.5	8.0	0.4535	17.525	7.948	0.179191	259.6
52252-05	6.045	3.480	1.664	7.768	3.404	15.0	69.0	253.5	7.5	0.6047	9.506	5.749	0.185447	177.8
52012-05	4.750	2.870	2.921	6.165	4.356	12.0	73.6	194.8	8.5	0.8065	6.466	5.214	0.198521	144.0
52031-05	7.976	5.995	1.677	10.944	4.674	21.9	67.5	909.3	9.1	0.4035	28.212	11.385	0.202748	348.7
52555-05	6.045	2.845	1.677	7.507	3.099	14.0	85.9	169.9	7.5	0.8065	6.353	5.124	0.219839	161.5
52222-05	6.693	4.102	1.677	8.714	3.728	17.0	78.2	377.5	8.0	0.6047	13.208	7.987	0.240379	221.2
52040-05	5.423	3.467	2.921	7.129	4.654	14.0	85.9	302.4	9.0	0.8065	9.435	7.609	0.272397	184.0
52001-05	6.706	3.429	1.677	8.434	3.391	16.0	98.1	264.1	8.0	0.8065	9.230	7.444	0.298375	201.5
52067-05	9.931	7.214	1.410	13.493	5.017	26.9	93.2	1481.8	10.2	0.4535	40.852	18.528	0.329518	498.4
52036-05	5.436	2.845	2.934	6.864	4.356	13.0	119.7	204.3	9.0	1.2100	6.353	7.688	0.411477	167.8
52103-05	8.001	4.699	1.689	10.323	4.038	19.9	122.7	561.1	9.1	0.8065	17.333	13.979	0.495380	298.2
52128-05	9.271	5.969	1.715	12.207	4.699	23.9	147.3	1010.5	10.2	0.8065	27.968	22.556	0.716143	414.1
52013-05	6.731	3.429	2.934	8.461	4.648	16.0	196.3	330.8	10.1	1.6130	9.230	14.887	0.952977	235.9
52022-05	8.001	4.699	2.947	10.323	5.296	19.9	245.4	685.1	11.1	1.6130	17.333	27.957	1.622713	339.0
52448-05	8.001	4.064	2.947	10.052	4.979	18.9	291.4	512.5	11.1	2.0165	12.965	26.143	1.897065	315.8
52042-05	9.271	5.969	2.985	12.207	5.969	23.9	294.5	1212.6	12.2	1.6130	27.968	45.112	2.387163	462.7

Magnetics Cased Toroids 0.5 mil Iron Alloy														
Part No.	OD cm 1	ID cm 2	HT cm 3	Wound OD cm 4	Wound HT cm 5	MPL cm 6	Wtfe grams 7	Wtcu grams 8	MLT cm 9	Ac cm sq 10	Wa cm sq 11	Ap cm 4th 12	Kg cm 5th 13	At cm sq 14
52120-05	8.649	4.686	2.947	10.987	5.290	20.9	322.2	713.1	11.6	2.0165	17.237	34.758	2.409828	372.0
52261-05	9.271	5.321	2.972	11.906	5.632	22.9	352.8	962.0	12.2	2.0165	22.225	44.817	2.969969	433.1
52100-05	9.957	5.918	4.318	12.878	7.277	24.9	575.4	1454.1	14.9	3.0248	27.492	83.157	6.764262	554.7
52101-05	12.535	7.798	4.394	16.374	8.293	31.9	883.4	2895.5	17.1	3.6295	47.734	173.250	14.745070	847.3
52112-05	13.830	9.665	5.703	18.591	10.535	36.9	1135.0	5264.2	20.2	4.0325	73.327	295.688	23.624240	1157.6
52516-05	15.761	10.909	4.407	21.132	9.861	41.9	1159.5	6530.9	19.7	3.6295	93.418	339.062	25.038460	1355.4
52426-05	15.761	9.639	5.703	20.510	10.522	39.9	1963.2	5636.5	21.7	6.4519	72.933	470.559	55.877440	1338.1

Magnetics Cased Toroids 1 mil Iron Alloy														
Part No.	OD cm 1	ID cm 2	HT cm 3	Wound OD cm 4	Wound HT cm 5	MPL cm 6	Wtfe grams 7	Wtcu grams 8	MLT cm 9	Ac cm sq 10	Wa cm sq 11	Ap cm 4th 12	Kg cm 5th 13	At cm sq 14
52402-1	1.384	0.686	0.661	1.732	1.004	3.3	0.5	2.8	2.2	0.0194	0.369	0.007	0.000026	10.2
52107-1	1.689	1.003	0.661	2.184	1.162	4.2	0.6	6.8	2.4	0.0194	0.790	0.015	0.000049	15.5
52403-1	1.854	1.169	0.661	2.429	1.245	4.7	0.7	9.7	2.5	0.0194	1.073	0.021	0.000064	18.8
52485-1	2.019	1.321	0.661	2.669	1.321	5.2	0.8	13.0	2.7	0.0194	1.370	0.027	0.000077	22.3
52153-1	1.537	0.686	0.661	1.895	1.004	3.5	1.0	3.0	2.3	0.0379	0.369	0.014	0.000093	11.6
52356-1	2.172	1.486	0.661	2.903	1.404	5.7	0.8	17.2	2.8	0.0194	1.733	0.034	0.000093	26.0
52154-1	1.689	0.851	0.661	2.119	1.086	4.0	1.2	4.9	2.4	0.0379	0.568	0.022	0.000136	14.3
52056-1	1.854	1.003	0.661	2.355	1.162	4.5	1.3	7.1	2.5	0.0379	0.790	0.030	0.000179	17.3
52057-1	2.172	1.321	0.661	2.823	1.321	5.5	1.6	13.6	2.8	0.0379	1.370	0.052	0.000282	24.2
52143-1	2.489	1.626	0.661	3.289	1.474	6.5	1.9	22.5	3.0	0.0379	2.075	0.079	0.000392	32.2
52155-1	1.697	0.846	0.978	2.126	1.401	4.0	2.3	5.8	2.9	0.0759	0.562	0.043	0.000443	16.4
52374-1	2.819	1.943	0.661	3.776	1.632	7.5	2.2	34.9	3.3	0.0379	2.963	0.112	0.000515	41.7
52086-1	3.137	2.261	0.661	4.253	1.791	8.5	2.5	50.9	3.6	0.0379	4.013	0.152	0.000648	52.3
52000-1	2.172	1.003	0.661	2.690	1.162	5.0	2.9	7.8	2.8	0.0759	0.790	0.060	0.000651	21.2
52063-1	2.489	1.321	0.661	3.150	1.321	6.0	3.5	14.8	3.0	0.0759	1.370	0.104	0.001035	28.7
52002-1	2.553	1.384	0.661	3.243	1.353	6.2	3.6	16.6	3.1	0.0759	1.503	0.114	0.001117	30.3
52632-1	2.172	1.003	0.813	2.690	1.314	5.0	4.3	8.5	3.0	0.1138	0.790	0.090	0.001347	22.5
52134-1	2.819	1.626	0.661	3.624	1.474	7.0	4.0	24.4	3.3	0.0759	2.075	0.157	0.001443	37.4
52459-1	3.137	1.943	0.661	4.094	1.632	8.0	4.6	37.6	3.6	0.0759	2.963	0.225	0.001914	47.3
52176-1	2.172	1.003	0.978	2.690	1.479	5.0	5.7	9.3	3.3	0.1509	0.790	0.119	0.002177	23.9
52011-1	3.467	2.248	0.661	4.573	1.785	9.0	5.2	54.0	3.8	0.0759	3.967	0.301	0.002385	58.5
52666-1	3.774	2.576	0.661	5.042	1.949	10.0	5.8	75.5	4.1	0.0759	5.209	0.395	0.002943	70.8
52033-1	2.489	1.321	0.978	3.150	1.638	6.0	6.9	17.3	3.6	0.1509	1.370	0.207	0.003508	31.8
52076-1	2.832	1.296	0.813	3.503	1.461	6.5	8.4	16.7	3.6	0.1703	1.318	0.225	0.004288	35.3
52296-1	2.553	1.257	0.978	3.192	1.606	6.0	8.3	15.9	3.6	0.1818	1.240	0.225	0.004544	32.1
52454-1	4.747	3.503	0.661	6.479	2.412	13.0	7.5	166.3	4.9	0.0759	9.632	0.731	0.004570	115.0
52061-1	2.819	1.626	0.978	3.624	1.791	7.0	8.0	28.2	3.8	0.1509	2.075	0.313	0.004947	41.0
52691-1	5.070	3.820	0.661	6.962	2.571	14.0	8.1	208.3	5.1	0.0759	11.454	0.869	0.005160	132.3
52654-1	5.387	4.138	0.661	7.441	2.730	15.0	8.7	256.5	5.4	0.0759	13.441	1.020	0.005769	150.7
52106-1	3.137	1.626	0.813	3.954	1.626	7.5	9.7	28.1	3.8	0.1703	2.075	0.353	0.006318	44.7

Magnetics Cased Toroids 1 mil Iron Alloy														
Part No.	OD cm 1	ID cm 2	HT cm 3	Wound OD cm 4	Wound HT cm 5	MPL cm 6	Wtfe grams 7	Wtcu grams 8	MLT cm 9	Ac cm sq 10	Wa cm sq 11	Ap cm 4th 12	Kg cm 5th 13	At cm sq 14
52007-1	2.832	1.296	0.978	3.503	1.626	6.5	11.2	18.0	3.8	0.2268	1.318	0.299	0.007080	37.2
52004-1	3.467	2.248	0.978	4.573	2.102	9.0	10.3	61.2	4.3	0.1509	3.967	0.599	0.008327	63.0
52748-1	7.328	6.007	0.661	10.335	3.664	20.9	12.1	697.0	6.9	0.0759	28.325	2.149	0.009429	286.6
52168-1	2.819	1.626	1.296	3.624	2.109	7.0	12.1	31.9	4.3	0.2268	2.075	0.471	0.009862	44.6
52115-1	3.772	2.578	0.978	5.041	2.267	10.0	11.5	85.0	4.6	0.1509	5.217	0.787	0.010368	75.8
52084-1	3.137	1.626	0.978	3.954	1.791	7.5	12.9	30.1	4.1	0.2268	2.075	0.471	0.010477	46.8
52167-1	3.137	1.308	0.978	3.836	1.632	7.0	16.1	19.5	4.1	0.3027	1.343	0.406	0.012076	42.8
52039-1	4.089	2.896	0.978	5.516	2.426	11.0	12.6	113.2	4.8	0.1509	6.583	0.993	0.012397	89.8
52392-1	4.089	2.578	0.813	5.358	2.102	10.5	13.6	84.8	4.6	0.1703	5.217	0.888	0.013237	80.4
52285-1	3.467	1.931	0.978	4.427	1.943	8.5	14.7	45.1	4.3	0.2268	2.927	0.664	0.013878	57.8
52094-1	2.819	1.321	1.296	3.499	1.956	6.5	16.8	21.1	4.3	0.3406	1.370	0.466	0.014683	40.7
52318-1	3.454	1.626	0.978	4.290	1.791	8.0	18.4	31.9	4.3	0.3027	2.075	0.628	0.017569	53.0
52029-1	3.785	2.248	0.978	4.895	2.102	9.5	16.4	64.8	4.6	0.2268	3.967	0.900	0.017766	69.9
52034-1	3.137	1.626	1.296	3.954	2.109	7.5	19.4	33.8	4.6	0.3406	2.075	0.707	0.021011	50.7
52228-1	3.772	2.578	1.296	5.041	2.585	10.0	17.3	94.4	5.1	0.2268	5.217	1.183	0.021078	80.8
52393-1	5.448	3.759	0.826	7.299	2.705	14.5	18.8	224.0	5.7	0.1703	11.092	1.889	0.022654	145.6
52133-1	3.188	1.384	1.296	3.917	1.988	7.2	24.8	24.7	4.6	0.4535	1.503	0.682	0.026753	48.5
52391-1	3.452	2.263	1.677	4.565	2.808	9.0	20.7	77.8	5.4	0.3027	4.020	1.217	0.027052	73.0
52181-1	3.467	1.931	1.296	4.427	2.261	8.5	22.0	50.4	4.8	0.3406	2.927	0.997	0.028020	62.2
52032-1	4.102	2.248	0.978	5.222	2.102	10.0	23.0	68.4	4.8	0.3027	3.967	1.201	0.029990	77.3
52188-1	3.467	1.613	1.296	4.299	2.102	8.0	27.6	35.2	4.8	0.4535	2.042	0.926	0.034666	57.4
52553-1	3.785	2.248	1.296	4.895	2.420	9.5	24.6	72.0	5.1	0.3406	3.967	1.351	0.036081	74.8
52018-1	5.448	3.759	0.991	7.299	2.870	14.5	25.0	234.4	5.9	0.2268	11.092	2.515	0.038385	149.4
52030-1	4.750	2.870	0.991	6.165	2.426	12.0	27.6	123.8	5.4	0.3027	6.466	1.957	0.043989	106.6
52504-1	4.102	2.566	1.296	5.365	2.579	10.5	27.2	98.4	5.4	0.3406	5.168	1.760	0.044785	88.6
52383-1	3.785	1.931	1.296	4.759	2.261	9.0	31.0	53.1	5.1	0.4535	2.927	1.327	0.047206	69.3
52323-1	7.455	6.007	0.978	10.454	3.981	21.1	29.3	758.3	7.5	0.1818	28.325	5.149	0.049723	302.3
52481-1	3.467	1.626	1.639	4.303	2.452	8.0	36.9	39.8	5.4	0.6053	2.075	1.256	0.056366	62.2
52026-1	4.102	2.248	1.296	5.222	2.420	10.0	34.5	75.5	5.4	0.4535	3.967	1.799	0.060948	82.5
52315-1	4.750	3.188	1.308	6.318	2.902	12.5	32.4	167.2	5.9	0.3406	7.978	2.717	0.062823	120.2

Magnetics Cased Toroids 1 mil Iron Alloy														
Part No.	OD cm 1	ID cm 2	HT cm 3	Wound OD cm 4	Wound HT cm 5	MPL cm 6	Wtfe grams 7	Wtcu grams 8	MLT cm 9	Ac cm sq 10	Wa cm sq 11	Ap cm 4th 12	Kg cm 5th 13	At cm sq 14
52514-1	4.102	1.613	1.321	4.983	2.127	9.0	46.6	39.2	5.4	0.6803	2.042	1.389	0.070076	72.3
52098-1	4.750	2.870	1.308	6.165	2.743	12.0	41.4	135.5	5.9	0.4535	6.466	2.932	0.090280	112.8
52038-1	4.102	2.248	1.639	5.222	2.763	10.0	46.0	83.3	5.9	0.6053	3.967	2.401	0.098471	88.1
52091-1	4.432	2.553	1.651	5.696	2.927	11.0	50.7	112.6	6.2	0.6053	5.116	3.097	0.121192	103.3
52139-1	5.397	3.493	1.308	7.115	3.054	14.0	48.3	218.3	6.4	0.4535	9.577	4.344	0.122932	147.7
52035-1	4.750	2.870	1.651	6.165	3.086	12.0	55.2	148.1	6.4	0.6053	6.466	3.914	0.147109	119.4
52559-1	5.067	3.188	1.664	6.636	3.258	13.0	59.9	190.5	6.7	0.6053	7.978	4.829	0.174100	137.0
52425-1	5.385	2.870	1.321	6.821	2.756	13.0	67.3	147.6	6.4	0.6803	6.466	4.399	0.186402	132.1
52190-1	4.762	2.223	1.664	5.907	2.775	11.0	75.9	89.3	6.5	0.9071	3.879	3.518	0.197260	106.3
52055-1	5.410	3.480	1.664	7.122	3.404	14.0	64.5	236.3	7.0	0.6053	9.506	5.754	0.199311	155.7
52345-1	6.045	4.115	1.677	8.070	3.734	16.0	73.7	355.4	7.5	0.6053	13.292	8.046	0.259087	196.9
52451-1	4.102	2.248	2.909	5.222	4.033	10.0	92.0	111.9	7.9	1.2097	3.967	4.799	0.292601	108.9
52169-1	5.436	2.845	1.664	6.864	3.086	13.0	89.7	158.4	7.0	0.9071	6.353	5.763	0.298249	140.5
52017-1	6.693	4.737	1.677	9.028	4.045	18.0	82.9	503.4	8.0	0.6053	17.614	10.662	0.321188	242.6
52066-1	7.341	4.725	1.359	9.665	3.721	18.9	98.3	501.5	8.0	0.6803	17.525	11.922	0.403179	259.6
52252-1	6.045	3.480	1.664	7.768	3.404	15.0	103.5	253.5	7.5	0.9071	9.506	8.623	0.417255	177.8
52012-1	4.750	2.870	2.921	6.165	4.356	12.0	110.4	194.8	8.5	1.2097	6.466	7.822	0.446672	144.0
52031-1	7.976	5.995	1.677	10.944	4.674	21.9	101.3	909.3	9.1	0.6053	28.212	17.077	0.456184	348.7
52555-1	6.045	2.845	1.677	7.507	3.099	14.0	128.9	169.9	7.5	1.2097	6.353	7.686	0.494638	161.5
52222-1	6.693	4.102	1.677	8.714	3.728	17.0	117.3	377.5	8.0	0.9071	13.208	11.981	0.540853	221.2
52040-1	5.423	3.467	2.921	7.129	4.654	14.0	128.9	302.4	9.0	1.2097	9.435	11.414	0.612892	184.0
52001-1	6.706	3.429	1.677	8.434	3.391	16.0	147.2	264.1	8.0	1.2097	9.230	11.165	0.671345	201.5
52067-1	9.931	7.214	1.410	13.493	5.017	26.9	139.7	1481.8	10.2	0.6803	40.852	27.792	0.741414	498.4
52036-1	5.436	2.845	2.934	6.864	4.356	13.0	179.5	204.3	9.0	1.8150	6.353	11.532	0.925823	167.8
52103-1	8.001	4.699	1.689	10.323	4.038	19.9	184.1	561.1	9.1	1.2097	17.333	20.968	1.114605	298.2
52128-1	9.271	5.969	1.715	12.207	4.699	23.9	220.9	1010.5	10.2	1.2097	27.968	33.834	1.611321	414.1
52013-1	6.731	3.429	2.934	8.461	4.648	16.0	294.4	330.8	10.1	2.4195	9.230	22.331	2.144198	235.9
52022-1	8.001	4.699	2.947	10.323	5.296	19.9	368.1	685.1	11.1	2.4195	17.333	41.936	3.651105	339.0
52448-1	8.001	4.064	2.947	10.052	4.979	18.9	437.1	512.5	11.1	3.0248	12.965	39.215	4.268396	315.8
52042-1	9.271	5.969	2.985	12.207	5.969	23.9	441.8	1212.6	12.2	2.4195	27.968	67.667	5.371117	462.7

Magnetics Cased Toroids 1 mil Iron Alloy														
Part No.	OD cm 1	ID cm 2	HT cm 3	Wound OD cm 4	Wound HT cm 5	MPL cm 6	Wtfe grams 7	Wtcu grams 8	MLT cm 9	Ac cm sq 10	Wa cm sq 11	Ap cm 4th 12	Kg cm 5th 13	At cm sq 14
52120-1	8.649	4.686	2.947	10.987	5.290	20.9	483.3	713.1	11.6	3.0248	17.237	52.137	5.422114	372.0
52261-1	9.271	5.321	2.972	11.906	5.632	22.9	529.2	962.0	12.2	3.0248	22.225	67.225	6.682430	433.1
52100-1	9.957	5.918	4.318	12.878	7.277	24.9	863.0	1454.1	14.9	4.5372	27.492	124.736	15.219590	554.7
52101-1	12.535	7.798	4.394	16.374	8.293	31.9	1325.1	2895.5	17.1	5.4443	47.734	259.874	33.176420	847.3
52112-1	13.830	9.665	5.703	18.591	10.535	36.9	1702.5	5264.2	20.2	6.0487	73.327	443.532	53.154540	1157.6
52516-1	15.761	10.909	4.407	21.132	9.861	41.9	1739.3	6530.9	19.7	5.4443	93.418	508.593	56.336530	1355.4
52426-1	15.761	9.639	5.703	20.510	10.522	39.9	2944.8	5636.5	21.7	9.6779	72.933	705.838	125.724200	1338.1

Magnetics Cased Toroids 2 mil Iron Alloy														
Part No.	OD cm 1	ID cm 2	HT cm 3	Wound OD cm 4	Wound HT cm 5	MPL cm 6	Wtfe grams 7	Wtcu grams 8	MLT cm 9	Ac cm sq 10	Wa cm sq 11	Ap cm 4th 12	Kg cm 5th 13	At cm sq 14
52402-2	1.384	0.686	0.661	1.732	1.004	3.3	0.5	2.8	2.2	0.022	0.369	0.008	0.000033	10.2
52107-2	1.689	1.003	0.661	2.184	1.162	4.2	0.7	6.8	2.4	0.022	0.790	0.017	0.000063	15.5
52403-2	1.854	1.169	0.661	2.429	1.245	4.7	0.8	9.7	2.5	0.022	1.073	0.024	0.000082	18.8
52485-2	2.019	1.321	0.661	2.669	1.321	5.2	0.9	13.0	2.7	0.022	1.370	0.030	0.000099	22.3
52153-2	1.537	0.686	0.661	1.895	1.004	3.5	1.1	3.0	2.3	0.043	0.369	0.016	0.000119	11.6
52356-2	2.172	1.486	0.661	2.903	1.404	5.7	1.0	17.2	2.8	0.022	1.733	0.038	0.000120	26.0
52154-2	1.689	0.851	0.661	2.119	1.086	4.0	1.3	4.9	2.4	0.043	0.568	0.024	0.000175	14.3
52056-2	1.854	1.003	0.661	2.355	1.162	4.5	1.5	7.1	2.5	0.043	0.790	0.034	0.000230	17.3
52057-2	2.172	1.321	0.661	2.823	1.321	5.5	1.8	13.6	2.8	0.043	1.370	0.059	0.000362	24.2
52143-2	2.489	1.626	0.661	3.289	1.474	6.5	2.1	22.5	3.0	0.043	2.075	0.089	0.000503	32.2
52155-2	1.697	0.846	0.978	2.126	1.401	4.0	2.6	5.8	2.9	0.086	0.562	0.048	0.000569	16.4
52374-2	2.819	1.943	0.661	3.776	1.632	7.5	2.5	34.9	3.3	0.043	2.963	0.127	0.000662	41.7
52086-2	3.137	2.261	0.661	4.253	1.791	8.5	2.8	50.9	3.6	0.043	4.013	0.173	0.000832	52.3
52000-2	2.172	1.003	0.661	2.690	1.162	5.0	3.3	7.8	2.8	0.086	0.790	0.068	0.000836	21.2
52063-2	2.489	1.321	0.661	3.150	1.321	6.0	3.9	14.8	3.0	0.086	1.370	0.118	0.001329	28.7
52002-2	2.553	1.384	0.661	3.243	1.353	6.2	4.1	16.6	3.1	0.086	1.503	0.129	0.001435	30.3
52632-2	2.172	1.003	0.813	2.690	1.314	5.0	4.9	8.5	3.0	0.129	0.790	0.102	0.001730	22.5
52134-2	2.819	1.626	0.661	3.624	1.474	7.0	4.6	24.4	3.3	0.086	2.075	0.178	0.001853	37.4
52459-2	3.137	1.943	0.661	4.094	1.632	8.0	5.2	37.6	3.6	0.086	2.963	0.255	0.002458	47.3
52176-2	2.172	1.003	0.978	2.690	1.479	5.0	6.5	9.3	3.3	0.171	0.790	0.135	0.002797	23.9
52011-2	3.467	2.248	0.661	4.573	1.785	9.0	5.9	54.0	3.8	0.086	3.967	0.341	0.003063	58.5
52666-2	3.774	2.576	0.661	5.042	1.949	10.0	6.5	75.5	4.1	0.086	5.209	0.448	0.003780	70.8
52033-2	2.489	1.321	0.978	3.150	1.638	6.0	7.8	17.3	3.6	0.171	1.370	0.234	0.004506	31.8
52076-2	2.832	1.296	0.813	3.503	1.461	6.5	9.5	16.7	3.6	0.193	1.318	0.254	0.005508	35.3
52296-2	2.553	1.257	0.978	3.192	1.606	6.0	9.4	15.9	3.6	0.206	1.240	0.255	0.005836	32.1
52454-2	4.747	3.503	0.661	6.479	2.412	13.0	8.5	166.3	4.9	0.086	9.632	0.828	0.005870	115.0
52061-2	2.819	1.626	0.978	3.624	1.791	7.0	9.1	28.2	3.8	0.171	2.075	0.355	0.006355	41.0
52691-2	5.070	3.820	0.661	6.962	2.571	14.0	9.2	208.3	5.1	0.086	11.454	0.985	0.006627	132.3
52654-2	5.387	4.138	0.661	7.441	2.730	15.0	9.8	256.5	5.4	0.086	13.441	1.156	0.007409	150.7
52106-2	3.137	1.626	0.813	3.954	1.626	7.5	11.0	28.1	3.8	0.193	2.075	0.401	0.008115	44.7

Magnetics Cased Toroids 2 mil Iron Alloy														
Part No.	OD cm 1	ID cm 2	HT cm 3	Wound OD cm 4	Wound HT cm 5	MPL cm 6	Wtfe grams 7	Wtcu grams 8	MLT cm 9	Ac cm sq 10	Wa cm sq 11	Ap cm 4th 12	Kg cm 5th 13	At cm sq 14
52007-2	2.832	1.296	0.978	3.503	1.626	6.5	12.7	18.0	3.8	0.257	1.318	0.339	0.009094	37.2
52004-2	3.467	2.248	0.978	4.573	2.102	9.0	11.7	61.2	4.3	0.171	3.967	0.678	0.010695	63.0
52748-2	7.328	6.007	0.661	10.335	3.664	20.9	13.7	697.0	6.9	0.086	28.325	2.436	0.012110	286.6
52168-2	2.819	1.626	1.296	3.624	2.109	7.0	13.7	31.9	4.3	0.257	2.075	0.533	0.012667	44.6
52115-2	3.772	2.578	0.978	5.041	2.267	10.0	13.0	85.0	4.6	0.171	5.217	0.892	0.013317	75.8
52084-2	3.137	1.626	0.978	3.954	1.791	7.5	14.7	30.1	4.1	0.257	2.075	0.533	0.013458	46.8
52167-2	3.137	1.308	0.978	3.836	1.632	7.0	18.3	19.5	4.1	0.343	1.343	0.461	0.015511	42.8
52039-2	4.089	2.896	0.978	5.516	2.426	11.0	14.3	113.2	4.8	0.171	6.583	1.126	0.015924	89.8
52392-2	4.089	2.578	0.813	5.358	2.102	10.5	15.4	84.8	4.6	0.193	5.217	1.007	0.017003	80.4
52285-2	3.467	1.931	0.978	4.427	1.943	8.5	16.6	45.1	4.3	0.257	2.927	0.752	0.017825	57.8
52094-2	2.819	1.321	1.296	3.499	1.956	6.5	19.1	21.1	4.3	0.386	1.370	0.529	0.018859	40.7
52318-2	3.454	1.626	0.978	4.290	1.791	8.0	20.9	31.9	4.3	0.343	2.075	0.712	0.022566	53.0
52029-2	3.785	2.248	0.978	4.895	2.102	9.5	18.6	64.8	4.6	0.257	3.967	1.019	0.022820	69.9
52034-2	3.137	1.626	1.296	3.954	2.109	7.5	22.0	33.8	4.6	0.386	2.075	0.801	0.026988	50.7
52228-2	3.772	2.578	1.296	5.041	2.585	10.0	19.6	94.4	5.1	0.257	5.217	1.341	0.027074	80.8
52393-2	5.448	3.759	0.826	7.299	2.705	14.5	21.3	224.0	5.7	0.193	11.092	2.141	0.029098	145.6
52133-2	3.188	1.384	1.296	3.917	1.988	7.2	28.2	24.7	4.6	0.514	1.503	0.773	0.034363	48.5
52391-2	3.452	2.263	1.677	4.565	2.808	9.0	23.5	77.8	5.4	0.343	4.020	1.379	0.034746	73.0
52181-2	3.467	1.931	1.296	4.427	2.261	8.5	24.9	50.4	4.8	0.386	2.927	1.130	0.035990	62.2
52032-2	4.102	2.248	0.978	5.222	2.102	10.0	26.1	68.4	4.8	0.343	3.967	1.361	0.038521	77.3
52188-2	3.467	1.613	1.296	4.299	2.102	8.0	31.3	35.2	4.8	0.514	2.042	1.050	0.044527	57.4
52553-2	3.785	2.248	1.296	4.895	2.420	9.5	27.9	72.0	5.1	0.386	3.967	1.531	0.046344	74.8
52018-2	5.448	3.759	0.991	7.299	2.870	14.5	28.4	234.4	5.9	0.257	11.092	2.851	0.049304	149.4
52030-2	4.750	2.870	0.991	6.165	2.426	12.0	31.3	123.8	5.4	0.343	6.466	2.218	0.056501	106.6
52504-2	4.102	2.566	1.296	5.365	2.579	10.5	30.8	98.4	5.4	0.386	5.168	1.995	0.057524	88.6
52383-2	3.785	1.931	1.296	4.759	2.261	9.0	35.2	53.1	5.1	0.514	2.927	1.504	0.060633	69.3
52323-2	7.455	6.007	0.978	10.454	3.981	21.1	33.2	758.3	7.5	0.206	28.325	5.835	0.063867	302.3
52481-2	3.467	1.626	1.639	4.303	2.452	8.0	41.8	39.8	5.4	0.686	2.075	1.424	0.072399	62.2
52026-2	4.102	2.248	1.296	5.222	2.420	10.0	39.1	75.5	5.4	0.514	3.967	2.039	0.078284	82.5
52315-2	4.750	3.188	1.308	6.318	2.902	12.5	36.7	167.2	5.9	0.386	7.978	3.079	0.080692	120.2

Magnetics Cased Toroids 2 mil Iron Alloy														
Part No.	OD cm 1	ID cm 2	HT cm 3	Wound OD cm 4	Wound HT cm 5	MPL cm 6	Wtfe grams 7	Wtcu grams 8	MLT cm 9	Ac cm sq 10	Wa cm sq 11	Ap cm 4th 12	Kg cm 5th 13	At cm sq 14
52514-2	4.102	1.613	1.321	4.983	2.127	9.0	52.8	39.2	5.4	0.771	2.042	1.575	0.090009	72.3
52098-2	4.750	2.870	1.308	6.165	2.743	12.0	46.9	135.5	5.9	0.514	6.466	3.323	0.115960	112.8
52038-2	4.102	2.248	1.639	5.222	2.763	10.0	52.2	83.3	5.9	0.686	3.967	2.721	0.126481	88.1
52091-2	4.432	2.553	1.651	5.696	2.927	11.0	57.4	112.6	6.2	0.686	5.116	3.510	0.155665	103.3
52139-2	5.397	3.493	1.308	7.115	3.054	14.0	54.7	218.3	6.4	0.514	9.577	4.923	0.157900	147.7
52035-2	4.750	2.870	1.651	6.165	3.086	12.0	62.6	148.1	6.4	0.686	6.466	4.435	0.188954	119.4
52559-2	5.067	3.188	1.664	6.636	3.258	13.0	67.8	190.5	6.7	0.686	7.978	5.473	0.223622	137.0
52425-2	5.385	2.870	1.321	6.821	2.756	13.0	76.2	147.6	6.4	0.771	6.466	4.985	0.239424	132.1
52190-2	4.762	2.223	1.664	5.907	2.775	11.0	86.0	89.3	6.5	1.028	3.879	3.988	0.253370	106.3
52055-2	5.410	3.480	1.664	7.122	3.404	14.0	73.1	236.3	7.0	0.686	9.506	6.521	0.256004	155.7
52345-2	6.045	4.115	1.677	8.070	3.734	16.0	83.5	355.4	7.5	0.686	13.292	9.118	0.332782	196.9
52451-2	4.102	2.248	2.909	5.222	4.033	10.0	104.3	111.9	7.9	1.371	3.967	5.438	0.375830	108.9
52169-2	5.436	2.845	1.664	6.864	3.086	13.0	101.7	158.4	7.0	1.028	6.353	6.531	0.383084	140.5
52017-2	6.693	4.737	1.677	9.028	4.045	18.0	94.0	503.4	8.0	0.686	17.614	12.084	0.412548	242.6
52066-2	7.341	4.725	1.359	9.665	3.721	18.9	111.4	501.5	8.0	0.771	17.525	13.512	0.517862	259.6
52252-2	6.045	3.480	1.664	7.768	3.404	15.0	117.3	253.5	7.5	1.028	9.506	9.773	0.535940	177.8
52012-2	4.750	2.870	2.921	6.165	4.356	12.0	125.1	194.8	8.5	1.371	6.466	8.864	0.573726	144.0
52031-2	7.976	5.995	1.677	10.944	4.674	21.9	114.8	909.3	9.1	0.686	28.212	19.354	0.585943	348.7
52555-2	6.045	2.845	1.677	7.507	3.099	14.0	146.0	169.9	7.5	1.371	6.353	8.711	0.635335	161.5
52222-2	6.693	4.102	1.677	8.714	3.728	17.0	133.0	377.5	8.0	1.028	13.208	13.578	0.694695	221.2
52040-2	5.423	3.467	2.921	7.129	4.654	14.0	146.0	302.4	9.0	1.371	9.435	12.936	0.787226	184.0
52001-2	6.706	3.429	1.677	8.434	3.391	16.0	166.9	264.1	8.0	1.371	9.230	12.654	0.862305	201.5
52067-2	9.931	7.214	1.410	13.493	5.017	26.9	158.4	1481.8	10.2	0.771	40.852	31.497	0.952306	498.4
52036-2	5.436	2.845	2.934	6.864	4.356	13.0	203.4	204.3	9.0	2.057	6.353	13.069	1.189168	167.8
52103-2	8.001	4.699	1.689	10.323	4.038	19.9	208.6	561.1	9.1	1.371	17.333	23.764	1.431648	298.2
52128-2	9.271	5.969	1.715	12.207	4.699	23.9	250.3	1010.5	10.2	1.371	27.968	38.345	2.069652	414.1
52013-2	6.731	3.429	2.934	8.461	4.648	16.0	333.7	330.8	10.1	2.742	9.230	25.308	2.754103	235.9
52022-2	8.001	4.699	2.947	10.323	5.296	19.9	417.2	685.1	11.1	2.742	17.333	47.527	4.689641	339.0
52448-2	8.001	4.064	2.947	10.052	4.979	18.9	495.4	512.5	11.1	3.428	12.965	44.443	5.482517	315.8
52042-2	9.271	5.969	2.985	12.207	5.969	23.9	500.7	1212.6	12.2	2.742	27.968	76.690	6.898902	462.7

Magnetics Cased Toroids 2 mil Iron Alloy														
Part No.	OD cm 1	ID cm 2	HT cm 3	Wound OD cm 4	Wound HT cm 5	MPL cm 6	Wtfe grams 7	Wtcu grams 8	MLT cm 9	Ac cm sq 10	Wa cm sq 11	Ap cm 4th 12	Kg cm 5th 13	At cm sq 14
52120-2	8.649	4.686	2.947	10.987	5.290	20.9	547.7	713.1	11.6	3.428	17.237	59.089	6.964404	372.0
52261-2	9.271	5.321	2.972	11.906	5.632	22.9	599.8	962.0	12.2	3.428	22.225	76.189	8.583210	433.1
52100-2	9.957	5.918	4.318	12.878	7.277	24.9	978.1	1454.1	14.9	5.142	27.492	141.367	19.548720	554.7
52101-2	12.535	7.798	4.394	16.374	8.293	31.9	1501.8	2895.5	17.1	6.170	47.734	294.524	42.613270	847.3
52112-2	13.830	9.665	5.703	18.591	10.535	36.9	1929.5	5264.2	20.2	6.855	73.327	502.670	68.274050	1157.6
52516-2	15.761	10.909	4.407	21.132	9.861	41.9	1971.2	6530.9	19.7	6.170	93.418	576.405	72.361150	1355.4
52426-2	15.761	9.639	5.703	20.510	10.522	39.9	3337.5	5636.5	21.7	10.968	72.933	799.950	161.485800	1338.1

Magnetics Cased Toroids 4 mil Iron Alloy

Part No.	OD cm 1	ID cm 2	HT cm 3	Wound OD cm 4	Wound HT cm 5	MPL cm 6	Wtfe grams 7	Wtcu grams 8	MLT cm 9	Ac cm sq 10	Wa cm sq 11	Ap cm 4th 12	Kg cm 5th 13	At cm sq 14
52402-4	1.384	0.686	0.661	1.732	1.004	3.3	0.6	2.8	2.2	0.0233	0.369	0.009	0.000037	10.2
52107-4	1.689	1.003	0.661	2.184	1.162	4.2	0.8	6.8	2.4	0.0233	0.790	0.018	0.000071	15.5
52403-4	1.854	1.169	0.661	2.429	1.245	4.7	0.8	9.7	2.5	0.0233	1.073	0.025	0.000092	18.8
52485-4	2.019	1.321	0.661	2.669	1.321	5.2	0.9	13.0	2.7	0.0233	1.370	0.032	0.000111	22.3
52153-4	1.537	0.686	0.661	1.895	1.004	3.5	1.2	3.0	2.3	0.0455	0.369	0.017	0.000134	11.6
52356-4	2.172	1.486	0.661	2.903	1.404	5.7	1.0	17.2	2.8	0.0233	1.733	0.040	0.000135	26.0
52154-4	1.689	0.851	0.661	2.119	1.086	4.0	1.4	4.9	2.4	0.0455	0.568	0.026	0.000196	14.3
52056-4	1.854	1.003	0.661	2.355	1.162	4.5	1.6	7.1	2.5	0.0455	0.790	0.036	0.000258	17.3
52057-4	2.172	1.321	0.661	2.823	1.321	5.5	1.9	13.6	2.8	0.0455	1.370	0.062	0.000406	24.2
52143-4	2.489	1.626	0.661	3.289	1.474	6.5	2.3	22.5	3.0	0.0455	2.075	0.094	0.000564	32.2
52155-4	1.697	0.846	0.978	2.126	1.401	4.0	2.8	5.8	2.9	0.0911	0.562	0.051	0.000638	16.4
52374-4	2.819	1.943	0.661	3.776	1.632	7.5	2.6	34.9	3.3	0.0455	2.963	0.135	0.000742	41.7
52086-4	3.137	2.261	0.661	4.253	1.791	8.5	2.9	50.9	3.6	0.0455	4.013	0.183	0.000933	52.3
52000-4	2.172	1.003	0.661	2.690	1.162	5.0	3.5	7.8	2.8	0.0911	0.790	0.072	0.000937	21.2
52063-4	2.489	1.321	0.661	3.150	1.321	6.0	4.2	14.8	3.0	0.0911	1.370	0.125	0.001490	28.7
52002-4	2.553	1.384	0.661	3.243	1.353	6.2	4.3	16.6	3.1	0.0911	1.503	0.137	0.001609	30.3
52632-4	2.172	1.003	0.813	2.690	1.314	5.0	5.2	8.5	3.0	0.1366	0.790	0.108	0.001939	22.5
52134-4	2.819	1.626	0.661	3.624	1.474	7.0	4.8	24.4	3.3	0.0911	2.075	0.189	0.002078	37.4
52459-4	3.137	1.943	0.661	4.094	1.632	8.0	5.5	37.6	3.6	0.0911	2.963	0.270	0.002755	47.3
52176-4	2.172	1.003	0.978	2.690	1.479	5.0	6.9	9.3	3.3	0.1811	0.790	0.143	0.003136	23.9
52011-4	3.467	2.248	0.661	4.573	1.785	9.0	6.2	54.0	3.8	0.0911	3.967	0.361	0.003434	58.5
52666-4	3.774	2.576	0.661	5.042	1.949	10.0	6.9	75.5	4.1	0.0911	5.209	0.474	0.004238	70.8
52033-4	2.489	1.321	0.978	3.150	1.638	6.0	8.3	17.3	3.6	0.1811	1.370	0.248	0.005051	31.8
52076-4	2.832	1.296	0.813	3.503	1.461	6.5	10.1	16.7	3.6	0.2044	1.318	0.269	0.006175	35.3
52296-4	2.553	1.257	0.978	3.192	1.606	6.0	10.0	15.9	3.6	0.2181	1.240	0.271	0.006543	32.1
52454-4	4.747	3.503	0.661	6.479	2.412	13.0	9.0	166.3	4.9	0.0911	9.632	0.877	0.006581	115.0
52061-4	2.819	1.626	0.978	3.624	1.791	7.0	9.6	28.2	3.8	0.1811	2.075	0.376	0.007124	41.0
52691-4	5.070	3.820	0.661	6.962	2.571	14.0	9.7	208.3	5.1	0.0911	11.454	1.043	0.007430	132.3
52654-4	5.387	4.138	0.661	7.441	2.730	15.0	10.4	256.5	5.4	0.0911	13.441	1.224	0.008307	150.7
52106-4	3.137	1.626	0.813	3.954	1.626	7.5	11.7	28.1	3.8	0.2044	2.075	0.424	0.009098	44.7

Magnetics Cased Toroids 4 mil Iron Alloy														
Part No.	OD cm 1	ID cm 2	HT cm 3	Wound OD cm 4	Wound HT cm 5	MPL cm 6	Wtfe grams 7	Wtcu grams 8	MLT cm 9	Ac cm sq 10	Wa cm sq 11	Ap cm 4th 12	Kg cm 5th 13	At cm sq 14
52007-4	2.832	1.296	0.978	3.503	1.626	6.5	13.5	18.0	3.8	0.2721	1.318	0.359	0.010195	37.2
52004-4	3.467	2.248	0.978	4.573	2.102	9.0	12.4	61.2	4.3	0.1811	3.967	0.718	0.011991	63.0
52748-4	7.328	6.007	0.661	10.335	3.664	20.9	14.5	697.0	6.9	0.0911	28.325	2.579	0.013577	286.6
52168-4	2.819	1.626	1.296	3.624	2.109	7.0	14.5	31.9	4.3	0.2721	2.075	0.565	0.014201	44.6
52115-4	3.772	2.578	0.978	5.041	2.267	10.0	13.8	85.0	4.6	0.1811	5.217	0.945	0.014930	75.8
52084-4	3.137	1.626	0.978	3.954	1.791	7.5	15.5	30.1	4.1	0.2721	2.075	0.565	0.015087	46.8
52167-4	3.137	1.308	0.978	3.836	1.632	7.0	19.3	19.5	4.1	0.3632	1.343	0.488	0.017390	42.8
52039-4	4.089	2.896	0.978	5.516	2.426	11.0	15.2	113.2	4.8	0.1811	6.583	1.192	0.017852	89.8
52392-4	4.089	2.578	0.813	5.358	2.102	10.5	16.3	84.8	4.6	0.2044	5.217	1.066	0.019062	80.4
52285-4	3.467	1.931	0.978	4.427	1.943	8.5	17.6	45.1	4.3	0.2721	2.927	0.796	0.019984	57.8
52094-4	2.819	1.321	1.296	3.499	1.956	6.5	20.2	21.1	4.3	0.4087	1.370	0.560	0.021143	40.7
52318-4	3.454	1.626	0.978	4.290	1.791	8.0	22.1	31.9	4.3	0.3632	2.075	0.754	0.025299	53.0
52029-4	3.785	2.248	0.978	4.895	2.102	9.5	19.7	64.8	4.6	0.2721	3.967	1.079	0.025584	69.9
52034-4	3.137	1.626	1.296	3.954	2.109	7.5	23.3	33.8	4.6	0.4087	2.075	0.848	0.030256	50.7
52228-4	3.772	2.578	1.296	5.041	2.585	10.0	20.7	94.4	5.1	0.2721	5.217	1.420	0.030353	80.8
52393-4	5.448	3.759	0.826	7.299	2.705	14.5	22.5	224.0	5.7	0.2044	11.092	2.267	0.032622	145.6
52133-4	3.188	1.384	1.296	3.917	1.988	7.2	29.8	24.7	4.6	0.5442	1.503	0.818	0.038525	48.5
52391-4	3.452	2.263	1.677	4.565	2.808	9.0	24.9	77.8	5.4	0.3632	4.020	1.460	0.038954	73.0
52181-4	3.467	1.931	1.296	4.427	2.261	8.5	26.4	50.4	4.8	0.4087	2.927	1.196	0.040348	62.2
52032-4	4.102	2.248	0.978	5.222	2.102	10.0	27.6	68.4	4.8	0.3632	3.967	1.441	0.043186	77.3
52188-4	3.467	1.613	1.296	4.299	2.102	8.0	33.1	35.2	4.8	0.5442	2.042	1.111	0.049919	57.4
52553-4	3.785	2.248	1.296	4.895	2.420	9.5	29.5	72.0	5.1	0.4087	3.967	1.621	0.051957	74.8
52018-4	5.448	3.759	0.991	7.299	2.870	14.5	30.0	234.4	5.9	0.2721	11.092	3.018	0.055275	149.4
52030-4	4.750	2.870	0.991	6.165	2.426	12.0	33.1	123.8	5.4	0.3632	6.466	2.348	0.063344	106.6
52504-4	4.102	2.566	1.296	5.365	2.579	10.5	32.7	98.4	5.4	0.4087	5.168	2.112	0.064491	88.6
52383-4	3.785	1.931	1.296	4.759	2.261	9.0	37.2	53.1	5.1	0.5442	2.927	1.593	0.067976	69.3
52323-4	7.455	6.007	0.978	10.454	3.981	21.1	35.2	758.3	7.5	0.2181	28.325	6.178	0.071601	302.3
52481-4	3.467	1.626	1.639	4.303	2.452	8.0	44.2	39.8	5.4	0.7264	2.075	1.507	0.081167	62.2
52026-4	4.102	2.248	1.296	5.222	2.420	10.0	41.4	75.5	5.4	0.5442	3.967	2.159	0.087765	82.5
52315-4	4.750	3.188	1.308	6.318	2.902	12.5	38.9	167.2	5.9	0.4087	7.978	3.261	0.090465	120.2

Magnetics Cased Toroids 4 mil Iron Alloy

Part No.	OD cm 1	ID cm 2	HT cm 3	Wound OD cm 4	Wound HT cm 5	MPL cm 6	Wtfe grams 7	Wtcu grams 8	MLT cm 9	Ac cm sq 10	Wa cm sq 11	Ap cm 4th 12	Kg cm 5th 13	At cm sq 14
52514-4	4.102	1.613	1.321	4.983	2.127	9.0	55.9	39.2	5.4	0.8164	2.042	1.667	0.100910	72.3
52098-4	4.750	2.870	1.308	6.165	2.743	12.0	49.7	135.5	5.9	0.5442	6.466	3.519	0.130003	112.8
52038-4	4.102	2.248	1.639	5.222	2.763	10.0	55.3	83.3	5.9	0.7264	3.967	2.881	0.141799	88.1
52091-4	4.432	2.553	1.651	5.696	2.927	11.0	60.8	112.6	6.2	0.7264	5.116	3.716	0.174517	103.3
52139-4	5.397	3.493	1.308	7.115	3.054	14.0	58.0	218.3	6.4	0.5442	9.577	5.212	0.177022	147.7
52035-4	4.750	2.870	1.651	6.165	3.086	12.0	66.3	148.1	6.4	0.7264	6.466	4.696	0.211837	119.4
52559-4	5.067	3.188	1.664	6.636	3.258	13.0	71.8	190.5	6.7	0.7264	7.978	5.795	0.250704	137.0
52425-4	5.385	2.870	1.321	6.821	2.756	13.0	80.7	147.6	6.4	0.8164	6.466	5.278	0.268419	132.1
52190-4	4.762	2.223	1.664	5.907	2.775	11.0	91.1	89.3	6.5	1.0885	3.879	4.222	0.284055	106.3
52055-4	5.410	3.480	1.664	7.122	3.404	14.0	77.4	236.3	7.0	0.7264	9.506	6.905	0.287007	155.7
52345-4	6.045	4.115	1.677	8.070	3.734	16.0	88.4	355.4	7.5	0.7264	13.292	9.655	0.373085	196.9
52451-4	4.102	2.248	2.909	5.222	4.033	10.0	110.4	111.9	7.9	1.4517	3.967	5.758	0.421346	108.9
52169-4	5.436	2.845	1.664	6.864	3.086	13.0	107.6	158.4	7.0	1.0885	6.353	6.916	0.429478	140.5
52017-4	6.693	4.737	1.677	9.028	4.045	18.0	99.5	503.4	8.0	0.7264	17.614	12.794	0.462511	242.6
52066-4	7.341	4.725	1.359	9.665	3.721	18.9	118.0	501.5	8.0	0.8164	17.525	14.307	0.580578	259.6
52252-4	6.045	3.480	1.664	7.768	3.404	15.0	124.2	253.5	7.5	1.0885	9.506	10.347	0.600847	177.8
52012-4	4.750	2.870	2.921	6.165	4.356	12.0	132.5	194.8	8.5	1.4517	6.466	9.386	0.643208	144.0
52031-4	7.976	5.995	1.677	10.944	4.674	21.9	121.5	909.3	9.1	0.7264	28.212	20.492	0.656904	348.7
52555-4	6.045	2.845	1.677	7.507	3.099	14.0	154.6	169.9	7.5	1.4517	6.353	9.223	0.712278	161.5
52222-4	6.693	4.102	1.677	8.714	3.728	17.0	140.8	377.5	8.0	1.0885	13.208	14.377	0.778828	221.2
52040-4	5.423	3.467	2.921	7.129	4.654	14.0	154.6	302.4	9.0	1.4517	9.435	13.697	0.882565	184.0
52001-4	6.706	3.429	1.677	8.434	3.391	16.0	176.7	264.1	8.0	1.4517	9.230	13.398	0.966736	201.5
52067-4	9.931	7.214	1.410	13.493	5.017	26.9	167.7	1481.8	10.2	0.8164	40.852	33.350	1.067637	498.4
52036-4	5.436	2.845	2.934	6.864	4.356	13.0	215.4	204.3	9.0	2.1781	6.353	13.838	1.333185	167.8
52103-4	8.001	4.699	1.689	10.323	4.038	19.9	220.9	561.1	9.1	1.4517	17.333	25.161	1.605030	298.2
52128-4	9.271	5.969	1.715	12.207	4.699	23.9	265.1	1010.5	10.2	1.4517	27.968	40.600	2.320301	414.1
52013-4	6.731	3.429	2.934	8.461	4.648	16.0	353.3	330.8	10.1	2.9034	9.230	26.797	3.087645	235.9
52022-4	8.001	4.699	2.947	10.323	5.296	19.9	441.7	685.1	11.1	2.9034	17.333	50.323	5.257590	339.0
52448-4	8.001	4.064	2.947	10.052	4.979	18.9	524.5	512.5	11.1	3.6297	12.965	47.058	6.146488	315.8
52042-4	9.271	5.969	2.985	12.207	5.969	23.9	530.1	1212.6	12.2	2.9034	27.968	81.201	7.734409	462.7

Magnetics Cased Toroids 4 mil Iron Alloy														
Part No.	OD cm 1	ID cm 2	HT cm 3	Wound OD cm 4	Wound HT cm 5	MPL cm 6	Wtfe grams 7	Wtcu grams 8	MLT cm 9	Ac cm sq 10	Wa cm sq 11	Ap cm 4th 12	Kg cm 5th 13	At cm sq 14
52120-4	8.649	4.686	2.947	10.987	5.290	20.9	579.9	713.1	11.6	3.6297	17.237	62.565	7.807843	372.0
52261-4	9.271	5.321	2.972	11.906	5.632	22.9	635.0	962.0	12.2	3.6297	22.225	80.671	9.622698	433.1
52100-4	9.957	5.918	4.318	12.878	7.277	24.9	1035.7	1454.1	14.9	5.4446	27.492	149.683	21.916210	554.7
52101-4	12.535	7.798	4.394	16.374	8.293	31.9	1590.1	2895.5	17.1	6.5331	47.734	311.849	47.774040	847.3
52112-4	13.830	9.665	5.703	18.591	10.535	36.9	2043.0	5264.2	20.2	7.2584	73.327	532.239	76.542520	1157.6
52516-4	15.761	10.909	4.407	21.132	9.861	41.9	2087.1	6530.9	19.7	6.5331	93.418	610.311	81.124600	1355.4
52426-4	15.761	9.639	5.703	20.510	10.522	39.9	3533.8	5636.5	21.7	11.6135	72.933	847.006	181.042900	1338.1

Magnetics Cased Toroids 12 mil Iron Alloy

Part No.	OD cm 1	ID cm 2	HT cm 3	Wound OD cm 4	Wound HT cm 5	MPL cm 6	Wtfe grams 7	Wtcu grams 8	MLT cm 9	Ac cm sq 10	Wa cm sq 11	Ap cm 4th 12	Kg cm 5th 13	At cm sq 14
52402-12	1.384	0.686	0.661	1.732	1.004	3.3	0.6	2.8	2.2	0.0246	0.369	0.009	0.000041	10.2
52107-12	1.689	1.003	0.661	2.184	1.162	4.2	0.8	6.8	2.4	0.0246	0.790	0.019	0.000079	15.5
52403-12	1.854	1.169	0.661	2.429	1.245	4.7	0.9	9.7	2.5	0.0246	1.073	0.026	0.000102	18.8
52485-12	2.019	1.321	0.661	2.669	1.321	5.2	1.0	13.0	2.7	0.0246	1.370	0.034	0.000124	22.3
52153-12	1.537	0.686	0.661	1.895	1.004	3.5	1.3	3.0	2.3	0.0481	0.369	0.018	0.000149	11.6
52356-12	2.172	1.486	0.661	2.903	1.404	5.7	1.1	17.2	2.8	0.0246	1.733	0.043	0.000150	26.0
52154-12	1.689	0.851	0.661	2.119	1.086	4.0	1.5	4.9	2.4	0.0481	0.568	0.027	0.000218	14.3
52056-12	1.854	1.003	0.661	2.355	1.162	4.5	1.6	7.1	2.5	0.0481	0.790	0.038	0.000287	17.3
52057-12	2.172	1.321	0.661	2.823	1.321	5.5	2.0	13.6	2.8	0.0481	1.370	0.066	0.000453	24.2
52143-12	2.489	1.626	0.661	3.289	1.474	6.5	2.4	22.5	3.0	0.0481	2.075	0.100	0.000629	32.2
52155-12	1.697	0.846	0.978	2.126	1.401	4.0	2.9	5.8	2.9	0.0961	0.562	0.054	0.000710	16.4
52374-12	2.819	1.943	0.661	3.776	1.632	7.5	2.7	34.9	3.3	0.0481	2.963	0.142	0.000826	41.7
52086-12	3.137	2.261	0.661	4.253	1.791	8.5	3.1	50.9	3.6	0.0481	4.013	0.193	0.001039	52.3
52000-12	2.172	1.003	0.661	2.690	1.162	5.0	3.7	7.8	2.8	0.0961	0.790	0.076	0.001044	21.2
52063-12	2.489	1.321	0.661	3.150	1.321	6.0	4.4	14.8	3.0	0.0961	1.370	0.132	0.001660	28.7
52002-12	2.553	1.384	0.661	3.243	1.353	6.2	4.5	16.6	3.1	0.0961	1.503	0.145	0.001792	30.3
52632-12	2.172	1.003	0.813	2.690	1.314	5.0	5.5	8.5	3.0	0.1442	0.790	0.114	0.002161	22.5
52134-12	2.819	1.626	0.661	3.624	1.474	7.0	5.1	24.4	3.3	0.0961	2.075	0.199	0.002315	37.4
52459-12	3.137	1.943	0.661	4.094	1.632	8.0	5.9	37.6	3.6	0.0961	2.963	0.285	0.003070	47.3
52176-12	2.172	1.003	0.978	2.690	1.479	5.0	7.3	9.3	3.3	0.1911	0.790	0.151	0.003494	23.9
52011-12	3.467	2.248	0.661	4.573	1.785	9.0	6.6	54.0	3.8	0.0961	3.967	0.381	0.003827	58.5
52666-12	3.774	2.576	0.661	5.042	1.949	10.0	7.3	75.5	4.1	0.0961	5.209	0.501	0.004722	70.8
52033-12	2.489	1.321	0.978	3.150	1.638	6.0	8.7	17.3	3.6	0.1911	1.370	0.262	0.005628	31.8
52076-12	2.832	1.296	0.813	3.503	1.461	6.5	10.7	16.7	3.6	0.2157	1.318	0.284	0.006880	35.3
52296-12	2.553	1.257	0.978	3.192	1.606	6.0	10.5	15.9	3.6	0.2302	1.240	0.286	0.007290	32.1
52454-12	4.747	3.503	0.661	6.479	2.412	13.0	9.5	166.3	4.9	0.0961	9.632	0.926	0.007332	115.0
52061-12	2.819	1.626	0.978	3.624	1.791	7.0	10.2	28.2	3.8	0.1911	2.075	0.397	0.007938	41.0
52691-12	5.070	3.820	0.661	6.962	2.571	14.0	10.2	208.3	5.1	0.0961	11.454	1.101	0.008279	132.3
52654-12	5.387	4.138	0.661	7.441	2.730	15.0	11.0	256.5	5.4	0.0961	13.441	1.292	0.009255	150.7
52106-12	3.137	1.626	0.813	3.954	1.626	7.5	12.3	28.1	3.8	0.2157	2.075	0.448	0.010137	44.7

Magnetics Cased Toroids 12 mil Iron Alloy

Part No.	OD cm 1	ID cm 2	HT cm 3	Wound OD cm 4	Wound HT cm 5	MPL cm 6	Wtfe grams 7	Wtcu grams 8	MLT cm 9	Ac cm sq 10	Wa cm sq 11	Ap cm 4th 12	Kg cm 5th 13	At cm sq 14
52007-12	2.832	1.296	0.978	3.503	1.626	6.5	14.2	18.0	3.8	0.2872	1.318	0.379	0.011359	37.2
52004-12	3.467	2.248	0.978	4.573	2.102	9.0	13.1	61.2	4.3	0.1911	3.967	0.758	0.013360	63.0
52748-12	7.328	6.007	0.661	10.335	3.664	20.9	15.4	697.0	6.9	0.0961	28.325	2.723	0.015128	286.6
52168-12	2.819	1.626	1.296	3.624	2.109	7.0	15.3	31.9	4.3	0.2872	2.075	0.596	0.015822	44.6
52115-12	3.772	2.578	0.978	5.041	2.267	10.0	14.5	85.0	4.6	0.1911	5.217	0.997	0.016635	75.8
52084-12	3.137	1.626	0.978	3.954	1.791	7.5	16.4	30.1	4.1	0.2872	2.075	0.596	0.016810	46.8
52167-12	3.137	1.308	0.978	3.836	1.632	7.0	20.4	19.5	4.1	0.3834	1.343	0.515	0.019376	42.8
52039-12	4.089	2.896	0.978	5.516	2.426	11.0	16.0	113.2	4.8	0.1911	6.583	1.258	0.019891	89.8
52392-12	4.089	2.578	0.813	5.358	2.102	10.5	17.2	84.8	4.6	0.2157	5.217	1.125	0.021239	80.4
52285-12	3.467	1.931	0.978	4.427	1.943	8.5	18.6	45.1	4.3	0.2872	2.927	0.841	0.022266	57.8
52094-12	2.819	1.321	1.296	3.499	1.956	6.5	21.3	21.1	4.3	0.4314	1.370	0.591	0.023558	40.7
52318-12	3.454	1.626	0.978	4.290	1.791	8.0	23.3	31.9	4.3	0.3834	2.075	0.796	0.028189	53.0
52029-12	3.785	2.248	0.978	4.895	2.102	9.5	20.8	64.8	4.6	0.2872	3.967	1.139	0.028505	69.9
52034-12	3.137	1.626	1.296	3.954	2.109	7.5	24.6	33.8	4.6	0.4314	2.075	0.895	0.033711	50.7
52228-12	3.772	2.578	1.296	5.041	2.585	10.0	21.9	94.4	5.1	0.2872	5.217	1.498	0.033819	80.8
52393-12	5.448	3.759	0.826	7.299	2.705	14.5	23.8	224.0	5.7	0.2157	11.092	2.393	0.036347	145.6
52133-12	3.188	1.384	1.296	3.917	1.988	7.2	31.5	24.7	4.6	0.5745	1.503	0.864	0.042924	48.5
52391-12	3.452	2.263	1.677	4.565	2.808	9.0	26.2	77.8	5.4	0.3834	4.020	1.541	0.043403	73.0
52181-12	3.467	1.931	1.296	4.427	2.261	8.5	27.9	50.4	4.8	0.4314	2.927	1.263	0.044956	62.2
52032-12	4.102	2.248	0.978	5.222	2.102	10.0	29.2	68.4	4.8	0.3834	3.967	1.521	0.048118	77.3
52188-12	3.467	1.613	1.296	4.299	2.102	8.0	35.0	35.2	4.8	0.5745	2.042	1.173	0.055620	57.4
52553-12	3.785	2.248	1.296	4.895	2.420	9.5	31.2	72.0	5.1	0.4314	3.967	1.711	0.057890	74.8
52018-12	5.448	3.759	0.991	7.299	2.870	14.5	31.7	234.4	5.9	0.2872	11.092	3.186	0.061587	149.4
52030-12	4.750	2.870	0.991	6.165	2.426	12.0	35.0	123.8	5.4	0.3834	6.466	2.479	0.070578	106.6
52504-12	4.102	2.566	1.296	5.365	2.579	10.5	34.5	98.4	5.4	0.4314	5.168	2.230	0.071855	88.6
52383-12	3.785	1.931	1.296	4.759	2.261	9.0	39.3	53.1	5.1	0.5745	2.927	1.681	0.075739	69.3
52323-12	7.455	6.007	0.978	10.454	3.981	21.1	37.1	758.3	7.5	0.2302	28.325	6.522	0.079778	302.3
52481-12	3.467	1.626	1.639	4.303	2.452	8.0	46.7	39.8	5.4	0.7667	2.075	1.591	0.090437	62.2
52026-12	4.102	2.248	1.296	5.222	2.420	10.0	43.7	75.5	5.4	0.5745	3.967	2.279	0.097788	82.5
52315-12	4.750	3.188	1.308	6.318	2.902	12.5	41.0	167.2	5.9	0.4314	7.978	3.442	0.100795	120.2

Magnetics Cased Toroids 12 mil Iron Alloy

Part No.	OD cm 1	ID cm 2	HT cm 3	Wound OD cm 4	Wound HT cm 5	MPL cm 6	Wtfe grams 7	Wtcu grams 8	MLT cm 9	Ac cm sq 10	Wa cm sq 11	Ap cm 4th 12	Kg cm 5th 13	At cm sq 14
52514-12	4.102	1.613	1.321	4.983	2.127	9.0	59.0	39.2	5.4	0.8617	2.042	1.760	0.112434	72.3
52098-12	4.750	2.870	1.308	6.165	2.743	12.0	52.4	135.5	5.9	0.5745	6.466	3.714	0.144849	112.8
52038-12	4.102	2.248	1.639	5.222	2.763	10.0	58.3	83.3	5.9	0.7667	3.967	3.041	0.157992	88.1
52091-12	4.432	2.553	1.651	5.696	2.927	11.0	64.2	112.6	6.2	0.7667	5.116	3.923	0.194446	103.3
52139-12	5.397	3.493	1.308	7.115	3.054	14.0	61.2	218.3	6.4	0.5745	9.577	5.502	0.197238	147.7
52035-12	4.750	2.870	1.651	6.165	3.086	12.0	70.0	148.1	6.4	0.7667	6.466	4.957	0.236029	119.4
52559-12	5.067	3.188	1.664	6.636	3.258	13.0	75.8	190.5	6.7	0.7667	7.978	6.117	0.279334	137.0
52425-12	5.385	2.870	1.321	6.821	2.756	13.0	85.2	147.6	6.4	0.8617	6.466	5.572	0.299072	132.1
52190-12	4.762	2.223	1.664	5.907	2.775	11.0	96.2	89.3	6.5	1.1490	3.879	4.457	0.316493	106.3
52055-12	5.410	3.480	1.664	7.122	3.404	14.0	81.7	236.3	7.0	0.7667	9.506	7.289	0.319783	155.7
52345-12	6.045	4.115	1.677	8.070	3.734	16.0	93.3	355.4	7.5	0.7667	13.292	10.191	0.415690	196.9
52451-12	4.102	2.248	2.909	5.222	4.033	10.0	116.6	111.9	7.9	1.5323	3.967	6.078	0.469463	108.9
52169-12	5.436	2.845	1.664	6.864	3.086	13.0	113.6	158.4	7.0	1.1490	6.353	7.300	0.478524	140.5
52017-12	6.693	4.737	1.677	9.028	4.045	18.0	105.0	503.4	8.0	0.7667	17.614	13.505	0.515328	242.6
52066-12	7.341	4.725	1.359	9.665	3.721	18.9	124.5	501.5	8.0	0.8617	17.525	15.102	0.646879	259.6
52252-12	6.045	3.480	1.664	7.768	3.404	15.0	131.1	253.5	7.5	1.1490	9.506	10.922	0.669462	177.8
52012-12	4.750	2.870	2.921	6.165	4.356	12.0	139.8	194.8	8.5	1.5323	6.466	9.907	0.716661	144.0
52031-12	7.976	5.995	1.677	10.944	4.674	21.9	128.3	909.3	9.1	0.7667	28.212	21.631	0.731921	348.7
52555-12	6.045	2.845	1.677	7.507	3.099	14.0	163.2	169.9	7.5	1.5323	6.353	9.735	0.793619	161.5
52222-12	6.693	4.102	1.677	8.714	3.728	17.0	148.6	377.5	8.0	1.1490	13.208	15.176	0.867768	221.2
52040-12	5.423	3.467	2.921	7.129	4.654	14.0	163.2	302.4	9.0	1.5323	9.435	14.458	0.983351	184.0
52001-12	6.706	3.429	1.677	8.434	3.391	16.0	186.5	264.1	8.0	1.5323	9.230	14.143	1.077135	201.5
52067-12	9.931	7.214	1.410	13.493	5.017	26.9	177.0	1481.8	10.2	0.8617	40.852	35.203	1.189558	498.4
52036-12	5.436	2.845	2.934	6.864	4.356	13.0	227.3	204.3	9.0	2.2991	6.353	14.607	1.485432	167.8
52103-12	8.001	4.699	1.689	10.323	4.038	19.9	233.1	561.1	9.1	1.5323	17.333	26.559	1.788321	298.2
52128-12	9.271	5.969	1.715	12.207	4.699	23.9	279.8	1010.5	10.2	1.5323	27.968	42.856	2.585274	414.1
52013-12	6.731	3.429	2.934	8.461	4.648	16.0	373.0	330.8	10.1	3.0647	9.230	28.285	3.440246	235.9
52022-12	8.001	4.699	2.947	10.323	5.296	19.9	466.3	685.1	11.1	3.0647	17.333	53.118	5.857994	339.0
52448-12	8.001	4.064	2.947	10.052	4.979	18.9	553.7	512.5	11.1	3.8314	12.965	49.672	6.848403	315.8
52042-12	9.271	5.969	2.985	12.207	5.969	23.9	559.6	1212.6	12.2	3.0647	27.968	85.712	8.617659	462.7

Magnetics Cased Toroids 12 mil Iron Alloy														
Part No.	OD cm 1	ID cm 2	HT cm 3	Wound OD cm 4	Wound HT cm 5	MPL cm 6	Wtfe grams 7	Wtcu grams 8	MLT cm 9	Ac cm sq 10	Wa cm sq 11	Ap cm 4th 12	Kg cm 5th 13	At cm sq 14
52120-12	8.649	4.686	2.947	10.987	5.290	20.9	612.1	713.1	11.6	3.8314	17.237	66.041	8.699479	372.0
52261-12	9.271	5.321	2.972	11.906	5.632	22.9	670.3	962.0	12.2	3.8314	22.225	85.152	10.721590	433.1
52100-12	9.957	5.918	4.318	12.878	7.277	24.9	1093.2	1454.1	14.9	5.7471	27.492	157.998	24.418990	554.7
52101-12	12.535	7.798	4.394	16.374	8.293	31.9	1678.5	2895.5	17.1	6.8961	47.734	329.174	53.229720	847.3
52112-12	13.830	9.665	5.703	18.591	10.535	36.9	2156.5	5264.2	20.2	7.6617	73.327	561.807	85.283490	1157.6
52516-12	15.761	10.909	4.407	21.132	9.861	41.9	2203.1	6530.9	19.7	6.8961	93.418	644.217	90.388830	1355.4
52426-12	15.761	9.639	5.703	20.510	10.522	39.9	3730.1	5636.5	21.7	12.2587	72.933	894.062	201.717600	1338.1

7

HIGH FLUX, MPP, SENDUST, KOOL Mμ, AND IRON POWDER CORES

POWDER-CORE MANUFACTURERS

Manufacturers address and phone number can be found from the Table of Contents.

Notes

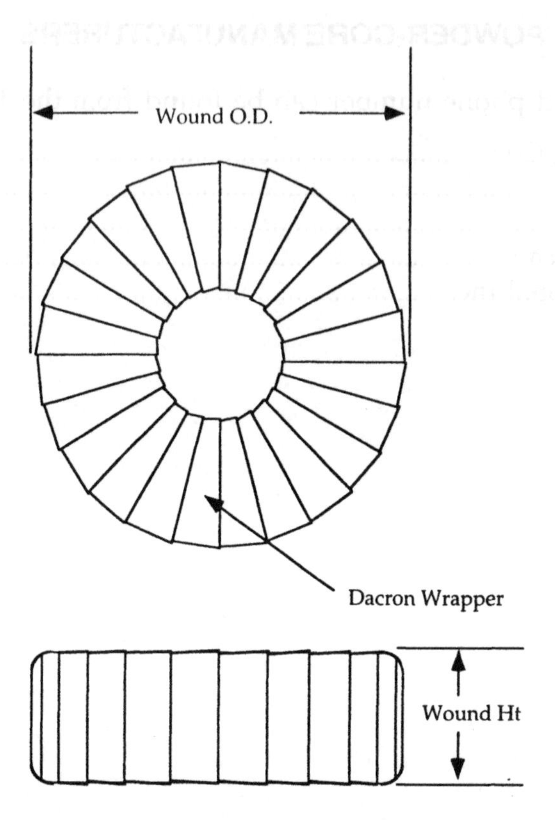

Figure 118 Toroidal dimensional outline.

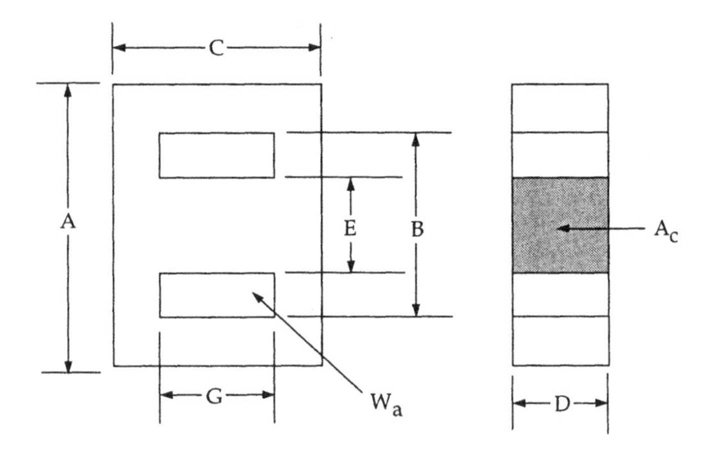

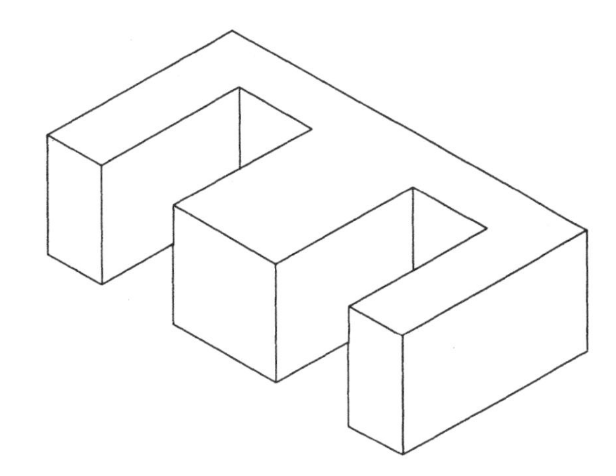

Figure 119

TOROIDAL POWDER-CORE TABLE DESCRIPTION

Definitions for the Following Tables

Information given is listed by column as

Col.	Dim.	Description	Units
1.	OD	Core outside diameter	cm
2.	ID	Core inside diameter	cm
3.	HT	Core height	cm
4.	OD	Finished wound (K_u = 0.4)	cm
5.	HT	Finished wound (K_u = 0.4)	cm
6.	MPL	Magnetic path length [MPL = 3.14(OD + ID)(0.5)]	cm
7.	W_{tfe}	Core weight [W_{tfe} = (MPL)(A_c)(Density)]	grams
8.	W_{tcu}	Copper weight [W_{tcu} = (MLT)(W_a)(K_u)(8.89)]	grams
9.	MLT	Mean length turn [MLT = (OD + 2HT)(0.8), core]	cm
10.	A_c	Effective iron area [A_c = (OD − ID)(HT)(SF)(0.5), core]	cm sq
11.	W_a	Window area [$W_a = 3.14(\text{ID})^2(0.25)$]	cm sq
12.	A_p	Area product [A_p = (A_c)(W_a)]	cm 4th
13.	K_g	Core geometry [K_g = (A_p)(A_c)(K_u)/MLT]	cm 5th
14.	A_t	Surface area	cm sq
15.	Perm	Core permeability	*B/H*
16.	AL	Core millihenrys per 1000 turns	mh

EE POWDER-CORE TABLE DESCRIPTION

Definitions for the Following Tables

Information given is listed by column as

Col.	Dim.	Description	Units
1.	Ht	Finished transformer height	cm
2.	W_{th}	Finished transformer width	cm
3.	Lt	Finished transformer length	cm
4.	G	Window length	cm
5.	MPL	Mean magnetic path length	cm
6.	W_{tfe}	Core weight	grams
7.	W_{tcu}	Copper weight [W_{tcu} = (MLT)(W_a)(K_u)(8.89)]	grams
8.	MLT	Mean length turn	cm
9.	A_c	Iron area (gross) [A_c = (D)(E)]	cm sq
10.	W_a	Window area [W_a = (F)(G)]	cm sq
11.	A_p	Area product (gross) [A_p = (A_c)(W_a)]	cm 4th
12.	K_g	Core geometry (gross) [K_g = (A_p)(A_c)(K_u)/(MLT)]	cm 5th
13.	A_t	Surface area	cm sq
14.	Perm	Core permeability	B/H
15.	AL	Core millihenrys per 1000 turns	mh

Arnold Engineering Co. High Flux Powder Cores																
Part No.	OD cm 1	ID cm 2	HT cm 3	Wound OD cm 4	Wound HT cm 5	MPL cm 6	Wtfe grams 7	Wtcu grams 8	MLT cm 9	Ac cm sq 10	Wa cm sq 11	Ap cm 4th 12	Kg cm 5th 13	At cm sq 14	Perm 15	AL 16
HF-014-200	0.407	0.127	0.203	0.422	0.267	0.8	0.1	0.0	0.7	0.012	0.013	0.000	0.0000010	0.8	200	42
HF-014-173	0.407	0.127	0.203	0.422	0.267	0.8	0.1	0.0	0.7	0.012	0.013	0.000	0.0000010	0.8	173	36
HF-014-160	0.407	0.127	0.203	0.422	0.267	0.8	0.1	0.0	0.7	0.012	0.013	0.000	0.0000010	0.8	160	33
HF-014-147	0.407	0.127	0.203	0.422	0.267	0.8	0.1	0.0	0.7	0.012	0.013	0.000	0.0000010	0.8	147	31
HF-014-125	0.407	0.127	0.203	0.422	0.267	0.8	0.1	0.0	0.7	0.012	0.013	0.000	0.0000010	0.8	125	26
HF-014-60	0.407	0.127	0.203	0.422	0.267	0.8	0.1	0.0	0.7	0.012	0.013	0.000	0.0000010	0.8	60	12
HF-014-26	0.407	0.127	0.203	0.422	0.267	0.8	0.1	0.0	0.7	0.012	0.013	0.000	0.0000010	0.8	26	5
HF-014-14	0.407	0.127	0.203	0.422	0.267	0.8	0.1	0.0	0.7	0.012	0.013	0.000	0.0000010	0.8	14	3
HF-015-200	0.445	0.170	0.305	0.469	0.390	1.0	0.2	0.1	0.8	0.020	0.023	0.000	0.0000040	1.1	200	56
HF-015-173	0.445	0.170	0.305	0.469	0.390	1.0	0.2	0.1	0.8	0.020	0.023	0.000	0.0000040	1.1	173	48
HF-015-160	0.445	0.170	0.305	0.469	0.390	1.0	0.2	0.1	0.8	0.020	0.023	0.000	0.0000040	1.1	160	45
HF-015-147	0.445	0.170	0.305	0.469	0.390	1.0	0.2	0.1	0.8	0.020	0.023	0.000	0.0000040	1.1	147	41
HF-015-125	0.445	0.170	0.305	0.469	0.390	1.0	0.2	0.1	0.8	0.020	0.023	0.000	0.0000040	1.1	125	35
HF-015-60	0.445	0.170	0.305	0.469	0.390	1.0	0.2	0.1	0.8	0.020	0.023	0.000	0.0000040	1.1	60	17
HF-015-26	0.445	0.170	0.305	0.469	0.390	1.0	0.2	0.1	0.8	0.020	0.023	0.000	0.0000040	1.1	26	7
HF-015-14	0.445	0.170	0.305	0.469	0.390	1.0	0.2	0.1	0.8	0.020	0.023	0.000	0.0000040	1.1	14	4
HF-018-200	0.516	0.185	0.305	0.540	0.398	1.1	0.2	0.1	0.9	0.028	0.027	0.001	0.0000090	1.4	200	67
HF-018-173	0.516	0.185	0.305	0.540	0.398	1.1	0.2	0.1	0.9	0.028	0.027	0.001	0.0000090	1.4	173	57
HF-018-160	0.516	0.185	0.305	0.540	0.398	1.1	0.2	0.1	0.9	0.028	0.027	0.001	0.0000090	1.4	160	53
HF-018-147	0.516	0.185	0.305	0.540	0.398	1.1	0.2	0.1	0.9	0.028	0.027	0.001	0.0000090	1.4	147	49
HF-018-125	0.516	0.185	0.305	0.540	0.398	1.1	0.2	0.1	0.9	0.028	0.027	0.001	0.0000090	1.4	125	42
HF-018-60	0.516	0.185	0.305	0.540	0.398	1.1	0.2	0.1	0.9	0.028	0.027	0.001	0.0000090	1.4	60	20
HF-018-26	0.516	0.185	0.305	0.540	0.398	1.1	0.2	0.1	0.9	0.028	0.027	0.001	0.0000090	1.4	26	9
HF-018-14	0.516	0.185	0.305	0.540	0.398	1.1	0.2	0.1	0.9	0.028	0.027	0.001	0.0000090	1.4	14	5
HF-027-200	0.711	0.216	0.305	0.735	0.413	1.5	0.6	0.1	1.1	0.048	0.037	0.002	0.0000320	2.2	200	86
HF-027-160	0.711	0.216	0.305	0.735	0.413	1.5	0.6	0.1	1.1	0.048	0.037	0.002	0.0000320	2.2	160	69
HF-027-173	0.711	0.216	0.305	0.735	0.413	1.5	0.6	0.1	1.1	0.048	0.037	0.002	0.0000320	2.2	173	75
HF-027-147	0.711	0.216	0.305	0.735	0.413	1.5	0.6	0.1	1.1	0.048	0.037	0.002	0.0000320	2.2	147	64
HF-027-125	0.711	0.216	0.305	0.735	0.413	1.5	0.6	0.1	1.1	0.048	0.037	0.002	0.0000320	2.2	125	54
HF-027-60	0.711	0.216	0.305	0.735	0.413	1.5	0.6	0.1	1.1	0.048	0.037	0.002	0.0000320	2.2	60	26

Arnold Engineering Co. High Flux Powder Cores

Part No.	OD cm 1	ID cm 2	HT cm 3	Wound OD cm 4	Wound HT cm 5	MPL cm 6	Wtfe grams 7	Wtcu grams 8	MLT cm 9	Ac cm sq 10	Wa cm sq 11	Ap cm 4th 12	Kg cm 5th 13	At cm sq 14	Perm 15	AL 16
HF-027-26	0.711	0.216	0.305	0.735	0.413	1.5	0.6	0.1	1.1	0.048	0.037	0.002	0.0000320	2.2	26	11
HF-027-14	0.711	0.216	0.305	0.735	0.413	1.5	0.6	0.1	1.1	0.048	0.037	0.002	0.0000320	2.2	14	6
HF-025-200	0.686	0.228	0.330	0.714	0.444	1.4	0.6	0.2	1.1	0.048	0.041	0.002	0.0000350	2.2	200	83
HF-025-173	0.686	0.228	0.330	0.714	0.444	1.4	0.6	0.2	1.1	0.048	0.041	0.002	0.0000350	2.2	173	69
HF-025-160	0.686	0.228	0.330	0.714	0.444	1.4	0.6	0.2	1.1	0.048	0.041	0.002	0.0000350	2.2	160	64
HF-025-147	0.686	0.228	0.330	0.714	0.444	1.4	0.6	0.2	1.1	0.048	0.041	0.002	0.0000350	2.2	147	58
HF-025-125	0.686	0.228	0.330	0.714	0.444	1.4	0.6	0.2	1.1	0.048	0.041	0.002	0.0000350	2.2	125	52
HF-025-60	0.686	0.228	0.330	0.714	0.444	1.4	0.6	0.2	1.1	0.048	0.041	0.002	0.0000350	2.2	60	24
HF-025-26	0.686	0.228	0.330	0.714	0.444	1.4	0.6	0.2	1.1	0.048	0.041	0.002	0.0000350	2.2	26	10
HF-025-14	0.686	0.228	0.330	0.714	0.444	1.4	0.6	0.2	1.1	0.048	0.041	0.002	0.0000350	2.2	14	6
HF-026-200	0.711	0.216	0.529	0.735	0.637	1.5	1.0	0.2	1.4	0.090	0.037	0.003	0.0000840	2.8	200	165
HF-026-173	0.711	0.216	0.529	0.735	0.637	1.5	1.0	0.2	1.4	0.090	0.037	0.003	0.0000840	2.8	173	144
HF-026-160	0.711	0.216	0.529	0.735	0.637	1.5	1.0	0.2	1.4	0.090	0.037	0.003	0.0000840	2.8	160	132
HF-026-147	0.711	0.216	0.529	0.735	0.637	1.5	1.0	0.2	1.4	0.090	0.037	0.003	0.0000840	2.8	147	122
HF-026-125	0.711	0.216	0.529	0.735	0.637	1.5	1.0	0.2	1.4	0.090	0.037	0.003	0.0000840	2.8	125	103
HF-026-60	0.711	0.216	0.529	0.735	0.637	1.5	1.0	0.2	1.4	0.090	0.037	0.003	0.0000840	2.8	60	50
HF-026-26	0.711	0.216	0.529	0.735	0.637	1.5	1.0	0.2	1.4	0.090	0.037	0.003	0.0000840	2.8	26	21
HF-026-14	0.711	0.216	0.529	0.735	0.637	1.5	1.0	0.2	1.4	0.090	0.037	0.003	0.0000840	2.8	14	12
HF-031-200	0.838	0.345	0.369	0.890	0.542	1.9	0.9	0.4	1.3	0.058	0.093	0.005	0.0001000	3.4	200	83
HF-031-173	0.838	0.345	0.369	0.890	0.542	1.9	0.9	0.4	1.3	0.058	0.093	0.005	0.0001000	3.4	173	73
HF-031-160	0.838	0.345	0.369	0.890	0.542	1.9	0.9	0.4	1.3	0.058	0.093	0.005	0.0001000	3.4	160	66
HF-031-147	0.838	0.345	0.369	0.890	0.542	1.9	0.9	0.4	1.3	0.058	0.093	0.005	0.0001000	3.4	147	62
HF-031-125	0.838	0.345	0.369	0.890	0.542	1.9	0.9	0.4	1.3	0.058	0.093	0.005	0.0001000	3.4	125	52
HF-031-60	0.838	0.345	0.369	0.890	0.542	1.9	0.9	0.4	1.3	0.058	0.093	0.005	0.0001000	3.4	60	25
HF-031-26	0.838	0.345	0.369	0.890	0.542	1.9	0.9	0.4	1.3	0.058	0.093	0.005	0.0001000	3.4	26	11
HF-031-14	0.838	0.345	0.369	0.890	0.542	1.9	0.9	0.4	1.3	0.058	0.093	0.005	0.0001000	3.4	14	6
HF-028-200	0.755	0.345	0.559	0.812	0.732	1.7	1.1	0.5	1.5	0.077	0.093	0.007	0.0001480	3.5	200	112
HF-028-173	0.755	0.345	0.559	0.812	0.732	1.7	1.1	0.5	1.5	0.077	0.093	0.007	0.0001480	3.5	173	95
HF-028-160	0.755	0.345	0.559	0.812	0.732	1.7	1.1	0.5	1.5	0.077	0.093	0.007	0.0001480	3.5	160	89
HF-028-147	0.755	0.345	0.559	0.812	0.732	1.7	1.1	0.5	1.5	0.077	0.093	0.007	0.0001480	3.5	147	81

Arnold Engineering Co. High Flux Powder Cores																
Part No.	OD cm 1	ID cm 2	HT cm 3	Wound OD cm 4	Wound HT cm 5	MPL cm 6	Wtfe grams 7	Wtcu grams 8	MLT cm 9	Ac cm sq 10	Wa cm sq 11	Ap cm 4th 12	Kg cm 5th 13	At cm sq 14	Perm 15	AL 16
HF-028-125	0.755	0.345	0.559	0.812	0.732	1.7	1.1	0.5	1.5	0.077	0.093	0.007	0.0001480	3.5	125	70
HF-028-60	0.755	0.345	0.559	0.812	0.732	1.7	1.1	0.5	1.5	0.077	0.093	0.007	0.0001480	3.5	60	33
HF-028-26	0.755	0.345	0.559	0.812	0.732	1.7	1.1	0.5	1.5	0.077	0.093	0.007	0.0001480	3.5	26	14
HF-028-14	0.755	0.345	0.559	0.812	0.732	1.7	1.1	0.5	1.5	0.077	0.093	0.007	0.0001480	3.5	14	8
HF-039-200	1.016	0.427	0.369	1.081	0.583	2.3	1.3	0.7	1.4	0.074	0.143	0.011	0.0002230	4.7	200	106
HF-039-173	1.016	0.427	0.369	1.081	0.583	2.3	1.3	0.7	1.4	0.074	0.143	0.011	0.0002230	4.7	173	92
HF-039-160	1.016	0.427	0.369	1.081	0.583	2.3	1.3	0.7	1.4	0.074	0.143	0.011	0.0002230	4.7	160	84
HF-039-147	1.016	0.427	0.369	1.081	0.583	2.3	1.3	0.7	1.4	0.074	0.143	0.011	0.0002230	4.7	147	78
HF-039-125	1.016	0.427	0.369	1.081	0.583	2.3	1.3	0.7	1.4	0.074	0.143	0.011	0.0002230	4.7	125	66
HF-039-60	1.016	0.427	0.369	1.081	0.583	2.3	1.3	0.7	1.4	0.074	0.143	0.011	0.0002230	4.7	60	32
HF-039-26	1.016	0.427	0.369	1.081	0.583	2.3	1.3	0.7	1.4	0.074	0.143	0.011	0.0002230	4.7	26	14
HF-039-14	1.016	0.427	0.369	1.081	0.583	2.3	1.3	0.7	1.4	0.074	0.143	0.011	0.0002230	4.7	14	7
HF-038-200	1.016	0.427	0.447	1.081	0.661	2.3	1.7	0.8	1.5	0.092	0.143	0.013	0.0003170	5.0	200	85
HF-038-173	1.016	0.427	0.447	1.081	0.661	2.3	1.7	0.8	1.5	0.092	0.143	0.013	0.0003170	5.0	173	74
HF-038-160	1.016	0.427	0.447	1.081	0.661	2.3	1.7	0.8	1.5	0.092	0.143	0.013	0.0003170	5.0	160	68
HF-038-147	1.016	0.427	0.447	1.081	0.661	2.3	1.7	0.8	1.5	0.092	0.143	0.013	0.0003170	5.0	147	63
HF-038-125	1.016	0.427	0.447	1.081	0.661	2.3	1.7	0.8	1.5	0.092	0.143	0.013	0.0003170	5.0	125	53
HF-038-60	1.016	0.427	0.447	1.081	0.661	2.3	1.7	0.8	1.5	0.092	0.143	0.013	0.0003170	5.0	60	25
HF-038-26	1.016	0.427	0.447	1.081	0.661	2.3	1.7	0.8	1.5	0.092	0.143	0.013	0.0003170	5.0	26	11
HF-038-14	1.016	0.427	0.447	1.081	0.661	2.3	1.7	0.8	1.5	0.092	0.143	0.013	0.0003170	5.0	14	6
HF-040-200	1.067	0.457	0.447	1.138	0.676	2.4	1.8	0.9	1.6	0.096	0.164	0.016	0.0003850	5.5	200	105
HF-040-173	1.067	0.457	0.447	1.138	0.676	2.4	1.8	0.9	1.6	0.096	0.164	0.016	0.0003850	5.5	173	92
HF-040-160	1.067	0.457	0.447	1.138	0.676	2.4	1.8	0.9	1.6	0.096	0.164	0.016	0.0003850	5.5	160	84
HF-040-147	1.067	0.457	0.447	1.138	0.676	2.4	1.8	0.9	1.6	0.096	0.164	0.016	0.0003850	5.5	147	78
HF-040-125	1.067	0.457	0.447	1.138	0.676	2.4	1.8	0.9	1.6	0.096	0.164	0.016	0.0003850	5.5	125	66
HF-040-60	1.067	0.457	0.447	1.138	0.676	2.4	1.8	0.9	1.6	0.096	0.164	0.016	0.0003850	5.5	60	32
HF-040-26	1.067	0.457	0.447	1.138	0.676	2.4	1.8	0.9	1.6	0.096	0.164	0.016	0.0003850	5.5	26	14
HF-040-14	1.067	0.457	0.447	1.138	0.676	2.4	1.8	0.9	1.6	0.096	0.164	0.016	0.0003850	5.5	14	7
HF-044-200	1.169	0.584	0.447	1.274	0.739	2.8	2.0	1.6	1.7	0.091	0.268	0.024	0.0005370	6.8	200	85
HF-044-173	1.169	0.584	0.447	1.274	0.739	2.8	2.0	1.6	1.7	0.091	0.268	0.024	0.0005370	6.8	173	73

Arnold Engineering Co. High Flux Powder Cores																
Part No.	OD cm 1	ID cm 2	HT cm 3	Wound OD cm 4	Wound HT cm 5	MPL cm 6	Wtfe grams 7	Wtcu grams 8	MLT cm 9	Ac cm sq 10	Wa cm sq 11	Ap cm 4th 12	Kg cm 5th 13	At cm sq 14	Perm 15	AL 16
HF-044-160	1.169	0.584	0.447	1.274	0.739	2.8	2.0	1.6	1.7	0.091	0.268	0.024	0.0005370	6.8	160	68
HF-044-147	1.169	0.584	0.447	1.274	0.739	2.8	2.0	1.6	1.7	0.091	0.268	0.024	0.0005370	6.8	147	62
HF-044-125	1.169	0.584	0.447	1.274	0.739	2.8	2.0	1.6	1.7	0.091	0.268	0.024	0.0005370	6.8	125	53
HF-044-60	1.169	0.584	0.447	1.274	0.739	2.8	2.0	1.6	1.7	0.091	0.268	0.024	0.0005370	6.8	60	26
HF-044-26	1.169	0.584	0.447	1.274	0.739	2.8	2.0	1.6	1.7	0.091	0.268	0.024	0.0005370	6.8	26	11
HF-044-14	1.169	0.584	0.447	1.274	0.739	2.8	2.0	1.6	1.7	0.091	0.268	0.024	0.0005370	6.8	14	6
HF-050-200	1.321	0.711	0.526	1.457	0.882	3.2	2.9	2.7	1.9	0.113	0.397	0.045	0.0010680	9.0	200	90
HF-050-173	1.321	0.711	0.526	1.457	0.882	3.2	2.9	2.7	1.9	0.113	0.397	0.045	0.0010680	9.0	173	79
HF-050-160	1.321	0.711	0.526	1.457	0.882	3.2	2.9	2.7	1.9	0.113	0.397	0.045	0.0010680	9.0	160	72
HF-050-147	1.321	0.711	0.526	1.457	0.882	3.2	2.9	2.7	1.9	0.113	0.397	0.045	0.0010680	9.0	147	67
HF-050-125	1.321	0.711	0.526	1.457	0.882	3.2	2.9	2.7	1.9	0.113	0.397	0.045	0.0010680	9.0	125	56
HF-050-60	1.321	0.711	0.526	1.457	0.882	3.2	2.9	2.7	1.9	0.113	0.397	0.045	0.0010680	9.0	60	27
HF-050-26	1.321	0.711	0.526	1.457	0.882	3.2	2.9	2.7	1.9	0.113	0.397	0.045	0.0010680	9.0	26	12
HF-050-14	1.321	0.711	0.526	1.457	0.882	3.2	2.9	2.7	1.9	0.113	0.397	0.045	0.0010680	9.0	14	6
HF-065-200	1.715	0.965	0.686	1.908	1.169	4.2	6.6	6.4	2.5	0.196	0.731	0.143	0.0045490	15.6	200	115
HF-065-173	1.715	0.965	0.686	1.908	1.169	4.2	6.6	6.4	2.5	0.196	0.731	0.143	0.0045490	15.6	173	104
HF-065-160	1.715	0.965	0.686	1.908	1.169	4.2	6.6	6.4	2.5	0.196	0.731	0.143	0.0045490	15.6	160	92
HF-065-147	1.715	0.965	0.686	1.908	1.169	4.2	6.6	6.4	2.5	0.196	0.731	0.143	0.0045490	15.6	147	88
HF-065-125	1.715	0.965	0.686	1.908	1.169	4.2	6.6	6.4	2.5	0.196	0.731	0.143	0.0045490	15.6	125	72
HF-065-60	1.715	0.965	0.686	1.908	1.169	4.2	6.6	6.4	2.5	0.196	0.731	0.143	0.0045490	15.6	60	36
HF-065-26	1.715	0.965	0.686	1.908	1.169	4.2	6.6	6.4	2.5	0.196	0.731	0.143	0.0045490	15.6	26	15
HF-065-14	1.715	0.965	0.686	1.908	1.169	4.2	6.6	6.4	2.5	0.196	0.731	0.143	0.0045490	15.6	14	8
HF-068-200	1.778	0.914	0.686	1.946	1.143	4.2	7.8	5.9	2.5	0.232	0.656	0.152	0.0056030	15.9	200	142
HF-068-173	1.778	0.914	0.686	1.946	1.143	4.2	7.8	5.9	2.5	0.232	0.656	0.152	0.0056030	15.9	173	123
HF-068-160	1.778	0.914	0.686	1.946	1.143	4.2	7.8	5.9	2.5	0.232	0.656	0.152	0.0056030	15.9	160	114
HF-068-147	1.778	0.914	0.686	1.946	1.143	4.2	7.8	5.9	2.5	0.232	0.656	0.152	0.0056030	15.9	147	105
HF-068-125	1.778	0.914	0.686	1.946	1.143	4.2	7.8	5.9	2.5	0.232	0.656	0.152	0.0056030	15.9	125	89
HF-068-60	1.778	0.914	0.686	1.946	1.143	4.2	7.8	5.9	2.5	0.232	0.656	0.152	0.0056030	15.9	60	43
HF-068-26	1.778	0.914	0.686	1.946	1.143	4.2	7.8	5.9	2.5	0.232	0.656	0.152	0.0056030	15.9	26	19
HF-068-14	1.778	0.914	0.686	1.946	1.143	4.2	7.8	5.9	2.5	0.232	0.656	0.152	0.0056030	15.9	14	10

Arnold Engineering Co. High Flux Powder Cores																
Part No.	OD cm 1	ID cm 2	HT cm 3	Wound OD cm 4	Wound HT cm 5	MPL cm 6	Wtfe grams 7	Wtcu grams 8	MLT cm 9	Ac cm sq 10	Wa cm sq 11	Ap cm 4th 12	Kg cm 5th 13	At cm sq 14	Perm 15	AL 16
HF-080-200	2.081	1.219	0.686	2.333	1.296	5.2	9.6	11.5	2.8	0.232	1.166	0.271	0.0090910	22.2	200	109
HF-080-173	2.081	1.219	0.686	2.333	1.296	5.2	9.6	11.5	2.8	0.232	1.166	0.271	0.0090910	22.2	173	96
HF-080-160	2.081	1.219	0.686	2.333	1.296	5.2	9.6	11.5	2.8	0.232	1.166	0.271	0.0090910	22.2	160	87
HF-080-147	2.081	1.219	0.686	2.333	1.296	5.2	9.6	11.5	2.8	0.232	1.166	0.271	0.0090910	22.2	147	81
HF-080-125	2.081	1.219	0.686	2.333	1.296	5.2	9.6	11.5	2.8	0.232	1.166	0.271	0.0090910	22.2	125	68
HF-080-60	2.081	1.219	0.686	2.333	1.296	5.2	9.6	11.5	2.8	0.232	1.166	0.271	0.0090910	22.2	60	32
HF-080-26	2.081	1.219	0.686	2.333	1.296	5.2	9.6	11.5	2.8	0.232	1.166	0.271	0.0090910	22.2	26	14
HF-080-14	2.081	1.219	0.686	2.333	1.296	5.2	9.6	11.5	2.8	0.232	1.166	0.271	0.0090910	22.2	14	8
HF-090-200	2.341	1.346	0.813	2.615	1.486	5.8	15.1	16.0	3.2	0.327	1.422	0.465	0.0191670	28.2	200	144
HF-090-173	2.341	1.346	0.813	2.615	1.486	5.8	15.1	16.0	3.2	0.327	1.422	0.465	0.0191670	28.2	173	124
HF-090-160	2.341	1.346	0.813	2.615	1.486	5.8	15.1	16.0	3.2	0.327	1.422	0.465	0.0191670	28.2	160	115
HF-090-147	2.341	1.346	0.813	2.615	1.486	5.8	15.1	16.0	3.2	0.327	1.422	0.465	0.0191670	28.2	147	106
HF-090-125	2.341	1.346	0.813	2.615	1.486	5.8	15.1	16.0	3.2	0.327	1.422	0.465	0.0191670	28.2	125	90
HF-090-60	2.341	1.346	0.813	2.615	1.486	5.8	15.1	16.0	3.2	0.327	1.422	0.465	0.0191670	28.2	60	43
HF-090-26	2.341	1.346	0.813	2.615	1.486	5.8	15.1	16.0	3.2	0.327	1.422	0.465	0.0191670	28.2	26	19
HF-090-14	2.341	1.346	0.813	2.615	1.486	5.8	15.1	16.0	3.2	0.327	1.422	0.465	0.0191670	28.2	14	10
HF-092-200	2.411	1.389	0.940	2.695	1.635	6.0	18.1	18.5	3.4	0.380	1.515	0.576	0.0254830	31.0	200	168
HF-092-173	2.411	1.389	0.940	2.695	1.635	6.0	18.1	18.5	3.4	0.380	1.515	0.576	0.0254830	31.0	173	146
HF-092-160	2.411	1.389	0.940	2.695	1.635	6.0	18.1	18.5	3.4	0.380	1.515	0.576	0.0254830	31.0	160	135
HF-092-147	2.411	1.389	0.940	2.695	1.635	6.0	18.1	18.5	3.4	0.380	1.515	0.576	0.0254830	31.0	147	124
HF-092-125	2.411	1.389	0.940	2.695	1.635	6.0	18.1	18.5	3.4	0.380	1.515	0.576	0.0254830	31.0	125	105
HF-092-60	2.411	1.389	0.940	2.695	1.635	6.0	18.1	18.5	3.4	0.380	1.515	0.576	0.0254830	31.0	60	51
HF-092-26	2.411	1.389	0.940	2.695	1.635	6.0	18.1	18.5	3.4	0.380	1.515	0.576	0.0254830	31.0	26	22
HF-092-14	2.411	1.389	0.940	2.695	1.635	6.0	18.1	18.5	3.4	0.380	1.515	0.576	0.0254830	31.0	14	12
HF-106-200	2.741	1.422	1.169	3.005	1.880	6.5	33.4	22.9	4.1	0.639	1.587	1.014	0.0638060	39.1	200	251
HF-106-173	2.741	1.422	1.169	3.005	1.880	6.5	33.4	22.9	4.1	0.639	1.587	1.014	0.0638060	39.1	173	217
HF-106-160	2.741	1.422	1.169	3.005	1.880	6.5	33.4	22.9	4.1	0.639	1.587	1.014	0.0638060	39.1	160	201
HF-106-147	2.741	1.422	1.169	3.005	1.880	6.5	33.4	22.9	4.1	0.639	1.587	1.014	0.0638060	39.1	147	185
HF-106-125	2.741	1.422	1.169	3.005	1.880	6.5	33.4	22.9	4.1	0.639	1.587	1.014	0.0638060	39.1	125	157
HF-106-60	2.741	1.422	1.169	3.005	1.880	6.5	33.4	22.9	4.1	0.639	1.587	1.014	0.0638060	39.1	60	75

Arnold Engineering Co. High Flux Powder Cores

Part No.	OD cm 1	ID cm 2	HT cm 3	Wound OD cm 4	Wound HT cm 5	MPL cm 6	Wtfe grams 7	Wtcu grams 8	MLT cm 9	Ac cm sq 10	Wa cm sq 11	Ap cm 4th 12	Kg cm 5th 13	At cm sq 14	Perm 15	AL 16
HF-106-26	2.741	1.422	1.169	3.005	1.880	6.5	33.4	22.9	4.1	0.639	1.587	1.014	0.0638060	39.1	26	32
HF-106-14	2.741	1.422	1.169	3.005	1.880	6.5	33.4	22.9	4.1	0.639	1.587	1.014	0.0638060	39.1	14	18
HF-135-200	3.481	2.289	0.940	4.006	2.085	9.0	33.1	62.7	4.3	0.458	4.113	1.884	0.0804670	63.5	200	126
HF-135-173	3.481	2.289	0.940	4.006	2.085	9.0	33.1	62.7	4.3	0.458	4.113	1.884	0.0804670	63.5	173	109
HF-135-160	3.481	2.289	0.940	4.006	2.085	9.0	33.1	62.7	4.3	0.458	4.113	1.884	0.0804670	63.5	160	101
HF-135-147	3.481	2.289	0.940	4.006	2.085	9.0	33.1	62.7	4.3	0.458	4.113	1.884	0.0804670	63.5	147	93
HF-135-125	3.481	2.289	0.940	4.006	2.085	9.0	33.1	62.7	4.3	0.458	4.113	1.884	0.0804670	63.5	125	79
HF-135-60	3.481	2.289	0.940	4.006	2.085	9.0	33.1	62.7	4.3	0.458	4.113	1.884	0.0804670	63.5	60	38
HF-135-26	3.481	2.289	0.940	4.006	2.085	9.0	33.1	62.7	4.3	0.458	4.113	1.884	0.0804670	63.5	26	16
HF-135-14	3.481	2.289	0.940	4.006	2.085	9.0	33.1	62.7	4.3	0.458	4.113	1.884	0.0804670	63.5	14	9
HF-131-200	3.351	1.943	0.927	3.750	1.899	8.3	36.3	43.9	4.2	0.546	2.964	1.618	0.0848690	54.9	200	174
HF-131-173	3.351	1.943	0.927	3.750	1.899	8.3	36.3	43.9	4.2	0.546	2.964	1.618	0.0848690	54.9	173	151
HF-131-160	3.351	1.943	0.927	3.750	1.899	8.3	36.3	43.9	4.2	0.546	2.964	1.618	0.0848690	54.9	160	140
HF-131-147	3.351	1.943	0.927	3.750	1.899	8.3	36.3	43.9	4.2	0.546	2.964	1.618	0.0848690	54.9	147	129
HF-131-125	3.351	1.943	0.927	3.750	1.899	8.3	36.3	43.9	4.2	0.546	2.964	1.618	0.0848690	54.9	125	109
HF-131-60	3.351	1.943	0.927	3.750	1.899	8.3	36.3	43.9	4.2	0.546	2.964	1.618	0.0848690	54.9	60	52
HF-131-26	3.351	1.943	0.927	3.750	1.899	8.3	36.3	43.9	4.2	0.546	2.964	1.618	0.0848690	54.9	26	23
HF-131-14	3.351	1.943	0.927	3.750	1.899	8.3	36.3	43.9	4.2	0.546	2.964	1.618	0.0848690	54.9	14	12
HF-130-200	3.351	1.943	1.118	3.750	2.090	8.3	44.2	47.1	4.5	0.666	2.964	1.974	0.1176400	57.5	200	203
HF-130-173	3.351	1.943	1.118	3.750	2.090	8.3	44.2	47.1	4.5	0.666	2.964	1.974	0.1176400	57.5	173	176
HF-130-160	3.351	1.943	1.118	3.750	2.090	8.3	44.2	47.1	4.5	0.666	2.964	1.974	0.1176400	57.5	160	163
HF-130-147	3.351	1.943	1.118	3.750	2.090	8.3	44.2	47.1	4.5	0.666	2.964	1.974	0.1176400	57.5	147	150
HF-130-125	3.351	1.943	1.118	3.750	2.090	8.3	44.2	47.1	4.5	0.666	2.964	1.974	0.1176400	57.5	125	127
HF-130-60	3.351	1.943	1.118	3.750	2.090	8.3	44.2	47.1	4.5	0.666	2.964	1.974	0.1176400	57.5	60	61
HF-130-26	3.351	1.943	1.118	3.750	2.090	8.3	44.2	47.1	4.5	0.666	2.964	1.974	0.1176400	57.5	26	28
HF-130-14	3.351	1.943	1.118	3.750	2.090	8.3	44.2	47.1	4.5	0.666	2.964	1.974	0.1176400	57.5	14	14
HF-141-200	3.631	2.189	1.097	4.097	2.192	9.2	49.3	62.3	4.7	0.670	3.762	2.520	0.1449390	67.3	200	187
HF-141-173	3.631	2.189	1.097	4.097	2.192	9.2	49.3	62.3	4.7	0.670	3.762	2.520	0.1449390	67.3	173	162
HF-141-160	3.631	2.189	1.097	4.097	2.192	9.2	49.3	62.3	4.7	0.670	3.762	2.520	0.1449390	67.3	160	150
HF-141-147	3.631	2.189	1.097	4.097	2.192	9.2	49.3	62.3	4.7	0.670	3.762	2.520	0.1449390	67.3	147	138

Arnold Engineering Co. High Flux Powder Cores																
Part No.	OD cm 1	ID cm 2	HT cm 3	Wound OD cm 4	Wound HT cm 5	MPL cm 6	Wtfe grams 7	Wtcu grams 8	MLT cm 9	Ac cm sq 10	Wa cm sq 11	Ap cm 4th 12	Kg cm 5th 13	At cm sq 14	Perm 15	AL 16
HF-141-125	3.631	2.189	1.097	4.097	2.192	9.2	49.3	62.3	4.7	0.670	3.762	2.520	0.1449390	67.3	125	117
HF-141-60	3.631	2.189	1.097	4.097	2.192	9.2	49.3	62.3	4.7	0.670	3.762	2.520	0.1449390	67.3	60	56
HF-141-26	3.631	2.189	1.097	4.097	2.192	9.2	49.3	62.3	4.7	0.670	3.762	2.520	0.1449390	67.3	26	24
HF-141-14	3.631	2.189	1.097	4.097	2.192	9.2	49.3	62.3	4.7	0.670	3.762	2.520	0.1449390	67.3	14	13
HF-157-200	4.041	2.359	1.499	4.528	2.679	10.1	85.7	87.5	5.6	1.060	4.368	4.631	0.3486550	86.4	200	269
HF-157-173	4.041	2.359	1.499	4.528	2.679	10.1	85.7	87.5	5.6	1.060	4.368	4.631	0.3486550	86.4	173	233
HF-157-160	4.041	2.359	1.499	4.528	2.679	10.1	85.7	87.5	5.6	1.060	4.368	4.631	0.3486550	86.4	160	215
HF-157-147	4.041	2.359	1.499	4.528	2.679	10.1	85.7	87.5	5.6	1.060	4.368	4.631	0.3486550	86.4	147	198
HF-157-125	4.041	2.359	1.499	4.528	2.679	10.1	85.7	87.5	5.6	1.060	4.368	4.631	0.3486550	86.4	125	168
HF-157-60	4.041	2.359	1.499	4.528	2.679	10.1	85.7	87.5	5.6	1.060	4.368	4.631	0.3486550	86.4	60	81
HF-157-26	4.041	2.359	1.499	4.528	2.679	10.1	85.7	87.5	5.6	1.060	4.368	4.631	0.3486550	86.4	26	35
HF-157-14	4.041	2.359	1.499	4.528	2.679	10.1	85.7	87.5	5.6	1.060	4.368	4.631	0.3486550	86.4	14	19
HF-185-200	4.721	2.819	1.575	5.315	2.985	11.9	125.9	139.7	6.3	1.320	6.238	8.234	0.6904750	115.9	200	285
HF-185-173	4.721	2.819	1.575	5.315	2.985	11.9	125.9	139.7	6.3	1.320	6.238	8.234	0.6904750	115.9	173	246
HF-185-160	4.721	2.819	1.575	5.315	2.985	11.9	125.9	139.7	6.3	1.320	6.238	8.234	0.6904750	115.9	160	228
HF-185-147	4.721	2.819	1.575	5.315	2.985	11.9	125.9	139.7	6.3	1.320	6.238	8.234	0.6904750	115.9	147	210
HF-185-125	4.721	2.819	1.575	5.315	2.985	11.9	125.9	139.7	6.3	1.320	6.238	8.234	0.6904750	115.9	125	178
HF-185-60	4.721	2.819	1.575	5.315	2.985	11.9	125.9	139.7	6.3	1.320	6.238	8.234	0.6904750	115.9	60	86
HF-185-26	4.721	2.819	1.575	5.315	2.985	11.9	125.9	139.7	6.3	1.320	6.238	8.234	0.6904750	115.9	26	37
HF-185-14	4.721	2.819	1.575	5.315	2.985	11.9	125.9	139.7	6.3	1.320	6.238	8.234	0.6904750	115.9	14	20
HF-200-200	5.131	3.149	1.397	5.810	2.972	13.0	129.3	175.5	6.3	1.240	7.784	9.652	0.7551430	132.5	200	243
HF-200-173	5.131	3.149	1.397	5.810	2.972	13.0	129.3	175.5	6.3	1.240	7.784	9.652	0.7551430	132.5	173	210
HF-200-160	5.131	3.149	1.397	5.810	2.972	13.0	129.3	175.5	6.3	1.240	7.784	9.652	0.7551430	132.5	160	195
HF-200-147	5.131	3.149	1.397	5.810	2.972	13.0	129.3	175.5	6.3	1.240	7.784	9.652	0.7551430	132.5	147	179
HF-200-125	5.131	3.149	1.397	5.810	2.972	13.0	129.3	175.5	6.3	1.240	7.784	9.652	0.7551430	132.5	125	152
HF-200-60	5.131	3.149	1.397	5.810	2.972	13.0	129.3	175.5	6.3	1.240	7.784	9.652	0.7551430	132.5	60	73
HF-200-26	5.131	3.149	1.397	5.810	2.972	13.0	129.3	175.5	6.3	1.240	7.784	9.652	0.7551430	132.5	26	32
HF-200-14	5.131	3.149	1.397	5.810	2.972	13.0	129.3	175.5	6.3	1.240	7.784	9.652	0.7551430	132.5	14	17
HF-184-200	4.721	2.359	1.854	5.144	3.034	11.2	174.7	104.7	6.7	1.950	4.368	8.518	0.9853460	111.3	200	450
HF-184-173	4.721	2.359	1.854	5.144	3.034	11.2	174.7	104.7	6.7	1.950	4.368	8.518	0.9853460	111.3	173	390

Arnold Engineering Co. High Flux Powder Cores																
Part No.	OD cm 1	ID cm 2	HT cm 3	Wound OD cm 4	Wound HT cm 5	MPL cm 6	Wtfe grams 7	Wtcu grams 8	MLT cm 9	Ac cm sq 10	Wa cm sq 11	Ap cm 4th 12	Kg cm 5th 13	At cm sq 14	Perm 15	AL 16
HF-184-160	4.721	2.359	1.854	5.144	3.034	11.2	174.7	104.7	6.7	1.950	4.368	8.518	0.9853460	111.3	160	360
HF-184-147	4.721	2.359	1.854	5.144	3.034	11.2	174.7	104.7	6.7	1.950	4.368	8.518	0.9853460	111.3	147	330
HF-184-125	4.721	2.359	1.854	5.144	3.034	11.2	174.7	104.7	6.7	1.950	4.368	8.518	0.9853460	111.3	125	281
HF-184-60	4.721	2.359	1.854	5.144	3.034	11.2	174.7	104.7	6.7	1.950	4.368	8.518	0.9853460	111.3	60	135
HF-184-26	4.721	2.359	1.854	5.144	3.034	11.2	174.7	104.7	6.7	1.950	4.368	8.518	0.9853460	111.3	26	59
HF-184-14	4.721	2.359	1.854	5.144	3.034	11.2	174.7	104.7	6.7	1.950	4.368	8.518	0.9853460	111.3	14	32
HF-225-200	5.771	3.509	1.448	6.523	3.203	14.6	168.5	238.3	6.9	1.440	9.666	13.919	1.1562790	163.7	200	250
HF-225-173	5.771	3.509	1.448	6.523	3.203	14.6	168.5	238.3	6.9	1.440	9.666	13.919	1.1562790	163.7	173	218
HF-225-160	5.771	3.509	1.448	6.523	3.203	14.6	168.5	238.3	6.9	1.440	9.666	13.919	1.1562790	163.7	160	200
HF-225-147	5.771	3.509	1.448	6.523	3.203	14.6	168.5	238.3	6.9	1.440	9.666	13.919	1.1562790	163.7	147	185
HF-225-125	5.771	3.509	1.448	6.523	3.203	14.6	168.5	238.3	6.9	1.440	9.666	13.919	1.1562790	163.7	125	156
HF-225-60	5.771	3.509	1.448	6.523	3.203	14.6	168.5	238.3	6.9	1.440	9.666	13.919	1.1562790	163.7	60	75
HF-225-26	5.771	3.509	1.448	6.523	3.203	14.6	168.5	238.3	6.9	1.440	9.666	13.919	1.1562790	163.7	26	33
HF-225-14	5.771	3.509	1.448	6.523	3.203	14.6	168.5	238.3	6.9	1.440	9.666	13.919	1.1562790	163.7	14	18
HF-300-200	7.831	4.869	1.321	8.894	3.756	20.0	283.2	554.5	8.4	1.770	18.610	32.940	2.7835220	284.9	200	227
HF-300-173	7.831	4.869	1.321	8.894	3.756	20.0	283.2	554.5	8.4	1.770	18.610	32.940	2.7835220	284.9	173	197
HF-300-160	7.831	4.869	1.321	8.894	3.756	20.0	283.2	554.5	8.4	1.770	18.610	32.940	2.7835220	284.9	160	182
HF-300-147	7.831	4.869	1.321	8.894	3.756	20.0	283.2	554.5	8.4	1.770	18.610	32.940	2.7835220	284.9	147	167
HF-300-125	7.831	4.869	1.321	8.894	3.756	20.0	283.2	554.5	8.4	1.770	18.610	32.940	2.7835220	284.9	125	142
HF-300-60	7.831	4.869	1.321	8.894	3.756	20.0	283.2	554.5	8.4	1.770	18.610	32.940	2.7835220	284.9	60	68
HF-300-26	7.831	4.869	1.321	8.894	3.756	20.0	283.2	554.5	8.4	1.770	18.610	32.940	2.7835220	284.9	26	30
HF-300-14	7.831	4.869	1.321	8.894	3.756	20.0	283.2	554.5	8.4	1.770	18.610	32.940	2.7835220	284.9	14	16
HF-400-200	10.211	5.669	1.702	11.330	4.537	25.0	712.0	977.1	10.9	3.560	25.228	89.812	11.7418100	452.1	200	365
HF-400-173	10.211	5.669	1.702	11.330	4.537	25.0	712.0	977.1	10.9	3.560	25.228	89.812	11.7418100	452.1	173	316
HF-400-160	10.211	5.669	1.702	11.330	4.537	25.0	712.0	977.1	10.9	3.560	25.228	89.812	11.7418100	452.1	160	292
HF-400-147	10.211	5.669	1.702	11.330	4.537	25.0	712.0	977.1	10.9	3.560	25.228	89.812	11.7418100	452.1	147	268
HF-400-125	10.211	5.669	1.702	11.330	4.537	25.0	712.0	977.1	10.9	3.560	25.228	89.812	11.7418100	452.1	125	228
HF-400-60	10.211	5.669	1.702	11.330	4.537	25.0	712.0	977.1	10.9	3.560	25.228	89.812	11.7418100	452.1	60	112
HF-400-26	10.211	5.669	1.702	11.330	4.537	25.0	712.0	977.1	10.9	3.560	25.228	89.812	11.7418100	452.1	26	47
HF-400-14	10.211	5.669	1.702	11.330	4.537	25.0	712.0	977.1	10.9	3.560	25.228	89.812	11.7418100	452.1	14	26

Arnold Engineering Co. High Flux Powder Cores

Part No.	OD cm 1	ID cm 2	HT cm 3	Wound OD cm 4	Wound HT cm 5	MPL cm 6	Wtfe grams 7	Wtcu grams 8	MLT cm 9	Ac cm sq 10	Wa cm sq 11	Ap cm 4th 12	Kg cm 5th 13	At cm sq 14	Perm 15	AL 16
HF-520-200	13.301	7.809	2.081	14.922	5.986	33.2	1418.3	2378.1	14.0	5.340	47.870	255.624	39.0835700	784.8	200	414
HF-520-173	13.301	7.809	2.081	14.922	5.986	33.2	1418.3	2378.1	14.0	5.340	47.870	255.624	39.0835700	784.8	173	358
HF-520-160	13.301	7.809	2.081	14.922	5.986	33.2	1418.3	2378.1	14.0	5.340	47.870	255.624	39.0835700	784.8	160	332
HF-520-147	13.301	7.809	2.081	14.922	5.986	33.2	1418.3	2378.1	14.0	5.340	47.870	255.624	39.0835700	784.8	147	304
HF-520-125	13.301	7.809	2.081	14.922	5.986	33.2	1418.3	2378.1	14.0	5.340	47.870	255.624	39.0835700	784.8	125	259
HF-520-60	13.301	7.809	2.081	14.922	5.986	33.2	1418.3	2378.1	14.0	5.340	47.870	255.624	39.0835700	784.8	60	124
HF-520-26	13.301	7.809	2.081	14.922	5.986	33.2	1418.3	2378.1	14.0	5.340	47.870	255.624	39.0835700	784.8	26	54
HF-520-14	13.301	7.809	2.081	14.922	5.986	33.2	1418.3	2378.1	14.0	5.340	47.870	255.624	39.0835700	784.8	14	26
HF-521-200	13.301	7.809	2.591	14.922	6.496	33.2	1771.6	2517.0	14.8	6.670	47.870	319.291	57.6115800	812.3	200	414
HF-521-173	13.301	7.809	2.591	14.922	6.496	33.2	1771.6	2517.0	14.8	6.670	47.870	319.291	57.6115800	812.3	173	258
HF-521-160	13.301	7.809	2.591	14.922	6.496	33.2	1771.6	2517.0	14.8	6.670	47.870	319.291	57.6115800	812.3	160	332
HF-521-147	13.301	7.809	2.591	14.922	6.496	33.2	1771.6	2517.0	14.8	6.670	47.870	319.291	57.6115800	812.3	147	305
HF-521-125	13.301	7.809	2.591	14.922	6.496	33.2	1771.6	2517.0	14.8	6.670	47.870	319.291	57.6115800	812.3	125	259
HF-521-60	13.301	7.809	2.591	14.922	6.496	33.2	1771.6	2517.0	14.8	6.670	47.870	319.291	57.6115800	812.3	60	124
HF-521-26	13.301	7.809	2.591	14.922	6.496	33.2	1771.6	2517.0	14.8	6.670	47.870	319.291	57.6115800	812.3	26	54
HF-521-14	13.301	7.809	2.591	14.922	6.496	33.2	1771.6	2517.0	14.8	6.670	47.870	319.291	57.6115800	812.3	14	29

Arnold Engineering Co. MPP Powder Cores

Part No.	OD cm 1	ID cm 2	HT cm 3	Wound OD cm 4	Wound HT cm 5	MPL cm 6	Wtfe grams 7	Wtcu grams 8	MLT cm 9	Ac cm sq 10	Wa cm sq 11	Ap cm 4th 12	Kg cm 5th 13	At cm sq 14	Perm 15	AL 16
A-481033-1	0.419	0.114	0.216	0.430	0.273	0.8	0.1	0.0	0.7	0.0138	0.010	0.000	0.000001	0.8	160	33
A-479026-1	0.419	0.114	0.216	0.430	0.273	0.8	0.1	0.0	0.7	0.0139	0.010	0.000	0.000001	0.8	125	26
A-483052-1	0.419	0.114	0.216	0.430	0.273	0.8	0.1	0.0	0.7	0.0139	0.010	0.000	0.000001	0.8	250	52
A-482036-1	0.419	0.114	0.216	0.430	0.273	0.8	0.1	0.0	0.7	0.0139	0.010	0.000	0.000001	0.8	173	36
A-522043-1	0.419	0.114	0.216	0.430	0.273	0.8	0.1	0.0	0.7	0.0140	0.010	0.000	0.000001	0.8	205	43
A-480031-1	0.419	0.114	0.216	0.430	0.273	0.8	0.1	0.0	0.7	0.0141	0.010	0.000	0.000001	0.8	147	31
A-468007-1	0.457	0.157	0.318	0.477	0.396	1.0	0.2	0.1	0.9	0.0207	0.019	0.000	0.000004	1.2	26	7
A-473048-1	0.457	0.157	0.318	0.477	0.396	1.0	0.2	0.1	0.9	0.0213	0.019	0.000	0.000004	1.2	173	48
A-474057-1	0.457	0.157	0.318	0.477	0.396	1.0	0.2	0.1	0.9	0.0214	0.019	0.000	0.000004	1.2	205	57
A-471041-1	0.457	0.157	0.318	0.477	0.396	1.0	0.2	0.1	0.9	0.0214	0.019	0.000	0.000004	1.2	147	41
A-470035-1	0.457	0.157	0.318	0.477	0.396	1.0	0.2	0.1	0.9	0.0215	0.019	0.000	0.000004	1.2	125	35
A-475070-1	0.457	0.157	0.318	0.477	0.396	1.0	0.2	0.1	0.9	0.0215	0.019	0.000	0.000004	1.2	250	70
A-472045-1	0.457	0.157	0.318	0.477	0.396	1.0	0.2	0.1	0.9	0.0216	0.019	0.000	0.000004	1.2	160	45
A-469017-1	0.457	0.157	0.318	0.477	0.396	1.0	0.2	0.1	0.9	0.0218	0.019	0.000	0.000004	1.2	60	17
A-467004-1	0.457	0.157	0.318	0.477	0.396	1.0	0.2	0.1	0.9	0.0220	0.019	0.000	0.000004	1.2	14	4
A-356057-1	0.528	0.173	0.318	0.549	0.404	1.1	0.3	0.1	0.9	0.0289	0.023	0.001	0.000008	1.4	173	57
A-355053-1	0.528	0.173	0.318	0.549	0.404	1.1	0.3	0.1	0.9	0.0290	0.023	0.001	0.000008	1.4	160	53
A-357068-1	0.528	0.173	0.318	0.549	0.404	1.1	0.3	0.1	0.9	0.0291	0.023	0.001	0.000009	1.4	205	68
A-358083-1	0.528	0.173	0.318	0.549	0.404	1.1	0.3	0.1	0.9	0.0291	0.023	0.001	0.000009	1.4	250	83
A-352020-1	0.528	0.173	0.318	0.549	0.404	1.1	0.3	0.1	0.9	0.0292	0.023	0.001	0.000009	1.4	60	20
A-354049-1	0.528	0.173	0.318	0.549	0.404	1.1	0.3	0.1	0.9	0.0292	0.023	0.001	0.000009	1.4	147	49
A-353042-1	0.528	0.173	0.318	0.549	0.404	1.1	0.3	0.1	0.9	0.0294	0.023	0.001	0.000009	1.4	125	42
A-351009-1	0.528	0.173	0.318	0.549	0.404	1.1	0.3	0.1	0.9	0.0303	0.023	0.001	0.000009	1.4	26	9
A-350005-1	0.528	0.173	0.318	0.549	0.404	1.1	0.3	0.1	0.9	0.0313	0.023	0.001	0.000010	1.4	14	5
A-530010-2	0.699	0.216	0.343	0.723	0.451	1.4	0.5	0.1	1.1	0.0440	0.037	0.002	0.000026	2.3	26	10
A-528058-2	0.699	0.216	0.343	0.723	0.451	1.4	0.6	0.1	1.1	0.0451	0.037	0.002	0.000027	2.3	147	58
A-526069-2	0.699	0.216	0.343	0.723	0.451	1.4	0.6	0.1	1.1	0.0456	0.037	0.002	0.000027	2.3	173	69
A-529024-2	0.699	0.216	0.343	0.723	0.451	1.4	0.6	0.1	1.1	0.0457	0.037	0.002	0.000028	2.3	60	24
A-527064-2	0.699	0.216	0.343	0.723	0.451	1.4	0.6	0.1	1.1	0.0457	0.037	0.002	0.000028	2.3	160	64
A-512082-2	0.699	0.216	0.343	0.723	0.451	1.4	0.6	0.1	1.1	0.0457	0.037	0.002	0.000028	2.3	205	82

Arnold Engineering Co. MPP Powder Cores

Part No.	OD cm 1	ID cm 2	HT cm 3	Wound OD cm 4	Wound HT cm 5	MPL cm 6	Wtfe grams 7	Wtcu grams 8	MLT cm 9	Ac cm sq 10	Wa cm sq 11	Ap cm 4th 12	Kg cm 5th 13	At cm sq 14	Perm 15	AL 16
A-525100-2	0.699	0.216	0.343	0.723	0.451	1.4	0.6	0.1	1.1	0.0457	0.037	0.002	0.000028	2.3	250	100
A-521120-2	0.699	0.216	0.343	0.723	0.451	1.4	0.6	0.1	1.1	0.0457	0.037	0.002	0.000028	2.3	300	120
A-543140-2	0.699	0.216	0.343	0.723	0.451	1.4	0.6	0.1	1.1	0.0457	0.037	0.002	0.000028	2.3	350	140
A-461069-2	0.724	0.203	0.318	0.745	0.419	1.5	0.6	0.1	1.1	0.0500	0.032	0.002	0.000030	2.3	160	69
A-406151-2	0.724	0.203	0.318	0.745	0.419	1.5	0.6	0.1	1.1	0.0500	0.032	0.002	0.000030	2.3	350	151
A-520052-2	0.699	0.216	0.343	0.723	0.451	1.4	0.6	0.1	1.1	0.0475	0.037	0.002	0.000030	2.3	125	52
A-331054-2	0.724	0.203	0.318	0.745	0.419	1.5	0.6	0.1	1.1	0.0501	0.032	0.002	0.000030	2.3	125	54
A-362108-2	0.724	0.203	0.318	0.745	0.419	1.5	0.6	0.1	1.1	0.0501	0.032	0.002	0.000030	2.3	250	108
A-460026-2	0.724	0.203	0.318	0.745	0.419	1.5	0.6	0.1	1.1	0.0502	0.032	0.002	0.000030	2.3	60	26
A-384130-2	0.724	0.203	0.318	0.745	0.419	1.5	0.6	0.1	1.1	0.0502	0.032	0.002	0.000030	2.3	300	130
A-465075-2	0.724	0.203	0.318	0.745	0.419	1.5	0.6	0.1	1.1	0.0502	0.032	0.002	0.000030	2.3	173	75
A-462089-2	0.724	0.203	0.318	0.745	0.419	1.5	0.6	0.1	1.1	0.0503	0.032	0.002	0.000030	2.3	205	89
A-464064-2	0.724	0.203	0.318	0.745	0.419	1.5	0.6	0.1	1.1	0.0505	0.032	0.002	0.000030	2.3	147	64
A-531006-2	0.699	0.216	0.343	0.723	0.451	1.4	0.6	0.1	1.1	0.0490	0.037	0.002	0.000032	2.3	14	6
A-385247-2	0.724	0.203	0.541	0.745	0.643	1.5	1.2	0.2	1.4	0.0954	0.032	0.003	0.000082	2.9	300	247
W-134103-4	0.724	0.203	0.541	0.745	0.643	1.5	1.2	0.2	1.4	0.0955	0.032	0.003	0.000082	2.9	125	103
D-134103-4	0.724	0.203	0.541	0.745	0.643	1.5	1.2	0.2	1.4	0.0955	0.032	0.003	0.000082	2.9	125	103
A-134103-2	0.724	0.203	0.541	0.745	0.643	1.5	1.2	0.2	1.4	0.0955	0.032	0.003	0.000082	2.9	125	103
A-363206-2	0.724	0.203	0.541	0.745	0.643	1.5	1.2	0.2	1.4	0.0955	0.032	0.003	0.000082	2.9	250	206
W-200170-4	0.724	0.203	0.541	0.745	0.643	1.5	1.2	0.2	1.4	0.0961	0.032	0.003	0.000083	2.9	205	170
D-200170-4	0.724	0.203	0.541	0.745	0.643	1.5	1.2	0.2	1.4	0.0961	0.032	0.003	0.000083	2.9	205	170
A-200170-2	0.724	0.203	0.541	0.745	0.643	1.5	1.2	0.2	1.4	0.0961	0.032	0.003	0.000083	2.9	205	170
W-224122-4	0.724	0.203	0.541	0.745	0.643	1.5	1.2	0.2	1.4	0.0962	0.032	0.003	0.000083	2.9	147	122
D-224122-4	0.724	0.203	0.541	0.745	0.643	1.5	1.2	0.2	1.4	0.0962	0.032	0.003	0.000083	2.9	147	122
A-224122-2	0.724	0.203	0.541	0.745	0.643	1.5	1.2	0.2	1.4	0.0962	0.032	0.003	0.000083	2.9	147	122
W-222144-4	0.724	0.203	0.541	0.745	0.643	1.5	1.2	0.2	1.4	0.0965	0.032	0.003	0.000084	2.9	173	144
D-222144-4	0.724	0.203	0.541	0.745	0.643	1.5	1.2	0.2	1.4	0.0965	0.032	0.003	0.000084	2.9	173	144
A-222144-2	0.724	0.203	0.541	0.745	0.643	1.5	1.2	0.2	1.4	0.0965	0.032	0.003	0.000084	2.9	173	144
W-135050-4	0.724	0.203	0.541	0.745	0.643	1.5	1.2	0.2	1.4	0.0966	0.032	0.003	0.000084	2.9	60	50
D-135050-4	0.724	0.203	0.541	0.745	0.643	1.5	1.2	0.2	1.4	0.0966	0.032	0.003	0.000084	2.9	60	50

Arnold Engineering Co. MPP Powder Cores																
Part No.	OD cm 1	ID cm 2	HT cm 3	Wound OD cm 4	Wound HT cm 5	MPL cm 6	Wtfe grams 7	Wtcu grams 8	MLT cm 9	Ac cm sq 10	Wa cm sq 11	Ap cm 4th 12	Kg cm 5th 13	At cm sq 14	Perm 15	AL 16
A-135050-2	0.724	0.203	0.541	0.745	0.643	1.5	1.2	0.2	1.4	0.0966	0.032	0.003	0.000084	2.9	60	50
A-337296-1	0.724	0.203	0.541	0.745	0.643	1.5	1.2	0.2	1.4	0.0980	0.032	0.003	0.000086	2.9	350	296
W-338066-4	0.851	0.333	0.381	0.898	0.547	1.9	1.0	0.4	1.3	0.0610	0.087	0.005	0.000100	3.5	160	66
D-338066-4	0.851	0.333	0.381	0.898	0.547	1.9	1.0	0.4	1.3	0.0610	0.087	0.005	0.000100	3.5	160	66
A-338066-2	0.851	0.333	0.381	0.898	0.547	1.9	1.0	0.4	1.3	0.0610	0.087	0.005	0.000100	3.5	160	66
A-386124-2	0.851	0.333	0.381	0.898	0.547	1.9	1.0	0.4	1.3	0.0612	0.087	0.005	0.000101	3.5	300	124
A-407145-2	0.851	0.333	0.381	0.898	0.547	1.9	1.0	0.4	1.3	0.0613	0.087	0.005	0.000101	3.5	350	145
W-137052-4	0.851	0.333	0.381	0.898	0.547	1.9	1.0	0.4	1.3	0.0615	0.087	0.005	0.000102	3.5	125	52
D-137052-4	0.851	0.333	0.381	0.898	0.547	1.9	1.0	0.4	1.3	0.0615	0.087	0.005	0.000102	3.5	125	52
A-137052-2	0.851	0.333	0.381	0.898	0.547	1.9	1.0	0.4	1.3	0.0615	0.087	0.005	0.000102	3.5	125	52
A-364104-2	0.851	0.333	0.381	0.898	0.547	1.9	1.0	0.4	1.3	0.0615	0.087	0.005	0.000102	3.5	250	104
W-138025-4	0.851	0.333	0.381	0.898	0.547	1.9	1.0	0.4	1.3	0.0616	0.087	0.005	0.000102	3.5	60	25
D-138025-4	0.851	0.333	0.381	0.898	0.547	1.9	1.0	0.4	1.3	0.0616	0.087	0.005	0.000102	3.5	60	25
A-138025-2	0.851	0.333	0.381	0.898	0.547	1.9	1.0	0.4	1.3	0.0616	0.087	0.005	0.000102	3.5	60	25
W-201086-4	0.851	0.333	0.381	0.898	0.547	1.9	1.0	0.4	1.3	0.0621	0.087	0.005	0.000104	3.5	205	86
D-201086-4	0.851	0.333	0.381	0.898	0.547	1.9	1.0	0.4	1.3	0.0621	0.087	0.005	0.000104	3.5	205	86
A-201086-2	0.851	0.333	0.381	0.898	0.547	1.9	1.0	0.4	1.3	0.0621	0.087	0.005	0.000104	3.5	205	86
W-225062-4	0.851	0.333	0.381	0.898	0.547	1.9	1.0	0.4	1.3	0.0624	0.087	0.005	0.000105	3.5	147	62
D-225062-4	0.851	0.333	0.381	0.898	0.547	1.9	1.0	0.4	1.3	0.0624	0.087	0.005	0.000105	3.5	147	62
A-225062-2	0.851	0.333	0.381	0.898	0.547	1.9	1.0	0.4	1.3	0.0624	0.087	0.005	0.000105	3.5	147	62
W-223073-4	0.851	0.333	0.381	0.898	0.547	1.9	1.0	0.4	1.3	0.0624	0.087	0.005	0.000105	3.5	173	73
D-223073-4	0.851	0.333	0.381	0.898	0.547	1.9	1.0	0.4	1.3	0.0624	0.087	0.005	0.000105	3.5	173	73
A-223073-2	0.851	0.333	0.381	0.898	0.547	1.9	1.0	0.4	1.3	0.0624	0.087	0.005	0.000105	3.5	173	73
A-339011-2	0.851	0.333	0.381	0.898	0.547	1.9	1.0	0.4	1.3	0.0626	0.087	0.005	0.000106	3.5	26	11
A-340006-2	0.851	0.333	0.381	0.898	0.547	1.9	1.0	0.4	1.3	0.0634	0.087	0.006	0.000108	3.5	14	6
A-540014-2	0.767	0.333	0.572	0.819	0.738	1.7	1.1	0.5	1.5	0.0740	0.087	0.006	0.000125	3.6	26	14
A-535095-2	0.767	0.333	0.572	0.819	0.738	1.7	1.1	0.5	1.5	0.0755	0.087	0.007	0.000130	3.6	173	95
A-539033-2	0.767	0.333	0.572	0.819	0.738	1.7	1.1	0.5	1.5	0.0756	0.087	0.007	0.000130	3.6	60	33
A-537081-2	0.767	0.333	0.572	0.819	0.738	1.7	1.1	0.5	1.5	0.0758	0.087	0.007	0.000131	3.6	147	81
A-453113-2	0.767	0.333	0.572	0.819	0.738	1.7	1.1	0.5	1.5	0.0758	0.087	0.007	0.000131	3.6	205	113

Arnold Engineering Co. MPP Powder Cores																
Part No.	OD cm 1	ID cm 2	HT cm 3	Wound OD cm 4	Wound HT cm 5	MPL cm 6	Wtfe grams 7	Wtcu grams 8	MLT cm 9	Ac cm sq 10	Wa cm sq 11	Ap cm 4th 12	Kg cm 5th 13	At cm sq 14	Perm 15	AL 16
A-534138-2	0.767	0.333	0.572	0.819	0.738	1.7	1.1	0.5	1.5	0.0759	0.087	0.007	0.000131	3.6	250	138
A-533166-2	0.767	0.333	0.572	0.819	0.738	1.7	1.1	0.5	1.5	0.0761	0.087	0.007	0.000132	3.6	300	166
A-532194-2	0.767	0.333	0.572	0.819	0.738	1.7	1.1	0.5	1.5	0.0762	0.087	0.007	0.000132	3.6	350	194
A-536089-2	0.767	0.333	0.572	0.819	0.738	1.7	1.1	0.5	1.5	0.0765	0.087	0.007	0.000133	3.6	160	89
A-538070-2	0.767	0.333	0.572	0.819	0.738	1.7	1.1	0.5	1.5	0.0770	0.087	0.007	0.000135	3.6	125	70
A-541008-2	0.767	0.333	0.572	0.819	0.738	1.7	1.2	0.5	1.5	0.0786	0.087	0.007	0.000140	3.6	14	8
W-496084-4	1.029	0.414	0.381	1.090	0.588	2.3	1.4	0.7	1.4	0.0739	0.135	0.010	0.000205	4.8	205	84
D-496084-4	1.029	0.414	0.381	1.090	0.588	2.3	1.4	0.7	1.4	0.0739	0.135	0.010	0.000205	4.8	205	84
A-496084-2	1.029	0.414	0.381	1.090	0.588	2.3	1.4	0.7	1.4	0.0739	0.135	0.010	0.000205	4.8	205	84
W-500025-4	1.029	0.414	0.381	1.090	0.588	2.3	1.4	0.7	1.4	0.0751	0.135	0.010	0.000212	4.8	60	25
D-500025-4	1.029	0.414	0.381	1.090	0.588	2.3	1.4	0.7	1.4	0.0751	0.135	0.010	0.000212	4.8	60	25
A-500025-2	1.029	0.414	0.381	1.090	0.588	2.3	1.4	0.7	1.4	0.0751	0.135	0.010	0.000212	4.8	60	25
A-501011-2	1.029	0.414	0.381	1.090	0.588	2.3	1.5	0.7	1.4	0.0763	0.135	0.010	0.000219	4.8	26	11
W-250053-4	1.029	0.414	0.381	1.090	0.588	2.3	1.5	0.7	1.4	0.0765	0.135	0.010	0.000220	4.8	125	53
D-250053-4	1.029	0.414	0.381	1.090	0.588	2.3	1.5	0.7	1.4	0.0765	0.135	0.010	0.000220	4.8	125	53
A-250053-2	1.029	0.414	0.381	1.090	0.588	2.3	1.5	0.7	1.4	0.0765	0.135	0.010	0.000220	4.8	125	53
A-495106-2	1.029	0.414	0.381	1.090	0.588	2.3	1.5	0.7	1.4	0.0765	0.135	0.010	0.000220	4.8	250	106
W-498068-4	1.029	0.414	0.381	1.090	0.588	2.3	1.5	0.7	1.4	0.0766	0.135	0.010	0.000221	4.8	160	68
D-498068-4	1.029	0.414	0.381	1.090	0.588	2.3	1.5	0.7	1.4	0.0766	0.135	0.010	0.000221	4.8	160	68
A-498068-2	1.029	0.414	0.381	1.090	0.588	2.3	1.5	0.7	1.4	0.0766	0.135	0.010	0.000221	4.8	160	68
A-493149-2	1.029	0.414	0.381	1.090	0.588	2.3	1.5	0.7	1.4	0.0768	0.135	0.010	0.000221	4.8	350	149
A-494128-2	1.029	0.414	0.381	1.090	0.588	2.3	1.5	0.7	1.4	0.0769	0.135	0.010	0.000222	4.8	300	128
W-497074-4	1.029	0.414	0.381	1.090	0.588	2.3	1.5	0.7	1.4	0.0771	0.135	0.010	0.000224	4.8	173	74
D-497074-4	1.029	0.414	0.381	1.090	0.588	2.3	1.5	0.7	1.4	0.0771	0.135	0.010	0.000224	4.8	173	74
A-497074-2	1.029	0.414	0.381	1.090	0.588	2.3	1.5	0.7	1.4	0.0771	0.135	0.010	0.000224	4.8	173	74
A-502006-2	1.029	0.414	0.381	1.090	0.588	2.3	1.5	0.7	1.4	0.0773	0.135	0.010	0.000224	4.8	14	6
W-499063-4	1.029	0.414	0.381	1.090	0.588	2.3	1.5	0.7	1.4	0.0773	0.135	0.010	0.000224	4.8	147	63
D-499063-4	1.029	0.414	0.381	1.090	0.588	2.3	1.5	0.7	1.4	0.0773	0.135	0.010	0.000224	4.8	147	63
A-499063-2	1.029	0.414	0.381	1.090	0.588	2.3	1.5	0.7	1.4	0.0773	0.135	0.010	0.000224	4.8	147	63
A-249007-2	1.029	0.414	0.460	1.090	0.667	2.3	1.7	0.7	1.6	0.0902	0.135	0.012	0.000281	5.1	14	7

Arnold Engineering Co. MPP Powder Cores																
Part No.	OD cm 1	ID cm 2	HT cm 3	Wound OD cm 4	Wound HT cm 5	MPL cm 6	Wtfe grams 7	Wtcu grams 8	MLT cm 9	Ac cm sq 10	Wa cm sq 11	Ap cm 4th 12	Kg cm 5th 13	At cm sq 14	Perm 15	AL 16
W-240084-4	1.029	0.414	0.460	1.090	0.667	2.3	1.8	0.7	1.6	0.0947	0.135	0.013	0.000310	5.1	160	84
D-240084-4	1.029	0.414	0.460	1.090	0.667	2.3	1.8	0.7	1.6	0.0947	0.135	0.013	0.000310	5.1	160	84
A-240084-2	1.029	0.414	0.460	1.090	0.667	2.3	1.8	0.7	1.6	0.0947	0.135	0.013	0.000310	5.1	160	84
W-246066-4	1.029	0.414	0.460	1.090	0.667	2.3	1.8	0.7	1.6	0.0952	0.135	0.013	0.000313	5.1	125	66
D-246066-4	1.029	0.414	0.460	1.090	0.667	2.3	1.8	0.7	1.6	0.0952	0.135	0.013	0.000313	5.1	125	66
A-246066-2	1.029	0.414	0.460	1.090	0.667	2.3	1.8	0.7	1.6	0.0952	0.135	0.013	0.000313	5.1	125	66
A-365132-2	1.029	0.414	0.460	1.090	0.667	2.3	1.8	0.7	1.6	0.0952	0.135	0.013	0.000313	5.1	250	132
A-408185-2	1.029	0.414	0.460	1.090	0.667	2.3	1.8	0.7	1.6	0.0953	0.135	0.013	0.000314	5.1	350	185
A-387159-2	1.029	0.414	0.460	1.090	0.667	2.3	1.8	0.7	1.6	0.0956	0.135	0.013	0.000315	5.1	300	159
W-245078-4	1.029	0.414	0.460	1.090	0.667	2.3	1.8	0.7	1.6	0.0957	0.135	0.013	0.000316	5.1	147	78
D-245078-4	1.029	0.414	0.460	1.090	0.667	2.3	1.8	0.7	1.6	0.0957	0.135	0.013	0.000316	5.1	147	78
A-245078-2	1.029	0.414	0.460	1.090	0.667	2.3	1.8	0.7	1.6	0.0957	0.135	0.013	0.000316	5.1	147	78
W-202109-4	1.029	0.414	0.460	1.090	0.667	2.3	1.8	0.7	1.6	0.0959	0.135	0.013	0.000318	5.1	205	109
D-202109-4	1.029	0.414	0.460	1.090	0.667	2.3	1.8	0.7	1.6	0.0959	0.135	0.013	0.000318	5.1	205	109
A-202109-2	1.029	0.414	0.460	1.090	0.667	2.3	1.8	0.7	1.6	0.0959	0.135	0.013	0.000318	5.1	205	109
W-244092-4	1.029	0.414	0.460	1.090	0.667	2.3	1.8	0.7	1.6	0.0959	0.135	0.013	0.000318	5.1	173	92
D-244092-4	1.029	0.414	0.460	1.090	0.667	2.3	1.8	0.7	1.6	0.0959	0.135	0.013	0.000318	5.1	173	92
A-244092-2	1.029	0.414	0.460	1.090	0.667	2.3	1.8	0.7	1.6	0.0959	0.135	0.013	0.000318	5.1	173	92
W-247032-4	1.029	0.414	0.460	1.090	0.667	2.3	1.9	0.7	1.6	0.0962	0.135	0.013	0.000319	5.1	60	32
D-247032-4	1.029	0.414	0.460	1.090	0.667	2.3	1.9	0.7	1.6	0.0962	0.135	0.013	0.000319	5.1	60	32
A-247032-2	1.029	0.414	0.460	1.090	0.667	2.3	1.9	0.7	1.6	0.0962	0.135	0.013	0.000319	5.1	60	32
A-248014-2	1.029	0.414	0.460	1.090	0.667	2.3	1.9	0.7	1.6	0.0971	0.135	0.013	0.000326	5.1	26	14
A-342007-2	1.080	0.445	0.460	1.146	0.682	2.4	1.9	0.9	1.6	0.0953	0.155	0.015	0.000352	5.5	14	7
W-308084-4	1.080	0.445	0.460	1.146	0.682	2.4	2.0	0.9	1.6	0.1000	0.155	0.016	0.000388	5.5	160	84
D-308084-4	1.080	0.445	0.460	1.146	0.682	2.4	2.0	0.9	1.6	0.1000	0.155	0.016	0.000388	5.5	160	84
A-308084-2	1.080	0.445	0.460	1.146	0.682	2.4	2.0	0.9	1.6	0.1000	0.155	0.016	0.000388	5.5	160	84
W-309105-4	1.080	0.445	0.460	1.146	0.682	2.4	2.0	0.9	1.6	0.1000	0.155	0.016	0.000388	5.5	200	105
D-309105-4	1.080	0.445	0.460	1.146	0.682	2.4	2.0	0.9	1.6	0.1000	0.155	0.016	0.000388	5.5	200	105
A-309105-2	1.080	0.445	0.460	1.146	0.682	2.4	2.0	0.9	1.6	0.1000	0.155	0.016	0.000388	5.5	200	105
W-292066-4	1.080	0.445	0.460	1.146	0.682	2.4	2.0	0.9	1.6	0.1006	0.155	0.016	0.000392	5.5	125	66

Arnold Engineering Co. MPP Powder Cores

Part No.	OD cm 1	ID cm 2	HT cm 3	Wound OD cm 4	Wound HT cm 5	MPL cm 6	Wtfe grams 7	Wtcu grams 8	MLT cm 9	Ac cm sq 10	Wa cm sq 11	Ap cm 4th 12	Kg cm 5th 13	At cm sq 14	Perm 15	AL 16
D-292066-4	1.080	0.445	0.460	1.146	0.682	2.4	2.0	0.9	1.6	0.1006	0.155	0.016	0.000392	5.5	125	66
A-292066-2	1.080	0.445	0.460	1.146	0.682	2.4	2.0	0.9	1.6	0.1006	0.155	0.016	0.000392	5.5	125	66
A-366132-2	1.080	0.445	0.460	1.146	0.682	2.4	2.0	0.9	1.6	0.1006	0.155	0.016	0.000392	5.5	250	132
A-409185-2	1.080	0.445	0.460	1.146	0.682	2.4	2.0	0.9	1.6	0.1007	0.155	0.016	0.000393	5.5	350	185
A-388159-2	1.080	0.445	0.460	1.146	0.682	2.4	2.1	0.9	1.6	0.1010	0.155	0.016	0.000395	5.5	300	159
W-239078-4	1.080	0.445	0.460	1.146	0.682	2.4	2.1	0.9	1.6	0.1011	0.155	0.016	0.000396	5.5	147	78
D-239078-4	1.080	0.445	0.460	1.146	0.682	2.4	2.1	0.9	1.6	0.1011	0.155	0.016	0.000396	5.5	147	78
A-239078-2	1.080	0.445	0.460	1.146	0.682	2.4	2.1	0.9	1.6	0.1011	0.155	0.016	0.000396	5.5	147	78
W-238092-4	1.080	0.445	0.460	1.146	0.682	2.4	2.1	0.9	1.6	0.1013	0.155	0.016	0.000398	5.5	173	92
D-238092-4	1.080	0.445	0.460	1.146	0.682	2.4	2.1	0.9	1.6	0.1013	0.155	0.016	0.000398	5.5	173	92
A-238092-2	1.080	0.445	0.460	1.146	0.682	2.4	2.1	0.9	1.6	0.1013	0.155	0.016	0.000398	5.5	173	92
W-307032-4	1.080	0.445	0.460	1.146	0.682	2.4	2.1	0.9	1.6	0.1016	0.155	0.016	0.000400	5.5	60	32
D-307032-4	1.080	0.445	0.460	1.146	0.682	2.4	2.1	0.9	1.6	0.1016	0.155	0.016	0.000400	5.5	60	32
A-307032-2	1.080	0.445	0.460	1.146	0.682	2.4	2.1	0.9	1.6	0.1016	0.155	0.016	0.000400	5.5	60	32
A-341014-2	1.080	0.445	0.460	1.146	0.682	2.4	2.1	0.9	1.6	0.1026	0.155	0.016	0.000408	5.5	26	14
A-410148-2	1.181	0.572	0.460	1.281	0.745	2.8	2.2	1.5	1.7	0.0926	0.256	0.024	0.000524	6.9	350	148
A-256011-2	1.181	0.572	0.460	1.281	0.745	2.8	2.2	1.5	1.7	0.0927	0.256	0.024	0.000524	6.9	26	11
A-389127-2	1.181	0.572	0.460	1.281	0.745	2.8	2.2	1.5	1.7	0.0927	0.256	0.024	0.000525	6.9	300	127
W-253053-4	1.181	0.572	0.460	1.281	0.745	2.8	2.2	1.5	1.7	0.0929	0.256	0.024	0.000527	6.9	125	53
D-253053-4	1.181	0.572	0.460	1.281	0.745	2.8	2.2	1.5	1.7	0.0929	0.256	0.024	0.000527	6.9	125	53
A-253053-2	1.181	0.572	0.460	1.281	0.745	2.8	2.2	1.5	1.7	0.0929	0.256	0.024	0.000527	6.9	125	53
A-367106-2	1.181	0.572	0.460	1.281	0.745	2.8	2.2	1.5	1.7	0.0929	0.256	0.024	0.000527	6.9	250	106
W-251074-4	1.181	0.572	0.460	1.281	0.745	2.8	2.2	1.5	1.7	0.0937	0.256	0.024	0.000536	6.9	173	74
D-251074-4	1.181	0.572	0.460	1.281	0.745	2.8	2.2	1.5	1.7	0.0937	0.256	0.024	0.000536	6.9	173	74
A-251074-2	1.181	0.572	0.460	1.281	0.745	2.8	2.2	1.5	1.7	0.0937	0.256	0.024	0.000536	6.9	173	74
A-257006-2	1.181	0.572	0.460	1.281	0.745	2.8	2.2	1.5	1.7	0.0939	0.256	0.024	0.000538	6.9	14	6
W-252063-4	1.181	0.572	0.460	1.281	0.745	2.8	2.2	1.5	1.7	0.0939	0.256	0.024	0.000538	6.9	147	63
D-252063-4	1.181	0.572	0.460	1.281	0.745	2.8	2.2	1.5	1.7	0.0939	0.256	0.024	0.000538	6.9	147	63
A-252063-2	1.181	0.572	0.460	1.281	0.745	2.8	2.2	1.5	1.7	0.0939	0.256	0.024	0.000538	6.9	147	63
W-203088-4	1.181	0.572	0.460	1.281	0.745	2.8	2.2	1.5	1.7	0.0940	0.256	0.024	0.000540	6.9	205	88

Arnold Engineering Co. MPP Powder Cores

Part No.	OD cm 1	ID cm 2	HT cm 3	Wound OD cm 4	Wound HT cm 5	MPL cm 6	Wtfe grams 7	Wtcu grams 8	MLT cm 9	Ac cm sq 10	Wa cm sq 11	Ap cm 4th 12	Kg cm 5th 13	At cm sq 14	Perm 15	AL 16
D-203088-4	1.181	0.572	0.460	1.281	0.745	2.8	2.2	1.5	1.7	0.0940	0.256	0.024	0.000540	6.9	205	88
A-203088-2	1.181	0.572	0.460	1.281	0.745	2.8	2.2	1.5	1.7	0.0940	0.256	0.024	0.000540	6.9	205	88
W-255026-4	1.181	0.572	0.460	1.281	0.745	2.8	2.2	1.5	1.7	0.0949	0.256	0.024	0.000550	6.9	60	26
D-255026-4	1.181	0.572	0.460	1.281	0.745	2.8	2.2	1.5	1.7	0.0949	0.256	0.024	0.000550	6.9	60	26
A-255026-2	1.181	0.572	0.460	1.281	0.745	2.8	2.2	1.5	1.7	0.0949	0.256	0.024	0.000550	6.9	60	26
A-053006-2	1.334	0.699	0.538	1.464	0.888	3.2	3.0	2.6	1.9	0.1089	0.383	0.042	0.000941	9.1	14	6
A-390134-2	1.334	0.699	0.538	1.464	0.888	3.2	3.1	2.6	1.9	0.1135	0.383	0.043	0.001023	9.1	300	134
W-290056-4	1.334	0.699	0.538	1.464	0.888	3.2	3.1	2.6	1.9	0.1138	0.383	0.044	0.001029	9.1	125	56
D-313056-4	1.334	0.699	0.538	1.464	0.888	3.2	3.1	2.6	1.9	0.1138	0.383	0.044	0.001029	9.1	125	56
A-050056-2	1.334	0.699	0.538	1.464	0.888	3.2	3.1	2.6	1.9	0.1138	0.383	0.044	0.001029	9.1	125	56
A-368112-2	1.334	0.699	0.538	1.464	0.888	3.2	3.1	2.6	1.9	0.1138	0.383	0.044	0.001029	9.1	250	112
A-412157-2	1.334	0.699	0.538	1.464	0.888	3.2	3.1	2.6	1.9	0.1139	0.383	0.044	0.001031	9.1	350	157
W-311027-4	1.334	0.699	0.538	1.464	0.888	3.2	3.1	2.6	1.9	0.1143	0.383	0.044	0.001038	9.1	60	27
D-311027-4	1.334	0.699	0.538	1.464	0.888	3.2	3.1	2.6	1.9	0.1143	0.383	0.044	0.001038	9.1	60	27
A-051027-2	1.334	0.699	0.538	1.464	0.888	3.2	3.1	2.6	1.9	0.1143	0.383	0.044	0.001038	9.1	60	27
W-261045-4	1.334	0.699	0.538	1.464	0.888	3.2	3.1	2.6	1.9	0.1143	0.383	0.044	0.001038	9.1	100	45
D-261045-4	1.334	0.699	0.538	1.464	0.888	3.2	3.1	2.6	1.9	0.1143	0.383	0.044	0.001038	9.1	100	45
A-261045-2	1.334	0.699	0.538	1.464	0.888	3.2	3.1	2.6	1.9	0.1143	0.383	0.044	0.001038	9.1	100	45
W-294072-4	1.334	0.699	0.538	1.464	0.888	3.2	3.1	2.6	1.9	0.1143	0.383	0.044	0.001038	9.1	160	72
D-301072-4	1.334	0.699	0.538	1.464	0.888	3.2	3.1	2.6	1.9	0.1143	0.383	0.044	0.001038	9.1	160	72
A-301072-2	1.334	0.699	0.538	1.464	0.888	3.2	3.1	2.6	1.9	0.1143	0.383	0.044	0.001038	9.1	160	72
W-295090-4	1.334	0.699	0.538	1.464	0.888	3.2	3.1	2.6	1.9	0.1143	0.383	0.044	0.001038	9.1	200	90
D-315090-4	1.334	0.699	0.538	1.464	0.888	3.2	3.1	2.6	1.9	0.1143	0.383	0.044	0.001038	9.1	200	90
A-204093-2	1.334	0.699	0.538	1.464	0.888	3.2	3.1	2.6	1.9	0.1152	0.383	0.044	0.001055	9.1	205	93
W-143067-4	1.334	0.699	0.538	1.464	0.888	3.2	3.1	2.6	1.9	0.1158	0.383	0.044	0.001065	9.1	147	67
D-143067-4	1.334	0.699	0.538	1.464	0.888	3.2	3.1	2.6	1.9	0.1158	0.383	0.044	0.001065	9.1	147	67
A-143067-2	1.334	0.699	0.538	1.464	0.888	3.2	3.1	2.6	1.9	0.1158	0.383	0.044	0.001065	9.1	147	67
W-172079-4	1.334	0.699	0.538	1.464	0.888	3.2	3.1	2.6	1.9	0.1160	0.383	0.044	0.001069	9.1	173	79
D-172079-4	1.334	0.699	0.538	1.464	0.888	3.2	3.1	2.6	1.9	0.1160	0.383	0.044	0.001069	9.1	173	79
A-172079-2	1.334	0.699	0.538	1.464	0.888	3.2	3.1	2.6	1.9	0.1160	0.383	0.044	0.001069	9.1	173	79

Arnold Engineering Co. MPP Powder Cores																
Part No.	OD cm 1	ID cm 2	HT cm 3	Wound OD cm 4	Wound HT cm 5	MPL cm 6	Wtfe grams 7	Wtcu grams 8	MLT cm 9	Ac cm sq 10	Wa cm sq 11	Ap cm 4th 12	Kg cm 5th 13	At cm sq 14	Perm 15	AL 16
A-052012-2	1.334	0.699	0.538	1.464	0.888	3.2	3.2	2.6	1.9	0.1172	0.383	0.045	0.001092	9.1	26	12
A-268008-2	1.727	0.953	0.699	1.914	1.175	4.2	6.8	6.3	2.5	0.1914	0.712	0.136	0.004176	15.7	14	8
W-285092-4	1.727	0.953	0.699	1.914	1.175	4.2	6.9	6.3	2.5	0.1926	0.712	0.137	0.004228	15.7	160	92
D-285092-4	1.727	0.953	0.699	1.914	1.175	4.2	6.9	6.3	2.5	0.1926	0.712	0.137	0.004228	15.7	160	92
A-285092-2	1.727	0.953	0.699	1.914	1.175	4.2	6.9	6.3	2.5	0.1926	0.712	0.137	0.004228	15.7	160	92
W-316115-4	1.727	0.953	0.699	1.914	1.175	4.2	6.9	6.3	2.5	0.1926	0.712	0.137	0.004228	15.7	200	115
D-316115-4	1.727	0.953	0.699	1.914	1.175	4.2	6.9	6.3	2.5	0.1926	0.712	0.137	0.004228	15.7	200	115
W-281072-4	1.727	0.953	0.699	1.914	1.175	4.2	6.9	6.3	2.5	0.1929	0.712	0.137	0.004243	15.7	125	72
D-281072-4	1.727	0.953	0.699	1.914	1.175	4.2	6.9	6.3	2.5	0.1929	0.712	0.137	0.004243	15.7	125	72
A-281072-2	1.727	0.953	0.699	1.914	1.175	4.2	6.9	6.3	2.5	0.1929	0.712	0.137	0.004243	15.7	125	72
A-369144-2	1.727	0.953	0.699	1.914	1.175	4.2	6.9	6.3	2.5	0.1929	0.712	0.137	0.004243	15.7	250	144
A-391173-2	1.727	0.953	0.699	1.914	1.175	4.2	6.9	6.3	2.5	0.1932	0.712	0.138	0.004253	15.7	300	173
A-267015-2	1.727	0.953	0.699	1.914	1.175	4.2	6.9	6.3	2.5	0.1932	0.712	0.138	0.004257	15.7	26	15
W-264088-4	1.727	0.953	0.699	1.914	1.175	4.2	7.2	6.3	2.5	0.2005	0.712	0.143	0.004583	15.7	147	88
D-264088-4	1.727	0.953	0.699	1.914	1.175	4.2	7.2	6.3	2.5	0.2005	0.712	0.143	0.004583	15.7	147	88
A-264088-2	1.727	0.953	0.699	1.914	1.175	4.2	7.2	6.3	2.5	0.2005	0.712	0.143	0.004583	15.7	147	88
W-266036-4	1.727	0.953	0.699	1.914	1.175	4.2	7.2	6.3	2.5	0.2010	0.712	0.143	0.004604	15.7	60	36
D-266036-4	1.727	0.953	0.699	1.914	1.175	4.2	7.2	6.3	2.5	0.2010	0.712	0.143	0.004604	15.7	60	36
A-266036-2	1.727	0.953	0.699	1.914	1.175	4.2	7.2	6.3	2.5	0.2010	0.712	0.143	0.004604	15.7	60	36
A-262123-2	1.727	0.953	0.699	1.914	1.175	4.2	7.2	6.3	2.5	0.2010	0.712	0.143	0.004604	15.7	205	123
W-263104-4	1.727	0.953	0.699	1.914	1.175	4.2	7.2	6.3	2.5	0.2014	0.712	0.143	0.004622	15.7	173	104
D-263104-4	1.727	0.953	0.699	1.914	1.175	4.2	7.2	6.3	2.5	0.2014	0.712	0.143	0.004622	15.7	173	104
A-263104-2	1.727	0.953	0.699	1.914	1.175	4.2	7.2	6.3	2.5	0.2014	0.712	0.143	0.004622	15.7	173	104
W-194123-4	1.791	0.902	0.699	1.954	1.149	4.2	8.6	5.8	2.6	0.2393	0.638	0.153	0.005732	16.0	173	123
D-194123-4	1.791	0.902	0.699	1.954	1.149	4.2	8.6	5.8	2.6	0.2393	0.638	0.153	0.005732	16.0	173	123
A-194123-2	1.791	0.902	0.699	1.954	1.149	4.2	8.6	5.8	2.6	0.2393	0.638	0.153	0.005732	16.0	173	123
W-190089-4	1.791	0.902	0.699	1.954	1.149	4.2	8.6	5.8	2.6	0.2396	0.638	0.153	0.005748	16.0	125	89
D-190089-4	1.791	0.902	0.699	1.954	1.149	4.2	8.6	5.8	2.6	0.2396	0.638	0.153	0.005748	16.0	125	89
A-190089-2	1.791	0.902	0.699	1.954	1.149	4.2	8.6	5.8	2.6	0.2396	0.638	0.153	0.005748	16.0	125	89
A-370178-2	1.791	0.902	0.699	1.954	1.149	4.2	8.6	5.8	2.6	0.2396	0.638	0.153	0.005748	16.0	250	178

Arnold Engineering Co. MPP Powder Cores																
Part No.	OD cm 1	ID cm 2	HT cm 3	Wound OD cm 4	Wound HT cm 5	MPL cm 6	Wtfe grams 7	Wtcu grams 8	MLT cm 9	Ac cm sq 10	Wa cm sq 11	Ap cm 4th 12	Kg cm 5th 13	At cm sq 14	Perm 15	AL 16
W-205146-4	1.791	0.902	0.699	1.954	1.149	4.2	8.6	5.8	2.6	0.2397	0.638	0.153	0.005752	16.0	205	146
D-205146-4	1.791	0.902	0.699	1.954	1.149	4.2	8.6	5.8	2.6	0.2397	0.638	0.153	0.005752	16.0	205	146
A-205146-2	1.791	0.902	0.699	1.954	1.149	4.2	8.6	5.8	2.6	0.2397	0.638	0.153	0.005752	16.0	205	146
W-559114-4	1.791	0.902	0.699	1.954	1.149	4.2	8.6	5.8	2.6	0.2398	0.638	0.153	0.005756	16.0	160	114
D-559114-4	1.791	0.902	0.699	1.954	1.149	4.2	8.6	5.8	2.6	0.2398	0.638	0.153	0.005756	16.0	160	114
A-559114-2	1.791	0.902	0.699	1.954	1.149	4.2	8.6	5.8	2.6	0.2398	0.638	0.153	0.005756	16.0	160	114
A-392214-2	1.791	0.902	0.699	1.954	1.149	4.2	8.6	5.8	2.6	0.2401	0.638	0.153	0.005770	16.0	300	214
A-187010-2	1.791	0.902	0.699	1.954	1.149	4.2	8.6	5.8	2.6	0.2404	0.638	0.153	0.005785	16.0	14	10
W-193105-4	1.791	0.902	0.699	1.954	1.149	4.2	8.6	5.8	2.6	0.2404	0.638	0.153	0.005785	16.0	147	105
D-193105-4	1.791	0.902	0.699	1.954	1.149	4.2	8.6	5.8	2.6	0.2404	0.638	0.153	0.005785	16.0	147	105
A-193105-2	1.791	0.902	0.699	1.954	1.149	4.2	8.6	5.8	2.6	0.2404	0.638	0.153	0.005785	16.0	147	105
W-189043-4	1.791	0.902	0.699	1.954	1.149	4.2	8.7	5.8	2.6	0.2412	0.638	0.154	0.005824	16.0	60	43
D-189043-4	1.791	0.902	0.699	1.954	1.149	4.2	8.7	5.8	2.6	0.2412	0.638	0.154	0.005824	16.0	60	43
A-189043-2	1.791	0.902	0.699	1.954	1.149	4.2	8.7	5.8	2.6	0.2412	0.638	0.154	0.005824	16.0	60	43
A-188019-2	1.791	0.902	0.699	1.954	1.149	4.2	8.8	5.8	2.6	0.2459	0.638	0.157	0.006055	16.0	26	19
W-133032-4	2.096	1.207	0.699	2.342	1.302	5.2	9.7	11.4	2.8	0.2201	1.143	0.252	0.007927	22.4	60	32
E-403032-4	2.096	1.207	0.699	2.342	1.302	5.2	9.7	11.4	2.8	0.2201	1.143	0.252	0.007927	22.4	60	32
D-131032-4	2.096	1.207	0.699	2.342	1.302	5.2	9.7	11.4	2.8	0.2201	1.143	0.252	0.007927	22.4	60	32
B-055032-4	2.096	1.207	0.699	2.342	1.302	5.2	9.7	11.4	2.8	0.2201	1.143	0.252	0.007927	22.4	60	32
A-848032-2	2.096	1.207	0.699	2.342	1.302	5.2	9.7	11.4	2.8	0.2201	1.143	0.252	0.007927	22.4	60	32
D-608014-4	2.096	1.207	0.699	2.342	1.302	5.2	9.8	11.4	2.8	0.2223	1.143	0.254	0.008081	22.4	26	14
B-056014-4	2.096	1.207	0.699	2.342	1.302	5.2	9.8	11.4	2.8	0.2223	1.143	0.254	0.008081	22.4	26	14
W-511014-4	2.096	1.207	0.699	2.342	1.302	5.2	9.8	11.4	2.8	0.2223	1.143	0.254	0.008081	22.4	26	14
A-511014-2	2.096	1.207	0.699	2.342	1.302	5.2	9.8	11.4	2.8	0.2223	1.143	0.254	0.008081	22.4	26	14
A-393163-2	2.096	1.207	0.699	2.342	1.302	5.2	9.9	11.4	2.8	0.2243	1.143	0.256	0.008227	22.4	300	163
W-271087-4	2.096	1.207	0.699	2.342	1.302	5.2	9.9	11.4	2.8	0.2244	1.143	0.256	0.008240	22.4	160	87
D-271087-4	2.096	1.207	0.699	2.342	1.302	5.2	9.9	11.4	2.8	0.2244	1.143	0.256	0.008240	22.4	160	87
A-271087-2	2.096	1.207	0.699	2.342	1.302	5.2	9.9	11.4	2.8	0.2244	1.143	0.256	0.008240	22.4	160	87
W-114068-4	2.096	1.207	0.699	2.342	1.302	5.2	9.9	11.4	2.8	0.2245	1.143	0.257	0.008248	22.4	125	68
E-951068-4	2.096	1.207	0.699	2.342	1.302	5.2	9.9	11.4	2.8	0.2245	1.143	0.257	0.008248	22.4	125	68

Arnold Engineering Co. MPP Powder Cores

Part No.	OD cm 1	ID cm 2	HT cm 3	Wound OD cm 4	Wound HT cm 5	MPL cm 6	Wtfe grams 7	Wtcu grams 8	MLT cm 9	Ac cm sq 10	Wa cm sq 11	Ap cm 4th 12	Kg cm 5th 13	At cm sq 14	Perm 15	AL 16
D-130068-4	2.096	1.207	0.699	2.342	1.302	5.2	9.9	11.4	2.8	0.2245	1.143	0.257	0.008248	22.4	125	68
B-054068-4	2.096	1.207	0.699	2.342	1.302	5.2	9.9	11.4	2.8	0.2245	1.143	0.257	0.008248	22.4	125	68
A-206068-2	2.096	1.207	0.699	2.342	1.302	5.2	9.9	11.4	2.8	0.2245	1.143	0.257	0.008248	22.4	125	68
A-371136-2	2.096	1.207	0.699	2.342	1.302	5.2	9.9	11.4	2.8	0.2245	1.143	0.257	0.008248	22.4	250	136
W-317109-4	2.096	1.207	0.699	2.342	1.302	5.2	9.9	11.4	2.8	0.2249	1.143	0.257	0.008278	22.4	200	109
D-317109-4	2.096	1.207	0.699	2.342	1.302	5.2	9.9	11.4	2.8	0.2249	1.143	0.257	0.008278	22.4	200	109
W-144081-4	2.096	1.207	0.699	2.342	1.302	5.2	10.0	11.4	2.8	0.2274	1.143	0.260	0.008462	22.4	147	81
D-144081-4	2.096	1.207	0.699	2.342	1.302	5.2	10.0	11.4	2.8	0.2274	1.143	0.260	0.008462	22.4	147	81
A-144081-2	2.096	1.207	0.699	2.342	1.302	5.2	10.0	11.4	2.8	0.2274	1.143	0.260	0.008462	22.4	147	81
A-207113-2	2.096	1.207	0.699	2.342	1.302	5.2	10.0	11.4	2.8	0.2275	1.143	0.260	0.008468	22.4	205	113
W-241083-4	2.096	1.207	0.699	2.342	1.302	5.2	10.1	11.4	2.8	0.2284	1.143	0.261	0.008533	22.4	150	83
D-241083-4	2.096	1.207	0.699	2.342	1.302	5.2	10.1	11.4	2.8	0.2284	1.143	0.261	0.008533	22.4	150	83
A-241083-2	2.096	1.207	0.699	2.342	1.302	5.2	10.1	11.4	2.8	0.2284	1.143	0.261	0.008533	22.4	150	83
W-173096-4	2.096	1.207	0.699	2.342	1.302	5.2	10.1	11.4	2.8	0.2290	1.143	0.262	0.008582	22.4	173	96
D-173096-4	2.096	1.207	0.699	2.342	1.302	5.2	10.1	11.4	2.8	0.2290	1.143	0.262	0.008582	22.4	173	96
A-173096-2	2.096	1.207	0.699	2.342	1.302	5.2	10.1	11.4	2.8	0.2290	1.143	0.262	0.008582	22.4	173	96
D-058008-4	2.096	1.207	0.699	2.342	1.302	5.2	10.4	11.4	2.8	0.2359	1.143	0.270	0.009100	22.4	14	8
B-041008-4	2.096	1.207	0.699	2.342	1.302	5.2	10.4	11.4	2.8	0.2359	1.143	0.270	0.009100	22.4	14	8
W-057008-4	2.096	1.207	0.699	2.342	1.302	5.2	10.4	11.4	2.8	0.2359	1.143	0.270	0.009100	22.4	14	8
A-057008-2	2.096	1.207	0.699	2.342	1.302	5.2	10.4	11.4	2.8	0.2359	1.143	0.270	0.009100	22.4	14	8
W-062010-4	2.350	1.334	0.826	2.618	1.492	5.8	16.2	15.9	3.2	0.3288	1.396	0.459	0.018866	28.3	14	10
D-062010-4	2.350	1.334	0.826	2.618	1.492	5.8	16.2	15.9	3.2	0.3288	1.396	0.459	0.018866	28.3	14	10
B-062010-4	2.350	1.334	0.826	2.618	1.492	5.8	16.2	15.9	3.2	0.3288	1.396	0.459	0.018866	28.3	14	10
A-062010-2	2.350	1.334	0.826	2.618	1.492	5.8	16.2	15.9	3.2	0.3288	1.396	0.459	0.018866	28.3	14	10
W-059043-4	2.350	1.334	0.826	2.618	1.492	5.8	16.2	15.9	3.2	0.3299	1.396	0.461	0.018992	28.3	60	43
D-059043-4	2.350	1.334	0.826	2.618	1.492	5.8	16.2	15.9	3.2	0.3299	1.396	0.461	0.018992	28.3	60	43
A-059043-2	2.350	1.334	0.826	2.618	1.492	5.8	16.2	15.9	3.2	0.3299	1.396	0.461	0.018992	28.3	60	43
W-174124-4	2.350	1.334	0.826	2.618	1.492	5.8	16.2	15.9	3.2	0.3300	1.396	0.461	0.018997	28.3	173	124
D-174124-4	2.350	1.334	0.826	2.618	1.492	5.8	16.2	15.9	3.2	0.3300	1.396	0.461	0.018997	28.3	173	124
A-174124-2	2.350	1.334	0.826	2.618	1.492	5.8	16.2	15.9	3.2	0.3300	1.396	0.461	0.018997	28.3	173	124

Arnold Engineering Co. MPP Powder Cores																
Part No.	OD cm 1	ID cm 2	HT cm 3	Wound OD cm 4	Wound HT cm 5	MPL cm 6	Wtfe grams 7	Wtcu grams 8	MLT cm 9	Ac cm sq 10	Wa cm sq 11	Ap cm 4th 12	Kg cm 5th 13	At cm sq 14	Perm 15	AL 16
A-208147-2	2.350	1.334	0.826	2.618	1.492	5.8	16.2	15.9	3.2	0.3301	1.396	0.461	0.019013	28.3	205	147
D-283115-4	2.350	1.334	0.826	2.618	1.492	5.8	16.3	15.9	3.2	0.3309	1.396	0.462	0.019103	28.3	160	115
W-300115-4	2.350	1.334	0.826	2.618	1.492	5.8	16.3	15.9	3.2	0.3309	1.396	0.462	0.019103	28.3	160	115
A-300115-2	2.350	1.334	0.826	2.618	1.492	5.8	16.3	15.9	3.2	0.3309	1.396	0.462	0.019103	28.3	160	115
W-160090-4	2.350	1.334	0.826	2.618	1.492	5.8	16.3	15.9	3.2	0.3315	1.396	0.463	0.019169	28.3	125	90
D-159090-4	2.350	1.334	0.826	2.618	1.492	5.8	16.3	15.9	3.2	0.3315	1.396	0.463	0.019169	28.3	125	90
A-310090-2	2.350	1.334	0.826	2.618	1.492	5.8	16.3	15.9	3.2	0.3315	1.396	0.463	0.019169	28.3	125	90
W-318144-4	2.350	1.334	0.826	2.618	1.492	5.8	16.3	15.9	3.2	0.3315	1.396	0.463	0.019169	28.3	200	144
D-318144-4	2.350	1.334	0.826	2.618	1.492	5.8	16.3	15.9	3.2	0.3315	1.396	0.463	0.019169	28.3	200	144
A-372180-2	2.350	1.334	0.826	2.618	1.492	5.8	16.3	15.9	3.2	0.3315	1.396	0.463	0.019169	28.3	250	180
A-394216-2	2.350	1.334	0.826	2.618	1.492	5.8	16.3	15.9	3.2	0.3315	1.396	0.463	0.019169	28.3	300	216
W-147106-4	2.350	1.334	0.826	2.618	1.492	5.8	16.3	15.9	3.2	0.3320	1.396	0.463	0.019227	28.3	147	106
D-147106-4	2.350	1.334	0.826	2.618	1.492	5.8	16.3	15.9	3.2	0.3320	1.396	0.463	0.019227	28.3	147	106
A-147106-2	2.350	1.334	0.826	2.618	1.492	5.8	16.3	15.9	3.2	0.3320	1.396	0.463	0.019227	28.3	147	106
W-060019-4	2.350	1.334	0.826	2.618	1.492	5.8	16.5	15.9	3.2	0.3364	1.396	0.470	0.019747	28.3	26	19
D-060019-4	2.350	1.334	0.826	2.618	1.492	5.8	16.5	15.9	3.2	0.3364	1.396	0.470	0.019747	28.3	26	19
B-060019-4	2.350	1.334	0.826	2.618	1.492	5.8	16.5	15.9	3.2	0.3364	1.396	0.470	0.019747	28.3	26	19
A-060019-2	2.350	1.334	0.826	2.618	1.492	5.8	16.5	15.9	3.2	0.3364	1.396	0.470	0.019747	28.3	26	19
A-442105-2	2.421	1.377	0.953	2.698	1.641	6.0	20.2	18.3	3.5	0.3987	1.488	0.593	0.027339	31.1	125	105
A-447253-2	2.421	1.377	0.953	2.698	1.641	6.0	20.3	18.3	3.5	0.4003	1.488	0.596	0.027557	31.1	300	253
A-443124-2	2.421	1.377	0.953	2.698	1.641	6.0	20.3	18.3	3.5	0.4004	1.488	0.596	0.027570	31.1	147	124
A-444135-2	2.421	1.377	0.953	2.698	1.641	6.0	20.3	18.3	3.5	0.4005	1.488	0.596	0.027584	31.1	160	135
A-272173-2	2.421	1.377	0.953	2.698	1.641	6.0	20.3	18.3	3.5	0.4006	1.488	0.596	0.027594	31.1	205	173
A-445146-2	2.421	1.377	0.953	2.698	1.641	6.0	20.3	18.3	3.5	0.4006	1.488	0.596	0.027596	31.1	173	146
A-446211-2	2.421	1.377	0.953	2.698	1.641	6.0	20.3	18.3	3.5	0.4006	1.488	0.596	0.027600	31.1	250	211
A-440022-2	2.421	1.377	0.953	2.698	1.641	6.0	20.4	18.3	3.5	0.4016	1.488	0.598	0.027741	31.1	26	22
A-441051-2	2.421	1.377	0.953	2.698	1.641	6.0	20.5	18.3	3.5	0.4035	1.488	0.600	0.027994	31.1	60	51
A-439012-2	2.421	1.377	0.953	2.698	1.641	6.0	20.6	18.3	3.5	0.4069	1.488	0.605	0.028466	31.1	14	12
A-330170-2	2.756	1.410	0.927	3.014	1.632	6.5	28.5	20.5	3.7	0.5117	1.560	0.798	0.044296	36.6	173	170
W-105032-4	2.756	1.410	1.181	3.014	1.886	6.5	35.6	22.7	4.1	0.6409	1.560	1.000	0.062591	39.4	26	32

Arnold Engineering Co. MPP Powder Cores																
Part No.	OD cm 1	ID cm 2	HT cm 3	Wound OD cm 4	Wound HT cm 5	MPL cm 6	Wtfe grams 7	Wtcu grams 8	MLT cm 9	Ac cm sq 10	Wa cm sq 11	Ap cm 4th 12	Kg cm 5th 13	At cm sq 14	Perm 15	AL 16
D-067032-4	2.756	1.410	1.181	3.014	1.886	6.5	35.6	22.7	4.1	0.6409	1.560	1.000	0.062591	39.4	26	32
B-165032-4	2.756	1.410	1.181	3.014	1.886	6.5	35.6	22.7	4.1	0.6409	1.560	1.000	0.062591	39.4	26	32
A-066032-2	2.756	1.410	1.181	3.014	1.886	6.5	35.6	22.7	4.1	0.6409	1.560	1.000	0.062591	39.4	26	32
W-099075-4	2.756	1.410	1.181	3.014	1.886	6.5	36.2	22.7	4.1	0.6509	1.560	1.015	0.064562	39.4	60	75
D-269075-4	2.756	1.410	1.181	3.014	1.886	6.5	36.2	22.7	4.1	0.6509	1.560	1.015	0.064562	39.4	60	75
B-065075-4	2.756	1.410	1.181	3.014	1.886	6.5	36.2	22.7	4.1	0.6509	1.560	1.015	0.064562	39.4	60	75
A-395075-5	2.756	1.410	1.181	3.014	1.886	6.5	36.2	22.7	4.1	0.6509	1.560	1.015	0.064562	39.4	60	75
A-894075-2	2.756	1.410	1.181	3.014	1.886	6.5	36.2	22.7	4.1	0.6509	1.560	1.015	0.064562	39.4	60	75
A-209257-2	2.756	1.410	1.181	3.014	1.886	6.5	36.3	22.7	4.1	0.6528	1.560	1.018	0.064941	39.4	205	257
W-175217-4	2.756	1.410	1.181	3.014	1.886	6.5	36.3	22.7	4.1	0.6531	1.560	1.019	0.065011	39.4	173	217
D-175217-4	2.756	1.410	1.181	3.014	1.886	6.5	36.3	22.7	4.1	0.6531	1.560	1.019	0.065011	39.4	173	217
A-175217-2	2.756	1.410	1.181	3.014	1.886	6.5	36.3	22.7	4.1	0.6531	1.560	1.019	0.065011	39.4	173	217
W-319251-4	2.756	1.410	1.181	3.014	1.886	6.5	36.3	22.7	4.1	0.6535	1.560	1.019	0.065080	39.4	200	251
D-319251-4	2.756	1.410	1.181	3.014	1.886	6.5	36.3	22.7	4.1	0.6535	1.560	1.019	0.065080	39.4	200	251
W-098157-4	2.756	1.410	1.181	3.014	1.886	6.5	36.4	22.7	4.1	0.6540	1.560	1.020	0.065184	39.4	125	157
E-115157-4	2.756	1.410	1.181	3.014	1.886	6.5	36.4	22.7	4.1	0.6540	1.560	1.020	0.065184	39.4	125	157
D-671157-4	2.756	1.410	1.181	3.014	1.886	6.5	36.4	22.7	4.1	0.6540	1.560	1.020	0.065184	39.4	125	157
B-064157-4	2.756	1.410	1.181	3.014	1.886	6.5	36.4	22.7	4.1	0.6540	1.560	1.020	0.065184	39.4	125	157
A-930157-2	2.756	1.410	1.181	3.014	1.886	6.5	36.4	22.7	4.1	0.6540	1.560	1.020	0.065184	39.4	125	157
A-373314-2	2.756	1.410	1.181	3.014	1.886	6.5	36.4	22.7	4.1	0.6540	1.560	1.020	0.065184	39.4	250	314
W-297201-4	2.756	1.410	1.181	3.014	1.886	6.5	36.4	22.7	4.1	0.6541	1.560	1.020	0.065210	39.4	160	201
D-296201-4	2.756	1.410	1.181	3.014	1.886	6.5	36.4	22.7	4.1	0.6541	1.560	1.020	0.065210	39.4	160	201
A-302201-2	2.756	1.410	1.181	3.014	1.886	6.5	36.4	22.7	4.1	0.6541	1.560	1.020	0.065210	39.4	160	201
A-396377-2	2.756	1.410	1.181	3.014	1.886	6.5	36.4	22.7	4.1	0.6543	1.560	1.021	0.065253	39.4	300	377
W-145185-4	2.756	1.410	1.181	3.014	1.886	6.5	36.4	22.7	4.1	0.6553	1.560	1.022	0.065444	39.4	147	185
D-145185-4	2.756	1.410	1.181	3.014	1.886	6.5	36.4	22.7	4.1	0.6553	1.560	1.022	0.065444	39.4	147	185
A-145185-2	2.756	1.410	1.181	3.014	1.886	6.5	36.4	22.7	4.1	0.6553	1.560	1.022	0.065444	39.4	147	185
D-069018-4	2.756	1.410	1.181	3.014	1.886	6.5	37.2	22.7	4.1	0.6695	1.560	1.044	0.068304	39.4	14	18
W-068018-4	2.756	1.410	1.181	3.014	1.886	6.5	37.2	22.7	4.1	0.6695	1.560	1.044	0.068304	39.4	14	18
A-068018-2	2.756	1.410	1.181	3.014	1.886	6.5	37.2	22.7	4.1	0.6695	1.560	1.044	0.068304	39.4	14	18

Arnold Engineering Co. MPP Powder Cores

Part No.	OD cm 1	ID cm 2	HT cm 3	Wound OD cm 4	Wound HT cm 5	MPL cm 6	Wtfe grams 7	Wtcu grams 8	MLT cm 9	Ac cm sq 10	Wa cm sq 11	Ap cm 4th 12	Kg cm 5th 13	At cm sq 14	Perm 15	AL 16
A-346016-2	3.493	2.273	0.953	4.010	2.089	9.1	34.1	62.3	4.3	0.4435	4.057	1.799	0.073925	63.6	26	16
W-321126-4	3.493	2.273	0.953	4.010	2.089	9.1	35.0	62.3	4.3	0.4541	4.057	1.842	0.077478	63.6	200	126
D-321126-4	3.493	2.273	0.953	4.010	2.089	9.1	35.0	62.3	4.3	0.4541	4.057	1.842	0.077478	63.6	200	126
W-177109-4	3.493	2.273	0.953	4.010	2.089	9.1	35.0	62.3	4.3	0.4541	4.057	1.842	0.077493	63.6	173	109
D-177109-4	3.493	2.273	0.953	4.010	2.089	9.1	35.0	62.3	4.3	0.4541	4.057	1.842	0.077493	63.6	173	109
A-177109-2	3.493	2.273	0.953	4.010	2.089	9.1	35.0	62.3	4.3	0.4541	4.057	1.842	0.077493	63.6	173	109
W-304101-4	3.493	2.273	0.953	4.010	2.089	9.1	35.0	62.3	4.3	0.4550	4.057	1.846	0.077786	63.6	160	101
D-304101-4	3.493	2.273	0.953	4.010	2.089	9.1	35.0	62.3	4.3	0.4550	4.057	1.846	0.077786	63.6	160	101
A-304101-2	3.493	2.273	0.953	4.010	2.089	9.1	35.0	62.3	4.3	0.4550	4.057	1.846	0.077786	63.6	160	101
W-585079-4	3.493	2.273	0.953	4.010	2.089	9.1	35.1	62.3	4.3	0.4555	4.057	1.848	0.077971	63.6	125	79
D-585079-4	3.493	2.273	0.953	4.010	2.089	9.1	35.1	62.3	4.3	0.4555	4.057	1.848	0.077971	63.6	125	79
A-585079-2	3.493	2.273	0.953	4.010	2.089	9.1	35.1	62.3	4.3	0.4555	4.057	1.848	0.077971	63.6	125	79
W-149093-4	3.493	2.273	0.953	4.010	2.089	9.1	35.1	62.3	4.3	0.4560	4.057	1.850	0.078132	63.6	147	93
D-149093-4	3.493	2.273	0.953	4.010	2.089	9.1	35.1	62.3	4.3	0.4560	4.057	1.850	0.078132	63.6	147	93
A-149093-2	3.493	2.273	0.953	4.010	2.089	9.1	35.1	62.3	4.3	0.4560	4.057	1.850	0.078132	63.6	147	93
W-345038-4	3.493	2.273	0.953	4.010	2.089	9.1	35.1	62.3	4.3	0.4565	4.057	1.852	0.078300	63.6	60	38
D-345038-4	3.493	2.273	0.953	4.010	2.089	9.1	35.1	62.3	4.3	0.4565	4.057	1.852	0.078300	63.6	60	38
A-345038-2	3.493	2.273	0.953	4.010	2.089	9.1	35.1	62.3	4.3	0.4565	4.057	1.852	0.078300	63.6	60	38
A-212130-2	3.493	2.273	0.953	4.010	2.089	9.1	35.2	62.3	4.3	0.4570	4.057	1.854	0.078502	63.6	205	130
A-347009-2	3.493	2.273	0.953	4.010	2.089	9.1	35.7	62.3	4.3	0.4633	4.057	1.880	0.080673	63.6	14	9
A-197109-2	3.366	1.930	0.940	3.758	1.905	8.3	40.8	43.6	4.2	0.5773	2.925	1.689	0.092921	55.2	125	109
A-166151-2	3.366	1.930	0.940	3.758	1.905	8.3	40.9	43.6	4.2	0.5778	2.925	1.690	0.093098	55.2	173	151
A-162129-2	3.366	1.930	0.940	3.758	1.905	8.3	41.1	43.6	4.2	0.5809	2.925	1.699	0.094108	55.2	147	129
A-210180-2	3.366	1.930	0.940	3.758	1.905	8.3	41.1	43.6	4.2	0.5813	2.925	1.700	0.094214	55.2	205	180
D-344014-4	3.366	1.930	1.130	3.758	2.096	8.3	46.8	46.8	4.5	0.6620	2.925	1.936	0.113927	57.8	14	14
B-344014-4	3.366	1.930	1.130	3.758	2.096	8.3	46.8	46.8	4.5	0.6620	2.925	1.936	0.113927	57.8	14	14
A-344014-2	3.366	1.930	1.130	3.758	2.096	8.3	46.8	46.8	4.5	0.6620	2.925	1.936	0.113927	57.8	14	14
A-211208-2	3.366	1.930	1.130	3.758	2.096	8.3	47.5	46.8	4.5	0.6717	2.925	1.965	0.117286	57.8	205	208
W-320203-4	3.366	1.930	1.130	3.758	2.096	8.3	47.5	46.8	4.5	0.6719	2.925	1.966	0.117370	57.8	200	203
D-320203-4	3.366	1.930	1.130	3.758	2.096	8.3	47.5	46.8	4.5	0.6719	2.925	1.966	0.117370	57.8	200	203

Arnold Engineering Co. MPP Powder Cores																
Part No.	OD cm 1	ID cm 2	HT cm 3	Wound OD cm 4	Wound HT cm 5	MPL cm 6	Wtfe grams 7	Wtcu grams 8	MLT cm 9	Ac cm sq 10	Wa cm sq 11	Ap cm 4th 12	Kg cm 5th 13	At cm sq 14	Perm 15	AL 16
D-116127-4	3.366	1.930	1.130	3.758	2.096	8.3	47.6	46.8	4.5	0.6726	2.925	1.967	0.117602	57.8	125	127
B-070127-4	3.366	1.930	1.130	3.758	2.096	8.3	47.6	46.8	4.5	0.6726	2.925	1.967	0.117602	57.8	125	127
W-548127-4	3.366	1.930	1.130	3.758	2.096	8.3	47.6	46.8	4.5	0.6726	2.925	1.967	0.117602	57.8	125	127
A-548127-2	3.366	1.930	1.130	3.758	2.096	8.3	47.6	46.8	4.5	0.6726	2.925	1.967	0.117602	57.8	125	127
A-374254-2	3.366	1.930	1.130	3.758	2.096	8.3	47.6	46.8	4.5	0.6726	2.925	1.967	0.117602	57.8	250	254
W-291061-4	3.366	1.930	1.130	3.758	2.096	8.3	47.6	46.8	4.5	0.6730	2.925	1.969	0.117756	57.8	60	61
D-291061-4	3.366	1.930	1.130	3.758	2.096	8.3	47.6	46.8	4.5	0.6730	2.925	1.969	0.117756	57.8	60	61
A-291061-2	3.366	1.930	1.130	3.758	2.096	8.3	47.6	46.8	4.5	0.6730	2.925	1.969	0.117756	57.8	60	61
W-176176-4	3.366	1.930	1.130	3.758	2.096	8.3	47.6	46.8	4.5	0.6735	2.925	1.970	0.117912	57.8	173	176
D-176176-4	3.366	1.930	1.130	3.758	2.096	8.3	47.6	46.8	4.5	0.6735	2.925	1.970	0.117912	57.8	173	176
A-176176-2	3.366	1.930	1.130	3.758	2.096	8.3	47.6	46.8	4.5	0.6735	2.925	1.970	0.117912	57.8	173	176
W-303163-4	3.366	1.930	1.130	3.758	2.096	8.3	47.7	46.8	4.5	0.6744	2.925	1.973	0.118239	57.8	160	163
D-303163-4	3.366	1.930	1.130	3.758	2.096	8.3	47.7	46.8	4.5	0.6744	2.925	1.973	0.118239	57.8	160	163
A-303163-2	3.366	1.930	1.130	3.758	2.096	8.3	47.7	46.8	4.5	0.6744	2.925	1.973	0.118239	57.8	160	163
W-148150-4	3.366	1.930	1.130	3.758	2.096	8.3	47.8	46.8	4.5	0.6755	2.925	1.976	0.118624	57.8	147	150
D-148150-4	3.366	1.930	1.130	3.758	2.096	8.3	47.8	46.8	4.5	0.6755	2.925	1.976	0.118624	57.8	147	150
A-148150-2	3.366	1.930	1.130	3.758	2.096	8.3	47.8	46.8	4.5	0.6755	2.925	1.976	0.118624	57.8	147	150
D-013015-4	3.366	1.930	1.181	3.758	2.146	8.3	50.2	47.7	4.6	0.7093	2.925	2.075	0.128464	58.4	14	15
B-761015-4	3.366	1.930	1.181	3.758	2.146	8.3	50.2	47.7	4.6	0.7093	2.925	2.075	0.128464	58.4	14	15
W-074015-4	3.366	1.930	1.181	3.758	2.146	8.3	50.2	47.7	4.6	0.7093	2.925	2.075	0.128464	58.4	14	15
A-074015-2	3.366	1.930	1.181	3.758	2.146	8.3	50.2	47.7	4.6	0.7093	2.925	2.075	0.128464	58.4	14	15
D-270028-4	3.366	1.930	1.181	3.758	2.146	8.3	50.4	47.7	4.6	0.7129	2.925	2.085	0.129785	58.4	26	28
B-579028-4	3.366	1.930	1.181	3.758	2.146	8.3	50.4	47.7	4.6	0.7129	2.925	2.085	0.129785	58.4	26	28
W-073028-4	3.366	1.930	1.181	3.758	2.146	8.3	50.4	47.7	4.6	0.7129	2.925	2.085	0.129785	58.4	26	28
A-073028-2	3.366	1.930	1.181	3.758	2.146	8.3	50.4	47.7	4.6	0.7129	2.925	2.085	0.129785	58.4	26	28
D-381065-4	3.366	1.930	1.181	3.758	2.146	8.3	50.7	47.7	4.6	0.7172	2.925	2.098	0.131334	58.4	60	65
B-072065-4	3.366	1.930	1.181	3.758	2.146	8.3	50.7	47.7	4.6	0.7172	2.925	2.098	0.131334	58.4	60	65
W-071065-4	3.366	1.930	1.181	3.758	2.146	8.3	50.7	47.7	4.6	0.7172	2.925	2.098	0.131334	58.4	60	65
A-071065-2	3.366	1.930	1.181	3.758	2.146	8.3	50.7	47.7	4.6	0.7172	2.925	2.098	0.131334	58.4	60	65
D-298028-4	3.366	1.930	1.130	3.758	2.096	8.3	50.4	46.8	4.5	0.7129	2.925	2.085	0.132128	57.8	26	28

Arnold Engineering Co. MPP Powder Cores																
Part No.	OD cm 1	ID cm 2	HT cm 3	Wound OD cm 4	Wound HT cm 5	MPL cm 6	Wtfe grams 7	Wtcu grams 8	MLT cm 9	Ac cm sq 10	Wa cm sq 11	Ap cm 4th 12	Kg cm 5th 13	At cm sq 14	Perm 15	AL 16
B-298028-4	3.366	1.930	1.130	3.758	2.096	8.3	50.4	46.8	4.5	0.7129	2.925	2.085	0.132128	57.8	26	28
A-298028-2	3.366	1.930	1.130	3.758	2.096	8.3	50.4	46.8	4.5	0.7129	2.925	2.085	0.132128	57.8	26	28
B-079024-4	3.645	2.172	1.110	4.102	2.196	9.1	52.1	61.8	4.7	0.6711	3.702	2.485	0.142173	67.5	26	24
W-078024-4	3.645	2.172	1.110	4.102	2.196	9.1	52.1	61.8	4.7	0.6711	3.702	2.485	0.142173	67.5	26	24
D-078024-4	3.645	2.172	1.110	4.102	2.196	9.1	52.1	61.8	4.7	0.6711	3.702	2.485	0.142173	67.5	26	24
A-078024-2	3.645	2.172	1.110	4.102	2.196	9.1	52.1	61.8	4.7	0.6711	3.702	2.485	0.142173	67.5	26	24
B-081013-4	3.645	2.172	1.110	4.102	2.196	9.1	52.4	61.8	4.7	0.6751	3.702	2.500	0.143870	67.5	14	13
W-080013-4	3.645	2.172	1.110	4.102	2.196	9.1	52.4	61.8	4.7	0.6751	3.702	2.500	0.143870	67.5	14	13
D-080013-4	3.645	2.172	1.110	4.102	2.196	9.1	52.4	61.8	4.7	0.6751	3.702	2.500	0.143870	67.5	14	13
A-080013-2	3.645	2.172	1.110	4.102	2.196	9.1	52.4	61.8	4.7	0.6751	3.702	2.500	0.143870	67.5	14	13
B-077056-4	3.645	2.172	1.110	4.102	2.196	9.1	52.7	61.8	4.7	0.6786	3.702	2.512	0.145350	67.5	60	56
W-076056-4	3.645	2.172	1.110	4.102	2.196	9.1	52.7	61.8	4.7	0.6786	3.702	2.512	0.145350	67.5	60	56
D-076056-4	3.645	2.172	1.110	4.102	2.196	9.1	52.7	61.8	4.7	0.6786	3.702	2.512	0.145350	67.5	60	56
A-076056-2	3.645	2.172	1.110	4.102	2.196	9.1	52.7	61.8	4.7	0.6786	3.702	2.512	0.145350	67.5	60	56
W-322187-4	3.645	2.172	1.110	4.102	2.196	9.1	52.8	61.8	4.7	0.6798	3.702	2.517	0.145869	67.5	200	187
D-322187-4	3.645	2.172	1.110	4.102	2.196	9.1	52.8	61.8	4.7	0.6798	3.702	2.517	0.145869	67.5	200	187
D-293117-4	3.645	2.172	1.110	4.102	2.196	9.1	52.9	61.8	4.7	0.6805	3.702	2.520	0.146181	67.5	125	117
B-075117-4	3.645	2.172	1.110	4.102	2.196	9.1	52.9	61.8	4.7	0.6805	3.702	2.520	0.146181	67.5	125	117
W-324117-4	3.645	2.172	1.110	4.102	2.196	9.1	52.9	61.8	4.7	0.6805	3.702	2.520	0.146181	67.5	125	117
A-324117-2	3.645	2.172	1.110	4.102	2.196	9.1	52.9	61.8	4.7	0.6805	3.702	2.520	0.146181	67.5	125	117
W-178162-4	3.645	2.172	1.110	4.102	2.196	9.1	52.9	61.8	4.7	0.6808	3.702	2.521	0.146311	67.5	173	162
D-178162-4	3.645	2.172	1.110	4.102	2.196	9.1	52.9	61.8	4.7	0.6808	3.702	2.521	0.146311	67.5	173	162
A-178162-2	3.645	2.172	1.110	4.102	2.196	9.1	52.9	61.8	4.7	0.6808	3.702	2.521	0.146311	67.5	173	162
A-213192-2	3.645	2.172	1.110	4.102	2.196	9.1	52.9	61.8	4.7	0.6810	3.702	2.521	0.146364	67.5	205	192
D-284150-4	3.645	2.172	1.110	4.102	2.196	9.1	52.9	61.8	4.7	0.6816	3.702	2.524	0.146650	67.5	160	150
W-305150-4	3.645	2.172	1.110	4.102	2.196	9.1	52.9	61.8	4.7	0.6816	3.702	2.524	0.146650	67.5	160	150
A-305150-2	3.645	2.172	1.110	4.102	2.196	9.1	52.9	61.8	4.7	0.6816	3.702	2.524	0.146650	67.5	160	150
W-150138-4	3.645	2.172	1.110	4.102	2.196	9.1	53.0	61.8	4.7	0.6826	3.702	2.527	0.147050	67.5	147	138
D-150138-4	3.645	2.172	1.110	4.102	2.196	9.1	53.0	61.8	4.7	0.6826	3.702	2.527	0.147050	67.5	147	138
A-150138-2	3.645	2.172	1.110	4.102	2.196	9.1	53.0	61.8	4.7	0.6826	3.702	2.527	0.147050	67.5	147	138

Arnold Engineering Co. MPP Powder Cores																
Part No.	OD cm 1	ID cm 2	HT cm 3	Wound OD cm 4	Wound HT cm 5	MPL cm 6	Wtfe grams 7	Wtcu grams 8	MLT cm 9	Ac cm sq 10	Wa cm sq 11	Ap cm 4th 12	Kg cm 5th 13	At cm sq 14	Perm 15	AL 16
A-214275-2	4.051	2.350	1.511	4.534	2.686	10.1	91.7	87.2	5.7	1.0733	4.333	4.651	0.352840	86.6	205	275
W-306215-4	4.051	2.350	1.511	4.534	2.686	10.1	91.9	87.2	5.7	1.0751	4.333	4.659	0.354044	86.6	160	215
D-306215-4	4.051	2.350	1.511	4.534	2.686	10.1	91.9	87.2	5.7	1.0751	4.333	4.659	0.354044	86.6	160	215
A-306215-2	4.051	2.350	1.511	4.534	2.686	10.1	91.9	87.2	5.7	1.0751	4.333	4.659	0.354044	86.6	160	215
W-110168-4	4.051	2.350	1.511	4.534	2.686	10.1	91.9	87.2	5.7	1.0753	4.333	4.660	0.354176	86.6	125	168
D-082168-4	4.051	2.350	1.511	4.534	2.686	10.1	91.9	87.2	5.7	1.0753	4.333	4.660	0.354176	86.6	125	168
A-254168-2	4.051	2.350	1.511	4.534	2.686	10.1	91.9	87.2	5.7	1.0753	4.333	4.660	0.354176	86.6	125	168
W-323269-4	4.051	2.350	1.511	4.534	2.686	10.1	92.0	87.2	5.7	1.0761	4.333	4.663	0.354703	86.6	200	269
D-323269-4	4.051	2.350	1.511	4.534	2.686	10.1	92.0	87.2	5.7	1.0761	4.333	4.663	0.354703	86.6	200	269
W-085035-4	4.051	2.350	1.511	4.534	2.686	10.1	92.0	87.2	5.7	1.0771	4.333	4.667	0.355312	86.6	26	35
D-085035-4	4.051	2.350	1.511	4.534	2.686	10.1	92.0	87.2	5.7	1.0771	4.333	4.667	0.355312	86.6	26	35
A-085035-2	4.051	2.350	1.511	4.534	2.686	10.1	92.0	87.2	5.7	1.0771	4.333	4.667	0.355312	86.6	26	35
W-179233-4	4.051	2.350	1.511	4.534	2.686	10.1	92.1	87.2	5.7	1.0776	4.333	4.670	0.355664	86.6	173	233
D-179233-4	4.051	2.350	1.511	4.534	2.686	10.1	92.1	87.2	5.7	1.0776	4.333	4.670	0.355664	86.6	173	233
A-179233-2	4.051	2.350	1.511	4.534	2.686	10.1	92.1	87.2	5.7	1.0776	4.333	4.670	0.355664	86.6	173	233
W-151198-4	4.051	2.350	1.511	4.534	2.686	10.1	92.1	87.2	5.7	1.0777	4.333	4.670	0.355726	86.6	147	198
D-151198-4	4.051	2.350	1.511	4.534	2.686	10.1	92.1	87.2	5.7	1.0777	4.333	4.670	0.355726	86.6	147	198
A-151198-2	4.051	2.350	1.511	4.534	2.686	10.1	92.1	87.2	5.7	1.0777	4.333	4.670	0.355726	86.6	147	198
W-083081-4	4.051	2.350	1.511	4.534	2.686	10.1	92.3	87.2	5.7	1.0801	4.333	4.681	0.357345	86.6	60	81
D-083081-4	4.051	2.350	1.511	4.534	2.686	10.1	92.3	87.2	5.7	1.0801	4.333	4.681	0.357345	86.6	60	81
A-083081-2	4.051	2.350	1.511	4.534	2.686	10.1	92.3	87.2	5.7	1.0801	4.333	4.681	0.357345	86.6	60	81
W-086019-4	4.051	2.350	1.511	4.534	2.686	10.1	92.8	87.2	5.7	1.0859	4.333	4.705	0.361136	86.6	14	19
D-086019-4	4.051	2.350	1.511	4.534	2.686	10.1	92.8	87.2	5.7	1.0859	4.333	4.705	0.361136	86.6	14	19
A-086019-2	4.051	2.350	1.511	4.534	2.686	10.1	92.8	87.2	5.7	1.0859	4.333	4.705	0.361136	86.6	14	19
W-195246-4	4.737	2.807	1.588	5.324	2.991	11.8	135.1	139.2	6.3	1.3409	6.184	8.292	0.702616	116.4	173	246
D-195246-4	4.737	2.807	1.588	5.324	2.991	11.8	135.1	139.2	6.3	1.3409	6.184	8.292	0.702616	116.4	173	246
A-195246-2	4.737	2.807	1.588	5.324	2.991	11.8	135.1	139.2	6.3	1.3409	6.184	8.292	0.702616	116.4	173	246
W-287037-4	4.737	2.807	1.588	5.324	2.991	11.8	135.2	139.2	6.3	1.3419	6.184	8.298	0.703715	116.4	26	37
B-411037-4	4.737	2.807	1.588	5.324	2.991	11.8	135.2	139.2	6.3	1.3419	6.184	8.298	0.703715	116.4	26	37
D-091037-4	4.737	2.807	1.588	5.324	2.991	11.8	135.2	139.2	6.3	1.3419	6.184	8.298	0.703715	116.4	26	37

Arnold Engineering Co. MPP Powder Cores																
Part No.	OD cm 1	ID cm 2	HT cm 3	Wound OD cm 4	Wound HT cm 5	MPL cm 6	Wtfe grams 7	Wtcu grams 8	MLT cm 9	Ac cm sq 10	Wa cm sq 11	Ap cm 4th 12	Kg cm 5th 13	At cm sq 14	Perm 15	AL 16
A-091037-2	4.737	2.807	1.588	5.324	2.991	11.8	135.2	139.2	6.3	1.3419	6.184	8.298	0.703715	116.4	26	37
W-089178-4	4.737	2.807	1.588	5.324	2.991	11.8	135.2	139.2	6.3	1.3428	6.184	8.304	0.704629	116.4	125	178
D-089178-4	4.737	2.807	1.588	5.324	2.991	11.8	135.2	139.2	6.3	1.3428	6.184	8.304	0.704629	116.4	125	178
A-089178-2	4.737	2.807	1.588	5.324	2.991	11.8	135.2	139.2	6.3	1.3428	6.184	8.304	0.704629	116.4	125	178
W-216292-4	4.737	2.807	1.588	5.324	2.991	11.8	135.3	139.2	6.3	1.3432	6.184	8.306	0.705015	116.4	205	292
D-216292-4	4.737	2.807	1.588	5.324	2.991	11.8	135.3	139.2	6.3	1.3432	6.184	8.306	0.705015	116.4	205	292
A-216292-2	4.737	2.807	1.588	5.324	2.991	11.8	135.3	139.2	6.3	1.3432	6.184	8.306	0.705015	116.4	205	292
W-326228-4	4.737	2.807	1.588	5.324	2.991	11.8	135.3	139.2	6.3	1.3437	6.184	8.310	0.705619	116.4	160	228
D-326228-4	4.737	2.807	1.588	5.324	2.991	11.8	135.3	139.2	6.3	1.3437	6.184	8.310	0.705619	116.4	160	228
A-326228-2	4.737	2.807	1.588	5.324	2.991	11.8	135.3	139.2	6.3	1.3437	6.184	8.310	0.705619	116.4	160	228
B-957020-4	4.737	2.807	1.588	5.324	2.991	11.8	135.7	139.2	6.3	1.3471	6.184	8.330	0.709160	116.4	14	20
W-092020-4	4.737	2.807	1.588	5.324	2.991	11.8	135.7	139.2	6.3	1.3471	6.184	8.330	0.709160	116.4	14	20
D-092020-4	4.737	2.807	1.588	5.324	2.991	11.8	135.7	139.2	6.3	1.3471	6.184	8.330	0.709160	116.4	14	20
A-092020-2	4.737	2.807	1.588	5.324	2.991	11.8	135.7	139.2	6.3	1.3471	6.184	8.330	0.709160	116.4	14	20
W-153210-4	4.737	2.807	1.588	5.324	2.991	11.8	135.7	139.2	6.3	1.3471	6.184	8.330	0.709160	116.4	147	210
D-153210-4	4.737	2.807	1.588	5.324	2.991	11.8	135.7	139.2	6.3	1.3471	6.184	8.330	0.709160	116.4	147	210
A-153210-2	4.737	2.807	1.588	5.324	2.991	11.8	135.7	139.2	6.3	1.3471	6.184	8.330	0.709160	116.4	147	210
W-090086-4	4.737	2.807	1.588	5.324	2.991	11.8	136.1	139.2	6.3	1.3516	6.184	8.358	0.713895	116.4	60	86
D-090086-4	4.737	2.807	1.588	5.324	2.991	11.8	136.1	139.2	6.3	1.3516	6.184	8.358	0.713895	116.4	60	86
A-090086-2	4.737	2.807	1.588	5.324	2.991	11.8	136.1	139.2	6.3	1.3516	6.184	8.358	0.713895	116.4	60	86
W-181210-4	5.144	3.112	1.410	5.807	2.965	13.0	138.1	172.2	6.4	1.2526	7.600	9.519	0.748701	132.2	173	210
D-181210-4	5.144	3.112	1.410	5.807	2.965	13.0	138.1	172.2	6.4	1.2526	7.600	9.519	0.748701	132.2	173	210
A-181210-2	5.144	3.112	1.410	5.807	2.965	13.0	138.1	172.2	6.4	1.2526	7.600	9.519	0.748701	132.2	173	210
A-349017-2	5.144	3.112	1.410	5.807	2.965	13.0	138.1	172.2	6.4	1.2530	7.600	9.523	0.749211	132.2	14	17
W-217249-4	5.144	3.112	1.410	5.807	2.965	13.0	138.1	172.2	6.4	1.2534	7.600	9.525	0.749641	132.2	205	249
D-217249-4	5.144	3.112	1.410	5.807	2.965	13.0	138.1	172.2	6.4	1.2534	7.600	9.525	0.749641	132.2	205	249
A-217249-2	5.144	3.112	1.410	5.807	2.965	13.0	138.1	172.2	6.4	1.2534	7.600	9.525	0.749641	132.2	205	249
D-163152-4	5.144	3.112	1.410	5.807	2.965	13.0	138.3	172.2	6.4	1.2548	7.600	9.536	0.751328	132.2	125	152
W-715152-4	5.144	3.112	1.410	5.807	2.965	13.0	138.3	172.2	6.4	1.2548	7.600	9.536	0.751328	132.2	125	152
A-715152-2	5.144	3.112	1.410	5.807	2.965	13.0	138.3	172.2	6.4	1.2548	7.600	9.536	0.751328	132.2	125	152

Arnold Engineering Co. MPP Powder Cores																
Part No.	OD cm 1	ID cm 2	HT cm 3	Wound OD cm 4	Wound HT cm 5	MPL cm 6	Wtfe grams 7	Wtcu grams 8	MLT cm 9	Ac cm sq 10	Wa cm sq 11	Ap cm 4th 12	Kg cm 5th 13	At cm sq 14	Perm 15	AL 16
D-164073-4	5.144	3.112	1.410	5.807	2.965	13.0	138.4	172.2	6.4	1.2554	7.600	9.541	0.752152	132.2	60	73
W-106073-4	5.144	3.112	1.410	5.807	2.965	13.0	138.4	172.2	6.4	1.2554	7.600	9.541	0.752152	132.2	60	73
A-106073-2	5.144	3.112	1.410	5.807	2.965	13.0	138.4	172.2	6.4	1.2554	7.600	9.541	0.752152	132.2	60	73
W-154179-4	5.144	3.112	1.410	5.807	2.965	13.0	138.5	172.2	6.4	1.2565	7.600	9.549	0.753414	132.2	147	179
D-154179-4	5.144	3.112	1.410	5.807	2.965	13.0	138.5	172.2	6.4	1.2565	7.600	9.549	0.753414	132.2	147	179
A-154179-2	5.144	3.112	1.410	5.807	2.965	13.0	138.5	172.2	6.4	1.2565	7.600	9.549	0.753414	132.2	147	179
W-327195-4	5.144	3.112	1.410	5.807	2.965	13.0	138.6	172.2	6.4	1.2576	7.600	9.558	0.754730	132.2	160	195
D-327195-4	5.144	3.112	1.410	5.807	2.965	13.0	138.6	172.2	6.4	1.2576	7.600	9.558	0.754730	132.2	160	195
A-327195-2	5.144	3.112	1.410	5.807	2.965	13.0	138.6	172.2	6.4	1.2576	7.600	9.558	0.754730	132.2	160	195
A-348032-2	5.144	3.112	1.410	5.807	2.965	13.0	140.0	172.2	6.4	1.2700	7.600	9.652	0.769689	132.2	26	32
D-198330-4	4.737	2.350	1.867	5.156	3.042	11.1	188.2	104.4	6.8	1.9886	4.333	8.617	1.011463	111.8	147	330
W-152330-4	4.737	2.350	1.867	5.156	3.042	11.1	188.2	104.4	6.8	1.9886	4.333	8.617	1.011463	111.8	147	330
A-152330-2	4.737	2.350	1.867	5.156	3.042	11.1	188.2	104.4	6.8	1.9886	4.333	8.617	1.011463	111.8	147	330
W-108281-4	4.737	2.350	1.867	5.156	3.042	11.1	188.4	104.4	6.8	1.9913	4.333	8.629	1.014260	111.8	125	281
D-466281-4	4.737	2.350	1.867	5.156	3.042	11.1	188.4	104.4	6.8	1.9913	4.333	8.629	1.014260	111.8	125	281
A-438281-2	4.737	2.350	1.867	5.156	3.042	11.1	188.4	104.4	6.8	1.9913	4.333	8.629	1.014260	111.8	125	281
W-140135-4	4.737	2.350	1.867	5.156	3.042	11.1	188.6	104.4	6.8	1.9931	4.333	8.637	1.016066	111.8	60	135
D-139135-4	4.737	2.350	1.867	5.156	3.042	11.1	188.6	104.4	6.8	1.9931	4.333	8.637	1.016066	111.8	60	135
A-759135-2	4.737	2.350	1.867	5.156	3.042	11.1	188.6	104.4	6.8	1.9931	4.333	8.637	1.016066	111.8	60	135
W-325360-4	4.737	2.350	1.867	5.156	3.042	11.1	188.6	104.4	6.8	1.9931	4.333	8.637	1.016066	111.8	160	360
D-325360-4	4.737	2.350	1.867	5.156	3.042	11.1	188.6	104.4	6.8	1.9931	4.333	8.637	1.016066	111.8	160	360
A-325360-2	4.737	2.350	1.867	5.156	3.042	11.1	188.6	104.4	6.8	1.9931	4.333	8.637	1.016066	111.8	160	360
W-215462-4	4.737	2.350	1.867	5.156	3.042	11.1	188.9	104.4	6.8	1.9963	4.333	8.651	1.019373	111.8	205	462
D-215462-4	4.737	2.350	1.867	5.156	3.042	11.1	188.9	104.4	6.8	1.9963	4.333	8.651	1.019373	111.8	205	462
A-215462-2	4.737	2.350	1.867	5.156	3.042	11.1	188.9	104.4	6.8	1.9963	4.333	8.651	1.019373	111.8	205	462
W-180390-4	4.737	2.350	1.867	5.156	3.042	11.1	188.9	104.4	6.8	1.9969	4.333	8.653	1.019985	111.8	173	390
D-180390-4	4.737	2.350	1.867	5.156	3.042	11.1	188.9	104.4	6.8	1.9969	4.333	8.653	1.019985	111.8	173	390
A-180390-2	4.737	2.350	1.867	5.156	3.042	11.1	188.9	104.4	6.8	1.9969	4.333	8.653	1.019985	111.8	173	390
W-087059-4	4.737	2.350	1.867	5.156	3.042	11.1	190.2	104.4	6.8	2.0101	4.333	8.711	1.033509	111.8	26	59
D-087059-4	4.737	2.350	1.867	5.156	3.042	11.1	190.2	104.4	6.8	2.0101	4.333	8.711	1.033509	111.8	26	59

Arnold Engineering Co. MPP Powder Cores

Part No.	OD cm 1	ID cm 2	HT cm 3	Wound OD cm 4	Wound HT cm 5	MPL cm 6	Wtfe grams 7	Wtcu grams 8	MLT cm 9	Ac cm sq 10	Wa cm sq 11	Ap cm 4th 12	Kg cm 5th 13	At cm sq 14	Perm 15	AL 16
A-087059-2	4.737	2.350	1.867	5.156	3.042	11.1	190.2	104.4	6.8	2.0101	4.333	8.711	1.033509	111.8	26	59
W-088032-4	4.737	2.350	1.867	5.156	3.042	11.1	191.6	104.4	6.8	2.0247	4.333	8.774	1.048578	111.8	14	32
D-088032-4	4.737	2.350	1.867	5.156	3.042	11.1	191.6	104.4	6.8	2.0247	4.333	8.774	1.048578	111.8	14	32
A-088032-2	4.737	2.350	1.867	5.156	3.042	11.1	191.6	104.4	6.8	2.0247	4.333	8.774	1.048578	111.8	14	32
W-107156-4	5.779	3.493	1.461	6.523	3.207	14.6	179.0	237.0	7.0	1.4463	9.575	13.848	1.151120	163.8	125	156
D-927156-4	5.779	3.493	1.461	6.523	3.207	14.6	179.0	237.0	7.0	1.4463	9.575	13.848	1.151120	163.8	125	156
B-912156-4	5.779	3.493	1.461	6.523	3.207	14.6	179.0	237.0	7.0	1.4463	9.575	13.848	1.151120	163.8	125	156
A-109156-2	5.779	3.493	1.461	6.523	3.207	14.6	179.0	237.0	7.0	1.4463	9.575	13.848	1.151120	163.8	125	156
W-141075-4	5.779	3.493	1.461	6.523	3.207	14.6	179.3	237.0	7.0	1.4486	9.575	13.870	1.154812	163.8	60	75
D-093075-4	5.779	3.493	1.461	6.523	3.207	14.6	179.3	237.0	7.0	1.4486	9.575	13.870	1.154812	163.8	60	75
A-488075-2	5.779	3.493	1.461	6.523	3.207	14.6	179.3	237.0	7.0	1.4486	9.575	13.870	1.154812	163.8	60	75
W-328200-4	5.779	3.493	1.461	6.523	3.207	14.6	179.3	237.0	7.0	1.4486	9.575	13.870	1.154812	163.8	160	200
D-328200-4	5.779	3.493	1.461	6.523	3.207	14.6	179.3	237.0	7.0	1.4486	9.575	13.870	1.154812	163.8	160	200
A-328200-2	5.779	3.493	1.461	6.523	3.207	14.6	179.3	237.0	7.0	1.4486	9.575	13.870	1.154812	163.8	160	200
W-155185-4	5.779	3.493	1.461	6.523	3.207	14.6	180.5	237.0	7.0	1.4584	9.575	13.965	1.170578	163.8	147	185
D-155185-4	5.779	3.493	1.461	6.523	3.207	14.6	180.5	237.0	7.0	1.4584	9.575	13.965	1.170578	163.8	147	185
A-155185-2	5.779	3.493	1.461	6.523	3.207	14.6	180.5	237.0	7.0	1.4584	9.575	13.965	1.170578	163.8	147	185
W-182218-4	5.779	3.493	1.461	6.523	3.207	14.6	180.8	237.0	7.0	1.4603	9.575	13.983	1.173579	163.8	173	218
D-182218-4	5.779	3.493	1.461	6.523	3.207	14.6	180.8	237.0	7.0	1.4603	9.575	13.983	1.173579	163.8	173	218
A-182218-2	5.779	3.493	1.461	6.523	3.207	14.6	180.8	237.0	7.0	1.4603	9.575	13.983	1.173579	163.8	173	218
W-218259-4	5.779	3.493	1.461	6.523	3.207	14.6	181.2	237.0	7.0	1.4641	9.575	14.019	1.179732	163.8	205	259
D-218259-4	5.779	3.493	1.461	6.523	3.207	14.6	181.2	237.0	7.0	1.4641	9.575	14.019	1.179732	163.8	205	259
A-218259-2	5.779	3.493	1.461	6.523	3.207	14.6	181.2	237.0	7.0	1.4641	9.575	14.019	1.179732	163.8	205	259
D-095033-4	5.779	3.493	1.461	6.523	3.207	14.6	182.1	237.0	7.0	1.4709	9.575	14.084	1.190618	163.8	26	33
B-377033-4	5.779	3.493	1.461	6.523	3.207	14.6	182.1	237.0	7.0	1.4709	9.575	14.084	1.190618	163.8	26	33
W-094033-4	5.779	3.493	1.461	6.523	3.207	14.6	182.1	237.0	7.0	1.4709	9.575	14.084	1.190618	163.8	26	33
A-094033-2	5.779	3.493	1.461	6.523	3.207	14.6	182.1	237.0	7.0	1.4709	9.575	14.084	1.190618	163.8	26	33
D-097018-4	5.779	3.493	1.461	6.523	3.207	14.6	184.4	237.0	7.0	1.4900	9.575	14.267	1.221744	163.8	14	18
W-096018-4	5.779	3.493	1.461	6.523	3.207	14.6	184.4	237.0	7.0	1.4900	9.575	14.267	1.221744	163.8	14	18
A-096018-2	5.779	3.493	1.461	6.523	3.207	14.6	184.4	237.0	7.0	1.4900	9.575	14.267	1.221744	163.8	14	18

Arnold Engineering Co. MPP Powder Cores																
Part No.	OD cm 1	ID cm 2	HT cm 3	Wound OD cm 4	Wound HT cm 5	MPL cm 6	Wtfe grams 7	Wtcu grams 8	MLT cm 9	Ac cm sq 10	Wa cm sq 11	Ap cm 4th 12	Kg cm 5th 13	At cm sq 14	Perm 15	AL 16
A-123068-2	7.844	4.859	1.334	8.901	3.763	20.0	305.2	554.2	8.4	1.7995	18.534	33.352	2.855153	285.5	60	68
D-627142-4	7.844	4.859	1.334	8.901	3.763	20.0	305.9	554.2	8.4	1.8038	18.534	33.431	2.868604	285.5	125	142
A-866142-2	7.844	4.859	1.334	8.901	3.763	20.0	305.9	554.2	8.4	1.8038	18.534	33.431	2.868604	285.5	125	142
A-156167-2	7.844	4.859	1.334	8.901	3.763	20.0	305.9	554.2	8.4	1.8038	18.534	33.432	2.868879	285.5	147	167
A-219233-2	7.844	4.859	1.334	8.901	3.763	20.0	306.1	554.2	8.4	1.8047	18.534	33.448	2.871561	285.5	205	233
A-183197-2	7.844	4.859	1.334	8.901	3.763	20.0	306.6	554.2	8.4	1.8081	18.534	33.511	2.882400	285.5	173	197
A-335016-2	7.844	4.859	1.334	8.901	3.763	20.0	307.8	554.2	8.4	1.8146	18.534	33.633	2.903340	285.5	14	16
A-124030-2	7.844	4.859	1.334	8.901	3.763	20.0	310.7	554.2	8.4	1.8321	18.534	33.956	2.959442	285.5	26	30
B-673033-4	7.844	4.859	1.486	8.901	3.915	20.0	341.8	570.2	8.7	2.0153	18.534	37.352	3.480006	290.4	26	33
B-600040-4	10.224	5.652	1.422	11.335	4.248	24.9	647.1	932.1	10.5	3.0529	25.072	76.543	8.940644	440.6	26	40
A-126040-2	10.224	5.652	1.422	11.335	4.248	24.9	647.1	932.1	10.5	3.0529	25.072	76.543	8.940644	440.6	26	40
A-157268-2	10.224	5.652	1.715	11.335	4.540	24.9	766.8	973.8	10.9	3.6178	25.072	90.706	12.018143	452.6	147	268
A-542228-2	10.224	5.652	1.715	11.335	4.540	24.9	767.2	973.8	10.9	3.6195	25.072	90.750	12.029625	452.6	125	228
A-220374-2	10.224	5.652	1.715	11.335	4.540	24.9	767.3	973.8	10.9	3.6203	25.072	90.769	12.034773	452.6	205	374
A-184316-2	10.224	5.652	1.715	11.335	4.540	24.9	768.3	973.8	10.9	3.6246	25.072	90.879	12.063807	452.6	173	316
A-125112-2	10.224	5.652	1.715	11.335	4.540	24.9	785.1	973.8	10.9	3.7042	25.072	92.873	12.598997	452.6	60	112
A-430026-2	13.317	7.795	2.096	14.930	5.993	33.2	1381.5	2375.9	14.0	4.9011	47.701	233.790	32.722617	786.0	14	26
A-128124-2	13.317	7.795	2.096	14.930	5.993	33.2	1537.4	2375.9	14.0	5.4541	47.701	260.166	40.522691	786.0	60	124
A-158304-2	13.317	7.795	2.096	14.930	5.993	33.2	1538.4	2375.9	14.0	5.4576	47.701	260.337	40.576063	786.0	147	304
A-185358-2	13.317	7.795	2.096	14.930	5.993	33.2	1539.4	2375.9	14.0	5.4612	47.701	260.506	40.628543	786.0	173	358
A-127259-2	13.317	7.795	2.096	14.930	5.993	33.2	1541.4	2375.9	14.0	5.4681	47.701	260.837	40.732110	786.0	125	259
A-221425-2	13.317	7.795	2.096	14.930	5.993	33.2	1542.2	2375.9	14.0	5.4712	47.701	260.985	40.778152	786.0	205	425
B-694054-4	13.317	7.795	2.096	14.930	5.993	33.2	1545.0	2375.9	14.0	5.4811	47.701	261.457	40.925899	786.0	26	54
A-129054-2	13.317	7.795	2.096	14.930	5.993	33.2	1545.0	2375.9	14.0	5.4811	47.701	261.457	40.925899	786.0	26	54

Arnold Engineering Co. MSS Powder Cores																
Part No.	OD cm 1	ID cm 2	HT cm 3	Wound OD cm 4	Wound HT cm 5	MPL cm 6	Wtfe grams 7	Wtcu grams 8	MLT cm 9	Ac cm sq 10	Wa cm sq 11	Ap cm 4th 12	Kg cm 5th 13	At cm sq 14	Perm 15	AL 16
MS-014-125	0.407	0.127	0.203	0.422	0.267	0.8	0.1	0.0	0.7	0.012	0.013	0.000	0.000001	0.8	125	26
MS-014-90	0.407	0.127	0.203	0.422	0.267	0.8	0.1	0.0	0.7	0.012	0.013	0.000	0.000001	0.8	90	19
MS-014-75	0.407	0.127	0.203	0.422	0.267	0.8	0.1	0.0	0.7	0.012	0.013	0.000	0.000001	0.8	75	16
MS-014-60	0.407	0.127	0.203	0.422	0.267	0.8	0.1	0.0	0.7	0.012	0.013	0.000	0.000001	0.8	60	12
MS-015-125	0.445	0.170	0.305	0.469	0.390	1.0	0.1	0.1	0.8	0.020	0.023	0.000	0.000004	1.1	125	35
MS-015-90	0.445	0.170	0.305	0.469	0.390	1.0	0.1	0.1	0.8	0.020	0.023	0.000	0.000004	1.1	90	25
MS-015-75	0.445	0.170	0.305	0.469	0.390	1.0	0.1	0.1	0.8	0.020	0.023	0.000	0.000004	1.1	75	21
MS-015-60	0.445	0.170	0.305	0.469	0.390	1.0	0.1	0.1	0.8	0.020	0.023	0.000	0.000004	1.1	60	17
MS-018-125	0.516	0.185	0.305	0.540	0.398	1.1	0.2	0.1	0.9	0.028	0.027	0.001	0.000009	1.4	125	42
MS-018-90	0.516	0.185	0.305	0.540	0.398	1.1	0.2	0.1	0.9	0.028	0.027	0.001	0.000009	1.4	90	30
MS-018-75	0.516	0.185	0.305	0.540	0.398	1.1	0.2	0.1	0.9	0.028	0.027	0.001	0.000009	1.4	75	25
MS-018-60	0.516	0.185	0.305	0.540	0.398	1.1	0.2	0.1	0.9	0.028	0.027	0.001	0.000009	1.4	60	20
MS-027-125	0.711	0.216	0.305	0.735	0.413	1.5	0.4	0.1	1.1	0.048	0.037	0.002	0.000032	2.2	125	54
MS-027-90	0.711	0.216	0.305	0.735	0.413	1.5	0.4	0.1	1.1	0.048	0.037	0.002	0.000032	2.2	90	39
MS-027-75	0.711	0.216	0.305	0.735	0.413	1.5	0.4	0.1	1.1	0.048	0.037	0.002	0.000032	2.2	75	32
MS-027-60	0.711	0.216	0.305	0.735	0.413	1.5	0.4	0.1	1.1	0.048	0.037	0.002	0.000032	2.2	60	26
MS-025-125	0.686	0.228	0.330	0.714	0.444	1.4	0.4	0.2	1.1	0.048	0.041	0.002	0.000035	2.2	125	52
MS-025-90	0.686	0.228	0.330	0.714	0.444	1.4	0.4	0.2	1.1	0.048	0.041	0.002	0.000035	2.2	90	37
MS-025-75	0.686	0.228	0.330	0.714	0.444	1.4	0.4	0.2	1.1	0.048	0.041	0.002	0.000035	2.2	75	31
MS-025-60	0.686	0.228	0.330	0.714	0.444	1.4	0.4	0.2	1.1	0.048	0.041	0.002	0.000035	2.2	60	24
MS-026-125	0.711	0.216	0.529	0.735	0.637	1.5	0.8	0.2	1.4	0.090	0.037	0.003	0.000084	2.8	125	103
MS-026-90	0.711	0.216	0.529	0.735	0.637	1.5	0.8	0.2	1.4	0.090	0.037	0.003	0.000084	2.8	90	74
MS-026-75	0.711	0.216	0.529	0.735	0.637	1.5	0.8	0.2	1.4	0.090	0.037	0.003	0.000084	2.8	75	62
MS-026-60	0.711	0.216	0.529	0.735	0.637	1.5	0.8	0.2	1.4	0.090	0.037	0.003	0.000084	2.8	60	50
MS-031-125	0.838	0.345	0.369	0.890	0.542	1.9	0.7	0.4	1.3	0.058	0.093	0.005	0.000100	3.4	125	52
MS-031-90	0.838	0.345	0.369	0.890	0.542	1.9	0.7	0.4	1.3	0.058	0.093	0.005	0.000100	3.4	90	37
MS-031-75	0.838	0.345	0.369	0.890	0.542	1.9	0.7	0.4	1.3	0.058	0.093	0.005	0.000100	3.4	75	31
MS-031-60	0.838	0.345	0.369	0.890	0.542	1.9	0.7	0.4	1.3	0.058	0.093	0.005	0.000100	3.4	60	25
MS-028-125	0.755	0.345	0.559	0.812	0.732	1.7	0.8	0.5	1.5	0.077	0.093	0.007	0.000148	3.5	125	70
MS-028-90	0.755	0.345	0.559	0.812	0.732	1.7	0.8	0.5	1.5	0.077	0.093	0.007	0.000148	3.5	90	50

Arnold Engineering Co. MSS Powder Cores

Part No.	OD cm 1	ID cm 2	HT cm 3	Wound OD cm 4	Wound HT cm 5	MPL cm 6	Wtfe grams 7	Wtcu grams 8	MLT cm 9	Ac cm sq 10	Wa cm sq 11	Ap cm 4th 12	Kg cm 5th 13	At cm sq 14	Perm 15	AL 16
MS-028-75	0.755	0.345	0.559	0.812	0.732	1.7	0.8	0.5	1.5	0.077	0.093	0.007	0.000148	3.5	75	42
MS-028-60	0.755	0.345	0.559	0.812	0.732	1.7	0.8	0.5	1.5	0.077	0.093	0.007	0.000148	3.5	60	33
MS-039-125	1.016	0.427	0.369	1.081	0.583	2.3	1.0	0.7	1.4	0.074	0.143	0.011	0.000223	4.7	125	53
MS-039-90	1.016	0.427	0.369	1.081	0.583	2.3	1.0	0.7	1.4	0.074	0.143	0.011	0.000223	4.7	90	38
MS-039-75	1.016	0.427	0.369	1.081	0.583	2.3	1.0	0.7	1.4	0.074	0.143	0.011	0.000223	4.7	75	32
MS-039-60	1.016	0.427	0.369	1.081	0.583	2.3	1.0	0.7	1.4	0.074	0.143	0.011	0.000223	4.7	60	25
MS-038-125	1.016	0.427	0.447	1.081	0.661	2.3	1.3	0.8	1.5	0.092	0.143	0.013	0.000317	5.0	125	66
MS-038-90	1.016	0.427	0.447	1.081	0.661	2.3	1.3	0.8	1.5	0.092	0.143	0.013	0.000317	5.0	90	48
MS-038-75	1.016	0.427	0.447	1.081	0.661	2.3	1.3	0.8	1.5	0.092	0.143	0.013	0.000317	5.0	75	40
MS-038-60	1.016	0.427	0.447	1.081	0.661	2.3	1.3	0.8	1.5	0.092	0.143	0.013	0.000317	5.0	60	32
MS-040-125	1.067	0.457	0.447	1.138	0.676	2.4	1.4	0.9	1.6	0.096	0.164	0.016	0.000385	5.5	125	66
MS-040-90	1.067	0.457	0.447	1.138	0.676	2.4	1.4	0.9	1.6	0.096	0.164	0.016	0.000385	5.5	90	48
MS-040-75	1.067	0.457	0.447	1.138	0.676	2.4	1.4	0.9	1.6	0.096	0.164	0.016	0.000385	5.5	75	40
MS-040-60	1.067	0.457	0.447	1.138	0.676	2.4	1.4	0.9	1.6	0.096	0.164	0.016	0.000385	5.5	60	32
MS-044-125	1.169	0.584	0.447	1.274	0.739	2.8	1.5	1.6	1.7	0.091	0.268	0.024	0.000537	6.8	125	53
MS-044-90	1.169	0.584	0.447	1.274	0.739	2.8	1.5	1.6	1.7	0.091	0.268	0.024	0.000537	6.8	90	38
MS-044-75	1.169	0.584	0.447	1.274	0.739	2.8	1.5	1.6	1.7	0.091	0.268	0.024	0.000537	6.8	75	32
MS-044-60	1.169	0.584	0.447	1.274	0.739	2.8	1.5	1.6	1.7	0.091	0.268	0.024	0.000537	6.8	60	26
MS-050-125	1.321	0.711	0.526	1.457	0.882	3.2	2.2	2.7	1.9	0.113	0.397	0.045	0.001068	9.0	125	56
MS-050-90	1.321	0.711	0.526	1.457	0.882	3.2	2.2	2.7	1.9	0.113	0.397	0.045	0.001068	9.0	90	40
MS-050-75	1.321	0.711	0.526	1.457	0.882	3.2	2.2	2.7	1.9	0.113	0.397	0.045	0.001068	9.0	75	34
MS-050-60	1.321	0.711	0.526	1.457	0.882	3.2	2.2	2.7	1.9	0.113	0.397	0.045	0.001068	9.0	60	27
MS-050-125	1.715	0.965	0.686	1.908	1.169	4.2	5.1	6.4	2.5	0.196	0.731	0.143	0.004549	15.6	125	72
MS-050-90	1.715	0.965	0.686	1.908	1.169	4.2	5.1	6.4	2.5	0.196	0.731	0.143	0.004549	15.6	90	52
MS-050-75	1.715	0.965	0.686	1.908	1.169	4.2	5.1	6.4	2.5	0.196	0.731	0.143	0.004549	15.6	75	43
MS-050-60	1.715	0.965	0.686	1.908	1.169	4.2	5.1	6.4	2.5	0.196	0.731	0.143	0.004549	15.6	60	35
MS-068-125	1.778	0.914	0.686	1.946	1.143	4.2	6.0	5.9	2.5	0.232	0.656	0.152	0.005603	15.9	125	89
MS-068-90	1.778	0.914	0.686	1.946	1.143	4.2	6.0	5.9	2.5	0.232	0.656	0.152	0.005603	15.9	90	64
MS-068-75	1.778	0.914	0.686	1.946	1.143	4.2	6.0	5.9	2.5	0.232	0.656	0.152	0.005603	15.9	75	53
MS-068-60	1.778	0.914	0.686	1.946	1.143	4.2	6.0	5.9	2.5	0.232	0.656	0.152	0.005603	15.9	60	43

Arnold Engineering Co. MSS Powder Cores																
Part No.	OD cm 1	ID cm 2	HT cm 3	Wound OD cm 4	Wound HT cm 5	MPL cm 6	Wtfe grams 7	Wtcu grams 8	MLT cm 9	Ac cm sq 10	Wa cm sq 11	Ap cm 4th 12	Kg cm 5th 13	At cm sq 14	Perm 15	AL 16
MS-080-125	2.081	1.219	0.686	2.333	1.296	5.2	7.4	11.5	2.8	0.232	1.166	0.271	0.009091	22.2	125	68
MS-080-90	2.081	1.219	0.686	2.333	1.296	5.2	7.4	11.5	2.8	0.232	1.166	0.271	0.009091	22.2	90	49
MS-080-75	2.081	1.219	0.686	2.333	1.296	5.2	7.4	11.5	2.8	0.232	1.166	0.271	0.009091	22.2	75	41
MS-080-60	2.081	1.219	0.686	2.333	1.296	5.2	7.4	11.5	2.8	0.232	1.166	0.271	0.009091	22.2	60	33
MS-090-125	2.341	1.346	0.813	2.615	1.486	5.8	11.6	16.0	3.2	0.327	1.422	0.465	0.019167	28.2	125	90
MS-090-90	2.341	1.346	0.813	2.615	1.486	5.8	11.6	16.0	3.2	0.327	1.422	0.465	0.019167	28.2	90	65
MS-090-75	2.341	1.346	0.813	2.615	1.486	5.8	11.6	16.0	3.2	0.327	1.422	0.465	0.019167	28.2	75	54
MS-090-60	2.341	1.346	0.813	2.615	1.486	5.8	11.6	16.0	3.2	0.327	1.422	0.465	0.019167	28.2	60	43
MS-092-125	2.411	1.389	0.940	2.695	1.635	6.0	13.9	18.5	3.4	0.380	1.515	0.576	0.025483	31.0	125	105
MS-092-90	2.411	1.389	0.940	2.695	1.635	6.0	13.9	18.5	3.4	0.380	1.515	0.576	0.025483	31.0	90	76
MS-092-75	2.411	1.389	0.940	2.695	1.635	6.0	13.9	18.5	3.4	0.380	1.515	0.576	0.025483	31.0	75	63
MS-092-60	2.411	1.389	0.940	2.695	1.635	6.0	13.9	18.5	3.4	0.380	1.515	0.576	0.025483	31.0	60	50
MS-106-125	2.741	1.422	1.169	3.005	1.880	6.5	25.7	22.9	4.1	0.639	1.587	1.014	0.063806	39.1	125	157
MS-106-90	2.741	1.422	1.169	3.005	1.880	6.5	25.7	22.9	4.1	0.639	1.587	1.014	0.063806	39.1	90	113
MS-106-75	2.741	1.422	1.169	3.005	1.880	6.5	25.7	22.9	4.1	0.639	1.587	1.014	0.063806	39.1	75	94
MS-106-60	2.741	1.422	1.169	3.005	1.880	6.5	25.7	22.9	4.1	0.639	1.587	1.014	0.063806	39.1	60	75
MS-135-125	3.481	2.289	0.940	4.006	2.085	9.0	25.5	62.7	4.3	0.458	4.113	1.884	0.080467	63.5	125	79
MS-135-90	3.481	2.289	0.940	4.006	2.085	9.0	25.5	62.7	4.3	0.458	4.113	1.884	0.080467	63.5	90	57
MS-135-75	3.481	2.289	0.940	4.006	2.085	9.0	25.5	62.7	4.3	0.458	4.113	1.884	0.080467	63.5	75	47
MS-135-60	3.481	2.289	0.940	4.006	2.085	9.0	25.5	62.7	4.3	0.458	4.113	1.884	0.080467	63.5	60	38
MS-131-125	3.351	1.943	0.927	3.750	1.899	8.3	27.9	43.9	4.2	0.546	2.964	1.618	0.084869	54.9	125	109
MS-131-90	3.351	1.943	0.927	3.750	1.899	8.3	27.9	43.9	4.2	0.546	2.964	1.618	0.084869	54.9	90	78
MS-131-75	3.351	1.943	0.927	3.750	1.899	8.3	27.9	43.9	4.2	0.546	2.964	1.618	0.084869	54.9	75	65
MS-131-60	3.351	1.943	0.927	3.750	1.899	8.3	27.9	43.9	4.2	0.546	2.964	1.618	0.084869	54.9	60	52
MS-130-125	3.351	1.943	1.118	3.750	2.090	8.3	34.0	47.1	4.5	0.666	2.964	1.974	0.117640	57.5	125	127
MS-130-90	3.351	1.943	1.118	3.750	2.090	8.3	34.0	47.1	4.5	0.666	2.964	1.974	0.117640	57.5	90	91
MS-130-75	3.351	1.943	1.118	3.750	2.090	8.3	34.0	47.1	4.5	0.666	2.964	1.974	0.117640	57.5	75	76
MS-130-60	3.351	1.943	1.118	3.750	2.090	8.3	34.0	47.1	4.5	0.666	2.964	1.974	0.117640	57.5	60	61
MS-141-125	3.631	2.189	1.097	4.097	2.192	9.2	37.9	62.3	4.7	0.670	3.762	2.520	0.144939	67.3	125	117
MS-141-90	3.631	2.189	1.097	4.097	2.192	9.2	37.9	62.3	4.7	0.670	3.762	2.520	0.144939	67.3	90	84

Arnold Engineering Co. MSS Powder Cores																
Part No.	OD cm 1	ID cm 2	HT cm 3	Wound OD cm 4	Wound HT cm 5	MPL cm 6	Wtfe grams 7	Wtcu grams 8	MLT cm 9	Ac cm sq 10	Wa cm sq 11	Ap cm 4th 12	Kg cm 5th 13	At cm sq 14	Perm 15	AL 16
MS-141-75	3.631	2.189	1.097	4.097	2.192	9.2	37.9	62.3	4.7	0.670	3.762	2.520	0.144939	67.3	75	70
MS-141-60	3.631	2.189	1.097	4.097	2.192	9.2	37.9	62.3	4.7	0.670	3.762	2.520	0.144939	67.3	60	56
MS-157-125	4.041	2.359	1.499	4.528	2.679	10.1	65.9	87.5	5.6	1.060	4.368	4.631	0.348655	86.4	125	168
MS-157-90	4.041	2.359	1.499	4.528	2.679	10.1	65.9	87.5	5.6	1.060	4.368	4.631	0.348655	86.4	90	121
MS-157-75	4.041	2.359	1.499	4.528	2.679	10.1	65.9	87.5	5.6	1.060	4.368	4.631	0.348655	86.4	75	101
MS-157-60	4.041	2.359	1.499	4.528	2.679	10.1	65.9	87.5	5.6	1.060	4.368	4.631	0.348655	86.4	60	81
MS-185-125	4.721	2.819	1.575	5.315	2.985	11.9	96.8	139.7	6.3	1.320	6.238	8.234	0.690475	115.9	125	178
MS-185-90	4.721	2.819	1.575	5.315	2.985	11.9	96.8	139.7	6.3	1.320	6.238	8.234	0.690475	115.9	90	128
MS-185-75	4.721	2.819	1.575	5.315	2.985	11.9	96.8	139.7	6.3	1.320	6.238	8.234	0.690475	115.9	75	107
MS-185-60	4.721	2.819	1.575	5.315	2.985	11.9	96.8	139.7	6.3	1.320	6.238	8.234	0.690475	115.9	60	85
MS-200-125	5.131	3.149	1.397	5.810	2.972	13.0	99.4	175.5	6.3	1.240	7.784	9.652	0.755143	132.5	125	152
MS-200-90	5.131	3.149	1.397	5.810	2.972	13.0	99.4	175.5	6.3	1.240	7.784	9.652	0.755143	132.5	90	109
MS-200-75	5.131	3.149	1.397	5.810	2.972	13.0	99.4	175.5	6.3	1.240	7.784	9.652	0.755143	132.5	75	91
MS-200-60	5.131	3.149	1.397	5.810	2.972	13.0	99.4	175.5	6.3	1.240	7.784	9.652	0.755143	132.5	60	73
MS-184-125	4.721	2.359	1.854	5.144	3.034	11.2	134.3	104.7	6.7	1.950	4.368	8.518	0.985346	111.3	125	281
MS-184-90	4.721	2.359	1.854	5.144	3.034	11.2	134.3	104.7	6.7	1.950	4.368	8.518	0.985346	111.3	90	202
MS-184-75	4.721	2.359	1.854	5.144	3.034	11.2	134.3	104.7	6.7	1.950	4.368	8.518	0.985346	111.3	75	169
MS-184-60	4.721	2.359	1.854	5.144	3.034	11.2	134.3	104.7	6.7	1.950	4.368	8.518	0.985346	111.3	60	135
MS-225-125	5.771	3.509	1.448	6.523	3.203	14.6	129.6	238.3	6.9	1.440	9.666	13.919	1.156279	163.7	125	156
MS-225-90	5.771	3.509	1.448	6.523	3.203	14.6	129.6	238.3	6.9	1.440	9.666	13.919	1.156279	163.7	90	112
MS-225-75	5.771	3.509	1.448	6.523	3.203	14.6	129.6	238.3	6.9	1.440	9.666	13.919	1.156279	163.7	75	94
MS-225-60	5.771	3.509	1.448	6.523	3.203	14.6	129.6	238.3	6.9	1.440	9.666	13.919	1.156279	163.7	60	75
MS-300-125	7.831	4.869	1.321	8.894	3.756	20.0	217.7	554.5	8.4	1.770	18.610	32.940	2.783522	284.9	125	142
MS-300-90	7.831	4.869	1.321	8.894	3.756	20.0	217.7	554.5	8.4	1.770	18.610	32.940	2.783522	284.9	90	102
MS-300-75	7.831	4.869	1.321	8.894	3.756	20.0	217.7	554.5	8.4	1.770	18.610	32.940	2.783522	284.9	75	85
MS-300-60	7.831	4.869	1.321	8.894	3.756	20.0	217.7	554.5	8.4	1.770	18.610	32.940	2.783522	284.9	60	68
MS-400-125	10.211	5.669	1.702	11.330	4.537	25.0	547.4	977.1	10.9	3.560	25.228	89.812	11.741810	452.1	125	228
MS-400-90	10.211	5.669	1.702	11.330	4.537	25.0	547.4	977.1	10.9	3.560	25.228	89.812	11.741810	452.1	90	164
MS-400-75	10.211	5.669	1.702	11.330	4.537	25.0	547.4	977.1	10.9	3.560	25.228	89.812	11.741810	452.1	75	137
MS-400-60	10.211	5.669	1.702	11.330	4.537	25.0	547.4	977.1	10.9	3.560	25.228	89.812	11.741810	452.1	60	109

Arnold Engineering Co. MSS Powder Cores																
Part No.	OD cm 1	ID cm 2	HT cm 3	Wound OD cm 4	Wound HT cm 5	MPL cm 6	Wtfe grams 7	Wtcu grams 8	MLT cm 9	Ac cm sq 10	Wa cm sq 11	Ap cm 4th 12	Kg cm 5th 13	At cm sq 14	Perm 15	AL 16
MS-520-125	13.301	7.809	2.081	14.922	5.986	33.2	1090.3	2378.1	14.0	5.340	47.870	255.624	39.083570	784.8	125	256
MS-520-90	13.301	7.809	2.081	14.922	5.986	33.2	1090.3	2378.1	14.0	5.340	47.870	255.624	39.083570	784.8	90	184
MS-520-75	13.301	7.809	2.081	14.922	5.986	33.2	1090.3	2378.1	14.0	5.340	47.870	255.624	39.083570	784.8	75	154
MS-520-60	13.301	7.809	2.081	14.922	5.986	33.2	1090.3	2378.1	14.0	5.340	47.870	255.624	39.083570	784.8	60	123
MS-521-125	13.301	7.809	2.591	14.922	6.496	33.2	1361.9	2517.0	14.8	6.670	47.870	319.291	57.611580	812.3	125	259
MS-521-90	13.301	7.809	2.591	14.922	6.496	33.2	1361.9	2517.0	14.8	6.670	47.870	319.291	57.611580	812.3	90	186
MS-521-75	13.301	7.809	2.591	14.922	6.496	33.2	1361.9	2517.0	14.8	6.670	47.870	319.291	57.611580	812.3	75	155
MS-521-60	13.301	7.809	2.591	14.922	6.496	33.2	1361.9	2517.0	14.8	6.670	47.870	319.291	57.611580	812.3	60	124

Magnetics Kool Mμ Powder Cores																
Part No	OD cm 1	ID cm 2	HT cm 3	Wound OD cm 4	Wound HT cm 5	MPL cm 6	Wtfe grams 7	Wtcu grams 8	MLT cm 9	Ac cm sq 10	Wa cm sq 11	Ap cm 4th 12	Kg cm 5th 13	At cm sq 14	Perm 15	AL 16
77140	0.381	0.152	0.183	0.403	0.259	0.8	0.1	0.1	0.6	0.0137	0.018	0.000	0.000002	0.7	125	26
77444	0.381	0.152	0.183	0.403	0.259	0.8	0.1	0.1	0.6	0.0137	0.018	0.000	0.000002	0.7	90	19
77141	0.381	0.152	0.183	0.403	0.259	0.8	0.1	0.1	0.6	0.0137	0.018	0.000	0.000002	0.7	60	13
77445	0.381	0.152	0.183	0.403	0.259	0.8	0.1	0.1	0.6	0.0173	0.018	0.000	0.000004	0.7	75	16
77020	0.699	0.229	0.343	0.727	0.458	1.4	0.4	0.2	1.1	0.0470	0.041	0.002	0.000033	2.3	125	50
77824	0.699	0.229	0.343	0.727	0.458	1.4	0.4	0.2	1.1	0.0470	0.041	0.002	0.000033	2.3	90	36
77825	0.699	0.229	0.343	0.727	0.458	1.4	0.4	0.2	1.1	0.0470	0.041	0.002	0.000033	2.3	75	30
77021	0.699	0.229	0.343	0.727	0.458	1.4	0.4	0.2	1.1	0.0470	0.041	0.002	0.000033	2.3	60	24
77270	0.732	0.221	0.554	0.757	0.665	1.4	0.8	0.2	1.5	0.0920	0.038	0.004	0.000088	3.0	125	103
77874	0.732	0.221	0.554	0.757	0.665	1.4	0.8	0.2	1.5	0.0920	0.038	0.004	0.000088	3.0	90	74
77875	0.732	0.221	0.554	0.757	0.665	1.4	0.8	0.2	1.5	0.0920	0.038	0.004	0.000088	3.0	75	62
77271	0.732	0.221	0.554	0.757	0.665	1.4	0.8	0.2	1.5	0.0920	0.038	0.004	0.000088	3.0	60	50
77030	0.851	0.343	0.381	0.902	0.553	1.8	0.7	0.4	1.3	0.0615	0.092	0.006	0.000108	3.5	125	52
77834	0.851	0.343	0.381	0.902	0.553	1.8	0.7	0.4	1.3	0.0615	0.092	0.006	0.000108	3.5	90	37
77835	0.851	0.343	0.381	0.902	0.553	1.8	0.7	0.4	1.3	0.0615	0.092	0.006	0.000108	3.5	75	31
77031	0.851	0.343	0.381	0.902	0.553	1.8	0.7	0.4	1.3	0.0615	0.092	0.006	0.000108	3.5	60	25
77280	1.029	0.427	0.381	1.093	0.595	2.2	1.0	0.7	1.4	0.0752	0.143	0.011	0.000226	4.8	125	53
77884	1.029	0.427	0.381	1.093	0.595	2.2	1.0	0.7	1.4	0.0752	0.143	0.011	0.000226	4.8	90	38
77885	1.029	0.427	0.381	1.093	0.595	2.2	1.0	0.7	1.4	0.0752	0.143	0.011	0.000226	4.8	75	32
77281	1.029	0.427	0.381	1.093	0.595	2.2	1.0	0.7	1.4	0.0752	0.143	0.011	0.000226	4.8	60	25
77040	1.080	0.457	0.457	1.150	0.686	2.4	1.5	0.9	1.6	0.1000	0.164	0.016	0.000411	5.6	125	66
77844	1.080	0.457	0.457	1.150	0.686	2.4	1.5	0.9	1.6	0.1000	0.164	0.016	0.000411	5.6	90	48
77845	1.080	0.457	0.457	1.150	0.686	2.4	1.5	0.9	1.6	0.1000	0.164	0.016	0.000411	5.6	75	40
77041	1.080	0.457	0.457	1.150	0.686	2.4	1.5	0.9	1.6	0.1000	0.164	0.016	0.000411	5.6	60	32
77050	1.346	0.699	0.551	1.476	0.901	3.1	2.2	2.7	2.0	0.1140	0.384	0.044	0.001018	9.3	125	56
77054	1.346	0.699	0.551	1.476	0.901	3.1	2.2	2.7	2.0	0.1140	0.384	0.044	0.001018	9.3	90	40
77055	1.346	0.699	0.551	1.476	0.901	3.1	2.2	2.7	2.0	0.1140	0.384	0.044	0.001018	9.3	75	34
77051	1.346	0.699	0.551	1.476	0.901	3.1	2.2	2.7	2.0	0.1140	0.384	0.044	0.001018	9.3	60	27
77120	1.740	0.953	0.711	1.926	1.188	4.1	4.9	6.4	2.5	0.1920	0.713	0.137	0.004156	16.0	125	72
77224	1.740	0.953	0.711	1.926	1.188	4.1	4.9	6.4	2.5	0.1920	0.713	0.137	0.004156	16.0	90	52

Magnetics Kool Mμ Powder Cores																
Part No.	OD cm 1	ID cm 2	HT cm 3	Wound OD cm 4	Wound HT cm 5	MPL cm 6	Wtfe grams 7	Wtcu grams 8	MLT cm 9	Ac cm sq 10	Wa cm sq 11	Ap cm 4th 12	Kg cm 5th 13	At cm sq 14	Perm 15	AL 16
77225	1.740	0.953	0.711	1.926	1.188	4.1	4.9	6.4	2.5	0.1920	0.713	0.137	0.004156	16.0	75	43
77121	1.740	0.953	0.711	1.926	1.188	4.1	4.9	6.4	2.5	0.1920	0.713	0.137	0.004156	16.0	60	35
77206	2.110	1.207	0.711	2.355	1.315	5.1	7.1	11.5	2.8	0.2260	1.144	0.258	0.008269	22.7	125	68
77210	2.110	1.207	0.711	2.355	1.315	5.1	7.1	11.5	2.8	0.2260	1.144	0.258	0.008269	22.7	90	49
77211	2.110	1.207	0.711	2.355	1.315	5.1	7.1	11.5	2.8	0.2260	1.144	0.258	0.008269	22.7	75	41
77848	2.110	1.207	0.711	2.355	1.315	5.1	7.1	11.5	2.8	0.2260	1.144	0.258	0.008269	22.7	60	32
77310	2.360	1.339	0.838	2.630	1.508	5.7	11.5	16.2	3.2	0.3310	1.407	0.466	0.019103	28.7	125	90
77314	2.360	1.339	0.838	2.630	1.508	5.7	11.5	16.2	3.2	0.3310	1.407	0.466	0.019103	28.7	90	65
77315	2.360	1.339	0.838	2.630	1.508	5.7	11.5	16.2	3.2	0.3310	1.407	0.466	0.019103	28.7	75	54
77059	2.360	1.339	0.838	2.630	1.508	5.7	11.5	16.2	3.2	0.3310	1.407	0.466	0.019103	28.7	60	43
77130	1.190	0.589	0.472	1.295	0.767	2.7	15.0	1.7	1.7	0.9060	0.272	0.247	0.052376	7.1	125	53
77334	1.190	0.589	0.472	1.295	0.767	2.7	15.0	1.7	1.7	0.9060	0.272	0.247	0.052376	7.1	90	38
77335	1.190	0.589	0.472	1.295	0.767	2.7	15.0	1.7	1.7	0.9060	0.272	0.247	0.052376	7.1	75	32
77131	1.190	0.589	0.472	1.295	0.767	2.7	15.0	1.7	1.7	0.9060	0.272	0.247	0.052376	7.1	60	26
77930	2.770	1.410	1.199	3.027	1.904	6.4	25.5	22.9	4.1	0.6540	1.561	1.021	0.064582	39.8	125	157
77934	2.770	1.410	1.199	3.027	1.904	6.4	25.5	22.9	4.1	0.6540	1.561	1.021	0.064582	39.8	90	113
77935	2.770	1.410	1.199	3.027	1.904	6.4	25.5	22.9	4.1	0.6540	1.561	1.021	0.064582	39.8	75	94
77894	2.770	1.410	1.199	3.027	1.904	6.4	25.5	22.9	4.1	0.6540	1.561	1.021	0.064582	39.8	60	75
77585	3.520	2.260	0.983	4.028	2.113	9.0	25.0	62.6	4.4	0.4540	4.009	1.820	0.075320	64.4	125	79
77589	3.520	2.260	0.983	4.028	2.113	9.0	25.0	62.6	4.4	0.4540	4.009	1.820	0.075320	64.4	90	57
77590	3.520	2.260	0.983	4.028	2.113	9.0	25.0	62.6	4.4	0.4540	4.009	1.820	0.075320	64.4	75	47
77586	3.520	2.260	0.983	4.028	2.113	9.0	25.0	62.6	4.4	0.4540	4.009	1.820	0.075320	64.4	60	38
77548	3.380	1.930	1.161	3.770	2.126	8.2	33.7	47.4	4.6	0.6720	2.924	1.965	0.115789	58.5	125	127
77552	3.380	1.930	1.161	3.770	2.126	8.2	33.7	47.4	4.6	0.6720	2.924	1.965	0.115789	58.5	90	91
77553	3.380	1.930	1.161	3.770	2.126	8.2	33.7	47.4	4.6	0.6720	2.924	1.965	0.115789	58.5	75	76
77071	3.380	1.930	1.161	3.770	2.126	8.2	33.7	47.4	4.6	0.6720	2.924	1.965	0.115789	58.5	60	61
77324	3.670	2.150	1.128	4.116	2.203	9.0	37.4	61.2	4.7	0.6780	3.629	2.460	0.140739	67.9	125	117
77328	3.670	2.150	1.128	4.116	2.203	9.0	37.4	61.2	4.7	0.6780	3.629	2.460	0.140739	67.9	90	84
77329	3.670	2.150	1.128	4.116	2.203	9.0	37.4	61.2	4.7	0.6780	3.629	2.460	0.140739	67.9	75	70
77076	3.670	2.150	1.128	4.116	2.203	9.0	37.4	61.2	4.7	0.6780	3.629	2.460	0.140739	67.9	60	56

Magnetics Kool Mμ Powder Cores

Part No.	OD cm 1	ID cm 2	HT cm 3	Wound OD cm 4	Wound HT cm 5	MPL cm 6	Wtfe grams 7	Wtcu grams 8	MLT cm 9	Ac cm sq 10	Wa cm sq 11	Ap cm 4th 12	Kg cm 5th 13	At cm sq 14	Perm 15	AL 16
77254	4.070	2.330	1.537	4.543	2.702	9.8	64.9	86.6	5.7	1.0720	4.262	4.569	0.342768	87.2	125	168
77258	4.070	2.330	1.537	4.543	2.702	9.8	64.9	86.6	5.7	1.0720	4.262	4.569	0.342768	87.2	90	121
77259	4.070	2.330	1.537	4.543	2.702	9.8	64.9	86.6	5.7	1.0720	4.262	4.569	0.342768	87.2	75	101
77083	4.070	2.330	1.537	4.543	2.702	9.8	64.9	86.6	5.7	1.0720	4.262	4.569	0.342768	87.2	60	81
77089	4.760	2.790	1.613	5.338	3.008	11.6	95.8	138.8	6.4	1.3400	6.111	8.188	0.686955	117.1	125	178
77093	4.760	2.790	1.613	5.338	3.008	11.6	95.8	138.8	6.4	1.3400	6.111	8.188	0.686955	117.1	90	128
77094	4.760	2.790	1.613	5.338	3.008	11.6	95.8	138.8	6.4	1.3400	6.111	8.188	0.686955	117.1	75	107
77090	4.760	2.790	1.613	5.338	3.008	11.6	95.8	138.8	6.4	1.3400	6.111	8.188	0.686955	117.1	60	86
77715	5.170	3.090	1.435	5.822	2.980	12.7	97.9	171.4	6.4	1.2510	7.495	9.377	0.729483	133.0	125	152
77719	5.170	3.090	1.435	5.822	2.980	12.7	97.9	171.4	6.4	1.2510	7.495	9.377	0.729483	133.0	90	109
77720	5.170	3.090	1.435	5.822	2.980	12.7	97.9	171.4	6.4	1.2510	7.495	9.377	0.729483	133.0	75	91
77716	5.170	3.090	1.435	5.822	2.980	12.7	97.9	171.4	6.4	1.2510	7.495	9.377	0.729483	133.0	60	73
77438	4.760	2.330	1.892	5.170	3.057	10.7	131.4	103.6	6.8	1.9900	4.262	8.481	0.987635	112.6	125	281
77442	4.760	2.330	1.892	5.170	3.057	10.7	131.4	103.6	6.8	1.9900	4.262	8.481	0.987635	112.6	90	202
77443	4.760	2.330	1.892	5.170	3.057	10.7	131.4	103.6	6.8	1.9900	4.262	8.481	0.987635	112.6	75	169
77439	4.760	2.330	1.892	5.170	3.057	10.7	131.4	103.6	6.8	1.9900	4.262	8.481	0.987635	112.6	60	135
77109	5.800	3.470	1.486	6.532	3.221	14.3	127.0	235.9	7.0	1.4440	9.452	13.649	1.123400	164.6	125	156
77213	5.800	3.470	1.486	6.532	3.221	14.3	127.0	235.9	7.0	1.4440	9.452	13.649	1.123400	164.6	90	112
77214	5.800	3.470	1.486	6.532	3.221	14.3	127.0	235.9	7.0	1.4440	9.452	13.649	1.123400	164.6	75	94
77110	5.800	3.470	1.486	6.532	3.221	14.3	127.0	235.9	7.0	1.4440	9.452	13.649	1.123400	164.6	60	75

Magnetics High Flux Powder Cores																
Part No.	OD cm 1	ID cm 2	HT cm 3	Wound OD cm 4	Wound HT cm 5	MPL cm 6	Wtfe grams 7	Wtcu grams 8	MLT cm 9	Ac cm sq 10	Wa cm sq 11	Ap cm 4th 12	Kg cm 5th 13	At cm sq 14	Perm 15	AL 16
58018	0.699	0.229	0.343	0.727	0.458	1.4	0.5	0.2	1.1	0.0470	0.041	0.002	0.000033	2.3	160	64
58019	0.699	0.229	0.343	0.727	0.458	1.4	0.5	0.2	1.1	0.0470	0.041	0.002	0.000033	2.3	147	59
58020	0.699	0.229	0.343	0.727	0.458	1.4	0.5	0.2	1.1	0.0470	0.041	0.002	0.000033	2.3	125	50
58021	0.699	0.229	0.343	0.727	0.458	1.4	0.5	0.2	1.1	0.0470	0.041	0.002	0.000033	2.3	60	24
58022	0.699	0.229	0.343	0.727	0.458	1.4	0.5	0.2	1.1	0.0470	0.041	0.002	0.000033	2.3	26	10
58023	0.699	0.229	0.343	0.727	0.458	1.4	0.5	0.2	1.1	0.0470	0.041	0.002	0.000033	2.3	14	6
58238	0.724	0.229	0.318	0.750	0.433	1.4	0.5	0.2	1.1	0.0476	0.041	0.002	0.000034	2.3	160	69
58239	0.724	0.229	0.318	0.750	0.433	1.4	0.5	0.2	1.1	0.0476	0.041	0.002	0.000034	2.3	147	64
58240	0.724	0.229	0.318	0.750	0.433	1.4	0.5	0.2	1.1	0.0476	0.041	0.002	0.000034	2.3	125	54
58241	0.724	0.229	0.318	0.750	0.433	1.4	0.5	0.2	1.1	0.0476	0.041	0.002	0.000034	2.3	60	26
58242	0.724	0.229	0.318	0.750	0.433	1.4	0.5	0.2	1.1	0.0476	0.041	0.002	0.000034	2.3	26	11
58243	0.724	0.229	0.318	0.750	0.433	1.4	0.5	0.2	1.1	0.0476	0.041	0.002	0.000034	2.3	14	6
58268	0.732	0.221	0.554	0.757	0.665	1.4	1.0	0.2	1.5	0.0920	0.038	0.004	0.000088	3.0	160	132
58269	0.732	0.221	0.554	0.757	0.665	1.4	1.0	0.2	1.5	0.0920	0.038	0.004	0.000088	3.0	147	122
58270	0.732	0.221	0.554	0.757	0.665	1.4	1.0	0.2	1.5	0.0920	0.038	0.004	0.000088	3.0	125	103
58271	0.732	0.221	0.554	0.757	0.665	1.4	1.0	0.2	1.5	0.0920	0.038	0.004	0.000088	3.0	60	50
58272	0.732	0.221	0.554	0.757	0.665	1.4	1.0	0.2	1.5	0.0920	0.038	0.004	0.000088	3.0	26	21
58273	0.732	0.221	0.554	0.757	0.665	1.4	1.0	0.2	1.5	0.0920	0.038	0.004	0.000088	3.0	14	12
58028	0.851	0.343	0.381	0.902	0.553	1.8	0.9	0.4	1.3	0.0615	0.092	0.006	0.000108	3.5	160	66
58029	0.851	0.343	0.381	0.902	0.553	1.8	0.9	0.4	1.3	0.0615	0.092	0.006	0.000108	3.5	147	62
58030	0.851	0.343	0.381	0.902	0.553	1.8	0.9	0.4	1.3	0.0615	0.092	0.006	0.000108	3.5	125	25
58031	0.851	0.343	0.381	0.902	0.553	1.8	0.9	0.4	1.3	0.0615	0.092	0.006	0.000108	3.5	60	52
58032	0.851	0.343	0.381	0.902	0.553	1.8	0.9	0.4	1.3	0.0615	0.092	0.006	0.000108	3.5	26	11
58033	0.851	0.343	0.381	0.902	0.553	1.8	0.9	0.4	1.3	0.0615	0.092	0.006	0.000108	3.5	14	6
58408	0.762	0.345	0.572	0.818	0.745	1.7	1.0	0.5	1.5	0.0725	0.093	0.007	0.000129	3.6	160	89
58409	0.762	0.345	0.572	0.818	0.745	1.7	1.0	0.5	1.5	0.0725	0.093	0.007	0.000129	3.6	147	81
58410	0.762	0.345	0.572	0.818	0.745	1.7	1.0	0.5	1.5	0.0725	0.093	0.007	0.000129	3.6	125	70
58411	0.762	0.345	0.572	0.818	0.745	1.7	1.0	0.5	1.5	0.0725	0.093	0.007	0.000129	3.6	60	33
58278	1.029	0.427	0.381	1.093	0.595	2.2	1.3	0.7	1.4	0.0752	0.143	0.011	0.000226	4.8	160	68
58279	1.029	0.427	0.381	1.093	0.595	2.2	1.3	0.7	1.4	0.0752	0.143	0.011	0.000226	4.8	147	63

Magnetics High Flux Powder Cores																
Part No.	OD cm 1	ID cm 2	HT cm 3	Wound OD cm 4	Wound HT cm 5	MPL cm 6	Wtfe grams 7	Wtcu grams 8	MLT cm 9	Ac cm sq 10	Wa cm sq 11	Ap cm 4th 12	Kg cm 5th 13	At cm sq 14	Perm 15	AL 16
58280	1.029	0.427	0.381	1.093	0.595	2.2	1.3	0.7	1.4	0.0752	0.143	0.011	0.000226	4.8	125	53
58281	1.029	0.427	0.381	1.093	0.595	2.2	1.3	0.7	1.4	0.0752	0.143	0.011	0.000226	4.8	60	25
58282	1.029	0.427	0.381	1.093	0.595	2.2	1.3	0.7	1.4	0.0752	0.143	0.011	0.000226	4.8	26	11
58283	1.029	0.427	0.381	1.093	0.595	2.2	1.3	0.7	1.4	0.0752	0.143	0.011	0.000226	4.8	14	6
58288	1.029	0.427	0.457	1.093	0.671	2.2	1.6	0.8	1.6	0.0945	0.143	0.014	0.000329	5.1	160	84
58289	1.029	0.427	0.457	1.093	0.671	2.2	1.6	0.8	1.6	0.0945	0.143	0.014	0.000329	5.1	147	78
58290	1.029	0.427	0.457	1.093	0.671	2.2	1.6	0.8	1.6	0.0945	0.143	0.014	0.000329	5.1	125	66
58291	1.029	0.427	0.457	1.093	0.671	2.2	1.6	0.8	1.6	0.0945	0.143	0.014	0.000329	5.1	60	32
58292	1.029	0.427	0.457	1.093	0.671	2.2	1.6	0.8	1.6	0.0945	0.143	0.014	0.000329	5.1	26	14
58293	1.029	0.427	0.457	1.093	0.671	2.2	1.6	0.8	1.6	0.0945	0.143	0.014	0.000329	5.1	14	7
58038	1.080	0.457	0.457	1.150	0.686	2.4	1.9	0.9	1.6	0.1000	0.164	0.016	0.000411	5.6	160	84
58039	1.080	0.457	0.457	1.150	0.686	2.4	1.9	0.9	1.6	0.1000	0.164	0.016	0.000411	5.6	147	78
58041	1.080	0.457	0.457	1.150	0.686	2.4	1.9	0.9	1.6	0.1000	0.164	0.016	0.000411	5.6	125	66
58040	1.080	0.457	0.457	1.150	0.686	2.4	1.9	0.9	1.6	0.1000	0.164	0.016	0.000411	5.6	60	32
58042	1.080	0.457	0.457	1.150	0.686	2.4	1.9	0.9	1.6	0.1000	0.164	0.016	0.000411	5.6	26	14
58043	1.080	0.457	0.457	1.150	0.686	2.4	1.9	0.9	1.6	0.1000	0.164	0.016	0.000411	5.6	14	7
58128	1.190	0.589	0.472	1.295	0.767	2.7	1.9	1.7	1.7	0.0906	0.272	0.025	0.000524	7.1	160	68
58129	1.190	0.589	0.472	1.295	0.767	2.7	1.9	1.7	1.7	0.0906	0.272	0.025	0.000524	7.1	147	63
58130	1.190	0.589	0.472	1.295	0.767	2.7	1.9	1.7	1.7	0.0906	0.272	0.025	0.000524	7.1	125	53
58131	1.190	0.589	0.472	1.295	0.767	2.7	1.9	1.7	1.7	0.0906	0.272	0.025	0.000524	7.1	60	26
58133	1.190	0.589	0.472	1.295	0.767	2.7	1.9	1.7	1.7	0.0906	0.272	0.025	0.000524	7.1	14	6
58132	0.190	0.589	0.472	0.544	0.767	2.7	1.9	0.9	0.9	0.0906	0.272	0.025	0.000986	2.1	26	11
58048	1.346	0.699	0.551	1.476	0.901	3.1	2.8	2.7	2.0	0.1140	0.384	0.044	0.001018	9.3	160	72
58049	1.346	0.699	0.551	1.476	0.901	3.1	2.8	2.7	2.0	0.1140	0.384	0.044	0.001018	9.3	147	67
58050	1.346	0.699	0.551	1.476	0.901	3.1	2.8	2.7	2.0	0.1140	0.384	0.044	0.001018	9.3	125	56
58051	1.346	0.699	0.551	1.476	0.901	3.1	2.8	2.7	2.0	0.1140	0.384	0.044	0.001018	9.3	60	27
58052	1.346	0.699	0.551	1.476	0.901	3.1	2.8	2.7	2.0	0.1140	0.384	0.044	0.001018	9.3	26	12
58053	1.346	0.699	0.551	1.476	0.901	3.1	2.8	2.7	2.0	0.1140	0.384	0.044	0.001018	9.3	14	6
58118	1.740	0.953	0.711	1.926	1.188	4.1	6.3	6.4	2.5	0.1920	0.713	0.137	0.004156	16.0	160	92
58119	1.740	0.953	0.711	1.926	1.188	4.1	6.3	6.4	2.5	0.1920	0.713	0.137	0.004156	16.0	147	88

Magnetics High Flux Powder Cores																
Part No.	OD cm 1	ID cm 2	HT cm 3	Wound OD cm 4	Wound HT cm 5	MPL cm 6	Wtfe grams 7	Wtcu grams 8	MLT cm 9	Ac cm sq 10	Wa cm sq 11	Ap cm 4th 12	Kg cm 5th 13	At cm sq 14	Perm 15	AL 16
58120	1.740	0.953	0.711	1.926	1.188	4.1	6.3	6.4	2.5	0.1920	0.713	0.137	0.004156	16.0	125	72
58121	1.740	0.953	0.711	1.926	1.188	4.1	6.3	6.4	2.5	0.1920	0.713	0.137	0.004156	16.0	60	35
58122	1.740	0.953	0.711	1.926	1.188	4.1	6.3	6.4	2.5	0.1920	0.713	0.137	0.004156	16.0	26	15
58123	1.740	0.953	0.711	1.926	1.188	4.1	6.3	6.4	2.5	0.1920	0.713	0.137	0.004156	16.0	14	8
58378	1.803	0.902	0.711	1.965	1.162	4.1	7.7	5.9	2.6	0.2320	0.639	0.148	0.005330	16.3	160	114
58379	1.803	0.902	0.711	1.965	1.162	4.1	7.7	5.9	2.6	0.2320	0.639	0.148	0.005330	16.3	147	105
58380	1.803	0.902	0.711	1.965	1.162	4.1	7.7	5.9	2.6	0.2320	0.639	0.148	0.005330	16.3	125	89
58381	1.803	0.902	0.711	1.965	1.162	4.1	7.7	5.9	2.6	0.2320	0.639	0.148	0.005330	16.3	60	43
58382	1.803	0.902	0.711	1.965	1.162	4.1	7.7	5.9	2.6	0.2320	0.639	0.148	0.005330	16.3	26	19
58383	1.803	0.902	0.711	1.965	1.162	4.1	7.7	5.9	2.6	0.2320	0.639	0.148	0.005330	16.3	14	10
58204	2.110	1.207	0.711	2.355	1.315	5.1	9.2	11.5	2.8	0.2260	1.144	0.258	0.008269	22.7	160	87
58205	2.110	1.207	0.711	2.355	1.315	5.1	9.2	11.5	2.8	0.2260	1.144	0.258	0.008269	22.7	147	81
58206	2.110	1.207	0.711	2.355	1.315	5.1	9.2	11.5	2.8	0.2260	1.144	0.258	0.008269	22.7	125	68
58848	2.110	1.207	0.711	2.355	1.315	5.1	9.2	11.5	2.8	0.2260	1.144	0.258	0.008269	22.7	60	32
58208	2.110	1.207	0.711	2.355	1.315	5.1	9.2	11.5	2.8	0.2260	1.144	0.258	0.008269	22.7	26	14
58209	2.110	1.207	0.711	2.355	1.315	5.1	9.2	11.5	2.8	0.2260	1.144	0.258	0.008269	22.7	14	8
58308	2.360	1.339	0.838	2.630	1.508	5.7	15.0	16.2	3.2	0.3310	1.407	0.466	0.019103	28.7	160	115
58309	2.360	1.339	0.838	2.630	1.508	5.7	15.0	16.2	3.2	0.3310	1.407	0.466	0.019103	28.7	147	106
58310	2.360	1.339	0.838	2.630	1.508	5.7	15.0	16.2	3.2	0.3310	1.407	0.466	0.019103	28.7	125	90
58059	2.360	1.339	0.838	2.630	1.508	5.7	15.0	16.2	3.2	0.3310	1.407	0.466	0.019103	28.7	60	43
58312	2.360	1.339	0.838	2.630	1.508	5.7	15.0	16.2	3.2	0.3310	1.407	0.466	0.019103	28.7	26	19
58313	2.360	1.339	0.838	2.630	1.508	5.7	15.0	16.2	3.2	0.3310	1.407	0.466	0.019103	28.7	14	10
58348	2.430	1.377	0.970	2.707	1.659	5.9	18.3	18.5	3.5	0.3880	1.488	0.578	0.025638	31.4	160	135
58349	2.430	1.377	0.970	2.707	1.659	5.9	18.3	18.5	3.5	0.3880	1.488	0.578	0.025638	31.4	147	124
58350	2.430	1.377	0.970	2.707	1.659	5.9	18.3	18.5	3.5	0.3880	1.488	0.578	0.025638	31.4	125	105
58351	2.430	1.377	0.970	2.707	1.659	5.9	18.3	18.5	3.5	0.3880	1.488	0.578	0.025638	31.4	60	51
58352	2.430	1.377	0.970	2.707	1.659	5.9	18.3	18.5	3.5	0.3880	1.488	0.578	0.025638	31.4	26	22
58353	2.430	1.377	0.970	2.707	1.659	5.9	18.3	18.5	3.5	0.3880	1.488	0.578	0.025638	31.4	14	12
58928	2.770	1.410	1.199	3.027	1.904	6.4	33.2	22.9	4.1	0.6540	1.561	1.021	0.064582	39.8	160	201
58929	2.770	1.410	1.199	3.027	1.904	6.4	33.2	22.9	4.1	0.6540	1.561	1.021	0.064582	39.8	147	185

Magnetics High Flux Powder Cores																
Part No.	OD cm 1	ID cm 2	HT cm 3	Wound OD cm 4	Wound HT cm 5	MPL cm 6	Wtfe grams 7	Wtcu grams 8	MLT cm 9	Ac cm sq 10	Wa cm sq 11	Ap cm 4th 12	Kg cm 5th 13	At cm sq 14	Perm 15	AL 16
58930	2.770	1.410	1.199	3.027	1.904	6.4	33.2	22.9	4.1	0.6540	1.561	1.021	0.064582	39.8	125	157
58894	2.770	1.410	1.199	3.027	1.904	6.4	33.2	22.9	4.1	0.6540	1.561	1.021	0.064582	39.8	60	75
58932	2.770	1.410	1.199	3.027	1.904	6.4	33.2	22.9	4.1	0.6540	1.561	1.021	0.064582	39.8	26	32
58933	2.770	1.410	1.199	3.027	1.904	6.4	33.2	22.9	4.1	0.6540	1.561	1.021	0.064582	39.8	14	18
58583	3.520	2.260	0.983	4.028	2.113	9.0	32.5	62.6	4.4	0.4540	4.009	1.820	0.075320	64.4	160	101
58584	3.520	2.260	0.983	4.028	2.113	9.0	32.5	62.6	4.4	0.4540	4.009	1.820	0.075320	64.4	147	93
58585	3.520	2.260	0.983	4.028	2.113	9.0	32.5	62.6	4.4	0.4540	4.009	1.820	0.075320	64.4	125	79
58586	3.520	2.260	0.983	4.028	2.113	9.0	32.5	62.6	4.4	0.4540	4.009	1.820	0.075320	64.4	60	38
58587	3.520	2.260	0.983	4.028	2.113	9.0	32.5	62.6	4.4	0.4540	4.009	1.820	0.075320	64.4	26	16
58588	3.520	2.260	0.983	4.028	2.113	9.0	32.5	62.6	4.4	0.4540	4.009	1.820	0.075320	64.4	14	9
58546	3.380	1.930	1.161	3.770	2.126	8.2	43.8	47.4	4.6	0.6720	2.924	1.965	0.115789	58.5	160	163
58547	3.380	1.930	1.161	3.770	2.126	8.2	43.8	47.4	4.6	0.6720	2.924	1.965	0.115789	58.5	147	150
58548	3.380	1.930	1.161	3.770	2.126	8.2	43.8	47.4	4.6	0.6720	2.924	1.965	0.115789	58.5	125	127
58071	3.380	1.930	1.161	3.770	2.126	8.2	43.8	47.4	4.6	0.6720	2.924	1.965	0.115789	58.5	60	61
58550	3.380	1.930	1.161	3.770	2.126	8.2	43.8	47.4	4.6	0.6720	2.924	1.965	0.115789	58.5	26	28
58551	3.380	1.930	1.161	3.770	2.126	8.2	43.8	47.4	4.6	0.6720	2.924	1.965	0.115789	58.5	14	14
58322	3.670	2.150	1.128	4.116	2.203	9.0	48.7	61.2	4.7	0.6780	3.629	2.460	0.140739	67.9	160	150
58323	3.670	2.150	1.128	4.116	2.203	9.0	48.7	61.2	4.7	0.6780	3.629	2.460	0.140739	67.9	147	138
58324	3.670	2.150	1.128	4.116	2.203	9.0	48.7	61.2	4.7	0.6780	3.629	2.460	0.140739	67.9	125	117
58076	3.670	2.150	1.128	4.116	2.203	9.0	48.7	61.2	4.7	0.6780	3.629	2.460	0.140739	67.9	60	56
58326	3.670	2.150	1.128	4.116	2.203	9.0	48.7	61.2	4.7	0.6780	3.629	2.460	0.140739	67.9	26	24
58327	3.670	2.150	1.128	4.116	2.203	9.0	48.7	61.2	4.7	0.6780	3.629	2.460	0.140739	67.9	14	13
58252	4.070	2.330	1.537	4.543	2.702	9.8	84.4	86.6	5.7	1.0720	4.262	4.569	0.342768	87.2	160	215
58253	4.070	2.330	1.537	4.543	2.702	9.8	84.4	86.6	5.7	1.0720	4.262	4.569	0.342768	87.2	147	198
58254	4.070	2.330	1.537	4.543	2.702	9.8	84.4	86.6	5.7	1.0720	4.262	4.569	0.342768	87.2	125	168
58083	4.070	2.330	1.537	4.543	2.702	9.8	84.4	86.6	5.7	1.0720	4.262	4.569	0.342768	87.2	60	81
58256	4.070	2.330	1.537	4.543	2.702	9.8	84.4	86.6	5.7	1.0720	4.262	4.569	0.342768	87.2	26	35
58257	4.070	2.330	1.537	4.543	2.702	9.8	84.4	86.6	5.7	1.0720	4.262	4.569	0.342768	87.2	14	19
58089	4.760	2.790	1.613	5.338	3.008	11.6	124.7	138.8	6.4	1.3400	6.111	8.188	0.686955	117.1	125	178
58090	4.760	2.790	1.613	5.338	3.008	11.6	124.7	138.8	6.4	1.3400	6.111	8.188	0.686955	117.1	60	86

Magnetics High Flux Powder Cores

Part No.	OD cm 1	ID cm 2	HT cm 3	Wound OD cm 4	Wound HT cm 5	MPL cm 6	Wtfe grams 7	Wtcu grams 8	MLT cm 9	Ac cm sq 10	Wa cm sq 11	Ap cm 4th 12	Kg cm 5th 13	At cm sq 14	Perm 15	AL 16
58091	4.760	2.790	1.613	5.338	3.008	11.6	124.7	138.8	6.4	1.3400	6.111	8.188	0.686955	117.1	26	37
58092	4.760	2.790	1.613	5.338	3.008	11.6	124.7	138.8	6.4	1.3400	6.111	8.188	0.686955	117.1	14	20
58715	5.170	3.090	1.435	5.822	2.980	12.7	127.4	171.4	6.4	1.2510	7.495	9.377	0.729483	133.0	125	152
58716	5.170	3.090	1.435	5.822	2.980	12.7	127.4	171.4	6.4	1.2510	7.495	9.377	0.729483	133.0	60	73
58717	5.170	3.090	1.435	5.822	2.980	12.7	127.4	171.4	6.4	1.2510	7.495	9.377	0.729483	133.0	26	32
58718	5.170	3.090	1.435	5.822	2.980	12.7	127.4	171.4	6.4	1.2510	7.495	9.377	0.729483	133.0	14	17
58438	4.760	2.330	1.892	5.170	3.057	10.7	171.0	103.6	6.8	1.9900	4.262	8.481	0.987635	112.6	125	281
58439	4.760	2.330	1.892	5.170	3.057	10.7	171.0	103.6	6.8	1.9900	4.262	8.481	0.987635	112.6	60	135
58440	4.760	2.330	1.892	5.170	3.057	10.7	171.0	103.6	6.8	1.9900	4.262	8.481	0.987635	112.6	26	59
58441	4.760	2.330	1.892	5.170	3.057	10.7	171.0	103.6	6.8	1.9900	4.262	8.481	0.987635	112.6	14	32
58109	5.800	3.470	1.486	6.532	3.221	14.3	165.2	235.9	7.0	1.4440	9.452	13.649	1.123400	164.6	125	156
58110	5.800	3.470	1.486	6.532	3.221	14.3	165.2	235.9	7.0	1.4440	9.452	13.649	1.123400	164.6	60	75
58111	5.800	3.470	1.486	6.532	3.221	14.3	165.2	235.9	7.0	1.4440	9.452	13.649	1.123400	164.6	26	33
58112	5.800	3.470	1.486	6.532	3.221	14.3	165.2	235.9	7.0	1.4440	9.452	13.649	1.123400	164.6	14	18
58867	7.890	4.800	1.397	8.918	3.797	20.0	283.2	549.7	8.5	1.7700	18.086	32.013	2.651763	287.4	60	68
58868	7.890	4.800	1.397	8.918	3.797	20.0	283.2	549.7	8.5	1.7700	18.086	32.013	2.651763	287.4	26	30
58869	7.890	4.800	1.397	8.918	3.797	20.0	283.2	549.7	8.5	1.7700	18.086	32.013	2.651763	287.4	14	16

Magnetics MPP Powder Cores																
Part No.	OD cm 1	ID cm 2	HT cm 3	Wound OD cm 4	Wound HT cm 5	MPL cm 6	Wtfe grams 7	Wtcu grams 8	MLT cm 9	Ac cm sq 10	Wa cm sq 11	Ap cm 4th 12	Kg cm 5th 13	At cm sq 14	Perm 15	AL 16
55135	0.394	0.152	0.196	0.416	0.272	0.8	0.1	0.0	0.6	0.0137	0.018	0.000	0.000002	0.8	300	62
55137	0.394	0.152	0.196	0.416	0.272	0.8	0.1	0.0	0.6	0.0137	0.018	0.000	0.000002	0.8	200	42
55134	0.394	0.152	0.196	0.416	0.272	0.8	0.1	0.0	0.6	0.0137	0.018	0.000	0.000002	0.8	173	36
55138	0.394	0.152	0.196	0.416	0.272	0.8	0.1	0.0	0.6	0.0137	0.018	0.000	0.000002	0.8	160	33
55139	0.394	0.152	0.196	0.416	0.272	0.8	0.1	0.0	0.6	0.0137	0.018	0.000	0.000002	0.8	147	31
55140	0.394	0.152	0.196	0.416	0.272	0.8	0.1	0.0	0.6	0.0137	0.018	0.000	0.000002	0.8	125	26
55145	0.432	0.198	0.297	0.465	0.396	0.9	0.2	0.1	0.8	0.0211	0.031	0.001	0.000007	1.1	300	84
55147	0.432	0.198	0.297	0.465	0.396	0.9	0.2	0.1	0.8	0.0211	0.031	0.001	0.000007	1.1	200	56
55144	0.432	0.198	0.297	0.465	0.396	0.9	0.2	0.1	0.8	0.0211	0.031	0.001	0.000007	1.1	173	48
55148	0.432	0.198	0.297	0.465	0.396	0.9	0.2	0.1	0.8	0.0211	0.031	0.001	0.000007	1.1	160	45
55149	0.432	0.198	0.297	0.465	0.396	0.9	0.2	0.1	0.8	0.0211	0.031	0.001	0.000007	1.1	147	41
55150	0.432	0.198	0.297	0.465	0.396	0.9	0.2	0.1	0.8	0.0211	0.031	0.001	0.000007	1.1	125	35
55175	0.521	0.193	0.330	0.547	0.427	1.1	0.3	0.1	0.9	0.0285	0.029	0.001	0.000010	1.5	300	99
55177	0.521	0.193	0.330	0.547	0.427	1.1	0.3	0.1	0.9	0.0285	0.029	0.001	0.000010	1.5	200	67
55174	0.521	0.193	0.330	0.547	0.427	1.1	0.3	0.1	0.9	0.0285	0.029	0.001	0.000010	1.5	173	57
55178	0.521	0.193	0.330	0.547	0.427	1.1	0.3	0.1	0.9	0.0285	0.029	0.001	0.000010	1.5	160	53
55179	0.521	0.193	0.330	0.547	0.427	1.1	0.3	0.1	0.9	0.0285	0.029	0.001	0.000010	1.5	147	49
55180	0.521	0.193	0.330	0.547	0.427	1.1	0.3	0.1	0.9	0.0285	0.029	0.001	0.000010	1.5	125	42
55181	0.521	0.193	0.330	0.547	0.427	1.1	0.3	0.1	0.9	0.0285	0.029	0.001	0.000010	1.5	60	20
55016	0.699	0.229	0.343	0.727	0.458	1.4	0.5	0.2	1.1	0.0470	0.041	0.002	0.000033	2.3	550	220
55015	0.699	0.229	0.343	0.727	0.458	1.4	0.5	0.2	1.1	0.0470	0.041	0.002	0.000033	2.3	300	120
55017	0.699	0.229	0.343	0.727	0.458	1.4	0.5	0.2	1.1	0.0470	0.041	0.002	0.000033	2.3	200	80
55014	0.699	0.229	0.343	0.727	0.458	1.4	0.5	0.2	1.1	0.0470	0.041	0.002	0.000033	2.3	173	69
55018	0.699	0.229	0.343	0.727	0.458	1.4	0.5	0.2	1.1	0.0470	0.041	0.002	0.000033	2.3	160	64
55019	0.699	0.229	0.343	0.727	0.458	1.4	0.5	0.2	1.1	0.0470	0.041	0.002	0.000033	2.3	147	59
55020	0.699	0.229	0.343	0.727	0.458	1.4	0.5	0.2	1.1	0.0470	0.041	0.002	0.000033	2.3	125	50
55021	0.699	0.229	0.343	0.727	0.458	1.4	0.5	0.2	1.1	0.0470	0.041	0.002	0.000033	2.3	60	24
55022	0.699	0.229	0.343	0.727	0.458	1.4	0.5	0.2	1.1	0.0470	0.041	0.002	0.000033	2.3	26	10
55023	0.699	0.229	0.343	0.727	0.458	1.4	0.5	0.2	1.1	0.0470	0.041	0.002	0.000033	2.3	14	6
55236	0.724	0.229	0.318	0.750	0.433	1.4	0.6	0.2	1.1	0.0476	0.041	0.002	0.000034	2.3	550	242

Magnetics MPP Powder Cores																
Part No.	OD cm 1	ID cm 2	HT cm 3	Wound OD cm 4	Wound HT cm 5	MPL cm 6	Wtfe grams 7	Wtcu grams 8	MLT cm 9	Ac cm sq 10	Wa cm sq 11	Ap cm 4th 12	Kg cm 5th 13	At cm sq 14	Perm 15	AL 16
55235	0.724	0.229	0.318	0.750	0.433	1.4	0.6	0.2	1.1	0.0476	0.041	0.002	0.000034	2.3	300	130
55237	0.724	0.229	0.318	0.750	0.433	1.4	0.6	0.2	1.1	0.0476	0.041	0.002	0.000034	2.3	200	86
55234	0.724	0.229	0.318	0.750	0.433	1.4	0.6	0.2	1.1	0.0476	0.041	0.002	0.000034	2.3	173	75
55238	0.724	0.229	0.318	0.750	0.433	1.4	0.6	0.2	1.1	0.0476	0.041	0.002	0.000034	2.3	160	69
55239	0.724	0.229	0.318	0.750	0.433	1.4	0.6	0.2	1.1	0.0476	0.041	0.002	0.000034	2.3	147	64
55240	0.724	0.229	0.318	0.750	0.433	1.4	0.6	0.2	1.1	0.0476	0.041	0.002	0.000034	2.3	125	54
55241	0.724	0.229	0.318	0.750	0.433	1.4	0.6	0.2	1.1	0.0476	0.041	0.002	0.000034	2.3	60	26
55242	0.724	0.229	0.318	0.750	0.433	1.4	0.6	0.2	1.1	0.0476	0.041	0.002	0.000034	2.3	26	11
55243	0.724	0.229	0.318	0.750	0.433	1.4	0.6	0.2	1.1	0.0476	0.041	0.002	0.000034	2.3	14	6
55266	0.732	0.221	0.554	0.757	0.665	1.4	1.1	0.2	1.5	0.0920	0.038	0.004	0.000088	3.0	550	466
55265	0.732	0.221	0.554	0.757	0.665	1.4	1.1	0.2	1.5	0.0920	0.038	0.004	0.000088	3.0	300	247
55267	0.732	0.221	0.554	0.757	0.665	1.4	1.1	0.2	1.5	0.0920	0.038	0.004	0.000088	3.0	200	165
55264	0.732	0.221	0.554	0.757	0.665	1.4	1.1	0.2	1.5	0.0920	0.038	0.004	0.000088	3.0	173	144
55268	0.732	0.221	0.554	0.757	0.665	1.4	1.1	0.2	1.5	0.0920	0.038	0.004	0.000088	3.0	160	132
55269	0.732	0.221	0.554	0.757	0.665	1.4	1.1	0.2	1.5	0.0920	0.038	0.004	0.000088	3.0	147	122
55270	0.732	0.221	0.554	0.757	0.665	1.4	1.1	0.2	1.5	0.0920	0.038	0.004	0.000088	3.0	125	103
55271	0.732	0.221	0.554	0.757	0.665	1.4	1.1	0.2	1.5	0.0920	0.038	0.004	0.000088	3.0	60	50
55272	0.732	0.221	0.554	0.757	0.665	1.4	1.1	0.2	1.5	0.0920	0.038	0.004	0.000088	3.0	26	21
55273	0.732	0.221	0.554	0.757	0.665	1.4	1.1	0.2	1.5	0.0920	0.038	0.004	0.000088	3.0	14	12
55026	0.851	0.343	0.381	0.902	0.553	1.8	0.9	0.4	1.3	0.0615	0.092	0.006	0.000108	3.5	550	229
55025	0.851	0.343	0.381	0.902	0.553	1.8	0.9	0.4	1.3	0.0615	0.092	0.006	0.000108	3.5	300	124
55027	0.851	0.343	0.381	0.902	0.553	1.8	0.9	0.4	1.3	0.0615	0.092	0.006	0.000108	3.5	200	83
55024	0.851	0.343	0.381	0.902	0.553	1.8	0.9	0.4	1.3	0.0615	0.092	0.006	0.000108	3.5	173	73
55028	0.851	0.343	0.381	0.902	0.553	1.8	0.9	0.4	1.3	0.0615	0.092	0.006	0.000108	3.5	160	66
55029	0.851	0.343	0.381	0.902	0.553	1.8	0.9	0.4	1.3	0.0615	0.092	0.006	0.000108	3.5	147	62
55030	0.851	0.343	0.381	0.902	0.553	1.8	0.9	0.4	1.3	0.0615	0.092	0.006	0.000108	3.5	125	52
55031	0.851	0.343	0.381	0.902	0.553	1.8	0.9	0.4	1.3	0.0615	0.092	0.006	0.000108	3.5	60	25
55032	0.851	0.343	0.381	0.902	0.553	1.8	0.9	0.4	1.3	0.0615	0.092	0.006	0.000108	3.5	26	11
55033	0.851	0.343	0.381	0.902	0.553	1.8	0.9	0.4	1.3	0.0615	0.092	0.006	0.000108	3.5	14	6
55405	0.762	0.345	0.572	0.818	0.745	1.7	1.0	0.5	1.5	0.0725	0.093	0.007	0.000129	3.6	300	166

Magnetics MPP Powder Cores

Part No.	OD cm 1	ID cm 2	HT cm 3	Wound OD cm 4	Wound HT cm 5	MPL cm 6	Wtfe grams 7	Wtcu grams 8	MLT cm 9	Ac cm sq 10	Wa cm sq 11	Ap cm 4th 12	Kg cm 5th 13	At cm sq 14	Perm 15	AL 16
55407	0.762	0.345	0.572	0.818	0.745	1.7	1.0	0.5	1.5	0.0725	0.093	0.007	0.000129	3.6	200	112
55404	0.762	0.345	0.572	0.818	0.745	1.7	1.0	0.5	1.5	0.0725	0.093	0.007	0.000129	3.6	173	95
55408	0.762	0.345	0.572	0.818	0.745	1.7	1.0	0.5	1.5	0.0725	0.093	0.007	0.000129	3.6	160	89
55409	0.762	0.345	0.572	0.818	0.745	1.7	1.0	0.5	1.5	0.0725	0.093	0.007	0.000129	3.6	147	81
55410	0.762	0.345	0.572	0.818	0.745	1.7	1.0	0.5	1.5	0.0725	0.093	0.007	0.000129	3.6	125	70
55411	0.762	0.345	0.572	0.818	0.745	1.7	1.0	0.5	1.5	0.0725	0.093	0.007	0.000129	3.6	60	33
55276	1.029	0.427	0.381	1.093	0.595	2.2	1.4	0.7	1.4	0.0752	0.143	0.011	0.000226	4.8	550	232
55275	1.029	0.427	0.381	1.093	0.595	2.2	1.4	0.7	1.4	0.0752	0.143	0.011	0.000226	4.8	300	128
55277	1.029	0.427	0.381	1.093	0.595	2.2	1.4	0.7	1.4	0.0752	0.143	0.011	0.000226	4.8	200	84
55274	1.029	0.427	0.381	1.093	0.595	2.2	1.4	0.7	1.4	0.0752	0.143	0.011	0.000226	4.8	173	74
55278	1.029	0.427	0.381	1.093	0.595	2.2	1.4	0.7	1.4	0.0752	0.143	0.011	0.000226	4.8	160	68
55279	1.029	0.427	0.381	1.093	0.595	2.2	1.4	0.7	1.4	0.0752	0.143	0.011	0.000226	4.8	147	63
55280	1.029	0.427	0.381	1.093	0.595	2.2	1.4	0.7	1.4	0.0752	0.143	0.011	0.000226	4.8	125	53
55281	1.029	0.427	0.381	1.093	0.595	2.2	1.4	0.7	1.4	0.0752	0.143	0.011	0.000226	4.8	60	25
55282	1.029	0.427	0.381	1.093	0.595	2.2	1.4	0.7	1.4	0.0752	0.143	0.011	0.000226	4.8	26	11
55283	1.029	0.427	0.381	1.093	0.595	2.2	1.4	0.7	1.4	0.0752	0.143	0.011	0.000226	4.8	14	6
55286	1.029	0.427	0.457	1.093	0.671	2.2	1.8	0.8	1.6	0.0945	0.143	0.014	0.000329	5.1	550	290
55285	1.029	0.427	0.457	1.093	0.671	2.2	1.8	0.8	1.6	0.0945	0.143	0.014	0.000329	5.1	300	159
55287	1.029	0.427	0.457	1.093	0.671	2.2	1.8	0.8	1.6	0.0945	0.143	0.014	0.000329	5.1	200	105
55284	1.029	0.427	0.457	1.093	0.671	2.2	1.8	0.8	1.6	0.0945	0.143	0.014	0.000329	5.1	173	92
55288	1.029	0.427	0.457	1.093	0.671	2.2	1.8	0.8	1.6	0.0945	0.143	0.014	0.000329	5.1	160	84
55289	1.029	0.427	0.457	1.093	0.671	2.2	1.8	0.8	1.6	0.0945	0.143	0.014	0.000329	5.1	147	78
55290	1.029	0.427	0.457	1.093	0.671	2.2	1.8	0.8	1.6	0.0945	0.143	0.014	0.000329	5.1	125	66
55291	1.029	0.427	0.457	1.093	0.671	2.2	1.8	0.8	1.6	0.0945	0.143	0.014	0.000329	5.1	60	32
55292	1.029	0.427	0.457	1.093	0.671	2.2	1.8	0.8	1.6	0.0945	0.143	0.014	0.000329	5.1	26	14
55293	1.029	0.427	0.457	1.093	0.671	2.2	1.8	0.8	1.6	0.0945	0.143	0.014	0.000329	5.1	14	7
55036	1.080	0.457	0.457	1.150	0.686	2.4	2.0	0.9	1.6	0.1000	0.164	0.016	0.000411	5.6	550	290
55035	1.080	0.457	0.457	1.150	0.686	2.4	2.0	0.9	1.6	0.1000	0.164	0.016	0.000411	5.6	300	159
55037	1.080	0.457	0.457	1.150	0.686	2.4	2.0	0.9	1.6	0.1000	0.164	0.016	0.000411	5.6	200	105
55034	1.080	0.457	0.457	1.150	0.686	2.4	2.0	0.9	1.6	0.1000	0.164	0.016	0.000411	5.6	173	92

Magnetics MPP Powder Cores

Part No.	OD cm 1	ID cm 2	HT cm 3	Wound OD cm 4	Wound HT cm 5	MPL cm 6	Wtfe grams 7	Wtcu grams 8	MLT cm 9	Ac cm sq 10	Wa cm sq 11	Ap cm 4th 12	Kg cm 5th 13	At cm sq 14	Perm 15	AL 16
55038	1.080	0.457	0.457	1.150	0.686	2.4	2.0	0.9	1.6	0.1000	0.164	0.016	0.000411	5.6	160	84
55039	1.080	0.457	0.457	1.150	0.686	2.4	2.0	0.9	1.6	0.1000	0.164	0.016	0.000411	5.6	147	78
55040	1.080	0.457	0.457	1.150	0.686	2.4	2.0	0.9	1.6	0.1000	0.164	0.016	0.000411	5.6	125	66
55041	1.080	0.457	0.457	1.150	0.686	2.4	2.0	0.9	1.6	0.1000	0.164	0.016	0.000411	5.6	60	32
55042	1.080	0.457	0.457	1.150	0.686	2.4	2.0	0.9	1.6	0.1000	0.164	0.016	0.000411	5.6	26	14
55043	1.080	0.457	0.457	1.150	0.686	2.4	2.0	0.9	1.6	0.1000	0.164	0.016	0.000411	5.6	14	7
55125	1.190	0.589	0.472	1.295	0.767	2.7	2.1	1.7	1.7	0.0906	0.272	0.025	0.000524	7.1	300	127
55127	1.190	0.589	0.472	1.295	0.767	2.7	2.1	1.7	1.7	0.0906	0.272	0.025	0.000524	7.1	200	85
55124	1.190	0.589	0.472	1.295	0.767	2.7	2.1	1.7	1.7	0.0906	0.272	0.025	0.000524	7.1	173	74
55128	1.190	0.589	0.472	1.295	0.767	2.7	2.1	1.7	1.7	0.0906	0.272	0.025	0.000524	7.1	160	68
55129	1.190	0.589	0.472	1.295	0.767	2.7	2.1	1.7	1.7	0.0906	0.272	0.025	0.000524	7.1	147	63
55130	1.190	0.589	0.472	1.295	0.767	2.7	2.1	1.7	1.7	0.0906	0.272	0.025	0.000524	7.1	125	53
55131	1.190	0.589	0.472	1.295	0.767	2.7	2.1	1.7	1.7	0.0906	0.272	0.025	0.000524	7.1	60	26
55132	1.190	0.589	0.472	1.295	0.767	2.7	2.1	1.7	1.7	0.0906	0.272	0.025	0.000524	7.1	26	11
55133	1.190	0.589	0.472	1.295	0.767	2.7	2.1	1.7	1.7	0.0906	0.272	0.025	0.000524	7.1	14	6
55046	1.346	0.699	0.551	1.476	0.901	3.1	3.0	2.7	2.0	0.1140	0.384	0.044	0.001018	9.3	550	255
55045	1.346	0.699	0.551	1.476	0.901	3.1	3.0	2.7	2.0	0.1140	0.384	0.044	0.001018	9.3	300	134
55047	1.346	0.699	0.551	1.476	0.901	3.1	3.0	2.7	2.0	0.1140	0.384	0.044	0.001018	9.3	200	90
55044	1.346	0.699	0.551	1.476	0.901	3.1	3.0	2.7	2.0	0.1140	0.384	0.044	0.001018	9.3	173	79
55048	1.346	0.699	0.551	1.476	0.901	3.1	3.0	2.7	2.0	0.1140	0.384	0.044	0.001018	9.3	160	72
55049	1.346	0.699	0.551	1.476	0.901	3.1	3.0	2.7	2.0	0.1140	0.384	0.044	0.001018	9.3	147	67
55050	1.346	0.699	0.551	1.476	0.901	3.1	3.0	2.7	2.0	0.1140	0.384	0.044	0.001018	9.3	125	56
55051	1.346	0.699	0.551	1.476	0.901	3.1	3.0	2.7	2.0	0.1140	0.384	0.044	0.001018	9.3	60	27
55052	1.346	0.699	0.551	1.476	0.901	3.1	3.0	2.7	2.0	0.1140	0.384	0.044	0.001018	9.3	26	12
55053	1.346	0.699	0.551	1.476	0.901	3.1	3.0	2.7	2.0	0.1140	0.384	0.044	0.001018	9.3	14	6
55116	1.740	0.953	0.711	1.926	1.188	4.1	6.7	6.4	2.5	0.1920	0.713	0.137	0.004156	16.0	550	317
55115	1.740	0.953	0.711	1.926	1.188	4.1	6.7	6.4	2.5	0.1920	0.713	0.137	0.004156	16.0	300	173
55117	1.740	0.953	0.711	1.926	1.188	4.1	6.7	6.4	2.5	0.1920	0.713	0.137	0.004156	16.0	200	115
55114	1.740	0.953	0.711	1.926	1.188	4.1	6.7	6.4	2.5	0.1920	0.713	0.137	0.004156	16.0	173	104
55118	1.740	0.953	0.711	1.926	1.188	4.1	6.7	6.4	2.5	0.1920	0.713	0.137	0.004156	16.0	160	92

Magnetics MPP Powder Cores																
Part No.	OD cm 1	ID cm 2	HT cm 3	Wound OD cm 4	Wound HT cm 5	MPL cm 6	Wtfe grams 7	Wtcu grams 8	MLT cm 9	Ac cm sq 10	Wa cm sq 11	Ap cm 4th 12	Kg cm 5th 13	At cm sq 14	Perm 15	AL 16
55119	1.740	0.953	0.711	1.926	1.188	4.1	6.7	6.4	2.5	0.1920	0.713	0.137	0.004156	16.0	147	88
55120	1.740	0.953	0.711	1.926	1.188	4.1	6.7	6.4	2.5	0.1920	0.713	0.137	0.004156	16.0	125	72
55121	1.740	0.953	0.711	1.926	1.188	4.1	6.7	6.4	2.5	0.1920	0.713	0.137	0.004156	16.0	60	35
55122	1.740	0.953	0.711	1.926	1.188	4.1	6.7	6.4	2.5	0.1920	0.713	0.137	0.004156	16.0	26	15
55123	1.740	0.953	0.711	1.926	1.188	4.1	6.7	6.4	2.5	0.1920	0.713	0.137	0.004156	16.0	14	8
55375	1.803	0.902	0.711	1.965	1.162	4.1	8.2	5.9	2.6	0.2320	0.639	0.148	0.005330	16.3	300	214
55377	1.803	0.902	0.711	1.965	1.162	4.1	8.2	5.9	2.6	0.2320	0.639	0.148	0.005330	16.3	200	142
55374	1.803	0.902	0.711	1.965	1.162	4.1	8.2	5.9	2.6	0.2320	0.639	0.148	0.005330	16.3	173	123
55378	1.803	0.902	0.711	1.965	1.162	4.1	8.2	5.9	2.6	0.2320	0.639	0.148	0.005330	16.3	160	114
55379	1.803	0.902	0.711	1.965	1.162	4.1	8.2	5.9	2.6	0.2320	0.639	0.148	0.005330	16.3	147	105
55380	1.803	0.902	0.711	1.965	1.162	4.1	8.2	5.9	2.6	0.2320	0.639	0.148	0.005330	16.3	125	89
55381	1.803	0.902	0.711	1.965	1.162	4.1	8.2	5.9	2.6	0.2320	0.639	0.148	0.005330	16.3	60	43
55382	1.803	0.902	0.711	1.965	1.162	4.1	8.2	5.9	2.6	0.2320	0.639	0.148	0.005330	16.3	26	19
55383	1.803	0.902	0.711	1.965	1.162	4.1	8.2	5.9	2.6	0.2320	0.639	0.148	0.005330	16.3	14	10
55202	2.110	1.207	0.711	2.355	1.315	5.1	9.8	11.5	2.8	0.2260	1.144	0.258	0.008269	22.7	550	320
55201	2.110	1.207	0.711	2.355	1.315	5.1	9.8	11.5	2.8	0.2260	1.144	0.258	0.008269	22.7	300	163
55203	2.110	1.207	0.711	2.355	1.315	5.1	9.8	11.5	2.8	0.2260	1.144	0.258	0.008269	22.7	200	109
55200	2.110	1.207	0.711	2.355	1.315	5.1	9.8	11.5	2.8	0.2260	1.144	0.258	0.008269	22.7	173	96
55204	2.110	1.207	0.711	2.355	1.315	5.1	9.8	11.5	2.8	0.2260	1.144	0.258	0.008269	22.7	160	87
55205	2.110	1.207	0.711	2.355	1.315	5.1	9.8	11.5	2.8	0.2260	1.144	0.258	0.008269	22.7	147	81
55206	2.110	1.207	0.711	2.355	1.315	5.1	9.8	11.5	2.8	0.2260	1.144	0.258	0.008269	22.7	125	68
55848	2.110	1.207	0.711	2.355	1.315	5.1	9.8	11.5	2.8	0.2260	1.144	0.258	0.008269	22.7	60	32
55208	2.110	1.207	0.711	2.355	1.315	5.1	9.8	11.5	2.8	0.2260	1.144	0.258	0.008269	22.7	26	14
55209	2.110	1.207	0.711	2.355	1.315	5.1	9.8	11.5	2.8	0.2260	1.144	0.258	0.008269	22.7	14	8
55306	2.360	1.339	0.838	2.630	1.508	5.7	16.0	16.2	3.2	0.3310	1.407	0.466	0.019103	28.7	550	396
55305	2.360	1.339	0.838	2.630	1.508	5.7	16.0	16.2	3.2	0.3310	1.407	0.466	0.019103	28.7	300	216
55307	2.360	1.339	0.838	2.630	1.508	5.7	16.0	16.2	3.2	0.3310	1.407	0.466	0.019103	28.7	200	144
55304	2.360	1.339	0.838	2.630	1.508	5.7	16.0	16.2	3.2	0.3310	1.407	0.466	0.019103	28.7	173	124
55308	2.360	1.339	0.838	2.630	1.508	5.7	16.0	16.2	3.2	0.3310	1.407	0.466	0.019103	28.7	160	115
55309	2.360	1.339	0.838	2.630	1.508	5.7	16.0	16.2	3.2	0.3310	1.407	0.466	0.019103	28.7	147	106

Magnetics MPP Powder Cores

Part No.	OD cm 1	ID cm 2	HT cm 3	Wound OD cm 4	Wound HT cm 5	MPL cm 6	Wtfe grams 7	Wtcu grams 8	MLT cm 9	Ac cm sq 10	Wa cm sq 11	Ap cm 4th 12	Kg cm 5th 13	At cm sq 14	Perm 15	AL 16
55310	2.360	1.339	0.838	2.630	1.508	5.7	16.0	16.2	3.2	0.3310	1.407	0.466	0.019103	28.7	125	90
55059	2.360	1.339	0.838	2.630	1.508	5.7	16.0	16.2	3.2	0.3310	1.407	0.466	0.019103	28.7	60	43
55312	2.360	1.339	0.838	2.630	1.508	5.7	16.0	16.2	3.2	0.3310	1.407	0.466	0.019103	28.7	26	19
55313	2.360	1.339	0.838	2.630	1.508	5.7	16.0	16.2	3.2	0.3310	1.407	0.466	0.019103	28.7	14	10
55345	2.430	1.377	0.970	2.707	1.659	5.9	19.4	18.5	3.5	0.3880	1.488	0.578	0.025638	31.4	300	253
55347	2.430	1.377	0.970	2.707	1.659	5.9	19.4	18.5	3.5	0.3880	1.488	0.578	0.025638	31.4	200	169
55344	2.430	1.377	0.970	2.707	1.659	5.9	19.4	18.5	3.5	0.3880	1.488	0.578	0.025638	31.4	173	146
55348	2.430	1.377	0.970	2.707	1.659	5.9	19.4	18.5	3.5	0.3880	1.488	0.578	0.025638	31.4	160	135
55349	2.430	1.377	0.970	2.707	1.659	5.9	19.4	18.5	3.5	0.3880	1.488	0.578	0.025638	31.4	147	124
55350	2.430	1.377	0.970	2.707	1.659	5.9	19.4	18.5	3.5	0.3880	1.488	0.578	0.025638	31.4	125	105
55351	2.430	1.377	0.970	2.707	1.659	5.9	19.4	18.5	3.5	0.3880	1.488	0.578	0.025638	31.4	60	51
55352	2.430	1.377	0.970	2.707	1.659	5.9	19.4	18.5	3.5	0.3880	1.488	0.578	0.025638	31.4	26	22
55353	2.430	1.377	0.970	2.707	1.659	5.9	19.4	18.5	3.5	0.3880	1.488	0.578	0.025638	31.4	14	12
55926	2.770	1.410	1.199	3.027	1.904	6.4	35.3	22.9	4.1	0.6540	1.561	1.021	0.064582	39.8	550	740
55925	2.770	1.410	1.199	3.027	1.904	6.4	35.3	22.9	4.1	0.6540	1.561	1.021	0.064582	39.8	300	377
55927	2.770	1.410	1.199	3.027	1.904	6.4	35.3	22.9	4.1	0.6540	1.561	1.021	0.064582	39.8	200	251
55924	2.770	1.410	1.199	3.027	1.904	6.4	35.3	22.9	4.1	0.6540	1.561	1.021	0.064582	39.8	173	217
55928	2.770	1.410	1.199	3.027	1.904	6.4	35.3	22.9	4.1	0.6540	1.561	1.021	0.064582	39.8	160	201
55929	2.770	1.410	1.199	3.027	1.904	6.4	35.3	22.9	4.1	0.6540	1.561	1.021	0.064582	39.8	147	185
55930	2.770	1.410	1.199	3.027	1.904	6.4	35.3	22.9	4.1	0.6540	1.561	1.021	0.064582	39.8	125	157
55894	2.770	1.410	1.199	3.027	1.904	6.4	35.3	22.9	4.1	0.6540	1.561	1.021	0.064582	39.8	60	75
55932	2.770	1.410	1.199	3.027	1.904	6.4	35.3	22.9	4.1	0.6540	1.561	1.021	0.064582	39.8	26	32
55933	2.770	1.410	1.199	3.027	1.904	6.4	35.3	22.9	4.1	0.6540	1.561	1.021	0.064582	39.8	14	18
55581	3.520	2.260	0.983	4.028	2.113	9.0	34.5	62.6	4.4	0.4540	4.009	1.820	0.075320	64.4	550	348
55580	3.520	2.260	0.983	4.028	2.113	9.0	34.5	62.6	4.4	0.4540	4.009	1.820	0.075320	64.4	300	190
55582	3.520	2.260	0.983	4.028	2.113	9.0	34.5	62.6	4.4	0.4540	4.009	1.820	0.075320	64.4	200	126
55579	3.520	2.260	0.983	4.028	2.113	9.0	34.5	62.6	4.4	0.4540	4.009	1.820	0.075320	64.4	173	109
55583	3.520	2.260	0.983	4.028	2.113	9.0	34.5	62.6	4.4	0.4540	4.009	1.820	0.075320	64.4	160	101
55584	3.520	2.260	0.983	4.028	2.113	9.0	34.5	62.6	4.4	0.4540	4.009	1.820	0.075320	64.4	147	93
55585	3.520	2.260	0.983	4.028	2.113	9.0	34.5	62.6	4.4	0.4540	4.009	1.820	0.075320	64.4	125	79

Magnetics MPP Powder Cores

Part No.	OD cm 1	ID cm 2	HT cm 3	Wound OD cm 4	Wound HT cm 5	MPL cm 6	Wtfe grams 7	Wtcu grams 8	MLT cm 9	Ac cm sq 10	Wa cm sq 11	Ap cm 4th 12	Kg cm 5th 13	At cm sq 14	Perm 15	AL 16
55586	3.520	2.260	0.983	4.028	2.113	9.0	34.5	62.6	4.4	0.4540	4.009	1.820	0.075320	64.4	60	38
55587	3.520	2.260	0.983	4.028	2.113	9.0	34.5	62.6	4.4	0.4540	4.009	1.820	0.075320	64.4	26	16
55588	3.520	2.260	0.983	4.028	2.113	9.0	34.5	62.6	4.4	0.4540	4.009	1.820	0.075320	64.4	14	9
55544	3.380	1.930	1.161	3.770	2.126	8.2	46.6	47.4	4.6	0.6720	2.924	1.965	0.115789	58.5	550	559
55543	3.380	1.930	1.161	3.770	2.126	8.2	46.6	47.4	4.6	0.6720	2.924	1.965	0.115789	58.5	300	305
55545	3.380	1.930	1.161	3.770	2.126	8.2	46.6	47.4	4.6	0.6720	2.924	1.965	0.115789	58.5	200	203
55542	3.380	1.930	1.161	3.770	2.126	8.2	46.6	47.4	4.6	0.6720	2.924	1.965	0.115789	58.5	173	176
55546	3.380	1.930	1.161	3.770	2.126	8.2	46.6	47.4	4.6	0.6720	2.924	1.965	0.115789	58.5	160	163
55547	3.380	1.930	1.161	3.770	2.126	8.2	46.6	47.4	4.6	0.6720	2.924	1.965	0.115789	58.5	147	150
55548	3.380	1.930	1.161	3.770	2.126	8.2	46.6	47.4	4.6	0.6720	2.924	1.965	0.115789	58.5	125	127
55071	3.380	1.930	1.161	3.770	2.126	8.2	46.6	47.4	4.6	0.6720	2.924	1.965	0.115789	58.5	60	61
55550	3.380	1.930	1.161	3.770	2.126	8.2	46.6	47.4	4.6	0.6720	2.924	1.965	0.115789	58.5	26	28
55551	3.380	1.930	1.161	3.770	2.126	8.2	46.6	47.4	4.6	0.6720	2.924	1.965	0.115789	58.5	14	14
55320	3.670	2.150	1.128	4.116	2.203	9.0	51.8	61.2	4.7	0.6780	3.629	2.460	0.140739	67.9	550	515
55319	3.670	2.150	1.128	4.116	2.203	9.0	51.8	61.2	4.7	0.6780	3.629	2.460	0.140739	67.9	300	281
55321	3.670	2.150	1.128	4.116	2.203	9.0	51.8	61.2	4.7	0.6780	3.629	2.460	0.140739	67.9	200	187
55318	3.670	2.150	1.128	4.116	2.203	9.0	51.8	61.2	4.7	0.6780	3.629	2.460	0.140739	67.9	173	162
55322	3.670	2.150	1.128	4.116	2.203	9.0	51.8	61.2	4.7	0.6780	3.629	2.460	0.140739	67.9	160	150
55323	3.670	2.150	1.128	4.116	2.203	9.0	51.8	61.2	4.7	0.6780	3.629	2.460	0.140739	67.9	147	138
55324	3.670	2.150	1.128	4.116	2.203	9.0	51.8	61.2	4.7	0.6780	3.629	2.460	0.140739	67.9	125	117
55076	3.670	2.150	1.128	4.116	2.203	9.0	51.8	61.2	4.7	0.6780	3.629	2.460	0.140739	67.9	60	56
55326	3.670	2.150	1.128	4.116	2.203	9.0	51.8	61.2	4.7	0.6780	3.629	2.460	0.140739	67.9	26	24
55327	3.670	2.150	1.128	4.116	2.203	9.0	51.8	61.2	4.7	0.6780	3.629	2.460	0.140739	67.9	14	13
55250	4.070	2.330	1.537	4.543	2.702	9.8	89.7	86.6	5.7	1.0720	4.262	4.569	0.342768	87.2	550	740
55249	4.070	2.330	1.537	4.543	2.702	9.8	89.7	86.6	5.7	1.0720	4.262	4.569	0.342768	87.2	300	403
55251	4.070	2.330	1.537	4.543	2.702	9.8	89.7	86.6	5.7	1.0720	4.262	4.569	0.342768	87.2	200	269
55248	4.070	2.330	1.537	4.543	2.702	9.8	89.7	86.6	5.7	1.0720	4.262	4.569	0.342768	87.2	173	233
55252	4.070	2.330	1.537	4.543	2.702	9.8	89.7	86.6	5.7	1.0720	4.262	4.569	0.342768	87.2	160	215
55253	4.070	2.330	1.537	4.543	2.702	9.8	89.7	86.6	5.7	1.0720	4.262	4.569	0.342768	87.2	147	198
55254	4.070	2.330	1.537	4.543	2.702	9.8	89.7	86.6	5.7	1.0720	4.262	4.569	0.342768	87.2	125	168

Magnetics MPP Powder Cores																
Part No.	OD cm 1	ID cm 2	HT cm 3	Wound OD cm 4	Wound HT cm 5	MPL cm 6	Wtfe grams 7	Wtcu grams 8	MLT cm 9	Ac cm sq 10	Wa cm sq 11	Ap cm 4th 12	Kg cm 5th 13	At cm sq 14	Perm 15	AL 16
55083	4.070	2.330	1.537	4.543	2.702	9.8	89.7	86.6	5.7	1.0720	4.262	4.569	0.342768	87.2	60	81
55256	4.070	2.330	1.537	4.543	2.702	9.8	89.7	86.6	5.7	1.0720	4.262	4.569	0.342768	87.2	26	35
55257	4.070	2.330	1.537	4.543	2.702	9.8	89.7	86.6	5.7	1.0720	4.262	4.569	0.342768	87.2	14	19
55084	4.760	2.790	1.613	5.338	3.008	11.6	132.5	138.8	6.4	1.3400	6.111	8.188	0.686955	117.1	300	427
55086	4.760	2.790	1.613	5.338	3.008	11.6	132.5	138.8	6.4	1.3400	6.111	8.188	0.686955	117.1	200	285
55082	4.760	2.790	1.613	5.338	3.008	11.6	132.5	138.8	6.4	1.3400	6.111	8.188	0.686955	117.1	173	246
55087	4.760	2.790	1.613	5.338	3.008	11.6	132.5	138.8	6.4	1.3400	6.111	8.188	0.686955	117.1	160	228
55088	4.760	2.790	1.613	5.338	3.008	11.6	132.5	138.8	6.4	1.3400	6.111	8.188	0.686955	117.1	147	210
55089	4.760	2.790	1.613	5.338	3.008	11.6	132.5	138.8	6.4	1.3400	6.111	8.188	0.686955	117.1	125	178
55090	4.760	2.790	1.613	5.338	3.008	11.6	132.5	138.8	6.4	1.3400	6.111	8.188	0.686955	117.1	60	86
55091	4.760	2.790	1.613	5.338	3.008	11.6	132.5	138.8	6.4	1.3400	6.111	8.188	0.686955	117.1	26	27
55092	4.760	2.790	1.613	5.338	3.008	11.6	132.5	138.8	6.4	1.3400	6.111	8.188	0.686955	117.1	14	20
55710	5.170	3.090	1.435	5.822	2.980	12.7	135.4	171.4	6.4	1.2510	7.495	9.377	0.729483	133.0	300	365
55712	5.170	3.090	1.435	5.822	2.980	12.7	135.4	171.4	6.4	1.2510	7.495	9.377	0.729483	133.0	200	243
55709	5.170	3.090	1.435	5.822	2.980	12.7	135.4	171.4	6.4	1.2510	7.495	9.377	0.729483	133.0	173	210
55713	5.170	3.090	1.435	5.822	2.980	12.7	135.4	171.4	6.4	1.2510	7.495	9.377	0.729483	133.0	160	195
55714	5.170	3.090	1.435	5.822	2.980	12.7	135.4	171.4	6.4	1.2510	7.495	9.377	0.729483	133.0	147	179
55715	5.170	3.090	1.435	5.822	2.980	12.7	135.4	171.4	6.4	1.2510	7.495	9.377	0.729483	133.0	125	152
55716	5.170	3.090	1.435	5.822	2.980	12.7	135.4	171.4	6.4	1.2510	7.495	9.377	0.729483	133.0	60	73
55717	5.170	3.090	1.435	5.822	2.980	12.7	135.4	171.4	6.4	1.2510	7.495	9.377	0.729483	133.0	26	32
55718	5.170	3.090	1.435	5.822	2.980	12.7	135.4	171.4	6.4	1.2510	7.495	9.377	0.729483	133.0	14	17
55433	4.760	2.330	1.892	5.170	3.057	10.7	181.7	103.6	6.8	1.9900	4.262	8.481	0.987635	112.6	300	674
55435	4.760	2.330	1.892	5.170	3.057	10.7	181.7	103.6	6.8	1.9900	4.262	8.481	0.987635	112.6	200	450
55432	4.760	2.330	1.892	5.170	3.057	10.7	181.7	103.6	6.8	1.9900	4.262	8.481	0.987635	112.6	173	390
55436	4.760	2.330	1.892	5.170	3.057	10.7	181.7	103.6	6.8	1.9900	4.262	8.481	0.987635	112.6	160	360
55437	4.760	2.330	1.892	5.170	3.057	10.7	181.7	103.6	6.8	1.9900	4.262	8.481	0.987635	112.6	147	330
55438	4.760	2.330	1.892	5.170	3.057	10.7	181.7	103.6	6.8	1.9900	4.262	8.481	0.987635	112.6	125	281
55439	4.760	2.330	1.892	5.170	3.057	10.7	181.7	103.6	6.8	1.9900	4.262	8.481	0.987635	112.6	60	135
55440	4.760	2.330	1.892	5.170	3.057	10.7	181.7	103.6	6.8	1.9900	4.262	8.481	0.987635	112.6	26	59
55441	4.760	2.330	1.892	5.170	3.057	10.7	181.7	103.6	6.8	1.9900	4.262	8.481	0.987635	112.6	14	32

Magnetics MPP Powder Cores																
Part No.	OD cm 1	ID cm 2	HT cm 3	Wound OD cm 4	Wound HT cm 5	MPL cm 6	Wtfe grams 7	Wtcu grams 8	MLT cm 9	Ac cm sq 10	Wa cm sq 11	Ap cm 4th 12	Kg cm 5th 13	At cm sq 14	Perm 15	AL 16
55104	5.800	3.470	1.486	6.532	3.221	14.3	175.5	235.9	7.0	1.4440	9.452	13.649	1.123400	164.6	300	374
55106	5.800	3.470	1.486	6.532	3.221	14.3	175.5	235.9	7.0	1.4440	9.452	13.649	1.123400	164.6	200	250
55103	5.800	3.470	1.486	6.532	3.221	14.3	175.5	235.9	7.0	1.4440	9.452	13.649	1.123400	164.6	173	218
55107	5.800	3.470	1.486	6.532	3.221	14.3	175.5	235.9	7.0	1.4440	9.452	13.649	1.123400	164.6	160	200
55108	5.800	3.470	1.486	6.532	3.221	14.3	175.5	235.9	7.0	1.4440	9.452	13.649	1.123400	164.6	147	185
55109	5.800	3.470	1.486	6.532	3.221	14.3	175.5	235.9	7.0	1.4440	9.452	13.649	1.123400	164.6	125	156
55110	5.800	3.470	1.486	6.532	3.221	14.3	175.5	235.9	7.0	1.4440	9.452	13.649	1.123400	164.6	60	75
55111	5.800	3.470	1.486	6.532	3.221	14.3	175.5	235.9	7.0	1.4440	9.452	13.649	1.123400	164.6	26	33
55112	5.800	3.470	1.486	6.532	3.221	14.3	175.5	235.9	7.0	1.4440	9.452	13.649	1.123400	164.6	14	18
55866	7.890	4.800	1.397	8.918	3.797	20.0	300.9	549.7	8.5	1.7700	18.086	32.013	2.651763	287.4	125	142
55867	7.890	4.800	1.397	8.918	3.797	20.0	300.9	549.7	8.5	1.7700	18.086	32.013	2.651763	287.4	60	68
55868	7.890	4.800	1.397	8.918	3.797	20.0	300.9	549.7	8.5	1.7700	18.086	32.013	2.651763	287.4	26	30
55869	7.890	4.800	1.397	8.918	3.797	20.0	300.9	549.7	8.5	1.7700	18.086	32.013	2.651763	287.4	14	16

Micrometals Toroidal Iron Powder Cores																
Part No.	OD cm 1	ID cm 2	HT cm 3	Wound OD cm 4	Wound HT cm 5	MPL cm 6	Wtfe grams 7	Wtcu grams 8	MLT cm 9	Ac cm sq 10	Wa cm sq 11	Ap cm 4th 12	Kg cm 5th 13	At cm sq 14	Perm 15	AL 16
T16-52	0.406	0.198	0.152	0.441	0.251	0.9	0.1	0.1	0.6	0.015	0.031	0.000	0.000005	0.8	75	14
T16-26	0.406	0.198	0.152	0.441	0.251	0.9	0.1	0.1	0.6	0.015	0.031	0.000	0.000005	0.8	75	15
T16-18	0.406	0.198	0.152	0.441	0.251	0.9	0.1	0.1	0.6	0.015	0.031	0.000	0.000005	0.8	55	10
T16-8/90	0.406	0.198	0.152	0.441	0.251	0.9	0.1	0.1	0.6	0.015	0.031	0.000	0.000005	0.8	35	6
T20-52	0.508	0.224	0.178	0.543	0.290	1.2	0.2	0.1	0.7	0.023	0.039	0.001	0.000012	1.2	75	18
T20-26	0.508	0.224	0.178	0.543	0.290	1.2	0.2	0.1	0.7	0.023	0.039	0.001	0.000012	1.2	75	19
T20-18	0.508	0.224	0.178	0.543	0.290	1.2	0.2	0.1	0.7	0.023	0.039	0.001	0.000012	1.2	55	13
T20-8/90	0.508	0.224	0.178	0.543	0.290	1.2	0.2	0.1	0.7	0.023	0.039	0.001	0.000012	1.2	35	8
T25-52	0.648	0.305	0.244	0.700	0.397	1.5	0.4	0.2	0.9	0.037	0.073	0.003	0.000044	2.0	75	23
T25-26	0.648	0.305	0.244	0.700	0.397	1.5	0.4	0.2	0.9	0.037	0.073	0.003	0.000044	2.0	75	25
T25-18	0.648	0.305	0.244	0.700	0.397	1.5	0.4	0.2	0.9	0.037	0.073	0.003	0.000044	2.0	55	17
T25-8/90	0.648	0.305	0.244	0.700	0.397	1.5	0.4	0.2	0.9	0.037	0.073	0.003	0.000044	2.0	35	10
T26-52	0.673	0.267	0.483	0.711	0.617	1.5	0.9	0.3	1.3	0.090	0.056	0.005	0.000138	2.6	75	56
T26-26	0.673	0.267	0.483	0.711	0.617	1.5	0.9	0.3	1.3	0.090	0.056	0.005	0.000138	2.6	75	57
T26-18	0.673	0.267	0.483	0.711	0.617	1.5	0.9	0.3	1.3	0.090	0.056	0.005	0.000138	2.6	55	42
T26-8/90	0.673	0.267	0.483	0.711	0.617	1.5	0.9	0.3	1.3	0.090	0.056	0.005	0.000138	2.6	35	24
T30-52	0.780	0.384	0.325	0.848	0.517	1.8	0.8	0.5	1.1	0.060	0.116	0.007	0.000146	3.1	75	31
T30-26	0.780	0.384	0.325	0.848	0.517	1.8	0.8	0.5	1.1	0.060	0.116	0.007	0.000146	3.1	75	34
T30-18	0.780	0.384	0.325	0.848	0.517	1.8	0.8	0.5	1.1	0.060	0.116	0.007	0.000146	3.1	55	22
T30-8/90	0.780	0.384	0.325	0.848	0.517	1.8	0.8	0.5	1.1	0.060	0.116	0.007	0.000146	3.1	35	14
T37-52	0.953	0.521	0.325	1.055	0.586	2.3	1.0	1.0	1.3	0.064	0.213	0.014	0.000272	4.5	75	26
T37-26	0.953	0.521	0.325	1.055	0.586	2.3	1.0	1.0	1.3	0.064	0.213	0.014	0.000272	4.5	75	29
T37-18	0.953	0.521	0.325	1.055	0.586	2.3	1.0	1.0	1.3	0.064	0.213	0.014	0.000272	4.5	55	19
T37-8/90	0.953	0.521	0.325	1.055	0.586	2.3	1.0	1.0	1.3	0.064	0.213	0.014	0.000272	4.5	35	12
T38-52	0.953	0.445	0.483	1.028	0.706	2.2	1.7	0.8	1.5	0.114	0.155	0.018	0.000526	4.8	75	49
T38-26	0.953	0.445	0.483	1.028	0.706	2.2	1.7	0.8	1.5	0.114	0.155	0.018	0.000526	4.8	75	49
T38-18	0.953	0.445	0.483	1.028	0.706	2.2	1.7	0.8	1.5	0.114	0.155	0.018	0.000526	4.8	55	36
T38-8/90	0.953	0.445	0.483	1.028	0.706	2.2	1.7	0.8	1.5	0.114	0.155	0.018	0.000526	4.8	35	20
T44-52	1.120	0.582	0.404	1.228	0.695	2.7	1.9	1.5	1.5	0.099	0.266	0.026	0.000676	6.2	75	35
T44-26	1.120	0.582	0.404	1.228	0.695	2.7	1.9	1.5	1.5	0.099	0.266	0.026	0.000676	6.2	75	37

Micrometals Toroidal Iron Powder Cores																
Part No.	OD cm 1	ID cm 2	HT cm 3	Wound OD cm 4	Wound HT cm 5	MPL cm 6	Wtfe grams 7	Wtcu grams 8	MLT cm 9	Ac cm sq 10	Wa cm sq 11	Ap cm 4th 12	Kg cm 5th 13	At cm sq 14	Perm 15	AL 16
T44-18	1.120	0.582	0.404	1.228	0.695	2.7	1.9	1.5	1.5	0.099	0.266	0.026	0.000676	6.2	55	26
T44-8/90	1.120	0.582	0.404	1.228	0.695	2.7	1.9	1.5	1.5	0.099	0.266	0.026	0.000676	6.2	35	18
T50-52	1.270	0.770	0.483	1.435	0.868	3.2	2.5	3.0	1.8	0.112	0.465	0.052	0.001306	8.8	75	33
T50-26	1.270	0.770	0.483	1.435	0.868	3.2	2.5	3.0	1.8	0.112	0.465	0.052	0.001306	8.8	75	33
T50-18	1.270	0.770	0.483	1.435	0.868	3.2	2.5	3.0	1.8	0.112	0.465	0.052	0.001306	8.8	55	24
T50-8/90	1.270	0.770	0.483	1.435	0.868	3.2	2.5	3.0	1.8	0.112	0.465	0.052	0.001306	8.8	35	18
T51-52C	1.270	0.508	0.635	1.344	0.889	2.8	4.4	1.5	2.0	0.223	0.203	0.045	0.001983	8.1	75	75
T51-26C	1.270	0.508	0.635	1.344	0.889	2.8	4.4	1.5	2.0	0.223	0.203	0.045	0.001983	8.1	75	83
T51-8C/90	1.270	0.508	0.635	1.344	0.889	2.8	4.4	1.5	2.0	0.223	0.203	0.045	0.001983	8.1	35	37
T50-52B	1.270	0.770	0.635	1.435	1.020	3.2	3.3	3.4	2.0	0.148	0.465	0.069	0.002007	9.6	75	44
T50-26B	1.270	0.770	0.635	1.435	1.020	3.2	3.3	3.4	2.0	0.148	0.465	0.069	0.002007	9.6	75	44
T50-18B	1.270	0.770	0.635	1.435	1.020	3.2	3.3	3.4	2.0	0.148	0.465	0.069	0.002007	9.6	55	32
T50-8B/90	1.270	0.770	0.635	1.435	1.020	3.2	3.3	3.4	2.0	0.148	0.465	0.069	0.002007	9.6	35	23
T50-52D	1.270	0.770	0.953	1.435	1.338	3.2	5.0	4.2	2.5	0.223	0.465	0.104	0.003644	11.2	75	66
T50-26D	1.270	0.770	0.953	1.435	1.338	3.2	5.0	4.2	2.5	0.223	0.465	0.104	0.003644	11.2	75	72
T60-52	1.520	0.853	0.594	1.690	1.021	3.7	4.9	4.4	2.2	0.187	0.571	0.107	0.003688	12.2	75	47
T60-26	1.520	0.853	0.594	1.690	1.021	3.7	4.9	4.4	2.2	0.187	0.571	0.107	0.003688	12.2	75	50
T60-18	1.520	0.853	0.594	1.690	1.021	3.7	4.9	4.4	2.2	0.187	0.571	0.107	0.003688	12.2	55	35
T60-8/90	1.520	0.853	0.594	1.690	1.021	3.7	4.9	4.4	2.2	0.187	0.571	0.107	0.003688	12.2	35	19
T68-52	1.750	0.940	0.483	1.930	0.953	4.2	5.3	5.4	2.2	0.179	0.694	0.124	0.004091	14.4	75	40
T68-26	1.750	0.940	0.483	1.930	0.953	4.2	5.3	5.4	2.2	0.179	0.694	0.124	0.004091	14.4	75	44
T68-18	1.750	0.940	0.483	1.930	0.953	4.2	5.3	5.4	2.2	0.179	0.694	0.124	0.004091	14.4	55	29
T68-8/90	1.750	0.940	0.483	1.930	0.953	4.2	5.3	5.4	2.2	0.179	0.694	0.124	0.004091	14.4	35	20
T68-52A	1.750	0.940	0.635	1.930	1.105	4.2	7.2	6.0	2.4	0.242	0.694	0.168	0.006725	15.4	75	54
T68-26A	1.750	0.940	0.635	1.930	1.105	4.2	7.2	6.0	2.4	0.242	0.694	0.168	0.006725	15.4	75	58
T68-18A	1.750	0.940	0.635	1.930	1.105	4.2	7.2	6.0	2.4	0.242	0.694	0.168	0.006725	15.4	55	40
T68-8A/90	1.750	0.940	0.635	1.930	1.105	4.2	7.2	6.0	2.4	0.242	0.694	0.168	0.006725	15.4	35	26
T72-52	1.830	0.711	0.660	1.930	1.016	4.0	9.8	3.6	2.5	0.349	0.397	0.138	0.007672	14.8	75	82
T72-26	1.830	0.711	0.660	1.930	1.016	4.0	9.8	3.6	2.5	0.349	0.397	0.138	0.007672	14.8	75	90
T72-18	1.830	0.711	0.660	1.930	1.016	4.0	9.8	3.6	2.5	0.349	0.397	0.138	0.007672	14.8	55	60

Micrometals Toroidal Iron Powder Cores

Part No.	OD cm 1	ID cm 2	HT cm 3	Wound OD cm 4	Wound HT cm 5	MPL cm 6	Wtfe grams 7	Wtcu grams 8	MLT cm 9	Ac cm sq 10	Wa cm sq 11	Ap cm 4th 12	Kg cm 5th 13	At cm sq 14	Perm 15	AL 16
T72-8/90	1.830	0.711	0.660	1.930	1.016	4.0	9.8	3.6	2.5	0.349	0.397	0.138	0.007672	14.8	35	36
T80-52	2.020	1.260	0.635	2.296	1.265	5.1	8.3	11.7	2.6	0.231	1.246	0.288	0.010107	21.4	75	42
T80-26	2.020	1.260	0.635	2.296	1.265	5.1	8.3	11.7	2.6	0.231	1.246	0.288	0.010107	21.4	75	46
T80-18	2.020	1.260	0.635	2.296	1.265	5.1	8.3	11.7	2.6	0.231	1.246	0.288	0.010107	21.4	55	31
T80-8/90	2.020	1.260	0.635	2.296	1.265	5.1	8.3	11.7	2.6	0.231	1.246	0.288	0.010107	21.4	35	18
T68-52D	1.750	0.940	0.953	1.930	1.423	4.2	10.6	7.2	2.9	0.358	0.694	0.248	0.012158	17.7	75	80
T68-26D	1.750	0.940	0.953	1.930	1.423	4.2	10.6	7.2	2.9	0.358	0.694	0.248	0.012158	17.7	75	87
T80-52B	2.020	1.260	0.953	2.296	1.583	5.1	12.5	13.9	3.1	0.347	1.246	0.432	0.019111	24.1	75	63
T80-26B	2.020	1.260	0.953	2.296	1.583	5.1	12.5	13.9	3.1	0.347	1.246	0.432	0.019111	24.1	75	71
T80-18B	2.020	1.260	0.953	2.296	1.583	5.1	12.5	13.9	3.1	0.347	1.246	0.432	0.019111	24.1	55	47
T80-8B/90	2.020	1.260	0.953	2.296	1.583	5.1	12.5	13.9	3.1	0.347	1.246	0.432	0.019111	24.1	35	30
T94-52	2.390	1.420	0.792	2.688	1.452	6.0	15.1	17.4	3.1	0.362	1.583	0.573	0.026772	29.1	75	57
T94-26	2.390	1.420	0.792	2.688	1.452	6.0	15.1	17.4	3.1	0.362	1.583	0.573	0.026772	29.1	75	60
T94-18	2.390	1.420	0.792	2.688	1.452	6.0	15.1	17.4	3.1	0.362	1.583	0.573	0.026772	29.1	55	42
T94-8/90	2.390	1.420	0.792	2.688	1.452	6.0	15.1	17.4	3.1	0.362	1.583	0.573	0.026772	29.1	35	25
T80-52D	2.020	1.260	1.270	2.296	1.900	5.1	16.3	16.2	3.6	0.453	1.246	0.565	0.028042	26.7	75	83
T80-26D	2.020	1.260	1.270	2.296	1.900	5.1	16.3	16.2	3.6	0.453	1.246	0.565	0.028042	26.7	75	92
T90-52	2.290	1.400	0.953	2.591	1.653	5.8	16.0	18.4	3.4	0.395	1.539	0.608	0.028606	29.4	75	64
T90-26	2.290	1.400	0.953	2.591	1.653	5.8	16.0	18.4	3.4	0.395	1.539	0.608	0.028606	29.4	75	70
T90-18	2.290	1.400	0.953	2.591	1.653	5.8	16.0	18.4	3.4	0.395	1.539	0.608	0.028606	29.4	55	47
T90-8/90	2.290	1.400	0.953	2.591	1.653	5.8	16.0	18.4	3.4	0.395	1.539	0.608	0.028606	29.4	35	30
T106-52A	2.690	1.450	0.792	2.969	1.517	6.5	20.9	20.1	3.4	0.461	1.650	0.761	0.041034	34.6	75	67
T106-26A	2.690	1.450	0.792	2.969	1.517	6.5	20.9	20.1	3.4	0.461	1.650	0.761	0.041034	34.6	75	67
T106-18A	2.690	1.450	0.792	2.969	1.517	6.5	20.9	20.1	3.4	0.461	1.650	0.761	0.041034	34.6	55	49
T124-26	3.160	1.800	0.711	3.523	1.611	7.8	24.9	33.2	3.7	0.459	2.543	1.167	0.058473	46.3	75	58
T106-52	2.690	1.450	1.110	2.969	1.835	6.5	29.9	23.1	3.9	0.659	1.650	1.088	0.072990	38.0	75	95
T106-26	2.690	1.450	1.110	2.969	1.835	6.5	29.9	23.1	3.9	0.659	1.650	1.088	0.072990	38.0	75	93
T106-18	2.690	1.450	1.110	2.969	1.835	6.5	29.9	23.1	3.9	0.659	1.650	1.088	0.072990	38.0	55	70
T106-8/90	2.690	1.450	1.110	2.969	1.835	6.5	29.9	23.1	3.9	0.659	1.650	1.088	0.072990	38.0	35	45
T106-52B	2.690	1.450	1.460	2.969	2.185	6.5	39.0	26.3	4.5	0.858	1.650	1.416	0.108290	41.7	75	124

Micrometals Toroidal Iron Powder Cores

Part No.	OD cm 1	ID cm 2	HT cm 3	Wound OD cm 4	Wound HT cm 5	MPL cm 6	Wtfe grams 7	Wtcu grams 8	MLT cm 9	Ac cm sq 10	Wa cm sq 11	Ap cm 4th 12	Kg cm 5th 13	At cm sq 14	Perm 15	AL 16
T106-26B	2.690	1.450	1.460	2.969	2.185	6.5	39.0	26.3	4.5	0.858	1.650	1.416	0.108290	41.7	75	124
T106-18B	2.690	1.450	1.460	2.969	2.185	6.5	39.0	26.3	4.5	0.858	1.650	1.416	0.108290	41.7	55	91
T130-52	3.300	1.980	1.110	3.719	2.100	8.3	40.5	48.3	4.4	0.698	3.078	2.148	0.135813	56.9	75	79
T130-26	3.300	1.980	1.110	3.719	2.100	8.3	40.5	48.3	4.4	0.698	3.078	2.148	0.135813	56.9	75	81
T130-18	3.300	1.980	1.110	3.719	2.100	8.3	40.5	48.3	4.4	0.698	3.078	2.148	0.135813	56.9	55	58
T130-8/90	3.300	1.980	1.110	3.719	2.100	8.3	40.5	48.3	4.4	0.698	3.078	2.148	0.135813	56.9	35	35
T132-52	3.300	1.780	1.110	3.643	2.000	8.0	44.9	39.1	4.4	0.805	2.487	2.002	0.145993	53.8	75	95
T132-26	3.300	1.780	1.110	3.643	2.000	8.0	44.9	39.1	4.4	0.805	2.487	2.002	0.145993	53.8	75	103
T131-52	3.300	1.630	1.110	3.590	1.925	7.7	47.8	32.8	4.4	0.885	2.086	1.846	0.147966	51.7	75	108
T131-26	3.300	1.630	1.110	3.590	1.925	7.7	47.8	32.8	4.4	0.885	2.086	1.846	0.147966	51.7	75	116
T131-18	3.300	1.630	1.110	3.590	1.925	7.7	47.8	32.8	4.4	0.885	2.086	1.846	0.147966	51.7	55	79
T131-8/90	3.300	1.630	1.110	3.590	1.925	7.7	47.8	32.8	4.4	0.885	2.086	1.846	0.147966	51.7	35	53
T150-52	3.840	2.150	1.110	4.268	2.185	9.4	58.2	62.6	4.8	0.887	3.629	3.219	0.235554	71.5	75	89
T150-26	3.840	2.150	1.110	4.268	2.185	9.4	58.2	62.6	4.8	0.887	3.629	3.219	0.235554	71.5	75	96
T157-52	3.990	2.410	1.450	4.503	2.655	10.1	74.9	89.4	5.5	1.060	4.559	4.833	0.371763	85.3	75	99
T157-26	3.990	2.410	1.450	4.503	2.655	10.1	74.9	89.4	5.5	1.060	4.559	4.833	0.371763	85.3	75	100
T157-18	3.990	2.410	1.450	4.503	2.655	10.1	74.9	89.4	5.5	1.060	4.559	4.833	0.371763	85.3	55	73
T157-8/90	3.990	2.410	1.450	4.503	2.655	10.1	74.9	89.4	5.5	1.060	4.559	4.833	0.371763	85.3	35	42
T175-52	4.450	2.720	1.650	5.035	3.010	11.2	105.1	128.0	6.2	1.340	5.808	7.782	0.672799	107.4	75	105
T175-26	4.450	2.720	1.650	5.035	3.010	11.2	105.1	128.0	6.2	1.340	5.808	7.782	0.672799	107.4	75	105
T175-18	4.450	2.720	1.650	5.035	3.010	11.2	105.1	128.0	6.2	1.340	5.808	7.782	0.672799	107.4	55	82
T200-52	5.080	3.180	1.400	5.778	2.990	13.0	115.6	178.0	6.3	1.270	7.938	10.082	0.812410	131.7	75	92
T200-26	5.080	3.180	1.400	5.778	2.990	13.0	115.6	178.0	6.3	1.270	7.938	10.082	0.812410	131.7	75	92
T200-18	5.080	3.180	1.400	5.778	2.990	13.0	115.6	178.0	6.3	1.270	7.938	10.082	0.812410	131.7	55	67
T200-8/90	5.080	3.180	1.400	5.778	2.990	13.0	115.6	178.0	6.3	1.270	7.938	10.082	0.812410	131.7	35	43
T184-52	4.670	2.410	1.800	5.115	3.005	11.2	147.4	107.3	6.6	1.880	4.559	8.572	0.974280	109.8	75	159
T184-26	4.670	2.410	1.800	5.115	3.005	11.2	147.4	107.3	6.6	1.880	4.559	8.572	0.974280	109.8	75	169
T184-18	4.670	2.410	1.800	5.115	3.005	11.2	147.4	107.3	6.6	1.880	4.559	8.572	0.974280	109.8	55	116
T184-8/90	4.670	2.410	1.800	5.115	3.005	11.2	147.4	107.3	6.6	1.880	4.559	8.572	0.974280	109.8	35	72
T225-52	5.720	3.570	1.400	6.498	3.180	14.6	145.1	241.1	6.8	1.420	9.949	14.127	1.177272	162.3	75	92

Micrometals Toroidal Iron Powder Cores																
Part No.	OD cm 1	ID cm 2	HT cm 3	Wound OD cm 4	Wound HT cm 5	MPL cm 6	Wtfe grams 7	Wtcu grams 8	MLT cm 9	Ac cm sq 10	Wa cm sq 11	Ap cm 4th 12	Kg cm 5th 13	At cm sq 14	Perm 15	AL 16
T225-26	5.720	3.570	1.400	6.498	3.180	14.6	145.1	241.1	6.8	1.420	9.949	14.127	1.177272	162.3	75	98
T225-18	5.720	3.570	1.400	6.498	3.180	14.6	145.1	241.1	6.8	1.420	9.949	14.127	1.177272	162.3	55	67
T225-8/90	5.720	3.570	1.400	6.498	3.180	14.6	145.1	241.1	6.8	1.420	9.949	14.127	1.177272	162.3	35	43
T201-52	5.080	2.410	2.220	5.492	3.425	11.8	232.1	123.5	7.6	2.810	4.559	12.812	1.890817	130.6	75	224
T201-26	5.080	2.410	2.220	5.492	3.425	11.8	232.1	123.5	7.6	2.810	4.559	12.812	1.890817	130.6	75	224
T201-18	5.080	2.410	2.220	5.492	3.425	11.8	232.1	123.5	7.6	2.810	4.559	12.812	1.890817	130.6	55	164
T200-52B	5.080	3.180	2.540	5.778	4.130	13.0	211.1	229.4	8.1	2.320	7.938	18.417	2.102694	155.5	75	155
T200-26B	5.080	3.180	2.540	5.778	4.130	13.0	211.1	229.4	8.1	2.320	7.938	18.417	2.102694	155.5	75	160
T200-18B	5.080	3.180	2.540	5.778	4.130	13.0	211.1	229.4	8.1	2.320	7.938	18.417	2.102694	155.5	55	120
T200-8B/90	5.080	3.180	2.540	5.778	4.130	13.0	211.1	229.4	8.1	2.320	7.938	18.417	2.102694	155.5	35	79
T300-52	7.720	4.900	1.270	8.810	3.720	19.8	232.8	550.1	8.2	1.680	18.848	31.664	2.592406	279.5	75	80
T300-26	7.720	4.900	1.270	8.810	3.720	19.8	232.8	550.1	8.2	1.680	18.848	31.664	2.592406	279.5	75	80
T300-18	7.720	4.900	1.270	8.810	3.720	19.8	232.8	550.1	8.2	1.680	18.848	31.664	2.592406	279.5	55	58
T300-8/90	7.720	4.900	1.270	8.810	3.720	19.8	232.8	550.1	8.2	1.680	18.848	31.664	2.592406	279.5	35	37
T225-52B	5.720	3.570	2.540	6.502	4.325	14.6	264.7	307.4	8.6	2.590	10.005	25.912	3.107076	189.3	75	155
T225-26B	5.720	3.570	2.540	6.502	4.325	14.6	264.7	307.4	8.6	2.590	10.005	25.912	3.107076	189.3	75	160
T250-52	6.350	3.180	2.540	6.922	4.130	15.0	403.2	258.1	9.1	3.840	7.938	30.483	5.120473	202.7	75	242
T250-26	6.350	3.180	2.540	6.922	4.130	15.0	403.2	258.1	9.1	3.840	7.938	30.483	5.120473	202.7	75	242
T250-18	6.350	3.180	2.540	6.922	4.130	15.0	403.2	258.1	9.1	3.840	7.938	30.483	5.120473	202.7	55	177
T300-26D	7.720	4.900	2.540	8.810	4.990	19.8	468.5	686.3	10.2	3.380	18.848	63.706	8.411149	319.9	75	160
T400-52	10.200	5.720	1.650	11.339	4.510	25.0	605.5	986.4	10.8	3.460	25.684	88.866	11.388070	451.6	75	131
T400-26	10.200	5.720	1.650	11.339	4.510	25.0	605.5	986.4	10.8	3.460	25.684	88.866	11.388070	451.6	75	131
T400-18	10.200	5.720	1.650	11.339	4.510	25.0	605.5	986.4	10.8	3.460	25.684	88.866	11.388070	451.6	55	96
T400-8/90	10.200	5.720	1.650	11.339	4.510	25.0	605.5	986.4	10.8	3.460	25.684	88.866	11.388070	451.6	35	60
T400-26D	10.200	5.720	3.300	11.339	6.160	25.0	1198.8	1227.5	13.4	6.850	25.684	175.935	35.867710	519.2	75	262
T520-52	13.200	7.820	2.030	14.836	5.940	33.1	1214.1	2357.1	13.8	5.240	48.005	251.544	38.183420	775.2	75	149
T520-26	13.200	7.820	2.030	14.836	5.940	33.1	1214.1	2357.1	13.8	5.240	48.005	251.544	38.183420	775.2	75	149

Micrometals EE Iron Powder Cores

Part No.	HT cm 1	Wth cm 2	Lt cm 3	G cm 4	MPL cm 5	Wtfe grams 6	Wtcu grams 7	MLT cm 8	Ac cm sq 9	Wa cm sq 10	Ap cm 4th 11	Kg cm 5th 12	At cm sq 13	Perm 14	AL 15
E49-52	1.11	0.95	1.27	0.793	2.9	2.0	2.4	2.7	0.101	0.252	0.025	0.000384	6.5	75	38
E49-26	1.11	0.95	1.27	0.793	2.9	2.0	2.4	2.7	0.101	0.252	0.025	0.000384	6.5	75	38
E49-18	1.11	0.95	1.27	0.793	2.9	2.0	2.4	2.7	0.101	0.252	0.025	0.000384	6.5	55	29
E49-8	1.11	0.95	1.27	0.793	2.9	2.0	2.4	2.7	0.101	0.252	0.025	0.000384	6.5	35	21
E75-52	1.61	1.43	1.91	1.160	4.2	6.6	7.5	3.8	0.226	0.554	0.125	0.002972	14.4	75	59
E75-26	1.61	1.43	1.91	1.160	4.2	6.6	7.5	3.8	0.226	0.554	0.125	0.002972	14.4	75	64
E75-18	1.61	1.43	1.91	1.160	4.2	6.6	7.5	3.8	0.226	0.554	0.125	0.002972	14.4	55	46
E75-8	1.61	1.43	1.91	1.160	4.2	6.6	7.5	3.8	0.226	0.554	0.125	0.002972	14.4	35	34
E100-52	1.91	1.91	2.54	1.270	5.1	14.4	14.3	4.9	0.403	0.810	0.326	0.010626	23.5	75	85
E100-26	1.91	1.91	2.54	1.270	5.1	14.4	14.3	4.9	0.403	0.810	0.326	0.010626	23.5	75	92
E100-18	1.91	1.91	2.54	1.270	5.1	14.4	14.3	4.9	0.403	0.810	0.326	0.010626	23.5	55	65
E100-8	1.91	1.91	2.54	1.270	5.1	14.4	14.3	4.9	0.403	0.810	0.326	0.010626	23.5	35	48
E137-52	2.91	2.54	3.49	1.960	7.4	47.0	37.1	6.7	0.907	1.555	1.411	0.076252	46.8	75	131
E137-26	2.91	2.54	3.49	1.960	7.4	47.0	37.1	6.7	0.907	1.555	1.411	0.076252	46.8	75	134
E137-18	2.91	2.54	3.49	1.960	7.4	47.0	37.1	6.7	0.907	1.555	1.411	0.076252	46.8	55	100
E137-8	2.91	2.54	3.49	1.960	7.4	47.0	37.1	6.7	0.907	1.555	1.411	0.076252	46.8	35	67
E162-52	3.41	2.86	4.13	2.140	8.4	95.2	49.5	8.2	1.610	1.701	2.739	0.215532	63.2	75	199
E162-26	3.41	2.86	4.13	2.140	8.4	95.2	49.5	8.2	1.610	1.701	2.739	0.215532	63.2	75	210
E162-18	3.41	2.86	4.13	2.140	8.4	95.2	49.5	8.2	1.610	1.701	2.739	0.215532	63.2	55	149
E162-8	3.41	2.86	4.13	2.140	8.4	95.2	49.5	8.2	1.610	1.701	2.739	0.215532	63.2	35	105
E168-52	4.22	3.07	4.28	3.070	10.4	129.5	91.3	8.9	1.810	2.870	5.196	0.420572	84.4	75	179
E168-26	4.22	3.07	4.28	3.070	10.4	129.5	91.3	8.9	1.810	2.870	5.196	0.420572	84.4	75	195
E168-18	4.22	3.07	4.28	3.070	10.4	129.5	91.3	8.9	1.810	2.870	5.196	0.420572	84.4	55	135
E168-8	4.22	3.07	4.28	3.070	10.4	129.5	91.3	8.9	1.810	2.870	5.196	0.420572	84.4	35	97
E187-52	3.94	3.18	4.74	2.420	9.5	163.1	65.2	9.4	2.480	1.948	4.831	0.509005	82.4	75	265
E187-26	3.94	3.18	4.74	2.420	9.5	163.1	65.2	9.4	2.480	1.948	4.831	0.509005	82.4	75	265
E187-18	3.94	3.18	4.74	2.420	9.5	163.1	65.2	9.4	2.480	1.948	4.831	0.509005	82.4	55	213
E187-8	3.94	3.18	4.74	2.420	9.5	163.1	65.2	9.4	2.480	1.948	4.831	0.509005	82.4	35	144
E168-52A	4.22	3.07	4.28	3.070	10.4	172.2	101.5	9.9	2.410	2.870	6.918	0.670637	92.9	75	230
E168-26A	4.22	3.07	4.28	3.070	10.4	172.2	101.5	9.9	2.410	2.870	6.918	0.670637	92.9	75	232

Micrometals EE Iron Powder Cores															
Part No.	HT cm 1	Wth cm 2	Lt cm 3	G cm 4	MPL cm 5	Wtfe grams 6	Wtcu grams 7	MLT cm 8	Ac cm sq 9	Wa cm sq 10	Ap cm 4th 11	Kg cm 5th 12	At cm sq 13	Perm 14	AL 15
E168-18A	4.22	3.07	4.28	3.070	10.4	172.2	101.5	9.9	2.410	2.870	6.918	0.670637	92.9	55	170
E168-8A	4.22	3.07	4.28	3.070	10.4	172.2	101.5	9.9	2.410	2.870	6.918	0.670637	92.9	35	116
E225-52	4.76	3.81	5.69	2.900	11.5	285.6	110.7	11.2	3.580	2.784	9.967	1.276322	119.3	75	325
E225-26	4.76	3.81	5.69	2.900	11.5	285.6	110.7	11.2	3.580	2.784	9.967	1.276322	119.3	75	325
E225-18	4.76	3.81	5.69	2.900	11.5	285.6	110.7	11.2	3.580	2.784	9.967	1.276322	119.3	55	240
E225-8	4.76	3.81	5.69	2.900	11.5	285.6	110.7	11.2	3.580	2.784	9.967	1.276322	119.3	35	156
E220-52	5.54	3.86	5.61	3.830	13.2	333.9	167.9	11.6	3.600	4.079	14.684	1.827264	141.3	75	262
E220-26	5.54	3.86	5.61	3.830	13.2	333.9	167.9	11.6	3.600	4.079	14.684	1.827264	141.3	75	275
E220-18	5.54	3.86	5.61	3.830	13.2	333.9	167.9	11.6	3.600	4.079	14.684	1.827264	141.3	55	196
E220-8	5.54	3.86	5.61	3.830	13.2	333.9	167.9	11.6	3.600	4.079	14.684	1.827264	141.3	35	143
E305-52	7.75	5.38	7.75	5.380	18.5	728.0	426.5	14.8	5.620	8.097	45.505	6.905384	258.4	75	287
E305-26	7.75	5.38	7.75	5.380	18.5	728.0	426.5	14.8	5.620	8.097	45.505	6.905384	258.4	75	287
E305-18	7.75	5.38	7.75	5.380	18.5	728.0	426.5	14.8	5.620	8.097	45.505	6.905384	258.4	55	222
E305-8	7.75	5.38	7.75	5.380	18.5	728.0	426.5	14.8	5.620	8.097	45.505	6.905384	258.4	35	156
E305-52A	7.75	5.38	7.75	5.380	18.5	973.0	472.0	16.4	7.490	8.097	60.646	11.083210	282.9	75	382
E305-26A	7.75	5.38	7.75	5.380	18.5	973.0	472.0	16.4	7.490	8.097	60.646	11.083210	282.9	75	382
E450-26	9.24	7.93	11.40	5.720	22.9	1960.0	972.6	21.5	12.200	12.698	154.921	35.100000	476.0	75	540

MMG North America High Flux Powder Cores																
Part No.	OD cm 1	ID cm 2	HT cm 3	Wound OD cm 4	Wound HT cm 5	MPL cm 6	Wtfe grams 7	Wtcu grams 8	MLT cm 9	Ac cm sq 10	Wa cm sq 11	Ap cm 4th 12	Kg cm 5th 13	At cm sq 14	Perm 15	AL 16
G-60ZH	1.270	0.762	0.475	1.431	0.856	3.2	2.8	2.9	1.8	0.11	0.456	0.050	0.001242	8.7	125	61
G-60SH	1.270	0.762	0.475	1.431	0.856	3.2	2.8	2.9	1.8	0.11	0.456	0.050	0.001242	8.7	140	55
G-60NH	1.270	0.762	0.475	1.431	0.856	3.2	2.8	2.9	1.8	0.11	0.456	0.050	0.001242	8.7	60	27
G-62ZH	1.664	1.016	0.635	1.883	1.143	4.2	6.7	6.8	2.3	0.20	0.810	0.162	0.005524	15.1	140	81
G-62SH	1.664	1.016	0.635	1.883	1.143	4.2	6.7	6.8	2.3	0.20	0.810	0.162	0.005524	15.1	140	74
G-62NH	1.664	1.016	0.635	1.883	1.143	4.2	6.7	6.8	2.3	0.20	0.810	0.162	0.005524	15.1	60	35
G-29ZH	2.032	1.270	0.635	2.310	1.270	5.2	9.5	11.9	2.6	0.23	1.266	0.291	0.010142	21.7	125	76
G-29SH	2.032	1.270	0.635	2.310	1.270	5.2	9.5	11.9	2.6	0.23	1.266	0.291	0.010142	21.7	140	69
G-29NH	2.032	1.270	0.635	2.310	1.270	5.2	9.5	11.9	2.6	0.23	1.266	0.291	0.010142	21.7	60	33
G-57ZH	2.692	1.473	1.118	2.979	1.855	6.5	34.0	23.9	3.9	0.65	1.703	1.107	0.073013	38.4	125	171
G-57SH	2.692	1.473	1.118	2.979	1.855	6.5	34.0	23.9	3.9	0.65	1.703	1.107	0.073013	38.4	140	155
G-57NH	2.692	1.473	1.118	2.979	1.855	6.5	34.0	23.9	3.9	0.65	1.703	1.107	0.073013	38.4	60	75
G-56ZH	3.302	1.994	1.118	3.726	2.115	8.3	46.6	49.2	4.4	0.70	3.121	2.185	0.138081	57.3	125	145
G-56SH	3.302	1.994	1.118	3.726	2.115	8.3	46.6	49.2	4.4	0.70	3.121	2.185	0.138081	57.3	140	132
G-56NH	3.302	1.994	1.118	3.726	2.115	8.3	46.6	49.2	4.4	0.70	3.121	2.185	0.138081	57.3	60	63
G-48ZH	3.988	2.413	1.448	4.503	2.655	10.1	87.6	89.5	5.5	1.09	4.571	4.982	0.394427	85.2	125	187
G-48SH	3.988	2.413	1.448	4.503	2.655	10.1	87.6	89.5	5.5	1.09	4.571	4.982	0.394427	85.2	140	170
G-48NH	3.988	2.413	1.448	4.503	2.655	10.1	87.6	89.5	5.5	1.09	4.571	4.982	0.394427	85.2	60	82
G-1ZH	5.050	3.000	1.700	5.679	3.200	12.6	168.9	169.8	6.8	1.67	7.065	11.799	1.165892	132.6	125	228
G-1SH	5.050	3.000	1.700	5.679	3.200	12.6	168.9	169.8	6.8	1.67	7.065	11.799	1.165892	132.6	140	207
G-1NH	5.050	3.000	1.700	5.679	3.200	12.6	168.9	169.8	6.8	1.67	7.065	11.799	1.165892	132.6	60	100
G-6ZH	7.000	4.600	2.000	8.054	4.300	18.2	319.2	519.8	8.8	2.19	16.611	36.377	3.621186	259.8	125	207
G-6SH	7.000	4.600	2.000	8.054	4.300	18.2	319.2	519.8	8.8	2.19	16.611	36.377	3.621186	259.8	140	188
G-6NH	7.000	4.600	2.000	8.054	4.300	18.2	319.2	519.8	8.8	2.19	16.611	36.377	3.621186	259.8	60	91
G-9ZH	8.000	4.800	2.900	9.016	5.300	20.1	696.6	710.0	11.0	4.33	18.086	78.314	12.286240	341.3	125	372
G-9SH	8.000	4.800	2.900	9.016	5.300	20.1	696.6	710.0	11.0	4.33	18.086	78.314	12.286240	341.3	140	339
G-9NH	8.000	4.800	2.900	9.016	5.300	20.1	696.6	710.0	11.0	4.33	18.086	78.314	12.286240	341.3	60	163

MMG North America MPP Powder Cores

Part No.	OD cm 1	ID cm 2	HT cm 3	Wound OD cm 4	Wound HT cm 5	MPL cm 6	Wtfe grams 7	Wtcu grams 8	MLT cm 9	Ac cm sq 10	Wa cm sq 11	Ap cm 4th 12	Kg cm 5th 13	At cm sq 14	Perm 15	AL 16
G-60Z	1.270	0.762	0.475	1.431	0.856	3.2	2.8	2.9	1.8	0.11	0.456	0.050	0.001242	8.7	125	61
G-60S	1.270	0.762	0.475	1.431	0.856	3.2	2.8	2.9	1.8	0.11	0.456	0.050	0.001242	8.7	140	55
G-60N	1.270	0.762	0.475	1.431	0.856	3.2	2.8	2.9	1.8	0.11	0.456	0.050	0.001242	8.7	60	27
G-62Z	1.664	1.016	0.635	1.883	1.143	4.2	6.7	6.8	2.3	0.20	0.810	0.162	0.005524	15.1	140	81
G-62S	1.664	1.016	0.635	1.883	1.143	4.2	6.7	6.8	2.3	0.20	0.810	0.162	0.005524	15.1	140	74
G-62N	1.664	1.016	0.635	1.883	1.143	4.2	6.7	6.8	2.3	0.20	0.810	0.162	0.005524	15.1	60	35
G-29Z	2.032	1.270	0.635	2.310	1.270	5.2	9.5	11.9	2.6	0.23	1.266	0.291	0.010142	21.7	125	76
G-29S	2.032	1.270	0.635	2.310	1.270	5.2	9.5	11.9	2.6	0.23	1.266	0.291	0.010142	21.7	140	69
G-29N	2.032	1.270	0.635	2.310	1.270	5.2	9.5	11.9	2.6	0.23	1.266	0.291	0.010142	21.7	60	33
G-57Z	2.692	1.473	1.118	2.979	1.855	6.5	34.0	23.9	3.9	0.65	1.703	1.107	0.073013	38.4	125	171
G-57S	2.692	1.473	1.118	2.979	1.855	6.5	34.0	23.9	3.9	0.65	1.703	1.107	0.073013	38.4	140	155
G-57N	2.692	1.473	1.118	2.979	1.855	6.5	34.0	23.9	3.9	0.65	1.703	1.107	0.073013	38.4	60	75
G-56Z	3.302	1.994	1.118	3.726	2.115	8.3	46.6	49.2	4.4	0.70	3.121	2.185	0.138081	57.3	125	145
G-56S	3.302	1.994	1.118	3.726	2.115	8.3	46.6	49.2	4.4	0.70	3.121	2.185	0.138081	57.3	140	132
G-56N	3.302	1.994	1.118	3.726	2.115	8.3	46.6	49.2	4.4	0.70	3.121	2.185	0.138081	57.3	60	63
G-48Z	3.988	2.413	1.448	4.503	2.655	10.1	87.6	89.5	5.5	1.09	4.571	4.982	0.394427	85.2	125	187
G-48S	3.988	2.413	1.448	4.503	2.655	10.1	87.6	89.5	5.5	1.09	4.571	4.982	0.394427	85.2	140	170
G-48N	3.988	2.413	1.448	4.503	2.655	10.1	87.6	89.5	5.5	1.09	4.571	4.982	0.394427	85.2	60	82
G-1Z	5.050	3.000	1.700	5.679	3.200	12.6	168.9	169.8	6.8	1.67	7.065	11.799	1.165892	132.6	125	228
G-1S	5.050	3.000	1.700	5.679	3.200	12.6	168.9	169.8	6.8	1.67	7.065	11.799	1.165892	132.6	140	207
G-1N	5.050	3.000	1.700	5.679	3.200	12.6	168.9	169.8	6.8	1.67	7.065	11.799	1.165892	132.6	60	100
G-6Z	7.000	4.600	2.000	8.054	4.300	18.2	319.2	519.8	8.8	2.19	16.611	36.377	3.621186	259.8	125	207
G-6S	7.000	4.600	2.000	8.054	4.300	18.2	319.2	519.8	8.8	2.19	16.611	36.377	3.621186	259.8	140	188
G-6N	7.000	4.600	2.000	8.054	4.300	18.2	319.2	519.8	8.8	2.19	16.611	36.377	3.621186	259.8	60	91
G-9Z	8.000	4.800	2.900	9.016	5.300	20.1	696.6	710.0	11.0	4.33	18.086	78.314	12.286240	341.3	125	372
G-9S	8.000	4.800	2.900	9.016	5.300	20.1	696.6	710.0	11.0	4.33	18.086	78.314	12.286240	341.3	140	339
G-9N	8.000	4.800	2.900	9.016	5.300	20.1	696.6	710.0	11.0	4.33	18.086	78.314	12.286240	341.3	60	163

Pyroferric Toroidal Iron Powder Cores

Part No.	OD cm 1	ID cm 2	HT cm 3	Wound OD cm 4	Wound HT cm 5	MPL cm 6	Wtfe grams 7	Wtcu grams 8	MLT cm 9	Ac cm sq 10	Wa cm sq 11	Ap cm 4th 12	Kg cm 5th 13	At cm sq 14	Perm 15	AL 16.0
PT-310E	0.787	0.396	0.317	0.858	0.515	1.9	0.8	0.5	1.1	0.0620	0.123	0.0080	0.000167	3.1	75	30.5
PT-310C	0.787	0.396	0.317	0.858	0.515	1.9	0.8	0.5	1.1	0.0620	0.123	0.0080	0.000167	3.1	67	27.2
PT-310B	0.787	0.396	0.317	0.858	0.515	1.9	0.8	0.5	1.1	0.0620	0.123	0.0080	0.000167	3.1	60	24.4
PT-380E	0.965	0.533	0.317	1.070	0.584	2.4	1.1	1.0	1.3	0.0685	0.223	0.0150	0.000327	4.6	75	26.0
PT-380C	0.965	0.533	0.317	1.070	0.584	2.4	1.1	1.0	1.3	0.0685	0.223	0.0150	0.000327	4.6	67	23.2
PT-380B	0.965	0.533	0.317	1.070	0.584	2.4	1.1	1.0	1.3	0.0685	0.223	0.0150	0.000327	4.6	60	20.8
PT-437E	1.110	0.635	0.317	1.239	0.635	2.7	1.4	1.6	1.4	0.0753	0.317	0.0240	0.000515	6.0	75	24.7
PT-437C	1.110	0.635	0.317	1.239	0.635	2.7	1.4	1.6	1.4	0.0753	0.317	0.0240	0.000515	6.0	67	22.1
PT-437B	1.110	0.635	0.317	1.239	0.635	2.7	1.4	1.6	1.4	0.0753	0.317	0.0240	0.000515	6.0	60	19.8
PT-440E	1.118	0.584	0.406	1.227	0.698	2.7	2.0	1.5	1.5	0.1084	0.268	0.0290	0.000815	6.2	75	36.7
PT-440D	1.118	0.584	0.406	1.227	0.698	2.7	2.0	1.5	1.5	0.1084	0.268	0.0290	0.000815	6.2	72	35.2
PT-440C	1.118	0.584	0.406	1.227	0.698	2.7	2.0	1.5	1.5	0.1084	0.268	0.0290	0.000815	6.2	67	32.8
PT-440B	1.118	0.584	0.406	1.227	0.698	2.7	2.0	1.5	1.5	0.1084	0.268	0.0290	0.000815	6.2	60	29.4
PT-500E	1.270	0.762	0.475	1.431	0.856	3.2	2.7	2.9	1.8	0.1206	0.456	0.0550	0.001493	8.7	75	34.3
PT-500C	1.270	0.762	0.475	1.431	0.856	3.2	2.7	2.9	1.8	0.1206	0.456	0.0550	0.001493	8.7	67	30.6
PT-500B	1.270	0.762	0.475	1.431	0.856	3.2	2.7	2.9	1.8	0.1206	0.456	0.0550	0.001493	8.7	60	27.4
PT-501E	1.270	0.508	0.635	1.344	0.889	2.8	4.7	1.5	2.0	0.2419	0.203	0.0490	0.002333	8.1	75	78.0
PT-501C	1.270	0.508	0.635	1.344	0.889	2.8	4.7	1.5	2.0	0.2419	0.203	0.0490	0.002333	8.1	67	69.7
PT-501B	1.270	0.508	0.635	1.344	0.889	2.8	4.7	1.5	2.0	0.2419	0.203	0.0490	0.002333	8.1	60	62.3
PT-500AE	1.270	0.762	0.635	1.431	1.016	3.2	3.6	3.3	2.0	0.1613	0.456	0.0740	0.002334	9.5	75	46.5
PT-500AC	1.270	0.762	0.635	1.431	1.016	3.2	3.6	3.3	2.0	0.1613	0.456	0.0740	0.002334	9.5	67	41.5
PT-500AB	1.270	0.762	0.635	1.431	1.016	3.2	3.6	3.3	2.0	0.1613	0.456	0.0740	0.002334	9.5	60	37.2
PT-941E	2.388	1.600	6.735	2.761	7.535	6.3	11.0	90.7	12.7	0.2500	2.010	0.5020	0.003960	90.9	75	36.3
PT-500DE	1.270	0.762	0.953	1.431	1.334	3.2	5.4	4.1	2.5	0.2419	0.456	0.1100	0.004199	11.1	75	68.0
PT-500DC	1.270	0.762	0.953	1.431	1.334	3.2	5.4	4.1	2.5	0.2419	0.456	0.1100	0.004199	11.1	67	60.8
PT-500DB	1.270	0.762	0.953	1.431	1.334	3.2	5.4	4.1	2.5	0.2419	0.456	0.1100	0.004199	11.1	60	54.4
PT-680E	1.727	0.965	0.475	1.918	0.958	4.2	5.4	5.6	2.1	0.1810	0.731	0.1320	0.004473	14.3	75	42.1
PT-680C	1.727	0.965	0.475	1.918	0.958	4.2	5.4	5.6	2.1	0.1810	0.731	0.1320	0.004473	14.3	67	37.6
PT-680B	1.727	0.965	0.475	1.918	0.958	4.2	5.4	5.6	2.1	0.1810	0.731	0.1320	0.004473	14.3	60	33.7
PT-680AE	1.727	0.965	0.635	1.918	1.118	4.2	7.2	6.2	2.4	0.2419	0.731	0.1770	0.007136	15.4	75	56.4

Pyroferric Toroidal Iron Powder Cores

Part No.	OD cm 1	ID cm 2	HT cm 3	Wound OD cm 4	Wound HT cm 5	MPL cm 6	Wtfe grams 7	Wtcu grams 8	MLT cm 9	Ac cm sq 10	Wa cm sq 11	Ap cm 4th 12	Kg cm 5th 13	At cm sq 14	Perm 15	AL 16.0
PT-680AC	1.727	0.965	0.635	1.918	1.118	4.2	7.2	6.2	2.4	0.2419	0.731	0.1770	0.007136	15.4	67	50.4
PT-680AB	1.727	0.965	0.635	1.918	1.118	4.2	7.2	6.2	2.4	0.2419	0.731	0.1770	0.007136	15.4	60	45.1
PT-720E	1.829	0.711	0.660	1.930	1.016	4.0	10.3	3.6	2.5	0.3690	0.397	0.1460	0.008579	14.8	75	86.0
PT-720C	1.829	0.711	0.660	1.930	1.016	4.0	10.3	3.6	2.5	0.3690	0.397	0.1460	0.008579	14.8	67	77.7
PT-720B	1.829	0.711	0.660	1.930	1.016	4.0	10.3	3.6	2.5	0.3690	0.397	0.1460	0.008579	14.8	60	69.6
PT-800E	2.032	1.270	0.635	2.310	1.270	5.2	8.8	11.9	2.6	0.2419	1.266	0.3060	0.011219	21.7	75	45.4
PT-800C	2.032	1.270	0.635	2.310	1.270	5.2	8.8	11.9	2.6	0.2419	1.266	0.3060	0.011219	21.7	67	40.6
PT-800B	2.032	1.270	0.635	2.310	1.270	5.2	8.8	11.9	2.6	0.2419	1.266	0.3060	0.011219	21.7	60	36.4
PT-680DE	1.727	0.965	0.953	1.918	1.436	4.2	10.7	7.6	2.9	0.3629	0.731	0.2650	0.013250	17.6	75	81.5
PT-680DC	1.727	0.965	0.953	1.918	1.436	4.2	10.7	7.6	2.9	0.3629	0.731	0.2650	0.013250	17.6	67	72.7
PT-680DB	1.727	0.965	0.953	1.918	1.436	4.2	10.7	7.6	2.9	0.3629	0.731	0.2650	0.013250	17.6	60	65.2
PT-941C	2.388	1.600	0.635	2.761	1.435	6.3	11.0	20.9	2.9	0.2500	2.010	0.5020	0.017168	30.1	67	32.4
PT-941B	2.388	1.600	0.635	2.761	1.435	6.3	11.0	20.9	2.9	0.2500	2.010	0.5020	0.017168	30.1	60	29.0
PT-800AE	2.032	1.270	0.953	2.310	1.588	5.2	13.2	14.2	3.2	0.3629	1.266	0.4590	0.021171	24.3	75	65.7
PT-800AC	2.032	1.270	0.953	2.310	1.588	5.2	13.2	14.2	3.2	0.3629	1.266	0.4590	0.021171	24.3	67	61.2
PT-800AB	2.032	1.270	0.953	2.310	1.588	5.2	13.2	14.2	3.2	0.3629	1.266	0.4590	0.021171	24.3	60	52.6
PT-940E	2.388	1.422	0.792	2.687	1.503	6.0	16.0	17.9	3.2	0.3825	1.587	0.6070	0.029234	29.6	75	58.5
PT-940C	2.388	1.422	0.792	2.687	1.503	6.0	16.0	17.9	3.2	0.3825	1.587	0.6070	0.029234	29.6	67	52.2
PT-940B	2.388	1.422	0.792	2.687	1.503	6.0	16.0	17.9	3.2	0.3825	1.587	0.6070	0.029234	29.6	60	46.8
PT-800DE	2.032	1.270	1.270	2.310	1.905	5.2	17.6	16.5	3.7	0.4839	1.266	0.6130	0.032423	27.0	75	88.7
PT-800DC	2.032	1.270	1.270	2.310	1.905	5.2	17.6	16.5	3.7	0.4839	1.266	0.6130	0.032423	27.0	67	79.2
PT-800DB	2.032	1.270	1.270	2.310	1.905	5.2	17.6	16.5	3.7	0.4839	1.266	0.6130	0.032423	27.0	60	71.0
PT-900E	2.286	1.397	0.953	2.586	1.652	5.8	17.1	18.3	3.4	0.4234	1.532	0.6490	0.032758	29.3	75	68.5
PT-900C	2.286	1.397	0.953	2.586	1.652	5.8	17.1	18.3	3.4	0.4234	1.532	0.6490	0.032758	29.3	67	61.2
PT-900B	2.286	1.397	0.953	2.586	1.652	5.8	17.1	18.3	3.4	0.4234	1.532	0.6490	0.032758	29.3	60	54.8
PT-1060E	2.692	1.473	1.118	2.979	1.855	6.5	31.2	23.9	3.9	0.6814	1.703	1.1610	0.080238	38.4	75	96.9
PT-1060C	2.692	1.473	1.118	2.979	1.855	6.5	31.2	23.9	3.9	0.6814	1.703	1.1610	0.080238	38.4	67	86.5
PT-1060B	2.692	1.473	1.118	2.979	1.855	6.5	31.2	23.9	3.9	0.6814	1.703	1.1610	0.080238	38.4	60	77.5
PT-1060AE	2.692	1.473	1.460	2.979	2.197	6.5	40.8	27.2	4.5	0.8900	1.703	1.5160	0.120201	42.1	75	128.2
PT-1060AC	2.692	1.473	1.460	2.979	2.197	6.5	40.8	27.2	4.5	0.8900	1.703	1.5160	0.120201	42.1	67	114.5

Pyroferric Toroidal Iron Powder Cores																
Part No.	OD cm 1	ID cm 2	HT cm 3	Wound OD cm 4	Wound HT cm 5	MPL cm 6	Wtfe grams 7	Wtcu grams 8	MLT cm 9	Ac cm sq 10	Wa cm sq 11	Ap cm 4th 12	Kg cm 5th 13	At cm sq 14	Perm 15	AL 16.0
PT-1060AB	2.692	1.473	1.460	2.979	2.197	6.5	40.8	27.2	4.5	0.8900	1.703	1.5160	0.120201	42.1	60	102.6
PT-1300E	3.302	1.981	1.110	3.721	2.101	8.3	42.6	48.4	4.4	0.7331	3.081	2.2580	0.149913	57.0	75	82.6
PT-1300C	3.302	1.981	1.110	3.721	2.101	8.3	42.6	48.4	4.4	0.7331	3.081	2.2580	0.149913	57.0	67	73.8
PT-1300B	3.302	1.981	1.110	3.721	2.101	8.3	42.6	48.4	4.4	0.7331	3.081	2.2580	0.149913	57.0	60	66.1
PT-1301E	3.302	1.626	1.110	3.590	1.923	7.7	50.4	32.6	4.4	0.9303	2.075	1.9310	0.162641	51.7	75	115.1
PT-1301C	3.302	1.626	1.110	3.590	1.923	7.7	50.4	32.6	4.4	0.9303	2.075	1.9310	0.162641	51.7	67	102.9
PT-1301B	3.302	1.626	1.110	3.590	1.923	7.7	50.4	32.6	4.4	0.9303	2.075	1.9310	0.162641	51.7	60	92.1
PT-1510E	3.835	2.146	1.110	4.262	2.183	9.4	61.7	62.3	4.8	0.9374	3.615	3.3890	0.262322	71.3	75	94.0
PT-1510C	3.835	2.146	1.110	4.262	2.183	9.4	61.7	62.3	4.8	0.9374	3.615	3.3890	0.262322	71.3	67	84.0
PT-1510B	3.835	2.146	1.110	4.262	2.183	9.4	61.7	62.3	4.8	0.9374	3.615	3.3890	0.262322	71.3	60	75.2
PT-1570E	3.988	2.426	1.448	4.508	2.661	10.1	79.8	90.5	5.5	1.1310	4.620	5.2250	0.429246	85.5	75	101.2
PT-1570C	3.988	2.426	1.448	4.508	2.661	10.1	79.8	90.5	5.5	1.1310	4.620	5.2250	0.429246	85.5	67	90.4
PT-1570B	3.988	2.426	1.448	4.508	2.661	10.1	79.8	90.5	5.5	1.1310	4.620	5.2250	0.429246	85.5	60	81.0
PT-2000E	5.080	3.175	1.397	5.777	2.985	13.0	120.7	177.3	6.3	1.3300	7.913	10.5250	0.888863	131.5	75	93.0
PT-2000C	5.080	3.175	1.397	5.777	2.985	13.0	120.7	177.3	6.3	1.3300	7.913	10.5250	0.888863	131.5	67	83.0
PT-2000B	5.080	3.175	1.397	5.777	2.985	13.0	120.7	177.3	6.3	1.3300	7.913	10.5250	0.888863	131.5	60	77.5
PT-1840E	4.674	2.413	1.803	5.120	3.010	11.1	158.8	107.7	6.6	2.0380	4.571	9.3150	1.146390	110.1	75	170.4
PT-1840C	4.674	2.413	1.803	5.120	3.010	11.1	158.8	107.7	6.6	2.0380	4.571	9.3150	1.146390	110.1	67	152.2
PT-1840B	4.674	2.413	1.803	5.120	3.010	11.1	158.8	107.7	6.6	2.0380	4.571	9.3150	1.146390	110.1	60	142.0
PT-2000BE	5.080	3.175	2.540	5.777	4.128	13.0	219.6	228.7	8.1	2.4190	7.913	19.1420	2.278794	155.4	75	115.0
PT-2000BB	5.080	3.175	2.540	5.777	4.128	13.0	219.6	228.7	8.1	2.4190	7.913	19.1420	2.278794	155.4	60	138.0

Pyroferric EE Iron Powder Cores															
Part No.	Ht cm 1	Wth cm 2	Lt cm 3	G cm 4	MPL cm 5	Wtfe grams 6	Wtcu grams 7	MLT cm 8	Ac cm sq 9	Wa cm sq 10	Ap cm 4th 11	Kg cm 5th 12	At cm sq 13	Perm 14	AL 15
PE-76E	1.68	1.45	1.93	1.16	4.3	6.8	7.7	3.8	0.226	0.568	0.128	0.003035	15.1	75	60
PE-76C	1.68	1.45	1.93	1.16	4.3	6.8	7.7	3.8	0.226	0.568	0.128	0.003035	15.1	67	54
PE-100E	1.90	1.92	2.54	1.27	5.1	14.4	14.4	5.0	0.403	0.813	0.328	0.010608	23.4	75	90
PE-100C	1.90	1.92	2.54	1.27	5.1	14.4	14.4	5.0	0.403	0.813	0.328	0.010608	23.4	67	80
PE-137E	2.86	2.53	3.49	1.90	7.3	46.3	35.7	6.7	0.907	1.501	1.361	0.073845	46.1	75	129
PE-137C	2.86	2.53	3.49	1.90	7.3	46.3	35.7	6.7	0.907	1.501	1.361	0.073845	46.1	67	115
PE-162E	3.41	2.85	4.13	2.14	8.3	94.5	49.1	8.2	1.610	1.691	2.722	0.214588	63.1	75	205
PE-162C	3.41	2.85	4.13	2.14	8.3	94.5	49.1	8.2	1.610	1.691	2.722	0.214588	63.1	67	183
PE-168E	4.22	3.07	4.28	3.07	10.3	129.5	90.8	8.9	1.810	2.855	5.168	0.418122	84.2	75	190
PE-168C	4.22	3.07	4.28	3.07	10.3	129.5	90.8	8.9	1.810	2.855	5.168	0.418122	84.2	67	170
PE-168AE	4.22	3.07	4.28	3.07	10.3	172.2	101.0	9.9	2.410	2.855	6.881	0.666762	92.7	75	232
PE-168AC	4.22	3.07	4.28	3.07	10.3	172.2	101.0	9.9	2.410	2.855	6.881	0.666762	92.7	67	207

8

FERRITE CORES

FERRITE CORE MANUFACTURERS

Manufacturers' addresses and phone numbers can be found in the Table of Contents.

Notes

FERRITE MATERIAL CROSS-REFERENCE GUIDE

		Materials					
Permeability	μ_i	1500	2300	2500	3000	5000	10000
Curie Temp.	°C	>230	>230	>230	>250	>125	>125
Flux Density @15 Oe	B_m	0.48T	0.50T	0.50T	0.49T	0.43T	0.43T
Residual Flux @ 25°C	B_r	0.08T	0.12T	0.12T	0.10T	0.10T	0.07T
Coercivity, Oersted	H_c	0.2	0.18	0.18	0.2	0.1	0.04
Resistivity, Ω - m	ρ	20	6	5	2	1	0.15
Density, g/cm³	δ	4.7	4.8	4.8	4.8	4.8	4.8
Application		Power	Power	Power	Power	Filter	Filter
Manufacturer's		Materials Designation					
Magnetics		K	R	P	F	J	W
Thomson LCC			F2	B2	B1	A4	A2
Philips Components		3F4	3F3	3C85	3C81	3E2A	3E5
Fair-Rite			78	77		75	76
Siemens		N47	N67		N41	T35	T38
TDK Corp.		PC50	PC40	PC30		H5B	H5C2
MMG				F-44	F-5	F-10	
Ceramics Mag.		MN8CX	MN80	MN60	MN67	MN60LL	MC25
Tokin			2500B2	2500B	3100B	6000H	12001H
Ferrite International				TSF-05	TSF-10	TSF-15	

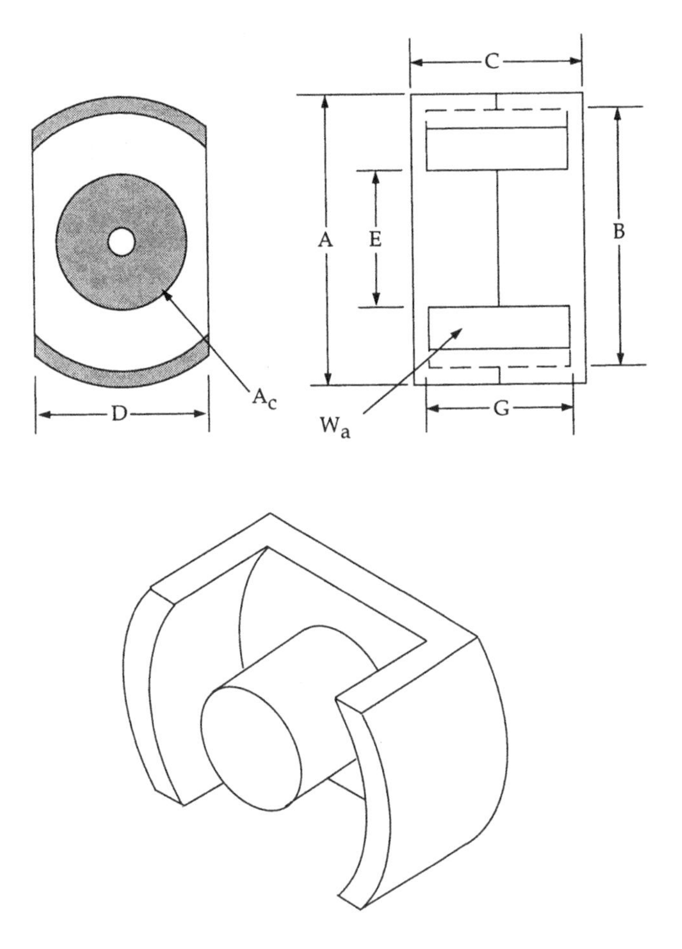

Figure 120 Outline of a DS ferrite core.

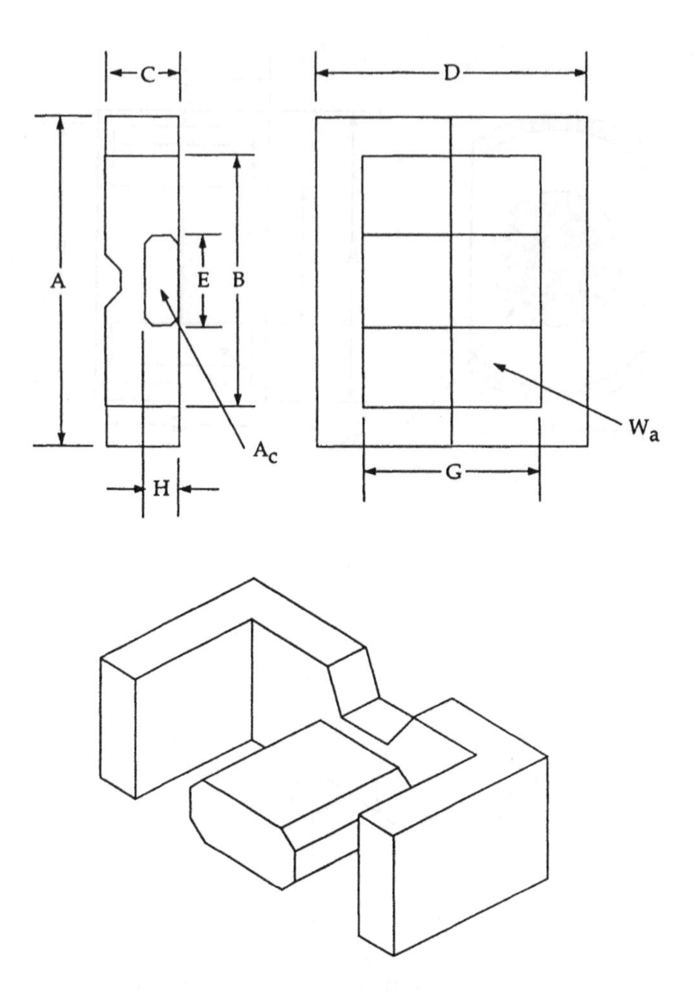

Figure 121 Outline of a ETD ferrite core.

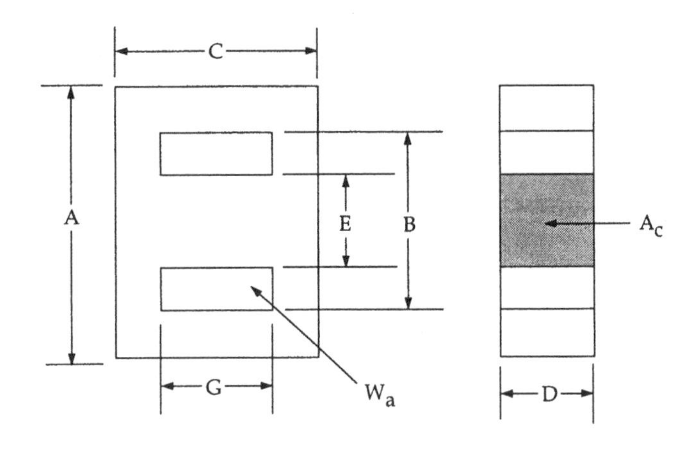

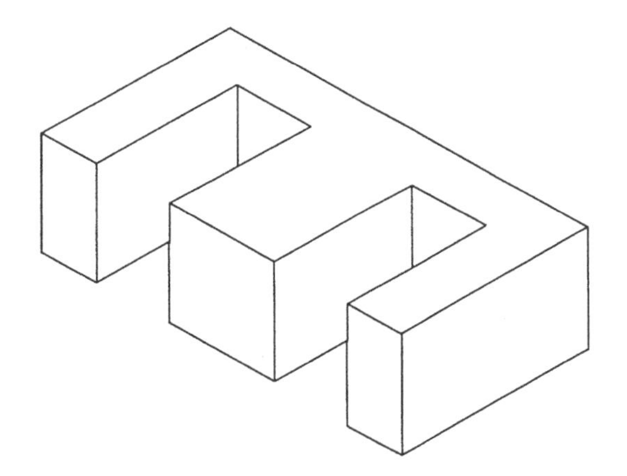

Figure 122 Outline of a EE & EI ferrite core.

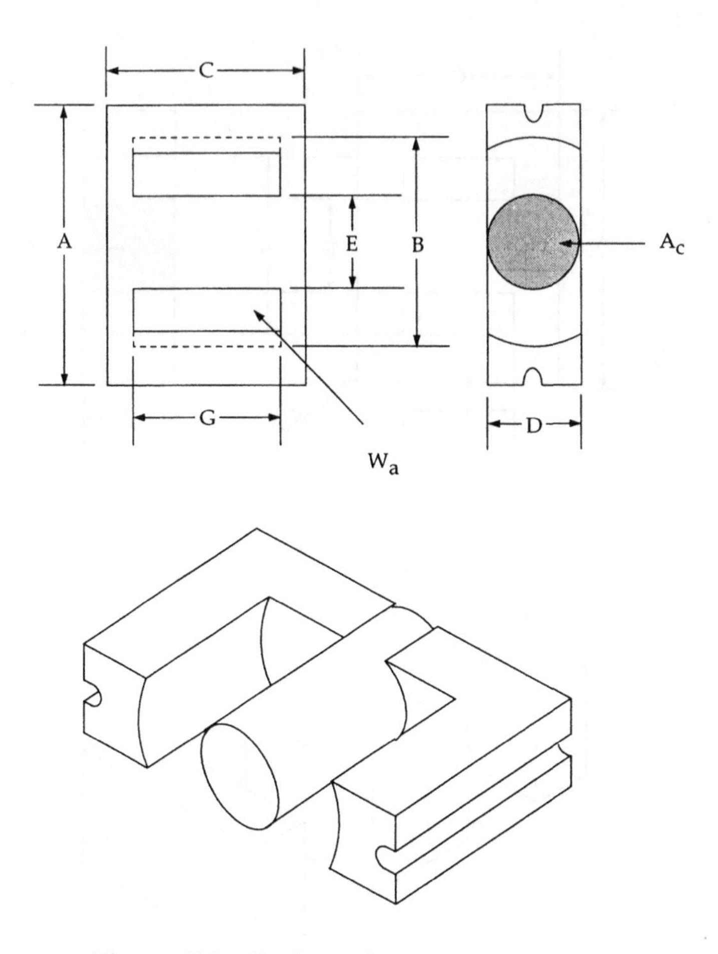

Figure 123 Outline of a EC ferrite core.

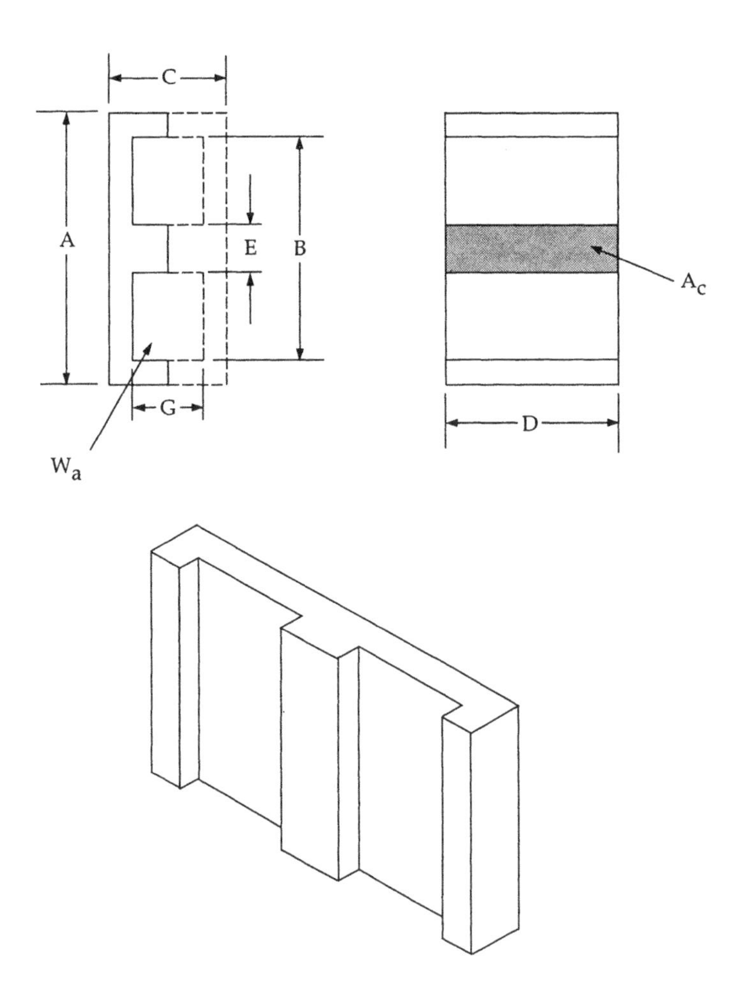

Figure 124 Outline of a EELP low-profile ferrite core.

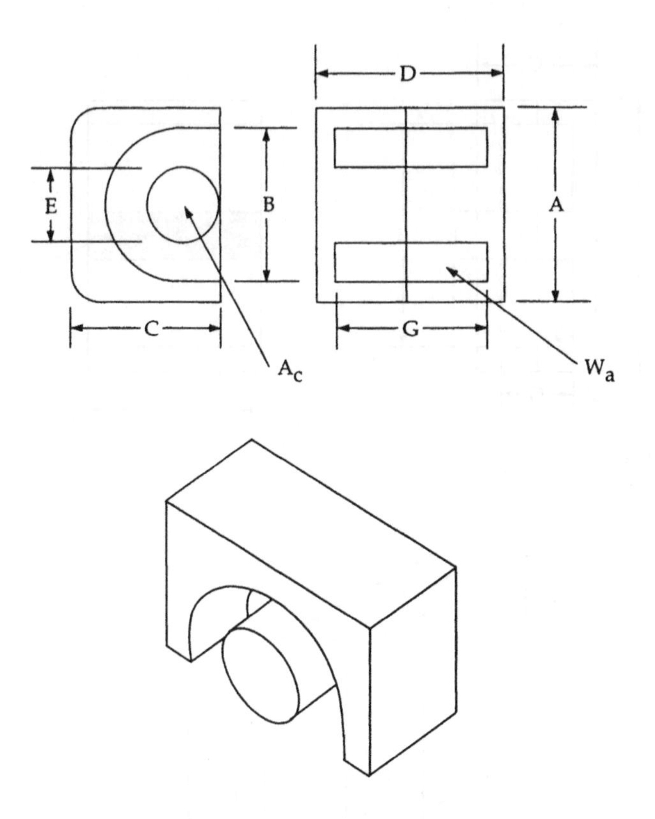

Figure 125 Outline of a EP ferrite core.

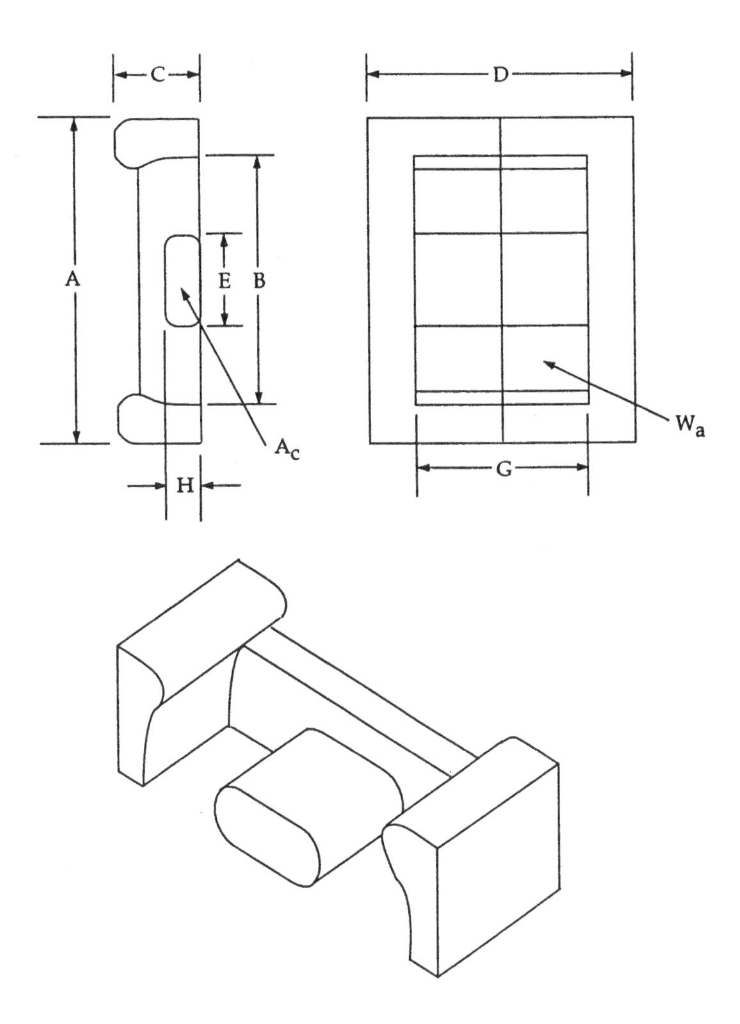

Figure 126 Outline of a EPC ferrite core.

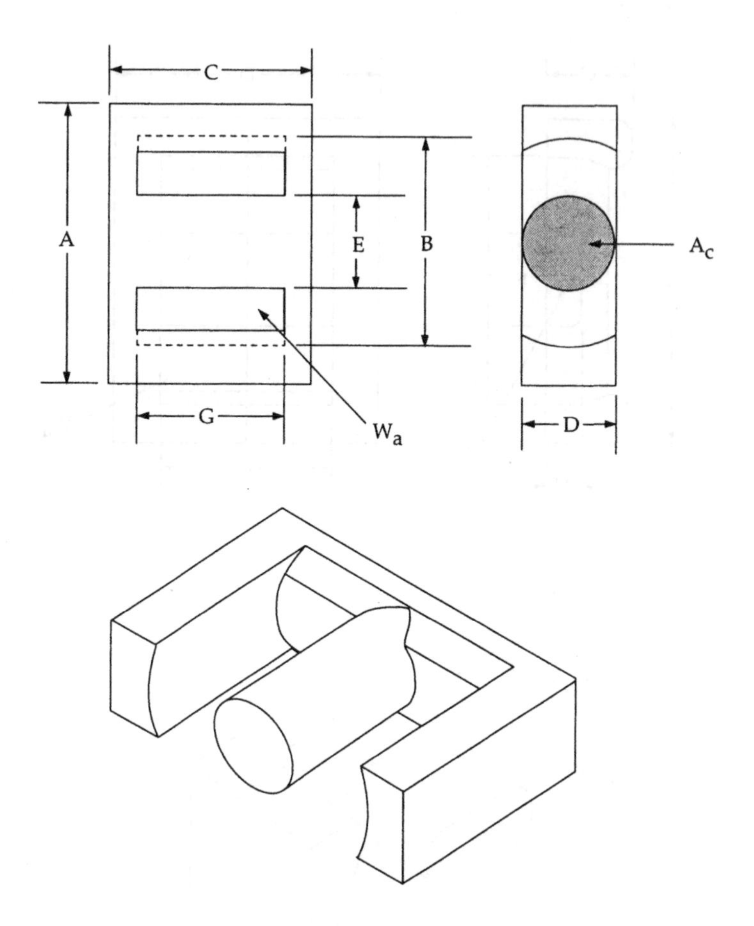

Figure 127 Outline of a ETD ferrite core.

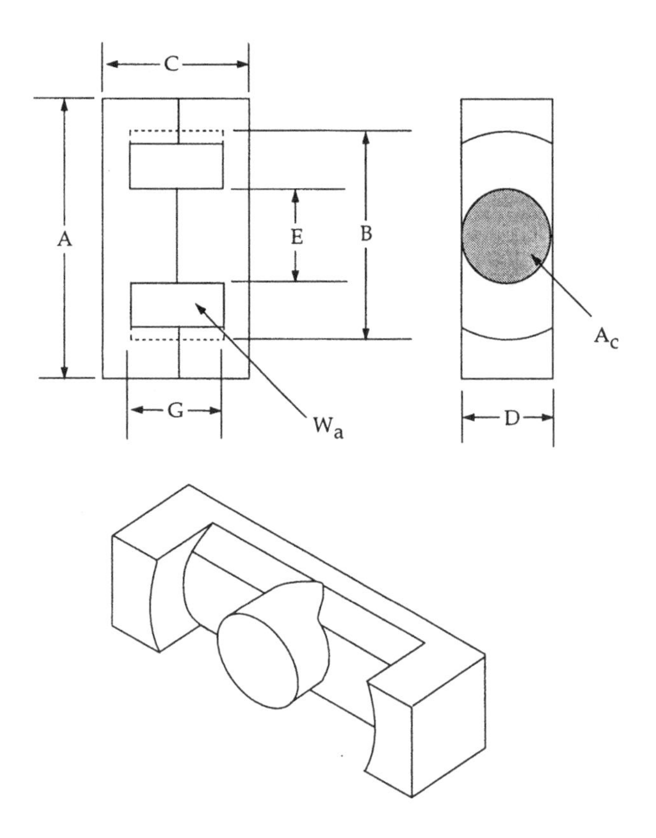

Figure 128 Outline of a ETDLP low-profile ferrite core.

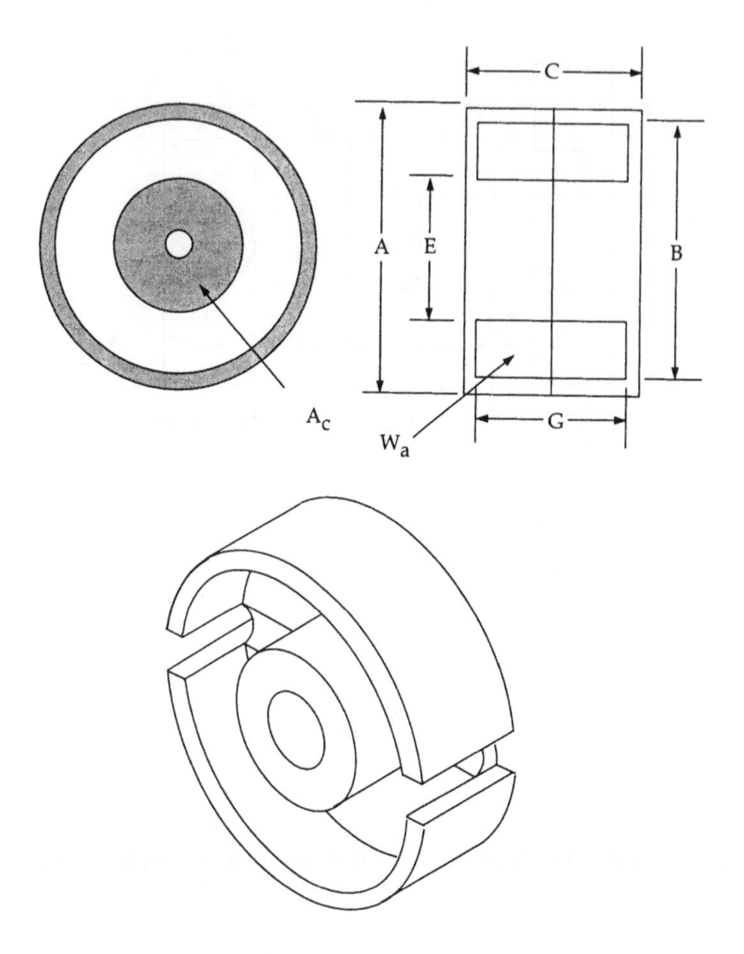

Figure 129 Outline of a POT ferrite core.

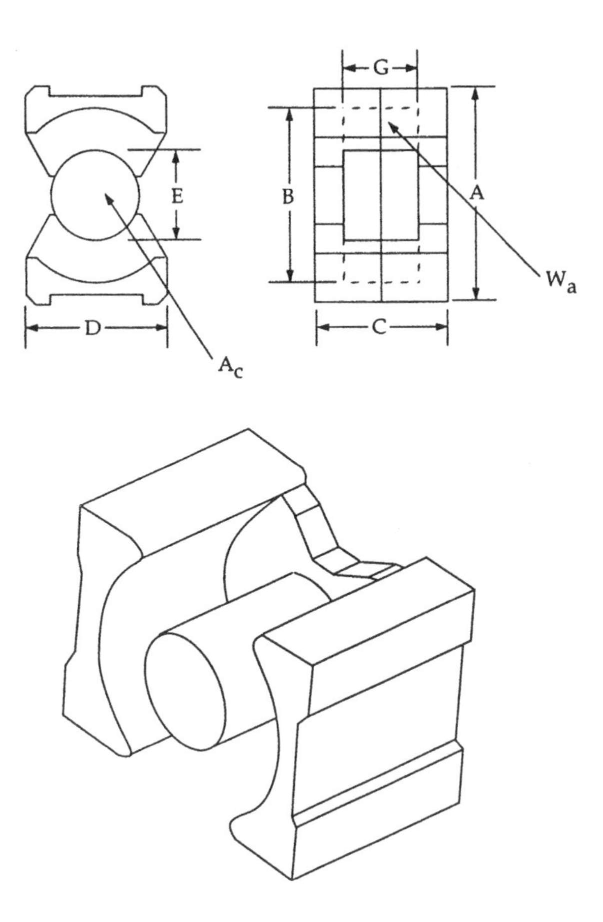

Figure 130 Outline of a PQ ferrite core.

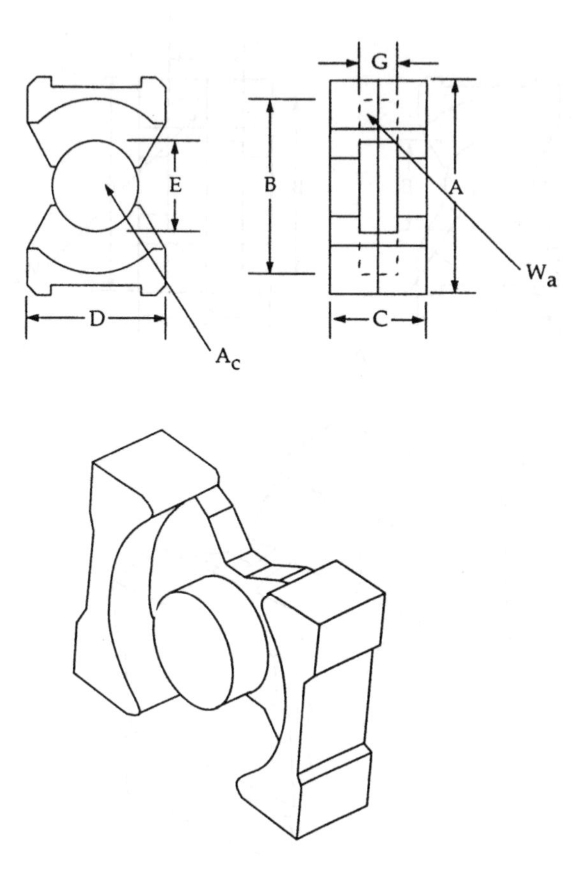

Figure 131 Outline of a PQLP low-profile ferrite core.

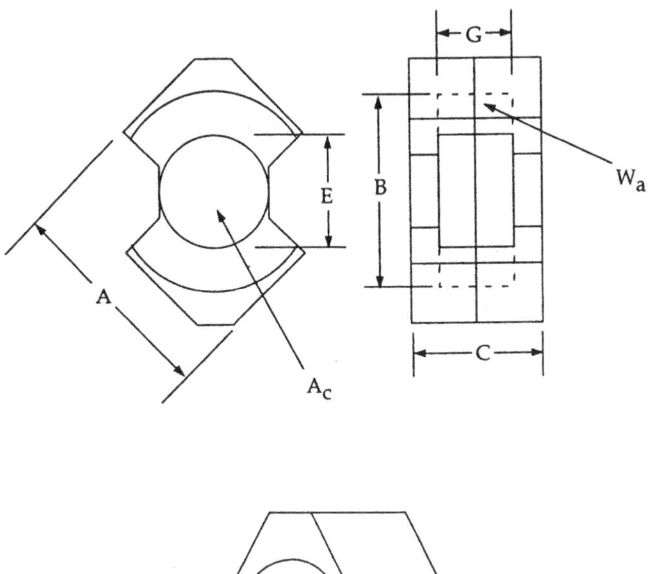

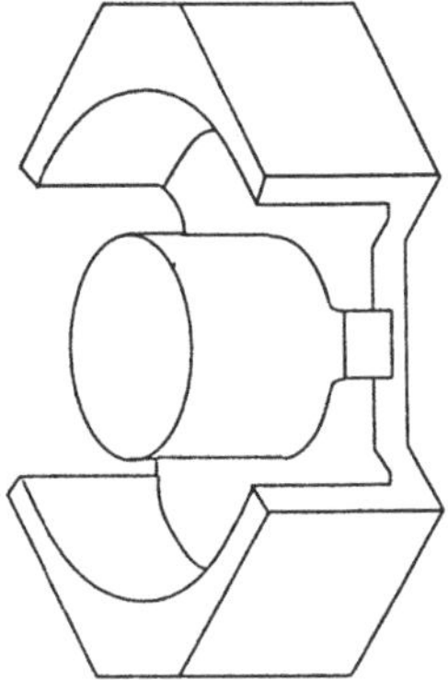

Figure 132 Outline of a RM ferrite core.

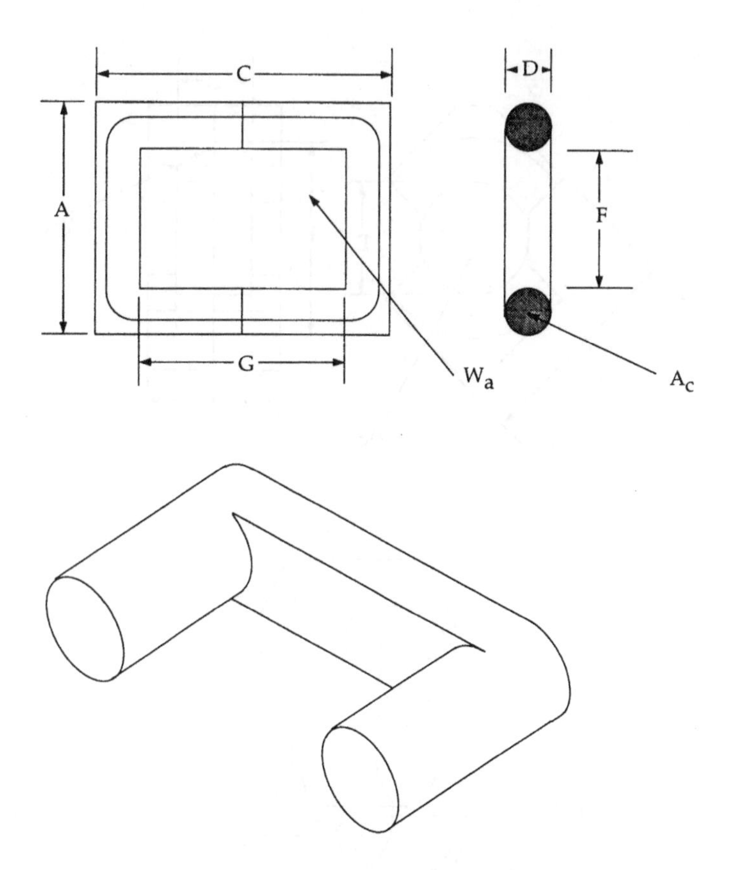

Figure 133 Outline of a RUI ferrite core.

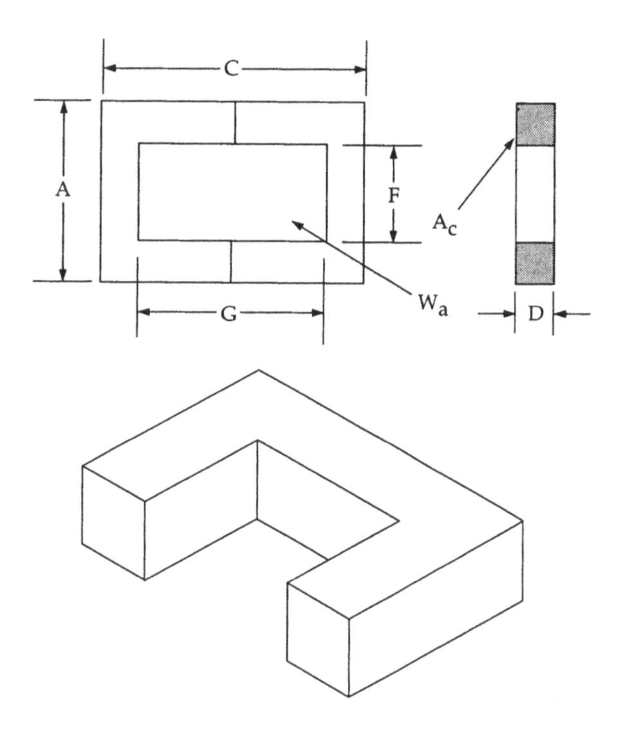

Figure 134 Outline of a SUI ferrite core.

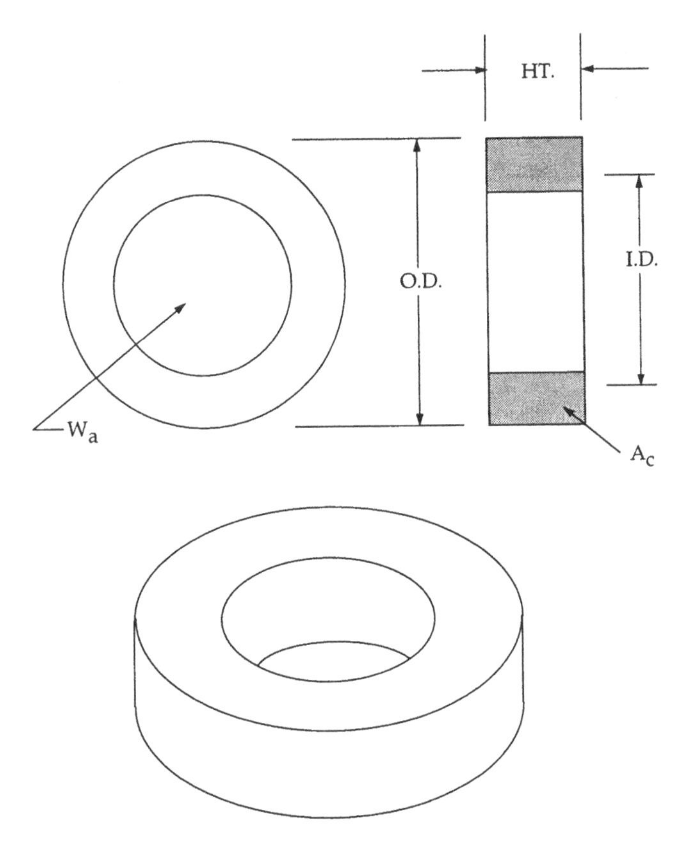

Figure 135 Outline of a toroidal ferrite core.

FERRITE CUT CORES TABLE DESCRIPTION SUCH AS EE, PC, EC, ETD, EP, PQ, RM, EPC, EFD, etc.

Definitions for the Following Tables

Information given as listed by column:

Col.	Dim.	Description	Units
1.	Ht	Finished transformer height	cm
2.	Wth	Finished transformer width	cm
3.	Lt	Finished transformer length	cm
4.	G	Window length	cm
5.	MPL	Mean Magnetic Path Length	cm
6.	W_{tfe}	Core weight	gram
7.	W_{tcu}	Copper weight $[W_{tcu} = (MLT)(W_a)(K_u)(8.89)]$	gram
8.	MLT	Mean Length Turn	cm
9.	A_c	Iron area (gross) $[A_c = (D)(E)]$	cm sq
10.	W_a	Window area $[W_a = (F)(G)]$	cm sq
11.	A_p	Area product (gross) $[A_p = (A_c)(W_a)]$	cm 4th
12.	K_g	Core geometry (gross) $[K_g = (A_p)(A_c)(K_u)/(MLT)]$	cm 5th
13.	A_t	Surface area	cm sq
14.	Perm	Core permeability	(B/H)
15.	AL	Core millihenrys per 1000 turns	mh

FERRITE CORE TABLE DESCRIPTION

Definitions for the Following Tables

Information given is listed by column:

Col.	Dim	Description	Units
1.	OD	Core outside diameter	cm
2.	ID	Core inside diameter	cm
3.	HT	Core height	cm
4.	OD	Finished wound (K_u = 0.4)	cm
5.	HT	Finished wound (K_u = 0.4)	cm
6.	MPL	Magnetic Path Length [MPL = 3.14(OD + ID)(0.5)]	cm
7.	W_{tfe}	Core weight [W_{tfe} = (MPL)(A_c)(Density)	grams
8.	W_{tcu}	Copper weight [W_{tcu} = (MLT)(W_a)(K_u)(8.89)]	grams
9.	MLT	Mean Length Turn [MLT = (OD + 2HT)(0.8), core]	cm
10.	A_c	Effective iron area [A_c = (OD − ID)(HT)(SF)(0.5), core]	cm sq
11.	W_a	Window area [$W_a = 3.14(ID)^2(0.25)$]	cm sq
12.	A_p	Area product [A_p = (A_c)(W_a)]	cm 4th
13.	K_g	Core geometry [K_g = (A_p)(A_c)(K_u)/(MLT)]	cm 5th
14.	A_t	Surface area	cm sq
15.	Perm	Core permeability	(B/H)
16.	AL	Core millihenrys per 1000 turns	mh

Fair-Rite EE Cores

Part No.	Ht. cm 1	Wth. cm 2	Lt. cm 3	G cm 4	MPL cm 5	Wtfe grams 6	Wtcu grams 7	MLT cm 8	Ac cm sq 9	Wa cm sq 10	Ap cm 4th 11	Kg cm 5th 12	At cm sq 13	Perm 14	AL 15
EE-0190	1.16	0.945	1.27	0.82	2.8	0.2	2.4	2.7	0.101	0.246	0.025	0.000372	6.7	2300	715
EE-0200	1.16	1.260	1.27	0.82	2.8	3.2	2.9	3.3	0.202	0.246	0.050	0.001206	8.3	2300	1450
EE-0160	1.64	1.410	1.93	1.12	4.0	4.4	7.1	3.8	0.225	0.524	0.118	0.002779	14.5	2300	1250
EE-0120	1.64	1.885	1.93	1.12	4.0	8.8	8.9	4.8	0.450	0.524	0.236	0.008899	17.9	2300	2550
EE-0150	1.96	1.880	2.54	1.27	4.9	9.8	13.7	5.0	0.400	0.775	0.310	0.009989	23.7	2300	1950
EE-0340	3.20	1.880	2.54	2.54	4.9	14.2	27.3	5.0	0.400	1.549	0.620	0.019979	34.6	2300	1950
EE-0180	2.12	1.920	2.80	1.12	4.8	16.8	12.1	5.3	0.660	0.644	0.425	0.021198	27.2	2300	3200
EE-0140	1.96	2.490	2.54	1.27	4.9	20.2	17.6	6.4	0.800	0.775	0.620	0.031069	29.1	2300	3925
EE-0300	3.00	1.960	3.01	2.00	6.7	24.4	22.9	5.2	0.600	1.230	0.738	0.033807	38.1	2300	2100
EE-0170	2.12	2.270	2.80	1.12	4.8	24.6	13.7	6.0	0.960	0.644	0.618	0.039610	30.7	2300	4450
EE-3750	2.90	2.465	3.46	1.90	6.9	30.0	34.1	6.6	0.860	1.463	1.258	0.066020	45.6	2300	3100
EE-5000	3.30	2.780	4.08	2.03	7.6	58.6	44.7	7.9	1.500	1.583	2.375	0.179542	60.3	2300	4850
EE-0360	4.23	3.435	4.29	2.99	9.8	88.4	89.1	9.1	1.840	2.766	5.089	0.413295	85.9	2300	4800
EE-6250	3.97	3.160	4.71	2.40	8.9	102.0	63.9	9.4	2.370	1.920	4.550	0.460874	82.3	2300	6800

Fair-Rite EI Cores															
Part No.	Ht cm 1	Wth cm 2	Lt cm 3	G cm 4	MPL cm 5	Wtfe grams 6	Wtcu grams 7	MLT cm 8	Ac cm sq 9	Wa cm sq 10	Ap cm 4th 11	Kg cm 5th 12	At cm sq 13	Perm 14	AL 15
EI-0210	1.07	1.410	1.930	0.560	2.9	3.4	3.6	3.80	0.225	0.262	0.059	0.00139	10.7	2300	1475
EI-0200	1.31	1.880	2.540	0.635	3.6	7.2	6.8	5.00	0.400	0.387	0.155	0.00500	17.9	2300	2325
EI-0360	2.73	3.435	4.285	1.495	6.8	63.9	44.6	9.10	1.840	1.383	2.544	0.20665	62.7	2300	6400

Fair-Rite EP Cores															
Part No.	Ht cm 1	Wth cm 2	Lt cm 3	G cm 4	MPL cm 5	Wtfe grams 6	Wtcu grams 7	MLT cm 8	Ac cm sq 9	Wa cm sq 10	Ap cm 4th 11	Kg cm 5th 12	At cm sq 13	Perm 14	AL 15
EP-07	0.840	0.74	0.92	0.52	1.6	1.4	0.7	1.9	0.103	0.107	0.011	0.000236	3.5	2300	1100
EP-10	1.070	1.02	1.15	0.74	1.9	2.8	1.7	2.2	0.113	0.226	0.026	0.000535	5.7	2300	1050
EP-13	1.163	1.30	1.25	0.92	2.4	5.2	2.2	2.4	0.195	0.260	0.051	0.001638	7.7	2300	1600
EP-17	1.408	1.68	1.80	1.13	2.9	11.8	3.8	3.0	0.339	0.347	0.118	0.005252	13.7	2300	2500
EP-20	1.883	2.14	2.40	1.43	4.0	28.2	8.3	4.2	0.780	0.554	0.432	0.032086	23.8	2300	4200

Fair Rite ETD Cores

Part No.	Ht cm 1	Wth cm 2	Lt cm 3	G cm 4	MPL cm 5	Wtfe grams 6	Wtcu grams 7	MLT cm 8	Ac cm sq 9	Wa cm sq 10	Ap cm 4th 11	Kg cm 5th 12	At cm sq 13	Perm 14	AL 15
ETD-29	3.16	2.27	2.98	2.20	7.2	28	32.4	6.3	0.76	1.452	1.104	0.053415	41.7	2300	2350
ETD-34	3.46	2.63	3.42	2.42	7.9	42	47.8	7.2	0.97	1.876	1.819	0.098564	53.2	2300	2800
ETD-39	3.96	3.01	3.91	2.92	9.2	60	76.5	8.4	1.25	2.570	3.212	0.191848	69.5	2300	3150
ETD-44	4.46	3.33	4.40	3.30	10.3	100	102.4	9.4	1.73	3.053	5.281	0.387419	87.3	2300	3900
ETD-49	4.94	3.70	4.87	3.62	11.4	124	138.3	10.4	2.11	3.747	7.906	0.642931	107.2	2300	4500

Fair-Rite PQ Cores

Part No.	Ht cm 1	Wth cm 2	Lt cm 3	G cm 4	MPL cm 5	Wtfe grams 6	Wtcu grams 7	MLT cm 8	Ac cm sq 9	Wa cm sq 10	Ap cm 4th 11	Kg cm 5th 12	At cm sq 13	Perm 14	AL 15
PQ-201621	1.62	2.320	2.125	1.00	3.7	13	7.1	4.4	0.62	0.460	0.285	0.016194	17.4	2300	3800
PQ-202021	2.02	2.320	2.125	1.40	4.6	15	10.0	4.4	0.62	0.644	0.399	0.022671	20.2	2300	3150
PQ-262021	2.04	2.950	2.725	1.12	4.6	32	11.7	5.6	1.19	0.588	0.700	0.059725	29.2	2300	6100
PQ-262521	2.50	2.950	2.725	1.58	5.6	36	16.4	5.6	1.18	0.830	0.979	0.082845	33.5	2300	5200
PQ-322021	2.08	3.605	3.300	1.12	5.6	42	18.7	6.7	1.70	0.787	1.338	0.136408	37.4	2300	7250
PQ-323021	3.06	3.605	3.300	2.10	7.5	56	35.0	6.7	1.61	1.475	2.375	0.229401	48.2	2300	5100
PQ-353521	3.50	4.365	3.610	2.47	8.8	72	58.3	7.5	1.96	2.180	4.272	0.445678	62.2	2300	5000
PQ-404021	4.00	5.010	4.150	2.92	10.2	96	96.2	8.4	2.01	3.227	6.485	0.621719	78.8	2300	4400

Fair-Rite Toroidal Cores

Part No.	OD cm 1	ID cm 2	HT cm 3	Wound OD cm 4	Wound HT cm 5	MPL cm 6	Wtfe grams 7	Wtcu grams 8	MLT cm 9	Ac cm sq 10	Wa cm sq 11	Ap cm 4th 12	Kg cm 5th 13	At cm sq 14	Perm 15	AL 16
T-0008	0.395	0.215	0.135	0.502	0.243	0.9	0.1	0.1	0.5	0.0110	0.036	0.000	0.000003	0.8	2000	285
T-0021	0.495	0.220	0.135	0.610	0.245	1.0	0.1	0.1	0.6	0.0150	0.038	0.001	0.000006	1.1	2000	360
T-0085	0.395	0.215	0.255	0.502	0.363	0.9	0.1	0.1	0.7	0.0210	0.036	0.001	0.000009	1.0	2000	570
T-0001	0.595	0.305	0.165	0.749	0.318	1.3	0.1	0.2	0.7	0.0200	0.073	0.001	0.000016	1.6	2000	390
T-0103	0.495	0.220	0.255	0.610	0.365	1.0	0.2	0.1	0.8	0.0300	0.038	0.001	0.000017	1.3	2000	725
T-0009	0.595	0.305	0.305	0.749	0.458	1.3	0.3	0.3	1.0	0.0410	0.073	0.003	0.000051	2.0	2000	780
T-0104	0.775	0.305	0.490	0.942	0.643	1.5	0.9	0.4	1.4	0.0990	0.073	0.007	0.000204	3.3	2000	1650
T-0002	0.950	0.475	0.330	1.191	0.568	2.1	0.7	0.8	1.3	0.0720	0.177	0.013	0.000285	4.3	2000	880
T-0025	0.950	0.475	0.490	1.191	0.728	2.1	1.1	1.0	1.5	0.1090	0.177	0.019	0.000545	4.9	2000	1325
T-0003	1.270	0.715	0.490	1.625	0.848	3.0	1.8	2.6	1.8	0.1290	0.401	0.052	0.001484	8.5	2000	1100
T-0011	1.270	0.790	0.635	1.659	1.030	3.1	2.2	3.5	2.0	0.1500	0.490	0.073	0.002170	9.7	2000	1200
T-0051	1.600	0.960	0.475	2.074	0.955	3.9	2.6	5.2	2.0	0.1450	0.723	0.105	0.002982	13.0	2000	940
T-0109	1.590	0.890	0.470	2.032	0.915	3.7	3.3	4.5	2.0	0.1590	0.622	0.099	0.003107	12.3	2000	1075
T-0049	1.600	0.960	0.635	2.074	1.115	3.9	3.6	5.9	2.3	0.1990	0.723	0.144	0.004991	14.0	2000	1300
T-0019	1.270	0.790	1.270	1.659	1.665	3.1	4.5	5.3	3.0	0.2990	0.490	0.146	0.005748	13.0	2000	2400
T-0006	2.100	1.320	0.635	2.750	1.295	5.2	6.0	13.1	2.7	0.2430	1.368	0.332	0.011983	23.1	2000	1175
T-0018	2.210	1.370	0.635	2.885	1.320	5.4	6.8	14.6	2.8	0.2620	1.473	0.386	0.014531	25.0	2000	1200
T-0013	2.540	1.550	0.635	3.304	1.410	6.2	9.1	20.4	3.0	0.3080	1.886	0.581	0.023479	31.8	2000	1250
T-0005	2.100	1.320	1.190	2.750	1.850	5.2	11.5	17.4	3.6	0.4600	1.368	0.629	0.032302	27.8	2000	2200
T-0014	2.540	1.550	0.815	3.304	1.590	6.2	12.0	22.4	3.3	0.4100	1.886	0.773	0.038013	33.6	2000	1600
T-0076	2.210	1.370	1.270	2.885	1.955	5.4	13.5	19.9	3.8	0.5200	1.473	0.766	0.041937	30.8	2000	2425
T-0010	2.900	1.900	0.750	3.834	1.700	7.3	13.0	35.5	3.5	0.3700	2.834	1.049	0.044086	43.6	2000	1275
T-0016	3.110	1.905	0.790	4.049	1.743	7.6	17.0	38.0	3.8	0.4700	2.849	1.339	0.067089	47.9	2000	1550
T-0064	2.540	1.550	1.270	3.304	2.045	6.2	18.0	27.3	4.1	0.6200	1.886	1.169	0.071355	38.4	2000	2500
T-0017	3.175	1.905	0.950	4.115	1.903	7.6	21.5	41.1	4.1	0.5900	2.849	1.681	0.097701	51.2	2000	1950
T-0012	2.900	1.900	1.385	3.834	2.335	7.3	24.0	45.7	4.5	0.6800	2.834	1.927	0.115553	51.2	2000	2350
T-0110	3.600	2.300	1.015	4.731	2.165	8.9	27.0	66.5	4.5	0.6400	4.153	2.658	0.151059	67.3	2000	1800
T-0027	3.600	2.300	1.270	4.731	2.420	8.9	33.5	72.5	4.9	0.7900	4.153	3.281	0.211048	71.1	2000	2250
T-0112	3.600	2.300	1.525	4.731	2.675	8.9	40.0	78.6	5.3	0.9500	4.153	3.945	0.281787	74.9	2000	2675
T-0038	6.100	3.555	1.270	7.858	3.048	14.5	110.0	243.8	6.9	1.5800	9.921	15.675	1.433241	172.1	2000	2725
T-0111	7.365	3.885	1.270	9.312	3.213	16.7	205.0	333.9	7.9	2.1500	11.848	25.474	2.764676	230.1	2000	3225

Fair-Rite UU Cores															
Part No.	Ht cm 1	Wth cm 2	Lt cm 3	G cm 4	MPL cm 5	Wtfe grams 6	Wtcu grams 7	MLT cm 8	Ac cm sq 9	Wa cm sq 10	Ap cm 4th 11	Kg cm 5th 12	At cm sq 13	Perm 14	AL 15
UU-0020	0.89	0.635	1.120	0.26	2.1	0.1	0.5	2.2	0.124	0.060	0.007	0.000164	4.0	2300	1000
UU-0260	2.52	1.905	3.785	1.24	7.2	14.4	27.3	5.0	0.400	1.544	0.618	0.019849	36.7	2300	1400
UU-0270	2.64	1.905	3.785	1.37	7.3	15.6	30.2	5.0	0.400	1.706	0.682	0.021930	38.1	2300	1375
UU-0250	3.15	1.905	3.785	1.88	8.4	17.4	41.4	5.0	0.400	2.341	0.936	0.030094	43.9	2300	1175
UU-0240	3.78	1.905	3.785	2.51	9.6	19.2	55.3	5.0	0.400	3.125	1.250	0.040179	51.1	2300	1050

Ferrite International PC Cores															
Part No.	Ht cm 1	Wth cm 2	Lt cm 3	G cm 4	MPL cm 5	Wtfe grams 6	Wtcu grams 7	MLT cm 8	Ac cm sq 9	Wa cm sq 10	Ap cm 4th 11	Kg cm 5th 12	At cm sq 13	Perm 14	AL 15
PC14-08-00	0.836	1.400	1.400	0.579	2.0	3.3	1.8	2.9	0.25	0.171	0.043	0.001458	6.8	2200	2350
PC18-11-00	1.060	1.800	1.800	0.740	2.6	7.1	3.8	3.7	0.43	0.287	0.123	0.005717	11.1	2200	3443
PC22-13-00	1.340	2.162	2.162	0.940	3.1	13.3	6.7	4.5	0.64	0.421	0.270	0.015431	16.4	2200	4230
PC26-16-00	1.610	2.550	2.550	1.120	3.8	22.3	11.0	5.3	0.93	0.580	0.539	0.037709	23.1	2200	6300
PC30-19-00	1.880	3.000	3.000	1.320	4.5	36.9	17.7	6.2	1.36	0.798	1.085	0.094643	31.8	2200	7685
PC36-22-00	2.170	3.561	3.561	1.480	5.2	61.8	28.6	7.5	1.99	1.073	2.135	0.226390	44.2	2200	9760
PC42-29-00	2.962	4.242	4.242	2.052	6.8	104.0	59.8	8.7	2.66	1.939	5.158	0.632589	67.7	2200	9815
PC66-56-00	5.578	6.629	6.629	4.242	12.3	550.0	262.1	13.2	7.15	5.576	39.869	8.626670	185.1	2200	14610

Ferrite International ELP Cores															
Part No.	Ht cm 1	Wth cm 2	Lt cm 3	G cm 4	MPL cm 5	Wtfe grams 6	Wtcu grams 7	MLT cm 8	Ac cm sq 9	Wa cm sq 10	Ap cm 4th 11	Kg cm 5th 12	At cm sq 13	Perm 14	AL 15
ELP19-06-13	1.234	2.228	1.905	0.762	3.2	9.9	7.3	5.6	0.601	0.365	0.219	0.009414	16.9	2200	4680
ELP25-07-13	1.372	2.558	2.540	0.762	3.8	15.3	11.2	6.4	0.785	0.491	0.385	0.018782	23.7	2200	5215
ELP30-08-13	1.625	2.236	3.010	0.762	4.2	24.4	8.9	6.8	1.120	0.368	0.412	0.027147	26.4	2200	7255
ELP34-08-16	1.631	3.260	3.432	0.762	4.6	32.9	18.7	8.3	1.384	0.637	0.882	0.058997	38.0	2200	8380
ELP41-10-19	1.951	3.568	4.064	0.762	5.1	60.9	21.4	9.5	2.273	0.634	1.440	0.137559	50.5	2200	12415

Ferrite International Toroidal Cores																
Part No.	OD cm 1	ID cm 2	HT cm 3	Wound OD cm 4	Wound HT cm 5	MPL cm 6	Wtfe grams 7	Wtcu grams 8	MLT cm 9	Ac cm sq 10	Wa cm sq 11	Ap cm 4th 12	Kg cm 5th 13	At cm sq 14	Perm 15	AL 16
TR10-05-03	0.953	0.475	0.318	1.194	0.556	2.1	0.7	0.8	1.3	0.073	0.177	0.013	0.000297	4.3	2200	972
TR13-08-03	1.270	0.792	0.318	1.660	0.714	3.1	1.1	2.7	1.5	0.074	0.492	0.036	0.000707	8.0	2200	659
TR10-05-06	0.953	0.475	0.635	1.194	0.873	2.1	1.5	1.1	1.8	0.146	0.177	0.026	0.000849	5.5	2200	1944
TR13-07-05	1.270	0.714	0.478	1.624	0.835	3.0	1.8	2.5	1.8	0.129	0.400	0.052	0.001496	8.4	2200	1210
TR13-08-06	1.270	0.792	0.635	1.660	1.031	3.1	2.3	3.6	2.0	0.149	0.492	0.073	0.002152	9.7	2200	1318
TR13-07-06	1.270	0.714	0.635	1.624	0.992	3.0	2.5	2.9	2.0	0.172	0.400	0.069	0.002331	9.2	2200	1610
TR16-09-05	1.588	0.907	0.470	2.037	0.924	3.7	2.8	4.6	2.0	0.156	0.646	0.101	0.003108	12.4	2200	1158
TR16-09-10	1.588	0.907	0.953	2.037	1.407	3.7	5.7	6.4	2.8	0.316	0.646	0.204	0.009228	15.5	2200	2347
TR22-14-06	2.210	1.372	0.635	2.886	1.321	5.4	6.9	14.6	2.8	0.261	1.478	0.386	0.014463	25.0	2200	1333
TR22-14-07	2.210	1.372	0.699	2.886	1.385	5.4	7.5	15.2	2.9	0.287	1.478	0.424	0.016867	25.6	2200	1466
TR25-16-08	2.540	1.549	0.792	3.303	1.567	6.2	11.5	22.1	3.3	0.385	1.884	0.725	0.033849	33.4	2200	1724
TR22-14-13	2.210	1.372	1.270	2.886	1.956	5.4	13.7	20.0	3.8	0.522	1.478	0.771	0.042383	30.8	2200	2665
TR29-19-08	2.901	1.900	0.762	3.835	1.712	7.3	13.3	35.7	3.5	0.376	2.834	1.066	0.045270	43.7	2200	1419
TR32-19-06	3.175	1.880	0.635	4.103	1.575	7.6	14.8	35.1	3.6	0.402	2.775	1.115	0.050435	46.7	2200	1465
TR32-19-11	3.175	1.880	1.105	4.103	2.045	7.6	25.7	42.5	4.3	0.699	2.775	1.939	0.125871	52.8	2200	2549
TR44-10-16	4.445	1.005	1.588	5.209	2.091	8.9	75.3	17.2	6.1	1.750	0.793	1.388	0.159307	76.8	2200	5452
TR37-30-16	3.731	2.967	1.588	5.210	3.072	10.4	30.6	135.8	5.5	0.604	6.910	4.174	0.182498	92.9	2200	1602
TR48-35-10	4.763	3.493	0.953	6.489	2.700	12.8	37.1	181.7	5.3	0.600	9.578	5.747	0.258511	121.1	2200	1300
TR36-23-15	3.637	2.261	1.537	4.750	2.668	8.9	44.9	76.6	5.4	1.038	4.013	4.166	0.322143	75.2	2200	3216
TR51-32-19	5.080	3.175	1.905	6.643	3.493	12.5	108.0	200.1	7.1	1.781	7.913	14.094	1.411735	142.1	2200	3940

Ferrite International SPC Cores

Part No.	Ht cm 1	Wth cm 2	Lt cm 3	G cm 4	MPL cm 5	Wtfe grams 6	Wtcu grams 7	MLT cm 8	Ac cm sq 9	Wa cm sq 10	Ap cm 4th 11	Kg cm 5th 12	At cm sq 13	Perm 14	AL 15
DSC14-09-08	0.836	1.400	1.400	0.579	2.4	2.9	1.8	3.0	0.25	0.171	0.043	0.001420	6.8	2200	1980
DSC23-13-11	1.107	2.286	2.286	0.752	3.0	11.5	5.2	4.6	0.66	0.323	0.213	0.012356	16.2	2200	4530
DSC23-13-18	1.803	2.286	2.286	1.448	4.2	16.5	10.1	4.6	0.66	0.622	0.410	0.023792	21.1	2200	3220
DSC30-18-19	1.880	3.000	3.000	1.320	5.9	31.0	17.7	6.2	1.27	0.798	1.013	0.082531	31.8	2200	5510

Ferrite International PQ Cores

Part No.	Ht cm 1	Wth cm 2	Lt cm 3	G cm 4	MPL cm 5	Wtfe grams 6	Wtcu grams 7	MLT cm 8	Ac cm sq 9	Wa cm sq 10	Ap cm 4th 11	Kg cm 5th 12	At cm sq 13	Perm 14	AL 15
PQ20-16-00	1.621	2.317	2.126	1.031	3.7	14.1	7.4	4.4	0.62	0.473	0.293	0.016611	17.4	2200	3660
PQ20-20-00	2.022	2.317	2.126	1.433	4.5	16.9	10.2	4.4	0.62	0.657	0.407	0.023088	20.2	2200	3025
PQ26-20-00	2.017	2.951	2.725	1.148	4.6	32.3	12.0	5.6	1.19	0.603	0.718	0.061294	29.0	2200	6390
PQ26-25-00	2.474	2.951	2.725	1.610	5.6	37.9	16.8	5.6	1.18	0.846	0.998	0.084522	33.2	2200	5290
PQ32-20-00	2.057	3.605	3.302	1.148	5.6	45.7	19.1	6.7	1.70	0.806	1.371	0.139753	37.2	2200	7220
PQ32-30-00	3.033	3.603	3.302	2.129	7.5	60.8	35.4	6.7	1.61	1.493	2.405	0.232019	47.9	2200	5360
PQ35-35-00	3.475	4.366	3.612	2.499	8.8	77.7	58.9	7.5	1.96	2.205	4.323	0.450910	62.0	2200	5695
PQ40-40-00	3.973	5.009	4.148	2.951	10.2	105.6	97.3	8.4	2.01	3.261	6.554	0.628084	78.4	2200	5310

Ferrite International UU Cores															
Part No.	Ht cm 1	Wth cm 2	Lt cm 3	G cm 4	MPL cm 5	Wtfe grams 6	Wtcu grams 7	MLT cm 8	Ac cm sq 9	Wa cm sq 10	Ap cm 4th 11	Kg cm 5th 12	At cm sq 13	Perm 14	AL 15
UU27-25-09	5.080	2.746	3.429	1.880	11.7	37.8	39.6	7.6	0.65	1.456	0.946	0.032185	71.6	2200	1530
UU27-20-09	4.013	2.210	3.429	1.880	9.6	30.7	34.0	6.6	0.64	1.456	0.932	0.036290	53.8	2200	1835
UU27-16-09	3.175	1.791	3.429	1.880	7.9	25.1	28.7	5.5	0.63	1.456	0.917	0.041759	41.4	2200	2210
UU27-13-09	2.540	1.473	3.429	1.880	6.7	20.9	25.4	4.9	0.62	1.456	0.903	0.045684	33.0	2200	2560
UU39-19-10	3.759	2.946	5.944	1.981	10.8	44.6	104.4	7.3	0.78	4.001	3.120	0.132688	87.5	2200	1995
UU41-30-12	6.096	4.001	6.020	2.337	15.5	80.9	158.3	10.0	1.00	4.452	4.452	0.178070	141.7	2200	1800
UU4125-10	5.080	3.493	6.020	2.337	13.5	67.5	142.2	9.0	1.00	4.452	4.452	0.198209	117.8	2200	2050
UU41-22-10	4.445	3.175	6.020	2.337	12.2	62.2	132.2	8.3	1.00	4.452	4.452	0.213285	103.9	2200	2265
UU41-21-12	4.125	3.015	6.020	2.337	11.6	59.6	123.9	7.8	1.00	4.452	4.452	0.227453	97.2	2200	2385
UU69-24-14	4.841	5.601	11.255	2.769	17.7	124.1	534.7	12.5	1.38	11.990	16.546	0.728317	253.4	2200	2150
UU81-45-16	8.920	8.181	13.221	3.175	27.2	268.7	1023.6	17.8	1.89	16.129	30.484	1.291297	481.4	2200	1920
UU81-33-16	6.528	6.985	13.221	3.175	22.4	209.8	886.4	15.5	1.89	16.129	30.484	1.491220	380.4	2200	2332
UU101-57-25	11.039	7.361	14.724	6.121	29.8	964.0	1936.6	18.1	6.03	30.039	181.138	24.099036	607.8	2200	7045

Ferrite International PQLP Cores															
Part No.	Ht cm 1	Wth cm 2	Lt cm 3	G cm 4	MPL cm 5	Wtfe grams 6	Wtcu grams 7	MLT cm 8	Ac cm sq 9	Wa cm sq 10	Ap cm 4th 11	Kg cm 5th 12	At cm sq 13	Perm 14	AL 15
PQLP20-14-14	1.351	2.317	2.126	0.762	3.2	12.5	5.4	4.4	0.62	0.349	0.216	0.012196	15.5	2200	4285
PQLP26-16-19	1.630	2.951	2.725	0.762	3.9	28.0	7.9	5.6	1.19	0.400	0.477	0.040685	25.4	2200	6975
PQLP32-17-22	1.671	3.605	3.302	0.762	4.8	39.4	12.7	6.7	1.70	0.535	0.910	0.092763	32.9	2200	8050
PQLP35-17-26	1.737	4.366	3.612	0.762	5.3	44.9	18.0	7.5	1.96	0.672	1.318	0.137492	40.4	2200	8565
PQLP40-18-28	1.783	5.009	4.148	0.762	5.8	63.5	25.1	8.4	2.01	0.842	1.692	0.162182	48.0	2200	8490

Ferrite International ETDLP Cores															
Part No.	Ht cm 1	Wth cm 2	Lt cm 3	G cm 4	MPL cm 5	Wtfe grams 6	Wtcu grams 7	MLT cm 8	Ac cm sq 9	Wa cm sq 10	Ap cm 4th 11	Kg cm 5th 12	At cm sq 13	Perm 14	AL 15
ETD34-09-11	1.803	2.631	3.421	0.762	4.7	32.7	15.1	7.2	0.97	0.591	0.574	0.031070	33.1	2200	5240
ETD39-09-13	1.798	3.010	3.909	0.762	5.0	46.3	20.0	8.4	1.2	0.671	0.805	0.046150	39.6	2200	6245
ETD44-10-15	1.920	3.330	4.399	0.762	5.4	72.1	23.6	9.4	1.73	0.705	1.219	0.089398	48.0	2200	8050
ETD49-10-16	2.083	3.701	4.869	0.762	5.9	95.0	29.1	10.4	2.11	0.789	1.664	0.135304	58.2	2200	9065

Ferrite International ETD Cores															
Part No.	Ht cm 1	Wth cm 2	Lt cm 3	G cm 4	MPL cm 5	Wtfe grams 6	Wtcu grams 7	MLT cm 8	Ac cm sq 9	Wa cm sq 10	Ap cm 4th 11	Kg cm 5th 12	At cm sq 13	Perm 14	AL 15
ETD34-00-00	3.460	2.631	3.421	2.418	8.0	39.7	47.8	7.2	0.973	1.875	1.825	0.099134	53.2	2200	3070
ETD39-00-00	3.958	3.010	3.909	2.922	9.4	58.5	76.5	8.4	1.249	2.571	3.212	0.191673	69.5	2200	3360
ETD44-00-00	4.460	3.330	4.399	3.302	10.5	90.8	102.4	9.4	1.732	3.053	5.287	0.388241	87.3	2200	4160
ETD49-00-00	4.938	3.701	4.869	3.616	11.6	122.4	138.2	10.4	2.114	3.743	7.912	0.644410	107.2	2200	4600

Ferrite International EE Cores														
Part No.	Ht cm 1	Wth cm 2	Lt cm 3	G cm 4	MPL cm 5	Wtfe grams 6	Wtcu grams 7	MLT cm 8	Ac cm sq 9	Wa cm sq 10	Ap cm 4th 11	Kg cm 5th 12	At cm sq 13	Perm 14	AL 15
EE13-06-03	1.138	0.953	1.270	0.818	2.8	1.4	2.5	2.7	0.101	0.26	0.026	0.000396	6.7	2200	850
EE10-06-05	1.118	1.016	1.024	0.838	2.6	1.6	2.2	2.7	0.119	0.227	0.027	0.000476	6.1	2200	985
EE13-06-04	1.280	0.899	1.270	0.930	2.9	2.0	2.4	2.7	0.133	0.25	0.033	0.000654	7	2200	1055
EE12-08-04	1.550	0.965	1.232	1.118	3.2	2.7	2.8	2.6	0.145	0.298	0.043	0.000948	8.6	2200	1040
EE17-07-03	1.422	1.054	1.676	0.812	3.0	3.0	3.0	2.9	0.162	0.289	0.047	0.001037	10.3	2200	1255
EE13-06-06	1.138	1.270	1.270	0.818	2.8	2.8	3.1	3.3	0.202	0.260	0.052	0.001280	8.2	2200	1716
EE16-08-05	1.610	1.214	1.610	1.178	3.7	3.9	5.5	3.4	0.205	0.450	0.092	0.002213	11.6	2200	1280
EE19-08-05	1.610	1.433	1.905	1.138	4.0	4.5	7.4	3.8	0.225	0.545	0.123	0.002896	14.4	2200	1450
EE20-10-06	1.982	1.329	1.999	1.300	4.6	7.3	6.9	3.7	0.315	0.525	0.165	0.005636	16.9	2200	1540
EE19-08-09	1.610	1.829	1.905	1.138	4.0	8.3	8.9	4.6	0.412	0.545	0.225	0.008039	17.2	2200	2650
EE25-10-06	1.930	1.433	2.540	1.438	4.9	9.8	9.0	4.0	0.394	0.625	0.246	0.009607	20.6	2200	2100
EE25-13-07	2.510	1.792	2.505	1.790	5.8	15.1	17.0	5.0	0.517	0.960	0.496	0.020615	27.4	2200	2415
EE25-16-06	3.200	1.923	2.540	2.566	7.4	14.7	29.2	5.0	0.399	1.653	0.659	0.021173	35.1	2200	1450
EE25-10-13	1.930	2.558	2.540	1.320	4.9	19.5	19.5	6.4	0.787	0.850	0.669	0.032702	29.4	2200	4225
EE30-15-07	2.982	2.017	3.000	2.062	6.6	20.8	25.5	5.2	0.573	1.344	0.770	0.033944	38.4	2200	2330
EE28-11-11	2.108	2.278	2.799	1.138	4.8	24.6	14.1	6.0	0.967	0.660	0.638	0.041111	30.6	2200	5420
EE30-13-09	2.616	2.184	3.048	1.752	6.1	26.1	23.7	6.1	0.834	1.090	0.909	0.049532	36.0	2200	3700
EE30-13-11	2.626	2.035	3.010	1.626	5.8	32.5	17.3	6.2	1.083	0.785	0.851	0.059422	34.6	2200	4975
EE41-16-06	3.296	2.298	4.064	2.108	7.8	30.1	42.3	6.8	0.758	1.753	1.329	0.059425	52.3	2200	2705
EE34-14-09	2.824	2.604	3.432	1.956	6.9	28.7	39.3	6.8	0.813	1.635	1.329	0.063944	45.7	2200	3230
EE25-16-13	3.200	2.558	2.540	2.566	7.4	29.4	37.8	6.4	0.798	1.653	1.319	0.065360	42.4	2200	2900
EE32-16-09	3.220	2.328	3.211	2.322	7.5	31.0	36.7	6.3	0.822	1.640	1.348	0.070441	45.4	2200	2950
EE34-14-13	2.824	2.942	3.432	1.956	6.9	39.2	44.4	7.6	1.108	1.635	1.812	0.105145	50.0	2200	4410
EE30-21-11	4.252	1.994	3.000	3.252	9.0	50.0	32.8	6.1	1.098	1.511	1.659	0.119450	50.5	2200	3260
EE35-24-10	4.760	2.600	3.493	3.810	10.7	48.0	74.2	6.8	0.900	3.058	2.752	0.145160	70.1	2200	2320
EE40-17-11	3.398	2.802	4.001	2.052	7.7	51.5	46.8	7.4	1.274	1.778	2.265	0.155895	61.2	2200	4550
EE41-16-12	3.296	2.908	4.064	2.108	7.8	59.1	51.1	8.2	1.487	1.753	2.606	0.189086	61.3	2200	5300
EE42-21-15	4.180	3.370	4.199	3.032	9.7	89.1	87.9	8.9	1.815	2.768	5.024	0.408434	82.9	2200	5155
EE43-21-15	4.222	3.451	4.285	3.028	9.8	91.3	93.1	9.1	1.838	2.887	5.307	0.430246	85.8	2200	5180
EE47-21-16	3.926	3.226	4.686	2.438	8.9	107.6	68.3	9.5	2.346	2.028	4.759	0.471621	82.0	2200	7290

Ferrite International EE Cores

Part No.	Ht cm 1	Wth cm 2	Lt cm 3	G cm 4	MPL cm 5	Wtfe grams 6	Wtcu grams 7	MLT cm 8	Ac cm sq 9	Wa cm sq 10	Ap cm 4th 11	Kg cm 5th 12	At cm sq 13	Perm 14	AL 15
EE42-21-20	4.180	3.787	4.199	3.032	9.7	113.1	96.1	9.8	2.304	2.768	6.378	0.601950	89.9	2200	6555
EE43-21-20	4.222	3.868	4.285	3.028	9.8	116.0	101.7	9.9	2.334	2.887	6.739	0.635354	92.9	2200	6575
EE43-21-22	4.222	4.130	4.285	3.028	9.8	131.4	107.0	10.4	2.645	2.887	7.637	0.774945	97.4	2200	7435
EE42-33-20	6.518	3.767	4.229	5.298	14.3	167.5	166.8	9.7	2.350	4.811	11.305	1.090004	127.4	2200	4535
EE56-24-19	4.720	3.861	5.613	2.922	10.7	186.8	115.7	11.2	3.398	2.894	9.835	1.189449	117.9	2200	8815
EE59-25-19	4.938	3.884	5.872	2.922	10.9	207.1	117.2	11.2	3.642	2.947	10.732	1.397637	127.1	2200	9200
EE55-28-21	5.502	4.186	5.512	3.760	12.3	222.3	162.1	11.5	3.540	3.978	14.082	1.740310	138.5	2200	7930
EE106-24-17	4.810	8.450	10.574	2.854	15.4	264.8	619.2	18.2	3.366	9.575	32.230	2.386319	313.9	2200	6025
EE55-28-25	5.502	4.656	5.512	3.760	12.3	272.7	175.4	12.4	4.344	3.978	17.281	2.421906	148.9	2200	9735
EE72-28-19	5.588	5.334	7.239	3.606	13.7	256.8	299.2	13.6	3.690	6.182	22.813	2.473826	191.6	2200	7450
EE80-38-20	7.620	6.020	8.001	5.638	19.4	363.2	602.2	14.9	3.925	11.386	44.690	4.717403	278.0	2200	5880
EE76-51-25	10.160	5.080	7.620	7.620	21.7	648.9	507.8	14.8	5.624	9.677	54.426	8.297486	309.7	2200	7145
EE70-54-32	10.796	5.717	6.985	8.570	23.2	810.4	596.3	15.4	6.997	10.892	76.215	13.855820	331.5	2200	8275

Magnetics DS Cores

Part No.	Ht cm 1	Wth cm 2	Lt cm 3	G cm 4	MPL cm 5	Wtfe grams 6	Wtcu grams 7	MLT cm 8	Ac cm sq 9	Wa cm sq 10	Ap cm 4th 11	Kg cm 5th 12	At cm sq 13	Perm 14	AL 15
DS-42311	1.107	2.327	2.286	0.726	2.7	10	4.8	4.6	0.378	0.292	0.110	0.003615	16.2	2300	2580
DS-42318	1.803	2.327	2.286	1.387	4.0	13	9.1	4.6	0.407	0.557	0.227	0.008001	21.1	2300	2180
DS-42616	1.610	2.682	2.550	1.102	3.9	15	10.2	5.4	0.627	0.536	0.336	0.015699	23.1	2300	2870
DS-43019	1.880	3.180	3.000	1.300	4.6	22	16.7	6.3	0.960	0.747	0.717	0.043793	31.8	2300	3330
DS-43622	2.172	3.764	3.561	1.458	5.3	37	26.7	7.5	1.250	1.005	1.257	0.084228	44.2	2300	4020
DS-44229	2.957	4.630	4.239	2.042	7.2	78	56.0	8.6	1.780	1.828	3.255	0.269171	67.6	2300	4830

Magnetics EC Cores

Part No.	Ht cm 1	Wth cm 2	Lt cm 3	G cm 4	MPL cm 5	Wtfe grams 6	Wtcu grams 7	MLT cm 8	Ac cm sq 9	Wa cm sq 10	Ap cm 4th 11	Kg cm 5th 12	At cm sq 13	Perm 14	AL 15
EC-43517	3.459	2.275	3.449	2.383	7.6	36	35.3	6.3	0.709	1.578	1.119	0.050463	50.3	2300	1660
EC-44119	3.901	2.705	4.064	2.697	8.8	52	55.4	7.5	1.060	2.083	2.208	0.125223	67.6	2300	2210
EC-45224	4.841	3.302	5.220	3.099	10.3	111	97.8	9.0	1.410	3.040	4.287	0.267204	106.5	2300	2900
EC-47035	6.899	4.450	7.000	4.465	14.1	253	258.4	11.6	2.110	6.278	13.246	0.965788	201.9	2300	3310

Magnetics EE Cores															
Part No.	Ht cm 1	Wth cm 2	Lt cm 3	G cm 4	MPL cm 5	Wtfe grams 6	Wtcu grams 7	MLT cm 8	Ac cm sq 9	Wa cm sq 10	Ap cm 4th 11	Kg cm 5th 12	At cm sq 13	Perm 14	AL 15
EE-40904	0.813	0.505	0.886	0.406	1.5	0.5	0.4	1.7	0.036	0.064	0.002	0.000020	2.9	2300	370
EE-41203	1.138	0.955	1.270	0.792	2.8	1.3	2.4	2.7	0.100	0.253	0.025	0.000377	6.7	2300	440
EE-41208	1.549	0.965	1.232	1.092	3.2	2.5	2.7	2.6	0.115	0.291	0.033	0.000583	8.6	2300	630
EE-41707	1.422	1.118	1.676	0.787	3.0	3.0	3.2	3.0	0.126	0.300	0.038	0.000629	10.6	2300	760
EE-41205	1.138	1.273	1.270	0.792	2.8	2.6	3.0	3.3	0.200	0.253	0.051	0.001220	8.2	2300	1100
EE-41709	1.880	0.991	1.778	1.346	4.2	4.5	4.8	3.6	0.181	0.376	0.068	0.001380	12.3	2300	740
EE-41808	1.621	1.453	1.930	1.107	4.0	4.4	7.4	3.8	0.228	0.540	0.123	0.002917	14.7	2300	865
EE-42211	2.245	1.885	2.200	1.499	5.1	8.2	15.6	4.5	0.246	0.982	0.242	0.005315	24.4	2300	920
EE-42216	1.143	2.730	2.159	0.297	3.2	13.0	4.0	6.6	0.806	0.170	0.137	0.006693	20.6	2300	3590
EE-41810	1.621	1.930	1.930	1.118	4.0	8.5	9.3	4.8	0.454	0.546	0.248	0.009384	18.1	2300	1725
EE-42510	1.905	1.935	2.515	1.250	4.9	9.5	14.3	4.9	0.384	0.813	0.312	0.009704	23.5	2300	1325
EE-43524	4.763	2.565	3.454	0.312	10.7	46.0	6.1	6.7	0.831	0.255	0.212	0.010491	69.0	2300	1320
EE-42515	3.175	1.915	2.540	2.515	7.4	15.0	28.4	5.0	0.397	1.610	0.639	0.020467	34.8	2300	865
EE-43007	3.002	1.999	3.010	1.941	6.6	20.0	23.4	5.2	0.491	1.254	0.616	0.023077	38.5	2300	1545
EE-45114	2.794	1.905	5.080	1.539	6.4	37.0	30.4	8.8	0.781	0.977	0.763	0.027248	45.5	2300	2300
EE-42520	1.905	2.570	2.515	1.250	4.8	19.0	18.5	6.4	0.768	0.813	0.624	0.029914	29.1	2300	2650
EE-42810	2.108	2.278	2.799	1.087	4.8	23.0	13.5	6.0	0.860	0.631	0.543	0.031085	30.6	2300	3155
EE-44008	1.702	3.101	4.064	0.711	5.2	26.0	20.0	7.8	0.951	0.723	0.687	0.033647	41.3	2300	3150
EE-43618	1.270	3.810	3.556	0.483	4.2	28.0	15.5	8.9	1.350	0.490	0.662	0.040261	39.6	2300	5170
EE-43208	1.270	3.937	3.175	0.635	4.2	26.0	19.2	8.9	1.290	0.605	0.780	0.045070	38.2	2300	5025
EE-43009	2.680	2.141	3.048	1.765	6.2	26.0	22.8	6.1	0.807	1.060	0.856	0.045634	36.3	2300	2170
EE-43013	2.631	1.882	3.000	1.600	5.8	32.0	13.8	6.0	1.070	0.650	0.696	0.049958	33.5	2300	3070
EE-43515	2.830	2.530	3.454	1.930	6.9	33.0	36.4	6.6	0.821	1.540	1.264	0.062458	45.4	2300	2000
EE-42530	3.175	2.550	2.540	2.515	7.4	30.0	36.8	6.4	0.794	1.610	1.278	0.063145	42.0	2300	1730
EE-43520	4.125	2.540	3.493	3.124	9.4	42.0	59.2	6.7	0.905	2.480	2.244	0.121069	61.5	2300	1460
EE-44011	3.399	2.802	4.001	2.002	7.7	49.0	45.6	7.4	1.140	1.734	1.976	0.121700	61.2	2300	3000
EE-45015	3.175	3.678	5.017	1.580	7.7	70.0	62.7	9.5	1.420	1.862	2.644	0.158671	80.3	2300	3615
EE-44317	3.302	2.827	4.087	2.083	7.8	57.0	47.2	8.1	1.490	1.640	2.444	0.180038	60.9	2300	2925
EE-44308	1.702	5.486	4.318	0.838	5.8	64.0	48.3	12.0	2.270	1.128	2.561	0.193032	72.0	2300	6420
EE-44020	4.216	3.457	4.285	2.982	9.8	87.0	91.9	9.1	1.830	2.852	5.219	0.421320	85.9	2300	3450

Magnetics EE Cores															
Part No.	Ht cm 1	Wth cm 2	Lt cm 3	G cm 4	MPL cm 5	Wtfe grams 6	Wtcu grams 7	MLT cm 8	Ac cm sq 9	Wa cm sq 10	Ap cm 4th 11	Kg cm 5th 12	At cm sq 13	Perm 14	AL 15
EE-44721	3.942	3.203	4.712	2.413	8.9	103.0	66.3	9.4	2.360	1.974	4.658	0.465517	82.3	2300	4020
EE-45021	4.262	3.480	4.953	2.499	9.3	108.0	86.3	9.6	2.130	2.523	5.375	0.476026	94.6	2300	4600
EE-44924	4.755	3.137	4.907	3.018	10.4	132.0	78.8	9.3	2.440	2.376	5.797	0.606538	97.4	2300	4030
EE-44022	4.216	3.914	4.285	2.982	9.8	114.0	101.2	10.0	2.370	2.852	6.758	0.641915	93.6	2300	4150
EE-46016	4.460	4.460	5.999	2.758	11.0	135.0	162.1	11.4	2.400	3.997	9.593	0.807391	128.7	2300	4300
EE-45724	4.729	3.955	5.657	2.896	10.7	179.0	121.6	11.4	3.390	3.004	10.185	1.213144	120.1	2300	6070
EE-45528	5.512	4.196	5.486	3.708	12.3	212.0	160.9	11.4	3.460	3.956	13.688	1.656632	138.5	2300	3280
EE-45530	5.512	4.595	5.486	3.708	12.3	255.0	172.1	12.2	4.130	3.956	16.339	2.206450	147.2	2300	5640
EE-47228	5.588	5.334	7.239	3.556	13.7	264.0	295.1	13.6	3.630	6.097	22.131	2.360834	191.6	2300	3110
EE-48020	7.620	6.020	8.001	5.639	18.5	357.0	602.2	14.9	3.820	11.386	43.496	4.468523	278.0	2300	3505

Magnetics EI Cores															
Part No.	Ht cm 1	Wth cm 2	Lt cm 3	G cm 4	MPL cm 5	Wtfe grams 6	Wtcu grams 7	MLT cm 8	Ac cm sq 9	Wa cm sq 10	Ap cm 4th 11	Kg cm 5th 12	At cm sq 13	Perm 14	AL 15
EI-43618	1.016	3.861	3.556	0.241	3.7	25	7.8	9.0	1.35	0.245	0.331	0.019903	36.3	2300	5870
EI-43208	0.953	3.937	3.175	0.318	3.5	22	9.6	8.9	1.29	0.302	0.390	0.022535	33.7	2300	5930
EI-44308	1.257	5.486	4.318	0.419	4.9	54	24.2	12.0	2.27	0.564	1.281	0.096516	63.3	2300	7600
EI-44008	1.295	3.101	4.064	0.356	4.4	21	10.0	7.8	9.51	0.361	3.436	1.682348	35.5	2300	3690

Magnetics EP Cores															
Part No.	Ht cm 1	Wth cm 2	Lt cm 3	G cm 4	MPL cm 5	Wtfe grams 6	Wtcu grams 7	MLT cm 8	Ac cm sq 9	Wa cm sq 10	Ap cm 4th 11	Kg cm 5th 12	At cm sq 13	Perm 14	AL 15
EP-41010	1.052	1.026	1.151	0.721	1.9	2.8	1.6	2.1	0.078	0.207	0.016	0.000235	5.7	2300	780
EP-41313	1.139	1.285	1.250	0.899	2.4	5.1	2.0	2.4	0.140	0.234	0.033	0.000766	7.7	2300	1150
EP-41717	1.388	1.681	1.798	1.118	2.9	11.6	3.4	3.0	0.241	0.322	0.078	0.002513	13.7	2300	1790
EP-40707	0.824	0.742	0.919	0.498	1.6	1.4	0.6	1.9	0.780	0.094	0.073	0.012056	3.5	2300	810
EP-42120	1.850	2.139	2.400	1.397	4.0	27.6	7.4	4.2	0.780	0.497	0.387	0.028932	23.8	2300	3170

Magnetics ETD Cores															
Part No.	Ht cm 1	Wth cm 2	Lt cm 3	G cm 4	MPL cm 5	Wtfe grams 6	Wtcu grams 7	MLT cm 8	Ac cm sq 9	Wa cm sq 10	Ap cm 4th 11	Kg cm 5th 12	At cm sq 13	Perm 14	AL 15
ETD-43434	3.459	2.629	3.419	2.357	7.9	40	46.5	7.2	0.915	1.826	1.671	0.085426	53.1	2300	1750
ETD-43939	3.957	3.010	3.909	2.845	9.3	60	74.5	8.4	1.230	2.504	3.080	0.181017	69.5	2300	1930
ETD-44444	4.460	3.330	4.399	3.226	10.4	94	100.1	9.4	1.720	2.982	5.130	0.374089	87.3	2300	2390
ETD-44949	4.938	3.701	4.869	3.536	11.4	124	135.1	10.4	2.090	3.660	7.649	0.615966	107.2	2300	2760
ETD-47054	10.795	5.410	6.858	8.367	23.1	396	708.4	14.0	3.140	14.270	44.809	4.031674	311.8	2300	2440

Magnetics PC Cores

Part No.	Ht cm 1	Wth cm 2	Lt cm 3	G cm 4	MPL cm 5	Wtfe grams 6	Wtcu grams 7	MLT cm 8	Ac cm sq 9	Wa cm sq 10	Ap cm 4th 11	Kg cm 5th 12	At cm sq 13	Perm 14	AL 15
PC-40704	0.417	0.724	0.724	0.279	1.0	0.5	0.2	1.6	0.059	0.038	0.002	0.000033	1.8	2300	620
PC-40905	0.526	0.914	0.914	0.361	1.3	1.0	0.5	1.9	0.081	0.065	0.005	0.000088	2.8	2300	760
PC-41107	0.650	1.110	1.110	0.442	1.5	1.8	0.8	2.3	0.133	0.095	0.013	0.000291	4.2	2300	1150
PC-41408	0.836	1.405	1.405	0.559	2.0	3.2	1.6	2.9	0.197	0.157	0.031	0.000833	6.8	2300	1540
PC-41811	1.054	1.801	1.801	0.721	2.6	7.3	3.5	3.7	0.360	0.267	0.096	0.003730	11.1	2300	2300
PC-42213	1.341	2.162	2.162	0.919	3.1	13.0	6.2	4.4	0.509	0.391	0.199	0.009116	16.4	2300	3030
PC-42616	1.610	2.550	2.550	1.102	3.8	20.0	10.1	5.3	0.751	0.536	0.403	0.022857	23.1	2300	3910
PC-43019	1.880	3.000	3.000	1.300	4.5	34.0	16.5	6.2	1.150	0.747	0.859	0.063638	31.8	2300	5010
PC-43622	2.172	3.561	3.561	1.458	5.3	57.0	26.7	7.5	1.740	1.005	1.749	0.163206	44.2	2300	6530
PC-44229	2.957	4.239	4.239	2.042	6.9	104.0	56.0	8.6	2.130	1.828	3.895	0.385432	67.6	2300	6900
PC-44529	2.921	4.501	4.501	1.880	6.7	149.6	48.4	9.1	2.990	1.490	4.454	0.582745	73.1	2300	9660

Magnetics PQ Cores

Part No.	Ht cm 1	Wth cm 2	Lt cm 3	G cm 4	MPL cm 5	Wtfe grams 6	Wtcu grams 7	MLT cm 8	Ac cm sq 9	Wa cm sq 10	Ap cm 4th 11	Kg cm 5th 12	At cm sq 13	Perm 14	AL 15
PQ-42610	1.016	2.885	2.725	0.239	2.9	15	2.3	5.6	0.938	0.118	0.110	0.007375	19.8	2300	5800
PQ-42614	1.189	2.885	2.725	0.671	3.3	14	6.6	5.6	0.709	0.330	0.234	0.011834	21.4	2300	4210
PQ-42016	1.621	2.256	2.126	1.001	3.7	13	6.6	4.3	0.580	0.428	0.248	0.013270	17.4	2300	2690
PQ-42020	2.022	2.256	2.126	1.402	4.5	15	9.3	4.3	0.580	0.600	0.348	0.018591	20.2	2300	2210
PQ-43214	1.189	3.528	3.302	0.671	3.4	21	10.5	6.6	0.920	0.445	0.410	0.022740	27.6	2300	5150
PQ-42620	2.017	2.885	2.725	1.118	4.6	31	11.0	5.6	1.090	0.551	0.600	0.046617	29.0	2300	4170
PQ-42625	2.474	2.885	2.725	1.580	5.6	36	15.5	5.6	1.090	0.779	0.849	0.065900	33.2	2300	3450
PQ-43220	2.057	3.528	3.302	1.118	5.6	42	17.5	6.6	1.370	0.742	1.017	0.084044	37.2	2300	4980
PQ-43230	3.033	3.528	3.302	2.098	7.5	55	32.9	6.6	1.370	1.394	1.909	0.157774	47.9	2300	3500
PQ-43535	3.475	4.290	3.612	2.469	8.8	73	55.4	7.5	1.560	2.085	3.253	0.271479	62.0	2300	3610
PQ-44040	3.973	4.917	4.148	2.921	10.2	95	91.8	8.3	1.670	3.094	5.167	0.413744	78.4	2300	3200

Magnetics RM Cores

Part No.	Ht cm 1	Wth cm 2	Lt cm 3	G cm 4	MPL cm 5	Wtfe grams 6	Wtcu grams 7	MLT cm 8	Ac cm sq 9	Wa cm sq 10	Ap cm 4th 11	Kg cm 5th 12	At cm sq 13	Perm 14	AL 15
RM-41510	1.041	1.214	1.214	0.630	2.2	3.3	1.5	2.5	0.181	0.166	0.030	0.000857	8.0	2300	1290
RM-41812	1.240	1.450	1.450	0.798	2.7	5.1	2.6	3.1	0.320	0.239	0.076	0.003147	11.4	2300	1790
RM-42316	1.641	1.935	1.935	1.082	3.8	13.0	6.8	4.2	0.540	0.456	0.246	0.012756	20.2	2300	2025
RM-42819	1.859	2.413	2.413	1.240	4.4	23.0	11.8	5.2	0.900	0.638	0.574	0.039749	29.6	2300	3035
RM-43723	2.347	2.924	2.924	1.676	5.7	42.0	22.4	6.2	1.250	1.022	1.277	0.103477	44.5	2300	3450

Magnetics RS Cores

Part No.	Ht cm 1	Wth cm 2	Lt cm 3	G cm 4	MPL cm 5	Wtfe grams 6	Wtcu grams 7	MLT cm 8	Ac cm sq 9	Wa cm sq 10	Ap cm 4th 11	Kg cm 5th 12	At cm sq 13	Perm 14	AL 15
RS-41408	0.836	1.501	1.405	0.559	2.0	2.9	1.6	2.9	0.230	0.157	0.036	0.001135	6.8	2300	1320
RS-42311	1.107	2.327	2.286	0.726	2.7	11.7	4.8	4.6	0.378	0.292	0.110	0.003615	16.2	2300	2950
RS-42318	1.803	2.327	2.286	1.387	3.9	17.4	9.1	4.6	0.407	0.557	0.227	0.008001	21.1	2300	2300
RS-43019	1.880	3.180	3.000	1.300	4.6	31.0	16.7	6.3	0.960	0.747	0.717	0.043793	31.8	2300	4150

Magnetics Toroidal Cores

Part No.	OD cm 1	ID cm 2	HT cm 3	Wound OD cm 4	Wound HT cm 5	MPL cm 6	Wtfe grams 7	Wtcu grams 8	MLT cm 9	Ac cm sq 10	Wa cm sq 11	Ap cm 4th 12	Kg cm 5th 13	At cm sq 14	Perm 15	AL 16
TC-40200	0.254	0.127	0.127	0.277	0.191	0.6	0.0	0.0	0.4	0.008	0.013	0.000	0.000001	0.3	3000	525
TC-40502	0.394	0.224	0.127	0.439	0.239	0.9	0.0	0.1	0.5	0.011	0.039	0.000	0.000004	0.8	3000	44
TC-40401	0.483	0.229	0.127	0.522	0.241	1.0	0.1	0.1	0.6	0.015	0.041	0.001	0.000007	1.0	3000	570
TC-40503	0.394	0.224	0.254	0.439	0.366	0.9	0.1	0.1	0.7	0.021	0.039	0.001	0.000010	1.0	3000	885
TC-40601	0.584	0.305	0.152	0.641	0.305	1.3	0.1	0.2	0.7	0.021	0.073	0.002	0.000018	1.6	2300	448
TC-40402	0.483	0.229	0.254	0.522	0.368	1.0	0.2	0.1	0.8	0.031	0.041	0.001	0.000020	1.3	3000	1140
TC-40603	0.584	0.305	0.318	0.641	0.470	1.3	0.3	0.3	1.0	0.044	0.073	0.003	0.000058	1.9	2300	938
TC-40705	0.762	0.318	0.478	0.810	0.636	1.5	0.7	0.4	1.4	0.098	0.079	0.008	0.000221	3.2	2300	1920
TC-40301	0.351	1.829	0.127	1.622	1.041	0.8	0.0	4.5	0.5	0.010	2.625	0.027	0.000226	11.6	3000	495
TC-41003	0.953	0.475	0.318	1.037	0.555	2.1	0.7	0.8	1.3	0.070	0.177	0.012	0.000273	4.3	2300	1007
TC-41005	0.953	0.475	0.478	1.037	0.715	2.1	1.1	1.0	1.5	0.107	0.177	0.019	0.000531	4.9	2300	1518
TC-41303	1.270	0.792	0.318	1.443	0.714	3.1	1.1	2.7	1.5	0.072	0.493	0.035	0.000671	8.0	2300	685
TC-40907	0.953	0.559	0.711	1.069	0.991	2.3	1.5	1.7	1.9	0.135	0.245	0.033	0.000941	6.2	2300	1733
TC-41506	1.321	0.737	0.396	1.467	0.765	3.1	1.6	2.6	1.7	0.109	0.426	0.046	0.001197	8.5	2300	1022
TC-41407	1.270	0.714	0.478	1.412	0.834	3.0	1.8	2.5	1.8	0.126	0.400	0.050	0.001427	8.4	2300	1247
TC-41305	1.270	0.792	0.508	1.443	0.904	3.1	1.8	3.2	1.8	0.117	0.493	0.058	0.001476	9.0	2300	1094
TC-41450	1.400	0.899	0.500	1.602	0.950	3.5	2.0	4.3	1.9	0.120	0.635	0.076	0.001904	10.8	2300	993
TC-41206	1.270	0.516	0.635	1.346	0.893	2.5	2.6	1.5	2.0	0.221	0.209	0.046	0.002007	8.1	2300	2594
TC-41306	1.270	0.792	0.635	1.443	1.031	3.1	2.2	3.6	2.0	0.146	0.493	0.072	0.002069	9.7	2300	1366
TC-41406	1.270	0.714	0.635	1.412	0.992	3.0	2.4	2.9	2.0	0.169	0.400	0.068	0.002248	9.2	2300	1660
TC-41605	1.588	0.889	0.470	1.764	0.914	3.7	2.7	4.5	2.0	0.153	0.620	0.095	0.002873	12.3	2300	1265
TC-42206	2.210	1.372	0.635	2.509	1.321	5.4	6.6	14.6	2.8	0.250	1.477	0.369	0.013262	25.0	2300	1389
TC-41809	1.844	0.975	1.026	2.029	1.514	4.1	8.1	8.3	3.1	0.403	0.747	0.301	0.015564	19.6	2300	2806
TC-42207	2.210	1.372	0.792	2.509	1.478	5.4	8.3	15.9	3.0	0.315	1.477	0.465	0.019308	26.5	2300	1725
TC-43205	3.200	1.501	0.450	3.455	1.200	6.7	11.2	20.6	3.3	0.345	1.769	0.610	0.025679	39.7	2300	1481
TC-42507	2.540	1.549	0.792	2.873	1.567	6.2	11.2	22.1	3.3	0.374	1.885	0.705	0.031951	33.4	2300	1801
TC-42212	2.210	1.372	1.270	2.509	1.956	5.4	13.4	20.0	3.8	0.511	1.477	0.755	0.040594	30.8	2300	2778
TC-42908	2.901	1.900	0.749	3.335	1.699	7.3	12.7	35.5	3.5	0.358	2.834	1.014	0.041276	43.6	2300	1458
TC-43806	3.810	1.905	0.635	4.152	1.588	8.3	22.9	41.2	4.1	0.570	2.849	1.624	0.091099	59.6	2300	2024
TC-42915	2.901	1.900	1.524	3.335	2.474	7.4	26.7	48.0	4.8	0.740	2.834	2.097	0.130423	52.9	2300	2964

Magnetics Toroidal Cores

Part No.	OD cm 1	ID cm 2	HT cm 3	Wound OD cm 4	Wound HT cm 5	MPL cm 6	Wtfe grams 7	Wtcu grams 8	MLT cm 9	Ac cm sq 10	Wa cm sq 11	Ap cm 4th 12	Kg cm 5th 13	At cm sq 14	Perm 15	AL 16
TC-43113	3.099	1.900	1.300	3.509	2.250	7.6	27.3	45.9	4.6	0.745	2.834	2.111	0.137964	54.1	2300	2852
TC-43610	3.599	2.301	1.001	4.114	2.151	9.0	27.3	66.2	4.5	0.628	4.157	2.611	0.146366	67.1	2300	2033
TC-43615	3.599	2.301	1.499	4.114	2.649	9.0	41.2	78.0	5.3	0.946	4.157	3.933	0.281994	74.5	2300	3096
TC-43813	3.810	1.905	1.270	4.152	2.223	8.3	46.3	51.5	5.1	1.150	2.849	3.276	0.296655	69.1	2300	3850
TC-44416	4.445	1.905	1.588	4.741	2.540	8.9	80.4	61.8	6.1	1.870	2.849	5.327	0.653669	90.2	2300	5363
TC-44916	4.907	3.383	1.588	5.716	3.279	12.7	71.5	206.6	6.5	1.160	8.986	10.423	0.747993	135.5	2300	2714
TC-44715	4.689	2.700	1.501	5.239	2.851	11.0	75.8	125.2	6.2	1.420	5.723	8.126	0.750172	110.9	2300	3707
TC-43825	3.810	1.905	2.540	4.152	3.493	8.3	93.0	72.0	7.1	2.310	2.849	6.581	0.854972	88.2	2300	8061
TC-44925	4.907	3.180	1.905	5.627	3.495	12.3	87.1	196.9	7.0	1.460	7.939	11.590	0.970601	136.8	2300	3420
TC-46113	6.096	3.556	1.270	6.830	3.048	14.5	109.7	243.9	6.9	1.560	9.926	15.485	1.398620	172.0	2300	3148
TC-44932	4.907	3.383	3.175	5.716	4.867	12.7	145.4	287.8	9.0	2.360	8.986	21.206	2.222829	168.3	2300	5428
TC-47313	7.366	3.886	1.270	8.098	3.213	16.5	169.7	334.1	7.9	2.120	11.856	25.134	2.689449	230.1	2300	3702
TC-48613	8.573	5.555	1.270	9.830	4.047	21.5	195.0	765.8	8.9	1.870	24.223	45.298	3.811325	344.3	2300	2507

Magnetics UU Cores

Part No.	Ht cm 1	Wth cm 2	Lt cm 3	G cm 4	MPL cm 5	Wtfe grams 6	Wtcu grams 7	MLT cm 8	Ac cm sq 9	Wa cm sq 10	Ap cm 4th 11	Kg cm 5th 12	At cm sq 13	Perm 14	AL 15
UI-41106	0.602	1.349	1.803	0.211	2.5	1.5	1.7	3.2	0.115	0.152	0.017	0.000254	7.5	2300	710
UU-41106	0.838	1.349	1.803	4.216	2.9	1.8	34.1	3.2	0.115	3.031	0.349	0.005070	19.9	2300	770
UU-42515	3.175	1.905	3.810	1.854	8.3	17.0	41.4	4.9	0.404	2.355	0.951	0.031109	44.1	2300	830
UU-42220	4.115	1.580	3.160	2.794	9.6	19.0	41.8	4.4	0.397	2.661	1.057	0.038019	44.0	2300	670
UU-42512	2.591	2.550	3.820	1.270	6.9	29.0	37.1	6.4	0.800	1.626	1.301	0.064853	45.7	2300	1430
UU-42530	3.175	2.540	3.810	1.854	8.3	34.0	53.7	6.4	0.808	2.355	1.903	0.095908	53.0	2300	1570
UU-42515	3.175	1.905	3.810	1.854	6.9	29.0	41.4	4.9	0.800	2.355	1.884	0.121985	44.1	2300	1430
UU-44119	4.196	3.073	6.101	2.682	12.1	54.0	145.8	8.0	0.805	5.110	4.113	0.165008	103.0	2300	1220
UU-44121	4.125	3.073	6.033	2.179	11.3	55.0	117.5	8.0	0.988	4.152	4.102	0.203694	97.4	2300	1410
UU-44125	5.080	3.073	6.101	3.134	13.3	64.0	170.4	8.0	0.988	5.971	5.899	0.290455	115.9	2300	1200
UU-44130	6.096	3.073	6.101	4.150	15.3	75.0	225.7	8.0	0.988	7.906	7.812	0.384606	134.5	2300	1050
UU-49925	11.430	7.620	15.240	6.223	30.8	975.0	2107.1	18.7	6.450	31.613	203.903	28.066610	649.0	2500	3690

MMG North America EE Cores

Part No.	Ht cm 1	Wth cm 2	Lt cm 3	G cm 4	MPL cm 5	Wtfe grams 6	Wtcu grams 7	MLT cm 8	Ac cm sq 9	Wa cm sq 10	Ap cm 4th 11	Kg cm 5th 12	At cm sq 13	Perm 14	AL 15
EE-24x25	1.905	1.956	2.540	1.298	4.9	9.1	15.3	5.0	0.381	0.863	0.329	0.010011	23.7	1510	1480
EE-30x30x7	3.000	1.995	3.010	1.890	6.7	20.6	22.7	5.2	0.600	1.219	0.731	0.033480	38.4	1600	1800
EE-25x12.5	2.008	2.633	2.615	1.398	5.1	18.0	22.5	6.6	0.770	0.963	0.742	0.034794	31.3	1500	2845
EE-375	1.880	2.488	3.468	1.930	6.9	30.6	35.3	6.6	0.880	1.507	1.326	0.070826	33.8	1490	2890
EE-21	3.296	2.858	4.064	2.108	7.7	58.0	49.1	8.1	1.530	1.700	2.601	0.196038	60.8	1425	3585
EE-42x15	4.200	3.310	4.215	3.020	9.7	84.0	86.1	8.8	1.750	2.741	4.796	0.379890	82.4	1490	3500
EE-42x20	4.200	3.795	0.215	3.020	9.7	108.0	95.6	9.8	2.320	2.741	6.358	0.601629	26.6	1470	4560
EE-55x21	5.500	4.185	5.515	3.810	12.3	213.0	164.2	11.5	3.500	4.029	14.102	1.722946	138.5	1550	5570
EE-55x25	5.500	4.585	5.515	3.810	12.3	256.0	175.6	12.3	4.200	4.029	16.922	2.319128	147.3	1600	6875
EE-65x27	6.500	5.240	6.515	4.510	14.7	384.0	282.6	13.9	5.320	5.705	30.351	4.636562	201.2	1625	7430
EE-70x32	6.450	5.800	6.950	4.298	14.6	488.0	311.0	15.5	6.710	5.641	37.852	6.552274	221.9	1860	11125

MMG North America EFD Cores

Part No.	Ht cm 1	Wth cm 2	Lt cm 3	G cm 4	MPL cm 5	Wtfe grams 6	Wtcu grams 7	MLT cm 8	Ac cm sq 9	Wa cm sq 10	Ap cm 4th 11	Kg cm 5th 12	At cm sq 13	Perm 14	AL 15
EFD-15	0.750	1.5	1.50	1.10	3.4	2.8	3.0	2.7	0.122	0.314	0.038	0.000696	7.3	1220	675
EFD-20	0.990	2.0	2.01	1.54	4.7	7.0	6.8	3.8	0.290	0.501	0.145	0.004431	13.4	1350	1120
EFD-25	1.295	2.5	2.51	1.86	5.7	16.5	12.4	4.9	0.570	0.716	0.408	0.019149	21.7	1425	1790

MMG North America EP Cores

Part No.	Ht cm 1	Wth cm 2	Lt cm 3	G cm 4	MPL cm 5	Wtfe grams 6	Wtcu grams 7	MLT cm 8	Ac cm sq 9	Wa cm sq 10	Ap cm 4th 11	Kg cm 5th 12	At cm sq 13	Perm 14	AL 15
EP-7	0.840	0.740	0.92	0.52	1.6	1.4	0.7	1.8	0.085	0.107	0.009	0.000167	3.5	1115	920
EP-10	1.070	1.030	1.15	0.74	1.9	2.8	1.7	2.2	0.085	0.226	0.019	0.000303	5.7	1150	855
EP-13	1.163	1.285	1.25	0.92	2.4	5.1	2.2	2.4	0.149	0.260	0.039	0.000956	7.7	1220	1235
EP-17	1.419	1.680	1.80	1.15	2.9	11.6	3.9	3.0	0.253	0.367	0.093	0.003126	13.7	1320	1980
EP-20	1.883	2.140	2.40	1.43	4.0	27.6	8.3	4.2	0.600	0.554	0.332	0.018986	23.8	1265	3115

MMG North America ETD Cores

Part No.	Ht cm 1	Wth cm 2	Lt cm 3	G cm 4	MPL cm 5	Wtfe grams 6	Wtcu grams 7	MLT cm 8	Ac cm sq 9	Wa cm sq 10	Ap cm 4th 11	Kg cm 5th 12	At cm sq 13	Perm 14	AL 15
ETD-29	3.18	2.27	2.98	2.20	7.0	28	32.4	6.3	0.704	1.452	1.022	0.045834	41.9	1440	1950
ETD-34	3.46	2.63	3.42	2.36	7.9	40	46.6	7.2	0.916	1.829	1.675	0.085716	53.2	1450	2250
ETD-39	3.96	3.01	3.91	2.84	9.2	60	74.4	8.4	1.230	2.499	3.074	0.180669	69.5	1455	2470
ETD-44	4.46	3.33	4.40	3.22	10.3	94	99.9	9.4	1.720	2.979	5.123	0.373670	87.3	1480	3100
ETD-49	4.93	3.70	4.87	3.54	11.4	124	135.2	10.4	2.090	3.664	7.658	0.616860	107.0	1515	3525

MMG North America PC Cores

Part No.	Ht cm 1	Wth cm 2	Lt cm 3	G cm 4	MPL cm 5	Wtfe grams 6	Wtcu grams 7	MLT cm 8	Ac cm sq 9	Wa cm sq 10	Ap cm 4th 11	Kg cm 5th 12	At cm sq 13	Perm 14	AL 15
P-0905	0.53	0.915	0.915	0.374	1.2	1.0	0.5	2.0	0.098	0.072	0.007	0.000135	2.8	1340	1350
P-1107	0.65	1.110	1.110	0.454	1.6	1.8	0.9	2.3	0.159	0.104	0.017	0.000454	4.2	1495	1880
P-1408	0.84	1.400	1.400	0.580	2.0	3.2	1.8	2.9	0.250	0.171	0.043	0.001455	6.7	1569	2500
P-1811	1.06	1.790	1.790	0.740	2.6	7.3	3.8	3.7	0.430	0.285	0.123	0.005682	11.0	1460	3075
P-2213	1.24	2.160	2.160	0.940	3.2	13.0	6.7	4.5	0.630	0.421	0.265	0.014941	15.7	1840	4650
P-2616	1.61	2.550	2.550	1.120	3.8	20.0	10.9	5.3	0.940	0.577	0.542	0.038281	23.1	1910	6000
P-3019	1.88	3.000	3.000	2.095	4.5	34.0	28.1	6.2	1.360	1.267	1.724	0.150373	31.8	1970	7500

MMG North America RM Cores

Part No.	Ht cm 1	Wth cm 2	Lt cm 3	G cm 4	MPL cm 5	Wtfe grams 6	Wtcu grams 7	MLT cm 8	Ac cm sq 9	Wa cm sq 10	Ap cm 4th 11	Kg cm 5th 12	At cm sq 13	Perm 14	AL 15
RM4	1.04	0.965	0.965	0.72	2.2	1.7	0.8	1.9	0.13	0.121	0.016	0.000434	5.9	1160	860
RM5	1.04	1.205	1.205	0.65	2.2	3.0	1.6	2.5	0.13	0.182	0.024	0.000483	7.9	1160	1570
RM6	1.22	1.440	1.440	0.82	2.9	5.3	2.8	3.1	0.37	0.252	0.093	0.004471	11.2	1240	2000
RM7	1.34	1.685	1.685	0.86	3.0	7.2	4.5	3.6	0.43	0.344	0.148	0.006994	14.7	1320	2370
RM8	1.64	1.930	1.930	1.10	3.8	12.0	7.4	4.2	0.64	0.492	0.315	0.019189	20.1	1365	2906
RM10	1.86	2.405	2.405	1.27	4.4	22.0	13.0	5.2	0.98	0.695	0.681	0.050985	29.5	1360	3800
RM12	2.35	2.925	2.925	1.70	5.7	45.0	24.3	6.2	1.40	1.097	1.535	0.138201	44.6	1275	4000
RM14	2.89	3.410	3.410	2.11	7.0	74.0	40.1	7.2	2.00	1.567	3.133	0.348073	62.7	1360	4880

MMG North America RU Cores

Part No.	Ht cm 1	Wth cm 2	Lt cm 3	G cm 4	MPL cm 5	Wtfe grams 6	Wtcu grams 7	MLT cm 8	Ac cm sq 9	Wa cm sq 10	Ap cm 4th 11	Kg cm 5th 12	At cm sq 13	Perm 14	AL 15
UI-12.7	0.889	1.267	2.044	0.381	3.3	1.6	3.3	3.1	0.126	0.294	0.037	0.000601	9.5	1210	580
UU-12.7	1.270	1.267	2.044	0.762	3.8	2.4	6.5	3.1	0.126	0.588	0.074	0.001201	12.0	1210	500
UU-15	2.240	1.165	2.045	1.200	4.8	9.2	7.8	3.5	0.320	0.624	0.200	0.007262	17.0	1335	1120
UU-20	3.074	1.370	2.720	1.648	6.8	19.2	15.8	4.4	0.549	1.022	0.561	0.028244	29.1	1545	1566
UU-25x20	3.886	2.100	3.300	2.264	8.6	48.8	40.5	6.1	1.050	1.868	1.961	0.134961	50.5	1620	2490
UU-30x25	5.060	2.620	4.200	2.980	11.2	97.8	80.7	7.6	1.685	2.980	5.021	0.444258	82.7	1590	3000
UU-100	11.430	6.248	15.392	6.300	30.8	952.0	1828.9	16.4	6.452	31.361	202.344	31.843050	584.9	1185	3120

MMG North America Toroidal Cores

Part No.	OD cm 1	ID cm 2	HT cm 3	Wound OD cm 4	Wound HT cm 5	MPL cm 6	Wtfe grams 7	Wtcu grams 8	MLT cm 9	Ac cm sq 10	Wa cm sq 11	Ap cm 4th 12	Kg cm 5th 13	At cm sq 14	Perm 15	AL 16
T-6.4x3.2x3.0	0.635	0.318	0.300	0.796	0.459	1.4	0.3	0.3	1.0	0.046	0.079	0.004	0.000068	2.1	1900	788
T-6.4x3.2x4.0	0.635	0.318	0.396	0.796	0.555	1.4	0.4	0.3	1.1	0.060	0.079	0.005	0.000100	2.4	1900	1041
T-9.5x4.8x3.2	0.952	0.475	0.318	1.193	0.556	2.1	0.7	0.8	1.3	0.073	0.177	0.013	0.000297	4.3	1900	840
T-12.9x7.4x5	1.270	0.735	0.500	1.634	0.868	3.0	1.9	2.7	1.8	0.131	0.424	0.056	0.001603	8.6	1900	1039
T-12.9x7.9x6.4	1.270	0.790	0.635	1.659	1.030	3.1	2.3	3.5	2.0	0.150	0.490	0.073	0.002170	9.7	1900	1146
T-12.9x7.4x6.4	1.270	0.735	0.635	1.634	1.003	3.0	2.4	3.1	2.0	0.165	0.424	0.070	0.002273	9.3	1900	1320
T-12.7x6.4x6.4	1.270	0.635	0.635	1.592	0.953	2.8	2.6	2.3	2.0	0.194	0.317	0.061	0.002345	8.7	1900	1673
T-16.8x9.7x5.0	1.670	0.965	0.500	2.148	0.983	3.9	3.3	5.6	2.1	0.172	0.731	0.126	0.004050	13.9	1900	1050
T-16.9x9.8x7.0	1.690	0.980	0.700	2.175	1.190	4.0	4.7	6.6	2.5	0.242	0.754	0.182	0.007144	15.6	1900	1450
T-22x13.7x6.4	2.210	1.370	0.635	2.885	1.320	5.4	6.9	14.6	2.8	0.262	1.473	0.386	0.014531	25.0	1900	1154
T-20x11x12.5	2.000	1.100	1.250	2.548	1.800	4.6	12.2	12.2	3.6	0.546	0.950	0.519	0.031463	24.6	1900	2840
T-22x13.7x12.7	2.210	1.370	1.270	2.885	1.955	5.4	13.7	19.9	3.8	0.523	1.473	0.771	0.042422	30.8	1900	2308
T-25x15x10	2.500	1.500	1.000	3.240	1.750	6.0	14.3	22.6	3.6	0.489	1.766	0.864	0.046927	34.3	1900	1941
T-24x12.0x12.0	2.400	1.200	1.200	3.008	1.800	5.2	17.5	15.4	3.8	0.691	1.130	0.781	0.056223	31.2	1900	3161
T-31.5x19.6x9.6	3.150	1.960	0.952	4.115	1.932	7.7	20.9	43.4	4.0	0.556	3.016	1.677	0.092229	51.5	1900	1716
T-31.5x19.6x12.5	3.150	1.960	1.250	4.115	2.230	7.7	27.4	48.5	4.5	0.730	3.016	2.201	0.142216	55.4	1900	2254
T-38.1x25.4x12.7	3.810	2.540	1.270	5.059	2.540	9.7	37.4	91.5	5.1	0.795	5.065	4.026	0.252039	80.5	1900	1957
T-38.1x25.4x15.9	3.810	2.540	1.587	5.059	2.857	9.7	46.8	100.6	5.6	0.994	5.065	5.034	0.358241	85.6	1900	2445
T-38.1x25.4x19	3.810	2.540	1.905	5.059	3.175	9.7	56.2	109.8	6.1	1.193	5.065	6.042	0.472969	90.6	1900	2935
T-63.0x38x25	6.300	3.800	2.500	8.174	4.400	15.2	225.6	364.4	9.0	3.059	11.335	34.675	4.693398	217.8	1900	6319

Philips EC Cores

Part No.	Ht cm 1	Wth cm 2	Lt cm 3	G cm 4	MPL cm 5	Wtfe grams 6	Wtcu grams 7	MLT cm 8	Ac cm sq 9	Wa cm sq 10	Ap cm 4th 11	Kg cm 5th 12	At cm sq 13	Perm 14	AL 15
F-EC35	3.46	2.27	3.43	2.44	7.8	36	36.0	6.3	0.841	1.610	1.354	0.072543	50.0	2000	2100
F-EC41	3.90	2.71	4.06	2.78	8.9	52	57.3	7.5	1.213	2.155	2.613	0.169488	67.6	2000	2700
F-EC52	4.84	3.30	5.22	3.18	10.5	111	100.2	9.0	1.800	3.116	5.610	0.446519	106.4	2000	3600
F-EC70	6.90	4.45	7.00	4.54	14.4	253	262.7	11.6	2.790	6.379	17.797	1.715154	201.9	2000	3900

Philips EE Cores

Part No.	Ht cm 1	Wth cm 2	Lt cm 3	G cm 4	MPL cm 5	Wtfe grams 6	Wtcu grams 7	MLT cm 8	Ac cm sq 9	Wa cm sq 10	Ap cm 4th 11	Kg cm 5th 12	At cm sq 13	Perm 14	AL 15
F-814E250	1.140	1.270	1.270	0.820	2.8	2.6	3.0	3.3	0.2019	0.258	0.052	0.001270	8.2	2000	1465
F-813E187	1.620	1.430	1.910	1.140	4.0	4.2	7.4	3.8	0.2252	0.547	0.123	0.002925	14.5	2000	1170
F-813E343	1.620	1.830	1.910	1.140	4.0	7.5	8.9	4.6	0.4123	0.547	0.226	0.008097	17.3	2000	2150
F-812E250	1.940	1.930	2.540	1.320	4.9	9.1	15.1	5.0	0.3948	0.851	0.336	0.010631	23.8	2000	1730
F-782E727	2.980	2.010	3.000	2.020	6.6	19.0	24.8	5.2	0.5768	1.333	0.769	0.033856	38.3	2000	1900
F-E375	2.820	2.550	3.430	1.960	6.9	29.0	37.7	6.7	0.8129	1.588	1.291	0.062901	45.2	2000	2440
F-E21	3.300	2.860	4.060	2.100	7.8	60.0	49.1	8.1	1.4903	1.701	2.535	0.186301	60.9	2000	4060
F-783E608	4.200	3.310	4.220	3.060	9.8	77.0	87.4	8.8	1.7613	2.785	4.905	0.391521	82.5	2000	3900
F-E625	3.920	3.180	4.690	2.440	8.9	108.0	66.0	9.4	2.3419	1.976	4.629	0.461680	81.4	2000	5655
F-783E776	4.200	3.790	4.220	3.060	9.8	105.0	96.9	9.8	2.3303	2.785	6.489	0.618113	90.6	2000	5000
F-E75	4.720	3.810	5.610	2.920	10.7	182.0	111.8	11.2	3.3806	2.818	9.526	1.154430	117.2	2000	6900
F-E55	5.500	4.190	5.510	3.800	12.4	220.0	164.1	11.5	3.5226	4.028	14.189	1.745130	138.5	2000	6300
F-E65	6.497	5.083	6.485	4.364	14.7	758.0	252.5	13.7	5.3800	5.193	27.937	4.395911	197.5	2300	8400

Philips EFD Cores															
Part No.	Ht cm 1	Wth cm 2	Lt cm 3	G cm 4	MPL cm 5	Wtfe grams 6	Wtcu grams 7	MLT cm 8	Ac cm sq 9	Wa cm sq 10	Ap cm 4th 11	Kg cm 5th 12	At cm sq 13	Perm 14	AL 15
EFD-15	0.75	1.5	1.5	1.10	3.4	2.8	3.0	2.7	0.15	0.314	0.047	0.001053	7.3	1800	550
EFD-20	0.99	2.0	2.0	1.54	4.7	7.0	6.8	3.8	0.31	0.501	0.155	0.005063	13.3	1800	950
EFD-25	1.28	2.5	2.5	1.86	5.7	16.5	11.5	4.8	0.58	0.679	0.394	0.019111	21.6	1800	1600
EFD-30	1.30	3.0	3.0	2.24	6.8	22.3	17.0	5.5	0.69	0.874	0.603	0.030470	28.9	1800	1800

Philips EP Cores															
Part No.	Ht cm 1	Wth cm 2	Lt cm 3	G cm 4	MPL cm 5	Wtfe grams 6	Wtcu grams 7	MLT cm 8	Ac cm sq 9	Wa cm sq 10	Ap cm 4th 11	Kg cm 5th 12	At cm sq 13	Perm 14	AL 15
F-EP7	0.845	0.74	0.92	0.52	1.6	1.4	0.7	1.8	0.108	0.107	0.012	0.000270	3.5	2000	1120
F-EP10	1.065	1.02	1.15	0.74	1.9	2.8	1.7	2.2	0.113	0.226	0.026	0.000535	5.6	2000	1025
F-EP13	1.165	1.28	1.25	0.92	2.4	5.1	2.2	2.4	0.195	0.262	0.051	0.001658	7.7	2000	1475
F-EP17	1.415	1.68	1.80	1.14	3.0	11.6	3.9	3.0	0.337	0.359	0.121	0.005406	13.7	2000	2230
F-EP20	1.885	2.14	2.40	1.44	4.1	27.7	8.3	4.2	0.787	0.554	0.436	0.032619	23.9	2000	3950

Philips ETD Cores															
Part No.	Ht cm 1	Wth cm 2	Lt cm 3	G cm 4	MPL cm 5	Wtfe grams 6	Wtcu grams 7	MLT cm 8	Ac cm sq 9	Wa cm sq 10	Ap cm 4th 11	Kg cm 5th 12	At cm sq 13	Perm 14	AL 15
F-ETD29	3.36	2.27	2.98	2.20	7.1	30	32.4	6.3	0.761	1.452	1.105	0.053556	43.8	2000	1900
F-ETD34	3.46	2.63	3.42	2.42	7.9	40	47.8	7.2	0.974	1.876	1.827	0.099378	53.2	2000	2300
F-ETD39	3.96	3.01	3.91	2.92	9.2	60	76.5	8.4	1.252	2.570	3.217	0.192463	69.5	2000	2600
F-ETD44	4.46	3.33	4.40	3.30	10.3	94	102.1	9.5	1.742	3.036	5.289	0.389685	87.3	2000	3200
F-ETD49	4.94	3.70	4.87	3.62	11.4	124	137.9	10.4	2.110	3.729	7.867	0.638330	107.2	2000	3600
F-ETD54	5.52	4.12	5.45	4.04	12.7	396	186.9	11.7	2.800	4.505	12.613	1.210584	133.7	2000	4400
F-ETD59	6.20	4.47	5.98	4.40	13.9	499	232.7	12.9	3.677	5.082	18.687	2.134739	163.1	2000	5400

Philips PC Cores															
Part No.	Ht cm 1	Wth cm 2	Lt cm 3	G cm 4	MPL cm 5	Wtfe grams 6	Wtcu grams 7	MLT cm 8	Ac cm sq 9	Wa cm sq 10	Ap cm 4th 11	Kg cm 5th 12	At cm sq 13	Perm 14	AL 15
PC-704	0.42	0.72	0.72	0.28	1.0	0.5	0.2	1.5	0.071	0.042	0.003	0.000055	1.8	2300	1150
PC-905	0.52	0.91	0.91	0.38	1.3	1.3	0.5	1.9	0.101	0.072	0.007	0.000151	2.8	2300	1230
PC-1107	0.66	1.11	1.11	0.46	1.5	1.8	0.9	2.3	0.168	0.106	0.018	0.000513	4.2	2300	1800
PC-1408	0.84	1.40	1.40	0.58	2.0	3.2	1.8	2.9	0.252	0.171	0.043	0.001479	6.8	2300	2240
PC-1408T	0.84	1.40	1.40	0.58	2.1	3.2	1.8	3.0	0.293	0.171	0.050	0.001947	6.8	2300	2000
PC-1811T	1.06	1.79	1.79	0.74	2.7	7.3	3.7	3.7	0.406	0.285	0.116	0.005087	11.0	2000	2505
PC-1811	1.06	1.79	1.79	0.74	2.6	6.4	3.7	3.7	0.432	0.285	0.123	0.005759	11.0	2300	3680
PC-2311T	1.10	2.29	2.29	0.76	2.9	11.5	5.3	4.6	0.610	0.327	0.199	0.010676	16.1	2000	4115
PC-2213	1.34	2.15	2.15	0.94	3.1	12.0	6.7	4.5	0.635	0.423	0.269	0.015291	16.3	2300	4650
PC-2318T	1.80	2.29	2.29	1.44	4.2	16.5	10.0	4.6	0.622	0.619	0.385	0.021032	21.2	2000	3000
PC-2616T	1.62	2.55	2.55	1.12	4.0	20.0	10.9	5.3	0.884	0.577	0.510	0.033856	23.2	2000	3500
PC-2616	1.62	2.55	2.55	1.12	3.8	20.0	10.9	5.3	0.955	0.577	0.551	0.039513	23.2	2300	6000
PC-3019T	1.88	3.00	3.00	1.32	4.7	31.0	17.7	6.2	1.208	0.799	0.965	0.074751	31.8	2000	4350
PC-3019	1.88	3.00	3.00	1.32	4.5	34.0	17.7	6.2	1.381	0.799	1.103	0.097694	31.8	2300	7580
PC-3622	2.16	3.55	3.55	1.44	5.3	54.0	29.3	7.6	2.019	1.087	2.195	0.234161	43.9	2300	9660
PC-4229	2.96	4.24	4.24	2.06	6.8	104.0	60.0	8.7	2.665	1.947	5.188	0.637907	67.6	2300	10300
PC-6656	5.58	6.63	6.63	4.24	12.3	50.4	262.2	13.2	7.161	5.576	39.927	8.649361	185.2	2700	18200

Philips RM Cores

Part No.	Ht cm 1	Wth cm 2	Lt cm 3	G cm 4	MPL cm 5	Wtfe grams 6	Wtcu grams 7	MLT cm 8	Ac cm sq 9	Wa cm sq 10	Ap cm 4th 11	Kg cm 5th 12	At cm sq 13	Perm 14	AL 15
RM5	1.04	1.44	1.44	0.66	2.1	3.0	1.7	2.5	0.213	0.185	0.039	0.001317	10.1	2300	1960
RM5I	1.04	1.44	1.44	0.66	2.3	3.2	1.7	2.5	0.252	0.185	0.047	0.001843	10.1	2000	1800
RM6	1.24	1.76	1.76	0.82	2.6	4.5	2.5	3.0	0.320	0.230	0.073	0.003117	14.9	2300	2700
RM6SI	1.24	1.76	1.76	0.82	2.9	5.0	2.9	3.1	0.368	0.258	0.095	0.004474	14.9	2000	2350
RM8	1.64	2.27	2.27	1.10	3.6	12.5	7.3	4.2	0.523	0.490	0.256	0.012767	25.2	2300	3400
RM8I	1.64	2.27	2.27	1.10	3.8	13.5	7.3	4.2	0.632	0.490	0.309	0.018643	25.2	2000	3250
RM10	1.86	2.79	2.79	1.26	4.2	19.6	13.0	5.2	0.839	0.699	0.587	0.037639	36.3	2300	4765
RM10I	1.86	2.79	2.79	1.26	4.5	23.0	13.0	5.2	0.968	0.699	0.677	0.050104	36.3	2000	4400
RM12I	2.44	3.68	3.68	1.72	5.7	45.0	24.5	6.2	1.458	1.109	1.618	0.151652	63.0	2000	5500
RM14I	3.02	4.15	4.15	2.10	7.0	74.0	40.0	7.2	2.000	1.565	3.129	0.347969	84.6	2000	6250

Philips Toroidal Cores

Part No.	OD cm 1	ID cm 2	HT cm 3	Wound OD cm 4	Wound HT cm 5	MPL cm 6	Wtfe grams 7	Wtcu grams 8	MLT cm 9	Ac cm sq 10	Wa cm sq 11	Ap cm 4th 12	Kg cm 5th 13	At cm sq 14	Perm 15	AL 16
F-135T050	0.343	0.178	0.127	0.432	0.216	0.8	0.0	0.0	0.5	0.0096	0.025	0.000	0.000002	0.6	2300	375
F-213T050	0.483	0.229	0.127	0.600	0.242	1.0	0.0	0.1	0.6	0.0100	0.041	0.000	0.000003	1.0	2300	430
F-891T050	0.394	0.224	0.127	0.505	0.239	0.9	0.0	0.1	0.5	0.0100	0.039	0.000	0.000003	0.8	2300	325
F-1041T060	0.584	0.305	0.152	0.737	0.305	1.3	0.1	0.2	0.7	0.0200	0.073	0.001	0.000016	1.6	2300	450
F-266T125	0.953	0.475	0.318	1.194	0.556	2.1	0.7	0.8	1.3	0.0716	0.177	0.013	0.000286	4.3	2300	1000
F-768T188	1.270	0.714	0.478	1.624	0.835	2.9	1.8	2.5	1.8	0.1284	0.400	0.051	0.001482	8.4	2000	1000
F-204T250	1.270	0.792	0.635	1.660	1.031	3.1	2.2	3.6	2.0	0.1476	0.492	0.073	0.002112	9.7	2000	1200
F-331T185	1.588	0.907	0.470	2.037	0.924	3.7	2.8	4.6	2.0	0.1546	0.646	0.100	0.003053	12.4	2000	1050
F-846T250	2.210	1.372	0.630	2.886	1.316	5.4	6.8	14.6	2.8	0.2590	1.478	0.383	0.014283	25.0	2000	1220
F-846T500	2.210	1.372	1.270	2.886	1.956	5.4	13.7	20.0	3.8	0.5201	1.478	0.769	0.042075	30.8	2000	2440
F-502T300	2.901	1.900	0.762	3.835	1.712	7.3	13.3	35.7	3.5	0.3735	2.834	1.058	0.044670	43.7	2000	1290
F-500T400	3.599	2.301	1.016	4.731	2.167	9.0	28.1	66.6	4.5	0.6452	4.156	2.682	0.153630	67.3	2000	1820
F-500T600	3.599	2.301	1.524	4.731	2.675	9.0	42.2	78.6	5.3	0.9693	4.156	4.029	0.293740	74.9	2000	2720
F-528T500	3.886	1.956	1.270	4.875	2.248	8.5	48.4	54.9	5.1	1.1750	3.003	3.529	0.322636	71.7	2000	3480
F-400T750	5.080	3.175	1.905	6.643	3.493	12.5	107.5	200.1	7.1	1.7740	7.913	14.038	1.400660	142.1	2000	3555
F-144T500	7.366	3.886	1.270	9.313	3.213	16.5	170.8	334.1	7.9	2.1290	11.854	25.238	2.712054	230.1	2000	3220

Philips UU Cores

Part No.	Ht cm 1	Wth cm 2	Lt cm 3	G cm 4	MPL cm 5	Wtfe grams 6	Wtcu grams 7	MLT cm 8	Ac cm sq 9	Wa cm sq 10	Ap cm 4th 11	Kg cm 5th 12	At cm sq 13	Perm 14	AL 15
F-376UI	2.23	1.91	3.81	0.95	6.4	11	21.2	5.0	0.403	1.207	0.486	0.015828	33.6	2700	1750
F-376UU	3.18	1.91	3.81	1.90	8.4	16	42.5	5.0	0.403	2.413	0.972	0.031656	44.4	2700	1400
F-1F30	4.44	2.37	4.76	2.54	11.0	45	83.1	6.4	0.865	3.632	3.142	0.168983	75.0	2700	2300
F-1F30	4.12	3.07	6.06	2.22	11.3	53	119.3	8.0	1.039	4.174	4.336	0.224165	98.2	2000	1680
F-1F19	5.90	5.08	10.10	3.62	18.5	116	565.7	11.9	1.465	13.394	19.622	0.968142	248.0	2000	1520
F-1F10	5.40	5.31	10.61	2.54	17.3	160	434.9	12.4	2.039	9.855	20.095	1.320694	240.2	2700	3800
F-U64	8.10	4.80	6.80	5.30	21.0	306	505.5	11.2	2.903	12.720	36.926	3.836673	226.3	2000	2500
F-1F5UI	8.26	7.62	15.24	3.18	24.5	815	1076.7	18.7	6.452	16.154	104.228	14.351150	506.7	2000	4650
F-1F4UI	10.35	6.62	12.92	4.80	25.8	1121	1110.4	18.0	8.368	17.376	145.402	27.081410	495.2	2000	5325
F-1F5UU	11.44	7.62	15.24	6.36	30.7	1010	2153.5	18.7	6.452	32.309	208.456	28.702290	652.1	2000	3750
F-1F4UU	15.20	6.62	12.92	9.60	35.4	1512	2220.9	18.0	8.394	34.752	291.708	54.499920	684.0	2000	3900

Siemens EC Cores

Part No.	Ht cm 1	Wth cm 2	Lt cm 3	G cm 4	MPL cm 5	Wtfe grams 6	Wtcu grams 7	MLT cm 8	Ac cm sq 9	Wa cm sq 10	Ap cm 4th 11	Kg cm 5th 12	At cm sq 13	Perm 14	AL 15
B66337-EC35	3.49	2.275	3.45	2.38	7.7	369	34.9	6.4	0.843	1.541	1.299	0.068865	50.6	2000	2100
B66339-EC41	3.93	2.705	4.06	2.70	8.9	52	54.9	7.5	1.210	2.045	2.475	0.158719	67.9	2000	2700
B66341-EC52	4.87	3.300	5.22	3.10	10.5	110	96.9	9.1	1.800	2.984	5.371	0.423531	106.8	2000	3400
B66343-EC70	6.93	4.450	7.00	2.46	14.4	252	141.5	11.7	2.790	3.407	9.506	0.908502	202.4	2000	3900

Siemens EF Cores															
Part No.	Ht cm 1	Wth cm 2	Lt cm 3	G cm 4	MPL cm 5	Wtfe grams 6	Wtcu grams 7	MLT cm 8	Ac cm sq 9	Wa cm sq 10	Ap cm 4th 11	Kg cm 5th 12	At cm sq 13	Perm 14	AL 15
B66219	1.56	1.05	1.43	1.04	3.4	2.8	3.6	3.1	0.155	0.322	0.050	0.000999	9.7	2000	860
B66393	1.16	1.13	1.60	0.72	2.9	3.0	2.8	3.3	0.192	0.238	0.046	0.001054	8.7	2000	1100
B66307	1.64	1.13	1.60	1.14	3.8	3.6	4.4	3.3	0.201	0.376	0.076	0.001829	11.3	2100	990
B66379	1.60	1.43	1.90	1.14	4.0	4.4	7.4	3.8	0.225	0.542	0.122	0.002871	14.3	2100	1100
B66314	1.74	1.60	2.10	1.20	4.3	4.8	9.7	4.1	0.216	0.660	0.143	0.002979	17.4	2100	950
B66312	1.87	1.41	2.00	1.23	4.3	6.8	7.3	4.1	0.318	0.504	0.160	0.005030	16.5	2000	1300
B66311	2.04	1.41	2.00	1.40	4.5	7.3	8.3	4.1	0.318	0.574	0.183	0.005725	17.6	2000	1490
B66315	1.96	1.88	2.54	1.30	4.9	9.6	14.0	4.9	0.388	0.797	0.309	0.009718	23.6	2100	1600
B66317	2.56	1.75	2.50	1.74	5.8	16.0	15.4	5.0	0.525	0.870	0.457	0.019268	27.5	2100	1800
B66319	1.46	1.96	3.00	1.94	6.7	22.0	22.2	5.2	0.600	1.193	0.716	0.032793	22.7	2100	1850
B66403	2.50	2.31	2.80	1.73	5.9	25.0	23.2	6.1	0.830	1.073	0.890	0.048654	34.6	2100	3100
B66229	1.90	2.27	3.20	2.24	7.4	30.0	33.0	6.3	0.830	1.478	1.227	0.064867	30.3	2100	2250
B66389	2.30	2.58	3.60	2.40	8.1	50.0	42.7	7.0	1.200	1.716	2.059	0.141342	41.2	2100	3000
B66381	2.50	2.86	4.06	2.10	7.7	58.0	48.9	8.1	1.490	1.691	2.519	0.184524	49.8	2100	4000
B66325	3.04	3.25	4.20	2.96	9.7	88.0	80.2	8.8	1.780	2.560	4.558	0.368572	64.2	2100	3800
B66383	3.12	3.18	4.69	2.44	8.9	106.0	66.0	9.4	2.330	1.976	4.605	0.457000	68.8	2100	5400
B66329	4.00	3.73	4.20	2.96	9.7	116.0	88.9	9.8	2.400	2.560	6.145	0.604169	86.3	2100	5100
B66385	3.76	3.81	5.61	2.92	10.7	184.0	111.8	11.2	3.400	2.818	9.581	1.167718	99.1	2100	6700
B66335	4.20	4.13	5.50	3.70	12.0	215.0	152.7	11.4	3.540	3.756	13.294	1.646245	112.4	2000	5800
B66344	5.00	4.53	5.50	3.70	12.0	256.0	163.4	12.2	4.200	3.756	15.773	2.165802	136.2	2000	6800
B66387	5.48	5.16	6.50	4.44	14.7	394.0	265.3	13.9	5.350	5.372	28.742	4.429095	175.5	2000	7200
B66375	4.04	5.89	8.00	5.58	18.4	358.0	566.9	14.8	3.900	10.797	42.109	4.449419	175.2	2000	4150
B66371	6.40	5.80	7.05	4.38	14.9	514.0	313.6	15.5	6.830	5.694	38.890	6.859105	223.4	2000	8850

Siemens EFD Cores															
Part No.	Ht cm 1	Wth cm 2	Lt cm 3	G cm 4	MPL cm 5	Wtfe grams 6	Wtcu grams 7	MLT cm 8	Ac cm sq 9	Wa cm sq 10	Ap cm 4th 11	Kg cm 5th 12	At cm sq 13	Perm 14	AL 15
B66411-EFD10	0.425	1.04	1.05	0.75	2.3	0.8	0.8	1.8	0.072	0.116	0.008	0.000132	3.3	2200	500
B66417-EFD20	0.990	2.00	2.00	1.54	4.7	7.2	6.8	3.8	0.310	0.501	0.155	0.005063	13.3	2200	1200
B66421-EFD25	1.275	2.50	2.50	1.86	5.7	16.6	11.5	4.8	0.580	0.679	0.394	0.019111	21.6	2200	2000
B66423-EFD30	1.300	3.00	3.00	2.24	6.8	24.0	17.0	5.5	0.690	0.874	0.603	0.030470	28.9	2200	2050
B66413-EFD15	0.750	1.50	1.50	1.10	3.4	2.8	3.0	2.7	15.000	0.314	4.703	10.527990	7.3	2200	780

Siemens EP Cores															
Part No.	Ht cm 1	Wth cm 2	Lt cm 3	G cm 4	MPL cm 5	Wtfe grams 6	Wtcu grams 7	MLT cm 8	Ac cm sq 9	Wa cm sq 10	Ap cm 4th 11	Kg cm 5th 12	At cm sq 13	Perm 14	AL 15
B65839-EP7	0.840	0.75	0.94	0.50	1.6	1.4	0.6	1.9	0.103	0.095	0.010	0.000212	3.6	2100	1100
B65841-EP10	1.073	1.04	1.18	0.72	1.9	2.8	1.6	2.1	0.113	0.207	0.023	0.000493	5.9	2100	1100
B65843-EP13	1.160	1.30	1.28	0.90	24.2	4.5	2.0	2.4	0.195	0.234	0.046	0.001489	8.0	2100	1580
B65845EP17	1.408	1.70	1.85	1.12	28.5	12.0	3.3	3.0	0.339	0.316	0.107	0.004909	14.3	2100	2400
B65847-EP20	1.885	2.16	2.45	1.40	4.0	27.5	7.4	4.2	0.780	0.497	0.388	0.028940	24.7	2100	4000

Siemens ET Cores														
Part No.	Ht cm 1	Wth cm 2	Lt cm 3	G cm 4	MPL cm 5	Wtfe grams 6	Wtcu grams 7	MLT cm 8	Ac cm sq 9	Wa cm sq 10	Ap cm 4th 11	Kg cm 5th 12	At cm sq 13	Perm 14	AL 15
B66358	3.20	2.20	3.06	2.14	7.1	28	29.0	6.2	0.8	1.305	0.992	0.048307	42.3	2200	2200
B66361	3.50	2.56	3.40	2.36	7.9	40	43.3	7.1	1.0	1.711	1.661	0.090572	52.7	2200	2600
B66363	4.00	2.93	3.89	2.84	9.2	60	69.3	8.3	1.3	2.343	2.929	0.176038	68.9	2200	2700
B66365	4.50	3.25	4.38	3.22	10.3	94	93.1	9.4	1.7	2.785	4.819	0.354574	86.6	2200	3500
B66367	4.98	3.61	4.85	3.54	11.4	124	126.2	10.3	2.1	3.434	7.245	0.591752	106.3	2200	3800
B66395	5.56	4.01	5.45	3.96	12.7	180	169.8	11.6	2.8	4.118	11.532	1.114003	132.8	2200	4450
B66397	6.24	4.36	5.98	4.40	13.9	260	215.7	12.8	3.7	4.730	17.406	1.998068	162.2	2200	5300

Siemens PC Cores															
Part No.	Ht cm 1	Wth cm 2	Lt cm 3	G cm 4	MPL cm 5	Wtfe grams 6	Wtcu grams 7	MLT cm 8	Ac cm sq 9	Wa cm sq 10	Ap cm 4th 11	Kg cm 5th 12	At cm sq 13	Perm 14	AL 15
B65491	0.26	0.335	0.335	0.17	0.5	0.1	0.0	0.9	0.0137	0.011	0.000	0.000001	0.4	4300	500
B65495	0.41	0.465	0.465	0.27	0.8	0.2	0.1	1.1	0.0280	0.020	0.001	0.000006	0.9	4300	800
B65501	0.33	0.580	0.580	0.22	0.8	0.2	0.1	1.3	0.0470	0.022	0.001	0.000015	1.1	4300	1500
B65511	0.42	0.735	0.735	0.28	1.0	0.5	0.2	1.5	0.0700	0.039	0.003	0.000050	1.8	4300	2000
B65517	0.54	0.930	0.930	0.36	1.2	1.0	0.4	1.9	0.0980	0.065	0.006	0.000128	2.9	2100	1500
B65531	0.66	1.130	1.130	0.44	1.6	1.8	0.8	2.3	0.1590	0.095	0.015	0.000414	4.3	2100	2000
B65541	0.85	1.430	1.430	0.56	2.0	3.5	1.6	2.9	0.2500	0.157	0.039	0.001341	7.0	2100	2800
B65651	1.06	1.840	1.840	0.73	2.6	6.6	3.5	3.7	0.4300	0.266	0.115	0.005337	11.4	2100	3600
B65661	1.36	2.200	2.200	0.92	3.2	14.0	6.2	4.4	0.6300	0.391	0.246	0.013961	17.0	2100	4400
B65671	1.63	2.600	2.600	1.10	3.7	23.0	10.0	5.3	0.9300	0.534	0.496	0.034864	23.9	2100	5500
B65701	1.90	3.050	3.050	1.30	4.5	37.0	16.5	6.2	1.3600	0.748	1.017	0.089132	32.8	2100	6400
B65611	2.20	3.600	3.600	1.46	5.2	59.5	26.6	7.5	2.0200	1.000	2.020	0.218332	45.2	2100	8000
B65621	2.50	4.100	4.100	1.70	6.2	82.0	41.5	8.3	2.4200	1.403	3.394	0.394688	58.6	2100	8400

Siemens PM Cores

Part No.	Ht cm 1	Wth cm 2	Lt cm 3	G cm 4	MPL cm 5	Wtfe grams 6	Wtcu grams 7	MLT cm 8	Ac cm sq 9	Wa cm sq 10	Ap cm 4th 11	Kg cm 5th 12	At cm sq 13	Perm 14	AL 15
B65646-PM50/39	3.9	5.0	5.0	2.64	8.4	140	84.7	9.5	3.7	2.508	9.280	1.445414	128.0	2000	7400
B65684-PM62/49	4.9	6.2	6.2	3.34	10.9	280	164.7	11.9	5.7	3.891	22.179	4.248139	198.4	2000	9200
B65686-PM74/59	5.9	7.4	7.4	4.07	12.8	460	281.6	13.9	7.9	5.698	45.014	10.235180	284.2	2000	10000
B65713-PM87/70	7.0	8.7	8.7	4.80	14.6	770	475.8	15.8	9.1	8.496	77.314	17.867760	395.0	2000	12000
B65733-PM114/93	9.3	11.4	11.4	6.30	20.0	1940	1052.9	20.9	17.2	14.175	243.810	80.307860	684.0	2000	16000

Siemens RM Cores

Part No.	Ht cm 1	Wth cm 2	Lt cm 3	G cm 4	MPL cm 5	Wtfe grams 6	Wtcu grams 7	MLT cm 8	Ac cm sq 9	Wa cm sq 10	Ap cm 4th 11	Kg cm 5th 12	At cm sq 13	Perm 14	AL 15
B65817-RM3	0.75	0.75	0.75	0.44	1.5	0.6	0.4	1.5	0.084	0.066	0.006	0.000121	3.4	2300	800
B65805-RM5	1.05	1.23	1.23	0.63	2.1	3.0	1.5	2.5	0.208	0.167	0.035	0.001142	8.2	2100	2000
B65807-RM6	1.25	1.47	1.47	0.80	2.7	5.3	2.7	3.1	0.313	0.240	0.075	0.003022	11.7	2100	2400
B65809-R6	1.25	1.47	1.47	0.80	2.6	4.0	2.7	3.1	0.320	0.240	0.077	0.003159	11.7	2100	3000
B65819-RM7	1.35	1.72	1.72	0.84	3.0	7.2	4.0	3.6	0.400	0.315	0.126	0.005578	15.2	2100	2700
B65811-RM8	1.65	1.97	1.97	1.08	3.5	12.0	6.8	4.2	0.520	0.456	0.237	0.011831	20.8	2100	3300
B65813-RM10	1.87	2.47	2.47	1.24	4.2	22.0	11.8	5.2	0.830	0.639	0.530	0.033842	30.7	2100	4200
B65815-RM12	2.46	2.98	2.98	1.68	5.7	45.0	22.3	6.2	1.460	1.016	1.484	0.140742	47.1	2100	5300
B65887-RM14	3.02	3.48	3.48	2.08	7.0	74.0	37.0	7.1	2.000	1.456	2.912	0.325971	66.3	2100	6000
B65803-RM4	1.05	0.98	0.98	0.70	2.1	1.7	1.0	2.0	11.000	0.142	1.559	3.395394	6.0	2100	1100

Siemens RU Cores															
Part No.	Ht cm 1	Wth cm 2	Lt cm 3	G cm 4	MPL cm 5	Wtfe grams 6	Wtcu grams 7	MLT cm 8	Ac cm sq 9	Wa cm sq 10	Ap cm 4th 11	Kg cm 5th 12	At cm sq 13	Perm 14	AL 15
B67354	3.60	2.70	4.00	2.30	9.5	44.0	66.0	7.3	0.94	2.530	2.378	0.121909	64.1	2000	2000
B67359	6.16	2.62	4.87	4.10	14.1	86.0	146.3	7.4	1.17	5.535	6.476	0.407769	106.6	2100	1950
B67313	6.34	2.73	5.23	4.20	14.8	97.5	167.1	7.8	1.37	6.006	8.228	0.576378	117.2	2100	2000
B67317	7.14	3.04	5.47	4.86	16.4	125.0	220.7	8.5	1.43	7.290	10.425	0.700450	141.1	2100	2050
B67312	6.38	3.76	5.57	4.78	15.8	137.0	262.7	9.9	1.77	7.457	13.199	0.943209	151.0	2000	2450
B67368	6.80	3.23	5.96	4.80	16.3	140.0	253.1	9.0	1.66	7.872	13.068	0.959529	150.4	2100	2350
B67314	7.33	3.40	6.50	4.91	17.6	145.5	308.2	9.3	1.80	9.329	16.792	1.301301	170.9	2100	2300

Siemens SU Cores															
Part No.	Ht cm 1	Wth cm 2	Lt cm 3	G cm 4	MPL cm 5	Wtfe grams 6	Wtcu grams 7	MLT cm 8	Ac cm sq 9	Wa cm sq 10	Ap cm 4th 11	Kg cm 5th 12	At cm sq 13	Perm 14	AL 15
B67366	1.860	1.03	1.45	1.14	4.1	4.2	4.9	2.9	0.197	0.467	0.092	0.002483	11.0	2000	800
B67350	2.240	1.19	2.04	1.18	4.8	8.6	7.8	3.6	0.320	0.614	0.196	0.007051	17.1	2000	1200
B67364	2.340	1.33	2.32	1.14	5.3	9.6	10.1	3.8	0.320	0.752	0.241	0.008144	20.1	2000	1300
B67348	3.180	1.40	2.71	1.60	6.8	18.0	15.7	4.4	0.550	1.008	0.554	0.027802	29.7	2000	1600
B67318	3.400	2.10	3.00	2.20	8.1	27.0	38.2	5.4	0.665	1.980	1.317	0.064608	43.0	2000	1650
B67352	4.000	2.12	3.30	2.20	8.6	46.0	39.5	6.2	1.050	1.804	1.894	0.129247	51.4	2000	2500
B67355	4.540	2.50	3.48	2.60	9.8	65.0	57.4	6.9	1.310	2.340	3.065	0.232759	64.7	2000	2500
B67362	5.280	3.66	4.12	3.20	11.8	146.0	112.7	9.5	2.650	3.328	8.819	0.981887	103.7	2000	4400
B67345-3	15.200	5.06	12.76	9.60	35.4	800.0	1781.2	15.1	4.480	33.216	148.808	17.683010	593.4	2000	2900
B67345-10	15.200	5.46	12.76	9.60	35.4	1000.0	1875.7	15.9	5.600	33.216	186.010	26.237800	615.7	2000	3600
B67345-1	15.200	6.46	12.76	9.60	35.4	1500.0	2111.9	17.9	8.400	33.216	279.014	52.431650	671.6	2000	5400

Siemens Toroidal Cores

Part No.	OD cm 1	ID cm 2	HT cm 3	Wound OD.cm 4	Wound HT cm 5	MPL cm 6	Wtfe grams 7	Wtcu grams 8	MLT cm 9	Ac cm sq 10	Wa cm sq 11	Ap cm 4th 12	Kg cm 5th 13	At cm sq 14	Perm 15	AL 16
A-35	0.250	0.150	0.100	0.324	0.175	0.6	0.0	0.0	0.4	0.0049	0.018	0.000	0.000000	0.3	4300	440
A-61	0.394	0.224	0.130	0.505	0.242	0.9	0.0	0.1	0.5	0.0108	0.039	0.000	0.000004	0.8	4300	630
A-36	0.400	0.240	0.160	0.518	0.280	0.9	0.1	0.1	0.6	0.0125	0.045	0.001	0.000005	0.9	4300	700
A-56	0.584	0.305	0.152	0.737	0.305	1.3	0.1	0.2	0.7	0.0205	0.073	0.001	0.000017	1.6	4300	850
A-37	0.630	0.380	0.250	0.817	0.440	1.5	0.2	0.4	0.9	0.0306	0.113	0.003	0.000047	2.2	4300	1090
A-62	0.953	0.475	0.317	1.194	0.555	2.1	0.7	0.8	1.3	0.0728	0.177	0.013	0.000296	4.3	4300	1900
A-38	1.000	0.600	0.400	1.296	0.700	2.4	0.9	1.4	1.4	0.0783	0.283	0.022	0.000481	5.5	4300	1760
A-44	1.250	0.750	0.500	1.620	0.875	3.0	1.8	2.8	1.8	0.1223	0.442	0.054	0.001468	8.6	2100	1070
A-644	1.330	0.830	0.500	1.739	0.915	3.3	1.9	3.6	1.9	0.1227	0.541	0.066	0.001747	9.7	4300	2020
A-658	1.400	0.900	0.500	1.843	0.950	3.5	2.1	4.3	1.9	0.1230	0.636	0.078	0.002004	10.8	4300	1900
A-623	1.500	1.040	0.530	2.012	1.050	3.9	2.3	6.2	2.0	0.1205	0.849	0.102	0.002408	13.0	4300	1670
A-45	1.600	0.960	0.630	2.074	1.110	3.9	3.7	5.9	2.3	0.1973	0.723	0.143	0.004923	14.0	2100	1350
A-652	1.700	1.070	0.680	2.227	1.215	4.2	4.3	7.8	2.4	0.2104	0.899	0.189	0.006501	16.3	750	470
A-632	2.000	1.000	0.700	2.506	1.200	4.4	7.1	7.6	2.7	0.3363	0.785	0.264	0.013056	19.3	2100	2030
A-638	2.210	1.370	0.635	2.885	1.320	5.4	6.9	14.6	2.8	0.2617	1.473	0.386	0.014498	25.0	2100	1270
S-641	2.300	1.450	0.780	3.014	1.505	5.7	8.3	18.1	3.1	0.2994	1.650	0.494	0.019164	28.5	4300	2840
A-626	2.300	1.450	0.940	3.014	1.665	5.7	9.9	19.6	3.3	0.3601	1.650	0.594	0.025600	30.0	4300	3420
A-618	2.530	1.480	1.000	3.262	1.740	6.0	14.9	22.2	3.6	0.5126	1.719	0.881	0.049868	34.5	2100	2260
A-657	3.050	2.000	1.250	4.034	2.250	7.7	24.2	49.6	4.4	0.6466	3.140	2.030	0.118271	54.0	2000	2110
A-58	3.400	2.000	1.000	4.388	2.000	8.2	26.3	48.2	4.3	0.6608	3.140	2.075	0.126954	57.8	4300	4360
A-48	3.400	2.050	1.250	4.411	2.275	8.2	32.9	55.4	4.7	0.8260	3.299	2.725	0.190746	62.1	4300	5460
A-674	3.600	2.300	1.500	4.731	2.650	9.0	41.7	78.0	5.3	0.9589	4.153	3.982	0.289266	74.5	2100	2810
A-22	4.180	2.620	1.250	5.470	2.560	10.3	47.8	102.4	5.3	0.9575	5.389	5.160	0.369780	90.9	4300	5000
A-659	4.000	2.400	1.600	5.184	2.800	9.6	58.5	92.6	5.8	1.2530	4.522	5.666	0.492983	87.8	4300	7000
A-82	5.000	3.000	2.000	6.480	3.500	12.0	114.3	180.9	7.2	1.9570	7.065	13.826	1.503216	137.1	4300	8700
A-40	5.830	4.080	1.760	7.840	3.800	15.2	112.6	347.6	7.5	1.5240	13.067	19.915	1.622999	190.0	4300	5400
A-84	10.200	6.580	1.500	13.436	4.790	25.5	330.8	1276.3	10.6	2.6720	33.988	90.815	9.191586	485.5	4300	5500

Steward Toroidal Cores																
Part No.	OD cm 1	ID cm 2	HT cm 3	Wound OD cm 4	Wound HT cm 5	MPL cm 6	Wtfe grams 7	Wtcu grams 8	MLT cm 9	Ac cm sq 10	Wa cm sq 11	Ap cm 4th 12	Kg cm 5th 13	At cm sq 14	Perm 15	AL 16
T-0250-000	0.630	0.380	0.190	0.817	0.380	1.5	0.2	0.3	0.8	0.0237	0.113	0.003	0.000031	2.0	5000	974
T-0255-000	0.650	0.250	0.250	0.788	0.375	1.2	0.3	0.2	0.9	0.0460	0.049	0.002	0.000045	1.9	5000	2380
T-0300-100	0.762	0.175	0.318	0.893	0.406	1.1	0.4	0.1	1.1	0.0780	0.024	0.002	0.000052	2.4	5000	4665
T-0250-100	0.630	0.380	0.318	0.817	0.508	1.5	0.3	0.4	1.0	0.0393	0.113	0.004	0.000069	2.4	5000	1622
T-0255-100	0.650	0.250	0.400	0.788	0.525	1.2	0.4	0.2	1.2	0.0737	0.049	0.004	0.000092	2.3	5000	3813
T-0301-100	0.762	0.318	0.318	0.932	0.477	1.5	0.5	0.3	1.1	0.0662	0.079	0.005	0.000124	2.8	5000	2780
T-0302-000	0.762	0.280	0.478	0.920	0.618	1.4	0.7	0.3	1.4	0.1060	0.062	0.007	0.000201	3.1	5000	4791
T-0295-000	0.750	0.239	0.610	0.895	0.730	1.3	0.9	0.3	1.6	0.1397	0.045	0.006	0.000222	3.3	5000	6972
T-0301-000	0.762	0.318	0.478	0.932	0.637	1.5	0.7	0.4	1.4	0.0996	0.079	0.008	0.000229	3.2	5000	4180
T-0315-000	0.800	0.318	0.478	0.973	0.637	1.5	0.8	0.4	1.4	0.1074	0.079	0.009	0.000261	3.4	5000	4413
T-0354-000	0.900	0.600	0.400	1.195	0.700	2.3	0.7	1.4	1.4	0.0590	0.283	0.017	0.000289	4.9	5000	1617
T-0375-100	0.953	0.475	0.318	1.194	0.556	2.1	0.7	0.8	1.3	0.0728	0.177	0.013	0.000296	4.3	5000	2209
T-0325-000	0.825	0.445	0.478	1.047	0.701	1.9	0.8	0.8	1.4	0.0881	0.155	0.014	0.000339	4.0	5000	2957
T-0315-100	0.800	0.318	0.600	0.973	0.759	1.5	1.0	0.5	1.6	0.1348	0.079	0.011	0.000361	3.8	5000	5541
T-0433-000	1.100	0.500	0.300	1.359	0.550	2.3	0.9	0.9	1.4	0.0853	0.196	0.017	0.000420	5.2	5000	2361
T-0395-100	1.000	0.600	0.400	1.296	0.700	2.4	0.9	1.4	1.4	0.0792	0.283	0.022	0.000493	5.5	5000	2072
T-0394-200	1.000	0.500	0.400	1.253	0.650	2.2	1.0	1.0	1.4	0.0959	0.196	0.019	0.000501	5.0	5000	2765
T-0377-000	0.953	0.560	0.478	1.230	0.758	2.3	1.0	1.3	1.5	0.0918	0.246	0.023	0.000543	5.3	5000	2546
T-0375-300	0.953	0.475	0.478	1.194	0.716	2.1	1.1	1.0	1.5	0.1095	0.177	0.019	0.000556	4.9	5000	3322
T-0376-100	0.953	0.508	0.483	1.207	0.737	2.2	1.1	1.1	1.5	0.1038	0.203	0.021	0.000569	5.1	5000	3033
T-0354-100	0.900	0.600	0.650	1.195	0.950	2.3	1.1	1.8	1.8	0.0961	0.283	0.027	0.000593	5.8	5000	2637
T-0433-100	1.100	0.500	0.400	1.359	0.650	2.3	1.3	1.1	1.5	0.1135	0.196	0.022	0.000665	5.7	5000	3141
T-0394-000	1.050	0.500	0.500	1.306	0.750	2.2	1.3	1.1	1.6	0.1203	0.196	0.024	0.000693	5.8	5000	3467
T-0375-000	0.953	0.475	0.635	1.194	0.873	2.1	1.5	1.1	1.8	0.1456	0.177	0.026	0.000845	5.5	5000	4417
T-0376-000	0.953	0.508	0.635	1.207	0.889	2.2	1.4	1.3	1.8	0.1366	0.203	0.028	0.000850	5.7	5000	3991
T-0395-000	1.000	0.600	0.600	1.296	0.900	2.4	1.4	1.8	1.8	0.1186	0.283	0.034	0.000903	6.3	5000	3100
T-0472-000	1.200	0.600	0.400	1.504	0.700	2.6	1.5	1.6	1.6	0.1148	0.283	0.032	0.000931	6.9	5000	2765
T-0377-100	0.953	0.560	0.711	1.230	0.991	2.3	1.5	1.7	1.9	0.1367	0.246	0.034	0.000968	6.2	5000	3793
T-0520-000	1.321	0.737	0.395	1.687	0.764	3.1	1.7	2.6	1.7	0.1125	0.426	0.048	0.001278	8.5	5000	2313
T-0502-100	1.270	0.516	0.470	1.548	0.728	2.5	2.0	1.3	1.8	0.1657	0.209	0.035	0.001298	7.3	5000	4235
T-0501-000	1.270	0.714	0.478	1.624	0.835	3.0	1.8	2.5	1.8	0.1292	0.400	0.052	0.001501	8.4	5000	2752
T-0551-000	1.400	0.900	0.500	1.843	0.950	3.5	2.1	4.3	1.9	0.1232	0.636	0.078	0.002011	10.8	5000	2213

Steward Toroidal Cores																
Part No.	OD cm 1	ID cm 2	HT cm 3	Wound OD cm 4	Wound HT cm 5	MPL cm 6	Wtfe grams 7	Wtcu grams 8	MLT cm 9	Ac cm sq 10	Wa cm sq 11	Ap cm 4th 12	Kg cm 5th 13	At cm sq 14	Perm 15	AL 16
T-0502-000	1.270	0.516	0.635	1.548	0.893	2.5	2.7	1.5	2.0	0.2239	0.209	0.047	0.002063	8.1	5000	5724
T-0500-100	1.270	0.792	0.635	1.660	1.031	3.1	2.3	3.6	2.0	0.1488	0.492	0.073	0.002146	9.7	5000	2996
T-0570-000	1.450	0.850	0.550	1.870	0.975	3.4	2.7	4.1	2.0	0.1607	0.567	0.091	0.002872	11.2	5000	2930
T-0625-000	1.588	0.889	0.470	2.030	0.915	3.7	2.8	4.5	2.0	0.1596	0.620	0.099	0.003126	12.3	5000	2724

TDK Corp. EC Cores															
Part No.	Ht cm 1	Wth cm 2	Lt cm 3	G cm 4	MPL cm 5	Wtfe grams 6	Wtcu grams 7	MLT cm 8	Ac cm sq 9	Wa cm sq 10	Ap cm 4th 11	Kg cm 5th 12	At cm sq 13	Perm 14	AL 15
EC.35	3.460	2.275	3.45	2.31	7.7	36	34.2	6.3	0.843	1.530	1.290	0.069180	50.3	2300	2400
EC-41	3.900	2.705	4.06	2.78	8.9	57	57.1	7.5	1.210	2.148	2.599	0.168283	67.6	2300	3200
EC-52	4.840	3.300	5.22	3.18	10.5	111	100.2	9.0	1.800	3.116	5.610	0.446519	106.4	2300	4200
EC-70	6.900	4.450	7.00	4.55	14.4	256	263.2	11.6	2.790	6.393	17.836	1.718932	201.9	2300	4800
EIC-90	5.500	7.000	9.00	3.55	14.6	480	476.9	18.9	6.320	7.100	44.872	6.005740	254.0	2300	9000
EIC-120	6.525	9.500	12.00	3.55	18.0	705	936.0	22.8	7.530	11.538	86.877	11.470420	430.6	2300	8400
EC-90	9.000	7.000	9.00	7.10	21.6	698	953.8	18.9	6.240	14.200	88.608	11.709320	366.0	2300	6000
EC-120	10.100	9.500	12.00	7.10	25.0	780	1871.9	22.8	7.530	23.075	173.755	22.940840	584.3	2300	6300

TDK Corp. EE Cores

Part No.	Ht cm 1	Wth cm 2	Lt cm 3	G cm 4	MPL cm 5	Wtfe grams 6	Wtcu grams 7	MLT cm 8	Ac cm sq 9	Wa cm sq 10	Ap cm 4th 11	Kg cm 5th 12	At cm sq 13	Perm 14	AL 15
EE-8	0.800	0.775	0.830	0.600	1.9	0.7	1.0	2.1	0.070	0.125	0.009	0.000114	3.5	2300	610
EE-10	1.100	1.000	1.020	0.840	2.6	1.5	2.1	2.7	0.121	0.221	0.027	0.000483	6.0	2300	850
EE-12	1.390	0.780	1.210	0.890	2.7	2.4	1.5	2.4	0.140	0.176	0.025	0.000582	7.0	2300	650
EF-126	1.280	0.875	1.270	0.930	3.0	2.0	2.3	2.7	0.130	0.239	0.031	0.000607	7.0	2300	810
EE-125	1.480	1.125	1.240	0.980	3.1	2.4	3.2	2.9	0.150	0.314	0.047	0.000986	9.2	2300	720
SEE-1212	1.200	1.200	1.200	0.940	3.0	2.4	3.2	3.1	0.165	0.289	0.048	0.001011	7.9	2300	850
EE-13	1.200	1.340	1.300	0.920	3.0	2.7	3.9	3.3	0.171	0.334	0.057	0.001173	9.0	2300	1130
EE-129	1.138	1.285	1.290	0.818	2.8	2.7	3.2	3.3	0.192	0.270	0.052	0.001205	8.4	2300	900
SEE-13	1.600	1.340	1.300	1.320	3.8	3.2	5.7	3.3	0.171	0.479	0.082	0.001683	11.1	2300	940
EE-16	1.440	1.250	1.600	1.035	3.5	3.3	4.8	3.4	0.192	0.398	0.077	0.001740	11.0	2300	1140
EF-16	1.610	1.125	1.610	1.180	3.8	3.9	4.6	3.3	0.201	0.398	0.080	0.001964	11.2	2300	1100
EE-25432	3.170	1.880	2.540	0.254	7.3	15.0	2.8	4.9	0.404	0.158	0.064	0.002106	34.4	2300	1330
SEE-16	1.430	1.613	1.600	1.100	3.7	4.1	7.1	3.9	0.217	0.513	0.111	0.002498	13.2	2300	1240
EE-1916	1.620	1.405	1.929	1.143	3.9	4.8	7.1	3.8	0.224	0.531	0.119	0.002831	14.5	2300	1350
EE-19	1.590	1.465	1.910	1.120	3.9	4.8	7.4	3.8	0.230	0.540	0.124	0.002983	14.6	2300	1250
EE-1624	2.440	1.240	1.600	2.040	5.5	5.2	9.3	3.4	0.194	0.775	0.150	0.003472	16.6	2300	800
EE-2020	2.000	1.290	2.015	1.300	4.3	7.5	6.6	3.7	0.310	0.507	0.157	0.005336	16.9	2300	1460
EE-22	1.870	1.300	2.200	1.070	4.0	8.8	5.3	3.8	0.410	0.388	0.159	0.006781	17.1	2300	2180
EF-20	1.980	1.405	2.000	1.440	4.5	7.4	8.6	4.0	0.335	0.605	0.203	0.006793	17.2	2300	1570
EE-1927	2.710	1.475	2.000	2.230	6.2	7.9	14.9	3.8	0.245	1.087	0.267	0.006804	23.0	2300	670
EE-2519	1.892	1.849	2.540	1.282	4.9	9.1	13.5	4.9	0.400	0.782	0.313	0.010316	22.9	2300	2000
EE-254	1.932	1.850	2.540	1.296	4.9	10.0	13.6	4.9	0.403	0.787	0.317	0.010534	23.3	2300	2000
EE-2229	2.940	1.575	2.200	2.140	6.3	12.0	16.3	4.3	0.358	1.070	0.383	0.012822	26.8	2300	1300
EF-25	2.510	1.745	2.505	1.790	5.8	15.0	16.0	4.9	0.518	0.917	0.475	0.020065	27.1	2300	2000
EE-29	3.000	1.950	2.980	2.200	7.0	10.0	26.2	5.4	0.571	1.364	0.779	0.032974	37.2	2300	1970
EE-3030	3.000	1.960	3.010	1.990	6.7	22.0	23.0	5.2	0.597	1.249	0.745	0.034378	38.1	2300	2100
EE-2821	2.095	2.200	2.800	1.145	4.8	24.0	13.0	5.9	0.959	0.621	0.596	0.038852	29.9	2300	4400
EE-30	2.630	1.970	3.000	1.630	5.8	32.0	15.9	6.1	1.090	0.734	0.800	0.057136	34.1	2300	4690
EE-3528	2.780	2.540	3.440	1.990	7.0	28.0	38.0	6.6	0.777	1.612	1.252	0.058700	44.8	2300	2700
EF-32	3.220	2.265	3.210	2.320	7.4	32.0	34.5	6.2	0.832	1.566	1.303	0.069965	44.8	2300	2590

Part No.	Ht cm 1	Wth cm 2	Lt cm 3	G cm 4	MPL cm 5	Wtfe grams 6	Wtcu grams 7	MLT cm 8	Ac cm sq 9	Wa cm sq 10	Ap cm 4th 11	Kg cm 5th 12	At cm sq 13	Perm 14	AL 15
TDK Corp. EE Cores															
EE-2833	3.350	2.200	2.800	2.450	7.3	33.0	28.6	5.8	0.864	1.397	1.207	0.072422	42.5	2300	2850
EE-35	2.866	2.503	3.454	1.940	6.9	33.0	35.4	6.6	0.894	1.504	1.344	0.072547	45.6	2300	3170
EE-3328	2.750	2.650	3.300	1.850	6.8	20.0	32.9	7.3	1.110	1.277	1.417	0.086719	44.9	2300	3940
EE-3042	4.250	1.980	3.000	3.250	9.0	51.0	32.2	6.1	1.070	1.479	1.582	0.110715	50.3	2300	2890
EE-503	5.120	1.570	5.030	3.180	10.5	68.0	38.6	7.1	1.209	1.526	1.845	0.125324	75.7	2300	2900
EE-40	3.400	2.760	4.000	2.050	7.7	50.0	45.2	7.3	1.270	1.732	2.200	0.152232	60.8	2300	4150
EE-3548	4.850	2.450	3.500	3.670	10.5	57.0	63.2	6.7	1.033	2.661	2.749	0.169901	69.1	2300	2460
EE-3347	4.750	2.660	3.300	3.850	10.6	64.0	69.2	7.3	1.170	2.676	3.131	0.201523	68.8	2300	2800
EE-4133	3.352	2.801	4.128	2.082	7.8	64.0	45.9	8.1	1.613	1.594	2.571	0.204982	61.8	2300	5060
EE-623	6.200	1.670	6.230	3.740	12.6	102.0	58.9	8.4	1.530	1.982	3.033	0.222252	108.0	2300	3100
EE-4054	5.450	2.720	4.000	4.050	12.2	88.0	84.1	7.5	1.440	3.149	4.534	0.347807	87.8	2300	2990
EE-4438	3.760	3.430	4.400	2.360	8.7	85.0	72.7	9.0	1.910	2.277	4.350	0.370153	80.0	2300	5330
EE-42421	4.200	3.250	4.215	3.030	9.7	80.0	82.7	8.7	1.820	2.659	4.839	0.402916	81.7	2300	4700
EE-4215	4.216	3.451	4.285	3.036	9.8	91.0	93.3	9.1	1.832	2.893	5.301	0.428326	85.7	2300	4800
EE-4739	3.926	3.172	4.712	2.440	9.1	108.0	65.5	9.4	2.420	1.964	4.753	0.490346	81.7	2300	6660
EE-50	4.260	3.450	5.000	2.550	9.6	116.0	86.4	9.6	2.260	2.537	5.734	0.541532	94.9	2300	6110
EE-42422	4.200	3.725	4.215	3.030	9.7	116.0	91.6	9.7	2.350	2.659	6.248	0.605915	89.7	2300	6100
EE-4220	4.216	3.881	4.285	3.036	9.8	120.0	102.1	9.9	2.340	2.893	6.770	0.638273	93.1	2300	6000
EE-4460	6.050	3.480	4.400	4.650	13.3	125.0	148.3	9.1	1.837	4.604	8.457	0.686122	116.7	2300	3600
EE-60	4.460	4.400	6.000	2.810	11.0	135.0	160.4	11.3	2.470	3.990	9.856	0.861209	127.8	2300	5670
EE-5066	6.670	3.400	5.000	4.950	14.4	169.0	162.1	9.5	2.250	4.802	10.803	1.024146	134.7	2300	4000
EE-4265	6.528	3.760	4.191	5.300	14.3	177.0	164.6	9.8	2.375	4.744	11.266	1.096761	126.7	2300	4400
EE-5747	4.720	3.810	5.657	2.926	10.2	190.0	112.0	11.2	3.440	2.824	9.713	1.197811	118.0	2300	8530
EE-6565	6.500	3.795	6.520	4.510	14.7	214.0	218.0	11.1	2.660	5.536	14.726	1.415082	164.3	2300	4400
EE-6071	7.170	4.410	6.000	5.570	16.5	210.0	319.6	11.3	2.440	7.937	19.367	1.669427	184.4	2300	3700
EE-5555	5.500	4.125	5.515	3.760	12.3	234.0	156.1	11.4	3.540	3.863	13.676	1.704086	137.6	2300	7100

TDK Corp. EER Cores															
Part No.	Ht cm 1	Wth cm 2	Lt cm 3	G cm 4	MPL cm 5	Wtfe grams 6	Wtcu grams 7	MLT cm 8	Ac cm sq 9	Wa cm sq 10	Ap cm 4th 11	Kg cm 5th 12	At cm sq 13	Perm 14	AL 15
EER-255	1.86	1.98	2.550	1.240	4.8	11.0	14.5	5.3	0.448	0.763	0.342	0.011467	23.3	2300	1920
EER-22	2.94	1.55	2.200	2.140	6.2	12.0	15.0	4.5	0.375	0.947	0.355	0.011950	26.5	2300	1400
EIR-255	3.10	1.98	2.550	2.480	4.8	11.0	29.0	5.3	0.448	1.525	0.683	0.022934	34.5	2300	1920
EER-28	2.80	2.27	2.855	1.930	6.4	28.0	25.0	6.4	0.821	1.090	0.895	0.045638	37.2	2300	2870
EER-28L	3.38	2.27	2.855	2.506	7.6	33.0	32.4	6.4	0.814	1.416	1.153	0.058252	43.1	2300	2520
EIR-28	3.38	2.27	2.855	2.506	5.0	24.0	32.4	6.4	0.892	1.416	1.263	0.069951	43.1	2300	4000
EIR-30	4.28	2.00	3.000	3.300	5.8	37.0	30.3	6.5	1.230	1.320	1.624	0.123579	51.0	2300	4930
EER-33	3.40	2.62	3.300	2.500	7.7	51.0	42.4	7.8	1.290	1.525	1.967	0.129752	51.4	2300	4100
EER-35	4.14	2.56	3.500	2.940	9.1	52.0	53.6	7.2	1.070	2.102	2.249	0.134207	61.6	2300	2770
EIR-35	4.88	2.56	3.500	3.680	6.8	39.0	67.1	7.2	1.110	2.631	2.921	0.180781	70.6	2300	3860
EER-39	4.54	2.93	3.930	3.400	10.3	69.0	83.0	8.3	1.300	2.805	3.647	0.227947	76.9	2300	3200
EER-40	4.48	2.90	4.000	3.080	9.8	78.0	72.2	8.4	1.490	2.418	3.603	0.255824	76.7	2300	3620
EER-4215	4.32	3.04	4.215	3.190	9.9	90.0	79.7	9.0	1.760	2.504	4.407	0.346563	79.1	2300	4500
EER-42	4.48	2.94	4.200	3.080	9.9	102.0	68.4	9.0	1.940	2.141	4.153	0.358447	80.1	2300	4690
EIR-40	5.46	2.90	4.000	4.060	7.6	65.0	95.1	8.4	1.590	3.187	5.067	0.384006	90.2	2300	5000
EER-45	4.80	3.22	4.500	3.300	10.8	106.0	94.5	9.3	1.940	2.871	5.570	0.466762	92.8	2300	4670
EER-49	3.80	3.64	4.900	2.480	9.1	110.0	88.9	10.5	2.310	2.381	5.500	0.483858	87.4	2300	6250
EER-4220	4.24	3.41	4.215	3.050	9.9	116.0	80.7	10.3	2.400	2.211	5.307	0.496344	83.4	2300	5340
EER-48	4.20	4.01	4.800	3.000	10.0	134.0	116.8	11.4	2.610	2.880	7.517	0.688236	98.7	2300	6600
EER-49LS	4.78	3.63	4.900	3.080	10.4	137.0	109.7	10.5	2.490	2.941	7.324	0.695623	104.0	2300	6030
EER-50	5.20	3.62	5.000	3.540	11.3	142.0	127.7	10.3	2.380	3.487	8.299	0.766970	112.8	2300	5300
EER-49LZ	5.40	3.64	4.900	3.700	11.8	150.0	132.7	10.5	2.460	3.552	8.738	0.818681	114.7	2300	5300

TDK Corp. EI Cores

Part No.	Ht cm 1	Wth cm 2	Lt cm 3	G cm 4	MPL cm 5	Wtfe grams 6	Wtcu grams 7	MLT cm 8	Ac cm sq 9	Wa cm sq 10	Ap cm 4th 11	Kg cm 5th 12	At cm sq 13	Perm 14	AL 15
EI-125	0.890	1.125	1.24	0.510	2.1	1.9	1.7	2.9	0.144	0.163	0.024	0.000473	6.4	2300	1200
EI-16	1.420	1.240	1.60	1.020	3.5	3.3	4.6	3.4	0.198	0.388	0.077	0.001808	10.8	2300	1100
EI-19	1.585	1.475	2.00	1.115	4.0	5.1	7.4	3.8	0.240	0.544	0.130	0.003254	15.1	2300	1400
EI-22B	1.930	1.560	2.20	1.130	4.3	8.3	8.5	4.2	0.348	0.565	0.197	0.006489	19.1	2300	1660
EI-2219	1.870	1.575	2.20	1.070	4.2	8.5	8.1	4.3	0.370	0.535	0.198	0.006848	18.7	2300	2000
EI-22	1.905	1.300	2.20	1.055	3.9	9.8	5.2	3.8	0.420	0.382	0.161	0.007016	17.4	2300	2400
EI-254	1.905	1.880	2.54	1.270	4.8	10.0	13.8	4.9	0.400	0.791	0.316	0.010320	23.2	2300	1860
EI-25	1.825	1.925	2.53	1.235	4.7	9.8	13.8	5.0	0.410	0.772	0.316	0.010338	22.8	2300	2140
EI-28	2.025	2.200	2.80	1.225	4.8	22.0	14.3	5.8	0.860	0.698	0.600	0.035877	29.3	2300	4300
EI-30	2.675	1.980	3.00	1.625	5.8	34.0	16.1	6.1	1.110	0.739	0.821	0.059574	34.6	2300	4690
EI-35	2.885	2.450	3.50	1.815	6.7	36.0	31.3	6.7	1.014	1.316	1.334	0.080962	45.7	2300	3800
EI-33	2.830	2.620	3.30	1.905	6.7	41.0	33.0	7.2	1.180	1.286	1.517	0.099366	45.6	2300	4400
EI-3329	2.875	2.660	3.30	1.925	6.8	41.0	34.6	7.3	1.185	1.338	1.585	0.103362	46.4	2300	4400
EE-40	3.475	2.720	4.00	2.025	7.7	60.0	42.0	7.5	1.480	1.574	2.330	0.183699	61.2	2300	4860
EI-4437	3.725	3.480	4.40	2.325	8.7	85.0	74.1	9.1	1.920	2.302	4.419	0.374762	80.1	2300	5460
EI-50	4.235	3.400	5.00	2.475	9.4	115.0	81.0	9.5	2.300	2.401	5.522	0.535085	93.8	2300	6110
EI-60	4.435	4.410	6.00	2.785	10.9	139.0	159.8	11.3	2.470	3.969	9.803	0.855365	127.5	2300	5670
EI-70	6.440	5.570	7.00	4.280	14.6	523.0	277.9	15.2	6.950	5.157	35.844	6.576563	218.5	2300	8400

TDK Corp. EP Cores															
Part No.	Ht cm 1	Wth cm 2	Lt cm 3	G cm 4	MPL cm 5	Wtfe grams 6	Wtcu grams 7	MLT cm 8	Ac cm sq 9	Wa cm sq 10	Ap cm 4th 11	Kg cm 5th 12	At cm sq 13	Perm 14	AL 15
EP-10	0.647	1.020	1.15	0.74	1.9	2.8	-0.3	0.8	0.113	-0.087	-0.010	-0.000540	5.7	2300	800
EP-7	0.840	0.740	0.92	0.52	1.6	1.4	0.7	1.9	0.103	0.107	0.011	0.000236	3.5	2300	830
EP-13	1.163	1.285	1.25	0.93	2.4	5.1	2.2	2.4	0.195	0.262	0.051	0.001652	7.7	2300	1170
EP-17	1.416	1.680	1.80	1.13	2.9	12.0	3.8	3.0	0.339	0.357	0.121	0.005445	13.7	2300	1840
LP-23-8	1.210	2.340	1.65	1.74	4.4	9.6	6.5	3.1	0.313	0.592	0.185	0.007488	14.7	2300	1600
EP-20	1.883	2.140	2.40	1.43	4.0	28.0	8.3	4.2	0.780	0.554	0.432	0.032086	23.8	2300	3200
LP-22-13	1.810	2.240	2.50	1.64	4.9	21.0	13.9	4.6	0.679	0.853	0.579	0.034400	23.4	2300	3310
LP-32-13	1.810	3.180	2.50	2.41	6.4	30.0	20.4	4.6	0.703	1.253	0.881	0.054188	30.6	2300	2630
EP-30	2.773	2.985	3.10	2.34	6.3	75.0	24.3	6.3	1.790	1.082	1.937	0.219388	46.6	2300	5200

TDK Corp. EPC Cores															
Part No.	Ht cm 1	Wth cm 2	Lt cm 3	G cm 4	MPL cm 5	Wtfe grams 6	Wtcu grams 7	MLT cm 8	Ac cm sq 9	Wa cm sq 10	Ap cm 4th 11	Kg cm 5th 12	At cm sq 13	Perm 14	AL 15
EPC-10	0.470	0.81	1.02	0.53	1.8	1.1	0.5	1.9	0.0939	0.069	0.006	0.000128	2.9	2300	1000
EPC-13	0.705	1.32	1.33	0.90	3.1	2.1	2.0	2.5	0.1250	0.221	0.028	0.000549	5.9	2300	870
EPC-17	0.930	1.71	1.76	1.21	4.0	4.5	4.9	3.4	0.2280	0.399	0.091	0.002428	10.2	2300	1150
EPC-19	0.965	1.95	1.91	1.45	4.6	5.3	6.9	3.7	0.2270	0.529	0.120	0.002981	12.1	2300	940
EPC-25B	0.975	2.28	2.51	1.75	4.6	11.0	9.2	4.6	0.3330	0.569	0.189	0.005544	17.7	2300	1560
EPC-25	1.258	2.50	2.51	1.80	5.9	13.0	14.4	4.9	0.4640	0.824	0.382	0.014385	20.6	2300	1560
EPC-27	1.230	3.20	2.71	2.40	7.3	18.0	18.8	5.1	0.5460	1.032	0.563	0.024036	26.8	2300	1540
EPC-30	1.230	3.50	3.01	2.60	8.2	23.0	21.9	5.5	0.6100	1.118	0.682	0.030146	31.5	2300	1570

TDK Corp. ETD Cores

Part No.	Ht cm 1	Wth cm 2	Lt cm 3	G cm 4	MPL cm 5	Wtfe grams 6	Wtcu grams 7	MLT cm 8	Ac cm sq 9	Wa cm sq 10	Ap cm 4th 11	Kg cm 5th 12	At cm sq 13	Perm 14	AL 15
ETD-19	2.73	1.490	1.96	1.88	5.5	14.0	11.4	4.5	0.413	0.705	0.291	0.010582	22.5	2300	1720
ETD-24	2.89	1.860	2.44	2.02	6.2	20.0	19.6	5.4	0.563	1.020	0.574	0.023979	30.5	2300	2125
ETD-29	3.16	2.270	2.98	2.20	7.1	28.0	32.4	6.3	0.736	1.452	1.069	0.050095	41.7	2300	2500
ETD-34	3.46	2.638	3.42	2.42	7.9	40.0	47.9	7.2	0.971	1.876	1.821	0.098547	53.3	2300	2780
ETD-39	3.96	3.018	3.91	2.92	9.2	60.0	76.6	8.4	1.250	2.570	3.212	0.191482	69.7	2300	3150
ETD-44	4.46	3.340	4.40	3.30	10.3	94.0	102.6	9.5	1.750	3.053	5.342	0.395590	87.5	2300	4000
ETD-49	4.94	3.710	4.87	3.62	11.4	124.0	138.5	10.4	2.130	3.747	7.980	0.653917	107.4	2300	4440

TDK Corp. PC Cores

Part No.	Ht cm 1	Wth cm 2	Lt cm 3	G cm 4	MPL cm 5	Wtfe grams 6	Wtcu grams 7	MLT cm 8	Ac cm sq 9	Wa cm sq 10	Ap cm 4th 11	Kg cm 5th 12	At cm sq 13	Perm 14	AL 15
C-1408	0.832	1.40	1.40	0.58	2.0	3	1.8	2.9	0.200	0.171	0.034	0.000931	6.7	2500	1310
C-231152	1.100	2.28	2.28	0.75	3.3	12	5.2	4.5	0.690	0.324	0.224	0.013582	16.0	2500	3750
C-231712	1.685	2.30	2.30	1.15	3.6	16	7.4	4.0	0.620	0.529	0.328	0.020543	20.5	2500	2770
C-231852	1.780	2.28	2.28	1.41	3.9	17	9.9	4.5	0.637	0.610	0.388	0.021762	20.9	2500	2730
C-301952	1.880	3.00	3.00	1.32	4.7	29	17.7	6.2	1.320	0.799	1.054	0.089254	31.8	2500	4870

TDK Corp. PCC Cores															
Part No.	Ht cm 1	Wth cm 2	Lt cm 3	G cm 4	MPL cm 5	Wtfe grams 6	Wtcu grams 7	MLT cm 8	Ac cm sq 9	Wa cm sq 10	Ap cm 4th 11	Kg cm 5th 12	At cm sq 13	Perm 14	AL 15
P-9-5	0.54	0.93	0.93	0.36	1.2	0.8	0.4	1.9	0.0729	0.065	0.005	0.000071	2.9	2000	1300
P-11-7	0.66	1.13	1.13	0.44	1.6	1.8	0.8	2.3	0.1240	0.095	0.012	0.000252	4.3	2000	1800
P-14-8	0.85	1.42	1.42	0.56	2.0	3.2	1.6	2.9	0.1840	0.157	0.029	0.000726	7.0	2000	2050
P-18-11	1.06	1.82	1.82	0.73	2.6	6.7	3.6	3.7	0.3440	0.270	0.093	0.003448	11.3	2000	3100
P-18-14	1.40	1.80	1.80	1.06	3.0	8.3	5.1	3.7	0.3380	0.387	0.131	0.004788	13.0	2000	2600
P-22-13	1.36	2.16	2.16	0.92	3.2	12.7	6.2	4.4	0.4770	0.391	0.187	0.008004	16.5	2000	3700
P-23-17	1.72	2.30	2.30	1.13	3.7	20.0	6.5	4.7	0.6610	0.384	0.254	0.014152	20.7	2000	5100
P-26-16	1.63	2.55	2.55	1.10	3.8	21.1	10.0	5.3	0.7130	0.534	0.380	0.020492	23.3	2000	5000
P-28-23	2.32	2.80	2.80	1.63	4.9	40.0	15.0	5.6	0.9330	0.750	0.700	0.046424	32.7	2000	5800
P-30-19	1.90	3.00	3.00	1.30	4.5	35.3	16.7	6.3	1.0900	0.748	0.815	0.056539	32.0	2000	6500
P-34-28	2.82	3.40	3.40	2.01	5.9	65.0	31.0	6.7	1.1800	1.307	1.542	0.109004	48.3	2000	6000
P-41-25	2.50	4.10	4.10	1.70	6.2	90.0	41.5	8.3	2.0100	1.403	2.819	0.272280	58.6	2000	13200
P-42-29	2.94	4.24	4.24	2.03	6.9	104.0	55.6	8.6	2.0400	1.817	3.706	0.351399	67.4	2000	13200
P-45-29	2.92	4.50	4.50	1.88	6.7	139.1	52.3	9.3	2.8600	1.589	4.543	0.560929	73.1	2000	13200

TDK Corp. PQ Cores															
Part No.	Ht cm 1	Wth cm 2	Lt cm 3	G cm 4	MPL cm 5	Wtfe grams 6	Wtcu grams 7	MLT cm 8	Ac cm sq 9	Wa cm sq 10	Ap cm 4th 11	Kg cm 5th 12	At cm sq 13	Perm 14	AL 15
PQ-2016	1.620	2.320	2.05	1.03	3.7	13	7.4	4.4	0.62	0.474	0.294	0.016679	16.9	2300	3880
PQ-2020	2.020	2.320	2.05	1.43	4.5	15	10.2	4.4	0.62	0.658	0.408	0.023157	19.7	2300	3150
PQ-2620	2.015	2.950	2.65	1.15	4.6	31	12.0	5.6	1.19	0.604	0.718	0.061325	28.4	2300	6170
PQ-2625	2.475	2.950	2.65	1.61	5.6	36	16.8	5.6	1.18	0.845	0.997	0.084418	32.6	2300	5250
PQ-3220	2.055	3.605	3.20	1.15	5.6	42	19.2	6.7	1.70	0.808	1.373	0.140062	36.3	2300	7310
PQ-3230	3.035	3.605	3.20	2.13	7.5	55	35.5	6.7	1.61	1.496	2.409	0.232678	46.9	2300	5140
PQ-3535	3.475	4.365	3.51	2.50	8.8	73	59.0	7.5	1.96	2.206	4.324	0.451091	60.7	2300	4860
PQ-4040	3.975	5.010	4.05	2.95	10.2	95	97.2	8.4	2.01	3.260	6.552	0.628106	77.1	2300	4300
PQ-5050	4.995	5.600	5.00	3.61	11.3	195	158.5	10.3	3.28	4.332	14.209	1.812269	113.9	2300	6720

TDK Corp. RM Cores															
Part No.	Ht cm 1	Wth cm 2	Lt cm 3	G cm 4	MPL cm 5	Wtfe grams 6	Wtcu grams 7	MLT cm 8	Ac cm sq 9	Wa cm sq 10	Ap cm 4th 11	Kg cm 5th 12	At cm sq 13	Perm 14	AL 15
RM-4	1.04	0.963	0.963	0.720	2.3	1.7	1.1	2.0	0.140	0.157	0.022	0.000603	5.9	2300	680
RM-5	1.04	1.205	1.205	0.650	2.2	3.0	1.6	2.5	0.237	0.182	0.043	0.001606	7.9	2300	1250
RM-6	1.24	1.440	1.440	0.820	2.9	5.5	2.9	3.1	0.366	0.260	0.095	0.004449	11.3	2300	1600
R-6	1.24	1.445	1.445	0.820	2.8	5.6	2.9	3.1	0.381	0.258	0.098	0.004796	11.3	2300	1700
RM-7	1.34	1.685	1.685	0.865	3.2	7.7	4.5	3.6	0.460	0.345	0.159	0.008020	14.7	2300	1950
RM-8	1.64	1.935	1.935	1.100	3.8	13.0	7.3	4.2	0.640	0.490	0.313	0.019118	20.2	2300	1950
RM-10	1.86	2.415	2.415	1.270	4.4	23.0	13.0	5.2	0.980	0.695	0.681	0.050985	29.6	2300	3630
RM-12	2.35	2.925	2.925	1.710	5.7	42.0	24.4	6.2	1.400	1.103	1.544	0.139014	44.6	2300	4150
RM-14	2.88	3.420	3.420	2.110	7.0	70.0	39.8	7.2	1.780	1.556	2.770	0.274451	62.8	2300	4600

TDK Corp. SMD Cores															
Part No.	Ht cm 1	Wth cm 2	Lt cm 3	G cm 4	MPL cm 5	Wtfe grams 6	Wtcu grams 7	MLT cm 8	Ac cm sq 9	Wa cm sq 10	Ap cm 4th 11	Kg cm 5th 12	At cm sq 13	Perm 14	AL 15
EE-5-Z	0.53	0.445	0.525	0.400	1.3	0.2	0.3	1.5	0.0267	0.050	0.001	0.000010	1.4	2300	200
ER-95-5	0.49	0.913	0.935	0.335	1.4	0.6	0.7	2.7	0.0847	0.071	0.006	0.000074	2.9	2300	610
ER-11-3	0.39	1.037	1.083	0.210	1.3	0.8	0.5	3.2	0.1170	0.047	0.005	0.000081	3.1	2300	1040
ER-11-5	0.49	1.062	1.083	0.315	1.5	1.0	0.8	3.2	0.1190	0.074	0.009	0.000133	3.6	2300	870
EE-12-5	0.50	1.040	1.200	0.260	1.4	1.5	0.6	3.1	0.1750	0.057	0.010	0.000226	4.0	2300	2050
EEM-127	1.37	0.630	1.275	0.910	2.7	1.9	1.3	2.7	0.1200	0.137	0.016	0.000287	6.1	2300	820
ER-145-6	0.59	1.380	1.450	0.330	1.9	1.8	1.6	3.8	0.1760	0.117	0.021	0.000382	6.0	2300	1280

TDK Corp. Toroidal Cores

Part No.	OD cm 1	ID cm 2	HT cm 3	Wound OD cm 4	Wound HT cm 5	MPL cm 6	Wtfe grams 7	Wtcu grams 8	MLT cm 9	Ac cm sq 10	Wa cm sq 11	Ap cm 4th 12	Kg cm 5th 13	At cm sq 14	Perm 15	AL 16
T-3.05x1.27x1.27	0.305	0.127	0.127	0.373	0.191	0.6	0.0	0.0	0.4	0.0106	0.013	0.000	0.000001	0.4	2300	511
T-4x1x2	0.400	0.200	0.100	0.501	0.200	0.9	0.0	0.1	0.5	0.0096	0.031	0.000	0.000002	0.7	2300	319
T-3.94x1.27x2.23	0.394	0.223	0.127	0.505	0.239	0.9	0.0	0.1	0.5	0.0106	0.039	0.000	0.000003	0.8	2300	333
T-4.83x1.27x2.29	0.483	0.229	0.127	0.600	0.242	1.0	0.1	0.1	0.6	0.0154	0.041	0.001	0.000007	1.0	2300	436
T-5.84x1.5x3.05	0.584	0.305	0.152	0.737	0.305	1.3	0.1	0.2	0.7	0.0205	0.073	0.001	0.000017	1.6	2300	456
T-6x1.5x3	0.600	0.300	0.150	0.752	0.300	1.3	0.1	0.2	0.7	0.0216	0.071	0.002	0.000018	1.6	2300	476
T-8x2x4	0.800	0.400	0.200	1.003	0.400	1.7	0.3	0.4	1.0	0.0384	0.126	0.005	0.000077	2.8	2300	638
T-10x2.5x5	1.000	0.500	0.250	1.253	0.500	2.2	0.6	0.8	1.2	0.0601	0.196	0.012	0.000236	4.4	2300	796
T-9.52x3.18x4.75	0.952	0.475	0.318	1.193	0.556	2.1	0.7	0.8	1.3	0.0726	0.177	0.013	0.000294	4.3	2300	1013
T-12x3x6	1.200	0.600	0.300	1.504	0.600	2.6	1.1	1.4	1.4	0.0865	0.283	0.024	0.000587	6.4	2300	957
T-14x3.5x7	1.400	0.700	0.350	1.754	0.700	3.1	1.7	2.3	1.7	0.1180	0.385	0.045	0.001275	8.7	2300	1118
T-12.7x4.77x7.14	1.270	0.714	0.477	1.624	0.834	3.0	1.8	2.5	1.8	0.1290	0.400	0.052	0.001497	8.4	2300	1263
T-12.7x6.35x7.92	1.270	0.792	0.635	1.660	1.031	3.1	2.3	3.6	2.0	0.1490	0.492	0.073	0.002152	9.7	2300	1380
T-12.7x6.35x7.14	1.270	0.714	0.635	1.624	0.992	3.0	2.5	2.9	2.0	0.1720	0.400	0.069	0.002331	9.2	2300	1684
T-16x4x8	1.600	0.800	0.400	2.005	0.800	3.5	2.6	3.4	1.9	0.1540	0.502	0.077	0.002482	11.3	2300	1278
T-18x4.5x9	1.800	0.900	0.450	2.256	0.900	3.9	3.7	4.9	2.2	0.1950	0.636	0.124	0.004477	14.4	2300	1437
T-20x5x10	2.000	1.000	0.500	2.506	1.000	4.4	5.1	6.7	2.4	0.2400	0.785	0.188	0.007536	17.7	2300	1590
T-20x7.5x14.5	2.000	1.450	0.750	2.716	1.475	5.3	5.3	16.4	2.8	0.2040	1.650	0.337	0.009812	24.2	2300	1106
T-31x8x19	3.100	1.900	0.800	4.036	1.750	7.6	17.2	37.9	3.8	0.4710	2.834	1.335	0.066879	47.8	2300	1802
T-28x13x16	2.800	1.600	1.300	3.593	2.100	6.6	24.2	30.9	4.3	0.7600	2.010	1.527	0.107476	44.0	2300	3347
T-31x13x19	3.100	1.900	1.300	4.036	2.250	7.6	28.0	46.0	4.6	0.7650	2.834	2.168	0.145477	54.1	2300	2927
T-38x14x22	3.800	2.200	1.400	4.889	2.500	9.0	47.4	71.3	5.3	1.0900	3.799	4.141	0.341975	75.9	2300	3510
T-44.5x13x30	4.450	3.000	1.300	5.926	2.800	11.4	51.4	141.7	5.6	0.9300	7.065	6.570	0.433370	107.2	2300	2357
T-51x13x31	5.100	3.100	1.300	6.628	2.850	12.4	76.4	165.2	6.2	1.2700	7.544	9.581	0.790096	128.3	2300	2959
T-72x10x52	7.200	5.200	1.000	9.766	3.600	18.1	87.2	555.5	7.4	0.9910	21.226	21.035	1.132937	260.1	2300	1578
T-68x13.5x44	6.800	4.400	1.350	8.964	3.550	17.1	131.9	410.7	7.6	1.5950	15.198	24.240	2.034899	226.1	2300	2702
T-90x13.5x74	9.000	7.400	1.350	12.706	5.050	25.6	134.1	1430.8	9.4	1.0800	42.987	46.426	2.142717	454.9	2300	1219

TDK Corp. UU Cores															
Part No.	Ht cm 1	Wth cm 2	Lt cm 3	G cm 4	MPL cm 5	Wtfe grams 6	Wtcu grams 7	MLT cm 8	Ac cm sq 9	Wa cm sq 10	Ap cm 4th 11	Kg cm 5th 12	At cm sq 13	Perm 14	AL 15
UI-75	0.990	0.700	1.020	0.550	2.4	1.3	1.1	2.2	0.100	0.149	0.015	0.000273	4.4	2500	650
UU-75	1.540	0.700	1.020	1.100	3.4	1.8	2.3	2.2	0.100	0.297	0.030	0.000547	6.3	2500	500
UI-127	0.899	1.257	2.032	0.391	3.1	2.0	3.3	3.1	0.126	0.298	0.038	0.000610	9.5	2500	740
UU-913	1.340	0.980	1.380	0.940	3.5	1.8	4.1	2.6	0.100	0.451	0.045	0.000699	8.3	2500	570
UU-9	1.780	0.980	1.380	1.380	4.4	2.2	6.1	2.6	0.100	0.662	0.066	0.001026	10.3	2500	400
UU-105	1.560	1.050	1.600	1.050	4.0	2.6	5.7	2.8	0.126	0.578	0.073	0.001323	10.5	2500	540
UU-187	1.600	1.920	3.290	1.150	5.9	12.0	23.7	4.1	0.113	1.633	0.185	0.002041	26.7	2500	480
UI-13	1.543	1.290	1.910	0.850	4.0	4.9	6.3	3.4	0.230	0.519	0.119	0.003212	13.1	2500	1050
UI-21	1.415	1.750	3.220	0.435	4.7	7.6	7.6	4.4	0.310	0.487	0.151	0.004250	20.9	2500	1220
UU-157	1.940	1.250	2.220	1.200	5.0	6.5	9.8	3.5	0.247	0.780	0.193	0.005364	17.1	2500	750
UU-13	2.405	1.290	1.910	1.700	5.7	6.9	12.6	3.4	0.240	1.037	0.249	0.006995	18.6	2500	780
UU-1526	2.680	1.280	2.160	1.880	6.4	7.8	15.8	3.5	0.240	1.278	0.307	0.008475	21.8	2500	720
UU-21	1.850	1.750	3.220	0.870	5.5	8.9	15.3	4.4	0.310	0.974	0.302	0.008500	25.3	2500	1060
UI-17	1.800	1.670	2.400	0.800	4.6	12.0	8.9	4.4	0.490	0.560	0.274	0.012094	19.9	2500	1950
UI-254	2.235	1.925	3.830	0.965	6.4	19.0	21.9	5.0	0.389	1.245	0.484	0.015212	33.9	2500	1150
UU-18	3.330	1.550	2.650	2.380	8.0	14.0	29.4	4.1	0.333	2.023	0.674	0.021926	33.0	2500	830
UU-197	3.540	1.335	2.710	2.340	8.1	15.0	24.6	4.0	0.357	1.732	0.618	0.022125	32.3	2500	880
UU-17	2.600	1.700	2.430	1.600	6.4	15.0	18.5	4.5	0.480	1.168	0.561	0.024113	26.8	2500	1500
UU-25432	3.200	1.875	3.780	1.930	8.4	18.0	41.9	4.9	0.400	2.393	0.957	0.031101	44.1	2500	950
UU-22	3.100	1.890	3.190	2.000	7.9	18.0	36.7	4.7	0.440	2.180	0.959	0.035696	38.8	2500	1100
UU-2431	3.100	2.050	3.600	2.000	8.2	19.0	45.1	5.1	0.440	2.500	1.100	0.038182	44.3	2500	1070
UI-26	2.410	2.410	3.820	1.000	6.6	30.0	26.3	6.1	0.840	1.210	1.016	0.055915	41.6	2500	2460
UI-33	3.170	2.365	4.730	1.270	8.4	41.0	40.9	6.5	0.920	1.778	1.636	0.093096	56.7	2500	2230
UU-254	2.590	2.540	3.810	1.320	7.6	30.0	38.2	6.4	1.030	1.676	1.727	0.110950	45.8	2500	2550
UU-26	3.420	2.410	3.820	2.000	8.6	38.0	52.6	6.1	0.840	2.420	2.033	0.111830	54.2	2500	2010
UU-33	4.440	2.365	4.730	2.540	10.9	53.0	81.8	6.5	0.930	3.556	3.307	0.190262	74.7	2500	1780
UI-45	4.300	3.000	6.025	1.300	10.3	127.0	62.3	9.0	2.250	1.950	4.388	0.439336	95.8	2500	4520
UI-60	4.300	4.500	9.000	1.300	13.3	174.0	157.0	11.3	2.250	3.900	8.775	0.697782	161.1	2500	3620
UU-45	5.600	3.000	5.725	2.600	12.9	144.0	120.5	8.7	2.250	3.900	8.775	0.909012	114.1	2500	3650
UU-60	5.600	4.500	9.000	2.600	15.9	188.0	313.9	11.3	2.250	7.800	17.550	1.395565	196.2	2500	3060

TDK Corp. UU Cores

Part No.	Ht cm 1	Wth cm 2	Lt cm 3	G cm 4	MPL cm 5	Wtfe grams 6	Wtcu grams 7	MLT cm 8	Ac cm sq 9	Wa cm sq 10	Ap cm 4th 11	Kg cm 5th 12	At cm sq 13	Perm 14	AL 15
UI-70	4.900	5.200	10.400	1.300	15.1	261.0	206.6	13.1	3.250	4.420	14.365	1.420546	212.1	2500	4710
UI-55	6.000	3.500	7.025	2.000	13.3	292.0	117.2	11.0	4.000	3.000	12.000	1.747361	151.5	2500	6380
UU-70	6.200	5.200	10.400	2.600	17.7	302.0	413.2	13.1	3.240	8.840	28.642	2.823636	252.6	2500	4020
UU-55	8.000	3.500	7.025	4.000	17.3	371.0	234.4	11.0	4.000	6.000	24.000	3.494722	193.6	2500	5020
UI-101	8.240	7.640	15.220	3.190	24.4	825.0	1082.7	18.7	6.270	16.269	102.007	13.669920	506.3	2500	5810
UI-80	9.500	7.000	12.000	5.500	25.4	786.0	1321.2	16.9	6.030	22.000	132.660	18.946940	465.0	2500	5360
UU-101	11.460	7.640	15.220	6.380	31.0	1028.0	2165.4	18.7	6.310	32.538	205.315	27.689790	652.9	2500	4650
UI-100	10.580	7.000	14.000	4.500	26.2	1275.0	1209.0	18.9	8.910	18.000	160.380	30.262300	543.1	2500	7700
UU-80	15.000	7.000	12.000	11.000	36.2	1106.0	2642.4	16.9	5.990	44.000	263.560	37.392810	674.0	2500	3740
UU-100	15.090	7.000	14.000	9.000	3.5	1670.0	2418.0	18.9	8.950	36.000	322.200	61.069240	732.3	2500	5060

Thomson EC Cores

Part No.	Ht cm 1	Wth cm 2	Lt cm 3	G cm 4	MPL cm 5	Wtfe grams 6	Wtcu grams 7	MLT cm 8	Ac cm sq 9	Wa cm sq 10	Ap cm 4th 11	Kg cm 5th 12	At cm sq 13	Perm 14	AL 15
EC-3510A	3.46	2.275	3.45	2.45	7.7	36	36.3	6.3	0.84	1.623	1.363	0.072852	50.3	2500	2500
EC-4112A	3.90	2.705	4.06	2.78	8.9	56	57.1	7.5	1.21	2.148	2.599	0.168283	67.6	2500	3400
EC-5214A	4.84	3.300	5.22	3.18	10.5	110	100.2	9.0	1.80	3.116	5.610	0.446519	106.4	2500	4400
EC-7017A	6.90	4.450	7.00	4.55	14.4	252	263.2	11.6	2.79	6.393	17.836	1.718932	201.9	2500	5000

Thomson EE Cores															
Part No.	Ht cm 1	Wth cm 2	Lt cm 3	G cm 4	MPL cm 5	Wtfe grams 6	Wtcu grams 7	MLT cm 8	Ac cm sq 9	Wa cm sq 10	Ap cm 4th 11	Kg cm 5th 12	At cm sq 13	Perm 14	AL 15
E-1304A	1.28	0.920	1.260	0.93	3.0	1.5	2.5	2.7	0.124	0.263	0.033	0.000595	7.1	2500	840
E-1306A	1.20	1.383	1.300	0.93	3.1	26.0	4.3	3.4	0.163	0.357	0.058	0.001115	9.2	2500	1040
E-1605A	1.43	1.290	1.600	1.02	3.5	3.6	5.0	3.4	0.198	0.408	0.081	0.001858	11.1	2500	1150
E-1905A	1.58	1.490	1.915	1.12	4.0	4.6	7.8	3.9	0.218	0.566	0.123	0.002768	14.6	2500	1140
E-1605B	2.45	1.285	1.600	2.05	5.5	5.5	10.0	3.4	0.194	0.820	0.159	0.003595	17.0	2500	700
E-1907A	1.58	1.675	1.915	1.12	4.0	6.4	8.6	4.3	0.300	0.566	0.170	0.004787	15.9	2500	1600
E-2006A	1.99	1.435	2.000	1.43	4.6	7.4	8.9	4.1	0.322	0.618	0.199	0.006324	17.4	2500	1450
E-2005B	2.71	1.445	1.950	2.23	6.1	8.0	14.2	3.8	0.245	1.054	0.258	0.006655	22.3	2500	920
E-2506A	1.90	1.905	2.530	1.27	4.8	10.0	14.2	4.9	0.400	0.806	0.323	0.010457	23.3	2500	1950
E-2506B	1.98	1.905	2.530	1.35	5.0	10.5	15.0	4.9	0.400	0.857	0.343	0.011115	24.0	2500	2000
E-2206A	3.00	1.630	2.200	2.20	6.4	12.0	18.0	4.4	0.350	1.161	0.406	0.013029	27.7	2500	1250
E-2506C	3.20	1.904	2.540	2.57	7.4	15.0	28.6	4.9	0.403	1.628	0.656	0.021409	34.9	2500	1350
E-2507A	2.52	1.743	2.550	1.78	5.8	16.0	15.9	5.0	0.550	0.903	0.497	0.022048	27.5	2500	2200
E-3007B	3.00	2.000	3.010	2.00	6.6	22.0	24.1	5.2	0.600	1.295	0.777	0.035580	38.5	2500	2000
E-3008A	2.60	2.225	3.025	1.83	6.4	20.0	27.2	5.8	0.640	1.318	0.843	0.037164	36.2	2500	2250
E-3509A	2.88	2.595	3.490	1.98	7.0	30.0	39.7	6.7	0.830	1.663	1.380	0.068246	47.0	2500	3000
E-3510A	2.88	2.650	3.490	1.98	7.0	31.0	40.4	6.8	0.884	1.663	1.470	0.076167	47.7	2500	3000
E-3213A	2.80	2.657	3.190	1.93	6.6	37.0	33.8	7.1	1.130	1.338	1.512	0.096210	44.8	2500	4000
E-3510B	4.76	2.543	3.490	3.81	10.7	50.0	72.4	6.7	0.905	3.031	2.743	0.147894	69.1	2500	2100
E-3611A	3.56	2.700	3.590	2.46	8.1	54.0	48.4	7.0	1.160	1.937	2.247	0.148518	58.2	2500	3600
E-4113A	3.33	2.844	4.100	2.09	7.7	64.0	48.0	8.2	1.580	1.655	2.615	0.202688	61.5	2500	5000
E-3313A	2.66	2.665	3.300	3.81	10.4	64.0	68.5	7.2	1.180	2.657	3.136	0.204205	44.0	2500	2650
E-4012B	5.45	2.800	4.050	4.05	11.7	90.0	89.9	7.6	1.440	3.311	4.768	0.359685	89.7	2500	3250
E-4215A	4.20	3.310	4.215	3.03	9.7	90.0	86.4	8.8	1.800	2.750	4.950	0.403239	82.4	2500	5000
E-4215B	4.22	3.454	4.280	3.02	9.8	90.0	93.0	9.1	1.840	2.881	5.302	0.429949	85.8	2500	5000
E-4916A	4.12	3.240	4.885	2.42	9.1	120.0	68.6	9.5	2.560	2.034	5.208	0.562249	88.1	2500	7000
E-4220A	4.20	3.765	4.215	3.03	9.7	120.0	95.3	9.7	2.330	2.750	6.407	0.612584	90.1	2500	6500
E-4220B	4.22	3.867	4.280	3.02	9.8	120.0	101.5	9.9	2.330	2.881	6.714	0.631924	92.8	2500	6500
E-5521A	5.50	4.185	5.515	3.76	12.3	230.0	162.0	11.5	3.570	3.976	14.195	1.769029	138.5	2500	7200
E-5525A	5.50	4.585	5.515	3.76	12.3	270.0	173.3	12.3	4.200	3.976	16.700	2.288694	147.3	2500	8600

Thomson EE Cores

Part No.	Ht cm 1	Wth cm 2	Lt cm 3	G cm 4	MPL cm 5	Wtfe grams 6	Wtcu grams 7	MLT cm 8	Ac cm sq 9	Wa cm sq 10	Ap cm 4th 11	Kg cm 5th 12	At cm sq 13	Perm 14	AL 15
E-6527A	6.50	5.210	6.515	4.52	14.6	470.0	280.1	13.9	5.500	5.650	31.075	4.903177	200.4	2500	%10000
E-7032A	6.59	5.870	7.050	4.45	14.9	510.0	332.6	15.5	6.830	6.030	41.183	7.252927	229.6	2500	%11500

Thomson EFD Cores

Part No.	Ht cm 1	Wth cm 2	Lt cm 3	G cm 4	MPL cm 5	Wtfe grams 6	Wtcu grams 7	MLT cm 8	Ac cm sq 9	Wa cm sq 10	Ap cm 4th 11	Kg cm 5th 12	At cm sq 13	Perm 14	AL 15
EFD-1505	0.750	1.5	1.5	1.10	3.4	2.8	3.0	2.7	0.15	0.314	0.047	0.001053	7.3	1900	550
EFD-2007	0.990	2.0	2.0	1.54	4.7	7.0	6.8	3.8	0.31	0.501	0.155	0.005063	13.3	1900	950
EFD-2509	1.275	2.5	2.5	1.86	5.7	16.5	11.5	4.8	0.58	0.679	0.394	0.019111	21.6	1900	1600
EFD-3007	1.300	3.0	3.0	2.24	6.8	22.3	17.0	5.5	0.69	0.874	0.603	0.030470	28.9	1900	1800

Thomson EI Cores															
Part No.	Ht cm 1	Wth cm 2	Lt cm 3	G cm 4	MPL cm 5	Wtfe grams 6	Wtcu grams 7	MLT cm 8	Ac cm sq 9	Wa cm sq 10	Ap cm 4th 11	Kg cm 5th 12	At cm sq 13	Perm 14	AL 15
EI-2206A	1.900	1.630	2.200	1.100	4.3	8.5	9.0	4.4	0.363	0.580	0.211	0.007008	19.3	2500	1760
EI-2506C	1.918	1.904	2.540	1.283	4.8	10.0	14.3	4.9	0.403	0.814	0.328	0.010705	23.5	2500	1900
EI-3011B	2.680	2.035	3.025	1.630	5.7	34.0	17.4	6.2	1.100	0.791	0.870	0.061805	35.3	2500	4230
EI-3510B	2.855	2.543	3.490	1.905	6.9	32.0	36.2	6.7	0.905	1.515	1.371	0.073947	46.1	2500	3150
EI-3313A	2.830	2.665	3.300	1.905	6.7	42.0	34.2	7.2	1.200	1.329	1.594	0.105593	46.0	2500	4200
EI-4012B	2.475	2.800	4.050	2.025	7.7	60.0	44.9	7.6	1.480	1.655	2.450	0.189972	49.0	2500	5000
EI-4215B	2.707	3.454	4.280	1.511	6.8	65.0	46.5	9.1	1.840	1.441	2.651	0.214974	62.4	2500	7100

Thomson ETD Cores															
Part No.	Ht cm 1	Wth cm 2	Lt cm 3	G cm 4	MPL cm 5	Wtfe grams 6	Wtcu grams 7	MLT cm 8	Ac cm sq 9	Wa cm sq 10	Ap cm 4th 11	Kg cm 5th 12	At cm sq 13	Perm 14	AL 15
ETD-2910A	3.16	2.27	2.98	2.20	7.0	28	32.4	6.3	0.76	1.452	1.104	0.053415	41.7	2500	2100
ETD-3411A	3.46	2.63	3.42	2.42	7.9	40	47.8	7.2	0.97	1.876	1.819	0.098564	53.2	2500	2550
ETD-3913A	3.96	3.01	3.91	2.92	9.2	64	76.5	8.4	1.25	2.570	3.212	0.191848	69.5	2500	2800
ETD-4415A	4.46	3.33	4.40	3.30	10.3	94	102.4	9.4	1.73	3.053	5.281	0.387419	87.3	2500	3500
ETD-4916A	4.94	3.70	4.87	3.62	11.4	124	138.3	10.4	2.11	3.747	7.906	0.642931	107.2	2500	4000
ETD-5419A	5.52	4.12	5.45	4.04	12.7	180	186.9	11.7	2.80	4.505	12.613	1.210584	133.7	2500	4750

Thomson PC Cores

Part No.	Ht cm 1	Wth cm 2	Lt cm 3	G cm 4	MPL cm 5	Wtfe grams 6	Wtcu grams 7	MLT cm 8	Ac cm sq 9	Wa cm sq 10	Ap cm 4th 11	Kg cm 5th 12	At cm sq 13	Perm 14	AL 15
FP-0905	0.525	0.915	0.915	0.375	1.3	0.8	0.5	2.0	0.10	0.072	0.007	0.000141	2.8	2500	1200
FP-1107	0.645	1.110	1.110	0.455	2.0	1.7	0.9	2.3	0.16	0.105	0.017	0.000455	4.2	2500	1800
FP-1408	0.580	1.400	1.400	0.580	2.0	3.2	1.8	2.9	0.25	0.171	0.043	0.001455	5.6	2500	2500
FP-1811	1.055	1.790	1.790	0.740	2.6	6.5	3.8	3.7	0.43	0.285	0.123	0.005682	11.0	2500	3400
FP-2213	1.340	2.160	2.160	0.940	3.2	12.5	6.7	4.5	0.63	0.421	0.265	0.014941	16.4	2500	4500
FP-2616	1.610	2.550	2.550	1.120	3.7	21.0	10.9	5.3	0.93	0.577	0.536	0.037471	23.1	2500	5900
FP-3019	1.880	3.000	3.000	1.320	4.5	35.0	17.7	6.2	1.36	0.799	1.086	0.094745	31.8	2500	7200
FP-3622	2.180	3.560	3.560	1.500	5.2	58.0	29.0	7.5	2.02	1.088	2.197	0.236419	44.3	2500	9000
FP-7042	4.160	7.110	7.110	2.840	10.5	415.0	207.3	14.0	7.00	4.175	29.224	5.861282	172.2	2500	16000

Thomson RM Cores

Part No.	Ht cm 1	Wth cm 2	Lt cm 3	G cm 4	MPL cm 5	Wtfe grams 6	Wtcu grams 7	MLT cm 8	Ac cm sq 9	Wa cm sq 10	Ap cm 4th 11	Kg cm 5th 12	At cm sq 13	Perm 14	AL 15
RM-5	1.04	1.205	1.205	0.65	2.1	3.1	1.6	2.5	0.208	0.182	0.038	0.001237	7.9	2500	1600
RM-6	1.24	1.440	1.440	0.82	2.7	4.9	2.9	3.1	0.313	0.260	0.081	0.003254	11.3	2500	2500
RM-8	1.63	1.925	1.925	1.12	3.5	10.2	7.4	4.2	0.520	0.498	0.259	0.012850	20.0	2500	2800
RM-10	1.86	2.415	2.415	1.27	4.2	20.0	13.0	5.2	0.830	0.695	0.577	0.036572	29.6	2500	4300
RM-14	2.89	3.410	3.410	2.11	7.1	65.0	40.2	7.2	1.780	1.572	2.798	0.276939	62.7	2500	5400

Thomson RU Cores															
Part No.	Ht cm 1	Wth cm 2	Lt cm 3	G cm 4	MPL cm 5	Wtfe grams 6	Wtcu grams 7	MLT cm 8	Ac cm sq 9	Wa cm sq 10	Ap cm 4th 11	Kg cm 5th 12	At cm sq 13	Perm 14	AL 15
R-2810A	3.21	2.29	4.09	2.26	8.5	28	61.8	5.9	0.50	2.938	1.469	0.049637	54.8	2500	2500
R-3012C	4.48	2.75	4.57	3.30	11.3	45	122.1	6.7	0.66	5.115	3.376	0.132793	83.4	2500	1500
R-3110B	4.50	2.35	4.50	2.80	10.9	42	85.1	6.3	0.78	3.780	2.948	0.145381	73.6	2500	3000
R-3012D	5.18	2.70	4.50	3.95	13.0	54	140.4	6.7	0.68	5.925	4.029	0.164474	91.5	2500	1500
R-3511A	4.56	2.51	4.91	2.64	11.4	52	91.5	6.9	0.90	3.722	3.350	0.174495	81.5	2500	2500
R-3110C	5.10	2.35	4.50	3.40	12.1	52	103.3	6.3	0.78	4.590	3.580	0.176535	81.8	2500	2000
R-5211A	4.74	4.15	8.23	2.84	15.0	68	292.7	9.6	0.91	8.605	7.831	0.297999	162.5	2500	2200
R-3611A	5.84	2.67	5.15	4.12	14.2	70	162.3	7.1	0.96	6.448	6.190	0.335720	107.9	2500	2700
R-3513A	5.50	2.51	4.74	3.50	12.7	85	111.3	7.5	1.29	4.200	5.418	0.375160	93.9	2500	3000
R-5221B	5.50	4.15	8.23	3.50	16.4	74	360.7	9.6	0.93	10.605	9.863	0.383572	180.1	2500	2000
R-3513C	6.86	2.58	4.82	4.80	15.4	86	165.0	7.4	1.21	6.264	7.579	0.495278	115.5	2500	2800
R-3718B	5.10	3.29	5.18	3.43	12.7	100	159.0	8.7	1.49	5.111	7.615	0.518847	110.5	2500	4000
R-4718A	4.90	3.75	6.69	3.18	14.5	104	221.8	10.1	1.50	6.201	9.301	0.554789	139.1	2500	4000
R-4115A	6.74	2.84	5.46	4.36	15.4	152	179.6	8.4	1.65	6.017	9.928	0.780537	129.8	2500	3500
R-3915A	7.04	3.10	5.47	4.96	16.2	120	236.8	8.4	1.47	7.936	11.666	0.817588	141.3	2500	3000
R-5816A	5.68	4.33	8.53	3.30	16.6	140	360.2	11.0	1.69	9.174	15.504	0.949119	193.5	2500	4000
R-4022A	6.38	3.76	5.42	4.78	15.8	146	258.7	9.8	1.77	7.457	13.199	0.957710	147.9	2500	4000
R-4215A	7.26	2.96	5.67	4.80	16.8	120	214.7	8.7	1.78	6.960	12.389	1.016869	144.6	2500	3700
R-7016A	6.36	5.29	10.70	3.82	19.6	170	648.2	12.9	1.76	14.134	24.876	1.357881	279.0	2500	4100
R-5917A	7.16	4.35	8.55	4.38	19.1	192	471.3	11.4	2.08	11.607	24.143	1.759128	229.6	2500	4800
R-6420A	8.10	4.70	8.71	5.30	21.0	304	568.9	13.1	2.90	12.190	35.351	3.124593	273.2	2500	5000
R-5536A	7.50	5.60	7.52	5.10	18.9	420	524.8	14.5	4.33	10.200	44.166	5.287220	261.8	2500	12000

Thomson SU Cores

Part No.	Ht cm 1	Wth cm 2	Lt cm 3	G cm 4	MPL cm 5	Wtfe grams 6	Wtcu grams 7	MLT cm 8	Ac cm sq 9	Wa cm sq 10	Ap cm 4th 11	Kg cm 5th 12	At cm sq 13	Perm 14	AL 15
UP-1105A	1.56	1.00	1.550	1.05	4.0	2.7	5.1	2.7	0.128	0.525	0.067	0.001254	10.0	4000	1300
UP-1204A	1.84	0.74	1.545	1.01	4.0	3.5	3.2	2.6	0.165	0.348	0.057	0.001462	9.6	2500	950
UP-1606A	1.98	1.25	2.220	1.44	5.1	6.6	11.8	3.5	0.248	0.936	0.232	0.006489	17.9	2500	1200
UP-1506A	2.24	1.17	2.040	1.21	5.1	8.4	7.9	3.5	0.326	0.629	0.205	0.007611	17.0	2500	1600
UP-1706A	2.24	1.31	2.320	1.21	5.3	9.0	10.6	3.7	0.326	0.799	0.260	0.009091	19.6	2500	1540
UP-2007A	3.06	1.38	2.730	1.65	6.8	18.0	16.1	4.4	0.543	1.040	0.564	0.028073	29.1	2500	2000
UP-2507A	3.68	1.72	3.440	2.17	8.5	24.0	37.3	4.9	0.540	2.148	1.160	0.051324	43.8	2500	1250
UP-2513A	3.92	2.11	3.320	2.28	8.7	44.0	41.6	6.1	1.050	1.915	2.011	0.138305	51.1	2500	3000
UP-3126A	5.28	3.65	4.120	3.25	11.8	160.0	114.2	9.5	2.660	3.380	8.991	1.006885	103.7	2500	5600
UP-4628A	7.90	4.40	6.200	5.10	1.8	360.0	340.1	11.7	3.900	8.160	31.824	4.235959	204.1	2500	5600
UP-9316A	15.20	5.06	12.760	9.60	35.5	770.0	1781.2	15.1	4.500	33.216	149.472	17.841250	593.4	2500	3200
UP-9320A	15.20	5.46	12.760	9.60	35.5	960.0	1875.7	15.9	5.600	33.216	186.010	26.237800	615.7	2500	4000
UP-102A	11.42	7.62	15.240	6.34	30.8	1000.0	2146.7	18.7	6.450	32.207	207.736	28.594290	651.2	2500	5200
UP-141A	15.70	6.50	19.100	6.70	37.7	1600.0	2449.0	20.6	6.750	33.500	226.125	29.698290	872.1	2500	4000
UP-126A	18.20	9.00	19.600	12.60	48.0	2078.0	6648.5	21.2	5.600	88.200	493.920	52.192700	1237.0	2500	3000
UP-9330B	15.20	6.46	12.760	9.60	35.5	1450.0	2111.9	17.9	8.400	33.216	279.014	52.431650	671.6	2500	6000
UP-141B	15.70	8.00	19.100	6.70	37.7	3200.0	2806.4	23.6	13.500	33.500	452.250	103.665400	976.5	2500	9000

Thomson Toroidal Cores																
Part No.	OD cm 1	ID cm 2	HT cm 3	Wound OD cm 4	Wound HT cm 5	MPL cm 6	Wtfe grams 7	Wtcu grams 8	MLT cm 9	Ac cm sq 10	Wa cm sq 11	Ap cm 4th 12	Kg cm 5th 13	At cm sq 14	Perm 15	AL 16
T-0400A	0.400	0.240	0.160	0.518	0.280	1.0	0.1	0.1	0.6	0.013	0.045	0.001	0.000005	0.9	6000	960
T-0480A	0.484	0.228	0.128	0.601	0.242	1.1	0.1	0.1	0.6	0.016	0.041	0.001	0.000007	1.0	6000	1150
T-0500A	0.500	0.300	0.200	0.648	0.350	1.3	0.1	0.2	0.7	0.020	0.071	0.001	0.000016	1.4	6000	1220
T-0630A	0.630	0.380	0.250	0.817	0.440	1.6	0.2	0.4	0.9	0.032	0.113	0.004	0.000051	2.2	6000	1500
T-0800B	0.800	0.400	0.250	1.003	0.450	1.9	0.5	0.5	1.0	0.050	0.126	0.006	0.000121	3.0	6000	2100
T-0950A	0.952	0.475	0.317	1.193	0.555	2.2	0.8	0.8	1.3	0.076	0.177	0.013	0.000323	4.3	6000	2500
T-1000C	1.000	0.600	0.300	1.296	0.600	2.5	0.9	1.3	1.3	0.071	0.283	0.020	0.000445	5.1	6000	1800
T-1000A	1.000	0.600	0.400	1.296	0.700	2.5	1.0	1.4	1.4	0.080	0.283	0.023	0.000502	5.5	6000	2400
T-1300A	1.335	0.730	0.320	1.699	0.685	3.2	1.5	2.4	1.6	0.097	0.418	0.041	0.000996	8.2	6000	2200
T-1300B	1.302	0.720	0.320	1.660	0.680	3.3	1.7	2.2	1.6	0.107	0.407	0.044	0.001200	7.9	6000	2300
T-1270A	1.270	0.714	0.470	1.624	0.827	3.1	2.0	2.5	1.8	0.131	0.400	0.052	0.001554	8.4	6000	3100
T-1300C	1.320	0.720	0.500	1.679	0.860	3.2	2.3	2.7	1.9	0.150	0.407	0.061	0.001973	9.0	6000	3600
T-1400A	1.400	0.900	0.500	1.843	0.950	3.6	2.2	4.3	1.9	0.125	0.636	0.079	0.002070	10.8	6000	2400
T-1270C	1.270	0.792	0.635	1.660	1.031	3.2	2.4	3.6	2.0	0.152	0.492	0.075	0.002239	9.7	6000	3500
T-1270B	1.270	0.714	0.635	1.624	0.992	3.1	2.7	2.9	2.0	0.177	0.400	0.071	0.002468	9.2	6000	4200
T-1300D	1.335	0.730	0.600	1.699	0.965	3.2	2.8	3.0	2.0	0.181	0.418	0.076	0.002703	9.7	6000	3850
T-1600C	1.550	0.720	0.500	1.921	0.860	3.6	3.6	3.0	2.0	0.208	0.407	0.085	0.003452	11.0	6000	4600
T-1600B	1.590	1.100	0.625	2.132	1.175	4.2	3.1	7.7	2.3	0.150	0.950	0.142	0.003763	15.0	6000	2760
T-1400B	1.400	0.900	0.900	1.843	1.350	3.6	3.9	5.8	2.6	0.225	0.636	0.143	0.005030	13.1	6000	4470
T-1600A	1.600	0.960	0.630	2.074	1.110	4.0	3.9	5.9	2.3	0.200	0.723	0.145	0.005059	14.0	6000	3800
T-1250A	1.235	0.580	1.200	1.533	1.490	2.9	5.5	2.7	2.9	0.396	0.264	0.105	0.005696	10.9	6000	10100
T-2100A	2.060	1.400	0.500	2.749	1.200	5.4	4.3	13.4	2.4	0.165	1.539	0.254	0.006845	22.2	6000	2300
T-1900C	1.880	1.100	0.800	2.424	1.350	4.7	7.1	9.4	2.8	0.312	0.950	0.296	0.013285	19.5	6000	4800
T-2000B	2.000	1.000	0.700	2.506	1.200	4.7	8.0	7.6	2.7	0.350	0.785	0.275	0.014142	19.3	6000	5400
T-2210B	2.210	1.372	0.635	2.886	1.321	5.6	7.3	14.6	2.8	0.270	1.478	0.399	0.015477	25.0	6000	3500
T-2000C	2.000	1.000	0.800	2.506	1.300	4.7	9.1	8.0	2.9	0.400	0.785	0.314	0.017444	20.1	5000	5540
T-2000A	2.000	1.000	1.000	2.506	1.500	4.7	11.4	8.9	3.2	0.500	0.785	0.393	0.024531	21.7	6000	8000
T-1900A	1.900	1.140	1.500	2.462	2.070	4.8	13.2	14.2	3.9	0.570	1.020	0.582	0.033822	25.5	5000	6700
T-2540A	2.540	1.550	0.793	3.304	1.568	6.4	12.2	22.1	3.3	0.393	1.886	0.741	0.035299	33.4	5000	3910
T-2000D	2.000	1.050	1.500	2.527	2.025	4.8	16.6	12.3	4.0	0.712	0.865	0.616	0.043874	26.1	5000	9650

Thomson Toroidal Cores

Part No.	OD cm 1	ID cm 2	HT cm 3	Wound OD cm 4	Wound HT cm 5	MPL cm 6	Wtfe grams 7	Wtcu grams 8	MLT cm 9	Ac cm sq 10	Wa cm sq 11	Ap cm 4th 12	Kg cm 5th 13	At cm sq 14	Perm 15	AL 16
T-2210A	2.210	1.372	1.270	2.886	1.956	5.6	14.5	20.0	3.8	0.532	1.478	0.786	0.044023	30.8	6000	6900
T-2500A	2.500	1.500	1.000	3.240	1.750	6.3	15.3	22.6	3.6	0.500	1.766	0.883	0.049063	34.3	6000	6000
T-2600A	2.600	1.450	1.000	3.320	1.725	3.7	10.2	21.6	3.7	0.575	1.650	0.949	0.059314	35.3	6000	6500
T-2500B	2.500	1.500	1.500	3.240	2.250	6.3	22.8	27.6	4.4	0.750	1.766	1.325	0.090320	39.4	6000	8400
T-3800B	3.810	1.905	0.635	4.775	1.588	9.0	26.3	41.2	4.1	0.605	2.849	1.724	0.102631	59.6	2500	2175
T-2600B	2.600	1.450	1.495	3.320	2.220	6.4	26.5	26.2	4.5	0.860	1.650	1.419	0.109185	40.5	6000	9900
T-3150A	3.150	1.900	1.250	4.087	2.200	7.9	30.0	45.5	4.5	0.780	2.834	2.210	0.152577	54.5	6000	7500
T-2600C	2.600	1.450	2.000	3.320	2.725	6.4	35.5	31.0	5.3	1.150	1.650	1.898	0.165359	45.7	6000	13200
T-2800A	2.760	1.760	1.900	3.626	2.780	7.1	32.7	45.4	5.2	0.950	2.432	2.310	0.167266	52.3	6000	10250
T-3600A	3.600	2.300	1.500	4.731	2.650	9.3	43.8	78.0	5.3	0.975	4.153	4.049	0.299062	74.5	6000	7700
T-3800A	3.810	1.905	1.270	4.775	2.223	9.0	52.7	51.5	5.1	1.210	2.849	3.447	0.328418	69.1	2500	4000
T-5000A	5.000	1.500	1.000	5.941	1.750	10.2	86.6	35.2	5.6	1.750	1.766	3.091	0.386367	88.1	4000	6000
T-3600B	3.600	2.300	2.000	4.731	3.150	9.3	58.6	89.8	6.1	1.300	4.153	5.398	0.461709	81.9	2500	4500
T-4000A	4.000	2.400	1.600	5.184	2.800	10.0	62.1	92.6	5.8	1.280	4.522	5.788	0.514458	87.8	6000	9600
T-5200A	5.130	3.100	1.000	6.659	2.550	12.9	63.8	153.0	5.7	1.020	7.544	7.695	0.550394	122.9	2500	2500
T-5600A	5.540	3.235	1.800	7.139	3.418	13.8	138.5	213.6	7.3	2.070	8.215	17.005	1.925674	156.6	6000	11000
T-7500A	7.500	2.300	2.000	8.924	3.150	15.0	366.7	135.9	9.2	5.040	4.153	20.929	4.586259	213.3	2500	11100
T-6300A	6.300	3.800	2.500	8.174	4.400	16.0	244.4	364.4	9.0	3.150	11.335	35.707	4.976792	217.8	6000	15000
T-8000A	8.000	4.000	1.500	10.025	3.500	18.8	273.5	393.0	8.8	3.000	12.560	37.680	5.138182	268.0	2500	5000
T-100A	10.000	5.500	2.000	12.738	4.750	24.3	530.3	945.7	11.2	4.500	23.746	106.858	17.173630	444.7	2500	5300
T-152A	15.200	6.850	1.900	18.764	5.325	34.6	1330.7	1990.9	15.2	7.930	36.834	292.095	60.955590	866.5	2500	6600

Thomson UI Cores															
Part No.	Ht cm 1	Wth cm 2	Lt cm 3	G cm 4	MPL cm 5	Wtfe grams 6	Wtcu grams 7	MLT cm 8	Ac cm sq 9	Wa cm sq 10	Ap cm 4th 11	Kg cm 5th 12	At cm sq 13	Perm 14	AL 15
UI-9316A	10.35	5.06	12.76	4.8	25.5	580	890.6	15.1	4.56	16.608	75.732	9.160092	421.2	2500	4500
UI-9320A	10.35	5.46	12.76	4.8	25.5	750	937.9	15.9	5.70	16.608	94.666	13.591620	439.7	2500	5600
UI-102A	8.24	7.62	15.24	3.2	24.5	800	1073.3	18.7	6.45	16.104	103.868	14.297150	506.0	2500	6600
UI-9330A	10.35	6.46	12.76	4.8	25.5	1100	1056.0	17.9	8.56	16.608	142.165	27.224040	485.9	2500	8400
UI-126A	11.90	9.00	19.60	6.3	34.5	750	3324.3	21.2	6.00	44.100	264.600	29.957540	876.7	2500	4000

Tokin EE Cores															
Part No.	Ht cm 1	Wth cm 2	Lt cm 3	G cm 4	MPL cm 5	Wtfe grams 6	Wtcu grams 7	MLT cm 8	Ac cm sq 9	Wa cm sq 10	Ap cm 4th 11	Kg cm 5th 12	At cm sq 13	Perm 14	AL 15
FEE-12.5W	1.460	1.150	1.250	1.000	3.1	2.4	3.4	2.9	0.149	0.325	0.048	0.000986	9.2	2500	950
FEE-16	1.400	1.260	1.600	1.000	3.5	3.3	4.7	3.4	0.190	0.380	0.072	0.001595	10.8	2500	1200
FEE-20x8x5	1.600	1.490	2.020	1.160	4.0	4.3	8.1	4.0	0.216	0.574	0.124	0.002704	15.3	2500	1020
FEE-19x16x5	1.584	1.412	1.930	1.114	4.0	7.0	7.0	3.8	0.228	0.522	0.119	0.002882	14.3	2500	1450
FEE-19	1.560	1.490	1.910	1.100	4.0	4.8	7.4	3.9	0.230	0.534	0.123	0.002887	14.5	2500	1450
FEE-16W	2.400	1.280	1.600	2.000	5.5	5.2	9.5	3.4	0.190	0.780	0.148	0.003281	16.7	2500	700
FEE-19x12x5	2.460	1.480	1.910	2.000	5.8	6.5	13.3	3.9	0.223	0.960	0.214	0.004902	20.5	2300	960
FEE-19W	2.660	1.480	1.910	2.200	6.2	7.0	14.6	3.9	0.224	1.056	0.237	0.005441	21.9	2500	900
FEE-19.2	2.200	1.270	1.920	1.610	4.9	7.3	8.0	3.8	0.298	0.596	0.178	0.005613	17.2	2500	2620
FEE-22	1.840	1.300	2.200	1.040	4.0	8.8	5.1	3.9	0.410	0.364	0.149	0.006264	16.9	2500	2560
FEE-25x19x6	1.854	1.855	2.520	1.296	4.9	15.0	13.9	4.9	0.368	0.807	0.297	0.009006	22.4	2300	1900
FEE-25S	2.030	1.840	2.500	1.350	5.0	10.2	14.0	4.9	0.409	0.810	0.331	0.011170	23.7	2500	1860
FEE-25Z	1.990	1.940	2.530	1.350	5.0	10.3	15.2	5.1	0.414	0.837	0.347	0.011219	24.2	2500	2090
FEE-22SW	2.880	1.560	2.200	2.120	6.4	11.1	15.6	4.3	0.350	1.018	0.356	0.011555	26.2	2500	1380
FEE-22W	2.860	1.280	2.200	2.060	5.9	11.9	9.7	3.9	0.400	0.700	0.280	0.011566	23.9	2500	1700
FEE-25	2.020	1.840	2.530	1.320	5.0	10.6	13.7	4.9	0.424	0.792	0.336	0.011738	23.9	2500	2130
FEE-25L	3.120	1.840	2.540	2.514	7.4	14.7	25.9	4.9	0.400	1.483	0.593	0.019331	33.5	2500	1350
FEE-25W	3.060	1.920	2.530	2.440	7.2	15.0	26.9	5.1	0.417	1.488	0.621	0.020366	33.6	2500	1450
FEE-25SW	3.380	1.840	2.500	2.700	7.7	15.5	28.0	4.9	0.404	1.620	0.654	0.021798	35.5	2500	1380
FEE-25D	1.840	2.560	2.530	1.220	4.7	19.7	17.2	6.6	0.840	0.738	0.620	0.031719	28.4	2500	4500
FEE-29x30x7.1	3.000	1.950	2.980	2.200	7.0	20.1	26.2	5.4	0.570	1.364	0.777	0.032859	37.2	2500	2040
FEE-28S	2.100	2.210	2.800	1.240	4.9	21.5	14.3	5.9	0.870	0.688	0.599	0.035613	30.1	2500	4450
FEE-28x12.5x10.9	2.500	2.290	2.800	1.700	5.9	24.5	22.0	6.1	0.826	1.020	0.843	0.045845	34.4	2500	3500
FEE-29	2.720	2.200	2.900	1.800	6.2	26.5	23.1	6.0	0.850	1.080	0.918	0.051916	36.4	2500	3460
FEE-30	2.600	1.950	3.000	1.600	5.8	33.0	14.9	6.1	1.090	0.680	0.741	0.052611	33.7	2500	4750
FEE-34x28x9	2.794	2.528	3.440	1.940	6.9	36.0	36.8	6.6	0.816	1.560	1.273	0.062597	44.7	2500	2950
FEE-29M	3.000	2.310	2.980	2.200	7.0	30.2	29.7	6.1	0.860	1.364	1.173	0.065992	41.5	2500	3070
FEE-28W	3.300	2.210	2.800	2.400	7.3	32.1	27.7	5.9	0.877	1.332	1.168	0.070042	42.1	2500	3000
FEE-33M	2.700	2.620	3.300	1.920	6.6	37.6	33.2	7.2	1.130	1.296	1.464	0.091841	44.0	2500	4280
FEE-35M	3.100	2.450	3.500	1.900	7.0	37.0	32.7	6.7	1.060	1.378	1.460	0.092618	48.2	2500	3790

Tokin EE Cores															
Part No.	Ht cm 1	Wth cm 2	Lt cm 3	G cm 4	MPL cm 5	Wtfe grams 6	Wtcu grams 7	MLT cm 8	Ac cm sq 9	Wa cm sq 10	Ap cm 4th 11	Kg cm 5th 12	At cm sq 13	Perm 14	AL 15
FEE-30W	4.200	1.950	3.000	3.200	9.0	49.1	29.7	6.1	1.110	1.360	1.510	0.109119	49.5	2500	3000
FEE-33x32x13	3.220	2.600	3.300	2.320	7.6	43.6	38.9	7.2	1.150	1.508	1.734	0.110047	50.0	2500	3840
FEE-33x35x11.5	3.500	2.670	3.300	2.600	8.2	40.3	48.4	6.9	0.990	1.976	1.956	0.112362	53.9	2500	3000
FEE-31	4.100	2.150	3.150	3.060	8.9	48.8	36.9	6.5	1.100	1.607	1.767	0.120428	52.4	2500	3130
FEE-40	3.340	2.760	4.000	2.000	7.7	51.1	43.8	7.4	1.280	1.660	2.125	0.146732	59.9	2500	4200
FEE-35W	4.760	2.430	3.500	3.600	10.5	52.1	61.4	6.7	1.000	2.574	2.574	0.153372	67.6	2500	2360
FEE-40S	3.400	2.720	4.000	2.000	7.8	55.8	41.0	7.6	1.450	1.520	2.204	0.168324	60.2	2500	4750
FEE-33W	2.660	2.500	3.300	3.760	10.5	61.9	58.5	7.3	1.180	2.256	2.662	0.172312	42.1	2500	2830
FEE-35FT	4.400	2.670	3.500	3.200	9.7	58.8	60.0	7.2	1.210	2.352	2.846	0.191952	66.9	2500	3150
FEE-35SW	4.760	2.630	3.500	3.600	10.5	62.8	65.1	7.1	1.200	2.574	3.089	0.208436	70.9	2500	2850
FEE-40x44.6x11	4.460	2.680	4.000	3.060	9.7	70.5	60.6	7.5	1.450	2.264	3.283	0.252849	73.9	2500	3750
FEE-30DW	4.200	3.050	3.000	3.200	9.0	100.0	41.3	8.5	2.210	1.360	3.006	0.311027	65.4	2500	6100
FEE-40W	5.400	2.720	4.000	4.000	11.7	83.9	82.1	7.6	1.480	3.040	4.499	0.350722	87.0	2500	3070
FEE-42x15	4.200	3.270	4.200	3.100	9.9	97.5	85.8	8.9	1.760	2.713	4.774	0.377820	81.5	2500	4460
FEE-47x39x16	3.886	3.175	4.712	2.414	8.9	120.0	65.0	9.4	2.371	1.947	4.616	0.466308	81.1	2300	6650
FEE-50	4.200	3.450	5.000	2.500	9.6	110.0	83.8	9.7	2.260	2.438	5.509	0.515012	93.8	2500	5900
FEE-42x20	4.200	3.750	4.200	3.100	9.8	130.0	95.1	9.9	2.370	2.713	6.429	0.618369	89.5	2500	6000
FEE-44W	6.000	3.450	4.400	4.600	13.3	123.3	142.6	9.1	1.900	4.416	8.390	0.702095	115.3	2500	3500
FEE-60	4.400	4.400	6.000	2.760	1.1	140.0	156.7	11.4	2.470	3.864	9.544	0.826864	126.4	2500	5630
FEE-50W	6.600	3.400	5.000	4.900	14.2	161.8	158.8	9.6	2.300	4.655	10.707	1.027002	133.5	2500	4000
FEE-57x47x19	4.674	3.810	5.657	2.896	10.7	210.0	110.9	11.2	3.440	2.795	9.614	1.185530	117.1	2300	8000
FEE-55A	5.300	4.300	5.500	3.700	12.4	202.5	165.9	11.5	3.280	4.070	13.350	1.528065	137.1	2500	6720
FEE-60W	7.100	4.410	6.000	5.500	16.5	200.7	313.8	11.4	2.470	7.728	19.087	1.651347	182.8	2500	3700
FEE-72	5.400	5.200	7.130	3.500	13.4	240.0	277.0	13.5	3.580	5.775	20.675	2.194817	181.6	2500	6700
FEE-80	7.500	5.920	7.930	5.600	18.3	350.0	578.8	14.8	3.810	11.032	42.032	4.341705	270.1	2500	5200
FEE-83S	7.500	6.750	8.250	5.640	18.9	351.3	748.9	15.7	3.710	13.451	49.905	4.730091	302.0	2500	4900
FEE-83SS	8.500	6.770	8.300	6.640	21.1	360.2	887.2	15.7	3.420	15.903	54.388	4.742529	334.1	2500	4050
FEE-70	10.800	4.440	7.000	8.580	23.2	515.0	479.4	12.9	4.445	10.468	46.528	6.423552	287.6	2500	4820
FEE-78	9.900	6.240	7.730	8.000	22.9	540.0	823.5	15.5	4.810	14.960	71.958	8.943682	343.2	2500	5300

Tokin EI Cores															
Part No.	Ht cm 1	Wth cm 2	Lt cm 3	G cm 4	MPL cm 5	Wtfe grams 6	Wtcu grams 7	MLT cm 8	Ac cm sq 9	Wa cm sq 10	Ap cm 4th 11	Kg cm 5th 12	At cm sq 13	Perm 14	AL 15
FEI-12.5	0.88	1.15	1.25	0.500	2.2	1.9	1.7	2.9	0.152	0.163	0.025	0.000513	6.4	2300	1200
FEI-35S	2.93	2.67	3.50	1.800	6.7	41.5	33.8	7.2	0.120	1.323	0.159	0.001062	48.8	2300	3950
FEI-19N	1.17	1.48	1.91	0.710	3.2	3.5	4.7	3.9	0.221	0.341	0.075	0.001709	11.8	2500	1800
FEI-16	1.40	1.28	1.60	1.000	3.5	3.2	4.8	3.4	0.194	0.390	0.076	0.001710	10.9	2300	1100
FEI-19	1.56	1.49	1.91	1.100	4.0	4.4	7.4	3.9	0.227	0.534	0.121	0.002812	14.5	2300	1100
FEI-22S	1.84	1.60	2.20	1.060	4.2	7.7	8.3	4.4	0.362	0.530	0.192	0.006346	18.6	2300	1750
FEI-22	1.88	1.30	2.20	1.030	3.9	8.8	5.0	3.9	0.417	0.361	0.150	0.006418	17.2	2300	2200
FEI-25	1.80	1.94	2.53	1.220	4.7	11.0	13.8	5.1	0.412	0.756	0.312	0.010041	22.5	2300	1850
FEI-25S	1.99	1.86	2.50	1.350	4.9	10.0	14.3	4.9	0.460	0.824	0.379	0.014273	23.5	2500	2150
FEI-66	4.47	4.78	6.57	2.700	11.1	199.0	173.1	13.0	0.356	3.753	1.336	0.014666	142.0	2500	8000
FEI-25x21x7	2.54	1.72	2.54	1.775	5.8	15.5	15.1	4.9	0.536	0.861	0.461	0.020063	27.5	2500	2300
FEI-30S	2.65	1.55	3.00	1.600	5.8	21.1	12.9	5.3	0.727	0.680	0.494	0.026909	29.7	2500	3140
FEI-28	2.00	2.23	2.80	1.200	4.9	24.0	14.2	5.9	0.851	0.678	0.577	0.033390	29.2	2300	3750
FEI-30	2.65	1.98	3.00	1.600	5.8	35.0	15.5	6.2	1.110	0.704	0.781	0.056055	34.4	2300	4350
FEI-35	2.93	2.47	3.50	1.800	6.7	36.2	31.9	6.8	0.999	1.323	1.322	0.077944	46.2	2300	3400
FEI-33	2.83	2.62	3.30	1.880	6.7	41.5	32.1	7.3	1.181	1.241	1.465	0.095084	45.5	2300	4000
FEI-35T	2.90	2.69	3.50	1.800	6.8	41.1	34.1	7.1	1.212	1.341	1.625	0.110243	48.8	2500	4200
FEI-40	3.45	2.72	4.00	2.000	7.7	61.0	41.0	7.6	1.400	1.520	2.128	0.156916	60.8	2500	4350
FEI-44	3.70	3.45	4.40	2.300	8.6	82.7	71.3	9.1	1.900	2.208	4.195	0.351047	79.2	2500	5500
FEI-50	4.20	3.40	5.00	2.450	9.5	110.0	79.4	9.6	2.300	2.328	5.353	0.513501	93.2	2300	5500
FEI-60	4.40	4.41	6.00	2.750	10.9	140.0	156.9	11.4	2.470	3.864	9.543	0.825673	126.5	2300	5100

Tokin EP Cores															
Part No.	Ht cm 1	Wth cm 2	Lt cm 3	G cm 4	MPL cm 5	Wtfe grams 6	Wtcu grams 7	MLT cm 8	Ac cm sq 9	Wa cm sq 10	Ap cm 4th 11	Kg cm 5th 12	At cm sq 13	Perm 14	AL 15
FEP-7	0.850	0.75	0.92	0.50	1.6	1.4	0.7	1.9	0.10	0.100	0.010	0.000216	3.6	2500	550
FEP-10	1.083	1.04	1.15	0.72	1.9	2.8	1.7	2.2	0.11	0.214	0.024	0.000476	5.8	2500	850
FEP-13	1.175	1.30	1.25	0.90	2.4	5.1	2.1	2.4	0.20	0.248	0.050	0.001625	7.8	2500	1250
FEP-17	1.433	1.70	1.80	1.10	2.9	11.8	3.7	3.0	0.34	0.338	0.115	0.005143	14.0	2500	2500
FEP-20	1.905	2.16	2.40	1.40	4.0	27.6	7.9	4.2	0.78	0.525	0.410	0.030118	24.3	2500	4200

Tokin EQK Cores															
Part No.	Ht cm 1	Wth cm 2	Lt cm 3	G cm 4	MPL cm 5	Wtfe grams 6	Wtcu grams 7	MLT cm 8	Ac cm sq 9	Wa cm sq 10	Ap cm 4th 11	Kg cm 5th 12	At cm sq 13	Perm 14	AL 15
FQK-1623	0.87	3.03	1.65	1.72	4.4	9.6	15.4	7.5	0.313	0.576	0.180	0.002995	16.6	2300	1350
FQK-2522	1.29	3.28	2.50	1.61	4.9	21.0	24.8	8.5	0.679	0.821	0.558	0.017837	27.7	2300	3200
FQK-2532	1.29	4.22	2.50	2.38	6.4	30.0	44.8	10.4	0.703	1.214	0.853	0.023140	34.8	2300	2500

Tokin ETD Cores

Part No.	Ht cm 1	Wth cm 2	Lt cm 3	G cm 4	MPL cm 5	Wtfe grams 6	Wtcu grams 7	MLT cm 8	Ac cm sq 9	Wa cm sq 10	Ap cm 4th 11	Kg cm 5th 12	At cm sq 13	Perm 14	AL 15
EER-25.5	1.86	1.98	2.55	1.24	5.0	10.8	14.5	5.3	0.432	0.763	0.329	0.010662	23.3	2300	1850
EER-28	2.80	2.27	2.85	1.92	6.7	28.5	24.9	6.4	0.853	1.085	0.925	0.049010	37.1	2300	2870
EER-29	3.16	2.43	3.06	2.20	7.4	27.0	34.9	6.1	0.750	1.595	1.196	0.058406	45.2	2300	2500
EER-30	2.65	2.02	3.00	1.65	6.2	37.1	15.6	6.5	1.190	0.677	0.805	0.058995	34.8	2500	4830
EER-28L	3.38	2.27	2.85	2.50	7.8	33.2	32.4	6.4	0.847	1.413	1.196	0.062920	43.1	2300	2520
EER-34	3.46	2.55	3.50	2.36	8.3	41.5	43.0	7.1	0.966	1.699	1.641	0.089220	53.3	2300	2700
EER-35	3.36	2.56	3.50	2.16	7.9	45.0	39.4	7.2	1.101	1.544	1.700	0.104397	52.2	2300	3200
EER-35L	4.14	2.56	3.50	2.94	9.5	50.7	53.6	7.2	1.084	2.102	2.279	0.137742	61.6	2300	2770
EER-39	4.14	2.93	3.89	2.84	9.7	60.0	69.3	8.3	1.232	2.343	2.887	0.171005	70.8	2300	2950
EER-39L	4.44	2.87	3.90	3.40	10.6	70.0	79.1	8.2	1.319	2.703	3.565	0.228716	74.4	2300	3200
EER-40	4.48	2.90	4.00	3.08	10.2	78.1	72.2	8.4	1.528	2.418	3.694	0.269039	76.7	2300	3450
EER-42	4.24	3.05	4.20	3.00	10.2	87.6	74.2	9.1	1.825	2.295	4.188	0.336357	77.8	2300	4200
EER-44	4.50	3.25	4.38	3.22	10.9	88.6	93.1	9.4	1.720	2.785	4.791	0.350487	86.6	2300	3700
EER-43	4.30	3.19	4.30	3.06	10.3	103.5	81.7	9.4	2.000	2.433	4.865	0.412134	81.9	2500	5370
EER-42A	4.24	3.45	4.20	3.00	10.1	118.0	80.0	10.3	2.325	2.175	5.057	0.454628	83.8	2300	5340
EER-49	3.76	3.65	4.90	2.44	9.7	110.0	88.1	10.5	2.291	2.355	5.394	0.469992	86.8	2500	6520
EER-42L	5.08	2.98	4.20	3.58	11.4	135.9	75.5	9.5	2.390	2.238	5.348	0.538679	89.3	2500	5800
EER-49L	5.36	3.64	4.90	3.66	12.4	150.0	131.2	10.5	2.320	3.514	8.152	0.720277	114.0	2500	5390
EER-49LL	6.30	3.64	4.90	4.60	14.2	171.5	164.9	10.5	2.413	4.416	10.656	0.979299	130.1	2500	4240

Tokin PC Cores

Part No.	Ht cm 1	Wth cm 2	Lt cm 3	G cm 4	MPL cm 5	Wtfe grams 6	Wtcu grams 7	MLT cm 8	Ac cm sq 9	Wa cm sq 10	Ap cm 4th 11	Kg cm 5th 12	At cm sq 13	Perm 14	AL 15
FP-1107	0.66	1.13	1.13	0.44	1.6	1.6	0.3	2.9	0.159	0.029	0.005	0.000101	4.3	2500	1700
FP-0905	0.54	0.93	0.93	0.36	1.3	0.8	0.4	1.9	0.100	0.065	0.006	0.000133	2.9	2500	1100
FP-1408	0.84	1.41	1.41	0.56	2.0	2.9	1.6	2.9	0.250	0.157	0.039	0.001341	6.8	2500	2520
FP-1811	1.06	1.80	1.80	0.72	2.6	6.4	3.5	3.7	0.432	0.263	0.114	0.005313	11.1	2500	3450
FP-2213	1.36	2.20	2.20	0.92	3.2	12.4	6.2	4.4	0.633	0.391	0.248	0.014095	17.0	2500	4200
FP-2616	1.63	2.60	2.60	1.10	3.7	21.6	10.0	5.3	0.930	0.534	0.496	0.034864	23.9	2500	5300
FP-3019	1.90	3.05	3.05	1.30	4.5	36.2	16.5	6.2	1.360	0.748	1.017	0.089132	32.8	2500	6800
FP-3622	2.20	3.60	3.60	1.46	5.2	60.0	26.6	7.5	2.020	1.000	2.020	0.218332	45.2	2500	8800
FP-4229	2.97	4.31	4.31	2.03	6.9	115.0	55.6	8.6	2.650	1.817	4.815	0.592969	69.4	2500	9130

Tokin PQ Cores

Part No.	Ht cm 1	Wth cm 2	Lt cm 3	G cm 4	MPL cm 5	Wtfe grams 6	Wtcu grams 7	MLT cm 8	Ac cm sq 9	Wa cm sq 10	Ap cm 4th 11	Kg cm 5th 12	At cm sq 13	Perm 14	AL 15
FPQ-2016	1.64	2.30	2.090	1.00	3.7	13	6.6	4.3	0.62	0.430	0.267	0.015247	17.6	2300	3500
FPQ-2020	2.04	2.30	2.090	1.40	4.5	15	9.3	4.3	0.62	0.602	0.373	0.021346	20.4	2300	2950
FPQ-2620	2.04	2.93	2.695	1.12	4.6	13	11.0	5.6	1.19	0.552	0.656	0.055636	29.4	2300	5650
FPQ-2625	2.50	2.93	2.695	1.58	5.6	36	15.5	5.6	1.18	0.778	0.918	0.077174	33.7	2300	4600
FPQ-3220	2.08	3.58	3.250	1.12	5.6	42	17.6	6.6	1.70	0.745	1.266	0.129891	37.5	2300	7310
FPQ-3230	3.06	3.58	3.250	2.10	7.5	55	32.9	6.6	1.61	1.397	2.248	0.218441	48.3	2300	5140
FPQ-3535	3.50	4.34	3.570	2.47	8.8	73	55.5	7.5	1.96	2.087	4.091	0.428980	62.5	2300	4860
FPQ-4040	4.00	4.98	4.140	2.92	1.0	95	91.8	8.3	2.01	3.095	6.221	0.599768	79.7	2300	4300

Tokin RM Cores

Part No.	Ht cm 1	Wth cm 2	Lt cm 3	G cm 4	MPL cm 5	Wtfe grams 6	Wtcu grams 7	MLT cm 8	Ac cm sq 9	Wa cm sq 10	Ap cm 4th 11	Kg cm 5th 12	At cm sq 13	Perm 14	AL 15
RM-4	1.04	0.960	0.960	0.72	2.3	1.7	1.1	2.0	0.14	0.157	0.022	0.000603	5.8	4000	1240
RM-5	1.04	1.205	1.205	0.65	2.1	2.8	1.6	2.5	0.21	0.182	0.038	0.001261	7.9	2500	1350
RM-6	1.24	1.440	1.440	0.82	2.7	4.6	2.9	3.1	0.31	0.260	0.081	0.003192	11.3	2500	1720
RM-8	1.64	1.935	1.935	1.10	3.5	10.4	7.3	4.2	0.52	0.490	0.255	0.012621	20.2	2500	2200
RM-10	1.86	2.415	2.415	1.27	4.4	23.0	13.0	5.2	0.98	0.695	0.681	0.050985	29.6	2500	3900

Tokin Toroidal Cores

Part No.	OD cm 1	ID cm 2	HT cm 3	Wound OD cm 4	Wound HT cm 5	MPL cm 6	Wtfe grams 7	Wtcu grams 8	MLT cm 9	Ac cm sq 10	Wa cm sq 11	Ap cm 4th 12	Kg cm 5th 13	At cm sq 14	Perm 15	AL 16
FR-8.6x4.5x3.2	0.86	0.45	0.32	1.086	0.545	2.1	0.6	0.7	1.2	0.0648	0.159	0.010	0.000222	3.7	4000	1580
FR-8.2x4.5x4.1	0.82	0.45	0.41	1.044	0.635	2.0	0.7	0.7	1.3	0.0751	0.159	0.012	0.000273	3.8	4000	1900
FR-9.5x5x5	0.95	0.50	0.50	1.201	0.750	2.3	1.2	1.1	1.6	0.1104	0.196	0.022	0.000613	5.1	4000	2400
FR-10x5x5	0.99	0.47	0.46	1.231	0.695	2.3	1.3	0.9	1.5	0.1188	0.173	0.021	0.000641	5.1	4000	2470
FR-11x7x4	1.19	0.76	0.44	1.564	0.820	3.1	1.3	2.7	1.7	0.0881	0.453	0.040	0.000850	7.9	4000	1400
FR-12.7x7.6x6	1.27	0.76	0.60	1.645	0.980	3.2	2.3	3.2	2.0	0.1509	0.453	0.068	0.002090	9.3	4000	2400
FR-14x7x5	1.40	0.70	0.50	1.754	0.850	2.3	1.9	2.6	1.9	0.1729	0.385	0.067	0.002396	9.5	4000	2650
FR-16x7x4	1.60	0.70	0.40	1.968	0.750	3.6	3.1	2.6	1.9	0.1779	0.385	0.068	0.002536	10.7	4000	2470
FR-20x10x5	2.00	1.00	0.50	2.506	1.000	4.7	5.7	6.7	2.4	0.2479	0.785	0.195	0.008040	17.7	4000	2640
FR-19x10x10	1.85	0.98	1.03	2.341	1.520	4.4	9.6	8.4	3.1	0.4459	0.754	0.336	0.019169	19.8	4000	5970
FR-25x15x12	2.50	1.50	1.20	3.240	1.950	6.3	18.2	24.6	3.9	0.5979	1.766	1.056	0.064429	36.3	4000	5730
FR-38x19x13	3.81	1.90	1.27	4.773	2.220	9.0	52.4	51.2	5.1	1.2043	2.834	3.413	0.323625	69.0	4000	8570
FR-47x27x15	4.70	2.70	1.50	6.037	2.850	11.6	84.1	125.4	6.2	1.4914	5.723	8.535	0.826542	111.2	4000	8170
FR-68x44x15	6.80	4.40	1.50	8.964	3.700	17.6	151.9	423.7	7.8	1.7807	15.198	27.062	2.458671	230.3	4000	5000

9

LAMINATIONS

STEEL AND NICKEL LAMINATION MANUFACTURERS

Manufacturers' addresses and phone numbers can be found from the Table of Contents.

1. Magnetics
2. National Metal Products Corp.
3. Magnetic Metals
4. Temple Steel Co.
5. Thomas and Skinner Inc.

Notes

UU AND UI LAMINATION TABLE DESCRIPTION

Definitions for the Following Tables

Information given is listed by column:

Col.	Dim.	Description	Units
1.	D	Stack build	cm
2.	Ht	Finished transformer height	cm
3.	Wth	Finished transformer width (Wth = D + 2F)	cm
4.	Lt	Finished transformer length [Lt = 2(E + F)]	cm
5.	G	Window length	cm
6.	MPL	Mean Magnetic Path Length	cm
7.	W_{tfe}	Core weight [W_{tfe} = (MPL)(A_c)(7.63)]	gram
8.	W_{tcu}	Copper weight [W_{tcu} = (MLT)(W_a)(K_u)(8.89)]	gram
9.	MLT	Mean Length Turn [MLT = 2(D + 2J) + 2(E + 2J) + 3(3.14)(F), J = Bobbin Thickness]	cm
10.	A_c	Iron area (gross) [A_c = (D)(E)]	cm sq
11.	W_a	Window area [W_a = (F)(G)]	cm sq
12.	A_p	Area product (gross) [A_p = (A_c)(W_a)]	cm 4th
13.	K_g	Core geometry (gross) [K_g = (A_p)(A_c)(K_u)/(MLT)]	cm 5th
14.	A_t	Surface area	cm sq

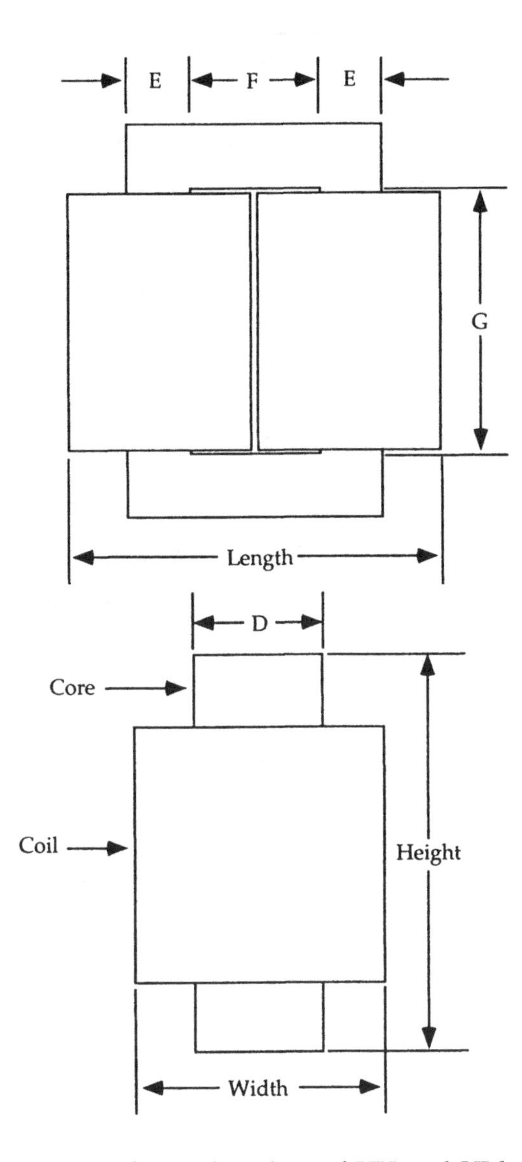

Figure 136 Mechanical outline of UU and UI laminations.

Standard UU and UI Laminations

Part No.	D cm 1	Ht cm 2	Wth cm 3	Lt cm 4	G cm 5	MPL cm 6	Wtfe grams 7	Wtcu grams 8	MLT cm 9	Ac cm sq 10	Wa cm sq 11	Ap cm 4th 12	Kg cm 5th 13	At cm sq 14
63DU - .25	0.040	1.429	0.357	0.953	0.794	3.2	0.2	1.2	1.3	0.0063	0.252	0.002	0.000003	3.6
63DU - .5	0.079	1.429	0.397	0.953	0.794	3.2	0.3	1.2	1.4	0.0126	0.252	0.003	0.000012	3.8
63DU - 1	0.159	1.429	0.476	0.953	0.794	3.2	0.7	1.4	1.5	0.0252	0.252	0.006	0.000042	4.2
124DU - .25	0.079	2.460	0.556	1.588	1.190	5.2	1.1	3.9	1.9	0.0252	0.567	0.014	0.000074	9.9
63DU - 1.5	0.238	1.429	0.556	0.953	0.794	3.2	1.0	1.5	1.7	0.0378	0.252	0.010	0.000085	4.6
63DU - 2	0.318	1.429	0.635	0.953	0.794	3.2	1.3	1.7	1.9	0.0504	0.252	0.013	0.000138	4.9
124DU - .5	0.159	2.460	0.635	1.588	1.190	5.2	2.2	4.3	2.1	0.0504	0.567	0.029	0.000273	10.5
18DU - .25	0.119	3.493	0.754	2.223	1.588	7.3	3.6	9.3	2.6	0.0567	1.008	0.057	0.000499	19.3
124DU - 1	0.318	2.460	0.794	1.588	1.190	5.2	4.5	4.9	2.4	0.1008	0.567	0.057	0.000950	11.8
26DU - .25	0.159	4.445	0.794	2.540	1.905	8.9	7.6	12.9	3.0	0.1008	1.210	0.122	0.001643	27.2
18DU - .5	0.238	3.493	0.873	2.223	1.588	7.3	7.1	10.2	2.8	0.1134	1.008	0.114	0.001830	20.7
124DU - 1.5	0.476	2.460	0.953	1.588	1.190	5.2	6.7	5.5	2.7	0.1512	0.567	0.086	0.001890	13.1
25DU - .25	0.159	4.604	1.111	3.175	2.064	9.9	8.0	24.4	3.5	0.1008	1.966	0.198	0.002289	36.8
124DU - 2	0.635	2.460	1.111	1.588	1.190	5.2	9.0	6.2	3.1	0.2016	0.567	0.114	0.003012	14.4
141L - .25	0.159	4.128	1.429	3.810	2.858	10.8	8.2	51.5	4.0	0.1008	3.629	0.366	0.003698	47.7
1DU - .25	0.159	6.350	1.111	3.175	3.810	13.3	11.2	45.0	3.5	0.1008	3.629	0.366	0.004226	51.8
26DU - .5	0.318	4.445	0.953	2.540	1.905	8.9	15.2	14.2	3.3	0.2016	1.210	0.244	0.005942	29.4
18DU - 1	0.476	3.493	1.111	2.223	1.588	7.3	14.2	11.9	3.3	0.2268	1.008	0.229	0.006267	23.4
25DU - .5	0.318	4.604	1.270	3.175	2.064	9.9	16.0	26.6	3.8	0.2016	1.966	0.396	0.008392	39.3
18DU - 1.5	0.714	3.493	1.349	2.223	1.588	7.3	21.3	13.6	3.8	0.3402	1.008	0.343	0.012327	26.1
39DU - .25	0.238	6.668	1.191	3.810	2.858	13.3	0.3	41.5	4.3	0.2268	2.722	0.617	0.013072	61.2
141L - .5	0.318	4.128	1.588	3.810	2.858	10.8	16.5	55.6	4.3	0.2016	3.629	0.732	0.013700	50.2
1DU - .5	0.318	6.350	1.270	3.175	3.810	13.3	22.3	49.1	3.8	0.2016	3.629	0.732	0.015493	54.8
108L - .25	0.258	5.556	1.132	3.810	3.334	12.7	25.8	45.1	4.4	0.2659	2.913	0.774	0.018898	55.8
18DU - 2	0.953	3.493	1.588	2.223	1.588	7.3	28.4	15.3	4.3	0.4536	1.008	0.457	0.019467	28.8
26DU - 1	0.635	4.445	1.270	2.540	1.905	8.9	30.4	17.0	3.9	0.4032	1.210	0.488	0.019943	33.9
37DU - .25	0.238	7.620	2.143	5.715	3.810	17.2	32.8	149.2	5.8	0.2268	7.258	1.646	0.025840	115.2
25DU - 1	0.635	4.604	1.588	3.175	2.064	9.9	32.0	31.1	4.4	0.4032	1.966	0.793	0.028771	44.3
250L - .25	0.258	7.461	1.132	3.810	5.239	16.5	33.5	70.9	4.4	0.2659	4.577	1.217	0.029698	74.6
101L - .25	0.278	5.080	1.865	5.398	2.858	13.3	30.5	91.6	5.7	0.3087	4.536	1.400	0.030455	79.8

Standard UU and UI Laminations

Part No.	D cm 1	Ht cm 2	Wth cm 3	Lt cm 4	G cm 5	MPL cm 6	Wtfe grams 7	Wtcu grams 8	MLT cm 9	Ac cm sq 10	Wa cm sq 11	Ap cm 4th 12	Kg cm 5th 13	At cm sq 14
26DU - 1.5	0.953	4.445	1.588	2.540	1.905	8.9	45.6	19.7	4.6	0.6048	1.210	0.732	0.038650	38.3
39DU - .5	0.476	6.668	1.429	3.810	2.858	13.3	0.6	46.1	4.8	0.4536	2.722	1.235	0.047057	66.2
141L - 1	0.635	4.128	1.905	3.810	2.858	10.8	32.9	63.8	4.9	0.4032	3.629	1.463	0.047759	55.2
1DU - 1	0.635	6.350	1.588	3.175	3.810	13.3	44.6	57.3	4.4	0.4032	3.629	1.463	0.053116	60.9
7L - .25	0.318	6.350	1.588	5.080	3.810	15.2	45.8	96.0	5.6	0.4032	4.839	1.951	0.056427	87.9
50UI - .25	0.318	6.350	1.588	5.080	3.810	15.2	45.5	96.0	5.6	0.4032	4.839	1.951	0.056427	87.9
25DU - 1.5	0.953	4.604	1.905	3.175	2.064	9.9	48.0	35.5	5.1	0.6048	1.966	1.189	0.056641	49.2
26DU - 2	1.270	4.445	1.905	2.540	1.905	8.9	60.8	23.3	5.4	0.8065	1.210	0.976	0.058115	42.7
108L - .5	0.516	5.556	1.389	3.810	3.334	12.7	51.5	50.5	4.9	0.5317	2.913	1.549	0.067596	60.6
4L - .25	0.318	6.350	2.223	6.350	3.810	16.5	48.8	169.7	6.6	0.4032	7.258	2.927	0.071805	117.7
25DU - 2	1.270	4.604	2.223	3.175	2.064	9.9	64.0	41.3	5.9	0.8065	1.966	1.585	0.086476	54.1
141L - 1.5	0.953	4.128	2.223	3.810	2.858	10.8	49.4	72.0	5.6	0.6048	3.629	2.195	0.095221	60.3
37DU - .5	0.476	7.620	2.381	5.715	3.810	17.2	65.5	161.5	6.3	0.4536	7.258	3.292	0.095490	121.6
1DU - 1.5	0.953	6.350	1.905	3.175	3.810	13.3	66.9	65.5	5.1	0.6048	3.629	2.195	0.104568	66.9
250L - .5	0.516	7.461	1.389	3.810	5.239	16.5	66.9	79.3	4.9	0.5317	4.577	2.434	0.106222	80.4
104L - .25	0.318	8.095	2.301	6.507	5.555	20.2	60.0	262.4	6.7	0.4032	11.020	4.443	0.107007	152.4
50DU - .25	0.318	10.160	2.858	7.620	5.080	22.9	75.8	347.4	7.6	0.4032	12.903	5.203	0.110844	204.8
101L - .5	0.556	5.080	2.143	5.398	2.858	13.3	61.0	100.6	6.2	0.6174	4.536	2.801	0.110961	85.6
105L - .25	0.318	9.366	2.223	6.350	6.826	22.5	67.5	304.0	6.6	0.4032	13.004	5.244	0.128651	169.5
102L - .25	0.357	8.255	1.945	6.033	5.398	19.7	74.8	197.2	6.5	0.5103	8.569	4.373	0.137916	137.0
141L - 2	1.270	4.128	2.540	3.810	2.858	10.8	65.8	82.7	6.4	0.8065	3.629	2.927	0.147237	65.3
39DU - 1	0.953	6.668	1.905	3.810	2.858	13.3	1.1	55.3	5.7	0.9073	2.722	2.469	0.156846	76.2
1DU - 2	1.270	6.350	2.223	3.175	3.810	13.3	89.2	76.3	5.9	0.8065	3.629	2.927	0.159649	73.0
106L - .25	0.357	8.255	2.580	7.303	5.398	21.0	79.5	318.6	7.5	0.5103	11.996	6.122	0.167311	175.4
60UI - .25	0.357	8.255	2.580	7.303	5.398	18.1	79.0	318.6	7.5	0.5103	11.996	6.122	0.167311	175.4
7L - .5	0.635	6.350	1.905	5.080	3.810	15.2	91.5	106.9	6.2	0.8065	4.839	3.902	0.202637	95.2
50UI - .5	0.635	6.350	1.905	5.080	3.810	15.2	91.0	106.9	6.2	0.8065	4.839	3.902	0.202637	95.2
108L - 1	1.031	5.556	1.905	3.810	3.334	12.7	103.0	61.2	5.9	1.0635	2.913	3.098	0.223163	70.3
4L - .5	0.635	6.350	2.540	6.350	3.810	16.5	97.5	186.1	7.2	0.8065	7.258	5.853	0.261920	125.8
107L - .25	0.397	9.525	2.461	7.303	6.350	23.2	107.8	355.0	7.6	0.6300	13.105	8.257	0.273181	195.7

Standard UU and UI Laminations

Part No.	D cm 1	Ht cm 2	Wth cm 3	Lt cm 4	G cm 5	MPL cm 6	Wtfe grams 7	Wtcu grams 8	MLT cm 9	Ac cm sq 10	Wa cm sq 11	Ap cm 4th 12	Kg cm 5th 13	At cm sq 14
39DU - 1.5	1.429	6.668	2.381	3.810	2.858	13.3	1.7	66.5	6.9	1.3609	2.722	3.704	0.293667	86.2
37DU - 1	0.953	7.620	2.858	5.715	3.810	17.2	131.0	186.1	7.2	0.9073	7.258	6.585	0.331493	134.3
250L - 1	1.031	7.461	1.905	3.810	5.239	16.5	133.8	96.1	5.9	1.0635	4.577	4.868	0.350684	92.0
101L - 1	1.111	5.080	2.699	5.398	2.858	13.3	122.0	118.5	7.3	1.2349	4.536	5.602	0.376698	97.3
104L - .5	0.635	8.095	2.619	6.507	5.555	20.2	120.0	287.3	7.3	0.8065	11.020	8.887	0.390959	161.7
50DU - .5	0.635	10.160	3.175	7.620	5.080	22.9	151.5	376.5	8.2	0.8065	12.903	10.406	0.409064	216.1
108L - 1.5	1.547	5.556	2.421	3.810	3.334	12.7	154.5	73.9	7.1	1.5952	2.913	4.647	0.415481	79.9
75UI - .25	0.476	9.525	2.381	7.620	5.715	22.9	154.0	316.0	8.2	0.9073	10.887	9.877	0.439207	197.8
39DU - 2	1.905	6.668	2.858	3.810	2.858	13.3	2.2	75.7	7.8	1.8145	2.722	4.939	0.458471	96.2
105L - .5	0.635	9.366	2.540	6.350	6.826	22.5	135.0	333.4	7.2	0.8065	13.004	10.487	0.469274	179.4
102L - .5	0.714	8.255	2.302	6.033	5.398	19.7	149.5	219.0	7.2	1.0207	8.569	8.746	0.496828	147.2
106L - .5	0.714	8.255	2.937	7.303	5.398	21.0	159.0	349.1	8.2	1.0207	11.996	12.244	0.610824	186.5
60UI - .5	0.714	8.255	2.937	7.303	5.398	18.1	158.0	349.1	8.2	1.0207	11.996	12.244	0.610824	186.5
37DU - 1.5	1.429	7.620	3.334	5.715	3.810	17.2	196.5	215.8	8.4	1.3609	7.258	9.877	0.643052	147.0
108L - 2	2.062	5.556	2.936	3.810	3.334	12.7	206.0	84.6	8.2	2.1269	2.913	6.195	0.645370	89.6
250L - 1.5	1.547	7.461	2.421	3.810	5.239	16.5	200.7	116.2	7.1	1.5952	4.577	7.302	0.652899	103.7
7L - 1	1.270	6.350	2.540	5.080	3.810	15.2	183.0	132.2	7.7	1.6129	4.839	7.804	0.655443	109.7
50UI - 1	1.270	6.350	2.540	5.080	3.810	15.2	182.0	132.2	7.7	1.6129	4.839	7.804	0.655443	109.7
101L - 1.5	1.667	5.080	3.254	5.398	2.858	13.3	183.0	139.6	8.7	1.8523	4.536	8.403	0.719185	108.9
75DU - .25	0.476	15.240	4.286	11.430	7.620	34.3	259.3	1151.3	11.2	0.9073	29.032	26.340	0.857115	460.9
4L - 1	1.270	6.350	3.175	6.350	3.810	16.5	195.0	224.0	8.7	1.6129	7.258	11.707	0.870227	141.9
107L - .5	0.794	9.525	2.858	7.303	6.350	23.2	215.5	391.9	8.4	1.2601	13.105	16.513	0.989599	209.1
250L - 2	2.062	7.461	2.936	3.810	5.239	16.5	267.6	132.9	8.2	2.1269	4.577	9.736	1.014153	115.3
37DU - 2	1.905	7.620	3.810	5.715	3.810	17.2	262.0	240.4	9.3	1.8145	7.258	13.170	1.026291	159.7
101L - 2	2.223	5.080	3.810	5.398	2.858	13.3	244.0	157.6	9.8	2.4698	4.536	11.203	1.133097	120.6
7L - 1.5	1.905	6.350	3.175	5.080	3.810	15.2	274.5	154.0	9.0	2.4193	4.839	11.707	1.265526	124.2
50UI - 1.5	1.905	6.350	3.175	5.080	3.810	15.2	273.0	154.0	9.0	2.4193	4.839	11.707	1.265526	124.2
104L - 1	1.270	8.095	3.254	6.507	5.555	20.2	240.0	344.9	8.8	1.6129	11.020	17.774	1.302679	180.2
50DU - 1	1.270	10.160	3.810	7.620	5.080	22.9	303.0	444.0	9.7	1.6129	12.903	20.812	1.387668	238.7
105L - 1	1.270	9.366	3.175	6.350	6.826	22.5	270.0	401.3	8.7	1.6129	13.004	20.974	1.559157	199.4

Standard UU and UI Laminations														
Part No.	D cm 1	Ht cm 2	Wth cm 3	Lt cm 4	G cm 5	MPL cm 6	Wtfe grams 7	Wtcu grams 8	MLT cm 9	Ac cm sq 10	Wa cm sq 11	Ap cm 4th 12	Kg cm 5th 13	At cm sq 14
75UI - .5	0.953	9.525	2.858	7.620	5.715	22.9	308.0	352.8	9.1	1.8145	10.887	19.755	1.573220	214.1
102L - 1	1.429	8.255	3.016	6.033	5.398	19.7	299.0	268.6	8.8	2.0413	8.569	17.491	1.620133	167.6
4L - 1.5	1.905	6.350	3.810	6.350	3.810	16.5	292.5	256.8	9.9	2.4193	7.258	17.560	1.708065	158.1
100UI - .25	0.635	12.700	3.175	10.160	7.620	30.5	364.3	739.6	10.7	1.6129	19.355	31.217	1.874238	351.6
7L - 2	2.540	6.350	3.810	5.080	3.810	15.2	366.0	175.9	10.2	3.2258	4.839	15.609	1.970298	138.7
50UI - 2	2.540	6.350	3.810	5.080	3.810	15.2	364.0	175.9	10.2	3.2258	4.839	15.609	1.970298	138.7
106L - 1	1.429	8.255	3.651	7.303	5.398	21.0	318.0	418.6	9.8	2.0413	11.996	24.488	2.037734	208.8
60UI - 1	1.429	8.255	3.651	7.303	5.398	18.1	316.0	418.6	9.8	2.0413	11.996	24.488	2.037734	208.8
104L - 1.5	1.905	8.095	3.889	6.507	5.555	20.2	360.0	394.7	10.1	2.4193	11.020	26.660	2.561465	198.8
4L - 2	2.540	6.350	4.445	6.350	3.810	16.5	390.0	289.6	11.2	3.2258	7.258	23.413	2.692814	174.2
50DU - 1.5	1.905	10.160	4.445	7.620	5.080	22.9	454.5	502.2	10.9	2.4193	12.903	31.217	2.759989	261.3
105L - 1.5	1.905	9.366	3.810	6.350	6.826	22.5	405.0	460.1	9.9	2.4193	13.004	31.461	3.060284	219.4
102L - 1.5	2.143	8.255	3.731	6.033	5.398	19.7	448.5	312.1	10.2	3.0620	8.569	26.237	3.136888	188.1
75DU - .5	0.953	15.240	4.763	11.430	7.620	34.3	518.5	1249.7	12.1	1.8145	29.032	52.679	3.158681	486.3
107L - 1	1.588	9.525	3.651	7.303	6.350	23.2	431.0	475.2	10.2	2.5202	13.105	33.026	3.264577	235.8
106L - 1.5	2.143	8.255	4.366	7.303	5.398	21.0	477.0	479.5	11.2	3.0620	11.996	36.731	4.002156	231.0
60UI - 1.5	2.143	8.255	4.366	7.303	5.398	18.1	474.0	479.5	11.2	3.0620	11.996	36.731	4.002156	231.0
104L - 2	2.540	8.095	4.524	6.507	5.555	20.2	480.0	444.5	11.3	3.2258	11.020	35.547	4.043843	217.3
50DU - 2	2.540	10.160	5.080	7.620	5.080	22.9	606.0	560.5	12.2	3.2258	12.903	41.623	4.396533	283.9
105L - 2	2.540	9.366	4.445	6.350	6.826	22.5	540.0	518.8	11.2	3.2258	13.004	41.948	4.824626	239.3
102L - 2	2.858	8.255	4.445	6.033	5.398	19.7	598.0	355.7	11.7	4.0827	8.569	34.982	4.894107	208.5
75UI - 1	1.905	9.525	3.810	7.620	5.715	22.9	616.0	434.3	11.2	3.6290	10.887	39.509	5.112141	246.8
125UI - .25	0.794	15.875	3.969	12.700	9.525	38.1	717.3	1433.5	13.3	2.5202	30.242	76.214	5.763487	549.4
106L - 2	2.858	8.255	5.080	7.303	5.398	21.0	636.0	540.5	12.7	4.0827	11.996	48.975	6.312607	253.2
60UI - 2	2.858	8.255	5.080	7.303	5.398	18.1	632.0	540.5	12.7	4.0827	11.996	48.975	6.312607	253.2
107L - 1.5	2.381	9.525	4.445	7.303	6.350	23.2	646.5	549.2	11.8	3.7802	13.105	49.539	6.355899	262.5
1125DU - .25	0.714	22.860	6.429	17.145	11.430	51.4	854.3	3838.4	16.5	2.0413	65.322	133.344	6.589073	1037.0
100UI - .5	1.270	12.700	3.810	10.160	7.620	30.5	728.5	840.8	12.2	3.2258	19.355	62.435	6.594800	380.6
125DU - .25	0.794	22.860	5.874	16.510	10.160	41.4	1043.8	2995.5	16.3	2.5202	51.613	130.072	8.033834	959.3
75UI - 1.5	2.858	9.525	4.763	7.620	5.715	22.9	924.0	508.1	13.1	5.4435	10.887	59.264	9.832691	279.4

Standard UU and UI Laminations														
Part No.	D cm 1	Ht cm 2	Wth cm 3	Lt cm 4	G cm 5	MPL cm 6	Wtfe grams 7	Wtcu grams 8	MLT cm 9	Ac cm sq 10	Wa cm sq 11	Ap cm 4th 12	Kg cm 5th 13	At cm sq 14
107L - 2	3.175	9.525	5.239	7.303	6.350	23.2	862.0	623.2	13.4	5.0403	13.105	66.052	9.958042	289.2
75DU - 1	1.905	15.240	5.715	11.430	7.620	34.3	1037.0	1467.0	14.2	3.6290	29.032	105.359	10.763039	537.1
150UI - .25	0.953	19.050	4.763	15.240	11.430	45.7	1237.5	2464.5	15.9	3.6290	43.548	158.038	14.414936	791.1
75UI - 2	3.810	9.525	5.715	7.620	5.715	22.9	1232.0	589.9	15.2	7.2581	10.887	79.019	15.056222	312.1
125UI - .5	1.588	15.875	4.763	12.700	9.525	38.1	1434.5	1625.8	15.1	5.0403	30.242	152.429	20.328089	594.8
75DU - 1.5	2.858	15.240	6.668	11.430	7.620	34.3	1555.5	1663.7	16.1	5.4435	29.032	158.038	21.354050	587.9
100UI - 1	2.540	12.700	5.080	10.160	7.620	30.5	1457.0	1015.6	14.8	6.4516	19.355	124.869	21.838393	438.7
1125DU - .5	1.429	22.860	7.144	17.145	11.430	51.4	1708.5	4216.7	18.2	4.0827	65.322	266.689	23.991524	1094.2
125DU - .5	1.588	22.860	6.668	16.510	10.160	41.4	2087.5	3323.6	18.1	5.0403	51.613	260.145	28.963258	1021.8
180UI - .25	1.143	20.574	5.715	18.288	11.430	50.3	1971.3	3533.7	19.0	5.2258	52.258	273.089	30.019058	1029.5
75DU - 2	3.810	15.240	7.620	11.430	7.620	34.3	2074.0	1881.8	18.2	7.2581	29.032	210.717	33.562017	638.7
100UI - 1.5	3.810	12.700	6.350	10.160	7.620	30.5	2185.5	1204.7	17.5	9.6774	19.355	187.304	41.422249	496.8
150UI - .5	1.905	19.050	5.715	15.240	11.430	45.7	2475.0	2790.5	18.0	7.2581	43.548	316.076	50.924133	856.4
100UI - 2	5.080	12.700	7.620	10.160	7.620	30.5	2914.0	1379.5	20.0	12.9032	19.355	249.739	64.307767	554.8
125UI - 1	3.175	15.875	6.350	12.700	9.525	38.1	2869.0	1967.2	18.3	10.0806	30.242	304.857	67.199280	685.5
1125DU - 1	2.858	22.860	8.573	17.145	11.430	51.4	3417.0	4880.5	21.0	8.1653	65.322	533.378	82.914382	1208.5
125DU - 1	3.175	22.860	8.255	16.510	10.160	41.4	4175.0	3906.3	21.3	10.0806	51.613	520.289	98.570537	1146.8
180UI - .5	2.286	20.574	6.858	18.288	11.430	50.3	3942.5	3995.7	21.5	10.4516	52.258	546.179	106.193385	1118.3
240UI - .25	1.524	27.432	7.620	24.384	15.240	67.1	4655.8	8397.4	25.4	9.2903	92.903	863.097	126.181607	1830.2
125UI - 1.5	4.763	15.875	7.938	12.700	9.525	38.1	4303.5	2331.0	21.7	15.1209	30.242	457.285	127.600332	776.2
1125DU - 1.5	4.286	22.860	10.001	17.145	11.430	51.4	5125.5	5592.5	24.1	12.2480	65.322	800.067	162.803878	1322.8
150UI - 1	3.810	19.050	7.620	15.240	11.430	45.7	4950.0	3412.7	22.0	14.5161	43.548	632.151	166.557784	987.1
125DU - 1.5	4.763	22.860	9.843	16.510	10.160	41.4	6262.5	4527.2	24.7	15.1209	51.613	780.434	191.366316	1271.8
125UI - 2	6.350	15.875	9.525	12.700	9.525	38.1	5738.0	2672.5	24.9	20.1613	30.242	609.714	197.862715	866.9
1125DU - 2	5.715	22.860	11.430	17.145	11.430	51.4	6834.0	6256.3	26.9	16.3306	65.322	1066.756	258.722216	1437.1
125DU - 2	6.350	22.860	11.430	16.510	10.160	41.4	8350.0	5109.9	27.8	20.1613	51.613	1040.579	301.410305	1396.8
150UI - 1.5	5.715	19.050	9.525	15.240	11.430	45.7	7425.0	4002.7	25.8	21.7741	43.548	948.227	319.515346	1117.7
180UI - 1	4.572	20.574	9.144	18.288	11.430	50.3	7885.0	4884.0	26.3	20.9032	52.258	1092.358	347.518560	1296.0
240UI - .5	3.048	27.432	9.144	24.384	15.240	67.1	9311.5	9404.4	28.5	18.5806	92.903	1726.195	450.684211	1988.1
150UI - 2	7.620	19.050	11.430	15.240	11.430	45.7	9900.0	4654.7	30.1	29.0322	43.548	1264.303	488.467147	1248.4

Standard UU and UI Laminations

Part No.	D cm 1	Ht cm 2	Wth cm 3	Lt cm 4	G cm 5	MPL cm 6	Wtfe grams 7	Wtcu grams 8	MLT cm 9	Ac cm sq 10	Wa cm sq 11	Ap cm 4th 12	Kg cm 5th 13	At cm sq 14
180UI - 1.5	6.858	20.574	11.430	18.288	11.430	50.3	11827.5	5807.9	31.3	31.3548	52.258	1638.536	657.526981	1473.7
180UI - 2	9.144	20.574	13.716	18.288	11.430	50.3	15770.0	6657.5	35.8	41.8064	52.258	2184.715	1019.760886	1651.4
240UI - 1	6.096	27.432	12.192	24.384	15.240	67.1	18623.0	11487.0	34.8	37.1612	92.903	3452.389	1475.896953	2304.0
240UI - 1.5	9.144	27.432	15.240	24.384	15.240	67.1	27934.5	13633.0	41.3	55.7418	92.903	5178.585	2798.029917	2619.9
240UI - 2	12.192	27.432	18.288	24.384	15.240	67.1	37246.0	15646.9	47.4	74.3224	92.903	6904.780	4334.042105	2935.7

EE AND EI LAMINATION TABLE DESCRIPTION

Definitions for the Following Tables

Information given is listed by column:

Col.	Dim.	Description	Units
1.	D	Stack build	cm
2.	Ht	Finished transformer height (Ht = G + 2E)	cm
3.	Wth	Finished transformer width (Wth = D + 2F)	cm
4.	Lt	Finished transformer length [Lt = 2(E + F)]	cm
5.	G	Window length	cm
6.	MPL	Mean Magnetic Path Length	cm
7.	W_{tfe}	Core weight [W_{tfe} = (MPL)(A_c)(7.63)]	gram
8.	W_{tcu}	Copper weight [W_{tcu} = (MLT)(W_a)(K_u)(8.89)]	gram
9.	MLT	Mean Length Turn [MLT = 2(D + 2J) + 2(E + 2J) + 3(3.14)(F), J = Bobbin Thickness]	cm
10.	A_c	Iron area (gross) [A_c = (D)(E)]	cm sq
11.	W_a	Window area [W_a = (F)(G)]	cm sq
12.	A_p	Area product (gross) [A_p = (A_c)(W_a)]	cm 4th
13.	K_g	Core geometry (gross) [K_g = (A_p)(A_c)(K_u)/(MLT)]	cm 5th
14.	A_t	Surface area	cm sq

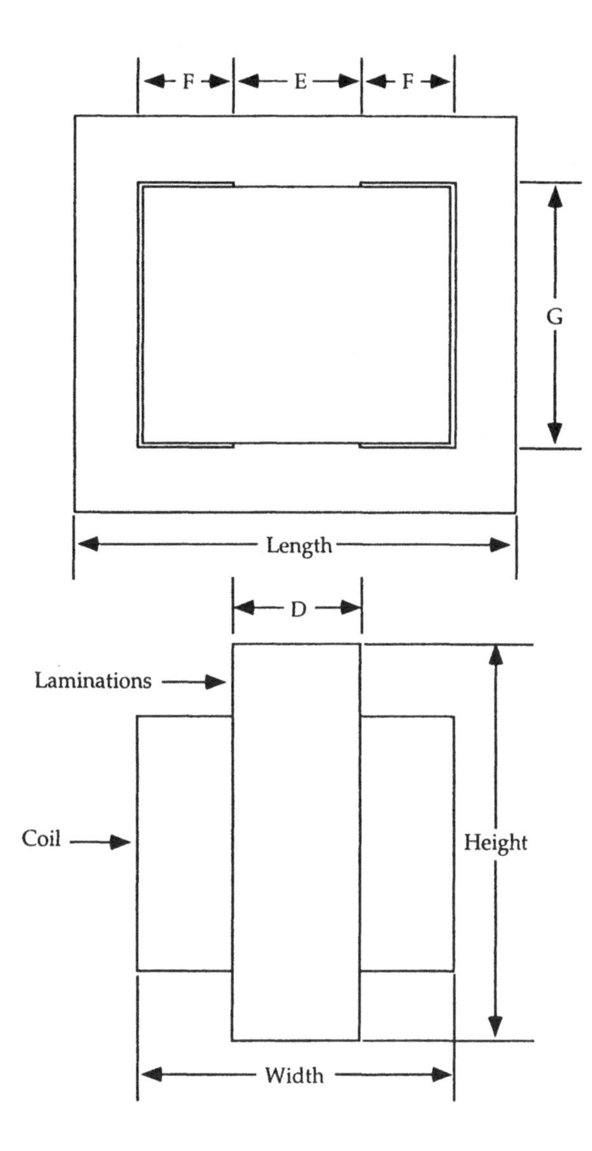

Figure 137 Mechanical outline of EE and EI laminations.

Standard EE and EI Laminations														
Part No.	D cm 1	Ht cm 2	Wth cm 3	Lt cm 4	G cm 5	MPL cm 6	Wtfe grams 7	Wtcu grams 8	MLT cm 9	Ac cm sq 10	Wa cm sq 11	Ap cm 4th 12	Kg cm 5th 13	At cm sq 14
094EI - .25	0.060	0.635	0.536	0.953	0.397	1.8	0.2	0.6	1.8	0.0142	0.095	0.001	0.000004	2.5
3031EE - .25	0.059	0.953	0.537	0.953	0.714	2.4	0.3	1.1	1.7	0.0139	0.170	0.002	0.000008	3.4
094EI - .5	0.119	0.635	0.595	0.953	0.397	1.8	0.4	0.6	1.9	0.0284	0.095	0.003	0.000016	2.6
3031EE - .5	0.118	0.953	0.596	0.953	0.714	2.4	0.5	1.1	1.9	0.0279	0.170	0.005	0.000028	3.6
2829EE - .25	0.079	1.111	0.714	1.270	0.794	2.9	0.5	2.0	2.2	0.0252	0.252	0.006	0.000029	5.4
3233EE - .25	0.089	1.054	0.851	1.473	0.699	2.9	0.7	2.4	2.5	0.0316	0.266	0.008	0.000043	6.3
094EI - 1	0.238	0.635	0.714	0.953	0.397	1.8	0.7	0.7	2.1	0.0567	0.095	0.005	0.000058	3.0
3031EE - 1	0.236	0.953	0.714	0.953	0.714	2.4	1.0	1.3	2.1	0.0558	0.170	0.010	0.000101	4.1
2829EE - .5	0.159	1.111	0.794	1.270	0.794	2.9	1.1	2.1	2.4	0.0504	0.252	0.013	0.000109	5.8
094EI - 1.5	0.357	0.635	0.833	0.953	0.397	1.8	1.1	0.8	2.3	0.0851	0.095	0.008	0.000117	3.4
186EE - .25	0.119	1.111	1.072	1.905	0.635	3.2	1.4	3.3	3.1	0.0567	0.302	0.017	0.000126	8.9
3233EE - .5	0.178	1.054	0.940	1.473	0.699	2.9	1.3	2.5	2.7	0.0632	0.266	0.017	0.000159	6.8
094EI - 2	0.476	0.635	0.953	0.953	0.397	1.8	1.5	0.9	2.6	0.1134	0.095	0.011	0.000188	3.8
3031EE - 1.5	0.354	0.953	0.832	0.953	0.714	2.4	1.5	1.4	2.3	0.0837	0.170	0.014	0.000204	4.5
187EE - .25	0.119	1.588	1.072	1.905	1.111	4.1	1.8	5.8	3.1	0.0567	0.529	0.030	0.000220	11.7
3031EE - 2	0.472	0.953	0.950	0.953	0.714	2.4	2.0	1.6	2.6	0.1116	0.170	0.019	0.000330	5.0
1867EE - .25	0.119	2.223	1.072	1.905	1.746	5.4	2.3	9.2	3.1	0.0567	0.832	0.047	0.000346	15.5
2829EE - 1	0.318	1.111	0.953	1.270	0.794	2.9	2.2	2.4	2.7	0.1008	0.252	0.025	0.000383	6.6
186EE - .5	0.238	1.111	1.191	1.905	0.635	3.2	2.7	3.6	3.3	0.1134	0.302	0.034	0.000467	9.6
3233EE - 1	0.356	1.054	1.118	1.473	0.699	2.9	2.7	2.9	3.0	0.1265	0.266	0.034	0.000562	7.7
2829EE - 1.5	0.476	1.111	1.111	1.270	0.794	2.9	3.3	2.7	3.0	0.1512	0.252	0.038	0.000770	7.3
187EE - .5	0.238	1.588	1.191	1.905	1.111	4.1	3.5	6.3	3.3	0.1134	0.529	0.060	0.000817	12.6
2425EE - .25	0.159	1.905	1.429	2.540	1.270	5.1	3.9	11.4	4.0	0.1008	0.806	0.081	0.000822	19.2
3233EE - 1.5	0.533	1.054	1.295	1.473	0.699	2.9	4.0	3.2	3.4	0.1897	0.266	0.050	0.001132	8.6
2829EE - 2	0.635	1.111	1.270	1.270	0.794	2.9	4.4	3.0	3.3	0.2016	0.252	0.051	0.001238	8.1
1867EE - .5	0.238	2.223	1.191	1.905	1.746	5.4	4.6	9.9	3.3	0.1134	0.832	0.094	0.001284	16.5
186EE - 1	0.476	1.111	1.429	1.905	0.635	3.2	5.5	4.1	3.8	0.2268	0.302	0.069	0.001634	11.0
3233EE - 2	0.711	1.054	1.473	1.473	0.699	2.9	5.4	3.5	3.7	0.2529	0.266	0.067	0.001821	9.5
187EE - 1	0.476	1.588	1.429	1.905	1.111	4.1	7.1	7.2	3.8	0.2268	0.529	0.120	0.002860	14.2
2425EE - .5	0.318	1.905	1.588	2.540	1.270	5.1	7.8	12.4	4.3	0.2016	0.806	0.163	0.003044	20.6

Standard EE and EI Laminations														
Part No.	D cm 1	Ht cm 2	Wth cm 3	Lt cm 4	G cm 5	MPL cm 6	Wtfe grams 7	Wtcu grams 8	MLT cm 9	Ac cm sq 10	Wa cm sq 11	Ap cm 4th 12	Kg cm 5th 13	At cm sq 14
186EE - 1.5	0.714	1.111	1.667	1.905	0.635	3.2	8.2	4.6	4.3	0.3402	0.302	0.103	0.003268	12.5
2638EE - .25	0.238	2.273	1.508	3.175	1.322	5.8	9.7	14.3	4.8	0.2268	0.840	0.190	0.003612	26.0
1867EE - 1	0.476	2.223	1.429	1.905	1.746	5.4	9.3	11.3	3.8	0.2268	0.832	0.189	0.004494	18.4
2627EE - .25	0.238	2.699	1.508	3.175	1.746	6.7	11.5	18.9	4.8	0.2268	1.109	0.252	0.004771	30.0
186EE - 2	0.953	1.111	1.905	1.905	0.635	3.2	10.9	5.1	4.8	0.4536	0.302	0.137	0.005229	13.9
187EE - 1.5	0.714	1.588	1.667	1.905	1.111	4.1	10.6	8.1	4.3	0.3402	0.529	0.180	0.005719	15.9
2728EE - .25	0.238	3.068	1.508	3.175	2.116	7.4	12.8	22.9	4.8	0.2268	1.344	0.305	0.005780	33.5
375EI - .25	0.238	2.858	1.826	3.493	1.905	7.3	12.4	28.4	5.3	0.2268	1.512	0.343	0.005891	37.1
1867EE - 1.5	0.714	2.223	1.667	1.905	1.746	5.4	13.9	12.7	4.3	0.3402	0.832	0.283	0.008987	20.4
187EE - 2	0.953	1.588	1.905	1.905	1.111	4.1	14.2	9.0	4.8	0.4536	0.529	0.240	0.009150	17.5
2425EE - 1	0.635	1.905	1.905	2.540	1.270	5.1	15.5	14.2	4.9	0.4032	0.806	0.325	0.010613	23.4
2638EE - .5	0.476	2.273	1.746	3.175	1.322	5.8	19.5	15.7	5.3	0.4536	0.840	0.381	0.013139	28.6
50EI - .25	0.318	3.175	1.588	3.810	1.905	7.6	23.0	24.0	5.6	0.4032	1.210	0.488	0.014107	39.9
1867EE - 2	0.953	2.223	1.905	1.905	1.746	5.4	18.5	14.1	4.8	0.4536	0.832	0.377	0.014378	22.4
2627EE - .5	0.476	2.699	1.746	3.175	1.746	6.7	22.9	20.7	5.3	0.4536	1.109	0.503	0.017354	32.8
21EI - .25	0.318	3.334	1.905	4.128	2.064	8.3	24.8	35.4	6.1	0.4032	1.638	0.661	0.017536	47.9
2728EE - .5	0.476	3.068	1.746	3.175	2.116	7.4	25.5	25.1	5.3	0.4536	1.344	0.609	0.021027	36.4
2425EE - 1.5	0.953	1.905	2.223	2.540	1.270	5.1	23.3	16.0	5.6	0.6048	0.806	0.488	0.021160	26.2
375EI - .5	0.476	2.858	2.064	3.493	1.905	7.3	24.9	31.0	5.8	0.4536	1.512	0.686	0.021616	40.1
56EI - .25	0.357	4.604	1.786	4.286	2.143	8.6	31.8	33.9	6.2	0.5103	1.531	0.781	0.025629	63.1
2425EE - 2	1.270	1.905	2.540	2.540	1.270	5.1	31.0	18.4	6.4	0.8065	0.806	0.650	0.032719	29.0
28EI - .25	0.318	4.604	2.858	5.080	3.016	11.4	37.2	103.1	7.6	0.4032	3.831	1.545	0.032907	89.2
625EI - .25	0.397	3.969	1.984	4.763	2.381	9.5	44.8	46.2	6.9	0.6300	1.890	1.191	0.043690	62.4
2638EE - 1	0.953	2.273	2.223	3.175	1.322	5.8	38.9	18.5	6.2	0.9073	0.840	0.762	0.044496	33.8
50EI - .5	0.635	3.175	1.905	3.810	1.905	7.6	46.0	26.7	6.2	0.8065	1.210	0.976	0.050659	44.4
2627EE - 1	0.953	2.699	2.223	3.175	1.746	6.7	45.8	24.5	6.2	0.9073	1.109	1.006	0.058773	38.4
21EI - .5	0.635	3.334	2.223	4.128	2.064	8.3	49.7	39.1	6.7	0.8065	1.638	1.321	0.063505	52.6
2728EE - 1	0.953	3.068	2.223	3.175	2.116	7.4	51.0	29.7	6.2	0.9073	1.344	1.219	0.071211	42.4
375EI - 1	0.953	2.858	2.540	3.493	1.905	7.3	49.7	36.1	6.7	0.9073	1.512	1.372	0.074191	46.2
2638EE - 1.5	1.429	2.273	2.699	3.175	1.322	5.8	58.4	22.0	7.4	1.3609	0.840	1.142	0.084449	39.0

Standard EE and EI Laminations

Part No.	D cm 1	Ht cm 2	Wth cm 3	Lt cm 4	G cm 5	MPL cm 6	Wtfe grams 7	Wtcu grams 8	MLT cm 9	Ac cm sq 10	Wa cm sq 11	Ap cm 4th 12	Kg cm 5th 13	At cm sq 14
56EI - .5	0.714	4.604	2.143	4.286	2.143	8.6	63.5	37.8	6.9	1.0207	1.531	1.563	0.091961	69.4
75EI - .25	0.476	4.763	2.381	5.715	2.858	11.4	78.0	79.0	8.2	0.9073	2.722	2.469	0.109802	89.8
2627EE - 1.5	1.429	2.699	2.699	3.175	1.746	6.7	68.7	29.0	7.4	1.3609	1.109	1.509	0.111544	44.0
28EI - .5	0.635	4.604	3.175	5.080	3.016	11.4	74.3	111.8	8.2	0.8065	3.831	3.089	0.121441	95.4
2638EE - 2	1.905	2.273	3.175	3.175	1.322	5.8	77.8	24.8	8.3	1.8145	0.840	1.523	0.132937	44.2
2728EE - 1.5	1.429	3.068	2.699	3.175	2.116	7.4	76.5	35.2	7.4	1.3609	1.344	1.828	0.135150	48.3
375EI - 1.5	1.429	2.858	3.016	3.493	1.905	7.3	74.6	42.3	7.9	1.3609	1.512	2.058	0.142462	52.2
625EI - .5	0.794	3.969	2.381	4.763	2.381	9.5	89.5	51.5	7.7	1.2601	1.890	2.382	0.156658	69.3
50EI - 1	1.270	3.175	2.540	3.810	1.905	7.6	92.0	33.0	7.7	1.6129	1.210	1.951	0.163861	53.2
2627EE - 2	1.905	2.699	3.175	3.175	1.746	6.7	91.6	32.8	8.3	1.8145	1.109	2.012	0.175589	49.6
21EI - 1	1.270	3.334	2.858	4.128	2.064	8.3	99.3	47.7	8.2	1.6129	1.638	2.642	0.208374	62.1
2728EE - 2	1.905	3.068	3.175	3.175	2.116	7.4	102.0	39.7	8.3	1.8145	1.344	2.438	0.212751	54.3
375EI - 2	1.905	2.858	3.493	3.493	1.905	7.3	99.4	47.4	8.8	1.8145	1.512	2.744	0.225901	58.3
87EI - .25	0.556	5.556	2.778	6.668	3.334	13.3	120.3	124.5	9.5	1.2349	3.705	4.575	0.239032	122.3
56EI - 1	1.429	4.604	2.858	4.286	2.143	8.6	127.0	46.6	8.6	2.0413	1.531	3.125	0.297902	82.1
50EI - 1.5	1.905	3.175	3.175	3.810	1.905	7.6	138.0	38.5	9.0	2.4193	1.210	2.927	0.316381	62.1
75EI - .5	0.953	4.763	2.858	5.715	2.858	11.4	156.0	88.2	9.1	1.8145	2.722	4.939	0.393305	99.8
21EI - 1.5	1.905	3.334	3.493	4.128	2.064	8.3	149.0	55.0	9.5	2.4193	1.638	3.963	0.405835	71.6
28EI - 1	1.270	4.604	3.810	5.080	3.016	11.4	148.6	131.8	9.7	1.6129	3.831	6.178	0.411964	107.7
100EI - .25	0.635	6.350	3.175	7.620	3.810	15.2	178.0	184.9	10.7	1.6129	4.839	7.804	0.468560	159.7
50EI - 2	2.540	3.175	3.810	3.810	1.905	7.6	184.0	44.0	10.2	3.2258	1.210	3.902	0.492574	71.0
625EI - 1	1.588	3.969	3.175	4.763	2.381	9.5	179.0	63.5	9.5	2.5202	1.890	4.763	0.508106	83.2
56EI - 1.5	2.143	4.604	3.572	4.286	2.143	8.6	190.5	54.4	10.0	3.0620	1.531	4.688	0.574465	94.8
21EI - 2	2.540	3.334	4.128	4.128	2.064	8.3	198.6	62.4	10.7	3.2258	1.638	5.284	0.636013	81.0
28EI - 1.5	1.905	4.604	4.445	5.080	3.016	11.4	222.9	149.1	10.9	2.4193	3.831	9.268	0.819372	120.0
112EI - .25	0.714	7.144	3.572	8.573	4.286	17.2	257.3	262.2	12.0	2.0413	6.124	12.501	0.847937	202.1
87EI - .5	1.111	5.556	3.334	6.668	3.334	13.3	240.5	139.2	10.6	2.4698	3.705	9.150	0.855558	135.8
56EI - 2	2.858	4.604	4.286	4.286	2.143	8.6	254.0	62.2	11.4	4.0827	1.531	6.251	0.893541	107.5
625EI - 1.5	2.381	3.969	3.969	4.763	2.381	9.5	268.5	74.2	11.0	3.7802	1.890	7.145	0.978815	97.0
75EI - 1	1.905	4.763	3.810	5.715	2.858	11.4	312.0	108.6	11.2	3.6290	2.722	9.877	1.278035	119.8

Standard EE and EI Laminations

Part No.	D cm 1	Ht cm 2	Wth cm 3	Lt cm 4	G cm 5	MPL cm 6	Wtfe grams 7	Wtcu grams 8	MLT cm 9	Ac cm sq 10	Wa cm sq 11	Ap cm 4th 12	Kg cm 5th 13	At cm sq 14
28EI - 2	2.540	4.604	5.080	5.080	3.016	11.4	297.2	166.4	12.2	3.2258	3.831	12.357	1.305221	132.3
125EI - .25	0.794	7.938	3.969	9.525	4.763	19.1	353.5	358.4	13.3	2.5202	7.560	19.054	1.440872	249.5
625EI - 2	3.175	3.969	4.763	4.763	2.381	9.5	358.0	84.9	12.6	5.0403	1.890	9.527	1.521315	110.9
100EI - .5	1.270	6.350	3.810	7.620	3.810	15.2	356.0	210.2	12.2	3.2258	4.839	15.609	1.648700	177.4
138EI - .25	0.873	8.731	4.366	10.478	5.239	21.0	470.0	475.7	14.6	3.0494	9.148	27.896	2.327013	301.9
75EI - 1.5	2.858	4.763	4.763	5.715	2.858	11.4	468.0	127.0	13.1	5.4435	2.722	14.816	2.458173	139.7
87EI - 1	2.223	5.556	4.445	6.668	3.334	13.3	481.0	171.1	13.0	4.9395	3.705	18.299	2.783890	163.0
112EI - .5	1.429	7.144	4.286	8.573	4.286	17.2	514.5	297.6	13.7	4.0827	6.124	25.002	2.987532	224.5
150EI - .25	0.953	9.525	4.763	11.430	5.715	22.9	614.3	616.1	15.9	3.6290	10.887	39.509	3.603734	359.3
75EI - 2	3.810	4.763	5.715	5.715	2.858	11.4	624.0	147.5	15.2	7.2581	2.722	19.755	3.764054	159.7
125EI - .5	1.588	7.938	4.763	9.525	4.763	19.1	707.0	406.4	15.1	5.0403	7.560	38.107	5.082021	277.2
87EI - 1.5	3.334	5.556	5.556	6.668	3.334	13.3	721.5	200.4	15.2	7.4093	3.705	27.449	5.348476	190.2
100EI - 1	2.540	6.350	5.080	7.620	3.810	15.2	712.0	253.9	14.8	6.4516	4.839	31.217	5.459598	212.9
36EI - .25	1.032	10.795	7.382	14.605	6.668	27.9	893.8	1558.0	20.7	4.2591	21.169	90.161	7.421701	585.5
175EI - .25	1.111	11.113	5.556	13.335	6.668	26.7	976.5	974.8	18.5	4.9395	14.819	73.196	7.817717	489.0
87EI - 2	4.445	5.556	6.668	6.668	3.334	13.3	962.0	232.4	17.6	9.8790	3.705	36.598	8.198326	217.3
138EI - .5	1.746	8.731	5.239	10.478	5.239	21.0	940.0	539.0	16.6	6.0988	9.148	55.793	8.214678	335.4
112EI - 1	2.858	7.144	5.715	8.573	4.286	17.2	1029.0	359.8	16.5	8.1653	6.124	50.004	9.883624	269.5
100EI - 1.5	3.810	6.350	6.350	7.620	3.810	15.2	1068.0	301.2	17.5	9.6774	4.839	46.826	10.355562	248.4
150EI - .5	1.905	9.525	5.715	11.430	5.715	22.9	1228.5	697.6	18.0	7.2581	10.887	79.019	12.731036	399.2
19EI - .25	1.111	12.065	10.001	17.780	7.620	33.0	1222.3	3068.7	25.5	4.9395	33.871	167.305	12.974537	867.9
100EI - 2	5.080	6.350	7.620	7.620	3.810	15.2	1424.0	344.9	20.0	12.9032	4.839	62.435	16.076942	283.9
125EI - 1	3.175	7.938	6.350	9.525	4.763	19.1	1414.0	491.8	18.3	10.0806	7.560	76.214	16.799823	332.7
112EI - 1.5	4.286	7.144	7.144	8.573	4.286	17.2	1543.5	426.6	19.6	12.2480	6.124	75.006	18.758227	314.4
212EI - .25	1.349	13.494	6.747	16.193	8.096	32.4	1782.0	1754.1	22.6	7.2833	21.850	159.137	20.535889	721.0
36EI - .5	2.064	10.795	8.414	14.605	6.668	27.9	1787.5	1728.4	23.0	8.5181	21.169	180.323	26.759823	637.9
138EI - 1	3.493	8.731	6.985	10.478	5.239	21.0	1880.0	652.6	20.1	12.1976	9.148	111.585	27.138271	402.5
225EI - .25	1.429	14.288	7.144	17.145	8.573	34.3	2099.0	2079.1	23.9	8.1653	24.496	200.017	27.370438	808.4
175EI - .5	2.223	11.113	6.668	13.335	6.668	26.7	1953.0	1102.5	20.9	9.8790	14.819	146.392	27.650039	543.3
112EI - 2	5.715	7.144	8.573	8.573	4.286	17.2	2058.0	488.8	22.4	16.3306	6.124	100.008	29.102814	359.3

Standard EE and EI Laminations														
Part No.	D cm 1	Ht cm 2	Wth cm 3	Lt cm 4	G cm 5	MPL cm 6	Wtfe grams 7	Wtcu grams 8	MLT cm 9	Ac cm sq 10	Wa cm sq 11	Ap cm 4th 12	Kg cm 5th 13	At cm sq 14
125EI - 1.5	4.763	7.938	7.938	9.525	4.763	19.1	2121.0	582.8	21.7	15.1209	7.560	114.321	31.900089	388.1
150EI - 1	3.810	9.525	7.620	11.430	5.715	22.9	2457.0	853.2	22.0	14.5161	10.887	158.038	41.639435	479.0
250EI - .25	1.588	15.875	7.938	19.050	9.525	38.1	2818.3	2844.7	26.5	10.0806	30.242	304.857	46.470449	998.0
19EI - .5	2.223	12.065	11.113	17.780	7.620	33.0	2444.5	3360.4	27.9	9.8790	33.871	334.611	47.391978	934.3
125EI - 2	6.350	7.938	9.525	9.525	4.763	19.1	2828.0	668.1	24.9	20.1613	7.560	152.429	49.465695	443.5
138EI - 1.5	5.239	8.731	8.731	10.478	5.239	21.0	2820.0	773.0	23.8	18.2963	9.148	167.378	51.551845	469.6
212EI - .5	2.699	13.494	8.096	16.193	8.096	32.4	3564.0	1963.8	25.3	14.5665	21.850	318.275	73.372510	801.2
138EI - 2	6.985	8.731	10.478	10.478	5.239	21.0	3760.0	899.6	27.7	24.3951	9.148	223.171	78.747745	536.7
150EI - 1.5	5.715	9.525	9.525	11.430	5.715	22.9	3685.5	1000.7	25.8	21.7741	10.887	237.057	79.878837	558.9
251EI - .25	1.588	20.320	11.748	22.860	13.970	50.8	3779.5	8185.1	32.4	10.0806	70.968	715.398	88.938903	1685.5
36EI - 1	4.128	10.795	10.478	14.605	6.668	27.9	3575.0	2054.8	27.3	17.0363	21.169	360.646	90.037618	742.7
175EI - 1	4.445	11.113	8.890	13.335	6.668	26.7	3906.0	1347.6	25.6	19.7580	14.819	292.785	90.477806	652.0
225EI - .5	2.858	14.288	8.573	17.145	8.573	34.3	4198.0	2328.0	26.7	16.3306	24.496	400.033	97.775934	898.2
300EI - .25	1.905	19.050	9.525	22.860	11.430	45.7	5035.5	4896.8	31.6	14.5161	43.548	632.151	116.078006	1437.1
150EI - 2	7.620	9.525	11.430	11.430	5.715	22.9	4914.0	1163.7	30.1	29.0322	10.887	316.076	122.116787	638.7
19EI - 1	4.445	12.065	13.335	17.780	7.620	33.0	4889.0	3920.9	32.6	19.7580	33.871	669.222	162.472022	1066.9
250EI - .5	3.175	15.875	9.525	19.050	9.525	38.1	5636.5	3186.1	29.6	20.1613	30.242	609.714	165.961994	1108.9
175EI - 1.5	6.668	11.113	11.113	13.335	6.668	26.7	5859.0	1603.0	30.4	29.6370	14.819	439.177	171.151247	760.7
36EI - 1.5	6.191	10.795	12.541	14.605	6.668	27.9	5362.5	2365.5	31.4	25.5544	21.169	540.969	175.974626	847.6
212EI - 1	5.398	13.494	10.795	16.193	8.096	32.4	7128.0	2399.3	30.9	29.1330	21.850	636.549	240.214294	961.4
175EI - 2	8.890	11.113	13.335	13.335	6.668	26.7	7812.0	1837.2	34.9	39.5160	14.819	585.569	265.476676	869.4
36EI - 2	8.255	10.795	14.605	14.605	6.668	27.9	7150.0	2706.3	36.0	34.0725	21.169	721.292	273.445097	952.4
19EI - 1.5	6.668	12.065	15.558	17.780	7.620	33.0	7333.5	4504.4	37.4	29.6370	33.871	1003.833	318.203102	1199.6
225EI - 1	5.715	14.288	11.430	17.145	8.573	34.3	8396.0	2843.9	32.6	32.6612	24.496	800.067	320.150914	1077.8
251EI - .5	3.175	20.320	13.335	22.860	13.970	50.8	7559.0	8986.4	35.6	20.1613	70.968	1430.796	324.035679	1822.6
300EI - .5	3.810	19.050	11.430	22.860	11.430	45.7	10071.0	5519.0	35.6	29.0322	43.548	1264.303	411.965429	1596.8
212EI - 1.5	8.096	13.494	13.494	16.193	8.096	32.4	10692.0	2849.8	36.7	43.6995	21.850	954.823	455.049972	1121.6
400EI - .25	2.540	25.400	12.700	30.480	15.240	61.0	11937.3	11551.5	42.0	25.8064	77.419	1997.911	491.514460	2554.8
19EI - 2	8.890	12.065	17.780	17.780	7.620	33.0	9778.0	5039.8	41.8	39.5160	33.871	1338.444	505.600776	1332.3
250EI - 1	6.350	15.875	12.700	19.050	9.525	38.1	11273.0	3869.0	36.0	40.3225	30.242	1219.427	546.679224	1330.6

Standard EE and EI Laminations														
Part No.	D cm 1	Ht cm 2	Wth cm 3	Lt cm 4	G cm 5	MPL cm 6	Wtfe grams 7	Wtcu grams 8	MLT cm 9	Ac cm sq 10	Wa cm sq 11	Ap cm 4th 12	Kg cm 5th 13	At cm sq 14
225EI - 1.5	8.573	14.288	14.288	17.145	8.573	34.3	12594.0	3376.6	38.8	48.9918	24.496	1200.100	606.705263	1257.5
212EI - 2	10.795	13.494	16.193	16.193	8.096	32.4	14256.0	3269.1	42.1	58.2660	21.850	1273.098	705.200000	1281.9
225EI - 2	11.430	14.288	17.145	17.145	8.573	34.3	16792.0	3874.4	44.5	65.3225	24.496	1600.134	940.000332	1437.1
250EI - 1.5	9.525	15.875	15.875	19.050	9.525	38.1	16909.5	4617.3	42.9	60.4837	30.242	1829.142	1030.693407	1552.4
251EI - 1	6.350	20.320	16.510	22.860	13.970	50.8	15118.0	10588.9	42.0	40.3225	70.968	2861.591	1099.987479	2096.8
300EI - 1	7.620	19.050	15.240	22.860	11.430	45.7	20142.0	6761.0	43.7	58.0644	43.548	2528.606	1345.157895	1916.1
250EI - 2	12.700	15.875	19.050	19.050	9.525	38.1	22546.0	5300.2	49.3	80.6450	30.242	2438.856	1596.262604	1774.2
400EI - .5	5.080	25.400	15.240	30.480	15.240	61.0	23874.5	13007.3	47.2	51.6128	77.419	3995.821	1746.012188	2838.7
251EI - 1.5	9.525	20.320	19.685	22.860	13.970	50.8	22677.0	12344.8	48.9	60.4837	70.968	4292.386	2122.930748	2371.0
300EI - 1.5	11.430	19.050	19.050	22.860	11.430	45.7	30213.0	7941.0	51.3	87.0966	43.548	3792.909	2576.858726	2235.5
251EI - 2	12.700	20.320	22.860	22.860	13.970	50.8	30236.0	13947.3	55.3	80.6450	70.968	5723.182	3340.469806	2645.2
300EI - 2	15.240	19.050	22.860	22.860	11.430	45.7	40284.0	9121.0	58.9	116.1288	43.548	5057.212	3988.414404	2554.8
400EI - 1	10.160	25.400	20.320	30.480	15.240	61.0	47749.0	15914.5	57.8	103.2256	77.419	7991.643	5708.230471	3406.4
400EI - 1.5	15.240	25.400	25.400	30.480	15.240	61.0	71623.5	18711.6	68.0	154.8384	77.419	11987.463	10923.621053	3974.2
400EI - 2	20.320	25.400	30.480	30.480	15.240	61.0	95498.0	21508.6	78.1	206.4512	77.419	15983.284	16894.337950	4541.9

THREE-PHASE LAMINATION TABLE DESCRIPTION

Definitions for the Following Tables

Information given is listed by column:

Col.	Dim.	Description	Units
1.	D	Stack build	cm
2.	Ht	Finished transformer height	cm
3.	Wth	Finished transformer width	cm
4.	Lt	Finished transformer length	cm
5.	G	Window length	cm
6.	W_{tfe}	Core weight	gram
7.	W_{tcu}	Copper weight [W_{tcu} = (MLT)(W_a)(K_u)(8.89)(1.5)]	gram
8.	MLT	Mean Length Turn [MLT = 2(D + 2J) + 2(E + 2J) + 3(3.14)(F)(0.5), J = Bobbin Thickness]	cm
9.	A_c	Iron area (gross) [A_c = (D)(E)]	cm sq
10.	W_a	Window area [W_z = (F)(G)]	cm sq
11.	A_p	Area product (gross) [A_p = (A_c)(W_a)(1.5)]	cm 4th
12.	K_g	Core geometry (gross) [K_g = (A_p)(A_c)(K_u)/(MLT)]	cm 5th
13.	A_t	Surface area	cm sq

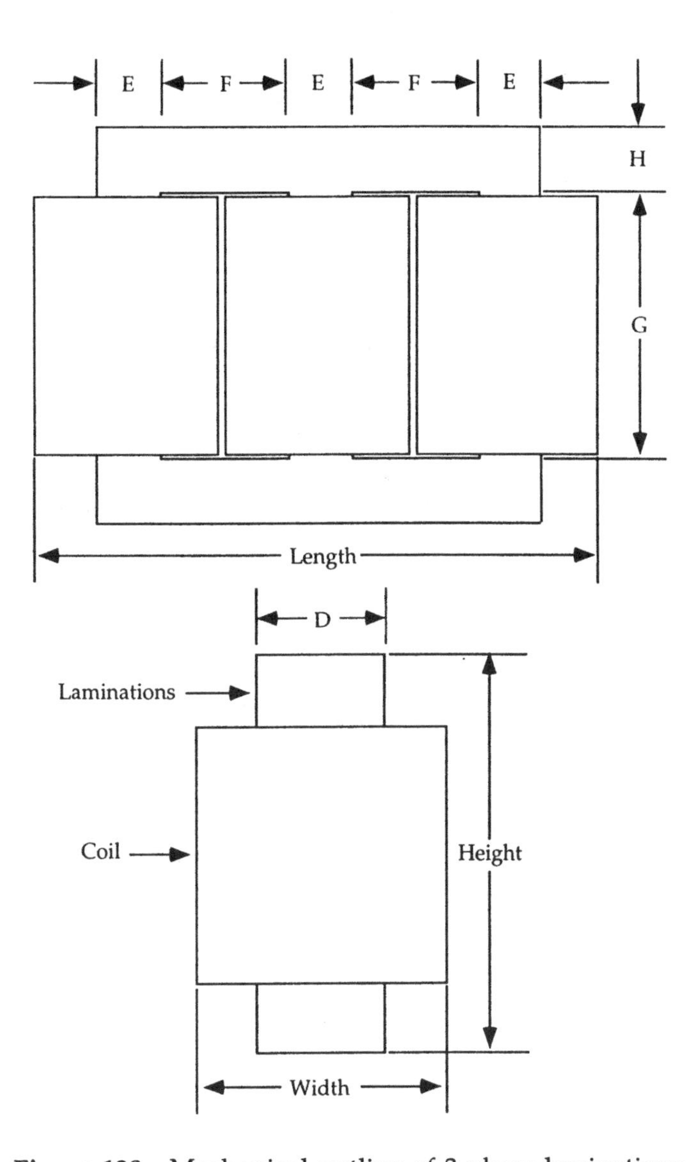

Figure 138 Mechanical outline of 3 phase laminations.

Standard 3 Phase Laminations

Part No.	D cm 1	Ht cm 2	Wth cm 3	Lt cm 4	G cm 5	Wtfe grams 6	Wtcu grams 7	MLT cm 8	Ac cm sq 9	Wa cm sq 10	Ap cm 4th 11	Kg cm 5th 12	At cm sq 13
25EI3P - .25	0.159	4.602	1.030	4.521	2.858	14.1	44.7	3.4	0.1008	2.490	0.376	0.004513	44.4
25EI3P - .5	0.318	4.602	1.189	4.521	2.858	28.3	48.9	3.7	0.2016	2.490	0.753	0.016495	47.3
375EI3P - .25	0.238	5.715	1.508	6.668	3.175	40.6	102.9	4.8	0.2268	4.032	1.372	0.026021	84.5
25EI3P - 1	0.635	4.602	1.506	4.521	2.858	56.5	57.3	4.3	0.4032	2.490	1.506	0.056273	53.1
50EI3P - .25	0.318	6.668	1.905	8.573	3.493	85.3	179.7	6.1	0.4032	5.544	3.353	0.089027	130.0
375EI3P - .5	0.476	5.715	1.746	6.668	3.175	81.3	113.1	5.3	0.4536	4.032	2.744	0.094659	90.4
25EI3P - 1.5	0.953	4.602	1.824	4.521	2.858	84.8	65.7	5.0	0.6048	2.490	2.259	0.110374	58.9
50LEI3P - .25	0.318	7.699	1.905	8.573	4.524	94.5	232.7	6.1	0.4032	7.181	4.344	0.115315	145.1
25EI3P - 2	1.270	4.602	2.141	4.521	2.858	113.0	76.8	5.8	0.8065	2.490	3.011	0.167902	64.7
562EI3P - .25	0.357	8.573	1.944	9.047	5.398	110.8	295.7	6.5	0.5094	8.569	6.548	0.206241	169.2
375EI3P - 1	0.953	5.715	2.223	6.668	3.175	162.5	133.6	6.2	0.9073	4.032	5.487	0.320578	102.2
50EI3P - .5	0.635	6.668	2.223	8.573	3.493	170.5	198.4	6.7	0.8065	5.544	6.707	0.322411	139.7
625EI3P - .25	0.397	9.604	2.381	10.714	5.634	185.8	446.6	7.5	0.6300	11.176	10.562	0.355315	226.3
50LEI3P - .5	0.635	7.699	2.223	8.573	4.524	189.0	257.0	6.7	0.8065	7.181	8.687	0.417610	155.4
375EI3P - 1.5	1.429	5.715	2.699	6.668	3.175	243.8	158.4	7.4	1.3609	4.032	8.231	0.608420	114.0
562EI3P - .5	0.714	8.573	2.301	9.047	5.398	221.5	328.3	7.2	1.0188	8.569	13.095	0.742990	181.8
375EI3P - 2	1.905	5.715	3.175	6.668	3.175	325.0	178.9	8.3	1.8145	4.032	10.975	0.957761	125.8
50EI3P - 1	1.270	6.668	2.858	8.573	3.493	341.0	241.9	8.2	1.6129	5.544	13.414	1.057896	159.1
625EI3P - .5	0.794	9.604	2.777	10.714	5.634	371.5	493.9	8.3	1.2601	11.176	21.124	1.285095	242.4
50LEI3P - 1	1.270	7.699	2.858	8.573	4.524	378.0	313.4	8.2	1.6129	7.181	17.374	1.370264	176.1
875EI3P - .25	0.556	11.669	3.334	15.001	6.111	458.8	935.4	10.3	1.2349	16.982	31.455	1.504556	398.2
50EI3P - 1.5	1.905	6.668	3.493	8.573	3.493	511.5	279.5	9.5	2.4193	5.544	20.121	2.060391	178.4
562EI3P - 1	1.427	8.573	3.015	9.047	5.398	443.0	402.7	8.8	2.0377	8.569	26.190	2.422965	206.9
50LEI3P - 1.5	1.905	7.699	3.493	8.573	4.524	567.0	362.0	9.5	2.4193	7.181	26.062	2.668769	196.8
50EI3P - 2	2.540	6.668	4.128	8.573	3.493	682.0	317.0	10.7	3.2258	5.544	26.827	3.228986	197.8
100EI3P - .25	0.635	13.716	4.445	19.050	7.620	724.0	1972.8	12.7	1.6129	29.032	70.239	3.557024	604.8
50LEI3P - 2	2.540	7.699	4.128	8.573	4.524	756.0	410.7	10.7	3.2258	7.181	34.749	4.182418	217.4
625EI3P - 1	1.588	9.604	3.571	10.714	5.634	743.0	600.4	10.1	2.5202	11.176	42.247	4.228147	274.7
562EI3P - 1.5	2.141	8.573	3.729	9.047	5.398	664.5	467.9	10.2	3.0565	8.569	39.285	4.691531	232.1
875EI3P - .5	1.111	11.669	3.890	15.001	6.111	917.5	1036.1	11.4	2.4698	16.982	62.911	5.433537	427.9

Standard 3 Phase Laminations

Part No.	D cm 1	Ht cm 2	Wth cm 3	Lt cm 4	G cm 5	Wtfe grams 6	Wtcu grams 7	MLT cm 8	Ac cm sq 9	Wa cm sq 10	Ap cm 4th 11	Kg cm 5th 12	At cm sq 13
120EI3P - .25	0.762	13.716	3.810	18.288	7.620	933.2	1587.4	12.8	2.3226	23.226	80.915	5.866757	578.3
562EI3P - 2	2.855	8.573	4.442	9.047	5.398	886.0	533.2	11.7	4.0754	8.569	52.380	7.319870	257.2
625EI3P - 1.5	2.381	9.604	4.365	10.714	5.634	1114.5	695.1	11.7	3.7802	11.176	63.371	8.218094	306.9
100EI3P - .5	1.270	13.716	5.080	19.050	7.620	1448.0	2200.5	14.2	3.2258	29.032	140.478	12.756199	646.5
625EI3P - 2	3.175	9.604	5.159	10.714	5.634	1486.0	789.7	13.2	5.0403	11.176	84.495	12.859180	339.2
875EI3P - 1	2.223	11.669	5.001	15.001	6.111	1835.0	1255.5	13.9	4.9395	16.982	125.822	17.935557	487.2
875EI3P - 1	2.223	11.669	5.001	15.001	6.111	1835.0	1255.5	13.9	4.9395	16.982	125.822	17.935557	487.2
150EI3P - .25	0.953	17.145	4.763	22.860	9.525	1830.7	3080.6	15.9	3.6290	36.290	197.547	18.018670	903.6
120EI3P - .5	1.524	13.716	4.572	18.288	7.620	1866.5	1801.0	14.5	4.6452	23.226	161.831	20.684044	627.1
875EI3P - 1.5	3.334	11.669	6.113	15.001	6.111	2752.5	1456.8	16.1	7.4093	16.982	188.732	34.778427	546.4
100EI3P - 1	2.540	13.716	6.350	19.050	7.620	2896.0	2593.8	16.7	6.4516	29.032	280.956	43.287147	729.7
180EI3P - .25	1.143	20.574	5.715	27.432	11.430	3162.5	5300.6	19.0	5.2258	52.258	409.634	45.028587	1301.2
875EI3P - 2	4.445	11.669	7.224	15.001	6.111	3670.0	1677.0	18.5	9.8790	16.982	251.643	53.711391	605.7
150EI3P - .5	1.905	17.145	5.715	22.860	9.525	3661.5	3488.1	18.0	7.2581	36.290	395.095	63.655147	979.8
120EI3P - 1	3.048	13.716	6.096	18.288	7.620	3733.0	2178.6	17.6	9.2903	23.226	323.662	68.395856	724.6
100EI3P - 1.5	3.810	13.716	7.620	19.050	7.620	4344.0	3019.4	19.5	9.6774	29.032	421.434	83.669119	812.9
120EI3P - 1.5	4.572	13.716	7.620	18.288	7.620	5599.5	2582.0	20.8	13.9355	23.226	485.492	129.848643	822.2
100EI3P - 2	5.080	13.716	8.890	19.050	7.620	5792.0	3412.7	22.0	12.9032	29.032	561.912	131.601108	896.1
180EI3P - .5	2.286	20.574	6.858	27.432	11.430	6325.0	5993.6	21.5	10.4516	52.258	819.268	159.290083	1411.0
240EI3P - .25	1.524	27.432	7.620	36.576	15.240	7535.3	12596.1	25.4	9.2903	92.903	1294.646	189.272465	2313.3
120EI3P - 2	6.096	13.716	9.144	18.288	7.620	7466.0	2959.6	23.9	18.5806	23.226	647.323	201.389363	919.7
150EI3P - 1	3.810	17.145	7.620	22.860	9.525	7323.0	4265.9	22.0	14.5161	36.290	790.189	208.197119	1132.3
150EI3P - 1.5	5.715	17.145	9.525	22.860	9.525	10984.5	5003.4	25.8	21.7741	36.290	1185.284	399.394127	1284.7
180EI3P - 1	4.572	20.574	9.144	27.432	11.430	12650.0	7326.0	26.3	20.9032	52.258	1638.536	521.277895	1630.4
150EI3P - 2	7.620	17.145	11.430	22.860	9.525	14646.0	5818.3	30.1	29.0322	36.290	1580.379	610.583823	1437.1
240EI3P - .5	3.048	27.432	9.144	36.576	15.240	15070.5	14106.5	28.5	18.5806	92.903	2589.293	676.026260	2508.4
180EI3P - 1.5	6.858	20.574	11.430	27.432	11.430	18975.0	8711.9	31.3	31.3548	52.258	2457.804	986.290416	1849.9
360EI3P - .25	2.286	41.148	11.430	54.864	22.860	25475.0	42172.9	37.8	20.9032	209.032	6554.144	1448.838781	5204.9
180EI3P - 2	9.144	20.574	13.716	27.432	11.430	25300.0	9986.3	35.8	41.8064	52.258	3277.073	1529.640997	2069.4
240EI3P - 1	6.096	27.432	12.192	36.576	15.240	30141.0	17230.5	34.8	37.1612	92.903	5178.585	2213.844875	2898.6

Standard 3 Phase Laminations

Part No.	D cm 1	Ht cm 2	Wth cm 3	Lt cm 4	G cm 5	Wtfe grams 6	Wtcu grams 7	MLT cm 8	Ac cm sq 9	Wa cm sq 10	Ap cm 4th 11	Kg cm 5th 12	At cm sq 13
240EI3P - 1.5	9.144	27.432	15.240	36.576	15.240	45211.5	20449.5	41.3	55.7418	92.903	7767.878	4197.044875	3288.8
360EI3P - .5	4.572	41.148	13.716	54.864	22.860	50950.0	47502.5	42.6	41.8064	209.032	13108.284	5145.140166	5643.9
240EI3P - 2	12.192	27.432	18.288	36.576	15.240	60282.0	23470.4	47.4	74.3224	92.903	10357.169	6501.062604	3679.0
360EI3P - 1	9.144	41.148	18.288	54.864	22.860	101900.0	58143.8	52.1	83.6127	209.032	26216.579	16813.961219	6521.8
360EI3P - 1.5	13.716	41.148	22.860	54.864	22.860	152850.0	68339.2	61.3	125.4191	209.032	39324.874	32187.445983	7399.7
360EI3P - 2	18.288	41.148	27.432	54.864	22.860	203800.0	78534.5	70.4	167.2255	209.032	52433.158	49793.562327	8277.7

10

C CORES

C CORE MANUFACTURERS

Manufacturers' addresses and phone numbers can be found from the Table of Contents.

Notes

C CORE TABLE DESCRIPTION

Definitions for the Following Tables

Information given is listed by column:

Col.	Dim.	Description	Units
1.	D	Strip width	cm
2.	E	Iron build	cm
3.	F	Window width	cm
4.	G	Window length	cm
5.	Ht	Finished transformer height (Ht = G + 2E)	cm
6.	Wth	Finished transformer width (Wth = D + F)	cm
7.	Lt	Finished transformer length [Lt = 2(E + F)]	cm
8.	MPL	Mean Magnetic Path Length	cm
9.	W_{tfe}	Core weight [$W_{tfe} = (MPL)(A_c)(7.63)$]	gram
10.	W_{tcu}	Copper weight [$W_{tcu} = (MLT)(W_a)(K_u)(8.89)$]	gram
11.	MLT	Mean Length Turn	cm
12.	A_c	Effective iron area [$A_c = (D)(E)SF$]	cm sq
13.	W_a	Window area [$W_a = (F)(G)$]	cm sq
14.	A_p	Area product [$A_p = (A_c)(W_a)$]	cm 4th
15.	K_g	Core geometry [$K_g = (A_p)(A_c)(K_u)/(MLT)$]	cm 5th
16.	A_t	Surface area	cm sq

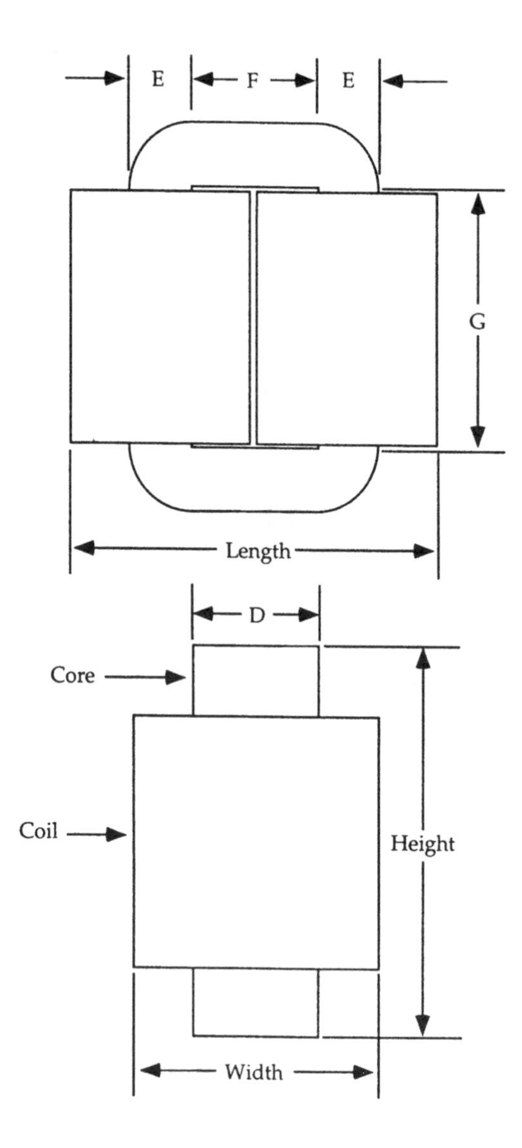

Figure 139 Mechanical outline of C cores.

National Arnold C Cores 1 mil Silicon Steel																
Part No.	D cm 1	E cm 2	F cm 3	G cm 4	Ht cm 5	Wth cm 6	Lt cm 7	MPL cm 8	Wtfe grams 9	Wtcu grams 10	MLT cm 11	Ac cm sq 12	Wa cm sq 13	Ap cm 4th 14	Kg cm 5th 15	At cm sq 16
CM-51-A	0.318	0.079	0.476	1.270	1.428	0.794	1.110	3.8	0.6	4.2	1.9	0.0209	0.605	0.013	0.000054	6.9
CM-1	0.476	0.159	0.159	0.476	0.794	0.635	0.636	1.9	0.9	0.5	1.9	0.0628	0.076	0.005	0.000062	2.6
CM-2-A	0.476	0.159	0.159	0.635	0.953	0.635	0.636	2.2	1.1	0.7	1.9	0.0628	0.101	0.006	0.000083	3.0
CM-51	0.318	0.159	0.318	0.953	1.271	0.636	0.954	3.2	1.0	2.0	1.9	0.0420	0.303	0.013	0.000115	4.9
CM-51-B	0.318	0.119	0.556	1.349	1.587	0.874	1.350	4.3	1.0	5.7	2.2	0.0314	0.750	0.024	0.000137	8.9
CM-7	0.318	0.159	0.635	1.588	1.906	0.953	1.588	5.1	1.6	8.5	2.4	0.0420	1.008	0.042	0.000301	11.9
CM-52	0.318	0.318	0.318	0.953	1.589	0.636	1.272	3.8	2.4	2.3	2.2	0.0839	0.303	0.025	0.000392	6.9
CM-53	0.318	0.318	0.476	0.953	1.589	0.794	1.588	4.1	2.6	3.9	2.4	0.0839	0.454	0.038	0.000527	8.9
CM-52-A	0.476	0.318	0.318	1.429	2.065	0.794	1.272	4.8	4.6	4.0	2.5	0.1256	0.454	0.057	0.001150	9.7
CM-54	0.635	0.318	0.397	0.953	1.589	1.032	1.430	4.0	5.1	4.0	2.9	0.1676	0.378	0.063	0.001447	9.8
CM-55	0.635	0.318	0.476	0.953	1.589	1.111	1.588	4.1	5.3	4.9	3.1	0.1676	0.454	0.076	0.001665	10.9
CM-53-A	0.556	0.357	0.357	1.667	2.381	0.913	1.428	5.5	6.9	5.9	2.8	0.1647	0.595	0.098	0.002312	12.7
CM-16	0.476	0.238	0.953	2.381	2.857	1.429	2.382	7.6	5.5	26.9	3.3	0.0940	2.269	0.213	0.002408	26.8
CM-3	0.635	0.318	0.635	1.270	1.906	1.270	1.906	5.1	6.5	9.5	3.3	0.1676	0.806	0.135	0.002737	15.3
CM-54-A	0.635	0.397	0.397	1.905	2.699	1.032	1.588	6.2	9.9	8.3	3.1	0.2092	0.756	0.158	0.004279	16.2
CM-52-B	0.953	0.397	0.318	1.429	2.223	1.271	1.430	5.1	12.2	5.8	3.6	0.3140	0.454	0.143	0.004969	14.6
CM-17	0.476	0.318	1.270	3.175	3.811	1.746	3.176	10.2	9.7	57.2	4.0	0.1256	4.032	0.507	0.006381	45.4
CM-7-A	0.635	0.476	0.635	1.588	2.540	1.270	2.222	6.4	12.2	13.0	3.6	0.2509	1.008	0.253	0.006999	21.0
CM-3-B	0.714	0.437	0.437	2.143	3.017	1.151	1.748	6.9	13.7	11.3	3.4	0.2590	0.936	0.243	0.007398	20.0
CM-21	0.635	0.635	0.635	1.270	2.540	1.270	2.540	6.4	16.2	11.3	3.9	0.3347	0.806	0.270	0.009159	22.6
CM-3-A	0.794	0.476	0.476	2.381	3.333	1.270	1.904	7.6	18.2	14.9	3.7	0.3137	1.133	0.356	0.012072	24.2
CM-7-D	0.953	0.476	0.635	1.588	2.540	1.588	2.222	6.4	18.2	15.3	4.3	0.3765	1.008	0.380	0.013413	24.0
CM-2	0.635	0.635	0.635	2.223	3.493	1.270	2.540	8.3	21.1	19.8	3.9	0.3347	1.412	0.472	0.016032	29.8
CM-4-B	0.953	0.635	0.635	1.310	2.580	1.588	2.540	6.4	24.6	13.6	4.6	0.5023	0.832	0.418	0.018325	26.1
CM-4-A	0.953	0.635	0.635	1.389	2.659	1.588	2.540	6.6	25.2	14.4	4.6	0.5023	0.882	0.443	0.019430	26.8
CM-3-D	0.873	0.556	0.556	2.619	3.731	1.429	2.224	8.6	26.4	21.4	4.1	0.4029	1.456	0.587	0.022841	31.1
CM-4	0.953	0.635	0.635	2.223	3.493	1.588	2.540	8.3	31.6	23.0	4.6	0.5023	1.412	0.709	0.031096	33.7
CM-18	0.953	0.556	0.794	2.540	3.652	1.747	2.700	8.9	29.8	33.5	4.7	0.4398	2.017	0.887	0.033393	38.4
CM-10	0.953	0.635	0.794	2.223	3.493	1.747	2.858	8.6	32.9	30.3	4.8	0.5023	1.765	0.887	0.036873	38.1
CM-8-A	0.953	0.635	0.635	2.858	4.128	1.588	2.540	9.5	36.5	29.6	4.6	0.5023	1.815	0.912	0.039979	38.9

National Arnold C Cores 1 mil Silicon Steel																
Part No.	D cm 1	E cm 2	F cm 3	G cm 4	Ht cm 5	Wth cm 6	Lt cm 7	MPL cm 8	Wtfe grams 9	Wtcu grams 10	MLT cm 11	Ac cm sq 12	Wa cm sq 13	Ap cm 4th 14	Kg cm 5th 15	At cm sq 16
CM-61	1.270	0.476	0.635	3.810	4.762	1.905	2.222	10.8	41.3	43.9	5.1	0.5018	2.419	1.214	0.047800	45.4
CM-33-A	1.072	0.714	0.714	3.175	4.603	1.786	2.856	10.6	51.5	41.1	5.1	0.6353	2.267	1.440	0.071745	48.9
CM-34	0.953	0.635	1.270	3.334	4.604	2.223	3.810	11.7	45.0	84.0	5.6	0.5023	4.234	2.127	0.076603	66.0
CM-8	0.953	0.953	0.635	2.858	4.764	1.588	3.176	10.8	62.1	33.7	5.2	0.7538	1.815	1.368	0.079069	50.6
CM-4-D	1.905	0.635	0.635	2.223	3.493	2.540	2.540	8.3	63.2	33.6	6.7	1.0040	1.412	1.417	0.085147	45.2
CM-9	0.953	0.953	0.794	2.858	4.764	1.747	3.494	11.1	63.9	44.1	5.5	0.7538	2.269	1.711	0.094353	56.1
CM-33	0.953	0.635	2.223	3.175	4.445	3.176	5.716	13.3	51.1	177.5	7.1	0.5023	7.058	3.545	0.100684	104.1
CM-490	0.476	0.476	3.810	16.510	17.462	4.286	8.572	42.5	61.0	1855.2	8.3	0.1881	62.903	11.829	0.107292	508.0
CM-12	0.953	0.635	2.699	3.175	4.445	3.652	6.668	14.3	54.8	238.3	7.8	0.5023	8.569	4.304	0.110563	126.7
CM-5	0.953	0.953	0.953	3.016	4.922	1.906	3.812	11.8	67.6	58.4	5.7	0.7538	2.874	2.167	0.114289	63.6
CM-4-E	1.191	0.794	0.794	3.493	5.081	1.985	3.176	11.8	70.4	55.5	5.6	0.7849	2.773	2.177	0.121509	60.0
CM-37	0.953	0.953	1.270	2.858	4.764	2.223	4.446	12.1	69.4	80.2	6.2	0.7538	3.630	2.736	0.132767	73.6
CM-24-A	1.588	0.635	1.270	2.858	4.128	2.858	3.810	10.8	68.9	91.0	7.0	0.8370	3.630	3.038	0.144301	70.4
CM-9-A	0.953	1.270	0.794	2.858	5.398	1.747	4.128	12.4	94.9	49.2	6.1	1.0046	2.269	2.280	0.150149	69.8
CM-492	0.635	0.635	3.175	9.843	11.113	3.810	7.620	28.6	73.0	881.6	7.9	0.3347	31.252	10.459	0.176506	296.0
CM-11-A	1.588	0.794	0.953	3.651	5.239	2.541	3.494	12.4	98.9	85.0	6.9	1.0465	3.479	3.641	0.221932	74.9
CM-34-A	1.905	0.635	1.270	3.334	4.604	3.175	3.810	11.7	90.0	115.7	7.7	1.0040	4.234	4.251	0.222256	82.1
CM-24-B	1.905	0.794	1.270	2.540	4.128	3.175	4.128	10.8	103.4	91.8	8.0	1.2554	3.226	4.050	0.254214	78.4
CM-11-B	1.588	0.794	1.270	3.810	5.398	2.858	4.128	13.3	106.5	126.7	7.4	1.0465	4.839	5.064	0.287780	90.9
CM-5-B	1.588	0.794	1.270	3.810	5.398	2.858	4.128	13.3	106.5	126.7	7.4	1.0465	4.839	5.064	0.287780	90.9
CM-17-A	1.270	1.111	1.270	3.175	5.397	2.540	4.762	13.3	119.1	105.6	7.4	1.1711	4.032	4.722	0.300394	91.7
CM-6	0.953	1.270	0.476	8.890	11.430	1.429	3.492	23.8	182.5	84.3	5.6	1.0046	4.232	4.251	0.304949	117.6
CM-11	1.905	0.953	0.953	3.016	4.922	2.858	3.812	11.8	135.1	79.9	7.8	1.5068	2.874	4.331	0.333809	80.2
CM-15	0.953	1.429	0.635	6.350	9.208	1.588	4.128	19.7	169.8	88.5	6.2	1.1303	4.032	4.558	0.334043	111.1
CM-107	1.588	0.794	1.588	3.810	5.398	3.176	4.764	14.0	111.6	169.2	7.9	1.0465	6.050	6.332	0.336997	105.9
CM-56	1.270	0.953	1.588	4.128	6.034	2.858	5.082	15.2	116.8	175.9	7.5	1.0046	6.555	6.585	0.350603	112.8
CM-5-A	2.540	0.635	1.270	3.810	5.080	3.810	3.810	12.7	129.7	154.0	9.0	1.3387	4.839	6.478	0.387476	100.0
CM-56-A	1.588	0.794	1.588	4.763	6.351	3.176	4.764	15.9	126.8	211.5	7.9	1.0465	7.564	7.916	0.421291	121.0
CM-24-H	1.270	0.635	4.445	6.350	7.620	5.715	10.160	24.1	123.2	1143.9	11.4	0.6694	28.226	18.893	0.443853	335.5
CM-70	0.953	0.953	3.334	5.715	7.621	4.287	8.574	21.9	126.0	640.6	9.5	0.7538	19.054	14.363	0.458075	244.1

National Arnold C Cores 1 mil Silicon Steel

Part No.	D cm 1	E cm 2	F cm 3	G cm 4	Ht cm 5	Wth cm 6	Lt cm 7	MPL cm 8	Wtfe grams 9	Wtcu grams 10	MLT cm 11	Ac cm sq 12	Wa cm sq 13	Ap cm 4th 14	Kg cm 5th 15	At cm sq 16
CM-47	1.588	0.794	1.905	5.715	7.303	3.493	5.398	18.4	147.1	323.8	8.4	1.0465	10.887	11.394	0.570315	155.5
CM-152	1.905	0.953	1.270	4.128	6.034	3.175	4.446	14.6	168.0	155.1	8.3	1.5068	5.243	7.900	0.572429	110.5
CM-32-A	1.588	0.794	2.223	6.668	8.256	3.811	6.034	21.0	167.3	467.1	8.9	1.0465	14.823	15.513	0.732750	194.4
CM-6-A	1.905	1.270	0.476	8.890	11.430	2.381	3.492	23.8	364.8	115.9	7.7	2.0081	4.232	8.497	0.885792	146.0
CM-92-A	1.270	1.270	3.810	3.810	6.350	5.080	10.160	20.3	207.6	602.4	11.7	1.3387	14.516	19.433	0.891705	258.1
CM-34-B	1.905	1.905	1.270	3.334	7.144	3.175	6.350	16.8	386.7	153.9	10.2	3.0121	4.234	12.754	1.503259	157.1
CM-14	2.540	1.270	1.588	3.969	6.509	4.128	5.716	16.2	330.8	240.3	10.7	2.6774	6.303	16.875	1.685702	159.2
CM-57	1.905	1.588	1.588	4.921	8.097	3.493	6.352	19.4	371.1	280.3	10.1	2.5109	7.815	19.621	1.953629	183.6
CM-414	1.588	1.588	1.588	6.668	9.844	3.176	6.352	22.9	365.1	355.9	9.5	2.0930	10.589	22.163	1.962851	207.8
CM-23	2.540	1.588	1.588	4.921	8.097	4.128	6.352	19.4	494.8	315.6	11.4	3.3478	7.815	26.162	3.084741	202.0
CM-96	2.540	2.540	3.810	8.573	13.653	6.350	12.700	34.9	1427.0	1945.5	16.7	5.3548	32.663	174.905	22.366695	604.1
CM-101-A	4.445	2.302	2.381	6.826	11.430	6.826	9.366	27.6	1789.9	1043.1	18.0	8.4929	16.253	138.032	25.981418	454.2
CM-97	3.810	2.223	2.858	9.525	13.971	6.668	10.162	33.7	1805.3	1681.4	17.4	7.0298	27.222	191.368	30.981037	555.0
CM-98-C	3.810	2.540	5.080	8.890	13.970	8.890	15.240	38.1	2335.0	3451.4	21.5	8.0322	45.161	362.746	54.228836	841.9
CM-199	4.128	3.175	3.334	11.748	18.098	7.462	13.018	42.9	3557.8	2877.0	20.7	10.8783	39.168	426.080	89.754934	850.9
CM-301-A	2.778	6.350	7.620	15.240	27.940	10.398	27.940	71.1	7945.1	12730.3	30.8	14.6414	116.129	1700.294	323.021292	2336.3

National Arnold C Cores 2 mil Silicon Steel																
Part No.	D cm 1	E cm 2	F cm 3	G cm 4	Ht cm 5	Wth cm 6	Lt cm 7	MPL cm 8	Wtfe grams 9	Wtcu grams 10	MLT cm 11	Ac cm sq 12	Wa cm sq 13	Ap cm 4th 14	Kg cm 5th 15	At cm sq 16
CL-1-H	0.32	0.16	0.32	0.95	1.27	0.64	0.95	3.2	1.1	2.0	1.9	0.045	0.303	0.014	0.00013	4.9
CL-1-F	0.64	0.16	0.24	0.71	1.03	0.87	0.79	2.5	1.7	1.4	2.4	0.090	0.170	0.015	0.00023	4.5
CL-1-R	0.48	0.16	0.32	1.11	1.43	0.79	0.95	3.5	1.8	2.7	2.2	0.067	0.353	0.024	0.00029	6.1
CL-1-L	0.48	0.24	0.24	0.71	1.19	0.71	0.95	2.9	2.2	1.3	2.2	0.101	0.170	0.017	0.00031	4.9
CL-1-K	0.64	0.16	0.24	1.43	1.75	0.87	0.79	4.0	2.7	2.9	2.4	0.090	0.340	0.031	0.00046	6.9
CL-1-C	0.64	0.24	0.24	0.71	1.19	0.87	0.95	2.9	2.9	1.5	2.5	0.135	0.170	0.023	0.00049	5.6
CL-26-B	0.32	0.32	0.48	0.95	1.59	0.79	1.59	4.1	2.8	3.9	2.4	0.090	0.454	0.041	0.00061	8.9
CL-163-D	0.48	0.24	0.64	1.27	1.75	1.11	1.75	4.8	3.7	8.1	2.8	0.101	0.806	0.081	0.00116	12.6
CL-1-P	0.48	0.32	0.32	1.43	2.07	0.79	1.27	4.8	4.9	4.0	2.5	0.135	0.454	0.061	0.00132	9.7
CL-26-A	0.64	0.32	0.48	0.95	1.59	1.11	1.59	4.1	5.7	4.9	3.1	0.180	0.454	0.082	0.00191	10.9
CL-163-A	0.48	0.32	0.64	1.27	1.91	1.11	1.91	5.1	5.2	8.6	3.0	0.135	0.806	0.109	0.00196	14.1
CL-40	0.48	0.32	0.48	1.75	2.38	0.95	1.59	5.7	5.9	8.1	2.7	0.135	0.831	0.112	0.00220	13.9
CL-1-A	0.64	0.32	0.64	0.95	1.59	1.27	1.91	4.4	6.1	7.1	3.3	0.180	0.605	0.109	0.00236	13.3
CL-163-F	0.56	0.36	0.36	1.67	2.38	0.91	1.43	5.5	7.4	5.9	2.8	0.177	0.595	0.105	0.00266	12.7
CL-1-D	0.64	0.32	0.60	1.27	1.91	1.23	1.83	5.0	6.9	8.7	3.2	0.180	0.756	0.136	0.00301	14.6
CL-1	0.64	0.32	0.64	1.27	1.91	1.27	1.91	5.1	7.0	9.5	3.3	0.180	0.806	0.145	0.00315	15.3
CL-163-B	0.48	0.48	0.40	1.59	2.54	0.87	1.75	5.9	9.0	6.6	2.9	0.202	0.630	0.127	0.00349	14.8
CL-153	0.64	0.32	0.64	1.43	2.07	1.27	1.91	5.4	7.4	10.7	3.3	0.180	0.907	0.163	0.00354	16.3
CL-2-K	0.64	0.32	0.64	1.59	2.22	1.27	1.91	5.7	7.8	11.9	3.3	0.180	1.008	0.181	0.00394	17.4
CL-162	0.64	0.32	2.54	0.79	1.43	3.18	5.72	7.9	10.9	45.2	6.3	0.180	2.017	0.362	0.00414	55.3
CL-163	0.64	0.40	0.64	1.27	2.06	1.27	2.06	5.4	9.2	9.9	3.5	0.224	0.806	0.181	0.00468	17.0
CL-147-D	0.64	0.40	0.40	1.91	2.70	1.03	1.59	6.2	10.6	8.3	3.1	0.224	0.756	0.170	0.00492	16.2
CL-2-H	0.48	0.48	0.64	1.59	2.54	1.11	2.22	6.4	9.8	11.9	3.3	0.202	1.008	0.203	0.00496	19.5
CL-4-A	0.64	0.32	0.64	2.22	2.86	1.27	1.91	7.0	9.6	16.6	3.3	0.180	1.412	0.254	0.00551	21.4
CL-147-B	0.56	0.52	0.40	1.75	2.78	0.95	1.83	6.4	12.4	7.8	3.2	0.255	0.693	0.177	0.00569	17.3
CL-3-J	0.64	0.48	0.48	1.43	2.38	1.11	1.90	5.7	11.7	8.2	3.4	0.269	0.680	0.183	0.00583	16.8
CL-1-E	0.95	0.32	0.64	1.27	1.91	1.59	1.91	5.1	10.5	11.3	3.9	0.270	0.806	0.218	0.00595	17.8
CL-147-A	0.64	0.48	0.64	1.27	2.22	1.27	2.22	5.7	11.7	10.4	3.6	0.269	0.806	0.217	0.00644	18.7
CL-147	0.64	0.32	0.79	2.54	3.18	1.43	2.22	7.9	10.9	25.5	3.6	0.180	2.017	0.362	0.00732	27.5
CL-2-F	0.64	0.32	0.64	3.18	3.81	1.27	1.91	8.9	12.2	23.7	3.3	0.180	2.016	0.362	0.00787	27.4

National Arnold C Cores 2 mil Silicon Steel

Part No.	D cm 1	E cm 2	F cm 3	G cm 4	Ht cm 5	Wth cm 6	Lt cm 7	MPL cm 8	Wtfe grams 9	Wtcu grams 10	MLT cm 11	Ac cm sq 12	Wa cm sq 13	Ap cm 4th 14	Kg cm 5th 15	At cm sq 16
CL-2	0.64	0.48	0.64	1.59	2.54	1.27	2.22	6.4	13.0	13.0	3.6	0.269	1.008	0.271	0.00805	21.0
CL-163-H	0.71	0.44	0.44	2.14	3.02	1.15	1.75	6.9	14.6	11.3	3.4	0.278	0.936	0.260	0.00851	20.0
CL-2-B	0.64	0.48	0.79	1.59	2.54	1.43	2.54	6.7	13.7	17.4	3.9	0.269	1.261	0.339	0.00942	24.4
CL-2-J	0.48	0.48	0.64	3.18	4.13	1.11	2.22	9.5	14.7	23.7	3.3	0.202	2.016	0.407	0.00991	30.0
CL-143	0.64	0.64	0.64	1.27	2.54	1.27	2.54	6.4	17.4	11.3	3.9	0.359	0.806	0.289	0.01053	22.6
CL-3-B	0.95	0.48	0.48	1.59	2.54	1.43	1.90	6.0	18.6	10.8	4.0	0.404	0.756	0.305	0.01228	20.6
CL-161	0.64	0.64	0.64	1.59	2.86	1.27	2.54	7.0	19.1	14.1	3.9	0.359	1.008	0.362	0.01317	25.0
CL-3-K	0.79	0.48	0.48	2.38	3.33	1.27	1.90	7.6	19.6	14.9	3.7	0.336	1.133	0.381	0.01388	24.2
CL-2-D	0.64	0.48	0.64	2.86	3.81	1.27	2.22	8.9	18.2	23.4	3.6	0.269	1.815	0.488	0.01448	29.8
CL-3-C	0.95	0.48	0.64	1.51	2.46	1.59	2.22	6.2	19.1	14.5	4.3	0.404	0.958	0.387	0.01465	23.4
CL-3-F	0.95	0.48	0.60	1.59	2.54	1.55	2.14	6.3	19.3	14.1	4.2	0.404	0.945	0.381	0.01467	23.1
CL-3	0.95	0.48	0.64	1.59	2.54	1.59	2.22	6.4	19.6	15.3	4.3	0.404	1.008	0.407	0.01542	24.0
CL-2-A	0.64	0.48	0.64	3.18	4.13	1.27	2.22	9.5	19.5	26.0	3.6	0.269	2.016	0.542	0.01609	32.0
CL-4-D	0.64	0.64	0.56	2.22	3.49	1.19	2.38	8.1	22.2	16.8	3.8	0.359	1.236	0.444	0.01666	27.8
CL-2-E	0.95	0.48	0.79	1.59	2.54	1.75	2.54	6.7	20.5	20.2	4.5	0.404	1.261	0.509	0.01822	27.6
CL-4	0.64	0.64	0.64	2.22	3.49	1.27	2.54	8.3	22.6	19.8	3.9	0.359	1.412	0.507	0.01843	29.8
CL-45	0.48	0.64	0.79	3.49	4.76	1.27	2.86	11.1	22.8	38.2	3.9	0.269	2.773	0.746	0.02071	42.5
CL-3-D	1.27	0.48	0.64	1.59	2.54	1.91	2.22	6.4	26.1	18.3	5.1	0.538	1.008	0.543	0.02291	27.0
CL-5-C	0.95	0.64	0.71	1.31	2.58	1.67	2.70	6.6	27.1	15.6	4.7	0.539	0.935	0.504	0.02307	27.9
CL-5-A	0.95	0.64	0.64	1.59	2.86	1.59	2.54	7.0	28.7	16.4	4.6	0.539	1.008	0.543	0.02554	28.4
CL-260	0.48	0.64	1.11	3.49	4.76	1.59	3.49	11.7	24.1	60.4	4.4	0.269	3.881	1.044	0.02568	53.8
CL-149-A	0.87	0.56	0.56	2.62	3.73	1.43	2.22	8.6	28.3	21.4	4.1	0.432	1.456	0.629	0.02626	31.1
CL-71	0.95	0.48	0.79	2.54	3.49	1.75	2.54	8.6	26.4	32.4	4.5	0.404	2.017	0.814	0.02914	35.8
CL-7-A	0.95	0.40	0.64	4.45	5.24	1.59	2.06	11.7	30.2	41.2	4.1	0.337	2.823	0.950	0.03119	42.8
CL-132-A	0.79	0.64	0.79	2.22	3.49	1.59	2.86	8.6	29.4	28.3	4.5	0.449	1.765	0.792	0.03150	36.1
CL-2-C	0.64	0.79	0.79	2.22	3.81	1.43	3.18	9.2	31.5	28.3	4.5	0.449	1.765	0.792	0.03150	39.1
CL-76	0.95	0.48	0.95	2.54	3.49	1.91	2.86	8.9	27.4	41.0	4.8	0.404	2.421	0.977	0.03314	40.5
CL-5	0.95	0.64	0.64	2.22	3.49	1.59	2.54	8.3	33.9	23.0	4.6	0.539	1.412	0.760	0.03576	33.7
CL-171	1.27	0.64	0.40	2.22	3.49	1.67	2.06	7.8	42.6	15.8	5.0	0.718	0.883	0.633	0.03607	30.9
CL-6-E	1.27	0.52	0.64	2.22	3.26	1.91	2.30	7.8	34.6	26.0	5.2	0.583	1.412	0.823	0.03710	33.5

National Arnold C Cores 2 mil Silicon Steel																
Part No.	D cm 1	E cm 2	F cm 3	G cm 4	Ht cm 5	Wth cm 6	Lt cm 7	MPL cm 8	Wtfe grams 9	Wtcu grams 10	MLT cm 11	Ac cm sq 12	Wa cm sq 13	Ap cm 4th 14	Kg cm 5th 15	At cm sq 16
CL-62	0.64	0.64	1.27	2.86	4.13	1.91	3.81	10.8	29.6	63.8	4.9	0.359	3.630	1.303	0.03784	55.2
CL-7	0.95	0.56	0.79	2.54	3.65	1.75	2.70	8.9	32.0	33.5	4.7	0.472	2.017	0.951	0.03840	38.4
CL-37	0.95	0.56	0.79	2.70	3.81	1.75	2.70	9.2	33.1	35.6	4.7	0.472	2.143	1.011	0.04080	39.8
CL-69	0.64	0.64	1.27	3.18	4.45	1.91	3.81	11.4	31.3	70.9	4.9	0.359	4.032	1.447	0.04203	58.9
CL-132-B	0.95	0.64	0.79	2.22	3.49	1.75	2.86	8.6	35.2	30.3	4.8	0.539	1.765	0.951	0.04240	38.1
CL-121-D	0.64	0.64	1.27	3.33	4.60	1.91	3.81	11.7	32.2	74.4	4.9	0.359	4.234	1.520	0.04414	60.7
CL-168	1.91	0.40	0.95	1.75	2.54	2.86	2.70	7.0	35.9	39.7	6.7	0.673	1.664	1.120	0.04495	40.6
CL-2-N	0.95	0.64	0.64	2.86	4.13	1.59	2.54	9.5	39.1	29.6	4.6	0.539	1.815	0.977	0.04597	38.9
CL-37-A	0.95	0.56	0.79	3.18	4.29	1.75	2.70	10.2	36.6	41.9	4.7	0.472	2.521	1.189	0.04799	44.0
CL-132	0.95	0.64	0.79	2.54	3.81	1.75	2.86	9.2	37.8	34.6	4.8	0.539	2.017	1.086	0.04844	41.0
CL-8-B	0.64	0.79	0.95	3.02	4.60	1.59	3.49	11.1	38.1	48.7	4.8	0.449	2.874	1.290	0.04861	51.8
CL-160	1.91	0.40	1.11	1.75	2.54	3.02	3.02	7.3	37.5	48.0	7.0	0.673	1.940	1.306	0.05054	45.3
CL-6	1.27	0.64	0.64	2.22	3.49	1.91	2.54	8.3	45.2	27.2	5.4	0.718	1.412	1.013	0.05372	37.5
CL-132-D	0.64	0.95	1.11	2.22	4.13	1.75	4.13	10.5	43.1	46.8	5.3	0.539	2.470	1.330	0.05378	54.5
CL-123-A	0.95	0.64	0.95	2.54	3.81	1.91	3.18	9.5	39.1	43.7	5.1	0.539	2.421	1.304	0.05529	46.0
CL-132-E	0.95	0.48	1.91	3.18	4.13	2.86	4.76	12.1	37.2	134.6	6.3	0.404	6.048	2.442	0.06303	82.9
CL-8-C	0.95	0.64	0.95	3.02	4.29	1.91	3.18	10.5	43.1	51.9	5.1	0.539	2.874	1.548	0.06565	50.8
CL-138	0.95	0.64	1.27	2.54	3.81	2.22	3.81	10.2	41.8	64.0	5.6	0.539	3.226	1.737	0.06710	56.5
CL-124	1.27	0.64	0.79	2.54	3.81	2.06	2.86	9.2	50.4	40.6	5.7	0.718	2.017	1.448	0.07336	45.3
CL-192	0.95	0.48	2.86	3.18	4.13	3.81	6.67	14.0	43.0	250.2	7.8	0.404	9.074	3.663	0.07631	126.4
CL-6-A	1.27	0.64	0.64	3.18	4.45	1.91	2.54	10.2	55.6	38.8	5.4	0.718	2.016	1.447	0.07672	46.0
CL-76-A	0.95	0.79	0.95	2.54	4.13	1.91	3.49	10.2	52.2	46.5	5.4	0.673	2.421	1.630	0.08135	51.8
CL-5-D	1.07	0.71	0.71	3.18	4.60	1.79	2.86	10.6	55.3	41.1	5.1	0.681	2.267	1.544	0.08249	48.9
CL-123	1.27	0.64	0.95	2.54	3.81	2.22	3.18	9.5	52.2	50.9	5.9	0.718	2.421	1.737	0.08434	50.4
CL-121	0.95	0.64	1.27	3.33	4.60	2.22	3.81	11.7	48.3	84.0	5.6	0.539	4.234	2.280	0.08808	66.0
CL-123-B	1.59	0.56	0.95	2.54	3.65	2.54	3.02	9.2	55.2	55.0	6.4	0.786	2.421	1.902	0.09353	51.7
CL-6-B	1.27	0.48	0.48	8.26	9.21	1.75	1.90	19.4	79.5	67.7	4.8	0.538	3.929	2.114	0.09386	72.0
CL-88	0.95	0.95	0.79	2.54	4.45	1.75	3.49	10.5	64.6	39.2	5.5	0.808	2.017	1.630	0.09642	52.8
CL-5-B	1.91	0.64	0.64	2.22	3.49	2.54	2.54	8.3	67.8	33.6	6.7	1.077	1.412	1.520	0.09790	45.2
CL-9-C	1.27	0.64	0.95	3.02	4.29	2.22	3.18	10.5	57.4	60.4	5.9	0.718	2.874	2.063	0.10014	55.6

National Arnold C Cores 2 mil Silicon Steel

Part No.	D cm 1	E cm 2	F cm 3	G cm 4	Ht cm 5	Wth cm 6	Lt cm 7	MPL cm 8	Wtfe grams 9	Wtcu grams 10	MLT cm 11	Ac cm sq 12	Wa cm sq 13	Ap cm 4th 14	Kg cm 5th 15	At cm sq 16
CL-9-E	1.27	0.64	0.95	3.18	4.45	2.22	3.18	10.8	59.1	63.6	5.9	0.718	3.026	2.172	0.10542	57.3
CL-42-A	2.54	0.32	1.27	3.49	4.13	3.81	3.18	10.8	59.2	131.2	8.3	0.719	4.436	3.189	0.11024	78.7
CL-216-D	0.64	0.64	1.91	6.99	8.26	2.54	5.08	20.3	55.6	281.0	5.9	0.359	13.306	4.775	0.11542	141.9
CL-62-B	1.27	0.64	1.27	2.86	4.13	2.54	3.81	10.8	59.1	82.8	6.4	0.718	3.630	2.605	0.11665	65.3
CL-208-A	1.27	0.64	1.19	3.02	4.29	2.46	3.65	11.0	60.0	80.3	6.3	0.718	3.592	2.578	0.11772	64.3
CL-123-D	1.59	0.64	0.95	2.54	3.81	2.54	3.18	9.5	65.2	56.4	6.6	0.898	2.421	2.172	0.11906	54.9
CL-490	0.48	0.48	3.81	16.51	17.46	4.29	8.57	42.5	65.5	1855.2	8.3	0.202	62.903	12.685	0.12337	508.0
CL-8	0.95	0.95	0.95	3.02	4.92	1.91	3.81	11.8	72.5	58.4	5.7	0.808	2.874	2.323	0.13141	63.6
CL-121-B	1.27	0.64	1.27	3.33	4.60	2.54	3.81	11.7	64.3	96.5	6.4	0.718	4.234	3.039	0.13608	71.4
CL-45-A	1.19	0.79	0.79	3.49	5.08	1.99	3.18	11.8	75.5	55.5	5.6	0.842	2.773	2.334	0.13971	60.0
CL-6-C	1.91	0.79	0.64	2.22	3.81	2.54	2.86	8.9	91.3	35.2	7.0	1.346	1.412	1.900	0.14612	51.6
CL-9-J	1.27	0.79	0.95	3.02	4.60	2.22	3.49	11.1	76.1	63.7	6.2	0.898	2.874	2.580	0.14858	62.1
CL-202-A	2.54	0.32	1.27	4.76	5.40	3.81	3.18	13.3	73.2	178.9	8.3	0.719	6.049	4.348	0.15033	96.4
CL-13	1.59	0.64	1.27	2.86	4.13	2.86	3.81	10.8	73.9	91.0	7.0	0.898	3.630	3.257	0.16592	70.4
CL-9-H	1.11	0.95	0.95	3.02	4.92	2.06	3.81	11.8	84.5	61.7	6.0	0.942	2.874	2.708	0.16924	66.3
CL-159-B	1.27	0.95	0.95	2.54	4.45	2.22	3.81	10.8	88.7	56.4	6.6	1.077	2.421	2.607	0.17152	63.3
CL-9-D	1.91	0.64	0.95	2.86	4.13	2.86	3.18	10.2	83.5	69.6	7.2	1.077	2.724	2.932	0.17577	63.1
CL-50	1.59	0.64	1.27	3.18	4.45	2.86	3.81	11.4	78.3	101.1	7.0	0.898	4.032	3.619	0.18432	74.6
CL-41	0.95	0.95	1.27	3.49	5.40	2.22	4.45	13.3	82.3	98.0	6.2	0.808	4.436	3.586	0.18657	82.1
CL-9-F	1.27	0.79	0.95	3.97	5.56	2.22	3.49	13.0	89.2	83.8	6.2	0.898	3.782	3.395	0.19553	73.0
CL-492	0.64	0.64	3.18	9.84	11.11	3.81	7.62	28.6	78.2	881.6	7.9	0.359	31.252	11.215	0.20295	296.0
CL-9	1.27	0.95	0.95	3.02	4.92	2.22	3.81	11.8	96.6	66.9	6.6	1.077	2.874	3.096	0.20366	69.1
CL-135-A	1.91	0.79	1.59	1.59	3.18	3.49	4.76	9.5	97.9	76.2	8.5	1.346	2.522	3.395	0.21508	75.6
CL-152-B	1.59	0.64	1.27	4.13	5.40	2.86	3.81	13.3	91.3	131.4	7.0	0.898	5.243	4.705	0.23965	87.3
CL-14-C	1.27	0.79	1.27	3.97	5.56	2.54	4.13	13.7	93.5	120.6	6.7	0.898	5.041	4.524	0.24130	87.0
CL-94	1.59	0.79	0.79	4.13	5.72	2.38	3.18	13.0	111.5	77.1	6.6	1.122	3.278	3.678	0.24945	73.6
CL-121-A	1.91	0.64	1.27	3.33	4.60	3.18	3.81	11.7	96.5	115.7	7.7	1.077	4.234	4.559	0.25555	82.1
CL-9-A	1.27	1.11	0.95	3.02	5.24	2.22	4.13	12.4	118.6	70.2	6.9	1.256	2.874	3.609	0.26405	76.4
CL-123-C	2.54	0.79	0.95	1.91	3.49	3.49	3.49	8.9	121.8	56.6	8.8	1.795	1.815	3.259	0.26670	67.2
CL-11-A	1.91	0.95	0.64	3.02	4.92	2.54	3.18	11.1	137.0	49.9	7.3	1.616	1.915	3.094	0.27318	67.6

National Arnold C Cores 2 mil Silicon Steel

Part No.	D cm 1	E cm 2	F cm 3	G cm 4	Ht cm 5	Wth cm 6	Lt cm 7	MPL cm 8	Wtfe grams 9	Wtcu grams 10	MLT cm 11	Ac cm sq 12	Wa cm sq 13	Ap cm 4th 14	Kg cm 5th 15	At cm sq 16
CL-86-B	0.95	1.27	1.27	3.18	5.72	2.22	5.08	14.0	114.8	98.2	6.8	1.077	4.032	4.343	0.27329	93.2
CL-192-A	0.95	0.48	8.26	8.26	9.21	9.21	17.46	34.9	107.6	3932.0	16.2	0.404	68.145	27.512	0.27381	781.2
CL-17-D	1.59	0.95	0.95	2.86	4.76	2.54	3.81	11.4	117.5	69.6	7.2	1.347	2.724	3.669	0.27503	72.6
CL-159	1.27	1.27	0.95	2.54	5.08	2.22	4.45	12.1	132.2	61.8	7.2	1.436	2.421	3.475	0.27772	77.8
CL-18-A	1.27	0.79	1.59	3.97	5.56	2.86	4.76	14.3	97.9	162.0	7.2	0.898	6.303	5.656	0.28089	101.9
CL-10	1.59	0.95	0.95	3.02	4.92	2.54	3.81	11.8	120.8	73.4	7.2	1.347	2.874	3.871	0.29024	74.6
CL-9-B	1.27	1.11	0.95	3.33	5.56	2.22	4.13	13.0	124.7	77.6	6.9	1.256	3.177	3.990	0.29189	80.5
CL-8-A	0.95	0.95	2.54	3.81	5.72	3.49	6.99	16.5	101.8	282.5	8.2	0.808	9.677	7.822	0.30814	149.2
CL-12	1.27	1.11	1.27	2.86	5.08	2.54	4.76	12.7	121.7	95.0	7.4	1.256	3.630	4.558	0.31091	87.1
CL-17-E	1.27	1.27	0.95	2.86	5.40	2.22	4.45	12.7	139.1	69.6	7.2	1.436	2.724	3.910	0.31249	82.1
CL-159-A	1.91	0.87	1.11	2.54	4.29	3.02	3.97	10.8	121.9	79.4	7.9	1.480	2.822	4.177	0.31270	76.0
CL-14-A	1.27	0.95	1.27	3.81	5.72	2.54	4.45	14.0	114.8	121.3	7.0	1.077	4.839	5.212	0.31864	92.8
CL-139-A	2.22	0.56	0.95	5.56	6.67	3.18	3.02	15.2	127.9	144.3	7.7	1.100	5.295	5.825	0.33448	97.5
CL-135	1.27	1.11	1.43	2.86	5.08	2.70	5.08	13.0	124.7	110.6	7.6	1.256	4.084	5.129	0.33836	93.8
CL-210-D	0.95	0.95	1.91	4.92	6.83	2.86	5.72	17.5	107.7	240.4	7.2	0.808	9.375	7.577	0.33976	135.2
CL-159-E	1.43	0.95	0.95	4.29	6.19	2.38	3.81	14.3	132.2	99.8	6.9	1.212	4.085	4.951	0.34946	87.6
CL-157-D	0.95	0.95	2.54	4.45	6.35	3.49	6.99	17.8	109.7	329.5	8.2	0.808	11.290	9.126	0.35949	162.5
CL-11	1.91	0.95	0.95	3.02	4.92	2.86	3.81	11.8	144.9	79.9	7.8	1.616	2.874	4.644	0.38382	80.2
CL-107	1.59	0.79	1.59	3.81	5.40	3.18	4.76	14.0	119.6	169.2	7.9	1.122	6.050	6.789	0.38748	105.9
CL-33	1.27	1.27	1.27	2.86	5.40	2.54	5.08	13.3	146.1	99.2	7.7	1.436	3.630	5.210	0.38945	95.2
CL-159-C	2.06	1.03	0.95	2.38	4.45	3.02	3.97	10.8	156.2	66.9	8.3	1.896	2.269	4.302	0.39318	78.2
CL-107-A	1.59	0.79	1.75	3.81	5.40	3.33	5.08	14.3	122.3	191.9	8.1	1.122	6.652	7.465	0.41301	113.6
CL-83	1.91	0.79	1.59	3.18	4.76	3.49	4.76	12.7	130.5	152.4	8.5	1.346	5.042	6.787	0.43002	101.9
CL-21-C	1.91	0.64	1.59	4.84	6.11	3.49	4.45	15.4	126.5	223.7	8.2	1.077	7.689	8.278	0.43575	120.0
CL-21-A	1.91	0.64	1.59	4.92	6.19	3.49	4.45	15.6	127.8	227.3	8.2	1.077	7.815	8.413	0.44286	121.3
CL-11-C	1.91	0.95	0.95	3.49	5.40	2.86	3.81	12.7	156.6	92.6	7.8	1.616	3.329	5.379	0.44452	86.5
CL-131-C	2.54	0.64	1.27	3.81	5.08	3.81	3.81	12.7	139.1	154.0	9.0	1.436	4.839	6.946	0.44552	100.0
CL-78	1.91	0.79	0.79	5.72	7.30	2.70	3.18	16.2	166.3	117.0	7.3	1.346	4.538	6.109	0.45354	97.9
CL-142	1.91	0.95	1.27	2.86	4.76	3.18	4.45	12.1	148.8	107.4	8.3	1.616	3.630	5.865	0.45569	91.2
CL-88-B	1.91	1.27	0.79	2.54	5.08	2.70	4.13	11.7	193.0	58.8	8.2	2.153	2.017	4.343	0.45586	83.6

National Arnold C Cores 2 mil Silicon Steel																
Part No.	D cm 1	E cm 2	F cm 3	G cm 4	Ht cm 5	Wth cm 6	Lt cm 7	MPL cm 8	Wtfe grams 9	Wtcu grams 10	MLT cm 11	Ac cm sq 12	Wa cm sq 13	Ap cm 4th 14	Kg cm 5th 15	At cm sq 16
CL-190	1.27	0.95	2.22	3.81	5.72	3.49	6.35	15.9	130.5	257.3	8.5	1.077	8.470	9.123	0.46008	140.0
CL-135-B	1.27	1.27	1.43	3.18	5.72	2.70	5.40	14.3	156.5	128.0	7.9	1.436	4.537	6.513	0.47149	107.2
CL-41-A	1.27	1.27	1.27	3.49	6.03	2.54	5.08	14.6	160.0	121.2	7.7	1.436	4.436	6.368	0.47598	104.8
CL-216-B	0.95	0.95	1.91	6.99	8.89	2.86	5.72	21.6	133.2	341.2	7.2	0.808	13.306	10.756	0.48227	170.6
CL-125	1.91	0.71	0.95	6.35	7.78	2.86	3.33	17.5	161.3	158.0	7.3	1.211	6.052	7.326	0.48313	109.9
CL-210-C	1.91	0.64	1.91	4.92	6.19	3.81	5.08	16.2	133.0	289.3	8.7	1.077	9.375	10.093	0.50080	139.1
CL-18	1.27	1.11	1.59	3.97	6.19	2.86	5.40	15.6	149.1	176.2	7.9	1.256	6.303	7.915	0.50560	119.0
CL-14-E	1.27	1.27	1.27	3.81	6.35	2.54	5.08	15.2	166.9	132.2	7.7	1.436	4.839	6.946	0.51918	109.7
CL-169	1.27	1.11	1.59	4.13	6.35	2.86	5.40	15.9	152.1	183.3	7.9	1.256	6.555	8.232	0.52586	121.6
CL-136	1.27	1.43	1.27	3.18	6.03	2.54	5.40	14.6	180.0	114.7	8.0	1.615	4.032	6.513	0.52599	108.7
CL-14	1.27	1.27	1.27	3.97	6.51	2.54	5.08	15.6	170.4	137.7	7.7	1.436	5.041	7.236	0.54084	112.1
CL-207	1.27	1.11	1.27	5.08	7.30	2.54	4.76	17.1	164.3	168.9	7.4	1.256	6.452	8.102	0.55263	119.5
CL-152	1.27	1.27	1.27	4.13	6.67	2.54	5.08	15.9	173.9	143.2	7.7	1.436	5.243	7.526	0.56251	114.5
CL-83-A	1.91	0.95	1.59	3.18	5.08	3.49	5.08	13.3	164.4	158.1	8.8	1.616	5.042	8.147	0.59715	110.5
CL-150	1.91	0.64	2.22	5.56	6.83	4.13	5.72	18.1	148.7	403.1	9.2	1.077	12.351	13.297	0.62392	170.3
CL-219	1.27	1.11	1.59	4.92	7.14	2.86	5.40	17.5	167.3	218.5	7.9	1.256	7.815	9.813	0.62688	134.7
CL-211	1.91	0.95	1.27	3.97	5.88	3.18	4.45	14.3	176.2	149.1	8.3	1.616	5.041	8.144	0.63283	108.1
CL-152-D	1.91	0.95	1.27	4.13	6.03	3.18	4.45	14.6	180.1	155.1	8.3	1.616	5.243	8.471	0.65818	110.5
CL-208-B	1.59	1.11	1.11	4.92	7.14	2.70	4.44	16.5	197.8	150.7	7.8	1.570	5.467	8.585	0.69570	116.2
CL-166-A	1.59	1.27	1.59	3.18	5.72	3.18	5.72	14.6	200.0	158.1	8.8	1.795	5.042	9.050	0.73691	121.8
CL-216-A	1.91	0.95	1.59	3.97	5.88	3.49	5.08	14.9	184.0	197.6	8.8	1.616	6.303	10.184	0.74648	124.2
CL-218-A	1.27	0.95	1.43	8.26	10.16	2.70	4.76	23.2	190.5	306.1	7.3	1.077	11.796	12.707	0.75025	166.5
CL-15	1.59	1.27	1.27	3.97	6.51	2.86	5.08	15.6	213.1	149.1	8.3	1.795	5.041	9.048	0.78094	119.5
CL-186-A	2.54	0.64	2.38	4.29	5.56	4.92	6.03	15.9	173.9	388.2	10.7	1.436	10.205	14.649	0.78639	169.0
CL-217	1.27	1.27	1.59	4.92	7.46	2.86	5.72	18.1	198.2	227.3	8.2	1.436	7.815	11.218	0.78731	144.5
CL-218	1.27	0.95	1.59	8.26	10.16	2.86	5.08	23.5	193.1	351.8	7.5	1.077	13.109	14.121	0.80615	178.3
CL-17-C	3.81	0.87	0.95	2.86	4.60	4.76	3.65	11.1	251.0	113.1	11.7	2.960	2.724	8.063	0.81752	105.6
CL-166	1.91	1.27	1.59	2.86	5.40	3.49	5.72	14.0	229.5	152.5	9.5	2.153	4.539	9.772	0.89056	123.2
CL-63	1.27	0.95	1.91	8.26	10.16	3.18	5.72	24.1	198.3	449.9	8.0	1.077	15.726	16.939	0.90725	202.5
CL-16-A	1.91	0.95	1.27	5.72	7.62	3.18	4.45	17.8	219.2	214.7	8.3	1.616	7.258	11.727	0.91122	134.7

National Arnold C Cores 2 mil Silicon Steel																
Part No.	D cm 1	E cm 2	F cm 3	G cm 4	Ht cm 5	Wth cm 6	Lt cm 7	MPL cm 8	Wtfe grams 9	Wtcu grams 10	MLT cm 11	Ac cm sq 12	Wa cm sq 13	Ap cm 4th 14	Kg cm 5th 15	At cm sq 16
CL-32	2.54	1.11	1.27	2.86	5.08	3.81	4.76	12.7	243.4	127.8	9.9	2.512	3.630	9.116	0.92469	112.1
CL-42-D	2.22	1.11	1.27	3.49	5.72	3.49	4.76	14.0	234.3	146.2	9.3	2.198	4.436	9.751	0.92486	116.3
CL-17-B	3.81	0.95	0.95	2.86	4.76	4.76	3.81	11.4	281.9	114.7	11.8	3.232	2.724	8.802	0.96105	110.8
CL-210	1.91	0.95	1.91	4.60	6.51	3.81	5.72	16.8	207.5	290.5	9.3	1.616	8.771	14.171	0.98326	153.1
CL-42	1.59	1.59	1.27	3.49	6.67	2.86	5.72	15.9	271.9	141.2	9.0	2.244	4.436	9.956	0.99823	130.9
CL-32-A	2.54	0.79	1.27	5.72	7.30	3.81	4.13	17.1	234.8	239.3	9.3	1.795	7.258	13.028	1.00901	139.3
CL-92-A	1.27	1.27	3.81	3.81	6.35	5.08	10.16	20.3	222.6	602.4	11.7	1.436	14.516	20.838	1.02529	258.1
CL-28	2.54	1.11	1.27	3.18	5.40	3.81	4.76	13.3	255.5	142.0	9.9	2.512	4.032	10.127	1.02725	117.5
CL-131-A	2.54	1.27	1.27	2.54	5.08	3.81	5.08	12.7	278.2	117.3	10.2	2.871	3.226	9.261	1.04045	116.1
CL-16	1.91	1.27	1.27	3.97	6.51	3.18	5.08	15.6	255.6	160.5	9.0	2.153	5.041	10.854	1.04426	126.8
CL-210-E	1.91	0.95	1.91	4.92	6.83	3.81	5.72	17.5	215.3	310.5	9.3	1.616	9.375	15.147	1.05097	159.1
CL-92	1.27	1.27	3.02	4.45	6.99	4.29	8.57	20.0	219.1	496.9	10.4	1.436	13.406	19.244	1.06013	222.5
CL-220	1.27	1.27	1.91	5.87	8.41	3.18	6.35	20.6	226.0	345.3	8.7	1.436	11.190	16.063	1.06273	181.3
CL-190-A	1.91	0.95	2.22	4.76	6.67	4.13	6.35	17.8	219.2	369.5	9.8	1.616	10.588	17.108	1.12664	175.3
CL-191-B	1.27	1.27	2.22	5.72	8.26	3.49	6.99	21.0	229.5	414.6	9.2	1.436	12.704	18.237	1.14093	199.2
CL-203	2.54	1.27	1.27	2.86	5.40	3.81	5.08	13.3	292.1	131.9	10.2	2.871	3.630	10.421	1.17071	121.8
CL-21-B	2.22	0.95	1.59	4.92	6.83	3.81	5.08	16.8	242.1	262.7	9.5	1.886	7.815	14.734	1.17552	148.1
CL-86	2.54	1.27	1.27	3.18	5.72	3.81	5.08	14.0	306.0	146.6	10.2	2.871	4.032	11.576	1.30056	127.4
CL-91	1.27	1.27	2.38	6.35	8.89	3.65	7.30	22.5	246.9	506.8	9.4	1.436	15.119	21.704	1.32207	223.9
CL-106-A	0.95	0.95	10.16	10.16	12.07	11.11	22.23	44.5	274.2	7404.2	20.2	0.808	103.226	83.438	1.33742	1221.1
CL-43	1.59	1.59	1.59	3.97	7.15	3.18	6.35	17.5	299.1	211.9	9.5	2.244	6.303	14.146	1.34337	156.3
CL-184	2.54	1.11	1.59	3.81	6.03	4.13	5.40	15.2	292.0	223.8	10.4	2.512	6.050	15.195	1.46739	145.4
CL-16-D	1.91	1.27	1.27	5.72	8.26	3.18	5.08	19.1	313.0	231.0	9.0	2.153	7.258	15.628	1.50363	155.6
CL-84-A	1.91	0.95	2.38	6.35	8.26	4.29	6.67	21.3	262.3	541.0	10.1	1.616	15.119	24.429	1.56912	219.9
CL-17	2.54	1.27	1.27	3.97	6.51	3.81	5.08	15.6	340.8	183.2	10.2	2.871	5.041	14.471	1.62580	141.5
CL-210-B	2.54	0.95	1.91	4.92	6.83	4.45	5.72	17.5	287.1	352.9	10.6	2.154	9.375	20.196	1.64421	175.0
CL-86-A	3.18	1.27	1.27	3.18	5.72	4.45	5.08	14.0	382.5	164.8	11.5	3.589	4.032	14.471	1.80755	141.1
CL-158	1.91	0.95	2.86	6.83	8.73	4.76	7.62	23.2	285.8	750.0	10.8	1.616	19.509	31.521	1.88441	267.1
CL-19	2.54	1.27	1.59	3.97	6.51	4.13	5.72	16.2	354.7	240.3	10.7	2.871	6.303	18.095	1.93823	159.2
CL-79	1.91	1.27	2.22	5.08	7.62	4.13	6.99	19.7	323.4	419.6	10.4	2.153	11.293	24.316	2.00449	204.5

National Arnold C Cores 2 mil Silicon Steel																
Part No.	D cm 1	E cm 2	F cm 3	G cm 4	Ht cm 5	Wth cm 6	Lt cm 7	MPL cm 8	Wtfe grams 9	Wtcu grams 10	MLT cm 11	Ac cm sq 12	Wa cm sq 13	Ap cm 4th 14	Kg cm 5th 15	At cm sq 16
CL-65	2.54	0.95	3.02	4.45	6.35	5.56	7.94	18.7	307.9	587.8	12.3	2.154	13.406	28.881	2.01867	236.6
CL-60-B	3.81	0.64	1.91	7.62	8.89	5.72	5.08	21.6	354.7	655.4	12.7	2.153	14.516	31.256	2.12027	240.3
CL-131-B	5.08	1.11	1.27	2.54	4.76	6.35	4.76	12.1	462.4	174.3	15.2	5.023	3.226	16.203	2.14299	155.0
CL-154	1.91	0.95	3.65	6.83	8.73	5.56	9.21	24.8	305.3	1068.4	12.1	1.616	24.922	40.268	2.15868	332.3
CL-21	1.91	1.59	1.59	4.92	8.10	3.49	6.35	19.4	397.9	280.3	10.1	2.692	7.815	21.040	2.24629	183.6
CL-112	2.54	1.27	1.75	4.45	6.99	4.29	6.03	17.5	382.5	302.7	11.0	2.871	7.761	22.281	2.33268	178.1
CL-209	1.91	1.11	3.02	6.51	8.73	4.92	8.25	23.5	337.7	794.1	11.4	1.884	19.631	36.978	2.44933	284.5
CL-98-B	1.27	1.27	5.48	8.89	11.43	6.75	13.49	33.8	370.4	2473.7	14.3	1.436	48.691	69.894	2.80906	589.2
CL-60-A	3.81	0.64	1.91	10.16	11.43	5.72	5.08	26.7	438.2	873.9	12.7	2.153	19.355	41.675	2.82703	295.2
CL-135-D	2.22	1.43	1.43	6.35	9.21	3.65	5.72	21.3	458.9	327.7	10.2	2.827	9.074	25.655	2.85685	197.9
CL-20	2.54	1.59	1.59	3.97	7.15	4.13	6.35	17.5	478.4	254.5	11.4	3.590	6.303	22.626	2.86069	182.0
CL-50-A	6.99	0.95	1.27	3.18	5.08	8.26	4.45	12.7	574.2	273.7	19.1	5.925	4.032	23.889	2.96615	192.8
CL-174	2.54	2.22	1.27	2.86	7.30	3.81	6.99	17.1	657.5	156.5	12.1	5.025	3.630	18.240	3.02320	188.4
CL-38	1.91	1.59	1.91	6.35	9.53	3.81	6.99	22.9	469.6	455.3	10.6	2.692	12.097	32.569	3.31371	234.7
CL-64	2.86	1.27	1.11	7.62	10.16	3.97	4.76	22.5	555.6	319.4	10.6	3.230	8.466	27.348	3.33117	203.9
CL-190-B	2.22	0.95	2.54	10.16	12.07	4.76	6.99	29.2	420.3	1004.7	10.9	1.886	25.806	48.658	3.35201	330.7
CL-23-A	3.18	1.27	1.59	4.92	7.46	4.76	5.72	18.1	495.6	333.2	12.0	3.589	7.815	28.044	3.35720	194.7
CL-157-A	2.54	1.27	2.54	4.92	7.46	5.08	7.62	20.0	438.2	543.0	12.2	2.871	12.499	35.885	3.37350	241.1
CL-178-A	3.81	1.27	1.59	3.97	6.51	5.40	5.72	16.2	532.1	301.9	13.5	4.306	6.303	27.143	3.47127	190.3
CL-22	2.54	1.59	1.59	4.92	8.10	4.13	6.35	19.4	530.6	315.6	11.4	3.590	7.815	28.053	3.54685	202.0
CL-191	2.54	1.27	2.22	5.72	8.26	4.76	6.99	21.0	459.1	529.4	11.7	2.871	12.704	36.474	3.57448	237.9
CL-210-A	2.86	1.43	1.91	4.92	7.78	4.76	6.67	19.4	537.1	405.8	12.2	3.635	9.375	34.075	4.06991	219.6
CL-139	2.22	1.59	3.02	4.45	7.62	5.24	9.21	21.3	510.0	618.1	13.0	3.142	13.406	42.119	4.08269	278.4
CL-156	2.54	1.27	2.54	6.03	8.57	5.08	7.62	22.2	486.9	665.7	12.2	2.871	15.324	43.994	4.13581	269.4
CL-173	1.91	1.59	2.54	6.51	9.69	4.45	8.26	24.5	502.3	680.9	11.6	2.692	16.533	44.513	4.13907	287.1
CL-182	1.91	1.91	1.91	5.87	9.68	3.81	7.62	23.2	571.2	446.4	11.2	3.230	11.190	36.142	4.16199	250.4
CL-156-A	2.54	1.27	2.62	6.03	8.57	5.16	7.78	22.4	490.3	693.3	12.3	2.871	15.800	45.362	4.22158	275.5
CL-85	2.22	1.59	3.49	4.45	7.62	5.72	10.16	22.2	532.8	757.2	13.7	3.142	15.526	48.781	4.47019	313.8
CL-165-D	2.38	1.59	1.59	7.30	10.48	3.97	6.35	24.1	619.7	455.2	11.0	3.365	11.597	39.026	4.75856	246.6
CL-24	2.54	1.59	1.91	5.87	9.05	4.45	6.99	21.9	600.1	471.7	11.9	3.590	11.190	40.170	4.86566	244.8

National Arnold C Cores 2 mil Silicon Steel

Part No.	D cm 1	E cm 2	F cm 3	G cm 4	Ht cm 5	Wth cm 6	Lt cm 7	MPL cm 8	Wtfe grams 9	Wtcu grams 10	MLT cm 11	Ac cm sq 12	Wa cm sq 13	Ap cm 4th 14	Kg cm 5th 15	At cm sq 16
CL-165	3.18	1.43	1.59	5.87	8.73	4.76	6.03	20.6	635.9	408.3	12.3	4.038	9.328	37.666	4.94251	227.9
CL-23	3.18	1.59	1.59	4.92	8.10	4.76	6.35	19.4	663.2	350.9	12.6	4.487	7.815	35.066	4.98456	220.3
CL-157-B	2.54	1.59	2.54	4.92	8.10	5.08	8.26	21.3	582.7	571.2	12.9	3.590	12.499	44.871	5.01339	267.6
CL-84	1.91	1.91	2.38	6.35	10.16	4.29	8.57	25.1	618.1	643.4	12.0	3.230	15.119	48.833	5.27227	298.5
CL-177	2.22	1.43	2.86	7.30	10.16	5.08	8.57	26.0	561.7	920.3	12.4	2.827	20.872	59.010	5.38218	332.0
CL-194	3.18	1.43	1.91	5.87	8.73	5.08	6.67	21.3	655.5	509.6	12.8	4.038	11.190	45.185	5.69874	251.1
CL-103-A	3.18	1.27	2.38	6.35	8.89	5.56	7.30	22.5	617.2	711.6	13.2	3.589	15.119	54.259	5.88445	285.6
CL-165-B	3.18	1.59	1.59	6.03	9.21	4.76	6.35	21.6	739.3	430.2	12.6	4.487	9.580	42.990	6.11092	245.1
CL-178	3.81	1.75	1.59	3.97	7.46	5.40	6.67	18.1	817.6	323.2	14.4	5.921	6.303	37.316	6.12787	229.9
CL-99	2.54	1.59	2.38	6.35	9.53	4.92	7.94	23.8	652.3	677.5	12.6	3.590	15.119	54.276	6.18438	292.9
CL-195	3.18	1.59	1.27	7.62	10.80	4.45	5.72	24.1	826.2	417.4	12.1	4.487	9.677	43.425	6.42690	254.1
CL-157	2.22	2.22	2.54	4.45	8.89	4.76	9.53	22.9	767.2	541.5	13.5	4.398	11.290	49.656	6.47683	299.7
CL-197	3.18	1.43	1.59	8.89	11.75	4.76	6.03	26.7	821.8	617.9	12.3	4.038	14.117	57.006	7.48024	293.0
CL-99-A	2.54	1.59	2.22	8.26	11.43	4.76	7.62	27.3	748.0	806.2	12.4	3.590	18.351	65.877	7.65691	327.5
CL-165-A	3.18	1.91	1.59	5.56	9.37	4.76	6.99	21.9	899.8	416.1	13.3	5.383	8.823	47.494	7.71167	262.4
CL-100	2.54	1.59	2.54	7.62	10.80	5.08	8.26	26.7	730.6	884.5	12.9	3.590	19.355	69.480	7.76306	339.6
CL-60	3.18	1.27	1.91	10.16	12.70	5.08	6.35	29.2	799.8	859.6	12.5	3.589	19.355	69.459	7.98364	335.5
CL-188-B	2.54	1.43	2.06	11.11	13.97	4.60	6.99	32.1	790.5	961.4	11.8	3.230	22.937	74.096	8.12323	364.6
CL-120	2.54	1.91	2.38	6.35	10.16	4.92	8.57	25.1	824.1	711.6	13.2	4.306	15.119	65.111	8.47360	322.3
CL-128	2.86	1.75	2.38	6.35	9.84	5.24	8.25	24.4	828.4	728.7	13.6	4.441	15.119	67.147	8.80062	318.8
CL-130-B	3.81	1.11	2.86	8.57	10.80	6.67	7.94	27.3	784.9	1319.6	15.1	3.767	24.502	92.305	9.18422	395.8
CL-120-A	2.54	1.91	2.38	6.99	10.80	4.92	8.57	26.4	865.9	782.8	13.2	4.306	16.631	71.622	9.32097	339.4
CL-65-A	7.62	0.95	3.02	4.45	6.35	10.64	7.94	18.7	923.8	1101.1	23.1	6.463	13.406	86.644	9.69797	381.8
CL-25-A	2.54	2.06	2.38	6.35	10.48	4.92	8.89	25.7	915.6	728.7	13.6	4.666	15.119	70.545	9.71375	337.6
CL-180	5.08	1.27	2.54	5.08	7.62	7.62	7.62	20.3	890.2	803.1	17.5	5.742	12.903	74.089	9.72165	322.6
CL-47	2.54	2.54	1.91	5.87	10.95	4.45	8.89	25.7	1126.7	547.5	13.8	5.742	11.190	64.252	10.72560	332.5
CL-49	2.86	1.98	2.38	6.35	10.32	5.24	8.73	25.4	978.0	754.3	14.0	5.047	15.119	76.300	10.97787	341.9
CL-25	2.54	2.22	2.38	6.35	10.80	4.92	9.21	26.4	1010.5	745.8	13.9	5.025	15.119	75.979	11.00969	353.4
CL-170	2.86	1.59	3.18	8.10	11.27	6.03	9.53	28.9	890.5	1324.0	14.5	4.039	25.705	103.829	11.58160	425.4
CL-98	2.54	1.59	5.08	7.62	10.80	7.62	13.34	31.8	869.7	2318.0	16.8	3.590	38.710	138.961	11.84937	591.2

National Arnold C Cores 2 mil Silicon Steel

Part No.	D cm 1	E cm 2	F cm 3	G cm 4	Ht cm 5	Wth cm 6	Lt cm 7	MPL cm 8	Wtfe grams 9	Wtcu grams 10	MLT cm 11	Ac cm sq 12	Wa cm sq 13	Ap cm 4th 14	Kg cm 5th 15	At cm sq 16
CL-47-D	3.81	1.91	1.91	5.87	9.68	5.72	7.62	23.2	1142.4	606.3	15.2	6.460	11.190	72.283	12.25783	316.3
CL-201	2.54	2.54	1.27	10.16	15.24	3.81	7.62	33.0	1446.6	585.6	12.8	5.742	12.903	74.089	13.33389	380.6
CL-209-A	2.54	2.22	3.02	6.51	10.96	5.56	10.48	27.9	1071.4	1038.0	14.9	5.025	19.631	98.653	13.33665	414.1
CL-102	2.54	2.54	2.38	6.35	11.43	4.92	9.84	27.6	1210.1	779.9	14.5	5.742	15.119	86.814	13.74533	386.0
CL-188	3.49	2.22	2.06	5.72	10.16	5.56	8.57	24.5	1289.2	641.0	15.3	6.911	11.796	81.518	14.74705	345.8
CL-188-A	3.33	2.22	2.06	6.35	10.80	5.40	8.57	25.7	1294.5	697.3	15.0	6.596	13.106	86.453	15.24509	357.5
CL-73	2.54	2.22	2.38	8.89	13.34	4.92	9.21	31.4	1205.3	1044.2	13.9	5.025	21.167	106.371	15.41357	425.1
CL-47-A	4.13	2.06	1.91	5.72	9.84	6.03	7.94	23.5	1359.4	626.8	16.2	7.583	10.887	82.556	15.46608	339.4
CL-164-A	2.54	2.22	3.18	7.30	11.75	5.72	10.80	29.8	1144.5	1246.6	15.1	5.025	23.187	116.522	15.49229	454.9
CL-53-A	3.49	1.98	2.38	6.83	10.79	5.87	8.73	26.4	1240.0	884.3	15.3	6.168	16.253	100.243	16.16404	380.0
CL-106	6.35	1.51	1.91	5.87	8.89	8.26	6.83	21.6	1403.9	776.8	19.5	8.523	11.190	95.366	16.65236	357.9
CL-141-B	1.27	2.54	5.72	15.24	20.32	6.99	16.51	52.1	1140.6	5327.3	17.2	2.871	87.097	250.051	16.69451	1069.4
CL-70	3.49	1.59	3.33	8.89	12.07	6.83	9.84	30.8	1160.2	1686.8	16.0	4.937	29.639	146.321	18.05372	494.4
CL-60-D	2.54	2.54	1.91	10.16	15.24	4.45	8.89	34.3	1502.3	947.0	13.8	5.742	19.355	111.134	18.55160	446.8
CL-82	3.81	1.43	2.54	11.91	14.76	6.35	7.94	34.6	1279.5	1643.4	15.3	4.846	30.241	146.537	18.58573	493.7
CL-98-H	1.91	1.75	5.08	16.83	20.32	6.99	13.65	50.8	1147.4	4829.0	15.9	2.960	85.486	253.061	18.86301	958.4
CL-53	3.49	2.22	2.38	6.83	11.27	5.87	9.21	27.3	1439.8	911.9	15.8	6.911	16.253	112.319	19.67819	405.8
CL-103	3.18	2.54	2.38	6.35	11.43	5.56	9.84	27.6	1512.7	848.2	15.8	7.177	15.119	108.518	19.74815	413.0
CL-164	2.54	2.54	3.18	7.62	12.70	5.72	11.43	31.8	1391.0	1355.2	15.8	5.742	24.194	138.917	20.25430	501.6
CL-189	4.13	1.98	2.38	6.83	10.79	6.51	8.73	26.4	1465.5	969.7	16.8	7.289	16.253	118.467	20.58656	404.8
CL-98-J	3.18	1.75	5.08	7.78	11.27	8.26	13.65	32.7	1231.1	2589.2	18.4	4.934	39.517	194.969	20.88247	648.3
CL-81	2.86	2.54	2.54	7.62	12.70	5.40	10.16	30.5	1502.5	1059.4	15.4	6.461	19.355	125.047	20.99574	453.2
CL-420	1.91	1.91	7.78	13.97	17.78	9.68	19.37	51.1	1259.7	7899.2	20.4	3.230	108.673	350.994	22.18387	1289.7
CL-227	3.18	1.91	3.65	8.89	12.70	6.83	11.11	32.7	1343.2	1904.4	16.5	5.383	32.457	174.720	22.80055	551.7
CL-52	3.97	2.22	2.38	6.83	11.27	6.35	9.21	27.3	1636.0	978.9	16.9	7.853	16.253	127.625	23.66684	425.3
CL-226	4.45	1.59	2.38	10.80	13.97	6.83	7.94	32.7	1567.6	1519.1	16.6	6.282	25.703	161.471	24.41356	490.7
CL-35	3.18	2.54	2.54	7.62	12.70	5.72	10.16	30.5	1669.2	1103.0	16.0	7.177	19.355	138.917	24.88650	467.7
CL-115	4.45	1.91	2.54	7.62	11.43	6.99	8.89	27.9	1606.6	1204.7	17.5	7.536	19.355	145.863	25.12059	448.4
CL-89-A	5.08	1.59	3.18	8.10	11.27	8.26	9.53	28.9	1582.8	1749.2	19.1	7.180	25.705	184.552	27.69584	517.8
CL-101	4.45	2.30	2.38	6.83	11.43	6.83	9.37	27.6	1919.3	1043.1	18.0	9.107	16.253	148.011	29.87354	454.2

National Arnold C Cores 2 mil Silicon Steel

Part No.	D cm 1	E cm 2	F cm 3	G cm 4	Ht cm 5	Wth cm 6	Lt cm 7	MPL cm 8	Wtfe grams 9	Wtcu grams 10	MLT cm 11	Ac cm sq 12	Wa cm sq 13	Ap cm 4th 14	Kg cm 5th 15	At cm sq 16
CL-198	6.35	1.59	3.18	6.35	9.53	9.53	9.53	25.4	1739.4	1539.2	21.5	8.975	20.161	180.939	30.25514	504.1
CL-36	3.18	2.54	3.49	7.62	12.70	6.67	12.07	32.4	1773.6	1658.4	17.5	7.177	26.617	191.039	31.30146	565.8
CL-27	5.08	2.22	2.38	6.35	10.80	7.46	9.21	26.4	2021.0	1030.1	19.2	10.051	15.119	151.959	31.88454	455.0
CL-54	5.08	1.91	1.91	10.16	13.97	6.99	7.62	31.8	2086.5	1223.5	17.8	8.613	19.355	166.701	32.30660	485.5
CL-130	3.81	2.54	2.54	7.62	12.70	6.35	10.16	30.5	2003.0	1204.7	17.5	8.613	19.355	166.701	32.81057	496.8
CL-130-A	3.81	2.38	2.86	7.94	12.70	6.67	10.48	31.1	1916.8	1426.7	17.7	8.074	22.687	183.167	33.44841	520.8
CL-36-A	3.81	1.91	3.49	10.48	14.29	7.30	10.80	35.6	1752.8	2307.5	17.7	6.460	36.600	236.421	34.45466	621.7
CL-104	3.18	3.18	2.54	7.62	13.97	5.72	11.43	33.0	2260.4	1190.4	17.3	8.972	19.355	173.647	36.02989	545.2
CL-98-A	2.54	2.54	5.08	10.16	15.24	7.62	15.24	40.6	1780.5	3440.1	18.7	5.742	51.613	296.357	36.31443	825.8
CL-39	4.45	2.86	1.91	7.62	13.34	6.35	9.53	30.5	2629.6	950.5	18.4	11.306	14.516	164.125	40.31224	500.9
CL-98-K	2.54	2.54	5.08	11.43	16.51	7.62	15.24	43.2	1891.8	3870.1	18.7	5.742	58.064	333.401	40.85374	883.9
CL-145	4.45	2.54	1.75	10.16	15.24	6.19	8.57	34.0	2604.6	1105.6	17.5	10.048	17.739	178.252	40.87671	520.6
CL-173-A	5.08	2.54	2.54	6.67	11.75	7.62	10.16	28.6	2503.9	1207.2	20.0	11.484	16.937	194.499	44.57426	521.0
CL-115-A	4.45	2.54	2.54	8.26	13.34	6.99	10.16	31.8	2434.2	1399.8	18.8	10.048	20.968	210.691	45.10759	547.6
CL-29-A	4.45	2.54	2.06	10.16	15.24	6.51	9.21	34.6	2653.4	1344.2	18.0	10.048	20.970	210.717	46.98329	557.0
CL-105	3.81	3.18	2.54	7.62	13.97	6.35	11.43	33.0	2712.4	1292.1	18.8	10.766	19.355	208.376	47.79846	577.4
CL-105-D	4.45	2.54	2.22	10.16	15.24	6.67	9.53	34.9	2677.7	1467.8	18.3	10.048	22.586	226.949	49.91147	575.5
CL-450	2.54	2.54	8.10	11.11	16.19	10.64	21.27	48.6	2128.2	7511.7	23.5	5.742	89.971	516.606	50.53616	1321.4
CL-144-A	5.08	2.54	2.54	7.62	12.70	7.62	10.16	30.5	2670.7	1379.5	20.0	11.484	19.355	222.268	50.93819	554.8
CL-29	4.45	2.86	2.54	7.62	13.34	6.99	10.80	31.8	2739.2	1335.9	19.4	11.306	19.355	218.833	50.98890	567.0
CL-450-A	2.54	2.54	8.10	12.38	17.46	10.64	21.27	51.1	2239.5	8370.1	23.5	5.742	100.253	575.644	56.31147	1402.4
CL-141-A	2.54	2.54	5.72	15.24	20.32	8.26	16.51	52.1	2281.2	6114.0	19.7	5.742	87.097	500.102	58.18578	1162.9
CL-98-C	3.81	2.54	5.08	8.89	13.97	8.89	15.24	38.1	2503.8	3451.4	21.5	8.613	45.161	388.968	62.35254	841.9
CL-144	5.08	2.86	2.54	7.62	13.34	7.62	10.80	31.8	3130.5	1423.3	20.7	12.922	19.355	250.095	62.50780	597.6
CL-289	2.54	2.54	6.67	15.24	20.32	9.21	18.42	54.0	2364.7	7674.2	21.2	5.742	101.620	583.496	63.10549	1326.3
CL-225-J	3.81	1.91	5.08	15.24	19.05	8.89	13.97	48.3	2378.6	5567.1	20.2	6.460	77.419	500.102	63.90180	1041.9
CL-68	5.08	2.54	3.49	7.62	12.70	8.57	12.07	32.4	2837.7	2038.7	21.5	11.484	26.617	305.662	65.18423	660.1
CL-225-M	2.54	3.18	5.08	12.70	19.05	7.62	16.51	48.3	2642.9	4591.5	20.0	7.177	64.516	463.057	66.42585	1041.9
CL-222-E	3.18	3.18	3.81	11.43	17.78	6.99	13.97	43.2	2955.9	2987.2	19.3	8.972	43.548	390.705	72.68766	843.5
CL-72-C	7.62	1.91	3.81	7.62	11.43	11.43	11.43	30.5	3004.6	2709.8	26.2	12.919	29.032	375.077	73.84628	725.8

National Arnold C Cores 2 mil Silicon Steel																
Part No.	D cm 1	E cm 2	F cm 3	G cm 4	Ht cm 5	Wth cm 6	Lt cm 7	MPL cm 8	Wtfe grams 9	Wtcu grams 10	MLT cm 11	Ac cm sq 12	Wa cm sq 13	Ap cm 4th 14	Kg cm 5th 15	At cm sq 16
CL-116	2.54	2.54	6.67	19.05	24.13	9.21	18.42	61.6	2698.6	9592.7	21.2	5.742	127.025	729.370	78.88187	1536.8
CL-176	3.81	3.49	3.18	10.80	17.78	6.99	13.34	41.9	3787.7	2487.1	20.4	11.844	34.274	405.957	94.24963	820.2
CL-127	3.81	2.54	5.72	12.70	17.78	9.53	16.51	47.0	3088.0	5804.2	22.5	8.613	72.581	625.128	95.76700	1124.2
CL-225-K	4.13	2.22	5.08	15.24	19.69	9.21	14.61	49.5	3086.6	5917.2	21.5	8.167	77.419	632.292	96.10320	1116.2
CL-109	5.08	2.54	2.54	14.61	19.69	7.62	10.16	44.5	3894.8	2644.1	20.0	11.484	37.097	426.013	97.63153	803.2
CL-191-A	6.35	3.33	2.22	7.62	14.29	8.57	11.11	33.0	4747.4	1413.5	23.5	18.842	16.939	319.171	102.51136	693.8
CL-199	4.13	3.18	3.33	11.75	18.10	7.46	13.02	42.9	3815.0	2877.0	20.7	11.665	39.168	456.881	103.20059	850.9
CL-144-B	5.08	2.54	2.54	16.19	21.27	7.62	10.16	47.6	4173.1	2931.6	20.0	11.484	41.130	472.333	108.24699	859.7
CL-222-F	4.45	3.49	3.81	9.53	16.51	8.26	14.61	40.6	4285.1	2926.0	22.7	13.819	36.290	501.476	122.24983	889.6
CL-294	1.27	4.29	20.00	29.53	38.10	21.27	48.58	116.2	4295.4	90576.8	43.1	4.845	590.649	2861.377	128.57448	6703.6
CL-75	8.89	2.22	4.45	8.89	13.34	13.34	13.34	35.6	4772.5	4274.7	30.4	17.589	39.516	695.032	160.74131	988.0
CL-185	5.08	3.81	3.97	9.05	16.67	9.05	15.56	41.3	5425.0	3170.8	24.8	17.226	35.915	618.672	171.69951	980.9
CL-67	4.13	3.81	4.45	11.91	19.53	8.57	16.51	47.9	5120.3	4454.6	23.7	13.998	52.922	740.785	175.22529	1127.1
CL-67-A	3.81	3.81	7.62	10.16	17.78	11.43	22.86	50.8	5007.6	7713.8	28.0	12.919	77.419	1000.204	184.47170	1509.7
CL-129	5.08	3.18	4.60	12.38	18.73	9.68	15.56	46.7	5112.1	4977.9	24.6	14.355	57.011	818.387	191.37662	1130.1
CL-56	4.13	3.81	5.08	11.91	19.53	9.21	17.78	49.2	5255.9	5305.4	24.7	13.998	60.482	846.612	192.16400	1226.5
CL-222	5.08	3.81	3.81	10.80	18.42	8.89	15.24	44.5	5842.2	3594.6	24.6	17.226	41.129	708.478	198.62036	1043.5
CL-151	3.81	3.18	6.67	16.51	22.86	10.48	19.69	59.1	4851.2	9886.6	25.3	10.766	110.089	1185.227	202.10490	1622.3
CL-98-E	5.72	3.18	5.08	10.16	16.51	10.80	16.51	43.2	5320.6	4876.8	26.6	16.149	51.613	833.503	202.62809	1129.0
CL-305-B	4.45	2.86	6.99	15.24	20.96	11.43	19.69	55.9	4820.8	9989.1	26.4	11.306	106.451	1203.581	206.27447	1594.5
CL-287	4.45	4.45	4.45	10.16	19.05	8.89	17.78	47.0	6304.7	4107.1	25.6	17.585	45.161	794.144	218.41519	1174.2
CL-222-C	4.76	5.08	3.81	8.89	19.05	8.57	17.78	45.7	7512.2	3189.8	26.5	21.535	33.871	729.392	237.23386	1154.1
CL-225	5.08	3.81	5.08	11.91	19.53	10.16	17.78	49.2	6468.1	5714.9	26.6	17.226	60.482	1041.857	270.16511	1297.6
CL-229-A	5.08	2.54	6.03	22.86	27.94	11.11	17.15	67.9	5953.5	12519.4	25.5	11.484	137.914	1583.788	284.99081	1837.6
CL-117	3.18	3.65	6.67	25.88	33.18	9.84	20.64	79.7	6273.1	15172.5	24.7	10.317	172.541	1780.075	297.05819	2234.1
CL-300-C	7.14	2.54	3.81	20.32	25.40	10.95	12.70	58.4	7198.7	7313.6	26.6	16.150	77.419	1250.299	304.03091	1402.4
CL-137	4.13	3.81	6.35	16.51	24.13	10.48	20.32	61.0	6510.6	9939.6	26.7	13.998	104.839	1467.491	308.18078	1718.6
CL-230	3.02	4.84	9.53	15.24	24.92	12.54	28.73	68.9	6832.5	16145.6	31.3	12.997	145.161	1886.671	313.58823	2409.2
CL-98-F	10.16	2.54	5.08	10.16	15.24	15.24	15.24	40.6	7121.9	6348.8	34.6	22.968	51.613	1185.427	314.83400	1290.3
CL-67-B	5.08	2.54	8.26	21.27	26.35	13.34	21.59	69.2	6064.8	18119.7	29.0	11.484	175.609	2016.663	319.25514	2248.8

National Arnold C Cores 2 mil Silicon Steel																
Part No.	D cm 1	E cm 2	F cm 3	G cm 4	Ht cm 5	Wth cm 6	Lt cm 7	MPL cm 8	Wtfe grams 9	Wtcu grams 10	MLT cm 11	Ac cm sq 12	Wa cm sq 13	Ap cm 4th 14	Kg cm 5th 15	At cm sq 16
CL-304	4.13	3.81	7.62	16.51	24.13	11.75	22.86	63.5	6781.9	12819.5	28.7	13.998	125.806	1760.989	344.08434	1975.0
CL-72-A	7.94	3.18	4.60	12.38	18.73	12.54	15.56	46.7	7988.1	6217.9	30.7	22.431	57.011	1278.810	374.10465	1326.1
CL-229	6.35	3.18	6.03	18.42	24.77	12.38	18.42	61.6	8433.0	11508.1	29.1	17.944	111.098	1993.483	491.18185	1828.3
CL-452-A	3.18	4.92	10.95	20.00	29.85	14.13	31.75	81.6	8657.5	26489.9	34.0	13.906	219.113	3046.877	498.48433	3204.5
CL-228	5.08	5.72	6.35	11.43	22.86	11.43	24.13	58.4	11517.4	8356.0	32.4	25.839	72.581	1875.383	598.69188	1887.1
CL-292-E	3.49	6.03	9.37	17.94	30.01	12.86	30.80	78.7	11268.2	20531.7	34.4	18.755	168.017	3151.188	687.93066	2959.9
CL-292-A	3.49	6.19	9.21	17.78	30.16	12.70	30.80	78.7	11563.0	20046.0	34.4	19.246	163.718	3150.986	704.50899	2950.3
CL-303-A	5.08	5.08	11.43	11.43	21.59	16.51	33.02	66.0	11573.1	18156.0	39.1	22.968	130.645	3000.612	705.37577	2764.5
CL-231	8.57	3.49	5.40	15.56	22.54	13.97	17.78	55.9	11364.1	10100.9	33.8	26.652	83.982	2238.247	705.47096	1777.7
CL-301-N	7.62	2.54	8.89	25.40	30.48	16.51	22.86	78.7	10349.0	28499.9	35.5	17.226	225.806	3889.683	755.10349	2974.2
CL-225-F	6.35	5.08	5.08	15.24	25.40	11.43	20.32	61.0	13353.6	8713.8	31.7	28.710	77.419	2222.676	806.43226	1871.0
CL-300-F	6.35	5.08	6.67	12.70	22.86	13.02	23.50	59.1	12936.5	10282.2	34.1	28.710	84.684	2431.234	817.69271	2010.2
CL-290	5.08	3.81	8.57	25.88	33.50	13.65	24.77	84.1	11058.5	25286.9	32.1	17.226	221.835	3821.278	821.37844	2988.7
CL-225-C	5.08	7.62	5.08	12.70	27.94	10.16	25.40	66.0	17359.6	7844.2	34.2	34.452	64.516	2222.676	895.82954	2193.5
CL-301-P	5.08	5.08	8.89	17.78	27.94	13.97	27.94	73.7	12908.4	19725.1	35.1	22.968	158.064	3630.370	950.39504	2761.3
CL-291	5.08	3.81	13.34	25.88	33.50	18.42	34.29	93.7	12310.2	48506.5	39.5	17.226	345.056	5943.864	1035.99893	4387.3
CL-301-K	6.35	5.08	7.62	15.24	25.40	13.97	25.40	66.0	14466.4	14717.4	35.6	28.710	116.129	3334.014	1074.29718	2400.0
CL-296	5.08	6.35	7.94	15.24	27.94	13.02	28.58	71.8	15718.5	15546.4	36.1	28.710	120.975	3473.150	1103.66913	2665.0
CL-308	5.72	5.72	8.89	13.97	25.40	14.61	29.21	68.6	15210.5	16620.0	37.6	29.069	124.193	3610.112	1115.40043	2672.6
CL-317	5.08	5.08	8.89	20.96	31.12	13.97	27.94	80.0	14021.2	23247.5	35.1	22.968	186.290	4278.651	1120.10844	3027.4
CL-301-L	6.35	5.08	9.21	15.24	25.40	15.56	28.58	69.2	15162.1	19228.3	38.5	28.710	140.330	4028.819	1200.70770	2757.0
CL-232	6.35	4.76	6.35	21.59	31.12	12.70	22.23	74.9	15389.9	16093.6	33.0	26.918	137.097	3690.377	1203.67635	2496.1
CL-305	7.62	5.08	6.99	15.24	25.40	14.61	24.13	64.8	17025.8	14226.5	37.6	34.452	106.451	3667.415	1344.75653	2388.7
CL-301-I	5.08	5.08	8.89	25.40	35.56	13.97	27.94	88.9	15579.1	28178.7	35.1	22.968	225.806	5186.244	1357.70720	3400.0
CL-292	6.35	5.08	9.21	17.78	27.94	15.56	28.58	74.3	16274.9	22433.0	38.5	28.710	163.718	4700.288	1400.82565	2981.2
CL-309	3.81	7.62	10.80	20.32	35.56	14.61	36.83	92.7	18277.7	31687.8	40.6	25.839	219.354	5667.823	1441.98905	4075.8
CL-288	4.05	6.19	10.48	26.67	39.05	14.53	33.34	99.1	16858.3	37507.3	37.7	22.304	279.448	6232.937	1473.29873	4188.0
CL-301-D	5.08	7.62	8.26	15.24	30.48	13.34	31.75	77.5	20364.2	17526.2	39.2	34.452	125.806	4334.218	1524.59835	3091.9
CL-225-E	5.72	5.72	11.43	16.51	27.94	17.15	34.29	78.7	17463.9	27929.9	41.6	29.069	188.709	5485.494	1532.44428	3527.4
CL-453	5.72	5.08	10.95	21.43	31.59	16.67	32.07	85.1	16775.4	33060.8	39.6	25.839	234.755	6065.759	1582.99095	3703.2

National Arnold C Cores 2 mil Silicon Steel																
Part No.	D cm 1	E cm 2	F cm 3	G cm 4	Ht cm 5	Wth cm 6	Lt cm 7	MPL cm 8	Wtfe grams 9	Wtcu grams 10	MLT cm 11	Ac cm sq 12	Wa cm sq 13	Ap cm 4th 14	Kg cm 5th 15	At cm sq 16
CL-300-E	7.62	6.35	6.67	12.70	25.40	14.29	26.04	64.1	21073.9	11932.4	39.6	43.064	84.684	3646.851	1585.36792	2453.7
CL-301-H	5.08	7.62	8.89	15.24	30.48	13.97	33.02	78.7	20698.0	19354.7	40.2	34.452	135.484	4667.619	1601.12999	3245.2
CL-313	6.35	4.29	11.43	25.40	33.97	17.78	31.43	90.8	16782.1	41329.6	40.0	24.222	290.322	7032.275	1701.97241	4069.5
CL-225-N	4.05	6.19	12.07	29.21	41.59	16.11	36.51	107.3	18263.0	50423.8	40.2	22.304	352.419	7860.500	1742.95496	4956.6
CL-294-A	5.08	4.45	20.00	29.53	38.42	25.08	48.90	116.8	17916.3	107686.2	51.3	20.097	590.649	11870.107	1861.10466	7425.9
CL-225-B	5.08	5.72	11.43	27.94	39.37	16.51	34.29	101.6	20030.3	45823.7	40.4	25.839	319.354	8251.684	2113.57251	4609.7
CL-300-A	5.08	7.62	8.89	20.32	35.56	13.97	33.02	88.9	23368.7	25806.2	40.2	34.452	180.645	6223.492	2134.83999	3722.6
CL-301	8.89	6.35	7.62	15.24	27.94	16.51	27.94	71.1	27263.5	18029.3	43.7	50.242	116.129	5834.524	2685.67312	3019.3
CL-316	7.62	5.08	9.53	25.40	35.56	17.15	29.21	90.2	23702.6	35763.8	41.6	34.452	241.935	8335.034	2763.07986	3911.3
CL-300-D	6.35	7.62	7.62	20.32	35.56	13.97	30.48	86.4	28376.3	22420.3	40.7	43.064	154.838	6668.027	2820.81564	3548.4
CL-301-C	6.35	6.35	7.62	27.94	40.64	13.97	27.94	96.5	26428.9	28905.0	38.2	35.887	212.903	7640.448	2872.67953	3800.0
CL-311	3.81	7.62	11.75	43.82	59.06	15.56	38.74	141.6	27917.5	77097.6	42.1	25.839	514.739	13300.155	3263.58237	6901.8
CL-301-A	5.08	7.62	15.24	25.40	40.64	20.32	45.72	111.8	29377.8	69022.2	50.1	34.452	387.096	13336.055	3665.11389	6296.8
CL-300	5.72	5.72	14.45	35.56	46.99	20.16	40.32	122.9	27252.1	84679.7	46.4	29.069	513.700	14932.476	3745.46973	6649.6
CL-307-A	6.35	7.62	12.70	20.32	35.56	19.05	40.64	96.5	31714.7	44686.2	48.7	43.064	258.064	11113.379	3931.33862	5019.3
CL-297	10.16	6.35	8.26	19.05	31.75	18.42	29.21	80.0	35053.1	26392.6	47.2	57.419	157.258	9029.620	4394.18680	3680.6
CL-306-A	6.35	10.16	8.26	19.05	39.37	14.61	36.83	95.3	41729.9	26168.9	46.8	57.419	157.258	9029.620	4431.74689	4454.8
CL-309-A	7.62	7.62	10.80	20.32	35.56	18.42	36.83	92.7	36555.4	37943.6	48.6	51.677	219.354	11335.647	4816.98863	4627.4
CL-301-F	5.08	10.16	13.97	25.40	45.72	19.05	48.26	119.4	41841.1	67164.4	53.2	45.935	354.838	16299.623	5626.48903	6858.1
CL-301-E	5.08	12.70	16.51	26.67	52.07	21.59	58.42	137.2	60091.0	97543.2	62.3	57.419	440.322	25282.937	9321.37368	9177.4
CL-301-J	7.62	10.16	13.97	25.40	45.72	21.59	48.26	119.4	62761.7	74079.1	58.7	68.903	354.838	24449.434	11477.92923	7335.5
CL-302-A	10.16	8.89	15.24	32.39	50.17	25.40	48.26	130.8	80232.6	110994.6	63.2	80.387	493.547	39674.763	20171.98894	8758.0
CL-501	11.43	10.64	9.37	41.91	63.18	20.80	40.00	145.1	119782.8	83823.4	60.1	108.197	392.529	42470.403	30607.57887	8549.8

National Arnold C Cores 4 mil Silicon Steel																
Part No.	D cm 1	E cm 2	F cm 3	G cm 4	Ht cm 5	Wth cm 6	Lt cm 7	MPL cm 8	Wtfe grams 9	Wtcu grams 10	MLT cm 11	Ac cm sq 12	Wa cm sq 13	Ap cm 4th 14	Kg cm 5th 15	At cm sq 16
CH-210-L	0.32	0.16	0.48	1.27	1.59	0.79	1.27	4.1	1.4	4.5	2.1	0.05	0.61	0.028	0.00024	8
CH-172-E	0.32	0.16	0.64	1.59	1.91	0.95	1.59	5.1	1.8	8.5	2.4	0.05	1.01	0.046	0.00035	12
CH-109-A	0.48	0.16	0.64	1.11	1.43	1.11	1.59	4.1	2.1	6.7	2.7	0.07	0.71	0.048	0.00049	10
CH-231-Q	0.32	0.24	0.56	1.59	2.06	0.87	1.59	5.2	2.7	7.5	2.4	0.07	0.88	0.060	0.00069	12
CH-121-J	0.64	0.16	0.32	1.91	2.22	0.95	0.95	5.1	3.5	5.4	2.5	0.09	0.61	0.055	0.00080	10
CH-215-L	0.32	0.24	0.64	1.75	2.22	0.95	1.75	5.7	3.0	9.9	2.5	0.07	1.11	0.076	0.00082	14
CH-246-P	0.48	0.16	0.79	1.91	2.22	1.27	1.91	6.0	3.1	15.7	2.9	0.07	1.51	0.103	0.00096	18
CH-123	0.32	0.32	0.95	1.27	1.91	1.27	2.54	5.7	4.0	13.7	3.2	0.09	1.21	0.110	0.00126	19
CH-121-L	0.48	0.32	0.32	1.43	2.07	0.79	1.27	4.8	5.0	4.0	2.5	0.14	0.45	0.062	0.00135	10
CH-121-M	0.56	0.36	0.36	1.67	2.38	0.91	1.43	5.5	7.5	5.9	2.8	0.18	0.60	0.106	0.00272	13
CH-139-J	0.48	0.24	0.95	2.38	2.86	1.43	2.38	7.6	5.9	26.9	3.3	0.10	2.27	0.231	0.00283	27
CH-121-A	0.64	0.32	0.64	1.27	1.91	1.27	1.91	5.1	7.0	9.5	3.3	0.18	0.81	0.147	0.00322	15
CH-172-C	0.64	0.32	0.64	1.59	2.22	1.27	1.91	5.7	7.9	11.9	3.3	0.18	1.01	0.183	0.00402	17
CH-172-H	0.32	0.64	0.64	1.59	2.86	0.95	2.54	7.0	9.7	11.9	3.3	0.18	1.01	0.183	0.00402	22
CH-139-K	0.48	0.32	0.95	2.38	3.02	1.43	2.54	7.9	8.3	28.2	3.5	0.14	2.27	0.309	0.00482	29
CH-109	0.64	0.48	0.64	0.95	1.91	1.27	2.22	5.1	10.5	7.8	3.6	0.27	0.61	0.165	0.00494	17
CH-447-D	0.64	0.40	0.40	1.91	2.70	1.03	1.59	6.2	10.7	8.3	3.1	0.23	0.76	0.172	0.00503	16
CH-447-A	0.56	0.52	0.40	1.75	2.78	0.95	1.83	6.4	12.5	7.8	3.2	0.26	0.69	0.179	0.00582	17
CH-121	0.95	0.32	0.64	1.27	1.91	1.59	1.91	5.1	10.6	11.3	3.9	0.27	0.81	0.220	0.00608	18
CH-210-H	0.64	0.48	0.48	1.59	2.54	1.11	1.90	6.0	12.5	9.1	3.4	0.27	0.76	0.206	0.00663	18
CH-210-F	0.48	0.64	0.48	1.59	2.86	0.95	2.22	6.7	13.8	9.1	3.4	0.27	0.76	0.206	0.00663	20
CH-121-E	0.56	0.56	0.64	1.27	2.38	1.19	2.38	6.0	12.8	10.4	3.6	0.28	0.81	0.224	0.00688	20
CH-238-A	0.95	0.28	0.64	1.91	2.46	1.59	1.83	6.2	11.3	16.6	3.9	0.24	1.21	0.288	0.00711	21
CH-210-J	0.64	0.32	0.79	2.54	3.18	1.43	2.22	7.9	11.0	25.5	3.6	0.18	2.02	0.367	0.00748	28
CH-385-D	0.48	0.32	1.27	3.18	3.81	1.75	3.18	10.2	10.6	57.2	4.0	0.14	4.03	0.549	0.00750	45
CH-172-F	0.32	0.95	0.64	1.59	3.49	0.95	3.18	8.3	17.2	14.2	3.9	0.27	1.01	0.275	0.00760	30
CH-447	0.64	0.48	0.64	1.59	2.54	1.27	2.22	6.4	13.2	13.0	3.6	0.27	1.01	0.274	0.00823	21
CH-193	0.64	0.48	0.95	1.27	2.22	1.59	2.86	6.4	13.2	17.8	4.1	0.27	1.21	0.329	0.00868	25
CH-238	0.95	0.32	0.64	1.91	2.54	1.59	1.91	6.4	13.2	17.0	3.9	0.27	1.21	0.330	0.00912	22
CH-121-H	0.64	0.60	0.64	1.27	2.46	1.27	2.46	6.2	16.1	11.1	3.9	0.34	0.81	0.274	0.00965	22

National Arnold C Cores 4 mil Silicon Steel

Part No.	D cm 1	E cm 2	F cm 3	G cm 4	Ht cm 5	Wth cm 6	Lt cm 7	MPL cm 8	Wtfe grams 9	Wtcu grams 10	MLT cm 11	Ac cm sq 12	Wa cm sq 13	Ap cm 4th 14	Kg cm 5th 15	At cm sq 16
CH-122	0.95	0.32	0.68	1.95	2.58	1.63	1.99	6.5	13.6	18.7	4.0	0.27	1.31	0.358	0.00974	23
CH-283-A	0.64	0.64	0.64	1.27	2.54	1.27	2.54	6.4	17.6	11.3	3.9	0.36	0.81	0.293	0.01077	23
CH-290-A	0.64	0.32	1.27	3.18	3.81	1.91	3.18	10.2	14.1	61.8	4.3	0.18	4.03	0.733	0.01237	48
CH-45-L	1.27	0.48	0.48	1.11	2.06	1.75	1.90	5.1	21.1	9.1	4.8	0.54	0.53	0.288	0.01292	20
CH-172-A	0.64	0.64	0.64	1.59	2.86	1.27	2.54	7.0	19.3	14.1	3.9	0.36	1.01	0.366	0.01347	25
CH-231-E	0.95	0.40	0.64	1.91	2.70	1.59	2.06	6.7	17.3	17.7	4.1	0.34	1.21	0.412	0.01367	24
CH-231	0.95	0.48	0.64	1.59	2.54	1.59	2.22	6.4	19.8	15.3	4.3	0.41	1.01	0.412	0.01577	24
CH-246-E	1.27	0.32	0.79	1.91	2.54	2.06	2.22	6.7	18.5	27.1	5.0	0.36	1.51	0.550	0.01589	29
CH-210-A	0.95	0.48	0.71	1.59	2.54	1.67	2.38	6.5	20.3	17.7	4.4	0.41	1.13	0.463	0.01723	26
CH-191-A	0.95	0.48	0.64	1.75	2.70	1.59	2.22	6.7	20.8	16.8	4.3	0.41	1.11	0.453	0.01734	25
CH-447-C	1.27	0.48	0.40	1.75	2.70	1.67	1.75	6.2	25.7	11.6	4.7	0.54	0.69	0.377	0.01738	23
CH-203-B	0.64	0.40	1.75	2.54	3.33	2.38	4.29	10.2	17.6	82.2	5.2	0.23	4.44	1.006	0.01752	59
CH-49-E	0.64	0.32	1.91	3.81	4.45	2.54	4.45	12.7	17.6	136.9	5.3	0.18	7.26	1.319	0.01808	80
CH-121-C	1.27	0.48	0.64	1.27	2.22	1.91	2.22	5.7	23.7	14.6	5.1	0.54	0.81	0.439	0.01873	24
CH-283	0.64	0.64	0.64	2.22	3.49	1.27	2.54	8.3	22.9	19.8	3.9	0.36	1.41	0.512	0.01885	30
CH-231-B	0.48	0.79	0.64	2.54	4.13	1.11	2.86	9.5	24.7	22.6	3.9	0.34	1.61	0.549	0.01892	35
CH-210-K	0.95	0.48	0.71	1.75	2.70	1.67	2.38	6.8	21.3	19.4	4.4	0.41	1.25	0.509	0.01895	27
CH-231-K	0.64	0.64	0.56	2.54	3.81	1.19	2.38	8.7	24.2	19.2	3.8	0.36	1.41	0.513	0.01947	30
CH-446-C	0.64	0.56	1.11	2.06	3.18	1.75	3.33	8.6	20.8	37.0	4.5	0.32	2.29	0.729	0.02043	39
CH-246-L	0.95	0.48	0.79	1.75	2.70	1.75	2.54	7.0	21.8	22.2	4.5	0.41	1.39	0.566	0.02048	29
CH-215-M	1.27	0.48	0.64	1.43	2.38	1.91	2.22	6.0	25.0	16.4	5.1	0.54	0.91	0.494	0.02108	26
CH-45-F	0.48	0.64	0.79	3.49	4.76	1.27	2.86	11.1	23.1	38.2	3.9	0.27	2.77	0.754	0.02118	43
CH-246-R	0.95	0.48	0.79	1.91	2.86	1.75	2.54	7.3	22.7	24.3	4.5	0.41	1.51	0.618	0.02235	30
CH-377-A	0.79	0.32	1.59	3.49	4.13	2.38	3.81	11.4	19.8	101.1	5.1	0.23	5.55	1.260	0.02236	65
CH-231-H	1.27	0.48	0.64	1.59	2.54	1.91	2.22	6.4	26.4	18.3	5.1	0.54	1.01	0.549	0.02343	27
CH-122-B	0.95	0.64	0.71	1.31	2.58	1.67	2.70	6.6	27.4	15.6	4.7	0.54	0.94	0.509	0.02359	28
CH-100-C	0.64	0.64	0.64	2.86	4.13	1.27	2.54	9.5	26.4	25.5	3.9	0.36	1.82	0.659	0.02424	35
CH-246-F	0.95	0.64	0.79	1.27	2.54	1.75	2.86	6.7	27.7	17.3	4.8	0.54	1.01	0.549	0.02477	29
CH-238-B	0.95	0.56	0.64	1.91	3.02	1.59	2.38	7.3	26.6	19.0	4.4	0.48	1.21	0.577	0.02488	29
CH-231-A	0.95	0.48	0.64	2.54	3.49	1.59	2.22	8.3	25.7	24.5	4.3	0.41	1.61	0.658	0.02523	31

National Arnold C Cores 4 mil Silicon Steel																
Part No.	D cm 1	E cm 2	F cm 3	G cm 4	Ht cm 5	Wth cm 6	Lt cm 7	MPL cm 8	Wtfe grams 9	Wtcu grams 10	MLT cm 11	Ac cm sq 12	Wa cm sq 13	Ap cm 4th 14	Kg cm 5th 15	At cm sq 16
CH-430-F	0.95	0.48	0.95	1.91	2.86	1.91	2.86	7.6	23.7	30.7	4.8	0.41	1.82	0.741	0.02542	35
CH-172	0.95	0.64	0.64	1.59	2.86	1.59	2.54	7.0	29.0	16.4	4.6	0.54	1.01	0.549	0.02612	28
CH-260	0.48	0.64	1.11	3.49	4.76	1.59	3.49	11.7	24.4	60.4	4.4	0.27	3.88	1.056	0.02626	54
CH-402-B	0.95	0.48	0.64	2.70	3.65	1.59	2.22	8.6	26.7	26.0	4.3	0.41	1.71	0.700	0.02681	33
CH-233	0.64	0.32	1.91	5.87	6.51	2.54	4.45	16.8	23.3	211.1	5.3	0.18	11.19	2.034	0.02787	109
CH-191-B	0.95	0.64	0.64	1.75	3.02	1.59	2.54	7.3	30.3	18.1	4.6	0.54	1.11	0.604	0.02872	30
CH-210-B	0.95	0.48	0.79	2.54	3.49	1.75	2.54	8.6	26.7	32.4	4.5	0.41	2.02	0.823	0.02980	36
CH-246-J	1.03	0.64	0.79	1.35	2.62	1.83	2.86	6.8	30.7	19.0	5.0	0.59	1.07	0.632	0.02988	31
CH-246-B	0.95	0.64	0.79	1.59	2.86	1.75	2.86	7.3	30.4	21.7	4.8	0.54	1.26	0.687	0.03097	32
CH-231-U	0.56	0.95	0.60	2.54	4.45	1.15	3.10	10.1	36.7	23.4	4.4	0.48	1.51	0.721	0.03153	40
CH-446-B	0.64	0.64	1.11	2.54	3.81	1.75	3.49	9.8	27.3	47.1	4.7	0.36	2.82	1.024	0.03168	47
CH-246-H	0.79	0.64	0.79	2.22	3.49	1.59	2.86	8.6	29.7	28.3	4.5	0.45	1.77	0.801	0.03222	36
CH-215-J	0.79	0.64	0.79	2.22	3.49	1.59	2.86	8.6	29.7	28.3	4.5	0.45	1.77	0.801	0.03222	36
CH-407-B	0.64	0.64	0.95	2.86	4.13	1.59	3.18	10.2	28.1	43.0	4.4	0.36	2.72	0.988	0.03229	45
CH-231-M	1.59	0.48	0.64	1.59	2.54	2.22	2.22	6.4	33.0	20.6	5.7	0.68	1.01	0.686	0.03256	30
CH-215-K	1.27	0.48	0.64	2.22	3.18	1.91	2.22	7.6	31.6	25.6	5.1	0.54	1.41	0.768	0.03279	32
CH-430	0.95	0.48	0.95	2.54	3.49	1.91	2.86	8.9	27.7	41.0	4.8	0.41	2.42	0.988	0.03389	41
CH-38-A	0.64	0.64	0.95	3.02	4.29	1.59	3.18	10.5	29.0	45.4	4.4	0.36	2.87	1.043	0.03407	46
CH-215-B	1.27	0.48	0.79	2.06	3.02	2.06	2.54	7.6	31.6	31.2	5.3	0.54	1.64	0.892	0.03629	35
CH-283-B	0.95	0.64	0.64	2.22	3.49	1.59	2.54	8.3	34.3	23.0	4.6	0.54	1.41	0.769	0.03656	34
CH-231-D	1.27	0.48	0.64	2.54	3.49	1.91	2.22	8.3	34.3	29.2	5.1	0.54	1.61	0.878	0.03747	35
CH-172-B	0.95	0.79	0.64	1.59	3.18	1.59	2.86	7.6	39.6	17.6	4.9	0.68	1.01	0.687	0.03819	33
CH-99	0.95	0.56	0.79	2.54	3.65	1.75	2.70	8.9	32.4	33.5	4.7	0.48	2.02	0.962	0.03926	38
CH-330-P	0.64	0.48	1.27	5.08	6.03	1.91	3.49	14.6	30.3	106.1	4.6	0.27	6.45	1.755	0.04130	74
CH-99-B	0.95	0.56	0.79	2.70	3.81	1.75	2.70	9.2	33.5	35.6	4.7	0.48	2.14	1.022	0.04172	40
CH-231-J	0.95	0.64	0.64	2.54	3.81	1.59	2.54	8.9	36.9	26.3	4.6	0.54	1.61	0.878	0.04178	36
CH-215-E	1.27	0.64	0.64	1.75	3.02	1.91	2.54	7.3	40.4	21.3	5.4	0.73	1.11	0.805	0.04314	33
CH-246-A	0.95	0.64	0.79	2.22	3.49	1.75	2.86	8.6	35.6	30.3	4.8	0.54	1.77	0.961	0.04336	38
CH-231-W	1.27	0.64	0.64	1.83	3.10	1.91	2.54	7.5	41.3	22.3	5.4	0.73	1.16	0.842	0.04512	34
CH-38	0.64	0.64	1.27	3.33	4.60	1.91	3.81	11.7	32.5	74.4	4.9	0.36	4.23	1.537	0.04514	61

National Arnold C Cores 4 mil Silicon Steel																
Part No.	D cm 1	E cm 2	F cm 3	G cm 4	Ht cm 5	Wth cm 6	Lt cm 7	MPL cm 8	Wtfe grams 9	Wtcu grams 10	MLT cm 11	Ac cm sq 12	Wa cm sq 13	Ap cm 4th 14	Kg cm 5th 15	At cm sq 16
CH-402-H	1.27	0.48	0.64	3.18	4.13	1.91	2.22	9.5	39.5	36.5	5.1	0.54	2.02	1.097	0.04684	40
CH-45-J	1.59	0.48	1.43	1.27	2.22	3.02	3.81	7.3	37.9	45.0	7.0	0.68	1.82	1.235	0.04814	48
CH-246	0.95	0.64	0.79	2.54	3.81	1.75	2.86	9.2	38.3	34.6	4.8	0.54	2.02	1.098	0.04954	41
CH-7-A	0.95	0.64	0.79	2.54	3.81	1.75	2.86	9.2	38.3	34.6	4.8	0.54	2.02	1.098	0.04954	41
CH-99-F	0.79	0.48	0.79	5.95	6.91	1.59	2.54	15.4	40.0	70.5	4.2	0.34	4.73	1.608	0.05215	62
CH-231-R	0.64	1.11	0.64	2.54	4.76	1.27	3.49	10.8	52.3	28.1	4.9	0.63	1.61	1.024	0.05311	49
CH-342-D	0.64	1.11	0.64	2.54	4.76	1.27	3.49	10.8	52.3	28.1	4.9	0.63	1.61	1.024	0.05311	49
CH-140-B	1.59	0.48	1.43	1.43	2.38	3.02	3.81	7.6	39.6	50.7	7.0	0.68	2.04	1.389	0.05416	50
CH-215	1.27	0.64	0.64	2.22	3.49	1.91	2.54	8.3	45.7	27.2	5.4	0.73	1.41	1.025	0.05493	38
CH-402-A	1.27	0.48	0.64	3.81	4.76	1.91	2.22	10.8	44.8	43.9	5.1	0.54	2.42	1.316	0.05620	45
CH-231-C	1.27	0.64	0.56	2.54	3.81	1.83	2.38	8.7	48.4	26.6	5.3	0.73	1.41	1.025	0.05624	38
CH-115	0.95	0.64	0.95	2.54	3.81	1.91	3.18	9.5	39.6	43.7	5.1	0.54	2.42	1.318	0.05654	46
CH-191	0.95	0.95	0.64	1.75	3.65	1.59	3.18	8.6	53.5	20.6	5.2	0.82	1.11	0.906	0.05680	40
CH-402	1.27	0.48	0.95	2.86	3.81	2.22	2.86	9.5	39.5	54.2	5.6	0.54	2.72	1.482	0.05763	48
CH-99-C	0.95	0.56	0.79	3.81	4.92	1.75	2.70	11.4	41.6	50.3	4.7	0.48	3.03	1.443	0.05889	50
CH-217-A	1.27	0.48	0.95	3.02	3.97	2.22	2.86	9.8	40.9	57.2	5.6	0.54	2.87	1.564	0.06081	49
CH-116	0.95	0.64	0.95	2.86	4.13	1.91	3.18	10.2	42.2	49.2	5.1	0.54	2.72	1.483	0.06361	49
CH-99-D	0.95	0.56	0.79	4.13	5.24	1.75	2.70	12.1	43.9	54.5	4.7	0.48	3.28	1.563	0.06381	53
CH-45-H	1.59	0.71	0.79	1.27	2.70	2.38	3.02	7.0	54.4	23.2	6.5	1.02	1.01	1.029	0.06503	39
CH-184	0.95	0.64	1.19	2.54	3.81	2.14	3.65	10.0	41.6	58.7	5.5	0.54	3.03	1.648	0.06581	54
CH-235	0.95	0.64	0.95	3.02	4.29	1.91	3.18	10.5	43.5	51.9	5.1	0.54	2.87	1.565	0.06713	51
CH-250-D	1.27	0.56	0.95	2.54	3.65	2.22	3.02	9.2	44.7	49.5	5.8	0.64	2.42	1.538	0.06794	48
CH-215-H	0.95	0.79	0.79	2.38	3.97	1.75	3.18	9.5	49.5	34.6	5.1	0.68	1.89	1.287	0.06812	45
CH-45-M	1.59	0.64	0.79	1.67	2.94	2.38	2.86	7.5	51.7	29.7	6.3	0.91	1.32	1.201	0.06921	40
CH-276	1.27	0.48	0.95	3.49	4.45	2.22	2.86	10.8	44.8	66.2	5.6	0.54	3.33	1.811	0.07043	54
CH-100-F	1.27	0.64	0.64	2.86	4.13	1.91	2.54	9.5	52.8	34.9	5.4	0.73	1.82	1.317	0.07062	43
CH-98-C	0.95	0.64	1.11	2.86	4.13	2.06	3.49	10.5	43.5	60.2	5.3	0.54	3.18	1.729	0.07071	55
CH-210-D	0.95	0.79	0.79	2.54	4.13	1.75	3.18	9.8	51.2	36.9	5.1	0.68	2.02	1.373	0.07267	47
CH-129-D	1.59	0.40	0.48	6.35	7.14	2.06	1.75	15.2	66.0	57.2	5.3	0.57	3.02	1.715	0.07309	60
CH-246-K	1.59	0.71	0.79	1.43	2.86	2.38	3.02	7.3	56.9	26.1	6.5	1.02	1.14	1.158	0.07317	41

National Arnold C Cores 4 mil Silicon Steel																
Part No.	D cm 1	E cm 2	F cm 3	G cm 4	Ht cm 5	Wth cm 6	Lt cm 7	MPL cm 8	Wtfe grams 9	Wtcu grams 10	MLT cm 11	Ac cm sq 12	Wa cm sq 13	Ap cm 4th 14	Kg cm 5th 15	At cm sq 16
CH-7-D	1.27	0.64	0.79	2.54	3.81	2.06	2.86	9.2	51.0	40.6	5.7	0.73	2.02	1.464	0.07502	45
CH-216-A	1.27	0.64	2.22	1.27	2.54	3.49	5.72	9.5	52.8	79.4	7.9	0.73	2.82	2.049	0.07523	75
CH-1-J	0.95	0.64	1.27	2.86	4.13	2.22	3.81	10.8	44.9	72.0	5.6	0.54	3.63	1.977	0.07721	60
CH-425	1.91	0.56	0.64	2.22	3.34	2.54	2.38	7.9	57.8	32.8	6.5	0.95	1.41	1.346	0.07861	42
CH-430-B	0.95	0.79	0.91	2.54	4.13	1.87	3.41	10.1	52.4	44.0	5.3	0.68	2.32	1.579	0.08063	51
CH-402-G	1.59	0.48	0.95	2.86	3.81	2.54	2.86	9.5	49.4	60.4	6.2	0.68	2.72	1.853	0.08091	52
CH-105-E	0.95	0.64	1.91	2.38	3.65	2.86	5.08	11.1	46.2	106.0	6.6	0.54	4.54	2.470	0.08186	77
CH-240-A	0.95	0.44	2.06	4.60	5.48	3.02	5.00	15.1	43.1	217.2	6.4	0.37	9.50	3.562	0.08307	111
CH-126	0.95	0.79	0.95	2.54	4.13	1.91	3.49	10.2	52.8	46.5	5.4	0.68	2.42	1.648	0.08319	52
CH-398	1.27	0.56	1.27	2.54	3.65	2.54	3.65	9.8	47.7	71.7	6.3	0.64	3.23	2.050	0.08333	58
CH-327-E	0.95	0.64	0.95	3.81	5.08	1.91	3.18	12.1	50.1	65.6	5.1	0.54	3.63	1.978	0.08480	59
CH-283-C	2.54	0.64	0.64	1.27	2.54	3.18	2.54	6.4	70.3	22.8	8.0	1.45	0.81	1.171	0.08545	42
CH-100-A	3.81	0.32	0.64	2.86	3.49	4.45	1.91	8.3	68.7	65.0	10.1	1.09	1.82	1.979	0.08572	60
CH-250	1.27	0.64	0.95	2.54	3.81	2.22	3.18	9.5	52.8	50.9	5.9	0.73	2.42	1.757	0.08624	50
CH-194-A	0.95	0.48	0.48	11.11	12.07	1.43	1.90	25.1	78.1	75.5	4.0	0.41	5.29	2.160	0.08788	84
CH-191-D	0.95	0.95	0.56	3.02	4.92	1.51	3.02	11.0	68.3	30.4	5.1	0.82	1.68	1.371	0.08800	49
CH-404-C	0.64	1.11	1.27	2.54	4.76	1.91	4.76	12.1	58.4	67.6	5.9	0.63	3.23	2.048	0.08826	70
CH-39	0.95	0.64	1.27	3.33	4.60	2.22	3.81	11.7	48.8	84.0	5.6	0.54	4.23	2.306	0.09007	66
CH-43-H	0.95	0.64	0.95	4.13	5.40	1.91	3.18	12.7	52.8	71.1	5.1	0.54	3.93	2.143	0.09188	62
CH-114-C	1.27	0.64	1.27	2.22	3.49	2.54	3.81	9.5	52.8	64.4	6.4	0.73	2.82	2.049	0.09278	57
CH-402-E	1.27	0.64	0.64	3.81	5.08	1.91	2.54	11.4	63.3	46.6	5.4	0.73	2.42	1.756	0.09415	52
CH-246-D	1.91	0.56	0.79	2.22	3.34	2.70	2.70	8.3	60.1	42.5	6.8	0.95	1.77	1.683	0.09468	47
CH-202-A	1.59	0.40	1.27	3.81	4.60	2.86	3.33	11.7	50.9	113.1	6.6	0.57	4.84	2.745	0.09481	72
CH-98-E	1.27	0.48	1.11	4.29	5.24	2.38	3.17	12.7	52.7	99.0	5.8	0.54	4.76	2.591	0.09647	69
CH-250-B	1.27	0.64	1.11	2.54	3.81	2.38	3.49	9.8	54.5	61.8	6.2	0.73	2.82	2.048	0.09650	56
CH-402-C	1.27	0.64	0.95	2.86	4.13	2.22	3.18	10.2	56.3	57.3	5.9	0.73	2.72	1.977	0.09704	54
CH-210	0.95	0.95	0.79	2.54	4.45	1.75	3.49	10.5	65.4	39.2	5.5	0.82	2.02	1.648	0.09860	53
CH-367	0.95	0.79	0.95	3.02	4.60	1.91	3.49	11.1	57.7	55.2	5.4	0.68	2.87	1.957	0.09878	57
CH-100-B	1.59	0.64	0.64	2.86	4.13	2.22	2.54	9.5	66.0	39.1	6.1	0.91	1.82	1.647	0.09881	47
CH-194	0.95	0.48	0.48	12.70	13.65	1.43	1.90	28.3	88.0	86.3	4.0	0.41	6.05	2.468	0.10043	95

National Arnold C Cores 4 mil Silicon Steel																
Part No.	D cm 1	E cm 2	F cm 3	G cm 4	Ht cm 5	Wth cm 6	Lt cm 7	MPL cm 8	Wtfe grams 9	Wtcu grams 10	MLT cm 11	Ac cm sq 12	Wa cm sq 13	Ap cm 4th 14	Kg cm 5th 15	At cm sq 16
CH-404-A	0.95	0.79	1.27	2.54	4.13	2.22	4.13	10.8	56.1	67.6	5.9	0.68	3.23	2.197	0.10150	63
CH-314-D	1.27	0.64	1.11	2.70	3.97	2.38	3.49	10.2	56.3	65.7	6.2	0.73	3.00	2.176	0.10254	58
CH-98-B	0.95	0.71	1.23	3.18	4.60	2.18	3.89	11.7	54.5	78.8	5.7	0.61	3.91	2.392	0.10327	66
CH-173-D	1.91	0.48	0.64	3.97	4.92	2.54	2.22	11.1	69.2	57.1	6.4	0.82	2.52	2.057	0.10546	56
CH-114	1.27	0.64	1.27	2.54	3.81	2.54	3.81	10.2	56.3	73.6	6.4	0.73	3.23	2.341	0.10601	61
CH-1-A	1.27	0.64	0.95	3.18	4.45	2.22	3.18	10.8	59.8	63.6	5.9	0.73	3.03	2.196	0.10781	57
CH-203	1.27	0.64	0.95	3.18	4.45	2.22	3.18	10.8	59.8	63.6	5.9	0.73	3.03	2.196	0.10781	57
CH-356-E	1.27	0.64	1.11	2.86	4.13	2.38	3.49	10.5	58.0	69.6	6.2	0.73	3.18	2.305	0.10858	60
CH-342-A	1.27	0.79	0.79	2.54	4.13	2.06	3.18	9.8	68.2	42.9	6.0	0.91	2.02	1.830	0.11106	51
CH-1-H	1.27	0.64	1.27	2.70	3.97	2.54	3.81	10.5	58.0	78.2	6.4	0.73	3.43	2.488	0.11265	63
CH-139-A	0.95	0.95	0.95	2.54	4.45	1.91	3.81	10.8	67.3	49.2	5.7	0.82	2.42	1.979	0.11317	58
CH-488-C	0.64	0.64	3.81	5.08	6.35	4.45	8.89	20.3	56.3	614.6	8.9	0.36	19.36	7.024	0.11418	229
CH-342-B	1.91	0.64	0.64	2.54	3.81	2.54	2.54	8.9	73.8	38.3	6.7	1.09	1.61	1.756	0.11439	48
CH-231-T	2.38	0.64	0.64	1.91	3.18	3.02	2.54	7.6	79.1	32.9	7.6	1.36	1.21	1.646	0.11732	47
CH-1	1.27	0.64	1.27	2.86	4.13	2.54	3.81	10.8	59.8	82.8	6.4	0.73	3.63	2.634	0.11928	65
CH-215-C	1.91	0.64	0.79	2.22	3.49	2.70	2.86	8.6	71.2	43.5	6.9	1.09	1.77	1.922	0.12068	50
CH-45-E	0.95	0.95	1.27	2.22	4.13	2.22	4.45	10.8	67.3	62.4	6.2	0.82	2.82	2.308	0.12142	65
CH-382	1.59	0.64	0.95	2.54	3.81	2.54	3.18	9.5	66.0	56.4	6.6	0.91	2.42	2.197	0.12175	55
CH-490	0.48	0.48	3.81	16.51	17.46	4.29	8.57	42.5	66.2	1855.2	8.3	0.20	62.90	12.827	0.12615	508
CH-199	1.27	0.79	0.95	2.54	4.13	2.22	3.49	10.2	70.4	53.6	6.2	0.91	2.42	2.197	0.12796	57
CH-215-D	1.91	0.64	0.79	2.38	3.65	2.70	2.86	8.9	73.8	46.6	6.9	1.09	1.89	2.058	0.12925	52
CH-2	1.91	0.48	1.27	2.86	3.81	3.18	3.49	10.2	63.3	95.0	7.4	0.82	3.63	2.962	0.13131	68
CH-104-C	0.95	0.79	1.59	2.86	4.45	2.54	4.76	12.1	62.7	103.2	6.4	0.68	4.54	3.091	0.13165	79
CH-139	0.95	0.95	0.95	3.02	4.92	1.91	3.81	11.8	73.3	58.4	5.7	0.82	2.87	2.349	0.13438	64
CH-105-D	0.95	0.95	1.91	1.91	3.81	2.86	5.72	11.4	71.3	93.1	7.2	0.82	3.63	2.966	0.13450	84
CH-236-A	0.95	0.56	1.75	5.24	6.35	2.70	4.60	16.2	58.9	200.6	6.2	0.48	9.15	4.362	0.13492	110
CH-7-H	1.91	0.64	0.79	2.54	3.81	2.70	2.86	9.2	76.5	49.7	6.9	1.09	2.02	2.196	0.13789	54
CH-40	1.27	0.64	1.27	3.33	4.60	2.54	3.81	11.7	65.1	96.5	6.4	0.73	4.23	3.073	0.13915	71
CH-111-B	1.91	0.64	0.95	2.22	3.49	2.86	3.18	8.9	73.9	54.1	7.2	1.09	2.12	2.306	0.13981	56
CH-3-A	1.27	0.64	1.63	2.86	4.13	2.90	4.52	11.5	63.7	115.3	7.0	0.73	4.65	3.375	0.14053	79

National Arnold C Cores 4 mil Silicon Steel

Part No.	D cm 1	E cm 2	F cm 3	G cm 4	Ht cm 5	Wth cm 6	Lt cm 7	MPL cm 8	Wtfe grams 9	Wtcu grams 10	MLT cm 11	Ac cm sq 12	Wa cm sq 13	Ap cm 4th 14	Kg cm 5th 15	At cm sq 16
CH-45-D	0.95	0.95	1.59	2.22	4.13	2.54	5.08	11.4	71.3	84.3	6.7	0.82	3.53	2.885	0.14053	77
CH-100-D	1.27	0.95	0.64	2.86	4.76	1.91	3.18	10.8	89.7	39.1	6.1	1.09	1.82	1.977	0.14235	56
CH-314-F	1.27	0.68	1.43	2.86	4.21	2.70	4.21	11.3	66.4	97.9	6.7	0.77	4.08	3.151	0.14424	73
CH-137	1.59	0.64	0.95	3.02	4.29	2.54	3.18	10.5	72.6	66.9	6.6	0.91	2.87	2.609	0.14457	60
CH-402-M	0.95	1.11	0.64	3.49	5.72	1.59	3.49	12.7	92.3	43.6	5.5	0.95	2.22	2.114	0.14560	64
CH-364-A	1.27	0.71	1.27	2.86	4.29	2.54	3.97	11.1	69.2	84.8	6.6	0.82	3.63	2.962	0.14718	69
CH-363	1.59	0.71	0.95	2.54	3.97	2.54	3.33	9.8	76.6	57.7	6.7	1.02	2.42	2.470	0.15030	58
CH-217-B	1.27	0.79	0.95	3.02	4.60	2.22	3.49	11.1	77.0	63.7	6.2	0.91	2.87	2.609	0.15194	62
CH-98	1.59	0.64	1.11	2.86	4.13	2.70	3.49	10.5	72.6	76.8	6.8	0.91	3.18	2.882	0.15388	64
CH-310-C	0.95	0.95	1.27	2.86	4.76	2.22	4.45	12.1	75.3	80.2	6.2	0.82	3.63	2.967	0.15611	74
CH-360	1.59	0.79	0.79	2.54	4.13	2.38	3.18	9.8	85.2	47.5	6.6	1.13	2.02	2.289	0.15696	56
CH-98-A	0.95	0.95	1.11	3.18	5.08	2.06	4.13	12.4	77.2	74.8	6.0	0.82	3.53	2.883	0.15806	72
CH-129-B	1.59	0.56	1.27	3.41	4.53	2.86	3.65	11.6	70.3	106.2	6.9	0.79	4.34	3.444	0.15890	74
CH-382-A	1.91	0.64	0.95	2.54	3.81	2.86	3.18	9.5	79.1	61.8	7.2	1.09	2.42	2.635	0.15975	59
CH-431-K	1.27	0.44	2.38	5.24	6.11	3.65	5.64	17.0	64.7	344.2	7.8	0.50	12.47	6.231	0.16042	146
CH-356-B	1.27	0.79	1.11	2.86	4.45	2.38	3.81	11.1	77.0	73.2	6.5	0.91	3.18	2.882	0.16143	66
CH-456	1.27	0.95	0.79	2.70	4.61	2.06	3.49	10.8	89.7	48.0	6.3	1.09	2.14	2.334	0.16143	60
CH-45-B	1.59	0.95	0.79	1.91	3.81	2.38	3.49	9.2	95.7	37.3	6.9	1.36	1.51	2.060	0.16181	55
CH-307	0.95	0.95	0.95	3.65	5.56	1.91	3.81	13.0	81.2	70.7	5.7	0.82	3.48	2.844	0.16267	71
CH-203-A	1.27	0.64	1.75	3.18	4.45	3.02	4.76	12.4	68.6	141.1	7.2	0.73	5.54	4.024	0.16316	89
CH-179-H	1.27	0.64	1.43	3.65	4.92	2.70	4.13	12.7	70.3	123.6	6.7	0.73	5.22	3.787	0.16503	82
CH-183-B	1.27	0.64	1.27	3.97	5.24	2.54	3.81	13.0	72.1	114.9	6.4	0.73	5.04	3.659	0.16565	79
CH-173-A	1.27	0.79	1.27	2.70	4.29	2.54	4.13	11.1	77.0	82.0	6.7	0.91	3.43	3.111	0.16780	70
CH-276-A	1.27	0.79	0.95	3.33	4.92	2.22	3.49	11.8	81.4	70.4	6.2	0.91	3.18	2.884	0.16796	66
CH-140	1.59	0.64	1.27	2.86	4.13	2.86	3.81	10.8	74.8	91.0	7.0	0.91	3.63	3.294	0.16967	70
CH-327-D	0.95	0.95	0.95	3.81	5.72	1.91	3.81	13.3	83.2	73.8	5.7	0.82	3.63	2.968	0.16976	73
CH-231-D	0.95	2.06	0.64	1.59	5.72	1.59	5.40	12.7	171.6	26.7	7.4	1.77	1.01	1.785	0.16993	87
CH-251	1.27	0.79	1.11	3.02	4.60	2.38	3.81	11.4	79.1	77.2	6.5	0.91	3.35	3.041	0.17035	68
CH-183-D	1.27	0.71	1.35	3.18	4.60	2.62	4.13	11.9	74.1	102.0	6.7	0.82	4.28	3.495	0.17046	76
CH-125	1.11	0.95	0.64	4.13	6.03	1.75	3.18	13.3	97.0	51.6	5.5	0.95	2.62	2.498	0.17207	66

National Arnold C Cores 4 mil Silicon Steel																
Part No.	D cm 1	E cm 2	F cm 3	G cm 4	Ht cm 5	Wth cm 6	Lt cm 7	MPL cm 8	Wtfe grams 9	Wtcu grams 10	MLT cm 11	Ac cm sq 12	Wa cm sq 13	Ap cm 4th 14	Kg cm 5th 15	At cm sq 16
CH-488-B	0.79	0.64	3.81	5.08	6.35	4.60	8.89	20.3	70.4	636.5	9.2	0.45	19.36	8.783	0.17238	234
CH-38-B	1.59	0.71	0.95	2.94	4.37	2.54	3.33	10.6	82.8	66.8	6.7	1.02	2.80	2.856	0.17379	63
CH-199-A	1.27	0.95	0.95	2.54	4.45	2.22	3.81	10.8	89.7	56.4	6.6	1.09	2.42	2.637	0.17539	63
CH-364	1.27	0.79	1.27	2.86	4.45	2.54	4.13	11.4	79.2	86.9	6.7	0.91	3.63	3.294	0.17769	72
CH-1-B	1.27	0.79	1.27	2.86	4.45	2.54	4.13	11.4	79.2	86.9	6.7	0.91	3.63	3.294	0.17769	72
CH-178-F	1.59	0.79	1.11	2.22	3.81	2.70	3.81	9.8	85.2	62.5	7.1	1.13	2.47	2.803	0.17877	63
CH-215-F	2.06	0.71	0.79	2.38	3.81	2.86	3.02	9.2	93.2	49.8	7.4	1.33	1.89	2.507	0.17951	58
CH-361-E	1.91	0.64	0.95	2.86	4.13	2.86	3.18	10.2	84.4	69.6	7.2	1.09	2.72	2.965	0.17975	63
CH-216	1.27	0.64	3.18	2.54	3.81	4.45	7.62	14.0	77.4	269.6	9.4	0.73	8.07	5.853	0.18073	144
CH-126-A	1.59	0.79	0.95	2.54	4.13	2.54	3.49	10.2	88.0	59.1	6.9	1.13	2.42	2.747	0.18154	62
CH-104-H	0.95	0.95	1.27	3.33	5.24	2.22	4.45	13.0	81.2	93.6	6.2	0.82	4.23	3.461	0.18211	80
CH-43	0.95	0.79	1.59	3.97	5.56	2.54	4.76	14.3	74.3	143.3	6.4	0.68	6.30	4.292	0.18283	95
CH-200-E	1.27	0.64	1.27	4.45	5.72	2.54	3.81	14.0	77.4	128.7	6.4	0.73	5.65	4.097	0.18552	86
CH-208-E	1.27	0.79	1.27	3.02	4.60	2.54	4.13	11.7	81.3	91.7	6.7	0.91	3.83	3.476	0.18751	74
CH-28-D	1.27	0.64	1.91	3.49	4.76	3.18	5.08	13.3	73.9	175.3	7.4	0.73	6.65	4.830	0.18925	101
CH-137-A	1.91	0.64	0.95	3.02	4.29	2.86	3.18	10.5	87.0	73.4	7.2	1.09	2.87	3.129	0.18968	65
CH-356	1.59	0.71	1.11	2.86	4.29	2.70	3.65	10.8	84.0	78.5	7.0	1.02	3.18	3.240	0.19013	68
CH-99-H	1.91	0.48	0.79	5.95	6.91	2.70	2.54	15.4	95.9	111.2	6.6	0.82	4.73	3.857	0.19031	83
CH-137-B	1.91	0.64	1.11	2.70	3.97	3.02	3.49	10.2	84.4	79.3	7.4	1.09	3.00	3.265	0.19128	67
CH-44-A	1.27	0.64	1.59	3.97	5.24	2.86	4.45	13.7	75.6	154.9	6.9	0.73	6.30	4.575	0.19217	94
CH-314-C	1.27	0.79	1.43	2.86	4.45	2.70	4.45	11.8	81.4	101.4	7.0	0.91	4.08	3.706	0.19278	79
CH-215-A	1.91	0.79	0.79	2.38	3.97	2.70	3.18	9.5	98.9	48.8	7.3	1.36	1.89	2.574	0.19323	59
CH-446	1.27	0.95	1.11	2.54	4.45	2.38	4.13	11.1	92.4	68.2	6.8	1.09	2.82	3.074	0.19701	69
CH-197	1.27	0.95	0.95	2.86	4.76	2.22	3.81	11.4	95.0	63.4	6.6	1.09	2.72	2.967	0.19735	67
CH-41	1.59	0.64	1.27	3.33	4.60	2.86	3.81	11.7	81.3	106.1	7.0	0.91	4.23	3.843	0.19793	77
CH-251-D	1.27	0.87	1.11	3.02	4.76	2.38	3.97	11.7	89.4	79.1	6.6	1.00	3.35	3.344	0.20104	72
CH-197-D	1.59	0.79	0.95	2.86	4.45	2.54	3.49	10.8	93.5	66.5	6.9	1.13	2.72	3.091	0.20427	65
CH-104-G	1.27	0.79	1.59	2.86	4.45	2.86	4.76	12.1	83.6	116.7	7.2	0.91	4.54	4.119	0.20683	85
CH-402-F	1.91	0.64	0.64	4.60	5.87	2.54	2.54	13.0	108.1	69.5	6.7	1.09	2.92	3.183	0.20735	69
CH-169-C	0.95	0.95	1.27	3.81	5.72	2.22	4.45	14.0	87.1	106.9	6.2	0.82	4.84	3.955	0.20810	86

National Arnold C Cores 4 mil Silicon Steel

Part No.	D cm 1	E cm 2	F cm 3	G cm 4	Ht cm 5	Wth cm 6	Lt cm 7	MPL cm 8	Wtfe grams 9	Wtcu grams 10	MLT cm 11	Ac cm sq 12	Wa cm sq 13	Ap cm 4th 14	Kg cm 5th 15	At cm sq 16
CH-217	1.27	0.95	0.95	3.02	4.92	2.22	3.81	11.8	97.7	66.9	6.6	1.09	2.87	3.131	0.20826	69
CH-445-A	2.22	0.64	1.11	2.38	3.65	3.33	3.49	9.5	92.3	75.9	8.1	1.27	2.65	3.361	0.21167	68
CH-108-A	1.59	0.79	0.95	3.02	4.60	2.54	3.49	11.1	96.2	70.2	6.9	1.13	2.87	3.262	0.21556	67
CH-402-N	0.95	1.43	0.64	3.49	6.35	1.59	4.13	14.0	130.7	48.7	6.2	1.23	2.22	2.719	0.21605	79
CH-114-B	1.27	0.95	1.27	2.54	4.45	2.54	4.45	11.4	95.0	80.8	7.0	1.09	3.23	3.514	0.21723	75
CH-327	2.22	0.48	1.27	3.81	4.76	3.49	3.49	12.1	87.7	137.6	8.0	0.95	4.84	4.608	0.21942	86
CH-261-A	1.91	0.64	0.95	3.49	4.76	2.86	3.18	11.4	95.0	85.0	7.2	1.09	3.33	3.624	0.21968	71
CH-3	1.91	0.64	1.27	2.86	4.13	3.18	3.81	10.8	89.7	99.2	7.7	1.09	3.63	3.952	0.22402	75
CH-380	2.22	0.64	0.95	2.86	4.13	3.18	3.18	10.2	98.5	75.7	7.8	1.27	2.72	3.460	0.22486	68
CH-445-B	1.59	0.95	1.11	2.06	3.97	2.70	4.13	10.2	105.6	60.6	7.4	1.36	2.29	3.123	0.22889	68
CH-197-B	1.91	0.56	1.27	3.81	4.92	3.18	3.65	12.4	90.1	129.5	7.5	0.95	4.84	4.613	0.23376	85
CH-251-B	1.27	0.95	1.11	3.02	4.92	2.38	4.13	12.1	100.3	81.0	6.8	1.09	3.35	3.650	0.23393	75
CH-104-A	2.54	0.48	1.59	2.86	3.81	4.13	4.13	10.8	89.6	147.4	9.1	1.09	4.54	4.939	0.23535	91
CH-330-H	1.27	0.64	1.27	5.72	6.99	2.54	3.81	16.5	91.4	165.5	6.4	0.73	7.26	5.268	0.23853	102
CH-391	0.95	0.95	1.59	3.81	5.72	2.54	5.08	14.6	91.1	144.4	6.7	0.82	6.05	4.945	0.24086	101
CH-201	1.27	0.95	0.95	3.49	5.40	2.22	3.81	12.7	105.6	77.5	6.6	1.09	3.33	3.626	0.24120	75
CH-464-A	2.54	0.64	0.95	2.54	3.81	3.49	3.18	9.5	105.5	72.8	8.5	1.45	2.42	3.514	0.24133	68
CH-2-B	1.27	1.11	0.95	2.70	4.92	2.22	4.13	11.7	113.8	62.8	6.9	1.27	2.57	3.266	0.24164	72
CH-252	1.59	0.79	1.11	3.02	4.60	2.70	3.81	11.4	99.0	84.8	7.1	1.13	3.35	3.802	0.24254	73
CH-176-A	0.95	1.27	1.43	2.54	5.08	2.38	5.40	13.0	108.2	91.6	7.1	1.09	3.63	3.954	0.24272	90
CH-187	1.27	0.95	1.27	2.86	4.76	2.54	4.45	12.1	100.3	91.0	7.0	1.09	3.63	3.954	0.24442	80
CH-335-B	1.59	0.64	1.27	4.13	5.40	2.86	3.81	13.3	92.3	131.4	7.0	0.91	5.24	4.758	0.24506	87
CH-342	1.27	1.27	0.79	2.54	5.08	2.06	4.13	11.7	130.1	49.7	6.9	1.45	2.02	2.928	0.24513	72
CH-385	1.91	0.64	1.27	3.18	4.45	3.18	3.81	11.4	94.9	110.1	7.7	1.09	4.03	4.390	0.24886	80
CH-316	0.95	0.95	1.59	3.97	5.88	2.54	5.08	14.9	93.1	150.5	6.7	0.82	6.30	5.152	0.25091	103
CH-174	0.95	1.27	1.27	2.86	5.40	2.22	5.08	13.3	110.8	88.4	6.8	1.09	3.63	3.954	0.25156	89
CH-446-A	1.91	0.79	1.11	2.38	3.97	3.02	3.81	10.2	105.5	72.9	7.8	1.36	2.65	3.601	0.25301	70
CH-180	1.59	0.79	1.27	2.86	4.45	2.86	4.13	11.4	99.0	95.1	7.4	1.13	3.63	4.119	0.25382	78
CH-457-A	2.22	0.56	1.27	3.33	4.45	3.49	3.65	11.4	97.0	122.9	8.2	1.11	4.23	4.710	0.25684	83
CH-140-D	1.59	0.79	1.43	2.70	4.29	3.02	4.45	11.4	99.0	104.4	7.6	1.13	3.86	4.377	0.26087	82

National Arnold C Cores 4 mil Silicon Steel

Part No.	D cm 1	E cm 2	F cm 3	G cm 4	Ht cm 5	Wth cm 6	Lt cm 7	MPL cm 8	Wtfe grams 9	Wtcu grams 10	MLT cm 11	Ac cm sq 12	Wa cm sq 13	Ap cm 4th 14	Kg cm 5th 15	At cm sq 16
CH-42	1.91	0.64	1.27	3.33	4.60	3.18	3.81	11.7	97.6	115.7	7.7	1.09	4.23	4.610	0.26133	82
CH-200-C	1.59	0.64	1.27	4.45	5.72	2.86	3.81	14.0	96.7	141.5	7.0	0.91	5.65	5.123	0.26388	92
CH-45-A	1.59	0.79	1.59	2.54	4.13	3.18	4.76	11.4	99.0	112.8	7.9	1.13	4.03	4.577	0.26416	86
CH-5-B	2.54	0.48	1.43	3.49	4.45	3.97	3.81	11.7	97.5	157.7	8.9	1.09	4.99	5.431	0.26612	94
CH-104-F	1.59	0.64	1.43	4.13	5.40	3.02	4.13	13.7	94.5	153.1	7.3	0.91	5.90	5.354	0.26631	95
CH-434	1.27	0.64	1.91	4.92	6.19	3.18	5.08	16.2	89.7	247.0	7.4	0.73	9.38	6.804	0.26662	125
CH-407-A	1.91	0.79	0.95	2.86	4.45	2.86	3.49	10.8	112.2	72.7	7.5	1.36	2.72	3.708	0.26912	70
CH-99-A	3.81	0.56	0.79	2.54	3.65	4.60	2.70	8.9	129.3	77.4	10.8	1.91	2.02	3.845	0.27164	75
CH-327-B	1.59	0.79	0.95	3.81	5.40	2.54	3.49	12.7	110.0	88.7	6.9	1.13	3.63	4.120	0.27231	77
CH-174-A	0.95	1.27	1.27	3.10	5.64	2.22	5.08	13.8	114.8	95.7	6.8	1.09	3.93	4.283	0.27251	92
CH-308-B	1.27	0.95	0.95	3.97	5.88	2.22	3.81	13.7	113.5	88.1	6.6	1.09	3.78	4.120	0.27407	81
CH-9	1.59	0.64	1.59	3.97	5.24	3.18	4.45	13.7	94.5	169.2	7.5	0.91	6.30	5.720	0.27513	100
CH-138-A	1.27	1.27	0.79	2.86	5.40	2.06	4.13	12.4	137.2	56.0	6.9	1.45	2.27	3.294	0.27582	76
CH-402-D	2.22	0.71	0.95	2.86	4.29	3.18	3.33	10.5	114.2	77.3	8.0	1.43	2.72	3.891	0.27866	72
CH-430-H	0.95	0.48	8.26	8.26	9.21	9.21	17.46	34.9	108.8	3932.0	16.2	0.41	68.15	27.821	0.28000	781
CH-197-A	1.59	0.95	0.95	2.86	4.76	2.54	3.81	11.4	118.8	69.6	7.2	1.36	2.72	3.710	0.28125	73
CH-1-E	2.06	0.95	0.87	2.06	3.97	2.94	3.65	9.7	130.8	51.3	8.0	1.77	1.80	3.190	0.28190	67
CH-405-D	1.59	0.79	1.27	3.18	4.76	2.86	4.13	12.1	104.5	105.6	7.4	1.13	4.03	4.576	0.28197	82
CH-269	1.27	0.95	1.27	3.33	5.24	2.54	4.45	13.0	108.2	106.1	7.0	1.09	4.23	4.612	0.28513	86
CH-44	1.27	0.79	1.59	3.97	5.56	2.86	4.76	14.3	99.0	162.0	7.2	0.91	6.30	5.720	0.28724	102
CH-430-A	4.45	0.48	0.95	2.54	3.49	5.40	2.86	8.9	129.2	104.6	12.2	1.90	2.42	4.609	0.28887	85
CH-428-B	1.91	0.64	1.59	3.18	4.45	3.49	4.45	12.1	100.2	146.7	8.2	1.09	5.04	5.489	0.29219	94
CH-4	2.22	0.56	1.27	3.81	4.92	3.49	3.65	12.4	105.1	140.4	8.2	1.11	4.84	5.383	0.29351	90
CH-45-C	1.59	0.95	0.79	3.49	5.40	2.38	3.49	12.4	128.7	68.4	6.9	1.36	2.77	3.778	0.29669	74
CH-108	1.59	0.95	0.95	3.02	4.92	2.54	3.81	11.8	122.1	73.4	7.2	1.36	2.87	3.915	0.29680	75
CH-104-B	1.59	0.79	1.59	2.86	4.45	3.18	4.76	12.1	104.5	126.9	7.9	1.13	4.54	5.150	0.29723	91
CH-105-F	1.91	0.64	1.91	2.86	4.13	3.81	5.08	12.1	100.2	168.0	8.7	1.09	5.44	5.927	0.29743	102
CH-124-D	1.27	0.95	1.11	3.97	5.88	2.38	4.13	14.0	116.1	106.6	6.8	1.09	4.41	4.803	0.30785	88
CH-103-C	0.95	0.95	1.59	4.92	6.83	2.54	5.08	16.8	105.0	186.5	6.7	0.82	7.82	6.388	0.31110	118
CH-404	1.59	0.95	1.27	2.54	4.45	2.86	4.45	11.4	118.8	88.1	7.7	1.36	3.23	4.394	0.31152	81

National Arnold C Cores 4 mil Silicon Steel																
Part No.	D cm 1	E cm 2	F cm 3	G cm 4	Ht cm 5	Wth cm 6	Lt cm 7	MPL cm 8	Wtfe grams 9	Wtcu grams 10	MLT cm 11	Ac cm sq 12	Wa cm sq 13	Ap cm 4th 14	Kg cm 5th 15	At cm sq 16
CH-456-C	2.22	0.64	0.79	4.60	5.87	3.02	2.86	13.3	129.3	98.4	7.6	1.27	3.66	4.644	0.31174	82
CH-118	1.27	0.95	1.07	4.13	6.03	2.34	4.05	14.2	118.1	106.0	6.7	1.09	4.43	4.820	0.31175	88
CH-261-C	1.27	1.11	0.95	3.49	5.72	2.22	4.13	13.3	129.2	81.3	6.9	1.27	3.33	4.227	0.31272	83
CH-270	1.27	1.11	1.27	2.86	5.08	2.54	4.76	12.7	123.1	95.0	7.4	1.27	3.63	4.609	0.31794	87
CH-1-C	1.27	1.11	1.27	2.86	5.08	2.54	4.76	12.7	123.1	95.0	7.4	1.27	3.63	4.609	0.31794	87
CH-169	1.27	0.95	1.27	3.81	5.72	2.54	4.45	14.0	116.1	121.3	7.0	1.09	4.84	5.271	0.32584	93
CH-139-H	1.91	0.95	0.95	2.54	4.45	2.86	3.81	10.8	134.6	67.3	7.8	1.63	2.42	3.955	0.33054	74
CH-127	1.27	0.95	1.59	3.33	5.24	2.86	5.08	13.7	113.5	142.1	7.5	1.09	5.29	5.767	0.33294	100
CH-361-J	1.91	0.87	0.95	3.02	4.76	2.86	3.65	11.4	130.5	78.3	7.7	1.50	2.87	4.302	0.33624	76
CH-425-A	2.22	1.11	0.64	2.22	4.45	2.86	3.49	10.2	172.3	41.5	8.3	2.22	1.41	3.138	0.33721	71
CH-139-B	2.22	0.95	0.95	2.06	3.97	3.18	3.81	9.8	143.2	59.1	8.5	1.91	1.97	3.750	0.33825	72
CH-304	1.59	0.79	1.27	3.81	5.40	2.86	4.13	13.3	115.5	126.7	7.4	1.13	4.84	5.491	0.33837	91
CH-1-D	3.81	0.48	1.27	2.86	3.81	5.08	3.49	10.2	126.5	146.9	11.4	1.63	3.63	5.924	0.33983	96
CH-149	1.59	0.71	1.59	3.97	5.40	3.18	4.60	14.0	108.8	172.7	7.7	1.02	6.30	6.432	0.34072	104
CH-261	1.59	0.95	0.95	3.49	5.40	2.54	3.81	12.7	132.0	85.1	7.2	1.36	3.33	4.534	0.34374	81
CH-49	0.95	0.95	1.91	4.92	6.83	2.86	5.72	17.5	108.9	240.4	7.2	0.82	9.38	7.663	0.34744	135
CH-139-D	2.86	0.79	0.87	2.22	3.81	3.73	3.33	9.4	146.0	64.1	9.3	2.04	1.94	3.964	0.34881	73
CH-176	0.95	1.27	1.27	3.97	6.51	2.22	5.08	15.6	129.3	122.7	6.8	1.09	5.04	5.491	0.34935	105
CH-160	1.59	0.95	1.27	2.86	4.76	2.86	4.45	12.1	125.4	99.2	7.7	1.36	3.63	4.944	0.35052	85
CH-99-E	1.59	0.95	0.79	4.13	6.03	2.38	3.49	13.7	141.9	80.8	6.9	1.36	3.28	4.464	0.35063	82
CH-2-A	1.91	0.95	0.95	2.70	4.61	2.86	3.81	11.1	138.6	71.5	7.8	1.63	2.57	4.203	0.35124	76
CH-202-B	1.59	0.79	1.27	3.97	5.56	2.86	4.13	13.7	118.2	132.0	7.4	1.13	5.04	5.720	0.35249	93
CH-124	2.22	0.64	1.11	3.97	5.24	3.33	3.49	12.7	123.1	126.5	8.1	1.27	4.41	5.602	0.35285	89
CH-328	1.91	0.87	0.95	3.18	4.92	2.86	3.65	11.7	134.2	82.4	7.7	1.50	3.03	4.529	0.35396	78
CH-5-J	1.27	0.95	1.43	3.81	5.72	2.70	4.76	14.3	118.8	141.3	7.3	1.09	5.44	5.931	0.35410	100
CH-374	1.91	0.64	1.59	3.97	5.24	3.49	4.45	13.7	113.4	183.4	8.2	1.09	6.30	6.862	0.36526	106
CH-107-D	2.54	0.48	1.59	4.45	5.40	4.13	4.13	14.0	116.0	229.2	9.1	1.09	7.06	7.681	0.36604	117
CH-173-C	1.91	0.95	0.64	3.97	5.88	2.54	3.18	13.0	162.3	65.6	7.3	1.63	2.52	4.118	0.36763	79
CH-5-H	1.59	0.79	1.43	3.81	5.40	3.02	4.45	13.7	118.2	147.4	7.6	1.13	5.44	6.178	0.36825	98
CH-43-C	1.59	0.95	0.95	3.81	5.72	2.54	3.81	13.3	138.6	92.8	7.2	1.36	3.63	4.945	0.37493	85

National Arnold C Cores 4 mil Silicon Steel																
Part No.	D cm 1	E cm 2	F cm 3	G cm 4	Ht cm 5	Wth cm 6	Lt cm 7	MPL cm 8	Wtfe grams 9	Wtcu grams 10	MLT cm 11	Ac cm sq 12	Wa cm sq 13	Ap cm 4th 14	Kg cm 5th 15	At cm sq 16
CH-173	1.27	1.27	1.27	2.70	5.24	2.54	5.08	13.0	144.2	93.6	7.7	1.45	3.43	4.976	0.37610	93
CH-5-E	2.22	0.64	1.43	3.49	4.76	3.65	4.13	12.4	120.0	152.1	8.6	1.27	4.99	6.341	0.37614	97
CH-431-C	1.27	0.64	3.49	5.08	6.35	4.76	8.26	19.7	109.0	624.8	9.9	0.73	17.74	12.879	0.37761	226
CH-200-B	1.27	0.95	1.27	4.45	6.35	2.54	4.45	15.2	126.7	141.5	7.0	1.09	5.65	6.149	0.38015	102
CH-179-E	1.59	0.79	1.43	3.97	5.56	3.02	4.45	14.0	121.0	153.6	7.6	1.13	5.67	6.436	0.38362	101
CH-5-F	1.27	0.95	1.43	4.13	6.03	2.70	4.76	14.9	124.1	153.1	7.3	1.09	5.90	6.426	0.38365	105
CH-214-A	1.91	0.71	1.27	3.97	5.40	3.18	3.97	13.3	124.5	140.5	7.8	1.22	5.04	6.171	0.38540	95
CH-103-D	1.59	0.95	1.59	2.70	4.61	3.18	5.08	12.4	128.7	124.7	8.2	1.36	4.29	5.838	0.38866	96
CH-183-A	1.27	1.27	1.23	2.86	5.40	2.50	5.00	13.3	146.8	95.2	7.6	1.45	3.52	5.103	0.38889	94
CH-405	1.59	0.95	1.27	3.18	5.08	2.86	4.45	12.7	132.0	110.2	7.7	1.36	4.03	5.492	0.38940	90
CH-254-A	1.91	0.79	1.27	3.33	4.92	3.18	4.13	12.4	128.6	120.5	8.0	1.36	4.23	5.764	0.39234	90
CH-361	1.91	0.95	0.95	3.02	4.92	2.86	3.81	11.8	146.5	79.9	7.8	1.63	2.87	4.696	0.39249	80
CH-251-C	2.54	0.64	1.43	3.02	4.29	3.97	4.13	11.4	126.6	141.0	9.2	1.45	4.31	6.256	0.39479	95
CH-138	1.27	1.27	1.27	2.86	5.40	2.54	5.08	13.3	147.7	99.2	7.7	1.45	3.63	5.269	0.39825	95
CH-318	2.54	0.64	1.27	3.33	4.60	3.81	3.81	11.7	130.1	134.8	9.0	1.45	4.23	6.146	0.39867	93
CH-310-E	1.91	0.87	1.27	2.86	4.60	3.18	4.29	11.7	134.2	105.3	8.2	1.50	3.63	5.433	0.39870	87
CH-169-A	1.27	1.07	1.27	3.81	5.95	2.54	4.68	14.4	135.1	125.4	7.3	1.23	4.84	5.929	0.39883	99
CH-139-C	2.06	1.03	0.95	2.38	4.45	3.02	3.97	10.8	157.9	66.9	8.3	1.92	2.27	4.350	0.40207	78
CH-253	1.59	0.95	1.27	3.33	5.24	2.86	4.45	13.0	135.3	115.7	7.7	1.36	4.23	5.767	0.40890	92
CH-380-A	2.54	0.79	0.95	2.86	4.45	3.49	3.49	10.8	149.5	85.0	8.8	1.82	2.72	4.944	0.40917	81
CH-366	2.22	1.11	0.79	2.22	4.45	3.02	3.81	10.5	177.7	53.5	8.5	2.22	1.77	3.923	0.40930	77
CH-45	1.59	0.79	1.59	3.97	5.56	3.18	4.76	14.3	123.7	176.3	7.9	1.13	6.30	7.152	0.41277	108
CH-356-D	1.59	1.11	1.11	2.86	5.08	2.70	4.44	12.4	150.0	87.5	7.8	1.59	3.18	5.042	0.41318	87
CH-314-E	1.59	0.79	1.43	4.29	5.87	3.02	4.45	14.6	126.5	165.9	7.6	1.13	6.13	6.950	0.41426	105
CH-360-A	1.91	1.19	0.79	2.54	4.92	2.70	3.97	11.4	178.1	57.7	8.0	2.04	2.02	4.118	0.41802	80
CH-179-A	1.91	0.71	1.43	3.97	5.40	3.33	4.29	13.7	127.5	163.2	8.1	1.22	5.67	6.943	0.42026	103
CH-1-G	1.27	1.11	1.27	3.81	6.03	2.54	4.76	14.6	141.5	126.7	7.4	1.27	4.84	6.145	0.42384	101
CH-173-B	1.91	0.95	1.27	2.70	4.61	3.18	4.45	11.8	146.5	101.4	8.3	1.63	3.43	5.601	0.44006	89
CH-334-B	1.91	0.87	0.95	3.97	5.72	2.86	3.65	13.3	152.3	103.0	7.7	1.50	3.78	5.661	0.44248	89
CH-327-A	2.22	0.79	0.95	3.81	5.40	3.18	3.49	12.7	154.0	105.1	8.1	1.59	3.63	5.768	0.45035	88

National Arnold C Cores 4 mil Silicon Steel

Part No.	D cm 1	E cm 2	F cm 3	G cm 4	Ht cm 5	Wth cm 6	Lt cm 7	MPL cm 8	Wtfe grams 9	Wtcu grams 10	MLT cm 11	Ac cm sq 12	Wa cm sq 13	Ap cm 4th 14	Kg cm 5th 15	At cm sq 16
CH-424	1.91	0.64	1.59	4.92	6.19	3.49	4.45	15.6	129.2	227.3	8.2	1.09	7.82	8.508	0.45287	121
CH-261-D	1.91	0.95	0.95	3.49	5.40	2.86	3.81	12.7	158.4	92.6	7.8	1.63	3.33	5.439	0.45456	87
CH-5	2.54	0.64	1.27	3.81	5.08	3.81	3.81	12.7	140.7	154.0	9.0	1.45	4.84	7.024	0.45559	100
CH-12-C	2.54	0.56	1.59	4.13	5.24	4.13	4.29	13.7	132.4	216.6	9.3	1.27	6.56	8.332	0.45582	117
CH-180-A	1.59	1.11	1.27	2.86	5.08	2.86	4.76	12.7	153.9	103.3	8.0	1.59	3.63	5.763	0.45757	93
CH-5-C	1.27	1.11	1.43	3.81	6.03	2.70	5.08	14.9	144.6	147.4	7.6	1.27	5.44	6.914	0.46127	109
CH-310	1.91	0.95	1.27	2.86	4.76	3.18	4.45	12.1	150.4	107.4	8.3	1.63	3.63	5.931	0.46599	91
CH-383	1.91	0.79	1.27	3.97	5.56	3.18	4.13	13.7	141.8	143.4	8.0	1.36	5.04	6.862	0.46706	99
CH-202	1.59	0.95	1.27	3.81	5.72	2.86	4.45	14.0	145.2	132.2	7.7	1.36	4.84	6.590	0.46728	99
CH-360-C	2.22	1.11	0.79	2.54	4.76	3.02	3.81	11.1	188.5	61.1	8.5	2.22	2.02	4.483	0.46767	81
CH-333-C	1.91	0.79	1.11	4.45	6.03	3.02	3.81	14.3	148.4	136.1	7.8	1.36	4.94	6.723	0.47233	98
CH-111-A	2.22	1.11	0.95	2.22	4.45	3.18	4.13	10.8	183.1	66.1	8.8	2.22	2.12	4.709	0.47728	83
CH-179-D	2.54	0.64	1.43	3.65	4.92	3.97	4.13	12.7	140.7	170.7	9.2	1.45	5.22	7.573	0.47791	105
CH-308-A	1.59	1.11	0.95	3.73	5.95	2.54	4.13	13.8	167.3	94.9	7.5	1.59	3.56	5.646	0.47797	92
CH-308	1.59	0.95	1.59	3.33	5.24	3.18	5.08	13.7	141.9	154.1	8.2	1.36	5.29	7.211	0.48010	107
CH-457	2.22	0.79	1.27	3.33	4.92	3.49	4.13	12.4	150.1	130.0	8.6	1.59	4.23	6.726	0.49491	96
CH-5-A	2.54	0.64	1.43	3.81	5.08	3.97	4.13	13.0	144.2	178.1	9.2	1.45	5.44	7.903	0.49872	108
CH-107-C	3.18	0.48	1.59	4.45	5.40	4.76	4.13	14.0	145.0	261.1	10.4	1.36	7.06	9.601	0.50212	129
CH-377	1.59	0.95	1.59	3.49	5.40	3.18	5.08	14.0	145.2	161.4	8.2	1.36	5.55	7.555	0.50299	109
CH-260-A	1.59	1.11	1.11	3.49	5.72	2.70	4.44	13.7	165.4	107.0	7.8	1.59	3.88	6.162	0.50498	96
CH-24-I	0.95	0.95	2.38	6.35	8.26	3.33	6.67	21.3	132.7	427.9	8.0	0.82	15.12	12.358	0.50773	192
CH-103	1.59	0.79	1.59	4.92	6.51	3.18	4.76	16.2	140.2	218.6	7.9	1.13	7.82	8.868	0.51178	124
CH-101	2.54	0.64	1.27	4.29	5.56	3.81	3.81	13.7	151.2	173.3	9.0	1.45	5.44	7.901	0.51251	107
CH-308-C	1.91	0.95	0.95	3.97	5.88	2.86	3.81	13.7	170.2	105.2	7.8	1.63	3.78	6.180	0.51651	93
CH-338-A	2.54	0.64	1.59	3.65	4.92	4.13	4.45	13.0	144.2	194.9	9.5	1.45	5.80	8.416	0.51706	113
CH-7-K	0.95	1.43	1.27	4.92	7.78	2.22	5.40	18.1	169.2	159.3	7.2	1.23	6.25	7.660	0.52406	128
CH-5-D	1.91	0.79	1.43	4.13	5.72	3.33	4.45	14.3	148.4	173.0	8.2	1.36	5.90	8.030	0.53005	110
CH-279	1.27	1.27	1.27	3.81	6.35	2.54	5.08	15.2	168.8	132.2	7.7	1.45	4.84	7.024	0.53091	110
CH-290	1.27	1.43	1.27	3.18	6.03	2.54	5.40	14.6	182.0	114.7	8.0	1.63	4.03	6.586	0.53787	109
CH-361-B	2.22	1.03	0.95	2.86	4.92	3.18	3.97	11.8	185.1	83.4	8.6	2.06	2.72	5.624	0.53917	88

National Arnold C Cores 4 mil Silicon Steel																
Part No.	D cm 1	E cm 2	F cm 3	G cm 4	Ht cm 5	Wth cm 6	Lt cm 7	MPL cm 8	Wtfe grams 9	Wtcu grams 10	MLT cm 11	Ac cm sq 12	Wa cm sq 13	Ap cm 4th 14	Kg cm 5th 15	At cm sq 16
CH-12-D	2.54	0.64	1.59	3.81	5.08	4.13	4.45	13.3	147.7	203.3	9.5	1.45	6.05	8.783	0.53957	116
CH-254	1.91	0.95	1.27	3.33	5.24	3.18	4.45	13.0	162.3	125.2	8.3	1.63	4.23	6.918	0.54360	98
CH-464-B	2.22	1.11	0.95	2.54	4.76	3.18	4.13	11.4	193.9	75.5	8.8	2.22	2.42	5.381	0.54534	87
CH-104	1.91	0.95	1.59	2.86	4.76	3.49	5.08	12.7	158.4	142.3	8.8	1.63	4.54	7.416	0.54967	105
CH-46	1.91	0.79	1.59	3.97	5.56	3.49	4.76	14.3	148.4	190.5	8.5	1.36	6.30	8.580	0.54971	115
CH-50	1.27	0.95	1.91	4.92	6.83	3.18	5.72	17.5	145.1	268.2	8.0	1.09	9.38	10.211	0.55306	143
CH-183	1.27	1.27	1.27	3.97	6.51	2.54	5.08	15.6	172.3	137.7	7.7	1.45	5.04	7.317	0.55307	112
CH-326	1.91	0.87	1.27	3.97	5.72	3.18	4.29	14.0	159.5	146.2	8.2	1.50	5.04	7.545	0.55369	104
CH-189	2.86	0.64	1.27	3.97	5.24	4.13	3.81	13.0	162.2	171.9	9.6	1.63	5.04	8.233	0.56102	108
CH-338	2.54	0.64	1.59	3.97	5.24	4.13	4.45	13.7	151.2	211.8	9.5	1.45	6.30	9.149	0.56209	119
CH-178-G	3.49	0.79	1.11	2.22	3.81	4.60	3.81	9.8	187.5	96.0	10.9	2.50	2.47	6.165	0.56333	92
CH-361-A	2.54	0.95	0.95	2.86	4.76	3.49	3.81	11.4	190.1	88.0	9.1	2.18	2.72	5.934	0.56883	89
CH-129-E	1.91	0.95	1.27	3.49	5.40	3.18	4.45	13.3	166.3	131.2	8.3	1.63	4.44	7.248	0.56952	101
CH-230	2.54	0.64	1.27	4.76	6.03	3.81	3.81	14.6	161.8	192.6	9.0	1.45	6.05	8.781	0.56955	115
CH-169-B	1.59	1.07	1.27	3.81	5.95	2.86	4.68	14.4	168.9	136.3	7.9	1.53	4.84	7.413	0.57350	106
CH-175	1.59	1.27	1.27	2.86	5.40	2.86	5.08	13.3	184.7	107.4	8.3	1.82	3.63	6.588	0.57505	102
CH-335	1.27	1.27	1.27	4.13	6.67	2.54	5.08	15.9	175.8	143.2	7.7	1.45	5.24	7.610	0.57522	115
CH-124-A	1.91	0.95	1.11	3.97	5.88	3.02	4.13	14.0	174.2	126.5	8.1	1.63	4.41	7.205	0.58363	100
CH-157	3.18	0.95	0.79	2.54	4.45	3.97	3.49	10.5	217.8	72.5	10.1	2.72	2.02	5.492	0.59169	88
CH-381-E	2.22	0.95	0.95	3.65	5.56	3.18	3.81	13.0	189.4	104.6	8.5	1.91	3.48	6.634	0.59833	95
CH-63-D	1.59	0.95	1.11	5.40	7.30	2.70	4.13	16.8	174.9	158.5	7.4	1.36	6.00	8.168	0.59861	114
CH-428	1.91	0.95	1.59	3.18	5.08	3.49	5.08	13.3	166.3	158.1	8.8	1.63	5.04	8.238	0.61064	111
CH-103-B	0.95	1.43	1.59	4.92	7.78	2.54	6.03	18.7	175.2	213.0	7.7	1.23	7.82	9.578	0.61260	146
CH-1-F	2.22	1.11	1.11	2.54	4.76	3.33	4.44	11.7	199.2	90.5	9.0	2.22	2.82	6.273	0.61827	94
CH-456-A	2.86	0.79	0.79	4.29	5.87	3.65	3.18	13.3	207.8	110.8	9.2	2.04	3.40	6.950	0.61995	98
CH-200-A	1.91	0.87	1.27	4.45	6.19	3.18	4.29	14.9	170.4	163.8	8.2	1.50	5.65	8.449	0.62010	111
CH-197-E	1.91	0.95	1.27	3.81	5.72	3.18	4.45	14.0	174.2	143.1	8.3	1.63	4.84	7.906	0.62121	106
CH-205-C	1.91	0.79	1.91	3.97	5.56	3.81	5.40	14.9	155.0	241.9	9.0	1.36	7.56	10.293	0.62296	131
CH-381-H	1.59	0.95	0.95	6.35	8.26	2.54	3.81	18.4	191.4	154.6	7.2	1.36	6.05	8.242	0.62488	117
CH-145-B	2.06	0.95	1.11	3.81	5.72	3.18	4.13	13.7	184.4	126.2	8.4	1.77	4.23	7.493	0.63273	101

National Arnold C Cores 4 mil Silicon Steel

Part No.	D cm 1	E cm 2	F cm 3	G cm 4	Ht cm 5	Wth cm 6	Lt cm 7	MPL cm 8	Wtfe grams 9	Wtcu grams 10	MLT cm 11	Ac cm sq 12	Wa cm sq 13	Ap cm 4th 14	Kg cm 5th 15	At cm sq 16
CH-207	1.27	1.27	1.75	3.65	6.19	3.02	6.03	15.9	175.8	191.1	8.4	1.45	6.38	9.253	0.63742	131
CH-361-C	1.91	1.27	0.95	3.02	5.56	2.86	4.45	13.0	216.3	86.4	8.5	2.18	2.87	6.258	0.64475	97
CH-330-C	2.54	0.64	1.27	5.40	6.67	3.81	3.81	15.9	175.8	218.2	9.0	1.45	6.86	9.951	0.64548	124
CH-245	1.91	0.95	1.27	3.97	5.88	3.18	4.45	14.3	178.2	149.1	8.3	1.63	5.04	8.236	0.64713	108
CH-158-A	3.18	0.64	0.79	5.87	7.14	3.97	2.86	15.9	219.8	157.1	9.5	1.81	4.66	8.463	0.64830	116
CH-171-A	1.27	1.27	1.59	3.97	6.51	2.86	5.72	16.2	179.4	183.4	8.2	1.45	6.30	9.149	0.64935	128
CH-186	3.18	0.64	1.27	3.97	5.24	4.45	3.81	13.0	180.2	183.2	10.2	1.81	5.04	9.146	0.64943	114
CH-105	1.91	0.95	1.91	3.02	4.92	3.81	5.72	13.7	170.2	190.3	9.3	1.63	5.75	9.388	0.65868	123
CH-314	2.86	0.64	1.43	4.29	5.56	4.29	4.13	14.0	174.1	214.3	9.8	1.63	6.13	10.004	0.66438	122
CH-314-A	2.86	0.79	1.43	2.86	4.45	4.29	4.45	11.8	183.1	147.5	10.2	2.04	4.08	8.341	0.67097	107
CH-335-C	1.91	0.95	1.27	4.13	6.03	3.18	4.45	14.6	182.1	155.1	8.3	1.63	5.24	8.566	0.67306	111
CH-346	3.18	0.64	1.11	4.60	5.87	4.29	3.49	14.0	193.4	181.4	10.0	1.81	5.12	9.281	0.67552	116
CH-456-B	2.54	0.95	0.79	3.97	5.88	3.33	3.49	13.3	221.7	99.1	8.8	2.18	3.15	6.865	0.67674	98
CH-129-J	2.22	0.95	1.27	3.33	5.24	3.49	4.45	13.0	189.4	134.8	9.0	1.91	4.23	8.073	0.68765	105
CH-345	2.54	0.79	1.27	3.81	5.40	3.81	4.13	13.3	184.7	159.5	9.3	1.82	4.84	8.783	0.68787	109
CH-356-A	1.91	1.27	1.11	2.86	5.40	3.02	4.76	13.0	216.3	98.3	8.7	2.18	3.18	6.914	0.69197	101
CH-47	2.22	0.79	1.59	3.97	5.56	3.81	4.76	14.3	173.2	204.7	9.1	1.59	6.30	10.012	0.69644	122
CH-11	2.54	0.71	1.59	3.97	5.40	4.13	4.60	14.0	174.0	215.4	9.6	1.63	6.30	10.287	0.69896	123
CH-31-F	1.27	0.95	1.27	8.26	10.16	2.54	4.45	22.9	190.0	262.7	7.0	1.09	10.48	11.420	0.70599	155
CH-330-D	1.27	1.27	1.27	5.08	7.62	2.54	5.08	17.8	196.9	176.2	7.7	1.45	6.45	9.365	0.70788	129
CH-103-A	1.59	0.95	1.59	4.92	6.83	3.18	5.08	16.8	174.9	227.4	8.2	1.36	7.82	10.644	0.70862	133
CH-178-E	3.81	0.79	1.11	2.54	4.13	4.92	3.81	10.5	217.7	118.1	11.8	2.72	2.82	7.683	0.71100	103
CH-381-D	2.22	0.79	0.95	6.03	7.62	3.18	3.49	17.1	207.8	166.4	8.1	1.59	5.75	9.133	0.71312	118
CH-277	2.54	0.79	1.27	3.97	5.56	3.81	4.13	13.7	189.1	166.2	9.3	1.82	5.04	9.149	0.71658	112
CH-98-D	2.54	0.95	1.11	3.18	5.08	3.65	4.13	12.4	205.9	117.1	9.3	2.18	3.53	7.685	0.71712	101
CH-310-A	2.54	0.95	1.27	2.86	4.76	3.81	4.45	12.1	200.6	123.8	9.6	2.18	3.63	7.907	0.71869	103
CH-104-J	2.86	0.79	1.59	2.86	4.45	4.45	4.76	12.1	188.1	167.9	10.4	2.04	4.54	9.269	0.72774	114
CH-159-A	1.27	1.27	0.79	7.62	10.16	2.06	4.13	21.9	242.6	149.2	6.9	1.45	6.05	8.783	0.73539	135
CH-430-D	3.81	0.87	0.95	2.54	4.29	4.76	3.65	10.5	239.3	100.5	11.7	2.99	2.42	7.246	0.74297	100
CH-327-C	2.54	0.95	0.95	3.81	5.72	3.49	3.81	13.3	221.7	117.4	9.1	2.18	3.63	7.910	0.75830	103

National Arnold C Cores 4 mil Silicon Steel																
Part No.	D cm 1	E cm 2	F cm 3	G cm 4	Ht cm 5	Wth cm 6	Lt cm 7	MPL cm 8	Wtfe grams 9	Wtcu grams 10	MLT cm 11	Ac cm sq 12	Wa cm sq 13	Ap cm 4th 14	Kg cm 5th 15	At cm sq 16
CH-279-A	1.59	1.27	1.27	3.81	6.35	2.86	5.08	15.2	211.1	143.1	8.3	1.82	4.84	8.783	0.76660	117
CH-351	1.91	1.27	1.27	2.86	5.40	3.18	5.08	13.3	221.6	115.5	9.0	2.18	3.63	7.903	0.76894	109
CH-3-B	1.91	1.27	1.27	2.86	5.40	3.18	5.08	13.3	221.6	115.5	9.0	2.18	3.63	7.903	0.76894	109
CH-405-A	1.59	1.43	1.27	3.18	6.03	2.86	5.40	14.6	227.6	123.8	8.6	2.04	4.03	8.235	0.77902	116
CH-381	2.22	1.11	0.95	3.65	5.87	3.18	4.13	13.7	231.5	108.5	8.8	2.22	3.48	7.734	0.78388	104
CH-282	1.59	1.27	1.59	3.33	5.87	3.18	5.72	14.9	206.7	166.0	8.8	1.82	5.29	9.610	0.79130	125
CH-12-E	1.91	0.95	1.59	4.13	6.03	3.49	5.08	15.2	190.0	205.5	8.8	1.63	6.56	10.711	0.79393	127
CH-177	1.59	1.27	1.27	3.97	6.51	2.86	5.08	15.6	215.5	149.1	8.3	1.82	5.04	9.149	0.79859	120
CH-51	1.59	0.95	1.91	4.92	6.83	3.49	5.72	17.5	181.5	289.4	8.7	1.36	9.38	12.768	0.80134	151
CH-49-A	0.95	1.59	1.91	4.92	8.10	2.86	6.99	20.0	207.9	282.7	8.5	1.36	9.38	12.768	0.82024	175
CH-330-L	1.91	0.95	1.27	5.08	6.99	3.18	4.45	16.5	205.9	190.8	8.3	1.63	6.45	10.541	0.82828	125
CH-243	1.59	1.59	1.27	2.86	6.03	2.86	5.72	14.6	253.0	115.6	9.0	2.27	3.63	8.238	0.83522	120
CH-361-H	3.81	0.87	0.95	2.86	4.60	4.76	3.65	11.1	253.8	113.1	11.7	2.99	2.72	8.153	0.83599	106
CH-464	2.54	1.27	0.95	2.54	5.08	3.49	4.45	12.1	267.3	83.7	9.7	2.90	2.42	7.028	0.83925	102
CH-48	2.54	0.79	1.59	3.97	5.56	4.13	4.76	14.3	197.9	219.0	9.8	1.82	6.30	11.440	0.85021	128
CH-208-F	2.86	0.95	1.27	2.86	4.76	4.13	4.45	12.1	225.7	132.0	10.2	2.45	3.63	8.897	0.85331	109
CH-353-A	1.75	1.03	1.59	4.45	6.51	3.33	5.24	16.2	200.4	217.3	8.7	1.62	7.06	11.447	0.85771	133
CH-107-E	3.18	0.64	1.59	4.45	5.72	4.76	4.45	14.6	202.2	269.1	10.7	1.81	7.06	12.808	0.86708	140
CH-435	1.27	0.95	2.54	6.51	8.42	3.81	6.99	21.9	182.1	531.6	9.0	1.09	16.53	18.009	0.86782	216
CH-430-C	3.81	0.95	0.95	2.54	4.45	4.76	3.81	10.8	269.2	101.9	11.8	3.27	2.42	7.910	0.87342	105
CH-348	2.54	0.95	1.43	3.18	5.08	3.97	4.76	13.0	216.4	158.7	9.8	2.18	4.54	9.884	0.87556	116
CH-408-A	2.22	1.11	1.11	3.65	5.87	3.33	4.44	14.0	236.9	130.1	9.0	2.22	4.06	9.016	0.88871	111
CH-124-B	2.54	0.95	1.11	3.97	5.88	3.65	4.13	14.0	232.2	146.4	9.3	2.18	4.41	9.606	0.89645	113
CH-129	2.22	1.11	1.27	3.33	5.56	3.49	4.76	13.7	231.5	139.6	9.3	2.22	4.23	9.412	0.90271	114
CH-445-C	4.13	0.95	1.11	2.06	3.97	5.24	4.13	10.2	274.5	103.7	12.7	3.54	2.29	8.119	0.90379	109
CH-430-E	3.18	1.11	0.95	2.54	4.76	4.13	4.13	11.4	276.9	91.9	10.7	3.17	2.42	7.685	0.91405	104
CH-179	3.49	0.71	1.27	3.97	5.40	4.76	3.97	13.3	228.4	197.5	11.0	2.24	5.04	11.314	0.92215	125
CH-57	1.27	1.11	2.22	5.72	7.94	3.49	6.67	20.3	196.9	400.3	8.9	1.27	12.70	16.133	0.92491	188
CH-330-B	1.91	0.95	1.27	5.72	7.62	3.18	4.45	17.8	221.7	214.7	8.3	1.63	7.26	11.859	0.93181	135
CH-224-J	1.11	1.59	1.43	5.24	8.42	2.54	6.03	19.7	238.5	214.3	8.0	1.59	7.49	11.887	0.93796	157

National Arnold C Cores 4 mil Silicon Steel

Part No.	D cm 1	E cm 2	F cm 3	G cm 4	Ht cm 5	Wth cm 6	Lt cm 7	MPL cm 8	Wtfe grams 9	Wtcu grams 10	MLT cm 11	Ac cm sq 12	Wa cm sq 13	Ap cm 4th 14	Kg cm 5th 15	At cm sq 16
CH-428-A	3.18	0.79	1.59	3.18	4.76	4.76	4.76	12.7	219.9	197.9	11.0	2.27	5.04	11.439	0.94044	126
CH-171	1.59	1.27	1.59	3.97	6.51	3.18	5.72	16.2	224.3	197.6	8.8	1.82	6.30	11.440	0.94201	136
CH-129-A	2.22	1.11	1.27	3.49	5.72	3.49	4.76	14.0	236.9	146.2	9.3	2.22	4.44	9.860	0.94576	116
CH-312-B	1.91	0.95	1.59	4.92	6.83	3.49	5.08	16.8	209.8	245.0	8.8	1.63	7.82	12.768	0.94645	141
CH-371	2.22	0.95	1.59	3.97	5.88	3.81	5.08	14.9	217.1	211.9	9.5	1.91	6.30	12.017	0.96954	131
CH-229-A	3.18	0.71	1.27	4.76	6.19	4.45	3.97	14.9	232.3	223.3	10.4	2.04	6.05	12.342	0.97033	132
CH-51-A	1.91	0.95	1.91	4.45	6.35	3.81	5.72	16.5	205.9	280.5	9.3	1.63	8.47	13.836	0.97076	150
CH-158	2.54	0.95	0.79	5.72	7.62	3.33	3.49	16.8	279.8	142.7	8.8	2.18	4.54	9.886	0.97444	121
CH-98-J	3.18	0.95	1.11	3.18	5.08	4.29	4.13	12.4	257.3	133.1	10.6	2.72	3.53	9.606	0.98635	112
CH-178-B	3.81	0.95	1.11	2.54	4.45	4.92	4.13	11.1	277.1	121.3	12.1	3.27	2.82	9.222	0.99732	113
CH-7	2.54	0.95	1.27	3.97	5.88	3.81	4.45	14.3	237.5	171.9	9.6	2.18	5.04	10.981	0.99807	121
CH-409-A	2.54	0.95	1.11	4.45	6.35	3.65	4.13	14.9	248.1	164.0	9.3	2.18	4.94	10.759	1.00396	121
CH-386-A	1.91	0.95	1.91	4.60	6.51	3.81	5.72	16.8	209.8	290.5	9.3	1.63	8.77	14.330	1.00548	153
CH-102-A	2.86	0.79	1.59	3.97	5.56	4.45	4.76	14.3	222.7	233.2	10.4	2.04	6.30	12.872	1.01063	135
CH-4-A	2.22	1.11	1.27	3.81	6.03	3.49	4.76	14.6	247.7	159.5	9.3	2.22	4.84	10.755	1.03159	122
CH-386	1.91	0.95	1.91	4.76	6.67	3.81	5.72	17.1	213.8	300.5	9.3	1.63	9.07	14.825	1.04021	156
CH-408	2.22	1.11	1.11	4.29	6.51	3.33	4.44	15.2	258.4	152.7	9.0	2.22	4.76	10.584	1.04328	121
CH-63-A	3.18	0.56	2.22	5.40	6.51	5.40	5.56	17.5	211.7	493.3	11.6	1.59	12.00	19.065	1.04808	193
CH-409	2.22	1.03	1.11	4.92	6.99	3.33	4.29	16.2	255.1	172.3	8.9	2.06	5.47	11.288	1.05198	126
CH-6	1.91	1.27	1.27	3.97	6.51	3.18	5.08	15.6	258.5	160.5	9.0	2.18	5.04	10.976	1.06785	127
CH-52	1.91	0.95	1.91	4.92	6.83	3.81	5.72	17.5	217.7	310.5	9.3	1.63	9.38	15.317	1.07472	159
CH-232-A	1.27	1.27	1.91	5.87	8.41	3.18	6.35	20.6	228.6	345.3	8.7	1.45	11.19	16.243	1.08674	181
CH-14	2.86	0.79	1.59	4.29	5.87	4.45	4.76	14.9	232.6	251.8	10.4	2.04	6.81	13.900	1.09135	141
CH-349	2.54	1.03	1.59	3.18	5.24	4.13	5.24	13.7	245.8	183.7	10.2	2.36	5.04	11.895	1.09559	128
CH-219	1.59	1.59	1.59	3.18	6.35	3.18	6.35	15.9	275.0	169.5	9.5	2.27	5.04	11.443	1.09892	141
CH-183-C	2.54	1.11	1.35	3.18	5.40	3.89	4.92	13.5	261.5	152.7	10.0	2.54	4.28	10.878	1.10201	121
CH-10-B	2.22	0.95	1.91	3.97	5.88	4.13	5.72	15.6	226.4	267.5	10.0	1.91	7.56	14.416	1.10491	148
CH-200	2.54	0.95	1.27	4.45	6.35	3.81	4.45	15.2	253.4	192.5	9.6	2.18	5.65	12.298	1.11776	129
CH-145	1.91	1.43	1.11	3.81	6.67	3.02	5.08	15.6	290.8	135.8	9.0	2.45	4.23	10.371	1.12673	126
CH-350-A	2.54	0.87	1.75	4.13	5.87	4.29	5.24	15.2	232.1	260.8	10.2	2.00	7.21	14.384	1.12845	145

National Arnold C Cores 4 mil Silicon Steel

Part No.	D cm 1	E cm 2	F cm 3	G cm 4	Ht cm 5	Wth cm 6	Lt cm 7	MPL cm 8	Wtfe grams 9	Wtcu grams 10	MLT cm 11	Ac cm sq 12	Wa cm sq 13	Ap cm 4th 14	Kg cm 5th 15	At cm sq 16
CH-388-D	2.22	0.95	1.91	4.13	6.03	4.13	5.72	15.9	231.0	278.3	10.0	1.91	7.86	14.994	1.14917	152
CH-59-A	1.91	0.95	2.22	4.76	6.67	4.13	6.35	17.8	221.7	369.5	9.8	1.63	10.59	17.300	1.15210	175
CH-202-C	1.59	1.59	1.27	3.97	7.15	2.86	5.72	16.8	291.4	160.5	9.0	2.27	5.04	11.440	1.15990	139
CH-112-A	1.91	1.11	1.59	4.60	6.83	3.49	5.40	16.8	244.6	237.4	9.1	1.90	7.31	13.926	1.16179	145
CH-404-B	3.18	1.11	1.27	2.54	4.76	4.45	4.76	12.1	292.2	128.2	11.2	3.17	3.23	10.241	1.16384	119
CH-482	1.27	1.27	2.22	5.72	8.26	3.49	6.99	21.0	232.1	414.6	9.2	1.45	12.70	18.442	1.16671	199
CH-384	3.18	0.79	1.59	3.97	5.56	4.76	4.76	14.3	247.4	247.4	11.0	2.27	6.30	14.300	1.17563	141
CH-63-E	2.54	0.95	1.11	5.24	7.15	3.65	4.13	16.5	274.5	193.3	9.3	2.18	5.82	12.680	1.18330	133
CH-178-C	2.86	0.95	1.27	3.97	5.88	4.13	4.45	14.3	267.3	183.3	10.2	2.45	5.04	12.356	1.18501	128
CH-100	2.54	0.95	1.59	3.97	5.88	4.13	5.08	14.9	248.1	226.1	10.1	2.18	6.30	13.731	1.18621	138
CH-208-A	2.86	1.11	1.27	3.02	5.24	4.13	4.76	13.0	283.8	143.6	10.5	2.86	3.83	10.946	1.18712	121
CH-243-A	2.22	1.43	1.27	2.86	5.72	3.49	5.40	14.0	304.8	127.9	9.9	2.86	3.63	10.377	1.19801	125
CH-362	2.54	0.79	1.91	4.92	6.51	4.45	5.40	16.8	233.1	342.3	10.3	1.82	9.38	17.016	1.20327	164
CH-7-B	2.54	1.27	0.95	3.65	6.19	3.49	4.45	14.3	316.5	120.3	9.7	2.90	3.48	10.101	1.20634	120
CH-129-F	3.18	0.95	1.27	3.49	5.40	4.45	4.45	13.3	277.1	171.3	10.9	2.72	4.44	12.080	1.21192	126
CH-379-B	2.70	0.95	1.11	4.92	6.83	3.81	4.13	15.9	280.4	187.7	9.7	2.31	5.47	12.656	1.21365	131
CH-401-B	2.86	1.19	0.95	3.49	5.88	3.81	4.29	13.7	319.2	120.8	10.2	3.06	3.33	10.198	1.22487	119
CH-7-C	2.54	0.95	1.27	4.92	6.83	3.81	4.45	16.2	269.2	213.1	9.6	2.18	6.25	13.615	1.23746	137
CH-312-A	1.91	1.11	1.59	4.92	7.14	3.49	5.40	17.5	253.8	253.8	9.1	1.90	7.82	14.885	1.24178	151
CH-273-B	1.27	1.27	1.59	7.62	10.16	2.86	5.72	23.5	260.2	352.0	8.2	1.45	12.10	17.565	1.24667	191
CH-223-B	2.54	1.11	1.27	3.81	6.03	3.81	4.76	14.6	283.0	170.4	9.9	2.54	4.84	12.289	1.26056	128
CH-255	1.91	1.27	1.59	3.97	6.51	3.49	5.72	16.2	269.0	211.8	9.5	2.18	6.30	13.724	1.26470	144
CH-183-E	2.54	1.11	1.35	3.65	5.87	3.89	4.92	14.4	279.9	175.6	10.0	2.54	4.93	12.509	1.26722	130
CH-224-H	2.22	1.11	1.43	4.29	6.51	3.65	5.08	15.9	269.2	207.3	9.5	2.22	6.13	13.614	1.27151	138
CH-376	2.54	1.11	1.11	4.29	6.51	3.65	4.44	15.2	295.3	163.5	9.7	2.54	4.76	12.094	1.27259	128
CH-372	2.22	1.11	1.59	3.97	6.19	3.81	5.40	15.6	263.9	219.0	9.8	2.22	6.30	14.010	1.27505	141
CH-379	2.54	1.03	1.11	4.92	6.99	3.65	4.29	16.2	291.5	184.6	9.5	2.36	5.47	12.898	1.28170	133
CH-102	2.86	0.91	1.59	3.97	5.80	4.45	5.00	14.8	264.6	238.5	10.6	2.35	6.30	14.802	1.30639	142
CH-317-H	1.27	1.27	4.45	4.45	6.99	5.72	11.43	22.9	253.2	890.0	12.7	1.45	19.76	28.681	1.31475	325
CH-330-N	2.54	0.95	1.27	5.24	7.15	3.81	4.45	16.8	279.8	226.8	9.6	2.18	6.65	14.495	1.31743	142

National Arnold C Cores 4 mil Silicon Steel																
Part No.	D cm 1	E cm 2	F cm 3	G cm 4	Ht cm 5	Wth cm 6	Lt cm 7	MPL cm 8	Wtfe grams 9	Wtcu grams 10	MLT cm 11	Ac cm sq 12	Wa cm sq 13	Ap cm 4th 14	Kg cm 5th 15	At cm sq 16
CH-24-A	1.27	1.27	2.30	6.35	8.89	3.57	7.14	22.4	247.9	483.5	9.3	1.45	14.62	21.219	1.32451	218
CH-387	2.22	0.95	1.91	4.76	6.67	4.13	5.72	17.1	249.5	321.1	10.0	1.91	9.07	17.300	1.32594	164
CH-370-A	1.59	1.27	1.91	4.92	7.46	3.49	6.35	18.7	259.4	310.5	9.3	1.82	9.38	17.016	1.32625	172
CH-58	1.59	1.11	2.22	5.72	7.94	3.81	6.67	20.3	246.2	429.0	9.5	1.59	12.70	20.173	1.34922	197
CH-29-D	1.27	1.27	2.38	6.35	8.89	3.65	7.30	22.5	249.7	506.8	9.4	1.45	15.12	21.947	1.35194	224
CH-403	2.22	1.27	1.27	3.97	6.51	3.49	5.08	15.6	301.6	171.9	9.6	2.54	5.04	12.808	1.35767	134
CH-53	2.22	0.95	1.91	4.92	6.83	4.13	5.72	17.5	254.1	331.7	10.0	1.91	9.38	17.874	1.36993	167
CH-161	1.59	1.59	1.59	3.97	7.15	3.18	6.35	17.5	302.5	211.9	9.5	2.27	6.30	14.305	1.37373	156
CH-57-A	1.27	1.75	2.22	3.97	7.46	3.49	7.94	19.4	294.9	317.8	10.1	2.00	8.82	17.608	1.38755	195
CH-336-A	3.18	0.95	4.92	1.59	3.49	8.10	11.75	16.8	349.7	461.0	16.6	2.72	7.82	21.281	1.39726	291
CH-388-C	2.54	0.95	1.91	4.13	6.03	4.45	5.72	15.9	263.9	296.0	10.6	2.18	7.86	17.132	1.41042	159
CH-401-A	3.18	1.19	0.95	3.49	5.88	4.13	4.29	13.7	354.6	128.3	10.8	3.40	3.33	11.329	1.42321	125
CH-208-B	2.86	1.27	1.27	2.86	5.40	4.13	5.08	13.3	332.4	140.1	10.9	3.27	3.63	11.857	1.42691	128
CH-179-F	1.91	1.43	1.43	3.97	6.83	3.33	5.72	16.5	308.7	192.0	9.5	2.45	5.67	13.896	1.43053	145
CH-335-A	3.18	0.95	1.27	4.13	6.03	4.45	4.45	14.6	303.5	202.4	10.9	2.72	5.24	14.277	1.43224	137
CH-65	1.27	1.27	2.54	6.51	9.05	3.81	7.62	23.2	256.7	568.8	9.7	1.45	16.53	23.999	1.44019	239
CH-330-A	1.91	1.27	1.27	5.40	7.94	3.18	5.08	18.4	306.0	218.2	9.0	2.18	6.86	14.927	1.45233	150
CH-292	3.18	0.79	1.59	4.92	6.51	4.76	4.76	16.2	280.3	306.8	11.0	2.27	7.82	17.730	1.45761	159
CH-478	2.54	0.95	1.59	4.92	6.83	4.13	5.08	16.8	279.8	280.3	10.1	2.18	7.82	17.024	1.47073	156
CH-124-E	2.54	1.27	1.11	3.97	6.51	3.65	4.76	15.2	337.6	156.4	10.0	2.90	4.41	12.802	1.49081	133
CH-112-C	1.91	1.27	1.59	4.76	7.30	3.49	5.72	17.8	295.4	254.2	9.5	2.18	7.56	16.469	1.51771	158
CH-205	3.18	0.91	1.59	3.97	5.80	4.76	5.00	14.8	293.9	252.8	11.3	2.61	6.30	16.443	1.52162	149
CH-381-C	3.18	0.95	0.95	5.72	7.62	4.13	3.81	17.1	356.3	200.7	10.4	2.72	5.45	14.832	1.55941	145
CH-103-E	1.59	1.51	1.59	4.92	7.94	3.18	6.19	19.1	313.3	258.2	9.3	2.16	7.82	16.842	1.56239	169
CH-300	2.54	0.95	1.91	4.60	6.51	4.45	5.72	16.8	279.8	330.1	10.6	2.18	8.77	19.107	1.57306	169
CH-107	2.86	0.95	1.59	4.45	6.35	4.45	5.08	15.9	297.0	269.2	10.7	2.45	7.06	17.303	1.58218	154
CH-317-A	1.27	1.27	5.08	5.08	7.62	6.35	12.70	25.4	281.3	1253.9	13.7	1.45	25.81	37.461	1.59192	400
CH-223	2.54	1.27	1.27	3.81	6.35	3.81	5.08	15.2	337.6	175.9	10.2	2.90	4.84	14.048	1.59594	139
CH-237-A	2.86	1.11	1.91	3.02	5.24	4.76	6.03	14.3	311.5	235.7	11.5	2.86	5.75	16.419	1.62681	154
CH-230-B	3.18	0.95	1.27	4.76	6.67	4.45	4.45	15.9	329.9	233.6	10.9	2.72	6.05	16.473	1.65256	148

National Arnold C Cores 4 mil Silicon Steel																
Part No.	D cm 1	E cm 2	F cm 3	G cm 4	Ht cm 5	Wth cm 6	Lt cm 7	MPL cm 8	Wtfe grams 9	Wtcu grams 10	MLT cm 11	Ac cm sq 12	Wa cm sq 13	Ap cm 4th 14	Kg cm 5th 15	At cm sq 16
CH-230-D	3.18	1.11	1.59	3.02	5.24	4.76	5.40	13.7	330.7	198.8	11.7	3.17	4.79	15.205	1.65407	144
CH-8	2.54	1.27	1.27	3.97	6.51	3.81	5.08	15.6	344.6	183.2	10.2	2.90	5.04	14.634	1.66254	142
CH-54	2.54	0.95	1.91	4.92	6.83	4.45	5.72	17.5	290.3	352.9	10.6	2.18	9.38	20.423	1.68137	175
CH-306	2.54	1.11	1.59	4.29	6.51	4.13	5.40	16.2	313.8	251.8	10.4	2.54	6.81	17.286	1.68802	154
CH-333	1.91	1.27	1.11	6.99	9.53	3.02	4.76	21.3	353.4	240.1	8.7	2.18	7.76	16.897	1.69118	166
CH-488-K	2.22	1.11	1.27	6.35	8.57	3.49	4.76	19.7	333.8	265.8	9.3	2.22	8.07	17.926	1.71931	164
CH-267-A	2.86	0.79	1.75	6.35	7.94	4.60	5.08	19.4	301.8	420.0	10.7	2.04	11.09	22.643	1.73639	189
CH-178	3.81	0.95	1.27	3.97	5.88	5.08	4.45	14.3	356.3	221.1	12.3	3.27	5.04	16.472	1.74540	147
CH-353	2.54	1.11	1.59	4.45	6.67	4.13	5.40	16.5	319.9	261.1	10.4	2.54	7.06	17.927	1.75064	158
CH-254-B	1.91	1.91	1.27	3.33	7.14	3.18	6.35	16.8	419.4	153.9	10.2	3.27	4.23	13.829	1.76751	157
CH-296	2.22	0.95	1.91	6.35	8.26	4.13	5.72	20.3	295.6	428.0	10.0	1.91	12.10	23.064	1.76774	195
CH-306-A	3.18	0.95	1.59	4.29	6.19	4.76	5.08	15.6	323.3	274.9	11.4	2.72	6.81	18.535	1.77767	158
CH-168	2.54	0.95	1.91	5.24	7.15	4.45	5.72	18.1	300.9	375.7	10.6	2.18	9.98	21.743	1.79002	182
CH-112-B	1.91	1.43	1.59	4.60	7.46	3.49	6.03	18.1	338.4	254.0	9.8	2.45	7.31	17.912	1.79692	166
CH-350-C	2.54	1.03	2.22	4.13	6.19	4.76	6.51	16.8	302.9	366.9	11.2	2.36	9.18	21.649	1.81720	183
CH-59	1.91	1.11	2.22	5.72	7.94	4.13	6.67	20.3	295.3	457.6	10.1	1.90	12.70	24.200	1.82014	207
CH-350-B	2.54	1.03	2.26	4.13	6.19	4.80	6.59	16.9	304.3	375.3	11.3	2.36	9.34	22.029	1.83907	186
CH-234	1.91	1.27	1.91	5.08	7.62	3.81	6.35	19.1	316.5	342.4	9.9	2.18	9.68	21.072	1.84471	184
CH-129-C	1.91	1.91	1.27	3.49	7.30	3.18	6.35	17.1	427.3	161.2	10.2	3.27	4.44	14.489	1.85181	160
CH-24-B	1.27	1.27	4.45	6.35	8.89	5.72	11.43	26.7	295.4	1271.4	12.7	1.45	28.23	40.973	1.87821	390
CH-369-A	2.54	1.11	1.75	4.45	6.67	4.29	5.71	16.8	326.1	294.0	10.7	2.54	7.76	19.711	1.88000	167
CH-240	2.54	0.95	2.06	5.24	7.15	4.60	6.03	18.4	306.2	416.6	10.8	2.18	10.81	23.557	1.89473	192
CH-107-F	2.86	1.27	1.59	3.18	5.72	4.45	5.72	14.6	364.1	203.6	11.4	3.27	5.04	16.470	1.89497	151
CH-12-A	2.54	1.27	1.59	3.81	6.35	4.13	5.72	15.9	351.7	230.7	10.7	2.90	6.05	17.565	1.90263	156
CH-209	2.54	1.11	1.59	4.92	7.14	4.13	5.40	17.5	338.4	289.1	10.4	2.54	7.82	19.847	1.93811	167
CH-272-A	1.27	1.27	2.22	9.53	12.07	3.49	6.99	28.6	316.5	691.1	9.2	1.45	21.17	30.736	1.94451	279
CH-224-F	2.22	1.43	1.43	4.29	7.14	3.65	5.72	17.1	374.0	221.2	10.2	2.86	6.13	17.511	1.97183	159
CH-141-B	3.18	0.64	2.86	6.67	7.94	6.03	6.99	21.6	298.9	861.7	12.7	1.81	19.06	34.579	1.97387	277
CH-12	2.54	1.27	1.59	3.97	6.51	4.13	5.72	16.2	358.7	240.3	10.7	2.90	6.30	18.298	1.98203	159
CH-20	3.18	0.79	1.91	5.87	7.46	5.08	5.40	18.7	324.3	459.1	11.5	2.27	11.19	25.388	1.99717	199

National Arnold C Cores 4 mil Silicon Steel

Part No.	D cm 1	E cm 2	F cm 3	G cm 4	Ht cm 5	Wth cm 6	Lt cm 7	MPL cm 8	Wtfe grams 9	Wtcu grams 10	MLT cm 11	Ac cm sq 12	Wa cm sq 13	Ap cm 4th 14	Kg cm 5th 15	At cm sq 16
CH-120	2.54	0.95	1.91	5.87	7.78	4.45	5.72	19.4	322.0	421.2	10.6	2.18	11.19	24.378	2.00698	194
CH-55	2.86	0.95	1.91	4.92	6.83	4.76	5.72	17.5	326.6	374.1	11.2	2.45	9.38	22.980	2.00807	183
CH-418-B	1.91	0.95	1.91	9.21	11.11	3.81	5.72	26.0	324.6	581.0	9.3	1.63	17.54	28.661	2.01097	241
CH-437-A	1.27	1.27	3.18	8.10	10.64	4.45	8.89	27.6	305.9	975.6	10.7	1.45	25.71	37.313	2.03001	330
CH-411	2.22	1.11	1.59	6.35	8.57	3.81	5.40	20.3	344.6	350.3	9.8	2.22	10.08	22.414	2.03995	185
CH-17	3.18	0.95	1.59	4.92	6.83	4.76	5.08	16.8	349.7	315.6	11.4	2.72	7.82	21.281	2.04104	171
CH-303	2.54	1.43	1.27	3.97	6.83	3.81	5.40	16.2	403.6	188.9	10.5	3.27	5.04	16.466	2.04139	152
CH-257	1.91	1.43	1.91	4.60	7.46	3.81	6.67	18.7	350.2	320.2	10.3	2.45	8.77	21.488	2.05113	185
CH-24	2.22	0.95	2.38	6.35	8.26	4.60	6.67	21.3	309.5	575.2	10.7	1.91	15.12	28.828	2.05510	229
CH-436	1.91	0.95	2.86	7.30	9.21	4.76	7.62	24.1	300.9	802.4	10.8	1.63	20.87	34.103	2.06166	279
CH-132	2.54	0.95	3.02	4.45	6.35	5.56	7.94	18.7	311.4	587.8	12.3	2.18	13.41	29.206	2.06428	237
CH-275	2.22	0.95	1.75	7.94	9.84	3.97	5.40	23.2	337.2	478.1	9.7	1.91	13.86	26.426	2.07749	214
CH-223-A	2.54	1.51	1.23	3.81	6.83	3.77	5.48	16.1	423.8	177.2	10.6	3.45	4.69	16.155	2.09461	153
CH-458	3.81	0.95	1.59	3.97	5.88	5.40	5.08	14.9	372.2	287.7	12.8	3.27	6.30	20.596	2.09754	166
CH-208	2.86	1.59	1.27	2.86	6.03	4.13	5.72	14.6	455.3	148.4	11.5	4.08	3.63	14.826	2.10751	150
CH-66	1.59	1.27	2.54	6.51	9.05	4.13	7.62	23.2	321.0	606.2	10.3	1.82	16.53	30.009	2.11284	250
CH-141-A	2.22	0.95	2.70	6.03	7.94	4.92	7.30	21.3	309.5	648.4	11.2	1.91	16.28	31.046	2.11460	245
CH-136	2.22	1.43	1.59	4.29	7.14	3.81	6.03	17.5	381.0	251.8	10.4	2.86	6.81	19.459	2.13866	169
CH-49-D	1.27	2.54	1.91	3.81	8.89	3.18	8.89	21.6	478.3	289.6	11.2	2.90	7.26	21.072	2.18118	232
CH-10-A	2.22	1.59	1.59	3.65	6.83	3.81	6.35	16.8	408.0	221.1	10.7	3.18	5.80	18.420	2.18305	167
CH-302-A	1.59	1.59	1.91	5.56	8.73	3.49	6.99	21.3	368.4	374.5	10.0	2.27	10.58	24.022	2.19151	208
CH-350	2.54	1.27	1.75	4.13	6.67	4.29	6.03	16.8	372.8	281.1	11.0	2.90	7.21	20.925	2.21528	172
CH-223-D	3.18	1.27	1.27	3.81	6.35	4.45	5.08	15.2	422.0	197.7	11.5	3.63	4.84	17.560	2.21808	153
CH-273-E	1.27	1.91	1.59	6.99	10.80	2.86	6.99	24.8	411.5	372.8	9.5	2.18	11.09	24.152	2.22574	228
CH-331	1.59	1.59	1.27	7.62	10.80	2.86	5.72	24.1	417.9	308.1	9.0	2.27	9.68	21.964	2.22686	202
CH-431-F	2.22	1.11	2.38	5.24	7.46	4.60	6.98	19.7	333.8	488.6	11.0	2.22	12.47	27.727	2.23825	216
CH-369	2.86	1.11	1.75	4.45	6.67	4.60	5.71	16.8	366.9	311.5	11.3	2.86	7.76	22.179	2.24609	175
CH-69-A	2.54	0.95	2.54	5.40	7.30	5.08	6.99	19.7	327.3	564.7	11.6	2.18	13.71	29.870	2.24744	228
CH-381-A	3.18	1.11	0.95	6.35	8.57	4.13	4.13	19.1	461.4	229.7	10.7	3.17	6.05	19.212	2.28513	167
CH-401	3.18	1.27	1.27	3.97	6.51	4.45	5.08	15.6	430.8	206.0	11.5	3.63	5.04	18.293	2.31064	156

National Arnold C Cores 4 mil Silicon Steel																
Part No.	D cm 1	E cm 2	F cm 3	G cm 4	Ht cm 5	Wth cm 6	Lt cm 7	MPL cm 8	Wtfe grams 9	Wtcu grams 10	MLT cm 11	Ac cm sq 12	Wa cm sq 13	Ap cm 4th 14	Kg cm 5th 15	At cm sq 16
CH-357	2.86	1.11	1.59	4.92	7.14	4.45	5.40	17.5	380.7	306.8	11.0	2.86	7.82	22.332	2.31241	175
CH-410-C	3.18	1.27	0.95	5.08	7.62	4.13	4.45	17.1	474.8	189.3	11.0	3.63	4.84	17.569	2.31970	158
CH-60	2.22	1.11	2.22	5.72	7.94	4.45	6.67	20.3	344.6	486.4	10.8	2.22	12.70	28.239	2.33211	216
CH-56	3.18	0.95	1.91	4.92	6.83	5.08	5.72	17.5	362.9	395.2	11.9	2.72	9.38	25.529	2.34569	191
CH-273-A	2.54	0.95	1.59	7.94	9.84	4.13	5.08	22.9	380.1	452.2	10.1	2.18	12.61	27.462	2.37242	211
CH-10	2.22	1.59	1.59	3.97	7.15	3.81	6.35	17.5	423.4	240.3	10.7	3.18	6.30	20.025	2.37320	174
CH-355-D	2.54	1.27	1.75	4.45	6.99	4.29	6.03	17.5	386.8	302.7	11.0	2.90	7.76	22.532	2.38539	178
CH-399-B	1.91	1.27	1.91	6.67	9.21	3.81	6.35	22.2	369.3	449.4	9.9	2.18	12.70	27.659	2.42136	216
CH-353-D	1.91	1.75	1.59	4.45	7.94	3.49	6.67	19.1	435.1	261.1	10.4	2.99	7.06	21.130	2.43209	186
CH-13-A	2.54	1.43	1.59	3.97	6.83	4.13	6.03	16.8	419.5	247.4	11.0	3.27	6.30	20.589	2.43710	170
CH-7-F	2.54	1.59	1.27	3.97	7.15	3.81	5.72	16.8	466.2	194.6	10.9	3.63	5.04	18.298	2.44710	164
CH-112	1.91	1.59	1.91	4.60	7.78	3.81	6.99	19.4	402.4	330.1	10.6	2.72	8.77	23.879	2.45687	197
CH-341	2.54	1.27	1.59	4.92	7.46	4.13	5.72	18.1	400.9	297.9	10.7	2.90	7.82	22.687	2.45744	178
CH-224-A	2.86	1.11	1.43	5.72	7.94	4.29	5.08	18.7	408.4	313.3	10.8	2.86	8.17	23.338	2.47254	180
CH-344	1.91	1.27	3.81	4.45	6.99	5.72	10.16	21.6	358.7	779.3	12.9	2.18	16.94	36.876	2.48208	299
CH-488-J	2.86	1.11	1.27	6.35	8.57	4.13	4.76	19.7	429.2	302.3	10.5	2.86	8.07	23.046	2.49942	180
CH-112-D	1.91	1.75	1.59	4.60	8.10	3.49	6.67	19.4	442.4	270.5	10.4	2.99	7.31	21.886	2.51909	189
CH-336	2.22	0.95	3.33	6.35	8.26	5.56	8.57	23.2	337.2	918.0	12.2	1.91	21.17	40.366	2.52458	303
CH-330-K	2.54	1.27	1.27	6.03	8.57	3.81	5.08	19.7	436.1	278.5	10.2	2.90	7.66	22.244	2.52712	178
CH-27-A	2.54	0.95	2.38	6.35	8.26	4.92	6.67	21.3	353.6	609.3	11.3	2.18	15.12	32.938	2.53290	239
CH-224-L	0.95	3.81	1.43	5.08	12.70	2.38	10.48	28.3	704.6	314.4	12.2	3.27	7.26	23.722	2.54635	333
CH-237	2.86	1.43	1.91	3.02	5.87	4.76	6.67	15.6	436.3	248.7	12.2	3.68	5.75	21.119	2.55075	176
CH-301	2.22	0.95	2.54	7.62	9.53	4.76	6.99	24.1	351.1	753.5	10.9	1.91	19.36	36.903	2.57082	271
CH-212	3.81	0.95	1.59	4.92	6.83	5.40	5.08	16.8	419.6	356.7	12.8	3.27	7.82	25.537	2.60066	186
CH-388-E	3.18	1.11	1.91	4.13	6.35	5.08	6.03	16.5	399.9	340.3	12.2	3.17	7.86	24.965	2.60480	186
CH-395	1.91	1.59	1.91	4.92	8.10	3.81	6.99	20.0	415.6	352.9	10.6	2.72	9.38	25.523	2.62603	204
CH-124-C	3.81	1.27	1.11	3.97	6.51	4.92	4.76	15.2	506.4	199.5	12.7	4.35	4.41	19.203	2.62967	162
CH-107-A	2.86	1.27	1.59	4.45	6.99	4.45	5.72	17.1	427.4	285.1	11.4	3.27	7.06	23.058	2.65295	177
CH-410-B	3.49	1.27	0.95	5.08	7.62	4.45	4.45	17.1	522.3	200.2	11.6	3.99	4.84	19.329	2.65411	165
CH-331-B	1.27	2.54	1.27	6.35	11.43	2.54	7.62	25.4	562.6	293.1	10.2	2.90	8.07	23.413	2.65990	245

National Arnold C Cores 4 mil Silicon Steel																
Part No.	D cm 1	E cm 2	F cm 3	G cm 4	Ht cm 5	Wth cm 6	Lt cm 7	MPL cm 8	Wtfe grams 9	Wtcu grams 10	MLT cm 11	Ac cm sq 12	Wa cm sq 13	Ap cm 4th 14	Kg cm 5th 15	At cm sq 16
CH-119	3.49	0.95	1.75	5.24	7.15	5.24	5.40	17.8	406.5	398.2	12.2	3.00	9.15	27.405	2.68285	195
CH-179-C	2.54	1.59	1.43	3.97	7.15	3.97	6.03	17.1	475.0	224.0	11.1	3.63	5.67	20.589	2.69159	173
CH-24-C	1.27	1.27	6.99	7.62	10.16	8.26	16.51	34.3	379.8	3152.2	16.7	1.45	53.23	77.263	2.69371	705
CH-465	1.43	1.43	3.18	7.14	10.00	4.60	9.21	26.4	369.6	912.1	11.3	1.84	22.68	41.686	2.70985	325
CH-107-J	2.54	1.43	1.59	4.45	7.30	4.13	6.03	17.8	443.2	277.1	11.0	3.27	7.06	23.058	2.72938	180
CH-135	2.22	1.59	1.75	4.29	7.46	3.97	6.67	18.4	446.4	292.0	11.0	3.18	7.48	23.775	2.75402	190
CH-317-B	1.27	1.27	7.62	7.62	10.16	8.89	17.78	35.6	393.9	3644.6	17.7	1.45	58.06	84.287	2.77262	781
CH-230-A	3.18	1.27	1.27	4.76	7.30	4.45	5.08	17.1	474.8	247.2	11.5	3.63	6.05	21.952	2.77289	171
CH-333-B	2.22	1.27	1.11	9.05	11.59	3.33	4.76	25.4	492.4	333.8	9.3	2.54	10.05	25.545	2.78022	208
CH-241	1.91	1.03	3.49	7.62	9.68	5.40	9.05	26.4	355.8	1132.6	12.0	1.77	26.62	47.095	2.78547	349
CH-352-A	3.18	1.27	1.75	3.81	6.35	4.92	6.03	16.2	448.3	289.5	12.2	3.63	6.65	24.141	2.86322	181
CH-67	1.91	1.27	2.54	6.51	9.05	4.45	7.62	23.2	385.1	643.5	10.9	2.18	16.53	35.999	2.86446	260
CH-61	2.54	1.11	2.22	5.72	7.94	4.76	6.67	20.3	393.8	515.0	11.4	2.54	12.70	32.266	2.87533	225
CH-224	2.86	1.43	1.43	4.29	7.14	4.29	5.72	17.1	480.9	248.8	11.4	3.68	6.13	22.512	2.89696	176
CH-388	3.81	0.95	1.91	4.76	6.67	5.72	5.72	17.1	427.6	430.2	13.3	3.27	9.07	29.651	2.90692	203
CH-13	2.54	1.59	1.59	3.97	7.15	4.13	6.35	17.5	483.8	254.5	11.4	3.63	6.30	22.880	2.92533	182
CH-205-D	2.86	1.59	1.27	3.97	7.15	4.13	5.72	16.8	524.5	206.0	11.5	4.08	5.04	20.589	2.92677	172
CH-16-A	2.86	1.27	1.59	4.92	7.46	4.45	5.72	18.1	451.1	315.6	11.4	3.27	7.82	25.528	2.93705	186
CH-410	2.86	1.11	1.59	6.35	8.57	4.45	5.40	20.3	443.1	395.8	11.0	2.86	10.08	28.817	2.98391	203
CH-411-B	2.54	1.27	1.59	6.03	8.57	4.13	5.72	20.3	450.2	365.2	10.7	2.90	9.58	27.814	3.01274	200
CH-57-B	2.22	1.43	2.22	4.76	7.62	4.45	7.30	19.7	429.5	429.3	11.4	2.86	10.59	30.272	3.03615	219
CH-7-J	2.54	1.83	1.27	3.97	7.62	3.81	6.19	17.8	566.3	203.2	11.3	4.17	5.04	21.041	3.09970	181
CH-312	1.91	1.91	1.59	4.92	8.73	3.49	6.99	20.6	514.3	297.9	10.7	3.27	7.82	25.523	3.11019	208
CH-75	1.59	1.43	2.86	7.30	10.16	4.45	8.57	26.0	405.7	826.0	11.1	2.04	20.87	42.627	3.12907	308
CH-205-E	3.18	1.27	1.91	3.97	6.51	5.08	6.35	16.8	466.0	335.8	12.5	3.63	7.56	27.439	3.18929	194
CH-331-A	1.27	2.54	1.27	7.62	12.70	2.54	7.62	27.9	618.9	351.8	10.2	2.90	9.68	28.096	3.19188	271
CH-107-B	2.86	1.43	1.59	4.45	7.30	4.45	6.03	17.8	498.7	293.1	11.7	3.68	7.06	25.945	3.26733	189
CH-28-B	3.81	1.27	1.91	3.18	5.72	5.72	6.35	15.2	506.4	300.4	14.0	4.35	6.05	26.340	3.28506	191
CH-224-B	2.54	1.59	1.43	4.92	8.10	3.97	6.03	19.1	527.7	277.8	11.1	3.63	7.03	25.528	3.33720	192
CH-388-F	2.54	1.59	1.91	3.97	7.15	4.45	6.99	18.1	501.3	318.7	11.9	3.63	7.56	27.448	3.36197	201

National Arnold C Cores 4 mil Silicon Steel																
Part No.	D cm 1	E cm 2	F cm 3	G cm 4	Ht cm 5	Wth cm 6	Lt cm 7	MPL cm 8	Wtfe grams 9	Wtcu grams 10	MLT cm 11	Ac cm sq 12	Wa cm sq 13	Ap cm 4th 14	Kg cm 5th 15	At cm sq 16
CH-232	2.54	1.27	1.91	5.87	8.41	4.45	6.35	20.6	457.2	446.4	11.2	2.90	11.19	32.487	3.36280	219
CH-370	2.86	1.27	1.91	4.92	7.46	4.76	6.35	18.7	466.9	395.2	11.9	3.27	9.38	30.624	3.37543	207
CH-332	2.22	1.11	2.54	7.62	9.84	4.76	7.30	24.8	420.0	775.2	11.3	2.22	19.36	43.021	3.39591	285
CH-479	3.18	1.27	1.59	4.92	7.46	4.76	5.72	18.1	501.1	333.2	12.0	3.63	7.82	28.359	3.43307	195
CH-62	2.86	1.11	2.22	5.72	7.94	5.08	6.67	20.3	443.1	543.8	12.0	2.86	12.70	36.306	3.44800	235
CH-429	2.54	1.27	3.02	4.45	6.99	5.56	8.57	20.0	443.1	618.0	13.0	2.90	13.41	38.921	3.48669	262
CH-7-E	2.54	1.91	1.91	3.02	6.83	4.45	7.62	17.5	580.2	255.2	12.5	4.35	5.75	25.021	3.48985	203
CH-162	3.81	1.27	1.59	3.97	6.51	5.40	5.72	16.2	538.1	301.9	13.5	4.35	6.30	27.448	3.54972	190
CH-236	3.49	1.11	1.75	5.24	7.46	5.24	5.71	18.4	490.7	408.5	12.6	3.49	9.15	31.948	3.55443	208
CH-26-F	3.18	0.95	2.38	6.35	8.26	5.56	6.67	21.3	442.0	677.5	12.6	2.72	15.12	41.173	3.55882	258
CH-317-F	1.59	1.59	5.08	5.08	8.26	6.67	13.34	26.7	461.9	1370.6	14.9	2.27	25.81	58.569	3.56001	444
CH-22	3.81	0.95	1.91	5.87	7.78	5.72	5.72	19.4	483.0	530.5	13.3	3.27	11.19	36.567	3.58498	229
CH-15	2.54	1.59	1.59	4.92	8.10	4.13	6.35	19.4	536.5	315.6	11.4	3.63	7.82	28.368	3.62700	202
CH-272	2.54	1.27	2.22	5.72	8.26	4.76	6.99	21.0	464.2	529.4	11.7	2.90	12.70	36.884	3.65526	238
CH-68	2.22	1.27	2.54	6.51	9.05	4.76	7.62	23.2	449.4	680.9	11.6	2.54	16.53	42.008	3.68641	271
CH-200-D	4.13	1.27	1.27	4.45	6.99	5.40	5.08	16.5	594.4	273.1	13.6	4.72	5.65	26.636	3.69471	188
CH-330-P	3.18	1.35	1.27	5.72	8.41	4.45	5.24	19.4	569.6	300.7	11.6	3.85	7.26	27.978	3.70300	196
CH-134	2.54	1.27	3.33	4.45	6.99	5.87	9.21	20.6	457.2	709.4	13.5	2.90	14.82	43.025	3.71138	285
CH-230-F	2.54	1.83	1.27	4.76	8.42	3.81	6.19	19.4	616.9	243.8	11.3	4.17	6.05	25.250	3.71979	197
CH-68-A	3.18	0.95	2.54	6.35	8.26	5.72	6.99	21.6	448.6	737.1	12.9	2.72	16.13	43.922	3.72273	270
CH-135-A	2.22	1.91	1.75	4.29	8.10	3.97	7.30	19.7	572.4	308.8	11.6	3.81	7.48	28.522	3.74677	214
CH-411-A	2.22	1.59	1.59	6.35	9.53	3.81	6.35	22.2	538.8	384.5	10.7	3.18	10.08	32.037	3.79688	222
CH-334-A	3.65	1.75	0.95	3.97	7.46	4.60	5.40	16.8	736.6	173.5	12.9	5.74	3.78	21.701	3.86102	186
CH-65-A	1.27	1.27	10.16	10.16	12.70	11.43	22.86	45.7	506.4	7943.1	21.6	1.45	103.23	149.843	4.02074	1290
CH-302	2.22	1.59	1.91	5.87	9.05	4.13	6.99	21.9	531.1	446.5	11.2	3.18	11.19	35.552	4.02650	235
CH-334	3.81	1.75	0.95	3.97	7.46	4.76	5.40	16.8	768.7	180.6	13.4	5.99	3.78	22.646	4.03988	190
CH-63	3.18	1.11	2.22	5.72	7.94	5.40	6.67	20.3	492.2	572.4	12.7	3.17	12.70	40.333	4.04237	244
CH-368-B	2.22	1.27	3.65	5.72	8.26	5.87	9.84	23.8	461.6	988.8	13.3	2.54	20.87	53.017	4.04350	338
CH-28-C	3.81	1.43	1.91	3.18	6.03	5.72	6.67	15.9	593.6	307.2	14.3	4.90	6.05	29.637	4.06652	204
CH-437	1.91	1.27	3.18	8.10	10.64	5.08	8.89	27.6	458.9	1091.6	11.9	2.18	25.71	55.970	4.08181	355

National Arnold C Cores 4 mil Silicon Steel

Part No.	D cm 1	E cm 2	F cm 3	G cm 4	Ht cm 5	Wth cm 6	Lt cm 7	MPL cm 8	Wtfe grams 9	Wtcu grams 10	MLT cm 11	Ac cm sq 12	Wa cm sq 13	Ap cm 4th 14	Kg cm 5th 15	At cm sq 16
CH-224-C	2.86	1.75	1.43	4.29	7.78	4.29	6.35	18.4	631.0	262.6	12.1	4.49	6.13	27.506	4.09743	200
CH-205-A	3.18	1.59	1.59	3.97	7.15	4.76	6.35	17.5	604.7	283.0	12.6	4.54	6.30	28.600	4.11111	199
CH-355	2.54	1.75	1.75	4.45	7.94	4.29	6.98	19.4	589.8	329.0	11.9	3.99	7.76	30.977	4.14855	214
CH-239	2.54	1.59	1.91	4.92	8.10	4.45	6.99	20.0	554.1	395.2	11.9	3.63	9.38	34.031	4.16837	223
CH-352	3.18	1.59	1.75	3.81	6.99	4.92	6.67	17.5	604.7	304.6	12.9	4.54	6.65	30.186	4.25548	205
CH-76	1.91	1.43	2.86	7.30	10.16	4.76	8.57	26.0	486.7	873.1	11.8	2.45	20.87	51.137	4.26032	320
CH-64-B	3.49	1.11	2.22	5.24	7.46	5.72	6.67	19.4	516.1	551.1	13.3	3.49	11.65	40.676	4.27078	241
CH-455-A	3.18	1.27	2.06	5.08	7.62	5.24	6.67	19.4	536.3	475.0	12.7	3.63	10.49	38.051	4.33606	230
CH-389	5.08	0.95	1.91	4.76	6.67	6.99	5.72	17.1	570.1	512.1	15.9	4.36	9.07	39.534	4.34089	235
CH-16	2.86	1.59	1.59	4.92	8.10	4.45	6.35	19.4	603.7	333.3	12.0	4.08	7.82	31.920	4.34851	211
CH-26-H	3.18	1.11	2.38	5.87	8.10	5.56	6.98	21.0	507.6	642.5	12.9	3.17	13.99	44.401	4.36469	260
CH-24-J	2.54	1.27	4.45	4.45	6.99	6.99	11.43	22.9	506.4	1068.4	15.2	2.90	19.76	57.362	4.38056	372
CH-488	1.91	1.43	3.81	6.35	9.21	5.72	10.48	26.0	486.7	1140.6	13.3	2.45	24.19	59.275	4.38157	374
CH-178-D	5.08	1.27	1.27	3.97	6.51	6.35	5.08	15.6	689.3	278.0	15.5	5.81	5.04	29.268	4.38284	200
CH-16-B	3.81	1.27	1.59	4.92	7.46	5.40	5.72	18.1	601.3	374.3	13.5	4.35	7.82	34.031	4.40115	211
CH-35	1.91	1.91	3.02	4.45	8.26	4.92	9.84	22.5	561.8	618.0	13.0	3.27	13.41	43.786	4.41284	295
CH-182	2.86	1.43	2.06	4.92	7.78	4.92	6.99	19.7	552.1	448.7	12.4	3.68	10.16	37.334	4.41864	230
CH-86	1.59	1.59	3.18	8.10	11.27	4.76	9.53	28.9	500.4	1091.8	11.9	2.27	25.71	58.339	4.43388	373
CH-439-A	2.54	1.27	4.13	4.76	7.30	6.67	10.80	22.9	506.4	1028.4	14.7	2.90	19.66	57.082	4.50670	357
CH-330-L	2.54	1.91	1.27	5.40	9.21	3.81	6.35	21.0	696.3	280.1	11.5	4.35	6.86	29.854	4.52530	216
CH-69	2.54	1.27	2.54	6.51	9.05	5.08	7.62	23.2	513.4	718.2	12.2	2.90	16.53	47.999	4.56295	282
CH-399-A	3.81	0.87	2.78	6.67	8.41	6.59	7.30	22.4	511.3	958.0	14.5	2.99	18.52	55.451	4.56545	311
CH-455	2.22	1.91	1.91	4.92	8.73	4.13	7.62	21.3	618.6	395.2	11.9	3.81	9.38	35.729	4.59480	239
CH-63-B	3.18	1.27	2.06	5.40	7.94	5.24	6.67	20.0	553.9	504.7	12.7	3.63	11.14	40.433	4.60750	238
CH-64	3.49	1.11	2.22	5.72	7.94	5.72	6.67	20.3	541.5	601.1	13.3	3.49	12.70	44.372	4.65881	253
CH-232-B	2.86	1.27	1.91	6.83	9.37	4.76	6.35	22.5	561.9	548.2	11.9	3.27	13.00	42.479	4.68212	249
CH-381-K	3.18	1.59	0.95	6.99	10.16	4.13	5.08	22.2	769.6	275.3	11.6	4.54	6.66	30.206	4.71417	217
CH-347	3.18	1.27	1.91	5.87	8.41	5.08	6.35	20.6	571.5	497.0	12.5	3.63	11.19	40.609	4.72005	238
CH-205-B	3.18	1.59	1.91	3.97	7.15	5.08	6.99	18.1	626.7	352.9	13.1	4.54	7.56	34.309	4.74477	219
CH-168-A	3.49	1.27	1.91	5.24	7.78	5.40	6.35	19.4	590.0	465.8	13.1	3.99	9.98	39.846	4.84840	232

National Arnold C Cores 4 mil Silicon Steel

Part No.	D cm 1	E cm 2	F cm 3	G cm 4	Ht cm 5	Wth cm 6	Lt cm 7	MPL cm 8	Wtfe grams 9	Wtcu grams 10	MLT cm 11	Ac cm sq 12	Wa cm sq 13	Ap cm 4th 14	Kg cm 5th 15	At cm sq 16
CH-224-D	5.08	1.27	1.43	3.97	6.51	6.51	5.40	15.9	703.4	317.8	15.8	5.81	5.67	32.932	4.85344	211
CH-141	2.86	1.19	2.86	6.03	8.42	5.72	8.10	22.5	527.0	808.9	13.2	3.06	17.24	52.822	4.90618	298
CH-431-J	2.54	1.27	3.49	5.72	8.26	6.03	9.53	23.5	520.5	973.4	13.7	2.90	19.96	57.956	4.90833	336
CH-412	3.18	1.35	1.59	6.35	9.05	4.76	5.87	21.3	625.6	435.6	12.1	3.85	10.08	38.871	4.93326	231
CH-15-A	2.54	1.91	1.59	4.92	8.73	4.13	6.99	20.6	685.7	333.2	12.0	4.35	7.82	34.031	4.94362	228
CH-397-B	2.54	1.59	2.54	4.76	7.94	5.08	8.26	21.0	580.5	552.9	12.9	3.63	12.10	43.918	4.96208	263
CH-392	2.86	1.43	1.91	5.87	8.73	4.76	6.67	21.3	596.6	484.4	12.2	3.68	11.19	41.131	4.96787	241
CH-49-C	2.86	1.59	1.91	4.92	8.10	4.76	6.99	20.0	623.4	416.4	12.5	4.08	9.38	38.292	5.00872	233
CH-368-A	2.54	1.27	3.65	5.72	8.26	6.19	9.84	23.8	527.5	1035.8	14.0	2.90	20.87	60.577	5.03919	350
CH-18	3.18	1.59	1.59	4.92	8.10	4.76	6.35	19.4	670.6	350.9	12.6	4.54	7.82	35.460	5.09720	220
CH-483-A	2.54	1.27	3.81	5.72	8.26	6.35	10.16	24.1	534.5	1100.2	14.2	2.90	21.77	63.215	5.16626	363
CH-390	2.54	1.27	2.54	7.62	10.16	5.08	7.62	25.4	562.6	840.8	12.2	2.90	19.36	56.191	5.34179	310
CH-397-A	2.54	1.59	2.86	4.76	7.94	5.40	8.89	21.6	598.1	646.3	13.4	3.63	13.61	49.416	5.37453	287
CH-181	1.91	1.91	2.38	6.35	10.16	4.29	8.57	25.1	625.1	643.4	12.0	3.27	15.12	49.382	5.39142	299
CH-154-A	3.49	1.27	1.91	5.87	8.41	5.40	6.35	20.6	628.7	522.3	13.1	3.99	11.19	44.676	5.43606	247
CH-77	2.22	1.43	2.86	7.30	10.16	5.08	8.57	26.0	568.0	920.3	12.4	2.86	20.87	59.673	5.50381	332
CH-399-D	7.30	0.64	2.78	6.11	7.38	10.08	6.83	20.3	647.1	1295.3	21.5	4.17	16.98	70.865	5.51459	373
CH-266-A	3.33	1.27	2.22	5.72	8.26	5.56	6.99	21.0	609.3	601.1	13.3	3.81	12.70	48.414	5.54611	262
CH-391-E	2.86	1.59	2.54	4.45	7.62	5.40	8.26	20.3	633.4	541.5	13.5	4.08	11.29	46.117	5.58644	265
CH-432	3.49	1.43	1.75	5.24	8.10	5.24	6.35	19.7	674.8	429.1	13.2	4.49	9.15	41.093	5.59691	234
CH-153	3.18	1.43	1.91	5.87	8.73	5.08	6.67	21.3	662.8	509.6	12.8	4.08	11.19	45.693	5.82752	251
CH-391	3.81	1.11	2.54	5.87	8.10	6.35	7.30	21.3	618.3	777.0	14.6	3.81	14.92	56.839	5.91395	291
CH-323	3.18	1.59	1.59	5.72	8.89	4.76	6.35	21.0	725.6	407.5	12.6	4.54	9.08	41.182	5.91963	238
CH-107-H	3.81	1.59	1.59	4.45	7.62	5.40	6.35	18.4	765.2	354.0	14.1	5.45	7.06	38.436	5.93527	228
CH-230-E	4.76	1.43	1.27	4.76	7.62	6.03	5.40	17.8	831.1	326.8	15.2	6.13	6.05	37.054	5.97566	225
CH-63-C	3.18	1.43	2.22	5.40	8.26	5.40	7.30	21.0	653.0	567.8	13.3	4.08	12.00	48.999	6.01476	263
CH-26	3.18	1.27	2.38	6.35	8.89	5.56	7.30	22.5	624.2	711.6	13.2	3.63	15.12	54.868	6.01743	286
CH-87	1.91	1.59	3.18	8.10	11.27	5.08	9.53	28.9	600.2	1149.8	12.6	2.72	25.71	69.985	6.05916	386
CH-141-E	1.91	1.91	2.86	6.35	10.16	4.76	9.53	26.0	648.8	820.6	12.7	3.27	18.15	59.275	6.09035	338
CH-24-D	1.59	1.59	6.99	7.62	10.80	8.57	17.15	35.6	615.8	3392.9	17.9	2.27	53.23	120.799	6.11750	761

National Arnold C Cores 4 mil Silicon Steel																
Part No.	D cm 1	E cm 2	F cm 3	G cm 4	Ht cm 5	Wth cm 6	Lt cm 7	MPL cm 8	Wtfe grams 9	Wtcu grams 10	MLT cm 11	Ac cm sq 12	Wa cm sq 13	Ap cm 4th 14	Kg cm 5th 15	At cm sq 16
CH-291-A	8.26	0.52	2.30	10.16	11.19	10.56	5.64	27.0	789.4	1860.7	22.4	3.83	23.39	89.662	6.14568	472
CH-24-F	2.54	1.27	4.45	6.35	8.89	6.99	11.43	26.7	590.8	1526.3	15.2	2.90	28.23	81.946	6.25795	442
CH-71	3.18	1.27	2.54	6.51	9.05	5.72	7.62	23.2	641.8	792.8	13.5	3.63	16.53	59.998	6.45820	303
CH-266	2.86	1.59	2.22	5.72	8.89	5.08	7.62	22.2	692.8	586.9	13.0	4.08	12.70	51.893	6.52700	275
CH-354	3.18	1.43	2.22	5.87	8.73	5.40	7.30	21.9	682.6	617.9	13.3	4.08	13.06	53.320	6.54514	275
CH-18-A	3.81	1.59	1.59	4.92	8.10	5.40	6.35	19.4	804.8	392.0	14.1	5.45	7.82	42.552	6.57085	239
CH-23-E	2.54	1.91	1.91	5.72	9.53	4.45	7.62	22.9	759.6	483.5	12.5	4.35	10.89	47.411	6.61289	269
CH-414-A	4.76	1.43	1.43	4.76	7.62	6.19	5.72	18.1	846.0	373.8	15.4	6.13	6.81	41.693	6.61511	236
CH-271	2.86	1.91	1.75	5.08	8.89	4.60	7.30	21.3	795.3	406.1	12.9	4.90	8.87	43.462	6.61626	252
CH-28-A	3.81	1.27	1.91	6.51	9.05	5.72	6.35	21.9	727.9	615.8	14.0	4.35	12.40	53.998	6.73462	272
CH-144	1.91	1.91	2.70	7.30	11.11	4.60	9.21	27.6	688.4	873.7	12.5	3.27	19.71	64.378	6.74717	351
CH-7-M	3.65	1.91	1.91	3.33	7.14	5.56	7.62	18.1	864.4	332.2	14.7	6.26	6.35	39.757	6.76677	244
CH-317-L	1.91	1.91	5.08	5.08	8.89	6.99	13.97	27.9	696.3	1487.0	16.2	3.27	25.81	84.287	6.79580	490
CH-23-A	2.54	1.91	1.91	5.87	9.68	4.45	7.62	23.2	770.1	497.0	12.5	4.35	11.19	48.730	6.79687	272
CH-141-D	2.86	1.43	2.86	6.03	8.89	5.72	8.57	23.5	659.0	838.1	13.7	3.68	17.24	63.377	6.81696	320
CH-78	2.54	1.43	2.86	7.30	10.16	5.40	8.57	26.0	649.0	967.3	13.0	3.27	20.87	68.182	6.83587	344
CH-26-A	3.18	1.43	2.38	5.87	8.73	5.56	7.62	22.2	692.5	674.1	13.6	4.08	13.99	57.110	6.88204	288
CH-415	3.18	1.59	1.59	6.67	9.84	4.76	6.35	22.9	791.6	475.5	12.6	4.54	10.59	48.049	6.90675	259
CH-333-A	3.18	1.75	1.11	7.62	11.11	4.29	5.71	24.4	930.6	367.1	12.2	4.99	8.47	42.238	6.91249	256
CH-471	3.18	1.43	1.91	6.99	9.84	5.08	6.67	23.5	732.0	606.0	12.8	4.08	13.31	54.335	6.92973	277
CH-205-F	3.33	1.91	1.91	3.97	7.78	5.24	7.62	19.4	844.7	378.5	14.1	5.72	7.56	43.219	7.01999	251
CH-153-A	3.18	1.59	1.91	5.87	9.05	5.08	6.99	21.9	758.6	522.3	13.1	4.54	11.19	50.777	7.02212	265
CH-488-A	3.81	1.59	1.27	6.35	9.53	5.08	5.72	21.6	897.1	390.2	13.6	5.45	8.07	43.913	7.02985	248
CH-390-D	2.54	1.91	2.54	4.92	8.73	5.08	8.89	22.5	749.0	599.4	13.5	4.35	12.50	54.433	7.03093	296
CH-289	2.86	1.75	1.91	5.87	9.37	4.76	7.30	22.5	772.4	509.6	12.8	4.49	11.19	50.255	7.04928	269
CH-413-A	2.86	1.91	1.59	5.87	9.68	4.45	6.99	22.5	842.9	418.8	12.6	4.90	9.33	45.707	7.09477	259
CH-390-C	2.54	1.91	2.54	5.08	8.89	5.08	8.89	22.9	759.6	618.8	13.5	4.35	12.90	56.191	7.25810	300
CH-381-J	3.33	2.22	0.95	5.72	10.16	4.29	6.35	22.2	1131.3	256.0	13.2	6.67	5.45	36.329	7.33317	254
CH-154	3.49	1.51	1.91	5.87	8.89	5.40	6.83	21.6	780.9	541.2	13.6	4.74	11.19	53.048	7.39618	268
CH-438	2.54	1.27	3.33	8.89	11.43	5.87	9.21	29.5	654.1	1418.9	13.5	2.90	29.64	86.049	7.42276	419

National Arnold C Cores 4 mil Silicon Steel

Part No.	D cm 1	E cm 2	F cm 3	G cm 4	Ht cm 5	Wth cm 6	Lt cm 7	MPL cm 8	Wtfe grams 9	Wtcu grams 10	MLT cm 11	Ac cm sq 12	Wa cm sq 13	Ap cm 4th 14	Kg cm 5th 15	At cm sq 16
CH-72	3.49	1.27	2.54	6.51	9.05	6.03	7.62	23.2	706.1	830.2	14.1	3.99	16.53	66.007	7.46462	313
CH-449	3.49	1.43	2.22	5.87	8.73	5.72	7.30	21.9	751.0	647.4	13.9	4.49	13.06	58.661	7.56052	286
CH-390-B	3.18	1.27	2.54	7.62	10.16	5.72	7.62	25.4	703.3	928.2	13.5	3.63	19.36	70.239	7.56052	332
CH-113	4.45	1.27	1.91	5.87	8.41	6.35	6.35	20.6	800.0	606.3	15.2	5.08	11.19	56.852	7.58280	275
CH-128-A	3.18	1.59	3.18	4.45	7.62	6.35	9.53	21.6	747.6	758.7	15.1	4.54	14.11	64.040	7.68835	323
CH-399	2.54	1.91	1.91	6.67	10.48	4.45	7.62	24.8	822.9	564.1	12.5	4.35	12.70	55.317	7.71562	292
CH-414-B	3.49	1.59	1.59	6.51	9.69	5.08	6.35	22.5	858.8	487.5	13.3	4.99	10.34	51.601	7.76893	266
CH-88	2.22	1.59	3.18	8.10	11.27	5.40	9.53	28.9	700.4	1207.9	13.2	3.18	25.71	81.667	7.85381	399
CH-418-C	2.54	1.59	1.91	9.37	12.54	4.45	6.99	28.9	800.3	752.2	11.9	3.63	17.84	64.770	7.93353	325
CH-15-B	2.54	2.54	1.59	4.92	10.00	4.13	8.26	23.2	1026.9	368.5	13.3	5.81	7.82	45.375	7.94698	284
CH-414	3.49	1.59	1.59	6.67	9.84	5.08	6.35	22.9	870.9	499.4	13.3	4.99	10.59	52.861	7.95870	270
CH-273-D	2.54	1.91	1.59	7.94	11.75	4.13	6.99	26.7	886.2	537.5	12.0	4.35	12.61	54.895	7.97449	295
CH-16-D	2.86	1.59	3.81	4.92	8.10	6.67	10.80	23.8	742.2	1032.2	15.5	4.08	18.75	76.583	8.08221	378
CH-19	2.86	1.91	1.91	5.87	9.68	4.76	7.62	23.2	866.6	522.3	13.1	4.90	11.19	54.831	8.18831	283
CH-433-C	1.91	1.27	5.08	12.70	15.24	6.99	12.70	40.6	675.2	3426.1	14.9	2.18	64.52	140.478	8.19305	726
CH-388-A	3.81	1.75	2.22	3.97	7.46	6.03	7.94	19.4	884.8	483.7	15.4	5.99	8.82	52.824	8.20489	273
CH-146	4.45	1.59	1.59	4.92	8.10	6.03	6.35	19.4	938.9	427.3	15.4	6.35	7.82	49.644	8.20491	257
CH-162-E	4.13	1.91	1.59	3.97	7.78	5.72	6.99	18.7	1011.7	344.6	15.4	7.08	6.30	44.608	8.21346	253
CH-79	2.86	1.43	2.86	7.30	10.16	5.72	8.57	26.0	730.2	1014.5	13.7	3.68	20.87	76.719	8.25199	356
CH-228	3.81	1.51	1.91	5.87	8.89	5.72	6.83	21.6	851.8	574.7	14.4	5.17	11.19	57.863	8.28655	278
CH-471-A	3.18	1.59	1.91	6.99	10.16	5.08	6.99	24.1	835.5	621.0	13.1	4.54	13.31	60.381	8.35027	292
CH-73	3.81	1.27	2.54	6.51	9.05	6.35	7.62	23.2	770.1	879.7	15.0	4.35	16.53	71.998	8.38124	324
CH-309	3.49	1.43	2.38	6.35	9.21	5.87	7.62	23.2	794.5	762.9	14.2	4.49	15.12	67.921	8.60106	311
CH-164	3.49	1.83	1.75	5.24	8.89	5.24	7.14	21.3	931.8	455.0	14.0	5.74	9.15	52.509	8.61995	270
CH-465-A	2.86	1.43	3.18	7.14	10.00	6.03	9.21	26.4	739.1	1142.7	14.2	3.68	22.68	83.372	8.65265	380
CH-293-A	2.86	1.91	1.91	6.43	10.24	4.76	7.62	24.3	908.1	571.6	13.1	4.90	12.25	60.012	8.96197	297
CH-389-A	3.18	2.06	1.91	4.76	8.89	5.08	7.94	21.6	971.7	454.2	14.1	5.90	9.07	53.515	8.96854	281
CH-248	3.81	1.59	1.91	5.87	9.05	5.72	6.99	21.9	910.3	581.1	14.6	5.45	11.19	60.932	9.08840	286
CH-165	2.54	1.43	3.33	8.89	11.75	5.87	9.53	30.2	751.8	1452.4	13.8	3.27	29.64	96.822	9.18085	436
CH-263	3.49	1.91	1.75	5.24	9.05	5.24	7.30	21.6	986.5	460.1	14.1	5.99	9.15	54.781	9.27716	277

National Arnold C Cores 4 mil Silicon Steel

Part No.	D cm 1	E cm 2	F cm 3	G cm 4	Ht cm 5	Wth cm 6	Lt cm 7	MPL cm 8	Wtfe grams 9	Wtcu grams 10	MLT cm 11	Ac cm sq 12	Wa cm sq 13	Ap cm 4th 14	Kg cm 5th 15	At cm sq 16
CH-325	2.54	1.98	2.38	6.35	10.32	4.92	8.73	25.4	878.9	720.1	13.4	4.54	15.12	68.573	9.28781	330
CH-388-B	5.08	1.59	1.91	3.97	7.15	6.99	6.99	18.1	1002.7	460.9	17.1	7.26	7.56	54.895	9.29965	273
CH-311-B	3.81	1.11	2.86	8.57	10.80	6.67	7.94	27.3	793.7	1319.6	15.1	3.81	24.50	93.342	9.39176	396
CH-74	4.13	1.27	2.54	6.51	9.05	6.67	7.62	23.2	834.4	917.1	15.6	4.72	16.53	78.007	9.43758	334
CH-355-B	4.45	1.75	1.75	4.45	7.94	6.19	6.98	19.4	1032.1	439.9	15.9	6.98	7.76	54.209	9.50224	271
CH-295-A	3.49	1.91	1.59	5.87	9.68	5.08	6.99	22.5	1030.1	461.0	13.9	5.99	9.33	55.863	9.62921	281
CH-431	3.81	1.59	2.38	5.24	8.42	6.19	7.94	21.6	897.1	680.9	15.4	5.45	12.47	67.924	9.63808	306
CH-80	3.18	1.43	2.86	7.30	10.16	6.03	8.57	26.0	811.2	1061.6	14.3	4.08	20.87	85.228	9.73266	368
CH-293	2.86	1.91	1.91	6.99	10.80	4.76	7.62	25.4	949.6	621.0	13.1	4.90	13.31	65.202	9.73703	311
CH-89	2.54	1.59	3.18	8.10	11.27	5.72	9.53	28.9	800.3	1265.9	13.8	3.63	25.71	93.313	9.78402	412
CH-30	3.81	1.27	2.54	7.62	10.16	6.35	7.62	25.4	844.0	1029.9	15.0	4.35	19.36	84.287	9.81181	355
CH-417-B	3.49	1.75	1.43	7.62	11.11	4.92	6.35	25.1	1050.4	516.1	13.3	5.49	10.89	59.769	9.84471	293
CH-24-E	1.91	1.91	6.99	6.35	10.16	8.89	17.78	34.3	854.5	3027.5	19.2	3.27	44.36	144.868	9.86028	752
CH-419-A	1.27	1.91	1.91	27.94	31.75	3.18	7.62	67.3	1118.3	1883.0	9.9	2.18	53.23	115.894	10.14591	705
CH-163-E	2.94	1.11	5.72	9.21	11.43	8.65	13.65	34.3	768.3	3307.8	17.7	2.94	52.62	154.540	10.26987	695
CH-295	3.18	2.06	1.59	6.35	10.48	4.76	7.30	24.1	1086.0	486.9	13.6	5.90	10.08	59.473	10.33245	296
CH-267	2.86	1.91	2.38	6.35	10.16	5.24	8.57	25.1	937.7	745.8	13.9	4.90	15.12	74.085	10.46763	334
CH-355-A	4.76	1.75	1.75	4.45	7.94	6.51	6.98	19.4	1105.9	457.4	16.6	7.48	7.76	58.088	10.49183	281
CH-106	3.81	1.59	2.38	5.72	8.89	6.19	7.94	22.5	936.6	742.8	15.4	5.45	13.61	74.096	10.51377	319
CH-419-E	3.18	1.59	1.91	8.89	12.07	5.08	6.99	27.9	967.4	790.4	13.1	4.54	16.94	76.848	10.62762	338
CH-82-A	3.81	0.95	2.86	13.02	14.92	6.67	7.62	35.6	886.7	1961.9	14.8	3.27	37.21	121.581	10.71701	506
CH-268	2.54	1.91	2.22	8.26	12.07	4.76	8.26	28.6	949.5	847.5	13.0	4.35	18.35	79.915	10.71798	359
CH-248-B	3.81	1.75	1.91	5.87	9.37	5.72	7.30	22.5	1029.7	593.6	14.9	5.99	11.19	66.995	10.75417	301
CH-390-A	2.54	1.91	2.54	7.62	11.43	5.08	8.89	27.9	928.4	928.2	13.5	4.35	19.36	84.287	10.88715	371
CH-21	3.49	1.91	1.91	5.87	9.68	5.40	7.62	23.2	1059.1	572.8	14.4	5.99	11.19	67.014	11.15203	305
CH-81	3.49	1.43	2.86	7.30	10.16	6.35	8.57	26.0	892.5	1108.8	14.9	4.49	20.87	93.764	11.27838	380
CH-329	2.22	2.22	2.54	7.62	12.07	4.76	9.53	29.2	991.3	928.3	13.5	4.45	19.36	86.082	11.35404	390
CH-204	3.49	1.91	1.91	6.03	9.84	5.40	7.62	23.5	1073.6	588.3	14.4	5.99	11.49	68.828	11.45390	310
CH-417	3.49	1.75	1.75	7.62	11.11	5.24	6.98	25.7	1077.0	654.2	13.8	5.49	13.31	73.027	11.59565	320
CH-322	3.18	2.22	1.59	6.35	10.80	4.76	7.62	24.8	1200.4	498.3	13.9	6.35	10.08	64.055	11.71142	312

National Arnold C Cores 4 mil Silicon Steel																
Part No.	D cm 1	E cm 2	F cm 3	G cm 4	Ht cm 5	Wth cm 6	Lt cm 7	MPL cm 8	Wtfe grams 9	Wtcu grams 10	MLT cm 11	Ac cm sq 12	Wa cm sq 13	Ap cm 4th 14	Kg cm 5th 15	At cm sq 16
CH-90	2.86	1.59	3.18	8.10	11.27	6.03	9.53	28.9	900.5	1324.0	14.5	4.08	25.71	104.995	11.84333	425
CH-37-A	3.81	1.27	3.49	7.62	10.16	7.30	9.53	27.3	907.3	1557.9	16.5	4.35	26.62	115.911	12.26665	446
CH-258	3.81	1.43	3.33	6.35	9.21	7.14	9.53	25.1	937.8	1244.3	16.5	4.90	21.17	103.738	12.30179	405
CH-431-A	3.81	1.59	3.49	5.08	8.26	7.30	10.16	23.5	976.3	1078.7	17.1	5.45	17.74	96.623	12.31015	392
CH-23	3.81	1.91	1.91	5.87	9.68	5.72	7.62	23.2	1155.2	606.3	15.2	6.53	11.19	73.096	12.53484	316
CH-82	3.81	1.43	2.86	7.30	10.16	6.67	8.57	26.0	973.5	1171.3	15.8	4.90	20.87	102.274	12.70243	391
CH-448	1.91	1.91	4.13	10.64	14.45	6.03	12.07	37.1	925.7	2296.5	14.7	3.27	43.91	143.400	12.73682	606
CH-314-B	3.18	3.18	1.43	4.29	10.64	4.60	9.21	24.1	1670.4	338.7	15.6	9.07	6.13	55.567	12.96675	342
CH-406	3.49	1.75	2.54	6.51	10.00	6.03	8.57	25.1	1050.4	886.2	15.1	5.49	16.53	90.747	13.21770	360
CH-35-D	3.81	1.91	3.02	4.45	8.26	6.83	9.84	22.5	1123.5	809.5	17.0	6.53	13.41	87.572	13.47478	364
CH-423	4.45	1.59	1.91	6.99	10.16	6.35	6.99	24.1	1169.7	751.1	15.9	6.35	13.31	84.533	13.53306	336
CH-291-B	3.53	2.30	2.30	4.45	9.05	5.83	9.21	22.7	1267.5	578.2	15.9	7.32	10.23	74.877	13.79261	337
CH-91	3.18	1.59	3.18	8.10	11.27	6.35	9.53	28.9	1000.4	1381.9	15.1	4.54	25.71	116.641	14.00335	439
CH-241-A	2.22	2.22	3.49	7.62	12.07	5.72	11.43	31.1	1056.0	1418.2	15.0	4.45	26.62	118.379	14.05492	482
CH-83	4.13	1.43	2.86	7.30	10.16	6.99	8.57	26.0	1054.7	1218.5	16.4	5.31	20.87	110.810	14.33366	403
CH-106-A	3.81	1.91	2.38	5.72	9.53	6.19	8.57	23.8	1186.8	773.4	16.0	6.53	13.61	88.887	14.53016	351
CH-416	4.13	1.75	1.75	7.62	11.11	5.87	6.98	25.7	1272.8	724.1	15.3	6.49	13.31	86.303	14.63095	343
CH-396	5.08	1.27	2.54	7.62	10.16	7.62	7.62	25.4	1125.3	1204.7	17.5	5.81	19.36	112.382	14.91201	400
CH-439	2.54	1.59	4.13	10.64	13.81	6.67	11.43	35.9	993.8	2395.8	15.3	3.63	43.91	159.384	15.08224	600
CH-419-B	1.27	2.54	1.91	26.67	31.75	3.18	8.89	67.3	1491.0	2026.9	11.2	2.90	50.81	147.502	15.26826	784
CH-449-B	4.45	1.75	2.22	5.87	9.37	6.67	7.94	23.2	1235.3	774.9	16.7	6.98	13.06	91.208	15.27015	348
CH-330-M	5.08	2.22	1.27	5.08	9.53	6.35	6.99	21.6	1674.4	399.6	17.4	10.16	6.45	65.571	15.30639	320
CH-23-D	4.45	1.91	1.91	5.72	9.53	6.35	7.62	22.9	1329.3	639.1	16.5	7.62	10.89	82.970	15.32236	334
CH-317	3.81	1.27	5.08	7.62	10.16	8.89	12.70	30.5	1012.8	2608.7	19.0	4.35	38.71	168.574	15.49442	613
CH-37	4.45	1.27	3.49	7.62	10.16	7.94	9.53	27.3	1058.5	1678.1	17.7	5.08	26.62	135.230	15.50032	471
CH-84	4.45	1.43	2.86	7.30	10.16	7.30	8.57	26.0	1135.7	1265.5	17.1	5.72	20.87	119.319	16.00166	415
CH-248-A	3.81	1.59	1.91	10.48	13.65	5.72	6.99	31.1	1292.9	1036.5	14.6	5.45	19.96	108.690	16.21182	403
CH-92	3.49	1.59	3.18	8.10	11.27	6.67	9.53	28.9	1100.6	1440.1	15.8	4.99	25.71	128.323	16.26469	452
CH-31-B	7.62	1.27	1.27	8.89	11.43	8.89	5.08	25.4	1687.9	842.7	21.0	8.71	11.29	98.335	16.32140	397
CH-85-A	5.72	1.27	2.86	6.51	9.05	8.57	8.26	23.8	1186.9	1274.9	19.3	6.53	18.60	121.518	16.47444	417

National Arnold C Cores 4 mil Silicon Steel																
Part No.	D cm 1	E cm 2	F cm 3	G cm 4	Ht cm 5	Wth cm 6	Lt cm 7	MPL cm 8	Wtfe grams 9	Wtcu grams 10	MLT cm 11	Ac cm sq 12	Wa cm sq 13	Ap cm 4th 14	Kg cm 5th 15	At cm sq 16
CH-321	3.18	1.59	3.18	9.53	12.70	6.35	9.53	31.8	1099.3	1625.9	15.1	4.54	30.24	137.229	16.47503	484
CH-426-A	1.27	1.27	34.29	34.29	36.83	35.56	71.12	142.2	1575.4	248876.4	59.5	1.45	1175.80	1706.809	16.64976	12568
CH-106-B	3.33	2.54	2.06	5.56	10.64	5.40	9.21	25.4	1477.1	636.0	15.6	7.62	11.47	87.400	17.08398	368
CH-393-D	3.81	1.59	2.54	8.89	12.07	6.35	8.26	29.2	1213.7	1252.6	15.6	5.45	22.58	122.957	17.16771	425
CH-128	5.08	1.27	3.18	7.62	10.16	8.26	8.89	26.7	1181.6	1591.7	18.5	5.81	24.19	140.478	17.63556	463
CH-128-B	3.49	1.91	2.06	8.89	12.70	5.56	7.94	29.5	1349.3	955.5	14.6	5.99	18.35	109.887	17.97503	400
CH-143	5.72	1.91	1.27	6.67	10.48	6.99	6.35	23.5	1756.6	543.5	18.0	9.80	8.47	82.976	18.01739	349
CH-329-A	2.22	2.22	6.35	6.99	11.43	8.57	17.15	35.6	1206.8	3070.8	19.5	4.45	44.36	197.270	18.02554	769
CH-337	3.18	1.59	3.49	9.84	13.02	6.67	10.16	33.0	1143.4	1909.5	15.6	4.54	34.38	156.014	18.13150	529
CH-311	3.49	1.75	2.86	8.26	11.75	6.35	9.21	29.2	1223.3	1306.5	15.6	5.49	23.59	129.498	18.25728	443
CH-93	3.81	1.59	3.18	8.10	11.27	6.99	9.53	28.9	1200.5	1517.0	16.6	5.45	25.71	139.969	18.36907	465
CH-106-C	3.49	2.54	2.06	5.56	10.64	5.56	9.21	25.4	1547.5	649.0	15.9	7.99	11.47	91.569	18.37761	375
CH-385-A	2.86	2.54	1.75	8.73	13.81	4.60	8.57	31.1	1551.0	766.8	14.1	6.53	15.24	99.597	18.40075	407
CH-36	3.49	1.59	3.33	8.89	12.07	6.83	9.84	30.8	1173.2	1686.8	16.0	4.99	29.64	147.965	18.46170	494
CH-264	4.45	1.59	2.54	7.62	10.80	6.99	8.26	26.7	1292.8	1161.1	16.9	6.35	19.36	122.957	18.52117	412
CH-25-B	3.49	2.22	2.38	6.35	10.80	5.87	9.21	26.4	1405.2	848.3	15.8	6.99	15.12	105.661	18.71964	392
CH-419	3.81	1.91	1.91	8.89	12.70	5.72	7.62	29.2	1455.9	917.6	15.2	6.53	16.94	110.627	18.97084	397
CH-313	4.76	1.43	2.86	7.94	10.80	7.62	8.57	27.3	1276.3	1426.9	17.7	6.13	22.69	138.972	19.25255	448
CH-391-J	3.81	2.54	2.54	4.45	9.53	6.35	10.16	24.1	1603.6	702.7	17.5	8.71	11.29	98.335	19.57202	392
CH-188	3.81	1.91	2.54	7.62	11.43	6.35	8.89	27.9	1392.6	1117.3	16.2	6.53	19.36	126.430	20.34949	423
CH-394	3.81	2.54	1.91	5.87	10.95	5.72	8.89	25.7	1709.1	656.8	16.5	8.71	11.19	97.461	20.56966	383
CH-465-B	5.08	1.43	3.18	7.14	10.00	8.26	9.21	26.4	1313.7	1517.9	18.8	6.53	22.68	148.192	20.57933	465
CH-94	4.13	1.59	3.18	8.10	11.27	7.30	9.53	28.9	1300.7	1575.2	17.2	5.90	25.71	151.652	20.76754	478
CH-85-B	4.76	1.43	2.86	8.57	11.43	7.62	8.57	28.6	1335.7	1541.0	17.7	6.13	24.50	150.090	20.79266	468
CH-355-E	8.89	1.59	1.67	4.84	8.02	10.56	6.51	19.4	1877.8	711.5	24.8	12.71	8.07	102.555	21.02556	390
CH-25	3.49	2.38	2.38	6.35	11.11	5.87	9.52	27.0	1541.2	865.3	16.1	7.49	15.12	113.171	21.05356	409
CH-294	2.86	1.91	3.97	9.05	12.86	6.83	11.75	33.7	1258.3	2090.1	16.4	4.90	35.92	175.987	21.07737	578
CH-133	3.81	2.54	1.27	8.57	13.65	5.08	7.62	29.8	1983.4	600.5	15.5	8.71	10.89	94.828	21.30051	398
CH-227-A	5.08	1.59	2.54	7.30	10.48	7.62	8.26	26.0	1442.4	1196.5	18.1	7.26	18.55	134.676	21.56135	426
CH-440	3.18	1.59	4.13	10.64	13.81	7.30	11.43	35.9	1242.3	2594.1	16.6	4.54	43.91	199.230	21.76468	632

National Arnold C Cores 4 mil Silicon Steel

Part No.	D cm 1	E cm 2	F cm 3	G cm 4	Ht cm 5	Wth cm 6	Lt cm 7	MPL cm 8	Wtfe grams 9	Wtcu grams 10	MLT cm 11	Ac cm sq 12	Wa cm sq 13	Ap cm 4th 14	Kg cm 5th 15	At cm sq 16
CH-128-C	5.08	1.43	3.18	7.62	10.48	8.26	9.21	27.3	1361.2	1619.0	18.8	6.53	24.19	158.066	21.95052	482
CH-461-E	2.54	2.54	3.49	7.62	12.70	6.03	12.07	32.4	1434.8	1538.2	16.3	5.81	26.62	154.548	22.08647	534
CH-431-E	3.81	2.22	3.49	5.08	9.53	7.30	11.43	26.0	1514.4	1158.9	18.4	7.62	17.74	135.260	22.45543	462
CH-394-A	3.18	3.18	1.91	5.87	12.22	5.08	10.16	28.3	1956.1	648.6	16.3	9.07	11.19	101.522	22.60434	427
CH-95	4.45	1.59	3.18	8.10	11.27	7.62	9.53	28.9	1400.5	1633.1	17.9	6.35	25.71	163.297	23.22513	491
CH-249-E	2.22	2.22	3.81	12.07	16.51	6.03	12.07	40.6	1379.2	2530.7	15.5	4.45	45.97	204.444	23.49289	676
CH-491	2.54	2.54	4.45	6.99	12.07	6.99	13.97	33.0	1462.9	1959.4	17.7	5.81	31.05	180.280	23.59401	611
CH-418-A	4.45	1.91	1.91	9.21	13.02	6.35	7.62	29.8	1735.5	1029.6	16.5	7.62	17.54	133.681	24.68736	432
CH-420	4.13	1.59	3.49	9.21	12.38	7.62	10.16	31.8	1429.4	2028.1	17.7	5.90	32.16	189.756	25.25407	551
CH-441	3.49	1.91	3.33	8.89	12.70	6.83	10.48	32.1	1465.3	1753.6	16.6	5.99	29.64	177.502	25.55574	532
CH-155-A	3.81	1.91	3.49	7.62	11.43	7.30	10.80	29.8	1487.6	1678.1	17.7	6.53	26.62	173.867	25.62297	518
CH-96	4.76	1.59	3.18	8.10	11.27	7.94	9.53	28.9	1500.7	1691.3	18.5	6.81	25.71	174.980	25.75047	505
CH-418	4.76	1.91	1.91	8.89	12.70	6.67	7.62	29.2	1820.0	1032.4	17.1	8.17	16.94	138.298	26.35177	436
CH-97	5.08	1.59	3.18	8.10	11.27	8.26	9.53	28.9	1600.6	1749.2	19.1	7.26	25.71	186.625	28.32171	518
CH-465-C	6.35	1.43	3.18	7.14	10.00	9.53	9.21	26.4	1642.2	1722.8	21.4	8.17	22.68	185.240	28.33128	514
CH-267-B	6.35	1.98	1.75	6.35	10.32	8.10	7.46	24.1	2087.4	789.2	20.0	11.34	11.09	125.712	28.48328	414
CH-84-B	3.49	2.86	2.54	6.03	11.75	6.03	10.80	28.6	1959.1	942.6	17.3	8.98	15.32	137.680	28.60506	468
CH-31-A	5.08	2.54	1.27	7.62	12.70	6.35	7.62	27.9	2475.7	621.1	18.0	11.61	9.68	112.382	28.92170	426
CH-324	3.49	1.91	3.97	9.05	12.86	7.46	11.75	33.7	1537.9	2252.3	17.6	5.99	35.92	215.089	29.21664	609
CH-148	2.86	2.06	4.13	10.64	14.76	6.99	12.38	37.8	1530.5	2643.7	16.9	5.31	43.91	233.095	29.23304	677
CH-491-A	3.81	1.91	4.45	7.62	11.43	8.26	12.70	31.8	1582.5	2315.5	19.2	6.53	33.87	221.253	30.07137	621
CH-419-C	4.13	2.54	1.91	7.62	12.70	6.03	8.89	29.2	2103.2	884.9	17.1	9.44	14.52	136.983	30.16190	448
CH-297	2.54	2.54	2.54	13.97	19.05	5.08	10.16	43.2	1913.0	1861.9	14.8	5.81	35.48	206.035	32.43002	632
CH-358-K	2.54	2.54	5.08	8.89	13.97	7.62	15.24	38.1	1687.9	3010.1	18.7	5.81	45.16	262.226	32.49319	768
CH-426	3.18	2.22	3.49	9.84	14.29	6.67	11.43	35.6	1723.7	2064.7	16.9	6.35	34.38	218.400	32.85935	608
CH-423-B	6.35	2.06	1.91	6.35	10.48	8.26	7.94	24.8	2229.0	878.7	20.4	11.80	12.10	142.690	32.95939	439
CH-147	3.81	1.91	3.49	9.84	13.65	7.30	10.80	34.3	1709.1	2167.7	17.7	6.53	34.38	224.589	33.09802	599
CH-460	4.13	2.06	3.49	7.62	11.75	7.62	11.11	30.5	1783.4	1768.4	18.7	7.67	26.62	204.101	33.50636	552
CH-31	3.81	2.54	2.54	7.62	12.70	6.35	10.16	30.5	2025.5	1204.7	17.5	8.71	19.36	168.574	33.55203	497
CH-451	3.81	2.54	3.33	6.35	11.43	7.14	11.75	29.5	1962.3	1411.6	18.8	8.71	21.17	184.391	34.26034	532

National Arnold C Cores 4 mil Silicon Steel

Part No.	D cm 1	E cm 2	F cm 3	G cm 4	Ht cm 5	Wth cm 6	Lt cm 7	MPL cm 8	Wtfe grams 9	Wtcu grams 10	MLT cm 11	Ac cm sq 12	Wa cm sq 13	Ap cm 4th 14	Kg cm 5th 15	At cm sq 16
CH-317-K	2.54	2.54	7.62	7.62	12.70	10.16	20.32	40.6	1800.5	4693.5	22.7	5.81	58.06	337.147	34.44797	1032
CH-442	3.49	1.91	4.13	10.64	14.45	7.62	12.07	37.1	1697.4	2792.3	17.9	5.99	43.91	262.938	35.21780	690
CH-227	5.72	1.91	2.54	7.30	11.11	8.26	8.89	27.3	2041.4	1322.1	20.0	9.80	18.55	181.756	35.54041	489
CH-32	4.13	2.54	2.54	7.62	12.70	6.67	10.16	30.5	2194.6	1248.5	18.1	9.44	19.36	182.644	38.00564	511
CH-324-A	3.49	1.91	4.76	10.64	14.45	8.26	13.34	38.4	1755.5	3401.5	18.9	5.99	50.66	303.386	38.48975	772
CH-155	5.08	1.91	3.49	7.62	11.43	8.57	10.80	29.8	1983.4	1918.5	20.3	8.71	26.62	231.822	39.84391	575
CH-34-C	5.08	2.22	2.54	7.62	12.07	7.62	9.53	29.2	2265.3	1335.9	19.4	10.16	19.36	196.714	41.20207	514
CH-311-C	5.08	2.06	2.86	7.94	12.07	7.94	9.84	29.8	2149.1	1580.5	19.6	9.44	22.69	214.086	41.24841	538
CH-84-A	4.45	2.86	2.54	6.03	11.75	6.99	10.80	28.6	2493.1	1057.7	19.4	11.43	15.32	175.204	41.28183	511
CH-385-E	4.45	2.54	1.75	10.16	15.24	6.19	8.57	34.0	2633.9	1105.6	17.5	10.16	17.74	180.254	41.80044	521
CH-97-A	5.08	1.59	3.18	12.07	15.24	8.26	9.53	36.8	2040.4	2606.8	19.1	7.26	38.31	278.117	42.20621	659
CH-358	3.49	2.54	3.33	8.89	13.97	6.83	11.75	34.6	2108.5	1887.5	17.9	7.99	29.64	236.669	42.21052	612
CH-33	4.45	2.54	2.54	7.62	12.70	6.99	10.16	30.5	2363.1	1292.1	18.8	10.16	19.36	196.669	42.57871	526
CH-198	4.45	2.22	3.49	7.62	12.07	7.94	11.43	31.1	2111.5	1858.5	19.6	8.89	26.62	236.705	42.88128	587
CH-130	6.51	1.75	3.18	7.46	10.95	9.68	9.84	28.3	2205.1	1913.1	22.7	10.23	23.69	242.294	43.64870	574
CH-422-J	2.54	2.54	8.89	9.21	14.29	11.43	22.86	46.4	2053.7	7197.3	24.7	5.81	81.86	475.310	44.64835	1322
CH-358-H	3.49	2.54	3.65	8.89	13.97	7.14	12.38	35.2	2147.1	2124.4	18.4	7.99	32.46	259.172	44.97407	648
CH-128-D	4.45	2.54	2.06	9.53	14.61	6.51	9.21	33.3	2584.7	1260.2	18.0	10.16	19.66	199.767	45.04221	537
CH-274	5.72	2.14	2.54	7.62	11.91	8.26	9.37	28.9	2429.9	1412.3	20.5	11.02	19.36	213.339	45.83924	531
CH-321-C	6.35	1.59	3.18	9.53	12.70	9.53	9.53	31.8	2198.7	2308.8	21.5	9.08	30.24	274.458	46.40827	625
CH-358-L	5.08	1.91	3.49	8.89	12.70	8.57	10.80	32.4	2152.2	2238.3	20.3	8.71	31.05	270.459	46.48457	624
CH-148-B	2.54	2.22	6.99	13.97	18.42	9.53	18.42	50.8	1969.8	7321.8	21.1	5.08	97.58	495.882	47.77080	1256
CH-443	3.81	2.54	3.33	8.89	13.97	7.14	11.75	34.6	2299.9	1976.2	18.8	8.71	29.64	258.148	47.96447	628
CH-393-E	5.08	2.22	2.54	8.89	13.34	7.62	9.53	31.8	2462.3	1558.5	19.4	10.16	22.58	229.499	48.06908	557
CH-399-C	6.35	2.54	1.91	6.67	11.75	8.26	8.89	27.3	3024.4	965.7	21.4	14.52	12.70	184.391	50.08021	511
CH-422-B	2.54	2.54	8.26	10.80	15.88	10.80	21.59	48.3	2138.1	7519.2	23.7	5.81	89.11	517.428	50.64681	1327
CH-373	5.08	1.98	3.65	8.73	12.70	8.73	11.27	32.7	2263.2	2343.7	20.7	9.07	31.88	289.150	50.74154	647
CH-475-A	3.49	2.54	3.18	11.11	16.19	6.67	11.43	38.7	2360.0	2215.6	17.7	7.99	35.28	281.741	50.95945	674
CH-34-D	5.08	1.98	3.65	8.81	12.78	8.73	11.27	32.9	2274.3	2365.2	20.7	9.07	32.17	291.800	51.20647	650
CH-450	2.54	2.54	8.10	11.11	16.19	10.64	21.27	48.6	2152.2	7511.7	23.5	5.81	89.97	522.410	51.67819	1321

National Arnold C Cores 4 mil Silicon Steel																
Part No.	D cm 1	E cm 2	F cm 3	G cm 4	Ht cm 5	Wth cm 6	Lt cm 7	MPL cm 8	Wtfe grams 9	Wtcu grams 10	MLT cm 11	Ac cm sq 12	Wa cm sq 13	Ap cm 4th 14	Kg cm 5th 15	At cm sq 16
CH-34	5.08	2.54	2.54	7.62	12.70	7.62	10.16	30.5	2700.7	1379.5	20.0	11.61	19.36	224.765	52.08930	555
CH-65-B	2.54	2.54	10.16	10.16	15.24	12.70	25.40	50.8	2250.6	9807.8	26.7	5.81	103.23	599.373	52.10073	1600
CH-358-B	3.49	2.54	4.92	8.89	13.97	8.41	14.92	37.8	2301.9	3173.6	20.4	7.99	43.75	349.325	54.69344	803
CH-167	2.54	3.33	3.81	11.43	18.10	6.35	14.29	43.8	2548.0	2839.7	18.3	7.62	43.55	331.904	55.17851	827
CH-148-A	2.54	2.54	5.08	15.24	20.32	7.62	15.24	50.8	2250.6	5160.2	18.7	5.81	77.42	449.530	55.70261	1058
CH-249-J	3.49	1.91	3.81	18.42	22.23	7.30	11.43	52.1	2379.3	4337.6	17.4	5.99	70.16	420.177	57.89441	942
CH-358-D	3.49	2.54	5.56	8.89	13.97	9.05	16.19	39.1	2379.3	3758.2	21.4	7.99	49.39	394.402	58.87384	885
CH-163	4.45	1.91	4.45	11.91	15.72	8.89	12.70	40.3	2344.6	3856.9	20.5	7.62	52.92	403.317	59.98956	837
CH-36-A	3.49	3.18	3.33	8.89	15.24	6.83	13.02	37.1	2829.1	2021.3	19.2	9.98	29.64	295.837	61.58644	698
CH-358-J	3.81	2.54	5.08	8.89	13.97	8.89	15.24	38.1	2531.9	3451.4	21.5	8.71	45.16	393.339	63.76159	842
CH-461	5.08	2.54	3.49	7.62	12.70	8.57	12.07	32.4	2869.6	2038.7	21.5	11.61	26.62	309.096	66.65727	660
CH-480	5.08	2.54	3.02	8.89	13.97	8.10	11.11	34.0	3010.1	1982.3	20.8	11.61	26.81	311.367	69.56569	655
CH-400	6.35	1.91	3.49	9.84	13.65	9.84	10.80	34.3	2848.6	2788.8	22.8	10.89	34.38	374.315	71.46329	723
CH-427	3.81	3.49	3.49	7.62	14.61	7.30	13.97	36.2	3308.1	1978.7	20.9	11.98	26.62	318.801	73.05914	728
CH-65-C	2.54	2.54	10.16	14.29	19.37	12.70	25.40	59.1	2616.4	13792.7	26.7	5.81	145.17	842.898	73.26922	1915
CH-486-A	2.22	2.22	15.24	20.32	24.77	17.46	34.93	80.0	2715.2	36809.9	33.4	4.45	309.68	1377.305	73.30215	3544
CH-415-B	5.08	3.81	1.59	8.26	15.88	6.67	10.80	34.9	4642.0	983.1	21.1	17.42	13.11	228.349	75.44504	650
CH-34-L	3.81	1.91	7.78	13.97	17.78	11.59	19.37	51.1	2547.8	9451.9	24.5	6.53	108.67	709.876	75.83433	1431
CH-249	4.45	2.54	3.81	10.16	15.24	8.26	12.70	38.1	2953.9	2858.7	20.8	10.16	38.71	393.339	76.98148	771
CH-358-F	5.08	2.54	3.49	8.89	13.97	8.57	12.07	34.9	3094.7	2378.5	21.5	11.61	31.05	360.612	77.76682	713
CH-249-C	5.72	2.06	3.81	10.16	14.29	9.53	11.75	36.2	2931.9	3077.3	22.4	10.62	38.71	410.948	78.05977	769
CH-444	3.81	2.54	4.45	11.91	16.99	8.26	13.97	42.9	2848.4	3856.9	20.5	8.71	52.92	460.934	78.35370	895
CH-241-B	2.22	2.22	20.96	20.96	25.40	23.18	46.36	92.7	3146.2	66205.8	42.4	4.45	439.11	1952.975	81.94435	5309
CH-463	3.81	3.81	4.45	6.35	13.97	8.26	16.51	36.8	3671.3	2312.0	23.0	13.06	28.23	368.755	83.65825	829
CH-473	5.72	2.06	4.13	10.64	14.76	9.84	12.38	37.8	3060.6	3568.3	22.9	10.62	43.91	466.108	86.60330	832
CH-475	5.08	2.54	3.18	10.80	15.88	8.26	11.43	38.1	3375.9	2564.4	21.0	11.61	34.27	398.021	87.87089	749
CH-393	5.72	3.18	2.54	7.62	13.97	8.26	11.43	33.0	4114.4	1554.3	22.6	16.33	19.36	316.076	91.42324	674
CH-34-A	7.62	2.54	2.54	7.62	12.70	10.16	10.16	30.5	4051.1	1756.7	25.5	17.42	19.36	337.147	92.03770	671
CH-422-C	2.54	2.54	8.89	19.69	24.77	11.43	22.86	67.3	2982.0	15386.5	24.7	5.81	175.00	1016.125	95.44990	2040
CH-473-B	6.35	1.83	4.13	12.70	16.35	10.48	11.91	41.0	3261.4	4408.8	23.6	10.44	52.43	547.092	96.56625	921

National Arnold C Cores 4 mil Silicon Steel

Part No.	D cm 1	E cm 2	F cm 3	G cm 4	Ht cm 5	Wth cm 6	Lt cm 7	MPL cm 8	Wtfe grams 9	Wtcu grams 10	MLT cm 11	Ac cm sq 12	Wa cm sq 13	Ap cm 4th 14	Kg cm 5th 15	At cm sq 16
CH-486	3.49	3.49	11.43	5.72	12.70	14.92	29.85	48.3	4043.6	7555.1	32.5	10.98	65.32	717.302	96.86863	1709
CH-481	6.35	2.46	3.33	8.89	13.81	9.68	11.59	34.3	3680.0	2495.0	23.7	14.06	29.64	416.865	99.06976	747
CH-473-D	6.35	1.79	4.76	12.70	16.27	11.11	13.10	42.1	3276.4	5239.5	24.4	10.21	60.49	617.422	103.49030	1011
CH-475-B	5.72	2.86	3.18	9.21	14.92	8.89	12.07	36.2	4060.0	2385.6	22.9	14.70	29.24	429.764	110.12598	767
CH-317-M	4.76	2.06	6.03	14.29	18.42	10.80	16.19	48.9	3301.0	7338.8	23.9	8.85	86.20	762.672	112.73878	1244
CH-473-E	5.72	2.86	4.13	8.57	14.29	9.84	13.97	36.8	4131.4	3076.0	24.4	14.70	35.39	520.228	125.14706	861
CH-488-E	7.94	2.70	3.81	6.35	11.75	11.75	13.02	31.1	4577.9	2449.5	28.5	19.28	24.19	466.504	126.37416	806
CH-311-A	7.62	2.70	2.86	8.57	13.97	10.48	11.11	33.7	4753.5	2295.0	26.3	18.51	24.50	453.519	127.47427	775
CH-249-A	6.35	1.91	3.81	16.51	20.32	10.16	11.43	48.3	4008.9	5122.3	22.9	10.89	62.90	684.831	130.23409	1052
CH-358-C	3.49	5.08	3.49	8.89	19.05	6.99	17.15	45.1	5493.8	2566.0	23.2	15.97	31.05	495.913	136.32359	1017
CH-321-B	7.62	2.54	3.18	9.53	14.61	10.80	11.43	35.6	4726.3	2852.0	26.5	17.42	30.24	526.793	138.40294	831
CH-453	7.62	2.22	3.49	11.27	15.72	11.11	11.43	38.4	4469.1	3694.0	26.4	15.25	39.37	600.203	138.71429	901
CH-163-D	6.83	2.54	5.72	7.62	12.70	12.54	16.51	36.8	4385.0	4478.6	28.9	15.60	43.55	679.538	146.65932	1036
CH-444-A	3.81	3.81	4.45	11.91	19.53	8.26	16.51	47.9	4779.0	4334.9	23.0	13.06	52.92	691.401	156.85593	1104
CH-373-A	5.08	3.81	3.65	8.81	16.43	8.73	14.92	40.2	5338.2	2783.0	24.3	17.42	32.17	560.361	160.49140	927
CH-474	5.08	3.18	4.45	11.43	17.78	9.53	15.24	44.5	4923.2	4391.1	24.3	14.52	50.81	737.510	176.19295	1058
CH-422-E	5.08	2.54	7.62	12.07	17.15	12.70	20.32	49.5	4388.7	9160.2	28.0	11.61	91.94	1067.634	176.99595	1494
CH-421-A	3.81	3.81	8.89	8.89	16.51	12.70	25.40	50.8	5063.8	8434.9	30.0	13.06	79.03	1032.514	179.77723	1632
CH-256	3.81	3.81	3.81	15.24	22.86	7.62	15.24	53.3	5317.0	4550.3	22.0	13.06	58.06	758.582	179.88237	1161
CH-262	3.81	3.81	3.97	17.46	25.08	7.78	15.56	58.1	5791.9	5493.1	22.3	13.06	69.31	905.508	212.31800	1292
CH-317-J	7.62	2.54	6.35	8.89	13.97	13.97	17.78	40.6	5401.4	6324.5	31.5	17.42	56.45	983.347	217.47608	1255
CH-339	7.62	5.08	1.27	10.16	20.32	8.89	12.70	43.2	11478.1	1312.7	28.6	34.84	12.90	449.530	218.95934	1052
CH-321-E	7.62	2.54	3.18	15.24	20.32	10.80	11.43	47.0	6245.4	4563.3	26.5	17.42	48.39	842.869	221.44470	1086
CH-278	6.35	2.86	3.81	13.97	19.69	10.16	13.34	47.0	5856.3	4847.9	25.6	16.33	53.23	869.360	221.75120	1109
CH-481-A	6.35	4.13	3.33	8.89	17.15	9.68	14.92	41.0	7372.9	2846.4	27.0	23.59	29.64	699.235	244.32776	1023
CH-470	5.40	3.97	3.33	12.38	20.32	8.73	14.61	47.3	6960.4	3638.6	24.8	19.28	41.29	796.064	247.73444	1098
CH-476	5.08	3.81	4.45	11.91	19.53	9.53	16.51	47.9	6371.9	4812.9	25.6	17.42	52.92	921.868	251.15992	1196
CH-486-C	3.18	3.18	15.24	20.32	26.67	18.42	36.83	83.8	5802.3	41003.4	37.2	9.07	309.68	2809.562	273.82908	3916
CH-433	5.08	3.81	5.08	11.91	19.53	10.16	17.78	49.2	6540.7	5714.9	26.6	17.42	60.48	1053.564	276.27035	1298
CH-256-A	7.62	2.54	4.21	15.24	20.32	11.83	13.49	49.1	6519.7	6415.9	28.1	17.42	64.12	1116.834	276.52888	1263

National Arnold C Cores 4 mil Silicon Steel																
Part No.	D cm 1	E cm 2	F cm 3	G cm 4	Ht cm 5	Wth cm 6	Lt cm 7	MPL cm 8	Wtfe grams 9	Wtcu grams 10	MLT cm 11	Ac cm sq 12	Wa cm sq 13	Ap cm 4th 14	Kg cm 5th 15	At cm sq 16
CH-185-B	5.08	3.81	6.35	12.07	19.69	11.43	20.32	52.1	6920.6	7782.2	28.6	17.42	76.61	1334.542	325.52295	1521
CH-247	7.94	3.18	3.81	12.38	18.73	11.75	13.97	45.1	7803.0	4936.4	29.4	22.68	47.18	1070.159	329.99566	1195
CH-321-F	7.62	3.49	3.18	13.02	20.00	10.80	13.34	46.4	8473.2	4178.1	28.4	23.96	41.33	990.111	333.74357	1165
CH-422-H	6.35	3.18	5.87	12.70	19.05	12.22	18.10	49.8	6901.3	7716.4	29.1	18.15	74.60	1353.623	337.75442	1449
CH-422	5.08	3.81	7.62	12.07	19.69	12.70	22.86	54.6	7258.2	9990.5	30.6	17.42	91.94	1601.450	365.14038	1748
CH-185-A	5.08	4.76	4.45	12.07	21.59	9.53	18.42	52.1	8652.0	5240.7	27.5	21.78	53.63	1167.847	370.17381	1388
CH-375	7.94	3.18	4.60	12.38	18.73	12.54	15.56	46.7	8077.9	6217.9	30.7	22.68	57.01	1293.179	382.55872	1326
CH-163-C	6.35	3.18	5.72	15.24	21.59	12.07	17.78	54.6	7560.6	8931.7	28.8	18.15	87.10	1580.379	397.74772	1573
CH-163-A	4.45	4.45	5.72	15.24	24.13	10.16	20.32	59.7	8098.6	8538.4	27.6	17.78	87.10	1548.771	399.59436	1681
CH-461-A	7.62	5.08	3.49	7.62	17.78	11.11	17.15	42.5	11309.5	3038.2	32.1	34.84	26.62	927.288	402.56014	1244
CH-156	7.30	3.18	6.19	12.07	18.42	13.49	18.73	49.2	7835.8	8470.9	31.9	20.87	74.69	1558.747	407.98408	1535
CH-185-F	8.89	3.65	2.22	16.19	23.50	11.11	11.75	51.4	11464.3	3813.0	29.8	29.21	36.00	1051.533	412.47348	1270
CH-156-A	8.57	3.33	4.60	11.11	17.78	13.18	15.88	44.8	8787.2	5869.1	32.3	25.72	51.16	1316.157	419.82401	1329
CH-495-B	5.08	3.81	8.89	12.70	20.32	13.97	25.40	58.4	7764.6	13069.6	32.6	17.42	112.90	1966.693	420.95226	2039
CH-433-D	6.35	4.45	5.08	11.91	20.80	11.43	19.05	51.8	10030.9	6534.3	30.4	25.40	60.48	1536.447	513.87198	1523
CH-433-A	7.62	3.81	5.08	11.91	19.53	12.70	17.78	49.2	9811.1	6893.5	32.1	26.13	60.48	1580.346	515.32923	1487
CH-321-H	7.62	3.49	3.18	21.27	28.26	10.80	13.34	62.9	11490.8	6827.5	28.4	23.96	67.54	1617.963	545.37771	1563
CH-422-A	6.35	4.45	5.08	12.70	21.59	11.43	19.05	53.3	10338.7	6922.4	30.2	25.40	64.52	1638.911	551.92021	1571
CH-185-C	6.99	4.45	4.45	13.34	22.23	11.43	17.78	53.3	11372.6	6545.7	31.1	27.94	59.27	1656.325	596.15544	1547
CH-262-A	7.62	3.81	3.97	17.46	25.08	11.59	15.56	58.1	11583.9	7469.8	30.3	26.13	69.31	1811.017	624.53557	1602
CH-452	7.62	3.18	6.35	18.42	24.77	13.97	19.05	62.2	10338.7	13628.8	32.8	21.77	116.94	2546.166	676.61019	2006
CH-495-A	5.08	5.08	8.89	12.70	22.86	13.97	27.94	63.5	11253.0	14089.4	35.1	23.23	112.90	2622.258	694.19444	2336
CH-497	2.54	7.62	6.99	29.21	44.45	9.53	29.21	102.9	13672.4	23140.6	31.9	17.42	204.03	3554.096	776.43524	3574
CH-469	3.81	7.62	7.62	13.97	29.21	11.43	30.48	73.7	14685.2	13491.0	35.6	26.13	106.45	2781.467	815.69143	2681
CH-476-A	7.14	6.03	4.45	11.91	23.97	11.59	20.96	56.8	16820.9	6501.8	34.5	38.79	52.92	2052.839	921.93672	1832
CH-319	4.45	6.51	7.62	16.19	29.21	12.07	28.26	73.7	14635.1	15220.0	34.7	26.04	123.39	3213.001	964.77851	2641
CH-485-E	10.16	2.46	7.94	23.50	28.42	18.10	20.80	72.7	12484.4	25812.4	38.9	22.50	186.50	4196.956	970.64853	2807
CH-185-D	8.26	5.72	4.45	12.07	23.50	12.70	20.32	55.9	18103.3	6891.0	36.1	42.46	53.63	2277.062	1070.25404	1865
CH-489-D	3.81	6.99	7.62	21.59	35.56	11.43	29.21	86.4	15782.3	20106.7	34.4	23.95	164.52	3940.411	1098.40739	3132
CH-485-F	5.08	5.08	7.94	23.50	33.66	13.02	26.04	83.2	14741.6	22282.8	33.6	23.23	186.50	4331.681	1197.74522	2984

National Arnold C Cores 4 mil Silicon Steel																
Part No.	D cm 1	E cm 2	F cm 3	G cm 4	Ht cm 5	Wth cm 6	Lt cm 7	MPL cm 8	Wtfe grams 9	Wtcu grams 10	MLT cm 11	Ac cm sq 12	Wa cm sq 13	Ap cm 4th 14	Kg cm 5th 15	At cm sq 16
CH-477	11.43	5.08	3.18	13.97	24.13	14.61	16.51	54.6	21774.5	6186.1	39.2	52.26	44.36	2317.889	1235.34749	1855
CH-485-H	5.08	5.08	8.26	24.13	34.29	13.34	26.67	85.1	15079.0	24151.5	34.1	23.23	199.19	4626.412	1260.56826	3119
CH-454	6.35	5.08	8.89	16.19	26.35	15.24	27.94	70.5	15613.8	19469.5	38.0	29.03	143.96	4179.353	1276.10072	2766
CH-484-A	10.16	3.81	6.67	16.51	24.13	16.83	20.96	61.6	16373.4	15512.1	39.6	34.84	110.09	3835.340	1348.83367	2326
CH-315-E	4.76	7.62	8.89	15.24	30.48	13.65	33.02	78.7	19624.5	19049.2	39.5	32.66	135.48	4425.525	1462.42589	3205
CH-530	8.89	5.72	3.81	17.78	29.21	12.70	19.05	66.0	23040.5	8770.2	36.4	45.73	67.74	3097.542	1556.12504	2165
CH-315-A	5.08	7.62	8.89	15.24	30.48	13.97	33.02	78.7	20930.6	19354.7	40.2	34.84	135.48	4720.064	1637.31258	3245
CH-422-F	10.16	5.08	7.62	12.07	22.23	17.78	25.40	59.7	21155.6	14273.2	43.7	46.45	91.94	4270.534	1817.45801	2513
CH-393-A	15.24	7.62	2.54	8.89	24.13	17.78	20.32	53.3	42536.3	4089.0	50.9	104.52	22.58	2360.032	1937.49041	2406
CH-265-A	8.89	6.35	3.81	20.32	33.02	12.70	20.32	73.7	28554.5	10372.8	37.7	50.81	77.42	3933.387	2121.58423	2503
CH-485-A	10.16	5.08	6.35	16.19	26.35	16.51	22.86	65.4	23181.5	15234.9	41.7	46.45	102.83	4776.403	2130.02301	2572
CH-486-B	5.08	5.72	11.43	27.94	39.37	16.51	34.29	101.6	20255.4	45823.7	40.4	26.13	319.35	8344.400	2161.33536	4610
CH-487-H	6.35	6.35	8.89	24.13	36.83	15.24	30.48	91.4	25319.2	30950.0	40.6	36.29	214.52	7784.828	2785.21459	3845
CH-485	10.16	6.67	5.40	15.88	29.21	15.56	24.13	69.2	32201.4	13208.9	43.3	60.97	85.69	5224.905	2939.76476	2782
CH-489	6.35	6.35	7.62	30.48	43.18	13.97	27.94	101.6	28132.5	31863.0	38.6	36.29	232.26	8428.686	3171.42450	4013
CH-489-A	3.81	8.57	14.61	31.12	48.26	18.42	46.36	125.7	28201.4	78393.7	48.5	29.40	454.44	13358.930	3238.05439	6957
CH-315	7.62	7.62	8.89	15.24	30.48	16.51	33.02	78.7	31395.9	21994.8	45.7	52.26	135.48	7080.097	3241.74947	3568
CH-485-B	10.16	5.00	7.94	23.50	33.50	18.10	25.88	82.9	28914.5	29181.5	44.0	45.73	186.50	8528.637	3545.46733	3565
CH-503	5.08	8.89	8.89	26.67	44.45	13.97	35.56	106.7	33083.8	36012.2	42.7	40.65	237.10	9636.798	3668.06994	4765
CH-320	8.26	7.30	8.89	17.15	31.75	17.15	32.39	81.3	33649.6	25088.9	46.3	54.26	152.42	8269.898	3877.39820	3736
CH-502	5.56	6.35	13.97	32.39	45.09	19.53	40.64	118.1	28614.7	74907.2	46.6	31.75	452.42	14365.435	3918.64459	6303
CH-487-E	5.08	10.16	11.43	20.32	40.64	16.51	43.18	104.1	36909.8	40668.6	49.2	46.45	232.26	10788.719	4071.00878	5348
CH-500-B	5.08	8.89	10.16	27.31	45.09	15.24	38.10	110.5	34265.4	44103.7	44.7	40.65	277.42	11275.709	4100.47695	5248
CH-487-F	8.89	6.35	8.89	21.59	34.29	17.78	30.48	86.4	33477.7	31159.3	45.7	50.81	191.94	9751.522	4340.88430	3942
CH-526-A	3.81	8.89	13.34	41.43	59.21	17.15	44.45	145.1	33748.6	92642.7	47.2	30.48	552.52	16842.988	4355.60720	7870
CH-486-F	5.08	8.89	11.43	27.94	45.72	16.51	40.64	114.3	35446.9	53034.9	46.7	40.65	319.35	12980.177	4518.78292	5755
CH-513	6.35	9.21	10.16	21.59	40.01	16.51	38.74	100.3	40285.2	37187.8	47.7	52.62	219.35	11543.245	5096.55727	4951
CH-487	10.16	5.08	12.70	25.40	35.56	22.86	35.56	96.5	34209.1	59230.2	51.6	46.45	322.58	14984.331	5392.04002	5265
CH-487-A	8.26	8.57	13.34	13.97	31.12	21.59	43.82	88.9	43204.4	36969.8	55.8	63.69	186.29	11865.385	5416.74223	5048
CH-487-B	7.62	8.89	8.89	20.32	38.10	16.51	35.56	94.0	43717.9	30958.1	48.2	60.97	180.65	11013.484	5573.10562	4510

National Arnold C Cores 4 mil Silicon Steel																
Part No.	D cm 1	E cm 2	F cm 3	G cm 4	Ht cm 5	Wth cm 6	Lt cm 7	MPL cm 8	Wtfe grams 9	Wtcu grams 10	MLT cm 11	Ac cm sq 12	Wa cm sq 13	Ap cm 4th 14	Kg cm 5th 15	At cm sq 16
CH-501	10.16	5.08	20.32	20.32	30.48	30.48	50.80	101.6	36009.6	93380.3	63.6	46.45	412.90	19179.944	5603.52183	7226
CH-493	10.16	6.35	7.62	27.62	40.32	17.78	27.94	95.9	42480.5	34579.9	46.2	58.06	210.49	12221.816	6144.25676	4294
CH-487-D	6.35	10.16	12.70	20.32	40.64	19.05	45.72	106.7	47262.6	49715.1	54.2	58.06	258.06	14984.331	6424.04214	5974
CH-492	7.62	8.26	12.22	21.59	38.10	19.84	40.96	100.6	43475.5	48949.2	52.2	56.61	263.92	14941.030	6486.89437	5451
CH-522	6.35	6.35	33.02	33.02	45.72	39.37	78.74	157.5	43605.4	301836.1	77.8	36.29	1090.32	39568.000	7378.00219	15323
CH-517	4.76	9.53	20.32	38.10	57.15	25.08	59.69	154.9	48269.9	168745.1	61.3	40.83	774.19	31610.892	8422.94613	11136
CH-484-B	11.43	8.89	6.35	20.96	38.74	17.78	30.48	90.2	62918.3	24522.6	51.8	91.45	133.06	12168.916	8589.32197	4371
CH-493-B	6.35	15.24	7.62	24.13	54.61	13.97	45.72	124.5	82709.5	36588.7	56.0	87.10	183.87	16014.504	9970.14877	6868
CH-487-J	7.62	11.43	11.75	20.32	43.18	19.37	46.36	109.9	65704.0	49032.0	57.8	78.39	238.72	18712.480	10157.92879	6397
CH-510	6.35	10.16	10.80	37.78	58.10	17.15	41.91	137.8	61048.0	74236.3	51.2	58.06	407.87	23682.581	10746.41149	7422
CH-486-D	10.16	10.16	8.89	20.32	40.64	19.05	38.10	99.1	70218.7	35853.0	55.8	92.90	180.65	16782.451	11173.97268	5374
CH-487-C	10.16	8.89	12.70	20.32	38.10	22.86	43.18	101.6	63016.8	54376.9	59.3	81.29	258.06	20978.064	11511.67103	6103
CH-487-K	7.62	10.16	13.97	25.40	45.72	21.59	48.26	119.4	63466.9	74079.1	58.7	69.68	354.84	24724.147	11737.30927	7336
CH-459	7.62	12.70	8.57	27.31	52.71	16.19	42.55	122.6	81444.2	46045.2	55.3	87.10	234.09	20388.074	12840.72938	6699
CH-528	6.35	10.16	13.97	40.64	60.96	20.32	48.26	149.9	66392.7	112591.1	55.8	58.06	567.74	32965.529	13728.96834	9187
CH-512	7.62	12.62	10.16	28.26	53.50	17.78	45.56	127.3	84083.8	58855.9	57.6	86.55	287.10	24849.999	14923.96872	7372
CH-486-E	10.16	10.16	8.89	27.94	48.26	19.05	38.10	114.3	81021.6	49297.8	55.8	92.90	248.39	23075.870	15364.21244	6245
CH-523	10.16	10.16	9.76	28.58	48.90	19.92	39.85	117.3	83159.5	56729.0	57.2	92.90	278.98	25917.879	16842.84776	6639
CH-500	11.43	11.43	10.16	30.48	53.34	21.59	43.18	127.0	113936.6	69252.1	62.9	117.58	309.68	36411.925	27231.79969	7845
CH-516	10.16	10.64	9.53	50.17	71.44	19.69	40.32	161.9	120157.3	98145.8	57.8	97.26	477.82	46470.821	31297.58175	9351

National Arnold C Cores 4 mil Silicon Steel																
Part No.	D cm 1	E cm 2	F cm 3	G cm 4	Ht cm 5	Wth cm 6	Lt cm 7	MPL cm 8	Wtfe grams 9	Wtcu grams 10	MLT cm 11	Ac cm sq 12	Wa cm sq 13	Ap cm 4th 14	Kg cm 5th 15	At cm sq 16
CZ-121-B	0.476	0.119	0.238	0.714	0.952	0.714	0.714	2.4	0.9	1.2	2.0	0.051	0.170	0.009	0.000090	3.5
CZ-121-F	0.476	0.318	0.318	1.270	1.906	0.794	1.272	4.4	4.6	3.6	2.5	0.136	0.404	0.055	0.001202	9.1
CZ-121-A	0.635	0.318	0.635	1.270	1.906	1.270	1.906	5.1	7.0	9.5	3.3	0.182	0.806	0.147	0.003218	15.3
CZ-447-B	0.635	0.318	0.318	3.175	3.811	0.953	1.272	8.3	11.5	10.1	2.8	0.182	1.010	0.183	0.004741	18.6
CZ-447-A	0.635	0.476	0.635	1.588	2.540	1.270	2.222	6.4	13.2	13.0	3.6	0.272	1.008	0.274	0.008230	21.0
CZ-121-D	1.270	0.318	0.635	1.270	1.906	1.905	1.906	5.1	14.1	13.7	4.8	0.364	0.806	0.293	0.008914	20.2
CZ-121-E	0.635	0.635	0.635	1.270	2.540	1.270	2.540	6.4	17.6	11.3	3.9	0.363	0.806	0.293	0.010769	22.6
CZ-2-A	1.270	0.318	0.635	1.588	2.224	1.905	1.906	5.7	15.9	17.1	4.8	0.364	1.008	0.367	0.011146	22.6
CZ-121-H	0.953	0.476	0.635	1.270	2.222	1.588	2.222	5.7	17.8	12.2	4.3	0.408	0.806	0.329	0.012613	21.6
CZ-45-E	1.270	0.476	0.476	1.111	2.063	1.746	1.904	5.1	21.1	9.1	4.8	0.544	0.529	0.288	0.012918	19.9
CZ-2-C	0.635	0.635	0.635	1.588	2.858	1.270	2.540	7.0	19.3	14.1	3.9	0.363	1.008	0.366	0.013466	25.0
CZ-2	0.953	0.476	0.635	1.588	2.540	1.588	2.222	6.4	19.8	15.3	4.3	0.408	1.008	0.412	0.015771	24.0
CZ-100-H	0.635	0.635	0.635	2.858	4.128	1.270	2.540	9.5	26.4	25.5	3.9	0.363	1.815	0.659	0.024235	34.7
CZ-4-B	0.953	0.635	0.794	1.270	2.540	1.747	2.858	6.7	27.7	17.3	4.8	0.545	1.008	0.549	0.024769	29.3
CZ-210-B	0.794	0.794	0.635	1.429	3.017	1.429	2.858	7.3	31.6	14.8	4.6	0.567	0.907	0.515	0.025508	30.0
CZ-114	1.270	0.318	1.270	2.223	2.859	2.540	3.176	8.3	22.9	58.0	5.8	0.364	2.823	1.026	0.025821	45.6
CZ-2-B	0.953	0.635	0.635	1.588	2.858	1.588	2.540	7.0	29.0	16.4	4.6	0.545	1.008	0.549	0.026118	28.4
CZ-210	0.794	0.794	0.714	1.429	3.017	1.508	3.016	7.5	32.3	17.1	4.7	0.567	1.020	0.579	0.027925	31.9
CZ-4-E	0.953	0.556	0.794	1.905	3.017	1.747	2.700	7.6	27.7	25.1	4.7	0.477	1.513	0.721	0.029447	32.7
CZ-4-C	0.953	0.635	0.794	1.588	2.858	1.747	2.858	7.3	30.4	21.7	4.8	0.545	1.261	0.687	0.030971	32.3
CZ-215-C	1.270	0.635	0.635	1.270	2.540	1.905	2.540	6.4	35.2	15.5	5.4	0.726	0.806	0.585	0.031382	29.0
CZ-4	0.953	0.476	0.953	2.540	3.492	1.906	2.858	8.9	27.7	41.0	4.8	0.408	2.421	0.988	0.033889	40.5
CZ-4-D	0.953	0.635	0.794	1.905	3.175	1.747	2.858	7.9	33.0	26.0	4.8	0.545	1.513	0.824	0.037153	35.2
CZ-114-A	1.270	0.476	1.270	1.588	2.540	2.540	3.492	7.6	31.6	43.7	6.1	0.544	2.017	1.097	0.039185	43.5
CZ-4-H	1.270	0.635	0.953	1.270	2.540	2.223	3.176	7.0	38.7	25.5	5.9	0.726	1.210	0.878	0.043122	36.7
CZ-215-F	1.270	0.635	0.635	1.905	3.175	1.905	2.540	7.6	42.2	23.3	5.4	0.726	1.210	0.878	0.047073	34.7
CZ-231-F	0.635	0.635	0.635	5.874	7.144	1.270	2.540	15.6	43.1	52.3	3.9	0.363	3.730	1.354	0.049809	57.7
CZ-231-D	0.635	1.111	0.635	2.540	4.762	1.270	3.492	10.8	52.3	28.1	4.9	0.635	1.613	1.024	0.053113	48.6
CZ-1-A	1.270	0.476	0.595	3.810	4.762	1.865	2.142	10.7	44.5	40.6	5.0	0.544	2.267	1.233	0.053319	43.9
CZ-215-B	1.588	0.635	0.635	1.588	2.858	2.223	2.540	7.0	48.4	21.7	6.1	0.908	1.008	0.915	0.054903	35.3

National Arnold C Cores 4 mil Silicon Steel

Part No.	D cm 1	E cm 2	F cm 3	G cm 4	Ht cm 5	Wth cm 6	Lt cm 7	MPL cm 8	Wtfe grams 9	Wtcu grams 10	MLT cm 11	Ac cm sq 12	Wa cm sq 13	Ap cm 4th 14	Kg cm 5th 15	At cm sq 16
CZ-215-A	1.270	0.635	0.635	2.223	3.493	1.905	2.540	8.3	45.7	27.2	5.4	0.726	1.412	1.025	0.054931	37.5
CZ-45	0.953	0.635	0.953	2.540	3.810	1.906	3.176	9.5	39.6	43.7	5.1	0.545	2.421	1.318	0.056536	46.0
CZ-44-C	0.953	0.953	0.635	1.746	3.652	1.588	3.176	8.6	53.5	20.6	5.2	0.817	1.109	0.906	0.056796	40.0
CZ-231-A	0.953	0.794	0.635	2.540	4.128	1.588	2.858	9.5	49.5	28.1	4.9	0.681	1.613	1.098	0.061077	41.7
CZ-231-C	1.270	0.635	0.635	2.540	3.810	1.905	2.540	8.9	49.2	31.1	5.4	0.726	1.613	1.171	0.062764	40.3
CZ-28-I	1.270	0.635	0.794	2.223	3.493	2.064	2.858	8.6	47.5	35.6	5.7	0.726	1.765	1.281	0.065659	42.1
CZ-129	1.588	0.397	0.476	6.350	7.144	2.064	1.746	15.2	66.0	57.2	5.3	0.567	3.023	1.715	0.073091	60.1
CZ-28-M	1.270	0.635	0.794	2.540	3.810	2.064	2.858	9.2	51.0	40.6	5.7	0.726	2.017	1.464	0.075022	45.3
CZ-216	0.953	0.476	1.111	5.080	6.032	2.064	3.174	14.3	44.5	100.6	5.0	0.408	5.644	2.304	0.075104	72.1
CZ-425	1.905	0.556	0.635	2.223	3.335	2.540	2.382	7.9	57.8	32.8	6.5	0.953	1.412	1.346	0.078612	42.1
CZ-215-J	1.270	0.635	1.270	1.905	3.175	2.540	3.810	8.9	49.2	55.2	6.4	0.726	2.419	1.756	0.079508	53.2
CZ-44-A	1.270	0.556	0.953	3.016	4.128	2.223	3.018	10.2	49.3	58.8	5.8	0.636	2.874	1.827	0.080666	52.4
CZ-44	0.953	0.794	0.953	2.540	4.128	1.906	3.494	10.2	52.8	46.5	5.4	0.681	2.421	1.648	0.083186	51.8
CZ-10-A	1.270	0.635	0.953	2.540	3.810	2.223	3.176	9.5	52.8	50.9	5.9	0.726	2.421	1.757	0.086244	50.4
CZ-44-F	1.349	0.556	0.794	3.651	4.763	2.143	2.700	11.1	57.2	58.4	5.7	0.675	2.899	1.957	0.093279	54.2
CZ-402-C	1.270	0.635	0.953	2.858	4.128	2.223	3.176	10.2	56.3	57.3	5.9	0.726	2.724	1.977	0.097042	53.9
CZ-4-F	1.270	1.032	0.953	1.270	3.334	2.223	3.970	8.6	77.2	28.9	6.7	1.180	1.210	1.428	0.100416	51.1
CZ-26	0.953	0.794	1.270	2.540	4.128	2.223	4.128	10.8	56.1	67.6	5.9	0.681	3.226	2.197	0.101498	62.7
CZ-46	1.270	0.635	1.270	2.540	3.810	2.540	3.810	10.2	56.3	73.6	6.4	0.726	3.226	2.341	0.106011	61.3
CZ-100-E	2.540	0.476	0.635	2.858	3.810	3.175	2.222	8.9	73.8	49.3	7.6	1.088	1.815	1.975	0.112549	52.8
CZ-45-D	0.953	0.953	0.953	2.540	4.446	1.906	3.812	10.8	67.3	49.2	5.7	0.817	2.421	1.979	0.113171	58.1
CZ-1	1.270	0.635	1.270	2.858	4.128	2.540	3.810	10.8	59.8	82.8	6.4	0.726	3.630	2.634	0.119283	65.3
CZ-215-D	1.905	0.635	0.794	2.381	3.651	2.699	2.858	8.9	73.8	46.6	6.9	1.089	1.891	2.058	0.129253	52.0
CZ-56	1.270	0.635	1.270	3.334	4.604	2.540	3.810	11.7	65.1	96.5	6.4	0.726	4.234	3.073	0.139150	71.4
CZ-215-E	1.270	0.953	0.635	2.858	4.764	1.905	3.176	10.8	89.7	39.1	6.1	1.089	1.815	1.977	0.142348	55.7
CZ-43-C	1.588	0.635	0.953	3.016	4.286	2.541	3.176	10.5	72.6	66.9	6.6	0.908	2.874	2.609	0.144565	60.3
CZ-231-E	0.635	2.223	0.635	2.540	6.986	1.270	5.716	15.2	147.7	40.8	7.1	1.270	1.613	2.049	0.146231	100.8
CZ-88	0.953	0.953	1.111	3.016	4.922	2.064	4.128	12.1	75.3	71.1	6.0	0.817	3.351	2.739	0.150143	69.5
CZ-45-C	1.905	0.635	0.953	2.540	3.810	2.858	3.176	9.5	79.1	61.8	7.2	1.089	2.421	2.635	0.159746	59.3
CZ-28	1.588	0.953	0.794	1.905	3.811	2.382	3.494	9.2	95.7	37.3	6.9	1.362	1.513	2.060	0.161809	55.4

National Arnold C Cores 4 mil Silicon Steel																
Part No.	D cm 1	E cm 2	F cm 3	G cm 4	Ht cm 5	Wth cm 6	Lt cm 7	MPL cm 8	Wtfe grams 9	Wtcu grams 10	MLT cm 11	Ac cm sq 12	Wa cm sq 13	Ap cm 4th 14	Kg cm 5th 15	At cm sq 16
CZ-456-H	1.588	0.635	0.794	3.969	5.239	2.382	2.858	12.1	83.6	70.6	6.3	0.908	3.151	2.860	0.164784	64.5
CZ-327	0.953	0.953	0.953	3.810	5.716	1.906	3.812	13.3	83.2	73.8	5.7	0.817	3.631	2.968	0.169756	72.6
CZ-28-C	2.064	0.714	0.794	2.381	3.809	2.858	3.016	9.2	93.2	49.8	7.4	1.326	1.891	2.507	0.179510	57.5
CZ-44-B	1.270	0.953	0.953	2.858	4.764	2.223	3.812	11.4	95.0	63.4	6.6	1.089	2.724	2.967	0.197350	67.2
CZ-55	1.588	0.635	1.270	3.334	4.604	2.858	3.810	11.7	81.3	106.1	7.0	0.908	4.234	3.843	0.197926	76.7
CZ-55-B	0.794	1.270	1.270	3.175	5.715	2.064	5.080	14.0	96.7	93.6	6.5	0.908	4.032	3.659	0.203439	89.7
CZ-30-C	1.270	0.635	1.270	4.921	6.191	2.540	3.810	14.9	82.6	142.5	6.4	0.726	6.250	4.536	0.205386	91.5
CZ-53-A	1.270	0.794	1.588	2.858	4.446	2.858	4.764	12.1	83.6	116.7	7.2	0.908	4.539	4.119	0.206832	84.9
CZ-217	1.270	0.953	0.953	3.016	4.922	2.223	3.812	11.8	97.7	66.9	6.6	1.089	2.874	3.131	0.208261	69.1
CZ-230	1.588	0.635	1.588	3.016	4.286	3.176	4.446	11.7	81.3	128.5	7.5	0.908	4.789	4.347	0.209070	85.5
CZ-30-A	1.270	0.714	1.270	4.128	5.556	2.540	3.968	13.7	85.0	122.5	6.6	0.816	5.243	4.278	0.212585	85.2
CZ-36	1.905	0.953	0.794	1.905	3.811	2.699	3.494	9.2	114.8	40.7	7.6	1.634	1.513	2.471	0.213357	60.0
CZ-43-H	1.588	0.794	0.953	3.016	4.604	2.541	3.494	11.1	96.2	70.2	6.9	1.135	2.874	3.262	0.215560	67.3
CZ-3	1.905	0.635	1.270	2.858	4.128	3.175	3.810	10.8	89.7	99.2	7.7	1.089	3.630	3.952	0.224016	75.4
CZ-58-B	1.270	0.794	1.588	3.175	4.763	2.858	4.764	12.7	88.0	129.6	7.2	0.908	5.042	4.576	0.229773	89.8
CZ-14-B	1.270	0.635	1.588	4.921	6.191	2.858	4.446	15.6	86.2	192.1	6.9	0.726	7.815	5.672	0.238261	107.8
CZ-45-A	2.540	0.635	0.953	2.540	3.810	3.493	3.176	9.5	105.5	72.8	8.5	1.452	2.421	3.514	0.241331	68.2
CZ-28-H	1.588	0.953	0.635	3.493	5.399	2.223	3.176	12.1	125.4	52.7	6.7	1.362	2.218	3.021	0.246137	67.6
CZ-25-C	0.953	0.953	2.858	2.858	4.764	3.811	7.622	15.2	95.1	252.9	8.7	0.817	8.168	6.677	0.250709	145.2
CZ-174	0.953	1.270	1.270	2.858	5.398	2.223	5.080	13.3	110.8	88.4	6.8	1.089	3.630	3.954	0.251563	88.5
CZ-44-E	2.064	0.794	0.953	2.381	3.969	3.017	3.494	9.8	110.8	63.1	7.8	1.475	2.269	3.347	0.252487	66.7
CZ-54	1.905	0.635	1.270	3.334	4.604	3.175	3.810	11.7	97.6	115.7	7.7	1.089	4.234	4.610	0.261326	82.1
CZ-457-A	1.588	0.794	0.794	4.286	5.874	2.382	3.176	13.3	115.5	80.1	6.6	1.135	3.403	3.862	0.264847	75.4
CZ-64	1.588	0.635	1.588	3.969	5.239	3.176	4.446	13.7	94.5	169.2	7.5	0.908	6.303	5.720	0.275132	100.0
CZ-44-H	2.223	0.714	0.953	2.858	4.286	3.176	3.334	10.5	114.2	77.3	8.0	1.429	2.724	3.891	0.278657	71.5
CZ-425-A	1.905	0.953	0.635	3.016	4.922	2.540	3.176	11.1	138.6	49.9	7.3	1.634	1.915	3.129	0.279356	67.6
CZ-44-K	1.588	0.953	0.953	2.858	4.764	2.541	3.812	11.4	118.8	69.6	7.2	1.362	2.724	3.710	0.281246	72.6
CZ-416	0.873	0.953	1.429	5.556	7.462	2.302	4.764	17.8	101.6	178.0	6.3	0.749	7.940	5.945	0.282468	116.5
CZ-10-B	1.270	1.270	0.953	2.540	5.080	2.223	4.446	12.1	133.6	61.8	7.2	1.452	2.421	3.514	0.283993	77.8
CZ-53	1.270	0.794	1.588	3.969	5.557	2.858	4.764	14.3	99.0	162.0	7.2	0.908	6.303	5.720	0.287235	101.9

National Arnold C Cores 4 mil Silicon Steel																
Part No.	D cm 1	E cm 2	F cm 3	G cm 4	Ht cm 5	Wth cm 6	Lt cm 7	MPL cm 8	Wtfe grams 9	Wtcu grams 10	MLT cm 11	Ac cm sq 12	Wa cm sq 13	Ap cm 4th 14	Kg cm 5th 15	At cm sq 16
CZ-5-C	2.223	0.556	1.270	3.810	4.922	3.493	3.652	12.4	105.1	140.4	8.2	1.112	4.839	5.383	0.293506	90.2
CZ-28-A	1.588	0.953	0.794	3.493	5.399	2.382	3.494	12.4	128.7	68.4	6.9	1.362	2.773	3.778	0.296692	74.0
CZ-43	1.588	0.953	0.953	3.016	4.922	2.541	3.812	11.8	122.1	73.4	7.2	1.362	2.874	3.915	0.296795	74.6
CZ-78-D	1.270	0.794	2.223	3.334	4.922	3.493	6.034	14.3	99.0	216.8	8.2	0.908	7.411	6.726	0.296827	121.8
CZ-25	1.270	1.032	1.429	2.858	4.922	2.699	4.922	12.7	114.3	108.3	7.5	1.180	4.084	4.817	0.304879	89.8
CZ-28-J	1.588	0.794	1.111	3.810	5.398	2.699	3.810	13.0	112.7	107.1	7.1	1.135	4.233	4.803	0.306389	83.8
CZ-32-B	0.953	1.270	1.429	3.493	6.033	2.382	5.398	14.9	124.0	126.0	7.1	1.089	4.991	5.437	0.333781	105.1
CZ-20-B	0.953	0.953	1.905	4.921	6.827	2.858	5.716	17.5	108.9	240.4	7.2	0.817	9.375	7.663	0.347439	135.2
CZ-43-E	1.905	0.953	0.953	2.699	4.605	2.858	3.812	11.1	138.6	71.5	7.8	1.634	2.572	4.203	0.351235	76.0
CZ-43-J	2.540	0.635	0.953	3.810	5.080	3.493	3.176	12.1	133.6	109.2	8.5	1.452	3.631	5.271	0.361997	85.1
CZ-25-D	1.588	1.032	1.111	2.858	4.922	2.699	4.286	12.1	135.8	85.7	7.6	1.475	3.175	4.683	0.363923	82.7
CZ-43-B	1.588	0.953	0.953	3.810	5.716	2.541	3.812	13.3	138.6	92.8	7.2	1.362	3.631	4.945	0.374930	84.7
CZ-100-F	3.810	0.635	0.635	3.334	4.604	4.445	2.540	10.5	174.1	80.6	10.7	2.177	2.117	4.610	0.375127	83.7
CZ-456-E	2.699	0.714	0.635	4.286	5.714	3.334	2.698	12.7	168.0	81.6	8.4	1.734	2.722	4.720	0.388416	83.3
CZ-361	1.905	0.953	0.953	3.016	4.922	2.858	3.812	11.8	146.5	79.9	7.8	1.634	2.874	4.696	0.392488	80.2
CZ-28-E	2.223	0.794	0.953	3.493	5.081	3.176	3.494	12.1	146.3	96.3	8.1	1.589	3.329	5.288	0.412884	83.9
CZ-5-A	2.540	0.635	1.270	3.810	5.080	3.810	3.810	12.7	140.7	154.0	9.0	1.452	4.839	7.024	0.455589	100.0
CZ-376-D	1.588	0.794	1.111	5.715	7.303	2.699	3.810	16.8	145.7	160.7	7.1	1.135	6.349	7.205	0.459584	108.6
CZ-15-A	1.588	0.953	1.032	4.445	6.351	2.620	3.970	14.8	153.5	119.2	7.3	1.362	4.587	6.248	0.465641	96.6
CZ-28-F	1.905	1.111	0.794	3.493	5.715	2.699	3.810	13.0	189.2	77.8	7.9	1.905	2.773	5.283	0.510380	87.9
CZ-231-H	3.175	0.635	0.635	5.874	7.144	3.810	2.540	15.6	215.4	122.4	9.2	1.815	3.730	6.768	0.532505	106.9
CZ-25-A	1.270	1.270	1.111	4.286	6.826	2.381	4.762	15.9	175.8	125.8	7.4	1.452	4.762	6.912	0.540014	108.9
CZ-254	1.905	0.953	1.270	3.334	5.240	3.175	4.446	13.0	162.3	125.2	8.3	1.634	4.234	6.918	0.543596	98.4
CZ-19	1.905	0.794	1.588	3.969	5.557	3.493	4.764	14.3	148.4	190.5	8.5	1.361	6.303	8.580	0.549707	115.0
CZ-50-A	1.270	0.953	1.905	4.921	6.827	3.175	5.716	17.5	145.1	268.2	8.0	1.089	9.375	10.211	0.553055	143.2
CZ-5-D	1.270	1.270	1.429	3.810	6.350	2.699	5.398	15.6	172.3	153.6	7.9	1.452	5.444	7.903	0.578576	117.5
CZ-209	2.858	0.953	1.349	1.905	3.811	4.207	4.604	10.3	193.0	94.6	10.3	2.451	2.570	6.299	0.596910	95.6
CZ-5	1.905	0.953	1.588	3.175	5.081	3.493	5.082	13.3	166.3	158.1	8.8	1.634	5.042	8.238	0.610641	110.5
CZ-30	2.540	0.714	1.270	4.128	5.556	3.810	3.968	13.7	170.0	169.8	9.1	1.632	5.243	8.557	0.613252	109.4
CZ-456-A	2.858	0.794	0.794	4.286	5.874	3.652	3.176	13.3	207.8	110.8	9.2	2.042	3.403	6.950	0.619948	98.4

National Arnold C Cores 4 mil Silicon Steel

Part No.	D cm 1	E cm 2	F cm 3	G cm 4	Ht cm 5	Wth cm 6	Lt cm 7	MPL cm 8	Wtfe grams 9	Wtcu grams 10	MLT cm 11	Ac cm sq 12	Wa cm sq 13	Ap cm 4th 14	Kg cm 5th 15	At cm sq 16
CZ-456-C	3.175	0.635	0.794	5.874	7.144	3.969	2.858	15.9	219.8	157.1	9.5	1.815	4.664	8.463	0.648298	116.2
CZ-80	2.858	0.794	1.429	2.858	4.446	4.287	4.446	11.8	183.1	147.5	10.2	2.042	4.084	8.341	0.670969	106.7
CZ-11-D	1.270	1.270	1.270	4.921	7.461	2.540	5.080	17.5	193.4	170.7	7.7	1.452	6.250	9.072	0.685722	126.6
CZ-310-A	2.540	0.953	1.270	2.858	4.764	3.810	4.446	12.1	200.6	123.8	9.6	2.179	3.630	7.907	0.718688	102.9
CZ-200-F	1.905	0.953	1.270	4.445	6.351	3.175	4.446	15.2	190.0	167.0	8.3	1.634	5.645	9.224	0.724741	115.4
CZ-50	1.588	1.270	1.588	3.334	5.874	3.176	5.716	14.9	206.7	166.0	8.8	1.815	5.294	9.610	0.791299	124.6
CZ-455	1.905	1.270	0.794	4.763	7.303	2.699	4.128	16.2	269.0	110.3	8.2	2.177	3.782	8.235	0.874152	113.9
CZ-254-A	2.223	1.111	1.270	3.334	5.556	3.493	4.762	13.7	231.5	139.6	9.3	2.223	4.234	9.412	0.902706	113.7
CZ-32-D	1.111	1.588	1.429	5.080	8.256	2.540	6.034	19.4	234.7	207.8	8.0	1.588	7.259	11.527	0.909493	154.1
CZ-49-F	1.905	1.270	1.111	3.969	6.509	3.016	4.762	15.2	253.2	136.5	8.7	2.177	4.410	9.601	0.960959	118.7
CZ-49-A	2.540	0.953	1.270	3.969	5.875	3.810	4.446	14.3	237.5	171.9	9.6	2.179	5.041	10.981	0.998067	121.2
CZ-21-B	2.540	0.794	1.905	4.128	5.716	4.445	5.398	15.2	211.1	287.1	10.3	1.815	7.864	14.274	1.009371	148.4
CZ-48	1.905	1.270	1.270	3.969	6.509	3.175	5.080	15.6	258.5	160.5	9.0	2.177	5.041	10.976	1.067854	126.8
CZ-20	1.905	0.953	1.905	4.921	6.827	3.810	5.716	17.5	217.7	310.5	9.3	1.634	9.375	15.317	1.074715	159.1
CZ-48-A	1.905	1.111	1.588	4.286	6.508	3.493	5.398	16.2	235.3	221.0	9.1	1.905	6.806	12.964	1.081545	139.3
CZ-14-A	2.858	0.794	1.588	4.286	5.874	4.446	4.764	14.9	232.6	251.8	10.4	2.042	6.806	13.900	1.091351	140.5
CZ-58	1.588	1.588	1.588	3.175	6.351	3.176	6.352	15.9	275.0	169.5	9.5	2.270	5.042	11.443	1.098915	141.2
CZ-183-C	2.540	1.111	1.349	3.175	5.397	3.889	4.920	13.5	261.5	152.7	10.0	2.540	4.283	10.878	1.102009	121.4
CZ-13-H	2.223	0.953	1.905	3.969	5.875	4.128	5.716	15.6	226.4	267.5	10.0	1.907	7.561	14.416	1.104907	148.3
CZ-75	3.175	0.794	1.588	3.810	5.398	4.763	4.764	14.0	241.9	237.5	11.0	2.269	6.050	13.727	1.128530	138.1
CZ-43-D	2.858	1.111	0.953	3.651	5.873	3.811	4.128	13.7	297.7	124.2	10.0	2.858	3.479	9.943	1.131809	116.2
CZ-416-A	1.111	1.588	1.429	6.350	9.526	2.540	6.034	21.9	265.4	259.7	8.0	1.588	9.074	14.408	1.136867	175.9
CZ-52	1.588	1.429	1.588	3.969	6.827	3.176	6.034	16.8	262.3	204.7	9.1	2.042	6.303	12.872	1.151134	145.9
CZ-456-D	3.810	0.794	0.794	5.556	7.144	4.604	3.176	15.9	329.8	176.8	11.3	2.723	4.411	12.011	1.160573	135.4
CZ-51-A	2.858	0.794	1.746	4.286	5.874	4.604	5.080	15.2	237.5	283.5	10.7	2.042	7.483	15.283	1.171996	149.5
CZ-2-E	1.905	0.476	1.270	26.194	27.146	3.175	3.492	56.8	353.9	871.1	7.4	0.816	33.266	27.149	1.203501	379.3
CZ-30-B	2.540	1.111	1.349	3.493	5.715	3.889	4.920	14.1	273.8	168.0	10.0	2.540	4.712	11.967	1.212383	127.0
CZ-31	2.858	0.913	1.588	3.969	5.795	4.446	5.002	14.8	264.6	238.5	10.6	2.348	6.303	14.802	1.306388	142.4
CZ-416-B	1.111	1.746	1.429	6.350	9.842	2.540	6.350	22.5	300.3	269.9	8.4	1.746	9.074	15.842	1.322434	187.3
CZ-381-E	3.810	0.794	0.953	5.556	7.144	4.763	3.494	16.2	336.4	216.9	11.5	2.723	5.295	14.416	1.362796	145.2

National Arnold C Cores 4 mil Silicon Steel																
Part No.	D cm 1	E cm 2	F cm 3	G cm 4	Ht cm 5	Wth cm 6	Lt cm 7	MPL cm 8	Wtfe grams 9	Wtcu grams 10	MLT cm 11	Ac cm sq 12	Wa cm sq 13	Ap cm 4th 14	Kg cm 5th 15	At cm sq 16
CZ-52-A	1.588	1.588	1.588	3.969	7.145	3.176	6.352	17.5	302.5	211.9	9.5	2.270	6.303	14.305	1.373731	156.3
CZ-49-D	1.905	1.588	1.111	3.969	7.145	3.016	5.398	16.5	343.0	146.4	9.3	2.723	4.410	12.006	1.400118	138.7
CZ-28-P	2.540	1.270	1.111	3.810	6.350	3.651	4.762	14.9	330.5	150.1	10.0	2.903	4.233	12.289	1.431083	130.3
CZ-11-A	3.175	0.953	0.794	6.191	8.097	3.969	3.494	17.8	369.5	176.7	10.1	2.723	4.916	13.386	1.442193	142.5
CZ-14	2.540	0.953	1.588	4.921	6.827	4.128	5.082	16.8	279.8	280.3	10.1	2.179	7.815	17.024	1.470731	155.6
CZ-456-B	3.810	0.953	0.794	5.239	7.145	4.604	3.494	15.9	395.9	171.4	11.6	3.268	4.160	13.593	1.533273	141.8
CZ-381-B	3.175	0.953	0.953	5.715	7.621	4.128	3.812	17.1	356.3	200.7	10.4	2.723	5.446	14.832	1.559405	145.2
CZ-34	2.858	0.953	1.588	4.445	6.351	4.446	5.082	15.9	297.0	269.2	10.7	2.451	7.059	17.303	1.582176	154.1
CZ-49	2.540	1.270	1.270	3.969	6.509	3.810	5.080	15.6	344.6	183.2	10.2	2.903	5.041	14.634	1.662544	141.5
CZ-43-F	3.334	1.191	0.953	3.810	6.192	4.287	4.288	14.3	389.7	144.0	11.2	3.574	3.631	12.976	1.662948	133.9
CZ-19-A	1.905	1.508	1.588	4.128	7.144	3.493	6.192	17.5	344.5	231.4	9.9	2.586	6.555	16.948	1.765643	162.5
CZ-12	3.175	0.953	1.588	4.286	6.192	4.763	5.082	15.6	323.3	274.9	11.4	2.723	6.806	18.535	1.777670	158.2
CZ-381-D	3.334	0.953	0.953	6.191	8.097	4.287	3.812	18.1	394.9	224.0	10.7	2.860	5.900	16.872	1.807246	156.6
CZ-99	2.223	1.667	1.191	3.651	6.985	3.414	5.716	16.4	416.1	158.6	10.3	3.335	4.348	14.502	1.886086	150.7
CZ-58-A	2.858	1.270	1.588	3.175	5.715	4.446	5.716	14.6	364.1	203.6	11.4	3.267	5.042	16.470	1.894966	150.8
CZ-5-B	2.858	1.270	1.270	3.810	6.350	4.128	5.080	15.2	379.9	186.8	10.9	3.267	4.839	15.807	1.902216	146.0
CZ-73	2.540	1.270	1.588	3.969	6.509	4.128	5.716	16.2	358.7	240.3	10.7	2.903	6.303	18.298	1.982028	159.2
CZ-63	1.905	1.270	2.381	4.763	7.303	4.286	7.302	19.4	321.8	431.4	10.7	2.177	11.341	24.693	2.010732	207.7
CZ-376-C	3.493	0.953	1.111	5.715	7.621	4.604	4.128	17.5	399.2	253.9	11.2	2.996	6.349	19.022	2.027342	162.6
CZ-17-A	3.175	0.953	1.588	4.921	6.827	4.763	5.082	16.8	349.7	315.6	11.4	2.723	7.815	21.281	2.041044	170.7
CZ-15	2.540	0.953	3.016	4.445	6.351	5.556	7.938	18.7	311.4	587.8	12.3	2.179	13.406	29.206	2.064283	236.6
CZ-38-A	3.493	1.270	0.953	4.128	6.668	4.446	4.446	15.2	464.3	162.7	11.6	3.993	3.934	15.706	2.156724	148.4
CZ-69	2.540	1.270	2.540	3.175	5.715	5.080	7.620	16.5	365.7	350.3	12.2	2.903	8.065	23.413	2.225745	196.8
CZ-381-A	3.175	1.111	0.953	6.350	8.572	4.128	4.128	19.1	461.4	229.7	10.7	3.175	6.052	19.212	2.285126	167.2
CZ-49-E	3.175	1.270	1.270	3.969	6.509	4.445	5.080	15.6	430.8	206.0	11.5	3.629	5.041	18.293	2.310643	156.3
CZ-22-B	2.223	1.111	2.223	5.715	7.937	4.446	6.668	20.3	344.6	486.4	10.8	2.223	12.704	28.239	2.332110	216.0
CZ-376-B	3.175	1.111	1.111	5.715	7.937	4.286	4.444	18.1	438.3	246.7	10.9	3.175	6.349	20.157	2.343141	166.8
CZ-381-C	4.128	0.953	0.953	6.191	8.097	5.081	3.812	18.1	489.0	261.7	12.5	3.541	5.900	20.890	2.371651	175.5
CZ-13-D	2.223	1.588	1.588	3.969	7.145	3.811	6.352	17.5	423.4	240.3	10.7	3.177	6.303	20.025	2.373197	173.5
CZ-41	3.810	0.953	1.588	4.921	6.827	5.398	5.082	16.8	419.6	356.7	12.8	3.268	7.815	25.537	2.600659	185.9

National Arnold C Cores 4 mil Silicon Steel

Part No.	D cm 1	E cm 2	F cm 3	G cm 4	Ht cm 5	Wth cm 6	Lt cm 7	MPL cm 8	Wtfe grams 9	Wtcu grams 10	MLT cm 11	Ac cm sq 12	Wa cm sq 13	Ap cm 4th 14	Kg cm 5th 15	At cm sq 16
CZ-352-A	3.175	1.270	1.746	3.810	6.350	4.921	6.032	16.2	448.3	289.5	12.2	3.629	6.652	24.141	2.863224	180.7
CZ-22-D	2.540	1.111	2.223	5.715	7.937	4.763	6.668	20.3	393.8	515.0	11.4	2.540	12.704	32.266	2.875327	225.2
CZ-20-A	3.810	0.953	1.905	4.763	6.669	5.715	5.716	17.1	427.6	430.2	13.3	3.268	9.074	29.651	2.906922	203.3
CZ-13-A	2.540	1.588	1.588	3.969	7.145	4.128	6.352	17.5	483.8	254.5	11.4	3.630	6.303	22.880	2.925333	182.0
CZ-74	2.540	1.270	1.588	6.668	9.208	4.128	5.716	21.6	478.3	403.7	10.7	2.903	10.589	30.742	3.329846	212.3
CZ-93	3.493	0.635	3.334	8.890	10.160	6.827	7.938	27.0	411.1	1485.9	14.1	1.996	29.639	59.167	3.351103	391.5
CZ-17-B	3.175	1.270	1.588	4.921	7.461	4.763	5.716	18.1	501.1	333.2	12.0	3.629	7.815	28.359	3.433069	194.7
CZ-6	2.540	1.270	3.016	4.445	6.985	5.556	8.572	20.0	443.1	618.0	13.0	2.903	13.406	38.921	3.486691	262.0
CZ-13-E	3.810	1.270	1.588	3.969	6.509	5.398	5.716	16.2	538.1	301.9	13.5	4.355	6.303	27.448	3.549715	190.3
CZ-18	2.540	1.588	1.588	4.921	8.097	4.128	6.352	19.4	536.5	315.6	11.4	3.630	7.815	28.368	3.627000	202.0
CZ-59	3.334	1.270	2.223	3.969	6.509	5.557	6.986	17.5	507.8	417.5	13.3	3.811	8.823	33.623	3.851706	218.3
CZ-78-E	3.175	1.111	2.223	5.715	7.937	5.398	6.668	20.3	492.2	572.4	12.7	3.175	12.704	40.333	4.042368	243.8
CZ-18-A	3.175	1.588	1.588	3.969	7.145	4.763	6.352	17.5	604.7	283.0	12.6	4.538	6.303	28.600	4.111113	199.2
CZ-68	2.858	1.588	1.588	4.921	8.097	4.446	6.352	19.4	603.7	333.3	12.0	4.085	7.815	31.920	4.348512	211.2
CZ-49-B	5.080	1.270	1.270	3.969	6.509	6.350	5.080	15.6	689.3	278.0	15.5	5.806	5.041	29.268	4.382841	200.4
CZ-68-C	2.858	1.429	2.064	4.921	7.779	4.922	6.986	19.7	552.1	448.7	12.4	3.676	10.157	37.334	4.418640	230.4
CZ-18-D	3.175	1.588	1.746	3.969	7.145	4.921	6.668	17.8	615.7	317.3	12.9	4.538	6.930	31.446	4.433065	209.1
CZ-9	2.540	1.270	3.651	5.080	7.620	6.191	9.842	22.5	499.3	920.7	14.0	2.903	18.547	53.846	4.479276	329.1
CZ-29	3.810	0.953	2.381	6.350	8.256	6.191	6.668	21.3	530.4	757.0	14.1	3.268	15.119	49.408	4.586760	276.7
CZ-71	3.493	0.794	4.128	6.985	8.573	7.621	9.844	25.4	483.8	1606.0	15.7	2.496	28.834	71.973	4.587916	423.3
CZ-78	3.493	1.111	2.223	5.715	7.937	5.716	6.668	20.3	541.5	601.1	13.3	3.493	12.704	44.372	4.658809	253.1
CZ-16-D	2.223	1.588	3.810	4.445	7.621	6.033	10.796	22.9	554.2	855.9	14.2	3.177	16.935	53.806	4.811446	338.4
CZ-49-C	5.080	1.270	1.429	3.969	6.509	6.509	5.398	15.9	703.4	317.8	15.8	5.806	5.672	32.932	4.853443	210.8
CZ-47-B	3.175	1.349	1.588	6.350	9.048	4.763	5.874	21.3	625.6	435.6	12.1	3.855	10.084	38.871	4.933264	231.3
CZ-18-E	3.175	1.588	2.064	3.969	7.145	5.239	7.304	18.4	637.7	389.6	13.4	4.538	8.192	37.173	5.044839	229.5
CZ-418	2.540	1.746	0.953	8.890	12.382	3.493	5.398	26.7	812.2	321.6	10.7	3.991	8.472	33.815	5.056831	244.6
CZ-68-B	2.858	1.746	1.588	4.921	8.413	4.446	6.668	20.0	685.4	342.1	12.3	4.491	7.815	35.096	5.121927	224.1
CZ-38-B	3.334	1.349	1.588	6.350	9.048	4.922	5.874	21.3	657.0	447.0	12.5	4.048	10.084	40.817	5.300988	236.1
CZ-73-A	5.080	1.270	1.588	3.969	6.509	6.668	5.716	16.2	717.4	358.8	16.0	5.806	6.303	36.597	5.309369	221.3
CZ-10	2.858	1.588	2.540	4.445	7.621	5.398	8.256	20.3	633.4	541.5	13.5	4.085	11.290	46.117	5.586436	265.0

National Arnold C Cores 4 mil Silicon Steel																
Part No.	D cm 1	E cm 2	F cm 3	G cm 4	Ht cm 5	Wth cm 6	Lt cm 7	MPL cm 8	Wtfe grams 9	Wtcu grams 10	MLT cm 11	Ac cm sq 12	Wa cm sq 13	Ap cm 4th 14	Kg cm 5th 15	At cm sq 16
CZ-40-B	3.493	1.429	1.746	5.239	8.097	5.239	6.350	19.7	674.8	429.1	13.2	4.492	9.147	41.093	5.596912	234.2
CZ-40	3.175	1.429	1.905	5.874	8.732	5.080	6.668	21.3	662.8	509.6	12.8	4.083	11.190	45.693	5.827517	251.1
CZ-8	3.810	1.111	2.540	5.874	8.096	6.350	7.302	21.3	618.3	777.0	14.6	3.810	14.920	56.839	5.913950	291.2
CZ-47	3.175	1.270	2.381	6.350	8.890	5.556	7.302	22.5	624.2	711.6	13.2	3.629	15.119	54.868	6.017425	285.6
CZ-41-A	5.080	1.270	1.588	4.921	7.461	6.668	5.716	18.1	801.8	444.9	16.0	5.806	7.815	45.375	6.582869	244.9
CZ-153-A	3.175	1.588	1.905	5.874	9.050	5.080	6.986	21.9	758.6	522.3	13.1	4.538	11.190	50.777	7.022117	265.2
CZ-488-A	3.810	1.588	1.270	6.350	9.526	5.080	5.716	21.6	897.1	390.2	13.6	5.445	8.065	43.913	7.029851	247.6
CZ-417-D	2.858	1.746	1.429	7.620	11.112	4.287	6.350	25.1	859.5	467.0	12.1	4.491	10.889	48.903	7.284752	270.9
CZ-59-A	3.810	1.588	2.223	4.286	7.462	6.033	7.622	19.4	804.8	511.7	15.1	5.445	9.528	51.881	7.482556	267.5
CZ-40-A	2.540	1.905	1.905	6.668	10.478	4.445	7.620	24.8	822.9	564.1	12.5	4.355	12.703	55.317	7.715615	291.5
CZ-47-A	3.810	1.270	2.381	6.350	8.890	6.191	7.302	22.5	749.0	791.1	14.7	4.355	15.119	65.842	7.794706	306.1
CZ-417-C	3.493	1.746	1.111	7.620	11.112	4.604	5.714	24.4	1023.8	386.2	12.8	5.489	8.466	46.468	7.951772	266.4
CZ-18-C	5.080	1.588	1.588	3.969	7.145	6.668	6.352	17.5	967.6	373.1	16.6	7.260	6.303	45.760	7.983936	250.6
CZ-388-A	3.810	1.746	2.223	3.969	7.461	6.033	7.938	19.4	884.8	483.7	15.4	5.987	8.823	52.824	8.204892	273.2
CZ-40-C	3.175	1.588	1.905	6.985	10.161	5.080	6.986	24.1	835.5	621.0	13.1	4.538	13.306	60.381	8.350270	292.0
CZ-9-B	1.588	1.588	9.525	9.525	12.701	11.113	22.226	44.5	769.8	7070.0	21.9	2.270	90.726	205.908	8.530026	1219.9
CZ-35-A	3.493	1.746	1.270	7.620	11.112	4.763	6.032	24.8	1037.1	450.1	13.1	5.489	9.677	53.118	8.916305	279.6
CZ-13-B	5.080	1.588	1.905	3.969	7.145	6.985	6.986	18.1	1002.7	460.9	17.1	7.260	7.561	54.895	9.299647	273.0
CZ-18-F	5.715	1.588	1.588	3.969	7.145	7.303	6.352	17.5	1088.5	401.5	17.9	8.168	6.303	51.480	9.388352	267.7
CZ-62	3.175	1.429	2.858	7.303	10.161	6.033	8.574	26.0	811.2	1061.6	14.3	4.083	20.872	85.228	9.732659	367.6
CZ-16-E	3.810	1.270	2.540	7.620	10.160	6.350	7.620	25.4	844.0	1029.9	15.0	4.355	19.355	84.287	9.811812	354.8
CZ-417-B	3.493	1.746	1.429	7.620	11.112	4.922	6.350	25.1	1050.4	516.1	13.3	5.489	10.889	59.769	9.844712	293.1
CZ-7	3.810	1.588	2.540	5.398	8.574	6.350	8.256	22.2	923.5	760.6	15.6	5.445	13.711	74.659	10.424219	323.0
CZ-70	2.540	2.858	2.064	4.445	10.161	4.604	9.844	24.5	1218.8	477.8	14.6	6.533	9.174	59.940	10.696566	337.1
CZ-60	3.810	1.588	3.651	4.286	7.462	7.461	10.478	22.2	923.4	965.1	17.3	5.445	15.648	85.208	10.700620	377.7
CZ-78-B	3.810	1.588	2.064	6.826	10.002	5.874	7.304	24.1	1002.6	744.1	14.9	5.445	14.089	76.717	11.250529	323.2
CZ-24	3.493	1.429	2.858	7.303	10.161	6.351	8.574	26.0	892.5	1108.8	14.9	4.492	20.872	93.764	11.278384	379.5
CZ-65	3.493	1.746	1.746	7.620	11.112	5.239	6.984	25.7	1077.0	654.2	13.8	5.489	13.305	73.027	11.595647	320.4
CZ-61	3.810	1.588	2.381	6.350	9.526	6.191	7.938	23.8	989.4	825.3	15.4	5.445	15.119	82.329	11.681965	337.2
CZ-79	4.445	1.588	2.064	5.874	9.050	6.509	7.304	22.2	1077.4	695.1	16.1	6.353	12.124	77.021	12.139515	318.9

National Arnold C Cores 4 mil Silicon Steel																
Part No.	D cm 1	E cm 2	F cm 3	G cm 4	Ht cm 5	Wth cm 6	Lt cm 7	MPL cm 8	Wtfe grams 9	Wtcu grams 10	MLT cm 11	Ac cm sq 12	Wa cm sq 13	Ap cm 4th 14	Kg cm 5th 15	At cm sq 16
CZ-7-A	5.080	1.270	2.540	6.350	8.890	7.620	7.620	22.9	1012.8	1003.9	17.5	5.806	16.129	93.652	12.426676	361.3
CZ-153-B	3.810	1.905	1.905	5.874	9.684	5.715	7.620	23.2	1155.2	606.3	15.2	6.532	11.190	73.096	12.534836	316.3
CZ-39	3.810	1.588	2.381	6.826	10.002	6.191	7.938	24.8	1029.0	887.2	15.4	5.445	16.253	88.500	12.557652	350.7
CZ-98	3.810	1.588	4.445	4.445	7.621	8.255	12.066	24.1	1002.6	1306.2	18.6	5.445	19.758	107.587	12.605060	452.5
CZ-61-B	3.810	1.588	2.659	6.350	9.526	6.469	8.494	24.4	1012.5	947.9	15.8	5.445	16.885	91.941	12.685237	361.2
CZ-9-C	1.746	1.905	8.890	8.890	12.700	10.636	21.590	43.2	986.3	6145.5	21.9	2.994	79.032	236.584	12.954833	1142.3
CZ-40-D	4.445	1.588	1.905	6.985	10.161	6.350	6.986	24.1	1169.7	751.1	15.9	6.353	13.306	84.533	13.533060	335.5
CZ-96	3.810	1.588	3.175	6.350	9.526	6.985	9.526	25.4	1055.4	1189.9	16.6	5.445	20.161	109.783	14.407557	407.3
CZ-16-B	3.810	1.588	2.540	7.620	10.796	6.350	8.256	26.7	1108.1	1073.7	15.6	5.445	19.355	105.392	14.715181	388.0
CZ-317	3.810	1.270	5.080	7.620	10.160	8.890	12.700	30.5	1012.8	2608.7	19.0	4.355	38.710	168.574	15.494416	612.9
CZ-248-A	3.810	1.588	1.905	10.478	13.654	5.715	6.986	31.1	1292.9	1036.5	14.6	5.445	19.961	108.690	16.211818	402.5
CZ-16	3.810	1.746	2.540	7.620	11.112	6.350	8.572	27.3	1247.3	1095.4	15.9	5.987	19.355	115.878	17.435871	405.0
CZ-23-B	4.128	1.826	2.858	5.874	9.526	6.986	9.368	24.8	1282.0	1027.5	17.2	6.784	16.788	113.888	17.956202	400.7
CZ-82	3.175	1.905	2.381	10.795	14.605	5.556	8.572	34.0	1411.0	1325.9	14.5	5.444	25.703	139.915	21.001567	471.6
CZ-227-A	5.080	1.588	2.540	7.303	10.479	7.620	8.256	26.0	1442.4	1196.5	18.1	7.260	18.550	134.676	21.561350	426.3
CZ-2-F	1.905	2.540	1.270	26.194	31.274	3.175	7.620	65.1	2162.7	1359.4	11.5	4.355	33.266	144.869	21.959180	697.8
CZ-97	5.080	1.588	3.175	8.096	11.272	8.255	9.526	28.9	1600.6	1749.2	19.1	7.260	25.705	186.625	28.321710	517.8
CZ-42	4.445	1.905	2.381	8.890	12.700	6.826	8.572	30.2	1753.9	1298.7	17.3	7.621	21.167	161.313	28.500048	471.8
CZ-33-A	5.080	1.905	2.540	7.620	11.430	7.620	8.890	27.9	1856.7	1292.1	18.8	8.710	19.355	168.574	31.282316	474.2
CZ-31-A	3.810	2.540	2.540	7.620	12.700	6.350	10.160	30.5	2025.5	1204.7	17.5	8.710	19.355	168.574	33.552025	496.8
CZ-31-C	4.128	2.540	2.540	7.620	12.700	6.668	10.160	30.5	2194.6	1248.5	18.1	9.437	19.355	182.644	38.005637	511.3
CZ-130	6.509	1.746	3.175	7.461	10.953	9.684	9.842	28.3	2205.1	1913.1	22.7	10.228	23.689	242.294	43.648701	574.0
CZ-31-B	5.080	2.540	2.540	7.620	12.700	7.620	10.160	30.5	2700.7	1379.5	20.0	11.613	19.355	224.765	52.089297	554.8
CZ-100	2.540	2.540	5.080	15.240	20.320	7.620	15.240	50.8	2250.6	5160.2	18.7	5.806	77.419	449.530	55.702610	1058.1
CZ-16-C	3.810	2.540	5.080	8.890	13.970	8.890	15.240	38.1	2531.9	3451.4	21.5	8.710	45.161	393.339	63.761587	841.9
CZ-13-C	6.350	2.223	3.493	7.620	12.066	9.843	11.432	31.1	3016.4	2199.5	23.2	12.704	26.617	338.150	73.947928	676.3
CZ-37-B	7.303	2.223	2.858	7.938	12.384	10.161	10.162	30.5	3398.4	1997.1	24.8	14.611	22.687	331.479	78.259284	659.0
CZ-393	5.715	3.175	2.540	7.620	13.970	8.255	11.430	33.0	4114.4	1554.3	22.6	16.331	19.355	316.076	91.423240	674.2
CZ-98-B	7.938	3.493	4.763	12.700	19.686	12.701	16.512	48.9	9310.4	6787.8	31.6	24.955	60.490	1509.512	477.494050	1436.5

National Arnold C Cores 12 mil Silicon Steel																
Part No.	D cm 1	E cm 2	F cm 3	G cm 4	Ht cm 5	Wth cm 6	Lt cm 7	MPL cm 8	Wtfe grams 9	Wtcu grams 10	MLT cm 11	Ac cm sq 12	Wa cm sq 13	Ap cm 4th 14	Kg cm 5th 15	At cm sq 16
CA-98	0.95	0.48	1.11	5.08	6.03	2.06	3.17	14.3	47.0	100.6	5.0	0.431	5.644	2.432	0.083680	72.1
CA-1-A	0.64	0.79	0.79	5.72	7.30	1.43	3.18	16.2	59.2	72.8	4.5	0.479	4.538	2.173	0.092280	71.3
CA-373-A	1.91	0.64	0.64	2.54	3.81	2.54	2.54	8.9	78.0	38.3	6.7	1.149	1.613	1.854	0.127454	48.4
CA-210	0.95	0.64	1.59	3.97	5.24	2.54	4.45	13.7	59.9	136.2	6.1	0.575	6.303	3.623	0.137111	87.7
CA-100	1.27	0.64	0.79	5.24	6.51	2.06	2.86	14.6	85.4	83.8	5.7	0.766	4.160	3.187	0.172410	71.8
CA-1	1.59	0.64	1.27	2.86	4.13	2.86	3.81	10.8	78.9	91.0	7.0	0.958	3.630	3.477	0.189044	70.4
CA-333-A	1.27	0.95	0.95	2.86	4.76	2.22	3.81	11.4	100.3	63.4	6.6	1.150	2.724	3.132	0.219887	67.2
CA-261-A	1.91	0.64	0.95	3.49	4.76	2.86	3.18	11.4	100.2	85.0	7.2	1.149	3.329	3.825	0.244769	70.8
CA-2	1.91	0.64	1.27	2.86	4.13	3.18	3.81	10.8	94.7	99.2	7.7	1.149	3.630	4.171	0.249598	75.4
CA-208-C	1.59	0.64	1.27	3.81	5.08	2.86	3.81	12.7	92.8	121.3	7.0	0.958	4.839	4.635	0.252014	83.1
CA-252-E	0.64	1.91	1.27	2.86	6.67	1.91	6.35	15.9	139.2	96.6	7.5	1.149	3.630	4.171	0.256271	114.9
CA-252	1.59	0.79	1.27	2.86	4.45	2.86	4.13	11.4	104.5	95.1	7.4	1.198	3.630	4.348	0.282806	77.6
CA-373	1.91	0.71	1.11	3.49	4.92	3.02	3.65	12.1	118.9	104.7	7.6	1.292	3.881	5.015	0.341467	81.3
CA-95-B	0.95	0.79	2.22	6.35	7.94	3.18	6.03	20.3	111.5	371.1	7.4	0.719	14.116	10.147	0.394711	170.4
CA-57	1.59	0.79	1.59	3.97	5.56	3.18	4.76	14.3	130.6	176.3	7.9	1.198	6.303	7.550	0.459910	108.4
CA-4-B	1.27	1.27	1.11	3.49	6.03	2.38	4.76	14.3	167.0	102.6	7.4	1.532	3.881	5.946	0.490358	97.6
CA-232-A	2.54	0.79	1.27	2.54	4.13	3.81	4.13	10.8	157.8	106.3	9.3	1.916	3.226	6.180	0.510950	88.9
CA-2-A	1.91	0.79	1.27	3.97	5.56	3.18	4.13	13.7	149.7	143.4	8.0	1.437	5.041	7.243	0.520401	99.3
CA-279-A	1.59	0.95	1.11	4.45	6.35	2.70	4.13	14.9	163.7	130.6	7.4	1.438	4.938	7.100	0.549214	100.5
CA-207	1.59	0.79	1.59	4.92	6.51	3.18	4.76	16.2	148.0	218.6	7.9	1.198	7.815	9.360	0.570224	123.5
CA-4	1.91	0.95	1.11	3.49	5.40	3.02	4.13	13.0	171.3	111.3	8.1	1.725	3.881	6.693	0.572289	93.6
CA-249	0.95	0.79	1.91	10.16	11.75	2.86	5.40	27.3	149.8	474.4	6.9	0.719	19.355	13.913	0.580395	212.7
CA-232	2.22	0.95	1.27	2.54	4.45	3.49	4.45	11.4	175.6	102.7	9.0	2.013	3.226	6.492	0.583710	92.0
CA-3-C	1.91	0.95	1.27	3.33	5.24	3.18	4.45	13.0	171.3	125.2	8.3	1.725	4.234	7.303	0.605674	98.4
CA-232-B	2.22	1.03	1.27	2.38	4.45	3.49	4.60	11.4	190.1	98.0	9.1	2.179	3.024	6.590	0.630521	93.7
CA-30	1.91	0.95	1.27	3.49	5.40	3.18	4.45	13.3	175.5	131.2	8.3	1.725	4.436	7.651	0.634558	100.8
CA-208	2.86	0.79	1.27	2.86	4.45	4.13	4.13	11.4	188.0	127.9	9.9	2.156	3.630	7.825	0.681153	99.4
CA-32-A	2.54	0.95	0.95	3.18	5.08	3.49	3.81	12.1	211.7	97.8	9.1	2.300	3.026	6.958	0.704082	93.6
CA-279	1.59	1.27	1.11	3.49	6.03	2.70	4.76	14.3	208.9	111.3	8.1	1.916	3.881	7.435	0.706233	104.4
CA-252-D	1.27	1.91	1.27	2.86	6.67	2.54	6.35	15.9	278.4	115.5	9.0	2.298	3.630	8.342	0.856752	131.5

National Arnold C Cores 12 mil Silicon Steel																
Part No.	D cm 1	E cm 2	F cm 3	G cm 4	Ht cm 5	Wth cm 6	Lt cm 7	MPL cm 8	Wtfe grams 9	Wtcu grams 10	MLT cm 11	Ac cm sq 12	Wa cm sq 13	Ap cm 4th 14	Kg cm 5th 15	At cm sq 16
CA-252-A	2.22	1.11	1.27	2.86	5.08	3.49	4.76	12.7	227.4	119.6	9.3	2.346	3.630	8.516	0.862194	105.9
CA-5	2.54	0.79	1.59	3.97	5.56	4.13	4.76	14.3	208.9	219.0	9.8	1.916	6.303	12.076	0.947305	128.1
CA-3	2.54	0.95	1.27	3.97	5.88	3.81	4.45	14.3	250.7	171.9	9.6	2.300	5.041	11.591	1.112043	121.2
CA-434	1.91	1.27	1.27	3.97	6.51	3.18	5.08	15.6	272.8	160.5	9.0	2.298	5.041	11.585	1.189801	126.8
CA-208-A	2.86	1.11	1.27	2.86	5.08	4.13	4.76	12.7	292.3	136.0	10.5	3.017	3.630	10.949	1.253399	118.3
CA-399	1.91	1.27	1.11	4.76	7.30	3.02	4.76	16.8	295.1	163.8	8.7	2.298	5.292	12.162	1.284892	131.0
CA-434-A	1.59	1.59	1.27	3.97	7.15	2.86	5.72	16.8	307.6	160.5	9.0	2.396	5.041	12.076	1.292355	139.1
CA-427	2.22	1.03	1.27	4.92	6.99	3.49	4.60	16.5	274.5	202.5	9.1	2.179	6.250	13.621	1.303148	134.8
CA-128	2.86	1.11	1.11	3.49	5.72	3.97	4.44	13.7	314.2	142.0	10.3	3.017	3.881	11.706	1.372605	121.6
CA-87	2.54	1.11	1.27	3.97	6.19	3.81	4.76	14.9	305.2	177.5	9.9	2.681	5.041	13.513	1.463125	131.1
CA-432	1.91	1.11	1.91	4.60	6.83	3.81	6.03	17.5	267.9	300.4	9.6	2.011	8.771	17.634	1.472621	163.4
CA-2-B	1.75	1.59	1.27	3.97	7.15	3.02	5.72	16.8	338.2	166.2	9.3	2.634	5.041	13.277	1.509060	143.1
CA-6-C	1.59	1.59	1.59	3.97	7.15	3.18	6.35	17.5	319.3	211.9	9.5	2.396	6.303	15.099	1.530607	156.3
CA-55-A	2.22	1.11	1.43	4.76	6.99	3.65	5.08	16.8	301.3	230.4	9.5	2.346	6.806	15.969	1.574387	146.4
CA-281	2.22	1.03	1.35	5.72	7.78	3.57	4.76	18.3	303.6	253.2	9.2	2.179	7.710	16.802	1.585963	152.5
CA-7	2.54	0.95	1.59	4.92	6.83	4.13	5.08	16.8	295.3	280.3	10.1	2.300	7.815	17.970	1.638685	155.6
CA-395	2.54	1.27	1.11	3.97	6.51	3.65	4.76	15.2	356.3	156.4	10.0	3.065	4.410	13.513	1.661052	133.0
CA-170-F	1.59	1.27	1.59	6.35	8.89	3.18	5.72	21.0	306.3	316.2	8.8	1.916	10.084	19.320	1.679234	178.3
CA-401-A	3.18	1.03	1.27	3.81	5.87	4.45	4.60	14.3	339.3	189.5	11.0	3.113	4.839	15.062	1.702405	136.8
CA-81-A	2.54	1.11	1.59	3.97	6.19	4.13	5.40	15.6	318.2	233.2	10.4	2.681	6.303	16.897	1.741683	148.4
CA-438-A	0.95	0.95	8.26	12.38	14.29	9.21	18.42	45.1	296.8	6245.1	17.2	0.863	102.222	88.197	1.771698	1065.7
CA-24	2.54	1.27	1.27	3.97	6.51	3.81	5.08	15.6	363.8	183.2	10.2	3.065	5.041	15.447	1.852402	141.5
CA-178-A	3.81	0.95	1.27	3.97	5.88	5.08	4.45	14.3	376.1	221.1	12.3	3.449	5.041	17.387	1.944719	147.4
CA-210-A	2.54	1.27	1.43	3.81	6.35	3.97	5.40	15.6	363.8	202.7	10.5	3.065	5.444	16.685	1.953121	147.3
CA-6-F	3.18	0.95	1.59	4.29	6.19	4.76	5.08	15.6	341.3	274.9	11.4	2.875	6.806	19.564	1.980676	158.2
CA-136	2.70	1.03	1.35	5.40	7.46	4.05	4.76	17.6	355.8	263.8	10.2	2.646	7.282	19.269	2.001849	158.9
CA-210-C	2.86	1.27	1.43	3.33	5.87	4.29	5.40	14.6	384.3	188.2	11.1	3.448	4.764	16.428	2.039950	145.5
CA-376-A	3.18	0.95	1.59	4.45	6.35	4.76	5.08	15.9	348.2	285.1	11.4	2.875	7.059	20.290	2.054154	161.4
CA-210-D	2.70	1.27	1.43	3.81	6.35	4.13	5.40	15.6	386.6	208.9	10.8	3.256	5.444	17.729	2.140302	151.0
CA-163	2.54	1.11	1.59	4.92	7.14	4.13	5.40	17.5	357.2	289.1	10.4	2.681	7.815	20.950	2.159441	166.5

National Arnold C Cores 12 mil Silicon Steel

Part No.	D cm 1	E cm 2	F cm 3	G cm 4	Ht cm 5	Wth cm 6	Lt cm 7	MPL cm 8	Wtfe grams 9	Wtcu grams 10	MLT cm 11	Ac cm sq 12	Wa cm sq 13	Ap cm 4th 14	Kg cm 5th 15	At cm sq 16
CA-81	2.54	1.27	1.59	3.97	6.51	4.13	5.72	16.2	378.7	240.3	10.7	3.065	6.303	19.315	2.208370	159.2
CA-8	3.18	0.95	1.59	4.92	6.83	4.76	5.08	16.8	369.1	315.6	11.4	2.875	7.815	22.463	2.274126	170.7
CA-200	2.22	0.95	2.38	6.35	8.26	4.60	6.67	21.3	326.7	575.2	10.7	2.013	15.119	30.429	2.289789	229.4
CA-281-A	2.86	1.03	1.35	5.72	7.78	4.21	4.76	18.3	390.3	288.0	10.5	2.802	7.710	21.602	2.304543	168.5
CA-205	3.81	0.95	1.59	3.97	5.88	5.40	5.08	14.9	392.8	287.7	12.8	3.449	6.303	21.741	2.337078	165.9
CA-358-E	2.54	1.11	1.59	5.40	7.62	4.13	5.40	18.4	376.7	317.1	10.4	2.681	8.572	22.980	2.368759	175.6
CA-241-A	1.59	1.59	1.59	6.35	9.53	3.18	6.35	22.2	406.3	339.0	9.5	2.396	10.084	24.157	2.448817	201.7
CA-401	3.18	1.27	1.27	3.97	6.51	4.45	5.08	15.6	454.7	206.0	11.5	3.831	5.041	19.309	2.574513	156.3
CA-118	1.59	1.59	1.91	5.87	9.05	3.49	6.99	21.9	400.5	396.0	10.0	2.396	11.190	26.807	2.581535	214.3
CA-118-A	1.59	1.59	1.75	6.35	9.53	3.33	6.67	22.5	412.1	382.5	9.7	2.396	11.087	26.561	2.623619	212.8
CA-6-D	2.22	1.59	1.59	3.97	7.15	3.81	6.35	17.5	446.9	240.3	10.7	3.354	6.303	21.137	2.644210	173.5
CA-402	2.86	1.27	1.27	4.76	7.30	4.13	5.08	17.1	451.1	233.6	10.9	3.448	6.049	20.858	2.649583	163.5
CA-227-B	2.86	1.27	1.59	4.13	6.67	4.45	5.72	16.5	434.4	264.7	11.4	3.448	6.555	22.604	2.745109	170.2
CA-326	1.91	1.43	1.59	6.35	9.21	3.49	6.03	21.6	426.1	350.3	9.8	2.586	10.084	26.078	2.761396	199.4
CA-416	1.91	1.91	1.59	3.97	7.78	3.49	6.99	18.7	492.8	240.3	10.7	3.448	6.303	21.729	2.794969	187.6
CA-25-A	2.86	1.43	1.75	3.33	6.19	4.60	6.35	15.9	470.0	246.8	11.9	3.880	5.821	22.585	2.939763	174.2
CA-161	2.54	1.11	1.91	5.87	8.10	4.45	6.03	20.0	409.1	433.8	10.9	2.681	11.190	29.999	2.951014	206.3
CA-69-A	1.59	2.22	1.59	4.45	8.89	3.18	7.62	21.0	536.3	269.2	10.7	3.354	7.059	23.672	2.961329	212.2
CA-438-B	0.95	0.95	12.70	19.05	20.96	13.65	27.31	67.3	443.1	20784.5	24.2	0.863	241.935	208.741	2.981938	2365.5
CA-10	3.18	0.95	1.91	5.87	7.78	5.08	5.72	19.4	424.8	471.7	11.9	2.875	11.190	32.165	3.119704	211.5
CA-431	3.18	1.27	1.91	3.49	6.03	5.08	6.35	15.9	464.0	295.5	12.5	3.831	6.654	25.490	3.127326	183.1
CA-326-A	1.91	1.59	1.83	5.56	8.73	3.73	6.83	21.1	463.0	377.4	10.5	2.874	10.145	29.156	3.204022	212.2
CA-6	2.54	1.59	1.59	3.97	7.15	4.13	6.35	17.5	510.7	254.5	11.4	3.832	6.303	24.151	3.259399	182.0
CA-241	1.91	1.59	1.59	6.35	9.53	3.49	6.35	22.2	487.4	361.7	10.1	2.874	10.084	28.980	3.302581	211.8
CA-61-F	1.59	1.27	3.49	7.62	10.16	5.08	9.53	27.3	399.2	1117.6	11.8	1.916	26.617	50.995	3.309729	358.1
CA-195-B	1.91	1.59	1.98	5.56	8.73	3.89	7.14	21.4	470.0	419.8	10.7	2.874	11.023	31.679	3.400619	223.0
CA-129	1.91	1.91	1.59	4.92	8.73	3.49	6.99	20.6	542.9	297.9	10.7	3.448	7.815	26.941	3.465367	207.6
CA-195-A	1.91	1.43	1.91	6.99	9.84	3.81	6.67	23.5	463.6	485.8	10.3	2.586	13.306	34.412	3.467255	235.3
CA-6-A	2.54	1.59	1.59	4.29	7.46	4.13	6.35	18.1	529.2	274.9	11.4	3.832	6.806	26.080	3.519724	188.7
CA-25-B	3.18	1.43	1.75	3.49	6.35	4.92	6.35	16.2	532.6	272.3	12.6	4.310	6.099	26.287	3.609179	185.7

National Arnold C Cores 12 mil Silicon Steel

Part No.	D cm 1	E cm 2	F cm 3	G cm 4	Ht cm 5	Wth cm 6	Lt cm 7	MPL cm 8	Wtfe grams 9	Wtcu grams 10	MLT cm 11	Ac cm sq 12	Wa cm sq 13	Ap cm 4th 14	Kg cm 5th 15	At cm sq 16
CA-69	2.54	1.59	1.59	4.45	7.62	4.13	6.35	18.4	538.5	285.1	11.4	3.832	7.059	27.048	3.650297	192.0
CA-62	3.18	1.11	0.95	9.21	11.43	4.13	4.13	24.8	633.2	333.1	10.7	3.351	8.775	29.406	3.692019	214.3
CA-432-A	1.91	1.91	1.91	4.60	8.41	3.81	7.62	20.6	542.9	349.9	11.2	3.448	8.771	30.237	3.716799	221.4
CA-204-A	2.54	1.59	1.91	3.97	7.15	4.45	6.99	18.1	529.2	318.7	11.9	3.832	7.561	28.972	3.745895	201.3
CA-195	2.54	1.27	1.91	5.87	8.41	4.45	6.35	20.6	482.6	446.4	11.2	3.065	11.190	34.292	3.746819	218.8
CA-9	3.18	1.27	1.59	4.92	7.46	4.76	5.72	18.1	529.0	333.2	12.0	3.831	7.815	29.935	3.825117	194.7
CA-354	2.54	1.75	1.59	3.97	7.46	4.13	6.67	18.1	581.8	261.6	11.7	4.213	6.303	26.554	3.833595	194.0
CA-271	3.81	1.11	1.59	4.92	7.14	5.40	5.40	17.5	535.8	365.5	13.2	4.021	7.815	31.424	3.843484	198.4
CA-61	1.27	1.75	3.49	7.62	11.11	4.76	10.48	29.2	469.5	1147.5	12.1	2.107	26.617	56.069	3.896820	389.7
CA-118-B	1.59	2.06	1.91	5.87	10.00	3.49	7.94	23.8	565.8	433.8	10.9	3.114	11.190	34.843	3.980306	252.7
CA-95-C	1.91	1.59	2.06	6.35	9.53	3.97	7.30	23.2	508.3	505.0	10.8	2.874	13.106	37.666	3.996442	246.5
CA-432-B	3.18	1.59	1.91	3.02	6.19	5.08	6.99	16.2	591.8	268.2	13.1	4.790	5.745	27.520	4.017239	196.2
CA-42-B	2.54	1.59	1.59	4.92	8.10	4.13	6.35	19.4	566.3	315.6	11.4	3.832	7.815	29.944	4.041194	202.0
CA-117-A	2.22	1.27	2.54	6.51	9.05	4.76	7.62	23.2	474.3	680.9	11.6	2.682	16.533	44.342	4.107386	270.9
CA-206	5.08	0.95	1.59	4.92	6.83	6.67	5.08	16.8	590.6	427.3	15.4	4.599	7.815	35.940	4.300358	216.1
CA-55-C	3.18	0.95	1.43	10.16	12.07	4.60	4.76	27.0	592.0	573.5	11.1	2.875	14.519	41.734	4.320051	259.0
CA-126-D	3.18	1.27	1.91	4.92	7.46	5.08	6.35	18.7	547.5	416.3	12.5	3.831	9.375	35.910	4.405833	215.7
CA-147	2.22	2.06	2.54	3.02	7.14	4.76	9.21	19.4	644.1	358.8	13.2	4.359	7.661	33.392	4.420703	245.4
CA-64	2.86	1.27	1.91	5.87	8.41	4.76	6.35	20.6	543.0	471.7	11.9	3.448	11.190	38.585	4.489231	228.1
CA-55-D	1.27	2.54	1.43	8.89	13.97	2.70	7.94	30.8	720.1	473.0	10.5	3.065	12.704	38.931	4.557282	311.0
CA-392	2.22	1.43	2.38	6.35	9.21	4.60	7.62	23.2	533.7	626.4	11.7	3.018	15.119	45.628	4.727717	268.1
CA-151-B	2.54	1.91	1.27	5.08	8.89	3.81	6.35	20.3	712.7	263.6	11.5	4.597	6.452	29.656	4.745043	209.7
CA-45	2.86	1.59	1.59	4.92	8.10	4.45	6.35	19.4	637.2	333.3	12.0	4.312	7.815	33.693	4.845102	211.2
CA-307-H	2.86	1.27	2.22	5.72	8.26	5.08	6.99	21.0	551.3	558.1	12.4	3.448	12.704	43.807	4.890839	247.6
CA-170	3.18	1.27	1.59	6.35	8.89	4.76	5.72	21.0	612.5	430.0	12.0	3.831	10.084	38.627	4.935886	224.6
CA-222	7.62	0.56	2.46	6.67	7.78	10.08	6.03	20.5	629.0	1250.6	21.4	4.025	16.410	66.048	4.961533	361.5
CA-11	3.81	0.95	2.38	6.35	8.26	6.19	6.67	21.3	559.9	757.0	14.1	3.449	15.119	52.152	5.110556	276.7
CA-307-B	3.49	1.11	2.22	5.72	7.94	5.72	6.67	20.3	571.6	601.1	13.3	3.687	12.704	46.837	5.190833	253.1
CA-125-J	3.18	1.27	1.91	5.87	8.41	5.08	6.35	20.6	603.2	497.0	12.5	3.831	11.190	42.865	5.259065	237.5
CA-314	3.49	1.59	1.59	3.97	7.15	5.08	6.35	17.5	702.2	297.3	13.3	5.270	6.303	33.213	5.278250	207.8

National Arnold C Cores 12 mil Silicon Steel

Part No.	D cm 1	E cm 2	F cm 3	G cm 4	Ht cm 5	Wth cm 6	Lt cm 7	MPL cm 8	Wtfe grams 9	Wtcu grams 10	MLT cm 11	Ac cm sq 12	Wa cm sq 13	Ap cm 4th 14	Kg cm 5th 15	At cm sq 16
CA-432-C	1.91	1.91	1.91	6.67	10.48	3.81	7.62	24.8	651.5	506.8	11.2	3.448	12.703	43.793	5.383062	268.6
CA-55-L	4.45	1.27	1.43	4.76	7.30	5.87	5.40	17.5	714.6	350.7	14.5	5.363	6.806	36.502	5.404017	213.5
CA-220	2.38	1.91	2.54	3.81	7.62	4.92	8.89	20.3	668.1	453.1	13.2	4.309	9.677	41.700	5.458351	259.3
CA-9-B	4.13	1.27	1.59	4.92	7.46	5.72	5.72	18.1	687.7	392.0	14.1	4.980	7.815	38.920	5.496926	219.8
CA-32	2.54	1.91	1.59	4.92	8.73	4.13	6.99	20.6	723.8	333.2	12.0	4.597	7.815	35.922	5.508169	227.5
CA-310	2.54	1.59	1.91	5.87	9.05	4.45	6.99	21.9	640.6	471.7	11.9	3.832	11.190	42.878	5.543812	244.8
CA-227-E	3.81	1.35	1.27	6.03	8.73	5.08	5.24	20.0	745.2	357.7	13.1	4.883	7.662	37.411	5.565739	219.7
CA-170-H	2.86	1.59	2.22	4.45	7.62	5.08	7.62	19.7	647.7	456.4	13.0	4.312	9.881	42.604	5.656282	242.8
CA-126	3.18	1.59	1.59	4.92	8.10	4.76	6.35	19.4	707.9	350.9	12.6	4.790	7.815	37.430	5.679287	220.3
CA-146	2.86	1.27	2.38	6.35	8.89	5.24	7.30	22.5	593.1	677.5	12.6	3.448	15.119	52.134	5.705936	275.3
CA-95-A	1.91	1.91	2.22	6.35	10.16	4.13	8.26	24.8	651.5	588.2	11.7	3.448	14.116	48.666	5.727205	285.9
CA-227-C	3.18	1.27	1.27	8.89	11.43	4.45	5.08	25.4	742.4	461.4	11.5	3.831	11.290	43.249	5.766546	250.0
CA-55-M	3.49	1.59	1.43	4.76	7.94	4.92	6.03	18.7	753.3	315.0	13.0	5.270	6.806	35.866	5.809290	215.2
CA-123	1.91	1.91	2.38	6.35	10.16	4.29	8.57	25.1	659.8	643.4	12.0	3.448	15.119	52.125	6.007103	298.5
CA-15-E	3.81	1.43	1.59	4.92	7.78	5.40	6.03	18.7	739.3	383.1	13.8	5.172	7.815	40.419	6.065277	224.8
CA-221-A	5.08	0.95	2.54	4.76	6.67	7.62	6.99	18.4	646.3	725.7	16.9	4.599	12.098	55.641	6.067710	281.9
CA-272	3.18	1.43	1.59	6.35	9.21	4.76	6.03	21.6	710.1	441.4	12.3	4.310	10.084	43.463	6.087724	238.2
CA-280	3.18	1.59	1.75	5.08	8.26	4.92	6.67	20.0	731.1	406.1	12.9	4.790	8.870	42.484	6.321919	234.8
CA-170-T	3.81	1.27	1.59	6.35	8.89	5.40	5.72	21.0	735.0	483.0	13.5	4.597	10.084	46.353	6.327735	243.2
CA-149	2.06	2.54	2.54	3.49	8.57	4.60	10.16	22.2	844.6	435.5	13.8	4.980	8.872	44.187	6.377165	295.1
CA-150	4.13	2.54	0.48	5.08	10.16	4.60	6.03	21.3	1616.7	128.1	14.9	9.961	2.418	24.086	6.441073	262.0
CA-125-B	3.18	1.43	1.91	5.87	8.73	5.08	6.67	21.3	699.6	509.6	12.8	4.310	11.190	48.231	6.493005	251.1
CA-127	2.22	1.91	1.91	6.35	10.16	4.13	7.62	24.1	740.7	509.9	11.9	4.023	12.097	48.666	6.606155	272.6
CA-358-H	3.81	1.43	1.59	5.40	8.26	5.40	6.03	19.7	777.0	420.3	13.8	5.172	8.572	44.337	6.653194	235.8
CA-138	3.18	1.27	2.38	6.35	8.89	5.56	7.30	22.5	658.9	711.6	13.2	3.831	15.119	57.917	6.704600	285.6
CA-170-C	3.18	1.59	2.38	4.29	7.46	5.56	7.94	19.7	719.4	503.4	13.9	4.790	10.205	48.880	6.750918	259.3
CA-257	5.08	1.27	1.59	4.92	7.46	6.67	5.72	18.1	846.3	444.9	16.0	6.129	7.815	47.896	7.334616	244.9
CA-168-A	3.81	1.59	2.22	3.81	6.99	6.03	7.62	18.4	807.7	454.8	15.1	5.748	8.470	48.681	7.411139	254.5
CA-168	5.08	1.27	2.22	3.81	6.35	7.30	6.99	17.1	801.8	512.2	17.0	6.129	8.470	51.911	7.483444	260.9
CA-95	1.91	1.91	2.54	7.62	11.43	4.45	8.89	27.9	735.0	840.8	12.2	3.448	19.355	66.727	7.532756	345.2

National Arnold C Cores 12 mil Silicon Steel																
Part No.	D cm 1	E cm 2	F cm 3	G cm 4	Ht cm 5	Wth cm 6	Lt cm 7	MPL cm 8	Wtfe grams 9	Wtcu grams 10	MLT cm 11	Ac cm sq 12	Wa cm sq 13	Ap cm 4th 14	Kg cm 5th 15	At cm sq 16
CA-420	1.91	1.91	5.08	5.08	8.89	6.99	13.97	27.9	735.0	1487.0	16.2	3.448	25.806	88.969	7.571868	490.3
CA-86	2.54	1.91	1.91	5.87	9.68	4.45	7.62	23.2	812.9	497.0	12.5	4.597	11.190	51.438	7.573054	272.4
CA-263	4.13	1.27	1.91	5.87	8.41	6.03	6.35	20.6	784.3	581.1	14.6	4.980	11.190	55.731	7.603006	265.6
CA-234-D	5.08	0.95	2.54	6.03	7.94	7.62	6.99	21.0	735.5	919.3	16.9	4.599	15.324	70.477	7.685596	319.0
CA-128-A	3.18	1.75	1.11	7.62	11.11	4.29	5.71	24.4	982.3	367.1	12.2	5.266	8.466	44.584	7.701873	255.7
CA-405	5.08	1.27	1.75	4.92	7.46	6.83	6.03	18.4	861.1	496.7	16.3	6.129	8.592	52.661	7.941333	256.5
CA-170-E	3.18	1.27	1.91	8.89	11.43	5.08	6.35	26.7	779.5	752.1	12.5	3.831	16.935	64.874	7.959328	306.5
CA-17	2.54	1.91	3.02	4.45	8.26	5.56	9.84	22.5	790.6	678.5	14.2	4.597	13.406	61.625	7.960999	317.6
CA-97	1.91	1.91	2.22	8.89	12.70	4.13	8.26	29.8	785.1	823.5	11.7	3.448	19.762	68.133	8.018087	348.8
CA-126-C	4.13	1.59	1.59	4.92	8.10	5.72	6.35	19.4	920.4	409.6	14.7	6.228	7.815	48.665	8.223558	247.9
CA-204	2.54	2.54	1.91	3.97	9.05	4.45	8.89	21.9	1024.5	369.9	13.8	6.129	7.561	46.341	8.257261	281.7
CA-42	3.18	1.98	1.59	4.92	8.89	4.76	7.14	21.0	956.8	372.9	13.4	5.984	7.815	46.764	8.341738	254.5
CA-126-B	3.18	1.43	3.49	4.92	7.78	6.67	9.84	22.5	741.4	935.2	15.3	4.310	17.189	74.089	8.348710	348.2
CA-61-E	3.18	1.91	1.75	4.92	8.73	4.92	7.30	21.0	918.7	412.8	13.5	5.746	8.592	49.370	8.399471	258.7
CA-176	3.18	1.27	2.54	7.62	10.16	5.72	7.62	25.4	742.4	928.2	13.5	3.831	19.355	74.141	8.423917	332.3
CA-333-B	5.72	0.95	1.11	11.43	13.34	6.83	4.13	28.9	1140.7	717.8	15.9	5.174	12.699	65.704	8.554419	340.1
CA-12	3.81	1.27	2.38	6.35	8.89	6.19	7.30	22.5	790.6	791.1	14.7	4.597	15.119	69.500	8.684842	306.1
CA-322	2.86	1.91	1.91	5.87	9.68	4.76	7.62	23.2	914.7	522.3	13.1	5.172	11.190	57.877	9.123389	283.4
CA-55-E	4.45	1.59	1.59	4.92	8.10	6.03	6.35	19.4	991.1	427.3	15.4	6.706	7.815	52.402	9.141889	257.0
CA-272-A	3.81	1.43	1.59	7.46	10.32	5.40	6.03	23.8	939.8	580.9	13.8	5.172	11.848	61.281	9.195902	282.9
CA-184	1.91	1.91	3.49	7.62	11.43	5.40	10.80	29.8	785.1	1297.8	13.7	3.448	26.617	91.763	9.228679	433.5
CA-170-B	3.18	1.91	2.38	4.29	8.10	5.56	8.57	21.0	918.7	526.4	14.5	5.746	10.205	58.637	9.290593	287.7
CA-170-K	3.18	1.59	1.91	6.99	10.16	5.08	6.99	24.1	881.9	621.0	13.1	4.790	13.306	63.735	9.303850	292.0
CA-18-F	3.81	1.59	1.91	5.40	8.57	5.72	6.99	21.0	919.1	534.0	14.6	5.748	10.283	59.105	9.305686	273.4
CA-17-A	2.54	1.91	3.18	5.08	8.89	5.72	10.16	24.1	846.3	830.7	14.5	4.597	16.129	74.141	9.412847	350.0
CA-125-H	2.54	2.22	1.91	5.72	10.16	4.45	8.26	24.1	987.7	508.1	13.1	5.364	10.887	58.399	9.547077	297.6
CA-18-A	3.18	2.22	1.59	4.92	9.37	4.76	7.62	21.9	1120.9	386.2	13.9	6.705	7.815	52.398	10.112334	276.4
CA-28	3.81	1.59	1.91	5.87	9.05	5.72	6.99	21.9	960.9	581.1	14.6	5.748	11.190	64.317	10.126269	285.5
CA-6-H	2.54	2.22	2.06	5.72	10.16	4.60	8.57	24.5	1000.7	561.0	13.4	5.364	11.796	63.274	10.150854	310.0
CA-125-E	3.18	1.91	1.91	5.72	9.53	5.08	7.62	22.9	1002.2	532.7	13.8	5.746	10.887	62.557	10.449938	290.3

National Arnold C Cores 12 mil Silicon Steel																
Part No.	D cm 1	E cm 2	F cm 3	G cm 4	Ht cm 5	Wth cm 6	Lt cm 7	MPL cm 8	Wtfe grams 9	Wtcu grams 10	MLT cm 11	Ac cm sq 12	Wa cm sq 13	Ap cm 4th 14	Kg cm 5th 15	At cm sq 16
CA-18-H	4.13	1.59	1.91	5.40	8.57	6.03	6.99	21.0	995.8	557.2	15.2	6.228	10.283	64.039	10.467989	283.3
CA-315	4.13	1.75	1.75	4.92	8.41	5.87	6.98	20.3	1061.5	467.6	15.3	6.847	8.592	58.831	10.527690	274.0
CA-397	2.54	2.54	2.86	3.81	8.89	5.40	10.80	23.5	1098.8	590.7	15.3	6.129	10.889	66.739	10.725446	346.4
CA-125	3.18	1.91	1.91	5.87	9.68	5.08	7.62	23.2	1016.2	547.5	13.8	5.746	11.190	64.297	10.740671	294.4
CA-111-B	3.18	1.43	2.86	7.30	10.16	6.03	8.57	26.0	856.3	1061.6	14.3	4.310	20.872	89.963	10.844104	367.6
CA-13	3.81	1.27	2.54	7.62	10.16	6.35	7.62	25.4	890.9	1029.9	15.0	4.597	19.355	88.969	10.932296	354.8
CA-125-F	4.13	1.59	1.91	5.72	8.89	6.03	6.99	21.6	1026.0	590.0	15.2	6.228	10.887	67.799	11.082726	291.6
CA-455-A	2.54	2.54	2.06	5.08	10.16	4.60	9.21	24.4	1143.3	522.3	14.0	6.129	10.485	64.264	11.246683	323.5
CA-154	1.91	3.49	0.95	9.53	16.51	2.86	8.89	34.9	1684.7	416.4	12.9	6.322	9.077	57.382	11.247485	412.2
CA-308-B	1.91	1.91	5.08	7.62	11.43	6.99	13.97	33.0	868.6	2230.4	16.2	3.448	38.710	133.454	11.357802	596.8
CA-248-A	3.18	1.43	1.91	10.48	13.34	5.08	6.67	30.5	1002.5	909.0	12.8	4.310	19.961	86.035	11.582176	359.3
CA-126-A	3.18	2.22	1.91	4.92	9.37	5.08	8.26	22.5	1153.4	479.9	14.4	6.705	9.375	62.857	11.711561	299.8
CA-55-J	4.45	1.83	1.59	4.92	8.57	6.03	6.83	20.3	1195.6	440.5	15.9	7.711	7.815	60.256	11.724521	279.7
CA-170-L	3.18	1.59	1.91	8.89	12.07	5.08	6.99	27.9	1021.2	790.4	13.1	4.790	16.935	81.118	11.841264	338.0
CA-152	2.54	1.91	1.27	12.70	16.51	3.81	6.35	35.6	1247.2	659.1	11.5	4.597	16.129	74.141	11.862609	364.5
CA-401-B	4.13	2.54	1.27	3.81	8.89	5.40	7.62	20.3	1544.3	277.8	16.1	9.961	4.839	48.198	11.893795	287.9
CA-6-B	5.08	1.59	2.06	4.29	7.46	7.14	7.30	19.1	1114.0	547.1	17.4	7.664	8.846	67.795	11.949121	293.8
CA-307	3.49	1.75	2.22	5.87	9.37	5.72	7.94	23.2	1024.6	676.8	14.6	5.794	13.058	75.655	12.028864	315.5
CA-322-A	3.49	1.91	1.91	5.87	9.68	5.40	7.62	23.2	1117.9	572.8	14.4	6.322	11.190	70.737	12.425564	305.4
CA-278	4.45	1.27	2.54	6.99	9.53	6.99	7.62	24.1	987.4	1024.2	16.2	5.363	17.742	95.148	12.572977	358.9
CA-425	5.08	1.43	1.91	5.87	8.73	6.99	6.67	21.3	1119.4	669.5	16.8	6.896	11.190	77.170	12.652513	309.8
CA-418	1.91	1.27	6.35	16.19	18.73	8.26	15.24	50.2	879.7	6189.5	16.9	2.298	102.826	236.332	12.835501	1067.4
CA-6-E	3.49	1.91	2.06	5.72	9.53	5.56	7.94	23.2	1117.9	614.3	14.6	6.322	11.796	74.566	12.874972	313.8
CA-94-F	3.81	1.59	2.22	6.67	9.84	6.03	7.62	24.1	1058.4	796.0	15.1	5.748	14.823	85.199	12.970466	332.6
CA-170-A	3.18	2.22	1.59	6.35	10.80	4.76	7.62	24.8	1267.1	498.3	13.9	6.705	10.084	67.613	13.048835	311.8
CA-260	1.59	1.59	5.72	16.19	19.37	7.30	14.61	50.2	917.0	5243.1	15.9	2.396	92.543	221.701	13.334215	989.4
CA-20	3.81	1.75	2.22	5.87	9.37	6.03	7.94	23.2	1117.6	715.9	15.4	6.320	13.058	82.521	13.529692	326.4
CA-13-A	4.45	1.27	2.54	7.62	10.16	6.99	7.62	25.4	1039.3	1117.3	16.2	5.363	19.355	103.798	13.715975	377.4
CA-94-H	5.08	1.27	2.22	6.99	9.53	7.30	6.99	23.5	1098.8	939.0	17.0	6.129	15.528	95.169	13.719648	351.7
CA-234-B	4.45	1.59	2.54	5.08	8.26	6.99	8.26	21.6	1104.7	774.1	16.9	6.706	12.903	86.525	13.757491	334.7

National Arnold C Cores 12 mil Silicon Steel																
Part No.	D cm 1	E cm 2	F cm 3	G cm 4	Ht cm 5	Wth cm 6	Lt cm 7	MPL cm 8	Wtfe grams 9	Wtcu grams 10	MLT cm 11	Ac cm sq 12	Wa cm sq 13	Ap cm 4th 14	Kg cm 5th 15	At cm sq 16
CA-234-C	5.08	1.27	2.54	6.35	8.89	7.62	7.62	22.9	1069.0	1003.9	17.5	6.129	16.129	98.855	13.845772	361.3
CA-36	3.81	1.91	1.91	5.87	9.68	5.72	7.62	23.2	1219.4	606.3	15.2	6.895	11.190	77.156	13.966283	316.3
CA-182-A	3.49	1.91	2.22	5.87	9.68	5.72	8.26	23.8	1148.6	691.6	14.9	6.322	13.058	82.545	14.013712	331.1
CA-234	4.13	1.75	2.54	5.08	8.57	6.67	8.57	22.2	1161.1	759.5	16.6	6.847	12.903	88.350	14.619323	340.1
CA-170-D	3.18	2.06	1.91	6.99	11.11	5.08	7.94	26.0	1236.7	666.1	14.1	6.226	13.306	82.840	14.654465	338.5
CA-170-J	3.49	1.91	1.91	6.99	10.80	5.40	7.62	25.4	1225.1	681.1	14.4	6.322	13.306	84.116	14.775718	334.3
CA-164	3.81	2.38	1.59	4.92	9.68	5.40	7.94	22.5	1482.3	436.0	15.7	8.618	7.815	67.346	14.795333	313.7
CA-173	3.81	1.59	1.91	8.73	11.91	5.72	6.99	27.6	1211.5	863.7	14.6	5.748	16.633	95.600	15.051490	358.1
CA-131	3.81	1.59	1.91	8.73	11.91	5.72	6.99	27.6	1211.5	863.7	14.6	5.748	16.633	95.600	15.051490	358.1
CA-182	3.49	1.91	2.38	6.03	9.84	5.87	8.57	24.4	1179.2	773.5	15.1	6.322	14.365	90.805	15.163482	348.8
CA-20-A	2.86	2.86	2.22	4.45	10.16	5.08	10.16	24.8	1466.4	545.7	15.5	7.760	9.881	76.676	15.324729	362.2
CA-132-B	3.81	1.59	1.91	8.89	12.07	5.72	6.99	27.9	1225.4	879.4	14.6	5.748	16.935	97.341	15.325592	362.1
CA-117	6.35	1.03	2.54	7.62	9.68	8.89	7.14	24.4	1161.3	1346.8	19.6	6.226	19.355	120.494	15.334186	416.6
CA-145	5.08	1.59	1.91	5.87	9.05	6.99	6.99	21.9	1281.2	682.1	17.1	7.664	11.190	85.756	15.334920	326.3
CA-94	3.18	2.22	2.22	5.87	10.32	5.40	8.89	25.1	1283.4	691.6	14.9	6.705	13.058	87.555	15.766396	351.4
CA-182-D	3.18	2.22	2.38	5.72	10.16	5.56	9.21	25.1	1283.3	732.7	15.1	6.705	13.607	91.239	16.160735	360.0
CA-318	3.81	1.27	6.35	6.35	8.89	10.16	15.24	30.5	1069.0	3003.3	20.9	4.597	40.323	185.353	16.271265	696.8
CA-38	3.81	1.59	2.54	7.62	10.80	6.35	8.26	26.7	1169.7	1073.7	15.6	5.748	19.355	111.247	16.395619	388.0
CA-14	5.08	1.27	2.54	7.62	10.16	7.62	7.62	25.4	1187.8	1204.7	17.5	6.129	19.355	118.626	16.614926	400.0
CA-94-K	6.35	1.03	2.06	9.68	11.75	8.41	6.19	27.6	1312.2	1322.9	18.6	6.226	19.988	124.435	16.648468	430.3
CA-18-E	3.49	2.38	1.91	5.40	10.16	5.40	8.57	24.1	1454.7	561.2	15.3	7.901	10.283	81.247	16.731380	340.1
CA-234-E	4.13	1.91	2.54	5.08	8.89	6.67	8.89	22.9	1303.0	774.1	16.9	7.471	12.903	96.395	17.075131	356.5
CA-170-M	4.13	1.59	1.91	8.89	12.07	6.03	6.99	27.9	1327.7	917.7	15.2	6.228	16.935	105.466	17.239796	374.3
CA-18	4.45	1.27	3.49	7.62	10.16	7.94	9.53	27.3	1117.3	1678.1	17.7	5.363	26.617	142.742	17.270414	470.6
CA-371-A	4.45	1.59	2.38	6.83	10.00	6.83	7.94	24.8	1267.1	960.6	16.6	6.706	16.253	108.986	17.589032	373.5
CA-330	2.54	2.38	2.54	7.62	12.38	5.08	9.84	29.8	1308.3	993.7	14.4	5.745	19.355	111.200	17.700319	421.1
CA-334	3.18	2.54	2.30	5.16	10.24	5.48	9.68	25.1	1466.2	661.0	15.7	7.661	11.876	90.985	17.813909	369.8
CA-94-D	3.18	2.22	2.22	6.67	11.11	5.40	8.89	26.7	1364.6	785.1	14.9	6.705	14.823	99.390	17.897571	374.1
CA-49	3.81	1.91	2.38	6.35	10.16	6.19	8.57	25.1	1319.6	859.4	16.0	6.895	15.119	104.250	17.988301	369.8
CA-93	3.81	1.91	2.38	6.35	10.16	6.19	8.57	25.1	1319.6	859.4	16.0	6.895	15.119	104.250	17.988301	369.8

National Arnold C Cores 12 mil Silicon Steel

Part No.	D cm 1	E cm 2	F cm 3	G cm 4	Ht cm 5	Wth cm 6	Lt cm 7	MPL cm 8	Wtfe grams 9	Wtcu grams 10	MLT cm 11	Ac cm sq 12	Wa cm sq 13	Ap cm 4th 14	Kg cm 5th 15	At cm sq 16
CA-248-B	3.81	1.59	1.91	10.48	13.65	5.72	6.99	31.1	1364.7	1036.5	14.6	5.748	19.961	114.729	18.063167	402.5
CA-94-B	3.18	2.22	2.06	7.46	11.91	5.24	8.57	27.9	1429.5	801.9	14.6	6.705	15.400	103.256	18.910646	382.1
CA-170-S	3.18	1.59	1.59	16.51	19.69	4.76	6.35	42.5	1555.0	1177.2	12.6	4.790	26.218	125.579	19.054061	478.0
CA-182-B	3.18	2.38	2.38	6.03	10.80	5.56	9.52	26.4	1444.0	789.6	15.5	7.182	14.365	103.162	19.171147	386.1
CA-267	3.81	1.75	2.54	7.62	11.11	6.35	8.57	27.3	1316.6	1095.4	15.9	6.320	19.355	122.316	19.427005	405.0
CA-55-H	4.45	1.75	1.59	8.89	12.38	6.03	6.67	27.9	1571.8	787.7	15.7	7.373	14.117	104.086	19.563043	372.8
CA-437-A	5.08	1.27	3.18	7.62	10.16	8.26	8.89	26.7	1247.2	1591.7	18.5	6.129	24.194	148.282	19.649497	462.9
CA-419	3.18	1.59	4.45	8.26	11.43	7.62	12.07	31.8	1160.4	2232.9	17.1	4.790	36.693	175.755	19.677378	577.5
CA-15-A	5.08	1.75	1.83	6.67	10.16	6.91	7.14	24.0	1541.2	750.5	17.3	8.426	12.176	102.595	19.948036	358.7
CA-94-A	3.18	2.22	2.22	7.62	12.07	5.40	8.89	28.6	1462.1	897.2	14.9	6.705	16.939	113.580	20.452833	401.3
CA-228-A	8.26	1.11	1.59	9.53	11.75	9.84	5.40	26.7	1773.0	1207.0	22.4	8.713	15.126	131.786	20.466315	450.2
CA-185	3.49	1.59	3.33	8.89	12.07	6.83	9.84	30.8	1238.4	1686.8	16.0	5.270	29.639	156.185	20.569980	494.4
CA-329-A	5.08	1.91	1.91	5.72	9.53	6.99	7.62	22.9	1603.6	688.2	17.8	9.194	10.887	100.091	20.705275	355.6
CA-130-D	3.97	1.75	2.54	7.78	11.27	6.51	8.57	27.6	1387.5	1140.6	16.2	6.583	19.759	130.079	21.100618	416.1
CA-329	5.08	1.91	1.91	5.87	9.68	6.99	7.62	23.2	1625.9	707.4	17.8	9.194	11.190	102.875	21.281327	360.3
CA-144	2.54	2.54	3.81	6.35	11.43	6.35	12.70	30.5	1425.4	1441.0	16.7	6.129	24.194	148.282	21.703698	519.4
CA-94-C	3.49	1.91	2.22	9.13	12.94	5.72	8.26	30.3	1462.5	1074.7	14.9	6.322	20.292	128.272	21.776841	422.0
CA-137	5.08	1.27	3.49	7.94	10.48	8.57	9.53	27.9	1306.7	1873.4	19.0	6.129	27.727	169.942	21.927944	507.1
CA-352	3.18	2.06	3.02	7.62	11.75	6.19	10.16	29.5	1402.6	1293.0	15.8	6.226	22.982	143.075	22.519727	460.2
CA-39	6.35	1.27	2.54	7.62	10.16	8.89	7.62	25.4	1484.8	1379.5	20.0	7.661	19.355	148.282	22.671002	445.2
CA-177	3.81	1.91	2.54	7.62	11.43	6.35	8.89	27.9	1469.9	1117.3	16.2	6.895	19.355	133.454	22.673346	422.6
CA-95-D	1.91	3.81	2.54	7.62	15.24	4.45	12.70	35.6	1870.8	1103.0	16.0	6.895	19.355	133.454	22.967625	558.1
CA-202-A	4.45	1.83	2.38	6.99	10.64	6.83	8.41	26.0	1531.8	1011.1	17.1	7.711	16.631	128.240	23.135520	404.3
CA-179-A	3.49	2.38	2.38	6.35	11.11	5.87	9.52	27.0	1626.8	865.3	16.1	7.901	15.119	119.458	23.457823	408.7
CA-234-A	5.08	1.75	2.54	6.03	9.53	7.62	8.57	24.1	1551.4	1005.7	18.5	8.426	15.324	129.122	23.580731	403.6
CA-431	5.08	1.91	2.22	5.87	9.68	7.30	8.26	23.8	1670.5	848.6	18.3	9.194	13.058	120.048	24.155400	388.1
CA-432-D	5.08	1.91	1.91	6.67	10.48	6.99	7.62	24.8	1737.3	803.0	17.8	9.194	12.703	116.781	24.157965	383.5
CA-132-D	3.18	2.54	1.91	8.26	13.34	5.08	8.89	30.5	1781.7	840.4	15.0	7.661	15.726	120.479	24.566789	424.2
CA-391	2.54	2.54	3.49	7.62	12.70	6.03	12.07	32.4	1514.5	1538.2	16.3	6.129	26.617	163.134	24.608693	534.3
CA-202-C	3.18	2.22	2.38	8.89	13.34	5.56	9.21	31.4	1608.2	1139.8	15.1	6.705	21.167	141.928	25.138921	453.8

National Arnold C Cores 12 mil Silicon Steel																
Part No.	D cm 1	E cm 2	F cm 3	G cm 4	Ht cm 5	Wth cm 6	Lt cm 7	MPL cm 8	Wtfe grams 9	Wtcu grams 10	MLT cm 11	Ac cm sq 12	Wa cm sq 13	Ap cm 4th 14	Kg cm 5th 15	At cm sq 16
CA-202-B	3.81	1.91	2.38	8.89	12.70	6.19	8.57	30.2	1586.8	1203.1	16.0	6.895	21.167	145.950	25.183621	444.8
CA-420-B	1.91	1.91	7.78	13.97	17.78	9.68	19.37	51.1	1344.7	7899.2	20.4	3.448	108.673	374.657	25.275778	1289.7
CA-170-N	5.08	2.22	1.59	6.35	10.80	6.67	7.62	24.8	2027.4	642.4	17.9	10.728	10.084	108.181	25.912964	382.0
CA-299	4.45	2.38	1.91	5.87	10.64	6.35	8.57	25.1	1924.2	694.7	17.5	10.054	11.190	112.508	25.916886	390.0
CA-16-H	2.54	2.54	3.81	7.62	12.70	6.35	12.70	33.0	1544.2	1729.2	16.7	6.129	29.032	177.939	26.044438	567.7
CA-111-C	3.89	1.98	2.86	7.30	11.27	6.75	9.68	28.3	1580.4	1265.4	17.0	7.330	20.872	152.991	26.310525	455.7
CA-171-A	6.35	1.11	2.54	11.43	13.65	8.89	7.30	32.4	1656.0	2036.5	19.7	6.702	29.032	194.577	26.444057	549.4
CA-329-E	7.62	1.11	1.91	12.07	14.29	9.53	6.03	32.4	1987.2	1771.0	21.7	8.043	22.984	184.848	27.443008	542.5
CA-244	3.81	1.75	3.33	8.89	12.38	7.14	10.16	31.4	1515.6	1808.9	17.2	6.320	29.639	187.310	27.588969	527.1
CA-179	5.08	1.91	2.38	6.35	10.16	7.46	8.57	25.1	1759.4	995.9	18.5	9.194	15.119	139.000	27.594273	417.4
CA-247	3.49	2.30	2.54	7.62	12.22	6.03	9.68	29.5	1721.0	1114.0	16.2	7.639	19.355	147.848	27.910623	454.3
CA-200-A	3.81	2.54	2.38	6.03	11.11	6.19	9.84	27.0	1893.1	881.4	17.3	9.194	14.365	132.061	28.146425	429.8
CA-177-A	3.81	1.91	2.54	9.53	13.34	6.35	8.89	31.8	1670.4	1396.6	16.2	6.895	24.194	166.818	28.341682	480.6
CA-413-F	2.54	3.02	2.86	7.62	13.65	5.40	11.75	33.0	1833.5	1255.1	16.2	7.278	21.778	158.491	28.467561	526.0
CA-385-B	3.49	2.30	2.06	9.13	13.73	5.56	8.73	31.6	1841.3	1034.3	15.4	7.639	18.840	143.917	28.483660	451.5
CA-290	3.81	1.91	3.49	7.62	11.43	7.30	10.80	29.8	1570.2	1678.1	17.7	6.895	26.617	183.526	28.549052	518.2
CA-55-B	4.45	2.54	1.59	6.83	11.91	6.03	8.26	27.0	2208.6	666.0	17.3	10.726	10.840	116.264	28.867709	407.6
CA-140	6.35	1.27	3.49	7.62	10.16	9.84	9.53	27.3	1596.2	2038.7	21.5	7.661	26.617	203.918	29.011471	545.6
CA-58	4.45	1.43	2.54	13.02	15.88	6.99	7.94	36.8	1695.8	1946.2	16.6	6.034	33.066	199.529	29.096982	555.7
CA-139	4.13	2.06	2.54	7.62	11.75	6.67	9.21	28.6	1764.8	1183.0	17.2	8.094	19.355	156.661	29.510350	453.9
CA-15	5.08	1.75	2.54	7.62	11.11	7.62	8.57	27.3	1755.4	1270.2	18.5	8.426	19.355	163.087	29.783718	455.0
CA-27-D	3.18	2.38	3.18	7.62	12.38	6.35	11.11	31.1	1704.9	1437.1	16.7	7.182	24.194	173.750	29.879421	513.1
CA-329-D	6.99	1.27	1.91	11.75	14.29	8.89	6.35	32.4	2082.5	1648.7	20.7	8.427	22.380	188.605	30.688995	529.0
CA-308-A	2.54	2.54	5.08	7.62	12.70	7.62	15.24	35.6	1662.9	2580.1	18.7	6.129	38.710	237.252	31.031856	709.7
CA-212	5.40	0.95	10.32	9.37	11.27	15.72	22.54	43.2	1610.2	10213.8	29.7	4.887	96.648	472.325	31.068395	1492.5
CA-70	3.49	2.22	3.49	7.30	11.75	6.99	11.43	30.5	1715.8	1589.6	17.5	7.377	25.509	188.175	31.684707	530.4
CA-142	5.08	0.95	7.82	12.07	13.97	12.90	17.54	43.6	1529.2	8437.8	25.2	4.599	94.324	433.814	31.724687	1243.5
CA-422-E	2.54	1.91	9.68	9.68	13.49	12.22	23.18	46.4	1625.9	8237.6	24.7	4.597	93.780	431.084	32.088111	1374.5
CA-170-R	2.86	3.18	1.91	8.89	15.24	4.76	10.16	34.3	2255.4	943.4	15.7	8.620	16.935	145.991	32.135837	503.3
CA-244-A	3.81	1.91	3.33	8.89	12.70	7.14	10.48	32.1	1687.1	1842.4	17.5	6.895	29.639	204.367	32.245091	546.5

National Arnold C Cores 12 mil Silicon Steel																
Part No.	D cm 1	E cm 2	F cm 3	G cm 4	Ht cm 5	Wth cm 6	Lt cm 7	MPL cm 8	Wtfe grams 9	Wtcu grams 10	MLT cm 11	Ac cm sq 12	Wa cm sq 13	Ap cm 4th 14	Kg cm 5th 15	At cm sq 16
CA-268	4.45	2.06	2.54	7.62	11.75	6.99	9.21	28.6	1900.3	1226.6	17.8	8.716	19.355	168.692	32.999491	467.2
CA-440-B	1.27	2.54	6.59	24.77	29.85	7.86	18.26	72.9	1703.8	10774.4	18.6	3.065	163.152	499.980	33.001598	1711.8
CA-340-A	3.49	2.22	3.49	7.62	12.07	6.99	11.43	31.1	1751.5	1658.6	17.5	7.377	26.617	196.343	33.060039	542.1
CA-181-B	7.62	1.11	2.38	12.07	14.29	10.00	6.98	33.3	2045.6	2289.9	22.4	8.043	28.727	231.036	33.156644	603.9
CA-420-A	1.91	3.49	3.81	9.53	16.51	5.72	14.61	40.6	1960.3	2243.6	17.4	6.322	36.290	229.407	33.365075	731.5
CA-27-E	3.18	2.54	3.18	7.62	12.70	6.35	11.43	31.8	1856.0	1464.5	17.0	7.661	24.194	185.353	33.368069	532.3
CA-421	2.54	2.54	4.45	8.89	13.97	6.99	13.97	36.8	1722.3	2493.7	17.7	6.129	39.516	242.195	33.457941	690.3
CA-111-A	3.81	2.54	2.86	6.35	11.43	6.67	10.80	28.6	2004.5	1161.8	18.0	9.194	18.148	166.847	34.081147	485.1
CA-211	5.72	3.18	1.59	3.81	10.16	7.30	9.53	23.5	3090.3	453.7	21.1	17.238	6.050	104.294	34.099132	440.8
CA-171	6.35	1.59	2.54	7.62	10.80	8.89	8.26	26.7	1949.5	1423.3	20.7	9.580	19.355	185.411	34.355637	484.7
CA-174	5.08	1.91	1.91	9.53	13.34	6.99	7.62	30.5	2138.1	1147.0	17.8	9.194	18.145	166.818	34.508791	466.9
CA-440-C	1.27	2.54	6.59	26.04	31.12	7.86	18.26	75.4	1763.2	11326.9	18.6	3.065	171.519	525.620	34.693988	1778.1
CA-385-D	3.97	2.22	2.06	9.84	14.29	6.03	8.57	32.7	2091.7	1187.7	16.4	8.382	20.316	170.287	34.727299	484.2
CA-48	5.08	1.91	2.54	7.62	11.43	7.62	8.89	27.9	1959.9	1292.1	18.8	9.194	19.355	177.939	34.854679	474.2
CA-128-C	5.08	2.86	1.11	7.62	13.34	6.19	7.94	28.9	3040.8	555.0	18.4	13.793	8.466	116.767	34.942586	449.7
CA-444	1.27	2.54	7.62	24.77	29.85	8.89	20.32	74.9	1752.0	13549.4	20.2	3.065	188.709	578.302	35.108231	1950.0
CA-302	3.18	2.54	3.49	7.62	12.70	6.67	12.07	32.4	1893.1	1658.4	17.5	·7.661	26.617	203.918	35.664138	565.8
CA-228	6.67	2.06	1.59	6.99	11.11	8.26	7.30	25.4	2534.1	835.1	21.2	13.075	11.092	145.026	35.821932	440.2
CA-300	5.08	2.22	2.38	6.35	10.80	7.46	9.21	26.4	2157.2	1030.1	19.2	10.728	15.119	162.203	36.328487	455.0
CA-27	3.18	2.70	3.18	7.62	13.02	6.35	11.75	32.4	2011.6	1491.9	17.3	8.141	24.194	196.956	36.985480	551.8
CA-181-C	6.99	1.27	2.38	11.75	14.29	9.37	7.30	33.3	2143.7	2135.0	21.5	8.427	27.972	235.731	37.021729	588.9
CA-340	3.49	2.38	3.49	7.62	12.38	6.99	11.75	31.8	1914.0	1688.5	17.8	7.901	26.617	210.298	37.254748	561.5
CA-214	4.13	2.06	3.49	7.62	11.75	7.62	11.11	30.5	1882.5	1768.4	18.7	8.094	26.617	215.440	37.332703	551.9
CA-111	3.81	2.54	2.54	7.62	12.70	6.35	10.16	30.5	2138.1	1204.7	17.5	9.194	19.355	177.939	37.383584	496.8
CA-444-A	1.27	2.54	7.62	26.67	31.75	8.89	20.32	78.7	1841.1	14591.7	20.2	3.065	203.225	622.786	37.808864	2061.3
CA-181	6.35	1.91	2.38	6.35	10.16	8.73	8.57	25.1	2199.3	1132.5	21.1	11.492	15.119	173.750	37.916949	465.0
CA-413-E	5.08	1.91	2.86	7.62	11.43	7.94	9.53	28.6	2004.5	1492.6	19.3	9.194	21.778	200.216	38.202441	506.9
CA-33	3.49	2.38	4.13	6.99	11.75	7.62	13.02	31.8	1914.0	1931.4	18.8	7.901	28.834	227.818	38.222444	604.7
CA-420-C	1.91	3.81	3.81	9.53	17.15	5.72	15.24	41.9	2204.9	2325.4	18.0	6.895	36.290	250.227	38.299184	776.6
CA-422-B	2.54	2.54	7.62	7.62	12.70	10.16	20.32	40.6	1900.5	4693.5	22.7	6.129	58.064	355.878	38.381843	1032.3

National Arnold C Cores 12 mil Silicon Steel

Part No.	D cm 1	E cm 2	F cm 3	G cm 4	Ht cm 5	Wth cm 6	Lt cm 7	MPL cm 8	Wtfe grams 9	Wtcu grams 10	MLT cm 11	Ac cm sq 12	Wa cm sq 13	Ap cm 4th 14	Kg cm 5th 15	At cm sq 16
CA-370	6.35	1.91	2.38	6.51	10.32	8.73	8.57	25.4	2227.2	1160.9	21.1	11.492	15.498	178.101	38.866365	470.5
CA-368-A	3.81	2.22	2.70	9.53	13.97	6.51	9.84	33.3	2046.8	1565.0	17.1	8.046	25.708	206.850	38.887914	537.1
CA-116-A	3.81	1.27	6.35	15.24	17.78	10.16	15.24	48.3	1692.6	7207.9	20.9	4.597	96.774	444.847	39.051036	1148.4
CA-202-D	5.72	1.75	2.38	8.89	12.38	8.10	8.25	29.5	2135.6	1466.0	19.5	9.480	21.167	200.653	39.064813	505.3
CA-247-A	3.49	2.54	2.54	9.21	14.29	6.03	10.16	33.7	2164.4	1385.7	16.7	8.429	23.388	197.131	39.888611	533.7
CA-368-E	7.14	1.67	2.70	6.67	10.00	9.84	8.73	25.4	2192.8	1476.8	23.1	11.314	17.997	203.610	39.930980	507.5
CA-189	5.08	2.70	3.02	4.13	9.53	8.10	11.43	25.1	2492.9	934.6	21.1	13.025	12.450	162.167	40.025917	492.0
CA-440	1.27	2.54	5.72	32.39	37.47	6.99	16.51	86.4	2019.3	11320.5	17.2	3.065	185.080	567.180	40.420332	1875.0
CA-188-A	3.18	2.70	2.54	9.84	15.24	5.72	10.48	35.6	2208.9	1453.0	16.3	8.141	25.001	203.531	40.551656	558.5
CA-371-B	6.35	1.91	2.38	6.83	10.64	8.73	8.57	26.0	2282.7	1217.4	21.1	11.492	16.253	186.775	40.759227	481.5
CA-317	2.54	2.54	6.35	8.89	13.97	8.89	17.78	40.6	1900.5	4162.9	20.7	6.129	56.452	345.992	40.903564	932.3
CA-229	4.45	3.18	1.59	6.67	13.02	6.03	9.53	29.2	2988.3	698.4	18.5	13.407	10.589	141.966	41.044979	479.7
CA-112	4.45	2.54	2.86	6.03	11.11	7.30	10.80	27.9	2286.7	1181.7	19.3	10.726	17.242	184.937	41.168310	501.9
CA-95-E	6.35	2.06	2.22	6.35	10.48	8.57	8.57	25.4	2413.2	1050.4	20.9	12.451	14.116	175.760	41.831046	469.6
CA-447	1.27	2.54	8.89	27.94	33.02	10.16	22.86	83.8	1959.9	19595.4	22.2	3.065	248.387	761.183	42.057643	2464.5
CA-240	4.13	2.54	2.54	7.62	12.70	6.67	10.16	30.5	2316.5	1248.5	18.1	9.961	19.355	192.791	42.345787	511.3
CA-435	4.13	2.54	4.45	5.08	10.16	8.57	13.97	29.2	2220.0	1696.7	21.1	9.961	22.581	224.922	42.410816	607.3
CA-440-A	1.27	2.54	6.59	32.39	37.47	7.86	18.26	88.1	2060.1	14089.6	18.6	3.065	213.352	653.821	43.155936	2109.8
CA-449	1.27	2.54	7.62	30.48	35.56	8.89	20.32	86.4	2019.3	16676.2	20.2	3.065	232.258	711.756	43.210131	2283.9
CA-449-A	1.27	2.54	7.62	30.96	36.04	8.89	20.32	87.3	2041.5	16936.7	20.2	3.065	235.885	722.871	43.884934	2311.7
CA-19	5.08	1.91	3.49	7.62	11.43	8.57	10.80	29.8	2093.6	1918.5	20.3	9.194	26.617	244.701	44.393990	574.7
CA-170-P	6.99	2.06	1.91	6.99	11.11	8.89	7.94	26.0	2720.8	1055.4	22.3	13.696	13.306	182.247	44.763238	483.7
CA-440-D	1.27	2.54	6.59	33.97	39.05	7.86	18.26	91.3	2134.4	14780.5	18.6	3.065	223.814	685.881	45.272089	2192.7
CA-371-D	6.99	1.91	2.38	6.83	10.64	9.37	8.57	26.0	2511.0	1313.9	22.7	12.641	16.253	205.452	45.695829	505.9
CA-55-R	5.08	2.54	1.59	8.89	13.97	6.67	8.26	31.1	2910.2	931.2	18.5	12.258	14.117	173.051	45.743572	494.8
CA-70-A	4.45	2.22	3.49	7.30	11.75	7.94	11.43	30.5	2183.4	1781.2	19.6	9.387	25.509	239.461	45.790596	574.5
CA-447-A	1.27	2.54	8.89	30.48	35.56	10.16	22.86	88.9	2078.7	21376.9	22.2	3.065	270.967	830.382	45.881065	2632.3
CA-130	5.08	2.22	2.54	7.62	12.07	7.62	9.53	29.2	2391.2	1335.9	19.4	10.728	19.355	207.642	45.907240	513.8
CA-444-B	1.27	2.54	7.62	32.39	37.47	8.89	20.32	90.2	2108.4	17718.5	20.2	3.065	246.774	756.240	45.910764	2395.2
CA-180-A	5.08	2.86	3.02	4.45	10.16	8.10	11.75	26.4	2773.5	1021.5	21.4	13.793	13.406	184.907	47.610021	524.5

National Arnold C Cores 12 mil Silicon Steel																
Part No.	D cm 1	E cm 2	F cm 3	G cm 4	Ht cm 5	Wth cm 6	Lt cm 7	MPL cm 8	Wtfe grams 9	Wtcu grams 10	MLT cm 11	Ac cm sq 12	Wa cm sq 13	Ap cm 4th 14	Kg cm 5th 15	At cm sq 16
CA-390	4.45	2.22	3.49	7.62	12.07	7.94	11.43	31.1	2228.8	1858.5	19.6	9.387	26.617	249.855	47.778220	586.8
CA-369	7.62	1.91	2.70	5.72	9.53	10.32	9.21	24.4	2572.4	1344.0	24.5	13.790	15.425	212.712	47.884984	520.9
CA-135	3.18	3.18	2.54	8.89	15.24	5.72	11.43	35.6	2598.3	1388.8	17.3	9.577	22.581	216.245	47.893541	588.7
CA-171-B	6.35	1.91	2.54	7.62	11.43	8.89	8.89	27.9	2449.9	1466.9	21.3	11.492	19.355	222.424	47.970298	525.8
CA-404	5.72	1.83	3.49	7.62	11.27	9.21	10.64	29.5	2233.7	2023.8	21.4	9.914	26.617	263.873	48.938003	592.3
CA-371	5.08	2.54	2.38	6.83	11.91	7.46	9.84	28.6	2672.5	1144.0	19.8	12.258	16.253	199.226	49.350373	510.5
CA-428	2.54	2.54	4.45	13.34	18.42	6.99	13.97	45.7	2138.1	3740.6	17.7	6.129	59.274	363.292	50.186911	876.6
CA-439	3.49	1.75	3.97	16.35	19.84	7.46	11.43	47.6	2105.3	3996.4	17.3	5.794	64.897	376.003	50.319618	-864.9
CA-16-F	3.81	2.54	3.81	7.62	12.70	7.62	12.70	33.0	2316.2	2012.9	19.5	9.194	29.032	266.908	50.340920	632.3
CA-234-H	7.62	1.91	2.54	6.35	10.16	10.16	8.89	25.4	2672.6	1391.1	24.3	13.790	16.129	222.424	50.586514	529.0
CA-43	2.86	2.86	3.49	10.64	16.35	6.35	12.70	39.7	2349.9	2315.1	17.5	7.760	37.152	288.287	51.062200	704.6
CA-421-A	2.54	2.54	7.62	10.16	15.24	10.16	20.32	45.7	2138.1	6258.0	22.7	6.129	77.419	474.504	51.175791	1187.1
CA-19-C	5.08	1.91	4.29	7.62	11.43	9.37	12.38	31.4	2204.8	2498.7	21.5	9.194	32.659	300.254	51.320393	663.8
CA-188	5.72	1.91	2.54	9.84	13.65	8.26	8.89	32.4	2555.7	1782.0	20.0	10.343	25.001	258.581	53.371668	576.2
CA-346	4.45	2.54	4.45	5.72	10.80	8.89	13.97	30.5	2494.4	1966.1	21.8	10.726	25.403	272.469	53.709917	651.6
CA-422-H	2.54	2.54	9.68	9.68	14.76	12.22	24.45	48.9	2286.6	8661.1	26.0	6.129	93.780	574.779	54.256059	1483.8
CA-159	2.86	2.86	5.08	8.89	14.61	7.94	15.88	39.4	2331.1	3214.4	20.0	7.760	45.161	350.440	54.344170	831.6
CA-130-A	5.08	2.22	3.18	7.62	12.07	8.26	10.80	30.5	2495.1	1755.6	20.4	10.728	24.194	259.553	54.580613	581.5
CA-317-E	2.54	2.54	2.70	20.32	25.40	5.24	10.48	56.2	2628.1	2926.4	15.0	6.129	54.844	336.138	54.918695	853.4
CA-103	3.81	2.54	3.18	9.53	14.61	6.99	11.43	35.6	2494.4	1989.6	18.5	9.194	30.242	278.030	55.264210	633.1
CA-26	3.49	3.49	3.18	6.35	13.34	6.67	13.34	33.0	2920.4	1402.7	19.6	11.591	20.161	233.689	55.378947	622.7
CA-135-A	5.08	2.30	2.54	8.89	13.49	7.62	9.68	32.1	2718.2	1571.2	19.6	11.110	22.581	250.858	56.969019	567.8
CA-34	5.08	1.98	3.65	8.81	12.78	8.73	11.27	32.9	2400.6	2365.2	20.7	9.575	32.169	308.011	57.054119	650.0
CA-16	5.08	2.54	2.54	7.62	12.70	7.62	10.16	30.5	2850.8	1379.5	20.0	12.258	19.355	237.252	58.037766	554.8
CA-19-D	5.08	2.22	3.49	7.62	12.07	8.57	11.43	31.1	2547.2	1978.7	20.9	10.728	26.617	285.549	58.613271	616.7
CA-185-A	3.49	2.54	3.33	11.11	16.19	6.83	11.75	39.1	2511.6	2359.5	17.9	8.429	37.051	312.286	58.791212	694.2
CA-16-K	6.35	1.75	2.54	10.95	14.45	8.89	8.57	34.0	2730.2	2077.3	21.0	10.533	27.823	293.054	58.805393	621.4
CA-437	5.08	2.38	3.02	7.62	12.38	8.10	10.79	30.8	2700.0	1673.1	20.5	11.491	22.982	264.078	59.286484	585.1
CA-445	1.27	3.18	8.89	26.99	33.34	10.16	24.13	84.5	2468.5	20011.3	23.5	3.831	239.923	919.059	60.039017	2550.8
CA-35-B	4.45	2.54	3.49	7.62	12.70	7.94	12.07	32.4	2650.4	1918.5	20.3	10.726	26.617	285.485	60.425153	628.7

Part No.	D cm 1	E cm 2	F cm 3	G cm 4	Ht cm 5	Wth cm 6	Lt cm 7	MPL cm 8	Wtfe grams 9	Wtcu grams 10	MLT cm 11	Ac cm sq 12	Wa cm sq 13	Ap cm 4th 14	Kg cm 5th 15	At cm sq 16
National Arnold C Cores 12 mil Silicon Steel																
CA-202	4.45	2.38	3.33	8.89	13.65	7.78	11.43	34.0	2606.2	2076.6	19.7	10.054	29.639	298.004	60.830052	638.8
CA-307-A	3.81	2.54	5.08	7.62	12.70	8.89	15.24	35.6	2494.4	2958.3	21.5	9.194	38.710	355.878	60.894003	780.6
CA-37	5.72	2.14	3.02	7.94	12.22	8.73	10.32	30.5	2705.8	1810.6	21.3	11.635	23.941	278.551	60.956183	594.2
CA-404-A	5.72	2.06	3.49	7.62	11.75	9.21	11.11	30.5	2606.3	2068.8	21.9	11.206	26.617	298.266	61.164861	624.5
CA-403	6.35	1.91	3.49	7.62	11.43	9.84	10.80	29.8	2617.0	2158.9	22.8	11.492	26.617	305.876	61.641428	631.1
CA-445-A	1.27	3.18	8.89	27.94	34.29	10.16	24.13	86.4	2524.1	20717.2	23.5	3.831	248.387	951.479	62.156890	2616.1
CA-19-A	5.08	1.91	3.81	10.16	13.97	8.89	11.43	35.6	2494.4	2858.7	20.8	9.194	38.710	355.878	63.016586	712.9
CA-276-A	5.40	2.70	2.54	6.83	12.22	7.94	10.48	29.5	3118.3	1294.6	21.0	13.841	17.338	239.971	63.271035	561.7
CA-413-C	5.08	2.54	2.86	7.62	12.70	7.94	10.80	31.1	2910.2	1590.9	20.5	12.258	21.778	266.955	63.716824	589.2
CA-55-P	5.08	3.14	1.59	8.89	15.16	6.67	9.45	33.5	3866.7	990.9	19.7	15.130	14.117	213.588	65.483715	574.7
CA-445-C	1.27	3.18	8.89	29.85	36.20	10.16	24.13	90.2	2635.5	22129.7	23.5	3.831	265.322	1016.353	66.394859	2746.8
CA-282	10.16	1.91	1.91	7.62	11.43	12.07	7.62	26.7	3741.6	1462.7	28.3	18.387	14.516	266.908	69.276025	604.8
CA-6-J	6.03	3.18	2.06	5.72	12.07	8.10	10.48	28.3	3923.4	942.6	22.5	18.197	11.796	214.648	69.524238	565.5
CA-43-A	2.86	2.86	5.72	10.64	16.35	8.57	17.15	44.1	2613.0	4541.9	21.0	7.760	60.785	471.675	69.674186	1004.4
CA-317-B	3.49	2.54	6.35	8.89	13.97	9.84	17.78	40.6	2613.6	4545.5	22.6	8.429	56.452	475.808	70.844108	992.8
CA-364	3.49	3.49	3.49	7.62	14.61	6.99	13.97	36.2	3201.3	1899.0	20.1	11.591	26.617	308.514	71.291437	709.8
CA-331	6.35	2.06	3.49	7.62	11.75	9.84	11.11	30.5	2895.8	2189.0	23.1	12.451	26.617	331.406	71.365667	653.5
CA-6-L	5.56	3.49	2.06	5.72	12.70	7.62	11.11	29.5	4154.1	929.3	22.2	18.437	11.796	217.476	72.392440	587.6
CA-311-A	2.54	2.54	5.08	17.78	22.86	7.62	15.24	55.9	2613.2	6020.2	18.7	6.129	90.322	553.588	72.407663	1174.2
CA-185-B	4.76	2.38	3.33	9.53	14.29	8.10	11.43	35.2	2897.0	2296.7	20.3	10.774	31.756	342.132	72.493886	679.6
CA-140-A	5.08	2.54	3.49	7.62	12.70	8.57	12.07	32.4	3029.0	2038.7	21.5	12.258	26.617	326.268	74.269366	660.1
CA-35	5.08	2.54	3.49	7.62	12.70	8.57	12.07	32.4	3029.0	2038.7	21.5	12.258	26.617	326.268	74.269366	660.1
CA-413-A	5.08	2.54	2.86	8.89	13.97	7.94	10.80	33.7	3147.8	1856.1	20.5	12.258	25.408	311.448	74.336295	636.8
CA-102	3.81	3.33	3.49	7.62	14.29	7.30	13.65	35.6	3274.3	1948.6	20.6	12.067	26.617	321.194	75.305645	705.1
CA-441-B	1.27	3.81	7.78	26.67	34.29	9.05	23.18	84.1	2951.0	16954.2	23.0	4.597	207.466	953.672	76.303049	2392.5
CA-441-C	1.27	3.81	7.78	26.83	34.45	9.05	23.18	84.5	2962.2	17055.3	23.0	4.597	208.703	959.358	76.757949	2402.7
CA-422	3.81	2.54	7.62	7.62	12.70	11.43	20.32	40.6	2850.8	5260.9	25.5	9.194	58.064	533.817	77.045155	1116.1
CA-332-E	5.08	2.22	3.18	10.80	15.24	8.26	10.80	36.8	3014.9	2487.1	20.4	10.728	34.274	367.700	77.322535	702.5
CA-332	5.08	2.54	3.02	8.89	13.97	8.10	11.11	34.0	3177.4	1982.3	20.8	12.258	26.812	328.666	77.509917	655.3
CA-43-B	2.86	2.86	6.99	10.64	16.35	9.84	19.69	46.7	2763.4	6077.9	23.0	7.760	74.292	576.491	77.777007	1193.5

National Arnold C Cores 12 mil Silicon Steel

Part No.	D cm 1	E cm 2	F cm 3	G cm 4	Ht cm 5	Wth cm 6	Lt cm 7	MPL cm 8	Wtfe grams 9	Wtcu grams 10	MLT cm 11	Ac cm sq 12	Wa cm sq 13	Ap cm 4th 14	Kg cm 5th 15	At cm sq 16
CA-307-J	5.08	3.81	2.22	5.72	13.34	7.30	12.07	31.1	4365.4	997.8	22.1	18.387	12.704	233.597	77.789512	625.1
CA-423	5.08	3.02	2.54	7.62	13.65	7.62	11.11	32.4	3596.4	1445.0	21.0	14.555	19.355	281.713	78.118440	619.5
CA-285	3.18	3.18	3.81	10.80	17.15	6.99	13.97	41.9	3062.3	2821.2	19.3	9.577	41.129	393.875	78.217561	816.9
CA-301	7.62	0.95	8.65	15.24	17.15	16.27	19.21	51.6	2715.9	14978.7	31.9	6.899	131.856	909.647	78.576527	1776.0
CA-155	7.62	0.95	8.65	15.24	17.15	16.27	19.21	51.6	2715.9	14978.7	31.9	6.899	131.856	909.647	78.576527	1776.0
CA-53	5.72	2.22	3.33	8.89	13.34	9.05	11.11	33.3	3070.2	2311.0	21.9	12.069	29.639	357.723	78.762414	679.6
CA-16-A	5.08	2.54	3.81	7.62	12.70	8.89	12.70	33.0	3088.3	2275.1	22.0	12.258	29.032	355.878	79.180044	696.8
CA-215	6.03	3.18	2.86	5.08	11.43	8.89	12.07	28.6	3967.6	1224.6	23.7	18.197	14.519	264.196	81.075526	621.0
CA-119-A	6.35	2.22	3.49	7.62	12.07	9.84	11.43	31.1	3184.0	2219.1	23.4	13.410	26.617	356.936	81.661657	676.3
CA-16-C	5.08	2.54	2.54	10.95	16.03	7.62	10.16	37.1	3474.4	1983.1	20.0	12.258	27.823	341.057	83.431193	673.4
CA-166	2.54	3.02	3.81	18.42	24.45	6.35	13.65	56.5	3138.1	4416.5	17.7	7.278	70.161	510.605	83.969010	1059.4
CA-426	5.08	3.81	1.59	8.26	15.88	6.67	10.80	34.9	4899.9	983.1	21.1	18.387	13.109	241.035	84.060675	650.1
CA-358-F	5.08	2.54	3.49	8.89	13.97	8.57	12.07	34.9	3266.6	2378.5	21.5	12.258	31.053	380.646	86.647594	712.6
CA-16-D	5.08	2.54	2.54	11.43	16.51	7.62	10.16	38.1	3563.4	2069.3	20.0	12.258	29.032	355.878	87.056649	690.3
CA-321	5.72	2.38	3.18	9.21	13.97	8.89	11.11	34.3	3382.1	2286.4	22.0	12.927	29.235	377.927	88.856247	695.9
CA-53-D	3.18	3.18	6.35	8.89	15.24	9.53	19.05	43.2	3155.1	4672.8	23.3	9.577	56.452	540.613	88.965428	1072.6
CA-16-B	5.08	2.54	2.54	11.75	16.83	7.62	10.16	38.7	3622.9	2126.9	20.0	12.258	29.840	365.779	89.478697	701.6
CA-185-D	5.08	2.54	3.33	9.53	14.61	8.41	11.75	35.9	3355.6	2404.2	21.3	12.258	31.756	389.271	89.649780	718.9
CA-441-D	1.27	3.81	10.16	27.94	35.56	11.43	27.94	91.4	3207.1	26971.5	26.7	4.597	283.870	1304.886	89.796881	3129.0
CA-398	4.45	2.86	2.86	10.80	16.51	7.30	11.43	38.7	3567.1	2184.2	19.9	12.069	30.852	372.342	90.283689	720.3
CA-436	2.54	2.54	11.43	15.24	20.32	13.97	27.94	63.5	2969.5	17785.8	28.7	6.129	174.193	1067.634	91.157663	2251.6
CA-53-H	3.18	3.18	7.62	8.26	14.61	10.80	21.59	44.5	3247.9	5652.8	25.3	9.577	62.903	602.397	91.311372	1218.5
CA-362	3.81	3.81	2.54	9.53	17.15	6.35	12.70	39.4	4142.5	1724.4	20.0	13.790	24.194	333.636	91.817559	737.1
CA-393	6.35	2.22	3.33	8.89	13.34	9.68	11.11	33.3	3411.4	2444.8	23.2	13.410	29.639	397.470	91.913800	710.7
CA-413-D	6.35	2.54	2.86	7.94	13.02	9.21	10.80	31.8	3712.2	1862.2	23.1	15.323	22.687	347.620	92.300069	661.6
CA-372	6.35	2.54	3.02	7.62	12.70	9.37	11.11	31.4	3674.7	1906.7	23.3	15.323	22.982	352.142	92.506619	667.0
CA-394	5.08	2.54	3.33	9.84	14.92	8.41	11.75	36.5	3415.1	2484.5	21.3	12.258	32.817	402.267	92.642812	731.7
CA-317-D	2.54	2.54	6.35	20.32	25.40	8.89	17.78	63.5	2969.5	9515.1	20.7	6.129	129.032	790.840	93.493862	1541.9
CA-53-E	3.18	3.18	6.99	8.89	15.24	10.16	20.32	44.5	3247.9	5360.2	24.3	9.577	62.097	594.674	93.842786	1164.5
CA-286	6.03	2.54	4.45	6.35	11.43	10.48	13.97	31.8	3526.6	2503.3	24.9	14.558	28.226	410.900	95.935426	761.3

National Arnold C Cores 12 mil Silicon Steel																
Part No.	D cm 1	E cm 2	F cm 3	G cm 4	Ht cm 5	Wth cm 6	Lt cm 7	MPL cm 8	Wtfe grams 9	Wtcu grams 10	MLT cm 11	Ac cm sq 12	Wa cm sq 13	Ap cm 4th 14	Kg cm 5th 15	At cm sq 16
CA-54	6.35	2.54	3.02	7.94	13.02	9.37	11.11	32.1	3749.1	1986.3	23.3	15.323	23.941	366.837	96.367131	680.0
CA-274-L	5.72	2.06	4.13	10.64	14.76	9.84	12.38	37.8	3230.6	3568.3	22.9	11.206	43.905	492.003	96.493178	831.9
CA-160	3.49	3.49	6.35	6.99	13.97	9.84	19.69	40.6	3594.4	3872.1	24.5	11.591	44.355	514.116	97.095484	1035.2
CA-332-C	5.08	2.54	3.18	10.80	15.88	8.26	11.43	38.1	3563.4	2564.4	21.0	12.258	34.274	420.134	97.905529	749.2
CA-413-B	6.35	2.54	2.86	8.57	13.65	9.21	10.80	33.0	3860.6	2046.0	23.5	15.323	24.502	375.428	97.985643	687.0
CA-54-F	6.35	2.54	3.18	7.94	13.02	9.53	11.43	32.4	3786.3	2149.2	24.0	15.323	25.203	386.177	98.699317	698.8
CA-358-B	5.08	2.82	3.33	8.89	14.53	8.41	12.30	35.7	3706.5	2302.5	21.8	13.600	29.639	403.084	100.370128	733.8
CA-187-A	5.72	3.18	2.54	7.62	13.97	8.26	11.43	33.0	4343.0	1554.3	22.6	17.238	19.355	333.636	101.863548	674.2
CA-111-D	7.62	2.54	2.54	7.62	12.70	10.16	10.16	30.5	4276.1	1756.7	25.5	18.387	19.355	355.878	102.548174	671.0
CA-53-A	5.72	1.91	6.51	9.68	13.49	12.22	16.83	40.0	3157.1	5889.5	26.3	10.343	63.033	651.934	102.648784	1096.3
CA-119	6.35	2.54	3.49	7.62	12.70	9.84	12.07	32.4	3786.3	2279.1	24.1	15.323	26.617	407.835	103.805169	723.0
CA-422-F	3.81	2.54	6.35	11.43	16.51	10.16	17.78	45.7	3207.1	6061.5	23.5	9.194	72.581	667.271	104.482783	1154.8
CA-276	5.08	2.54	2.86	12.70	17.78	7.94	10.80	41.3	3860.5	2651.5	20.5	12.258	36.297	444.925	106.194707	779.5
CA-417-A	5.08	2.54	2.54	13.97	19.05	7.62	10.16	43.2	4038.6	2529.1	20.0	12.258	35.484	434.962	106.402571	780.6
CA-27-A	6.35	2.70	3.18	7.62	13.02	9.53	11.75	32.4	4023.3	2056.1	23.9	16.282	24.194	393.912	107.345523	709.1
CA-196	5.72	2.86	3.49	7.62	13.34	9.21	12.70	33.7	3984.9	2219.1	23.4	15.517	26.617	413.005	109.332362	738.4
CA-332-A	5.08	2.54	4.92	8.89	13.97	10.00	14.92	37.8	3533.7	3699.7	23.8	12.258	43.748	536.261	110.562884	894.8
CA-442-A	3.81	1.91	7.30	19.37	23.18	11.11	18.42	61.0	3207.2	11926.4	23.7	6.895	141.445	975.281	113.441070	1666.9
CA-327	4.45	3.18	3.18	10.48	16.83	7.62	12.70	40.0	4092.5	2489.1	21.0	13.407	33.268	446.027	113.683935	796.8
CA-266	5.72	2.54	3.49	9.84	14.92	9.21	12.07	36.8	3875.5	2788.8	22.8	13.790	34.382	474.132	114.658881	786.2
CA-132	5.08	3.81	2.54	7.62	15.24	7.62	12.70	35.6	4988.8	1554.3	22.6	18.387	19.355	355.878	115.898081	735.5
CA-324	5.72	2.86	3.18	8.89	14.61	8.89	12.07	35.6	4210.3	2303.2	22.9	15.517	28.226	437.973	118.464555	754.1
CA-132-A	6.03	3.65	1.91	8.26	15.56	7.94	11.11	34.9	5575.9	1296.0	23.2	20.925	15.726	329.064	118.848210	713.5
CA-187	6.35	3.18	2.54	7.62	13.97	8.89	11.43	33.0	4825.5	1641.8	23.9	19.153	19.355	370.706	119.062014	706.5
CA-116-C	3.81	1.91	8.18	19.37	23.18	11.99	20.16	62.7	3299.1	14123.9	25.1	6.895	158.353	1091.866	120.061860	1849.0
CA-274-D	5.08	2.54	5.08	9.53	14.61	10.16	15.24	39.4	3682.2	4135.0	24.0	12.258	48.387	593.130	121.017470	948.4
CA-358-D	5.08	3.18	3.33	8.89	15.24	8.41	13.02	37.1	4343.0	2377.8	22.6	15.323	29.639	454.149	123.379507	787.6
CA-332-F	3.81	3.81	3.18	10.80	18.42	6.99	13.97	43.2	4543.4	2564.4	21.0	13.790	34.274	472.650	123.911686	870.2
CA-274-E	4.13	3.18	4.13	10.64	16.99	8.26	14.61	42.2	4011.7	3419.7	21.9	12.451	43.905	546.670	124.305185	913.0
CA-323-B	6.99	2.54	3.18	8.89	13.97	10.16	11.43	34.3	4409.8	2534.4	25.3	16.855	28.226	475.740	127.021918	771.0

National Arnold C Cores 12 mil Silicon Steel																
Part No.	D cm 1	E cm 2	F cm 3	G cm 4	Ht cm 5	Wth cm 6	Lt cm 7	MPL cm 8	Wtfe grams 9	Wtcu grams 10	MLT cm 11	Ac cm sq 12	Wa cm sq 13	Ap cm 4th 14	Kg cm 5th 15	At cm sq 16
CA-54-C	7.62	1.91	3.49	12.38	16.19	11.11	10.80	39.4	4142.7	3960.6	25.8	13.790	43.254	596.483	127.777434	896.3
CA-119-B	6.35	2.86	3.49	7.62	13.34	9.84	12.70	33.7	4427.6	2339.3	24.7	17.241	26.617	458.895	128.042545	771.5
CA-437-B	3.18	3.18	8.89	10.80	17.15	12.07	24.13	52.1	3804.7	9304.6	27.3	9.577	95.968	919.042	129.120812	1597.6
CA-424-A	3.18	3.18	10.16	10.16	16.51	13.34	26.67	53.3	3897.5	10740.2	29.3	9.577	103.226	988.550	129.421697	1774.2
CA-443-A	1.27	3.81	15.24	35.56	43.18	16.51	38.10	116.8	4098.0	66861.0	34.7	4.597	541.934	2491.145	132.022188	5509.7
CA-441	1.27	3.81	15.40	35.56	43.18	16.67	38.42	117.2	4109.1	68044.6	34.9	4.597	547.588	2517.135	132.446628	5568.7
CA-134	7.62	2.54	2.54	9.84	14.92	10.16	10.16	34.9	4899.9	2269.2	25.5	18.387	25.001	459.699	132.464787	761.3
CA-306	5.08	2.54	4.45	11.43	16.51	9.53	13.97	41.9	3919.8	4161.6	23.0	12.258	50.806	622.786	132.567919	951.6
CA-36-B	6.35	2.54	3.49	9.84	14.92	9.84	12.07	36.8	4306.1	2944.0	24.1	15.323	34.382	526.814	134.088488	820.4
CA-287-E	5.72	2.54	3.97	10.64	15.72	9.68	13.02	39.4	4142.5	3536.3	23.6	13.790	42.214	582.147	136.314001	885.1
CA-446	1.27	3.81	15.24	36.83	44.45	16.51	38.10	119.4	4187.1	69248.9	34.7	4.597	561.289	2580.115	136.737266	5648.4
CA-29	6.35	2.54	3.97	9.05	14.13	10.32	13.02	36.2	4231.7	3144.3	24.6	15.323	35.915	550.317	137.002201	847.5
CA-414-A	8.26	2.54	2.86	8.26	13.34	11.11	10.80	32.4	4922.2	2289.8	27.3	19.919	23.593	469.952	137.194235	766.2
CA-307-C	5.08	2.78	5.08	9.21	14.76	10.16	15.72	39.7	4059.8	4076.5	24.5	13.407	46.777	627.117	137.223139	970.5
CA-54-A	6.35	2.22	3.81	11.91	16.35	10.16	12.07	40.3	4126.0	3828.7	23.7	13.410	45.362	608.314	137.474576	904.3
CA-35-A	5.08	3.65	3.49	7.62	14.92	8.57	14.29	36.8	4951.4	2249.0	23.8	17.620	26.617	468.978	139.100495	825.2
CA-33-A	7.62	2.54	4.13	6.99	12.07	11.75	13.34	32.4	4543.5	2872.7	28.0	18.387	28.834	530.174	139.177704	834.7
CA-274-H	5.72	2.86	4.13	8.57	14.29	9.84	13.97	36.8	4360.9	3076.0	24.4	15.517	35.389	549.129	139.438543	861.3
CA-287	5.72	2.54	4.45	10.16	15.24	10.16	13.97	39.4	4142.5	3903.2	24.3	13.790	45.161	622.786	141.345897	929.0
CA-360-A	6.99	2.54	3.65	8.89	13.97	10.64	12.38	35.2	4532.2	3000.7	26.0	16.855	32.457	547.063	141.866528	832.3
CA-311	3.81	2.54	5.08	17.78	22.86	8.89	15.24	55.9	3919.8	6902.8	21.5	9.194	90.322	830.382	142.086006	1271.0
CA-446-A	1.27	3.81	15.24	39.37	46.99	16.51	38.10	124.5	4365.2	74024.7	34.7	4.597	599.999	2758.053	146.167422	5925.8
CA-441-A	1.27	3.81	15.40	39.53	47.15	16.67	38.42	125.1	4387.5	75639.4	34.9	4.597	608.707	2798.083	147.229549	6006.0
CA-307-E	6.99	2.86	3.49	7.78	13.50	10.48	12.70	34.0	4916.4	2549.5	26.4	18.965	27.172	515.317	148.153906	811.9
CA-443	1.27	3.81	15.24	40.01	47.63	16.51	38.10	125.7	4409.8	75218.6	34.7	4.597	609.676	2802.538	148.524962	5995.1
CA-287-C	5.08	3.06	4.92	8.89	15.00	10.00	15.95	39.8	4483.8	3860.2	24.8	14.748	43.748	645.202	153.391113	977.6
CA-421-C	6.67	2.54	4.45	8.89	13.97	11.11	13.97	36.8	4521.5	3739.3	26.6	16.090	39.516	635.809	153.774336	921.0
CA-332-B	5.08	3.81	3.02	8.89	16.51	8.10	13.65	39.1	5478.7	2224.5	23.3	18.387	26.812	492.998	155.411119	847.2
CA-450	2.86	2.86	8.26	19.69	25.40	11.11	22.23	67.3	3985.3	14446.4	25.0	7.760	162.500	1260.958	156.553401	1998.9
CA-133	5.08	2.54	5.08	12.70	17.78	10.16	15.24	45.7	4276.1	5513.3	24.0	12.258	64.516	790.840	161.356627	1109.7

National Arnold C Cores 12 mil Silicon Steel																
Part No.	D cm 1	E cm 2	F cm 3	G cm 4	Ht cm 5	Wth cm 6	Lt cm 7	MPL cm 8	Wtfe grams 9	Wtcu grams 10	MLT cm 11	Ac cm sq 12	Wa cm sq 13	Ap cm 4th 14	Kg cm 5th 15	At cm sq 16
CA-357	8.26	2.54	3.49	8.26	13.34	11.75	12.07	33.7	5115.2	2900.8	28.3	19.919	28.835	574.368	161.767529	847.6
CA-317-C	3.81	3.81	6.35	8.89	16.51	10.16	20.32	45.7	4810.7	5224.4	26.0	13.790	56.452	778.483	164.999835	1225.8
CA-414	9.53	2.54	2.86	8.26	13.34	12.38	10.80	32.4	5679.4	2502.9	29.8	22.984	23.593	542.253	167.103715	827.5
CA-274-C	4.13	3.81	4.13	10.64	18.26	8.26	15.88	44.8	5103.6	3617.9	23.2	14.941	43.905	656.004	169.189355	1017.4
CA-274-F	6.35	2.54	5.08	9.53	14.61	11.43	15.24	39.4	4602.8	4572.0	26.6	15.323	48.387	741.412	171.014556	1024.2
CA-274-A	5.72	2.86	4.13	10.64	16.35	9.84	13.97	41.0	4849.4	3816.2	24.4	15.517	43.905	681.271	172.992924	959.5
CA-274	6.35	2.62	4.13	10.64	15.87	10.48	13.49	40.0	4822.4	3939.9	25.2	15.799	43.905	693.667	173.716490	957.4
CA-287-A	5.72	2.86	4.45	10.16	15.88	10.16	14.61	40.6	4811.7	4005.3	24.9	15.517	45.161	700.757	174.390100	981.5
CA-275	10.16	2.54	5.08	5.08	10.16	15.24	15.24	30.5	5701.5	3174.4	34.6	24.516	25.806	632.672	179.357204	980.6
CA-317-A	6.35	2.54	6.35	8.89	13.97	12.70	17.78	40.6	4751.3	5734.3	28.6	15.323	56.452	864.981	185.590500	1174.2
CA-287-B	5.56	2.54	6.03	11.43	16.51	11.59	17.15	45.1	4612.0	6493.2	26.5	13.407	68.957	924.483	187.224983	1223.6
CA-441-F	1.27	3.81	15.24	50.80	58.42	16.51	38.10	147.3	5167.0	95515.7	34.7	4.597	774.192	3558.779	188.603126	7174.2
CA-19-F	7.62	2.54	3.81	10.16	15.24	11.43	12.70	38.1	5345.2	3787.8	27.5	18.387	38.710	711.756	190.235310	948.4
CA-53-H	5.08	3.18	6.35	8.89	15.24	11.43	19.05	43.2	5048.2	5479.3	27.3	15.323	56.452	864.981	194.225621	1203.2
CA-433	7.62	1.91	8.89	10.00	13.81	16.51	21.59	45.4	4777.2	10820.0	34.2	13.790	88.909	1226.080	197.619779	1629.8
CA-323	5.72	4.29	3.02	7.94	16.51	8.73	14.60	39.1	6933.6	2175.4	25.6	23.270	23.941	557.102	202.928246	922.1
CA-261	6.35	2.54	5.72	11.11	16.19	12.07	16.51	43.8	5122.6	6179.2	27.4	15.323	63.511	973.147	217.994136	1207.7
CA-316	3.18	3.18	7.62	20.32	26.67	10.80	21.59	68.6	5011.1	13914.6	25.3	9.577	154.838	1482.824	224.766454	2000.0
CA-459	7.14	2.78	4.13	10.64	16.19	11.27	13.81	40.6	5846.2	4299.9	27.5	18.854	43.905	827.781	226.669726	1032.0
CA-323-A	5.72	4.45	3.18	8.10	16.99	8.89	15.24	40.3	7424.7	2387.6	26.1	24.133	25.705	620.334	229.251290	977.8
CA-424-B	6.67	2.22	10.16	10.16	14.61	16.83	24.77	49.5	5321.9	12828.8	34.9	14.082	103.226	1453.604	234.275831	1867.9
CA-287-E	6.35	3.18	4.45	10.16	16.51	10.80	15.24	41.9	6124.7	4311.1	26.8	19.153	45.161	864.981	246.859497	1075.8
CA-316-A	4.45	3.18	7.62	12.70	19.05	12.07	21.59	53.3	5456.5	9642.3	28.0	13.407	96.774	1297.471	248.335069	1609.7
CA-337	3.18	3.18	12.70	17.78	24.13	15.88	31.75	73.7	5382.3	26696.2	33.2	9.577	225.806	2162.452	249.152434	2983.9
CA-274-B	6.35	3.18	5.08	9.53	15.88	11.43	16.51	41.9	6124.7	4790.5	27.8	19.153	48.387	926.765	255.021396	1135.5
CA-319	3.49	3.49	7.30	19.05	26.04	10.80	21.59	66.7	5897.0	12885.3	26.0	11.591	139.122	1612.564	287.052692	1948.6
CA-317-F	5.08	2.54	6.35	20.32	25.40	11.43	17.78	63.5	5939.1	11941.5	26.0	12.258	129.032	1581.679	297.989119	1761.3
CA-27-E	10.16	3.49	3.18	7.62	14.61	13.34	13.34	35.6	9148.0	2882.7	33.5	33.714	24.194	815.670	328.290401	1046.1
CA-330-B	7.94	3.33	4.76	14.76	21.43	12.70	16.19	52.4	10050.2	7811.4	31.2	25.142	70.321	1768.011	569.197823	1522.9

National Arnold C Cores 12 mil Silicon Steel

Part No.	D cm 1	E cm 2	F cm 3	G cm 4	Ht cm 5	Wth cm 6	Lt cm 7	MPL cm 8	Wtfe grams 9	Wtcu grams 10	MLT cm 11	Ac cm sq 12	Wa cm sq 13	Ap cm 4th 14	Kg cm 5th 15	At cm sq 16
CS-347-E	4.45	5.08	3.81	5.08	15.24	8.26	17.78	38.1	6236	1779	25.8	21.45	19.36	415.19	137.83	932
CS-348	6.350	2.858	3.493	9.843	15.559	9.843	12.702	38.1	5013	3071	25.1	17.24	34.38	592.77	162.76	872
CS-154-A	7.620	3.651	2.223	7.620	14.922	9.843	11.748	34.3	6915	1641	27.2	26.43	16.94	447.70	173.70	811
CS-335-D	6.350	2.223	3.810	16.034	20.480	10.160	12.066	48.6	4971	5288	24.3	13.41	61.09	819.23	180.52	1088
CS-309-A	3.175	5.080	5.080	10.160	20.320	8.255	20.320	50.8	5939	4606	25.1	15.32	51.61	790.84	193.16	1290
CS-300-A	8.573	2.858	3.493	7.620	13.336	12.066	12.702	33.7	5978	2798	29.6	23.28	26.62	619.54	195.13	887
CS-347-D	4.445	6.350	3.810	5.080	17.780	8.255	20.320	43.2	8834	1954	28.4	26.81	19.36	518.99	196.09	1158
CS-298-C	8.255	3.810	2.540	6.350	13.970	10.795	12.700	33.0	7528	1682	29.3	29.88	16.13	481.92	196.35	853
CS-299-A	5.715	4.286	3.016	7.938	16.510	8.731	14.604	39.1	6934	2175	25.6	23.27	23.94	557.10	202.93	922
CS-307-A	6.350	2.858	3.493	12.383	18.099	9.843	12.702	43.2	5681	3863	25.1	17.24	43.25	745.73	204.76	986
CS-306-B	6.350	2.540	4.445	12.700	17.780	10.795	13.970	44.5	5197	5134	25.6	15.32	56.45	864.98	207.29	1092
CS-298-E	9.525	2.540	2.540	11.430	16.510	12.065	10.160	38.1	6682	3028	29.3	22.98	29.03	667.27	209.13	927
CS-357-A	7.144	2.858	3.810	10.160	15.876	10.954	13.336	39.4	5827	3744	27.2	19.40	38.71	750.84	214.16	976
CS-306-D	7.620	2.540	4.445	10.160	15.240	12.065	13.970	39.4	5523	4579	28.5	18.39	45.16	830.38	214.18	1040
CS-154-I	3.810	1.905	13.970	27.940	31.750	17.780	31.750	91.4	4811	47440	34.2	6.90	390.32	2691.33	217.18	4061
CS-154-C	6.985	2.778	4.128	10.636	16.192	11.113	13.812	40.6	5716	4250	27.2	18.43	43.91	809.36	219.22	1022
CS-357-C	4.445	4.445	3.810	10.160	19.050	8.255	16.510	45.7	6548	3383	24.6	18.77	38.71	726.58	221.96	1081
CS-154	7.144	2.778	4.128	10.636	16.192	11.272	13.812	40.6	5846	4300	27.5	18.85	43.91	827.78	226.67	1032
CS-122-A	6.350	2.540	5.715	11.748	16.828	12.065	16.510	45.1	5271	6582	27.6	15.32	67.14	1028.75	228.71	1244
CS-303	3.810	3.493	8.255	12.383	19.369	12.065	23.496	55.2	5330	10317	28.4	12.64	102.22	1292.38	230.28	1714
CS-352-A	3.493	3.493	5.080	19.050	26.036	8.573	17.146	62.2	5504	7762	22.6	11.59	96.77	1121.71	230.57	1491
CS-301-H	5.715	3.175	3.810	12.700	19.050	9.525	13.970	45.7	6013	4229	24.6	17.24	48.39	834.09	234.00	1065
CS-290	6.350	3.175	3.175	12.700	19.050	9.525	12.700	44.5	6496	3563	24.9	19.15	40.32	772.30	238.09	1008
CS-317-B	6.985	2.858	2.858	14.764	20.480	9.843	11.432	46.7	6754	3810	25.4	18.97	42.20	800.24	239.10	1031
CS-162	8.890	3.175	1.905	12.383	18.733	10.795	10.160	41.3	8445	2377	28.3	26.81	23.59	632.54	239.42	956
CS-81-B	3.810	3.810	10.160	10.160	17.780	13.970	27.940	55.9	5880	11749	32.0	13.79	103.23	1423.51	245.33	1961
CS-344	7.144	2.699	4.445	11.748	17.146	11.589	14.288	43.2	6035	5177	27.9	18.32	52.22	956.54	251.38	1123
CS-301-A	6.985	2.858	3.810	12.700	18.416	10.795	13.336	44.5	6432	4626	26.9	18.97	48.39	917.66	258.94	1090
CS-122-B	6.350	2.540	6.350	12.700	17.780	12.700	17.780	48.3	5642	8307	29.0	15.32	80.65	1235.69	261.47	1406
CS-350	6.350	3.810	4.128	8.573	16.193	10.478	15.876	40.6	7127	3526	28.0	22.98	35.39	813.38	266.90	1060

National Arnold C Cores 12 mil Silicon Steel																
Part No.	D cm 1	E cm 2	F cm 3	G cm 4	Ht cm 5	Wth cm 6	Lt cm 7	MPL cm 8	Wtfe grams 9	Wtcu grams 10	MLT cm 11	Ac cm sq 12	Wa cm sq 13	Ap cm 4th 14	Kg cm 5th 15	At cm sq 16
CS-166	7.144	2.778	4.445	11.906	17.462	11.589	14.446	43.8	6303	5277	28.0	18.85	52.92	997.78	268.37	1145
CS-317-C	5.715	4.445	3.175	9.525	18.415	8.890	15.240	43.2	7951	2809	26.1	24.13	30.24	729.83	269.72	1047
CS-421	4.445	5.080	4.445	8.890	19.050	8.890	19.050	47.0	7691	3772	26.8	21.45	39.52	847.68	270.95	1223
CS-156-C	6.985	3.175	5.080	8.890	15.240	12.065	16.510	40.6	6533	4739	29.5	21.07	45.16	951.48	271.71	1140
CS-328	5.080	3.810	5.398	10.160	17.780	10.478	18.416	46.4	6503	5280	27.1	18.39	54.84	1008.41	273.97	1249
CS-343	6.350	3.175	5.080	10.160	16.510	11.430	16.510	43.2	6310	5072	27.6	19.15	51.61	988.55	274.07	1171
CS-185	5.080	3.810	6.985	8.890	16.510	12.065	21.590	47.0	6592	6528	29.6	18.39	62.10	1141.78	284.06	1419
CS-421-A	4.445	5.080	4.763	8.890	19.050	9.208	19.686	47.6	7795	4117	27.3	21.45	42.34	908.33	285.04	1270
CS-307-F	6.033	3.175	4.445	12.700	19.050	10.478	15.240	47.0	6524	5262	26.2	18.20	56.45	1027.25	285.27	1186
CS-317	7.938	2.858	2.858	14.764	20.480	10.796	11.432	46.7	7676	4096	27.3	21.55	42.20	909.42	287.23	1092
CS-307-C	6.350	3.056	4.445	12.700	18.812	10.795	15.002	46.5	6543	5341	26.6	18.44	56.45	1040.70	288.43	1186
CS-356-F	6.985	3.175	5.715	8.890	15.240	12.700	17.780	41.9	6737	5512	30.5	21.07	50.81	1070.41	295.68	1236
CS-299	8.573	3.810	3.016	7.938	15.558	11.589	13.652	37.1	8795	2615	30.7	31.03	23.94	742.89	300.18	1010
CS-298-F	5.715	2.540	3.493	25.876	30.956	9.208	12.066	68.9	7249	7331	22.8	13.79	90.39	1246.43	301.42	1468
CS-171	6.509	4.286	2.540	11.430	20.002	9.049	13.652	45.1	9117	2766	26.8	26.50	29.03	769.43	304.43	1068
CS-166-A	7.144	3.016	4.445	11.906	17.938	11.589	14.922	44.8	6992	5366	28.5	20.47	52.92	1083.26	311.04	1190
CS-156	7.144	2.778	5.080	12.700	18.256	12.224	15.716	46.7	6714	6661	29.0	18.85	64.52	1216.37	315.93	1292
CS-155	5.080	4.445	5.080	9.525	18.415	10.160	19.050	47.0	7691	4791	27.8	21.45	48.39	1037.98	319.90	1282
CS-81	4.445	3.810	10.160	10.160	17.780	14.605	27.940	55.9	6860	12215	33.3	16.09	103.23	1660.76	321.17	2019
CS-27	10.160	3.493	3.175	7.620	14.606	13.335	13.336	35.6	9148	2883	33.5	33.71	24.19	815.67	328.29	1046
CS-156-A	5.080	3.810	5.080	12.700	20.320	10.160	17.780	50.8	7127	6096	26.6	18.39	64.52	1186.26	328.35	1342
CS-347-C	6.350	5.080	3.334	7.620	17.780	9.684	16.828	42.2	9874	2612	28.9	30.65	25.41	778.54	330.10	1133
CS-274	6.350	3.810	4.128	10.636	18.256	10.478	15.876	44.8	7851	4374	28.0	22.98	43.91	1009.11	331.13	1169
CS-152-C	7.938	3.016	5.715	9.208	15.240	13.653	17.462	41.9	7273	6006	32.1	22.74	52.62	1196.87	339.25	1287
CS-280-A	7.144	3.572	4.128	10.478	17.622	11.272	15.400	43.5	8046	4480	29.1	24.24	43.25	1048.56	349.06	1169
CS-478	7.144	3.572	4.128	10.636	17.780	11.272	15.400	43.8	8105	4548	29.1	24.24	43.91	1064.38	354.33	1178
CS-328-A	8.255	2.858	5.398	10.478	16.194	13.653	16.512	43.2	7385	6419	31.9	22.41	56.56	1267.69	356.09	1304
CS-346	6.985	5.080	4.445	5.715	15.875	11.430	19.050	40.6	10453	2920	32.3	33.71	25.40	856.33	357.21	1223
CS-297-C	5.080	3.810	4.445	15.240	22.860	9.525	16.510	54.6	7661	6161	25.6	18.39	67.74	1245.57	358.20	1369
CS-337-A	7.620	4.445	4.445	6.350	15.240	12.065	17.780	39.4	9666	3244	32.3	32.18	28.23	908.23	361.64	1181

National Arnold C Cores 12 mil Silicon Steel

Part No.	D cm 1	E cm 2	F cm 3	G cm 4	Ht cm 5	Wth cm 6	Lt cm 7	MPL cm 8	Wtfe grams 9	Wtcu grams 10	MLT cm 11	Ac cm sq 12	Wa cm sq 13	Ap cm 4th 14	Kg cm 5th 15	At cm sq 16
CS-301	6.985	2.858	6.033	12.700	18.416	13.018	17.782	48.9	7076	8276	30.4	18.97	76.62	1453.08	362.91	1460
CS-307-B	6.350	3.493	4.445	12.700	19.686	10.795	15.876	48.3	7759	5517	27.5	21.07	56.45	1189.52	364.84	1269
CS-355	7.144	3.493	4.445	10.795	17.781	11.589	15.876	44.5	8040	5028	29.5	23.71	47.98	1137.52	366.03	1221
CS-328-B	8.255	2.858	5.398	10.795	16.511	13.653	16.512	43.8	7493	6614	31.9	22.41	58.27	1306.05	366.86	1324
CS-83	3.810	3.810	10.160	15.240	22.860	13.970	27.940	66.0	6949	17623	32.0	13.79	154.84	2135.27	367.99	2387
CS-76	3.810	5.080	9.366	9.843	20.003	13.176	28.892	58.7	8241	10917	33.3	18.39	92.19	1695.10	374.38	2064
CS-280-B	7.620	3.493	4.128	10.636	17.622	11.748	15.242	43.5	8393	4672	29.9	25.29	43.91	1110.19	375.26	1194
CS-318-A	8.573	2.619	3.810	16.034	21.272	12.383	12.858	50.2	8164	6426	29.6	21.33	61.09	1303.04	375.83	1313
CS-152-B	7.938	2.540	5.556	14.288	19.368	13.494	16.192	49.8	7285	8721	30.9	19.15	79.38	1520.56	377.09	1474
CS-149-E	7.620	1.588	10.160	25.400	28.576	17.780	23.496	77.5	6795	32654	35.6	11.50	258.06	2966.58	383.35	3066
CS-152-D	8.255	2.858	5.715	11.113	16.829	13.970	17.146	45.1	7711	7321	32.4	22.41	63.51	1423.48	393.71	1396
CS-156-F	7.144	3.175	5.080	12.700	19.050	12.224	16.510	48.3	7935	6844	29.8	21.55	64.52	1390.20	401.70	1369
CS-348-A	7.144	5.080	3.493	7.620	17.780	10.637	17.146	42.5	11192	2948	31.1	34.48	26.62	917.66	406.29	1211
CS-306-A	5.080	4.763	4.445	12.065	21.591	9.525	18.416	52.1	9133	5241	27.5	22.99	53.63	1232.73	412.45	1388
CS-81-A	4.445	4.445	10.160	10.160	19.050	14.605	29.210	58.4	8367	12681	34.5	18.77	103.23	1937.56	421.08	2161
CS-165	7.938	3.175	4.604	12.383	18.733	12.542	15.558	46.7	8527	6218	30.7	23.94	57.01	1365.02	426.25	1326
CS-79-E	8.573	2.699	5.080	13.811	19.209	13.653	15.558	48.6	8148	7918	31.7	21.98	70.16	1542.23	427.29	1437
CS-347	9.525	3.493	3.810	9.525	16.511	13.335	14.606	40.6	9801	4289	33.2	31.61	36.29	1147.04	436.36	1206
CS-348-B	7.620	5.080	3.493	7.620	17.780	11.113	17.146	42.5	11938	3038	32.1	36.77	26.62	978.80	448.53	1244
CS-333-A	7.620	2.858	5.715	14.605	20.321	13.335	17.146	52.1	8220	9244	31.1	20.69	83.47	1726.87	458.86	1565
CS-369	8.573	3.334	4.604	11.113	17.781	13.177	15.876	44.8	9275	5869	32.3	27.15	51.16	1389.28	467.77	1329
CS-334-A	7.620	3.889	4.604	10.160	17.938	12.224	16.986	45.1	9684	5233	31.5	28.15	46.78	1316.88	471.34	1320
CS-370-A	8.890	2.540	3.810	20.320	25.400	12.700	12.700	58.4	9562	8275	30.1	21.45	77.42	1660.76	474.10	1536
CS-354-A	4.128	4.128	5.080	22.860	31.116	9.208	18.416	72.4	8942	10449	25.3	16.19	116.13	1879.94	481.09	1891
CS-327-A	4.445	4.445	5.715	16.669	25.559	10.160	20.320	62.5	8958	9339	27.6	18.77	95.26	1788.11	486.97	1768
CS-242-A	6.350	3.175	6.033	16.510	22.860	12.383	18.416	57.8	8445	10533	29.7	19.15	99.61	1907.75	491.49	1711
CS-152	7.938	2.937	5.556	14.288	20.162	13.494	16.986	51.4	8692	8946	31.7	22.15	79.38	1758.22	491.55	1557
CS-78	6.350	3.810	7.620	10.160	17.780	13.970	22.860	50.8	8909	9112	33.1	22.98	77.42	1779.39	494.23	1716
CS-177-A	7.938	3.096	4.763	14.764	20.956	12.701	15.718	51.4	9163	7692	30.8	23.35	70.32	1641.80	498.43	1472
CS-154-H	7.620	4.128	4.128	10.636	18.892	11.748	16.512	46.0	10497	4870	31.2	29.88	43.91	1312.01	502.76	1319

National Arnold C Cores 12 mil Silicon Steel

Part No.	D cm 1	E cm 2	F cm 3	G cm 4	Ht cm 5	Wth cm 6	Lt cm 7	MPL cm 8	Wtfe grams 9	Wtcu grams 10	MLT cm 11	Ac cm sq 12	Wa cm sq 13	Ap cm 4th 14	Kg cm 5th 15	At cm sq 16
CS-167	5.080	3.810	8.573	13.970	21.590	13.653	24.766	60.3	8463	13652	32.1	18.39	119.77	2202.12	505.25	2074
CS-352	6.350	3.175	6.350	16.510	22.860	12.700	19.050	58.4	8537	11272	30.2	19.15	104.84	2007.99	508.80	1774
CS-307-D	6.668	4.128	4.445	12.700	20.956	11.113	17.146	50.8	10136	5979	29.8	26.15	56.45	1476.16	518.36	1419
CS-242-B	4.445	4.445	6.033	17.145	26.035	10.478	20.956	64.1	9185	10324	28.1	18.77	103.44	1941.50	519.35	1861
CS-317-A	7.938	3.175	4.763	14.764	21.114	12.701	15.876	51.8	9455	7732	30.9	23.94	70.32	1683.69	521.51	1489
CS-335-E	7.620	4.128	3.493	12.700	20.956	11.113	15.242	48.9	11149	4763	30.2	29.88	44.36	1325.63	524.75	1328
CS-152-A	7.620	3.453	6.033	11.430	18.336	13.653	18.972	48.7	9295	8051	32.8	25.00	68.96	1723.67	524.89	1548
CS-82-A	4.445	4.445	10.160	12.700	21.590	14.605	29.210	63.5	9094	15852	34.5	18.77	129.03	2421.95	526.36	2384
CS-155-A	7.620	4.128	5.080	9.525	17.781	12.700	18.416	45.7	10425	5624	32.7	29.88	48.39	1445.93	528.74	1407
CS-122-C	5.080	5.080	5.715	11.748	21.908	10.795	21.590	55.2	10334	7188	30.1	24.52	67.14	1646.01	536.11	1653
CS-327	4.445	4.445	6.668	16.828	25.718	11.113	22.226	64.8	9276	11597	29.1	18.77	112.21	2106.18	544.07	1972
CS-78-B	5.080	5.080	7.620	9.843	20.003	12.700	25.400	55.2	10334	8828	33.1	24.52	75.00	1838.80	544.78	1860
CS-317-D	6.112	2.699	4.763	30.480	35.878	10.875	14.924	81.3	9719	13379	25.9	15.67	145.18	2275.13	550.31	2073
CS-79-D	8.573	3.175	5.080	13.811	20.161	13.653	16.510	50.5	9960	8155	32.7	25.86	70.16	1814.22	574.07	1538
CS-356-B	11.748	3.016	4.445	10.795	16.827	16.193	14.922	42.5	10927	6437	37.7	33.66	47.98	1615.15	576.49	1423
CS-328-D	5.715	3.810	5.398	17.780	25.400	11.113	18.416	61.6	9722	9673	28.3	20.69	95.98	1985.32	579.62	1745
CS-154-L	4.445	3.810	12.700	16.510	24.130	17.145	33.020	73.7	9042	27785	37.3	16.09	209.68	3373.43	582.57	3166
CS-154-B	3.810	3.810	15.240	20.320	27.940	19.050	38.100	86.4	9087	44030	40.0	13.79	309.68	4270.53	589.17	4181
CS-144	7.303	3.175	5.318	17.463	23.813	12.621	16.986	58.3	9792	10079	30.5	22.03	92.87	2045.67	590.56	1704
CS-237-C	6.350	4.366	5.080	12.700	21.432	11.430	18.892	53.0	10656	6934	30.2	26.34	64.52	1699.22	592.30	1554
CS-481-A	6.350	4.763	3.810	13.970	23.496	10.160	17.146	54.6	11973	5569	29.4	28.73	53.23	1529.32	597.37	1486
CS-80	5.715	4.445	8.890	10.160	19.050	14.605	26.670	55.9	10290	11272	35.1	24.13	90.32	2179.75	599.59	2036
CS-335-A	8.573	3.413	3.810	16.034	22.860	12.383	14.446	53.3	11313	6771	31.2	27.80	61.09	1698.09	605.73	1480
CS-77	6.350	3.810	7.620	12.700	20.320	13.970	22.860	55.9	9800	11528	33.5	22.98	96.77	2224.24	610.42	1903
CS-237-B	6.350	4.445	5.080	12.700	21.590	11.430	19.050	53.3	10913	6970	30.4	26.81	64.52	1729.96	610.74	1571
CS-166-B	7.144	4.445	4.445	11.906	20.796	11.589	17.780	50.5	11620	5904	31.4	30.17	52.92	1596.52	614.07	1476
CS-253	6.350	5.715	6.826	6.826	18.256	13.176	25.082	50.2	13196	5975	36.1	34.48	46.59	1606.37	614.27	1746
CS-368	10.160	3.810	5.715	7.620	15.240	15.875	19.050	41.9	11759	5905	38.1	36.77	43.55	1601.45	617.83	1495
CS-318	8.573	3.413	4.128	15.399	22.225	12.701	15.082	52.7	11178	7159	31.7	27.80	63.57	1766.95	620.36	1505
CS-187	6.350	5.080	6.350	8.890	19.050	12.700	22.860	50.8	11878	6834	34.0	30.65	56.45	1729.96	622.87	1677

National Arnold C Cores 12 mil Silicon Steel

Part No.	D cm 1	E cm 2	F cm 3	G cm 4	Ht cm 5	Wth cm 6	Lt cm 7	MPL cm 8	Wtfe grams 9	Wtcu grams 10	MLT cm 11	Ac cm sq 12	Wa cm sq 13	Ap cm 4th 14	Kg cm 5th 15	At cm sq 16
CS-243-A	5.715	3.810	6.033	17.780	25.400	11.748	19.686	62.9	9922	11191	29.3	20.69	107.27	2218.86	625.79	1876
CS-154-F	8.890	4.128	4.128	10.636	18.892	13.018	16.512	46.0	12247	5267	33.7	34.86	43.91	1530.68	632.78	1409
CS-482	6.985	4.445	3.969	13.970	22.860	10.954	16.828	53.7	12076	5976	30.3	29.50	55.45	1635.46	636.67	1498
CS-177-B	7.938	3.572	4.763	14.764	21.908	12.701	16.670	53.3	10963	7930	31.7	26.94	70.32	1894.22	643.56	1574
CS-79-A	7.620	3.493	7.620	11.748	18.734	15.240	22.226	52.7	10169	11271	35.4	25.29	89.52	2263.58	646.64	1868
CS-335	8.573	3.572	3.810	16.034	23.178	12.383	14.764	54.0	11981	6840	31.5	29.09	61.09	1777.19	656.78	1515
CS-283-A	5.080	5.080	7.779	11.748	21.908	12.859	25.718	59.4	11106	10838	33.3	24.52	91.39	2240.47	658.82	2036
CS-318-B	8.255	3.334	4.763	16.510	23.178	13.018	16.194	55.9	11148	8913	31.9	26.15	78.64	2056.05	674.67	1649
CS-79	8.573	3.493	5.080	13.811	20.797	13.653	17.146	51.8	11234	8314	33.3	28.45	70.16	1995.92	681.56	1607
CS-122-D	6.350	4.445	6.350	12.700	21.590	12.700	21.590	55.9	11433	9399	32.8	26.81	80.65	2162.45	707.66	1803
CS-178-F	10.160	3.810	6.350	8.255	15.875	16.510	20.320	44.5	12472	7293	39.1	36.77	52.42	1927.67	724.73	1647
CS-154-D	5.715	5.715	5.080	11.748	23.178	10.795	21.590	56.5	13380	6717	31.7	31.03	59.68	1851.76	726.11	1735
CS-333	8.573	3.493	5.715	13.811	20.797	14.288	18.416	53.0	11509	9633	34.3	28.45	78.93	2245.41	744.49	1727
CS-311	7.620	3.175	6.350	18.415	24.765	13.970	19.050	62.2	10913	13629	32.8	22.98	116.94	2687.62	753.88	2006
CS-179	9.525	3.810	6.350	9.525	17.145	15.875	20.320	47.0	12361	8142	37.9	34.48	60.48	2085.22	759.62	1693
CS-161	8.573	3.413	5.398	15.399	22.225	13.971	17.622	55.2	11717	9950	33.7	27.80	83.12	2310.57	763.17	1749
CS-307-E	8.414	4.207	4.445	12.700	21.114	12.859	17.304	51.1	13116	6712	33.4	33.63	56.45	1898.34	763.68	1569
CS-304	8.573	3.493	6.033	13.811	20.797	14.606	19.052	53.7	11647	10317	34.8	28.45	83.32	2370.36	774.64	1788
CS-154-J	3.810	3.810	13.970	27.940	35.560	17.780	35.560	99.1	10423	52728	38.0	13.79	390.32	5382.65	781.58	4632
CS-297	7.620	4.445	4.445	13.811	22.701	12.065	17.780	54.3	13329	7057	32.3	32.18	61.39	1975.36	786.55	1626
CS-77-A	6.350	4.445	7.620	12.700	21.590	13.970	24.130	58.4	11952	11965	34.8	26.81	96.77	2594.94	800.50	2048
CS-319	9.525	3.572	4.286	15.399	22.543	13.811	15.716	53.7	13233	8012	34.1	32.32	66.00	2133.27	807.89	1643
CS-82	5.080	5.080	10.160	12.700	22.860	15.240	30.480	66.0	12353	17017	37.1	24.52	129.03	3163.36	836.44	2606
CS-338	10.160	3.810	2.540	20.320	27.940	12.700	12.700	61.0	17105	6083	33.1	36.77	51.61	1898.02	842.36	1665
CS-307-H	10.160	3.810	4.445	12.700	20.320	14.605	16.510	49.5	13898	7254	36.1	36.77	56.45	2075.95	845.08	1611
CS-255	8.573	3.572	5.398	16.034	23.178	13.971	17.940	57.2	12686	10459	34.0	29.09	86.55	2517.92	862.26	1826
CS-237-D	10.160	3.651	5.080	12.700	20.002	15.240	17.462	50.2	13488	8446	36.8	35.24	64.52	2273.51	870.52	1692
CS-318-C	8.255	3.889	4.763	16.510	24.288	13.018	17.304	58.1	13521	9223	33.0	30.50	78.64	2398.32	887.09	1775
CS-305	8.573	3.493	6.985	14.764	21.750	15.558	20.956	57.5	12474	13317	36.3	28.45	103.13	2933.77	919.31	2045
CS-237-J	6.350	5.080	6.668	12.700	22.860	13.018	23.496	59.1	13809	10282	34.1	30.65	84.68	2595.14	931.66	2010

National Arnold C Cores 12 mil Silicon Steel

Part No.	D cm 1	E cm 2	F cm 3	G cm 4	Ht cm 5	Wth cm 6	Lt cm 7	MPL cm 8	Wtfe grams 9	Wtcu grams 10	MLT cm 11	Ac cm sq 12	Wa cm sq 13	Ap cm 4th 14	Kg cm 5th 15	At cm sq 16
CS-316	7.620	3.810	6.350	16.510	24.130	13.970	20.320	61.0	12828	12692	34.0	27.58	104.84	2891.51	936.98	2029
CS-371	10.160	4.445	4.763	10.160	19.050	14.923	18.416	47.6	15590	6523	37.9	42.90	48.39	2076.17	940.00	1651
CS-153-A	4.445	4.445	14.605	19.050	27.940	19.050	38.100	85.1	12186	41084	41.5	18.77	278.23	5222.32	944.22	4126
CS-251	3.651	3.651	12.065	41.910	49.212	15.716	31.432	122.6	11841	61411	34.2	12.66	505.64	6403.13	949.64	5276
CS-316-D	6.350	4.445	7.620	15.240	24.130	13.970	24.130	63.5	12992	14358	34.8	26.81	116.13	3113.93	960.60	2242
CS-328-E	7.620	3.810	5.398	19.209	26.829	13.018	18.416	64.5	13564	12002	32.6	27.58	103.69	2859.84	969.26	2002
CS-145	4.445	4.445	7.620	27.940	36.830	12.065	24.130	88.9	12732	23136	30.6	18.77	212.90	3996.21	981.82	2977
CS-295	9.525	3.572	5.556	15.399	22.543	15.081	18.256	56.2	13859	10993	36.1	32.32	85.56	2765.38	989.49	1895
CS-480	5.080	5.080	6.350	20.320	30.480	11.430	22.860	73.7	13779	14272	31.1	24.52	129.03	3163.36	997.29	2355
CS-156-J	7.620	5.080	5.080	12.700	22.860	12.700	20.320	55.9	15679	7936	34.6	36.77	64.52	2372.52	1008.88	1819
CS-282	3.016	4.604	12.065	41.910	51.118	15.081	33.338	126.4	12719	62555	34.8	13.19	505.64	6670.15	1011.65	5511
CS-122-E	9.525	3.810	6.350	12.700	20.320	15.875	20.320	53.3	14031	10856	37.9	34.48	80.65	2780.30	1012.83	1923
CS-481	8.890	5.080	3.810	13.018	23.178	12.700	17.780	54.0	17669	6197	35.1	42.90	49.60	2127.94	1039.28	1710
CS-356-E	10.160	5.715	5.715	6.350	17.780	15.875	22.860	47.0	19777	5412	41.9	55.16	36.29	2001.81	1053.18	1842
CS-153-B	3.810	3.810	16.510	35.560	43.180	20.320	40.640	119.4	12561	87635	42.0	13.79	587.10	8096.22	1063.92	6413
CS-183	9.525	3.572	5.715	16.669	23.813	15.240	18.574	59.1	14564	12325	36.4	32.32	95.26	3079.11	1094.19	2013
CS-358	10.160	5.080	5.080	8.890	19.050	15.240	20.320	48.3	18055	6371	39.7	49.03	45.16	2214.35	1094.73	1768
CS-352-C	8.096	3.969	6.350	16.510	24.448	14.446	20.638	61.6	14347	13166	35.3	30.53	104.84	3200.34	1106.54	2110
CS-292-B	6.350	6.350	8.573	8.890	21.590	14.923	29.846	60.3	17632	10861	40.1	38.31	76.21	2919.48	1116.24	2388
CS-156-K	7.620	5.398	5.080	12.700	23.496	12.700	20.956	57.2	17040	8082	35.2	39.08	64.52	2521.04	1118.58	1895
CS-217	6.350	4.524	6.985	18.256	27.304	13.335	23.018	68.6	14280	15386	33.9	27.29	127.52	3480.10	1119.65	2346
CS-217-A	5.080	5.080	12.700	15.240	25.400	17.780	35.560	76.2	14254	28270	41.1	24.52	193.55	4745.04	1132.85	3458
CS-78-A	6.350	6.350	7.620	10.160	22.860	13.970	27.940	61.0	17817	10511	38.2	38.31	77.42	2965.65	1190.20	2310
CS-356-C	10.160	6.350	5.715	6.033	18.733	15.875	24.130	48.9	22866	5298	43.2	61.29	34.48	2113.20	1199.01	1975
CS-356-L	8.573	5.715	5.239	10.160	21.590	13.812	21.908	53.7	19056	7196	38.0	46.55	53.23	2477.51	1213.30	1908
CS-182-B	9.525	3.731	5.715	17.145	24.607	15.240	18.892	60.6	15622	12788	36.7	33.76	97.98	3308.02	1217.22	2085
CS-292-C	5.715	5.080	8.890	16.510	26.670	14.605	27.940	71.1	14967	18979	36.4	27.58	146.77	4048.11	1228.15	2724
CS-301-F	8.890	5.715	3.810	12.700	24.130	12.700	19.050	55.9	20579	6265	36.4	48.27	48.39	2335.45	1238.45	1842
CS-217-B	7.620	3.810	7.620	19.685	27.305	15.240	22.860	69.9	14699	19223	36.0	27.58	150.00	4137.08	1266.43	2545
CS-310	8.890	3.810	6.350	17.780	25.400	15.240	20.320	63.5	15590	14689	36.6	32.18	112.90	3632.92	1278.08	2232

National Arnold C Cores 12 mil Silicon Steel																
Part No.	D cm 1	E cm 2	F cm 3	G cm 4	Ht cm 5	Wth cm 6	Lt cm 7	MPL cm 8	Wtfe grams 9	Wtcu grams 10	MLT cm 11	Ac cm sq 12	Wa cm sq 13	Ap cm 4th 14	Kg cm 5th 15	At cm sq 16
CS-154-M	5.080	5.080	15.240	15.875	26.035	20.320	40.640	82.6	15442	38769	45.1	24.52	241.94	5931.30	1290.75	4206
CS-157	10.160	3.969	5.398	15.558	23.496	15.558	18.734	57.8	16891	11333	37.9	38.31	83.98	3217.25	1299.11	2023
CS-235	6.350	4.524	6.985	21.908	30.956	13.335	23.018	75.9	15801	18464	33.9	27.29	153.03	4176.28	1343.63	2612
CS-312-F	9.843	3.810	4.445	21.590	29.210	14.288	16.510	67.3	18297	12115	35.5	35.63	95.97	3419.01	1372.46	2136
CS-242	6.350	4.604	6.985	21.908	31.116	13.335	23.178	76.2	16148	18551	34.1	27.77	153.03	4250.13	1385.04	2633
CS-177	7.938	5.636	4.763	14.764	26.036	12.701	20.798	61.6	19976	8963	35.8	42.50	70.32	2988.76	1417.64	2058
CS-374	12.700	5.080	2.540	15.240	25.400	15.240	15.240	55.9	26132	5611	40.8	61.29	38.71	2372.52	1426.88	1910
CS-312-C	10.160	3.810	4.445	21.590	29.210	14.605	16.510	67.3	18886	12331	36.1	36.77	95.97	3529.12	1436.63	2165
CS-217-C	7.620	4.445	6.985	18.256	27.146	14.605	22.860	68.3	16759	16466	36.3	32.18	127.52	4103.20	1454.38	2453
CS-250-A	8.890	3.810	6.350	20.479	28.099	15.240	20.320	68.9	16915	16918	36.6	32.18	130.04	4184.40	1472.09	2424
CS-356	10.160	5.715	5.715	8.890	20.320	15.875	22.860	52.1	21915	7577	41.9	55.16	50.81	2802.54	1474.46	2039
CS-312-E	10.160	3.810	4.445	22.225	29.845	14.605	16.510	68.6	19243	12694	36.1	36.77	98.79	3632.92	1478.88	2204
CS-310-D	7.620	4.763	6.350	17.780	27.306	13.970	22.226	67.3	17708	14434	36.0	34.48	112.90	3892.82	1493.37	2356
CS-327-C	9.049	4.445	5.715	16.669	25.559	14.764	20.320	62.5	18236	12594	37.2	38.21	95.26	3640.17	1496.61	2190
CS-87	10.160	3.969	5.398	18.256	26.194	15.558	18.734	63.2	18468	13298	37.9	38.31	98.55	3775.17	1524.39	2208
CS-237-F	10.160	5.080	5.080	12.700	22.860	15.240	20.320	55.9	20906	9101	39.7	49.03	64.52	3163.36	1563.90	2039
CS-356-A	10.160	5.715	5.239	10.160	21.590	15.399	21.908	53.7	22584	7797	41.2	55.16	53.23	2936.13	1572.77	2046
CS-146	6.350	4.128	10.160	24.130	32.386	16.510	28.576	85.1	16168	33236	38.1	24.90	245.16	6105.03	1595.13	3528
CS-178-D	10.160	4.128	6.350	16.193	24.449	16.510	20.956	61.6	18726	14539	39.8	39.84	102.83	4096.93	1642.15	2314
CS-344-C	12.065	5.080	4.445	11.748	21.908	16.510	19.050	52.7	23415	7889	42.5	58.23	52.22	3040.54	1666.84	2007
CS-323	5.715	9.208	5.080	12.700	31.116	10.795	28.576	72.4	27613	8864	38.6	49.99	64.52	3225.32	1669.27	2693
CS-149-C	5.080	5.080	10.160	25.400	35.560	15.240	30.480	91.4	17105	34034	37.1	24.52	258.06	6326.72	1672.88	3768
CS-359	5.398	5.398	6.350	27.940	38.736	11.748	23.496	90.2	19045	20427	32.4	27.68	177.42	4911.22	1679.56	3008
CS-326	7.620	7.620	5.080	10.795	26.035	12.700	25.400	62.2	26191	7736	39.7	55.16	54.84	3024.96	1682.42	2319
CS-284-A	6.350	3.175	13.494	35.243	41.593	19.844	33.338	110.2	16101	70100	41.5	19.15	475.57	9108.66	1683.51	5404
CS-178-C	10.160	4.286	6.033	16.193	24.765	16.193	20.638	61.6	19442	13750	39.6	41.37	97.69	4041.38	1689.61	2286
CS-357-B	11.430	6.350	3.810	10.160	22.860	15.240	20.320	53.3	28062	5886	42.8	68.95	38.71	2669.08	1721.68	2052
CS-156-H	6.350	6.350	9.525	12.700	25.400	15.875	31.750	69.9	20416	17710	41.2	38.31	120.97	4633.83	1724.60	2944
CS-88	8.890	6.350	3.810	15.240	27.940	12.700	20.320	63.5	25984	7780	37.7	53.63	58.06	3113.93	1772.90	2168
CS-312-D	8.890	4.445	5.398	21.590	30.480	14.288	19.686	71.8	20553	15069	36.4	37.54	116.54	4375.05	1806.78	2442

National Arnold C Cores 12 mil Silicon Steel

Part No.	D cm 1	E cm 2	F cm 3	G cm 4	Ht cm 5	Wth cm 6	Lt cm 7	MPL cm 8	Wtfe grams 9	Wtcu grams 10	MLT cm 11	Ac cm sq 12	Wa cm sq 13	Ap cm 4th 14	Kg cm 5th 15	At cm sq 16
CS-178-A	10.160	4.445	7.620	13.970	22.860	17.780	24.130	61.0	19955	16046	42.4	42.90	106.45	4567.10	1848.98	2503
CS-178	10.160	4.207	6.033	18.574	26.988	16.193	20.480	66.0	20461	15709	39.4	40.61	112.06	4550.18	1874.74	2440
CS-238	7.144	4.842	7.779	20.955	30.639	14.923	25.242	76.8	19265	21680	37.4	32.86	163.01	5356.75	1882.64	2913
CS-312-H	10.160	5.398	6.985	10.795	21.591	17.145	24.766	57.2	22720	11610	43.3	52.10	75.40	3928.61	1890.94	2357
CS-305-A	10.160	3.969	6.191	20.955	28.893	16.351	20.320	70.2	20510	18081	39.2	38.31	129.73	4969.89	1943.06	2587
CS-339-A	10.160	4.445	8.255	13.970	22.860	18.415	25.400	62.2	20371	17792	43.4	42.90	115.32	4947.69	1957.03	2645
CS-182	9.525	4.921	5.715	17.145	26.987	15.240	21.272	65.4	22221	13617	39.1	44.53	97.98	4363.11	1988.55	2394
CS-312	10.160	3.810	6.985	21.590	29.210	17.145	21.590	72.4	20312	21516	40.1	36.77	150.81	5545.76	2033.18	2790
CS-287	8.890	4.445	6.350	21.590	30.480	15.240	21.590	73.7	21099	18455	37.9	37.54	137.10	5146.64	2041.51	2677
CS-322-A	10.160	5.715	8.890	8.890	20.320	19.050	29.210	58.4	24588	13187	46.9	55.16	79.03	4359.50	2049.94	2668
CS-182-A	8.890	4.366	7.620	19.685	28.417	16.510	23.972	72.1	20277	21171	39.7	36.87	150.00	5530.95	2055.29	2826
CS-202	10.160	4.604	5.715	18.415	27.623	15.875	20.638	66.7	22607	14864	39.7	44.44	105.24	4676.71	2093.06	2462
CS-178-K	15.240	3.334	8.255	13.970	20.638	23.495	23.178	57.8	21282	21047	51.3	48.27	115.32	5566.57	2094.10	2795
CS-310-C	9.525	4.763	6.350	17.780	27.306	15.875	22.226	67.3	22135	15964	39.8	43.10	112.90	4866.03	2109.80	2545
CS-339-B	9.525	5.080	8.890	12.700	22.860	18.415	27.940	63.5	22272	17819	44.4	45.97	112.90	5189.89	2150.06	2787
CS-293-L	5.715	5.715	8.890	24.130	35.560	14.605	29.210	88.9	21047	28707	37.6	31.03	214.52	6656.03	2195.12	3563
CS-178-E	10.160	5.080	6.350	15.240	25.400	16.510	22.860	63.5	23756	14338	41.7	49.03	96.77	4745.04	2233.59	2497
CS-158	7.144	5.001	7.779	23.813	33.815	14.923	25.560	83.2	21543	24846	37.7	33.94	185.24	6287.24	2262.98	3190
CS-178-P	7.620	6.668	6.350	15.240	28.576	13.970	26.036	69.9	25726	13683	39.8	48.27	96.77	4671.25	2268.32	2675
CS-359-A	5.715	5.834	6.350	30.480	42.148	12.065	24.368	97.0	23442	23321	33.9	31.67	193.55	6130.49	2292.31	3363
CS-237-H	10.160	6.350	5.080	12.700	25.400	15.240	22.860	61.0	28508	9684	42.2	61.29	64.52	3954.20	2296.56	2374
CS-246-A	10.160	3.969	5.398	27.940	35.878	15.558	18.734	82.6	24130	20353	37.9	38.31	150.82	5777.74	2333.01	2872
CS-203	10.160	4.604	5.715	20.638	29.846	15.875	20.638	71.1	24115	16658	39.7	44.44	117.95	5241.27	2345.73	2624
CS-87-A	10.160	5.080	5.398	18.256	28.416	15.558	20.956	67.6	25301	14077	40.2	49.03	98.55	4831.92	2359.12	2508
CS-178-B	10.160	5.080	6.350	16.193	26.353	16.510	22.860	65.4	24469	15235	41.7	49.03	102.83	5041.76	2373.27	2572
CS-284	5.080	5.080	13.335	31.750	41.910	18.415	36.830	110.5	20668	63342	42.1	24.52	423.39	10379.77	2419.39	5445
CS-236	8.573	5.080	6.985	20.638	30.798	15.558	24.130	75.6	23855	20243	39.5	41.37	144.16	5964.23	2499.56	2912
CS-334	7.620	7.620	6.985	12.700	27.940	14.605	29.210	69.9	29399	13458	42.7	55.16	88.71	4893.32	2530.76	2876
CS-237-E	12.383	3.969	5.080	24.448	32.386	17.463	18.098	74.9	26695	18503	41.9	46.69	124.20	5798.79	2585.00	2774
CS-178-H	7.620	7.620	6.350	13.970	29.210	13.970	27.940	71.1	29933	13143	41.7	55.16	88.71	4893.32	2591.32	2842

National Arnold C Cores 12 mil Silicon Steel

Part No.	D cm 1	E cm 2	F cm 3	G cm 4	Ht cm 5	Wth cm 6	Lt cm 7	MPL cm 8	Wtfe grams 9	Wtcu grams 10	MLT cm 11	Ac cm sq 12	Wa cm sq 13	Ap cm 4th 14	Kg cm 5th 15	At cm sq 16
CS-293	7.144	5.318	7.938	24.448	35.084	15.082	26.512	86.0	23695	26640	38.6	36.09	194.07	7004.35	2619.53	3381
CS-356-H	10.160	6.350	5.715	13.335	26.035	15.875	24.130	63.5	29695	11710	43.2	61.29	76.21	4670.90	2650.22	2559
CS-178-M	10.636	5.239	6.350	16.193	26.671	16.986	23.178	66.0	26674	15699	42.9	52.94	102.83	5443.16	2684.39	2664
CS-351-B	8.890	4.763	6.985	23.495	33.021	15.875	23.496	80.0	24558	23045	39.5	40.23	164.11	6601.58	2689.95	3080
CS-356-D	10.160	6.350	5.715	13.653	26.353	15.875	24.130	64.1	29993	11989	43.2	61.29	78.03	4782.28	2713.42	2584
CS-332	8.890	5.715	3.810	27.940	39.370	12.700	19.050	86.4	31804	13782	36.4	48.27	106.45	5137.99	2724.59	2810
CS-250	10.160	4.842	6.350	20.479	30.163	16.510	22.384	73.0	26040	19047	41.2	46.74	130.04	6077.49	2758.29	2840
CS-313-E	10.160	5.080	7.620	16.510	26.670	17.780	25.400	68.6	25657	19532	43.7	49.03	125.81	6168.55	2771.06	2897
CS-293-D	5.080	8.890	9.525	18.415	36.195	14.605	36.830	91.4	29933	27264	43.7	42.90	175.40	7525.33	2954.55	4122
CS-313-F	8.890	4.763	7.620	24.448	33.974	16.510	24.766	83.2	25532	26820	40.5	40.23	186.29	7493.84	2978.32	3332
CS-340-A	9.525	5.080	7.620	19.685	29.845	17.145	25.400	74.9	26280	22611	42.4	45.97	150.00	6895.13	2990.87	3101
CS-296	7.938	5.239	7.938	24.448	34.926	15.876	26.354	85.7	25842	27627	40.0	39.51	194.07	7667.21	3026.68	3454
CS-178-O	11.271	5.556	6.350	15.240	26.352	17.621	23.812	65.4	29688	15431	44.8	59.49	96.77	5757.14	3055.30	2741
CS-301-D	8.890	6.350	7.620	15.240	27.940	16.510	27.940	71.1	29102	18029	43.7	53.63	116.13	6227.86	3059.99	3019
CS-186	7.938	5.239	8.414	24.448	34.926	16.352	27.306	86.7	26129	29830	40.8	39.51	205.71	8126.98	3149.38	3590
CS-156-D	7.620	7.620	9.525	12.700	27.940	17.145	34.290	74.9	31537	20067	46.7	55.16	120.97	6672.71	3156.04	3469
CS-359-B	6.350	6.112	6.350	33.020	45.244	12.700	24.924	103.2	29029	26625	35.7	36.87	209.68	7730.93	3192.92	3727
CS-178-L	7.620	8.255	6.350	15.240	31.750	13.970	29.210	76.2	34744	14775	42.9	59.76	96.77	5783.02	3219.54	3139
CS-313	10.160	4.445	7.620	24.448	33.338	17.780	24.130	81.9	26815	28081	42.4	42.90	186.29	7992.59	3235.78	3382
CS-351-C	10.160	4.763	6.985	23.495	33.021	17.145	23.496	80.0	28066	24527	42.0	45.97	164.11	7544.66	3301.07	3224
CS-303-B	5.080	5.080	22.225	34.925	45.085	27.305	54.610	134.6	25182	154652	56.0	24.52	776.21	19029.58	3330.62	9465
CS-297-A	10.160	7.620	4.445	15.240	30.480	14.605	24.130	69.9	39198	10540	43.8	73.55	67.74	4982.29	3349.94	2795
CS-293-A	8.255	5.080	8.890	25.400	35.560	17.145	27.940	88.9	27023	33599	41.8	39.84	225.81	8995.80	3425.93	3803
CS-201	11.430	5.080	10.160	13.970	24.130	21.590	30.480	68.6	28864	25331	50.2	55.16	141.94	7829.31	3442.11	3416
CS-148	7.938	5.636	7.620	25.400	36.672	15.558	26.512	88.6	28727	27756	40.3	42.50	193.55	8226.11	3467.85	3567
CS-336-D	12.700	6.668	5.080	12.700	26.036	17.780	23.496	62.2	38200	10996	47.9	80.45	64.52	5190.28	3484.88	2714
CS-310-F	10.795	5.556	6.350	18.574	29.686	17.145	23.812	72.1	31333	18407	43.9	56.98	117.95	6720.28	3489.92	2966
CS-339	10.160	4.842	8.890	20.479	30.163	19.050	27.464	78.1	27852	29248	45.2	46.74	182.06	8508.49	3520.74	3508
CS-363	5.080	5.080	31.750	33.020	43.180	36.830	73.660	149.9	28033	264631	71.0	24.52	1048.39	25702.29	3550.79	13677
CS-336-C	10.160	8.255	5.080	12.700	29.210	15.240	26.670	68.6	41692	10558	46.0	79.68	64.52	5140.46	3559.87	2926

National Arnold C Cores 12 mil Silicon Steel																
Part No.	D cm 1	E cm 2	F cm 3	G cm 4	Ht cm 5	Wth cm 6	Lt cm 7	MPL cm 8	Wtfe grams 9	Wtcu grams 10	MLT cm 11	Ac cm sq 12	Wa cm sq 13	Ap cm 4th 14	Kg cm 5th 15	At cm sq 16
CS-248	6.350	6.350	7.620	30.480	43.180	13.970	27.940	101.6	29695	31533	38.2	38.31	232.26	8896.95	3570.61	4013
CS-336-A	10.160	6.191	5.080	20.638	33.020	15.240	22.542	76.2	34742	15619	41.9	59.76	104.84	6264.83	3574.37	2931
CS-293-B	8.255	5.080	9.525	25.400	35.560	17.780	29.210	90.2	27409	36856	42.8	39.84	241.94	9638.36	3585.22	3994
CS-250-B	10.160	5.556	6.350	20.955	32.067	16.510	23.812	76.8	31438	20166	42.6	53.63	133.06	7135.77	3591.64	3090
CS-202-A	8.890	6.668	5.715	20.638	33.974	14.605	24.766	79.4	34107	17324	41.3	56.31	117.95	6642.09	3622.33	3094
CS-149-D	7.620	5.398	10.160	25.400	36.196	17.780	31.116	92.7	27642	39647	43.2	39.08	258.06	10084.14	3648.33	4207
CS-186-A	8.255	5.080	8.573	27.940	38.100	16.828	27.306	93.3	28374	35217	41.3	39.84	239.53	9542.53	3677.89	3934
CS-186-B	7.938	5.874	8.414	24.448	36.196	16.352	28.576	89.2	30155	30759	42.0	44.30	205.71	9112.02	3839.52	3792
CS-148-C	8.573	5.636	7.620	25.400	36.672	16.193	26.512	88.6	31025	28630	41.6	45.90	193.55	8884.15	3921.37	3647
CS-325	10.160	5.080	7.938	23.495	33.655	18.098	26.036	83.2	31121	29286	44.2	49.03	186.50	9144.66	4061.56	3591
CS-345	10.160	4.445	6.985	34.290	43.180	17.145	22.860	100.3	32843	35255	41.4	42.90	239.52	10275.97	4260.41	3990
CS-285-A	7.620	7.620	13.335	13.970	29.210	20.955	41.910	85.1	35813	34866	52.6	55.16	186.29	10275.97	4307.91	4616
CS-190	11.430	6.985	4.445	19.050	33.020	15.875	22.860	74.9	43363	13558	45.0	75.85	84.68	6422.48	4327.62	3036
CS-205	7.620	7.620	5.715	25.400	40.640	13.335	26.670	92.7	39020	20993	40.7	55.16	145.16	8007.25	4344.29	3615
CS-360	6.668	6.985	6.985	33.020	46.990	13.653	27.940	108.0	36445	32387	39.5	44.25	230.65	10205.38	4574.09	4282
CS-351	7.620	7.620	6.985	23.495	38.735	14.605	29.210	91.4	38485	24897	42.7	55.16	164.11	9052.64	4681.91	3822
CS-340	10.160	6.350	7.620	19.685	32.385	17.780	27.940	80.0	37416	24643	46.2	61.29	150.00	9193.51	4878.61	3568
CS-336	10.160	5.715	5.080	33.020	44.450	15.240	21.590	99.1	41692	24421	40.9	55.16	167.74	9252.83	4986.58	3700
CS-314-B	10.160	6.826	6.350	20.479	34.131	16.510	26.352	81.0	40700	20882	45.2	65.88	130.04	8567.74	5000.11	3449
CS-341-A	10.160	6.350	8.255	19.685	32.385	18.415	29.210	81.3	38010	27272	47.2	61.29	162.50	9959.64	5173.52	3741
CS-365	11.430	7.620	5.080	17.780	33.020	16.510	25.400	76.2	48107	15189	47.3	82.74	90.32	7473.44	5230.23	3297
CS-336-B	11.430	5.398	6.668	26.353	37.149	18.098	24.132	87.6	39192	28332	45.3	58.61	175.72	10299.79	5326.01	3723
CS-342	15.240	5.715	3.810	25.400	36.830	19.050	19.050	81.3	51314	16899	49.1	82.74	96.77	8007.25	5396.58	3358
CS-312-B	10.160	5.080	6.985	34.290	44.450	17.145	24.130	102.9	38485	36336	42.7	49.03	239.52	11743.97	5398.96	4213
CS-310-E	8.255	8.890	6.350	20.955	38.735	14.605	30.480	90.2	47966	21518	45.5	69.72	133.06	9276.92	5688.91	3931
CS-246-L	10.160	5.080	10.160	27.940	38.100	20.320	30.480	96.5	36110	48097	47.6	49.03	283.87	13918.78	5729.34	4697
CS-148-A	7.938	7.938	7.620	25.400	41.276	15.558	31.116	97.8	44666	30924	44.9	59.86	193.55	11586.03	6174.34	4337
CS-360-A	7.303	7.620	7.620	33.020	48.260	14.923	30.480	111.8	45081	38496	43.0	52.87	251.61	13301.85	6537.73	4828
CS-149-F	7.620	7.620	10.160	25.400	40.640	17.780	35.560	101.6	42761	43725	47.6	55.16	258.06	14235.12	6592.00	4981
CS-351-A	7.620	9.525	6.985	23.495	42.545	14.605	33.020	99.1	52115	27121	46.5	68.95	164.11	11315.80	6715.73	4485

National Arnold C Cores 12 mil Silicon Steel

Part No.	D cm 1	E cm 2	F cm 3	G cm 4	Ht cm 5	Wth cm 6	Lt cm 7	MPL cm 8	Wtfe grams 9	Wtcu grams 10	MLT cm 11	Ac cm sq 12	Wa cm sq 13	Ap cm 4th 14	Kg cm 5th 15	At cm sq 16
CS-154-N	7.620	7.620	15.240	20.320	35.560	22.860	45.720	101.6	42761	61252	55.6	55.16	309.68	17082.14	6776.15	6039
CS-302	10.160	7.620	8.255	19.050	34.290	18.415	31.750	85.1	47750	27813	49.7	73.55	157.26	11566.03	6841.36	4106
CS-203-C	11.430	5.239	8.255	30.480	40.958	19.685	26.988	98.4	42722	42513	47.5	56.89	251.61	14313.65	6854.94	4540
CS-246	9.525	5.715	11.430	27.940	39.370	20.955	34.290	101.6	40089	56374	49.6	51.71	319.35	16514.96	6881.78	5265
CS-203-B	10.160	6.985	8.255	22.860	36.830	18.415	30.480	90.2	46384	32523	48.5	67.42	188.71	12722.63	7079.14	4263
CS-296-L	6.350	6.985	13.970	35.560	49.530	20.320	41.910	127.0	40831	88006	49.8	42.14	496.77	20932.54	7081.93	7087
CS-246-C	10.160	5.080	12.700	30.480	40.640	22.860	35.560	106.7	39911	71076	51.6	49.03	387.10	18980.15	7209.36	5858
CS-379	12.700	3.731	13.970	36.036	43.498	26.670	35.402	114.9	39476	100269	56.0	45.01	503.42	22661.34	7284.93	6871
CS-149	12.700	5.080	10.160	25.400	35.560	22.860	30.480	91.4	42761	48387	52.7	61.29	258.06	15816.79	7354.19	4774
CS-376	6.350	8.890	8.890	33.020	50.800	15.240	35.560	119.4	48849	47656	45.7	53.63	293.55	15742.65	7397.16	5613
CS-256	10.160	6.350	4.128	50.165	62.865	14.288	20.956	134.0	62658	29983	40.7	61.29	207.08	12692.04	7642.00	4820
CS-375	6.350	8.890	6.985	40.640	58.420	13.335	31.750	130.8	53526	43065	42.7	53.63	283.87	15223.66	7654.78	5618
CS-360-B	7.938	7.739	7.620	33.020	48.498	15.558	30.718	112.2	49978	39846	44.5	58.36	251.61	14684.24	7697.42	4973
CS-324	10.160	10.160	4.763	20.796	41.116	14.923	29.846	91.8	68656	17377	49.3	98.06	99.05	9713.40	7723.19	4185
CS-246-E	8.890	6.191	11.430	30.480	42.862	20.320	35.242	108.6	43319	61105	49.3	52.29	348.39	18215.76	7724.02	5629
CS-316-B	10.160	10.160	6.350	16.510	36.830	16.510	33.020	86.4	64617	19321	51.8	98.06	104.84	10280.92	7781.43	4223
CS-341	12.700	6.668	8.255	19.685	33.021	20.955	29.846	82.6	50673	30575	52.9	80.45	162.50	13073.01	7950.63	4166
CS-312-A	10.160	6.350	6.985	34.290	46.990	17.145	26.670	108.0	50482	38500	45.2	61.29	239.52	14679.96	7961.85	4677
CS-203-A	10.160	6.985	10.160	22.860	36.830	20.320	34.290	94.0	48344	42499	51.5	67.42	232.26	15658.63	8206.37	4848
CS-291-A	10.160	7.620	10.160	19.685	34.925	20.320	35.560	90.2	50601	37500	52.7	73.55	200.00	14709.62	8207.27	4729
CS-147	11.430	6.985	7.620	24.765	38.735	19.050	29.210	92.7	53652	33559	50.0	75.85	188.71	14312.96	8683.09	4426
CS-247	12.700	5.080	10.160	30.480	40.640	22.860	30.480	101.6	47513	58064	52.7	61.29	309.68	18980.15	8825.03	5316
CS-245	10.160	7.620	11.430	19.685	34.925	21.590	38.100	92.7	52026	43782	54.7	73.55	225.00	16548.32	8896.75	5118
CS-293-C	10.160	7.620	8.890	24.130	39.370	19.050	33.020	96.5	54164	38700	50.7	73.55	214.52	15777.25	9148.93	4816
CS-315-A	12.065	6.033	8.573	28.575	40.641	20.638	29.212	98.4	51931	44316	50.9	69.15	244.97	16939.61	9210.26	4844
CS-246-B	10.160	6.668	10.160	27.940	41.276	20.320	33.656	102.9	50517	51303	50.8	64.36	283.87	18269.77	9254.31	5282
CS-310-A	11.430	8.890	6.350	20.955	38.735	17.780	30.480	90.2	66414	24523	51.8	96.53	133.06	12844.97	9570.20	4371
CS-296-F	7.620	7.620	13.970	30.480	45.720	21.590	43.180	119.4	50245	81203	53.6	55.16	425.81	23487.94	9663.61	6936
CS-234	11.430	4.921	10.001	42.545	52.387	21.431	29.844	124.8	50872	75077	49.6	53.43	425.49	22736.06	9793.67	6258
CS-296-D	5.080	10.160	16.510	35.560	55.880	21.590	53.340	144.8	54164	119452	57.2	49.03	587.10	28786.57	9867.52	9336

National Arnold C Cores 12 mil Silicon Steel																
Part No.	D cm 1	E cm 2	F cm 3	G cm 4	Ht cm 5	Wth cm 6	Lt cm 7	MPL cm 8	Wtfe grams 9	Wtcu grams 10	MLT cm 11	Ac cm sq 12	Wa cm sq 13	Ap cm 4th 14	Kg cm 5th 15	At cm sq 16
CS-378	6.350	6.350	38.100	38.100	50.800	44.450	88.900	177.8	51967	444096	86.0	38.31	1451.61	55605.92	9903.46	19516
CS-315	13.335	5.477	9.843	28.575	39.529	23.178	30.640	98.7	52275	54303	54.3	69.38	281.26	19515.20	9975.73	5244
CS-149-A	13.335	5.874	10.160	26.035	37.783	23.495	32.068	95.9	54442	52284	55.6	74.41	264.52	19683.48	10540.31	5228
CS-320	10.160	10.160	7.620	20.320	40.640	17.780	35.560	96.5	72219	29633	53.8	98.06	154.84	15184.12	11066.79	4981
CS-320-B	10.160	9.525	7.620	24.130	43.180	17.780	34.290	101.6	71269	34359	52.5	91.94	183.87	16904.20	11829.57	5136
CS-315-B	9.525	8.255	9.843	28.575	45.085	19.368	36.196	109.9	62612	52239	52.2	74.70	281.26	21009.68	12019.02	5762
CS-329	10.160	7.620	13.335	24.130	39.370	23.495	41.910	105.4	59153	66036	57.7	73.55	321.77	23665.88	12063.94	6306
CS-372	13.335	7.620	5.080	33.020	48.260	18.415	25.400	106.7	78574	30482	51.1	96.53	167.74	16192.44	12235.15	4855
CS-293-P	10.160	8.573	9.525	25.400	42.546	19.685	36.196	104.1	65751	46144	53.6	82.75	241.94	20019.30	12353.80	5527
CS-296-B	7.620	7.620	16.510	35.560	50.800	24.130	48.260	134.6	56659	120287	57.6	55.16	587.10	32384.89	12401.88	8677
CS-360-D	8.890	9.009	8.255	34.290	52.308	17.145	34.528	121.1	70318	50303	50.0	76.09	283.06	21537.07	13116.00	5995
CS-313-A	13.335	6.350	7.620	35.243	47.943	20.955	27.940	111.1	68207	50183	52.5	80.44	268.55	21603.21	13228.20	5472
CS-296-A	5.080	12.700	15.240	35.560	60.960	20.320	55.880	152.4	71269	116211	60.3	61.29	541.93	33215.27	13503.65	10013
CS-246-F	8.890	8.573	11.430	30.480	47.626	20.320	40.006	118.1	65249	67007	54.1	72.40	348.39	25224.32	13506.53	6588
CS-329-B	10.478	5.874	13.335	41.910	53.658	23.813	38.418	134.0	59775	109018	54.9	58.47	558.87	32677.33	13932.17	7881
CS-148-D	14.923	7.303	7.620	25.400	40.006	22.543	29.846	95.3	75245	39665	57.6	103.53	193.55	20038.71	14399.64	5092
CS-360-K	8.890	8.890	8.890	38.100	55.880	17.780	35.560	129.5	74209	61106	50.7	75.08	338.71	25430.44	15053.86	6594
CS-322	12.700	7.620	8.573	29.845	45.085	21.273	32.386	107.3	75279	50329	55.3	91.94	255.86	23522.68	15638.00	5694
CS-292	13.970	9.208	8.890	19.685	38.101	22.860	36.196	94.0	87630	38290	61.5	122.20	175.00	21385.65	16989.71	5500
CS-293-H	15.240	7.620	8.890	24.130	39.370	24.130	33.020	96.5	81247	46451	60.9	110.32	214.52	23665.88	17150.49	5552
CS-329-F	12.700	7.620	13.335	24.130	39.370	26.035	41.910	105.4	73942	71848	62.8	91.94	321.77	29582.35	17324.91	6719
CS-329-H	10.160	7.620	10.160	41.910	57.150	20.320	35.560	134.6	75545	79838	52.7	73.55	425.81	31317.25	17473.55	7213
CS-252	6.985	10.160	12.065	44.450	64.770	19.050	44.450	153.7	79049	103835	54.4	67.42	536.29	36156.20	17907.88	8939
CS-293-K	8.890	8.890	8.890	49.530	67.310	17.780	35.560	152.4	87305	79437	50.7	75.08	440.32	33059.57	19570.02	7813
CS-296-J	10.160	7.620	16.510	35.560	50.800	26.670	48.260	134.6	75545	130893	62.7	73.55	587.10	43179.85	20261.37	9181
CS-367	5.715	11.430	36.513	36.513	59.373	42.228	95.886	191.8	90802	438204	92.4	62.06	1333.20	82733.44	22218.13	21160
CS-500	5.080	15.240	13.970	62.230	92.710	19.050	58.420	213.4	119732	195962	63.4	73.55	869.35	63939.39	29674.78	14887
CS-360-F	15.240	10.160	6.985	34.290	54.610	22.225	34.290	123.2	138262	53643	63.0	147.10	239.52	35231.91	32913.87	7129
CS-296-C	10.160	10.160	16.510	35.560	55.880	26.670	53.340	144.8	108329	141498	67.8	98.06	587.10	57573.13	33320.42	10445
CS-519	9.843	9.208	21.273	47.149	65.565	31.116	60.962	173.7	114099	259356	72.7	86.10	1003.00	86360.99	40903.49	14301

National Arnold C Cores 12 mil Silicon Steel																
Part No.	D cm 1	E cm 2	F cm 3	G cm 4	Ht cm 5	Wth cm 6	Lt cm 7	MPL cm 8	Wtfe grams 9	Wtcu grams 10	MLT cm 11	Ac cm sq 12	Wa cm sq 13	Ap cm 4th 14	Kg cm 5th 15	At cm sq 16
CS-373	10.160	10.160	35.560	35.560	55.880	45.720	91.440	182.9	136836	439252	97.7	98.06	1264.51	124003.67	49793.97	20800
CS-286-A	7.620	12.700	25.400	63.500	88.900	33.020	76.200	228.6	160355	468783	81.7	91.94	1612.90	148282.45	66715.89	21871

Magnetics C Cores 0.5 mil Iron Alloy

Part No.	D cm 1	E cm 2	F cm 3	G cm 4	Ht cm 5	Wth cm 6	Lt cm 7	MPL cm 8	Wtfe grams 9	Wtcu grams 10	MLT cm 11	Ac cm sq 12	Wa cm sq 13	Ap cm 4th 14	Kg cm 5th 15	At cm sq 16
MC1100-5	0.318	0.160	0.318	0.953	1.273	0.635	0.955	3.2	0.8	2.0	1.9	0.0305	0.302	0.009	0.000060	4.9
MC1505-5	0.318	0.160	0.795	1.113	1.433	1.113	1.910	4.5	1.2	8.2	2.6	0.0305	0.884	0.027	0.000126	11.9
MC1504-5	0.318	0.318	0.800	1.113	1.748	1.118	2.235	5.1	2.7	9.3	2.9	0.0605	0.890	0.054	0.000444	14.7
MC2340-5	0.635	0.318	0.478	0.953	1.588	1.113	1.590	4.1	4.4	5.0	3.1	0.1210	0.455	0.055	0.000869	10.9
MC1647-5	0.953	0.452	0.318	0.381	1.285	1.270	1.539	3.2	7.2	1.6	3.7	0.2584	0.121	0.031	0.000869	10.0
MC1606-5	0.635	0.318	0.478	1.270	1.905	1.113	1.590	4.8	5.0	6.6	3.1	0.1210	0.606	0.073	0.001159	12.6
MC0001-5	0.635	0.318	0.635	1.270	1.905	1.270	1.905	5.1	5.4	9.5	3.3	0.1210	0.806	0.098	0.001426	15.3
MC1582-5	0.953	0.318	0.635	1.270	1.905	1.588	1.905	5.1	8.0	11.3	3.9	0.1815	0.806	0.146	0.002692	17.7
MC1589-5	0.635	0.478	0.714	1.270	2.225	1.349	2.383	5.9	9.3	12.1	3.8	0.1819	0.906	0.165	0.003197	20.3
MC0147-5	0.635	0.318	0.800	2.540	3.175	1.435	2.235	8.0	8.4	25.8	3.6	0.1210	2.032	0.246	0.003333	27.7
MC0002-5	0.635	0.478	0.635	1.588	2.543	1.270	2.225	6.4	10.1	13.0	3.6	0.1819	1.008	0.183	0.003677	21.0
MC1645-5	0.953	0.478	0.318	1.588	2.543	1.270	1.590	5.7	13.6	6.8	3.8	0.2729	0.504	0.138	0.003987	17.4
MC1574-5	0.953	0.406	0.635	1.270	2.083	1.588	2.083	5.4	11.0	11.8	4.1	0.2323	0.806	0.187	0.004221	19.8
MC0143-5	0.635	0.635	0.635	1.270	2.540	1.270	2.540	6.4	13.4	11.3	3.9	0.2419	0.806	0.195	0.004786	22.6
MC1696-5	0.635	0.635	1.270	0.953	2.223	1.905	3.810	7.0	14.8	21.3	4.9	0.2419	1.210	0.293	0.005731	33.5
MC1576-5	0.478	0.795	0.478	2.383	3.973	0.955	2.545	8.9	17.7	15.0	3.7	0.2278	1.138	0.259	0.006377	29.6
MC1591-5	0.714	0.635	0.635	1.430	2.700	1.349	2.540	6.7	15.8	13.2	4.1	0.2719	0.908	0.247	0.006547	24.6
MC0003-5	0.953	0.478	0.635	1.588	2.543	1.588	2.225	6.4	15.1	15.3	4.3	0.2729	1.008	0.275	0.007041	24.0
MC1709-5	0.635	0.635	0.953	1.430	2.700	1.588	3.175	7.3	15.4	21.5	4.4	0.2419	1.362	0.330	0.007177	31.0
MC0004-5	0.635	0.635	0.635	2.223	3.493	1.270	2.540	8.3	17.4	19.8	3.9	0.2419	1.411	0.341	0.008376	29.8
MC1679-5	0.635	0.635	1.270	1.588	2.858	1.905	3.810	8.3	17.4	35.4	4.9	0.2419	2.016	0.488	0.009552	40.7
MC1705-5	0.635	0.635	0.953	2.223	3.493	1.588	3.175	8.9	18.8	33.4	4.4	0.2419	2.117	0.512	0.011154	38.5
MC1599-5	1.113	0.635	0.478	1.588	2.858	1.590	2.225	6.7	24.7	12.5	4.7	0.4239	0.758	0.321	0.011709	26.5
MC1700-5	1.430	0.478	0.478	1.981	2.936	1.908	1.910	6.8	24.4	17.4	5.2	0.4097	0.946	0.388	0.012281	27.9
MC1603-5	0.478	0.874	0.762	2.858	4.605	1.240	3.272	10.7	23.5	33.3	4.3	0.2503	2.177	0.545	0.012674	44.3
MC1693-5	0.635	0.635	0.953	2.540	3.810	1.588	3.175	9.5	20.1	38.2	4.4	0.2419	2.419	0.585	0.012748	41.5
MC8400-5	0.635	0.635	0.635	3.493	4.763	1.270	2.540	10.8	22.8	31.1	3.9	0.2419	2.218	0.537	0.013162	39.5
MC1142-5	1.270	0.318	1.113	2.858	3.493	2.383	2.860	9.2	19.5	62.5	5.5	0.2419	3.179	0.769	0.013460	47.4
MC1594-5	1.270	0.508	0.762	1.588	2.604	2.032	2.540	6.7	22.7	23.1	5.4	0.3871	1.210	0.468	0.013526	31.0
MC1614-5	1.270	0.556	0.699	1.448	2.560	1.969	2.510	6.5	24.1	19.3	5.4	0.4239	1.011	0.429	0.013566	29.7

Magnetics C Cores 0.5 mil Iron Alloy

Part No.	D cm 1	E cm 2	F cm 3	G cm 4	Ht cm 5	Wth cm 6	Lt cm 7	MPL cm 8	Wtfe grams 9	Wtcu grams 10	MLT cm 11	Ac cm sq 12	Wa cm sq 13	Ap cm 4th 14	Kg cm 5th 15	At cm sq 16
MC1669-5	1.270	0.478	0.889	1.588	2.543	2.159	2.733	6.9	21.8	27.6	5.5	0.3639	1.411	0.514	0.013593	33.3
MC1613-5	1.270	0.556	0.508	1.905	3.018	1.778	2.129	7.1	26.1	17.4	5.1	0.4239	0.968	0.410	0.013750	28.9
MC1685-5	0.635	0.478	1.270	3.810	4.765	1.905	3.495	12.1	19.2	79.6	4.6	0.1819	4.839	0.880	0.013846	59.9
MC1169-5	0.953	0.635	0.635	1.905	3.175	1.588	2.540	7.6	24.1	19.7	4.6	0.3629	1.210	0.439	0.013914	31.0
MC1699-5	1.430	0.516	0.516	1.880	2.911	1.946	2.062	6.9	26.5	18.3	5.3	0.4424	0.969	0.429	0.014292	29.2
MC1666-5	0.635	0.953	0.635	1.984	3.889	1.270	3.175	9.0	28.7	20.5	4.6	0.3629	1.260	0.457	0.014489	37.8
MC1605-5	1.113	0.635	0.478	1.984	3.254	1.590	2.225	7.5	27.6	15.7	4.7	0.4239	0.947	0.402	0.014632	29.5
MC1581-5	1.270	0.635	0.635	1.349	2.619	1.905	2.540	6.5	27.5	16.5	5.4	0.4839	0.856	0.414	0.014812	29.7
MC0076-5	0.953	0.478	0.953	2.540	3.495	1.905	2.860	8.9	21.2	41.0	4.8	0.2729	2.419	0.660	0.015130	40.6
MC1586-5	0.635	0.635	0.953	3.018	4.288	1.588	3.175	10.5	22.1	45.4	4.4	0.2419	2.874	0.695	0.015144	46.1
MC1638-5	0.953	0.478	1.113	2.383	3.338	2.065	3.180	8.9	21.2	47.3	5.0	0.2729	2.651	0.723	0.015746	43.9
MC0005-5	0.953	0.635	0.635	2.223	3.493	1.588	2.540	8.3	26.2	23.0	4.6	0.3629	1.411	0.512	0.016233	33.7
MC1695-5	0.635	0.635	0.635	4.445	5.715	1.270	2.540	12.7	26.8	39.6	3.9	0.2419	2.823	0.683	0.016752	46.8
MC1694-5	0.635	0.635	1.270	2.858	4.128	1.905	3.810	10.8	22.8	63.8	4.9	0.2419	3.629	0.878	0.017193	55.2
MC0007-5	0.953	0.556	0.787	2.540	3.653	1.740	2.687	8.9	24.6	33.2	4.7	0.3179	2.000	0.636	0.017343	38.2
MC1521-5	0.953	0.516	0.953	2.540	3.571	1.905	2.936	9.0	23.3	41.6	4.8	0.2947	2.419	0.713	0.017363	41.8
MC1684-5	0.635	0.478	1.524	4.445	5.400	2.159	4.003	13.8	22.0	121.1	5.0	0.1819	6.774	1.232	0.017846	78.0
MC1528-5	1.270	0.508	0.826	2.032	3.048	2.096	2.667	7.7	26.2	32.6	5.5	0.3871	1.677	0.649	0.018414	36.9
MC1677-5	0.635	0.953	0.889	1.984	3.889	1.524	3.683	9.6	30.3	31.2	5.0	0.3629	1.764	0.640	0.018660	44.9
MC1707-5	0.635	0.635	1.270	3.175	4.445	1.905	3.810	11.4	24.1	70.9	4.9	0.2419	4.032	0.976	0.019103	58.9
MC8100-5	0.635	0.635	1.270	3.335	4.605	1.905	3.810	11.8	24.8	74.4	4.9	0.2419	4.235	1.025	0.020066	60.7
MC1237-5	1.270	0.635	0.635	1.905	3.175	1.905	2.540	7.6	32.2	23.3	5.4	0.4839	1.210	0.585	0.020921	34.7
MC1656-5	0.953	0.635	1.588	1.588	2.858	2.540	4.445	8.9	28.2	54.4	6.1	0.3629	2.520	0.915	0.021852	54.4
MC1703-5	0.635	0.635	1.270	3.810	5.080	1.905	3.810	12.7	26.8	85.0	4.9	0.2419	4.839	1.171	0.022924	66.1
MC1691-5	0.635	0.673	1.308	3.531	4.877	1.943	3.962	12.4	27.7	83.4	5.1	0.2565	4.618	1.184	0.023926	66.0
MC0006-5	1.270	0.635	0.635	2.223	3.493	1.905	2.540	8.3	34.9	27.2	5.4	0.4839	1.411	0.683	0.024408	37.5
MC1200-5	0.953	0.635	0.953	2.540	3.810	1.905	3.175	9.5	30.2	43.7	5.1	0.3629	2.419	0.878	0.025096	46.0
MC1670-5	1.270	0.635	0.953	1.748	3.018	2.223	3.175	7.9	33.5	35.0	5.9	0.4839	1.665	0.805	0.026361	41.8
MC1698-5	0.635	0.635	1.524	4.445	5.715	2.159	4.318	14.5	30.6	128.7	5.3	0.2419	6.774	1.639	0.029697	84.9
MC1571-5	0.953	0.635	0.953	3.018	4.288	1.905	3.175	10.5	33.2	51.9	5.1	0.3629	2.874	1.043	0.029814	50.8

Magnetics C Cores 0.5 mil Iron Alloy

Part No.	D cm 1	E cm 2	F cm 3	G cm 4	Ht cm 5	Wth cm 6	Lt cm 7	MPL cm 8	Wtfe grams 9	Wtcu grams 10	MLT cm 11	Ac cm sq 12	Wa cm sq 13	Ap cm 4th 14	Kg cm 5th 15	At cm sq 16
MC1572-5	1.113	0.556	0.953	3.018	4.130	2.065	3.018	10.2	33.0	53.6	5.2	0.3713	2.874	1.067	0.030244	50.2
MC1604-5	0.635	0.635	1.588	4.445	5.715	2.223	4.445	14.6	30.8	136.5	5.4	0.2419	7.056	1.707	0.030368	87.9
MC1612-5	1.588	0.556	0.762	2.286	3.399	2.350	2.637	8.3	38.5	37.7	6.1	0.5298	1.742	0.923	0.032109	42.9
MC0124-5	1.270	0.635	0.787	2.540	3.810	2.057	2.845	9.2	38.8	40.2	5.7	0.4839	2.000	0.968	0.033126	45.1
MC1573-5	0.953	0.635	1.270	2.858	4.128	2.223	3.810	10.8	34.2	72.0	5.6	0.3629	3.629	1.317	0.034280	60.3
MC1577-5	0.635	0.635	2.223	4.445	5.715	2.858	5.715	15.9	33.5	226.1	6.4	0.2419	9.879	2.390	0.035931	119.4
MC1223-5	1.588	0.516	0.953	2.540	3.571	2.540	2.936	9.0	38.8	54.3	6.3	0.4911	2.419	1.188	0.036995	50.1
MC1371-5	1.588	0.635	0.787	2.223	3.493	2.375	2.845	8.6	45.2	39.1	6.3	0.6048	1.750	1.058	0.040717	46.0
MC1632-5	1.588	0.594	0.889	2.286	3.475	2.476	2.967	8.7	43.1	46.0	6.4	0.5661	2.032	1.151	0.040917	48.3
MC1598-5	0.635	0.635	1.588	6.350	7.620	2.223	4.445	18.4	38.9	195.0	5.4	0.2419	10.081	2.439	0.043383	113.3
MC1668-5	0.635	0.635	2.540	5.080	6.350	3.175	6.350	17.8	37.6	318.2	6.9	0.2419	12.903	3.122	0.043557	148.4
MC1490-5	0.953	0.635	1.588	3.175	4.445	2.540	4.445	12.1	38.2	108.9	6.1	0.3629	5.040	1.829	0.043704	76.6
MC1611-5	1.430	0.714	0.795	2.383	3.810	2.225	3.018	9.2	49.2	41.4	6.1	0.6124	1.894	1.160	0.046250	48.8
MC4400-5	1.270	0.635	1.270	2.540	3.810	2.540	3.810	10.2	42.9	73.6	6.4	0.4839	3.226	1.561	0.047116	61.3
MC1379-5	1.588	0.714	0.711	2.540	3.967	2.299	2.850	9.4	55.5	40.6	6.3	0.6798	1.806	1.228	0.052783	49.9
MC2350-5	1.588	0.635	0.953	2.540	3.810	2.540	3.175	9.5	50.3	56.3	6.5	0.6048	2.419	1.463	0.054063	54.8
MC1544-5	1.430	0.635	1.270	2.540	3.810	2.700	3.810	10.2	48.3	77.2	6.7	0.5448	3.226	1.758	0.056897	63.7
MC1317-5	1.588	0.516	0.953	3.970	5.001	2.540	2.936	11.9	51.1	84.8	6.3	0.4911	3.781	1.857	0.057823	65.8
MC0008-5	0.953	0.953	0.953	3.018	4.923	1.905	3.810	11.8	55.8	58.4	5.7	0.5444	2.874	1.565	0.059627	63.5
MC1692-5	1.270	0.635	1.588	2.858	4.128	2.858	4.445	11.4	48.3	111.5	6.9	0.4839	4.536	2.195	0.061477	77.6
MC1347-5	1.270	0.635	1.270	3.335	4.605	2.540	3.810	11.8	49.6	96.6	6.4	0.4839	4.235	2.049	0.061863	71.4
MC1583-5	1.270	0.635	1.270	3.493	4.763	2.540	3.810	12.1	51.0	101.1	6.4	0.4839	4.435	2.146	0.064784	73.4
MC1723-5	1.588	0.635	1.092	2.870	4.140	2.680	3.454	10.5	55.3	75.4	6.8	0.6048	3.135	1.896	0.067781	63.8
MC1363-5	0.953	0.795	1.588	3.335	4.925	2.540	4.765	13.0	51.7	120.4	6.4	0.4544	5.294	2.406	0.068358	86.1
MC1710-5	0.953	0.795	1.588	3.353	4.943	2.540	4.765	13.1	51.8	121.0	6.4	0.4544	5.323	2.418	0.068723	86.3
MC0013-5	1.588	0.635	1.270	2.858	4.128	2.858	3.810	10.8	57.0	90.9	7.0	0.6048	3.629	2.195	0.075358	70.4
MC1721-5	1.270	1.270	0.724	1.905	4.445	1.994	3.988	10.3	87.3	33.5	6.8	0.9677	1.379	1.335	0.075697	61.7
MC1547-5	0.953	0.953	1.113	3.653	5.558	2.065	4.130	13.3	63.4	86.2	6.0	0.5444	4.064	2.212	0.080749	77.4
MC1378-5	0.953	0.795	1.588	3.970	5.560	2.540	4.765	14.3	56.7	143.3	6.4	0.4544	6.302	2.864	0.081374	95.3
MC2300-5	1.270	0.635	1.270	4.445	5.715	2.540	3.810	14.0	59.0	128.7	6.4	0.4839	5.645	2.732	0.082453	85.5

Magnetics C Cores 0.5 mil Iron Alloy

Part No.	D cm 1	E cm 2	F cm 3	G cm 4	Ht cm 5	Wth cm 6	Lt cm 7	MPL cm 8	Wtfe grams 9	Wtcu grams 10	MLT cm 11	Ac cm sq 12	Wa cm sq 13	Ap cm 4th 14	Kg cm 5th 15	At cm sq 16
MC1380-5	1.270	0.635	1.664	3.810	5.080	2.934	4.597	13.5	57.0	158.5	7.0	0.4839	6.339	3.067	0.084443	95.0
MC1149-5	0.953	0.795	1.270	5.080	6.670	2.223	4.130	15.9	63.0	135.3	5.9	0.4544	6.452	2.931	0.090342	95.0
MC1543-5	0.953	0.953	1.270	3.810	5.715	2.223	4.445	14.0	66.4	106.9	6.2	0.5444	4.839	2.634	0.092327	86.3
MC0009-5	1.270	0.953	0.953	3.018	4.923	2.223	3.810	11.8	74.5	66.9	6.5	0.7258	2.874	2.086	0.092486	69.1
MC1184-5	1.588	0.795	0.953	3.018	4.608	2.540	3.495	11.1	73.5	70.2	6.9	0.7573	2.874	2.176	0.095985	67.3
MC1714-5	1.270	0.635	1.753	4.445	5.715	3.023	4.775	14.9	63.1	198.6	7.2	0.4839	7.790	3.769	0.101760	109.1
MC1219-5	0.953	0.953	1.905	3.335	5.240	2.858	5.715	14.3	67.9	162.9	7.2	0.5444	6.353	3.458	0.104460	108.0
MC1413-5	1.270	0.795	1.270	3.810	5.400	2.540	4.130	13.3	70.6	115.8	6.7	0.6058	4.839	2.931	0.105515	84.9
MC1178-5	1.588	0.635	1.588	3.493	4.763	3.175	4.445	12.7	67.1	148.8	7.5	0.6048	5.544	3.353	0.107524	92.7
MC1690-5	1.270	0.673	1.308	3.531	4.831	2.578	3.917	12.3	58.0	106.8	6.5	0.6194	4.618	2.860	0.108985	76.1
MC1151-5	1.270	0.795	1.270	3.970	5.560	2.540	4.130	13.7	72.2	120.7	6.7	0.6058	5.042	3.054	0.109947	87.1
MC1559-5	1.270	0.635	3.810	3.335	4.605	5.080	8.890	16.8	71.1	469.9	10.4	0.4839	12.706	6.148	0.114425	199.6
MC1362-5	1.588	0.635	1.588	3.810	5.080	3.175	4.445	13.3	70.4	162.3	7.5	0.6048	6.048	3.658	0.117299	97.6
MC1279-5	1.588	0.795	1.588	2.540	4.130	3.175	4.765	11.4	75.6	112.8	7.9	0.7573	4.032	3.053	0.117590	85.7
MC1164-5	1.588	0.953	0.953	2.700	4.605	2.540	3.810	11.1	88.0	65.7	7.2	0.9073	2.572	2.333	0.117875	70.6
MC1715-5	1.270	0.953	1.588	2.692	4.597	2.858	5.080	12.4	78.4	114.7	7.5	0.7258	4.274	3.102	0.119364	89.9
MC1713-5	1.270	0.635	2.540	4.293	5.563	3.810	6.350	16.2	68.5	325.9	8.4	0.4839	10.903	5.276	0.121477	148.5
MC1346-5	1.588	0.635	1.588	3.970	5.240	3.175	4.445	13.7	72.1	169.1	7.5	0.6048	6.302	3.812	0.122226	100.0
MC1155-5	1.270	0.795	1.588	3.970	5.560	2.858	4.765	14.3	75.6	162.0	7.2	0.6058	6.302	3.818	0.127959	101.9
MC1675-5	1.588	0.714	1.270	3.970	5.398	2.858	3.967	13.3	79.1	129.2	7.2	0.6798	5.042	3.428	0.129381	89.1
MC0010-5	1.588	0.953	0.953	3.018	4.923	2.540	3.810	11.8	93.1	73.4	7.2	0.9073	2.874	2.608	0.131736	74.6
MC1712-5	1.270	0.635	2.540	4.763	6.033	3.810	6.350	17.1	72.4	361.6	8.4	0.4839	12.097	5.853	0.134774	158.1
MC0012-5	1.270	1.113	1.270	2.858	5.083	2.540	4.765	12.7	94.0	95.1	7.4	0.8477	3.629	3.076	0.141608	87.2
MC1683-5	1.588	0.754	1.270	3.967	5.476	2.858	4.049	13.5	84.6	130.5	7.3	0.7185	5.039	3.621	0.142830	91.1
MC1300-5	0.953	1.270	1.270	3.970	6.510	2.223	5.080	15.6	98.6	122.8	6.8	0.7258	5.042	3.659	0.155168	104.8
MC1664-5	1.270	0.953	1.270	4.128	6.033	2.540	4.445	14.6	92.5	131.4	7.0	0.7258	5.242	3.805	0.156745	97.2
MC1264-5	1.270	1.270	0.953	3.175	5.715	2.223	4.445	13.3	112.7	77.3	7.2	0.9677	3.024	2.927	0.157708	86.3
MC1711-5	0.953	0.991	1.905	4.928	6.909	2.858	5.791	17.6	87.1	243.2	7.3	0.5661	9.387	5.314	0.165191	137.5
MC1349-5	1.588	0.953	1.270	3.335	5.240	2.858	4.445	13.0	103.1	115.7	7.7	0.9073	4.235	3.843	0.181532	92.3
MC1649-5	1.270	0.953	1.588	4.128	6.033	2.858	5.080	15.2	96.6	175.8	7.5	0.7258	6.552	4.756	0.182987	112.7

Magnetics C Cores 0.5 mil Iron Alloy																
Part No.	D cm 1	E cm 2	F cm 3	G cm 4	Ht cm 5	Wth cm 6	Lt cm 7	MPL cm 8	Wtfe grams 9	Wtcu grams 10	MLT cm 11	Ac cm sq 12	Wa cm sq 13	Ap cm 4th 14	Kg cm 5th 15	At cm sq 16
MC1560-5	0.635	0.635	8.890	15.240	16.510	9.525	19.050	50.8	107.3	8144.6	16.9	0.2419	135.484	32.778	0.187638	1261.3
MC1508-5	1.588	0.953	1.270	3.810	5.715	2.858	4.445	14.0	110.6	132.2	7.7	0.9073	4.839	4.390	0.207386	99.2
MC1414-5	1.270	0.953	1.588	5.080	6.985	2.858	5.080	17.1	108.6	216.4	7.5	0.7258	8.065	5.853	0.225215	127.8
MC0018-5	1.270	1.113	1.588	3.970	6.195	2.858	5.400	15.6	115.2	176.3	7.9	0.8477	6.302	5.343	0.230341	119.0
MC1221-5	1.588	0.953	1.905	3.335	5.240	3.493	5.715	14.3	113.2	196.1	8.7	0.9073	6.353	5.764	0.241019	121.9
MC1154-5	1.270	1.270	1.588	3.335	5.875	2.858	5.715	14.9	126.1	154.0	8.2	0.9677	5.294	5.124	0.242447	117.3
MC0014-5	1.270	1.270	1.270	3.970	6.510	2.540	5.080	15.6	131.5	137.7	7.7	0.9677	5.042	4.879	0.245870	112.1
MC1724-5	1.588	0.795	3.175	4.001	5.591	4.763	7.940	17.5	115.9	467.8	10.4	0.7573	12.702	9.618	0.281279	197.5
MC1359-5	1.270	1.270	1.588	3.970	6.510	2.858	5.715	16.2	136.8	183.3	8.2	0.9677	6.302	6.099	0.288610	128.1
MC1249-5	1.270	1.270	1.270	4.763	7.303	2.540	5.080	17.1	144.8	165.2	7.7	0.9677	6.048	5.853	0.294949	124.2
MC1608-5	1.430	1.270	1.905	3.175	5.715	3.335	6.350	15.2	145.0	193.5	9.0	1.0897	6.048	6.591	0.319230	133.7
MC1725-5	1.588	0.795	3.175	4.928	6.518	4.763	7.940	19.4	128.2	576.2	10.4	0.7573	15.645	11.847	0.346464	221.0
MC1617-5	0.953	1.753	2.057	3.810	7.315	3.010	7.620	18.7	163.9	252.2	9.0	1.0016	7.839	7.851	0.347643	172.5
MC1367-5	1.588	1.270	1.588	3.335	5.875	3.175	5.715	14.9	157.6	166.0	8.8	1.2097	5.294	6.404	0.351535	124.6
MC1716-5	1.270	1.270	1.549	4.928	7.468	2.819	5.639	18.0	152.4	220.5	8.1	0.9677	7.635	7.389	0.352201	142.4
MC1718-5	1.270	1.270	1.905	4.293	6.833	3.175	6.350	17.5	147.6	252.4	8.7	0.9677	8.177	7.914	0.352964	151.1
MC0015-5	1.588	1.270	1.270	3.970	6.510	2.858	5.080	15.6	164.3	149.1	8.3	1.2097	5.042	6.099	0.354840	119.5
MC1510-5	1.588	0.953	1.905	4.928	6.833	3.493	5.715	17.5	138.4	289.7	8.7	0.9073	9.387	8.516	0.356113	151.2
MC1720-5	1.270	1.270	2.692	3.658	6.198	3.962	7.925	17.8	150.2	347.2	9.9	0.9677	9.848	9.530	0.372064	182.8
MC1726-5	1.588	0.953	1.905	5.232	7.137	3.493	5.715	18.1	143.2	307.6	8.7	0.9073	9.968	9.043	0.378141	156.8
MC1727-5	1.588	0.953	1.905	5.410	7.315	3.493	5.715	18.4	146.1	318.1	8.7	0.9073	10.306	9.351	0.390990	160.1
MC1728-5	1.588	1.588	0.953	3.810	6.985	2.540	5.080	15.9	209.6	109.1	8.5	1.5121	3.629	5.487	0.392622	120.2
MC1634-5	0.953	1.270	3.302	5.715	8.255	4.255	9.144	23.1	146.5	673.5	10.0	0.7258	18.871	13.697	0.396172	265.5
MC1717-5	1.270	1.270	1.905	5.080	7.620	3.175	6.350	19.1	160.9	298.7	8.7	0.9677	9.677	9.365	0.417709	166.1
MC1235-5	1.430	1.588	1.588	3.493	6.668	3.018	6.350	16.5	196.3	180.1	9.1	1.3621	5.544	7.552	0.450398	143.1
MC1610-5	1.270	1.270	3.810	3.810	6.350	5.080	10.160	20.3	171.7	602.4	11.7	0.9677	14.516	14.048	0.465980	258.1
MC1635-5	0.953	1.270	3.810	6.985	9.525	4.763	10.160	26.7	169.0	1025.3	10.8	0.7258	26.613	19.316	0.517576	342.3
MC1609-5	1.430	1.430	1.994	4.923	7.783	3.424	6.848	19.6	209.4	330.1	9.5	1.2270	9.815	12.043	0.624884	184.0
MC1569-5	1.430	1.430	2.692	5.875	8.735	4.122	8.245	22.9	244.8	593.7	10.6	1.2270	15.818	19.408	0.902434	253.2
MC1530-5	1.588	1.588	1.588	5.875	9.050	3.175	6.350	21.3	280.8	313.4	9.5	1.5121	9.327	14.103	0.902592	192.6

Magnetics C Cores 0.5 mil Iron Alloy																
Part No.	D cm 1	E cm 2	F cm 3	G cm 4	Ht cm 5	Wth cm 6	Lt cm 7	MPL cm 8	Wtfe grams 9	Wtcu grams 10	MLT cm 11	Ac cm sq 12	Wa cm sq 13	Ap cm 4th 14	Kg cm 5th 15	At cm sq 16
MC1722-5	1.270	1.270	4.128	8.390	10.930	5.398	10.795	30.1	254.4	1498.4	12.2	0.9677	34.628	33.511	1.066059	428.6
MC1515-5	1.588	1.588	2.223	6.985	10.160	3.810	7.620	24.8	326.9	576.7	10.4	1.5121	15.524	23.474	1.359003	262.1
MC1719-5	1.270	1.270	3.810	11.278	13.818	5.080	10.160	35.3	297.8	1783.0	11.7	0.9677	42.968	41.582	1.379302	485.7

Magnetics C Cores 1 mil Iron Alloy

Part No.	D cm 1	E cm 2	F cm 3	G cm 4	Ht cm 5	Wth cm 6	Lt cm 7	MPL cm 8	Wtfe grams 9	Wtcu grams 10	MLT cm 11	Ac cm sq 12	Wa cm sq 13	Ap cm 4th 14	Kg cm 5th 15	At cm sq 16
MC1100-1	0.318	0.160	0.318	0.953	1.273	0.635	0.955	3.2	1.1	2.0	1.9	0.0406	0.302	0.012	0.00011	4.9
MC1505-1	0.318	0.160	0.795	1.113	1.433	1.113	1.910	4.5	1.6	8.2	2.6	0.0406	0.884	0.036	0.00022	11.9
MC1504-1	0.318	0.318	0.800	1.113	1.748	1.118	2.235	5.1	3.6	9.3	2.9	0.0806	0.890	0.072	0.00079	14.7
MC2340-1	0.635	0.318	0.478	0.953	1.588	1.113	1.590	4.1	5.8	5.0	3.1	0.1613	0.455	0.073	0.00155	10.9
MC1647-1	0.953	0.452	0.318	0.381	1.285	1.270	1.539	3.2	9.6	1.6	3.7	0.3445	0.121	0.042	0.00155	10.0
MC1606-1	0.635	0.318	0.478	1.270	1.905	1.113	1.590	4.8	6.7	6.6	3.1	0.1613	0.606	0.098	0.00206	12.6
MC0001-1	0.635	0.318	0.635	1.270	1.905	1.270	1.905	5.1	7.2	9.5	3.3	0.1613	0.806	0.130	0.00254	15.3
MC1582-1	0.953	0.318	0.635	1.270	1.905	1.588	1.905	5.1	10.7	11.3	3.9	0.2419	0.806	0.195	0.00479	17.7
MC1589-1	0.635	0.478	0.714	1.270	2.225	1.349	2.383	5.9	12.4	12.1	3.8	0.2426	0.906	0.220	0.00568	20.3
MC0147-1	0.635	0.318	0.800	2.540	3.175	1.435	2.235	8.0	11.2	25.8	3.6	0.1613	2.032	0.328	0.00593	27.7
MC0002-1	0.635	0.478	0.635	1.588	2.543	1.270	2.225	6.4	13.5	13.0	3.6	0.2426	1.008	0.245	0.00654	21.0
MC1645-1	0.953	0.478	0.318	1.588	2.543	1.270	1.590	5.7	18.2	6.8	3.8	0.3639	0.504	0.183	0.00709	17.4
MC1574-1	0.953	0.406	0.635	1.270	2.083	1.588	2.083	5.4	14.7	11.8	4.1	0.3097	0.806	0.250	0.00750	19.8
MC0143-1	0.635	0.635	0.635	1.270	2.540	1.270	2.540	6.4	17.9	11.3	3.9	0.3226	0.806	0.260	0.00851	22.6
MC1696-1	0.635	0.635	1.270	0.953	2.223	1.905	3.810	7.0	19.7	21.3	4.9	0.3226	1.210	0.390	0.01019	33.5
MC1576-1	0.478	0.795	0.478	2.383	3.973	0.955	2.545	8.9	23.6	15.0	3.7	0.3037	1.138	0.346	0.01134	29.6
MC1591-1	0.714	0.635	0.635	1.430	2.700	1.349	2.540	6.7	21.1	13.2	4.1	0.3626	0.908	0.329	0.01164	24.6
MC0003-1	0.953	0.478	0.635	1.588	2.543	1.588	2.225	6.4	20.2	15.3	4.3	0.3639	1.008	0.367	0.01252	24.0
MC1709-1	0.635	0.635	0.953	1.430	2.700	1.588	3.175	7.3	20.6	21.5	4.4	0.3226	1.362	0.439	0.01276	31.0
MC0004-1	0.635	0.635	0.635	2.223	3.493	1.270	2.540	8.3	23.2	19.8	3.9	0.3226	1.411	0.455	0.01489	29.8
MC1679-1	0.635	0.635	1.270	1.588	2.858	1.905	3.810	8.3	23.2	35.4	4.9	0.3226	2.016	0.650	0.01698	40.7
MC1705-1	0.635	0.635	0.953	2.223	3.493	1.588	3.175	8.9	25.0	33.4	4.4	0.3226	2.117	0.683	0.01983	38.5
MC1599-1	1.113	0.635	0.478	1.588	2.858	1.590	2.225	6.7	32.9	12.5	4.7	0.5652	0.758	0.428	0.02082	26.5
MC1700-1	1.430	0.478	0.478	1.981	2.936	1.908	1.910	6.8	32.6	17.4	5.2	0.5463	0.946	0.517	0.02183	27.9
MC1603-1	0.478	0.874	0.762	2.858	4.605	1.240	3.272	10.7	31.3	33.3	4.3	0.3338	2.177	0.727	0.02253	44.3
MC1693-1	0.635	0.635	0.953	2.540	3.810	1.588	3.175	9.5	26.8	38.2	4.4	0.3226	2.419	0.780	0.02266	41.5
MC8400-1	0.635	0.635	0.635	3.493	4.763	1.270	2.540	10.8	30.4	31.1	3.9	0.3226	2.218	0.715	0.02340	39.5
MC1142-1	1.270	0.318	1.113	2.858	3.493	2.383	2.860	9.2	25.9	62.5	5.5	0.3226	3.179	1.025	0.02393	47.4
MC1594-1	1.270	0.508	0.762	1.588	2.604	2.032	2.540	6.7	30.3	23.1	5.4	0.5161	1.210	0.624	0.02405	31.0
MC1614-1	1.270	0.556	0.699	1.448	2.560	1.969	2.510	6.5	32.2	19.3	5.4	0.5652	1.011	0.572	0.02412	29.7

Magnetics C Cores 1 mil Iron Alloy

Part No.	D cm 1	E cm 2	F cm 3	G cm 4	Ht cm 5	Wth cm 6	Lt cm 7	MPL cm 8	Wtfe grams 9	Wtcu grams 10	MLT cm 11	Ac cm sq 12	Wa cm sq 13	Ap cm 4th 14	Kg cm 5th 15	At cm sq 16
MC1669-1	1.270	0.478	0.889	1.588	2.543	2.159	2.733	6.9	29.1	27.6	5.5	0.4852	1.411	0.685	0.02417	33.3
MC1613-1	1.270	0.556	0.508	1.905	3.018	1.778	2.129	7.1	34.8	17.4	5.1	0.5652	0.968	0.547	0.02444	28.9
MC1685-1	0.635	0.478	1.270	3.810	4.765	1.905	3.495	12.1	25.6	79.6	4.6	0.2426	4.839	1.174	0.02462	59.9
MC1169-1	0.953	0.635	0.635	1.905	3.175	1.588	2.540	7.6	32.2	19.7	4.6	0.4839	1.210	0.585	0.02474	31.0
MC1699-1	1.430	0.516	0.516	1.880	2.911	1.946	2.062	6.9	35.3	18.3	5.3	0.5899	0.969	0.572	0.02541	29.2
MC1666-1	0.635	0.953	0.635	1.984	3.889	1.270	3.175	9.0	38.2	20.5	4.6	0.4839	1.260	0.610	0.02576	37.8
MC1605-1	1.113	0.635	0.478	1.984	3.254	1.590	2.225	7.5	36.8	15.7	4.7	0.5652	0.947	0.535	0.02601	29.5
MC1581-1	1.270	0.635	0.635	1.349	2.619	1.905	2.540	6.5	36.7	16.5	5.4	0.6452	0.856	0.553	0.02633	29.7
MC0076-1	0.953	0.478	0.953	2.540	3.495	1.905	2.860	8.9	28.3	41.0	4.8	0.3639	2.419	0.880	0.02690	40.6
MC1586-1	0.635	0.635	0.953	3.018	4.288	1.588	3.175	10.5	29.5	45.4	4.4	0.3226	2.874	0.927	0.02692	46.1
MC1638-1	0.953	0.478	1.113	2.383	3.338	2.065	3.180	8.9	28.3	47.3	5.0	0.3639	2.651	0.964	0.02799	43.9
MC0005-1	0.953	0.635	0.635	2.223	3.493	1.588	2.540	8.3	34.9	23.0	4.6	0.4839	1.411	0.683	0.02886	33.7
MC1695-1	0.635	0.635	0.635	4.445	5.715	1.270	2.540	12.7	35.8	39.6	3.9	0.3226	2.823	0.911	0.02978	46.8
MC1694-1	0.635	0.635	1.270	2.858	4.128	1.905	3.810	10.8	30.4	63.8	4.9	0.3226	3.629	1.171	0.03057	55.2
MC0007-1	0.953	0.556	0.787	2.540	3.653	1.740	2.687	8.9	32.9	33.2	4.7	0.4239	2.000	0.848	0.03083	38.2
MC1521-1	0.953	0.516	0.953	2.540	3.571	1.905	2.936	9.0	31.0	41.6	4.8	0.3929	2.419	0.951	0.03087	41.8
MC1684-1	0.635	0.478	1.524	4.445	5.400	2.159	4.003	13.8	29.3	121.1	5.0	0.2426	6.774	1.643	0.03173	78.0
MC1528-1	1.270	0.508	0.826	2.032	3.048	2.096	2.667	7.7	34.9	32.6	5.5	0.5161	1.677	0.866	0.03274	36.9
MC1677-1	0.635	0.953	0.889	1.984	3.889	1.524	3.683	9.6	40.4	31.2	5.0	0.4839	1.764	0.853	0.03317	44.9
MC1707-1	0.635	0.635	1.270	3.175	4.445	1.905	3.810	11.4	32.2	70.9	4.9	0.3226	4.032	1.301	0.03396	58.9
MC8100-1	0.635	0.635	1.270	3.335	4.605	1.905	3.810	11.8	33.1	74.4	4.9	0.3226	4.235	1.366	0.03567	60.7
MC1237-1	1.270	0.635	0.635	1.905	3.175	1.905	2.540	7.6	42.9	23.3	5.4	0.6452	1.210	0.780	0.03719	34.7
MC1656-1	0.953	0.635	1.588	1.588	2.858	2.540	4.445	8.9	37.6	54.4	6.1	0.4839	2.520	1.219	0.03885	54.4
MC1703-1	0.635	0.635	1.270	3.810	5.080	1.905	3.810	12.7	35.8	85.0	4.9	0.3226	4.839	1.561	0.04075	66.1
MC1691-1	0.635	0.673	1.308	3.531	4.877	1.943	3.962	12.4	36.9	83.4	5.1	0.3419	4.618	1.579	0.04254	66.0
MC0006-1	1.270	0.635	0.635	2.223	3.493	1.905	2.540	8.3	46.5	27.2	5.4	0.6452	1.411	0.911	0.04339	37.5
MC1200-1	0.953	0.635	0.953	2.540	3.810	1.905	3.175	9.5	40.2	43.7	5.1	0.4839	2.419	1.171	0.04462	46.0
MC1670-1	1.270	0.635	0.953	1.748	3.018	2.223	3.175	7.9	44.7	35.0	5.9	0.6452	1.665	1.074	0.04686	41.8
MC1698-1	0.635	0.635	1.524	4.445	5.715	2.159	4.318	14.5	40.8	128.7	5.3	0.3226	6.774	2.185	0.05280	84.9
MC1571-1	0.953	0.635	0.953	3.018	4.288	1.905	3.175	10.5	44.3	51.9	5.1	0.4839	2.874	1.391	0.05300	50.8

Magnetics C Cores 1 mil Iron Alloy

Part No.	D cm 1	E cm 2	F cm 3	G cm 4	Ht cm 5	Wth cm 6	Lt cm 7	MPL cm 8	Wtfe grams 9	Wtcu grams 10	MLT cm 11	Ac cm sq 12	Wa cm sq 13	Ap cm 4th 14	Kg cm 5th 15	At cm sq 16
MC1572-1	1.113	0.556	0.953	3.018	4.130	2.065	3.018	10.2	43.9	53.6	5.2	0.4951	2.874	1.423	0.05377	50.2
MC1604-1	0.635	0.635	1.588	4.445	5.715	2.223	4.445	14.6	41.1	136.5	5.4	0.3226	7.056	2.276	0.05399	87.9
MC1612-1	1.588	0.556	0.762	2.286	3.399	2.350	2.637	8.3	51.3	37.7	6.1	0.7065	1.742	1.231	0.05708	42.9
MC0124-1	1.270	0.635	0.787	2.540	3.810	2.057	2.845	9.2	51.8	40.2	5.7	0.6452	2.000	1.290	0.05889	45.1
MC1573-1	0.953	0.635	1.270	2.858	4.128	2.223	3.810	10.8	45.6	72.0	5.6	0.4839	3.629	1.756	0.06094	60.3
MC1577-1	0.635	0.635	2.223	4.445	5.715	2.858	5.715	15.9	44.7	226.1	6.4	0.3226	9.879	3.187	0.06388	119.4
MC1223-1	1.588	0.516	0.953	2.540	3.571	2.540	2.936	9.0	51.7	54.3	6.3	0.6548	2.419	1.584	0.06577	50.1
MC1371-1	1.588	0.635	0.787	2.223	3.493	2.375	2.845	8.6	60.3	39.1	6.3	0.8065	1.750	1.411	0.07239	46.0
MC1632-1	1.588	0.594	0.889	2.286	3.475	2.476	2.967	8.7	57.5	46.0	6.4	0.7548	2.032	1.534	0.07274	48.3
MC1598-1	0.635	0.635	1.588	6.350	7.620	2.223	4.445	18.4	51.9	195.0	5.4	0.3226	10.081	3.252	0.07713	113.3
MC1668-1	0.635	0.635	2.540	5.080	6.350	3.175	6.350	17.8	50.1	318.2	6.9	0.3226	12.903	4.162	0.07744	148.4
MC1490-1	0.953	0.635	1.588	3.175	4.445	2.540	4.445	12.1	51.0	108.9	6.1	0.4839	5.040	2.439	0.07770	76.6
MC1611-1	1.430	0.714	0.795	2.383	3.810	2.225	3.018	9.2	65.7	41.4	6.1	0.8165	1.894	1.547	0.08222	48.8
MC4400-1	1.270	0.635	1.270	2.540	3.810	2.540	3.810	10.2	57.2	73.6	6.4	0.6452	3.226	2.081	0.08376	61.3
MC1379-1	1.588	0.714	0.711	2.540	3.967	2.299	2.850	9.4	74.0	40.6	6.3	0.9064	1.806	1.637	0.09384	49.9
MC2350-1	1.588	0.635	0.953	2.540	3.810	2.540	3.175	9.5	67.1	56.3	6.5	0.8065	2.419	1.951	0.09611	54.8
MC1544-1	1.430	0.635	1.270	2.540	3.810	2.700	3.810	10.2	64.4	77.2	6.7	0.7265	3.226	2.343	0.10115	63.7
MC1317-1	1.588	0.516	0.953	3.970	5.001	2.540	2.936	11.9	68.1	84.8	6.3	0.6548	3.781	2.476	0.10280	65.8
MC0008-1	0.953	0.953	0.953	3.018	4.923	1.905	3.810	11.8	74.5	58.4	5.7	0.7258	2.874	2.086	0.10600	63.5
MC1692-1	1.270	0.635	1.588	2.858	4.128	2.858	4.445	11.4	64.4	111.5	6.9	0.6452	4.536	2.927	0.10929	77.6
MC1347-1	1.270	0.635	1.270	3.335	4.605	2.540	3.810	11.8	66.2	96.6	6.4	0.6452	4.235	2.733	0.10998	71.4
MC1583-1	1.270	0.635	1.270	3.493	4.763	2.540	3.810	12.1	68.0	101.1	6.4	0.6452	4.435	2.862	0.11517	73.4
MC1723-1	1.588	0.635	1.092	2.870	4.140	2.680	3.454	10.5	73.7	75.4	6.8	0.8065	3.135	2.528	0.12050	63.8
MC1363-1	0.953	0.795	1.588	3.335	4.925	2.540	4.765	13.0	68.9	120.4	6.4	0.6058	5.294	3.207	0.12153	86.1
MC1710-1	0.953	0.795	1.588	3.353	4.943	2.540	4.765	13.1	69.1	121.0	6.4	0.6058	5.323	3.224	0.12217	86.3
MC1690-1	1.270	0.650	1.308	3.531	4.831	2.578	3.917	12.3	64.2	107.0	6.5	0.6854	4.618	3.394	0.13347	76.1
MC0013-1	1.588	0.635	1.270	2.858	4.128	2.858	3.810	10.8	76.0	90.9	7.0	0.8065	3.629	2.927	0.13397	70.4
MC1721-1	1.270	1.270	0.724	1.905	4.445	1.994	3.988	10.3	116.5	33.5	6.8	1.2903	1.379	1.779	0.13457	61.7
MC1547-1	0.953	0.953	1.113	3.653	5.558	2.065	4.130	13.3	84.5	86.2	6.0	0.7258	4.064	2.949	0.14355	77.4
MC1378-1	0.953	0.795	1.588	3.970	5.560	2.540	4.765	14.3	75.6	143.3	6.4	0.6058	6.302	3.818	0.14467	95.3

Magnetics C Cores 1 mil Iron Alloy																
Part No.	D cm 1	E cm 2	F cm 3	G cm 4	Ht cm 5	Wth cm 6	Lt cm 7	MPL cm 8	Wtfe grams 9	Wtcu grams 10	MLT cm 11	Ac cm sq 12	Wa cm sq 13	Ap cm 4th 14	Kg cm 5th 15	At cm sq 16
MC2300-1	1.270	0.635	1.270	4.445	5.715	2.540	3.810	14.0	78.7	128.7	6.4	0.6452	5.645	3.642	0.14658	85.5
MC1380-1	1.270	0.635	1.664	3.810	5.080	2.934	4.597	13.5	76.0	158.5	7.0	0.6452	6.339	4.089	0.15012	95.0
MC1149-1	0.953	0.795	1.270	5.080	6.670	2.223	4.130	15.9	84.0	135.3	5.9	0.6058	6.452	3.908	0.16061	95.0
MC1543-1	0.953	0.953	1.270	3.810	5.715	2.223	4.445	14.0	88.5	106.9	6.2	0.7258	4.839	3.512	0.16414	86.3
MC0009-1	1.270	0.953	0.953	3.018	4.923	2.223	3.810	11.8	99.3	66.9	6.5	0.9677	2.874	2.781	0.16442	69.1
MC1184-1	1.588	0.795	0.953	3.018	4.608	2.540	3.495	11.1	98.0	70.2	6.9	1.0097	2.874	2.902	0.17064	67.3
MC1714-1	1.270	0.635	1.753	4.445	5.715	3.023	4.775	14.9	84.1	198.6	7.2	0.6452	7.790	5.026	0.18091	109.1
MC1219-1	0.953	0.953	1.905	3.335	5.240	2.858	5.715	14.3	90.5	162.9	7.2	0.7258	6.353	4.611	0.18571	108.0
MC1413-1	1.270	0.795	1.270	3.810	5.400	2.540	4.130	13.3	94.1	115.8	6.7	0.8077	4.839	3.908	0.18758	84.9
MC1178-1	1.588	0.635	1.588	3.493	4.763	3.175	4.445	12.7	89.4	148.8	7.5	0.8065	5.544	4.471	0.19115	92.7
MC1151-1	1.270	0.795	1.270	3.970	5.560	2.540	4.130	13.7	96.3	120.7	6.7	0.8077	5.042	4.073	0.19546	87.1
MC1559-1	1.270	0.635	3.810	3.335	4.605	5.080	8.890	16.8	94.8	469.9	10.4	0.6452	12.706	8.198	0.20342	199.6
MC1362-1	1.588	0.635	1.588	3.810	5.080	3.175	4.445	13.3	93.9	162.3	7.5	0.8065	6.048	4.878	0.20853	97.6
MC1279-1	1.588	0.795	1.588	2.540	4.130	3.175	4.765	11.4	100.8	112.8	7.9	1.0097	4.032	4.071	0.20905	85.7
MC1164-1	1.588	0.953	0.953	2.700	4.605	2.540	3.810	11.1	117.4	65.7	7.2	1.2097	2.572	3.111	0.20956	70.6
MC1715-1	1.270	0.953	1.588	2.692	4.597	2.858	5.080	12.4	104.5	114.7	7.5	0.9677	4.274	4.136	0.21220	89.9
MC1713-1	1.270	0.635	2.540	4.293	5.563	3.810	6.350	16.2	91.3	325.9	8.4	0.6452	10.903	7.034	0.21596	148.5
MC1346-1	1.588	0.635	1.588	3.970	5.240	3.175	4.445	13.7	96.1	169.1	7.5	0.8065	6.302	5.083	0.21729	100.0
MC1155-1	1.270	0.795	1.588	3.970	5.560	2.858	4.765	14.3	100.8	162.0	7.2	0.8077	6.302	5.091	0.22748	101.9
MC1675-1	1.588	0.714	1.270	3.970	5.398	2.858	3.967	13.3	105.5	129.2	7.2	0.9064	5.042	4.570	0.23001	89.1
MC0010-1	1.588	0.953	0.953	3.018	4.923	2.540	3.810	11.8	124.1	73.4	7.2	1.2097	2.874	3.477	0.23420	74.6
MC1712-1	1.270	0.635	2.540	4.763	6.033	3.810	6.350	17.1	96.6	361.6	8.4	0.6452	12.097	7.804	0.23960	158.1
MC0012-1	1.270	1.113	1.270	2.858	5.083	2.540	4.765	12.7	125.4	95.1	7.4	1.1303	3.629	4.102	0.25175	87.2
MC1683-1	1.588	0.754	1.270	3.967	5.476	2.858	4.049	13.5	112.8	130.5	7.3	0.9581	5.039	4.827	0.25392	91.1
MC1300-1	0.953	1.270	1.270	3.970	6.510	2.223	5.080	15.6	131.5	122.8	6.8	0.9677	5.042	4.879	0.27585	104.8
MC1664-1	1.270	0.953	1.270	4.128	6.033	2.540	4.445	14.6	123.4	131.4	7.0	0.9677	5.242	5.073	0.27866	97.2
MC1264-1	1.270	1.270	0.953	3.175	5.715	2.223	4.445	13.3	150.2	77.3	7.2	1.2903	3.024	3.902	0.28037	86.3
MC1711-1	0.953	0.991	1.905	4.928	6.909	2.858	5.791	17.6	116.2	243.2	7.3	0.7548	9.387	7.086	0.29367	137.5
MC1349-1	1.588	0.953	1.270	3.335	5.240	2.858	4.445	13.0	137.5	115.7	7.7	1.2097	4.235	5.124	0.32272	92.3
MC1649-1	1.270	0.953	1.588	4.128	6.033	2.858	5.080	15.2	128.8	175.8	7.5	0.9677	6.552	6.341	0.32531	112.7

Magnetics C Cores 1 mil Iron Alloy

Part No.	D cm 1	E cm 2	F cm 3	G cm 4	Ht cm 5	Wth cm 6	Lt cm 7	MPL cm 8	Wtfe grams 9	Wtcu grams 10	MLT cm 11	Ac cm sq 12	Wa cm sq 13	Ap cm 4th 14	Kg cm 5th 15	At cm sq 16
MC1560-1	0.635	0.635	8.890	15.240	16.510	9.525	19.050	50.8	143.1	8144.6	16.9	0.3226	135.484	43.704	0.33358	1261.3
MC1508-1	1.588	0.953	1.270	3.810	5.715	2.858	4.445	14.0	147.5	132.2	7.7	1.2097	4.839	5.853	0.36869	99.2
MC1414-1	1.270	0.953	1.588	5.080	6.985	2.858	5.080	17.1	144.8	216.4	7.5	0.9677	8.065	7.804	0.40038	127.8
MC0018-1	1.270	1.113	1.588	3.970	6.195	2.858	5.400	15.6	153.6	176.3	7.9	1.1303	6.302	7.124	0.40949	119.0
MC1221-1	1.588	0.953	1.905	3.335	5.240	3.493	5.715	14.3	150.9	196.1	8.7	1.2097	6.353	7.685	0.42848	121.9
MC1154-1	1.270	1.270	1.588	3.335	5.875	2.858	5.715	14.9	168.1	154.0	8.2	1.2903	5.294	6.831	0.43102	117.3
MC0014-1	1.270	1.270	1.270	3.970	6.510	2.540	5.080	15.6	175.3	137.7	7.7	1.2903	5.042	6.506	0.43710	112.1
MC1724-1	1.588	0.795	3.175	4.001	5.591	4.763	7.940	17.5	154.5	467.8	10.4	1.0097	12.702	12.824	0.50005	197.5
MC1359-1	1.270	1.270	1.588	3.970	6.510	2.858	5.715	16.2	182.4	183.3	8.2	1.2903	6.302	8.132	0.51308	128.1
MC1249-1	1.270	1.270	1.270	4.763	7.303	2.540	5.080	17.1	193.1	165.2	7.7	1.2903	6.048	7.804	0.52435	124.2
MC1608-1	1.430	1.270	1.905	3.175	5.715	3.335	6.350	15.2	193.3	193.5	9.0	1.4529	6.048	8.788	0.56752	133.7
MC1725-1	1.588	0.795	3.175	4.928	6.518	4.763	7.940	19.4	170.9	576.2	10.4	1.0097	15.645	15.797	0.61594	221.0
MC1617-1	0.953	1.753	2.057	3.810	7.315	3.010	7.620	18.7	218.5	252.2	9.0	1.3355	7.839	10.468	0.61803	172.5
MC1367-1	1.588	1.270	1.588	3.335	5.875	3.175	5.715	14.9	210.2	166.0	8.8	1.6129	5.294	8.539	0.62495	124.6
MC1716-1	1.270	1.270	1.549	4.928	7.468	2.819	5.639	18.0	203.1	220.5	8.1	1.2903	7.635	9.851	0.62614	142.4
MC1718-1	1.270	1.270	1.905	4.293	6.833	3.175	6.350	17.5	196.8	252.4	8.7	1.2903	8.177	10.551	0.62749	151.1
MC0015-1	1.588	1.270	1.270	3.970	6.510	2.858	5.080	15.6	219.1	149.1	8.3	1.6129	5.042	8.132	0.63083	119.5
MC1510-1	1.588	0.953	1.905	4.928	6.833	3.493	5.715	17.5	184.5	289.7	8.7	1.2097	9.387	11.355	0.63309	151.2
MC1720-1	1.270	1.270	2.692	3.658	6.198	3.962	7.925	17.8	200.3	347.2	9.9	1.2903	9.848	12.707	0.66145	182.8
MC1726-1	1.588	0.953	1.905	5.232	7.137	3.493	5.715	18.1	191.0	307.6	8.7	1.2097	9.968	12.058	0.67225	156.8
MC1727-1	1.588	0.953	1.905	5.410	7.315	3.493	5.715	18.4	194.7	318.1	8.7	1.2097	10.306	12.467	0.69509	160.1
MC1728-1	1.588	1.588	0.953	3.810	6.985	2.540	5.080	15.9	279.4	109.1	8.5	2.0161	3.629	7.317	0.69800	120.2
MC1634-1	0.953	1.270	3.302	5.715	8.255	4.255	9.144	23.1	195.3	673.5	10.0	0.9677	18.871	18.262	0.70431	265.5
MC1717-1	1.270	1.270	1.905	5.080	7.620	3.175	6.350	19.1	214.6	298.7	8.7	1.2903	9.677	12.487	0.74259	166.1
MC1235-1	1.430	1.588	1.588	3.493	6.668	3.018	6.350	16.5	261.8	180.1	9.1	1.8161	5.544	10.069	0.80071	143.1
MC1610-1	1.270	1.270	3.810	3.810	6.350	5.080	10.160	20.3	228.9	602.4	11.7	1.2903	14.516	18.730	0.82841	258.1
MC1635-1	0.953	1.270	3.810	6.985	9.525	4.763	10.160	26.7	225.3	1025.3	10.8	0.9677	26.613	25.754	0.92014	342.3
MC1609-1	1.430	1.430	1.994	4.923	7.783	3.424	6.848	19.6	279.3	330.1	9.5	1.6360	9.815	16.057	1.11091	184.0
MC1569-1	1.430	1.430	2.692	5.875	8.735	4.122	8.245	22.9	326.4	593.7	10.6	1.6360	15.818	25.878	1.60433	253.2
MC1530-1	1.588	1.588	1.588	5.875	9.050	3.175	6.350	21.3	374.5	313.4	9.5	2.0161	9.327	18.804	1.60461	192.6

Magnetics C Cores 1 mil Iron Alloy																
Part No.	D cm 1	E cm 2	F cm 3	G cm 4	Ht cm 5	Wth cm 6	Lt cm 7	MPL cm 8	Wtfe grams 9	Wtcu grams 10	MLT cm 11	Ac cm sq 12	Wa cm sq 13	Ap cm 4th 14	Kg cm 5th 15	At cm sq 16
MC1722-1	1.270	1.270	4.128	8.390	10.930	5.398	10.795	30.1	339.2	1498.4	12.2	1.2903	34.628	44.681	1.89522	428.6
MC1515-1	1.588	1.588	2.223	6.985	10.160	3.810	7.620	24.8	435.9	576.7	10.4	2.0161	15.524	31.299	2.41601	262.1
MC1719-1	1.270	1.270	3.810	11.278	13.818	5.080	10.160	35.3	397.1	1783.0	11.7	1.2903	42.968	55.442	2.45209	485.7

Magnetics C Cores 2 mil Iron Alloy

Part No.	D cm 1	E cm 2	F cm 3	G cm 4	Ht cm 5	Wth cm 6	Lt cm 7	MPL cm 8	Wtfe grams 9	Wtcu grams 10	MLT cm 11	Ac cm sq 12	Wa cm sq 13	Ap cm 4th 14	Kg cm 5th 15	At cm sq 16
MC1100-2	0.318	0.160	0.318	0.953	1.273	0.635	0.955	3.2	1.2	2.0	1.9	0.0432	0.302	0.013	0.000121	4.9
MC1505-2	0.318	0.160	0.795	1.113	1.433	1.113	1.910	4.5	1.7	8.2	2.6	0.0432	0.884	0.038	0.000253	11.9
MC1504-2	0.318	0.318	0.800	1.113	1.748	1.118	2.235	5.1	3.8	9.3	2.9	0.0857	0.890	0.076	0.000891	14.7
MC2340-2	0.635	0.318	0.478	0.953	1.588	1.113	1.590	4.1	6.2	5.0	3.1	0.1714	0.455	0.078	0.001745	10.9
MC1647-2	0.953	0.452	0.318	0.381	1.285	1.270	1.539	3.2	10.2	1.6	3.7	0.3660	0.121	0.044	0.001745	10.0
MC1606-2	0.635	0.318	0.478	1.270	1.905	1.113	1.590	4.8	7.1	6.6	3.1	0.1714	0.606	0.104	0.002326	12.6
MC0001-2	0.635	0.318	0.635	1.270	1.905	1.270	1.905	5.1	7.6	9.5	3.3	0.1714	0.806	0.138	0.002862	15.3
MC1582-2	0.953	0.318	0.635	1.270	1.905	1.588	1.905	5.1	11.4	11.3	3.9	0.2571	0.806	0.207	0.005403	17.7
MC1589-2	0.635	0.478	0.714	1.270	2.225	1.349	2.383	5.9	13.2	12.1	3.8	0.2577	0.906	0.234	0.006417	20.3
MC0147-2	0.635	0.318	0.800	2.540	3.175	1.435	2.235	8.0	11.9	25.8	3.6	0.1714	2.032	0.348	0.006689	27.7
MC0002-2	0.635	0.478	0.635	1.588	2.543	1.270	2.225	6.4	14.3	13.0	3.6	0.2577	1.008	0.260	0.007379	21.0
MC1645-2	0.953	0.478	0.318	1.588	2.543	1.270	1.590	5.7	19.3	6.8	3.8	0.3866	0.504	0.195	0.008001	17.4
MC1574-2	0.953	0.406	0.635	1.270	2.083	1.588	2.083	5.4	15.6	11.8	4.1	0.3290	0.806	0.265	0.008471	19.8
MC0143-2	0.635	0.635	0.635	1.270	2.540	1.270	2.540	6.4	19.0	11.3	3.9	0.3427	0.806	0.276	0.009606	22.6
MC1696-2	0.635	0.635	1.270	0.953	2.223	1.905	3.810	7.0	20.9	21.3	4.9	0.3427	1.210	0.415	0.011502	33.5
MC1576-2	0.478	0.795	0.478	2.383	3.973	0.955	2.545	8.9	25.1	15.0	3.7	0.3227	1.138	0.367	0.012798	29.6
MC1591-2	0.714	0.635	0.635	1.430	2.700	1.349	2.540	6.7	22.4	13.2	4.1	0.3852	0.908	0.350	0.013140	24.6
MC0003-2	0.953	0.478	0.635	1.588	2.543	1.588	2.225	6.4	21.4	15.3	4.3	0.3866	1.008	0.390	0.014131	24.0
MC1709-2	0.635	0.635	0.953	1.430	2.700	1.588	3.175	7.3	21.9	21.5	4.4	0.3427	1.362	0.467	0.014404	31.0
MC0004-2	0.635	0.635	0.635	2.223	3.493	1.270	2.540	8.3	24.7	19.8	3.9	0.3427	1.411	0.484	0.016810	29.8
MC1679-2	0.635	0.635	1.270	1.588	2.858	1.905	3.810	8.3	24.7	35.4	4.9	0.3427	2.016	0.691	0.019170	40.7
MC1705-2	0.635	0.635	0.953	2.223	3.493	1.588	3.175	8.9	26.6	33.4	4.4	0.3427	2.117	0.726	0.022386	38.5
MC1599-2	1.113	0.635	0.478	1.588	2.858	1.590	2.225	6.7	35.0	12.5	4.7	0.6005	0.758	0.455	0.023499	26.5
MC1700-2	1.430	0.478	0.478	1.981	2.936	1.908	1.910	6.8	34.6	17.4	5.2	0.5804	0.946	0.549	0.024647	27.9
MC1603-2	0.478	0.874	0.762	2.858	4.605	1.240	3.272	10.7	33.2	33.3	4.3	0.3547	2.177	0.772	0.025436	44.3
MC1693-2	0.635	0.635	0.953	2.540	3.810	1.588	3.175	9.5	28.5	38.2	4.4	0.3427	2.419	0.829	0.025584	41.5
MC8400-2	0.635	0.635	0.635	3.493	4.763	1.270	2.540	10.8	32.3	31.1	3.9	0.3427	2.218	0.760	0.026416	39.5
MC1142-2	1.270	0.318	1.113	2.858	3.493	2.383	2.860	9.2	27.6	62.5	5.5	0.3427	3.179	1.090	0.027014	47.4
MC1594-2	1.270	0.508	0.762	1.588	2.604	2.032	2.540	6.7	32.2	23.1	5.4	0.5484	1.210	0.663	0.027146	31.0
MC1614-2	1.270	0.556	0.699	1.448	2.560	1.969	2.510	6.5	34.2	19.3	5.4	0.6005	1.011	0.607	0.027227	29.7

Magnetics C Cores 2 mil Iron Alloy

Part No.	D cm 1	E cm 2	F cm 3	G cm 4	Ht cm 5	Wth cm 6	Lt cm 7	MPL cm 8	Wtfe grams 9	Wtcu grams 10	MLT cm 11	Ac cm sq 12	Wa cm sq 13	Ap cm 4th 14	Kg cm 5th 15	At cm sq 16
MC1669-2	1.270	0.478	0.889	1.588	2.543	2.159	2.733	6.9	30.9	27.6	5.5	0.5155	1.411	0.727	0.027280	33.3
MC1613-2	1.270	0.556	0.508	1.905	3.018	1.778	2.129	7.1	37.0	17.4	5.1	0.6005	0.968	0.581	0.027595	28.9
MC1685-2	0.635	0.478	1.270	3.810	4.765	1.905	3.495	12.1	27.2	79.6	4.6	0.2577	4.839	1.247	0.027788	59.9
MC1169-2	0.953	0.635	0.635	1.905	3.175	1.588	2.540	7.6	34.2	19.7	4.6	0.5141	1.210	0.622	0.027924	31.0
MC1699-2	1.430	0.516	0.516	1.880	2.911	1.946	2.062	6.9	37.5	18.3	5.3	0.6267	0.969	0.607	0.028684	29.2
MC1666-2	0.635	0.953	0.635	1.984	3.889	1.270	3.175	9.0	40.6	20.5	4.6	0.5141	1.260	0.648	0.029079	37.8
MC1605-2	1.113	0.635	0.478	1.984	3.254	1.590	2.225	7.5	39.1	15.7	4.7	0.6005	0.947	0.569	0.029365	29.5
MC1581-2	1.270	0.635	0.635	1.349	2.619	1.905	2.540	6.5	38.9	16.5	5.4	0.6855	0.856	0.587	0.029728	29.7
MC0076-2	0.953	0.478	0.953	2.540	3.495	1.905	2.860	8.9	30.0	41.0	4.8	0.3866	2.419	0.935	0.030366	40.6
MC1586-2	0.635	0.635	0.953	3.018	4.288	1.588	3.175	10.5	31.4	45.4	4.4	0.3427	2.874	0.985	0.030394	46.1
MC1638-2	0.953	0.478	1.113	2.383	3.338	2.065	3.180	8.9	30.0	47.3	5.0	0.3866	2.651	1.025	0.031602	43.9
MC0005-2	0.953	0.635	0.635	2.223	3.493	1.588	2.540	8.3	37.1	23.0	4.6	0.5141	1.411	0.726	0.032578	33.7
MC1695-2	0.635	0.635	0.635	4.445	5.715	1.270	2.540	12.7	38.0	39.6	3.9	0.3427	2.823	0.967	0.033620	46.8
MC1694-2	0.635	0.635	1.270	2.858	4.128	1.905	3.810	10.8	32.3	63.8	4.9	0.3427	3.629	1.244	0.034506	55.2
MC0007-2	0.953	0.556	0.787	2.540	3.653	1.740	2.687	8.9	34.9	33.2	4.7	0.4504	2.000	0.901	0.034807	38.2
MC1521-2	0.953	0.516	0.953	2.540	3.571	1.905	2.936	9.0	33.0	41.6	4.8	0.4175	2.419	1.010	0.034847	41.8
MC1684-2	0.635	0.478	1.524	4.445	5.400	2.159	4.003	13.8	31.2	121.1	5.0	0.2577	6.774	1.746	0.035817	78.0
MC1528-2	1.270	0.508	0.826	2.032	3.048	2.096	2.667	7.7	37.1	32.6	5.5	0.5484	1.677	0.920	0.036955	36.9
MC1677-2	0.635	0.953	0.889	1.984	3.889	1.524	3.683	9.6	42.9	31.2	5.0	0.5141	1.764	0.907	0.037449	44.9
MC1707-2	0.635	0.635	1.270	3.175	4.445	1.905	3.810	11.4	34.2	70.9	4.9	0.3427	4.032	1.382	0.038339	58.9
MC8100-2	0.635	0.635	1.270	3.335	4.605	1.905	3.810	11.8	35.2	74.4	4.9	0.3427	4.235	1.452	0.040272	60.7
MC1237-2	1.270	0.635	0.635	1.905	3.175	1.905	2.540	7.6	45.6	23.3	5.4	0.6855	1.210	0.829	0.041988	34.7
MC1656-2	0.953	0.635	1.588	1.588	2.858	2.540	4.445	8.9	39.9	54.4	6.1	0.5141	2.520	1.296	0.043856	54.4
MC1703-2	0.635	0.635	1.270	3.810	5.080	1.905	3.810	12.7	38.0	85.0	4.9	0.3427	4.839	1.658	0.046007	66.1
MC1691-2	0.635	0.673	1.308	3.531	4.877	1.943	3.962	12.4	39.2	83.4	5.1	0.3633	4.618	1.678	0.048018	66.0
MC0006-2	1.270	0.635	0.635	2.223	3.493	1.905	2.540	8.3	49.4	27.2	5.4	0.6855	1.411	0.967	0.048986	37.5
MC1200-2	0.953	0.635	0.953	2.540	3.810	1.905	3.175	9.5	42.8	43.7	5.1	0.5141	2.419	1.244	0.050367	46.0
MC1670-2	1.270	0.635	0.953	1.748	3.018	2.223	3.175	7.9	47.5	35.0	5.9	0.6855	1.665	1.141	0.052905	41.8
MC1698-2	0.635	0.635	1.524	4.445	5.715	2.159	4.318	14.5	43.3	128.7	5.3	0.3427	6.774	2.322	0.059601	84.9
MC1571-2	0.953	0.635	0.953	3.018	4.288	1.905	3.175	10.5	47.0	51.9	5.1	0.5141	2.874	1.478	0.059836	50.8

Magnetics C Cores 2 mil Iron Alloy																
Part No.	D cm 1	E cm 2	F cm 3	G cm 4	Ht cm 5	Wth cm 6	Lt cm 7	MPL cm 8	Wtfe grams 9	Wtcu grams 10	MLT cm 11	Ac cm sq 12	Wa cm sq 13	Ap cm 4th 14	Kg cm 5th 15	At cm sq 16
MC1572-2	1.113	0.556	0.953	3.018	4.130	2.065	3.018	10.2	46.7	53.6	5.2	0.5260	2.874	1.512	0.060698	50.2
MC1604-2	0.635	0.635	1.588	4.445	5.715	2.223	4.445	14.6	43.7	136.5	5.4	0.3427	7.056	2.419	0.060947	87.9
MC1612-2	1.588	0.556	0.762	2.286	3.399	2.350	2.637	8.3	54.5	37.7	6.1	0.7506	1.742	1.308	0.064441	42.9
MC0124-2	1.270	0.635	0.787	2.540	3.810	2.057	2.845	9.2	55.0	40.2	5.7	0.6855	2.000	1.371	0.066483	45.1
MC1573-2	0.953	0.635	1.270	2.858	4.128	2.223	3.810	10.8	48.5	72.0	5.6	0.5141	3.629	1.866	0.068797	60.3
MC1577-2	0.635	0.635	2.223	4.445	5.715	2.858	5.715	15.9	47.5	226.1	6.4	0.3427	9.879	3.386	0.072111	119.4
MC1223-2	1.588	0.516	0.953	2.540	3.571	2.540	2.936	9.0	55.0	54.3	6.3	0.6958	2.419	1.683	0.074247	50.1
MC1371-2	1.588	0.635	0.787	2.223	3.493	2.375	2.845	8.6	64.0	39.1	6.3	0.8569	1.750	1.499	0.081717	46.0
MC1632-2	1.588	0.594	0.889	2.286	3.475	2.476	2.967	8.7	61.1	46.0	6.4	0.8020	2.032	1.630	0.082118	48.3
MC1598-2	0.635	0.635	1.588	6.350	7.620	2.223	4.445	18.4	55.1	195.0	5.4	0.3427	10.081	3.455	0.087067	113.3
MC1668-2	0.635	0.635	2.540	5.080	6.350	3.175	6.350	17.8	53.2	318.2	6.9	0.3427	12.903	4.422	0.087417	148.4
MC1490-2	0.953	0.635	1.588	3.175	4.445	2.540	4.445	12.1	54.2	108.9	6.1	0.5141	5.040	2.591	0.087712	76.6
MC1611-2	1.430	0.714	0.795	2.383	3.810	2.225	3.018	9.2	69.8	41.4	6.1	0.8676	1.894	1.643	0.092821	48.8
MC4400-2	1.270	0.635	1.270	2.540	3.810	2.540	3.810	10.2	60.8	73.6	6.4	0.6855	3.226	2.211	0.094559	61.3
MC1379-2	1.588	0.714	0.711	2.540	3.967	2.299	2.850	9.4	78.7	40.6	6.3	0.9631	1.806	1.740	0.105932	49.9
MC2350-2	1.588	0.635	0.953	2.540	3.810	2.540	3.175	9.5	71.3	56.3	6.5	0.8569	2.419	2.073	0.108501	54.8
MC1544-2	1.430	0.635	1.270	2.540	3.810	2.700	3.810	10.2	68.5	77.2	6.7	0.7719	3.226	2.490	0.114190	63.7
MC1317-2	1.588	0.516	0.953	3.970	5.001	2.540	2.936	11.9	72.3	84.8	6.3	0.6958	3.781	2.631	0.116048	65.8
MC0008-2	0.953	0.953	0.953	3.018	4.923	1.905	3.810	11.8	79.1	58.4	5.7	0.7712	2.874	2.216	0.119668	63.5
MC1692-2	1.270	0.635	1.588	2.858	4.128	2.858	4.445	11.4	68.4	111.5	6.9	0.6855	4.536	3.110	0.123382	77.6
MC1347-2	1.270	0.635	1.270	3.335	4.605	2.540	3.810	11.8	70.3	96.6	6.4	0.6855	4.235	2.903	0.124156	71.4
MC1583-2	1.270	0.635	1.270	3.493	4.763	2.540	3.810	12.1	72.2	101.1	6.4	0.6855	4.435	3.040	0.130019	73.4
MC1723-2	1.588	0.635	1.092	2.870	4.140	2.680	3.454	10.5	78.3	75.4	6.8	0.8569	3.135	2.686	0.136033	63.8
MC1363-2	0.953	0.795	1.588	3.335	4.925	2.540	4.765	13.0	73.2	120.4	6.4	0.6437	5.294	3.408	0.137192	86.1
MC1710-2	0.953	0.795	1.588	3.353	4.943	2.540	4.765	13.1	73.4	121.0	6.4	0.6437	5.323	3.426	0.137923	86.3
MC0013-2	1.588	0.635	1.270	2.858	4.128	2.858	3.810	10.8	80.8	90.9	7.0	0.8569	3.629	3.110	0.151239	70.4
MC1721-2	1.270	1.270	0.724	1.905	4.445	1.994	3.988	10.3	123.7	33.5	6.8	1.3710	1.379	1.891	0.151920	61.7
MC1690-2	1.270	0.650	1.308	3.531	4.831	2.578	3.917	12.3	68.8	107.0	6.5	0.7350	4.618	3.394	0.153470	76.1
MC1547-2	0.953	0.953	1.113	3.653	5.558	2.065	4.130	13.3	89.8	86.2	6.0	0.7712	4.064	3.134	0.162059	77.4
MC1378-2	0.953	0.795	1.588	3.970	5.560	2.540	4.765	14.3	80.3	143.3	6.4	0.6437	6.302	4.057	0.163313	95.3

Magnetics C Cores 2 mil Iron Alloy

Part No.	D cm 1	E cm 2	F cm 3	G cm 4	Ht cm 5	Wth cm 6	Lt cm 7	MPL cm 8	Wtfe grams 9	Wtcu grams 10	MLT cm 11	Ac cm sq 12	Wa cm sq 13	Ap cm 4th 14	Kg cm 5th 15	At cm sq 16
MC2300-2	1.270	0.635	1.270	4.445	5.715	2.540	3.810	14.0	83.6	128.7	6.4	0.6855	5.645	3.870	0.165478	85.5
MC1380-2	1.270	0.635	1.664	3.810	5.080	2.934	4.597	13.5	80.7	158.5	7.0	0.6855	6.339	4.345	0.169472	95.0
MC1149-2	0.953	0.795	1.270	5.080	6.670	2.223	4.130	15.9	89.2	135.3	5.9	0.6437	6.452	4.153	0.181311	95.0
MC1543-2	0.953	0.953	1.270	3.810	5.715	2.223	4.445	14.0	94.1	106.9	6.2	0.7712	4.839	3.731	0.185294	86.3
MC0009-2	1.270	0.953	0.953	3.018	4.923	2.223	3.810	11.8	105.5	66.9	6.5	1.0282	2.874	2.955	0.185615	69.1
MC1184-2	1.588	0.795	0.953	3.018	4.608	2.540	3.495	11.1	104.1	70.2	6.9	1.0728	2.874	3.083	0.192636	67.3
MC1714-2	1.270	0.635	1.753	4.445	5.715	3.023	4.775	14.9	89.4	198.6	7.2	0.6855	7.790	5.340	0.204227	109.1
MC1219-2	0.953	0.953	1.905	3.335	5.240	2.858	5.715	14.3	96.2	162.9	7.2	0.7712	6.353	4.899	0.209645	108.0
MC1413-2	1.270	0.795	1.270	3.810	5.400	2.540	4.130	13.3	99.9	115.8	6.7	0.8582	4.839	4.153	0.211763	84.9
MC1178-2	1.588	0.635	1.588	3.493	4.763	3.175	4.445	12.7	95.0	148.8	7.5	0.8569	5.544	4.751	0.215795	92.7
MC1151-2	1.270	0.795	1.270	3.970	5.560	2.540	4.130	13.7	102.3	120.7	6.7	0.8582	5.042	4.327	0.220657	87.1
MC1559-2	1.270	0.635	3.810	3.335	4.605	5.080	8.890	16.8	100.7	469.9	10.4	0.6855	12.706	8.710	0.229644	199.6
MC1362-2	1.588	0.635	1.588	3.810	5.080	3.175	4.445	13.3	99.8	162.3	7.5	0.8569	6.048	5.183	0.235413	97.6
MC1279-2	1.588	0.795	1.588	2.540	4.130	3.175	4.765	11.4	107.1	112.8	7.9	1.0728	4.032	4.326	0.235997	85.7
MC1164-2	1.588	0.953	0.953	2.700	4.605	2.540	3.810	11.1	124.7	65.7	7.2	1.2853	2.572	3.305	0.236568	70.6
MC1715-2	1.270	0.953	1.588	2.692	4.597	2.858	5.080	12.4	111.0	114.7	7.5	1.0282	4.274	4.395	0.239556	89.9
MC1713-2	1.270	0.635	2.540	4.293	5.563	3.810	6.350	16.2	97.0	325.9	8.4	0.6855	10.903	7.474	0.243797	148.5
MC1346-2	1.588	0.635	1.588	3.970	5.240	3.175	4.445	13.7	102.1	169.1	7.5	0.8569	6.302	5.400	0.245301	100.0
MC1155-2	1.270	0.795	1.588	3.970	5.560	2.858	4.765	14.3	107.1	162.0	7.2	0.8582	6.302	5.409	0.256806	101.9
MC1675-2	1.588	0.714	1.270	3.970	5.398	2.858	3.967	13.3	112.1	129.2	7.2	0.9631	5.042	4.856	0.259660	89.1
MC0010-2	1.588	0.953	0.953	3.018	4.923	2.540	3.810	11.8	131.8	73.4	7.2	1.2853	2.874	3.694	0.264386	74.6
MC1712-2	1.270	0.635	2.540	4.763	6.033	3.810	6.350	17.1	102.6	361.6	8.4	0.6855	12.097	8.292	0.270485	158.1
MC0012-2	1.270	1.113	1.270	2.858	5.083	2.540	4.765	12.7	133.2	95.1	7.4	1.2010	3.629	4.358	0.284200	87.2
MC1683-2	1.588	0.754	1.270	3.967	5.476	2.858	4.049	13.5	119.9	130.5	7.3	1.0179	5.039	5.129	0.286652	91.1
MC1300-2	0.953	1.270	1.270	3.970	6.510	2.223	5.080	15.6	139.7	122.8	6.8	1.0282	5.042	5.184	0.311414	104.8
MC1664-2	1.270	0.953	1.270	4.128	6.033	2.540	4.445	14.6	131.1	131.4	7.0	1.0282	5.242	5.390	0.314578	97.2
MC1264-2	1.270	1.270	0.953	3.175	5.715	2.223	4.445	13.3	159.6	77.3	7.2	1.3710	3.024	4.146	0.316512	86.3
MC1711-2	0.953	0.991	1.905	4.928	6.909	2.858	5.791	17.6	123.4	243.2	7.3	0.8020	9.387	7.529	0.331530	137.5
MC1349-2	1.588	0.953	1.270	3.335	5.240	2.858	4.445	13.0	146.1	115.7	7.7	1.2853	4.235	5.444	0.364325	92.3
MC1649-2	1.270	0.953	1.588	4.128	6.033	2.858	5.080	15.2	136.8	175.8	7.5	1.0282	6.552	6.737	0.367245	112.7

Magnetics C Cores 2 mil Iron Alloy																
Part No.	D cm 1	E cm 2	F cm 3	G cm 4	Ht cm 5	Wth cm 6	Lt cm 7	MPL cm 8	Wtfe grams 9	Wtcu grams 10	MLT cm 11	Ac cm sq 12	Wa cm sq 13	Ap cm 4th 14	Kg cm 5th 15	At cm sq 16
MC1560-2	0.635	0.635	8.890	15.240	16.510	9.525	19.050	50.8	152.0	8144.6	16.9	0.3427	135.484	46.436	0.376579	1261.3
MC1508-2	1.588	0.953	1.270	3.810	5.715	2.858	4.445	14.0	156.8	132.2	7.7	1.2853	4.839	6.219	0.416213	99.2
MC1414-2	1.270	0.953	1.588	5.080	6.985	2.858	5.080	17.1	153.9	216.4	7.5	1.0282	8.065	8.292	0.451993	127.8
MC0018-2	1.270	1.113	1.588	3.970	6.195	2.858	5.400	15.6	163.2	176.3	7.9	1.2010	6.302	7.569	0.462281	119.0
MC1221-2	1.588	0.953	1.905	3.335	5.240	3.493	5.715	14.3	160.3	196.1	8.7	1.2853	6.353	8.166	0.483712	121.9
MC1154-2	1.270	1.270	1.588	3.335	5.875	2.858	5.715	14.9	178.6	154.0	8.2	1.3710	5.294	7.258	0.486577	117.3
MC0014-2	1.270	1.270	1.270	3.970	6.510	2.540	5.080	15.6	186.2	137.7	7.7	1.3710	5.042	6.912	0.493447	112.1
MC1724-2	1.588	0.795	3.175	4.001	5.591	4.763	7.940	17.5	164.2	467.8	10.4	1.0728	12.702	13.626	0.564511	197.5
MC1359-2	1.270	1.270	1.588	3.970	6.510	2.858	5.715	16.2	193.8	183.3	8.2	1.3710	6.302	8.640	0.579223	128.1
MC1249-2	1.270	1.270	1.270	4.763	7.303	2.540	5.080	17.1	205.2	165.2	7.7	1.3710	6.048	8.292	0.591947	124.2
MC1608-2	1.430	1.270	1.905	3.175	5.715	3.335	6.350	15.2	205.4	193.5	9.0	1.5437	6.048	9.337	0.640677	133.7
MC1725-2	1.588	0.795	3.175	4.928	6.518	4.763	7.940	19.4	181.6	576.2	10.4	1.0728	15.645	16.784	0.695334	221.0
MC1617-2	0.953	1.753	2.057	3.810	7.315	3.010	7.620	18.7	232.2	252.2	9.0	1.4189	7.839	11.123	0.697701	172.5
MC1367-2	1.588	1.270	1.588	3.335	5.875	3.175	5.715	14.9	223.3	166.0	8.8	1.7137	5.294	9.073	0.705512	124.6
MC1716-2	1.270	1.270	1.549	4.928	7.468	2.819	5.639	18.0	215.8	220.5	8.1	1.3710	7.635	10.467	0.706848	142.4
MC1718-2	1.270	1.270	1.905	4.293	6.833	3.175	6.350	17.5	209.2	252.4	8.7	1.3710	8.177	11.211	0.708379	151.1
MC0015-2	1.588	1.270	1.270	3.970	6.510	2.858	5.080	15.6	232.8	149.1	8.3	1.7137	5.042	8.640	0.712144	119.5
MC1510-2	1.588	0.953	1.905	4.928	6.833	3.493	5.715	17.5	196.1	289.7	8.7	1.2853	9.387	12.065	0.714700	151.2
MC1720-2	1.270	1.270	2.692	3.658	6.198	3.962	7.925	17.8	212.8	347.2	9.9	1.3710	9.848	13.501	0.746712	182.8
MC1726-2	1.588	0.953	1.905	5.232	7.137	3.493	5.715	18.1	202.9	307.6	8.7	1.2853	9.968	12.811	0.758908	156.8
MC1727-2	1.588	0.953	1.905	5.410	7.315	3.493	5.715	18.4	206.9	318.1	8.7	1.2853	10.306	13.247	0.784696	160.1
MC1728-2	1.588	1.588	0.953	3.810	6.985	2.540	5.080	15.9	296.9	109.1	8.5	2.1421	3.629	7.774	0.787971	120.2
MC1634-2	0.953	1.270	3.302	5.715	8.255	4.255	9.144	23.1	207.5	673.5	10.0	1.0282	18.871	19.404	0.795094	265.5
MC1717-2	1.270	1.270	1.905	5.080	7.620	3.175	6.350	19.1	228.0	298.7	8.7	1.3710	9.677	13.267	0.838319	166.1
MC1235-2	1.430	1.588	1.588	3.493	6.668	3.018	6.350	16.5	278.1	180.1	9.1	1.9296	5.544	10.699	0.903923	143.1
MC1610-2	1.270	1.270	3.810	3.810	6.350	5.080	10.160	20.3	243.2	602.4	11.7	1.3710	14.516	19.901	0.935197	258.1
MC1635-2	0.953	1.270	3.810	6.985	9.525	4.763	10.160	26.7	239.4	1025.3	10.8	1.0282	26.613	27.364	1.038747	342.3
MC1609-2	1.430	1.430	1.994	4.923	7.783	3.424	6.848	19.6	296.7	330.1	9.5	1.7382	9.815	17.061	1.254108	184.0
MC1569-2	1.430	1.430	2.692	5.875	8.735	4.122	8.245	22.9	346.8	593.7	10.6	1.7382	15.818	27.495	1.811135	253.2
MC1530-2	1.588	1.588	1.588	5.875	9.050	3.175	6.350	21.3	397.9	313.4	9.5	2.1421	9.327	19.979	1.811452	192.6

Magnetics C Cores 2 mil Iron Alloy

Part No.	D cm 1	E cm 2	F cm 3	G cm 4	Ht cm 5	Wth cm 6	Lt cm 7	MPL cm 8	Wtfe grams 9	Wtcu grams 10	MLT cm 11	Ac cm sq 12	Wa cm sq 13	Ap cm 4th 14	Kg cm 5th 15	At cm sq 16
MC1722-2	1.270	1.270	4.128	8.390	10.930	5.398	10.795	30.1	360.4	1498.4	12.2	1.3710	34.628	47.474	2.139521	428.6
MC1515-2	1.588	1.588	2.223	6.985	10.160	3.810	7.620	24.8	463.1	576.7	10.4	2.1421	15.524	33.255	2.727444	262.1
MC1719-2	1.270	1.270	3.810	11.278	13.818	5.080	10.160	35.3	422.0	1783.0	11.7	1.3710	42.968	58.907	2.768182	485.7

Magnetics C Cores 4 mil Iron Alloy																
Part No.	D cm 1	E cm 2	F cm 3	G cm 4	Ht cm 5	Wth cm 6	Lt cm 7	MPL cm 8	Wtfe grams 9	Wtcu grams 10	MLT cm 11	Ac cm sq 12	Wa cm sq 13	Ap cm 4th 14	Kg cm 5th 15	At cm sq 16
MC1100-4	0.318	0.160	0.318	0.953	1.273	0.635	0.955	3.2	1.3	2.0	1.9	0.0457	0.302	0.014	0.000136	4.9
MC1505-4	0.318	0.160	0.795	1.113	1.433	1.113	1.910	4.5	1.8	8.2	2.6	0.0457	0.884	0.040	0.000283	11.9
MC1504-4	0.318	0.318	0.800	1.113	1.748	1.118	2.235	5.1	4.0	9.3	2.9	0.0907	0.890	0.081	0.000999	14.7
MC2340-4	0.635	0.318	0.478	0.953	1.588	1.113	1.590	4.1	6.5	5.0	3.1	0.1815	0.455	0.083	0.001956	10.9
MC1647-4	0.953	0.452	0.318	0.381	1.285	1.270	1.539	3.2	10.8	1.6	3.7	0.3876	0.121	0.047	0.001956	10.0
MC1606-4	0.635	0.318	0.478	1.270	1.905	1.113	1.590	4.8	7.5	6.6	3.1	0.1815	0.606	0.110	0.002608	12.6
MC0001-4	0.635	0.318	0.635	1.270	1.905	1.270	1.905	5.1	8.0	9.5	3.3	0.1815	0.806	0.146	0.003209	15.3
MC1582-4	0.953	0.318	0.635	1.270	1.905	1.588	1.905	5.1	12.1	11.3	3.9	0.2722	0.806	0.219	0.006058	17.7
MC1589-4	0.635	0.478	0.714	1.270	2.225	1.349	2.383	5.9	14.0	12.1	3.8	0.2729	0.906	0.247	0.007194	20.3
MC0147-4	0.635	0.318	0.800	2.540	3.175	1.435	2.235	8.0	12.6	25.8	3.6	0.1815	2.032	0.369	0.007499	27.7
MC0002-4	0.635	0.478	0.635	1.588	2.543	1.270	2.225	6.4	15.1	13.0	3.6	0.2729	1.008	0.275	0.008273	21.0
MC1645-4	0.953	0.478	0.318	1.588	2.543	1.270	1.590	5.7	20.4	6.8	3.8	0.4094	0.504	0.206	0.008970	17.4
MC1574-4	0.953	0.406	0.635	1.270	2.083	1.588	2.083	5.4	16.5	11.8	4.1	0.3484	0.806	0.281	0.009497	19.8
MC0143-4	0.635	0.635	0.635	1.270	2.540	1.270	2.540	6.4	20.1	11.3	3.9	0.3629	0.806	0.293	0.010769	22.6
MC1696-4	0.635	0.635	1.270	0.953	2.223	1.905	3.810	7.0	22.1	21.3	4.9	0.3629	1.210	0.439	0.012895	33.5
MC1576-4	0.478	0.795	0.478	2.383	3.973	0.955	2.545	8.9	26.5	15.0	3.7	0.3417	1.138	0.389	0.014348	29.6
MC1591-4	0.714	0.635	0.635	1.430	2.700	1.349	2.540	6.7	23.8	13.2	4.1	0.4079	0.908	0.370	0.014732	24.6
MC0003-4	0.953	0.478	0.635	1.588	2.543	1.588	2.225	6.4	22.7	15.3	4.3	0.4094	1.008	0.413	0.015843	24.0
MC1709-4	0.635	0.635	0.953	1.430	2.700	1.588	3.175	7.3	23.1	21.5	4.4	0.3629	1.362	0.494	0.016148	31.0
MC0004-4	0.635	0.635	0.635	2.223	3.493	1.270	2.540	8.3	26.2	19.8	3.9	0.3629	1.411	0.512	0.018846	29.8
MC1679-4	0.635	0.635	1.270	1.588	2.858	1.905	3.810	8.3	26.2	35.4	4.9	0.3629	2.016	0.732	0.021491	40.7
MC1705-4	0.635	0.635	0.953	2.223	3.493	1.588	3.175	8.9	28.2	33.4	4.4	0.3629	2.117	0.768	0.025097	38.5
MC1599-4	1.113	0.635	0.478	1.588	2.858	1.590	2.225	6.7	37.0	12.5	4.7	0.6358	0.758	0.482	0.026345	26.5
MC1700-4	1.430	0.478	0.478	1.981	2.936	1.908	1.910	6.8	36.6	17.4	5.2	0.6146	0.946	0.581	0.027632	27.9
MC1603-4	0.478	0.874	0.762	2.858	4.605	1.240	3.272	10.7	35.2	33.3	4.3	0.3755	2.177	0.818	0.028516	44.3
MC1693-4	0.635	0.635	0.953	2.540	3.810	1.588	3.175	9.5	30.2	38.2	4.4	0.3629	2.419	0.878	0.028683	41.5
MC8400-4	0.635	0.635	0.635	3.493	4.763	1.270	2.540	10.8	34.2	31.1	3.9	0.3629	2.218	0.805	0.029615	39.5
MC1142-4	1.270	0.318	1.113	2.858	3.493	2.383	2.860	9.2	29.2	62.5	5.5	0.3629	3.179	1.154	0.030286	47.4
MC1594-4	1.270	0.508	0.762	1.588	2.604	2.032	2.540	6.7	34.1	23.1	5.4	0.5806	1.210	0.702	0.030434	31.0
MC1614-4	1.270	0.556	0.699	1.448	2.560	1.969	2.510	6.5	36.2	19.3	5.4	0.6358	1.011	0.643	0.030524	29.7

Magnetics C Cores 4 mil Iron Alloy																
Part No.	D cm 1	E cm 2	F cm 3	G cm 4	Ht cm 5	Wth cm 6	Lt cm 7	MPL cm 8	Wtfe grams 9	Wtcu grams 10	MLT cm 11	Ac cm sq 12	Wa cm sq 13	Ap cm 4th 14	Kg cm 5th 15	At cm sq 16
MC1669-4	1.270	0.478	0.889	1.588	2.543	2.159	2.733	6.9	32.7	27.6	5.5	0.5458	1.411	0.770	0.030583	33.3
MC1613-4	1.270	0.556	0.508	1.905	3.018	1.778	2.129	7.1	39.1	17.4	5.1	0.6358	0.968	0.615	0.030937	28.9
MC1685-4	0.635	0.478	1.270	3.810	4.765	1.905	3.495	12.1	28.8	79.6	4.6	0.2729	4.839	1.320	0.031154	59.9
MC1169-4	0.953	0.635	0.635	1.905	3.175	1.588	2.540	7.6	36.2	19.7	4.6	0.5444	1.210	0.658	0.031306	31.0
MC1699-4	1.430	0.516	0.516	1.880	2.911	1.946	2.062	6.9	39.7	18.3	5.3	0.6636	0.969	0.643	0.032158	29.2
MC1666-4	0.635	0.953	0.635	1.984	3.889	1.270	3.175	9.0	43.0	20.5	4.6	0.5444	1.260	0.686	0.032600	37.8
MC1605-4	1.113	0.635	0.478	1.984	3.254	1.590	2.225	7.5	41.4	15.7	4.7	0.6358	0.947	0.602	0.032921	29.5
MC1581-4	1.270	0.635	0.635	1.349	2.619	1.905	2.540	6.5	41.2	16.5	5.4	0.7258	0.856	0.622	0.033328	29.7
MC0076-4	0.953	0.478	0.953	2.540	3.495	1.905	2.860	8.9	31.8	41.0	4.8	0.4094	2.419	0.990	0.034043	40.6
MC1586-4	0.635	0.635	0.953	3.018	4.288	1.588	3.175	10.5	33.2	45.4	4.4	0.3629	2.874	1.043	0.034075	46.1
MC1638-4	0.953	0.478	1.113	2.383	3.338	2.065	3.180	8.9	31.8	47.3	5.0	0.4094	2.651	1.085	0.035429	43.9
MC0005-4	0.953	0.635	0.635	2.223	3.493	1.588	2.540	8.3	39.2	23.0	4.6	0.5444	1.411	0.768	0.036524	33.7
MC1695-4	0.635	0.635	0.635	4.445	5.715	1.270	2.540	12.7	40.2	39.6	3.9	0.3629	2.823	1.024	0.037692	46.8
MC1694-4	0.635	0.635	1.270	2.858	4.128	1.905	3.810	10.8	34.2	63.8	4.9	0.3629	3.629	1.317	0.038684	55.2
MC0007-4	0.953	0.556	0.787	2.540	3.653	1.740	2.687	8.9	37.0	33.2	4.7	0.4769	2.000	0.954	0.039022	38.2
MC1521-4	0.953	0.516	0.953	2.540	3.571	1.905	2.936	9.0	34.9	41.6	4.8	0.4420	2.419	1.069	0.039068	41.8
MC1684-4	0.635	0.478	1.524	4.445	5.400	2.159	4.003	13.8	33.0	121.1	5.0	0.2729	6.774	1.849	0.040154	78.0
MC1528-4	1.270	0.508	0.826	2.032	3.048	2.096	2.667	7.7	39.3	32.6	5.5	0.5806	1.677	0.974	0.041431	36.9
MC1677-4	0.635	0.953	0.889	1.984	3.889	1.524	3.683	9.6	45.4	31.2	5.0	0.5444	1.764	0.960	0.041985	44.9
MC1707-4	0.635	0.635	1.270	3.175	4.445	1.905	3.810	11.4	36.2	70.9	4.9	0.3629	4.032	1.463	0.042983	58.9
MC8100-4	0.635	0.635	1.270	3.335	4.605	1.905	3.810	11.8	37.2	74.4	4.9	0.3629	4.235	1.537	0.045149	60.7
MC1237-4	1.270	0.635	0.635	1.905	3.175	1.905	2.540	7.6	48.3	23.3	5.4	0.7258	1.210	0.878	0.047073	34.7
MC1656-4	0.953	0.635	1.588	1.588	2.858	2.540	4.445	8.9	42.2	54.4	6.1	0.5444	2.520	1.372	0.049167	54.4
MC1703-4	0.635	0.635	1.270	3.810	5.080	1.905	3.810	12.7	40.2	85.0	4.9	0.3629	4.839	1.756	0.051579	66.1
MC1691-4	0.635	0.673	1.308	3.531	4.877	1.943	3.962	12.4	41.5	83.4	5.1	0.3847	4.618	1.777	0.053834	66.0
MC0006-4	1.270	0.635	0.635	2.223	3.493	1.905	2.540	8.3	52.3	27.2	5.4	0.7258	1.411	1.024	0.054919	37.5
MC1200-4	0.953	0.635	0.953	2.540	3.810	1.905	3.175	9.5	45.3	43.7	5.1	0.5444	2.419	1.317	0.056467	46.0
MC1670-4	1.270	0.635	0.953	1.748	3.018	2.223	3.175	7.9	50.3	35.0	5.9	0.7258	1.665	1.208	0.059313	41.8
MC1698-4	0.635	0.635	1.524	4.445	5.715	2.159	4.318	14.5	45.9	128.7	5.3	0.3629	6.774	2.458	0.066819	84.9
MC1571-4	0.953	0.635	0.953	3.018	4.288	1.905	3.175	10.5	49.8	51.9	5.1	0.5444	2.874	1.565	0.067082	50.8

Magnetics C Cores 4 mil Iron Alloy

Part No.	D cm 1	E cm 2	F cm 3	G cm 4	Ht cm 5	Wth cm 6	Lt cm 7	MPL cm 8	Wtfe grams 9	Wtcu grams 10	MLT cm 11	Ac cm sq 12	Wa cm sq 13	Ap cm 4th 14	Kg cm 5th 15	At cm sq 16
MC1572-4	1.113	0.556	0.953	3.018	4.130	2.065	3.018	10.2	49.4	53.6	5.2	0.5570	2.874	1.601	0.068048	50.2
MC1604-4	0.635	0.635	1.588	4.445	5.715	2.223	4.445	14.6	46.3	136.5	5.4	0.3629	7.056	2.561	0.068328	87.9
MC1612-4	1.588	0.556	0.762	2.286	3.399	2.350	2.637	8.3	57.7	37.7	6.1	0.7948	1.742	1.384	0.072245	42.9
MC0124-4	1.270	0.635	0.787	2.540	3.810	2.057	2.845	9.2	58.3	40.2	5.7	0.7258	2.000	1.452	0.074534	45.1
MC1573-4	0.953	0.635	1.270	2.858	4.128	2.223	3.810	10.8	51.3	72.0	5.6	0.5444	3.629	1.975	0.077129	60.3
MC1577-4	0.635	0.635	2.223	4.445	5.715	2.858	5.715	15.9	50.3	226.1	6.4	0.3629	9.879	3.585	0.080844	119.4
MC1223-4	1.588	0.516	0.953	2.540	3.571	2.540	2.936	9.0	58.2	54.3	6.3	0.7367	2.419	1.782	0.083239	50.1
MC1371-4	1.588	0.635	0.787	2.223	3.493	2.375	2.845	8.6	67.8	39.1	6.3	0.9073	1.750	1.588	0.091614	46.0
MC1632-4	1.588	0.594	0.889	2.286	3.475	2.476	2.967	8.7	64.7	46.0	6.4	0.8492	2.032	1.726	0.092063	48.3
MC1598-4	0.635	0.635	1.588	6.350	7.620	2.223	4.445	18.4	58.3	195.0	5.4	0.3629	10.081	3.658	0.097611	113.3
MC1668-4	0.635	0.635	2.540	5.080	6.350	3.175	6.350	17.8	56.3	318.2	6.9	0.3629	12.903	4.683	0.098003	148.4
MC1490-4	0.953	0.635	1.588	3.175	4.445	2.540	4.445	12.1	57.3	108.9	6.1	0.5444	5.040	2.744	0.098335	76.6
MC1611-4	1.430	0.714	0.795	2.383	3.810	2.225	3.018	9.2	73.9	41.4	6.1	0.9186	1.894	1.740	0.104062	48.8
MC4400-4	1.270	0.635	1.270	2.540	3.810	2.540	3.810	10.2	64.4	73.6	6.4	0.7258	3.226	2.341	0.106011	61.3
MC1379-4	1.588	0.714	0.711	2.540	3.967	2.299	2.850	9.4	83.3	40.6	6.3	1.0198	1.806	1.842	0.118762	49.9
MC2350-4	1.588	0.635	0.953	2.540	3.810	2.540	3.175	9.5	75.4	56.3	6.5	0.9073	2.419	2.195	0.121642	54.8
MC1544-4	1.430	0.635	1.270	2.540	3.810	2.700	3.810	10.2	72.5	77.2	6.7	0.8173	3.226	2.636	0.128019	63.7
MC1317-4	1.588	0.516	0.953	3.970	5.001	2.540	2.936	11.9	76.6	84.8	6.3	0.7367	3.781	2.786	0.130102	65.8
MC0008-4	0.953	0.953	0.953	3.018	4.923	1.905	3.810	11.8	83.8	58.4	5.7	0.8165	2.874	2.347	0.134160	63.5
MC1692-4	1.270	0.635	1.588	2.858	4.128	2.858	4.445	11.4	72.4	111.5	6.9	0.7258	4.536	3.292	0.138324	77.6
MC1347-4	1.270	0.635	1.270	3.335	4.605	2.540	3.810	11.8	74.5	96.6	6.4	0.7258	4.235	3.074	0.139192	71.4
MC1583-4	1.270	0.635	1.270	3.493	4.763	2.540	3.810	12.1	76.4	101.1	6.4	0.7258	4.435	3.219	0.145765	73.4
MC1723-4	1.588	0.635	1.092	2.870	4.140	2.680	3.454	10.5	82.9	75.4	6.8	0.9073	3.135	2.844	0.152507	63.8
MC1363-4	0.953	0.795	1.588	3.335	4.925	2.540	4.765	13.0	77.5	120.4	6.4	0.6815	5.294	3.608	0.153806	86.1
MC1710-4	0.953	0.795	1.588	3.353	4.943	2.540	4.765	13.1	77.7	121.0	6.4	0.6815	5.323	3.627	0.154626	86.3
MC1690-4	1.270	0.650	1.308	3.531	4.831	2.578	3.917	12.3	69.6	107.0	6.5	0.7432	4.618	3.432	0.156938	76.1
MC0013-4	1.588	0.635	1.270	2.858	4.128	2.858	3.810	10.8	85.5	90.9	7.0	0.9073	3.629	3.292	0.169556	70.4
MC1721-4	1.270	1.270	0.724	1.905	4.445	1.994	3.988	10.3	131.0	33.5	6.8	1.4516	1.379	2.002	0.170318	61.7
MC1547-4	0.953	0.953	1.113	3.653	5.558	2.065	4.130	13.3	95.1	86.2	6.0	0.8165	4.064	3.318	0.181685	77.4
MC1378-4	0.953	0.795	1.588	3.970	5.560	2.540	4.765	14.3	85.1	143.3	6.4	0.6815	6.302	4.295	0.183092	95.3

Magnetics C Cores 4 mil Iron Alloy

Part No.	D cm 1	E cm 2	F cm 3	G cm 4	Ht cm 5	Wth cm 6	Lt cm 7	MPL cm 8	Wtfe grams 9	Wtcu grams 10	MLT cm 11	Ac cm sq 12	Wa cm sq 13	Ap cm 4th 14	Kg cm 5th 15	At cm sq 16
MC2300-4	1.270	0.635	1.270	4.445	5.715	2.540	3.810	14.0	88.5	128.7	6.4	0.7258	5.645	4.097	0.185519	85.5
MC1380-4	1.270	0.635	1.664	3.810	5.080	2.934	4.597	13.5	85.5	158.5	7.0	0.7258	6.339	4.601	0.189996	95.0
MC1149-4	0.953	0.795	1.270	5.080	6.670	2.223	4.130	15.9	94.5	135.3	5.9	0.6815	6.452	4.397	0.203269	95.0
MC1543-4	0.953	0.953	1.270	3.810	5.715	2.223	4.445	14.0	99.6	106.9	6.2	0.8165	4.839	3.951	0.207735	86.3
MC0009-4	1.270	0.953	0.953	3.018	4.923	2.223	3.810	11.8	111.7	66.9	6.5	1.0887	2.874	3.129	0.208095	69.1
MC1184-4	1.588	0.795	0.953	3.018	4.608	2.540	3.495	11.1	110.3	70.2	6.9	1.1359	2.874	3.265	0.215965	67.3
MC1714-4	1.270	0.635	1.753	4.445	5.715	3.023	4.775	14.9	94.6	198.6	7.2	0.7258	7.790	5.654	0.228961	109.1
MC1219-4	0.953	0.953	1.905	3.335	5.240	2.858	5.715	14.3	101.9	162.9	7.2	0.8165	6.353	5.188	0.235035	108.0
MC1413-4	1.270	0.795	1.270	3.810	5.400	2.540	4.130	13.3	105.8	115.8	6.7	0.9087	4.839	4.397	0.237409	84.9
MC1178-4	1.588	0.635	1.588	3.493	4.763	3.175	4.445	12.7	100.6	148.8	7.5	0.9073	5.544	5.030	0.241930	92.7
MC1151-4	1.270	0.795	1.270	3.970	5.560	2.540	4.130	13.7	108.4	120.7	6.7	0.9087	5.042	4.582	0.247380	87.1
MC1559-4	1.270	0.635	3.810	3.335	4.605	5.080	8.890	16.8	106.6	469.9	10.4	0.7258	12.706	9.222	0.257456	199.6
MC1362-4	1.588	0.635	1.588	3.810	5.080	3.175	4.445	13.3	105.6	162.3	7.5	0.9073	6.048	5.487	0.263923	97.6
MC1279-4	1.588	0.795	1.588	2.540	4.130	3.175	4.765	11.4	113.4	112.8	7.9	1.1359	4.032	4.580	0.264578	85.7
MC1164-4	1.588	0.953	0.953	2.700	4.605	2.540	3.810	11.1	132.1	65.7	7.2	1.3609	2.572	3.500	0.265218	70.6
MC1715-4	1.270	0.953	1.588	2.692	4.597	2.858	5.080	12.4	117.6	114.7	7.5	1.0887	4.274	4.653	0.268568	89.9
MC1713-4	1.270	0.635	2.540	4.293	5.563	3.810	6.350	16.2	102.7	325.9	8.4	0.7258	10.903	7.914	0.273322	148.5
MC1346-4	1.588	0.635	1.588	3.970	5.240	3.175	4.445	13.7	108.2	169.1	7.5	0.9073	6.302	5.718	0.275008	100.0
MC1155-4	1.270	0.795	1.588	3.970	5.560	2.858	4.765	14.3	113.4	162.0	7.2	0.9087	6.302	5.727	0.287907	101.9
MC1675-4	1.588	0.714	1.270	3.970	5.398	2.858	3.967	13.3	118.7	129.2	7.2	1.0198	5.042	5.142	0.291107	89.1
MC0010-4	1.588	0.953	0.953	3.018	4.923	2.540	3.810	11.8	139.6	73.4	7.2	1.3609	2.874	3.911	0.296405	74.6
MC1712-4	1.270	0.635	2.540	4.763	6.033	3.810	6.350	17.1	108.6	361.6	8.4	0.7258	12.097	8.780	0.303242	158.1
MC0012-4	1.270	1.113	1.270	2.858	5.083	2.540	4.765	12.7	141.0	95.1	7.4	1.2716	3.629	4.615	0.318618	87.2
MC1683-4	1.588	0.754	1.270	3.967	5.476	2.858	4.049	13.5	127.0	130.5	7.3	1.0778	5.039	5.431	0.321368	91.1
MC1300-4	0.953	1.270	1.270	3.970	6.510	2.223	5.080	15.6	147.9	122.8	6.8	1.0887	5.042	5.489	0.349128	104.8
MC1664-4	1.270	0.953	1.270	4.128	6.033	2.540	4.445	14.6	138.8	131.4	7.0	1.0887	5.242	5.707	0.352675	97.2
MC1264-4	1.270	1.270	0.953	3.175	5.715	2.223	4.445	13.3	169.0	77.3	7.2	1.4516	3.024	4.390	0.354844	86.3
MC1711-4	0.953	0.991	1.905	4.928	6.909	2.858	5.791	17.6	130.7	243.2	7.3	0.8492	9.387	7.971	0.371680	137.5
MC1349-4	1.588	0.953	1.270	3.335	5.240	2.858	4.445	13.0	154.7	115.7	7.7	1.3609	4.235	5.764	0.408447	92.3
MC1649-4	1.270	0.953	1.588	4.128	6.033	2.858	5.080	15.2	144.8	175.8	7.5	1.0887	6.552	7.134	0.411720	112.7

Magnetics C Cores 4 mil Iron Alloy																
Part No.	D cm 1	E cm 2	F cm 3	G cm 4	Ht cm 5	Wth cm 6	Lt cm 7	MPL cm 8	Wtfe grams 9	Wtcu grams 10	MLT cm 11	Ac cm sq 12	Wa cm sq 13	Ap cm 4th 14	Kg cm 5th 15	At cm sq 16
MC1560-4	0.635	0.635	8.890	15.240	16.510	9.525	19.050	50.8	160.9	8144.6	16.9	0.3629	135.484	49.167	0.422186	1261.3
MC1508-4	1.588	0.953	1.270	3.810	5.715	2.858	4.445	14.0	166.0	132.2	7.7	1.3609	4.839	6.585	0.466619	99.2
MC1414-4	1.270	0.953	1.588	5.080	6.985	2.858	5.080	17.1	163.0	216.4	7.5	1.0887	8.065	8.780	0.506733	127.8
MC0018-4	1.270	1.113	1.588	3.970	6.195	2.858	5.400	15.6	172.8	176.3	7.9	1.2716	6.302	8.014	0.518266	119.0
MC1221-4	1.588	0.953	1.905	3.335	5.240	3.493	5.715	14.3	169.8	196.1	8.7	1.3609	6.353	8.646	0.542293	121.9
MC1154-4	1.270	1.270	1.588	3.335	5.875	2.858	5.715	14.9	189.1	154.0	8.2	1.4516	5.294	7.685	0.545505	117.3
MC0014-4	1.270	1.270	1.270	3.970	6.510	2.540	5.080	15.6	197.2	137.7	7.7	1.4516	5.042	7.319	0.553207	112.1
MC1724-4	1.588	0.795	3.175	4.001	5.591	4.763	7.940	17.5	173.8	467.8	10.4	1.1359	12.702	14.428	0.632877	197.5
MC1359-4	1.270	1.270	1.588	3.970	6.510	2.858	5.715	16.2	205.2	183.3	8.2	1.4516	6.302	9.149	0.649371	128.1
MC1249-4	1.270	1.270	1.270	4.763	7.303	2.540	5.080	17.1	217.3	165.2	7.7	1.4516	6.048	8.780	0.663636	124.2
MC1608-4	1.430	1.270	1.905	3.175	5.715	3.335	6.350	15.2	217.5	193.5	9.0	1.6345	6.048	9.886	0.718268	133.7
MC1725-4	1.588	0.795	3.175	4.928	6.518	4.763	7.940	19.4	192.2	576.2	10.4	1.1359	15.645	17.771	0.779544	221.0
MC1617-4	0.953	1.753	2.057	3.810	7.315	3.010	7.620	18.7	245.9	252.2	9.0	1.5024	7.839	11.777	0.782198	172.5
MC1367-4	1.588	1.270	1.588	3.335	5.875	3.175	5.715	14.9	236.4	166.0	8.8	1.8145	5.294	9.607	0.790954	124.6
MC1716-4	1.270	1.270	1.549	4.928	7.468	2.819	5.639	18.0	228.5	220.5	8.1	1.4516	7.635	11.083	0.792452	142.4
MC1718-4	1.270	1.270	1.905	4.293	6.833	3.175	6.350	17.5	221.5	252.4	8.7	1.4516	8.177	11.870	0.794169	151.1
MC0015-4	1.588	1.270	1.270	3.970	6.510	2.858	5.080	15.6	246.5	149.1	8.3	1.8145	5.042	9.149	0.798389	119.5
MC1510-4	1.588	0.953	1.905	4.928	6.833	3.493	5.715	17.5	207.6	289.7	8.7	1.3609	9.387	12.775	0.801255	151.2
MC1720-4	1.270	1.270	2.692	3.658	6.198	3.962	7.925	17.8	225.3	347.2	9.9	1.4516	9.848	14.295	0.837144	182.8
MC1726-4	1.588	0.953	1.905	5.232	7.137	3.493	5.715	18.1	214.9	307.6	8.7	1.3609	9.968	13.565	0.850817	156.8
MC1727-4	1.588	0.953	1.905	5.410	7.315	3.493	5.715	18.4	219.1	318.1	8.7	1.3609	10.306	14.026	0.879728	160.1
MC1728-4	1.588	1.588	0.953	3.810	6.985	2.540	5.080	15.9	314.3	109.1	8.5	2.2681	3.629	8.231	0.883400	120.2
MC1634-4	0.953	1.270	3.302	5.715	8.255	4.255	9.144	23.1	219.7	673.5	10.0	1.0887	18.871	20.545	0.891386	265.5
MC1717-4	1.270	1.270	1.905	5.080	7.620	3.175	6.350	19.1	241.4	298.7	8.7	1.4516	9.677	14.048	0.939845	166.1
MC1235-4	1.430	1.588	1.588	3.493	6.668	3.018	6.350	16.5	294.5	180.1	9.1	2.0431	5.544	11.328	1.013395	143.1
MC1610-4	1.270	1.270	3.810	3.810	6.350	5.080	10.160	20.3	257.5	602.4	11.7	1.4516	14.516	21.072	1.048456	258.1
MC1635-4	0.953	1.270	3.810	6.985	9.525	4.763	10.160	26.7	253.5	1025.3	10.8	1.0887	26.613	28.974	1.164547	342.3
MC1609-4	1.430	1.430	1.994	4.923	7.783	3.424	6.848	19.6	314.2	330.1	9.5	1.8405	9.815	18.064	1.405989	184.0
MC1569-4	1.430	1.430	2.692	5.875	8.735	4.122	8.245	22.9	367.2	593.7	10.6	1.8405	15.818	29.112	2.030477	253.2
MC1530-4	1.588	1.588	1.588	5.875	9.050	3.175	6.350	21.3	421.3	313.4	9.5	2.2681	9.327	21.154	2.030832	192.6

Magnetics C Cores 4 mil Iron Alloy

Part No.	D cm 1	E cm 2	F cm 3	G cm 4	Ht cm 5	Wth cm 6	Lt cm 7	MPL cm 8	Wtfe grams 9	Wtcu grams 10	MLT cm 11	Ac cm sq 12	Wa cm sq 13	Ap cm 4th 14	Kg cm 5th 15	At cm sq 16
MC1722-4	1.270	1.270	4.128	8.390	10.930	5.398	10.795	30.1	381.6	1498.4	12.2	1.4516	34.628	50.267	2.398633	428.6
MC1515-4	1.588	1.588	2.223	6.985	10.160	3.810	7.620	24.8	490.4	576.7	10.4	2.2681	15.524	35.211	3.057757	262.1
MC1719-4	1.270	1.270	3.810	11.278	13.818	5.080	10.160	35.3	446.8	1783.0	11.7	1.4516	42.968	62.372	3.103429	485.7

11

CASELESS TOROIDS

CASELESS TOROIDAL CORE MANUFACTURERS

Manufacturers' addresses and phone numbers can be found from the Table of Contents.

Notes

CASELESS TOROIDAL CORE TABLE DESCRIPTION

Definitions for the Following Tables

Information given is listed by column:

Col.	Dim.	Description	Units
1.	OD	Core outside diameter	cm
2.	ID	Core inside diameter	cm
3.	HT	Core height	cm
4.	OD	Finished wound (K_u = 0.4)	cm
5.	HT	Finished wound (K_u = 0.4)	cm
6.	MPL	Magnetic Path Length [MPL = 3.14(OD + ID)(0.5)]	cm
7.	W_{tfe}	Core weight [W_{tfe} = (MPL)(A_c)(7.63)SF]	gram
8.	W_{tcu}	Copper weight [W_{tcu} = (MLT)(W_a)(K_u)(8.89)]	gram
9.	MLT	Mean Length Turn [MLT = (OD + 2HT)(0.8), core]	cm
10.	A_c	Effective iron area [A_c = (OD − ID)(HT)(SF)(0.5), core]	cm sq
11.	W_a	Window area [$W_a = 3.14(ID)^2(0.25)$]	cm sq
12.	A_p	Area product [A_p = (A_c)(W_a)]	cm 4th
13.	K_g	Core geometry [K_g = (A_p)(A_c)(K_u)/(MLT)]	cm 5th
14.	A_t	Surface area	cm sq

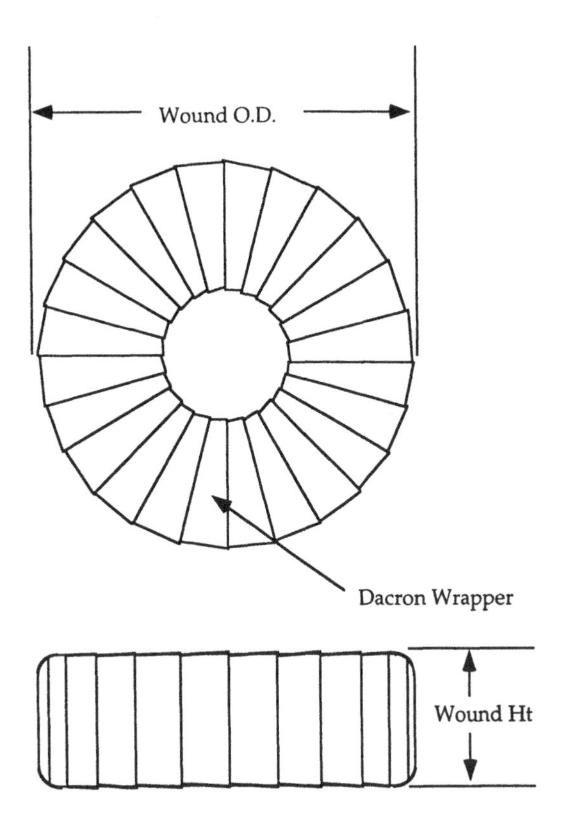

Figure 140 Mechanical outline of caseless toroids.

Magnetics Caseless Toroids 1 mil Iron Alloy														
Part No.	OD cm 1	ID cm 2	HT cm 3	Wound OD cm 4	Wound HT cm 5	MPL cm 6	Wtfe grams 7	Wtcu grams 8	MLT cm 9	Ac cm sq 10	Wa cm sq 11	Ap cm 4th 12	Kg cm 5th 13	At cm sq 14
53402-1	1.151	0.915	0.369	1.397	0.826	3.3	0.6	3.5	1.5	0.0194	0.657	0.013	0.000066	8.2
53107-1	1.468	1.232	0.369	1.815	0.985	4.2	0.7	7.5	1.8	0.0194	1.191	0.023	0.000102	13.3
53403-1	1.626	1.392	0.369	2.024	1.065	4.7	0.8	10.2	1.9	0.0194	1.521	0.030	0.000121	16.3
53485-1	1.783	1.550	0.369	2.232	1.144	5.2	0.9	13.5	2.0	0.0194	1.886	0.037	0.000141	19.6
53356-1	1.943	1.707	0.369	2.442	1.222	5.7	1.0	17.4	2.1	0.0194	2.287	0.044	0.000161	23.2
53153-1	1.308	0.915	0.369	1.530	0.826	3.5	1.2	3.8	1.6	0.0379	0.657	0.025	0.000231	9.4
53154-1	1.468	1.075	0.369	1.738	0.906	4.0	1.3	5.7	1.8	0.0379	0.907	0.034	0.000296	12.0
53056-1	1.626	1.232	0.369	1.945	0.985	4.5	1.5	8.0	1.9	0.0379	1.191	0.045	0.000363	14.8
53057-1	1.943	1.550	0.369	2.362	1.144	5.5	1.8	14.4	2.1	0.0379	1.886	0.072	0.000506	21.3
53143-1	2.261	1.867	0.369	2.780	1.302	6.5	2.1	23.3	2.4	0.0379	2.736	0.104	0.000657	29.1
53374-1	2.578	2.185	0.369	3.198	1.461	7.5	2.5	35.3	2.7	0.0379	3.747	0.142	0.000814	38.1
53155-1	1.468	1.075	0.686	1.738	1.223	4.0	2.6	7.3	2.3	0.0759	0.907	0.069	0.000920	14.0
53086-1	2.896	2.502	0.369	3.617	1.620	8.5	2.8	50.8	2.9	0.0379	4.914	0.186	0.000973	48.3
53000-1	1.943	1.232	0.369	2.217	0.985	5.0	3.3	9.1	2.1	0.0759	1.191	0.090	0.001280	18.1
53063-1	2.261	1.550	0.369	2.630	1.144	6.0	4.0	16.1	2.4	0.0759	1.886	0.143	0.001811	25.2
53002-1	2.324	1.613	0.369	2.711	1.175	6.2	4.1	17.8	2.4	0.0759	2.042	0.155	0.001920	26.8
53134-1	2.578	1.867	0.369	3.043	1.302	7.0	4.6	25.8	2.7	0.0759	2.736	0.208	0.002376	33.5
53632-1	1.943	1.232	0.529	2.217	1.145	5.0	5.0	10.2	2.4	0.1138	1.191	0.136	0.002572	19.4
53459-1	2.896	2.185	0.369	3.459	1.461	8.0	5.3	38.7	2.9	0.0759	3.747	0.284	0.002969	43.1
53011-1	3.213	2.502	0.369	3.876	1.620	9.0	5.9	55.2	3.2	0.0759	4.914	0.373	0.003581	53.9
53176-1	1.943	1.232	0.686	2.217	1.302	5.0	6.6	11.2	2.7	0.1509	1.191	0.180	0.004091	20.6
53666-1	3.531	2.820	0.369	4.293	1.779	10.0	6.6	75.8	3.4	0.0759	6.242	0.474	0.004210	65.8
53033-1	2.261	1.550	0.686	2.630	1.461	6.0	7.9	19.5	2.9	0.1509	1.886	0.285	0.005909	28.2
53454-1	4.483	3.772	0.369	5.547	2.255	13.0	8.6	165.9	4.2	0.0759	11.168	0.848	0.006159	109.0
53691-1	4.801	4.090	0.369	5.966	2.414	14.0	9.2	206.9	4.4	0.0759	13.131	0.996	0.006826	125.9
53654-1	5.118	4.407	0.369	6.384	2.572	15.0	9.9	254.0	4.7	0.0759	15.245	1.157	0.007496	143.9
53076-1	2.578	1.550	0.529	2.907	1.304	6.5	9.6	19.5	2.9	0.1703	1.886	0.321	0.007521	31.2
53296-1	2.324	1.486	0.686	2.657	1.429	6.0	9.5	18.2	3.0	0.1818	1.733	0.315	0.007748	28.4
53061-1	2.578	1.867	0.686	3.043	1.619	7.0	9.2	30.7	3.2	0.1509	2.736	0.413	0.007885	37.0
53106-1	2.896	1.867	0.529	3.317	1.462	7.5	11.1	30.8	3.2	0.1703	2.736	0.466	0.010035	40.4

Magnetics Caseless Toroids 1 mil Iron Alloy														
Part No.	OD cm 1	ID cm 2	HT cm 3	Wound OD cm 4	Wound HT cm 5	MPL cm 6	Wtfe grams 7	Wtcu grams 8	MLT cm 9	Ac cm sq 10	Wa cm sq 11	Ap cm 4th 12	Kg cm 5th 13	At cm sq 14
53748-1	7.023	6.312	0.369	8.900	3.525	20.9	13.9	690.5	6.2	0.0759	31.274	2.373	0.011603	277.7
53004-1	3.213	2.502	0.686	3.876	1.937	9.0	11.8	64.1	3.7	0.1509	4.914	0.741	0.012200	58.3
53007-1	2.578	1.550	0.686	2.907	1.461	6.5	12.8	21.2	3.2	0.2268	1.886	0.428	0.012276	32.9
53115-1	3.531	2.820	0.686	4.293	2.096	10.0	13.1	87.1	3.9	0.1509	6.242	0.942	0.014493	70.8
53168-1	2.578	1.867	1.004	3.043	1.937	7.0	13.8	35.7	3.7	0.2268	2.736	0.620	0.015341	40.5
53084-1	2.896	1.867	0.686	3.317	1.619	7.5	14.8	33.2	3.4	0.2268	2.736	0.620	0.016484	42.2
53039-1	3.848	3.137	0.686	4.710	2.254	11.0	14.5	114.7	4.2	0.1509	7.725	1.166	0.016846	84.4
53392-1	3.848	2.820	0.529	4.557	1.939	10.5	15.6	87.1	3.9	0.1703	6.242	1.063	0.018451	75.0
53167-1	2.896	1.550	0.686	3.192	1.461	7.0	18.4	22.9	3.4	0.3027	1.886	0.571	0.020237	38.0
53285-1	3.213	2.185	0.686	3.729	1.778	8.5	16.8	48.9	3.7	0.2268	3.747	0.850	0.021017	52.8
53094-1	2.578	1.550	1.004	2.907	1.779	6.5	19.3	24.6	3.7	0.3406	1.886	0.642	0.023852	36.2
53029-1	3.531	2.502	0.686	4.143	1.937	9.5	18.7	68.5	3.9	0.2268	4.914	1.114	0.025770	64.6
53318-1	3.213	1.867	0.686	3.597	1.619	8.0	21.1	35.7	3.7	0.3027	2.736	0.828	0.027332	47.9
53228-1	3.531	2.820	1.004	4.293	2.414	10.0	19.7	98.4	4.4	0.2268	6.242	1.416	0.028978	75.7
53393-1	5.118	4.090	0.529	6.224	2.574	14.5	21.5	230.7	4.9	0.1703	13.131	2.236	0.030832	138.3
53034-1	2.896	1.867	1.004	3.317	1.937	7.5	22.2	38.2	3.9	0.3406	2.736	0.932	0.032362	46.0
53391-1	3.213	2.502	1.321	3.876	2.572	9.0	23.7	81.8	4.7	0.3027	4.914	1.487	0.038439	67.2
53181-1	3.213	2.185	1.004	3.729	2.096	8.5	25.2	55.7	4.2	0.3406	3.747	1.276	0.041635	57.1
53133-1	2.959	1.613	1.004	3.272	1.810	7.2	28.4	28.9	4.0	0.4535	2.042	0.926	0.042288	43.6
53032-1	3.848	2.502	0.686	4.417	1.937	10.0	26.3	73.0	4.2	0.3027	4.914	1.487	0.043115	71.4
53553-1	3.531	2.502	1.004	4.143	2.255	9.5	28.2	77.4	4.4	0.3406	4.914	1.674	0.051458	69.4
53018-1	5.118	4.090	0.686	6.224	2.731	14.5	28.6	242.4	5.2	0.2268	13.131	2.978	0.052025	141.8
53188-1	3.213	1.867	1.004	3.597	1.937	8.0	31.6	40.6	4.2	0.4535	2.736	1.241	0.053900	52.0
53030-1	4.483	3.137	0.686	5.242	2.254	12.0	31.6	128.7	4.7	0.3027	7.725	2.338	0.060427	99.7
53323-1	7.150	6.312	0.686	9.000	3.842	21.1	33.5	758.2	6.8	0.1818	31.274	5.685	0.060628	293.0
53504-1	3.848	2.820	1.004	4.557	2.414	10.5	31.1	104.0	4.7	0.3406	6.242	2.126	0.061831	82.9
53383-1	3.531	2.185	1.004	4.006	2.096	9.0	35.5	59.0	4.4	0.4535	3.747	1.700	0.069587	63.6
53315-1	4.483	3.455	1.004	5.390	2.731	12.5	37.0	173.0	5.2	0.3406	9.370	3.191	0.083734	113.5
53481-1	3.213	1.867	1.321	3.597	2.254	8.0	42.2	45.6	4.7	0.6053	2.736	1.656	0.085611	56.1
53026-1	3.848	2.502	1.004	4.417	2.255	10.0	39.5	81.9	4.7	0.4535	4.914	2.229	0.086304	76.4

Magnetics Caseless Toroids 1 mil Iron Alloy														
Part No.	OD cm 1	ID cm 2	HT cm 3	Wound OD cm 4	Wound HT cm 5	MPL cm 6	Wtfe grams 7	Wtcu grams 8	MLT cm 9	Ac cm sq 10	Wa cm sq 11	Ap cm 4th 12	Kg cm 5th 13	At cm sq 14
53514-1	3.848	1.867	1.004	4.174	1.937	9.0	53.3	45.6	4.7	0.6803	2.736	1.861	0.108123	65.4
53098-1	4.483	3.137	1.004	5.242	2.572	12.0	47.4	142.6	5.2	0.4535	7.725	3.503	0.122400	105.7
53038-1	3.848	2.502	1.321	4.417	2.572	10.0	52.7	90.7	5.2	0.6053	4.914	2.974	0.138711	81.5
53139-1	5.118	3.772	1.004	6.071	2.890	14.0	55.3	226.4	5.7	0.4535	11.168	5.065	0.161200	139.9
53091-1	4.166	2.820	1.321	4.829	2.731	11.0	58.0	120.9	5.4	0.6053	6.242	3.778	0.167981	96.0
53035-1	4.483	3.137	1.321	5.242	2.889	12.0	63.2	156.6	5.7	0.6053	7.725	4.676	0.198623	111.7
53559-1	4.801	3.455	1.321	5.657	3.048	13.0	68.5	198.4	6.0	0.6053	9.370	5.672	0.230640	128.7
53425-1	5.118	3.137	1.004	5.795	2.572	13.0	77.0	156.6	5.7	0.6803	7.725	5.255	0.250858	123.5
53055-1	5.118	3.772	1.321	6.071	3.207	14.0	73.8	246.5	6.2	0.6053	11.168	6.760	0.263675	146.9
53190-1	4.483	2.502	1.321	4.979	2.572	11.0	86.9	99.6	5.7	0.9071	4.914	4.457	0.283730	97.7
53345-1	5.753	4.407	1.321	6.903	3.524	16.0	84.3	364.1	6.7	0.6053	15.245	9.228	0.332702	186.8
53451-1	3.848	2.502	2.591	4.417	3.842	10.0	105.3	126.2	7.2	1.2097	4.914	5.944	0.398187	101.8
53017-1	6.388	5.042	1.278	7.737	3.799	18.0	94.9	507.7	7.2	0.6053	19.955	12.079	0.408757	230.5
53169-1	5.118	3.137	1.321	5.795	2.889	13.0	102.6	170.5	6.2	0.9071	7.725	7.007	0.409533	130.2
53066-1	7.023	5.042	1.004	8.270	3.525	18.9	112.5	512.7	7.2	0.6803	19.955	13.576	0.511354	247.3
53252-1	5.753	3.772	1.321	6.616	3.207	15.0	118.5	266.7	6.7	0.9071	11.168	10.131	0.547328	167.5
53031-1	7.658	6.312	1.321	9.409	4.477	21.9	115.9	916.4	8.2	0.6053	31.274	18.931	0.556276	335.9
53012-1	4.483	3.137	2.591	5.242	4.159	12.0	126.3	212.4	7.7	1.2097	7.725	9.345	0.584836	135.8
53555-1	5.752	3.137	1.321	6.362	2.889	14.0	147.4	184.4	6.7	1.2097	7.725	9.345	0.673394	150.4
53222-1	6.388	4.407	1.448	7.442	3.651	17.0	134.2	402.6	7.4	0.9071	15.245	13.829	0.675581	213.1
53040-1	5.118	3.772	2.591	6.071	4.477	14.0	147.4	327.2	8.2	1.2097	11.168	13.511	0.793445	174.7
53067-1	9.563	7.582	1.004	11.600	4.795	26.9	159.9	1485.4	9.3	0.6803	45.126	30.700	0.902508	480.3
53001-1	6.388	3.772	1.321	7.175	3.207	16.0	168.4	286.9	7.2	1.2097	11.168	13.511	0.905041	190.0
53036-1	5.118	3.137	2.591	5.795	4.159	13.0	205.2	226.3	8.2	1.8150	7.725	14.020	1.235354	156.7
53103-1	7.658	5.042	1.321	8.816	3.842	19.9	210.6	584.7	8.2	1.2097	19.955	24.141	1.417702	283.6
53128-1	8.928	6.312	1.321	10.469	4.477	23.9	252.7	1029.4	9.3	1.2097	31.274	37.834	1.977968	396.8
53013-1	6.388	3.772	2.591	7.175	4.477	16.0	336.9	367.6	9.3	2.4195	11.168	27.022	2.825395	222.9
53022-1	7.658	5.042	2.591	8.816	5.112	19.9	421.2	728.9	10.3	2.4195	19.955	48.281	4.548985	324.1
53448-1	7.658	4.407	2.591	8.557	4.794	18.9	500.1	556.9	10.3	3.0248	15.245	46.114	5.431744	300.1
53042-1	8.928	6.312	2.591	10.469	5.747	23.9	505.4	1255.3	11.3	2.4195	31.274	75.668	6.487591	444.8

Magnetics Caseless Toroids 1 mil Iron Alloy														
Part No.	OD cm 1	ID cm 2	HT cm 3	Wound OD cm 4	Wound HT cm 5	MPL cm 6	Wtfe grams 7	Wtcu grams 8	MLT cm 9	Ac cm sq 10	Wa cm sq 11	Ap cm 4th 12	Kg cm 5th 13	At cm sq 14
53120-1	8.293	5.042	2.591	9.372	5.112	20.9	552.9	764.9	10.8	3.0248	19.955	60.360	6.774807	355.4
53261-1	8.928	5.677	2.591	10.192	5.429	22.9	605.5	1015.5	11.3	3.0248	25.298	76.522	8.202237	415.5
53100-1	9.563	6.312	3.861	11.015	7.017	24.9	987.5	1537.8	13.8	4.5372	31.274	141.898	18.623810	531.0
53101-1	12.108	8.217	3.861	14.044	7.969	31.9	1516.2	2989.9	15.9	5.4443	53.001	288.552	39.611030	813.7
53112-1	13.378	10.122	5.131	15.994	10.192	36.9	1948.0	5408.6	18.9	6.0487	80.426	486.468	62.236190	1119.8
53516-1	15.278	11.392	3.861	18.187	9.557	41.9	1990.0	6665.6	18.4	5.4443	101.874	554.626	65.642710	1314.4
53426-1	15.278	10.122	5.131	17.614	10.192	39.9	3369.4	5843.3	20.4	9.6779	80.426	778.349	147.471900	1292.4

Magnetics Caseless Toroids 2 mil Iron Alloy														
Part No.	OD cm 1	ID cm 2	HT cm 3	Wound OD cm 4	Wound HT cm 5	MPL cm 6	Wtfe grams 7	Wtcu grams 8	MLT cm 9	Ac cm sq 10	Wa cm sq 11	Ap cm 4th 12	Kg cm 5th 13	At cm sq 14
53402-2	1.151	0.915	0.369	1.397	0.826	3.3	0.6	3.5	1.5	0.022	0.657	0.014	0.000084	8.2
53107-2	1.468	1.232	0.369	1.815	0.985	4.2	0.8	7.5	1.8	0.022	1.191	0.026	0.000131	13.3
53403-2	1.626	1.392	0.369	2.024	1.065	4.7	0.9	10.2	1.9	0.022	1.521	0.033	0.000156	16.3
53485-2	1.783	1.550	0.369	2.232	1.144	5.2	1.0	13.5	2.0	0.022	1.886	0.041	0.000181	19.6
53356-2	1.943	1.707	0.369	2.442	1.222	5.7	1.1	17.4	2.1	0.022	2.287	0.050	0.000206	23.2
53153-2	1.308	0.915	0.369	1.530	0.826	3.5	1.3	3.8	1.6	0.043	0.657	0.028	0.000297	9.4
53154-2	1.468	1.075	0.369	1.738	0.906	4.0	1.5	5.7	1.8	0.043	0.907	0.039	0.000380	12.0
53056-2	1.626	1.232	0.369	1.945	0.985	4.5	1.7	8.0	1.9	0.043	1.191	0.051	0.000466	14.8
53057-2	1.943	1.550	0.369	2.362	1.144	5.5	2.1	14.4	2.1	0.043	1.886	0.081	0.000650	21.3
53143-2	2.261	1.867	0.369	2.780	1.302	6.5	2.4	23.3	2.4	0.043	2.736	0.118	0.000844	29.1
53374-2	2.578	2.185	0.369	3.198	1.461	7.5	2.8	35.3	2.7	0.043	3.747	0.161	0.001045	38.1
53155-2	1.468	1.075	0.686	1.738	1.223	4.0	3.0	7.3	2.3	0.086	0.907	0.078	0.001181	14.0
53086-2	2.896	2.502	0.369	3.617	1.620	8.5	3.2	50.8	2.9	0.043	4.914	0.211	0.001250	48.3
53000-2	1.943	1.232	0.369	2.217	0.985	5.0	3.7	9.1	2.1	0.086	1.191	0.102	0.001643	18.1
53063-2	2.261	1.550	0.369	2.630	1.144	6.0	4.5	16.1	2.4	0.086	1.886	0.162	0.002326	25.2
53002-2	2.324	1.613	0.369	2.711	1.175	6.2	4.6	17.8	2.4	0.086	2.042	0.176	0.002467	26.8
53134-2	2.578	1.867	0.369	3.043	1.302	7.0	5.2	25.8	2.7	0.086	2.736	0.235	0.003052	33.5
53632-2	1.943	1.232	0.529	2.217	1.145	5.0	5.6	10.2	2.4	0.129	1.191	0.154	0.003303	19.4
53459-2	2.896	2.185	0.369	3.459	1.461	8.0	6.0	38.7	2.9	0.086	3.747	0.322	0.003814	43.1
53011-2	3.213	2.502	0.369	3.876	1.620	9.0	6.7	55.2	3.2	0.086	4.914	0.423	0.004600	53.9
53176-2	1.943	1.232	0.686	2.217	1.302	5.0	7.4	11.2	2.7	0.171	1.191	0.204	0.005255	20.6
53666-2	3.531	2.820	0.369	4.293	1.779	10.0	7.5	75.8	3.4	0.086	6.242	0.537	0.005408	65.8
53033-2	2.261	1.550	0.686	2.630	1.461	6.0	8.9	19.5	2.9	0.171	1.886	0.322	0.007590	28.2
53454-2	4.483	3.772	0.369	5.547	2.255	13.0	9.7	165.9	4.2	0.086	11.168	0.961	0.007911	109.0
53691-2	4.801	4.090	0.369	5.966	2.414	14.0	10.5	206.9	4.4	0.086	13.131	1.129	0.008768	125.9
53654-2	5.118	4.407	0.369	6.384	2.572	15.0	11.2	254.0	4.7	0.086	15.245	1.311	0.009628	143.9
53076-2	2.578	1.550	0.529	2.907	1.304	6.5	10.9	19.5	2.9	0.193	1.886	0.364	0.009660	31.2
53296-2	2.324	1.486	0.686	2.657	1.429	6.0	10.8	18.2	3.0	0.206	1.733	0.357	0.009951	28.4
53061-2	2.578	1.867	0.686	3.043	1.619	7.0	10.4	30.7	3.2	0.171	2.736	0.468	0.010128	37.0
53106-2	2.896	1.867	0.529	3.317	1.462	7.5	12.6	30.8	3.2	0.193	2.736	0.528	0.012889	40.4

Magnetics Caseless Toroids 2 mil Iron Alloy														
Part No.	OD cm 1	ID cm 2	HT cm 3	Wound OD cm 4	Wound HT cm 5	MPL cm 6	Wtfe grams 7	Wtcu grams 8	MLT cm 9	Ac cm sq 10	Wa cm sq 11	Ap cm 4th 12	Kg cm 5th 13	At cm sq 14
53748-2	7.023	6.312	0.369	8.900	3.525	20.9	15.7	690.5	6.2	0.086	31.274	2.690	0.014903	277.7
53004-2	3.213	2.502	0.686	3.876	1.937	9.0	13.4	64.1	3.7	0.171	4.914	0.840	0.015670	58.3
53007-2	2.578	1.550	0.686	2.907	1.461	6.5	14.5	21.2	3.2	0.257	1.886	0.485	0.015768	32.9
53115-2	3.531	2.820	0.686	4.293	2.096	10.0	14.9	87.1	3.9	0.171	6.242	1.067	0.018616	70.8
53168-2	2.578	1.867	1.004	3.043	1.937	7.0	15.7	35.7	3.7	0.257	2.736	0.703	0.019704	40.5
53084-2	2.896	1.867	0.686	3.317	1.619	7.5	16.8	33.2	3.4	0.257	2.736	0.703	0.021173	42.2
53039-2	3.848	3.137	0.686	4.710	2.254	11.0	16.4	114.7	4.2	0.171	7.725	1.321	0.021638	84.4
53392-2	3.848	2.820	0.529	4.557	1.939	10.5	17.6	87.1	3.9	0.193	6.242	1.205	0.023700	75.0
53167-2	2.896	1.550	0.686	3.192	1.461	7.0	20.9	22.9	3.4	0.343	1.886	0.647	0.025993	38.0
53285-2	3.213	2.185	0.686	3.729	1.778	8.5	19.0	48.9	3.7	0.257	3.747	0.963	0.026995	52.8
53094-2	2.578	1.550	1.004	2.907	1.779	6.5	21.8	24.6	3.7	0.386	1.886	0.728	0.030636	36.2
53029-2	3.531	2.502	0.686	4.143	1.937	9.5	21.2	68.5	3.9	0.257	4.914	1.263	0.033100	64.6
53318-2	3.213	1.867	0.686	3.597	1.619	8.0	23.9	35.7	3.7	0.343	2.736	0.938	0.035106	47.9
53228-2	3.531	2.820	1.004	4.293	2.414	10.0	22.4	98.4	4.4	0.257	6.242	1.604	0.037221	75.7
53393-2	5.118	4.090	0.529	6.224	2.574	14.5	24.4	230.7	4.9	0.193	13.131	2.534	0.039602	138.3
53034-2	2.896	1.867	1.004	3.317	1.937	7.5	25.2	38.2	3.9	0.386	2.736	1.056	0.041568	46.0
53391-2	3.213	2.502	1.321	3.876	2.572	9.0	26.9	81.8	4.7	0.343	4.914	1.685	0.049372	67.2
53181-2	3.213	2.185	1.004	3.729	2.096	8.5	28.5	55.7	4.2	0.386	3.747	1.447	0.053477	57.1
53133-2	2.959	1.613	1.004	3.272	1.810	7.2	32.2	28.9	4.0	0.514	2.042	1.050	0.054317	43.6
53032-2	3.848	2.502	0.686	4.417	1.937	10.0	29.9	73.0	4.2	0.343	4.914	1.685	0.055379	71.4
53553-2	3.531	2.502	1.004	4.143	2.255	9.5	31.9	77.4	4.4	0.386	4.914	1.897	0.066095	69.4
53018-2	5.118	4.090	0.686	6.224	2.731	14.5	32.4	242.4	5.2	0.257	13.131	3.375	0.066823	141.8
53188-2	3.213	1.867	1.004	3.597	1.937	8.0	35.8	40.6	4.2	0.514	2.736	1.406	0.069231	52.0
53030-2	4.483	3.137	0.686	5.242	2.254	12.0	35.8	128.7	4.7	0.343	7.725	2.650	0.077615	99.7
53323-2	7.150	6.312	0.686	9.000	3.842	21.1	38.0	758.2	6.8	0.206	31.274	6.443	0.077874	293.0
53504-2	3.848	2.820	1.004	4.557	2.414	10.5	35.3	104.0	4.7	0.386	6.242	2.410	0.079419	82.9
53383-2	3.531	2.185	1.004	4.006	2.096	9.0	40.3	59.0	4.4	0.514	3.747	1.926	0.089381	63.6
53315-2	4.483	3.455	1.004	5.390	2.731	12.5	42.0	173.0	5.2	0.386	9.370	3.617	0.107551	113.5
53481-2	3.213	1.867	1.321	3.597	2.254	8.0	47.8	45.6	4.7	0.686	2.736	1.877	0.109963	56.1
53026-2	3.848	2.502	1.004	4.417	2.255	10.0	44.7	81.9	4.7	0.514	4.914	2.526	0.110853	76.4

Magnetics Caseless Toroids 2 mil Iron Alloy

Part No.	OD cm 1	ID cm 2	HT cm 3	Wound OD cm 4	Wound HT cm 5	MPL cm 6	Wtfe grams 7	Wtcu grams 8	MLT cm 9	Ac cm sq 10	Wa cm sq 11	Ap cm 4th 12	Kg cm 5th 13	At cm sq 14
53514-2	3.848	1.867	1.004	4.174	1.937	9.0	60.4	45.6	4.7	0.771	2.736	2.109	0.138878	65.4
53098-2	4.483	3.137	1.004	5.242	2.572	12.0	53.7	142.6	5.2	0.514	7.725	3.970	0.157216	105.7
53038-2	3.848	2.502	1.321	4.417	2.572	10.0	59.7	90.7	5.2	0.686	4.914	3.371	0.178166	81.5
53139-2	5.118	3.772	1.004	6.071	2.890	14.0	62.6	226.4	5.7	0.514	11.168	5.741	0.207052	139.9
53091-2	4.166	2.820	1.321	4.829	2.731	11.0	65.7	120.9	5.4	0.686	6.242	4.282	0.215763	96.0
53035-2	4.483	3.137	1.321	5.242	2.889	12.0	71.6	156.6	5.7	0.686	7.725	5.299	0.255120	111.7
53559-2	4.801	3.455	1.321	5.657	3.048	13.0	77.6	198.4	6.0	0.686	9.370	6.428	0.296244	128.7
53425-2	5.118	3.137	1.004	5.795	2.572	13.0	87.2	156.6	5.7	0.771	7.725	5.956	0.322214	123.5
53055-2	5.118	3.772	1.321	6.071	3.207	14.0	83.6	246.5	6.2	0.686	11.168	7.662	0.338677	146.9
53190-2	4.483	2.502	1.321	4.979	2.572	11.0	98.5	99.6	5.7	1.028	4.914	5.051	0.364436	97.7
53345-2	5.753	4.407	1.321	6.903	3.524	16.0	95.5	364.1	6.7	0.686	15.245	10.459	0.427337	186.8
53451-2	3.848	2.502	2.591	4.417	3.842	10.0	119.3	126.2	7.2	1.371	4.914	6.737	0.511450	101.8
53017-2	6.388	5.042	1.278	7.737	3.799	18.0	107.5	507.7	7.2	0.686	19.955	13.690	0.525026	230.5
53169-2	5.118	3.137	1.321	5.795	2.889	13.0	116.3	170.5	6.2	1.028	7.725	7.941	0.526022	130.2
53066-2	7.023	5.042	1.004	8.270	3.525	18.9	127.5	512.7	7.2	0.771	19.955	15.386	0.656806	247.3
53252-2	5.753	3.772	1.321	6.616	3.207	15.0	134.3	266.7	6.7	1.028	11.168	11.481	0.703012	167.5
53031-2	7.658	6.312	1.321	9.409	4.477	21.9	131.3	916.4	8.2	0.686	31.274	21.455	0.714505	335.9
53012-2	4.483	3.137	2.591	5.242	4.159	12.0	143.2	212.4	7.7	1.371	7.725	10.591	0.751189	135.8
53555-2	5.752	3.137	1.321	6.362	2.889	14.0	167.1	184.4	6.7	1.371	7.725	10.591	0.864937	150.4
53222-2	6.388	4.407	1.448	7.442	3.651	17.0	152.1	402.6	7.4	1.028	15.245	15.673	0.867747	213.1
53040-2	5.118	3.772	2.591	6.071	4.477	14.0	167.1	327.2	8.2	1.371	11.168	15.312	1.019136	174.7
53067-2	9.563	7.582	1.004	11.600	4.795	26.9	181.2	1485.4	9.3	0.771	45.126	34.793	1.159222	480.3
53001-2	6.388	3.772	1.321	7.175	3.207	16.0	190.9	286.9	7.2	1.371	11.168	15.312	1.162474	190.0
53036-2	5.118	3.137	2.591	5.795	4.159	13.0	232.6	226.3	8.2	2.057	7.725	15.890	1.586744	156.7
53103-2	7.658	5.042	1.321	8.816	3.842	19.9	238.7	584.7	8.2	1.371	19.955	27.359	1.820959	283.6
53128-2	8.928	6.312	1.321	10.469	4.477	23.9	286.4	1029.4	9.3	1.371	31.274	42.878	2.540589	396.8
53013-2	6.388	3.772	2.591	7.175	4.477	16.0	381.8	367.6	9.3	2.742	11.168	30.624	3.629064	222.9
53022-2	7.658	5.042	2.591	8.816	5.112	19.9	477.3	728.9	10.3	2.742	19.955	54.719	5.842919	324.1
53448-2	7.658	4.407	2.591	8.557	4.794	18.9	566.8	556.9	10.3	3.428	15.245	52.262	6.976774	300.1
53042-2	8.928	6.312	2.591	10.469	5.747	23.9	572.8	1255.3	11.3	2.742	31.274	85.757	8.332951	444.8

Magnetics Caseless Toroids 2 mil Iron Alloy														
Part No.	OD cm 1	ID cm 2	HT cm 3	Wound OD cm 4	Wound HT cm 5	MPL cm 6	Wtfe grams 7	Wtcu grams 8	MLT cm 9	Ac cm sq 10	Wa cm sq 11	Ap cm 4th 12	Kg cm 5th 13	At cm sq 14
53120-2	8.293	5.042	2.591	9.372	5.112	20.9	626.7	764.9	10.8	3.428	19.955	68.408	8.701864	355.4
53261-2	8.928	5.677	2.591	10.192	5.429	22.9	686.2	1015.5	11.3	3.428	25.298	86.725	10.535320	415.5
53100-2	9.563	6.312	3.861	11.015	7.017	24.9	1119.1	1537.8	13.8	5.142	31.274	160.817	23.921250	531.0
53101-2	12.108	8.217	3.861	14.044	7.969	31.9	1718.3	2989.9	15.9	6.170	53.001	327.026	50.878170	813.7
53112-2	13.378	10.122	5.131	15.994	10.192	36.9	2207.7	5408.6	18.9	6.855	80.426	551.331	79.938930	1119.8
53516-2	15.278	11.392	3.861	18.187	9.557	41.9	2255.3	6665.6	18.4	6.170	101.874	628.576	84.314420	1314.4
53426-2	15.278	10.122	5.131	17.614	10.192	39.9	3818.6	5843.3	20.4	10.968	80.426	882.129	189.419500	1292.4

Magnetics Inc. Caseless Toroids 4 mil Iron Alloy														
Part No.	OD cm 1	ID cm 2	HT cm 3	Wound OD cm 4	Wound HT cm 5	MPL cm 6	Wtfe grams 7	Wtcu grams 8	MLT cm 9	Ac cm sq 10	Wa cm sq 11	Ap cm 4th 12	Kg cm 5th 13	At cm sq 14
53402-4	1.151	0.915	0.369	1.397	0.826	3.3	0.7	3.5	1.5	0.0233	0.657	0.015	0.000094	8.2
53107-4	1.468	1.232	0.369	1.815	0.985	4.2	0.9	7.5	1.8	0.0233	1.191	0.028	0.000147	13.3
53403-4	1.626	1.392	0.369	2.024	1.065	4.7	1.0	10.2	1.9	0.0233	1.521	0.035	0.000175	16.3
53485-4	1.783	1.550	0.369	2.232	1.144	5.2	1.1	13.5	2.0	0.0233	1.886	0.044	0.000203	19.6
53356-4	1.943	1.707	0.369	2.442	1.222	5.7	1.2	17.4	2.1	0.0233	2.287	0.053	0.000231	23.2
53153-4	1.308	0.915	0.369	1.530	0.826	3.5	1.4	3.8	1.6	0.0455	0.657	0.030	0.000333	9.4
53154-4	1.468	1.075	0.369	1.738	0.906	4.0	1.6	5.7	1.8	0.0455	0.907	0.041	0.000426	12.0
53056-4	1.626	1.232	0.369	1.945	0.985	4.5	1.8	8.0	1.9	0.0455	1.191	0.054	0.000522	14.8
53057-4	1.943	1.550	0.369	2.362	1.144	5.5	2.2	14.4	2.1	0.0455	1.886	0.086	0.000729	21.3
53143-4	2.261	1.867	0.369	2.780	1.302	6.5	2.6	23.3	2.4	0.0455	2.736	0.125	0.000946	29.1
53374-4	2.578	2.185	0.369	3.198	1.461	7.5	3.0	35.3	2.7	0.0455	3.747	0.171	0.001171	38.1
53155-4	1.468	1.075	0.686	1.738	1.223	4.0	3.2	7.3	2.3	0.0911	0.907	0.083	0.001324	14.0
53086-4	2.896	2.502	0.369	3.617	1.620	8.5	3.4	50.8	2.9	0.0455	4.914	0.224	0.001402	48.3
53000-4	1.943	1.232	0.369	2.217	0.985	5.0	4.0	9.1	2.1	0.0911	1.191	0.108	0.001843	18.1
53063-4	2.261	1.550	0.369	2.630	1.144	6.0	4.8	16.1	2.4	0.0911	1.886	0.172	0.002607	25.2
53002-4	2.324	1.613	0.369	2.711	1.175	6.2	4.9	17.8	2.4	0.0911	2.042	0.186	0.002765	26.8
53134-4	2.578	1.867	0.369	3.043	1.302	7.0	5.5	25.8	2.7	0.0911	2.736	0.249	0.003421	33.5
53632-4	1.943	1.232	0.529	2.217	1.145	5.0	6.0	10.2	2.4	0.1366	1.191	0.163	0.003704	19.4
53459-4	2.896	2.185	0.369	3.459	1.461	8.0	6.3	38.7	2.9	0.0911	3.747	0.341	0.004276	43.1
53011-4	3.213	2.502	0.369	3.876	1.620	9.0	7.1	55.2	3.2	0.0911	4.914	0.447	0.005157	53.9
53176-4	1.943	1.232	0.686	2.217	1.302	5.0	7.9	11.2	2.7	0.1811	1.191	0.216	0.005891	20.6
53666-4	3.531	2.820	0.369	4.293	1.779	10.0	7.9	75.8	3.4	0.0911	6.242	0.568	0.006063	65.8
53033-4	2.261	1.550	0.686	2.630	1.461	6.0	9.5	19.5	2.9	0.1811	1.886	0.341	0.008509	28.2
53454-4	4.483	3.772	0.369	5.547	2.255	13.0	10.3	165.9	4.2	0.0911	11.168	1.017	0.008869	109.0
53691-4	4.801	4.090	0.369	5.966	2.414	14.0	11.1	206.9	4.4	0.0911	13.131	1.196	0.009829	125.9
53654-4	5.118	4.407	0.369	6.384	2.572	15.0	11.9	254.0	4.7	0.0911	15.245	1.388	0.010794	143.9
53076-4	2.578	1.550	0.529	2.907	1.304	6.5	11.6	19.5	2.9	0.2044	1.886	0.385	0.010830	31.2
53296-4	2.324	1.486	0.686	2.657	1.429	6.0	11.4	18.2	3.0	0.2181	1.733	0.378	0.011156	28.4
53061-4	2.578	1.867	0.686	3.043	1.619	7.0	11.0	30.7	3.2	0.1811	2.736	0.495	0.011355	37.0
53106-4	2.896	1.867	0.529	3.317	1.462	7.5	13.3	30.8	3.2	0.2044	2.736	0.559	0.014450	40.4

Magnetics Inc. Caseless Toroids 4 mil Iron Alloy														
Part No.	OD cm 1	ID cm 2	HT cm 3	Wound OD cm 4	Wound HT cm 5	MPL cm 6	Wtfe grams 7	Wtcu grams 8	MLT cm 9	Ac cm sq 10	Wa cm sq 11	Ap cm 4th 12	Kg cm 5th 13	At cm sq 14
53748-4	7.023	6.312	0.369	8.900	3.525	20.9	16.6	690.5	6.2	0.0911	31.274	2.848	0.016708	277.7
53004-4	3.213	2.502	0.686	3.876	1.937	9.0	14.2	64.1	3.7	0.1811	4.914	0.890	0.017568	58.3
53007-4	2.578	1.550	0.686	2.907	1.461	6.5	15.4	21.2	3.2	0.2721	1.886	0.513	0.017677	32.9
53115-4	3.531	2.820	0.686	4.293	2.096	10.0	15.8	87.1	3.9	0.1811	6.242	1.130	0.020871	70.8
53168-4	2.578	1.867	1.004	3.043	1.937	7.0	16.6	35.7	3.7	0.2721	2.736	0.745	0.022091	40.5
53084-4	2.896	1.867	0.686	3.317	1.619	7.5	17.8	33.2	3.4	0.2721	2.736	0.745	0.023737	42.2
53039-4	3.848	3.137	0.686	4.710	2.254	11.0	17.3	114.7	4.2	0.1811	7.725	1.399	0.024258	84.4
53392-4	3.848	2.820	0.529	4.557	1.939	10.5	18.7	87.1	3.9	0.2044	6.242	1.276	0.026570	75.0
53167-4	2.896	1.550	0.686	3.192	1.461	7.0	22.1	22.9	3.4	0.3632	1.886	0.685	0.029141	38.0
53285-4	3.213	2.185	0.686	3.729	1.778	8.5	20.1	48.9	3.7	0.2721	3.747	1.020	0.030264	52.8
53094-4	2.578	1.550	1.004	2.907	1.779	6.5	23.1	24.6	3.7	0.4087	1.886	0.771	0.034347	36.2
53029-4	3.531	2.502	0.686	4.143	1.937	9.5	22.5	68.5	3.9	0.2721	4.914	1.337	0.037109	64.6
53318-4	3.213	1.867	0.686	3.597	1.619	8.0	25.3	35.7	3.7	0.3632	2.736	0.994	0.039358	47.9
53228-4	3.531	2.820	1.004	4.293	2.414	10.0	23.7	98.4	4.4	0.2721	6.242	1.699	0.041729	75.7
53393-4	5.118	4.090	0.529	6.224	2.574	14.5	25.8	230.7	4.9	0.2044	13.131	2.683	0.044398	138.3
53034-4	2.896	1.867	1.004	3.317	1.937	7.5	26.7	38.2	3.9	0.4087	2.736	1.118	0.046602	46.0
53391-4	3.213	2.502	1.321	3.876	2.572	9.0	28.4	81.8	4.7	0.3632	4.914	1.785	0.055352	67.2
53181-4	3.213	2.185	1.004	3.729	2.096	8.5	30.2	55.7	4.2	0.4087	3.747	1.532	0.059954	57.1
53133-4	2.959	1.613	1.004	3.272	1.810	7.2	34.1	28.9	4.0	0.5442	2.042	1.111	0.060895	43.6
53032-4	3.848	2.502	0.686	4.417	1.937	10.0	31.6	73.0	4.2	0.3632	4.914	1.785	0.062086	71.4
53553-4	3.531	2.502	1.004	4.143	2.255	9.5	33.8	77.4	4.4	0.4087	4.914	2.008	0.074099	69.4
53018-4	5.118	4.090	0.686	6.224	2.731	14.5	34.4	242.4	5.2	0.2721	13.131	3.573	0.074916	141.8
53188-4	3.213	1.867	1.004	3.597	1.937	8.0	37.9	40.6	4.2	0.5442	2.736	1.489	0.077615	52.0
53030-4	4.483	3.137	0.686	5.242	2.254	12.0	37.9	128.7	4.7	0.3632	7.725	2.805	0.087015	99.7
53323-4	7.150	6.312	0.686	9.000	3.842	21.1	40.3	758.2	6.8	0.2181	31.274	6.822	0.087305	293.0
53504-4	3.848	2.820	1.004	4.557	2.414	10.5	37.4	104.0	4.7	0.4087	6.242	2.551	0.089037	82.9
53383-4	3.531	2.185	1.004	4.006	2.096	9.0	42.6	59.0	4.4	0.5442	3.747	2.040	0.100205	63.6
53315-4	4.483	3.455	1.004	5.390	2.731	12.5	44.5	173.0	5.2	0.4087	9.370	3.830	0.120577	113.5
53481-4	3.213	1.867	1.321	3.597	2.254	8.0	50.6	45.6	4.7	0.7264	2.736	1.987	0.123280	56.1
53026-4	3.848	2.502	1.004	4.417	2.255	10.0	47.4	81.9	4.7	0.5442	4.914	2.674	0.124278	76.4

Magnetics Inc. Caseless Toroids 4 mil Iron Alloy														
Part No.	OD cm 1	ID cm 2	HT cm 3	Wound OD cm 4	Wound HT cm 5	MPL cm 6	Wtfe grams 7	Wtcu grams 8	MLT cm 9	Ac cm sq 10	Wa cm sq 11	Ap cm 4th 12	Kg cm 5th 13	At cm sq 14
53514-4	3.848	1.867	1.004	4.174	1.937	9.0	63.9	45.6	4.7	0.8164	2.736	2.234	0.155697	65.4
53098-4	4.483	3.137	1.004	5.242	2.572	12.0	56.8	142.6	5.2	0.5442	7.725	4.204	0.176256	105.7
53038-4	3.848	2.502	1.321	4.417	2.572	10.0	63.2	90.7	5.2	0.7264	4.914	3.569	0.199743	81.5
53139-4	5.118	3.772	1.004	6.071	2.890	14.0	66.3	226.4	5.7	0.5442	11.168	6.078	0.232128	139.9
53091-4	4.166	2.820	1.321	4.829	2.731	11.0	69.6	120.9	5.4	0.7264	6.242	4.534	0.241893	96.0
53035-4	4.483	3.137	1.321	5.242	2.889	12.0	75.8	156.6	5.7	0.7264	7.725	5.611	0.286017	111.7
53559-4	4.801	3.455	1.321	5.657	3.048	13.0	82.2	198.4	6.0	0.7264	9.370	6.806	0.332122	128.7
53425-4	5.118	3.137	1.004	5.795	2.572	13.0	92.4	156.6	5.7	0.8164	7.725	6.306	0.361236	123.5
53055-4	5.118	3.772	1.321	6.071	3.207	14.0	88.5	246.5	6.2	0.7264	11.168	8.112	0.379693	146.9
53190-4	4.483	2.502	1.321	4.979	2.572	11.0	104.2	99.6	5.7	1.0885	4.914	5.349	0.408571	97.7
53345-4	5.753	4.407	1.321	6.903	3.524	16.0	101.1	364.1	6.7	0.7264	15.245	11.074	0.479091	186.8
53451-4	3.848	2.502	2.591	4.417	3.842	10.0	126.4	126.2	7.2	1.4517	4.914	7.133	0.573390	101.8
53017-4	6.388	5.042	1.278	7.737	3.799	18.0	113.8	507.7	7.2	0.7264	19.955	14.495	0.588610	230.5
53169-4	5.118	3.137	1.321	5.795	2.889	13.0	123.2	170.5	6.2	1.0885	7.725	8.408	0.589727	130.2
53066-4	7.023	5.042	1.004	8.270	3.525	18.9	135.0	512.7	7.2	0.8164	19.955	16.291	0.736350	247.3
53252-4	5.753	3.772	1.321	6.616	3.207	15.0	142.2	266.7	6.7	1.0885	11.168	12.157	0.788152	167.5
53031-4	7.658	6.312	1.321	9.409	4.477	21.9	139.1	916.4	8.2	0.7264	31.274	22.717	0.801037	335.9
53012-4	4.483	3.137	2.591	5.242	4.159	12.0	151.6	212.4	7.7	1.4517	7.725	11.214	0.842164	135.8
53555-4	5.752	3.137	1.321	6.362	2.889	14.0	176.9	184.4	6.7	1.4517	7.725	11.214	0.969687	150.4
53222-4	6.388	4.407	1.448	7.442	3.651	17.0	161.1	402.6	7.4	1.0885	15.245	16.594	0.972837	213.1
53040-4	5.118	3.772	2.591	6.071	4.477	14.0	176.9	327.2	8.2	1.4517	11.168	16.213	1.142561	174.7
53067-4	9.563	7.582	1.004	11.600	4.795	26.9	191.9	1485.4	9.3	0.8164	45.126	36.840	1.299612	480.3
53001-4	6.388	3.772	1.321	7.175	3.207	16.0	202.1	286.9	7.2	1.4517	11.168	16.213	1.303258	190.0
53036-4	5.118	3.137	2.591	5.795	4.159	13.0	246.2	226.3	8.2	2.1781	7.725	16.824	1.778910	156.7
53103-4	7.658	5.042	1.321	8.816	3.842	19.9	252.7	584.7	8.2	1.4517	19.955	28.969	2.041490	283.6
53128-4	8.928	6.312	1.321	10.469	4.477	23.9	303.3	1029.4	9.3	1.4517	31.274	45.401	2.848273	396.8
53013-4	6.388	3.772	2.591	7.175	4.477	16.0	404.3	367.6	9.3	2.9034	11.168	32.426	4.068569	222.9
53022-4	7.658	5.042	2.591	8.816	5.112	19.9	505.4	728.9	10.3	2.9034	19.955	57.938	6.550538	324.1
53448-4	7.658	4.407	2.591	8.557	4.794	18.9	600.2	556.9	10.3	3.6297	15.245	55.336	7.821710	300.1
53042-4	8.928	6.312	2.591	10.469	5.747	23.9	606.5	1255.3	11.3	2.9034	31.274	90.801	9.342130	444.8

Magnetics Inc. Caseless Toroids 4 mil Iron Alloy														
Part No.	OD cm 1	ID cm 2	HT cm 3	Wound OD cm 4	Wound HT cm 5	MPL cm 6	Wtfe grams 7	Wtcu grams 8	MLT cm 9	Ac cm sq 10	Wa cm sq 11	Ap cm 4th 12	Kg cm 5th 13	At cm sq 14
53120-4	8.293	5.042	2.591	9.372	5.112	20.9	663.5	764.9	10.8	3.6297	19.955	72.432	9.755722	355.4
53261-4	8.928	5.677	2.591	10.192	5.429	22.9	726.6	1015.5	11.3	3.6297	25.298	91.826	11.811220	415.5
53100-4	9.563	6.312	3.861	11.015	7.017	24.9	1185.0	1537.8	13.8	5.4446	31.274	170.277	26.818290	531.0
53101-4	12.108	8.217	3.861	14.044	7.969	31.9	1819.4	2989.9	15.9	6.5331	53.001	346.262	57.039880	813.7
53112-4	13.378	10.122	5.131	15.994	10.192	36.9	2337.6	5408.6	18.9	7.2584	80.426	583.762	89.620110	1119.8
53516-4	15.278	11.392	3.861	18.187	9.557	41.9	2388.0	6665.6	18.4	6.5331	101.874	665.551	94.525500	1314.4
53426-4	15.278	10.122	5.131	17.614	10.192	39.9	4043.3	5843.3	20.4	11.6135	80.426	934.019	212.359500	1292.4

Magnetics Caseless Toroids 12 mil Iron Alloy														
Part No.	OD cm 1	ID cm 2	HT cm 3	Wound OD cm 4	Wound HT cm 5	MPL cm 6	Wtfe grams 7	Wtcu grams 8	MLT cm 9	Ac cm sq 10	Wa cm sq 11	Ap cm 4th 12	Kg cm 5th 13	At cm sq 14
53402-12	1.151	0.915	0.369	1.397	0.826	3.3	0.7	3.5	1.5	0.0246	0.657	0.016	0.000105	8.2
53107-12	1.468	1.232	0.369	1.815	0.985	4.2	0.9	7.5	1.8	0.0246	1.191	0.029	0.000163	13.3
53403-12	1.626	1.392	0.369	2.024	1.065	4.7	1.0	10.2	1.9	0.0246	1.521	0.037	0.000195	16.3
53485-12	1.783	1.550	0.369	2.232	1.144	5.2	1.1	13.5	2.0	0.0246	1.886	0.046	0.000226	19.6
53356-12	1.943	1.707	0.369	2.442	1.222	5.7	1.2	17.4	2.1	0.0246	2.287	0.056	0.000258	23.2
53153-12	1.308	0.915	0.369	1.530	0.826	3.5	1.5	3.8	1.6	0.0481	0.657	0.032	0.000371	9.4
53154-12	1.468	1.075	0.369	1.738	0.906	4.0	1.7	5.7	1.8	0.0481	0.907	0.044	0.000475	12.0
53056-12	1.626	1.232	0.369	1.945	0.985	4.5	1.9	8.0	1.9	0.0481	1.191	0.057	0.000582	14.8
53057-12	1.943	1.550	0.369	2.362	1.144	5.5	2.3	14.4	2.1	0.0481	1.886	0.091	0.000812	21.3
53143-12	2.261	1.867	0.369	2.780	1.302	6.5	2.7	23.3	2.4	0.0481	2.736	0.131	0.001054	29.1
53374-12	2.578	2.185	0.369	3.198	1.461	7.5	3.1	35.3	2.7	0.0481	3.747	0.180	0.001305	38.1
53155-12	1.468	1.075	0.686	1.738	1.223	4.0	3.3	7.3	2.3	0.0961	0.907	0.087	0.001475	14.0
53086-12	2.896	2.502	0.369	3.617	1.620	8.5	3.6	50.8	2.9	0.0481	4.914	0.236	0.001562	48.3
53000-12	1.943	1.232	0.369	2.217	0.985	5.0	4.2	9.1	2.1	0.0961	1.191	0.115	0.002053	18.1
53063-12	2.261	1.550	0.369	2.630	1.144	6.0	5.0	16.1	2.4	0.0961	1.886	0.181	0.002905	25.2
53002-12	2.324	1.613	0.369	2.711	1.175	6.2	5.2	17.8	2.4	0.0961	2.042	0.196	0.003081	26.8
53134-12	2.578	1.867	0.369	3.043	1.302	7.0	5.9	25.8	2.7	0.0961	2.736	0.263	0.003812	33.5
53632-12	1.943	1.232	0.529	2.217	1.145	5.0	6.3	10.2	2.4	0.1442	1.191	0.172	0.004126	19.4
53459-12	2.896	2.185	0.369	3.459	1.461	8.0	6.7	38.7	2.9	0.0961	3.747	0.360	0.004764	43.1
53011-12	3.213	2.502	0.369	3.876	1.620	9.0	7.5	55.2	3.2	0.0961	4.914	0.472	0.005746	53.9
53176-12	1.943	1.232	0.686	2.217	1.302	5.0	8.3	11.2	2.7	0.1911	1.191	0.228	0.006564	20.6
53666-12	3.531	2.820	0.369	4.293	1.779	10.0	8.4	75.8	3.4	0.0961	6.242	0.600	0.006755	65.8
53033-12	2.261	1.550	0.686	2.630	1.461	6.0	10.0	19.5	2.9	0.1911	1.886	0.360	0.009481	28.2
53454-12	4.483	3.772	0.369	5.547	2.255	13.0	10.9	165.9	4.2	0.0961	11.168	1.074	0.009882	109.0
53691-12	4.801	4.090	0.369	5.966	2.414	14.0	11.7	206.9	4.4	0.0961	13.131	1.262	0.010952	125.9
53654-12	5.118	4.407	0.369	6.384	2.572	15.0	12.6	254.0	4.7	0.0961	15.245	1.465	0.012027	143.9
53076-12	2.578	1.550	0.529	2.907	1.304	6.5	12.2	19.5	2.9	0.2157	1.886	0.407	0.012067	31.2
53296-12	2.324	1.486	0.686	2.657	1.429	6.0	12.0	18.2	3.0	0.2302	1.733	0.399	0.012430	28.4
53061-12	2.578	1.867	0.686	3.043	1.619	7.0	11.6	30.7	3.2	0.1911	2.736	0.523	0.012651	37.0
53106-12	2.896	1.867	0.529	3.317	1.462	7.5	14.1	30.8	3.2	0.2157	2.736	0.590	0.016100	40.4

Magnetics Caseless Toroids 12 mil Iron Alloy

Part No.	OD cm 1	ID cm 2	HT cm 3	Wound OD cm 4	Wound HT cm 5	MPL cm 6	Wtfe grams 7	Wtcu grams 8	MLT cm 9	Ac cm sq 10	Wa cm sq 11	Ap cm 4th 12	Kg cm 5th 13	At cm sq 14
53748-12	7.023	6.312	0.369	8.900	3.525	20.9	17.6	690.5	6.2	0.0961	31.274	3.006	0.018616	277.7
53004-12	3.213	2.502	0.686	3.876	1.937	9.0	15.0	64.1	3.7	0.1911	4.914	0.939	0.019575	58.3
53007-12	2.578	1.550	0.686	2.907	1.461	6.5	16.2	21.2	3.2	0.2872	1.886	0.542	0.019696	32.9
53115-12	3.531	2.820	0.686	4.293	2.096	10.0	16.6	87.1	3.9	0.1911	6.242	1.193	0.023254	70.8
53168-12	2.578	1.867	1.004	3.043	1.937	7.0	17.5	35.7	3.7	0.2872	2.736	0.786	0.024614	40.5
53084-12	2.896	1.867	0.686	3.317	1.619	7.5	18.8	33.2	3.4	0.2872	2.736	0.786	0.026448	42.2
53039-12	3.848	3.137	0.686	4.710	2.254	11.0	18.3	114.7	4.2	0.1911	7.725	1.476	0.027028	84.4
53392-12	3.848	2.820	0.529	4.557	1.939	10.5	19.7	87.1	3.9	0.2157	6.242	1.347	0.029604	75.0
53167-12	2.896	1.550	0.686	3.192	1.461	7.0	23.4	22.9	3.4	0.3834	1.886	0.723	0.032469	38.0
53285-12	3.213	2.185	0.686	3.729	1.778	8.5	21.2	48.9	3.7	0.2872	3.747	1.076	0.033720	52.8
53094-12	2.578	1.550	1.004	2.907	1.779	6.5	24.4	24.6	3.7	0.4314	1.886	0.814	0.038269	36.2
53029-12	3.531	2.502	0.686	4.143	1.937	9.5	23.7	68.5	3.9	0.2872	4.914	1.411	0.041347	64.6
53318-12	3.213	1.867	0.686	3.597	1.619	8.0	26.7	35.7	3.7	0.3834	2.736	1.049	0.043852	47.9
53228-12	3.531	2.820	1.004	4.293	2.414	10.0	25.0	98.4	4.4	0.2872	6.242	1.793	0.046494	75.7
53393-12	5.118	4.090	0.529	6.224	2.574	14.5	27.2	230.7	4.9	0.2157	13.131	2.832	0.049468	138.3
53034-12	2.896	1.867	1.004	3.317	1.937	7.5	28.2	38.2	3.9	0.4314	2.736	1.180	0.051923	46.0
53391-12	3.213	2.502	1.321	3.876	2.572	9.0	30.0	81.8	4.7	0.3834	4.914	1.884	0.061673	67.2
53181-12	3.213	2.185	1.004	3.729	2.096	8.5	31.9	55.7	4.2	0.4314	3.747	1.617	0.066800	57.1
53133-12	2.959	1.613	1.004	3.272	1.810	7.2	36.0	28.9	4.0	0.5745	2.042	1.173	0.067849	43.6
53032-12	3.848	2.502	0.686	4.417	1.937	10.0	33.4	73.0	4.2	0.3834	4.914	1.884	0.069176	71.4
53553-12	3.531	2.502	1.004	4.143	2.255	9.5	35.7	77.4	4.4	0.4314	4.914	2.120	0.082561	69.4
53018-12	5.118	4.090	0.686	6.224	2.731	14.5	36.3	242.4	5.2	0.2872	13.131	3.772	0.083471	141.8
53188-12	3.213	1.867	1.004	3.597	1.937	8.0	40.0	40.6	4.2	0.5745	2.736	1.572	0.086479	52.0
53030-12	4.483	3.137	0.686	5.242	2.254	12.0	40.0	128.7	4.7	0.3834	7.725	2.961	0.096952	99.7
53323-12	7.150	6.312	0.686	9.000	3.842	21.1	42.5	758.2	6.8	0.2302	31.274	7.201	0.097275	293.0
53504-12	3.848	2.820	1.004	4.557	2.414	10.5	39.4	104.0	4.7	0.4314	6.242	2.693	0.099205	82.9
53383-12	3.531	2.185	1.004	4.006	2.096	9.0	45.0	59.0	4.4	0.5745	3.747	2.153	0.111648	63.6
53315-12	4.483	3.455	1.004	5.390	2.731	12.5	46.9	173.0	5.2	0.4314	9.370	4.042	0.134346	113.5
53481-12	3.213	1.867	1.321	3.597	2.254	8.0	53.4	45.6	4.7	0.7667	2.736	2.098	0.137359	56.1
53026-12	3.848	2.502	1.004	4.417	2.255	10.0	50.0	81.9	4.7	0.5745	4.914	2.823	0.138471	76.4

Magnetics Caseless Toroids 12 mil Iron Alloy														
Part No.	OD cm 1	ID cm 2	HT cm 3	Wound OD cm 4	Wound HT cm 5	MPL cm 6	Wtfe grams 7	Wtcu grams 8	MLT cm 9	Ac cm sq 10	Wa cm sq 11	Ap cm 4th 12	Kg cm 5th 13	At cm sq 14
53514-12	3.848	1.867	1.004	4.174	1.937	9.0	67.5	45.6	4.7	0.8617	2.736	2.358	0.173477	65.4
53098-12	4.483	3.137	1.004	5.242	2.572	12.0	60.0	142.6	5.2	0.5745	7.725	4.438	0.196384	105.7
53038-12	3.848	2.502	1.321	4.417	2.572	10.0	66.7	90.7	5.2	0.7667	4.914	3.767	0.222553	81.5
53139-12	5.118	3.772	1.004	6.071	2.890	14.0	70.0	226.4	5.7	0.5745	11.168	6.416	0.258636	139.9
53091-12	4.166	2.820	1.321	4.829	2.731	11.0	73.4	120.9	5.4	0.7667	6.242	4.786	0.269517	96.0
53035-12	4.483	3.137	1.321	5.242	2.889	12.0	80.1	156.6	5.7	0.7667	7.725	5.923	0.318679	111.7
53559-12	4.801	3.455	1.321	5.657	3.048	13.0	86.7	198.4	6.0	0.7667	9.370	7.184	0.370049	128.7
53425-12	5.118	3.137	1.004	5.795	2.572	13.0	97.5	156.6	5.7	0.8617	7.725	6.656	0.402488	123.5
53055-12	5.118	3.772	1.321	6.071	3.207	14.0	93.4	246.5	6.2	0.7667	11.168	8.563	0.423053	146.9
53190-12	4.483	2.502	1.321	4.979	2.572	11.0	110.0	99.6	5.7	1.1490	4.914	5.646	0.455229	97.7
53345-12	5.753	4.407	1.321	6.903	3.524	16.0	106.8	364.1	6.7	0.7667	15.245	11.689	0.533802	186.8
53451-12	3.848	2.502	2.591	4.417	3.842	10.0	133.4	126.2	7.2	1.5323	4.914	7.529	0.638870	101.8
53017-12	6.388	5.042	1.278	7.737	3.799	18.0	120.1	507.7	7.2	0.7667	19.955	15.300	0.655828	230.5
53169-12	5.118	3.137	1.321	5.795	2.889	13.0	130.0	170.5	6.2	1.1490	7.725	8.875	0.657073	130.2
53066-12	7.023	5.042	1.004	8.270	3.525	18.9	142.5	512.7	7.2	0.8617	19.955	17.196	0.820439	247.3
53252-12	5.753	3.772	1.321	6.616	3.207	15.0	150.1	266.7	6.7	1.1490	11.168	12.832	0.878157	167.5
53031-12	7.658	6.312	1.321	9.409	4.477	21.9	146.8	916.4	8.2	0.7667	31.274	23.979	0.892513	335.9
53012-12	4.483	3.137	2.591	5.242	4.159	12.0	160.0	212.4	7.7	1.5323	7.725	11.837	0.938337	135.8
53555-12	5.752	3.137	1.321	6.362	2.889	14.0	186.7	184.4	6.7	1.5323	7.725	11.837	1.080422	150.4
53222-12	6.388	4.407	1.448	7.442	3.651	17.0	170.0	402.6	7.4	1.1490	15.245	17.516	1.083933	213.1
53040-12	5.118	3.772	2.591	6.071	4.477	14.0	186.7	327.2	8.2	1.5323	11.168	17.114	1.273038	174.7
53067-12	9.563	7.582	1.004	11.600	4.795	26.9	202.5	1485.4	9.3	0.8617	45.126	38.886	1.448025	480.3
53001-12	6.388	3.772	1.321	7.175	3.207	16.0	213.4	286.9	7.2	1.5323	11.168	17.114	1.452087	190.0
53036-12	5.118	3.137	2.591	5.795	4.159	13.0	259.9	226.3	8.2	2.2991	7.725	17.759	1.982056	156.7
53103-12	7.658	5.042	1.321	8.816	3.842	19.9	266.7	584.7	8.2	1.5323	19.955	30.578	2.274623	283.6
53128-12	8.928	6.312	1.321	10.469	4.477	23.9	320.1	1029.4	9.3	1.5323	31.274	47.923	3.173539	396.8
53013-12	6.388	3.772	2.591	7.175	4.477	16.0	426.7	367.6	9.3	3.0647	11.168	34.227	4.533190	222.9
53022-12	7.658	5.042	2.591	8.816	5.112	19.9	533.5	728.9	10.3	3.0647	19.955	61.156	7.298593	324.1
53448-12	7.658	4.407	2.591	8.557	4.794	18.9	633.5	556.9	10.3	3.8314	15.245	58.411	8.714931	300.1
53042-12	8.928	6.312	2.591	10.469	5.747	23.9	640.2	1255.3	11.3	3.0647	31.274	95.846	10.408980	444.8

Magnetics Caseless Toroids 12 mil Iron Alloy

Part No.	OD cm 1	ID cm 2	HT cm 3	Wound OD cm 4	Wound HT cm 5	MPL cm 6	Wtfe grams 7	Wtcu grams 8	MLT cm 9	Ac cm sq 10	Wa cm sq 11	Ap cm 4th 12	Kg cm 5th 13	At cm sq 14
53120-12	8.293	5.042	2.591	9.372	5.112	20.9	700.4	764.9	10.8	3.8314	19.955	76.457	10.869800	355.4
53261-12	8.928	5.677	2.591	10.192	5.429	22.9	767.0	1015.5	11.3	3.8314	25.298	96.928	13.160030	415.5
53100-12	9.563	6.312	3.861	11.015	7.017	24.9	1250.8	1537.8	13.8	5.7471	31.274	179.737	29.880870	531.0
53101-12	12.108	8.217	3.861	14.044	7.969	31.9	1920.5	2989.9	15.9	6.8961	53.001	365.499	63.553700	813.7
53112-12	13.378	10.122	5.131	15.994	10.192	36.9	2467.4	5408.6	18.9	7.6617	80.426	616.193	99.854510	1119.8
53516-12	15.278	11.392	3.861	18.187	9.557	41.9	2520.7	6665.6	18.4	6.8961	101.874	702.527	105.320100	1314.4
53426-12	15.278	10.122	5.131	17.614	10.192	39.9	4267.9	5843.3	20.4	12.2587	80.426	985.909	236.610500	1292.4

National Arnold Caseless Toroids 2 mil Silicon Steel														
Part No.	OD cm 1	ID cm 2	HT cm 3	Wound OD cm 4	Wound HT cm 5	MPL cm 6	Wtfe grams 7	Wtcu grams 8	MLT cm 9	Ac cm sq 10	Wa cm sq 11	Ap cm 4th 12	Kg cm 5th 13	At cm sq 14
CRL-200	0.794	0.397	0.318	0.995	0.517	1.9	0.8	0.5	1.1	0.0562	0.124	0.007	0.000137	3.2
CRL-202	0.794	0.476	0.476	1.029	0.714	2.0	1.0	0.9	1.4	0.0674	0.178	0.012	0.000231	4.0
CRL-1-E	0.953	0.635	0.635	1.265	0.953	2.5	1.7	2.0	1.8	0.0899	0.317	0.028	0.000575	6.3
CRL-1-K	1.429	0.953	0.476	1.898	0.953	3.7	2.9	4.8	1.9	0.1008	0.713	0.072	0.001523	11.3
CRL-1-L	1.588	1.111	0.476	2.135	1.032	4.2	3.3	7.0	2.0	0.1010	0.970	0.098	0.001948	14.1
CRL-85-A	1.905	1.270	0.318	2.530	0.953	5.0	3.4	9.2	2.0	0.0899	1.267	0.114	0.002013	17.6
CRL-1-B	1.588	1.111	0.635	2.135	1.191	4.2	4.4	7.9	2.3	0.1348	0.970	0.131	0.003082	15.1
CRL-1-C	1.270	0.794	0.953	1.661	1.350	3.2	5.0	4.5	2.5	0.2019	0.495	0.100	0.003177	11.4
CRL-1-M	1.270	0.794	0.953	1.661	1.350	3.2	5.0	4.5	2.5	0.2019	0.495	0.100	0.003177	11.4
CRL-1-W	1.588	0.953	0.635	2.058	1.112	4.0	5.5	5.8	2.3	0.1794	0.713	0.128	0.004018	13.8
CRL-84	1.905	1.111	0.476	2.454	1.032	4.7	6.1	7.9	2.3	0.1682	0.970	0.163	0.004800	17.4
CRL-1-A	1.746	1.111	0.635	2.293	1.191	4.5	6.1	8.3	2.4	0.1794	0.970	0.174	0.005175	16.8
CRL-85-C	1.905	1.270	0.635	2.530	1.270	5.0	6.8	11.4	2.5	0.1794	1.267	0.227	0.006424	20.1
CRL-2	1.588	0.953	0.953	2.058	1.430	4.0	8.2	7.1	2.8	0.2693	0.713	0.192	0.007403	15.9
CRL-33	2.064	1.429	0.635	2.768	1.350	5.5	7.5	15.2	2.7	0.1794	1.604	0.288	0.007745	23.8
CRL-1-F	1.746	0.794	0.953	2.158	1.350	4.0	12.3	5.1	2.9	0.4037	0.495	0.200	0.011051	16.5
CRL-33-D	2.064	1.429	0.794	2.768	1.509	5.5	9.4	16.7	2.9	0.2244	1.604	0.360	0.011055	25.1
CRL-1-J	1.588	0.953	1.270	2.058	1.747	4.0	10.9	8.4	3.3	0.3589	0.713	0.256	0.011129	17.9
CRL-5-B	3.493	2.540	0.318	4.747	1.588	9.5	9.8	59.5	3.3	0.1349	5.068	0.683	0.011161	59.1
CRL-1-D	1.746	1.111	1.270	2.293	1.826	4.5	12.3	11.8	3.4	0.3589	0.970	0.348	0.014567	21.4
CRL-4-E	2.858	1.905	0.476	3.795	1.429	7.5	11.5	30.9	3.0	0.2019	2.851	0.575	0.015244	39.6
CRL-2-B	1.905	0.953	0.953	2.388	1.430	4.5	13.8	7.7	3.0	0.4037	0.713	0.288	0.015256	19.7
CRL-36	2.381	1.429	0.635	3.086	1.350	6.0	12.3	16.7	2.9	0.2690	1.604	0.432	0.015897	28.0
CRL-53	2.262	1.389	0.754	2.946	1.449	5.7	12.8	16.3	3.0	0.2929	1.515	0.444	0.017245	27.0
CRL-85-B	1.905	1.270	1.270	2.530	1.905	5.0	13.7	16.0	3.6	0.3589	1.267	0.455	0.018354	25.2
CRL-33-A	2.540	1.588	0.635	3.322	1.429	6.5	13.3	21.5	3.0	0.2690	1.981	0.533	0.018812	32.2
CRL-9	2.540	1.429	0.635	3.249	1.350	6.2	14.9	17.4	3.0	0.3139	1.604	0.504	0.020747	30.3
CRL-3-W	2.540	1.270	0.635	3.183	1.270	6.0	16.4	13.7	3.0	0.3589	1.267	0.455	0.021413	28.6
CRL-2-A	2.223	1.429	0.953	2.926	1.668	5.7	14.7	18.8	3.3	0.3367	1.604	0.540	0.022023	28.8
CRL-33-B	2.064	1.429	1.270	2.768	1.985	5.5	15.0	21.0	3.7	0.3589	1.604	0.576	0.022435	29.3

National Arnold Caseless Toroids 2 mil Silicon Steel

Part No.	OD cm 1	ID cm 2	HT cm 3	Wound OD cm 4	Wound HT cm 5	MPL cm 6	Wtfe grams 7	Wtcu grams 8	MLT cm 9	Ac cm sq 10	Wa cm sq 11	Ap cm 4th 12	Kg cm 5th 13	At cm sq 14
CRL-34	2.699	1.429	0.635	3.415	1.350	6.5	17.8	18.1	3.2	0.3589	1.604	0.576	0.026024	32.8
CRL-1-H	1.984	1.111	1.270	2.536	1.826	4.9	18.3	12.5	3.6	0.4934	0.970	0.478	0.026084	24.6
CRL-53-A	2.262	1.389	0.992	2.946	1.687	5.7	16.9	18.3	3.4	0.3854	1.515	0.584	0.026504	29.2
CRL-5-A	3.493	2.540	0.635	4.747	1.905	9.5	19.5	68.7	3.8	0.2693	5.068	1.365	0.038579	63.8
CRL-86-A	3.889	2.619	0.476	5.177	1.786	10.2	21.0	74.2	3.9	0.2690	5.388	1.449	0.040271	71.1
CRL-4	3.175	1.905	0.635	4.115	1.588	8.0	21.9	36.0	3.6	0.3589	2.851	1.023	0.041296	47.1
CRL-3-B	2.540	1.270	0.953	3.183	1.588	6.0	24.6	16.0	3.6	0.5386	1.267	0.682	0.041330	31.8
CRL-54-A	3.016	1.588	0.714	3.812	1.508	7.2	25.0	25.0	3.6	0.4537	1.981	0.899	0.045879	40.9
CRL-50-E	2.858	1.905	0.953	3.795	1.906	7.5	23.1	38.6	3.8	0.4042	2.851	1.152	0.048868	45.3
CRL-7	2.699	1.429	0.953	3.415	1.668	6.5	26.6	21.0	3.7	0.5386	1.604	0.864	0.050520	36.2
CRL-46-B	4.445	3.175	0.476	6.011	2.064	12.0	24.6	121.6	4.3	0.2690	7.918	2.130	0.053088	95.7
CRL-47-B	3.810	3.493	2.223	5.595	3.970	11.5	27.5	225.1	6.6	0.3136	9.584	3.005	0.057077	118.9
CRL-1-N	2.699	1.111	1.111	3.295	1.667	6.0	35.9	13.6	3.9	0.7851	0.970	0.761	0.060721	34.3
CRL-35	3.810	2.540	0.635	5.059	1.905	10.0	27.3	73.2	4.1	0.3589	5.068	1.819	0.064238	70.5
CRL-46-A	2.937	1.667	0.953	3.764	1.787	7.2	29.7	30.1	3.9	0.5386	2.183	1.176	0.065371	43.3
CRL-86	3.889	2.619	0.635	5.177	1.945	10.2	28.0	79.1	4.1	0.3589	5.388	1.934	0.067251	73.7
CRL-12	3.493	2.540	0.953	4.747	2.223	9.5	29.2	77.8	4.3	0.4042	5.068	2.048	0.076659	68.5
CRL-4-D	3.810	1.905	0.635	4.775	1.588	9.0	36.9	41.2	4.1	0.5383	2.851	1.534	0.081302	59.6
CRL-33-C	3.175	1.270	0.953	3.864	1.588	7.0	43.0	18.3	4.1	0.8079	1.267	1.024	0.081371	42.7
CRL-50-A	3.175	1.905	0.953	4.115	1.906	8.0	32.8	41.2	4.1	0.5386	2.851	1.535	0.081371	51.2
CRL-54	3.175	1.588	0.953	3.979	1.747	7.5	38.4	28.6	4.1	0.6730	1.981	1.333	0.088293	46.7
CRL-4-B	2.937	1.905	1.270	3.874	2.223	7.6	33.9	44.4	4.4	0.5832	2.851	1.663	0.088522	50.6
CRL-50-F	3.612	1.905	0.794	4.567	1.747	8.7	39.9	42.2	4.2	0.6031	2.851	1.719	0.099708	57.8
CRL-39	5.080	3.810	0.635	6.966	2.540	14.0	38.2	206.0	5.1	0.3589	11.402	4.092	0.115629	131.7
CRL-3-C	2.540	1.270	1.905	3.183	2.540	6.0	49.2	22.9	5.1	1.0766	1.267	1.364	0.115629	41.3
CRL-12-A	3.493	2.540	1.270	4.747	2.540	9.5	38.9	87.0	4.8	0.5386	5.068	2.729	0.121833	73.2
CRL-4-C	3.175	1.905	1.270	4.115	2.223	8.0	43.7	46.3	4.6	0.7177	2.851	2.046	0.128477	55.3
CRL-5	3.810	2.540	0.953	5.059	2.223	10.0	41.0	82.4	4.6	0.5386	5.068	2.729	0.128589	75.5
CRL-7-W	4.763	3.175	0.635	6.325	2.223	12.5	42.7	135.9	4.8	0.4487	7.918	3.553	0.132141	106.9
CRL-23-A	4.286	3.651	1.905	6.125	3.731	12.5	51.2	241.2	6.5	0.5383	10.471	5.636	0.187382	130.6

National Arnold Caseless Toroids 2 mil Silicon Steel														
Part No.	OD cm 1	ID cm 2	HT cm 3	Wound OD cm 4	Wound HT cm 5	MPL cm 6	Wtfe grams 7	Wtcu grams 8	MLT cm 9	Ac cm sq 10	Wa cm sq 11	Ap cm 4th 12	Kg cm 5th 13	At cm sq 14
CRL-40-A	3.810	2.580	1.270	5.079	2.560	10.0	53.2	94.5	5.1	0.6951	5.229	3.635	0.198939	81.3
CRL-40	3.810	2.540	1.270	5.059	2.540	10.0	54.6	91.5	5.1	0.7177	5.068	3.637	0.205563	80.5
CRL-1	3.175	4.445	0.635	5.738	2.858	12.0	-32.8	196.3	3.6	-0.3589	15.520	-5.570	0.224834	103.2
CRL-12-B	3.493	2.540	1.905	4.747	3.175	9.5	58.4	105.3	5.8	0.8079	5.068	4.094	0.226453	82.7
CRL-50-C	3.175	1.905	1.905	4.115	2.858	8.0	65.6	56.6	5.6	1.0766	2.851	3.069	0.236514	63.5
CRL-3	5.715	3.810	0.635	7.589	2.540	15.0	61.5	226.6	5.6	0.5383	11.402	6.138	0.236514	150.9
CRL-150	3.556	2.302	1.588	4.688	2.739	9.2	62.2	79.7	5.4	0.8862	4.163	3.689	0.242772	74.8
CRL-11	5.080	4.128	1.270	7.144	3.334	14.5	59.4	290.2	6.1	0.5380	13.385	7.202	0.254239	154.9
CRL-13-A	8.255	6.985	0.635	11.769	4.128	23.9	65.6	1038.5	7.6	0.3589	38.325	13.754	0.259095	370.0
CRL-46	4.445	3.175	1.270	6.011	2.858	12.0	65.6	157.3	5.6	0.7177	7.918	5.683	0.291993	110.7
CRL-35-A	3.810	2.540	1.588	5.059	2.858	10.0	68.3	100.7	5.6	0.8975	5.068	4.548	0.292135	85.6
CRL-45	4.763	1.905	0.953	5.797	1.906	10.5	96.9	54.1	5.3	1.2120	2.851	3.455	0.313961	87.4
CRL-41	4.445	2.858	1.270	5.851	2.699	11.5	78.5	127.5	5.6	0.8969	6.416	5.755	0.369450	103.3
CRL-6	5.080	3.175	0.953	6.643	2.541	13.0	79.9	157.4	5.6	0.8079	7.918	6.397	0.369889	122.3
CRL-47-A	4.445	3.493	1.905	6.183	3.652	12.5	76.8	225.1	6.6	0.8070	9.584	7.735	0.378077	130.9
CRL-47-C	5.080	3.493	1.111	6.799	2.858	13.5	80.6	199.1	5.8	0.7846	9.584	7.520	0.403993	133.6
CRL-6-W	4.445	2.540	1.270	5.703	2.540	11.0	90.1	100.7	5.6	1.0766	5.068	5.456	0.420470	96.6
CRL-47	5.080	3.493	1.270	6.799	3.017	13.5	92.2	207.8	6.1	0.8969	9.584	8.596	0.505871	137.0
CRL-11-A	5.715	4.128	1.111	7.752	3.175	15.5	92.6	302.2	6.3	0.7846	13.385	10.502	0.519088	171.6
CRL-25	6.350	5.080	1.270	8.884	3.810	18.0	98.3	512.7	7.1	0.7177	20.271	14.549	0.587323	230.2
CRL-50-D	4.128	1.905	1.905	5.112	2.858	9.5	136.3	64.4	6.4	1.8845	2.851	5.372	0.637652	86.9
CRL-10	5.715	4.763	2.223	8.105	4.605	16.5	118.3	515.1	8.1	0.9418	17.820	16.782	0.777703	220.3
CRL-9-W	5.715	3.810	1.270	7.589	3.175	15.0	122.9	267.8	6.6	1.0766	11.402	12.276	0.800510	166.1
CRL-26	7.620	6.350	1.270	10.807	4.445	21.9	120.2	915.5	8.1	0.7177	31.673	22.733	0.802980	334.2
CRL-45-C	4.128	2.223	2.540	5.238	3.652	10.0	163.9	101.7	7.4	2.1532	3.882	8.358	0.977253	103.1
CRL-38	6.350	5.715	4.128	9.259	6.986	19.0	168.7	1066.0	11.7	1.1665	25.655	29.926	1.194990	337.7
CRL-6-A	4.445	2.540	2.540	5.703	3.810	11.0	180.3	137.3	7.6	2.1532	5.068	10.912	1.233378	119.3
CRL-42-B	10.160	8.890	1.270	14.660	5.715	29.9	163.9	2242.9	10.2	0.7177	62.080	44.557	1.259073	600.5
CRL-19-A	6.350	3.810	1.270	8.229	3.175	16.0	174.8	288.4	7.1	1.4355	11.402	16.368	1.321476	188.4
CRL-45-D	4.128	2.143	3.175	5.205	4.247	9.9	210.8	107.5	8.4	2.8046	3.607	10.117	1.353975	111.9

National Arnold Caseless Toroids 2 mil Silicon Steel

Part No.	OD cm 1	ID cm 2	HT cm 3	Wound OD cm 4	Wound HT cm 5	MPL cm 6	Wtfe grams 7	Wtcu grams 8	MLT cm 9	Ac cm sq 10	Wa cm sq 11	Ap cm 4th 12	Kg cm 5th 13	At cm sq 14
CRL-45-B	4.763	2.223	2.540	5.908	3.652	11.0	240.4	108.7	7.9	2.8710	3.882	11.144	1.625259	122.5
CRL-6-C	5.080	3.175	2.540	6.643	4.128	13.0	213.1	228.9	8.1	2.1532	7.918	17.050	1.806705	155.4
CRL-8-W	5.715	3.175	1.905	7.293	3.493	14.0	229.5	214.6	7.6	2.1532	7.918	17.050	1.927153	163.5
CRL-6-B	5.080	2.540	2.540	6.366	3.810	12.0	262.2	146.5	8.1	2.8710	5.068	14.549	2.055629	139.8
CRL-37	7.620	5.080	1.270	10.119	3.810	20.0	218.5	585.9	8.1	1.4355	20.271	29.099	2.055629	281.8
CRL-50	18.733	16.193	0.635	26.911	8.732	54.9	300.5	11720.5	16.0	0.7177	205.969	147.832	2.652226	1874.8
CRL-40-B	6.350	2.540	2.540	7.728	3.810	14.0	458.9	164.8	9.1	4.3064	5.068	21.824	4.111259	186.2
CRL-10-W	7.620	5.080	1.905	10.119	4.445	20.0	327.8	659.1	9.1	2.1532	20.271	43.648	4.111259	302.0
CRL-10-B	7.620	5.080	2.223	10.119	4.763	20.0	382.5	695.8	9.7	2.5127	20.271	50.934	5.303306	312.1
CRL-10-A	7.620	5.080	2.858	10.119	5.398	20.0	491.8	769.0	10.7	3.2304	20.271	65.483	7.931036	332.3
CRL-48	8.573	6.033	2.540	11.546	5.557	22.9	502.6	1110.4	10.9	2.8710	28.590	82.081	8.629977	410.7
CRL-14	8.890	6.350	2.540	12.021	5.715	23.9	524.5	1258.8	11.2	2.8710	31.673	90.933	9.343770	442.6
CRL-42	12.065	8.890	1.588	16.459	6.033	32.9	563.6	2691.6	12.2	2.2436	62.080	139.285	10.252145	737.1
CRL-20-B	10.160	7.620	2.858	13.932	6.668	27.9	688.5	2059.9	12.7	3.2304	45.610	147.337	14.989838	596.5
CRL-28	12.700	10.160	2.540	17.768	7.620	35.9	786.7	4101.3	14.2	2.8710	81.084	232.788	18.794325	920.8
CRL-132	33.020	30.480	1.270	48.615	16.510	99.8	1092.6	73822.6	28.4	1.4355	729.753	1047.547	21.143616	6230.9
CRL-24	10.160	5.080	2.540	12.732	5.080	23.9	1048.9	878.8	12.2	5.7419	20.271	116.394	21.926713	457.6
CRL-130	42.545	40.005	1.270	63.097	21.273	129.7	1420.4	161234.7	36.1	1.4355	1257.114	1804.564	28.728144	10465.1
CRL-43	15.240	10.160	2.540	20.237	7.620	39.9	1748.2	4687.2	16.3	5.7419	81.084	465.576	65.780139	1127.2
CRL-21	15.240	10.160	2.540	20.237	7.620	39.9	1748.2	4687.2	16.3	5.7419	81.084	465.576	65.780139	1127.2
CRL-17	12.700	7.620	3.810	16.459	7.620	31.9	2097.8	2636.5	16.3	8.6129	45.610	392.830	83.252988	819.1

National Arnold Caseless Toroids 4 mil Silicon Steel														
Part No.	OD cm 1	ID cm 2	HT cm 3	Wound OD cm 4	Wound HT cm 5	MPL cm 6	Wtfe grams 7	Wtcu grams 8	MLT cm 9	Ac cm sq 10	Wa cm sq 11	Ap cm 4th 12	Kg cm 5th 13	At cm sq 14
CRH-100-A	0.794	0.397	0.318	0.995	0.517	1.9	0.8	0.5	1.1	0.0568	0.124	0.007	0.000140	3.2
CRH-86-F	1.270	0.953	0.318	1.742	0.795	3.5	1.2	3.9	1.5	0.0454	0.713	0.032	0.000385	9.1
CRH-86-D	1.588	1.270	0.318	2.221	0.953	4.5	1.6	8.0	1.8	0.0455	1.267	0.058	0.000590	14.4
CRH-89-E	1.429	0.953	0.318	1.898	0.795	3.7	1.9	4.2	1.7	0.0681	0.713	0.049	0.000801	10.4
CRH-1-D	1.270	0.794	0.476	1.661	0.873	3.2	2.5	3.1	1.8	0.1020	0.495	0.050	0.001158	8.9
CRH-89-B	1.588	0.953	0.318	2.058	0.795	4.0	2.8	4.5	1.8	0.0909	0.713	0.065	0.001324	11.8
CRH-1-F	1.389	0.913	0.476	1.838	0.933	3.6	2.8	4.4	1.9	0.1020	0.655	0.067	0.001454	10.7
CRH-89-D	1.429	0.953	0.476	1.898	0.953	3.7	2.9	4.8	1.9	0.1020	0.713	0.073	0.001557	11.3
CRH-86-J	1.905	1.270	0.318	2.530	0.953	5.0	3.5	9.2	2.0	0.0909	1.267	0.115	0.002058	17.6
CRH-86-A	1.905	1.270	0.318	2.530	0.953	5.0	3.5	9.2	2.0	0.0909	1.267	0.115	0.002058	17.6
CRH-86-B	1.746	1.270	0.476	2.373	1.111	4.7	3.7	9.7	2.2	0.1020	1.267	0.129	0.002441	17.1
CRH-89-C	1.588	0.953	0.476	2.058	0.953	4.0	4.1	5.2	2.0	0.1360	0.713	0.097	0.002598	12.8
CRH-1-J	1.429	0.794	0.635	1.824	1.032	3.5	4.8	3.8	2.2	0.1815	0.495	0.090	0.003020	11.1
CRH-1-M	1.429	0.794	0.635	1.824	1.032	3.5	4.8	3.8	2.2	0.1815	0.495	0.090	0.003020	11.1
CRH-1-A	1.270	0.794	0.953	1.661	1.350	3.2	5.1	4.5	2.5	0.2041	0.495	0.101	0.003249	11.4
CRH-44-C	1.746	1.111	0.476	2.293	1.032	4.5	4.7	7.4	2.2	0.1360	0.970	0.132	0.003324	15.7
CRH-89-A	1.588	0.953	0.635	2.058	1.112	4.0	5.5	5.8	2.3	0.1815	0.713	0.129	0.004109	13.8
CRH-86-K	1.984	1.270	0.476	2.609	1.111	5.1	6.0	10.6	2.3	0.1529	1.267	0.194	0.005047	19.8
CRH-2-A	1.746	1.111	0.635	2.293	1.191	4.5	6.2	8.3	2.4	0.1815	0.970	0.176	0.005292	16.8
CRH-102-A	2.223	1.588	0.476	3.006	1.270	6.0	6.2	17.9	2.5	0.1360	1.981	0.269	0.005771	26.2
CRH-86-C	2.064	1.270	0.476	2.690	1.111	5.2	6.8	10.9	2.4	0.1701	1.267	0.215	0.006075	20.7
CRH-100	1.667	0.913	0.714	2.122	1.171	4.1	7.5	5.8	2.5	0.2423	0.655	0.159	0.006208	14.9
CRH-16-B	1.905	1.270	0.635	2.530	1.270	5.0	6.9	11.4	2.5	0.1815	1.267	0.230	0.006569	20.1
CRH-2-C	2.064	1.111	0.476	2.619	1.032	5.0	7.8	8.3	2.4	0.2041	0.970	0.198	0.006698	19.2
CRH-91-A	1.588	0.953	0.953	2.058	1.430	4.0	8.3	7.1	2.8	0.2723	0.713	0.194	0.007571	15.9
CRH-2	2.223	1.588	0.635	3.006	1.429	6.0	8.3	19.7	2.8	0.1815	1.981	0.359	0.009336	27.7
CRH-100-B	1.667	0.913	0.953	2.122	1.410	4.1	10.0	6.7	2.9	0.3234	0.655	0.212	0.009580	16.5
CRH-118-A	2.064	1.270	0.635	2.690	1.270	5.2	9.1	12.0	2.7	0.2269	1.267	0.287	0.009781	22.1
CRH-1-K	1.905	0.794	0.794	2.329	1.191	4.2	12.8	4.9	2.8	0.3970	0.495	0.197	0.011170	17.2
CRH-1	1.588	0.953	1.270	2.058	1.747	4.0	11.1	8.4	3.3	0.3629	0.713	0.259	0.011380	17.9

National Arnold Caseless Toroids 4 mil Silicon Steel

Part No.	OD cm 1	ID cm 2	HT cm 3	Wound OD cm 4	Wound HT cm 5	MPL cm 6	Wtfe grams 7	Wtcu grams 8	MLT cm 9	Ac cm sq 10	Wa cm sq 11	Ap cm 4th 12	Kg cm 5th 13	At cm sq 14
CRH-85	2.540	1.588	0.476	3.322	1.270	6.5	10.1	19.7	2.8	0.2039	1.981	0.404	0.011794	30.6
CRH-1-C	2.223	1.429	0.635	2.926	1.350	5.7	9.9	15.9	2.8	0.2269	1.604	0.364	0.011819	25.8
CRH-86-E	1.746	1.111	1.111	2.293	1.667	4.5	10.9	10.9	3.2	0.3175	0.970	0.308	0.012313	20.2
CRH-3-A	2.540	1.905	0.635	3.483	1.588	7.0	9.7	30.9	3.0	0.1815	2.851	0.517	0.012317	36.4
CRH-16-L	1.905	1.270	0.953	2.530	1.588	5.0	10.4	13.7	3.0	0.2723	1.267	0.345	0.012327	22.7
CRH-17-B	3.175	1.905	0.318	4.115	1.271	8.0	11.1	30.9	3.0	0.1817	2.851	0.518	0.012352	43.0
CRH-116	1.270	0.635	2.540	1.592	2.858	3.0	16.6	5.7	5.1	0.7258	0.317	0.230	0.013138	18.3
CRH-100-C	1.826	0.913	0.953	2.288	1.410	4.3	12.9	7.0	3.0	0.3915	0.655	0.256	0.013448	18.3
CRH-101	2.302	1.548	0.714	3.063	1.488	6.0	11.2	20.0	3.0	0.2423	1.882	0.456	0.014809	29.0
CRH-19-C	3.651	3.175	0.635	5.256	2.223	10.7	11.1	110.9	3.9	0.1360	7.918	1.077	0.014885	80.1
CRH-86-H	1.746	1.111	1.270	2.293	1.826	4.5	12.4	11.8	3.4	0.3629	0.970	0.352	0.014896	21.4
CRH-3-C	2.858	1.905	0.476	3.795	1.429	7.5	11.7	30.9	3.0	0.2041	2.851	0.582	0.015589	39.6
CRH-44-E	2.699	1.270	0.476	3.352	1.111	6.2	14.6	13.2	2.9	0.3061	1.267	0.388	0.016256	29.3
CRH-124	1.984	1.191	0.953	2.572	1.549	5.0	12.9	12.3	3.1	0.3401	1.114	0.379	0.016563	22.9
CRH-2-H	2.223	1.588	0.953	3.006	1.747	6.0	12.4	23.3	3.3	0.2723	1.981	0.539	0.017788	30.7
CRH-118	2.064	1.270	0.953	2.690	1.588	5.2	13.6	14.3	3.2	0.3405	1.267	0.431	0.018501	24.8
CRH-16-C	1.905	1.270	1.270	2.530	1.905	5.0	13.8	16.0	3.6	0.3629	1.267	0.460	0.018769	25.2
CRH-16-F	2.540	1.588	0.635	3.322	1.429	6.5	13.5	21.5	3.0	0.2720	1.981	0.539	0.019237	32.2
CRH-85-C	2.540	1.588	0.635	3.322	1.429	6.5	13.5	21.5	3.0	0.2720	1.981	0.539	0.019237	32.2
CRH-16	2.540	1.270	0.635	3.183	1.270	6.0	16.6	13.7	3.0	0.3629	1.267	0.460	0.021897	28.6
CRH-77-F	3.493	2.858	0.635	4.923	2.064	10.0	13.8	86.9	3.8	0.1815	6.416	1.164	0.022176	70.0
CRH-85-D	2.699	1.746	0.635	3.558	1.508	7.0	14.5	27.0	3.2	0.2723	2.395	0.652	0.022371	36.7
CRH-85-E	2.540	1.905	0.953	3.483	1.906	7.0	14.5	36.1	3.6	0.2723	2.851	0.776	0.023774	39.9
CRH-16-H	3.175	1.270	0.476	3.864	1.111	7.0	21.7	14.9	3.3	0.4081	1.267	0.517	0.025557	36.9
CRH-8-D	2.858	1.905	0.635	3.795	1.588	7.5	15.5	33.5	3.3	0.2723	2.851	0.776	0.025605	41.5
CRH-16-P	2.223	1.270	0.953	2.852	1.588	5.5	17.1	14.9	3.3	0.4087	1.267	0.518	0.025626	27.0
CRH-44-L	2.223	1.111	0.953	2.786	1.509	5.2	19.1	11.4	3.3	0.4769	0.970	0.462	0.026701	25.4
CRH-2-B	2.223	1.588	1.270	3.006	2.064	6.0	16.6	26.8	3.8	0.3629	1.981	0.719	0.027385	33.7
CRH-19-A	3.651	3.175	0.953	5.256	2.541	10.7	16.7	125.2	4.4	0.2041	7.918	1.616	0.029688	85.3
CRH-85-L	2.858	1.588	0.635	3.647	1.429	7.0	19.3	23.3	3.3	0.3629	1.981	0.719	0.031598	37.3

National Arnold Caseless Toroids 4 mil Silicon Steel														
Part No.	OD cm 1	ID cm 2	HT cm 3	Wound OD cm 4	Wound HT cm 5	MPL cm 6	Wtfe grams 7	Wtcu grams 8	MLT cm 9	Ac cm sq 10	Wa cm sq 11	Ap cm 4th 12	Kg cm 5th 13	At cm sq 14
CRH-85-J	2.381	1.746	1.270	3.244	2.143	6.5	18.0	33.5	3.9	0.3629	2.395	0.869	0.032043	38.3
CRH-118-B	2.381	1.270	0.953	3.016	1.588	5.7	20.9	15.5	3.4	0.4765	1.267	0.604	0.033544	29.3
CRH-85-H	2.540	1.588	0.953	3.322	1.747	6.5	20.2	25.1	3.6	0.4083	1.981	0.809	0.037131	35.5
CRH-71-B	3.810	2.540	0.476	5.059	1.746	10.0	20.7	68.7	3.8	0.2720	5.068	1.379	0.039377	67.9
CRH-16-M	2.223	1.270	1.270	2.852	1.905	5.5	22.8	17.2	3.8	0.5446	1.267	0.690	0.039451	29.8
CRH-71-C	3.493	2.540	0.635	4.747	1.905	9.5	19.7	68.7	3.8	0.2723	5.068	1.380	0.039451	63.8
CRH-59-D	3.572	1.905	0.476	4.525	1.429	8.6	23.4	36.7	3.6	0.3571	2.851	1.018	0.040169	52.4
CRH-96-D	2.540	1.429	0.953	3.249	1.668	6.2	22.7	20.3	3.6	0.4765	1.604	0.764	0.040950	33.6
CRH-17	3.175	1.905	0.635	4.115	1.588	8.0	22.1	36.0	3.6	0.3629	2.851	1.034	0.042229	47.1
CRH-16-D	2.540	1.270	0.953	3.183	1.588	6.0	24.9	16.0	3.6	0.5446	1.267	0.690	0.042264	31.8
CRH-3-W	2.858	1.905	0.953	3.795	1.906	7.5	23.3	38.6	3.8	0.4087	2.851	1.165	0.049972	45.3
CRH-16-K	1.905	0.953	2.223	2.388	2.700	4.5	32.6	12.9	5.1	0.9523	0.713	0.679	0.050938	29.2
CRH-99-B	3.334	1.905	0.635	4.278	1.588	8.2	25.6	37.3	3.7	0.4083	2.851	1.164	0.051619	50.1
CRH-96-C	2.699	1.429	0.953	3.415	1.668	6.5	26.9	21.0	3.7	0.5446	1.604	0.874	0.051662	36.2
CRH-85-B	2.540	1.588	1.270	3.322	2.064	6.5	26.9	28.6	4.1	0.5441	1.981	1.078	0.057711	38.9
CRH-22	6.668	6.033	0.635	9.742	3.652	20.0	27.6	645.6	6.4	0.1815	28.590	5.188	0.059291	260.7
CRH-17-A	3.175	1.905	0.794	4.115	1.747	8.0	27.6	38.6	3.8	0.4538	2.851	1.294	0.061617	49.1
CRH-116-B	4.445	0.476	1.270	5.134	1.508	7.7	133.8	3.5	5.6	2.2683	0.178	0.404	0.065548	65.7
CRH-71	3.810	2.540	0.635	5.059	1.905	10.0	27.6	73.2	4.1	0.3629	5.068	1.839	0.065690	70.5
CRH-16-A	2.540	1.270	1.270	3.183	1.905	6.0	33.1	18.3	4.1	0.7258	1.267	0.920	0.065690	34.9
CRH-41-C	4.445	3.810	0.953	6.366	2.858	13.0	26.9	206.0	5.1	0.2723	11.402	3.105	0.066571	120.8
CRH-101-A	2.937	1.667	0.953	3.764	1.787	7.2	30.1	30.1	3.9	0.5446	2.183	1.189	0.066848	43.3
CRH-102	2.937	1.865	1.032	3.855	1.965	7.5	28.7	38.9	4.0	0.4978	2.732	1.360	0.067700	47.1
CRH-102-D	3.175	1.945	0.873	4.133	1.846	8.0	29.7	41.6	3.9	0.4832	2.972	1.436	0.070496	50.8
CRH-102-C	2.937	1.905	1.111	3.874	2.064	7.6	29.9	41.8	4.1	0.5159	2.851	1.471	0.073545	48.7
CRH-67-B	4.445	2.699	0.476	5.775	1.826	11.2	32.0	87.9	4.3	0.3740	5.722	2.140	0.074148	85.5
CRH-13-D	5.080	4.128	0.635	7.144	2.699	14.5	30.0	241.8	5.1	0.2720	13.385	3.641	0.077995	140.7
CRH-67-A	4.445	2.540	0.476	5.703	1.746	11.0	34.2	77.8	4.3	0.4081	5.068	2.068	0.078174	82.3
CRH-71-D	3.493	2.540	0.953	4.747	2.223	9.5	29.6	77.8	4.3	0.4087	5.068	2.071	0.078391	68.5
CRH-44-A	2.699	1.270	1.270	3.352	1.905	6.2	38.9	18.9	4.2	0.8167	1.267	1.035	0.080644	37.7

National Arnold Caseless Toroids 4 mil Silicon Steel														
Part No.	OD cm 1	ID cm 2	HT cm 3	Wound OD cm 4	Wound HT cm 5	MPL cm 6	Wtfe grams 7	Wtcu grams 8	MLT cm 9	Ac cm sq 10	Wa cm sq 11	Ap cm 4th 12	Kg cm 5th 13	At cm sq 14
CRH-8-H	3.810	3.175	1.270	5.403	2.858	11.0	30.4	143.0	5.1	0.3629	7.918	2.874	0.082113	94.3
CRH-7-H	3.810	1.905	0.635	4.775	1.588	9.0	37.3	41.2	4.1	0.5444	2.851	1.552	0.083139	59.6
CRH-44	3.175	1.905	0.953	4.115	1.906	8.0	33.2	41.2	4.1	0.5446	2.851	1.553	0.083210	51.2
CRH-17-D	3.493	1.905	0.794	4.442	1.747	8.5	36.7	41.2	4.1	0.5674	2.851	1.617	0.090308	55.3
CRH-2-D	2.540	1.588	1.746	3.322	2.540	6.5	37.0	34.0	4.8	0.7480	1.981	1.482	0.091864	43.8
CRH-95-D	3.493	2.223	0.873	4.587	1.985	9.0	34.2	57.9	4.2	0.4989	3.882	1.937	0.092216	61.6
CRH-97-B	5.556	5.080	1.270	8.150	3.810	16.7	34.7	466.9	6.5	0.2720	20.271	5.514	0.092644	201.8
CRH-44-K	2.858	1.270	1.270	3.522	1.905	6.5	44.9	19.5	4.3	0.9075	1.267	1.150	0.096655	40.5
CRH-95-B	2.858	1.588	1.270	3.647	2.064	7.0	38.7	30.4	4.3	0.7258	1.981	1.438	0.096655	44.5
CRH-95-C	3.334	2.223	1.072	4.427	2.184	8.7	35.7	60.5	4.4	0.5359	3.882	2.080	0.101769	61.1
CRH-85-A	2.540	1.588	1.905	3.322	2.699	6.5	40.4	35.8	5.1	0.8161	1.981	1.617	0.103880	45.5
CRH-87-A	4.763	3.493	0.635	6.489	2.382	13.0	35.9	164.5	4.8	0.3629	9.584	3.478	0.104607	114.6
CRH-4-H	2.858	1.905	1.588	3.795	2.541	7.5	38.9	48.9	4.8	0.6810	2.851	1.941	0.109550	52.9
CRH-113	3.572	3.175	2.858	5.184	4.446	10.6	41.3	209.2	7.4	0.5106	7.918	4.043	0.111125	114.6
CRH-4-A	3.493	2.143	0.953	4.549	2.025	8.9	39.1	55.4	4.3	0.5789	3.607	2.088	0.111976	61.4
CRH-73-B	6.668	5.715	0.635	9.550	3.493	19.5	40.4	579.4	6.4	0.2723	25.655	6.986	0.119838	247.9
CRH-46	3.493	1.905	0.953	4.442	1.906	8.5	44.1	43.8	4.3	0.6810	2.851	1.941	0.122435	57.6
CRH-97-D	5.636	5.080	1.270	8.222	3.810	16.8	40.8	471.5	6.5	0.3178	20.271	6.441	0.125166	204.5
CRH-7	3.175	1.905	1.270	4.115	2.223	8.0	44.2	46.3	4.6	0.7258	2.851	2.069	0.131380	55.3
CRH-45	3.810	2.540	0.953	5.059	2.223	10.0	41.5	82.4	4.6	0.5446	5.068	2.760	0.131495	75.5
CRH-12-W	3.810	2.540	0.953	5.059	2.223	10.0	41.5	82.4	4.6	0.5446	5.068	2.760	0.131495	75.5
CRH-3-E	2.858	1.588	1.588	3.647	2.382	7.0	48.4	34.0	4.8	0.9075	1.981	1.798	0.135190	48.2
CRH-96-A	2.699	1.588	1.905	3.484	2.699	6.7	48.9	36.7	5.2	0.9524	1.981	1.887	0.138021	48.6
CRH-199-B	2.858	1.905	1.905	3.795	2.858	7.5	46.6	54.1	5.3	0.8170	2.851	2.329	0.142663	56.7
CRH-44-B	3.651	1.905	0.953	4.607	1.906	8.7	49.9	45.1	4.4	0.7488	2.851	2.134	0.143802	60.9
CRH-96-B	2.699	1.429	1.905	3.415	2.620	6.5	53.9	29.7	5.2	1.0887	1.604	1.746	0.146046	46.4
CRH-85-F	2.540	1.588	2.540	3.322	3.334	6.5	53.8	42.9	6.1	1.0881	1.981	2.155	0.153896	52.1
CRH-77-H	4.128	2.858	0.953	5.535	2.382	11.0	45.6	110.1	4.8	0.5446	6.416	3.494	0.157708	89.5
CRH-4	4.128	2.858	0.953	5.535	2.382	11.0	45.6	110.1	4.8	0.5446	6.416	3.494	0.157708	89.5
CRH-4-J	3.810	1.905	0.953	4.775	1.906	9.0	56.0	46.4	4.6	0.8170	2.851	2.329	0.166424	64.4

National Arnold Caseless Toroids 4 mil Silicon Steel														
Part No.	OD cm 1	ID cm 2	HT cm 3	Wound OD cm 4	Wound HT cm 5	MPL cm 6	Wtfe grams 7	Wtcu grams 8	MLT cm 9	Ac cm sq 10	Wa cm sq 11	Ap cm 4th 12	Kg cm 5th 13	At cm sq 14
CRH-3-D	2.858	1.588	1.905	3.647	2.699	7.0	58.0	37.6	5.3	1.0887	1.981	2.157	0.176053	51.8
CRH-97-C	5.755	5.080	1.270	8.331	3.810	17.0	50.1	478.3	6.6	0.3858	20.271	7.820	0.181831	208.6
CRH-95-A	3.334	2.223	1.588	4.427	2.700	8.7	52.9	71.9	5.2	0.7939	3.882	3.082	0.187918	68.3
CRH-23-A	5.080	2.540	0.635	6.366	1.905	12.0	66.3	91.5	5.1	0.7258	5.068	3.678	0.210208	101.7
CRH-12-A	3.810	2.540	1.270	5.059	2.540	10.0	55.2	91.5	5.1	0.7258	5.068	3.678	0.210208	80.5
CRH-9	3.810	2.540	1.270	5.059	2.540	10.0	55.2	91.5	5.1	0.7258	5.068	3.678	0.210208	80.5
CRH-64-A	6.985	6.350	1.270	10.224	4.445	20.9	58.0	858.2	7.6	0.3629	31.673	11.494	0.218967	306.8
CRH-85-K	3.651	1.588	1.270	4.487	2.064	8.2	74.0	34.9	5.0	1.1790	1.981	2.335	0.222376	60.7
CRH-41-A	4.286	2.540	0.953	5.540	2.223	10.7	61.3	89.3	5.0	0.7488	5.068	3.795	0.229431	86.9
CRH-7-D	3.175	1.905	1.905	4.115	2.858	8.0	66.3	56.6	5.6	1.0887	2.851	3.103	0.241859	63.5
CRH-77-C	4.128	2.858	1.270	5.535	2.699	11.0	60.8	121.7	5.3	0.7258	6.416	4.657	0.253446	95.0
CRH-3-B	2.858	1.588	2.540	3.647	3.334	7.0	77.4	44.7	6.4	1.4516	1.981	2.875	0.262909	59.1
CRH-4-W	3.810	1.905	1.270	4.775	2.223	9.0	74.6	51.5	5.1	1.0887	2.851	3.103	0.266045	69.1
CRH-41	4.445	2.540	0.953	5.703	2.223	11.0	68.4	91.6	5.1	0.8170	5.068	4.140	0.266282	90.9
CRH-68-B	3.572	2.540	1.905	4.824	3.175	9.6	64.8	106.4	5.9	0.8847	5.068	4.483	0.268648	84.6
CRH-8	4.445	3.175	1.270	6.011	2.858	12.0	66.3	157.3	5.6	0.7258	7.918	5.747	0.298591	110.7
CRH-68	3.810	2.540	1.588	5.059	2.858	10.0	69.1	100.7	5.6	0.9075	5.068	4.599	0.298737	85.6
CRH-99-A	3.334	1.905	1.905	4.278	2.858	8.2	76.9	57.9	5.7	1.2250	2.851	3.492	0.299395	67.1
CRH-80	6.350	4.445	0.635	8.540	2.858	17.0	70.4	336.4	6.1	0.5444	15.520	8.448	0.301764	191.1
CRH-7-F	3.175	1.905	2.223	4.115	3.176	8.0	77.4	61.8	6.1	1.2704	2.851	3.622	0.301860	67.6
CRH-22-A	5.715	4.445	0.953	7.924	3.176	16.0	66.3	336.5	6.1	0.5446	15.520	8.453	0.302041	177.6
CRH-103	3.889	1.865	1.349	4.843	2.282	9.0	84.7	51.2	5.3	1.2287	2.732	3.357	0.313081	71.5
CRH-81-B	3.969	2.540	1.429	5.218	2.699	10.2	71.7	98.4	5.5	0.9189	5.068	4.657	0.313406	87.0
CRH-87-B	4.763	3.493	1.270	6.489	3.017	13.0	71.8	199.1	5.8	0.7258	9.584	6.956	0.345662	127.6
CRH-7-A	3.175	1.905	2.540	4.115	3.493	8.0	88.4	66.9	6.6	1.4516	2.851	4.138	0.363822	71.7
CRH-41-B	4.286	2.540	1.270	5.540	2.540	10.7	81.6	98.4	5.5	0.9978	5.068	5.057	0.369605	92.4
CRH-123	3.929	2.421	1.588	5.121	2.799	10.0	82.0	93.1	5.7	1.0776	4.604	4.961	0.376245	86.2
CRH-18	5.080	3.175	0.953	6.643	2.541	13.0	80.8	157.4	5.6	0.8170	7.918	6.469	0.378247	122.3
CRH-68-A	3.810	2.540	1.905	5.059	3.175	10.0	82.9	109.9	6.1	1.0887	5.068	5.517	0.394141	90.6
CRH-109-A	4.128	2.381	1.429	5.307	2.620	10.2	87.7	88.5	5.6	1.1234	4.453	5.003	0.402237	87.9

National Arnold Caseless Toroids 4 mil Silicon Steel

Part No.	OD cm 1	ID cm 2	HT cm 3	Wound OD cm 4	Wound HT cm 5	MPL cm 6	Wtfe grams 7	Wtcu grams 8	MLT cm 9	Ac cm sq 10	Wa cm sq 11	Ap cm 4th 12	Kg cm 5th 13	At cm sq 14
CRH-114	4.207	2.262	1.349	5.337	2.480	10.2	91.6	78.9	5.5	1.1807	4.019	4.745	0.405719	86.3
CRH-4-E	4.128	2.858	1.905	5.535	3.334	11.0	91.2	144.9	6.4	1.0887	6.416	6.985	0.479018	106.0
CRH-87-D	4.128	3.493	3.810	5.885	5.557	12.0	99.5	320.3	9.4	1.0887	9.584	10.434	0.483473	157.1
CRH-69	5.953	4.763	1.270	8.329	3.652	16.8	87.4	430.5	6.8	0.6801	17.820	12.119	0.485224	204.4
CRH-81-A	3.969	2.540	1.905	5.218	3.175	10.2	95.6	112.1	6.2	1.2250	5.068	6.208	0.488809	94.8
CRH-98-A	5.715	3.810	0.953	7.589	2.858	15.0	93.3	247.2	6.1	0.8170	11.402	9.315	0.499292	158.5
CRH-68-F	4.683	2.540	1.270	5.950	2.540	11.3	106.0	104.1	5.8	1.2247	5.068	6.207	0.526190	103.0
CRH-111	5.874	1.905	0.953	7.016	1.906	12.2	158.7	63.1	6.2	1.7021	2.851	4.852	0.530762	119.3
CRH-63	6.033	4.763	1.270	8.405	3.652	17.0	93.9	434.6	6.9	0.7258	17.820	12.934	0.547500	207.3
CRH-95-K	3.969	2.223	1.905	5.073	3.017	9.7	111.1	85.9	6.2	1.4968	3.882	5.810	0.558952	88.5
CRH-56	6.033	4.128	0.953	8.065	3.017	16.0	99.5	302.3	6.4	0.8170	13.385	10.935	0.562640	178.5
CRH-44-F	3.929	2.778	2.540	5.298	3.929	10.5	105.8	155.4	7.2	1.3156	6.062	7.975	0.582301	109.4
CRH-59-B	3.572	1.905	2.540	4.525	3.493	8.6	125.1	70.2	6.9	1.9054	2.851	5.431	0.598072	81.8
CRH-58-A	6.350	5.080	1.270	8.884	3.810	18.0	99.4	512.7	7.1	0.7258	20.271	14.713	0.600595	230.2
CRH-68-D	3.810	2.540	2.540	5.059	3.810	10.0	110.5	128.2	7.1	1.4516	5.068	7.356	0.600595	100.7
CRH-5-A	5.080	3.175	1.270	6.643	2.858	13.0	107.7	171.6	6.1	1.0887	7.918	8.621	0.615845	128.9
CRH-59-L	4.445	1.905	1.588	5.452	2.541	10.0	138.2	61.8	6.1	1.8151	2.851	5.174	0.616152	90.2
CRH-67	4.445	2.540	1.588	5.703	2.858	11.0	114.0	109.9	6.1	1.3613	5.068	6.899	0.616152	102.3
CRH-27-A	13.018	12.065	0.953	19.196	6.986	39.4	122.9	4854.4	11.9	0.4087	114.341	46.730	0.639856	999.6
CRH-4-F	4.128	2.143	1.905	5.205	2.977	9.9	127.9	81.5	6.4	1.7016	3.607	6.138	0.657939	91.2
CRH-59-N	4.366	1.905	1.746	5.367	2.699	9.9	145.3	63.7	6.3	1.9336	2.851	5.512	0.678158	90.7
CRH-4-C	4.128	2.143	1.984	5.205	3.056	9.9	133.2	83.1	6.5	1.7722	3.607	6.393	0.699712	92.5
CRH-79	5.080	2.540	1.270	6.366	2.540	12.0	132.6	109.9	6.1	1.4516	5.068	7.356	0.700694	114.4
CRH-59-C	4.286	2.540	1.905	5.540	3.175	10.7	122.5	116.7	6.5	1.4968	5.068	7.585	0.701159	103.4
CRH-43-D	5.080	3.810	1.905	6.966	3.810	14.0	116.0	288.4	7.1	1.0887	11.402	12.414	0.760128	159.5
CRH-59-A	3.810	1.905	2.540	4.775	3.493	9.0	149.2	72.1	7.1	2.1774	2.851	6.207	0.760128	88.2
CRH-98-C	5.715	3.810	1.270	7.589	3.175	15.0	124.3	267.8	6.6	1.0887	11.402	12.414	0.818600	166.1
CRH-4-L	4.366	1.905	1.984	5.367	2.937	9.9	165.2	67.6	6.7	2.1972	2.851	6.263	0.825628	94.7
CRH-59-J	4.683	2.778	1.746	6.055	3.135	11.7	133.9	141.0	6.5	1.4968	6.062	9.073	0.830609	117.2
CRH-88-A	8.255	7.303	1.588	11.959	5.240	24.4	126.9	1362.3	9.1	0.6803	41.894	28.500	0.848075	421.3

National Arnold Caseless Toroids 4 mil Silicon Steel														
Part No.	OD cm 1	ID cm 2	HT cm 3	Wound OD cm 4	Wound HT cm 5	MPL cm 6	Wtfe grams 7	Wtcu grams 8	MLT cm 9	Ac cm sq 10	Wa cm sq 11	Ap cm 4th 12	Kg cm 5th 13	At cm sq 14
CRH-5-F	5.874	3.969	1.270	7.827	3.255	15.5	128.5	296.2	6.7	1.0887	12.374	13.472	0.871562	176.2
CRH-8-C	4.445	3.175	2.540	6.011	4.128	12.0	132.6	214.6	7.6	1.4516	7.918	11.494	0.875868	134.6
CRH-22-B	5.398	4.445	2.540	7.625	4.763	15.5	128.5	462.6	8.4	1.0893	15.520	16.906	0.878738	205.3
CRH-10	5.080	3.175	1.588	6.643	3.176	13.0	134.7	186.0	6.6	1.3613	7.918	10.779	0.888689	135.5
CRH-68-E	4.128	2.540	2.540	5.379	3.810	10.5	145.1	132.7	7.4	1.8151	5.068	9.198	0.906592	109.8
CRH-119	3.294	2.302	5.715	4.428	6.866	8.8	171.1	174.4	11.8	2.5512	4.163	10.619	0.919987	126.2
CRH-115-A	4.921	3.016	1.746	6.407	3.254	12.5	142.4	171.0	6.7	1.4968	7.145	10.695	0.951331	129.9
CRH-115	4.921	3.016	1.746	6.407	3.254	12.5	142.4	171.0	6.7	1.4968	7.145	10.695	0.951331	129.9
CRH-5-H	5.874	3.969	1.349	7.827	3.334	15.5	136.4	301.7	6.9	1.1564	12.374	14.310	0.965240	178.1
CRH-59-E	4.763	1.905	1.905	5.797	2.858	10.5	195.8	69.5	6.9	2.4500	2.851	6.984	0.997959	104.8
CRH-95-E	4.128	2.223	2.540	5.238	3.652	10.0	165.8	101.7	7.4	2.1774	3.882	8.452	0.999338	103.1
CRH-76	4.128	2.858	3.175	5.535	4.604	11.0	151.9	191.2	8.4	1.8145	6.416	11.642	1.008050	128.1
CRH-5	5.715	3.175	1.270	7.293	2.858	14.0	154.7	186.0	6.6	1.4516	7.918	11.494	1.010617	149.0
CRH-80-B	6.350	4.445	1.270	8.540	3.493	17.0	140.9	392.5	7.1	1.0887	15.520	16.897	1.034619	208.1
CRH-120	5.715	2.858	1.270	7.162	2.699	13.5	167.8	150.7	6.6	1.6328	6.416	10.476	1.036040	141.2
CRH-34-C	8.890	7.620	1.270	12.732	5.080	25.9	143.6	1483.0	9.1	0.7258	45.610	33.104	1.051041	457.6
CRH-68-C	3.810	2.540	3.810	5.059	5.080	10.0	165.7	164.8	9.1	2.1774	5.068	11.035	1.051041	120.9
CRH-4-D	5.278	2.143	1.588	6.434	2.660	11.7	199.3	86.8	6.8	2.2403	3.607	8.081	1.070778	118.7
CRH-95-F	5.239	2.223	1.588	6.419	2.700	11.7	192.8	92.9	6.7	2.1552	3.882	8.366	1.071346	119.1
CRH-59	4.763	2.540	1.905	6.033	3.175	11.5	166.8	123.6	6.9	1.9057	5.068	9.657	1.073358	117.3
CRH-59-H	4.445	2.223	2.223	5.571	3.335	10.5	177.7	98.2	7.1	2.2228	3.882	8.628	1.078541	107.0
CRH-202	4.128	2.143	2.699	5.205	3.771	9.9	181.2	97.8	7.6	2.4109	3.607	8.697	1.100530	104.2
CRH-78-E	5.715	3.572	1.429	7.473	3.215	14.6	153.4	244.4	6.9	1.3781	10.022	13.811	1.110044	163.1
CRH-5-E	5.874	3.175	1.270	7.459	2.858	14.2	167.3	189.5	6.7	1.5425	7.918	12.214	1.119539	154.3
CRH-98-D	5.398	3.810	1.905	7.276	3.810	14.5	150.3	298.7	7.4	1.3613	11.402	15.522	1.147405	170.1
CRH-4-B	5.080	1.905	1.905	6.142	2.858	11.0	227.9	72.1	7.1	2.7218	2.851	7.759	1.187700	114.3
CRH-112	5.358	3.413	1.746	7.037	3.453	13.8	160.7	230.4	7.1	1.5282	9.150	13.983	1.207253	154.0
CRH-32	8.890	4.128	0.635	11.019	2.699	20.5	212.3	386.9	8.1	1.3607	13.385	18.214	1.219700	284.0
CRH-37-A	5.715	4.445	2.223	7.924	4.446	16.0	154.7	448.6	8.1	1.2704	15.520	19.717	1.232635	209.2
CRH-104	5.874	3.135	1.349	7.442	2.917	14.2	179.6	188.3	6.9	1.6627	7.720	12.836	1.244922	155.1

National Arnold Caseless Toroids 4 mil Silicon Steel														
Part No.	OD cm 1	ID cm 2	HT cm 3	Wound OD cm 4	Wound HT cm 5	MPL cm 6	Wtfe grams 7	Wtcu grams 8	MLT cm 9	Ac cm sq 10	Wa cm sq 11	Ap cm 4th 12	Kg cm 5th 13	At cm sq 14
CRH-28-B	4.128	2.223	3.016	5.238	4.128	10.0	196.8	112.2	8.1	2.5855	3.882	10.036	1.276965	111.0
CRH-50-C	10.160	8.890	1.270	14.660	5.715	29.9	165.7	2242.9	10.2	0.7258	62.080	45.058	1.287526	600.5
CRH-117	4.128	2.858	3.810	5.535	5.239	11.0	182.3	214.4	9.4	2.1774	6.416	13.970	1.294670	139.2
CRH-4-K	4.445	2.143	2.540	5.539	3.612	10.3	207.8	97.7	7.6	2.6312	3.607	9.492	1.310988	111.0
CRH-6-A	6.350	3.810	1.270	8.229	3.175	16.0	176.8	288.4	7.1	1.4516	11.402	16.552	1.351339	188.4
CRH-39	6.350	3.810	1.270	8.229	3.175	16.0	176.8	288.4	7.1	1.4516	11.402	16.552	1.351339	188.4
CRH-6-C	5.398	3.493	1.905	7.116	3.652	14.0	174.0	251.1	7.4	1.6331	9.584	15.651	1.387883	161.1
CRH-95-J	4.445	2.143	2.699	5.539	3.771	10.3	220.8	101.0	7.9	2.7959	3.607	10.086	1.432434	113.8
CRH-5-B	6.350	3.175	1.270	7.958	2.858	15.0	207.2	200.3	7.1	1.8145	7.918	14.368	1.466297	170.8
CRH-59-F	4.286	2.540	3.175	5.540	4.445	10.7	204.1	153.3	8.5	2.4946	5.068	12.642	1.482539	125.5
CRH-5-I	6.033	2.858	1.429	7.499	2.858	14.0	217.6	162.3	7.1	2.0417	6.416	13.100	1.504066	155.6
CRH-28-A	4.763	2.858	2.540	6.173	3.969	12.0	198.9	179.7	7.9	2.1774	6.416	13.970	1.545239	136.8
CRH-43-B	5.080	3.810	3.175	6.966	5.080	14.0	193.4	370.8	9.1	1.8145	11.402	20.690	1.642252	187.3
CRH-3-F	10.478	1.429	1.270	12.133	1.985	18.7	738.1	59.4	10.4	5.1715	1.604	8.295	1.647669	306.7
CRH-7-E	6.985	4.445	1.270	9.172	3.493	18.0	198.9	420.5	7.6	1.4516	15.520	22.529	1.716701	232.7
CRH-5-C	6.350	3.175	1.429	7.958	3.017	15.0	233.1	207.4	7.4	2.0417	7.918	16.167	1.792319	174.8
CRH-43-A	5.080	3.175	2.540	6.643	4.128	13.0	215.5	228.9	8.1	2.1774	7.918	17.241	1.847534	155.4
CRH-58	6.350	5.080	2.540	8.884	5.080	18.0	198.9	659.1	9.1	1.4516	20.271	29.425	1.868518	265.6
CRH-59-K	4.286	2.540	3.810	5.540	5.080	10.7	244.9	171.6	9.5	2.9935	5.068	15.170	1.907134	136.6
CRH-44-J	4.088	2.778	5.080	5.455	6.469	10.8	246.5	245.7	11.4	2.9947	6.062	18.153	1.907752	157.5
CRH-44-H	3.850	2.778	6.985	5.221	8.374	10.4	267.7	307.3	14.3	3.3696	6.062	20.426	1.931170	180.1
CRH-16-W	5.715	3.175	1.905	7.293	3.493	14.0	232.0	214.6	7.6	2.1774	7.918	17.241	1.970703	163.5
CRH-43	6.350	3.810	1.588	8.229	3.493	16.0	221.1	309.0	7.6	1.8151	11.402	20.696	1.971737	196.6
CRH-37-B	6.668	4.445	1.588	8.854	3.811	17.5	211.6	434.6	7.9	1.5886	15.520	24.654	1.989267	229.0
CRH-80-A	6.350	4.445	1.905	8.540	4.128	17.0	211.3	448.6	8.1	1.6331	15.520	25.345	2.036906	225.2
CRH-97-A	7.620	5.080	1.270	10.119	3.810	20.0	221.0	585.9	8.1	1.4516	20.271	29.425	2.102083	281.8
CRH-5-D	6.350	3.175	1.588	7.958	3.176	15.0	259.0	214.6	7.6	2.2689	7.918	17.966	2.139472	178.8
CRH-78-F	7.303	3.175	1.270	8.974	2.858	16.5	296.3	221.7	7.9	2.3592	7.918	18.681	2.238659	207.0
CRH-98-B	5.715	3.810	2.540	7.589	4.445	15.0	248.6	350.2	8.6	2.1774	11.402	24.828	2.503952	196.3
CRH-8-W	7.620	4.445	1.270	9.818	3.493	19.0	262.4	448.6	8.1	1.8145	15.520	28.161	2.514699	259.0

National Arnold Caseless Toroids 4 mil Silicon Steel														
Part No.	OD cm 1	ID cm 2	HT cm 3	Wound OD cm 4	Wound HT cm 5	MPL cm 6	Wtfe grams 7	Wtcu grams 8	MLT cm 9	Ac cm sq 10	Wa cm sq 11	Ap cm 4th 12	Kg cm 5th 13	At cm sq 14
CRH-90-A	6.350	3.969	1.905	8.304	3.890	16.2	252.5	357.6	8.1	2.0411	12.374	25.257	2.536988	209.7
CRH-33	8.890	4.128	0.953	11.019	3.017	20.5	318.7	411.1	8.6	2.0422	13.385	27.335	2.585368	295.0
CRH-6-W	6.350	3.810	1.905	8.229	3.810	16.0	265.2	329.6	8.1	2.1774	11.402	24.828	2.660449	204.8
CRH-6	9.049	7.303	1.588	12.696	5.240	25.7	244.6	1457.0	9.8	1.2477	41.894	52.270	2.667377	462.0
CRH-78-J	6.350	3.493	1.905	8.089	3.652	15.5	289.0	277.0	8.1	2.4492	9.584	23.473	2.829144	195.5
CRH-78	5.715	3.493	2.540	7.436	4.287	14.5	280.3	294.3	8.6	2.5397	9.584	24.341	2.863329	186.9
CRH-63-C	6.985	4.445	1.746	9.172	3.969	18.0	273.4	462.6	8.4	1.9957	15.520	30.973	2.949876	246.4
CRH-78-A	5.715	3.175	2.540	7.293	4.128	14.0	309.4	243.2	8.6	2.9032	7.918	22.989	3.091298	178.0
CRH-11-B	7.144	4.604	1.746	9.409	4.048	18.5	281.0	503.8	8.5	1.9957	16.650	33.228	3.117378	258.6
CRH-72-A	6.350	4.445	2.540	8.540	4.763	17.0	281.8	504.6	9.1	2.1774	15.520	33.793	3.218814	242.2
CRH-6-B	6.668	3.493	1.905	8.420	3.652	16.0	331.5	285.7	8.4	2.7218	9.584	26.085	3.387953	207.9
CRH-8-B	8.255	4.445	1.270	10.475	3.493	20.0	331.5	476.6	8.6	2.1774	15.520	33.793	3.408156	287.1
CRH-63-A	6.985	4.445	1.905	9.172	4.128	18.0	298.3	476.6	8.6	2.1774	15.520	33.793	3.408156	250.9
CRH-37	6.668	4.445	2.223	8.854	4.446	17.5	296.2	490.7	8.9	2.2238	15.520	34.513	3.452806	246.7
CRH-11-W	6.668	4.763	2.540	9.017	4.922	18.0	298.4	595.6	9.4	2.1774	17.820	38.802	3.595803	267.0
CRH-27-B	14.923	12.383	0.873	21.133	7.065	42.9	326.6	5711.6	13.3	0.9978	120.448	120.187	3.597322	1170.0
CRH-13-A	5.715	4.128	3.810	7.752	5.874	15.5	321.0	507.8	10.7	2.7209	13.385	36.420	3.715621	237.3
CRH-9-A	6.985	3.810	1.905	8.884	3.810	17.0	352.2	350.2	8.6	2.7218	11.402	31.035	3.912424	230.2
CRH-65-A	6.985	5.080	2.540	9.493	5.080	19.0	314.9	695.7	9.7	2.1774	20.271	44.138	3.982894	292.9
CRH-29	10.160	5.080	0.953	12.732	3.493	23.9	398.0	695.8	9.7	2.1786	20.271	44.161	3.986746	394.2
CRH-9-D	7.620	5.080	1.905	10.119	4.445	20.0	331.5	659.1	9.1	2.1774	20.271	44.138	4.204166	302.0
CRH-34	8.890	4.128	1.270	11.019	3.334	20.5	424.7	435.2	9.1	2.7215	13.385	36.428	4.336712	306.0
CRH-34-B	6.985	4.128	2.223	9.024	4.287	17.5	380.7	435.3	9.1	2.8580	13.385	38.255	4.782289	249.3
CRH-43-C	5.715	3.175	3.493	7.293	5.081	14.0	425.4	286.1	10.2	3.9925	7.918	31.614	4.968844	199.9
CRH-34-A	7.620	4.128	1.905	9.679	3.969	18.5	421.5	435.2	9.1	2.9935	13.385	40.069	5.247021	267.7
CRH-7-W	6.985	4.445	2.540	9.172	4.763	18.0	397.8	532.7	9.7	2.9032	15.520	45.058	5.421161	269.2
CRH-65-D	7.620	5.080	2.223	10.119	4.763	20.0	386.8	695.8	9.7	2.5409	20.271	51.506	5.423151	312.1
CRH-84-C	12.700	10.160	1.270	17.768	6.350	35.9	397.8	3515.4	12.2	1.4516	81.084	117.702	5.605554	849.9
CRH-84-C	12.700	10.160	1.270	17.768	6.350	35.9	397.8	3515.4	12.2	1.4516	81.084	117.702	5.605554	849.9
CRH-6-D	5.398	3.493	5.080	7.116	6.827	14.0	464.1	424.2	12.4	4.3548	9.584	41.736	5.841199	232.0

National Arnold Caseless Toroids 4 mil Silicon Steel														
Part No.	OD cm 1	ID cm 2	HT cm 3	Wound OD cm 4	Wound HT cm 5	MPL cm 6	Wtfe grams 7	Wtcu grams 8	MLT cm 9	Ac cm sq 10	Wa cm sq 11	Ap cm 4th 12	Kg cm 5th 13	At cm sq 14
CRH-79-D	6.350	2.540	3.175	7.728	4.445	14.0	580.1	183.1	10.2	5.4435	5.068	27.586	5.912108	201.6
CRH-57-A	8.890	6.350	1.905	12.021	5.080	23.9	397.8	1144.3	10.2	2.1774	31.673	68.966	5.912108	418.6
CRH-9-W	8.255	5.080	1.905	10.757	4.445	20.9	435.1	695.7	9.7	2.7218	20.271	55.173	6.223271	331.8
CRH-7-B	6.985	3.810	2.540	8.884	4.445	17.0	469.6	391.4	9.7	3.6290	11.402	41.380	6.223271	247.9
CRH-8-J	7.620	4.445	2.223	9.818	4.446	19.0	459.3	532.7	9.7	3.1761	15.520	49.293	6.487656	288.4
CRH-13-C	6.985	4.445	2.858	9.172	5.081	18.0	447.6	560.8	10.2	3.2667	15.520	50.699	6.519867	278.4
CRH-31	10.160	5.080	1.270	12.732	3.810	23.9	530.4	732.4	10.2	2.9032	20.271	58.851	6.726665	406.8
CRH-65	7.620	5.080	2.540	10.119	5.080	20.0	442.0	732.4	10.2	2.9032	20.271	58.851	6.726665	322.2
CRH-8-A	7.620	2.540	2.540	9.121	3.810	16.0	707.1	183.1	10.2	5.8064	5.068	29.425	6.726665	239.7
CRH-125-A	8.731	6.509	2.381	11.951	5.636	23.9	434.9	1277.4	10.8	2.3808	33.279	79.230	6.989855	435.7
CRH-79-E	5.715	2.540	4.445	7.042	5.715	13.0	628.4	210.6	11.7	6.3508	5.068	32.184	6.997422	204.2
CRH-106	8.414	5.040	1.984	10.901	4.504	21.1	485.8	702.8	9.9	3.0123	19.953	60.104	7.311138	340.7
CRH-26-B	15.240	12.700	1.270	21.613	7.620	43.9	486.2	6408.2	14.2	1.4516	126.693	183.909	7.507438	1250.6
CRH-29-A	6.350	3.810	3.810	8.229	5.715	16.0	530.4	453.2	11.2	4.3548	11.402	49.655	7.739487	254.0
CRH-78-B	5.715	3.493	5.080	7.436	6.827	14.5	560.6	432.8	12.7	5.0795	9.584	48.681	7.788256	246.2
CRH-65-C	7.620	5.080	2.858	10.119	5.398	20.0	497.3	769.0	10.7	3.2667	20.271	66.219	8.110263	332.3
CRH-11-C	7.898	4.564	2.381	10.157	4.663	19.6	533.6	589.3	10.1	3.5722	16.362	58.449	8.246103	310.7
CRH-126-A	7.898	4.564	2.381	10.157	4.663	19.6	533.6	589.3	10.1	3.5722	16.362	58.449	8.246103	310.7
CRH-9-B	6.985	3.810	3.175	8.884	5.080	17.0	587.0	432.6	10.7	4.5363	11.402	51.724	8.797779	265.6
CRH-75	8.573	6.033	2.540	11.546	5.557	22.9	508.3	1110.4	10.9	2.9032	28.590	83.003	8.824999	410.7
CRH-35	8.890	6.350	2.540	12.021	5.715	23.9	530.4	1258.8	11.2	2.9032	31.673	91.955	9.554922	442.6
CRH-57	8.890	6.350	2.540	12.021	5.715	23.9	530.4	1258.8	11.2	2.9032	31.673	91.955	9.554922	442.6
CRH-71-A	7.303	3.969	3.175	9.282	5.160	17.7	643.6	480.6	10.9	4.7635	12.374	58.943	10.282388	285.6
CRH-15-C	9.208	6.668	2.540	12.499	5.874	24.9	552.5	1419.6	11.4	2.9032	34.925	101.395	10.301390	475.8
CRH-7-C	8.255	4.445	2.540	10.475	4.763	20.0	662.9	588.8	10.7	4.3548	15.520	67.587	11.035935	328.9
CRH-93-A	8.890	4.921	2.223	11.338	4.684	21.7	657.3	721.7	10.7	3.9704	19.022	75.524	11.242515	368.5
CRH-17-W	8.890	4.445	2.223	11.141	4.446	20.9	710.8	588.8	10.7	4.4466	15.520	69.010	11.504867	350.4
CRH-12-B	9.208	6.350	2.540	12.334	5.715	24.4	609.2	1287.4	11.4	3.2667	31.673	103.467	11.827940	460.2
CRH-10-W	10.160	6.350	1.905	13.286	5.080	25.9	646.4	1258.8	11.2	3.2661	31.673	103.449	12.092948	489.0
CRH-66	10.160	7.620	2.540	13.932	6.350	27.9	618.7	1977.4	12.2	2.9032	45.610	132.415	12.612497	582.5

National Arnold Caseless Toroids 4 mil Silicon Steel

Part No.	OD cm 1	ID cm 2	HT cm 3	Wound OD cm 4	Wound HT cm 5	MPL cm 6	Wtfe grams 7	Wtcu grams 8	MLT cm 9	Ac cm sq 10	Wa cm sq 11	Ap cm 4th 12	Kg cm 5th 13	At cm sq 14
CRH-74	10.160	5.080	1.905	12.732	4.445	23.9	795.5	805.6	11.2	4.3548	20.271	88.276	13.759087	432.2
CRH-20-C	9.525	7.620	3.810	13.326	7.620	26.9	671.2	2224.6	13.7	3.2661	45.610	148.966	14.189059	597.6
CRH-12	9.525	6.350	2.540	12.648	5.715	24.9	690.6	1316.0	11.7	3.6290	31.673	114.943	14.280454	478.1
CRH-13	6.985	4.445	5.080	9.172	7.303	18.0	795.5	757.0	13.7	5.8064	15.520	90.116	15.259564	342.4
CRH-32-A	7.620	4.128	3.810	9.679	5.874	18.5	843.1	580.3	12.2	5.9870	13.385	80.138	15.741063	325.6
CRH-78-H	7.620	3.493	3.810	9.428	5.557	17.5	942.5	415.5	12.2	7.0757	9.584	67.813	15.742450	304.1
CRH-21	9.525	4.445	2.540	11.814	4.763	21.9	972.3	644.8	11.7	5.8064	15.520	90.116	17.913401	395.8
CRH-15	8.890	6.350	3.810	12.021	6.985	23.9	795.5	1487.6	13.2	4.3548	31.673	137.932	18.191101	490.6
CRH-21-A	8.255	4.445	3.810	10.475	6.033	20.0	994.4	700.9	12.7	6.5322	15.520	101.380	20.857916	370.7
CRH-62-A	8.890	5.715	3.810	11.701	6.668	22.9	953.0	1205.0	13.2	5.4435	25.655	139.656	23.023112	459.9
CRH-66-B	12.700	7.620	1.905	16.459	5.715	31.9	1060.7	2142.2	13.2	4.3548	45.610	198.622	26.195185	720.7
CRH-60	8.890	5.080	3.810	11.407	6.350	21.9	1093.9	952.1	13.2	6.5322	20.271	132.415	26.195185	431.7
CRH-70-A	8.890	3.810	3.810	10.905	5.715	20.0	1325.9	535.5	13.2	8.7097	11.402	99.311	26.195185	382.4
CRH-72	9.525	6.350	3.810	12.648	6.985	24.9	1035.8	1544.8	13,7	5.4435	31.673	172.415	27.370869	528.6
CRH-32-B	9.049	4.128	3.810	11.189	5.874	20.7	1332.6	634.7	13.3	8.4371	13.385	112.932	28.580363	402.9
CRH-18-W	11.430	6.350	2.540	14.587	5.715	27.9	1237.5	1487.6	13.2	5.8064	31.673	183.909	32.339735	595.8
CRH-62-B	10.795	5.715	2.858	13.657	5.716	25.9	1293.0	1205.0	13.2	6.5334	25.655	167.617	33.162876	537.9
CRH-14	8.890	6.033	5.715	11.858	8.732	23.4	1314.3	1652.7	16.3	7.3475	28.590	210.064	37.978440	545.9
CRH-79-F	8.890	2.540	6.350	10.532	7.620	18.0	2486.0	311.3	17.3	18.1451	5.068	91.955	38.641227	426.1
CRH-62	10.160	5.715	3.810	12.997	6.668	24.9	1450.2	1297.7	14.2	7.6210	25.655	195.519	41.902064	537.3
CRH-66-A	12.700	7.620	2.540	16.459	6.350	31.9	1414.3	2307.0	14.2	5.8064	45.610	264.829	43.242846	753.5
CRH-38-A	9.525	3.810	5.080	11.592	6.985	20.9	2088.3	638.5	15.7	13.0645	11.402	148.966	49.432850	465.2
CRH-21-C	11.430	7.620	3.810	15.178	7.620	29.9	1491.6	2471.7	15.2	6.5322	45.610	297.933	51.080611	724.8
CRH-3	9.843	5.556	5.080	12.600	7.858	24.2	1808.9	1379.8	16.0	9.8001	24.248	237.630	58.210986	560.1
CRH-24	11.430	5.715	3.810	14.324	6.668	26.9	2013.7	1390.4	15.2	9.7984	25.655	251.381	64.648899	622.0
CRH-20-D	12.700	8.890	3.810	17.079	8.255	33.9	1690.5	3588.6	16.3	6.5322	62.080	405.520	65.180988	900.7
CRH-84-B	15.240	10.160	2.540	20.237	7.620	39.9	1767.8	4687.2	16.3	5.8064	81.084	470.808	67.266649	1127.2
CRH-20-A	12.700	7.620	3.810	16.459	7.620	31.9	2121.4	2636.5	16.3	8.7097	45.610	397.244	85.134352	819.1
CRH-82-A	13.335	10.160	5.080	18.373	10.160	36.9	2044.1	5419.5	18.8	7.2581	81.084	588.510	90.900877	1116.1
CRH-13-F	15.240	8.890	2.858	19.635	7.303	37.9	2362.1	3700.9	16.8	8.1667	62.080	506.989	98.788923	1055.6

National Arnold Caseless Toroids 4 mil Silicon Steel

Part No.	OD cm 1	ID cm 2	HT cm 3	Wound OD cm 4	Wound HT cm 5	MPL cm 6	Wtfe grams 7	Wtcu grams 8	MLT cm 9	Ac cm sq 10	Wa cm sq 11	Ap cm 4th 12	Kg cm 5th 13	At cm sq 14
CRH-94-A	13.653	8.890	4.128	18.025	8.573	35.4	2390.8	3869.2	17.5	8.8477	62.080	549.266	110.907963	995.3
CRH-84-A	15.240	10.160	3.810	20.237	8.890	39.9	2651.8	5273.0	18.3	8.7097	81.084	706.212	134.533297	1207.9
CRH-92-A	14.605	9.525	5.080	19.290	9.843	37.9	3358.9	5020.7	19.8	11.6129	71.265	827.592	194.038410	1180.3
CRH-11	13.970	7.620	5.080	17.768	8.890	33.9	3756.7	3130.9	19.3	14.5161	45.610	662.073	199.144684	991.6
CRH-82-B	15.240	10.160	5.080	20.237	10.160	39.9	3535.7	5858.9	20.3	11.6129	81.084	941.615	215.253276	1288.6
CRH-11-A	15.240	5.080	5.080	18.242	7.620	31.9	5657.1	1464.7	20.3	23.2258	20.271	470.808	215.253276	958.9
CRH-92-B	14.605	7.620	5.080	18.431	8.890	34.9	4253.9	3213.3	19.8	15.9677	45.610	728.281	234.786476	1047.8
CRH-25-A	15.240	7.620	5.080	19.099	8.890	35.9	4773.2	3295.7	20.3	17.4193	45.610	794.488	272.429927	1105.8
CRH-92-C	15.875	9.525	5.080	20.574	9.843	39.9	4419.6	5278.2	20.8	14.5161	71.265	1034.490	288.395502	1300.4
CRH-84-D	16.510	10.160	5.080	21.515	10.160	41.9	4640.6	6151.9	21.3	14.5161	81.084	1177.019	320.317375	1413.1
CRH-25	16.510	7.620	5.080	20.447	8.890	37.9	5878.1	3460.4	21.3	20.3225	45.610	926.903	353.149906	1227.2
CRH-66-C	17.780	7.620	5.080	21.810	8.890	39.9	7071.4	3625.2	22.4	23.2258	45.610	1059.317	440.290792	1355.6
CRH-53-E	21.590	15.240	5.398	29.100	13.018	57.9	6809.6	16808.4	25.9	15.4248	182.438	2814.072	670.142335	2519.0
CRH-53-D	21.590	15.240	5.715	29.100	13.335	57.9	7209.5	17137.4	26.4	16.3306	182.438	2979.330	736.739587	2548.0
CRH-40-C	24.765	6.350	5.080	29.173	8.255	48.9	15700.7	3146.9	27.9	42.0967	31.673	1333.342	803.568914	2092.4
CRH-40-D	19.050	6.350	7.620	22.802	10.795	39.9	13258.8	3089.7	27.4	43.5483	31.673	1379.319	875.867822	1589.2
CRH-53-B	24.130	15.240	6.350	31.629	13.970	61.9	11988.2	19114.8	29.5	25.4032	182.438	4634.513	1598.307763	2958.1
CRH-53-C	29.210	15.240	6.985	36.861	14.605	69.8	23396.3	22410.4	34.5	43.9112	182.438	8011.087	4073.372719	3823.7

National Arnold Caseless Toroids 4 mil Silicon Steel														
Part No.	OD cm 1	ID cm 2	HT cm 3	Wound OD cm 4	Wound HT cm 5	MPL cm 6	Wtfe grams 7	Wtcu grams 8	MLT cm 9	Ac cm sq 10	Wa cm sq 11	Ap cm 4th 12	Kg cm 5th 13	At cm sq 14
CRZ-10-L	0.953	0.476	0.318	1.194	0.556	2.2	1.2	0.8	1.3	0.0683	0.178	0.012	0.000261	4.3
CRZ-49	1.270	0.635	0.318	1.592	0.636	3.0	2.1	1.7	1.5	0.0909	0.317	0.029	0.000686	7.2
CRZ-10-C	1.588	0.953	0.318	2.058	0.795	4.0	2.8	4.5	1.8	0.0909	0.713	0.065	0.001324	11.8
CRZ-10-A	1.905	1.270	0.318	2.530	0.953	5.0	3.5	9.2	2.0	0.0909	1.267	0.115	0.002058	17.6
CRZ-9-B	1.746	1.270	0.476	2.373	1.111	4.7	3.7	9.7	2.2	0.1020	1.267	0.129	0.002441	17.1
CRZ-9-C	1.429	1.111	0.953	1.981	1.509	4.0	4.2	9.2	2.7	0.1364	0.970	0.132	0.002703	15.5
CRZ-3-B	2.223	1.588	0.318	3.006	1.112	6.0	4.2	16.1	2.3	0.0909	1.981	0.180	0.002860	24.7
CRZ-10-B	2.223	1.270	0.318	2.852	0.953	5.5	5.7	10.3	2.3	0.1364	1.267	0.173	0.004121	21.3
CRZ-3-C	2.223	1.588	0.476	3.006	1.270	6.0	6.2	17.9	2.5	0.1360	1.981	0.269	0.005771	26.2
CRZ-86-C	2.064	1.270	0.476	2.690	1.111	5.2	6.8	10.9	2.4	0.1701	1.267	0.215	0.006075	20.7
CRZ-9	1.905	1.270	0.635	2.530	1.270	5.0	6.9	11.4	2.5	0.1815	1.267	0.230	0.006569	20.1
CRZ-48	1.429	0.873	1.151	1.859	1.588	3.6	7.9	6.4	3.0	0.2880	0.599	0.172	0.006653	14.7
CRZ-3	2.223	1.588	0.635	3.006	1.429	6.0	8.3	19.7	2.8	0.1815	1.981	0.359	0.009336	27.7
CRZ-7-B	2.381	1.429	0.476	3.086	1.191	6.0	9.3	15.2	2.7	0.2039	1.604	0.327	0.010006	26.5
CRZ-10-J	2.223	1.270	0.635	2.852	1.270	5.5	11.4	12.6	2.8	0.2723	1.267	0.345	0.013449	24.1
CRZ-8-A	2.858	1.905	0.476	3.795	1.429	7.5	11.7	30.9	3.0	0.2041	2.851	0.582	0.015589	39.6
CRZ-3-A	2.223	1.588	0.953	3.006	1.747	6.0	12.4	23.3	3.3	0.2723	1.981	0.539	0.017788	30.7
CRZ-9-A	1.905	1.270	1.270	2.530	1.905	5.0	13.8	16.0	3.6	0.3629	1.267	0.460	0.018769	25.2
CRZ-14-A	2.540	1.588	0.635	3.322	1.429	6.5	13.5	21.5	3.0	0.2720	1.981	0.539	0.019237	32.2
CRZ-10-K	2.699	1.270	0.556	3.352	1.191	6.2	17.0	13.7	3.0	0.3575	1.267	0.453	0.021248	30.2
CRZ-10-H	3.175	1.270	0.476	3.864	1.111	7.0	21.7	14.9	3.3	0.4081	1.267	0.517	0.025557	36.9
CRZ-8	2.858	1.905	0.635	3.795	1.588	7.5	15.5	33.5	3.3	0.2723	2.851	0.776	0.025605	41.5
CRZ-7	2.699	1.429	0.635	3.415	1.350	6.5	18.0	18.1	3.2	0.3629	1.604	0.582	0.026612	32.8
CRZ-9-D	2.223	1.588	1.270	3.006	2.064	6.0	16.6	26.8	3.8	0.3629	1.981	0.719	0.027385	33.7
CRZ-13	3.493	2.540	0.635	4.747	1.905	9.5	19.7	68.7	3.8	0.2723	5.068	1.380	0.039451	63.8
CRZ-4	2.858	1.905	0.953	3.795	1.906	7.5	23.3	38.6	3.8	0.4087	2.851	1.165	0.049972	45.3
CRZ-7-A	2.699	1.429	0.953	3.415	1.668	6.5	26.9	21.0	3.7	0.5446	1.604	0.874	0.051662	36.2
CRZ-6-D	4.445	2.699	0.476	5.775	1.826	11.2	32.0	87.9	4.3	0.3740	5.722	2.140	0.074148	85.5
CRZ-11-K	3.175	1.905	0.953	4.115	1.906	8.0	33.2	41.2	4.1	0.5446	2.851	1.553	0.083210	51.2
CRZ-4-C	2.858	1.905	1.429	3.795	2.382	7.5	35.0	46.4	4.6	0.6128	2.851	1.747	0.093646	51.0

National Arnold Caseless Toroids 4 mil Silicon Steel														
Part No.	OD cm 1	ID cm 2	HT cm 3	Wound OD cm 4	Wound HT cm 5	MPL cm 6	Wtfe grams 7	Wtcu grams 8	MLT cm 9	Ac cm sq 10	Wa cm sq 11	Ap cm 4th 12	Kg cm 5th 13	At cm sq 14
CRZ-26	3.572	1.389	0.992	4.334	1.687	7.8	57.9	24.0	4.4	0.9745	1.515	1.477	0.129513	52.5
CRZ-5	3.810	2.540	0.953	5.059	2.223	10.0	41.5	82.4	4.6	0.5446	5.068	2.760	0.131495	75.5
CRZ-11-B	3.810	1.905	0.953	4.775	1.906	9.0	56.0	46.4	4.6	0.8170	2.851	2.329	0.166424	64.4
CRZ-16-A	3.493	2.143	1.270	4.549	2.342	8.9	52.1	61.9	4.8	0.7715	3.607	2.783	0.177962	65.9
CRZ-19	3.810	2.540	1.270	5.059	2.540	10.0	55.2	91.5	5.1	0.7258	5.068	3.678	0.210208	80.5
CRZ-4-H	3.493	2.223	1.588	4.587	2.700	9.0	62.2	73.6	5.3	0.9075	3.882	3.523	0.239700	71.9
CRZ-4-H	3.493	2.223	1.588	4.587	2.700	9.0	62.2	73.6	5.3	0.9075	3.882	3.523	0.239700	71.9
CRZ-11-E	3.810	1.905	1.270	4.775	2.223	9.0	74.6	51.5	5.1	1.0887	2.851	3.103	0.266045	69.1
CRZ-41	4.445	2.540	0.953	5.703	2.223	11.0	68.4	91.6	5.1	0.8170	5.068	4.140	0.266282	90.9
CRZ-6	4.445	3.175	1.270	6.011	2.858	12.0	66.3	157.3	5.6	0.7258	7.918	5.747	0.298591	110.7
CRZ-68	3.810	2.540	1.588	5.059	2.858	10.0	69.1	100.7	5.6	0.9075	5.068	4.599	0.298737	85.6
CRZ-4-B	4.763	2.858	0.953	6.173	2.382	12.0	74.6	121.7	5.3	0.8170	6.416	5.242	0.321056	106.0
CRZ-11-A	3.175	1.905	2.540	4.115	3.493	8.0	88.4	66.9	6.6	1.4516	2.851	4.138	0.363822	71.7
CRZ-10-E	3.810	1.270	1.905	4.560	2.540	8.0	132.6	27.5	6.1	2.1774	1.267	2.759	0.394141	69.0
CRZ-4-L	6.985	2.223	0.635	8.332	1.747	14.5	150.2	91.2	6.6	1.3607	3.882	5.282	0.435341	154.7
CRZ-4-L	6.985	2.223	0.635	8.332	1.747	14.5	150.2	91.2	6.6	1.3607	3.882	5.282	0.435341	154.7
CRZ-4-E	4.128	2.858	1.905	5.535	3.334	11.0	91.2	144.9	6.4	1.0887	6.416	6.985	0.479018	106.0
CRZ-27	3.810	1.826	1.984	4.744	2.897	8.9	119.7	58.0	6.2	1.7713	2.619	4.639	0.528252	78.5
CRZ-11-C	5.080	2.540	1.111	6.366	2.381	12.0	116.0	105.3	5.8	1.2699	5.068	6.435	0.559580	111.2
CRZ-11-H	4.128	1.905	1.905	5.112	2.858	9.5	137.8	64.4	6.4	1.9057	2.851	5.432	0.652062	86.9
CRZ-11	4.445	2.540	1.905	5.703	3.175	11.0	136.7	119.0	6.6	1.6331	5.068	8.276	0.818600	107.9
CRZ-22	7.620	6.350	1.270	10.807	4.445	21.9	121.5	915.5	8.1	0.7258	31.673	22.989	0.821126	334.2
CRZ-6-A	4.445	3.175	2.540	6.011	4.128	12.0	132.6	214.6	7.6	1.4516	7.918	11.494	0.875868	134.6
CRZ-78-B	5.715	3.175	1.270	7.293	2.858	14.0	154.7	186.0	6.6	1.4516	7.918	11.494	1.010617	149.0
CRZ-10	4.445	2.540	2.223	5.703	3.493	11.0	159.6	128.2	7.1	1.9057	5.068	9.657	1.034968	113.6
CRZ-10-D	4.763	2.540	1.905	6.033	3.175	11.5	166.8	123.6	6.9	1.9057	5.068	9.657	1.073358	117.3
CRZ-4-A	4.763	2.223	1.905	5.908	3.017	11.0	182.3	94.7	6.9	2.1774	3.882	8.452	1.073358	110.8
CRZ-6-E	6.350	3.175	1.111	7.958	2.699	15.0	181.2	193.1	6.9	1.5873	7.918	12.569	1.163756	166.8
CRZ-25	4.128	2.064	2.858	5.173	3.890	9.7	197.0	93.7	7.9	2.6545	3.346	8.883	1.197658	105.2
CRZ-25	4.128	2.064	2.858	5.173	3.890	9.7	197.0	93.7	7.9	2.6545	3.346	8.883	1.197658	105.2

National Arnold Caseless Toroids 4 mil Silicon Steel														
Part No.	OD cm 1	ID cm 2	HT cm 3	Wound OD cm 4	Wound HT cm 5	MPL cm 6	Wtfe grams 7	Wtcu grams 8	MLT cm 9	Ac cm sq 10	Wa cm sq 11	Ap cm 4th 12	Kg cm 5th 13	At cm sq 14
CRZ-20	6.350	3.810	1.270	8.229	3.175	16.0	176.8	288.4	7.1	1.4516	11.402	16.552	1.351339	188.4
CRZ-6-C	5.398	3.493	1.905	7.116	3.652	14.0	174.0	251.1	7.4	1.6331	9.584	15.651	1.387883	161.1
CRZ-11-D	4.445	1.905	3.175	5.452	4.128	10.0	276.2	87.5	8.6	3.6290	2.851	10.345	1.738855	117.3
CRZ-29-A	7.303	3.175	1.111	8.974	2.699	16.5	259.2	214.6	7.6	2.0638	7.918	16.342	1.770399	202.5
CRZ-2	6.350	5.080	2.540	8.884	5.080	18.0	198.9	659.1	9.1	1.4516	20.271	29.425	1.868518	265.6
CRZ-29	7.303	4.763	1.270	9.646	3.652	19.0	209.9	499.0	7.9	1.4516	17.820	25.868	1.907435	256.7
CRZ-21	7.620	5.080	1.270	10.119	3.810	20.0	221.0	585.9	8.1	1.4516	20.271	29.425	2.102083	281.8
CRZ-78-A	5.715	3.175	2.540	7.293	4.128	14.0	309.4	243.2	8.6	2.9032	7.918	22.989	3.091298	178.0
CRZ-12	6.350	3.810	2.540	8.229	4.445	16.0	353.6	370.8	9.1	2.9032	11.402	33.104	4.204166	221.2
CRZ-12-B	6.033	3.969	3.175	7.985	5.160	15.7	353.6	435.9	9.9	2.9489	12.374	36.490	4.344946	229.5
CRZ-18	8.573	6.033	2.540	11.546	5.557	22.9	508.3	1110.4	10.9	2.9032	28.590	83.003	8.824999	410.7
CRZ-15-C	7.303	3.493	2.858	9.090	4.605	17.0	634.1	355.0	10.4	4.9000	9.584	46.962	8.837617	261.2
CRZ-15-A	8.890	6.350	2.858	12.021	6.033	23.9	596.8	1316.1	11.7	3.2667	31.673	103.467	11.570424	454.6

National Arnold Caseless Toroids 12 mil Silicon Steel														
Part No.	OD cm 1	ID cm 2	HT cm 3	Wound OD cm 4	Wound HT cm 5	MPL cm 6	Wtfe grams 7	Wtcu grams 8	MLT cm 9	Ac cm sq 10	Wa cm sq 11	Ap cm 4th 12	Kg cm 5th 13	At cm sq 14
CRA-16-J	1.826	1.270	1.270	2.451	1.905	4.9	12.4	15.7	3.5	0.3354	1.267	0.425	0.016	24
CRA-16-C	1.905	1.270	1.270	2.530	1.905	5.0	14.6	16.0	3.6	0.3831	1.267	0.485	0.021	25
CRA-16	2.540	1.270	0.635	3.183	1.270	6.0	17.5	13.7	3.0	0.3831	1.267	0.485	0.024	29
CRA-16-F	2.223	1.270	1.111	2.852	1.746	5.5	21.1	16.0	3.6	0.5029	1.267	0.637	0.036	28
CRA-21-B	3.175	1.905	0.635	4.115	1.588	8.0	23.3	36.0	3.6	0.3831	2.851	1.092	0.047	47
CRA-16-B	2.540	1.270	0.953	3.183	1.588	6.0	26.3	16.0	3.6	0.5749	1.267	0.728	0.047	32
CRA-7	2.858	1.905	0.953	3.795	1.906	7.5	24.6	38.6	3.8	0.4314	2.851	1.230	0.056	45
CRA-85-B	2.540	1.588	1.270	3.322	2.064	6.5	28.4	28.6	4.1	0.5743	1.981	1.138	0.064	39
CRA-70-D	3.810	2.540	0.635	5.059	1.905	10.0	29.2	73.2	4.1	0.3831	5.068	1.941	0.073	71
CRA-16-D	2.540	1.270	1.270	3.183	1.905	6.0	35.0	18.3	4.1	0.7661	1.267	0.971	0.073	35
CRA-16-E	2.699	1.270	1.270	3.352	1.905	6.2	41.0	18.9	4.2	0.8620	1.267	1.092	0.090	38
CRA-7-A	3.175	1.905	0.953	4.115	1.906	8.0	35.0	41.2	4.1	0.5749	2.851	1.639	0.093	51
CRA-16-H	2.818	1.270	1.270	3.479	1.905	6.4	45.8	19.3	4.3	0.9338	1.267	1.183	0.103	40
CRA-85-A	2.540	1.588	1.905	3.322	2.699	6.5	42.6	35.8	5.1	0.8614	1.981	1.706	0.116	46
CRA-21-D	3.493	2.223	0.953	4.587	2.065	9.0	39.4	59.6	4.3	0.5749	3.882	2.232	0.119	63
CRA-21	3.334	2.540	1.588	4.593	2.858	9.2	42.2	93.9	5.2	0.5989	5.068	3.035	0.140	74
CRA-7-B	3.175	1.905	1.270	4.115	2.223	8.0	46.7	46.3	4.6	0.7661	2.851	2.184	0.146	55
CRA-70-B	3.810	2.540	0.953	5.059	2.223	10.0	43.8	82.4	4.6	0.5749	5.068	2.913	0.147	76
CRA-85	2.699	1.588	1.905	3.484	2.699	6.7	51.7	36.7	5.2	1.0053	1.981	1.991	0.154	49
CRA-95	3.572	1.032	1.270	4.234	1.786	7.2	84.6	14.5	4.9	1.5323	0.837	1.282	0.161	52
CRA-70	3.810	2.540	1.270	5.059	2.540	10.0	58.3	91.5	5.1	0.7661	5.068	3.883	0.234	81
CRA-70-A	4.128	2.858	1.270	5.535	2.699	11.0	64.2	121.7	5.3	0.7661	6.416	4.916	0.282	95
CRA-6	5.874	4.763	0.953	8.254	3.335	16.7	64.1	394.4	6.2	0.5029	17.820	8.962	0.290	193
CRA-32	3.493	1.905	1.588	4.442	2.541	8.5	77.5	54.1	5.3	1.1978	2.851	3.415	0.307	66
CRA-16-K	4.366	1.270	1.270	5.178	1.905	8.9	126.2	24.9	5.5	1.8677	1.267	2.366	0.320	73
CRA-21-F	4.604	2.381	0.953	5.801	2.144	11.0	84.3	82.5	5.2	1.0063	4.453	4.481	0.346	92
CRA-70-E	4.445	2.540	1.270	5.703	2.540	11.0	96.2	100.7	5.6	1.1492	5.068	5.824	0.479	97
CRA-53	5.398	3.493	0.953	7.116	2.700	14.0	91.9	199.1	5.8	0.8623	9.584	8.265	0.488	140
CRA-67	5.398	4.128	1.270	7.446	3.334	15.0	87.5	302.3	6.4	0.7661	13.385	10.255	0.495	165
CRA-32-A	4.128	1.905	1.588	5.112	2.541	9.5	121.3	59.2	5.8	1.6768	2.851	4.780	0.549	82

National Arnold Caseless Toroids 12 mil Silicon Steel														
Part No.	OD cm 1	ID cm 2	HT cm 3	Wound OD cm 4	Wound HT cm 5	MPL cm 6	Wtfe grams 7	Wtcu grams 8	MLT cm 9	Ac cm sq 10	Wa cm sq 11	Ap cm 4th 12	Kg cm 5th 13	At cm sq 14
CRA-22-A	5.239	3.651	1.270	7.037	3.096	14.0	102.1	231.7	6.2	0.9580	10.471	10.030	0.618	146
CRA-8-H	5.080	3.175	1.270	6.643	2.858	13.0	113.7	171.6	6.1	1.1492	7.918	9.100	0.686	129
CRA-15	3.175	1.905	3.810	4.115	4.763	8.0	140.0	87.5	8.6	2.2984	2.851	6.552	0.697	88
CRA-12-D	4.842	1.905	1.429	5.883	2.382	10.6	161.2	62.4	6.2	1.9936	2.851	5.683	0.736	98
CRA-21-A	4.604	2.699	1.588	5.938	2.938	11.5	125.8	126.6	6.2	1.4369	5.722	8.222	0.759	110
CRA-12	5.080	2.540	1.270	6.366	2.540	12.0	140.0	109.9	6.1	1.5323	5.068	7.765	0.781	114
CRA-13-D	4.286	2.540	1.905	5.540	3.175	10.7	129.3	116.7	6.5	1.5799	5.068	8.007	0.781	103
CRA-32-B	3.493	1.905	3.175	4.442	4.128	8.5	155.0	79.8	7.9	2.3949	2.851	6.827	0.831	89
CRA-12-B	4.763	1.905	1.588	5.797	2.541	10.5	172.3	64.4	6.4	2.1558	2.851	6.145	0.834	99
CRA-7-C	3.175	1.905	4.445	4.115	5.398	8.0	163.3	97.8	9.7	2.6814	2.851	7.644	0.849	96
CRA-70-C	4.445	2.540	1.905	5.703	3.175	11.0	144.3	119.0	6.6	1.7238	5.068	8.736	0.912	108
CRA-41-E	7.620	6.350	1.270	10.807	4.445	21.9	128.3	915.5	8.1	0.7661	31.673	24.266	0.915	334
CRA-22	5.080	3.334	1.588	6.720	3.255	13.2	132.8	205.1	6.6	1.3170	8.731	11.499	0.917	140
CRA-75-A	8.255	7.303	1.588	11.959	5.240	24.4	133.9	1362.3	9.1	0.7181	41.894	30.084	0.945	421
CRA-60-B	5.874	3.969	1.270	7.827	3.255	15.5	135.6	296.2	6.7	1.1492	12.374	14.220	0.971	176
CRA-8	5.080	3.175	1.588	6.643	3.176	13.0	142.2	186.0	6.6	1.4369	7.918	11.378	0.990	136
CRA-30-B	6.668	5.080	1.270	9.187	3.810	18.5	134.9	531.0	7.4	0.9580	20.271	19.419	1.010	242
CRA-35-E	5.318	2.143	1.429	6.477	2.501	11.7	192.7	83.9	6.5	2.1551	3.607	7.774	1.025	117
CRA-107-P	5.715	3.175	1.270	7.293	2.858	14.0	163.3	186.0	6.6	1.5323	7.918	12.133	1.126	149
CRA-12-A	5.080	2.540	1.588	6.366	2.858	12.0	175.0	119.0	6.6	1.9159	5.068	9.709	1.127	121
CRA-32-D	3.810	1.905	3.175	4.775	4.128	9.0	196.8	82.4	8.1	2.8730	2.851	8.190	1.158	98
CRA-3	4.763	2.540	1.905	6.033	3.175	11.5	176.1	123.6	6.9	2.0115	5.068	10.194	1.196	117
CRA-32-C	5.556	1.905	1.588	6.665	2.541	11.7	246.3	70.8	7.0	2.7539	2.851	7.850	1.238	123
CRA-35-K	5.477	1.984	1.588	6.601	2.580	11.7	235.6	76.1	6.9	2.6348	3.092	8.147	1.240	122
CRA-8-F	5.080	3.175	1.905	6.643	3.493	13.0	170.6	200.3	7.1	1.7238	7.918	13.650	1.323	142
CRA-64	7.620	6.350	1.588	10.807	4.763	21.9	160.4	972.8	8.6	0.9580	31.673	30.342	1.346	345
CRA-78	6.350	3.810	1.270	8.229	3.175	16.0	186.6	288.4	7.1	1.5323	11.402	17.471	1.506	188
CRA-8-B	4.921	3.175	2.540	6.483	4.128	12.7	204.4	225.3	8.0	2.1065	7.918	16.680	1.757	150
CRA-49	6.350	5.715	5.080	9.259	7.938	19.0	221.6	1205.0	13.2	1.5323	25.655	39.311	1.824	365
CRA-8-A	5.080	3.175	2.540	6.643	4.128	13.0	227.4	228.9	8.1	2.2984	7.918	18.199	2.059	155

National Arnold Caseless Toroids 12 mil Silicon Steel

Part No.	OD cm 1	ID cm 2	HT cm 3	Wound OD cm 4	Wound HT cm 5	MPL cm 6	Wtfe grams 7	Wtcu grams 8	MLT cm 9	Ac cm sq 10	Wa cm sq 11	Ap cm 4th 12	Kg cm 5th 13	At cm sq 14
CRA-51	6.350	5.080	2.540	8.884	5.080	18.0	209.9	659.1	9.1	1.5323	20.271	31.060	2.082	266
CRA-59	7.461	5.080	1.270	9.961	3.810	19.7	215.9	576.7	8.0	1.4363	20.271	29.116	2.091	275
CRA-107	5.715	3.175	1.905	7.293	3.493	14.0	244.9	214.6	7.6	2.2984	7.918	18.199	2.196	164
CRA-23	7.144	5.080	1.588	9.648	4.128	19.2	228.1	595.1	8.3	1.5569	20.271	31.559	2.381	271
CRA-21-E	5.318	2.223	2.540	6.504	3.652	11.8	337.5	114.8	8.3	3.7341	3.882	14.495	2.603	141
CRA-13-C	5.239	2.540	2.540	6.534	3.810	12.2	303.6	148.8	8.3	3.2563	5.068	16.502	2.604	145
CRA-62-D	7.620	4.445	1.270	9.818	3.493	19.0	277.0	448.6	8.1	1.9153	15.520	29.726	2.802	259
CRA-38	7.620	4.128	1.270	9.679	3.334	18.5	296.6	386.9	8.1	2.1065	13.385	28.197	2.923	248
CRA-13	6.350	2.540	1.905	7.728	3.175	14.0	367.4	146.5	8.1	3.4476	5.068	17.471	2.964	171
CRA-107-D	6.350	2.858	1.905	7.838	3.334	14.5	348.8	185.4	8.1	3.1598	6.416	20.274	3.153	179
CRA-107-E	6.350	3.175	1.905	7.958	3.493	15.0	328.0	228.9	8.1	2.8730	7.918	22.749	3.216	187
CRA-41-B	8.890	6.350	1.270	12.021	4.445	23.9	279.9	1029.9	9.1	1.5323	31.673	48.532	3.253	395
CRA-107-M	6.033	2.858	2.223	7.499	3.652	14.0	357.3	191.3	8.4	3.3526	6.416	21.510	3.441	174
CRA-107-B	5.715	3.175	2.540	7.293	4.128	14.0	326.6	243.2	8.6	3.0645	7.918	24.266	3.444	178
CRA-42	7.620	5.080	1.588	10.119	4.128	20.0	291.7	622.6	8.6	1.9159	20.271	38.838	3.446	292
CRA-50	9.208	6.668	1.270	12.499	4.604	24.9	291.6	1167.2	9.4	1.5323	34.925	53.514	3.490	426
CRA-8-J	7.303	3.175	1.588	8.974	3.176	16.5	391.1	236.1	8.4	3.1138	7.918	24.656	3.663	216
CRA-8-D	5.080	3.175	3.810	6.643	5.398	13.0	341.1	286.1	10.2	3.4476	7.918	27.299	3.705	182
CRA-41-D	8.255	6.350	1.905	11.407	5.080	22.9	301.8	1087.1	9.7	1.7238	31.673	54.598	3.900	386
CRA-11-B	6.668	4.763	2.540	9.017	4.922	18.0	314.9	595.6	9.4	2.2984	17.820	40.957	4.006	267
CRA-42-C	7.938	5.159	1.588	10.475	4.168	20.6	329.1	661.0	8.9	2.0962	20.906	43.824	4.133	309
CRA-81	5.715	4.128	3.810	7.752	5.874	15.5	338.9	507.8	10.7	2.8721	13.385	38.443	4.140	237
CRA-30	7.620	5.080	1.905	10.119	4.445	20.0	349.9	659.1	9.1	2.2984	20.271	46.590	4.684	302
CRA-14-D	8.890	3.810	1.270	10.905	3.175	20.0	466.5	370.8	9.1	3.0645	11.402	34.943	4.684	295
CRA-107-C	6.350	3.810	2.540	8.229	4.445	16.0	373.2	370.8	9.1	3.0645	11.402	34.943	4.684	221
CRA-65-D	10.795	8.255	1.270	14.890	5.398	29.9	349.9	2030.6	10.7	1.5323	53.528	82.018	4.712	600
CRA-11	10.160	8.890	2.540	14.660	6.985	29.9	349.9	2691.5	12.2	1.5323	62.080	95.122	4.782	659
CRA-8-C	5.398	3.175	3.810	6.967	5.398	13.5	413.4	293.2	10.4	4.0231	7.918	31.856	4.922	194
CRA-107-F	6.350	3.175	2.540	7.958	4.128	15.0	437.4	257.5	9.1	3.8306	7.918	30.332	5.083	203
CRA-107-A	8.255	3.175	1.588	10.006	3.176	18.0	525.0	257.5	9.1	3.8318	7.918	30.342	5.086	257

National Arnold Caseless Toroids 12 mil Silicon Steel														
Part No.	OD cm 1	ID cm 2	HT cm 3	Wound OD cm 4	Wound HT cm 5	MPL cm 6	Wtfe grams 7	Wtcu grams 8	MLT cm 9	Ac cm sq 10	Wa cm sq 11	Ap cm 4th 12	Kg cm 5th 13	At cm sq 14
CRA-9-G	5.398	3.334	4.128	7.040	5.795	13.7	423.6	339.1	10.9	4.0471	8.731	35.336	5.237	206
CRA-107-N	7.620	3.175	1.905	9.316	3.493	17.0	520.5	257.5	9.1	4.0222	7.918	31.849	5.604	238
CRA-9	8.573	5.080	1.588	11.081	4.128	21.4	431.2	677.5	9.4	2.6348	20.271	53.409	5.989	336
CRA-7-W	6.985	4.445	2.540	9.172	4.763	18.0	419.9	532.7	9.7	3.0645	15.520	47.561	6.040	269
CRA-35-C	5.278	2.143	4.763	6.434	5.835	11.7	630.9	151.9	11.8	7.0927	3.607	25.586	6.129	183
CRA-1-A	10.160	7.620	1.588	13.932	5.398	27.9	408.3	1730.3	10.7	1.9159	45.610	87.384	6.277	541
CRA-13-E	6.350	2.540	3.175	7.728	4.445	14.0	612.3	183.1	10.2	5.7460	5.068	29.119	6.587	202
CRA-41	8.890	6.350	1.905	12.021	5.080	23.9	419.9	1144.3	10.2	2.2984	31.673	72.797	6.587	419
CRA-14-F	7.303	2.540	2.540	8.771	3.810	15.5	678.0	178.5	9.9	5.7466	5.068	29.122	6.757	226
CRA-79	6.985	3.810	2.540	8.884	4.445	17.0	495.7	391.4	9.7	3.8306	11.402	43.678	6.934	248
CRA-35-B	5.715	3.493	4.445	7.436	6.192	14.5	517.8	398.2	11.7	4.6915	9.584	44.963	7.222	231
CRA-33	7.620	5.080	2.540	10.119	5.080	20.0	466.5	732.4	10.2	3.0645	20.271	62.120	7.495	322
CRA-14	7.620	2.540	2.540	9.121	3.810	16.0	746.4	183.1	10.2	6.1290	5.068	31.060	7.495	240
CRA-79-E	7.303	3.810	2.540	9.216	4.445	17.5	561.4	401.7	9.9	4.2143	11.402	48.053	8.177	262
CRA-107-J	6.350	3.175	3.493	7.958	5.081	15.0	601.5	300.4	10.7	5.2679	7.918	41.713	8.239	226
CRA-23-B	10.478	7.303	1.588	14.075	5.240	27.9	510.4	1627.3	10.9	2.3949	41.894	100.331	8.799	543
CRA-14-E	7.620	2.540	2.858	9.121	4.128	16.0	839.9	192.3	10.7	6.8964	5.068	34.949	9.036	249
CRA-78-D	6.350	3.810	4.128	8.229	6.033	16.0	606.5	473.8	11.7	4.9804	11.402	56.789	9.682	262
CRA-63	8.573	6.033	2.540	11.546	5.557	22.9	536.5	1110.4	10.9	3.0645	28.590	87.614	9.833	411
CRA-34-A	7.303	3.493	2.858	9.090	4.605	17.0	669.3	355.0	10.4	5.1723	9.584	49.571	9.847	261
CRA-53-J	17.780	15.240	1.270	25.465	8.890	51.9	606.5	10546.1	16.3	1.5323	182.438	279.542	10.540	1729
CRA-14-A	8.890	6.350	2.540	12.021	5.715	23.9	559.8	1258.8	11.2	3.0645	31.673	97.063	10.646	443
CRA-23-E	10.319	7.303	1.905	13.918	5.557	27.7	576.5	1683.9	11.3	2.7291	41.894	114.332	11.042	547
CRA-63-A	8.573	5.715	2.540	11.384	5.398	22.4	590.6	996.5	10.9	3.4482	25.655	88.464	11.171	396
CRA-69-B	7.303	5.398	4.445	9.972	7.144	20.0	612.3	1054.4	13.0	4.0222	22.888	92.060	11.433	380
CRA-35-A	7.303	3.969	3.175	9.282	5.160	17.7	679.4	480.6	10.9	5.0281	12.374	62.217	11.457	286
CRA-35	7.303	3.493	3.175	9.090	4.922	17.0	743.6	372.2	10.9	5.7460	9.584	55.069	11.588	270
CRA-38-A	7.620	4.128	3.175	9.679	5.239	18.5	741.6	532.0	11.2	5.2664	13.385	70.492	13.287	306
CRA-107-L	8.890	3.175	2.540	10.701	4.128	19.0	997.2	314.7	11.2	6.8951	7.918	54.598	13.474	319
CRA-69-A	7.303	5.080	4.445	9.805	6.985	19.5	696.7	933.8	13.0	4.6936	20.271	95.143	13.789	366

National Arnold Caseless Toroids 12 mil Silicon Steel

Part No.	OD cm 1	ID cm 2	HT cm 3	Wound OD cm 4	Wound HT cm 5	MPL cm 6	Wtfe grams 7	Wtcu grams 8	MLT cm 9	Ac cm sq 10	Wa cm sq 11	Ap cm 4th 12	Kg cm 5th 13	At cm sq 14
CRA-69-C	7.303	5.398	5.080	9.972	7.779	20.0	699.8	1137.1	14.0	4.5968	22.888	105.212	13.847	400
CRA-66	10.160	7.620	2.540	13.932	6.350	27.9	653.1	1977.4	12.2	3.0645	45.610	139.771	14.053	583
CRA-107-H	5.715	3.175	6.985	7.293	8.573	14.0	898.0	443.4	15.7	8.4274	7.918	66.731	14.284	280
CRA-73-D	6.350	4.128	6.033	8.380	8.097	16.5	799.7	701.2	14.7	6.3675	13.385	85.231	14.735	323
CRA-107-K	7.620	3.175	3.493	9.316	5.081	17.0	954.3	329.0	11.7	7.3750	7.918	58.398	14.743	285
CRA-73-A	10.160	5.080	1.905	12.732	4.445	23.9	839.7	805.6	11.2	4.5968	20.271	93.181	15.330	432
CRA-62-B	8.255	4.445	3.175	10.475	5.398	20.0	874.7	644.8	11.7	5.7460	15.520	89.177	17.542	350
CRA-68	11.430	8.890	2.540	15.848	6.985	31.9	746.4	2915.7	13.2	3.0645	62.080	190.244	17.656	742
CRA-79-D	7.620	3.810	3.810	9.549	5.715	18.0	944.7	494.3	12.2	6.8951	11.402	78.621	17.786	315
CRA-73	6.985	4.128	5.080	9.024	7.144	17.5	918.3	652.9	13.7	6.8939	13.385	92.277	18.552	330
CRA-62	10.160	5.080	2.223	12.732	4.763	23.9	979.9	842.3	11.7	5.3641	20.271	108.735	19.967	445
CRA-15-A	8.890	6.350	3.810	12.021	6.985	23.9	839.7	1487.6	13.2	4.5968	31.673	145.595	20.268	491
CRA-54	6.509	3.334	6.350	8.189	8.017	15.5	1129.9	477.1	15.4	9.5766	8.731	83.616	20.843	311
CRA-65	12.700	10.160	2.540	17.768	7.620	35.9	839.7	4101.3	14.2	3.0645	81.084	248.482	21.414	921
CRA-82-B	25.400	23.495	1.905	37.427	13.653	76.8	1010.3	36031.3	23.4	1.7238	433.608	747.447	22.055	3804
CRA-2	9.525	6.350	3.175	12.648	6.350	24.9	911.2	1430.4	12.7	4.7883	31.673	151.661	22.872	503
CRA-77-E	8.255	3.810	3.810	10.224	5.715	19.0	1163.4	514.9	12.7	8.0443	11.402	91.725	23.240	348
CRA-77-E	8.255	3.810	3.810	10.224	5.715	19.0	1163.4	514.9	12.7	8.0443	11.402	91.725	23.240	348
CRA-62-A	8.255	4.445	3.810	10.475	6.033	20.0	1049.7	700.9	12.7	6.8951	15.520	107.012	23.240	371
CRA-10-B	10.160	5.080	2.540	12.732	5.080	23.9	1119.6	878.8	12.2	6.1290	20.271	124.241	24.983	458
CRA-73-B	12.065	5.080	1.905	14.768	4.445	26.9	1299.0	915.5	12.7	6.3206	20.271	128.123	25.506	549
CRA-82-A	10.160	7.620	3.810	13.932	7.620	27.9	979.7	2307.0	14.2	4.5968	45.610	209.657	27.102	638
CRA-48	11.430	7.620	2.540	15.178	6.350	29.9	1049.7	2142.2	13.2	4.5968	45.610	209.657	29.187	664
CRA-77	8.890	5.080	3.810	11.407	6.350	21.9	1154.6	952.1	13.2	6.8951	20.271	139.771	29.187	432
CRA-77-A	8.890	3.810	3.810	10.905	5.715	20.0	1399.5	535.5	13.2	9.1935	11.402	104.828	29.187	382
CRA-2-A	9.525	6.350	3.810	12.648	6.985	24.9	1093.4	1544.8	13.7	5.7460	31.673	181.994	30.497	529
CRA-65-C	10.795	8.255	3.810	14.890	7.938	29.9	1049.7	2804.2	14.7	4.5968	53.528	246.055	30.710	719
CRA-60-A	9.049	3.969	3.810	11.132	5.795	20.5	1434.6	586.8	13.3	9.1935	12.374	113.760	31.371	397
CRA-35-J	7.303	3.493	6.350	9.090	8.097	17.0	1487.2	545.4	16.0	11.4919	9.584	110.138	31.638	361
CRA-79-A	10.160	3.810	3.175	12.285	5.080	21.9	1603.6	535.5	13.2	9.5766	11.402	109.196	31.670	433

National Arnold Caseless Toroids 12 mil Silicon Steel

Part No.	OD cm 1	ID cm 2	HT cm 3	Wound OD cm 4	Wound HT cm 5	MPL cm 6	Wtfe grams 7	Wtcu grams 8	MLT cm 9	Ac cm sq 10	Wa cm sq 11	Ap cm 4th 12	Kg cm 5th 13	At cm sq 14
CRA-57-D	10.478	5.398	2.858	13.195	5.557	24.9	1312.4	1054.4	13.0	6.8964	22.888	157.845	33.610	504
CRA-68-D	11.430	8.890	3.810	15.848	8.255	31.9	1119.6	3364.3	15.2	4.5968	62.080	285.366	34.429	805
CRA-54-A	9.366	4.286	3.810	11.586	5.953	21.4	1504.5	697.3	13.6	9.1935	14.429	132.658	35.900	427
CRA-55-A	11.430	6.350	2.540	14.587	5.715	27.9	1306.2	1487.6	13.2	6.1290	31.673	194.126	36.033	596
CRA-20-H	10.478	7.938	4.763	14.412	8.732	28.9	1268.5	2816.7	16.0	5.7466	49.496	284.431	40.854	721
CRA-1	10.160	7.620	5.080	13.932	8.890	27.9	1306.2	2636.5	16.3	6.1290	45.610	279.542	42.158	694
CRA-20-E	12.700	10.795	5.080	18.135	10.478	36.9	1294.6	5952.8	18.3	4.5968	91.536	420.769	42.305	1113
CRA-16-A	10.160	3.810	3.810	12.285	5.715	21.9	1924.4	576.7	14.2	11.4919	11.402	131.035	42.347	457
CRA-36	10.160	6.350	3.810	13.286	6.985	25.9	1364.6	1602.1	14.2	6.8951	31.673	218.392	42.347	569
CRA-68-B	13.018	8.890	2.540	17.393	6.985	34.4	1307.9	3196.2	14.5	4.9804	62.080	309.184	42.543	856
CRA-8-E	8.255	3.175	6.350	10.006	7.938	18.0	2099.3	472.0	16.8	15.3226	7.918	121.329	44.359	407
CRA-57	10.160	5.715	3.810	12.997	6.668	24.9	1530.8	1297.7	14.2	8.0443	25.655	206.381	46.687	537
CRA-29	8.255	4.763	6.350	10.612	8.732	20.5	1643.6	1062.3	16.8	10.5327	17.820	187.693	47.171	468
CRA-65-A	12.065	8.255	3.175	16.128	7.303	31.9	1399.5	2804.2	14.7	5.7460	53.528	307.569	47.985	778
CRA-25-B	12.700	7.620	2.540	16.459	6.350	31.9	1492.8	2307.0	14.2	6.1290	45.610	279.542	48.181	754
CRA-61-A	10.160	5.080	3.810	12.732	6.350	23.9	1679.5	1025.3	14.2	9.1935	20.271	186.361	48.181	508
CRA-84-A	10.795	5.715	3.493	13.657	6.351	25.9	1668.0	1297.7	14.2	8.4286	25.655	216.239	51.251	565
CRA-27	14.288	10.160	2.540	19.297	7.620	38.4	1459.5	4467.6	15.5	4.9804	81.084	403.832	51.922	1046
CRA-35-F	8.573	3.493	6.350	10.455	8.097	19.0	2216.1	580.0	17.0	15.3226	9.584	146.850	52.887	437
CRA-92	12.700	10.478	5.080	17.950	10.319	36.4	1489.6	5608.3	18.3	5.3617	86.239	462.386	54.225	1088
CRA-29-A	8.890	4.763	5.715	11.270	8.097	21.4	1833.5	1030.1	16.3	11.2033	17.820	199.642	55.035	486
CRA-20	13.970	8.890	2.540	18.344	6.985	35.9	1679.5	3364.3	15.2	6.1290	62.080	380.488	61.208	931
CRA-37	11.113	7.620	4.445	14.863	8.255	29.4	1656.0	2595.4	16.0	7.3750	45.610	336.372	62.010	732
CRA-74	12.700	6.350	2.858	15.915	6.033	29.9	1968.5	1659.4	14.7	8.6204	31.673	273.038	63.904	699
CRA-77-C	8.890	4.445	6.350	11.141	8.573	20.9	2143.1	953.2	17.3	13.4072	15.520	208.079	64.608	495
CRA-61-B	9.843	5.556	5.080	12.600	7.858	24.2	1909.4	1379.8	16.0	10.3445	24.248	250.831	64.859	560
CRA-67-A	12.065	7.620	3.493	15.815	7.303	30.9	1740.2	2471.9	15.2	7.3750	45.610	336.372	65.108	755
CRA-31	10.160	6.350	5.080	13.286	8.255	25.9	1819.4	1830.9	16.3	9.1935	31.673	291.190	65.873	622
CRA-65-B	12.700	10.160	5.080	17.768	10.160	35.9	1679.5	5273.0	18.3	6.1290	81.084	496.964	66.621	1063
CRA-35-H	9.525	3.493	6.350	11.493	8.097	20.5	2839.1	606.0	17.8	18.1940	9.584	174.370	71.372	500

National Arnold Caseless Toroids 12 mil Silicon Steel

Part No.	OD cm 1	ID cm 2	HT cm 3	Wound OD cm 4	Wound HT cm 5	MPL cm 6	Wtfe grams 7	Wtcu grams 8	MLT cm 9	Ac cm sq 10	Wa cm sq 11	Ap cm 4th 12	Kg cm 5th 13	At cm sq 14
CRA-69	11.430	5.715	3.810	14.324	6.668	26.9	2125.6	1390.4	15.2	10.3427	25.655	265.347	72.032	622
CRA-66-E	11.430	8.255	5.080	15.504	9.208	30.9	1807.7	3287.6	17.3	7.6613	53.528	410.092	72.761	826
CRA-9-B	10.160	5.080	5.080	12.732	7.620	23.9	2239.3	1171.8	16.3	12.2580	20.271	248.482	74.948	559
CRA-41-A	12.700	6.350	3.175	15.915	6.350	29.9	2186.8	1716.5	15.2	9.5766	31.673	303.323	76.241	715
CRA-20-F	12.700	10.795	7.620	18.135	13.018	36.9	1941.9	7275.6	22.4	6.8951	91.536	631.154	77.879	1258
CRA-110-A	12.700	9.843	5.080	17.590	10.002	35.4	1862.9	4949.1	18.3	6.8939	76.103	524.649	79.110	1038
CRA-66-D	13.335	8.255	3.175	17.400	7.303	33.9	1982.7	2997.6	15.7	7.6613	53.528	410.092	79.803	874
CRA-77-D	8.890	5.080	7.620	11.407	10.160	21.9	2309.2	1391.5	19.3	13.7903	20.271	279.542	79.879	568
CRA-140-B	13.335	9.525	3.810	18.032	8.573	35.9	1889.4	4248.3	16.8	6.8951	71.265	491.383	80.844	996
CRA-84-D	12.065	8.890	5.080	16.459	9.525	32.9	1924.4	3925.0	17.8	7.6613	62.080	475.610	81.975	918
CRA-20-D	15.875	10.795	2.540	21.187	7.938	41.9	1959.4	5456.7	16.8	6.1290	91.536	561.025	82.046	1233
CRA-63-B	10.160	6.033	5.715	13.138	8.732	25.4	2174.6	1756.0	17.3	11.2033	28.590	320.300	83.103	631
CRA-66-F	12.700	8.255	3.810	16.760	7.938	32.9	2020.6	3094.3	16.3	8.0443	53.528	430.597	85.233	859
CRA-40	13.970	10.160	3.810	18.987	8.890	37.9	1994.4	4980.1	17.3	6.8951	81.084	559.084	89.277	1096
CRA-79-C	9.525	3.810	7.303	11.592	9.208	20.9	3168.9	782.8	19.3	19.8249	11.402	226.051	92.857	546
CRA-40-A	13.970	12.065	7.620	20.062	13.653	40.9	2151.8	9501.3	23.4	6.8951	114.341	788.396	93.052	1492
CRA-56-A	12.700	5.080	3.810	15.456	6.350	27.9	2939.0	1171.8	16.3	13.7903	20.271	279.542	94.856	683
CRA-86-A	9.843	2.540	9.525	11.599	10.795	19.5	4904.4	416.5	23.1	33.0415	5.068	167.446	95.744	604
CRA-18-B	9.843	4.763	6.668	12.273	9.050	22.9	2817.0	1175.0	18.5	16.0899	17.820	286.722	99.515	585
CRA-92-A	12.700	10.478	7.620	17.950	12.859	36.4	2234.4	6854.6	22.4	8.0425	86.239	693.578	99.823	1231
CRA-42-A	13.335	10.160	5.080	18.373	10.160	36.9	2157.6	5419.5	18.8	7.6613	81.084	621.205	101.282	1116
CRA-56-C	12.700	6.350	3.810	15.915	6.985	29.9	2624.1	1830.9	16.3	11.4919	31.673	363.987	102.926	747
CRA-19-B	10.001	4.921	6.668	12.502	9.129	23.4	2877.9	1262.8	18.7	16.0899	19.022	306.059	105.508	604
CRA-75-C	11.430	6.350	5.080	14.587	8.255	27.9	2612.5	1945.4	17.3	12.2580	31.673	388.253	110.218	712
CRA-91-A	13.335	5.715	3.810	16.357	6.668	29.9	3149.0	1529.4	16.8	13.7903	25.655	353.795	116.415	763
CRA-17	11.748	6.668	5.080	15.054	8.414	28.9	2705.9	2176.7	17.5	12.2580	34.925	428.113	119.770	754
CRA-20-B	13.970	8.890	3.810	18.344	8.255	35.9	2519.2	3812.9	17.3	9.1935	62.080	570.732	121.515	1004
CRA-89-A	16.510	7.620	2.540	20.447	6.350	37.9	3102.3	2801.3	17.3	10.7258	45.610	489.199	121.515	1064
CRA-66-A	13.335	8.255	4.128	17.400	8.256	33.9	2577.8	3287.8	17.3	9.9609	53.528	533.184	122.991	926
CRA-66-C	11.430	7.620	6.350	15.178	10.160	29.9	2624.1	3130.9	19.3	11.4919	45.610	524.141	124.811	846

National Arnold Caseless Toroids 12 mil Silicon Steel														
Part No.	OD cm 1	ID cm 2	HT cm 3	Wound OD cm 4	Wound HT cm 5	MPL cm 6	Wtfe grams 7	Wtcu grams 8	MLT cm 9	Ac cm sq 10	Wa cm sq 11	Ap cm 4th 12	Kg cm 5th 13	At cm sq 14
CRA-94	13.176	10.636	7.620	18.488	12.938	37.4	2624.1	7183.2	22.7	9.1935	88.859	816.931	132.152	1288
CRA-28	15.240	10.160	3.810	20.237	8.890	39.9	2799.1	5273.0	18.3	9.1935	81.084	745.446	149.897	1208
CRA-87-A	12.700	7.620	5.080	16.459	8.890	31.9	2985.7	2966.1	18.3	12.2580	45.610	559.084	149.897	885
CRA-48-E	10.795	7.620	9.208	14.550	13.018	28.9	3065.3	3790.1	23.4	13.8868	45.610	633.372	150.551	927
CRA-20-C	13.970	7.938	4.128	17.905	8.097	34.4	3106.0	3129.5	17.8	11.8275	49.496	585.414	155.764	959
CRA-75	12.700	6.350	5.080	15.915	8.255	29.9	3498.9	2059.8	18.3	15.3226	31.673	485.316	162.648	810
CRA-120-B	20.955	15.240	2.540	28.480	10.160	56.9	2991.5	13512.2	20.8	6.8951	182.438	1257.939	166.577	2182
CRA-47-D	15.240	9.525	3.810	19.929	8.573	38.9	3070.3	4634.5	18.3	10.3427	71.265	737.074	166.740	1160
CRA-99-A	13.970	8.255	4.445	18.047	8.573	34.9	3214.6	3481.0	18.3	12.0665	53.528	645.895	170.466	997
CRA-31-A	10.160	6.350	10.160	13.286	13.335	25.9	3638.8	2746.4	24.4	18.3871	31.673	582.379	175.660	833
CRA-47-B	15.240	8.890	3.810	19.635	8.255	37.9	3323.9	4037.2	18.3	11.4919	62.080	713.415	179.320	1114
CRA-88-A	19.050	7.620	2.540	23.185	6.350	41.9	4408.6	3130.9	19.3	13.7903	45.610	628.970	179.728	1306
CRA-46-A	13.970	6.985	4.445	17.507	7.938	32.9	3704.4	2492.3	18.3	14.7480	38.325	565.211	182.321	918
CRA-48-A	15.240	7.620	3.810	19.099	7.620	35.9	3778.8	2966.1	18.3	13.7903	45.610	628.970	189.713	1030
CRA-91-B	15.240	5.715	4.128	18.427	6.986	32.9	4691.2	1714.8	18.8	18.6766	25.655	479.156	190.437	937
CRA-47-E	13.970	8.890	5.080	18.344	9.525	35.9	3358.9	4261.5	19.3	12.2580	62.080	760.976	193.288	1077
CRA-18-A	12.700	4.763	6.350	15.356	8.732	27.4	5011.2	1287.6	20.3	23.9400	17.820	426.610	201.044	791
CRA-57-A	14.605	7.620	4.445	18.431	8.255	34.9	3928.9	3048.5	18.8	14.7480	45.610	672.648	211.113	1011
CRA-110-B	13.335	9.843	7.620	18.201	12.542	36.4	3511.6	6186.4	22.9	12.6393	76.103	961.887	212.731	1237
CRA-145	12.700	5.398	6.350	15.563	9.049	28.4	4777.9	1653.9	20.3	22.0247	22.888	504.105	218.558	823
CRA-72	13.970	6.985	5.080	17.507	8.573	32.9	4233.6	2630.8	19.3	16.8548	38.325	645.956	225.600	952
CRA-43-C	16.193	11.430	4.763	21.826	10.478	43.4	3568.0	7508.3	20.6	10.7759	102.622	1105.843	231.667	1466
CRA-28-A	15.875	10.160	4.445	20.873	9.525	40.9	3765.6	5712.5	19.8	12.0665	81.084	978.397	238.357	1308
CRA-43-A	15.240	11.430	6.350	20.898	12.065	41.9	3673.8	8156.7	22.4	11.4919	102.622	1179.318	242.531	1477
CRA-201-B	19.685	17.145	6.985	28.355	15.558	57.9	3720.5	22106.6	26.9	8.4274	230.899	1945.875	243.629	2648
CRA-122	10.795	5.080	10.160	13.406	12.700	24.9	5248.3	1794.3	24.9	27.5806	20.271	559.084	247.788	817
CRA-83-A	11.748	8.573	11.430	15.982	15.717	31.9	4198.8	5683.8	27.7	17.2379	57.731	995.166	247.841	1190
CRA-123-C	17.780	11.430	3.810	23.402	9.525	45.9	4023.7	7415.2	20.3	11.4919	102.622	1179.318	266.784	1560
CRA-76-A	13.970	6.350	5.715	17.265	8.890	31.9	5038.4	2288.6	20.3	20.6854	31.673	655.177	266.784	950
CRA-87-C	12.700	7.620	7.620	16.459	11.430	31.9	4478.5	3625.2	22.4	18.3871	45.610	838.626	275.946	1016

National Arnold Caseless Toroids 12 mil Silicon Steel														
Part No.	OD cm 1	ID cm 2	HT cm 3	Wound OD cm 4	Wound HT cm 5	MPL cm 6	Wtfe grams 7	Wtcu grams 8	MLT cm 9	Ac cm sq 10	Wa cm sq 11	Ap cm 4th 12	Kg cm 5th 13	At cm sq 14
CRA-56-B	12.700	5.080	7.620	15.456	10.160	27.9	5878.1	1611.2	22.4	27.5806	20.271	559.084	275.946	868
CRA-47-C	15.240	8.890	5.080	19.635	9.525	37.9	4431.9	4485.8	20.3	15.3226	62.080	951.220	286.912	1193
CRA-110-D	13.653	9.843	8.573	18.510	13.495	36.9	4369.6	6667.9	24.6	15.5150	76.103	1180.735	297.398	1322
CRA-64-A	15.875	8.255	4.763	20.022	8.891	37.9	4986.4	3868.0	20.3	17.2397	53.528	922.804	313.154	1188
CRA-47-F	16.510	8.890	4.445	20.949	8.890	39.9	4898.4	4485.8	20.3	16.0887	62.080	998.781	316.320	1274
CRA-144	18.733	16.193	8.890	26.911	16.987	54.9	4490.3	21394.4	29.2	10.7258	205.969	2209.174	324.475	2572
CRA-40-C	13.970	10.160	8.890	18.987	13.970	37.9	4653.5	7323.7	25.4	16.0887	81.084	1304.530	330.522	1399
CRA-52-B	17.780	12.700	5.080	24.043	11.430	47.9	4478.5	10070.1	22.4	12.2580	126.693	1553.011	340.674	1771
CRA-9-C	15.240	5.080	6.350	18.242	8.890	31.9	7464.2	1611.2	22.4	30.6451	20.271	621.205	340.674	1032
CRA-87-B	13.335	7.620	7.620	17.110	11.430	32.9	5195.8	3707.6	22.9	20.6854	45.610	943.454	341.483	1074
CRA-45-B	15.875	8.255	5.080	20.022	9.208	37.9	5318.3	3964.5	20.8	18.3871	53.528	984.221	347.550	1208
CRA-45-A	16.510	7.620	5.080	20.447	8.890	37.9	6204.6	3460.4	21.3	21.4516	45.610	978.397	393.479	1227
CRA-132-B	40.958	40.323	25.400	61.898	45.562	127.7	7464.3	333386.1	73.4	7.6613	1277.179	9784.822	408.489	14870
CRA-9-D	16.510	5.080	6.350	19.649	8.890	33.9	8922.1	1684.4	23.4	34.4757	20.271	698.855	412.420	1155
CRA-83	14.605	8.573	7.620	18.841	11.907	36.4	6065.8	4901.6	23.9	21.8328	57.731	1260.439	461.031	1262
CRA-83-B	14.288	8.573	8.255	18.517	12.542	35.9	6140.8	5058.1	24.6	22.4092	57.731	1293.715	470.666	1268
CRA-9-F	16.510	5.080	7.144	19.649	9.684	33.9	10037.7	1776.0	24.6	38.7866	20.271	786.240	495.089	1204
CRA-80-A	16.510	9.525	6.033	21.224	10.796	40.9	6246.7	5793.3	22.9	20.0167	71.265	1426.493	499.610	1427
CRA-75-B	19.050	6.350	5.080	22.802	8.255	39.9	9330.3	2631.9	23.4	30.6451	31.673	970.632	509.160	1407
CRA-50-A	16.510	8.255	6.033	20.690	10.161	38.9	7022.4	4351.4	22.9	23.6561	53.528	1266.264	524.127	1332
CRA-123-A	18.733	11.430	5.080	24.366	10.795	47.4	6371.4	8435.0	23.1	17.6221	102.622	1808.412	551.484	1758
CRA-112	18.733	15.558	10.160	26.536	17.939	53.9	6298.1	21123.2	31.2	15.3226	190.131	2913.297	571.520	2600
CRA-112	18.733	15.558	10.160	26.536	17.939	53.9	6298.1	21123.2	31.2	15.3226	190.131	2913.297	571.520	2600
CRA-119-B	20.320	13.970	5.080	27.196	12.065	53.9	6298.0	13292.5	24.4	15.3226	153.299	2348.930	590.413	2192
CRA-92-B	13.653	10.478	17.463	18.852	22.702	37.9	7617.8	11918.0	38.9	26.3364	86.239	2271.220	615.654	1902
CRA-143-A	21.590	16.510	5.715	29.780	13.970	59.9	6298.0	20112.6	26.4	13.7903	214.112	2952.663	616.567	2699
CRA-90-A	19.050	8.890	5.080	23.629	9.525	43.9	8210.7	5158.6	23.4	24.5161	62.080	1521.951	638.690	1583
CRA-123	15.875	11.430	10.160	21.515	15.875	42.9	7021.0	10566.7	29.0	21.4516	102.622	2201.394	652.346	1799
CRA-123-E	15.240	11.430	12.700	20.898	18.415	41.9	7347.6	11864.4	32.5	22.9838	102.622	2358.636	666.960	1894
CRA-110-C	15.558	9.843	9.525	20.401	14.447	39.9	7872.7	7492.5	27.7	25.8568	76.103	1967.777	735.096	1579

National Arnold Caseless Toroids 12 mil Silicon Steel														
Part No.	OD cm 1	ID cm 2	HT cm 3	Wound OD cm 4	Wound HT cm 5	MPL cm 6	Wtfe grams 7	Wtcu grams 8	MLT cm 9	Ac cm sq 10	Wa cm sq 11	Ap cm 4th 12	Kg cm 5th 13	At cm sq 14
CRA-44-A	16.510	8.255	7.620	20.690	11.748	38.9	8869.6	4834.8	25.4	29.8790	53.528	1599.359	752.554	1435
CRA-128	17.145	12.700	10.160	23.425	16.510	46.9	7674.2	13503.0	30.0	21.4516	126.693	2717.770	778.065	2076
CRA-140-A	14.288	9.208	12.700	18.817	17.304	36.9	8630.9	7519.5	31.8	30.6451	66.600	2040.976	787.970	1578
CRA-40-B	15.558	12.065	15.240	21.552	21.273	43.4	8372.4	14975.1	36.8	25.2858	114.341	2891.199	793.978	2169
CRA-90-B	20.320	10.160	5.080	25.465	10.160	47.9	8957.1	7030.7	24.4	24.5161	81.084	1987.855	799.449	1831
CRA-137	17.780	13.335	10.160	24.381	16.828	48.9	8000.7	15139.4	30.5	21.4516	139.679	2996.342	843.520	2222
CRA-52-A	21.590	12.700	5.080	27.865	11.430	53.9	8817.1	11443.2	25.4	21.4516	126.693	2717.770	918.117	2219
CRA-49-A	19.050	8.890	6.350	23.629	10.795	43.9	10263.3	5607.2	25.4	30.6451	62.080	1902.439	918.117	1678
CRA-123-D	17.780	11.430	8.255	23.402	13.970	45.9	8718.0	10010.5	27.4	24.8991	102.622	2555.189	927.705	1886
CRA-40-D	15.875	12.065	15.240	21.856	21.273	43.9	9237.0	15078.2	37.1	27.5806	114.341	3153.584	938.170	2210
CRA-44-C	21.590	11.430	5.080	27.314	10.795	51.9	9703.5	9269.0	25.4	24.5161	102.622	2515.879	971.330	2097
CRA-127-A	22.225	19.685	15.240	32.212	25.083	65.8	9237.0	45637.4	42.2	18.3871	304.381	5596.665	976.247	4166
CRA-87-D	12.700	7.620	20.320	16.459	24.130	31.9	11942.8	6920.9	42.7	49.0322	45.610	2236.337	1027.863	1672
CRA-111-A	17.145	13.018	13.335	23.597	19.844	47.4	9451.4	16592.4	35.1	26.1409	133.117	3479.812	1038.064	2345
CRA-111	18.098	13.018	10.160	24.521	16.669	48.9	9144.0	14548.6	30.7	24.5161	133.117	3263.516	1041.291	2227
CRA-27-B	20.320	10.160	6.033	25.465	11.113	47.9	10637.4	7470.4	25.9	29.1153	81.084	2360.773	1061.176	1907
CRA-120-A	23.495	15.240	5.080	30.990	12.700	60.9	9248.7	17467.0	26.9	19.9193	182.438	3634.047	1075.438	2744
CRA-51-A	19.050	9.525	7.303	23.873	12.066	44.9	11317.4	6823.2	26.9	33.0415	71.265	2354.703	1155.855	1799
CRA-127-C	22.860	19.685	12.700	32.794	22.543	66.8	9767.7	41788.4	38.6	19.1532	304.381	5829.859	1156.863	4010
CRA-93	13.176	4.048	25.400	15.680	27.424	27.1	22737.2	2342.6	51.2	110.1293	12.871	1417.523	1220.074	1736
CRA-127-D	22.860	19.685	13.335	32.794	23.178	66.8	10256.0	42888.1	39.6	20.1108	304.381	6121.352	1242.738	4075
CRA-44-B	18.098	10.160	8.890	23.142	13.970	44.4	11354.0	8275.9	28.7	33.5202	81.084	2717.941	1269.663	1856
CRA-43-D	16.193	11.430	15.240	21.826	20.955	43.4	11416.4	13625.6	37.3	34.4794	102.622	3538.326	1306.957	2184
CRA-119-A	19.368	13.970	10.160	26.262	17.145	52.4	10410.2	17308.1	31.8	26.0507	153.299	3993.551	1310.660	2497
CRA-27-A	20.320	10.160	6.985	25.465	12.065	47.9	12316.0	7909.6	27.4	33.7096	81.084	2733.300	1343.518	1983
CRA-140	17.304	9.208	10.160	21.911	14.764	41.7	12416.6	7128.4	30.1	39.0713	66.600	2602.164	1351.131	1770
CRA-48-D	15.240	7.620	15.240	19.099	19.050	35.9	15115.1	5932.2	36.6	55.1612	45.610	2515.879	1517.704	1715
CRA-143	21.590	16.510	10.160	29.780	18.415	59.9	11196.4	25527.6	33.5	24.5161	214.112	5249.179	1535.305	3114
CRA-44-D	17.780	10.160	10.795	22.814	15.875	43.9	13085.7	9081.4	31.5	39.0725	81.084	3168.143	1572.102	1954
CRA-51-C	33.655	25.400	4.128	46.237	16.828	92.8	11458.0	60421.8	33.5	16.1864	506.773	8202.835	1584.004	5800

National Arnold Caseless Toroids 12 mil Silicon Steel

Part No.	OD cm 1	ID cm 2	HT cm 3	Wound OD cm 4	Wound HT cm 5	MPL cm 6	Wtfe grams 7	Wtcu grams 8	MLT cm 9	Ac cm sq 10	Wa cm sq 11	Ap cm 4th 12	Kg cm 5th 13	At cm sq 14
CRA-127-B	22.860	19.685	16.510	32.794	26.353	66.8	12698.0	48386.6	44.7	24.8991	304.381	7578.817	1688.494	4402
CRA-144-A	21.273	16.193	11.430	29.301	19.527	58.9	12386.3	25859.3	35.3	27.5806	205.969	5680.733	1775.066	3145
CRA-53-B	24.130	15.240	6.350	31.629	13.970	61.9	12654.2	19114.8	29.5	26.8145	182.438	4891.986	1780.831	2958
CRA-23-D	20.955	7.303	8.890	25.172	12.542	44.4	19526.9	4616.4	31.0	57.6490	41.894	2415.130	1797.209	1986
CRA-53-D	24.130	10.795	6.350	29.759	11.748	54.9	16838.3	9590.6	29.5	40.2217	91.536	3681.729	2010.391	2488
CRA-51-E	34.290	25.400	4.445	46.850	17.145	93.8	13429.8	62251.2	34.5	18.7701	506.773	9512.195	2067.451	5968
CRA-124	18.415	12.700	15.240	24.667	21.590	48.9	15430.0	17622.6	39.1	41.3709	126.693	5241.414	2217.424	2628
CRA-139	32.385	22.225	5.080	43.324	16.193	85.8	16048.1	46960.2	34.0	24.5161	387.998	9512.195	2740.648	5150
CRA-120	20.955	15.240	15.240	28.480	22.860	56.9	17949.2	26694.8	41.1	41.3709	182.438	7547.636	3035.408	3318
CRA-200	35.560	25.400	5.080	48.086	17.780	95.8	17914.2	65913.1	36.6	24.5161	506.773	12424.092	3331.037	6315
CRA-112-A	21.590	15.558	15.240	29.267	23.019	58.4	19443.6	28163.9	41.7	43.6656	190.131	8302.208	3481.096	3460
CRA-18-E	18.415	4.763	25.400	21.702	27.782	36.4	45761.5	3508.8	55.4	164.7114	17.820	2935.154	3492.403	2633
CRA-136	21.590	15.875	17.780	29.435	25.718	58.9	21675.4	32184.1	45.7	48.2660	197.958	9554.660	4034.694	3737
CRA-200-D	36.830	25.400	5.080	49.334	17.780	97.8	20573.3	67744.0	37.6	27.5806	506.773	13977.103	4101.902	6575
CRA-200-K	36.830	25.400	5.080	49.334	17.780	97.8	20573.3	67744.0	37.6	27.5806	506.773	13977.103	4101.902	6575
CRA-113-A	21.273	16.828	24.130	29.654	32.544	59.9	23268.0	44000.1	55.6	50.9475	222.439	11332.713	4151.792	4411
CRA-50-C	40.640	30.480	5.080	55.729	20.320	111.7	20899.9	105460.9	40.6	24.5161	729.753	17890.692	4317.024	8432
CRA-133-B	25.400	20.320	16.510	35.535	26.670	71.8	21832.9	53902.2	46.7	39.8386	324.335	12921.055	4405.659	4958
CRA-201-A	23.495	17.145	15.240	31.963	23.813	63.8	22392.7	35454.0	43.2	45.9677	230.899	10613.863	4519.633	3994
CRA-141-A	31.750	22.225	8.255	42.698	19.368	84.8	24164.0	53268.3	38.6	37.3487	387.998	14491.235	5607.429	5459
CRA-127	26.035	19.685	15.240	35.788	25.083	71.8	25191.8	48936.5	45.2	45.9677	304.381	13991.663	5690.205	4829
CRA-200-B	39.370	27.940	5.715	53.143	19.685	105.7	25034.4	88616.5	40.6	31.0282	613.196	19026.332	5810.553	7719
CRA-133	25.400	20.320	20.320	35.535	30.480	71.8	26871.3	60933.0	52.8	49.0322	324.335	15902.838	5903.623	5384
CRA-141	34.925	22.225	6.350	45.859	17.463	89.8	26241.5	52567.4	38.1	38.3064	387.998	14862.805	5977.325	5816
CRA-133-A	26.670	20.320	15.240	36.745	25.400	73.8	25891.6	52730.5	45.7	45.9677	324.335	14908.910	5995.867	5051
CRA-27-C	20.320	10.160	20.320	25.465	25.400	47.9	35828.3	14061.5	48.8	98.0643	81.084	7951.419	6395.591	3049
CRA-136-C	21.590	15.875	30.480	29.435	38.418	58.9	37157.9	46488.2	66.0	82.7418	197.958	16379.418	8208.734	4911
CRA-136-B	30.163	22.860	13.970	41.492	25.400	83.3	30800.4	67849.9	46.5	48.4609	410.486	19892.527	8295.694	6012
CRA-200-C	40.640	27.940	6.350	54.393	20.320	107.7	31489.8	93047.3	42.7	38.3064	613.196	23489.299	8434.476	8115
CRA-15-C	38.418	30.480	10.160	53.604	25.400	108.2	31637.8	121940.2	47.0	38.3088	729.753	27955.967	9116.409	8787

National Arnold Caseless Toroids 12 mil Silicon Steel														
Part No.	OD cm 1	ID cm 2	HT cm 3	Wound OD cm 4	Wound HT cm 5	MPL cm 6	Wtfe grams 7	Wtcu grams 8	MLT cm 9	Ac cm sq 10	Wa cm sq 11	Ap cm 4th 12	Kg cm 5th 13	At cm sq 14
CRA-50-D	30.480	15.240	10.160	38.197	17.780	71.8	40306.9	26365.2	40.6	73.5482	182.438	13418.019	9713.304	4423
CRA-200-E	40.640	27.940	6.985	54.393	20.955	107.7	34638.7	95262.7	43.7	42.1370	613.196	25838.228	9968.374	8224
CRA-200-H	40.640	27.940	6.985	54.393	20.955	107.7	34638.7	95262.7	43.7	42.1370	613.196	25838.228	9968.374	8224
CRA-200-J	33.973	29.210	21.590	48.710	36.195	99.3	36993.7	147100.4	61.7	48.8458	670.208	32736.793	10362.873	9261
CRA-88-B	22.860	7.620	25.400	27.362	29.210	47.9	67178.1	9557.4	58.9	183.8706	45.610	8386.262	10466.922	3685
CRA-121	34.290	29.210	20.320	49.003	34.925	99.8	37321.2	142862.0	59.9	49.0322	670.208	32861.723	10751.910	9144
CRA-115-A	38.418	30.480	12.700	53.604	27.940	108.2	39547.2	132486.3	51.1	47.8860	729.753	34944.959	13110.516	9214
CRA-135-A	41.593	33.655	12.700	58.407	29.528	118.2	43192.1	169561.2	53.6	47.8860	889.704	42604.336	15226.595	10771
CRA-136-A	32.385	22.860	15.240	43.650	26.670	86.8	45660.1	73410.7	50.3	68.9515	410.486	28303.634	15521.970	6647
CRA-118	31.750	24.130	20.320	43.711	32.385	87.8	49264.0	94187.0	57.9	73.5482	457.363	33638.229	17088.220	7445
CRA-53-H	23.813	15.240	33.655	31.310	41.275	61.4	64155.1	47292.9	72.9	137.0490	182.438	25003.002	18802.265	5597
CRA-151	45.720	33.020	8.890	62.015	25.400	123.7	50616.9	154712.6	50.8	53.6289	856.447	45930.314	19395.224	10984
CRA-132	47.308	40.323	15.240	67.621	35.402	137.7	53113.3	282628.7	62.2	50.5644	1277.179	64579.823	20989.362	14696
CRA-134	33.338	25.400	21.590	45.932	34.290	92.3	57316.2	110313.6	61.2	81.4062	506.773	41254.466	21944.956	8258
CRA-142	45.720	35.560	12.383	63.393	30.163	127.7	58223.4	199170.2	56.4	59.7604	993.275	59358.495	25163.046	12313
CRA-126-A	36.830	29.210	21.590	51.383	36.195	103.7	61859.9	152547.6	64.0	78.1450	670.208	52373.371	25576.286	9985
CRA-126	38.418	29.210	17.780	52.898	32.385	106.2	63040.3	141046.9	59.2	77.7662	670.208	52119.469	27394.166	9772
CRA-130	40.005	32.385	21.590	56.185	37.783	113.7	67807.9	194953.4	66.5	78.1450	823.823	64377.664	30238.582	11622
CRA-135	41.593	33.655	21.590	58.407	38.418	118.2	73426.5	214562.9	67.8	81.4062	889.704	72427.372	34775.431	12402
CRA-146	58.420	41.910	8.890	79.093	29.845	157.6	83844.4	299080.6	61.0	69.7176	1379.690	96188.678	44002.913	17234
CRA-132-A	47.308	40.323	25.400	67.621	45.562	137.7	88522.2	356457.7	78.5	84.2740	1277.179	107633.038	46227.980	16853
CRA-201	36.830	17.145	22.860	45.667	31.433	84.8	138292.5	54223.8	66.0	213.7496	230.899	49354.462	63897.609	7781
CRA-117	48.260	22.860	12.700	59.987	24.130	111.7	130624.2	86016.6	58.9	153.2255	410.486	62896.965	65418.266	10195
CRA-138	76.200	68.580	25.400	111.104	59.690	227.4	159548.1	1334739.8	101.6	91.9353	3694.376	339643.610	122934.005	40204

12

E CORES

E CORE MANUFACTURERS

Manufacturers' addresses and phone numbers can be found from the Table of Contents.

Notes

E CORE TABLE DESCRIPTION

Definitions for the Following Tables

Information given is listed by column:

Col.	Dim.	Description	Units
1.	D	Strip width	cm
2.	Ht	Finished transformer height (Ht = G + 2E)	cm
3.	Wth	Finished transformer width (Wth = D + F)	cm
4.	Lt	Finished transformer length [Lt = 3(E + F)]	cm
5.	G	Window length	cm
6.	W_{tfe}	Core weight	gram
7.	W_{tcu}	Copper weight [W_{tcu} = (MLT) $(W_a)(K_u)(8.89)(1.5)$]	gram
8.	MLT	Mean Length Turn [MLT = 2(D + 2J) + 2(E + 2J) + 3(3.14)(F)(0.5), J = Bobbin Thickness]	cm
9.	A_c	Effective iron area [A_c = (D)(E)SF]	cm sq
10.	W_a	Window area [W_a = (F)(G)]	cm sq
11.	A_p	Area product [$A_p = (A_c)(W_a)(1.5)$]	cm 4th
12.	K_g	Core geometry [$K_g = (A_p)(A_c)(K_u)/(\text{MLT})$]	cm 5th
13.	A_t	Surface area	cm sq

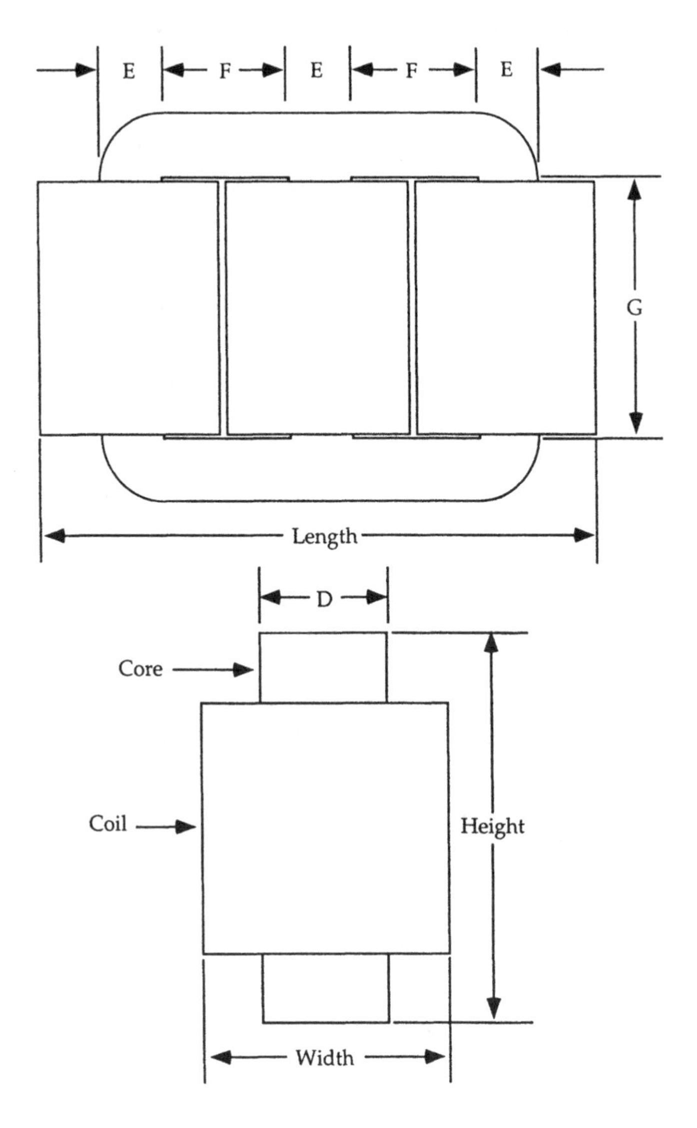

Figure 141 Mechanical outline of E cores.

Magnetics E Cores 0.5 mil Iron Alloy													
Part No.	D cm 1	Ht cm 2	Wth cm 3	Lt cm 4	G cm 5	Wtfe grams 6	Wtcu grams 7	MLT cm 8	Ac cm sq 9	Wa cm sq 10	Ap cm 4th 11	Kg cm 5th 12	At cm sq 13
ME1237-05	1.02	3.18	1.65	3.81	1.91	72.2	30.4	4.7	0.387	1.210	0.703	0.023	44
ME2160-05	0.95	3.81	2.07	5.24	2.54	71.1	80.3	5.3	0.363	2.826	1.538	0.042	72
ME1573-05	0.95	5.08	2.54	6.20	4.13	58.1	201.3	5.8	0.273	6.552	2.682	0.051	114
ME0043-05	0.95	4.76	2.22	6.67	2.86	129.8	120.2	6.2	0.544	3.629	2.963	0.104	105
ME0053-05	1.91	3.81	3.02	5.24	2.54	140.7	112.1	7.4	0.726	2.826	3.076	0.120	89
ME1555-05	1.27	5.08	2.54	6.67	3.18	186.1	151.6	7.0	0.726	4.032	4.390	0.181	118
ME2180-05	1.91	5.24	3.18	5.72	3.97	190.7	206.6	7.7	0.726	5.042	5.489	0.207	123
ME2540-05	1.59	4.92	2.86	6.67	3.02	222.5	157.0	7.7	0.907	3.832	5.215	0.246	122
ME0025-05	1.27	6.03	2.86	7.62	4.13	217.9	263.7	7.5	0.726	6.552	7.133	0.274	158
ME1151-05	1.91	4.76	3.18	6.67	2.86	258.8	161.0	8.3	1.089	3.629	5.926	0.310	126
ME1572-05	3.49	4.13	4.68	5.48	2.86	281.5	194.9	10.7	1.331	3.404	6.794	0.337	129
ME1559-05	1.91	5.88	3.18	6.67	3.97	304.2	223.7	8.3	1.089	5.042	8.234	0.431	148
ME1551-05	2.54	6.35	4.92	9.05	5.08	408.6	690.7	10.7	0.968	12.103	17.569	0.636	255
ME3800-05	1.91	6.99	3.81	8.57	5.08	390.4	480.8	9.3	1.089	9.677	15.804	0.739	224
ME6200-05	1.91	6.03	3.49	8.57	3.49	435.8	279.5	9.5	1.452	5.544	12.073	0.742	189
ME1571-05	3.81	6.19	5.24	6.20	4.92	444.5	448.7	12.0	1.452	7.039	15.328	0.745	199
ME1600-05	3.81	7.62	5.08	5.72	6.35	503.9	503.3	11.7	1.452	8.065	17.560	0.871	216
ME0067-05	1.59	7.62	3.49	9.53	5.08	454.0	480.8	9.3	1.210	9.677	17.560	0.912	246
ME1561-05	2.86	5.72	4.13	7.62	3.18	599.3	233.5	10.9	2.177	4.032	13.170	1.057	184
ME0078-05	2.54	6.51	3.81	7.62	3.97	585.7	274.9	10.2	1.935	5.042	14.638	1.109	194
ME1810-05	2.54	6.99	4.45	8.57	5.08	508.5	546.3	10.6	1.452	9.677	21.072	1.156	244
ME0090-05	2.54	7.62	4.45	9.53	5.08	721.9	579.1	11.2	1.935	9.677	28.096	1.939	278
ME1558-05	2.07	7.78	4.29	11.43	4.61	780.9	622.5	11.4	1.967	10.235	30.196	2.084	314
ME3001-05	1.91	8.26	4.45	12.38	5.08	780.9	797.1	11.6	1.815	12.903	35.120	2.201	356
ME3002-05	3.81	6.67	5.72	9.53	4.13	994.3	585.8	14.0	2.903	7.863	34.241	2.847	293
ME2250-05	3.81	7.62	5.72	9.53	5.08	1094.1	721.0	14.0	2.903	9.677	42.143	3.504	322
ME1440-05	2.54	10.16	5.08	11.43	7.62	971.6	1261.1	12.2	1.935	19.355	56.191	3.561	426
ME1570-05	2.54	8.89	5.08	14.29	4.45	1539.1	850.4	14.1	3.387	11.290	57.362	5.504	444
ME1570-05	2.54	8.89	5.08	14.29	4.45	1539.1	850.4	14.1	3.387	11.290	57.362	5.504	444
ME1560-05	2.54	10.03	5.99	15.12	6.86	1307.5	1803.5	14.3	2.419	23.673	85.909	5.821	561

Magnetics E Cores 0.5 mil Iron Alloy													
Part No.	D cm 1	Ht cm 2	Wth cm 3	Lt cm 4	G cm 5	Wtfe grams 6	Wtcu grams 7	MLT cm 8	Ac cm sq 9	Wa cm sq 10	Ap cm 4th 11	Kg cm 5th 12	At cm sq 13
ME1567-05	3.18	13.34	7.30	15.24	11.43	1480.0	3861.0	15.3	1.815	47.177	128.406	6.074	792
ME1566-05	3.81	10.16	6.35	11.43	7.62	1452.8	1544.8	15.0	2.903	19.355	84.287	6.541	481
ME1562-05	3.18	10.80	5.72	12.38	7.62	1593.5	1457.8	14.1	3.024	19.355	87.799	7.521	500
ME9200-05	3.81	9.53	5.72	11.43	5.72	1947.7	884.8	15.2	4.355	10.887	71.117	8.130	428
ME1565-05	3.81	11.28	9.14	21.11	7.87	2651.4	4528.5	20.2	3.890	42.000	245.089	18.868	996

Magnetics E Cores 1 mil Iron Alloy

Part No.	D cm 1	Ht cm 2	Wth cm 3	Lt cm 4	G cm 5	Wtfe grams 6	Wtcu grams 7	MLT cm 8	Ac cm sq 9	Wa cm sq 10	Ap cm 4th 11	Kg cm 5th 12	At cm sq 13
ME1237-1	1.02	3.18	1.65	3.81	1.91	72.2	30.4	4.7	0.516	1.210	0.937	0.041	44
ME2160-1	0.95	3.81	2.07	5.24	2.54	71.1	80.3	5.3	0.484	2.826	2.051	0.074	72
ME1573-1	0.95	5.08	2.54	6.20	4.13	58.1	201.3	5.8	0.364	6.552	3.576	0.090	114
ME0043-1	0.95	4.76	2.22	6.67	2.86	129.8	120.2	6.2	0.726	3.629	3.951	0.185	105
ME0053-1	1.91	3.81	3.02	5.24	2.54	140.7	112.1	7.4	0.968	2.826	4.102	0.214	89
ME1555-1	1.27	5.08	2.54	6.67	3.18	186.1	151.6	7.0	0.968	4.032	5.854	0.322	118
ME2180-1	1.91	5.24	3.18	5.72	3.97	190.7	206.6	7.7	0.968	5.042	7.319	0.369	123
ME2540-1	1.59	4.92	2.86	6.67	3.02	222.5	157.0	7.7	1.210	3.832	6.954	0.438	122
ME0025-1	1.27	6.03	2.86	7.62	4.13	217.9	263.7	7.5	0.968	6.552	9.511	0.488	158
ME1151-1	1.91	4.76	3.18	6.67	2.86	258.8	161.0	8.3	1.452	3.629	7.902	0.552	126
ME1572-1	3.49	4.13	4.68	5.48	2.86	281.5	194.9	10.7	1.774	3.404	9.059	0.599	129
ME1559-1	1.91	5.88	3.18	6.67	3.97	304.2	223.7	8.3	1.452	5.042	10.978	0.766	148
ME1551-1	2.54	6.35	4.92	9.05	5.08	408.6	690.7	10.7	1.290	12.103	23.426	1.130	255
ME3800-1	1.91	6.99	3.81	8.57	5.08	390.4	480.8	9.3	1.452	9.677	21.072	1.314	224
ME6200-1	1.91	6.03	3.49	8.57	3.49	435.8	279.5	9.5	1.936	5.544	16.097	1.319	189
ME1571-1	3.81	6.19	5.24	6.20	4.92	444.5	448.7	12.0	1.936	7.039	20.437	1.324	199
ME1600-1	3.81	7.62	5.08	5.72	6.35	503.9	503.3	11.7	1.936	8.065	23.413	1.549	216
ME0067-1	1.59	7.62	3.49	9.53	5.08	454.0	480.8	9.3	1.613	9.677	23.413	1.622	246
ME1561-1	2.86	5.72	4.13	7.62	3.18	599.3	233.5	10.9	2.903	4.032	17.560	1.878	184
ME0078-1	2.54	6.51	3.81	7.62	3.97	585.7	274.9	10.2	2.581	5.042	19.517	1.971	194
ME1810-1	2.54	6.99	4.45	8.57	5.08	508.5	546.3	10.6	1.936	9.677	28.096	2.055	244
ME0090-1	2.54	7.62	4.45	9.53	5.08	721.9	579.1	11.2	2.581	9.677	37.461	3.447	278
ME1558-1	2.07	7.78	4.29	11.43	4.61	780.9	622.5	11.4	2.623	10.235	40.262	3.704	314
ME3001-1	1.91	8.26	4.45	12.38	5.08	780.9	797.1	11.6	2.419	12.903	46.826	3.913	356
ME3002-1	3.81	6.67	5.72	9.53	4.13	994.3	585.8	14.0	3.871	7.863	45.655	5.061	293
ME2250-1	3.81	7.62	5.72	9.53	5.08	1094.1	721.0	14.0	3.871	9.677	56.191	6.229	322
ME1440-1	2.54	10.16	5.08	11.43	7.62	971.6	1261.1	12.2	2.581	19.355	74.922	6.331	426
ME1570-1	2.54	8.89	5.08	14.29	4.45	1539.1	850.4	14.1	4.516	11.290	76.482	9.784	444
ME1570-1	2.54	8.89	5.08	14.29	4.45	1539.1	850.4	14.1	4.516	11.290	76.482	9.784	444
ME1560-1	2.54	10.03	5.99	15.12	6.86	1307.5	1803.5	14.3	3.226	23.673	114.546	10.348	561

Magnetics E Cores 1 mil Iron Alloy													
Part No.	D cm 1	Ht cm 2	Wth cm 3	Lt cm 4	G cm 5	Wtfe grams 6	Wtcu grams 7	MLT cm 8	Ac cm sq 9	Wa cm sq 10	Ap cm 4th 11	Kg cm 5th 12	At cm sq 13
ME1567-1	3.18	13.34	7.30	15.24	11.43	1480.0	3861.0	15.3	2.419	47.177	171.208	10.799	792
ME1566-1	3.81	10.16	6.35	11.43	7.62	1452.8	1544.8	15.0	3.871	19.355	112.382	11.629	481
ME1562-1	3.18	10.80	5.72	12.38	7.62	1593.5	1457.8	14.1	4.032	19.355	117.065	13.371	500
ME9200-1	3.81	9.53	5.72	11.43	5.72	1947.7	884.8	15.2	5.806	10.887	94.822	14.454	428
ME1565-1	3.81	11.28	9.14	21.11	7.87	2651.4	4528.5	20.2	5.187	42.000	326.786	33.542	996

Magnetics E Cores 2 mil Iron Alloy													
Part No.	D cm 1	Ht cm 2	Wth cm 3	Lt cm 4	G cm 5	Wtfe grams 6	Wtcu grams 7	MLT cm 8	Ac cm sq 9	Wa cm sq 10	Ap cm 4th 11	Kg cm 5th 12	At cm sq 13
ME1237-2	1.016	3.175	1.651	3.81	1.905	72.2	30.4	4.7	0.548	1.210	0.995	0.046	44
ME2160-2	0.953	3.81	2.065	5.243	2.54	71.1	80.3	5.3	0.514	2.826	2.179	0.084	72
ME1573-2	0.953	5.083	2.54	6.195	4.128	58.1	201.3	5.8	0.387	6.552	3.800	0.102	114
ME0043-2	0.953	4.763	2.223	6.668	2.858	129.8	120.2	6.2	0.771	3.629	4.198	0.209	105
ME0053-2	1.905	3.81	3.018	5.243	2.54	140.7	112.1	7.4	1.028	2.826	4.358	0.241	89
ME1555-2	1.27	5.08	2.54	6.668	3.175	186.1	151.6	7	1.028	4.032	6.219	0.363	118
ME2180-2	1.905	5.24	3.175	5.715	3.97	190.7	206.6	7.7	1.028	5.042	7.777	0.417	123
ME2540-2	1.588	4.923	2.858	6.668	3.018	222.5	157	7.7	1.285	3.832	7.388	0.495	122
ME0025-2	1.27	6.033	2.858	7.62	4.128	217.9	263.7	7.5	1.028	6.552	10.106	0.551	158
ME1151-2	1.905	4.763	3.175	6.668	2.858	258.8	161	8.3	1.542	3.629	8.395	0.623	126
ME1572-2	3.493	4.128	4.684	5.479	2.858	281.5	194.9	10.7	1.885	3.404	9.625	0.677	129
ME1559-2	1.905	5.875	3.175	6.668	3.97	304.2	223.7	8.3	1.542	5.042	11.665	0.866	148
ME1551-2	2.54	6.35	4.923	9.053	5.08	408.6	690.7	10.7	1.371	12.103	24.890	1.277	255
ME3800-2	1.905	6.985	3.81	8.573	5.08	390.4	480.8	9.3	1.542	9.677	22.389	1.484	224
ME6200-2	1.905	6.033	3.493	8.573	3.493	435.8	279.5	9.5	2.056	5.544	17.103	1.490	189
ME1571-2	3.81	6.193	5.24	6.195	4.923	444.5	448.7	12	2.056	7.039	21.714	1.496	199
ME1600-2	3.81	7.62	5.08	5.715	6.35	503.9	503.3	11.7	2.056	8.065	24.876	1.750	216
ME0067-2	1.588	7.62	3.493	9.525	5.08	454	480.8	9.3	1.714	9.677	24.876	1.832	246
ME1561-2	2.858	5.715	4.128	7.62	3.175	599.3	233.5	10.9	3.085	4.032	18.658	2.122	184
ME0078-2	2.54	6.51	3.81	7.62	3.97	585.7	274.9	10.2	2.742	5.042	20.737	2.227	194
ME1810-2	2.54	6.985	4.445	8.573	5.08	508.5	546.3	10.6	2.056	9.677	29.852	2.322	244
ME0090-2	2.54	7.62	4.445	9.525	5.08	721.9	579.1	11.2	2.742	9.677	39.802	3.894	278
ME1558-2	2.065	7.78	4.288	11.43	4.605	780.9	622.5	11.4	2.786	10.235	42.778	4.185	314
ME3001-2	1.905	8.255	4.445	12.383	5.08	780.9	797.1	11.6	2.571	12.903	49.753	4.420	356
ME3002-2	3.81	6.668	5.715	9.525	4.128	994.3	585.8	14	4.113	7.863	48.509	5.718	293
ME2250-2	3.81	7.62	5.715	9.525	5.08	1094.1	721	14	4.113	9.677	59.703	7.037	322
ME1440-2	2.54	10.16	5.08	11.43	7.62	971.6	1261.1	12.2	2.742	19.355	79.604	7.152	426
ME1570-2	2.54	8.89	5.08	14.288	4.445	1539.1	850.4	14.1	4.798	11.290	81.263	11.053	444
ME1570-2	2.54	8.89	5.08	14.288	4.445	1539.1	850.4	14.1	4.798	11.290	81.263	11.053	444
ME1560-2	2.54	10.033	5.992	15.118	6.858	1307.5	1803.5	14.3	3.427	23.673	121.705	11.691	561

Magnetics E Cores 2 mil Iron Alloy													
Part No.	D cm 1	Ht cm 2	Wth cm 3	Lt cm 4	G cm 5	Wtfe grams 6	Wtcu grams 7	MLT cm 8	Ac cm sq 9	Wa cm sq 10	Ap cm 4th 11	Kg cm 5th 12	At cm sq 13
ME1567-2	3.175	13.335	7.303	15.24	11.43	1480	3861	15.3	2.571	47.177	181.909	12.199	792
ME1566-2	3.81	10.16	6.35	11.43	7.62	1452.8	1544.8	15	4.113	19.355	119.406	13.137	481
ME1562-2	3.175	10.795	5.715	12.383	7.62	1593.5	1457.8	14.1	4.284	19.355	124.381	15.105	500
ME9200-2	3.81	9.525	5.715	11.43	5.715	1947.7	884.8	15.2	6.169	10.887	100.749	16.328	428
ME1565-2	3.81	11.278	9.144	21.107	7.874	2651.4	4528.5	20.2	5.511	42.000	347.210	37.892	996

Magnetics E Cores 4 mil Iron Alloy													
Part No.	D cm 1	Ht cm 2	Wth cm 3	Lt cm 4	G cm 5	Wtfe grams 6	Wtcu grams 7	MLT cm 8	Ac cm sq 9	Wa cm sq 10	Ap cm 4th 11	Kg cm 5th 12	At cm sq 13
ME1237-4	1.02	3.18	1.65	3.81	1.91	72.2	30.4	4.7	0.581	1.210	1.054	0.052	44
ME2160-4	0.95	3.81	2.07	5.24	2.54	71.1	80.3	5.3	0.544	2.826	2.308	0.094	72
ME1573-4	0.95	5.08	2.54	6.20	4.13	58.1	201.3	5.8	0.409	6.552	4.023	0.114	114
ME0043-4	0.95	4.76	2.22	6.67	2.86	129.8	120.2	6.2	0.817	3.629	4.445	0.234	105
ME0053-4	1.91	3.81	3.02	5.24	2.54	140.7	112.1	7.4	1.089	2.826	4.614	0.270	89
ME1555-4	1.27	5.08	2.54	6.67	3.18	186.1	151.6	7.0	1.089	4.032	6.585	0.407	118
ME2180-4	1.91	5.24	3.18	5.72	3.97	190.7	206.6	7.7	1.089	5.042	8.234	0.467	123
ME2540-4	1.59	4.92	2.86	6.67	3.02	222.5	157.0	7.7	1.361	3.832	7.823	0.554	122
ME0025-4	1.27	6.03	2.86	7.62	4.13	217.9	263.7	7.5	1.089	6.552	10.700	0.618	158
ME1151-4	1.91	4.76	3.18	6.67	2.86	258.8	161.0	8.3	1.633	3.629	8.889	0.698	126
ME1572-4	3.49	4.13	4.68	5.48	2.86	281.5	194.9	10.7	1.996	3.404	10.192	0.758	129
ME1559-4	1.91	5.88	3.18	6.67	3.97	304.2	223.7	8.3	1.633	5.042	12.351	0.970	148
ME1551-4	2.54	6.35	4.92	9.05	5.08	408.6	690.7	10.7	1.452	12.103	26.354	1.430	255
ME3800-4	1.91	6.99	3.81	8.57	5.08	390.4	480.8	9.3	1.633	9.677	23.706	1.663	224
ME6200-4	1.91	6.03	3.49	8.57	3.49	435.8	279.5	9.5	2.177	5.544	18.109	1.669	189
ME1571-4	3.81	6.19	5.24	6.20	4.92	444.5	448.7	12.0	2.177	7.039	22.991	1.676	199
ME1600-4	3.81	7.62	5.08	5.72	6.35	503.9	503.3	11.7	2.177	8.065	26.339	1.961	216
ME0067-4	1.59	7.62	3.49	9.53	5.08	454.0	480.8	9.3	1.814	9.677	26.339	2.053	246
ME1561-4	2.86	5.72	4.13	7.62	3.18	599.3	233.5	10.9	3.266	4.032	19.755	2.377	184
ME0078-4	2.54	6.51	3.81	7.62	3.97	585.7	274.9	10.2	2.903	5.042	21.956	2.494	194
ME1810-4	2.54	6.99	4.45	8.57	5.08	508.5	546.3	10.6	2.177	9.677	31.608	2.601	244
ME0090-4	2.54	7.62	4.45	9.53	5.08	721.9	579.1	11.2	2.903	9.677	42.143	4.362	278
ME1558-4	2.07	7.78	4.29	11.43	4.61	780.9	622.5	11.4	2.950	10.235	45.294	4.688	314
ME3001-4	1.91	8.26	4.45	12.38	5.08	780.9	797.1	11.6	2.722	12.903	52.680	4.952	356
ME3002-4	3.81	6.67	5.72	9.53	4.13	994.3	585.8	14.0	4.355	7.863	51.362	6.406	293
ME2250-4	3.81	7.62	5.72	9.53	5.08	1094.1	721.0	14.0	4.355	9.677	63.215	7.884	322
ME1440-4	2.54	10.16	5.08	11.43	7.62	971.6	1261.1	12.2	2.903	19.355	84.287	8.013	426
ME1570-4	2.54	8.89	5.08	14.29	4.45	1539.1	850.4	14.1	5.081	11.290	86.043	12.383	444
ME1570-4	2.54	8.89	5.08	14.29	4.45	1539.1	850.4	14.1	5.081	11.290	86.043	12.383	444
ME1560-4	2.54	10.03	5.99	15.12	6.86	1307.5	1803.5	14.3	3.629	23.673	128.864	13.097	561

Magnetics E Cores 4 mil Iron Alloy													
Part No.	D cm 1	Ht cm 2	Wth cm 3	Lt cm 4	G cm 5	Wtfe grams 6	Wtcu grams 7	MLT cm 8	Ac cm sq 9	Wa cm sq 10	Ap cm 4th 11	Kg cm 5th 12	At cm sq 13
ME1567-4	3.18	13.34	7.30	15.24	11.43	1480.0	3861.0	15.3	2.722	47.177	192.609	13.667	792
ME1566-4	3.81	10.16	6.35	11.43	7.62	1452.8	1544.8	15.0	4.355	19.355	126.430	14.718	481
ME1562-4	3.18	10.80	5.72	12.38	7.62	1593.5	1457.8	14.1	4.536	19.355	131.698	16.923	500
ME9200-4	3.81	9.53	5.72	11.43	5.72	1947.7	884.8	15.2	6.532	10.887	106.675	18.293	428
ME1565-4	3.81	11.28	9.14	21.11	7.87	2651.4	4528.5	20.2	5.836	42.000	367.634	42.452	996

National Arnold E Cores 4 mil Silicon Steel													
Part No.	D cm 1	Ht cm 2	Wth cm 3	Lt cm 4	G cm 5	Wtfe grams 6	Wtcu grams 7	MLT cm 8	Ac cm sq 9	Wa cm sq 10	Ap cm 4th 11	Kg cm 5th 12	At cm sq 13
CTH-53-S	0.64	2.86	1.27	3.33	1.91	18	23	3.6	0.272	1.210	0.494	0.0148	32
CTH-43-E	0.64	2.86	1.27	3.81	1.59	29	21	3.9	0.363	1.008	0.549	0.0202	36
CTH-53-A	0.64	3.18	1.27	3.81	1.91	32	26	3.9	0.363	1.210	0.658	0.0242	39
CTH-53-R	0.64	3.25	1.35	4.05	1.98	34	31	4.1	0.363	1.417	0.771	0.0275	42
CTH-53-D	0.95	2.54	2.22	5.24	1.59	37	57	5.3	0.408	2.017	1.235	0.0383	56
CTH-53-B	0.64	4.13	1.51	4.52	2.86	43	58	4.3	0.363	2.495	1.358	0.0457	59
CTH-53-E	0.95	3.18	1.59	3.81	1.91	47	30	4.6	0.545	1.210	0.988	0.0470	43
CTH-43-B	1.27	2.86	1.91	3.81	1.59	57	29	5.4	0.726	1.008	1.098	0.0589	44
CTH-53-F	1.27	3.02	1.91	3.81	1.75	60	32	5.4	0.726	1.109	1.207	0.0647	46
CTH-43-A	0.95	3.49	1.59	4.76	1.59	75	28	5.2	0.817	1.008	1.236	0.0775	55
CTH-503	1.27	3.33	1.83	3.81	1.91	68	31	5.4	0.816	1.059	1.297	0.0777	48
CTH-53-K	0.95	4.13	1.83	4.52	2.86	64	66	5.0	0.545	2.495	2.038	0.0896	65
CTH-97	0.95	3.81	2.06	5.24	2.54	63	80	5.3	0.545	2.822	2.305	0.0943	72
CTH-57	1.11	3.81	1.98	4.52	2.54	68	62	5.3	0.635	2.217	2.112	0.1018	63
CTH-250-A	1.27	3.81	2.06	4.29	2.54	76	61	5.7	0.726	2.017	2.196	0.1125	62
CTH-504	1.59	4.29	1.87	3.69	2.38	108	22	6.1	1.362	0.662	1.352	0.1203	59
CTH-53-N	1.27	3.18	2.22	5.72	1.27	101	42	6.6	1.089	1.210	1.978	0.1315	69
CTH-69-C	1.27	3.81	2.38	5.24	2.54	83	93	6.2	0.726	2.822	3.072	0.1447	77
CTH-97-B	1.91	7.78	2.70	3.34	7.14	101	191	6.3	0.545	5.672	4.639	0.1606	110
CTH-124	0.95	3.81	2.38	7.15	1.91	98	94	6.5	0.817	2.722	3.338	0.1688	96
CTH-69-A	1.11	4.13	2.22	6.19	2.22	131	83	6.3	0.953	2.470	3.530	0.2143	89
CTH-504-A	1.59	4.29	2.14	4.53	2.38	145	46	6.6	1.362	1.324	2.705	0.2245	72
CTH-43	0.95	4.76	2.22	6.67	2.86	106	120	6.2	0.817	3.630	4.450	0.2342	105
CTH-503-A	1.27	4.45	2.22	5.48	2.70	133	88	6.4	0.998	2.572	3.850	0.2405	86
CTH-6-A	1.27	4.45	2.22	5.72	2.54	1333	85	6.6	1.089	2.421	3.955	0.2631	89
CTH-53	1.91	3.81	3.02	5.24	2.54	124	112	7.4	1.089	2.822	4.608	0.2700	89
CTH-69	1.27	4.45	2.38	6.19	2.54	140	102	6.8	1.089	2.822	4.611	0.2955	97
CTH-96-A	1.91	4.13	3.02	5.24	2.86	131	126	7.4	1.089	3.175	5.185	0.3038	94
CTH-102	1.27	4.76	2.70	6.67	3.18	129	169	7.0	0.908	4.537	6.176	0.3212	116
CTH-500-A	1.59	4.13	2.54	5.72	2.22	156	81	7.2	1.362	2.119	4.328	0.3281	90

National Arnold E Cores 4 mil Silicon Steel													
Part No.	D cm 1	Ht cm 2	Wth cm 3	Lt cm 4	G cm 5	Wtfe grams 6	Wtcu grams 7	MLT cm 8	Ac cm sq 9	Wa cm sq 10	Ap cm 4th 11	Kg cm 5th 12	At cm sq 13
CTH-6-B	1.27	5.24	2.22	5.72	3.33	151	111	6.6	1.089	3.177	5.191	0.3453	101
CTH-3-A	1.43	4.84	2.38	5.72	2.94	160	103	6.9	1.226	2.799	5.146	0.3673	98
CTH-96-B	1.91	4.76	3.02	5.24	3.49	147	154	7.4	1.089	3.881	6.337	0.3713	105
CTH-31-C	1.27	5.72	3.33	8.10	4.45	145	375	7.7	0.726	9.174	9.988	0.3786	174
CTH-4-L	2.54	4.45	3.33	4.29	3.18	95	110	8.2	1.452	2.521	5.489	0.3885	92
CTH-84-C	1.27	5.08	2.54	6.67	3.18	159	152	7.0	1.089	4.032	6.588	0.4073	118
CTH-84-B	1.27	5.08	2.54	6.67	3.18	159	152	7.0	1.089	4.032	6.588	0.4073	118
CTH-114	1.27	5.08	2.62	6.91	3.18	161	164	7.2	1.089	4.283	6.998	0.4252	123
CTH-500	1.91	4.13	2.86	5.72	2.22	189	88	7.8	1.634	2.119	5.192	0.4339	96
CTH-4-J	2.54	4.45	3.49	4.76	3.18	180	136	8.5	1.452	3.026	6.588	0.4525	102
CTH-96	1.91	4.45	3.02	5.72	2.86	177	131	7.8	1.361	3.175	6.484	0.4555	105
CTH-25-D	1.27	6.03	2.54	6.67	4.13	182	197	7.0	1.089	5.243	8.566	0.5296	135
CTH-41-A	1.59	5.08	2.86	6.67	3.18	200	165	7.7	1.362	4.032	8.238	0.5841	125
CTH-4-P	2.54	4.45	3.49	5.24	2.86	226	127	8.8	1.815	2.724	7.416	0.6137	108
CTH-25	1.27	6.03	2.86	7.62	4.13	193	264	7.5	1.089	6.555	10.711	0.6184	158
CTH-41-N	1.43	5.40	2.70	7.14	3.18	222	165	7.7	1.429	4.032	8.642	0.6430	134
CTH-83-C	1.91	5.08	2.74	5.83	2.86	249	101	7.9	1.905	2.381	6.802	0.6521	112
CTH-41-K	1.59	5.08	3.18	7.62	3.18	203	220	8.2	1.362	5.042	10.301	0.6858	146
CTH-135	2.22	4.13	3.49	6.67	2.22	240	135	9.0	1.907	2.823	8.074	0.6878	121
CTH-83-B	1.91	4.76	3.18	6.67	2.86	217	161	8.3	1.634	3.630	8.896	0.6990	127
CTH-55	1.59	4.92	3.49	8.57	3.02	221	266	8.7	1.362	5.745	11.738	0.7367	164
CTH-84-A	1.91	5.08	3.18	6.67	3.18	241	179	8.3	1.634	4.032	9.883	0.7765	133
CTH-40	1.91	4.61	3.49	7.62	2.70	227	202	8.8	1.634	4.286	10.504	0.7786	144
CTH-87-B	1.91	4.45	3.65	8.10	2.54	241	214	9.1	1.634	4.435	10.869	0.7836	150
CTH-77-E	1.59	5.72	3.18	7.62	3.81	232	264	8.2	1.362	6.050	12.361	0.8230	160
CTH-89-A	1.75	5.95	3.65	7.98	4.45	219	388	8.6	1.185	8.468	15.049	0.8295	185
CTH-133	1.91	4.61	3.65	8.10	2.70	239	228	9.1	1.634	4.712	11.550	0.8327	154
CTH-41-Q	1.27	6.03	2.86	8.57	3.49	257	242	8.2	1.452	5.547	12.078	0.8572	171
CTH-74	1.91	5.88	3.18	6.67	3.97	306	224	8.3	1.634	5.041	12.354	0.9707	148
CTH-41-H	2.22	5.08	3.49	6.67	3.18	284	193	9.0	1.907	4.032	11.532	0.9823	140

National Arnold E Cores 4 mil Silicon Steel

Part No.	D cm 1	Ht cm 2	Wth cm 3	Lt cm 4	G cm 5	Wtfe grams 6	Wtcu grams 7	MLT cm 8	Ac cm sq 9	Wa cm sq 10	Ap cm 4th 11	Kg cm 5th 12	At cm sq 13
CTH-83	1.91	5.40	3.18	7.62	2.86	328	173	9.0	2.177	3.630	11.855	1.1534	152
CTH-3	3.18	4.76	4.13	5.72	2.86	353	151	10.4	2.723	2.724	11.126	1.1698	134
CTH-41-L	2.22	5.72	3.49	6.67	3.81	307	231	9.0	1.907	4.839	13.839	1.1787	153
CTH-41-B	2.54	5.08	3.81	6.67	3.18	320	206	9.6	2.179	4.032	13.177	1.1976	148
CTH-78-C	1.91	5.72	3.18	7.62	3.18	348	193	9.0	2.177	4.032	13.170	1.2813	159
CTH-78-B	1.91	5.72	3.18	7.62	3.18	348	193	9.0	2.177	4.032	13.170	1.2813	159
CTH-41-C	1.91	5.24	3.49	8.57	2.70	344	216	9.5	2.177	4.286	13.999	1.2900	170
CTH-4-N	3.18	5.08	4.13	5.72	3.18	458	167	10.4	2.723	3.026	12.360	1.2995	140
CTH-89	1.27	7.62	3.18	9.53	5.08	323	448	8.7	1.452	9.677	21.072	1.4098	235
CTH-60	1.27	6.35	3.49	11.43	3.18	367	370	9.8	1.815	7.058	19.216	1.4216	241
CTH-74-E	2.54	5.88	3.81	6.67	3.97	356	258	9.6	2.179	5.041	16.472	1.4971	164
CTH-74-C	2.70	5.56	3.97	6.67	3.65	365	245	9.9	2.315	4.637	16.101	1.5050	162
CTH-41-P	1.91	5.72	3.49	8.57	3.18	419	254	9.5	2.177	5.042	16.467	1.5175	182
CTH-4-B	2.22	6.03	3.49	7.14	3.81	374	239	9.3	2.223	4.839	16.133	1.5474	167
CTH-41-U	3.18	5.08	4.45	6.67	3.18	399	234	10.9	2.723	4.032	16.471	1.6524	163
CTH-84	1.59	6.67	3.49	9.53	4.13	367	391	9.3	1.815	7.864	21.410	1.6688	221
CTH-41	1.91	6.03	3.49	8.57	3.49	390	280	9.5	2.177	5.547	18.117	1.6695	189
CTH-83-D	2.86	5.08	4.45	7.62	3.18	384	288	10.7	2.451	5.042	18.539	1.6952	178
CTH-56-D	1.91	6.35	3.49	8.34	3.97	199	312	9.3	2.042	6.303	19.305	1.6968	193
CTH-4-E	3.18	5.40	4.13	6.19	3.18	458	172	10.7	3.175	3.026	14.409	1.7138	154
CTH-67-B	1.43	7.62	3.33	9.53	5.08	358	464	9.0	1.633	9.677	23.710	1.7218	240
CTH-4-A	2.22	6.03	3.73	7.86	3.81	387	296	9.6	2.223	5.745	19.156	1.7662	185
CTH-74-D	2.70	6.19	3.97	6.67	4.29	215	288	9.9	2.315	5.443	18.901	1.7668	175
CTH-56	1.91	6.51	3.49	8.57	3.97	413	318	9.5	2.177	6.303	20.586	1.8971	201
CTH-4-K	3.18	5.72	4.45	6.67	3.81	439	280	10.9	2.723	4.839	19.765	1.9829	177
CTH-67	1.59	7.62	3.49	9.53	5.08	407	481	9.3	1.815	9.677	26.348	2.0537	246
CTH-1-H	3.81	5.08	5.08	6.67	3.18	481	265	12.3	3.268	4.032	19.765	2.0943	178
CTH-87	1.91	6.67	3.65	9.05	4.13	432	373	9.7	2.177	7.207	23.541	2.1139	218
CTH-55-B	3.18	4.92	5.08	8.57	3.02	442	363	11.9	2.723	5.745	23.469	2.1565	207
CTH-83-A	1.91	6.67	3.18	9.53	2.86	585	198	10.2	3.266	3.630	17.782	2.2727	211

National Arnold E Cores 4 mil Silicon Steel													
Part No.	D cm 1	Ht cm 2	Wth cm 3	Lt cm 4	G cm 5	Wtfe grams 6	Wtcu grams 7	MLT cm 8	Ac cm sq 9	Wa cm sq 10	Ap cm 4th 11	Kg cm 5th 12	At cm sq 13
CTH-66-E	2.22	6.03	3.97	9.05	3.49	462	336	10.3	2.541	6.099	23.244	2.2858	211
CTH-118	1.91	6.67	3.97	10.00	4.13	453	464	10.2	2.177	8.520	27.828	2.3766	245
CTH-23	1.91	6.67	3.97	10.00	4.13	453	464	10.2	2.177	8.520	27.828	2.3766	245
CTH-41-M	2.54	5.72	4.13	8.57	3.18	559	288	10.7	2.903	5.042	21.957	2.3783	200
CTH-78-A	2.54	6.35	3.81	7.62	3.81	504	264	10.2	2.903	4.839	21.072	2.3939	190
CTH-4-F	2.22	6.67	3.49	8.10	3.81	516	256	9.9	2.859	4.839	20.751	2.3956	197
CTH-78	2.54	6.51	3.81	7.62	3.97	517	275	10.2	2.903	5.041	21.951	2.4938	194
CTH-83-L	3.18	5.40	4.45	7.62	2.86	557	223	11.5	3.629	3.630	19.758	2.4958	185
CTH-64-C	2.22	7.30	4.45	9.53	5.40	429	669	10.5	1.907	12.000	34.319	2.5047	272
CTH-41-R	2.54	6.03	4.13	8.57	3.49	513	317	10.7	2.903	5.547	24.156	2.6165	208
CTH-41-W	3.18	6.19	4.76	7.62	4.29	494	412	11.4	2.723	6.806	27.802	2.6665	214
CTH-66	2.22	6.67	3.97	9.05	4.13	506	397	10.3	2.541	7.207	27.470	2.7014	228
CTH-50	2.06	6.99	3.33	11.43	1.91	914	152	11.8	4.718	2.419	17.123	2.7364	257
CTH-41-T	3.49	6.03	5.08	7.62	4.13	535	419	12.0	2.996	6.555	29.459	2.9436	219
CTH-66-C	2.22	6.03	4.76	11.43	3.49	522	548	11.6	2.541	8.872	33.815	2.9674	279
CTH-77	2.54	6.51	4.13	8.57	3.97	545	360	10.7	2.903	6.303	27.448	2.9730	220
CTH-45-C	3.49	6.99	4.76	6.67	5.08	562	396	11.5	2.996	6.452	28.993	3.0229	214
CTH-25-B	2.22	6.99	3.81	9.05	4.13	565	364	10.4	2.859	6.555	28.112	3.0897	231
CTH-4-C	2.54	6.83	3.77	8.21	3.81	1254	266	10.6	3.447	4.686	24.233	3.1419	211
CTH-83-J	3.18	5.72	4.76	8.57	3.18	619	323	12.0	3.629	5.042	27.446	3.3225	218
CTH-65	2.54	6.67	4.29	9.05	4.13	570	422	11.0	2.903	7.207	31.387	3.3229	238
CTH-65-C	2.22	6.83	3.89	9.77	3.65	623	352	10.8	3.177	6.086	29.005	3.3982	241
CTH-24-D	1.91	10.16	2.70	8.10	6.35	944	255	9.5	3.266	5.042	24.701	3.4060	251
CTH-296-A	1.59	8.81	3.57	10.72	5.64	562	601	10.1	2.270	11.182	38.067	3.4301	303
CTH-31-D	2.54	6.35	4.60	10.00	3.81	577	481	11.5	2.903	7.864	34.246	3.4677	257
CTH-66-B	2.22	6.67	4.76	11.43	4.13	559	648	11.6	2.541	10.485	39.962	3.5069	299
CTH-351	2.54	6.99	4.29	9.05	4.45	590	454	11.0	2.903	7.761	33.798	3.5781	246
CTH-74-A	4.45	6.35	5.72	6.67	4.45	668	410	13.6	3.813	5.645	32.283	3.6184	224
CTH-41-D	3.18	6.03	4.76	8.57	3.49	646	355	12.0	3.629	5.547	30.195	3.6553	227
CTH-77-A	2.54	6.83	4.13	9.05	3.97	631	371	11.0	3.267	6.303	30.884	3.6556	237

National Arnold E Cores 4 mil Silicon Steel													
Part No.	D cm 1	Ht cm 2	Wth cm 3	Lt cm 4	G cm 5	Wtfe grams 6	Wtcu grams 7	MLT cm 8	Ac cm sq 9	Wa cm sq 10	Ap cm 4th 11	Kg cm 5th 12	At cm sq 13
CTH-31	1.91	6.99	3.97	11.91	3.18	689	401	11.5	3.266	6.553	32.105	3.6573	285
CTH-89-D	2.22	7.94	4.13	9.53	5.40	574	581	10.6	2.541	10.283	39.193	3.7633	276
CTH-127	2.54	7.30	3.81	8.57	4.13	701	304	10.9	3.630	5.243	28.547	3.8177	230
CTH-65-A	2.54	7.30	4.29	9.05	4.76	605	487	11.0	2.903	8.316	36.216	3.8341	255
CTH-74-F	5.08	5.88	6.35	6.67	3.97	719	400	14.9	4.357	5.041	32.944	3.8597	228
CTH-66-A	2.22	8.89	3.81	8.57	6.35	61	543	10.1	2.541	10.084	38.433	3.8724	269
CTH-65-B	2.54	6.99	4.21	9.29	4.13	696	410	11.2	3.267	6.881	33.719	3.9469	248
CTH-47-A	2.22	7.62	3.81	9.53	4.45	668	404	10.7	3.177	7.059	33.639	3.9867	256
CTH-41-E	3.18	6.35	4.76	8.57	3.81	219	387	12.0	3.629	6.050	32.935	3.9870	235
CTH-39	2.22	9.21	3.81	8.57	6.67	623	570	10.1	2.541	10.589	40.357	4.0663	277
CTH-88	2.54	7.78	4.29	9.05	5.24	635	535	11.0	2.903	9.147	39.835	4.2172	267
CTH-24-B	2.54	7.62	4.92	10.48	5.40	594	799	11.6	2.540	12.853	48.964	4.2704	317
CTH-41-J	3.18	6.67	4.76	8.57	4.13	694	419	12.0	3.629	6.555	35.684	4.3198	243
CTH-22	3.02	7.78	5.24	9.53	5.87	607	838	12.0	2.587	13.058	50.668	4.3558	313
CTH-90	2.54	7.62	4.45	9.53	5.08	647	579	11.2	2.903	9.677	42.143	4.3624	278
CTH-47-F	4.45	6.35	6.03	7.62	4.45	707	531	14.1	3.813	7.059	40.366	4.3643	253
CTH-28	3.18	7.94	5.40	9.53	6.03	647	884	12.4	2.723	13.411	54.783	4.8303	323
CTH-69-B	4.13	5.40	5.80	9.29	2.54	926	329	14.5	5.309	4.234	33.719	4.9223	251
CTH-29	2.54	6.35	4.76	12.38	2.54	873	391	13.0	4.355	5.646	36.884	4.9468	302
CTH-70-D	3.18	6.67	5.08	9.53	4.13	727	524	12.5	3.629	7.864	42.807	4.9756	272
CTH-47-B	2.22	7.54	4.45	11.43	4.37	561	607	11.7	3.177	9.706	46.254	5.0154	313
CTH-4-D	2.86	7.62	4.09	8.69	4.29	858	326	11.6	4.288	5.272	33.907	5.0181	249
CTH-347-C	2.54	7.15	4.45	10.48	3.97	767	478	11.9	3.630	7.561	41.171	5.0430	282
CTH-49	2.86	7.62	4.45	9.05	4.76	780	471	11.7	3.676	7.564	41.702	5.2516	268
CTH-505-D	3.49	7.46	4.76	7.86	4.76	839	396	12.3	4.241	6.049	38.479	5.3129	250
CTH-77-F	3.81	6.51	5.40	8.57	3.97	818	453	13.5	4.355	6.303	41.171	5.3246	258
CTH-47-H	2.22	8.26	3.81	10.48	4.45	851	428	11.4	3.811	7.059	40.354	5.4170	292
CTH-47-E	2.86	8.10	4.13	8.57	4.92	879	383	11.5	4.085	6.250	38.292	5.4432	260
CTH-32	1.91	8.57	3.97	11.91	4.76	807	601	11.5	3.266	9.831	48.163	5.4866	335
CTH-505-C	2.86	7.94	4.21	8.81	4.76	862	398	11.6	4.085	6.425	39.368	5.5364	264

National Arnold E Cores 4 mil Silicon Steel													
Part No.	D cm 1	Ht cm 2	Wth cm 3	Lt cm 4	G cm 5	Wtfe grams 6	Wtcu grams 7	MLT cm 8	Ac cm sq 9	Wa cm sq 10	Ap cm 4th 11	Kg cm 5th 12	At cm sq 13
CTH-347-A	3.18	7.14	5.08	9.53	4.60	766	584	12.5	3.629	8.771	47.743	5.5493	286
CTH-30-B	3.18	6.83	4.76	9.53	3.65	870	391	12.6	4.538	5.798	39.463	5.6726	266
CTH-505-B	2.86	8.10	4.21	8.81	4.92	876	411	11.6	4.085	6.638	40.674	5.7200	268
CTH-47	2.86	8.10	4.45	9.53	4.92	907	500	12.0	4.085	7.815	47.880	6.5228	291
CTH-505-E	3.81	7.62	5.08	8.10	4.76	996	429	13.3	4.900	6.049	44.461	6.5581	269
CTH-90-D	3.33	7.62	5.24	9.53	5.08	846	661	12.8	3.811	9.677	55.317	6.5840	306
CTH-125	3.49	7.30	5.40	9.53	4.76	857	635	13.1	3.993	9.074	54.339	6.6118	301
CTH-89-C	2.54	7.62	4.45	11.43	3.81	957	484	12.5	4.355	7.258	47.411	6.6129	315
CTH-348	2.54	8.89	5.08	11.43	6.35	778	1051	12.2	2.903	16.129	70.239	6.6772	384
CTH-47-D	1.91	9.53	3.49	12.38	4.45	1092	452	12.0	4.355	7.059	46.109	6.6981	357
CTH-80	2.54	7.94	4.29	10.95	4.13	968	471	12.2	4.355	7.207	47.081	6.7008	309
CTH-95-A	2.54	7.38	5.08	12.50	4.13	926	723	12.9	3.719	10.485	58.496	6.7307	354
CTH-90-E	3.18	8.26	5.08	9.53	5.72	852	725	12.5	3.629	10.887	59.264	6.8884	319
CTH-413-A	1.91	9.53	4.29	12.86	5.72	901	869	12.0	3.266	13.607	66.665	7.2784	401
CTH-30-A	3.49	7.78	5.08	9.05	4.92	969	540	12.9	4.492	7.815	52.658	7.3096	294
CTH-376-A	2.86	8.61	4.01	9.88	4.33	1206	330	12.4	5.512	4.979	41.170	7.3105	299
CTH-296-B	2.54	8.81	4.52	10.72	5.64	905	715	12.0	3.630	11.182	60.888	7.3808	340
CTH-90-A	2.54	9.21	5.40	12.38	6.67	828	1293	12.7	2.903	19.057	82.991	7.5797	432
CTH-88-A	3.81	7.78	5.56	9.05	5.24	952	669	13.7	4.355	9.147	59.752	7.5879	310
CTH-17-B	4.13	7.62	5.72	8.57	5.08	1008	607	14.1	4.718	8.067	57.094	7.6394	300
CTH-30	3.18	8.10	4.76	9.53	4.92	1011	526	12.6	4.538	7.815	53.190	7.6458	302
CTH-89-B	2.54	8.26	4.45	11.43	4.45	1019	564	12.5	4.355	8.468	55.313	7.7150	335
CTH-11	2.54	8.89	5.72	13.34	6.35	834	1421	13.2	2.903	20.161	87.799	7.7168	459
CTH-82-B	3.49	7.15	6.35	12.15	4.76	904	1050	14.5	3.744	13.613	76.452	7.9166	391
CTH-80-A	3.49	7.30	5.24	10.00	4.13	1048	519	13.5	4.992	7.207	53.972	7.9767	305
CTH-17-C	4.13	7.14	5.72	9.05	4.29	1077	524	14.4	5.309	6.806	54.201	7.9804	296
CTH-347-B	3.18	7.94	5.72	11.43	5.40	882	986	13.5	3.629	13.711	74.636	8.0338	377
CTH-58-B	3.18	8.42	4.76	9.53	5.24	1027	560	12.6	4.538	8.320	56.627	8.1399	311
CTH-505	3.18	8.42	4.52	9.53	4.76	1150	436	12.7	5.218	6.425	50.289	8.2463	303
CTH-94-A	4.13	5.79	6.67	12.50	2.54	1282	561	16.3	6.045	6.452	58.496	8.6697	356

National Arnold E Cores 4 mil Silicon Steel

Part No.	D cm 1	Ht cm 2	Wth cm 3	Lt cm 4	G cm 5	Wtfe grams 6	Wtcu grams 7	MLT cm 8	Ac cm sq 9	Wa cm sq 10	Ap cm 4th 11	Kg cm 5th 12	At cm sq 13
CTH-58	3.18	8.42	4.92	10.00	5.24	1051	628	12.9	4.538	9.147	62.262	8.7773	327
CTH-45-D	4.13	8.10	5.56	8.57	5.24	1139	566	14.2	5.309	7.487	59.619	8.9327	308
CTH-58-A	3.18	8.26	5.08	10.48	5.08	1125	678	13.1	4.538	9.677	65.870	9.1094	339
CTH-61	3.18	7.78	5.40	11.43	4.60	1066	744	13.6	4.538	10.235	69.663	9.2809	357
CTH-33-C	2.86	8.89	5.40	11.91	6.03	977	1077	13.2	3.676	15.324	84.488	9.4322	407
CTH-95	2.54	8.26	5.08	13.34	4.45	1104	812	13.5	4.355	11.290	73.751	9.5263	401
CTH-6	1.91	10.16	5.08	15.24	6.35	1037	1421	13.2	3.266	20.161	98.774	9.7665	519
CTH-64-D	2.22	10.80	4.13	11.43	6.99	1105	841	11.9	3.811	13.306	76.073	9.7830	401
CTH-95-C	2.06	9.68	4.76	14.29	5.56	1101	1048	13.1	3.834	14.996	86.242	10.0954	461
CTH-70-H	4.05	8.26	6.11	10.00	5.72	1109	924	14.7	4.627	11.796	81.866	10.3123	367
CTH-24	2.22	10.16	4.60	12.86	6.35	1102	1016	12.6	3.811	15.119	86.437	10.4567	437
CTH-17	3.18	8.73	4.76	10.48	4.92	1276	553	13.3	5.444	7.815	63.808	10.4770	342
CTH-61-A	3.18	9.05	5.08	10.48	5.87	1138	783	13.1	4.538	11.190	76.165	10.5332	364
CTH-33-D	3.18	8.41	5.72	12.14	5.40	1103	1021	14.0	4.309	13.711	88.623	10.9409	409
CTH-90-I	4.45	8.26	6.35	9.53	5.72	1246	885	15.2	5.081	10.887	82.970	11.0663	364
CTH-33-A	3.18	9.37	5.72	11.67	6.67	1070	1233	13.6	3.855	16.937	97.931	11.0673	432
CTH-94	2.54	8.26	5.72	15.24	4.45	1185	1090	14.5	4.355	14.113	92.189	11.0881	472
CTH-90-J	4.45	7.62	6.67	10.48	5.08	1122	948	15.7	5.081	11.293	86.062	11.1146	379
CTH-71-M	2.54	9.53	4.76	12.38	5.72	1175	880	13.0	4.355	12.704	82.989	11.1302	411
CTH-90-B	2.54	9.84	5.40	13.34	6.67	1089	1357	13.4	3.630	19.057	103.771	11.2862	477
CTH-67-C	3.65	8.73	5.56	10.24	5.72	1210	808	13.9	4.955	10.887	80.920	11.5248	367
CTH-64-A	2.22	10.16	4.45	13.34	5.72	1257	880	13.0	4.448	12.704	84.756	11.6075	441
CTH-91-A	2.54	10.80	5.08	12.38	7.62	1133	1327	12.9	3.630	19.355	105.392	11.9077	471
CTH-33	2.54	9.53	5.08	13.34	5.72	1217	1044	13.5	4.355	14.516	94.823	12.2480	448
CTH-111	2.54	10.16	4.76	12.38	6.35	1239	978	13.0	4.355	14.116	92.209	12.3669	433
CTH-71-L	4.45	8.26	6.67	10.48	5.72	1233	1066	15.7	5.081	12.704	96.820	12.5039	400
CTH-24-C	2.22	11.43	4.60	12.86	7.62	1205	1220	12.6	3.811	18.143	103.725	12.5480	481
CTH-85	2.54	9.84	5.08	13.34	6.03	1249	1102	13.5	4.355	15.324	100.099	12.9296	459
CTH-70-B	3.18	8.57	5.08	12.38	4.13	1522	604	14.4	6.352	7.864	74.929	13.2261	391
CTH-82-C	2.54	9.53	5.40	14.29	5.72	1260	1218	14.0	4.355	16.333	106.694	13.2895	486

National Arnold E Cores 4 mil Silicon Steel													
Part No.	D cm 1	Ht cm 2	Wth cm 3	Lt cm 4	G cm 5	Wtfe grams 6	Wtcu grams 7	MLT cm 8	Ac cm sq 9	Wa cm sq 10	Ap cm 4th 11	Kg cm 5th 12	At cm sq 13
CTH-71-H	2.86	9.84	4.92	11.91	6.03	1336	888	13.4	4.900	12.452	91.524	13.4127	417
CTH-71	2.86	9.53	5.08	12.38	5.72	1327	923	13.6	4.900	12.704	93.378	13.4338	425
CTH-17-A	3.81	8.73	5.40	10.48	4.92	1538	614	14.7	6.532	7.815	76.570	13.5740	366
CTH-413	2.86	9.53	5.24	12.86	5.72	1351	1007	13.9	4.900	13.607	100.015	14.1313	443
CTH-45-B	3.49	9.37	5.00	10.72	5.24	1586	594	14.1	6.489	7.900	76.894	14.1646	377
CTH-70-E	3.18	8.41	5.56	13.34	4.29	1483	807	14.8	5.898	10.205	90.281	14.3676	427
CTH-409	2.54	10.16	5.40	14.29	6.35	1323	1354	14.0	4.355	18.148	118.549	14.7661	511
CTH-71-K	5.08	8.26	7.30	10.48	5.72	1417	1152	17.0	5.806	12.704	110.651	15.1120	424
CTH-90-K	2.54	10.48	5.40	14.29	6.67	1355	1422	14.0	4.355	19.057	124.486	15.5055	523
CTH-91-E	2.86	10.80	5.72	13.34	7.62	1306	1625	14.0	4.085	21.778	133.433	15.5866	528
CTH-35	2.54	10.16	5.72	15.24	6.35	1364	1558	14.5	4.355	20.161	131.698	15.8402	552
CTH-70-A	3.18	8.89	5.56	13.34	4.76	1548	897	14.8	5.898	11.341	100.329	15.9666	445
CTH-120	2.86	10.80	5.40	12.86	7.30	1390	1366	13.8	4.491	18.550	124.961	16.2624	498
CTH-86-C	2.86	8.89	6.35	16.19	5.08	1461	1478	15.6	4.900	17.744	130.423	16.3677	553
CTH-24-E	2.54	10.80	4.92	13.81	6.35	1533	1119	13.9	5.082	15.119	115.250	16.8877	498
CTH-91	2.22	12.07	4.76	14.29	7.62	1494	1393	13.5	4.448	19.355	129.122	17.0311	551
CTH-66-F	3.81	10.16	5.40	10.48	6.35	1740	793	14.7	6.532	10.084	98.805	17.5157	412
CTH-86-D	2.54	12.38	6.03	15.24	9.21	1362	2462	14.3	3.630	32.164	175.139	17.7246	667
CTH-70	3.18	9.21	5.56	13.81	4.76	1712	916	15.1	6.352	11.341	108.058	18.1324	468
CTH-61-C	3.81	9.53	5.72	11.43	5.72	1706	885	15.2	6.532	10.887	106.676	18.2933	428
CTH-85-B	3.18	9.84	5.72	13.34	6.03	1557	1206	14.8	5.444	15.324	125.124	18.4637	489
CTH-411	3.18	10.16	5.56	12.86	6.35	1588	1170	14.5	5.444	15.119	123.454	18.5308	481
CTH-82-F	2.54	10.16	6.67	18.10	6.35	1490	2234	16.0	4.355	26.213	171.228	18.6663	682
CTH-70-F	2.54	11.43	4.60	13.81	6.35	1773	979	14.0	5.806	13.106	114.152	18.9262	506
CTH-121	3.18	9.68	6.03	14.29	5.87	1590	1366	15.3	5.444	16.788	137.078	19.5657	523
CTH-82-K	3.18	9.68	6.03	14.29	5.87	1590	1366	15.3	5.444	16.788	137.078	19.5657	523
CTH-57-D	5.08	8.81	6.99	10.48	5.64	1780	982	17.1	7.260	10.737	116.927	19.8083	430
CTH-73-D	2.22	12.70	5.08	15.24	8.26	1603	1760	14.0	4.448	23.593	157.395	20.0192	620
CTH-112	2.70	10.96	5.24	14.29	6.51	1673	1273	14.4	5.400	16.533	133.913	20.0312	532
CTH-51	3.18	12.07	5.08	11.43	8.26	1731	1154	13.8	5.444	15.726	128.406	20.3209	486

National Arnold E Cores 4 mil Silicon Steel													
Part No.	D cm 1	Ht cm 2	Wth cm 3	Lt cm 4	G cm 5	Wtfe grams 6	Wtcu grams 7	MLT cm 8	Ac cm sq 9	Wa cm sq 10	Ap cm 4th 11	Kg cm 5th 12	At cm sq 13
CTH-98	3.81	8.89	6.35	13.34	5.08	1731	1117	16.2	6.532	12.903	126.430	20.3495	481
CTH-410-B	3.81	10.16	5.44	11.55	5.72	2020	766	15.4	7.623	9.298	106.317	21.0002	442
CTH-105	2.54	13.02	4.76	13.34	8.57	1777	1385	13.6	5.082	19.058	145.271	21.6744	559
CTH-48	5.40	8.81	7.94	11.67	6.11	1744	1515	18.3	6.554	15.524	152.614	21.8647	503
CTH-35-A	2.54	10.16	6.67	19.05	5.72	1732	2091	16.6	5.082	23.592	179.830	22.0008	703
CTH-4	3.18	10.16	6.35	15.24	6.35	1624	1694	15.8	5.444	20.161	164.623	22.7549	584
CTH-71-D	4.13	9.53	6.35	12.38	5.72	1909	1110	16.4	7.078	12.704	134.873	23.3215	480
CTH-33-B	3.49	9.84	6.03	13.81	5.72	1861	1216	15.7	6.489	14.516	141.284	23.3417	515
CTH-51-A	3.81	11.11	5.72	11.43	7.30	1937	1131	15.2	6.532	13.912	136.317	23.3764	483
CTH-502	2.86	11.11	5.56	14.77	6.67	1823	1441	15.0	5.718	17.997	154.360	23.5251	568
CTH-24-A	3.81	10.16	6.19	12.86	6.35	1898	1289	16.0	6.532	15.119	148.145	24.2169	510
CTH-109	3.81	13.02	6.35	11.91	10.16	1771	2104	15.3	4.900	25.806	189.679	24.3278	598
CTH-67-A	3.81	10.32	5.72	12.62	5.72	2071	931	16.0	7.894	10.887	128.907	25.3893	488
CTH-33-E	3.49	9.53	6.99	16.19	5.72	1850	1798	16.9	5.989	19.962	179.326	25.4366	615
CTH-410-A	2.86	11.43	5.24	14.76	6.35	2059	1221	15.1	6.533	15.119	148.171	25.5725	564
CTH-34	2.54	11.43	5.72	17.15	6.35	1970	1694	15.8	5.806	20.161	175.598	25.8900	654
CTH-235	5.40	8.89	7.46	11.43	5.40	2165	1090	18.3	8.482	11.141	141.760	26.2197	478
CTH-92	2.54	12.70	5.08	15.24	7.62	2026	1523	14.8	5.806	19.355	168.574	26.5337	619
CTH-236-A	3.49	10.95	5.28	13.22	5.72	2355	851	15.6	8.233	10.207	126.057	26.5507	507
CTH-82-L	3.18	10.16	7.30	18.10	6.35	1859	2412	17.2	5.444	26.213	214.036	27.0187	718
CTH-131	6.35	10.16	8.57	10.48	7.62	2087	1747	19.3	7.258	16.939	184.419	27.6867	544
CTH-410-F	2.86	12.86	5.08	14.29	7.78	2247	1374	14.9	6.533	17.293	169.470	29.7356	598
CTH-64	3.81	10.16	6.99	15.24	6.35	2054	1853	17.2	6.532	20.161	197.547	29.9564	616
CTH-36-A	3.18	10.48	6.11	16.43	5.40	2396	1408	16.6	7.258	15.854	172.603	30.0980	614
CTH-72	2.86	11.43	6.03	17.15	6.35	2224	1762	16.4	6.533	20.161	197.582	31.5065	672
CTH-2-A	3.49	10.80	6.03	15.24	5.72	2437	1290	16.7	7.985	14.516	173.867	33.3295	592
CTH-73	3.18	14.13	6.03	14.29	10.32	2143	2400	15.3	5.444	29.492	240.809	34.3716	703
CTH-45	2.86	10.80	6.67	20.00	5.08	2540	1861	18.0	7.351	19.355	213.426	34.8240	756
CTH-85-A	5.08	9.84	7.62	13.34	6.03	2498	1535	18.8	8.710	15.324	200.198	37.1508	577
CTH-36	3.18	11.43	6.35	17.15	6.35	2463	1831	17.0	7.258	20.161	219.497	37.4351	690

National Arnold E Cores 4 mil Silicon Steel													
Part No.	D cm 1	Ht cm 2	Wth cm 3	Lt cm 4	G cm 5	Wtfe grams 6	Wtcu grams 7	MLT cm 8	Ac cm sq 9	Wa cm sq 10	Ap cm 4th 11	Kg cm 5th 12	At cm sq 13
CTH-2	3.81	10.80	6.35	15.24	5.72	2661	1355	17.5	8.710	14.516	189.645	37.7460	608
CTH-79	4.45	10.16	7.62	15.24	6.35	2392	1990	18.5	7.621	20.161	230.472	37.9750	648
CTH-68-A	3.18	12.70	5.72	15.24	7.62	2532	1655	16.0	7.258	19.355	210.717	38.1733	655
CTH-27-B	4.13	11.43	7.78	16.19	7.94	2328	2828	18.3	6.487	28.982	281.994	39.9916	749
CTH-502-C	3.33	12.38	6.03	15.72	7.30	2645	1745	16.6	7.622	19.711	225.339	41.4002	674
CTH-82	2.54	13.65	6.67	20.00	8.57	2433	3256	17.2	5.806	35.389	308.229	41.5031	911
CTH-107	4.13	10.48	6.67	16.19	4.76	3167	1212	18.8	10.618	12.098	192.686	43.5869	637
CTH-119	2.86	13.97	6.03	17.15	8.89	2605	2467	16.4	6.533	28.226	276.615	44.1091	790
CTH-28-A	4.45	11.11	6.67	14.29	6.03	3079	1307	18.3	10.161	13.411	204.415	45.4607	611
CTH-71-A	5.72	9.84	7.94	13.34	5.40	3052	1292	20.2	11.434	12.000	205.808	46.6395	591
CTH-64-B	3.81	11.75	7.62	17.62	7.62	2604	2872	18.5	7.078	29.032	308.211	47.0481	799
CTH-129	4.13	13.18	5.87	13.34	7.78	3356	1247	17.2	10.027	13.582	204.289	47.6078	625
CTH-36-B	5.08	12.38	8.26	14.29	9.21	2655	2984	19.1	7.260	29.235	318.388	48.3176	754
CTH-73-B	3.18	14.13	8.26	20.96	10.32	2511	5241	18.7	5.444	52.421	428.030	49.7235	1094
CTH-38	4.45	9.53	8.26	19.05	4.45	3232	1876	20.8	10.161	16.935	258.129	50.5191	757
CTH-91-C	4.45	12.07	6.99	14.29	7.62	2990	1873	18.1	8.893	19.355	258.187	50.6308	668
CTH-82-A	3.81	12.38	6.67	16.19	7.30	3040	2004	18.0	8.710	20.872	272.682	52.7680	724
CTH-412	5.08	12.22	7.78	13.81	8.41	3017	2304	19.0	8.710	22.709	296.687	54.3338	702
CTH-68-D	3.65	13.97	6.19	15.24	8.89	3150	2045	17.0	8.346	22.581	282.693	55.5880	736
CTH-26-D	3.81	11.43	7.62	19.05	6.35	3123	2516	19.5	8.710	24.194	316.076	56.4767	823
CTH-73-E	4.45	14.13	7.30	14.29	10.32	5286	2832	18.0	7.621	29.492	337.132	57.0852	775
CTH-73-J	3.18	12.70	8.26	22.86	7.62	3094	4132	20.0	7.258	38.710	421.434	61.1343	1065
CTH-81-A	3.18	15.24	7.46	19.53	10.80	2931	4475	18.1	6.352	46.267	440.851	61.7742	1038
CTH-79-B	4.45	11.75	6.99	16.19	6.03	3721	1587	19.4	11.433	15.324	262.806	61.9227	713
CTH-64-E	4.45	13.02	8.26	17.15	9.21	1654	3649	19.5	7.621	35.082	401.043	62.7013	886
CTH-106	3.49	15.88	6.03	15.24	10.80	3361	2437	16.7	7.985	27.419	328.415	62.9557	808
CTH-62	5.72	11.75	7.62	13.34	6.67	4049	1377	20.3	13.065	12.703	248.928	64.0281	657
CTH-71-J	5.72	11.75	7.46	13.81	6.03	4430	1163	20.7	14.700	10.534	232.268	65.9679	666
CTH-128-B	4.13	12.70	7.30	17.15	7.62	3484	2470	19.1	9.437	24.194	342.457	67.5482	807
CTH-513	4.76	13.34	7.46	14.77	8.89	3429	2435	19.0	9.529	23.994	342.972	68.7142	765

National Arnold E Cores 4 mil Silicon Steel

Part No.	D cm 1	Ht cm 2	Wth cm 3	Lt cm 4	G cm 5	Wtfe grams 6	Wtcu grams 7	MLT cm 8	Ac cm sq 9	Wa cm sq 10	Ap cm 4th 11	Kg cm 5th 12	At cm sq 13
CTH-73-A	5.08	14.13	7.94	14.29	10.32	3751	3032	19.3	8.710	29.492	385.294	69.6471	811
CTH-497	5.72	11.43	8.41	14.77	6.99	3724	2105	20.9	11.434	18.853	323.339	70.6577	730
CTH-104-A	3.81	15.88	6.35	15.24	10.80	3674	2560	17.5	8.710	27.419	358.219	71.2981	827
CTH-132	3.18	13.34	9.21	25.72	8.26	3410	5714	21.5	7.258	49.802	542.203	73.1821	1283
CTH-104-B	3.81	16.51	6.35	15.24	11.43	3798	2711	17.5	8.710	29.032	379.291	75.4921	855
CTH-128-C	4.45	12.70	7.62	17.15	7.62	3742	2551	19.8	10.161	24.194	368.755	75.8094	826
CTH-68-B	6.35	12.70	8.89	13.34	8.89	3832	2542	21.1	10.887	22.581	368.755	76.0865	763
CTH-68	5.08	12.70	7.62	15.24	7.62	4053	2069	20.0	11.613	19.355	337.147	78.1339	761
CTH-73-L	5.72	13.65	8.57	14.29	9.84	3755	3083	20.5	9.798	28.131	413.461	78.8830	826
CTH-101	4.45	13.97	7.30	16.19	8.89	3032	2612	19.3	10.161	25.408	387.261	81.6696	835
CTH-113	4.13	17.78	7.94	17.15	13.97	3583	5356	18.9	7.078	53.226	565.054	84.8008	1106
CTH-108-A	6.35	13.02	9.21	14.29	9.21	4016	3062	21.8	10.887	26.316	429.764	85.7995	831
CTH-101-A	4.60	13.97	7.46	16.19	8.89	4066	2655	19.6	10.525	25.408	401.113	86.1947	845
CTH-93	5.08	11.75	8.57	18.10	6.67	4136	2676	21.5	11.613	23.291	405.719	87.4942	866
CTH-73-H	2.54	25.40	6.03	18.10	20.32	3880	6153	16.3	5.806	70.978	618.192	88.3459	1373
CTH-130	3.81	13.34	8.89	22.86	8.26	3199	4807	21.5	8.710	41.935	547.865	88.8108	1150
CTH-90-H	5.72	11.75	8.57	16.19	6.67	4399	2217	21.8	13.065	19.057	373.458	89.4700	802
CTH-79-A	6.35	10.16	11.43	20.96	6.35	4044	4354	25.3	10.887	32.258	526.793	90.6699	1060
CTH-358-A	3.49	15.24	6.83	19.53	8.89	4272	3032	19.2	9.981	29.639	443.755	92.3797	985
CTH-99	3.49	17.94	6.99	18.10	12.86	3968	4350	18.2	7.985	44.916	537.987	94.6321	1082
CTH-108	6.35	13.97	9.21	14.29	10.16	4255	3379	21.8	10.887	29.037	474.197	94.6701	876
CTH-34-E	5.08	13.34	7.62	16.19	7.62	4729	2135	20.7	13.067	19.355	379.357	95.8805	824
CTH-63	4.45	13.97	8.26	19.05	8.89	4111	3752	20.8	10.161	33.871	516.257	101.0382	1000
CTH-73-C	5.08	14.13	10.16	20.96	10.32	4009	6364	22.8	8.710	52.421	684.847	104.8219	1228
CTH-79-C	4.45	14.45	7.62	19.05	8.10	5142	2885	21.0	12.702	25.705	489.738	118.2552	980
CTH-26-B	6.35	11.43	10.16	19.05	6.35	4686	3172	24.6	14.516	24.194	526.793	124.4539	977
CTH-86	5.08	14.92	8.57	18.10	9.84	4980	3950	21.5	11.613	34.382	598.904	129.1550	1035
CTH-75	4.13	17.94	7.94	19.05	12.86	4800	5262	20.1	9.437	48.993	693.489	130.0145	1193
CTH-86-E	5.08	15.24	8.89	19.05	10.16	5175	4550	22.0	11.613	38.710	674.295	142.1293	1113
CTH-86-A	5.08	18.10	7.94	16.19	13.02	5598	4077	20.5	11.613	37.205	648.094	146.5455	1073

National Arnold E Cores 4 mil Silicon Steel

Part No.	D cm 1	Ht cm 2	Wth cm 3	Lt cm 4	G cm 5	Wtfe grams 6	Wtcu grams 7	MLT cm 8	Ac cm sq 9	Wa cm sq 10	Ap cm 4th 11	Kg cm 5th 12	At cm sq 13
CTH-75-D	4.13	17.94	8.73	21.43	12.86	5210	6752	21.4	9.437	59.203	838.011	147.9491	1363
CTH-86-F	5.72	14.92	9.21	18.10	9.84	5602	4183	22.8	13.065	34.382	673.767	154.3607	1077
CTH-31-A	3.81	17.78	6.35	22.86	7.62	7884	2332	22.6	17.419	19.355	505.721	156.0290	1226
CTH-73-F	5.08	15.40	10.16	22.86	10.32	6193	6720	24.0	11.613	52.421	913.130	176.5021	1378
CTH-75-B	5.08	17.94	8.89	19.05	12.86	6421	5759	22.0	11.613	48.993	853.421	179.8859	1264
CTH-128-A	6.35	13.97	9.53	19.05	7.62	7155	3207	24.9	18.145	24.194	658.491	192.3226	1081
CTH-76	6.35	14.92	10.16	19.05	9.84	6360	4916	24.6	14.516	37.502	816.570	192.9134	1182
CTH-498	4.45	14.29	11.11	29.53	7.94	6445	7489	26.5	12.702	52.931	1008.454	193.1624	1648
CTH-103	4.13	20.48	9.53	23.81	15.40	5803	10032	22.6	9.437	83.124	1176.610	196.2837	1710
CTH-75-C	4.13	19.05	10.48	26.67	13.97	5753	11414	24.1	9.437	88.710	1255.675	196.4938	1845
CTH-358	3.49	19.05	6.83	25.24	8.89	8077	3634	23.0	15.970	29.639	710.008	197.2967	1431
CTH-360	4.76	17.46	8.10	20.48	10.48	7130	4204	22.6	14.973	34.934	784.616	208.2829	1237
CTH-27-A	6.35	14.29	10.00	20.48	7.94	7558	3957	25.6	18.145	28.982	788.813	223.6593	1188
CTH-82-J	7.62	13.65	11.75	20.00	8.57	7302	5289	28.0	17.419	35.389	924.687	229.9668	1253
CTH-475	5.72	18.42	10.48	21.91	13.34	7144	8403	24.8	13.065	63.515	1244.679	262.2344	1555
CTH-358-B	4.45	19.05	8.89	24.77	11.43	8322	6587	24.3	15.242	50.806	1161.578	291.3791	1587
CTH-27-C	6.35	15.56	10.00	22.38	7.94	9621	4154	26.9	21.774	28.982	946.576	306.8458	1344
CTH-495	5.72	18.42	11.75	26.67	12.70	8580	11212	27.4	14.700	76.619	1689.465	362.1130	1904
CTH-501-J	4.45	20.00	8.89	26.67	11.11	10045	6739	25.6	17.782	49.397	1317.590	366.4517	1739
CTH-476-A	6.35	17.78	10.80	22.86	11.43	9447	7275	26.8	18.145	50.806	1382.831	373.8793	1577
CTH-496	4.76	19.69	9.53	26.67	11.43	10063	7572	26.1	17.696	54.441	1445.043	392.2511	1776
CTH-476-C	6.35	20.96	10.32	21.43	14.61	10482	8069	26.1	18.145	57.967	1577.734	438.7911	1672
CTH-476-D	6.35	17.46	10.80	24.77	9.84	11099	6513	27.9	21.774	43.752	1428.998	445.9901	1641
CTH-477	5.08	22.86	7.94	23.81	12.70	16578	4961	25.6	23.226	36.297	1264.524	458.4860	1714
CTH-501-A	5.72	20.96	12.38	28.58	15.24	9730	15411	28.4	14.700	101.620	2240.747	463.4312	2272
CTH-501-K	5.72	20.00	10.16	26.67	11.11	12900	7408	28.1	22.863	49.397	1694.045	551.0394	1857
CTH-501-E	4.45	24.92	9.84	29.53	16.03	12563	12498	27.1	17.782	86.552	2308.618	606.5911	2352
CTH-502-A	8.89	19.69	11.95	23.46	10.16	19549	5518	33.3	38.109	31.049	1774.856	811.9776	1838
CTH-501	6.99	29.85	13.65	34.29	20.32	25602	25426	35.2	29.943	135.494	6085.553	2071.7833	3544
CTH-501	6.99	29.85	13.65	34.29	20.32	25602	25426	35.2	29.943	135.494	6085.553	2071.7833	3544

National Arnold E Cores 4 mil Silicon Steel													
Part No.	D cm 1	Ht cm 2	Wth cm 3	Lt cm 4	G cm 5	Wtfe grams 6	Wtcu grams 7	MLT cm 8	Ac cm sq 9	Wa cm sq 10	Ap cm 4th 11	Kg cm 5th 12	At cm sq 13
CTH-508	5.08	53.34	17.15	55.25	40.64	44977	111462	42.6	29.032	490.322	21352.672	5818.3333	9004
CTH-494	7.62	39.05	17.78	50.96	25.40	54372	63401	46.1	46.813	258.064	18121.012	7366.9855	6626

National Arnold E Cores 4 mil Silicon Steel													
Part No.	D cm 1	Ht cm 2	Wth cm 3	Lt cm 4	G cm 5	Wtfe grams 6	Wtcu grams 7	MLT cm 8	Ac cm sq 9	Wa cm sq 10	Ap cm 4th 11	Kg cm 5th 12	At cm sq 13
CTZ-54	0.95	1.91	1.91	4.29	0.95	13.6	23.1	4.8	0.408	0.908	0.556	0.019	36
CTZ-53-A	0.64	3.18	1.27	3.81	1.91	30.8	25.5	3.9	0.363	1.210	0.658	0.024	39
CTZ-53-C	0.95	2.54	2.22	5.24	1.59	37.1	56.6	5.3	0.408	2.017	1.235	0.038	56
CTZ-53-D	0.95	3.97	1.83	4.29	2.86	52.5	63.8	4.8	0.477	2.495	1.785	0.071	60
CTZ-53-B	1.59	3.97	2.30	3.33	3.18	15.4	68.9	5.7	0.567	2.267	1.929	0.077	58
CTZ-99	1.43	3.97	2.06	3.57	2.86	73.4	54.0	5.6	0.715	1.815	1.947	0.100	57
CTZ-51	1.27	3.49	2.54	5.72	2.22	84.7	96.6	6.4	0.726	2.823	3.074	0.139	80
CTZ-68	1.59	3.81	2.38	5.24	1.91	140.9	56.0	6.9	1.362	1.513	3.090	0.243	77
CTZ-96-A	1.91	3.81	3.02	5.24	2.54	123.7	111.9	7.4	1.089	2.822	4.608	0.270	89
CTZ-106	1.11	4.61	2.38	6.67	2.70	127.3	119.4	6.5	0.953	3.428	4.899	0.286	105
CTZ-96-B	1.27	4.45	2.38	6.19	2.54	137.3	102.3	6.8	1.089	2.822	4.611	0.296	97
CTZ-500	1.91	4.13	2.86	5.72	2.22	188.9	88.4	7.8	1.634	2.119	5.192	0.434	96
CTZ-96	1.91	4.45	3.02	5.72	2.86	174.9	131.3	7.8	1.361	3.175	6.484	0.456	105
CTZ-77-M	1.27	5.72	2.54	6.67	3.81	174.0	181.9	7.0	1.089	4.839	7.906	0.489	130
CTZ-25-A	1.11	6.03	2.70	7.62	4.13	169.4	245.8	7.0	0.953	6.555	9.370	0.508	154
CTZ-25	1.27	6.03	2.86	7.62	4.13	194.3	263.9	7.5	1.089	6.555	10.711	0.618	158
CTZ-83-C	1.91	5.08	2.74	5.83	2.86	248.7	100.9	7.9	1.905	2.381	6.802	0.652	112
CTZ-83	1.91	4.76	3.18	6.67	2.86	228.3	161.0	8.3	1.634	3.630	8.896	0.699	127
CTZ-103	2.22	5.08	3.33	5.72	3.49	220.6	173.6	8.4	1.589	3.881	9.247	0.701	123
CTZ-107	2.22	5.72	3.33	5.72	4.13	242.4	205.2	8.4	1.589	4.586	10.928	0.828	135
CTZ-77-L	2.38	4.92	3.25	5.72	2.86	288.1	117.2	8.8	2.212	2.495	8.277	0.832	118
CTZ-97-B	2.54	5.08	3.65	5.72	3.49	260.9	186.7	9.0	1.815	3.881	10.566	0.850	130
CTZ-80	2.86	3.81	4.21	6.91	1.91	304.4	141.8	10.3	2.451	2.570	9.449	0.895	133
CTZ-3-B	0.95	8.26	2.86	9.53	5.72	415.9	455.6	7.8	1.089	10.887	17.789	0.988	240
CTZ-77-C	1.91	5.40	3.18	7.62	2.86	332.5	173.3	9.0	2.177	3.630	11.855	1.153	152
CTZ-97	2.86	5.08	3.97	6.19	3.18	345.6	187.7	10.0	2.451	3.527	12.970	1.275	144
CTZ-12-A	1.27	7.62	3.18	9.53	5.08	322.1	448.0	8.7	1.452	9.677	21.072	1.410	235
CTZ-56	1.59	6.51	3.18	8.57	3.97	342.9	296.4	8.8	1.815	6.303	17.160	1.413	191
CTZ-97-A	3.33	5.08	4.45	6.19	3.18	405.9	205.6	10.9	2.860	3.527	15.130	1.584	155
CTZ-78-C	2.22	5.72	3.49	7.62	3.18	404.5	206.2	9.6	2.541	4.032	15.368	1.629	167

National Arnold E Cores 4 mil Silicon Steel													
Part No.	D cm 1	Ht cm 2	Wth cm 3	Lt cm 4	G cm 5	Wtfe grams 6	Wtcu grams 7	MLT cm 8	Ac cm sq 9	Wa cm sq 10	Ap cm 4th 11	Kg cm 5th 12	At cm sq 13
CTZ-70	2.54	9.37	3.33	5.00	7.62	458.9	280.1	8.7	1.996	6.050	18.112	1.666	184
CTZ-84	1.59	6.67	3.49	9.53	4.13	367.4	390.7	9.3	1.815	7.864	21.410	1.669	221
CTZ-41-C	1.91	6.03	3.49	8.57	3.49	389.6	279.6	9.5	2.177	5.547	18.117	1.670	189
CTZ-104	3.81	4.45	5.08	6.67	2.54	219.7	212.3	12.3	3.268	3.226	15.812	1.675	163
CTZ-78-A	2.54	5.72	3.81	7.62	3.18	464.3	219.9	10.2	2.903	4.032	17.560	1.995	176
CTZ-1-H	3.81	5.08	5.08	6.67	3.18	27.6	265.3	12.3	3.268	4.032	19.765	2.094	178
CTZ-87	1.91	6.67	3.65	9.05	4.13	431.7	372.9	9.7	2.177	7.207	23.541	2.114	218
CTZ-105	2.54	5.40	4.21	9.29	2.54	537.3	252.1	11.2	3.267	4.234	20.748	2.429	205
CTZ-78	2.54	6.51	3.81	7.62	3.97	517.3	274.8	10.2	2.903	5.041	21.951	2.494	194
CTZ-41-A	2.54	6.03	4.13	8.57	3.49	513.2	317.2	10.7	2.903	5.547	24.156	2.616	208
CTZ-78-B	2.54	5.72	4.45	9.53	3.18	520.5	361.9	11.2	2.903	6.048	26.340	2.726	225
CTZ-81	2.54	6.51	3.97	8.10	3.97	530.5	316.8	10.5	2.903	5.672	24.699	2.739	207
CTZ-41	2.86	5.72	4.45	8.57	3.18	555.8	305.4	11.4	3.267	5.042	24.706	2.842	209
CTZ-77-H	2.22	6.99	3.49	8.57	3.81	596.1	263.9	10.2	3.177	4.839	23.060	2.866	212
CTZ-25-B	2.22	6.99	3.81	9.05	4.13	565.3	363.8	10.4	2.859	6.555	28.112	3.090	231
CTZ-47	1.27	9.53	2.86	12.38	4.45	727.1	403.7	10.7	2.903	7.059	30.739	3.330	329
CTZ-77-B	2.22	6.99	3.81	9.53	3.81	627.0	346.1	10.7	3.177	6.050	28.834	3.417	239
CTZ-351	2.54	6.99	4.29	9.05	4.45	590.3	454.1	11.0	2.903	7.761	33.798	3.578	246
CTZ-79	2.22	6.99	3.97	10.00	3.81	641.0	389.3	11.0	3.177	6.652	31.702	3.672	252
CTZ-25-C	2.86	6.67	4.45	8.57	4.13	617.9	397.1	11.4	3.267	6.555	32.121	3.696	234
CTZ-75	2.06	8.26	3.49	9.05	5.08	652.8	393.2	10.2	2.950	7.259	32.121	3.732	252
CTZ-41-B	2.54	6.67	4.13	9.53	3.49	693.1	336.0	11.4	3.630	5.547	30.204	3.862	241
CTZ-6-A	2.22	6.35	4.76	12.38	3.18	673.2	525.6	12.2	3.177	8.065	38.433	3.998	304
CTZ-2	2.54	6.35	5.08	11.43	3.81	618.8	630.6	12.2	2.903	9.677	42.143	4.006	300
CTZ-6-F	2.22	8.81	4.45	10.72	6.11	696.3	814.7	11.2	2.699	13.587	55.006	5.282	339
CTZ-6-E	2.54	6.99	5.08	11.91	4.13	743.8	701.0	12.5	3.267	10.485	51.378	5.356	329
CTZ-63	2.54	6.99	4.76	11.43	3.81	790.9	558.1	12.4	3.630	8.470	46.119	5.421	307
CTZ-71-D	2.54	8.26	4.76	10.48	5.72	708.0	794.1	11.7	2.903	12.704	55.326	5.483	329
CTZ-31	1.91	8.57	3.97	11.91	4.76	802.7	601.4	11.5	3.266	9.831	48.163	5.487	335
CTZ-32	3.18	7.94	4.76	8.57	5.40	782.8	548.3	12.0	3.629	8.572	46.662	5.649	277

National Arnold E Cores 4 mil Silicon Steel

Part No.	D cm 1	Ht cm 2	Wth cm 3	Lt cm 4	G cm 5	Wtfe grams 6	Wtcu grams 7	MLT cm 8	Ac cm sq 9	Wa cm sq 10	Ap cm 4th 11	Kg cm 5th 12	At cm sq 13
CTZ-98-A	2.54	7.30	5.40	12.86	4.45	792.8	883.1	13.0	3.267	12.704	62.249	6.241	373
CTZ-6-B	2.54	7.30	5.08	12.38	4.13	852.5	718.8	12.9	3.630	10.485	57.094	6.451	349
CTZ-48	2.86	8.10	4.45	9.53	4.92	908.3	499.9	12.0	4.085	7.815	47.880	6.523	291
CTZ-350-A	2.54	7.94	4.29	10.95	4.13	967.6	470.5	12.2	4.355	7.207	47.081	6.701	309
CTZ-6-C	2.54	7.30	5.40	13.34	4.13	887.4	840.2	13.4	3.630	11.798	64.242	6.987	381
CTZ-63-A	3.49	6.67	5.72	10.96	3.81	943.1	629.9	13.9	4.492	8.470	57.073	7.356	322
CTZ-15	2.22	8.73	4.13	11.91	4.60	1006.6	569.5	12.2	4.129	8.771	54.327	7.372	347
CTZ-350-B	2.54	7.94	4.60	11.91	4.13	1009.7	578.9	12.7	4.355	8.520	55.656	7.611	340
CTZ-30	3.18	8.10	4.76	9.53	4.92	1010.6	526.3	12.6	4.538	7.815	53.190	7.646	302
CTZ-5-C	2.54	8.89	4.45	11.43	5.08	954.0	644.7	12.5	4.355	9.677	63.215	8.817	355
CTZ-5-H	3.49	7.46	5.72	10.96	4.60	1024.7	761.1	13.9	4.492	10.235	68.967	8.889	349
CTZ-5	3.18	7.78	5.40	11.43	4.60	1073.6	743.8	13.6	4.538	10.235	69.663	9.281	357
CTZ-350-D	2.86	9.68	4.60	10.24	6.35	1121.2	733.3	12.4	4.288	11.087	71.310	9.864	359
CTZ-61	2.86	8.26	4.92	11.91	4.45	1171.0	654.5	13.4	4.900	9.174	67.433	9.882	364
CTZ-349	3.49	9.21	6.35	11.91	6.99	1003.8	1523.0	14.3	3.493	19.963	104.586	10.216	462
CTZ-62	2.86	8.26	5.08	12.38	4.45	601.1	718.1	13.6	4.900	9.881	72.628	10.448	380
CTZ-9-B	3.18	9.05	5.24	11.19	5.72	1206.8	851.4	13.5	4.764	11.796	84.283	11.867	387
CTZ-91-B	2.22	9.84	5.40	15.24	6.03	1227.6	1414.9	13.8	3.811	19.155	109.508	12.055	522
CTZ-5-D	2.54	9.53	5.08	13.34	5.72	1217.2	1044.2	13.5	4.355	14.516	94.823	12.248	448
CTZ-71-A	3.18	9.05	5.40	11.67	5.72	1242.6	934.0	13.8	4.764	12.704	90.776	12.550	405
CTZ-85	2.22	9.84	5.72	16.19	6.03	1195.9	1612.8	14.3	3.811	21.073	120.476	12.801	563
CTZ-8-B	2.86	8.89	6.67	16.19	5.72	1243.5	1798.1	15.5	4.085	21.774	133.410	14.079	574
CTZ-45-B	3.49	9.37	5.00	10.72	5.24	1581.0	593.7	14.1	6.489	7.900	76.894	14.165	377
CTZ-5-B	3.49	8.89	6.03	12.38	5.72	1335.9	1142.7	14.8	4.992	14.516	108.701	14.708	445
CTZ-46	3.49	9.37	5.16	11.19	5.24	819.0	668.0	14.3	6.489	8.733	85.001	15.386	394
CTZ-35	2.54	10.16	5.72	15.24	6.35	715.7	1557.5	14.5	4.355	20.161	131.698	15.840	552
CTZ-14	2.54	9.68	7.14	19.05	6.19	721.2	2494.7	16.4	3.991	28.503	170.651	16.604	715
CTZ-236-D	4.45	8.81	6.43	10.72	5.64	1578.7	954.1	16.0	6.353	11.182	106.554	16.926	415
CTZ-7-H	3.18	9.53	5.72	13.34	5.72	1532.0	1142.5	14.8	5.444	14.516	118.528	17.490	477
CTZ-102	2.54	9.68	7.78	20.96	6.19	761.0	3011.2	17.4	3.991	32.435	194.187	17.812	809

National Arnold E Cores 4 mil Silicon Steel													
Part No.	D cm 1	Ht cm 2	Wth cm 3	Lt cm 4	G cm 5	Wtfe grams 6	Wtcu grams 7	MLT cm 8	Ac cm sq 9	Wa cm sq 10	Ap cm 4th 11	Kg cm 5th 12	At cm sq 13
CTZ-236-C	3.81	9.21	6.51	12.86	6.03	1503.1	1376.6	15.8	5.445	16.283	132.998	18.277	490
CTZ-73-D	2.86	10.16	6.03	15.24	6.35	1534.3	1625.9	15.1	4.900	20.161	148.186	19.211	568
CTZ-3-A	2.86	10.80	4.76	13.34	5.72	1870.9	835.9	14.4	6.533	10.887	106.694	19.370	479
CTZ-73-A	2.22	12.70	5.08	15.24	8.26	1585.0	1760.2	14.0	4.448	23.593	157.395	20.019	620
CTZ-73-E	2.86	10.48	6.03	15.24	6.67	1594.6	1707.3	15.1	4.900	21.171	155.607	20.173	581
CTZ-7-D	3.49	9.84	5.72	12.86	5.72	1391.2	1030.9	15.2	6.489	12.704	123.651	21.097	476
CTZ-7-E	2.86	11.27	5.24	13.81	6.83	1788.4	1257.7	14.5	5.718	16.253	139.399	21.976	532
CTZ-10-A	2.54	10.16	6.67	19.05	5.72	1731.8	2090.8	16.6	5.082	23.592	179.830	22.001	703
CTZ-7-F	3.49	9.84	6.03	13.81	5.72	1861.4	1216.4	15.7	6.489	14.516	141.284	23.342	515
CTZ-8-D	3.81	10.16	6.19	12.86	6.35	1889.0	1289.1	16.0	6.532	15.119	148.145	24.217	510
CTZ-4-A	3.81	9.53	7.62	16.19	6.35	1736.8	2270.4	17.6	5.445	24.194	197.610	24.464	652
CTZ-4-E	2.22	13.97	5.40	16.19	9.53	1775.8	2336.5	14.5	4.448	30.242	201.754	24.779	722
CTZ-7-J	3.81	10.32	5.72	12.62	5.72	2187.5	930.9	16.0	7.894	10.887	128.907	25.389	488
CTZ-8	2.54	11.43	5.72	17.15	6.35	1969.6	1694.1	15.8	5.806	20.161	175.598	25.890	654
CTZ-236-Z	3.49	10.95	5.28	13.22	5.72	2354.7	851.3	15.6	8.233	10.207	126.057	26.551	507
CTZ-71	3.49	9.84	6.83	16.19	5.72	2010.4	1723.3	17.0	6.489	19.054	185.449	28.386	619
CTZ-7-G	3.18	10.80	5.72	15.24	5.72	2217.0	1240.9	16.0	7.258	14.516	158.038	28.630	575
CTZ-7-A	3.81	10.48	6.35	13.81	6.35	2420.8	1424.0	16.6	7.078	16.129	171.228	29.287	556
CTZ-64	3.81	10.16	6.99	15.24	6.35	2047.1	1853.0	17.2	6.532	20.161	197.547	29.956	616
CTZ-7-K	3.81	10.32	6.35	14.53	5.72	2340.2	1318.4	17.0	7.894	14.516	171.876	31.870	569
CTZ-7-C	3.49	10.80	6.03	15.24	5.72	2291.3	1290.1	16.7	7.985	14.516	173.867	33.330	592
CTZ-45	2.86	10.80	6.67	20.00	5.08	2514.2	1860.5	18.0	7.351	19.355	213.426	34.824	756
CTZ-7	3.81	10.80	6.35	15.24	5.72	2672.7	1355.3	17.5	8.710	14.516	189.645	37.746	608
CTZ-8-C	2.54	12.07	7.62	23.81	6.35	2679.9	3334.5	19.4	6.533	32.258	316.131	42.631	1005
CTZ-4-B	3.49	10.80	7.30	19.05	5.72	2756.1	2166.7	18.7	7.985	21.774	260.800	44.651	770
CTZ-7-B	3.81	11.75	7.62	17.62	7.62	2588.0	2871.9	18.5	7.078	29.032	308.211	47.048	799
CTZ-4-C	3.49	12.70	6.67	17.15	7.62	2949.5	2278.8	17.7	7.985	24.194	289.778	52.413	769
CTZ-5-F	4.13	13.34	6.03	14.29	7.62	3636.7	1376.6	17.8	10.618	14.516	231.199	55.231	671
CTZ-17	1.91	20.32	7.30	26.67	13.34	3629.0	7632.6	19.9	5.989	71.982	646.626	77.922	1620
CTZ-22-A	5.08	12.22	7.30	16.19	5.87	4230.1	1449.9	20.8	14.516	13.058	284.325	79.309	755

National Arnold E Cores 4 mil Silicon Steel													
Part No.	D cm 1	Ht cm 2	Wth cm 3	Lt cm 4	G cm 5	Wtfe grams 6	Wtcu grams 7	MLT cm 8	Ac cm sq 9	Wa cm sq 10	Ap cm 4th 11	Kg cm 5th 12	At cm sq 13
CTZ-7-M	5.40	14.61	8.57	19.05	8.26	6295.8	3208.0	22.9	15.425	26.210	606.417	163.053	1053
CTZ-360	4.76	17.46	8.10	20.48	10.48	7129.8	4204.2	22.6	14.973	34.934	784.616	208.283	1237
CTZ-7-L	6.35	14.61	9.53	19.05	8.26	6849.4	3474.2	24.9	18.145	26.210	713.365	208.350	1117

National Arnold E Cores 12 mil Silicon Steel													
Part No.	D cm 1	Ht cm 2	Wth cm 3	Lt cm 4	G cm 5	Wtfe grams 6	Wtcu grams 7	MLT cm 8	Ac cm sq 9	Wa cm sq 10	Ap cm 4th 11	Kg cm 5th 12	At cm sq 13
CTS-290	6.35	20.32	10.16	24.77	11.43	14822	6546	28.2	26.8	43.5	1752	667	1787
CTS-331-A	7.62	16.35	13.34	28.58	8.73	14434	8796	33.0	27.6	49.9	2064	689	1958
CTS-316-B	5.08	25.56	11.43	29.53	18.57	12714	17572	27.9	16.9	117.9	2982	720	2591
CTS-306-B	5.08	22.86	9.53	28.58	12.7	17615	8466	28.1	24.5	56.5	2076	724	2106
CTS-332	7.62	17.30	13.34	28.58	9.68	15035	9756	33.0	27.6	55.3	2290	764	2038
CTS-305-E	7.62	20.00	12.07	24.77	12.38	15683	9118	31.1	27.6	55.0	2277	809	1936
CTS-333	6.99	20.00	12.70	28.58	12.38	15330	11996	31.8	25.3	70.8	2684	854	2203
CTS-305-D	6.35	24.45	10.16	26.19	14.61	19134	8647	29.1	29.7	55.6	2478	1010	2160
CTS-275	9.84	21.59	14.13	22.38	15.24	17903	11839	34.0	29.7	65.3	2909	1017	2100
CTS-287-A	5.72	23.50	10.80	30.48	13.34	18235	10978	30.4	27.6	67.7	2803	1018	2391
CTS-156-A	5.72	23.50	10.80	31.43	12.7	19317	10674	31.0	29.3	64.5	2836	1072	2444
CTS-478-A	7.14	21.75	11.27	29.05	10.64	22239	7751	33.1	37.7	43.9	2483	1132	2225
CTS-310-A	7.94	18.42	14.29	32.39	9.53	19658	11599	36.0	33.5	60.5	3041	1134	2419
CTS-479	6.99	21.59	11.43	30.48	10.16	22524	8093	33.6	37.9	45.2	2569	1160	2303
CTS-255-B	9.84	21.59	15.08	25.24	15.24	18627	15109	35.5	29.7	79.8	3556	1190	2370
CTS-305-C	5.72	26.19	10.48	31.43	14.76	20719	11686	31.2	31.0	70.3	3273	1304	2637
CTS-310-B	7.62	19.69	13.97	34.29	9.53	22532	11803	36.6	36.8	60.5	3336	1341	2600
CTS-35-B	6.35	24.13	12.70	34.29	13.97	21893	16110	34.0	30.6	88.7	4078	1468	2881
CTS-255-D	8.89	20.40	14.61	30.48	11.51	20713	12931	36.9	37.5	65.8	3704	1509	2527
CTS-156	7.14	23.81	12.22	31.91	12.7	26711	11904	34.6	37.7	64.5	3649	1591	2656
CTS-255-A	7.62	25.08	13.34	30.48	16.19	23168	16941	34.3	32.2	92.5	4467	1675	2808
CTS-301-A	7.62	26.35	12.86	30.01	16.83	25246	16086	34.2	34.5	88.2	4560	1838	2831
CTS-255-C	7.62	24.13	13.65	33.34	13.97	25915	16224	36.1	36.8	84.3	4649	1895	2933
CTS-316-A	7.62	26.35	13.34	31.43	16.83	25733	17931	35.0	34.5	96.2	4974	1963	2980
CTS-152-B	7.94	25.08	13.49	32.86	14.29	28917	15502	36.6	40.7	79.4	4847	2156	2973
CTS-305-B	7.94	27.15	12.70	32.86	14.76	34060	13860	37.0	46.7	70.3	4925	2489	3073
CTS-331	9.53	24.13	15.24	32.39	13.97	32063	16778	39.4	46.0	79.8	5505	2569	3053
CTS-255-F	9.37	24.29	15.40	33.34	14.13	32014	17996	39.6	45.2	85.2	5779	2640	3149
CTS-298	5.08	41.91	10.48	37.15	27.94	36520	26886	33.4	33.7	150.8	7626	3077	4469
CTS-186-E	10.16	27.94	14.92	29.53	17.78	36928	17696	39.2	49.0	84.7	6229	3118	3172

National Arnold E Cores 12 mil Silicon Steel													
Part No.	D cm 1	Ht cm 2	Wth cm 3	Lt cm 4	G cm 5	Wtfe grams 6	Wtcu grams 7	MLT cm 8	Ac cm sq 9	Wa cm sq 10	Ap cm 4th 11	Kg cm 5th 12	At cm sq 13
CTS-186-C	7.62	32.39	12.70	37.15	17.78	45126	18808	39.0	52.9	90.3	7163	3880	3876
CTS-315	7.30	31.12	13.65	41.91	15.88	45555	22063	41.0	52.9	100.8	7994	4120	4215
CTS-255	8.57	30.32	13.97	37.63	16.03	45134	18986	41.1	58.2	86.6	7554	4275	3872
CTS-150	8.26	42.55	13.97	34.29	31.12	49837	36165	38.1	44.8	177.8	11955	5621	4803
CTS-217-C	9.53	31.12	15.88	41.91	15.88	59089	24452	45.5	69.0	100.8	10426	6323	4540
CTS-302-B	10.16	32.54	15.56	37.63	18.26	59967	23285	44.3	69.0	98.5	10193	6346	4324
CTS-302-A	10.16	31.43	15.56	40.01	15.56	64225	20556	45.9	76.6	84.0	9652	6446	4395

National Arnold E Cores 12 mil Silicon Steel													
Part No.	D cm 1	Ht cm 2	Wth cm 3	Lt cm 4	G cm 5	Wtfe grams 6	Wtcu grams 7	MLT cm 8	Ac cm sq 9	Wa cm sq 10	Ap cm 4th 11	Kg cm 5th 12	At cm sq 13
CTA-36	1.27	5.08	2.70	7.15	3.18	178	177	7.3	1.15	4.54	7.83	0.493	128
CTA-24	1.59	5.64	3.18	9.53	2.46	397	197	9.5	2.40	3.91	14.04	1.424	184
CTA-89-A	1.27	7.62	3.18	9.53	5.08	339	448	8.7	1.53	9.68	22.24	1.571	235
CTA-2	0.95	10.16	4.13	15.24	6.35	532	1195	11.1	1.72	20.16	52.16	3.239	471
CTA-25	1.91	6.67	3.81	11.43	2.86	686	326	11.2	3.45	5.44	28.16	3.461	261
CTA-5-F	2.86	8.26	5.40	11.43	5.72	878	995	12.9	3.45	14.52	75.08	8.058	375
CTA-22	3.18	8.10	5.08	10.00	5.24	1073	682	12.8	4.31	9.98	64.53	8.687	324
CTA-29	3.18	7.78	5.40	11.43	4.60	1117	744	13.6	4.79	10.24	73.53	10.341	357
CTA-5-C	2.54	8.26	5.08	13.34	4.45	1163	812	13.5	4.60	11.29	77.85	10.614	401
CTA-73	2.86	9.21	6.03	13.34	6.67	1020	1564	13.8	3.45	21.17	109.50	10.906	485
CTA-5-A	3.81	9.53	5.08	8.57	6.35	1394	585	13.6	5.75	8.07	69.53	11.749	331
CTA-10-B	2.54	10.16	4.45	11.43	6.35	1261	806	12.5	4.60	12.10	83.41	12.280	395
CTA-5-E	2.54	9.53	5.08	13.34	5.72	1463	1044	13.5	4.60	14.52	100.09	13.647	448
CTA-17	3.18	9.84	5.08	10.95	6.35	1422	867	13.4	5.27	12.10	95.56	14.977	400
CTA-16-C	3.49	10.16	6.67	13.34	7.62	1318	1951	15.1	4.21	24.19	152.94	17.052	552
CTA-5-D	4.76	7.62	6.51	10.95	3.81	1853	599	16.9	8.62	6.65	86.01	17.555	382
CTA-401	2.54	8.89	6.03	19.05	3.18	2007	999	16.9	6.90	11.09	114.72	18.739	596
CTA-9	3.18	10.00	4.76	12.38	4.92	1996	606	14.5	7.66	7.82	89.80	18.939	429
CTA-14	3.18	9.21	5.56	13.81	4.76	1803	916	15.1	6.71	11.34	114.06	20.203	468
CTA-21	3.81	9.21	6.99	14.29	6.03	1688	1696	16.6	5.75	19.16	165.15	22.877	551
CTA-10-A	3.18	10.95	5.08	13.34	5.87	2222	897	15.0	7.66	11.19	128.59	26.221	500
CTA-5-B	2.86	10.80	6.03	16.19	6.35	1969	1694	15.8	6.04	20.16	182.53	27.971	619
CTA-410-C	2.86	11.75	5.56	15.72	6.67	2300	1502	15.6	6.90	18.00	186.17	32.833	620
CTA-10-E	3.81	10.95	5.72	13.34	5.87	2664	985	16.5	9.19	11.19	154.31	34.378	531
CTA-9-A	5.08	10.00	6.67	12.38	4.92	3206	773	18.5	12.26	7.82	143.69	37.982	514
CTA-71	4.45	10.16	6.67	13.34	5.72	2677	1196	17.6	9.39	12.70	178.89	38.074	545
CTA-12	3.81	11.43	6.19	14.76	6.35	2900	1392	17.3	9.19	15.12	208.50	44.438	613
CTA-12-B	5.72	11.75	6.99	11.43	6.67	3985	873	19.3	13.79	8.47	175.17	50.014	567
CTA-5-H	3.81	11.43	6.67	16.19	6.35	3025	1743	18.0	9.19	18.15	250.27	51.122	681
CTA-28	2.86	15.24	6.67	18.10	10.80	3370	3675	16.8	6.04	41.13	372.36	53.665	929

National Arnold E Cores 12 mil Silicon Steel

Part No.	D cm 1	Ht cm 2	Wth cm 3	Lt cm 4	G cm 5	Wtfe grams 6	Wtcu grams 7	MLT cm 8	Ac cm sq 9	Wa cm sq 10	Ap cm 4th 11	Kg cm 5th 12	At cm sq 13
CTA-5	3.81	11.43	6.99	17.15	6.35	3130	1990	18.5	9.19	20.16	278.03	55.264	727
CTA-14-A	6.35	9.21	8.73	13.81	4.76	3601	1300	21.5	13.41	11.34	228.12	56.936	614
CTA-12-C	5.72	11.75	7.30	12.38	6.67	4113	1119	19.8	13.79	10.59	219.03	60.962	611
CTA-12-A	3.81	12.70	6.19	16.67	6.35	3923	1494	18.5	11.49	15.12	260.63	64.674	727
CTA-18	5.08	10.95	7.62	15.24	5.87	3787	1595	20.0	12.26	14.92	274.33	67.109	682
CTA-123-C	3.81	13.65	7.62	18.57	8.89	3535	3465	19.2	8.62	33.87	437.85	78.696	926
CTA-8-A	3.49	15.56	6.99	18.10	10.48	3736	3545	18.2	8.43	36.60	462.73	85.916	962
CTA-20	5.72	11.43	8.26	15.24	6.35	4420	1834	21.3	13.79	16.13	333.64	86.347	737
CTA-15-A	5.08	12.70	7.62	15.24	7.62	4282	2069	20.0	12.26	19.36	355.88	87.057	761
CTA-123-B	3.81	13.65	8.26	20.48	8.89	3702	4253	20.2	8.62	39.52	510.83	87.275	1038
CTA-307-B	3.81	13.73	6.87	18.34	7.62	4335	2403	19.3	11.06	23.29	386.37	88.364	869
CTA-1	4.45	12.70	7.94	18.10	7.62	4053	2878	20.3	10.73	26.62	428.23	90.638	878
CTA-11-A	6.99	11.59	9.84	14.29	7.78	4249	2785	23.5	12.64	22.23	421.56	90.772	797
CTA-4	3.49	13.97	6.99	20.00	7.62	4256	2758	19.4	10.54	26.62	420.64	91.245	945
CTA-16-B	5.08	12.38	8.26	16.67	7.62	4168	2674	20.7	11.49	24.19	417.00	92.490	832
CTA-15-J	6.35	12.07	8.89	14.29	7.62	4498	2266	21.9	13.41	19.36	389.33	95.144	768
CTA-3	4.45	14.92	7.94	18.10	9.84	4598	3717	20.3	10.73	34.38	553.15	117.080	994
CTA-15-B	5.08	13.97	7.62	17.15	7.62	5749	2200	21.3	15.32	19.36	444.85	127.921	889
CTA-27	5.08	13.97	8.10	17.62	8.26	2787	2846	21.4	13.79	24.90	515.10	132.628	935
CTA-3-B	4.45	14.92	8.89	20.96	9.84	4911	5079	21.8	10.73	43.75	703.91	138.758	1173
CTA-125	6.35	11.43	10.16	19.05	6.35	5508	3145	24.4	15.32	24.19	556.06	139.850	977
CTA-7	5.72	10.80	11.43	24.77	5.72	2772	4582	26.3	13.79	32.66	675.61	141.710	1231
CTA-3-A	5.08	14.92	8.57	18.10	9.84	5261	3950	21.5	12.26	34.38	632.18	143.904	1035
CTA-15	5.08	14.61	7.62	18.10	7.62	6544	2266	21.9	16.86	19.36	489.40	150.342	956
CTA-124	5.08	14.92	8.89	19.05	9.84	5377	4408	22.0	12.26	37.50	689.55	153.419	1095
CTA-15-D	7.62	12.70	10.16	15.24	7.62	6423	2635	25.5	18.39	19.36	533.82	153.822	903
CTA-309-A	3.81	15.88	8.26	22.86	9.53	5480	4915	21.8	11.49	42.34	729.83	154.142	1252
CTA-3-C	5.08	15.24	8.89	19.05	10.16	5472	4550	22.0	12.26	38.71	711.76	158.360	1113
CTA-16	5.08	15.24	7.62	19.05	7.62	7375	2332	22.6	18.39	19.36	533.82	173.847	1026
CTA-3-D	4.45	15.24	8.57	21.91	8.89	6695	4412	22.5	13.41	36.70	738.03	175.621	1200

National Arnold E Cores 12 mil Silicon Steel													
Part No.	D cm 1	Ht cm 2	Wth cm 3	Lt cm 4	G cm 5	Wtfe grams 6	Wtcu grams 7	MLT cm 8	Ac cm sq 9	Wa cm sq 10	Ap cm 4th 11	Kg cm 5th 12	At cm sq 13
CTA-4-A	5.40	13.97	8.89	20.00	7.62	6578	3329	23.4	16.28	26.62	650.05	180.566	1074
CTA-3-E	5.08	16.51	9.21	20.00	11.43	5943	5672	22.5	12.26	47.18	867.56	188.749	1249
CTA-16-A	5.08	15.24	8.26	20.96	7.62	7733	3043	23.6	18.39	24.19	667.27	208.121	1140
CTA-6-A	5.08	16.19	8.57	20.00	9.84	6985	4183	22.8	15.32	34.38	790.22	212.331	1180
CTA-128-A	6.35	13.97	9.53	19.05	7.62	7559	3207	24.9	19.15	24.19	695.07	214.285	1081
CTA-15-C	5.08	16.51	7.62	20.96	7.62	9162	2463	23.9	21.45	19.36	622.79	224.027	1173
CTA-19	5.08	15.24	8.57	21.91	7.62	7910	3419	24.1	18.39	26.62	734.10	224.219	1198
CTA-15-H	6.03	15.24	8.57	19.05	7.62	8761	2528	24.5	21.84	19.36	633.96	226.109	1091
CTA-358-A	5.08	16.51	8.57	21.91	8.89	8439	3989	24.1	18.39	31.05	856.45	261.589	1276
CTA-19-A	5.08	16.19	9.84	23.81	9.84	7610	6203	24.8	15.32	46.88	1077.53	266.257	1438
CTA-237-B	6.35	16.51	10.80	20.96	11.43	7588	6931	25.6	15.32	50.81	1167.72	279.848	1411
CTA-237-C	4.45	19.05	8.89	24.77	11.43	8793	6587	24.3	16.09	50.81	1226.11	324.654	1587
CTA-15-E	3.18	29.21	9.21	27.62	22.86	6719	16758	22.8	9.58	137.91	1981.13	333.145	2507
CTA-19-F	6.99	16.03	11.75	22.39	10.64	8815	7583	28.1	17.91	50.66	1360.95	347.439	1518
CTA-19-D	5.08	17.78	8.57	25.72	7.62	11817	3779	26.6	24.52	26.62	978.80	360.578	1518
CTA-6	6.35	17.46	9.84	21.91	9.84	10969	4882	26.6	22.98	34.38	1185.33	409.368	1434
CTA-237-A	6.35	17.78	10.80	22.86	11.43	9969	7275	26.8	19.15	50.81	1459.66	416.575	1577
CTA-358-F	5.08	19.05	8.57	25.72	8.89	12529	4409	26.6	24.52	31.05	1141.94	420.674	1606
CTA-307	5.08	21.59	10.16	24.77	15.24	9621	10448	25.3	15.32	77.42	1779.39	431.036	1882
CTA-350-A	6.35	17.46	10.48	24.77	9.21	12420	5728	28.3	24.90	38.01	1419.82	500.572	1614
CTA-308-I	6.67	20.40	11.75	23.81	14.68	10881	11238	28.2	18.10	74.60	2025.73	519.400	1894
CTA-308-E	5.72	20.32	9.53	24.77	11.43	13341	6299	27.1	24.13	43.55	1576.43	561.168	1730
CTA-308-B	7.62	20.32	10.48	20.00	12.70	15349	5530	28.6	27.58	36.30	1501.62	579.989	1571
CTA-308-A	6.67	20.32	11.75	24.77	13.97	12024	10931	28.9	20.11	70.97	2140.99	596.453	1937
CTA-313	6.35	17.78	11.43	28.58	8.89	14304	7415	30.8	26.81	45.16	1816.46	632.942	1895
CTA-308-D	6.35	20.32	10.16	24.77	11.43	14827	6594	28.4	26.81	43.55	1751.59	661.806	1787
CTA-308-F	7.62	20.64	10.80	21.19	12.86	8073	6363	29.2	28.15	40.83	1724.09	664.467	1679
CTA-308-J	7.94	20.40	13.02	23.81	14.68	12937	12248	30.8	21.55	74.60	2411.55	675.358	2007
CTA-311	7.62	17.78	12.70	26.67	10.16	14801	8824	32.1	27.58	51.61	2135.27	734.964	1923
CTA-308-H	8.57	20.40	13.65	23.81	14.68	8507	12754	32.1	23.28	74.60	2604.46	756.519	2063

13

PC BOARD MOUNTS/HEADERS AND TRANSFORMER MOUNTING BRACKETS

INTRODUCTION

Mounting and packaging for magnetic devices have become more important in recent times as companies gained the ability to view the total cost of a component as a combination of its price, availability, quality, durability and how efficiently it can be put on the circuit board. Today, magnetic components in small and medium quantities are being mounted and packaged to improve durability in transit, to speed inspection time by allowing plug in testing, and to reduce production and rework costs.

More and more, the magnetic design engineer using a toroid must consider packaging and production options in addition to core material and size. Often the best magnetic package is a compromise between the optimum magnetic design, "manufacturability," durability, and the availability of hardware. This section (paper, application note) will help a magnetic design engineer with the packaging aspects of designing on a toroidal core.

Current technology relies heavily on soldering a component to the printed-circuit board (PCB). For a toroidal core and winding, a header to toroid mount is used to facilitate connecting the windings to the PCB by providing a terminal as an intermediate connection or by positioning the winding for direct connection to the PCB. This can be done using through hole soldering or surface-mount soldering.

THROUGH-HOLE TOROID MOUNTS

Through-hole headers and mounts connect components to a printed-circuit board by inserting a terminal or lead through a hole in the board and soldering it to the opposite side. Through-hole headers and mounts have two basic configurations; horizontal or vertical.

Horizontal Toroid Mounts

Horizontal through-hole headers or toroid mounts are widely used as a platform or holder for mounting wound toroids on their side. They are usually molded from plastic (see the subsection entitled Selecting the Best Plastic for Your Application) with the size, shape, and number of termination points specific to the wound toroid. They are most often either a platform, as in Fig. 142, or cup shape like Fig. 143. The molding of either configuration will typically include standoffs which allow the printed-circuit-board cleaning solutions to easily flow under the component. The minimum standoff is usually 0.015 in.

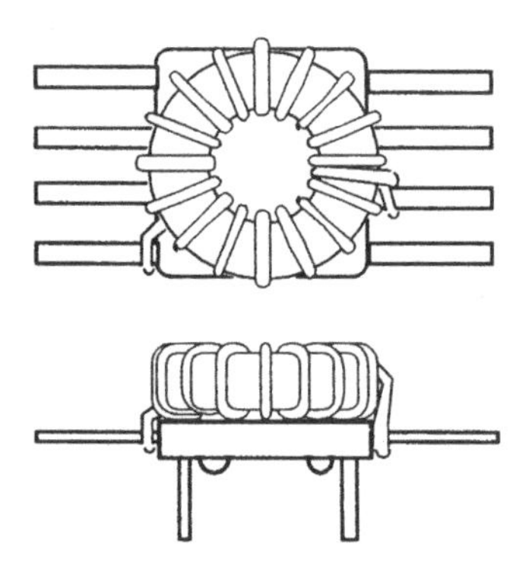

Figure 142 Horizontal platform/through hole.

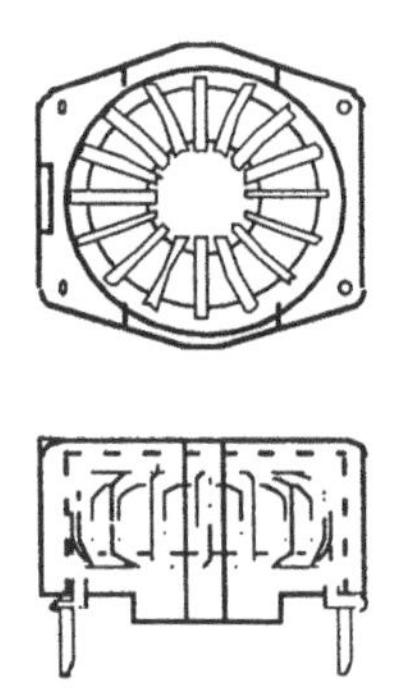

Figure 143 Horizontal cup/through hole.

The leads from the toroidal winding are attached to the mount's terminals, usually by soldering. If the winding's wire is large and stiff enough, the wire can be "self-leaded" and positioned through the header or mount into the printed-circuit board, as in Fig. 144. Once the toroid is attached to the mount, this component is ready for insertion into a printed-circuit board. Magnetic components are heavy and the mechanical characteristics of the solder connection are as important as electrical integrity. Printed-circuit boards with unplated through holes using heavy components may require a clinched terminal as described in MIL-STD-2000 and shown in Fig. 145. Printed-circuit boards with printed through holes offer good mechanical integrity without clinching, providing a successful intermetallic bond is created during the board solder process as shown in Fig. 146.

The toroids can be attached to the mount with either adhesives or mechanical means. Cup-shaped toroid mounts can be filled with a potting or encapsulation compound to both adhere and protect the wound toroid. Horizontal mounting offers both a low profile and a low center of gravity in applications that will experience shock and vibration. As the toroid's diameter gets larger, horizontal mounting begins to use up valuable circuit-board real estate. If there is room in the enclosure, vertical mounting is used to save board space.

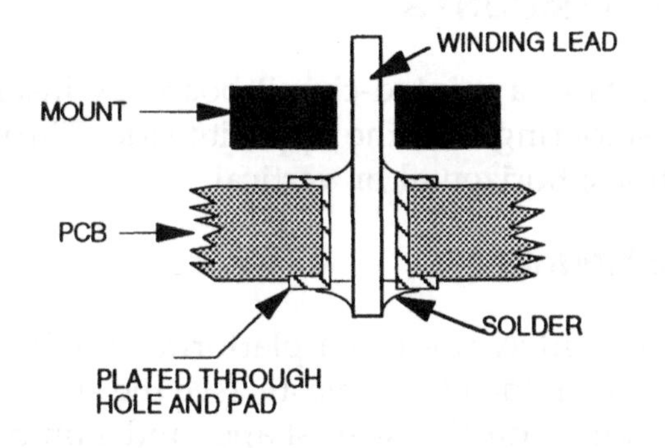

Figure 144 Self-lead/plated through hole.

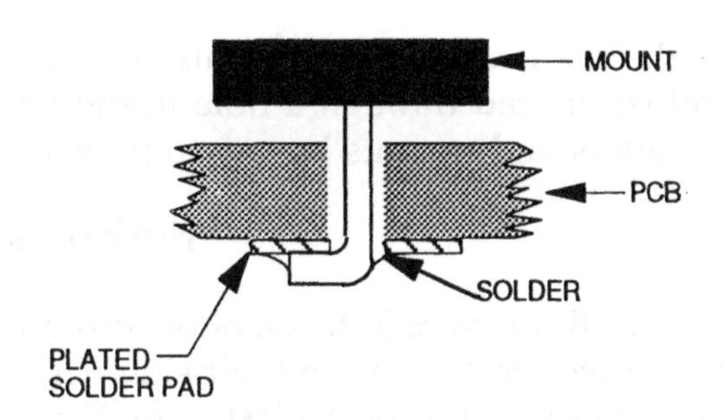

Figure 145 Clinched terminal.

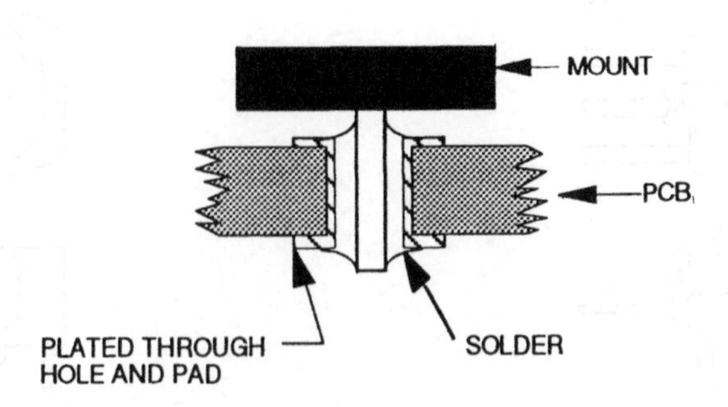

Figure 146 Plated through hole.

Vertical Toroid Mounts

Vertical through-hole headers and toroid mounts stand up wound toroids to save circuit-board real estate. As with horizontal mounts, vertical mounts are usually molded from plastic (see the subsection entitled Selecting the Best Plastic for Your Application) with the size, shape, and number of termination points specific to the application. The molding of either configuration will typically include standoffs which allow the printed-circuit-board cleaning solutions to flow easily under the component. The minimum standoff is usually 0.015 in. Vertical toroid mounts come in many configurations, several of which are shown in Figs. 147–152. Much of their structure is devoted to supporting the vertical toroid and creating a stable base for connection to the printed-circuit board.

The leads from the toroidal winding are attached to the mount's terminals, usually by soldering as shown in Fig. 147. If the winding's wire is large and stiff enough, the wire can be "self-leaded" and positioned through the header or mount into the printed circuit board, as in Figs. 144, 148, and 149. The advantage of self-leading mounts is that the expense and vulnerability of an additional intermediate solder connection is avoided. The toroids can be attached to the mount with either adhesives or mechanical means, or by encapsulation. Cup-shaped toroid mounts (Figs. 149–151) can be filled with a potting or encapsulation compound to both adhere and protect the wound toroid.

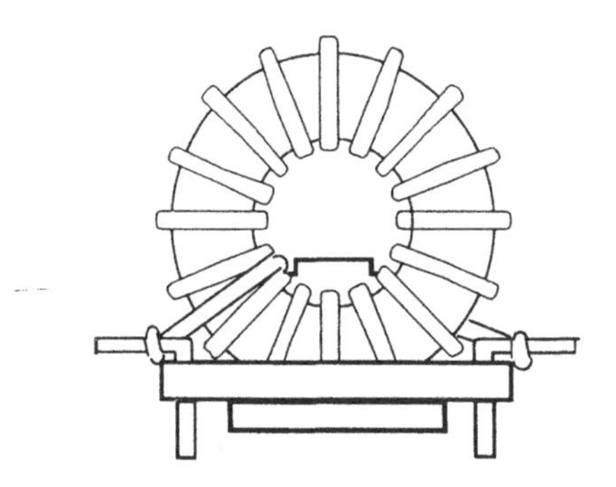

Figure 147 Vertical mount/through hole.

Figure 148 Verticle mount/self-leaded.

Figure 149 Vertical mount/self-leaded.

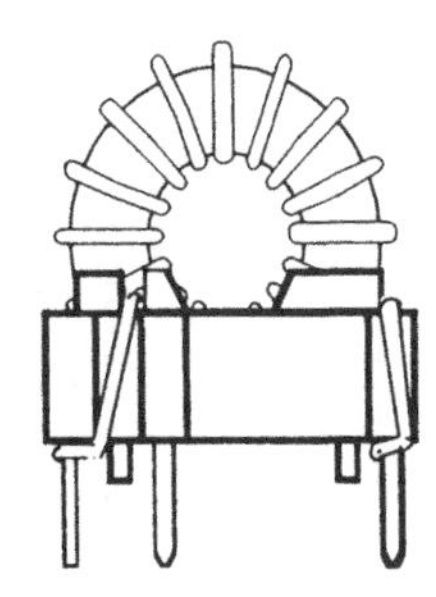

Figure 150 Vertical cup/through hole.

Vertical mounting saves circuit-board real estate when a toroid's diameter gets larger, but it creates a component height issue. Vertical mounting also raises the component's center of gravity, making it vulnerable to shock and vibration.

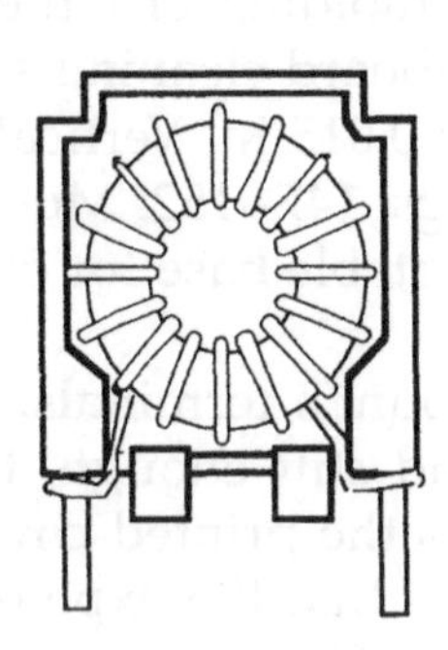

Figure 151 Vertical cup/through hole.

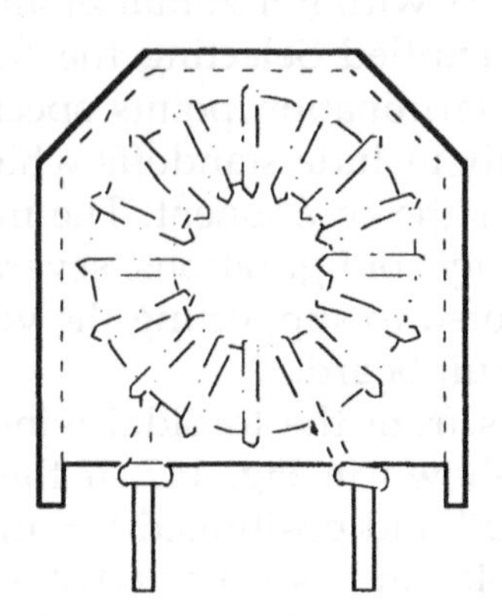

Figure 152 Vertical cup/through hole.

SURFACE-MOUNT TOROID MOUNTS

Surface-mount components are a direct response to cost-reduction efforts that center around improved circuit-board production. Automatic or robotic pick-and-place equipment can pick up and place surface-mount-style components on a printed-circuit board faster and more accurately than previous technology would allow. Instead of a pin or terminal passing through a printed-circuit board and being soldered on the opposite side, surface-mount components utilize a flat solderable surface that is soldered to a flat solderable pat on the face of the printed-circuit board (Fig. 152). The pad on the circuit board is usually coated with a pastelike formulation of solder and flux. With careful placement, surface-mount-style components on solder paste will stay in position until elevated temperatures, usually from an infrared oven, melt the solder paste and solder the mount's flat terminals to the circuit board's pad.

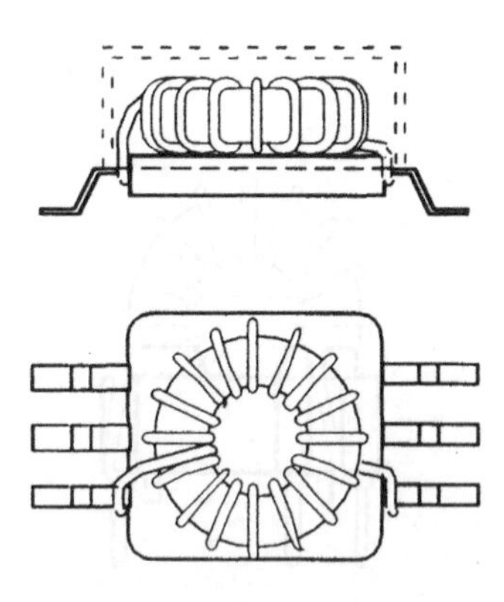

Figure 153 Gull wing toroid platform.

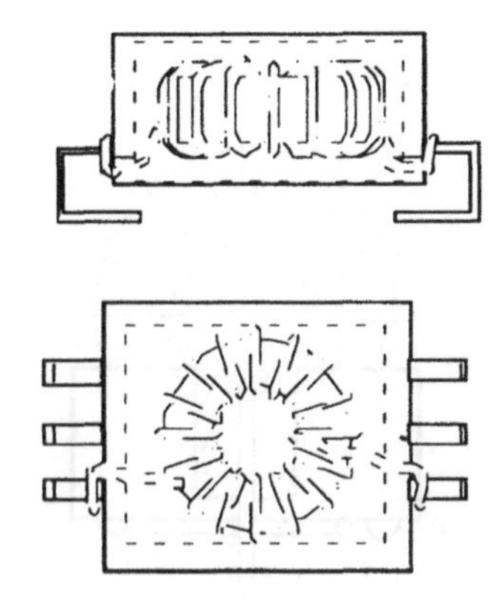

Figure 154 "J" lead toroid cup.

Because the mount's leads lay on the printed-circuit board, it is important that all the mount's leads are flat and on the same plane. If one or several of the leads are out of position or not on the same plane as the others, the solder connection can be defective. The industry specification for lead "co-planarity" allows a tolerance of 0.004 in. from the plane of the printed-circuit board. This makes the handling and packaging of surface-mount devices critical to the overall cost and performance of the technology.

A wound toroid must be incorporated in a header or mount in order to utilize surface-mount technology. There are several styles of surface mount headers and mounts for toroids which are usually molded from plastic (see the subsection entitled Selecting the Best Plastic for your Application) with the size, shape, and number of termination points specific to the wound toroid.

The most familiar styles are either gull wing (Figs. 153 and 155) or "J" lead (Figs. 154 and 156) horizontal surface mount devices. The gull wing style is widely used because it is relatively inexpensive to mold and form. The gull wings are flexible to withstand thermal expansion and contraction and it is easy to inspect the integrity of the gull wing lead to circuit-board solder connection. The "J" lead also has wide acceptance because it uses up less board real estate than the gull wing. However, the "J" lead to board solder connections are hidden from inspection and the leads are more difficult to form. Once the toroids are attached to the mount with either adhesives or mechanical means, or by encapsulation, the winding leads are connected, usually by soldering, to the mount's terminations.

Surface-mount components need to be shaped so the robot can pick them up from a package, tube, or tray and place them on the circuit board. Two common methods are either a mechanical grabber or a vacuum pickup. In either case the mount needs to be designed with a shape a robot can grab or a flat horizontal surface that allows vacuum pickup. Figure 153 shows how a cup can be shaped on the device which creates a flat horizontal surface for vacuum pickup.

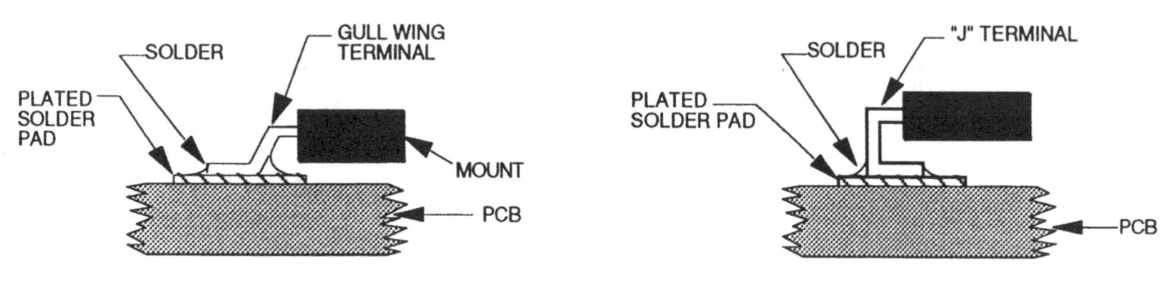

Figure 155 Gull wing.

Figure 156 "J" lead.

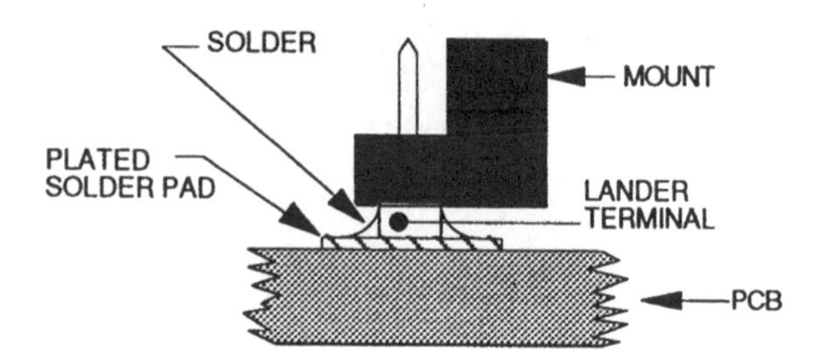

Figure 157 Lunar Lander.

Figure 154 shows an inverted-cup design with "J" lead terminals which also allows vacuum pickup. Both the gull wing and "J" lead are subject to co-planarity problems if packaging for shipment and production handling is not carefully considered.

New surface-mount techniques are being introduced to the industry constantly. One style is the "Lunar Lander" which is shown in Figs. 157 and 158. The "Lunar Lander" incorporates a lead style that is more rigidly supported by the plastic molding. This style is very robust and will tolerate handling and shipping with little or no effect on the co-planarity. This style solders well with the mount to board connections mostly visible for inspection. Figure 157 shows the Lunar Lander lead style incorporated with a cup-shaped mount. This can be filled with a potting or encapsulation compound to both adhere and protect the wound toroid. The cup shape also offers a flat surface important to pick-and-place vacuum pickup as well as a convenient location for a part number marking.

A self-leaded surface-mount header/toroid mount is a relatively recent development. In this style of toroid mount, the toroid's winding becomes the surface-mount connection to the circuit board. The winding terminations are first tinned, then channeled and gripped by the design and characteristics of the plastic molded mount, making a suitable surface-mount connection point. This concept improves reliability and reduced cost by eliminating an intermediate solder connection.

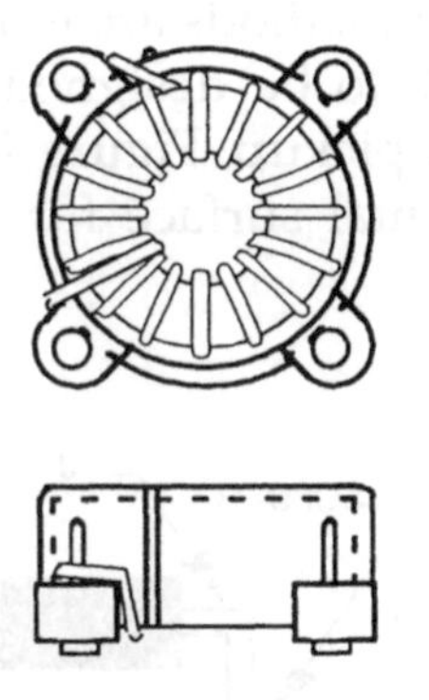

Figure 158 Lunar Lander surface mount.

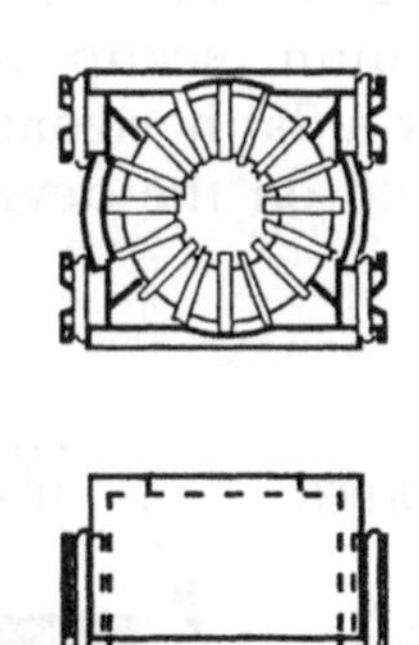

Figure 159 Self-lead surface mount.

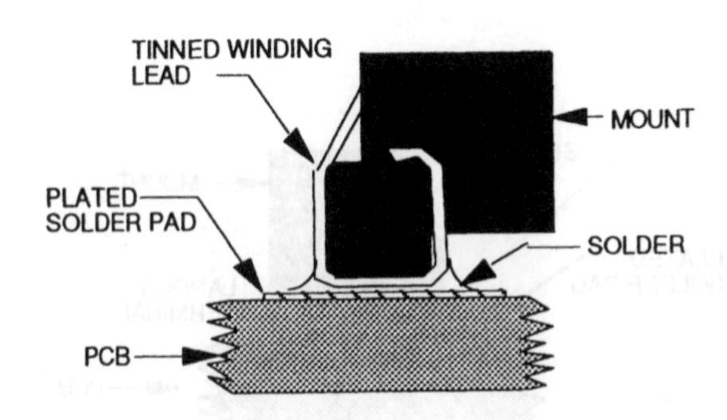

Figure 160 Self-lead suface mount.

Figure 160 shows how the tinned winding lead is held in position by the mount, on a solder paste pad for example, allowing the surface-mount connection to be made during infrared reflow. Although this can offer a cost-effective connection method, the mount is usually specific to a particular wire size and will be subject to co-planarity problems if the wire is not positioned correctly. As with the other surface-mount styles, a flat surface for vacuum pickup eneds to be incorporated into the toroid mount's design.

Selecting the Best Plastic for Your Application

Plastics used in molding toroid mounts and headers come in two broad categories: thermoset and thermoplastic. Thermoset plastics include epoxies, phenolics, and diallyl phthalate (DAP), which are known for their environmental stability and ability to tolerate over 750°F without melting. Thermoplastics include nylon, polypropylene, polycarbonate, polyester (Valox, Rynite), LCP (Vectra), and PPS (Ryton), which will begin to melt if they experience temperatures much above 500°F for an extended period. The chemistry that gives thermoplastics a lower melting point also make it less expensive to mold, giving it a cost advantage over the thermoset plastics.

Thermoplastics are widely used in applications that do not experience temperatures above 500°F, except for a few seconds during the winding lead to terminal and component to PCB soldering process. Thermoset plastics, on the other hand, are popular in magnetic applications when it is used in conjunction with self-stripping magnetic wire. The unstripped and untinned magnetic wire is wrapped around a terminal molded into a thermoset header or toroid mount and then dipped into a 750°F solder pot. The high-temperature solder will burn off the wire's insulation, tin the wire, and solder it to the terminal in a cost-effective way, without melting the mount.

There are trade-offs between the two plastic types that must be considered. The mount will be less expensive if molded from thermoplastic but will require pretinning the winding leads and careful heat management while soldering the leads to the mount and soldering the mount to the circuit board. The thermoset mount will be more expensive, but with self-stripping magnetic wire, several terminations can be soldered at once and the wire will not need to be stripped or pretinned, possibly making the overall cost of producing the component lower.

C CORE MOUNTING BRACKET MANUFACTURER

Manufacturer's address and phone number can be found from the Table of Contents.

1. Hallmark Metals

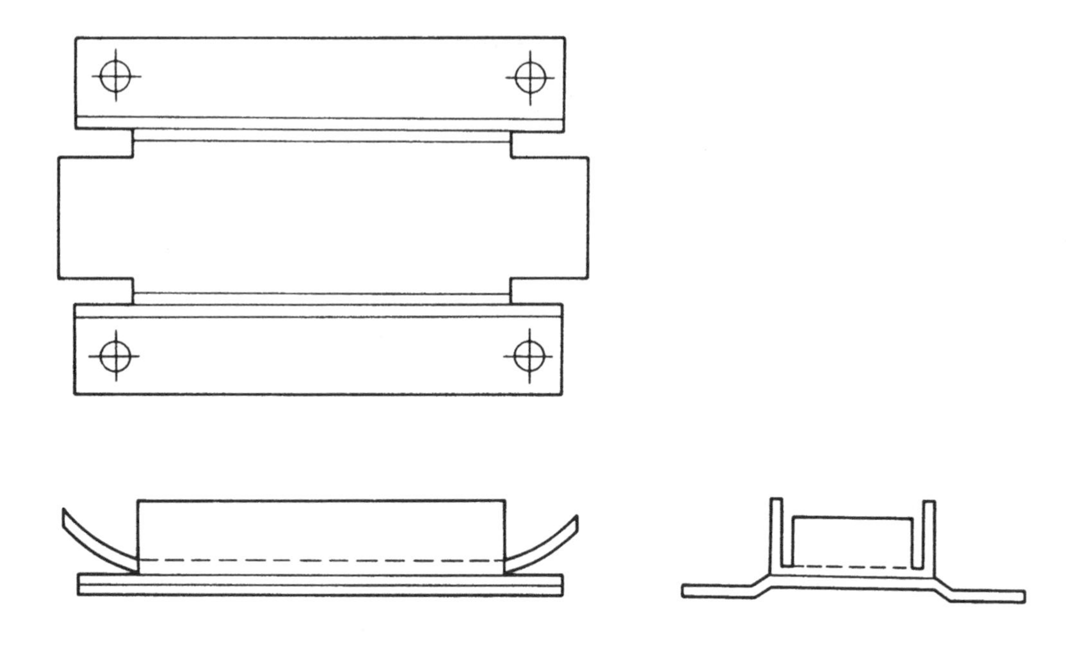

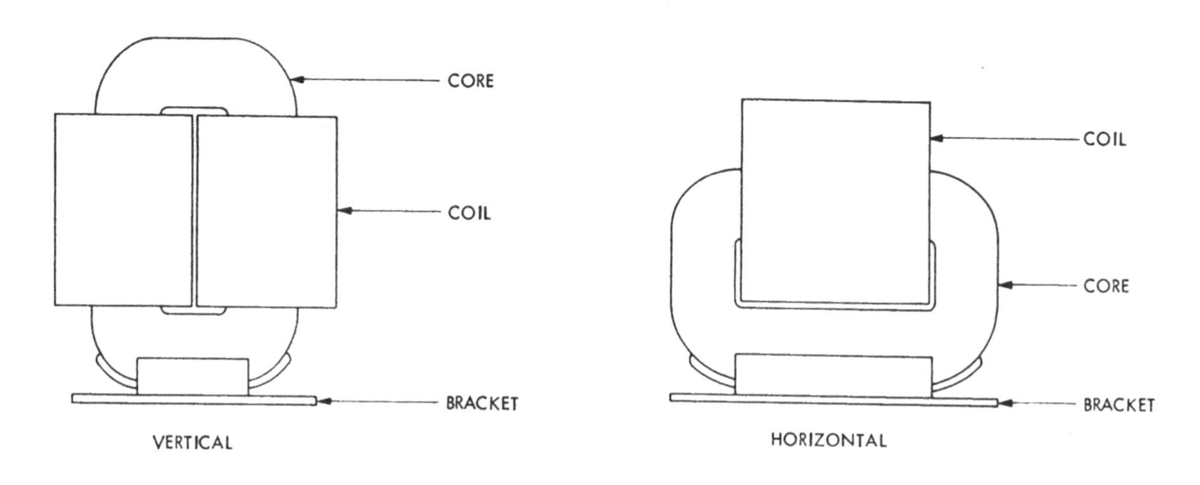

Figure 161 C core mounting bracket outline.

C CORE BANDING DATA

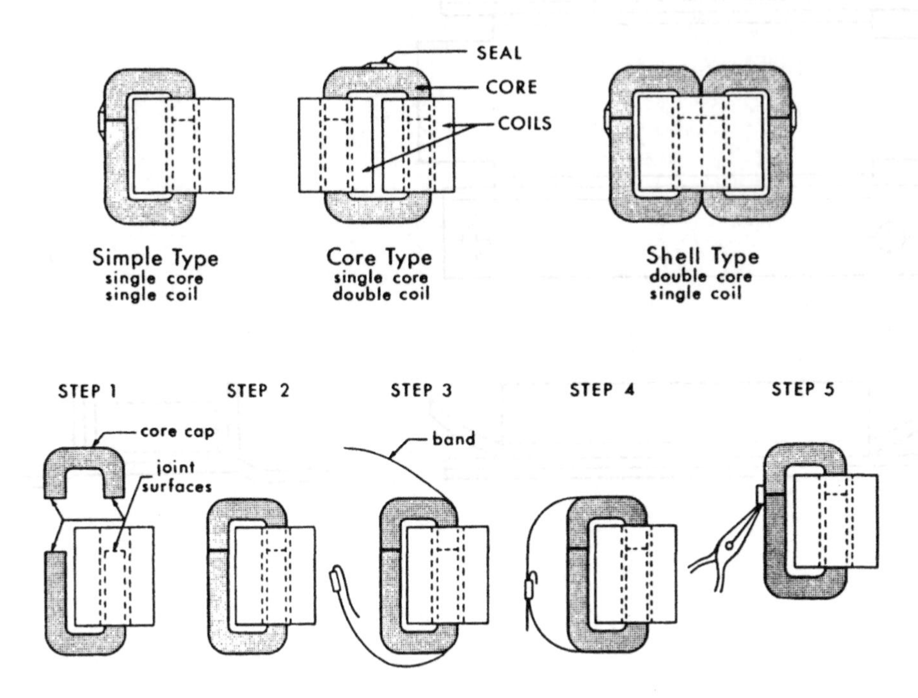

Core Strip Width (in.)	Core Cross-Section (D × E)-in²	Band Size (in.)	No. Bands Required	Seal Dimension (in.)	Banding Force (lb.)
Any	.188 or less	3/16 x .006	1	3/16 x 1/4	37.5
3/8 or larger	.188 to .375	3/8 x .006	1	3/8 x 3/8	75
3/8 to 1-1/2	.375 to .75	3/8 x .012	1	3/8 x 3/8	150
1-5/8 or larger		3/8 x .006	2	3/8 x 3/8	75
1/2 to 1-1/8	.75 to 1.5	3/8 x .012	1	3/8 x 3/8	150
1-1/4 or larger		3/8 x .012	2	3/8 x 3/8	150
3/4 or larger	1.5 to 3.0	3/4 x .023	1	7/8 x 1-7/8	600
3/4 or larger	3.0 to 4.25	3/4 x .035	1	7/8 x 1-7/8	900
2 or larger	4.25 to 6.0	3/4 x .023	2	7/8 x 1-7/8	600
3-1/4 or larger	6.0 to 9.0	3/4 x .023	3	7/8 x 1-7/8	600
3-1/4 or larger	9.0 to 13.5	3/4 x .035	3	7/8 x 1-7/8	900

Figure 162 Banding data.

LAMINATION BRACKET MANUFACTURERS AND DISTRIBUTORS

Manufacturers' addresses and phone numbers can be found from the Table of Contents.

1. Bahrs Die and Stamping Co.
2. Transformer Equipment and Material
3. RAM Sales and Distributing

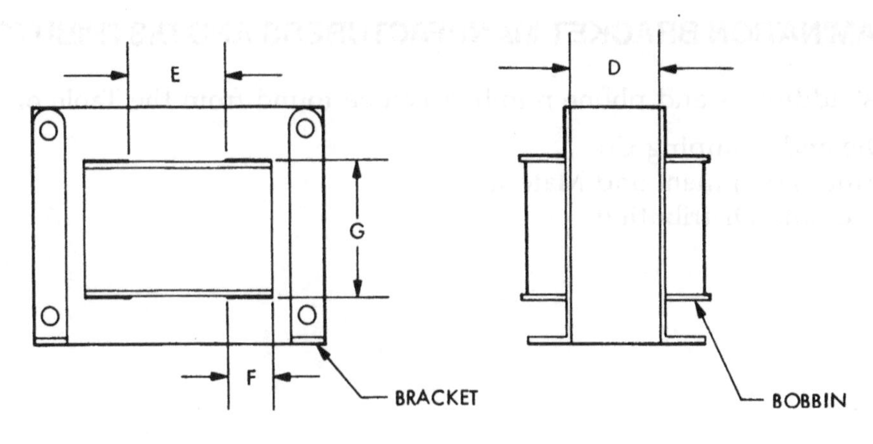

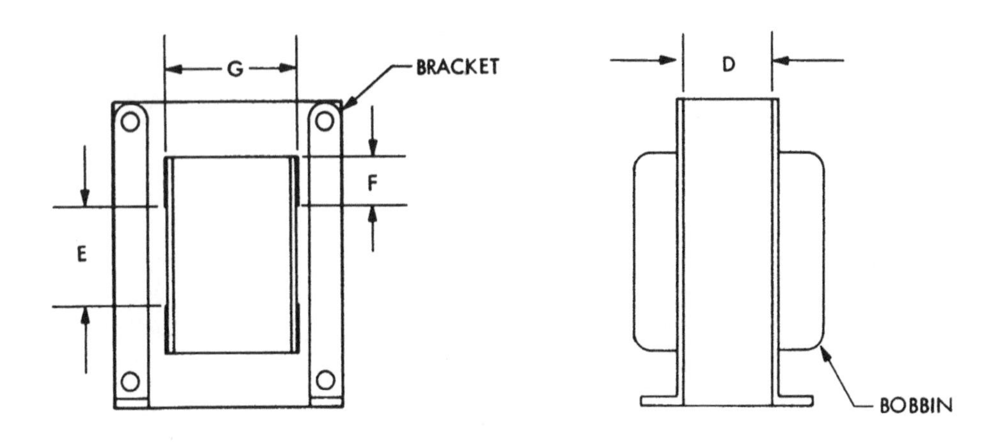

Figure 163 Transformer L brackets.

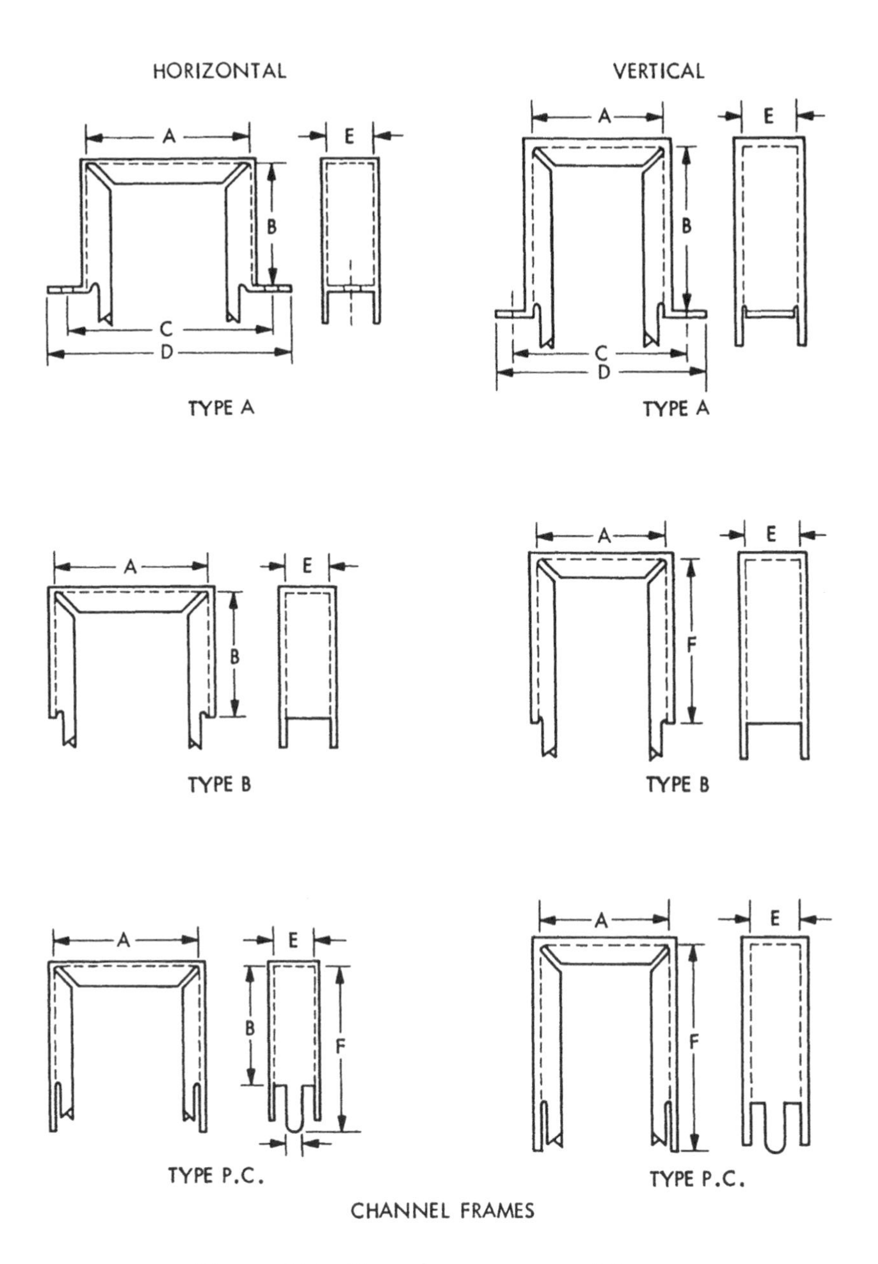

Figure 164 Transformer channel frames.

Table 13 Delbert Blinn Toroid Retainers

Part No.	A	B	C	D	R
100-1	0.250	0.180	0.190	4-40	0.343
100-2	0.343	0.210	0.250	4-40	0.250
100-3	0.484	0.250	0.250	6-32	0.250
100-4	0.615	0.250	0.250	6-32	0.250
100-5	0.812	0.250	0.255	6-32	0.250
100-6	1.000	0.375	0.375	10-32	0.187
100-7	0.850	0.218	0.410	6-32	0.250
100-8	1.750	0.750	0.562	10-32	0.218

Made of glass filled nylon.
Available with screw clearance holes.

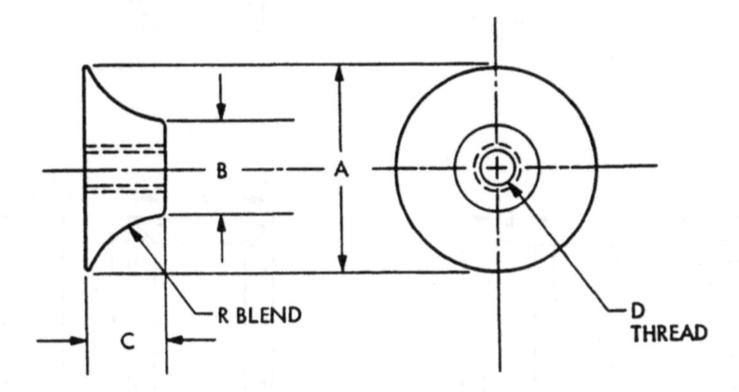

Figure 165 Toroid retainer outline.

INDEX

T - #0035 - 311024 - C0 - 276/219/38 - PB - 9780367400965 - Gloss Lamination